Shaw M. C.

December 1992.

GEOLOGICAL SOCIETY SPECIAL PUBLICATION NO 50

Classic Petroleum Provinces

EDITED BY

J. BROOKS
Brooks Associates
Glasgow, UK

1990

Published by

The Geological Society

London

Geological Society Special Publications
Series Editor K. COE

THE GEOLOGICAL SOCIETY

The Geological Society of London was founded in 1807 for the purposes of 'investigating the mineral structures of the earth'. It received its Royal Charter in 1825. The Society promotes all aspects of geological science by means of meetings, special lectures and courses, discussions, specialist groups, publications and library services.

It is expected that candidates for Fellowship will be graduates in geology or another earth science, or have equivalent qualifications or experience. All Fellows are entitled to receive for their subscription one of the Society's three journals: *The Quarterly Journal of Engineering Geology*, the *Journal of the Geological Society* or *Marine and Petroleum Geology*. On payment of an additional sum on the annual subscription, members may obtain copies of another journal.

Membership of the specialist groups is open to all Fellows without additional charge. Enquiries concerning Fellowship of the Society and membership of the specialist groups should be directed to the Executive Secretary, The Geological Society, Burlington House, Piccadilly, London W1V 0JU.

Published by the Geological Society from:
The Geological Society Publishing House
Unit 7
Brassmill Enterprise Centre
Brassmill Lane
Bath
Avon BA1 3JN
UK
(*Orders:* Tel. 0225 445046)

First published 1990

© The Geological Society 1990. All rights reserved. No reproduction, copy or transmission of this publication may be made without written permission. No paragraph of this publication may be reproduced, copied or transmitted save with the written permission or in accordance with the provisions of the Copyright Act 1956 (as amended) or under the terms of any licence permitting limited copying issued by the Copyright Licensing Agency, 33–34 Alfred Place, London WC1E 7DP. Users registered with Copyright Clearance Center: this publication is registered with CCC, 27 Congress St., Salem, MA 01970, USA. 0305–8719/90 $03.00.

British Library Cataloguing in Publication Data
Classic petroleum provinces.
  1. Natural gas & petroleum
  I. Brooks, J. (James), *1938*– II. Geological Society of London III. Series
  553.2'8

ISBN 0–903317–48–6

Printed in Great Britain at the Alden Press, Oxford

Contents

Acknowledgements	vii
BROOKS, J. Classic Petroleum provinces	1
JAMES, K. H. The Venezuelan hydrocarbon habitat	9
AYMARD, R., PIMENTEL, L., EITZ, P., LOPEZ, P., CHAOUCH, A., NAVARRO, J., MIJARES, J. & PEREIRA, J. G. Geological integration and evaluation of Northern Monagas, Eastern Venezuelan Basin	37
CHIGNE, N. & HERNANDEZ, L. Main aspects of petroleum exploration in the Apure area of Southwestern Venezuela 1985–1987	55
ROBERTO, M., MOMPART, L., PUCHE, E. & SCHERER, F. New oil discoveries in the Ceuta area, SE Lake Maracaibo, Venezuela	77
DASHWOOD, M. F. & ABBOTTS, I. L. Aspects of the petroleum geology of the Oriente Basin, Ecuador	89
MOHRIAK, W. U., MELLO, M. R., DEWEY, J. F. & MAXWELL, J. R. Petroleum geology of the Campos Basin, offshore Brazil	119
HUBBARD, R. J., EDRICH, S. P. & RATTEY, R. P. Geological evolution and hydrocarbon habitat of the 'Arctic Alaska microplate'	143
CREANEY, S. & ALLAN, J. Hydrocarbon generation and migration in the Western Canada sedimentary basin	189
LOPEZ, J. A. Structural styles of growth faults in the US Gulf Coast Basin	203
FAILS, T. G. The Northern Gulf Coast Basin: a classic petroleum province	221
CRAIG, D. H. Yates and other Guadalupian (Kazanian) oil fields, US Permian Basin	249
SASSEN, R. Geochemistry of carbonate source rocks and crude oil in Jurrassic salt basins of the Gulf Coast	265
HUSSEINI, M. I. & HUSSEINI, S. I. Origin of the Infracambrian salt basins of the Middle East	279
STONELEY, R. The Middle East Basin: a summary overview	293
ALSHARHAN, A. S. Geology and reservoir characteristics of Lower Cretaceous Kharaib Formation in Giant Zakum Field, Abu Dhabi, United Arab Emirates	299
GRANTHAM, P. J., LIJMBACH, G. W. M. & POSTHUMA, J. Geochemistry of crude oils in Oman	317
PAUL, S. K. People's Democratic Republic of Yeman: a future oil province	329
ZAKI, R., MCDOWELL, H., THREADGOLD, I. & OLDFIELD, O. Sub-salt imaging in the Gulf of Suez	341
HELMY, H. Southern Gulf of Suez, Egypt: structural geology of the B-trend oil fields	353
DOUST, H. Petroleum geology of the Niger Delta (abstract)	365
WEBER, K. J. Niger Delta reservoir geology: historical growth of the sedimentological model and its application to field development (abstract)	367
SELLA, M., TURCI, C. & RIVA, A. Petroleum geology of the 'Fossa Bradanica' (foredeep of the Southern Appenine thrust belt)	369
VAN DER BAAN, D. Zechstein reservoirs in The Netherlands	379
GLENNIE, K. W. & PROVAN, D. M. J. Lower Permian Rotliegend reservoir of the Southern North Sea gas province	399

CONTENTS

FRASER, A. J., NASH, D. F., STEELE, R. P. & EBDON, C. C. A regional assessment of the intra-carboniferous play of Northern England   417

PEGRUM, R. M. & SPENCER, A. M. Hydrocarbon plays in the northern North Sea   441

HEUM, O. R. & LARSEN, R. M. Haltenbanken hydrocarbon province (off-shore Mid-Norway) (abstract)   471

KONTOROVICH, A. E., MANDEL'BAUM, M. M., SURKOV, V. S., TROFIMUK, A. A. & ZOLOTOV, A. N. Lena-Tunguska upper Proterozoic–Palaeozoic petroleum superprovince   473

NESTEROV, I. I., SALMANOV, F. K., KONTOROVICH, A. E., KULAKHMETOV, N. K., SURKOV, V. S., TROFIMUK, A. A. & SHPILMAN, V. I. West Siberian oil and gas superprovince   491

DOLAN, P. Pakistan: a history of petroleum exploration and future potential   503

RAHMANIAN, V. D., MOORE, P. S., MUDGE, W. J. & SPRING, D. E. Sequence stratigraphy and the habitat of hydrocarbons, Gippsland Basin, Australia   525

LAWRENCE, S. R. Aspects of the petroleum geology of the Junggar Basin, Northwest China   545

Index   559

Acknowledgements

On behalf of the Geological Society Petroleum Group and PESGB, we thank the companies and individuals concerned with this volume for their generous support and their time and effort in presenting and publishing these extensive studies on classic petroleum provinces. The authors clearly show that these highly productive hydrocarbon provinces are not only classic petroleum provinces, but also classic geological provinces.

Many papers in this collection were first presented at a Geological Society (in association with the Petroleum Exploration Society of Great Britain) special two-day meeting held at the Royal Commonwealth Society and the Geological Society, Burlington House, London in May 1988. The meeting was truly international in a geographical, exploration and explorationist sense. The conference delegates (and hopefully the readers of the proceedings) are appreciative of the efforts of all the speakers who literally did come from the 'four corners of the world'. The effectiveness of the five sessions of the conference owed much to their chairmen: Richard Hardman; Ian Vann; John Brooks; David Roberts and Jim Brooks, whom we thank. The staff of the Geological Society and PESGB are also thanked for assistance, administration and organisation before and during the meeting.

Compiling the volume has been made relatively easy by the excellent response by most of the authors in providing good manuscript copy soon after the conference. The other authors are thanked for their special efforts — against all sorts of difficulties, work loads, distractions, in-house politics, etc, — that they finally made the publication. Alas, a very few did not make it!

The papers in this volume have all been refereed and some amended and corrected by the authors. The editor is most greatful to all those referees who undertook the onerous task of refereering the papers. The volume has benefited from their comments and suggestions.

This volume has been published 'in-house' by the Geological Society's Publishing House and we thank the staff — particularly David Ogden (Staff Editor) — for all their efforts and patience in the preparation of the book. Much work has contributed to the preparation of the volume, especially by the authors and we hope that readers will benefit in many ways from reading (and re-reading) the volume.

There is much oil and gas still to be produced and found in the classic petroleum provinces and we hope these proceedings will interest and assist you and your company. Our main hope is that the volume will inform everyone of some of the recent advances in petroleum geology and petroleum exploration. If the volume does this it will have been worthwhile.

JIM BROOKS
Glasgow, 1989

Classic petroleum provinces

JIM BROOKS

Brooks Associates, Glasgow G43 2EE, UK

In recent years there has been a proliferation of international meetings and publications with themes such as 'Frontier Areas for Exploration', 'Future Petroleum Provinces' and 'Future Potential for Exploration in Frontier Basins' in the hope that new exploitable hydrocarbon reserves will be discovered. Many of these geologically interesting areas, usually with limited hydrocarbon potential, have been promoted and often over emphasized in efforts to obtain investment monies into the regions. With the 1986 oil-price crash, there has been and will continue to be only limited funds available to explore in these frontier areas: deep water, jungles, mountain terrain, ice-covered lands and remote continental interiors.

The oil industry's present and likely future existence will continue to be focussed in the classic petroleum provinces of the world. Which company would choose not to explore in the mature petroleum provinces (Middle East, North and off-shore West Africa, West and East Texas, Mexico, Venezuela, Siberia, Western Canada, Gulf Coast, Niger Delta, North Sea) if they had the opportunity?

The concept of a petroleum province, composed of one or several sedimentary basins having common geological features and a comparable history was first suggested by Perrodon (1980). It generally includes several petroleum zones.

The classic petroleum provinces are still the world's major producers of hydrocarbons and will continue to be so well into, and through, the 21st century. The world's hydrocarbon supplies, oil company profits and major exploration and production activities will continue to be principally from these mature provinces.

In recent years much new data, new understandings, new interpretations, application of modern concepts, re-evaluation of these mature exploration areas, fields and basins has resulted in the discovery of major new hydrocarbon accumulations. Many of the basins are not only classic petroleum provinces, but also classic geological provinces.

Many basins have been major petroleum producing provinces for a great number of years (e.g. The Middle East and Texas Basins), but others have only recently fully revealed their hydrocarbon treasures (e.g. The North Sea and North Slope Alaska). Other regions, such as Venezuela and Siberia, still continue to surprise with new giant discoveries of oil and gas.

Nehring (1978) and Perrodon (1978) pointed out that some 20 giant petroleum provinces (each containing more than 10×10^9 bbl or 1.6×10^9 m^3 oil) contain over 85% of the total known world oil (i.e. production plus known recoverable reserves of conventional oil). However, these provinces cover only 20% of the total surface of sedimentary basins (Table 1). Among them, the Middle East (the Arabian–Iranian) province contains over 500×10^9 bbl (80×10^9 m^3) conventional oil and accounts for more than half of the world's past production plus known reserves. If heavy oil (API < 20°) and tar sands are included in the world reserve calculations, Western Canada and Eastern Venezuela provinces contain a similar amount of oil to the Middle East. Together these three provinces account for over 75% of the world reserves.

Currently, there are over 500 known giant oil and gas fields in the world containing a total of some 900×10^9 bbl of oil and 3500 TCF of natural gas. These giant fields, distributed throughout the classic petroleum provinces, contain over 1.5×10^{12} bbl of oil equivalent. A nice portfolio to have!

Of these reserves, some 270×10^9 bbl of oil (30% of reserves) and 198 TCF gas (6% of reserves) had been produced by mid-1984 (Carmalt & St John 1986). The remaining reserves, together with recent reported major hydrocarbon discoveries (particularly in Venezuela and Siberia) mean that over 70% of the oil (>600×10^9 bbl) and 95% (>4,000 TCF) of the gas remains to be recovered from giant fields in the major petroleum provinces.

Petroleum-rich provinces occur in a variety of geological settings. Various attempts have been made (see Bally & Snelson 1980; Klemme 1971, 1975; Kingston *et al.* 1983a, b; St John 1980; St John *et al.* 1984; Demaison 1984, 1989) to classify petroliferous basins. These proposed systems attempt to classify worldwide sedimentary basins into specific, as well as general, categories. Hydrocarbon plays/accumulations and basin classification have important relationships in some cases. Basin classification can be useful in that similarities and differences in

From BROOKS, J. (ed.), 1990, *Classic Petroleum Provinces*, Geological Society Special Publication No 50, pp 1–8.

Table 1. *Giant oil provinces of the world (Adapted and modified from Nehring 1978; Perrodon 1978; Bois et al. 1980 and Tissot & Welte 1984)*

Petroleum province	Surface (10^3 miles^2)	Production plus Proven Reserves Oil (10^9 bbl)	Production plus Proven Reserves Gas (TCF)	Oil richness (10^3 bbl/miles^2)	Gas richness (MCF/miles^2)	Total richness (Oil Equivalent) (10^3 bbl/miles^2)	Number of giant fields
Middle East (Arabian–Iraq–Iran)	900	500	750	80	120	110	79
Tampico-Reforma-Campeche	100	70	70	100	110	125	4
West Siberia	1300	60	700	10	70	20	14
Maracaibo	25	60	35	250	230	300	10
US Mid-Continent plus West Texas	400	50	230	20	90	35	16
Gulf Coast plus Mississippi plus East Texas	250	46	360	30	230	70	17
Volga–Ural	200	36	60	25	80	40	8
Sirte–Libya	80	34	32	65	70	75	15
North Sea	200	20	90	15	70	25	14
North Caucasus plus Kopet Dag	60	20	35	55	100	70	9
Niger Delta	60	19	53	50	140	70	6
California	25	19	32	120	210	160	12
North China	–	16	7	–	–	1	
West Canada	500	14	92	5	30	10	8
Orinoco plus Trinidad	50	13	10	40	30	45	8
Sumatra	85	13	18	25	30	30	4
Sahara, Algeria	25	10	125	10	90	20	4
North Slope, Alaska	70	10	10	20	60	30	1
Total 18 Giant Provinces	4550+	1000+	2750+	55 average	103 average	73 average	230
Total World		1100+ 154×10^9 metric tons	3700+				280

Table 2. *Basin classification as defined by Bally & Snelson (Bally & Snelson 1980; St John et al. 1984)*

1. Basins located on the rigid lithosphere, no associated with formation of megasutures
    11. Related to formation of oceanic crust
        111. Rifts
        112. Oceanic transform fault associated basins
        113-OC. Oceanic abyssal plains
        114. 'Atlantic-type' passive margins (shelf, slope & rise) which straddle continental and oceanic crust
            1141. Overlying earlier rift systems
            1142. Overlying earlier transform systems
            1143. Overlying earlier backarc basins of types 321 and 322
    12. Located on pre-Mesozoic continental lithosphere
        121. Cratonic basins
            1211. Located on earlier rifted grabens
            1212. Located on former backarc basins of type 321
2. Perisutural basins on rigid lithosphere associated with formation of compressional megasuture
    21-OC. Deep sea trench of moat on oceanic crust adjacent to B-subduction margin
    22. Foredeep and underlying platform sediments, or moat on continental crust adjacent to A-subduction margin
        221. Ramp with buried grabens, but with little or no block faulting
        222. Dominated by block faulting
    23. 'Chinese-type' basins associated with distal block faulting related to compressional or megasuture and without associated A-subduction margin
3. Episutural basins located and mostly contained in compressional megasuture
    31. Associated with B-subduction zone
        311. Forearc basin
        312. Circum-Pacific backarc basin
            3121-OC. Backarc basins floored by oceanic crust and associated with B-subduction (marginal sea *sensu stricto*)
            3122. Backarc basins floored by continental or intermediate crust, associated with B-subduction
    32. Backarc basins, associated with continental collision and on concave side of A-subduction arc
        321. On continental crust of 'Pannonian-type' basin
        322. On transitional and oceanic crust or 'W Mediterranean-type' basins
    33. Basins related to episutural megashear systems
        331. 'Great basin-type' basin
        332. 'California-type' basin
4. Folded belt
    41. Related to A-subduction
    42. Related to B-subduction
5. Plateau basalts

hydrocarbon-bearing basins can be compared and contrasted. Basin classifications attempt to catalogue and categorize prolific hydrocarbon plays with specific tectonic and depositional events.

The Bally & Snelson (1980) (see also St John *et al.* 1984) basin classification (Table 2) divides basins into three major categories depending upon their tectonic environment.

(i) Basins on rigid lithosphere (Type-1 Basins): basins located on the rigid lithosphere, no association with formation of megasutures. Examples are rifts, oceanic abyssal plains and 'Atlantic-type' margins.

(ii) Perisutural basins on rigid lithosphere (Type-2 Basins): associated with the formation of compressional megasuture. Examples are deep-sea trenches, foredeeps and 'Chinesetype' basins.

(iii) Episutural basins (Type-3 Basins): located and mostly contained in compressional megasuture. Examples are forearc and backarc basins and 'Basin and Range' basins.

Another two basin types have also been proposed: folded-belt (Type-4 Basins) and plateau basalts (Type-5 Basins).

The major classification divisions reflect the geological characteristics of the basin associated with megasutures, which are major compressional plate boundaries (Carmalt & St John 1986).

A number of interesting and potentially useful grouping and relationships of the major petroleum provinces, giant oil fields and hydrocarbon exploration success/failure with basin classification can be made.

Success in finding hydrocarbons using basin classification

Figure 1 uses the Bally and Snelson Classification to evaluate the most and least successful exploration in various types of basins. The relationship is normalised to the basin style which is the most successful for discovering hydrocarbons. Also reported is the percentage success of finding oil in each basin style.

The most successful exploration is in 'Atlantic-type passive margins — overlying earlier back arc basins' (Basin Type 1143). In these basins there has been a 85% exploration success in finding some hydrocarbons — a good striking rate!

There is a > 50% success for finding hydrocarbons in the following types of basins:

Basin Type 1143:	'Atlantic-type' passive margins overlying backarc basins.
Basin Type 23:	'Chinese-type' basins associated with distal block faulting related to compressional or megasuture and without associated A-subduction zone.
Basin Type 221:	Foredeep and underlying platform sediments — ramp with buried graben, but with little or no block faulting.
Basin Type 41:	Folded belt: related to A-subduction.
Basin Type 222:	Foredeep and underlying platform sediments: dominated by block faulting.
Basin Type 1211:	Cratonic basins located on earlier rifted graben.

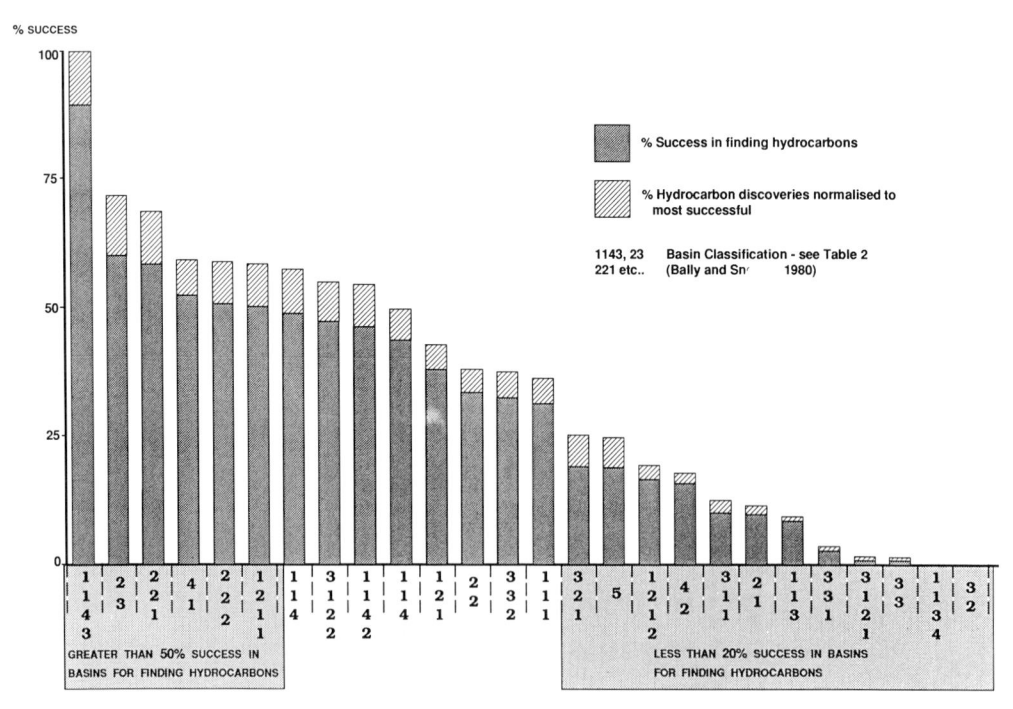

Fig. 1. Hydrocarbon exploration success and failure in different types of basin (using Bally & Snelson's basin classification).

There is a < 20% success for finding hydrocarbons in the following types of basins:
Basin Type 321:   Backarc basins on continental crust of 'Pannonian-type' basin.
Basin Type 5:   Plateau basalts.
Basin Type 1212:   Basins on Pre-Mesozoic continental lithosphere located on former backarc basins.
Basin Type 42:   Folded belt related to B-subduction.
Basin Type 311:   Forearc basins associated with B-subduction.
Basin Type 21:   Deep sea trench on oceanic crust adjacent to B-subduction margin.
Basin Type 113:   Basins located on oceanic abyssal plains.
Basin Type 331:   'Great basin-type' basins.
Basin Type 3121:   Backarc basins floored by oceanic crust and associated with B-subduction.
Basin Type 33:   Basins related to episutural megashear systems.

Distribution of giant oil and gas fields within different basin types

Carmalt & St John (1986) have related the number and distribution of giant fields to the major basin divisions of Bally's classification. Fig. 2 relates the number of giants fields by Bally's classification.

The largest group of giant fields (211 fields) is found in the 'Foredeep and underlying platform sediments' (see Table 2). This group of fields includes the majority of the Middle East Hydrocarbon Province — but not the megafields in the Zagros Foldbelt. The group also includes most of the Venezuelan petroleum province; the Permian; North Slope

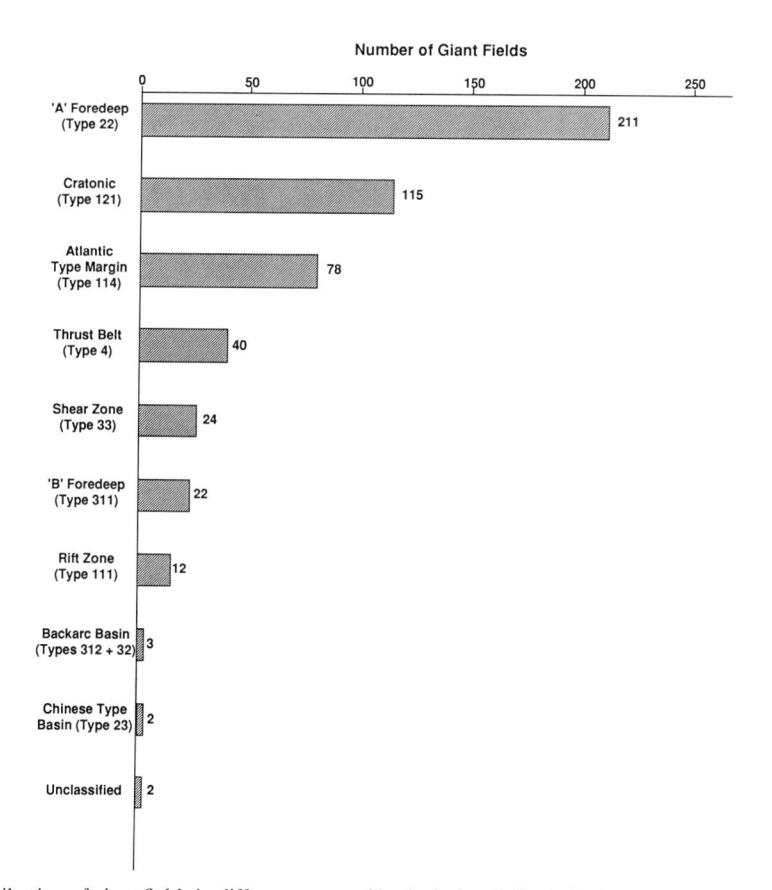

Fig. 2. Distribution of giant fields in different types of basin (using Bally & Snelson's basin classification — see Carmalt & St John 1986).

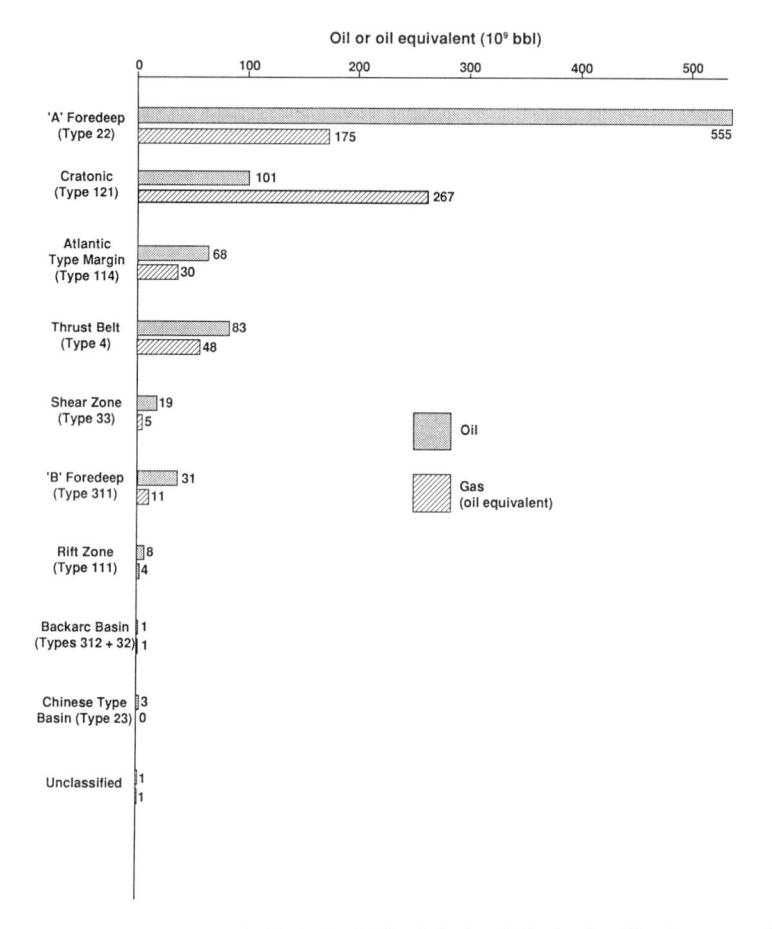

Fig. 3. Oil and gas reserves in giant fields (using Bally & Snelson's basin classification — see Carmalt & St John 1986).

of Alaska and Alberta provinces of North America; and a number of the mega-provinces of the USSR.

Some 115 giant fields been discovered in 'Cratonic basins on the Pre-Mesozoic continental crust'. The major petroleum discoveries of Western Siberia and the North Sea are found in this basin type.

'Atlantic-margin' type basins contain about 78 giant oil fields; and includes the Niger Delta, US Gulf Coast and the Campeche province in Mexico.

Forty giant fields in the Zagros foldbelt in Iraq and Iran account for the significant occurrence of giant fields in 'Thrust Belts'.

Figure 3 shows the distribution of hydrocarbons in giant oil and gas fields using the Bally and Snelson classification (data taken from Carmalt & St John 1986). The amount of gas in the giant fields and types of basin is given in oil equivalent.

The most prolific petroleum provinces are those on the continental crust, but in a mobile zone associated with collision. There are, however, large differences in 'oil richness' between the giant petroleum provinces.

Middle East, Sirte-Libya, Maracaibo, Reforma Campeche, California and North Caucasus petroleum provinces are very rich and have $> 50 \times 10^3$ bbl per mile^2. If heavy oil is included in the richness classification then Western Canada and East Venezuela are included in this category.

All other giant petroleum provinces, including North Sea, Gulf Coast, Western Siberia, Ural-Volga and Niger Delta, show moderate richness from 5 to 50×10^3 barrels per mile^2.

The non-giant petroleum provinces usually

show only moderate richness often $< 5 \times 10^3$ barrels per mile^2.

The known recoverable conventional oil reserves are very much concentrated in giant fields (Nehring 1978). There are between 20 000 and 30 000 fields smaller than 100×10^6 bbl and all these fields only account for about 10% of known oil reserves. However, the major petroleum provinces contain about 920 fields larger than 100×10^6 barrels, which contain about 90% of known oil reserves. Furthermore, 280 giant fields contain 75% of the known oil reserves, and 50% of the world's reserves are contained in only 35 mega-fields within the classic petroleum provinces.

Tissot & Welte (1984) point out that there are some differences in the degree of areal concentration of oil reserves and also in the role of giant fields from one petroleum province to another. A concentrated distribution of oil reserves is defined (Perrodon 1980); "as the occurrence of the major fraction of reserves in a small number of fields". There are large differences in size distribution between fields in major petroleum provinces.

In North Slope, Alaska, the Prudhoe Bay giant field contains the bulk of the known oil reserves in that area.

The Romashkino field contains > 35% of the known oil reserves in the Ural–Volga province of the USSR.

In the Western Siberia petroleum province, the Samotlor field contains 40% of the known oil reserves and the Urengory field has about 30% of the gas reserves in the province.

In the Algerian Sahara province, the Hassi Messaoud field contains > 65% of the oil reserves and the Hassi R'Mel field has > 30% of the known gas reserves.

Concentrated distribution of oil and gas reserves in a petroleum-rich province is controlled by the basin style and special geological situations. Often a stratigraphic or structural feature will have favoured large drainage areas such as a regional uplift, a regional unconformity with weathered beds or with a basal sandstone acting as major migration pathway. Tissot & Welte (1984) illustrate an extreme case of concentrated distribution with the huge heavy oil accumulations of Eastern Venezuela and Western Canada, where the possible conjunction of a regional unconformity for migration and an *in situ* degradation is responsible for sealing off and trapping the vast reserves.

It is essential that oil and gas exploration in frontier provinces continues and many of these regions are under-explored or still remain to be fully evaluated and tested. However many explorationists predict that as much as 50% of the undiscovered oil and gas lies in the classic petroleum provinces. Many of the sedimentary basins of the Middle East and Siberia have not been fully evaluated and explored. Recent discoveries of giant oil fields in Venezuela (this volume), support the hypothesis of the vast hydrocarbon potential still remaining in many of the classic petroleum provinces.

The exploration potential of the mature North Sea oil and gas province has also been generally underestimated and is substantially greater than most commentaries have suggested (Brooks & Nicholson 1988). There has been a lack of appreciation of the rapid and fundamental improvements taking place in our understanding of sedimentary basins. Recently, new concepts in structural geology, basin development, petroleum geochemistry and reservoir geology have been applied in the evaluation and re-evaluation of North Sea oil and gas fields and discoveries. North Sea sedimentary basins have good potential for further significant discoveries. It is concluded that the UK's known remaining oil and gas reserves should last for up to 30 and 50 years respectively.

Exploration drilling is likely to continue to be more selective in the future, with more emphasis on high gain for the risk involved. Such economics and philosophy will tend to concentrate exploration in the mature petroleum provinces. The challenge to companies is to maintain commitment to exploration teams taking imaginative overviews of basins rather than pursuing routine local mapping and remapping exercises.

There should be many future hydrocarbon discoveries made in the classical petroleum provinces. A reliable assessment of future discoveries must be based on thorough, sustained and imaginative mapping controlled by basin models which reflect the vast data base and experience gained over the years by the industry. As new wells are drilled and regions re-evaluated, new ideas are developing to replace our old hypotheses.

In the classic petroleum provinces, there is no fixed number of prospects known whose exhaustion will steadily diminish the chance of a giant field. In many of the mature petroleum provinces, new concepts, interpretation and modelling have barely started towards the highly detailed knowledge of the basins required for successful exploration.

Innovation in exploration thinking and introduction of new techniques can maintain the classical petroleum provinces oil and gas production and reserves well into the next century.

References

BALLY, A. W. & SNELSON, S. 1980. 'Facts and principles of world petroleum occurrence: realms of subsidence' In: MIALL, A. D. (ed.) Facts and Principles of World Petroleum Occurrence Canadian Society of Petroleum Geologists Memoir **6**, 9−94.

BROOKS, J. & NICHOLSON, J. 1988. 'Undeveloped Oil and Gas Potential on the UK Continental Shelf and some comments on the future exploration' In OPEC and Future World Oil Supply; Dallas Conference, 1987, ISEM, Southern Methodist University, 1−8.

CARMALT, S. W. & ST JOHN, B. 1986. 'Giant Oil and Gas Fields' In: HALBOUTY, M. T. (ed.) Future Petroleum Provinces of the World AAPG, Tulsa, OK, 11−53.

DEMAISON, G. 1984. 'The Generative Basin Concept' In: DEMAISON, G. & MURRIS, R. J. (eds) Petroleum Geochemistry and Basin Evaluation AAPG Memoir **35**, 1−14.

—— 1989. 'Genetic Classification of Petroleum Basins' Geological Society/AAPG Video Basin Development and Petroleum Exploration.

KINGSTON, D. R., DISHROON, C. P. & WILLIAMS, P. A. 1983a. 'Global Basin Classification System' AAPG Bulletin, **67**, 2175−2193.

——, —— & —— 1983b. 'Hydrocarbon Plays and Global Basin Classification', AAPG Bulletin, **67**, 2194−2198.

KLEMME, H. D. 1971. 'The Giants and the Supergiants' Oil and Gas Journal **69(9)**, 85−90; **69(10)**, 103−110; **69(11)**, 96−100.

—— 1975. 'Giant Oil Fields related to their geological setting − a possible economic guide to exploration'. Bulletin Canadian Petroleum Geologists, **23**, 30−66.

NEHRING, R. 1978. Giant Oil Fields and World Oil Resources. Rand Corporation Report R-2284 CIA.

PERRODON, A. 1978. 'Coup d'oeil sur les provinces geantes d'hydrocarbures'. Revue del'Institut Francois du Petrole, **33**, 493−513.

ST JOHN, B. 1980. Sedimentary Basins of the world and giant hydrocarbon accumulations AAPG Special Publication.

——, BALLY, A. W. & KLEMME, H. D. 1984. Sedimentary Provinces of the world − hydrocarbon productive and non-productive AAPG Special Publication.

TISSOT, B. P. & WELTE, D. H. 1984. Petroleum Formation and Occurrence. Springer, New York.

The Venezuelan hydrocarbon habitat

K. H. JAMES

Shell UK Exploration and Production, Shell-Mex House, Straud, London WC2R 0DX, UK. (Shell Internationale Petroleum Moatschappij B.V., Postbus 162, 2501 AN The Hague, Netherlands)

Abstract: Venezuela has produced some 40 billion barrels of oil. Remaining recoverable reserves are currently in the order of 60 billion barrels of oil and 92 trillion cubic feet of gas. The Orinoco Heavy Oil Belt, the world's largest known oil accumulation, has some 1.2 trillion barrels of oil-in-place. Only part of the 267 billion barrel, ultimate recoverable of this belt is included in the current reserves.

Most of Venezuela's hydrocarbon reserves are located in the Maracaibo, Eastern Venezuela, and Barinas−Apure Basins, and most are derived from Upper Cretaceous, marine source rocks. Tertiary, terrestrial source rocks have also contributed hydrocarbons in the Maracaibo and Eastern Venezuela basins, and Tertiary marine and terrestrial source rocks have generated hydrocarbons in the Falcon and offshore Venezuelan basins.

The hydrocarbon habitat has been profoundly influenced by Mesozoic−Recent plate-tectonics. Many of the structural traps are the consequence of Caribbean−South America dextral relative movements. Some of these structures, mainly in the northwest, were enhanced by stress generated in the Pacific. The northern part of the country has suffered considerable shortening and inversion. The resultant mountain chains limit the modern hydrocarbon basins. Reconstruction of late Cretaceous source rock palaeogeography, of Late Cretaceous to Palaeocene kitchens (now metamorphosed), and of Eocene to Recent kitchens, shows that the latter are remnants of a much larger petroliferous province and provides an explanation for the great abundance of hydrocarbons in the area.

Significant volumes of hydrocarbons remain to be discovered in extrapolations of traditional and developing plays.

This paper summarizes the structural and stratigraphic traps where abundant hydrocarbon reserves are located in Venezuela. It also describes the evolution of a plate-tectonic setting which allowed the deposition of widespread, prolific source rocks and the diachronous and continuing evolution of kitchen areas. The latter two subjects explain why Venezuela has such enormous hydrocarbon reserves. The regional overview highlights the main hydrocarbon plays and suggests areas for their possible extrapolation.

Regional setting

The main Venezuelan hydrocarbon provinces are the Maracaibo, Barinas−Apure, and Eastern Venezuela Basins (Fig. 1). They lie north and northwest of the Guayana Shield. Coastal and Interior Ranges form the northern limit to the Eastern Venezuela Basin, and the Merida Andes and the Perija Mountains form the northwestern limits to the Barinas−Apure and Maracaibo Basins. The El Baul High, a basement swell, separates the Barinas−Apure and Eastern Venezuela Basins. These provinces contain sedimentary sequences of Late Cretaceous to Recent age and their hydrocarbons are derived principally from Upper Cretaceous, marine source rocks. The Maracaibo Basin is traditionally the most important producing province, but the Eastern Venezuela Basin, with the world's largest known hydrocarbon accumulation, will assume greater importance in the future.

Further north lie the onshore Falcon Basin and the offshore Carupano, Cariaco, Bonaire, La Vela Bay, and Gulf of Venezuela basins. These are Tertiary to Recent in age and any hydrocarbons present are derived thermally from Tertiary source rocks or are biogenic in origin. Only the Falcon Basin has produced to date.

Hydrocarbon reserves and accumulations

Venezuela has produced more than 40 billion barrels of oil, most of which has come from the Maracaibo Basin. Official recoverable reserves are in the region of 60 billion barrels of oil and 92 trillion cubic feet of gas. Of these, 20×10^9 barrels are located in the Maracaibo Basin, 40×10^9 barrels are in the Eastern Venezuela Basin and 700×10^6 barrels are in the Barinas−Apure Basin. The reserves in the Eastern Venezuela Basin only partly include the 267×10^9 billion barrels currently estimated to be

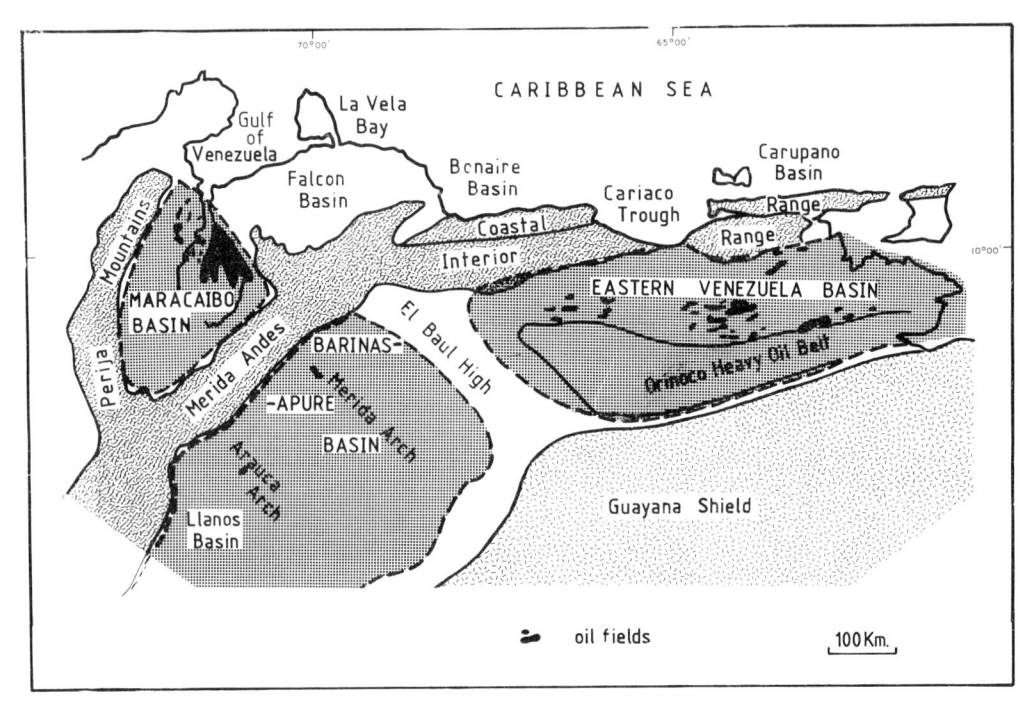

Fig. 1. The principal petroliferous basins (shaded) of Venezuela contain hydrocarbons derived from Upper Cretaceous source rocks. Other basins, further north, contain lesser amounts of hydrocarbons, derived from Tertiary source rocks or from biogenic activity.

ultimately recoverable from the 1.2 trillion barrel STOIIP of the Orinoco Heavy Oil Belt. Most of the gas reserves are thermal in origin and they lie principally in the northern part of the Eastern Venezuela Basin (66 TCF) and in the Maracaibo Basin (16 TCF). Ten TCF, in the Carupano Basin, may be of biogenic origin. Six billion barrels of recoverable oil have been discovered during the last five years, and there are good reasons to anticipate further, large discoveries.

Gonzalez de Juana *et al.* (1980) provide the most comprehensive, published compilation of Venezuelan hydrocarbon accumulations. These are shown, in updated form, in Fig. 2. The present paper does not treat these accumulations in detail, but focuses upon the main plays, summarized in Figs 3 & 4, and upon their plate-tectonic and structural settings.

Maracaibo Basin

In general, this basin contains heavy oil at shallow levels and lighter oil and gas at deeper levels. In the northeastern, Cabimas to Barua–Motatan fields, production comes from Tertiary sandstones. In the central and western lake fields it comes from Tertiary sandstones and Cretaceous limestones, and in the western, Mara to Alturitas fields it comes from Tertiary sandstones, Cretaceous limestones, and, locally, basement. Stratigraphic trapping (sand pinchout and truncation, heavy oil plugs) is important in the shallow, northeastern fields. Elsewhere, the main traps are structural closures associated with north–south trending, sinistral, strike-slip faults. Fracture porosity and permeability, generated by faulting, allow production from otherwise tight Cretaceous limestones and basement.

Most of the hydrocarbons originate from Upper Cretaceous, marine source rocks. Minor terrestrial contributions (?Palaeocene source) occur in the southwest. Crudes shallower than 5000 feet, in the northeast, are strongly biodegraded.

The Maracaibo Basin is at a mature stage of exploration in its northern half, but in the south, the North Andean Foredeep, adjacent to the Merida Andes, is poorly explored. Recoverable reserves in this basin recently increased by the discovery of 1.2×10^9 bbls of light crude trapped against E–W to NW–SE trending normal faults that interrupt the northward, monoclinal rise of the area between the traditional Ceuta and Barua highs (Roberto *et al.* 1990). There are

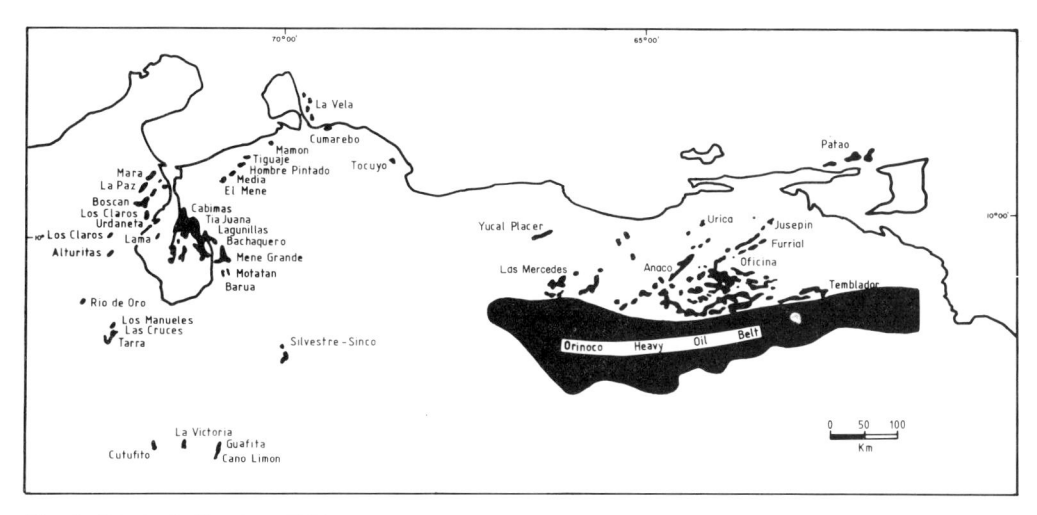

Fig. 2. Principal oil and gas fields.

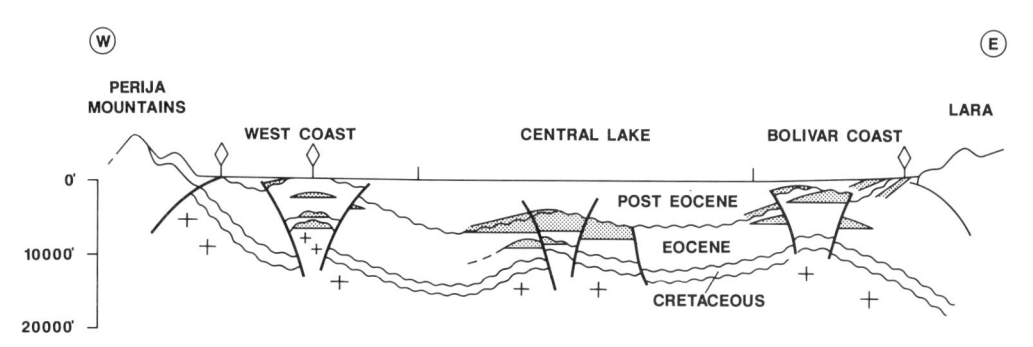

Fig. 3. Schematic cross section of the Maracaibo Basin, showing the main hydrocarbon plays.

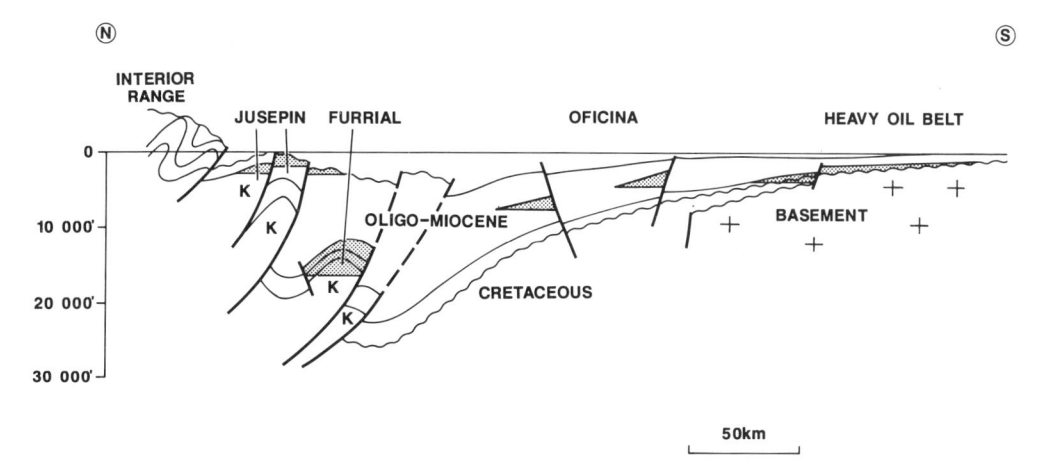

Fig. 4. Schematic cross section of the Eastern Venezuela Basin, showing the main hydrocarbon plays.

good expectations of further discoveries in the same area.

Eastern Venezuela Basin

This basin contains light oil and gas in the north and centre, and heavy oil or tar in the south. The reservoirs are provided by Tertiary and Cretaceous sandstones. The basin continues via the Gulf of Paria and Trinidad to the continental shelf east of Trinidad (Perez & Tarache 1985). A recent review of the basin and its prospectivity, excluding the Orinoco Heavy Oil Belt, was provided by Acosta de Moreno (1988) while the latest synopsis of the Orinoco Oil Belt, the world's largest known hydrocarbon accumulation, is provided in papers by Isea (1987), Kiser (1981, 1987), Martinez (1987), Santos & Frontado (1987), and Vega & de Rojas (1987).

The Jusepin and El Furrial trends, in the north of the basin, are associated with the roughly east-west trending thrust front of the Interior Ranges. The deeper of these two, the El Furrial trend, is the location of recent and continuing giant discoveries (Aymard & Pereira 1990; Carnevalli 1988). An element of stratigraphic trapping is provided in the Yucal–Placer Field by southward shale-out of Tertiary sandstones (Aymard *et al.* 1985).

Fields of the Anaco area are located in a narrow zone of en echelon structures associated with the northeast trending, dextral, strike-slip Anaco Fault (Fig. 5). In the Las Mercedes and Oficina fields trapping is provided by two sets of normal faults. One set (ENE–WSW) parallels the southern hinge line of the basin; the other (NW–SE) is the extensional strain resulting from dextral, strike-slip movement along the Caribbean–South American plate boundary (see later). Recent studies of the latter areas suggest that stratigraphic trapping may also be important (Rabasso-Vidal 1985; Zeuss 1985). Combination traps (faulting and sand shale-out) are responsible for the fields of the Temblador area.

The Orinoco Heavy Oil Belt is dominated by stratigraphic trapping, formed by onlap of sands onto basement and by heavy oil plugs. The belt, 460 km long and up to 40 km wide, includes six main producing areas, two of which were yielding 80 000 barrels per day at the end of 1985. Most of the oil is trapped in Miocene sandstones. Although it lies at shallow levels, there are no surface indications. Zamora *et al.* (1988) suggest that the belt may extend below the poorly explored, swampy, southern part of the Orinoco delta, where a further 25 to 226 × 10^9 bbls of oil-in-place may be trapped.

Most of the hydrocarbons in this basin are derived from Upper Cretaceous, marine source

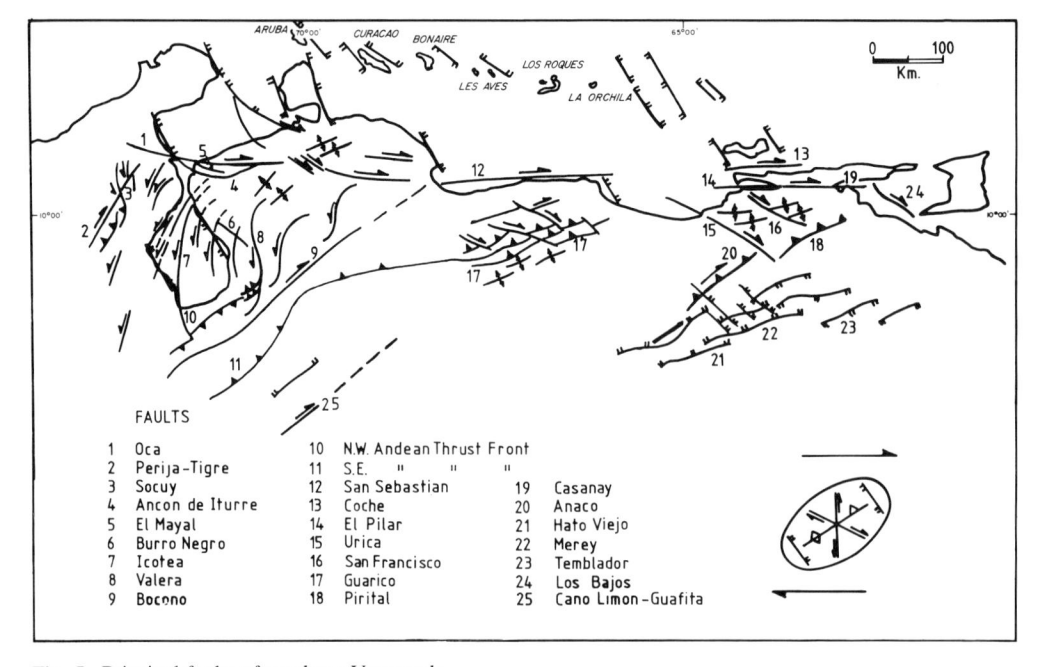

Fig. 5. Principal faults of northern Venezuela.

FAULTS

1	Oca	10	N.W. Andean Thrust Front		
2	Perija-Tigre	11	S.E. " " "		
3	Socuy	12	San Sebastian	19	Casanay
4	Ancon de Iturre	13	Coche	20	Anaco
5	El Mayal	14	El Pilar	21	Hato Viejo
6	Burro Negro	15	Urica	22	Merey
7	Icotea	16	San Francisco	23	Temblador
8	Valera	17	Guarico	24	Los Bajos
9	Bocono	18	Pirital	25	Cano Limon-Guafita

rocks, but Miocene, terrestrial source rocks have provided some of the oils in the basin centre (Cassani et al. 1988). As in the Maracaibo Basin, shallow oils, in the Heavy Oil Belt and in the Temblador and southern Oficina fields, are biodegraded.

Barinas−Apure Basin

This is the least explored of Venezuela's main hydrocarbon provinces. The most comprehensive structural analysis is based upon aeromagnetic data calibrated by seismic and well data (Young 1988). It shows that many structural elements parallel the adjacent Merida Andes and so probably result from similar stress (see later). In addition, there are important east−west trending structures, in the north and centre of the area, which, according to this paper, would be related to the plate boundary further north. This province is contiguous with the better known Llanos Basin of Colombia, which has the same general form (see, for example, Vasquez (1988)).

The Silvestre−Sinco fields are traditionally ascribed to anticlinal trapping. Their location above the hinge line of the basin suggests that normal faults may also play a trapping role, as they do in the Las Mercedes and Oficina fields of the Eastern Venezuela Basin. Reservoirs are provided by Tertiary and Cretaceous sandstones.

Important new discoveries in the same reservoirs along the Colombia−Venezuela border are located in en echelon structures developed where the dextral, strike-slip Cano Limon−Guafita fault system (Fig. 5) crosses a basement swell — the Arauca Arch (Chigne & Hernandez 1990). Hydrocarbons in this basin are derived from Upper Cretaceous marine source rocks (Cassani et al. 1988).

Mesozoic−Recent geological history: a summary

The sedimentary section of interest to the hydrocarbon geologist in Venezuela ranges in age from the Jurassic to the Late Tertiary. It rests upon Precambrian to Mesozoic, igneous or metamorphic basement.

Jurassic red beds and volcanics accumulated in grabens that formed in association with the opening of the Central Atlantic. In the early Cretaceous, a broad, northward tilted peneplain formed along the northern and northwestern flanks of the Guayana Shield. It became the site of mid−late Cretaceous, sandy deposition in the south, and lagoonal or marine, carbonate and shale deposition in the north. Venezuela's most important source rocks formed in this setting.

Further to the north lay an oceanic province. Late Cretaceous to early Palaeogene plate convergence resulted in southward obduction of oceanic material that came to lie in flysch troughs south of a thrust front. Mesozoic sediments flooring the troughs became metamorphosed and later resurfaced in northern uplifts that migrated southwards through time, along with their complementary foreland basins.

By the end of the Eocene the Caribbean−South American Plate boundary had become dominated by east−west, dextral transpression. In northcentral and northeastern parts of Venezuela, Oligocene to Recent sedimentation followed a general pattern of regression towards the east, with craton-derived paralic or continental, sandy deposits in the south and west and marine to continental shales and sands in the north, adjacent to the transpressional uplifts of the Coastal/Interior Ranges. In contrast, northwestern Venezuela has been thrust northwards, since at least the late Eocene, crossing the dextral plate boundary and developing major pull-apart basins in the process. These basins have been the site of lagoonal to marine, mainly clastic deposition.

Plate tectonics

The plate tectonics of the Caribbean region are complicated and the literature contains many models for the area's development. The subject receives detailed discussion here, for a major thesis of this paper is that the hydrocarbon habitat of northern South America cannot be properly understood without reference to its plate-tectonic setting.

Mesozoic plate tectonics

Figure 6 shows northern South America in relation to the neighbouring Caribbean, Atlantic, Cocos and Nazca plates.

The origin of the Caribbean Plate is still being debated. Many authors (e.g., Sykes et al. 1982; Pindell & Dewey 1982; Burke et al. 1984; Duncan & Hargreaves 1984; Ghosh et al. 1984; Mattson 1984; Speed 1986; Bouysse 1988, Ross & Scotese 1988) derive the plate from the Pacific area, often via a complicated series of movements. The Caribbean would thus be a non-subducted remnant of the pre-Tertiary, eastern Pacific, Farallon Plate. Donelly (1975), however, concluded that the Caribbean can best be explained by the growth of Mesozoic basins between the Americas as they separated from

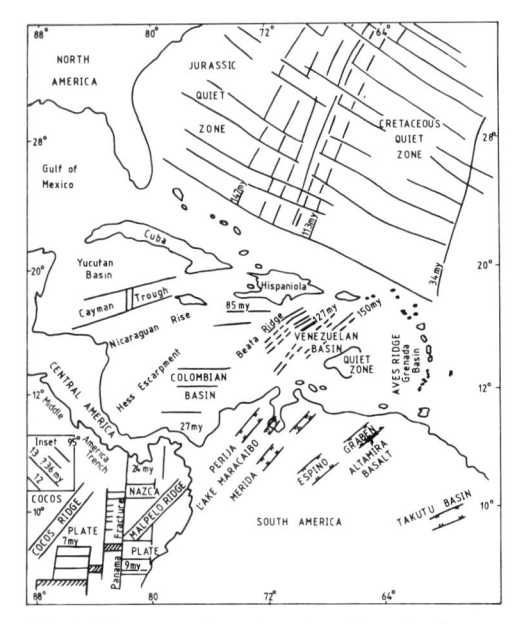

Fig. 6. Plate-tectonic setting of northern South America. Magnetic anomalies and spreading ridges are shown on the adjacent oceanic plates; Jurassic grabens are shown in South America.

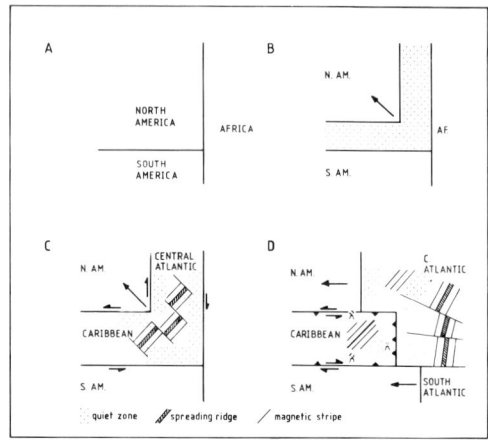

Fig. 7. Schematic illustration of the evolution of the (eastern) Caribbean Plate from A, Gondwana, through B and C, opening of the Central Atlantic and Caribbean oceanic areas, to D, isolation of the Caribbean Plate when the South Atlantic opened.

Africa at different times and rates. Similarly, Fox & Heezen (1975) suggested that a segment of the mid-Mesozoic, mid-oceanic ridge must have passed through the Caribbean region, into the Pacific, when North America moved northwest relative to Africa and South America.

Because the Central Atlantic began to open in the early Jurassic (Pitman & Talwani 1972) and the South Atlantic in the early Cretaceous (Larson & Ladd 1973), North America must have moved sinistrally relative to South America during the intervening period (Fig. 7). The extensional strain resulting from such relative displacement would have trended NE–SW, which is the direction of a symmetrical set of magnetic anomalies, attributed to the interval 153–127 Ma (Fig. 6), mapped in the Venezuela Basin (Ghosh et al. 1984). These considerations were argued by James (1985) as indicating in situ (between North and South America) formation of the eastern Caribbean Plate. The Colombian Basin, west of the NE–SW trending Beata Ridge, has magnetic anomalies that trend E–W and age northwards from 27 to 85 Ma (Christofferson 1976), continuing the sequence of anomalies on the Nazca Plate (Fig. 6). This part of the Caribbean, therefore, does seem to have had Pacific origins.

The stress field responsible for the opening of the Venezuela Basin also generated late Jurassic – early Cretaceous graben systems on the northern flank of the South America Plate (James 1985) in the Perija, Lake Maracaibo, Merida, Espino and Takutu areas (Crawford et al. 1985; Feo-Codecido et al. 1984; Maze 1984). Intermediate to basic, andesitic lavas and pyroclastics in the Perija area (Moticska 1975) are of mid–late Jurassic age (Odreman & Benedetto 1977). A middle–upper Jurassic basaltic body, trending N60°E in the Espino Graben, has been interpreted as a mantle-originated, rift extrusive (Moticska 1985). We shall see later that these extensional faults have been reactivated as dextral wrenches, from the Palaeogene onwards, with important consequences for northwestern Venezuela.

Interaction of the Caribbean and South American Plates

Maresch (1974) proposed that northern South America was the site of an Atlantic, rift-type margin in the mid-Mesozoic. A southward transgressive sedimentary sequence accumulated above the granitic basement of the continental margin while rift-associated, basic volcanics (now seen on the islands of Aruba, Curacao and Margarita, off the Venezuelan coast) formed further to the north. An island arc is supposed to have approached from the northwest, colliding obliquely (west to east) with the continent during the late Cretaceous, causing uplift in the north and generating a

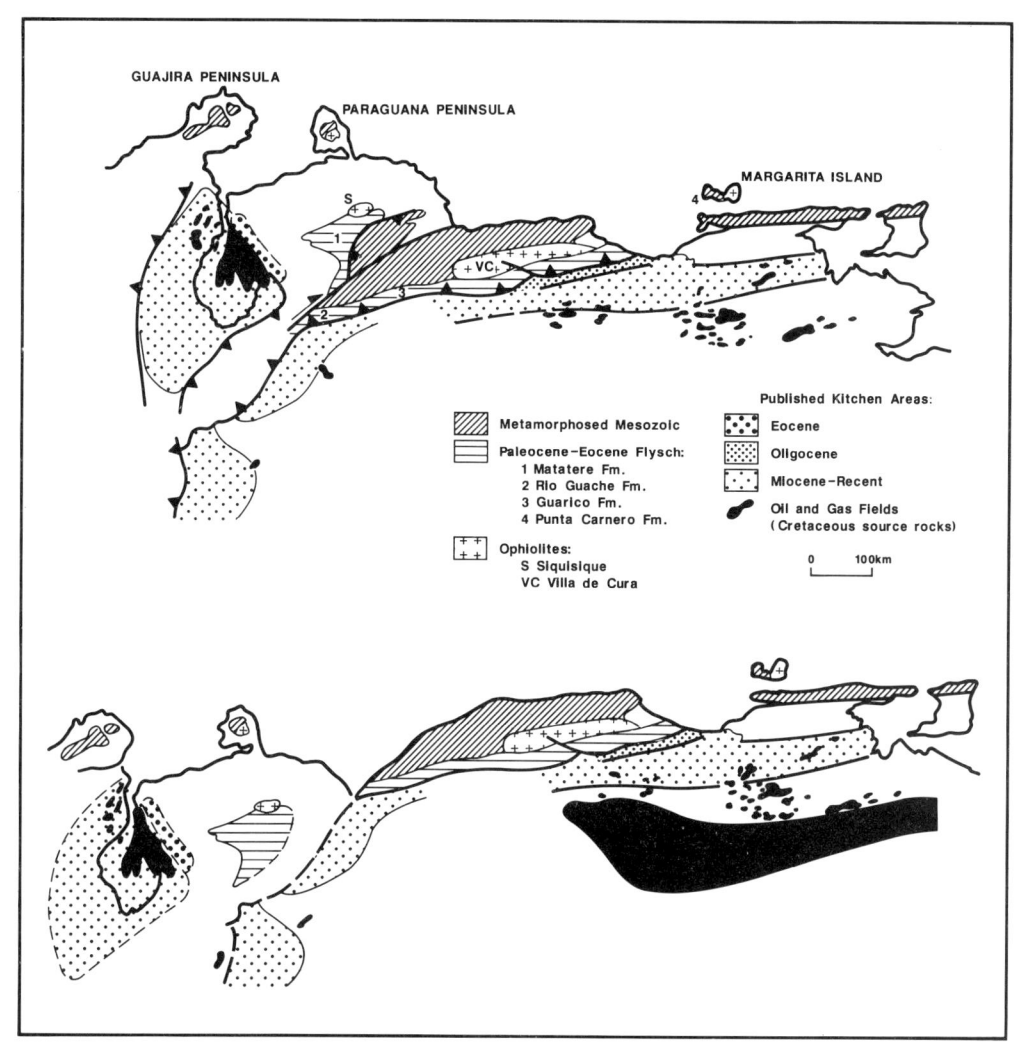

Fig. 8. Above, present-day disposition of metamorphosed Mesozoic rocks, Paleocene flysch basins, with associated ophiolitic nappes, the main oil fields, and the identified hydrocarbon kitchens. Below, the same elements in simple, east–west disposition, after restoration of displacements that affected the northwestern part of the country.

southward migrating, late Cretaceous to Eocene flysch trough to the south. Restriction of the sea between the approaching arc and the continent was suggested by Maresch as the reason for the euxinic character of part of the upper Cretaceous Caracas Group (see section on Cretaceous source rocks). High-pressure, low-temperature metamorphism of this unit, which forms part of the Venezuelan Coastal Ranges, has been attributed to obduction (Beets et al. 1984) during which the island–arc volcanics of the Villa de Cura Nappe (Fig. 8) came to rest on Palaeocene–Eocene flysch deposits of the Interior Ranges. The Siquisique ophiolites, further west, in the southern part of the Falcon Basin, were similarly emplaced (Bellizia et al. 1972). The vertical succession of these allochthonous units is interpreted to progress from continental margin, through oceanic crust, to island arc volcanics (Stephan et al. 1980, Wadge & Macdonald 1985).

According to Bell (1972), the Caribbean Plate assumed dextral movement relative to South America in the Eocene. Eocene to Recent volcanic activity in the Lesser Antilles results from subduction of the Americas Plate below the

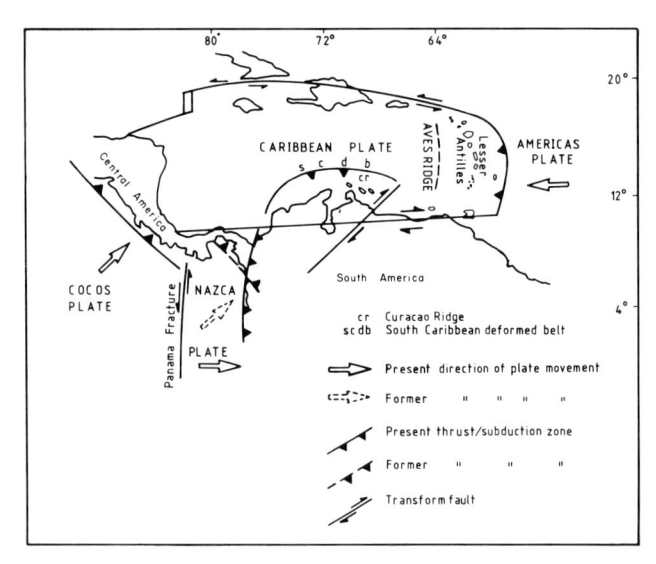

Fig. 9. Relationship of the Caribbean, South American and southeastern Pacific Plates; southern limit of the Caribbean Plate as indicated in Fig. 11. Northeasterly directed compression, arising from spreading in the Pacific area, generated the Central American isthmus and reactivated NE–SW trending, Mesozoic extensional faults in northwestern South America as Tertiary dextral wrenches. The northernmost part of the latter area has decoupled and is overriding the Caribbean Plate.

eastern edge of the Caribbean Plate. These volcanics therefore provide an indirect record of the relative movements of North and South America past the Caribbean Plate, sinistral in the north and dextral in the south (Fig. 9). The Aves Ridge, parallel to and west of the Lesser Antilles, is an abandoned island arc of late Cretaceous age (Bouysse et al. 1985). It surely indicates that the relative plate movements described above have operated since at least that time. Extending the argument further, the movements logically began when the Caribbean Plate assumed its independent identity as the South Atlantic began to open in the Early Cretaceous.

Biju-Duval et al. (1982) believe that recent, slight convergence between North and South America is responsible for transpression along the northern and southern margins of the Caribbean Plate. If such convergence had been more severe in the past, then the Cretaceous–Palaeogene island arc activity, collision, obduction and metamorphism along the northern part of Venezuela would have a simple explanation. Sibuet & Mascle (1978), using the fit of the trends of equatorial fracture zones and magnetic lineations, deduced that the early phase of South Atlantic opening (127–79 Ma) involved rotation of the South American and African plates about a pole located 18 to 24°N and 9 to 16°W. So South America apparently began a somewhat northwesterly movement at the time when Caribbean Plate generation ceased, and it continued moving in this direction until the late Santonian. Thereafter, the pole of rotation shifted many degrees to the northwest and South America assumed a more westerly drift. The early rotation would account for the Cretaceous compressive interaction between the Caribbean and South American Plate and it would explain why that compression migrated eastwards through time.

Tertiary–Recent plate tectonics

Modern earthquake activity in the Venezuelan Caribbean region is mainly associated with the Antilles Arc in the east (Fig. 10). Seismicity also occurs along the straight, east–west coastlines of northcentral and northeastern Venezuela where a major fault system exists (see Fig. 5: the San Sebastian, Coche, El Pilar, and Casanay Faults). The movement along this system is dextral strike-slip (transpressive) at a rate of one to two centimetres per annum (Molnar & Sykes 1969; Minster & Jordan 1978) and it is generally accepted that Caribbean–South American Plate boundary is here well defined.

In northwestern Venezuela and northern Colombia the coastline bulges convex to the north and, as Pennington (1981) noted, seis-

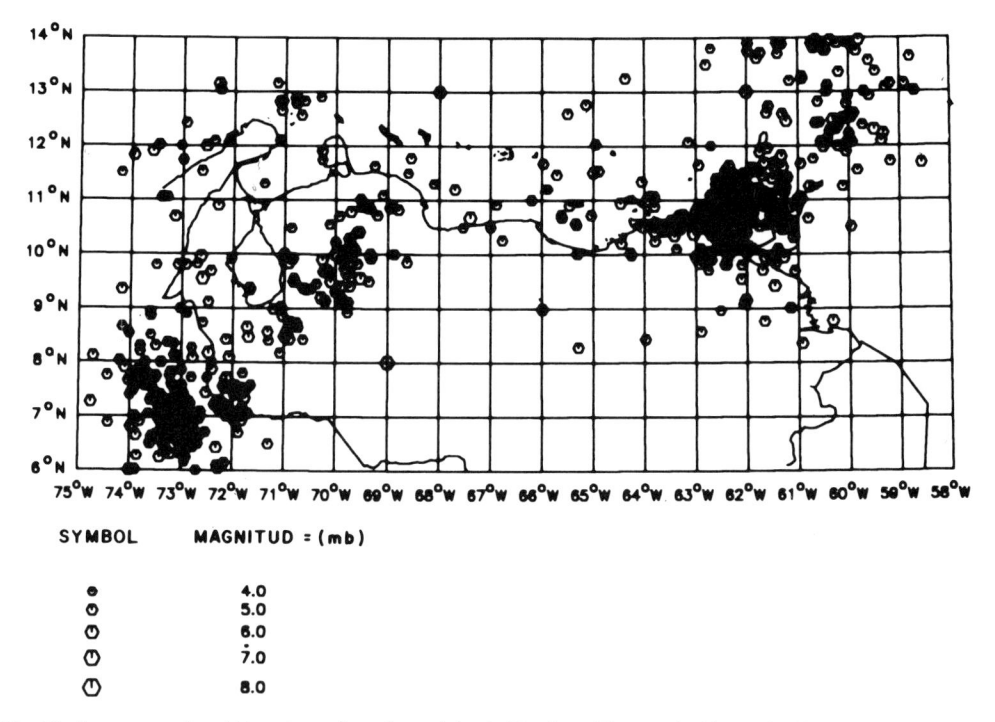

Fig. 10. Instrumental and historic earthquake activity in Northern Venezuela (shown by kind permission of E. Gajardo, Intevep, Venezuela).

micity fails to define the plate boundary between the Andean region and the Caribbean. Earthquake activity is concentrated along the dextral, Bonoco Fault (Fig. 5) of the Merida Andes, where the rate of displacement has been estimated at 0.9 cm a year. Neotectonic studies of this northwestern region indicate that the Bocono Fault is the most active element, though earthquake records and neotectonics both show that other faults, such as the Oca–Ancon system, are also active (Soulas 1987). Many authors regard the N55°E trending Bocono Fault as the Caribbean–South American plate boundary in this area and Eva *et al.* (1989) make the statement 'It is clear that geologically speaking the southern Caribbean–northern South America boundary extends in the west as far south as northern Ecuador'. This idea seems to be geometrically unsound for any plate moving northeastwards along the fault would have to make a turn of some 37° in order to comply with the approximately easterly movement of the Caribbean Plate at the fault's northern limit. Instead, it is held here that the area northwest of the Bocono Fault is a northward moving part of the South American plate that is progressively crossing the east–west plate boundary to override the Caribbean Plate. The same conclusion has been reached before, for example, by Burke *et al.* (1984) and Dewey & Pindell (1985). Seismic sections presented by Silver *et al.* (1975), Ladd *et al.* (1984) and Lehner *et al.* (1984), show reflectors at the southern limit of the Venezuela Basin dipping below deformed sediments of the Curacao Ridge. This ridge dies out in the east at the point where it is intercepted by the extrapolated Bocono Fault (Fig. 9); to the west the deformation continues around the northern offshore of Colombia as the South Caribbean Basin (Case 1974a).

To support this model of overriding, we have to find indications that the Caribbean–South American Plate boundary extends westward from its obvious expression along the east–west coastal segments of Venezuela. Figure 11 shows a series of Late Tertiary–Recent basins that line up across the northern margins of South America. Schubert (1979, 1982) discussed the origins of two of these, the Lake Valencia Basin and the Cariaco Trough, in relation to dextral wrench faults and more recently (1985, 1986) he discussed a series of pull-apart basins associated with dextral displacements along the Bocono and El Pilar faults. It is suggested here that all the basins shown in Fig. 11, the lake of the Maracaibo Basin included, are pull-apart fea-

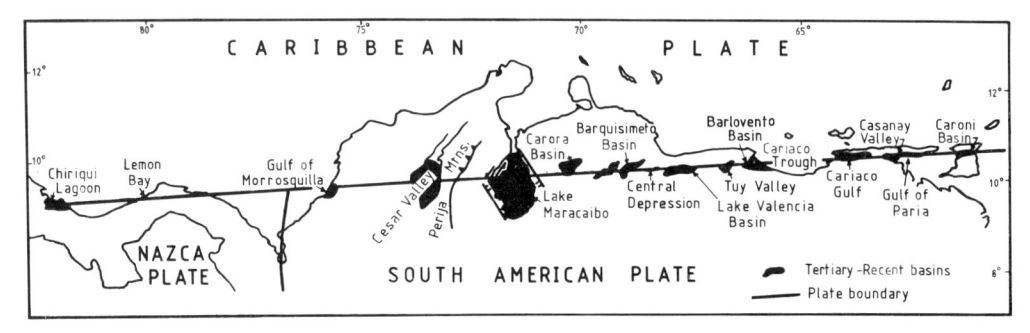

Fig. 11. Late Tertiary to Recent pull-apart basins along the north of South America and Panama indicate the southern boundary to the Caribbean Plate.

tures, and that they are the most recent expression of the plate boundary, as shown in Fig. 9. Older (Eocene to Recent) pull-apart basins in the offshore region (see p. 11) are expressions of earlier plate–plate interaction.

Support for the model also comes from comparison of the geology of northwestern Venezuela with that of the north-central and northeastern areas (Figs 12–14). This comparison is most striking when Fig. 12 is superposed upon Fig. 14. Metamorphic Mesozoic rocks are at or near the surface in the north of these areas while to the south, sedimentary sections thicken and become stratigraphically wider in range. The Cariaco Trough, Margarita Island and the Carupano Basin are analogues of the Urumaco Trough, Paraguana Peninsula, and La Vela Bay. The mainland parts of both areas are structurally characterized by east–west dextral faults, by NW–SE dextral synthetics to these faults, and by folds generated by the latter faults. These strong similarities suggest that the Falcon–Paraguana area originally was a westward continuation of the northcentral and northeastern parts of Venezuela and that it has been offset to the north by displacement along the Bocono Fault.

Finally, seismic refraction studies recently carried out in the Maracaibo Basin indicate that northward crustal thinning begins just north of the boundary suggested in Figs 9 & 11 (Gajardo & Nicolle 1986, Gajardo et al. 1990).

Cause and timing of northeastward movement of Northwestern Venezuela

The driving force for the northeasterly movement of northwestern Venezuela lies in the eastern Pacific area where the Cocos Plate and the northern part of the Nazca Plate are moving N70°E in response to spreading along both the N–S trending East Pacific Rise and the E–W trending Cocos–Nazca plate boundary (Figs 6 & 9). The movement is recorded by the Cocos and Malpelo Ridges, which formed as the Cocos Plate passed over the Galapagos Hot Spot (the Malpelo Ridge, originally contiguous with the Cocos Ridge, has been offset dextrally by the Panama Fracture). This northeasterly plate movement has generated compression against Central America and against northwestern South America. In the former area compression is marked by subduction, vulcanicity, and the Middle America Trench. In the latter area, compression reactivated NE–SW trending Jurassic normal faults (the Jurassic grabens described earlier) as dextral strike-slip faults (James 1985).

Interpretations of magnetic anomalies generated when the Farallon Plate split into the Cocos and Nazca Plates conclude that northeasterly plate movement in the eastern Pacific commenced 25 to 27 million years ago (Hey 1979, Lonsdale & Klitgord 1978). However, older anomalies may have been consumed by subduction at the eastern margins of the plates, so the preserved record may provide only a minimum date for such movement. West of the Cocos Ridge there is a set of NW–SE trending magnetic anomalies (Corvalan 1981; anomalies 12, 13, etc.; see the inset in Fig. 6) that also indicate northeastward plate movement, driven by the palaeo-East Pacific Rise. Their ages are early Oligocene-? Late Eocene. Moreover, andesitic volcanics in southern Central America record subduction since the Eocene (Malfait & Dinkleman 1972). So it seems likely that northeastward compression against northwestern South America began in the Eocene or even earlier. It is relevant to note here that a regional unconformity (the 'Post-Eocene Unconformity') truncates the Eocene section in the Maracaibo Basin. The greatest erosional vacuity (10 000′) is in the north of this basin (Fig. 15),

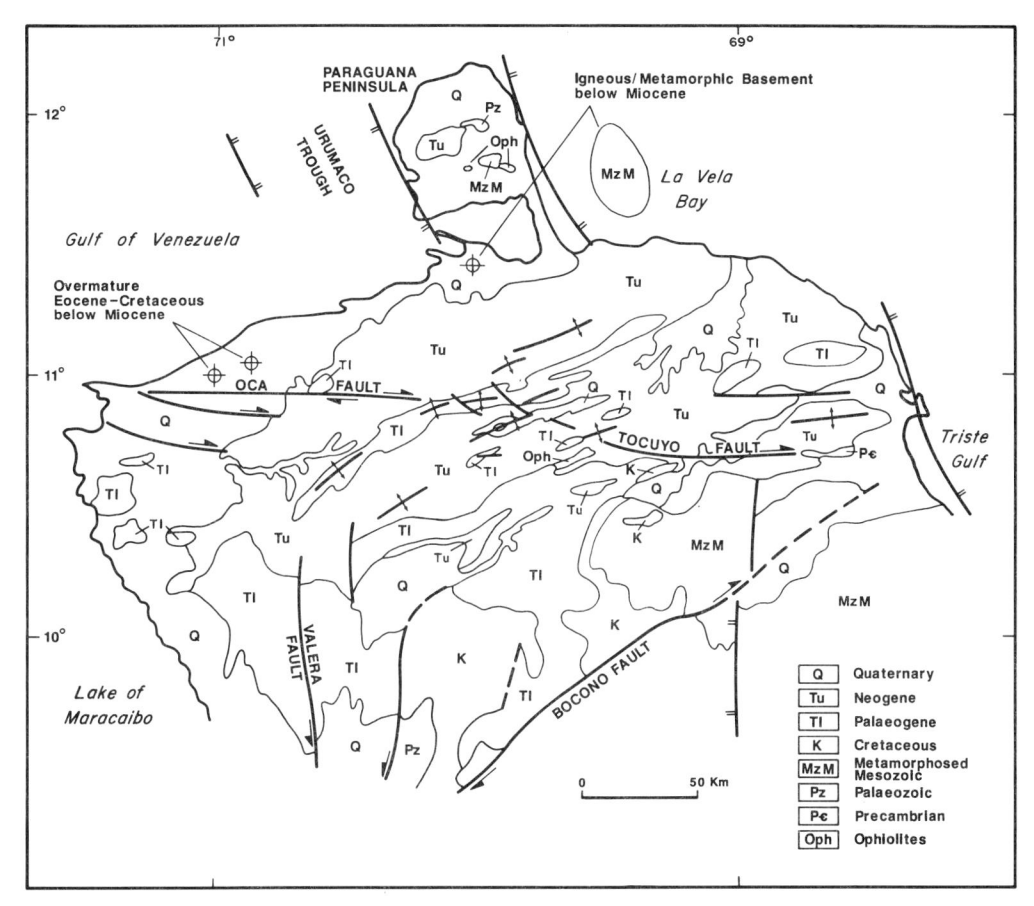

Fig. 12. Simplified geology of northwestern Venezuela.

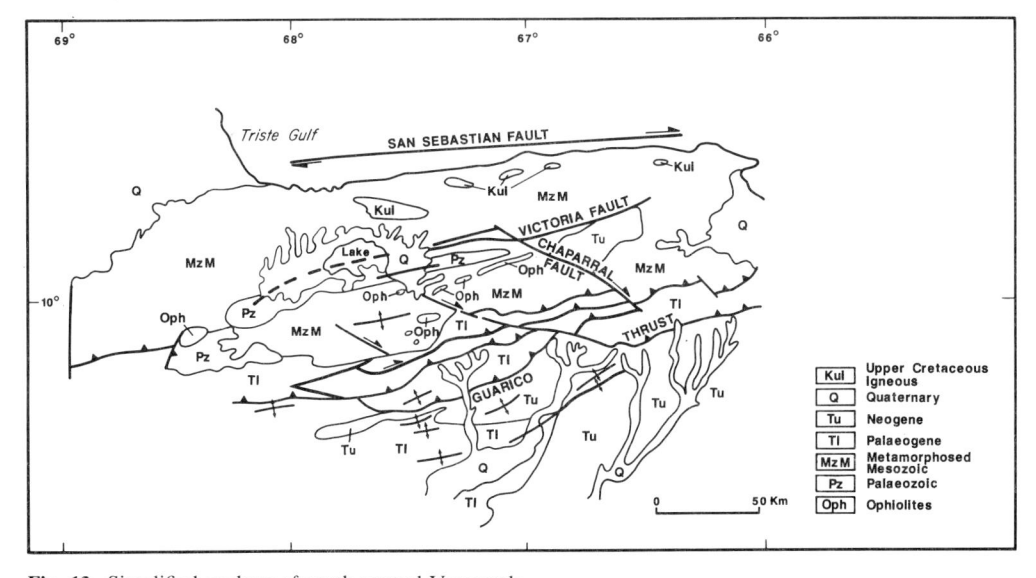

Fig. 13. Simplified geology of north-central Venezuela.

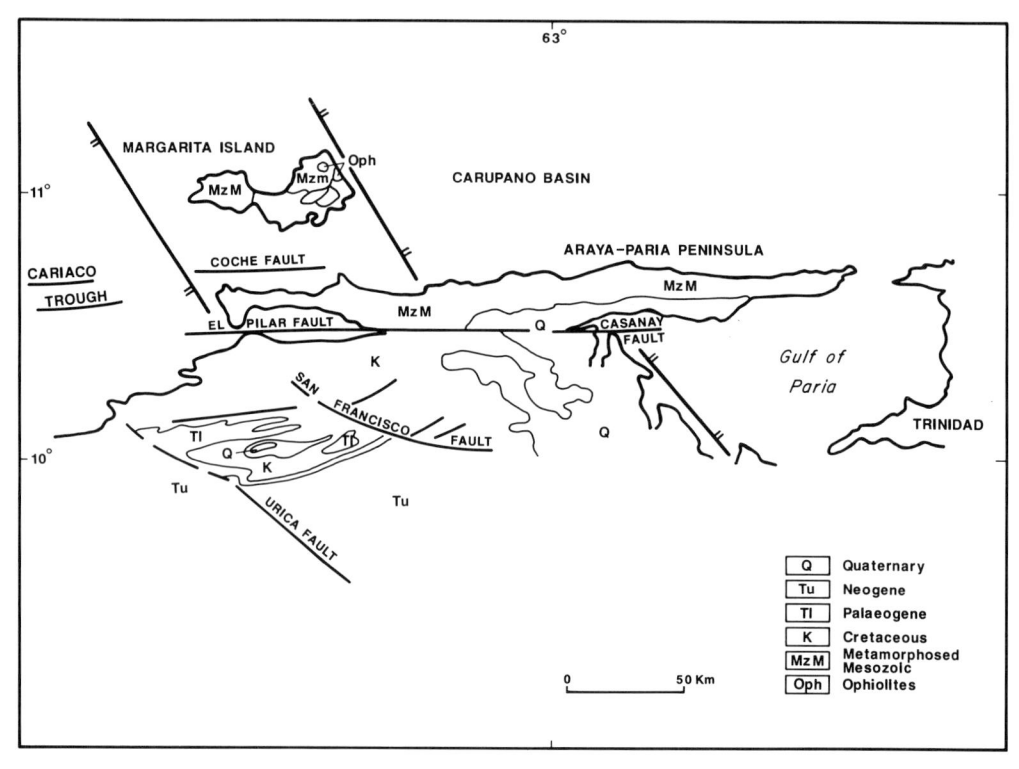

Fig. 14. Simplified geology of northeastern Venezuela.

and yet further north an estimated 13000' of Eocene were removed from the Guajira area (Fig. 8). Late Palaeocene–Eocene fluviatile sediments on Bonaire contain Precambrian exotics (Preim *et al.* 1986). Uplift obviously occurred along the northern margin of northwestern South America during the Palaeocene–Eocene.

Results and implications of the northeastward movement of Northwestern Venezuela

The northward movement of northwestern Venezuela resulted in compression and uplift in the north of the area as it first interacted with the Caribbean Plate. The subsequent, progressive passage of the area over the east–west, dextral, strike-slip plate-boundary resulted in the development of major pull-apart basins.

Compression resulted in enhanced activity along sinistral antithetics to the plate–plate dextral fault system. These are the N–S to NE–SW trending sinistral faults that dominate the Maracaibo and south Falcon basins. They are referred to in more detail in the later section on present-day structures.

The rhomboid shapes of Lake Maracaibo and the Gulf of Venezuela suggest pull-apart origins. Their southwestern and northeastern coastlines are defined by normal faults that result from east–west, dextral strike slip. Silver *et al.* (1975) and Muessig (1984) attribute the Neogene–Recent basins between the Netherlands Antilles, and the Oligo–Miocene Falcon Basin, to the same origins. Sedimentation above the Post-Eocene Unconformity in the Lake Maracaibo area commenced mostly in the Miocene (local, relief-filling Oligocene deposits only). The Gulf of Venezuela remains uncalibrated by the drill, but seismic correlation from the onshore Falcon Basin suggests an Oligocene–Recent, post-Eocene fill. The Falcon Basin is an inverted Oligo–Miocene pull-apart basin where extension was severe enough to allow basaltic intrusion during the Oligocene (Muessig 1984) (Fig. 20).

The implication of these northeasterly movements is that structural restoration is necessary if we are to understand correctly the pre-Eocene palaeogeography. There are various estimates of the amount of movement that has occurred along the Bocono Fault. Schubert (1985) estimates right lateral displacements of only some

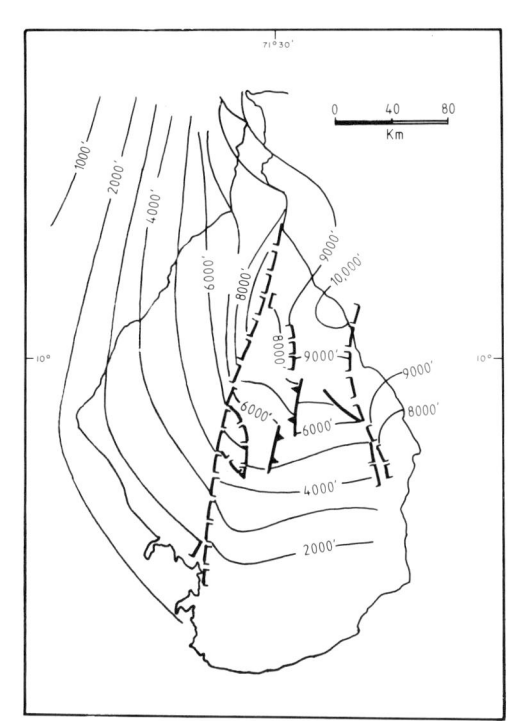

Fig. 15. Isopach map of missing overburden to the 'post-Eocene unconformity' of the Maracaibo Basin.

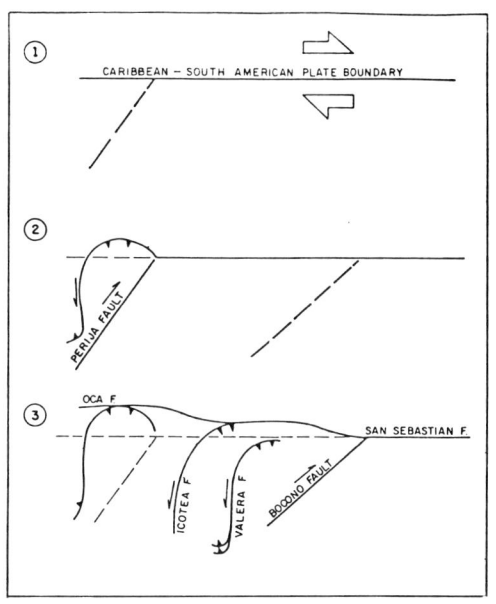

Fig. 16. Northeastward displacement of northwestern Venezuela, along faults such as the Perija-Tigre and Bocono (Fig. 5), resulted in the development of thrust-sinistral fault systems that dominate the northwestern basins.

few kilometers from analyses of pull-apart dimensions. In contrast, Dewey & Pindell (1985) estimated as much as 290 km displacement (all of which was assigned to the last nine million years, i.e. 3.2 cm pa).

The restoration shown in Fig. 8 (James 1985) shows a simplified east−west geology for northern Venezuela. Metamorphosed Mesozoic rocks of the Guajira and Paraguana peninsulas are brought into line with those of northcentral and northeastern Venezuela. Similarly, Palaeocene−Eocene flysch deposits of the southern Falcon Basin align with those of the Interior Ranges of the north central part of the country (see also their associated, allochthonous ophiolites). Finally, the oil fields and presently known kitchen areas form a single, east−west belt. The restoration indicates some 250 km of offset that developed between the late Eocene and today (say 40 Ma, which implies a rate of 0.6 cm pa; cf. the modern rate of 0.9 cm pa). Later, we shall see the implications of such a reconstruction for an understanding of the hydrocarbon habitat.

In detail, not all of this displacement occurred along the Bocono Fault. Some lateral movement has probably taken place along the northwestern, frontal-fault system of the Merida Andes. This major structural feature separates basement at 5000 m above sea level in the Andes from basement at 11 000 m below sea level in the adjacent foredeep of the southern Maracaibo Basin. Some earlier movements probably occurred along the Perija−Tigre Fault system. Thus the northeastward movements shifted their locus progressively eastward (as expressed in Fig. 16) following in the wake of the oblique collision between the Caribbean and South American plates.

Summary of plate tectonics

The part of the Caribbean Plate north of Venezuela (the Venezuelan Basin) formed in situ (between North and South America) during the mid-Mesozoic. Northern Venezuela at this time was a passive margin. When the South Atlantic began to open in the Neocomian, the newly created Caribbean Plate assumed a dextral, strike-slip relationship with the South American Plate. An element of convergence between these plates in the Late Cretaceous−Palaeogene resulted in island arc activity subparallel to the plate boundary, and culminated in arc−continent collision, obduction, crustal shortening, and flysch basin subsidence. Since the Late Eocene, northwestern Venezuela has been thrust northwards to override the Caribbean Plate, driven by spreading in the eastern Pacific. Major pull-apart basins developed in this northwestern area as it slipped over the dextral, strike-slip plate boundary.

Present-day structures

Figure 5 illustrates the main fault and fold systems of northern Venezuela, excluding the internal detail of the mountainous areas. The latter — the Perija Mountains, Merida Andes, and Coastal and Interior Ranges (Fig. 1) — are young, active transpressional uplifts associated with strike-slip faults. The stress–strain ellipse of Harding (1974) shows that many of the faults may be explained by the east–west, dextral, Caribbean–South America, strike-slip plate interaction. The major, east–west dextral fault system (the Oca, San Sebastian, Coche, El Pilar and Casanay faults), which crosses the north of the country, has already been mentioned as a function of this boundary (note that the Oca Fault is displaced to the north relative to the other faults in this series). NNW–SSE trending dextral faults, with associated NE–SW trending folds on the northern flank of the Eastern Venezuela Basin and in the northern part of the Falcon Basin, are synthetic responses to the above fault system. Roughly N–S trending sinistral faults, which dominate the Maracaibo Basin and the south of the Falcon Basin, are antithetics to the same regional system. Normal faults trending NW–SE control the margins of Lake Maracaibo, the Gulf of Venezuela, the basins between the islands of the Netherlands and Venezuelan Lesser Antilles, the Cariaco Basin, and the coastlines of the Paraguana Peninsula, the Falcon Basin and Margarita Island; they are extensional strain elements of the system.

The Maracaibo, Barinas–Apure, and Eastern Venezuela basins are asymmetric with foredeeps adjacent to the Merida Andes and the Coastal/Interior Ranges (Fig. 1). Reverse faulting and folding are commonly associated with these transpressional uplifts. They are well documented along the northern flank of the Eastern Venezuela Basin where important hydrocarbon plays are developing in overthrust/fold settings, and it is likely that they will become better known on the flanks of the Perija and Merida uplifts when exploration focuses on these areas.

Normal faults, mostly down to the north, are common above and parallel to the southern hinge line of the Eastern Venezuela Basin. A second set of normal faults in the same area trends NW–SE, parallel to the extensional strain resulting from the regional E–W dextral stress field. These normal fault systems form a conjugate set which influenced Tertiary sedimentation and resulted in combined structural/stratigraphic traps for oil migrating southwards from the northern foredeep. They form the main traps of the northern part of the Orinoco Heavy Oil Belt (Funes 1985) and of the Oficina area (Fig. 2). Normal faults are also important in the same structural setting in the Sylvestre–Sinco fields of the Barinas–Apure Basin. These faults again trap oil migrating out of the adjacent foredeep, but in this area the Merida basement arch provides an additional focus to migration.

Many of these normal faults are antithetic with respect to basin depocentres. In the central part of the Colombian Llanos Basin, where such faults provide one of the main oil plays, fault offset is commonly greater at shallower levels, possibly indicating oblique slip (McCollough & Padfield 1985).

In addition to all the above, there are important, NE–SW trending, dextral wrench faults (the Perija, Bocono, Cano Limon-Guafita, and Anaco faults). These are especially well expressed in the west where transpression along the Perija and Bocono faults has resulted in uplift of the Perija Mountains and the Merida Andes. En echelon fold structures generated by compression along the Cano Limon-Guafita and Anaco faults form important oil plays in the Barinas–Apure and Eastern Venezuela basins. As explained earlier, all these faults probably result from reactivation of Jurassic structures by Tertiary–Recent spreading activity in the eastern Pacific.

The Maracaibo Basin and the southern part of the Falcon Basin are dominated by a southward opening fan of sinistral wrench faults trending NE–SW to N–S. Of these, the Icotea Fault, which extends over 200 km through the middle of Lake Maracaibo, is the best documented example. Krause (1971) showed that this fault has some 16 km of strike-slip displacement in the central lake area. Further south, the front of the Merida Andes is displaced some 20 km by the same fault.

The prevalence of such faults in these areas is explained (Fig. 16) as a function of convergent interaction between this part of Venezuela and the Caribbean Plate (James 1985). Transpression along these faults has generated en echelon folds sub-parallel to the fault directions. Compression is greatest in the north of the basin, where the faults converge. Resultant fracture porosity and permeability in otherwise tight basement and Cretaceous limestones allows production from these structures. Synsedimentary and younger fault activity also generated Tertiary-level stratigraphic and fault traps on the fold flanks. One may surmise that many of the Maracaibo Basin's giant oil fields would not exist but for these sinistral wrench faults. They are best known, obviously, from the better explored and most productive areas; their northern and southern extremities are poorly known. Fortunately, the Valera Fault, east of the Maracaibo Basin, provides us with a well exposed model of what to expect at these extremities.

The Valera Fault is 220–240 km long. North of the town of Valera, it trends N–S and has demonstrable sinistral displacement (Soulas 1985). To the southwest of Valera the fault splays westward as a series of north or south yielding thrusts that transform the sinistral movement and resolve the space problem created where a southward moving block, west of the Valera Fault, meets a northeastward moving block on the northwestern side of the dextral Bocono Fault. Curvature to the east at the northern ends of most of the Maracaibo–Falcon sinistral faults suggests that they transform into thrusts in these areas also.

It is suggested here that these associated thrusts form important hydrocarbon plays remaining to be investigated in this area. For example, the Socuy Fault, which offsets the Perija and Tigre Faults (referred to as separate elements, but originally a single, dextral fault) partly transforms into thrusts along the western margin of the Maracaibo Basin. Oil seeps are common along this thrust front. Oil fields are present in folds associated with basinward continuations of the Socuy Fault system but the Perija Mountain thrust front has never been drilled. It offers the possibility of finding oil in Tertiary reservoirs below overthrusted Cretaceous and basement. Similarly, thrusts associated with the two faults shown east of the Valera Fault in the southern part of the Falcon Basin have superposed metamorphosed Cretaceous upon Palaeocene–Eocene flysch, creating a potential sub-thrust gas play.

Source rocks

Both Cretaceous and Tertiary source rocks are present in Venezuela. The former are marine; the latter have both marine and terrestrial components. The latest regional review is provided by Cassani et al. (1988).

Cretaceous source rocks

Maracaibo Basin. The upper Cretaceous (Cenomanian–Turonian–Coniacian) La Luna Formation (Fig. 17) of the Maracaibo Basin is one of the world's best known source rocks. Research on this unit, mostly geochemical, continues (e.g. Talukdar et al. 1985a,b; Vierma 1985a,b), as does the debate on the environment of its deposition. Hedberg (1931) concluded that the La Luna limestones were source rocks and he suggested that they were deep-water deposits accumulated far from land influence. Liddle (1931), in a discussion following Hedberg's paper, proposed that they had formed in fairly shallow water.

The La Luna is a limestone–calcareous shale, dark grey to black, organic rich and finely laminated. Nodules are common, nucleated by pelecypods or cephalopods. Planktonic (often dwarfed) and pelagic fossils are abundant, but

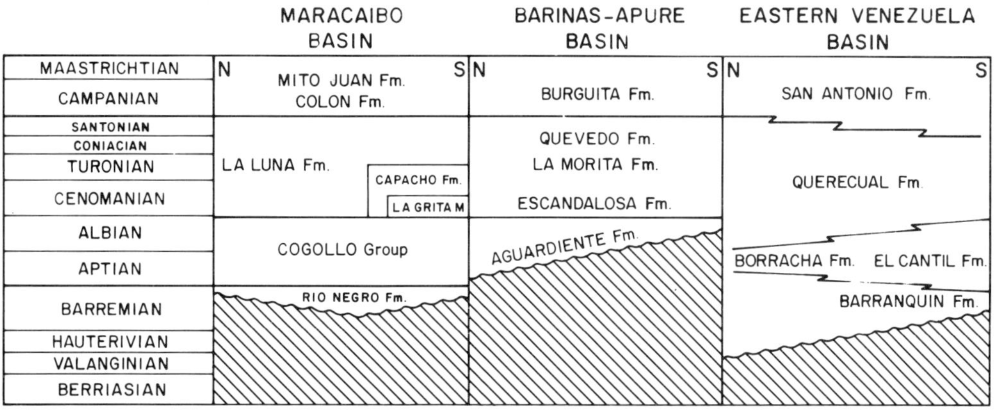

Fig. 17. Cretaceous source rocks of Venezuela.

benthonics are rare (dwarfed buliminids and rotalids). Chert is common and increases upwards and northwards in the sequence. The unit becomes sandy towards the south.

The high content of kerogen type II, absence of benthonic fossils, and abundance of planktonics are usually argued as indicating a deep, restricted-marine origin for the La Luna. However, planktonic fossils alone are not depth indicators and the absence of benthonics and abundance of organic matter indicate only euxinic bottom conditions, a relative paucity of clastic influx and high organic productivity/ preservation. Moreover, the underlying Cogollo Group, which is demonstrably the product of a shallow carbonate platform (Chacartegui 1985), with grainstone and ooid bar deposits, contains lagoonal, organic-rich sediments, with oysters, which are geochemically similar to the La Luna. Thus the La Luna could have formed in a rather shallow marine environment.

Upper Cretaceous source rocks are popularly attributed to upwelling of nutrient-rich water, following the model provided today by localities along the western coasts of South America and Africa. Macellari & De Vries (1987) explain the Late Cretaceous units of northwestern South America in these terms. However, Cretaceous organic-rich units are very widespread ('global anoxic events/crises'). For example, Turonian–Coniacian sediments from Deep Sea Drilling sites in the Venezuelan Basin have high organic carbon contents (Hay 1985). It is perhaps unreal to imagine that upwelling was that widely active. Late Cretaceous high sea-levels, transgressing over broad continental platforms, would have reworked widespread soil (nutrient) in shallow (limited circulation, photic zone) clear (carbonate deposits common) marine conditions, perhaps triggering high biologic productivity in the newly created ecological niches.

Source rocks also exist in the Coniacian-Maastrichtian Mito Juan section, also a shallow marine deposit, in the area west of Lake Maracaibo (Quijada & Caldera, 1985). In the south of the Maracaibo Basin, the La Grita Member of the Capacho Formation, lateral equivalent of the lower La Luna, has the same facies as the La Luna.

The La Luna has an average thickness of about 110 m in the Maracaibo Basin. Core and electric log studies indicate a net source rock thickness of between one and two thirds of the gross (Blaser & White 1984) and the oil yield is estimated to be 290×10^6 bbls/km^3 (Talukdar et al. 1985a). Since the kitchen area below the present-day basin is some 50 000 km^2, the La Luna here could have generated about one trillion barrels of oil. The hydrocarbon generating potentials of the Cogollo, Mito Juan and La Grita have not been determined. In one well, geochemical logs indicted Cogollo source-rock quality comparable with that of the La Luna over a 130 metre interval; however, this development appears to be restricted to a trough in the west of the Maracaibo Basin. Gallango et al. (1985) reported that they have identified small amounts of crudes derived from both the Capacho and Cogollo units.

Eastern Venezuela Basin. In the Eastern Venezuela Basin the Querecual Formation, age equivalent of the La Luna, also consists of black limestones and shales rich in planktonic foraminifera and with common nodules. Chert is present and increases in abundance upwards and northwards. The unit becomes sandy in the south, where it thins as a function of facies change into the overlying, northward prograding, sandy San Antonio Formation. Campos et al. (1985) interpret the environment of deposition of the Querecual as deep marine in the north and shallow marine in the south. The Querecual overlies coastal limestones and sandstones of the El Cantil Formation, and reef limestones of the Borracha Formation (cf. the Cogollo Group of the Maracaibo Basin).

Talukdar et al. (1985b) have determined that this source rock, of type II kerogen, has a generative capacity of 56 to 152×10^6 bbls/km^3. The thickness ranges from 353 to 700 m. Arnstein et al. (1982), indicate a kitchen area in the present day basin of about 28 000 km^2, capable of yielding between 0.6×10^{12} and 2.0×10^{12} bbls, assuming that the net source-rock thickness is one to two thirds of the gross as it is in the La Luna Formation.

Barinas–Apure Basin. The Upper Cretaceous Quevedo, La Morita and Escandalosa Formations have been identified as source-rock bearing units in the Barinas–Apure Basin (Chigne 1985, 1990). These are sandy equivalents of the La Luna, deposited in coastal settings in the south and in more open-marine conditions in the north. The organic material is type-II kerogen in the north, but kerogen type III becomes important in the south and over the Merida and Arauca Arches. There are no published data on the net source rock thicknesses of these units but the generative capacity has been estimated to be between 70 and 180×10^6 bbls/km^3 (Talukdar & De Toni 1988). Chigne's (1985) map indicates a kitchen area of some 40 000 km^2.

Former extent of Cretaceous source rocks. The original extent of these source-rock units is unknown. Outcrops in the Perija Mountains and the Merida Andes show that they extended across these areas. Gabela (1985) states that the La Luna is a source rock in the Colombian middle Magdalena Basin. Macellari & De Vries (1987) describe the La Luna in a regional context along with a Colombian equivalent, the Villeta Formation. The latter varies in thickness and age from a few hundred meters of Albian–Campanian age to over 5000 m of sequence dating back to the Barremian.

Palaeocene–Eocene flysch deposits (Fig. 8) of the Interior Ranges of north central Venezuela (Guarico Fm.), of the Falcon Basin (Matatere Fm.), and of the Portuguesa Basin (Rio Guacha Fm.) contain allochthonous Mesozoic elements such as Late Jurassic–Neocomian, calpionellid-bearing reef limestones, Aptian–Albian rudistid limestones and conglomerates, Aptian calpionellid pelagic limestones, and reworked La Luna and Querecual (Pierce 1960; Peirson et al. 1966; Gonzales Silva and Picard 1972). Olistostromes of the La Luna in the Rio Guacha Fm. are up to 1.5 km long. Olistostromes in the Guarico Fm. are considered to have come from the north. Olistoliths in the Palaeocene part of the Matatere Fm. contain Turonian ammonites, while those in the Eocene part contain middle Albian species (Renz et al. 1955). The Matatere Fm. also contains olistoliths of the Cogollo Group, and of volcanic material. The latter becomes more abundant northwards, towards Siquisique, where a middle Jurassic ophiolite complex crops out (Bartok et al. 1985). Also in the Siquisique area there are olistoliths of the Barquisimeto Formation, again a flysch deposit, but distinguished by incipient metamorphism (Bellizia et al. 1972) and a Maastrichtian age (Bushman 1959).

It is apparent that the La Luna/Querecual facies originally extended north of its present distribution, and that it probably was a diachronous, southward younging deposit. The age relationships of the La Luna olistoliths reported by Renz et al. (1955), and the presence of olistoliths of metamorphosed, Maastrichtian flysch within the Palaeocene–Eocene flysch, document progressive uplift and erosion in the north, through inversion of southward-migrating flysch basins.

Further support for the model of an original, more northerly extent of the Cretaceous source rocks can be found in metamorphic units of northern and offshore Venezuela. Several graphitic, metamorphic units and black limestones or marbles occur in the Northern Ranges of Venezuela. The Las Mercedes Formation of the Caracas area, for example, is a 500 m (minimum) thick calcareous schist, graphitic throughout, with layers of black limestone (Dengo 1953). Like the La Luna, the Las Mercedes contains nodules (of black limestone) and has a northward increasing silica content. Lenses of volcanic ash are present in the northernmost outcrops (Wehrmann 1972). No age-diagnostic fossils have been reported, but Furrer & Urbani (1973) have identified specimens of Ophthalmidiidae (shallow-water foraminifera). The conformably overlying Chuspita Formation, which contains meta conglomerates and quartzites, is interpreted as a shallow-water deposit. Calcareous graphitic phyllites and graphitic marbles similar to those of the Las Mercedes Formation form fifty percent of this unit (Gonzales de Juana 1982), which contains uncoiled ammonites identified as upper Albian genera by Macsotay (1972).

In eastern Venezuela, graphitic marbles and phyllites of the Carupano Formation and of the Yacua Member of the Cariaquito Formation are considered to be lithostratigraphic equivalents of the Las Mercedes Formation (Vignali 1979). Like the latter, the northernmost outcrops of the Carupano contain volcanics (the El Copey Member). Castro & Mederos (1985) describe a Barremian–Santonian unit, the Mejillones Group, of cherty brown, euxinic limestones from the offshore Carupano Basin. The unit has interbedded radiolarites and basalts and its thickness is at least 1100 m. Finally, on Margarita Island, the Cenomanian metasedimentary Los Robles Group contains thin lenses of black limestones, black marbles, and graphitic schists (Gonzalez de Juana & Vignali 1972) and there is a graphitic unit in the upper part of the pre-Cenomanian Juan Griego Group (Vignali 1979).

All these data, stratigraphically summarized in Fig. 18, point to the original, northerly existance of coeval or older, thicker equivalents of the La Luna-Querecual facies, in the vicinity of volcanic activity. It makes sense to link all the above mentioned units into one, large palaeogeographic province, probably in a back-arc setting, that extended right across the northern flank of South America in the Late Cretaceous (Fig. 19). The areal extent (more than half a million square kilometres) of this province must have been considerably larger than that of its present day remnants, for northern Venezuela has suffered much uplift, erosion and crustal shortening.

Maturation of the Cretaceous source rocks. The La Luna reached maturity for oil generation during the Eocene in the area of the Guajira

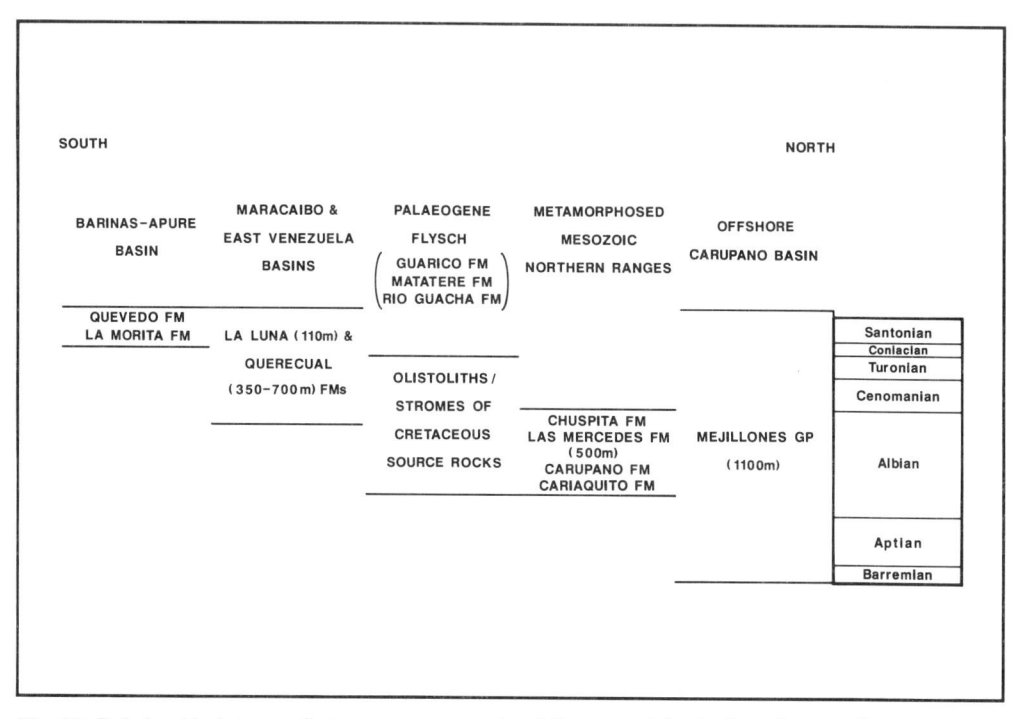

Fig. 18. Relationship between Cretaceous source rocks of the present day hydrocarbon provinces, allochthonous equivalents within Paleocene flysch, age-equivalent graphitic schists and black marbles of the northern ranges, and age-equivalent, organic rich units in the offshore region.

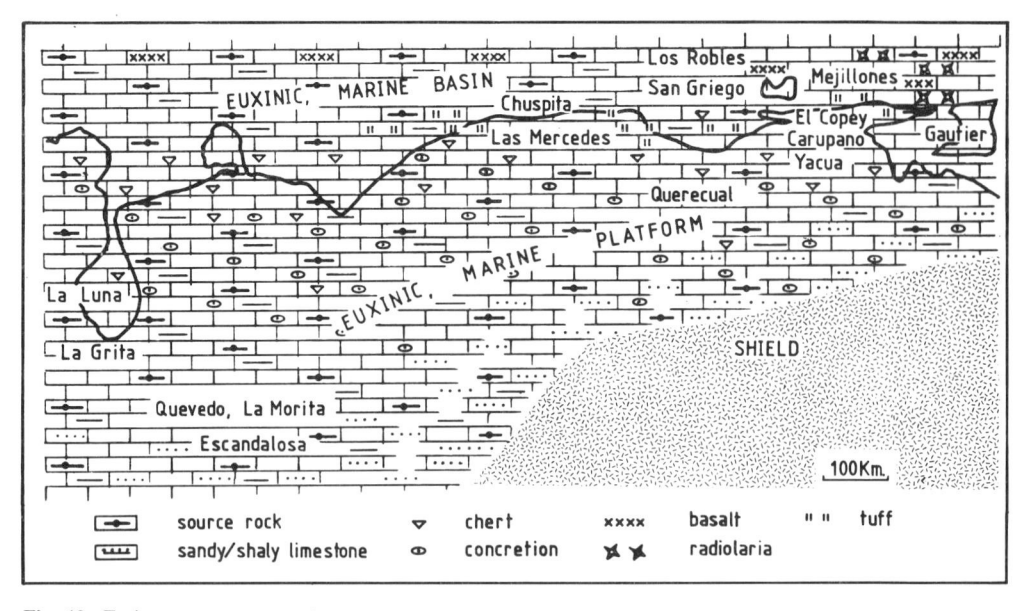

Fig. 19. Facies map, superposed upon a restored Venezuelan geography (see Fig. 8), of Upper Cretaceous, organic rich units.

Peninsula and in the northeast of the Maracaibo Basin (Talukdar et al. 1985b) (Fig. 8). In the latter area it is now mature for gas generation. On the Guajira Peninsula and in Tablazo Bay, at the entrance to Lake Maracaibo (Lew & Albarracin 1988), and also in the north of the Falcon Basin, the La Luna is now overmature (note, here, that the presence/absence of the La Luna below most of the Falcon Basin is untested). In the west and south of the Maracaibo Basin, maturity for oil generation was attained in the Miocene to Recent (Talukdar et al. 1985b). Concretions at the type locality on the Perija Mountains contain oil (Hedberg, 1931) but further to the south the La Luna at outcrop is post mature for oil generation as it is also at outcrop in the Merida Andes. The La Grita, a lateral equivalent of the La Luna in the south of the Maracaibo Basin, is mature for oil generation.

The Querecual Formation of the Eastern Venezuela Basin generated oil in the mid Oligocene-mid Miocene in the northern part of the basin, where it is now gas prone, and it has been producing oil further south since the Miocene (Arnstein et al. 1982). The Eastern Basin is now characterized by gas or medium gravity oil on its northern flank and by heavy oil (mostly degraded, Fiorillo 1985) in the south.

Generation in the Barinas–Apure Basin began in the depocentre adjacent to the Andes in the mid-Miocene (Chigne 1985). To the southeast the source rocks are immature.

The general picture, therefore, is that the Cretaceous source rocks are immature towards the craton and are mature to overmature below Tertiary depocentres. The oldest known of these lay beneath the northeast of the Maracaibo Basin and it may have extended below much of the Falcon Basin. Extrapolation of the trend of northward increasing maturity leads towards the graphitic metasediments of the north and suggests a regional model of earlier maturation below more northern basins, the oldest of which reached the metamorphic stage and then became inverted. The presence of mature/overmature La Luna in the Perija mountains and Merida Andes is also important, showing that these areas too, used to be part of the kitchen areas before Late Tertiary mountain building.

The same regional picture emerges in Colombia, where upper Cretaceous source rocks are immature to mature in the Llanos Basin, and are mature to overmature at outcrop in the adjacent Eastern Cordillera (Palmer & Russel 1988). McCollough & Padfield (1985) surmised that the oils of this basin originated in the Eastern Cordillera region, and that some must have migrated as much as 240 km. Likewise, in the Oriente Basin of Ecuador, the Cretaceous Napo Formation, identified as the most likely source rock for the oils present, is immature to just mature, leading Dashwood & Abbots (1990) to conclude that generation occurred below a kitchen that reached the metamorphic stage and then became inverted in the area of the adjacent Eastern Cordillera.

Summarizing the foregoing, Cretaceous source rocks originally extended over a very wide area. They probably were diachronous, younging southwards. They matured below Late Cretaceous to Tertiary depocentres which migrated southwards (Figs 20 & 21). The present-day basins are remnants of a much broader petroliferous province.

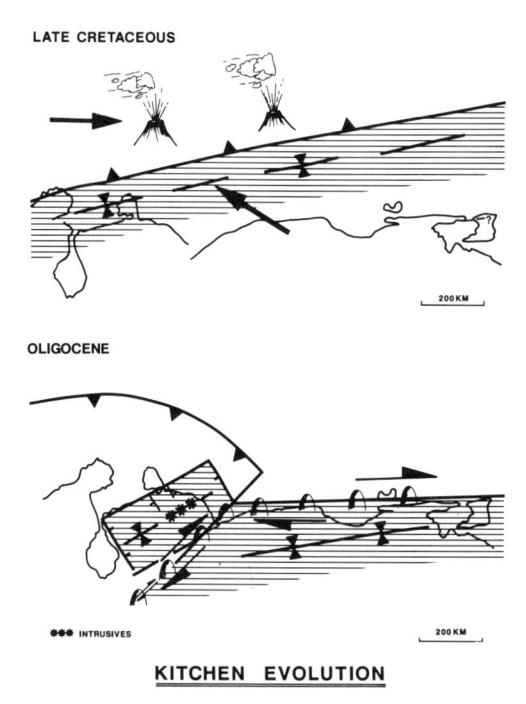

Fig. 20. The area below which Upper Cretaceous source rocks generated hydrocarbons in Venezuela migrated southwards through time. The Late Cretaceous kitchen lay below a flysch basin, south of an approaching island arc, that resulted from convergence between the Caribbean and South American plates. By late Palaeogene times, that plate–plate relationship had become one of dextral strike-slip and foreland basins formed adjacent to transpressional uplifts in the northcentral and northeastern part of the country. However, since the late Eocene, northwestern Venezuela has been progressively overriding the plate boundary (Fig. 9), and is characterised by pull-apart basins as a consequence.

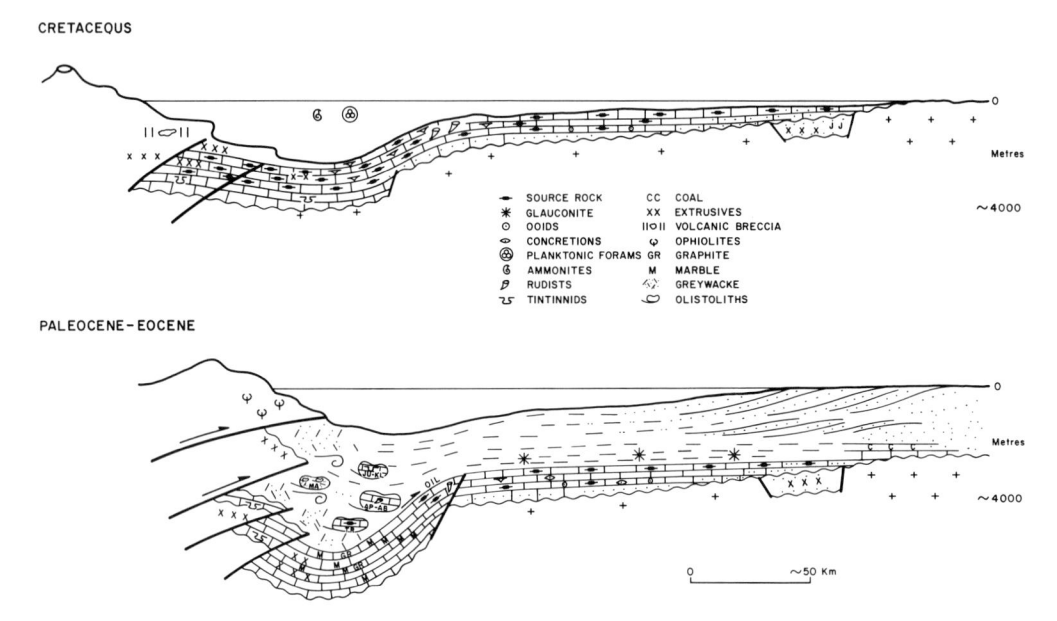

Fig. 21. Schematic, north–south, cross-sections of Venezuela illustrating, above, the Cretaceous development of organic rich units on a passive margin, south of an advancing island arc, and, below, the Neogene development of southward migrating flysch basins, in front of the arc-continent collision zone, where Cretaceous source rocks generated hydrocarbons and then became metamorphosed.

Tertiary source rocks

Tertiary source rocks (Fig. 22) are less well studied, especially quantitatively. They take second place in importance a long way behind the Cretaceous source rocks. Gallango *et al.* (1985), for example, conclude that 95% of the Maracaibo Basin crude has a Cretaceous origin.
Maracaibo Basin. Source rocks characterized by land-plant material are present in the south and west of the Maracaibo Basin in the Palaeocene Orocue and Marcelina formations, the Eocene Mirador and Carbonera formations, and the Miocene Lagunillas and La Rosa formations (Blaser & White 1984). The Carbonera Formation at outcrop in the Merida Andes is mature for oil generation. All these units should be mature in the north Andean foredeep, in the south of the Maracaibo Basin, where little exploration has been carried out. Talukdar *et al.* (1985a) reported that three fields (Los Manueles, West Tarra, Las Cruces, Fig. 2) in the southwest of the basin produce some crudes derived from terrestrial sources.
Eastern Venezuela Basin. In the Eastern Venezuela Basin, there are source rocks in the Merecure Group, and in the La Pascua and Roblecito formations, of Oligocene age, and in the Miocene Oficina and Chaguaramos forma-

tions. Arnstein *et al.* (1982) reported that these are capable of generating 1.7 to 7 kg hydrocarbons per ton. The organic material is mainly terrestrial and some waxy crudes in the Las Mercedes and Oficina areas probably are derived from these units. Geochemical correlation of source rocks and crudes indicate, however, that most of the oil there comes from Cretaceous source rocks. Aymard *et al.* (1985) reported overmature Tertiary source rocks in the 8 Tcf Yucal–Placer gas field in the north of the Eastern Basin, but higher organic carbon percentages occur in the overmature Cretaceous of the same area.

Offshore Basins and the Falcon Basin. Talukdar & Bolivar (1982) described Eocene and Oligocene, marine and terrestrial source rocks in the marine shales of the Cariaco Basin where oil, gas and condensate, of mainly marine origins, have been discovered. Their calculations estimate recoverable reserves of 52–157 million barrels in this area. The Patao High, a 40 km long, basement uplift in the Carupano Basin (Pereira 1985), is the site of six large accumulations of dry gas and condensate-associated gas in Mio–Pliocene sandstones and calcarenites (Kiser 1981, Bellizia *et al.* 1981). Source rock

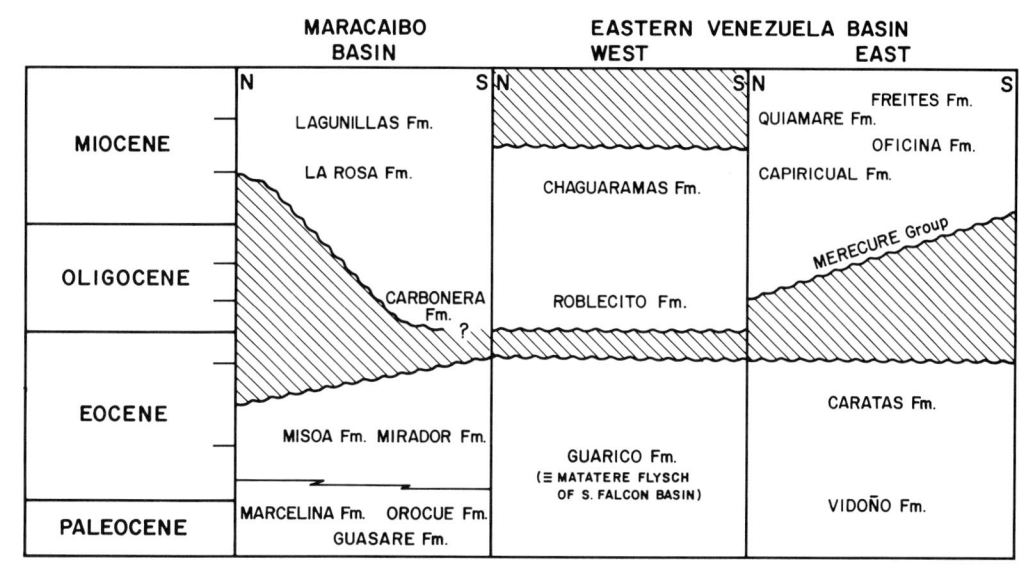

Fig. 22. Tertiary source rocks of Venezuela.

descriptions have not been published but the sedimentary section ranges from Eocene to Recent (Pereira 1985), as in the neighbouring Cariaco Trough. The Bonaire Basin likewise contains an Eocene to Recent section. One well encountered minor gas shows in this area. The sequence in La Vela Bay ranges only from Oligocene to Recent. Source rocks of terrestrial origin have been identified in the Oligocene. Rather small oil and gas accumulations here are associated with a basement high. The Gulf of Venezuela remains undrilled at this time. Logic indicates an analogy between it and La Vela Bay.

In the Falcon Basin, well data show that the Cretaceous is overmature and geochemistry points to source rocks of terrestrial origins for a string of smallish oil fields (El Mene–Mamon, Fig. 2). Likely source rock sections occur in the Eocene–Oligocene sequence and generation probably occurred in the Pliocene–Recent.

The general picture to emerge from the foregoing is that the Tertiary section locally includes mixed or terrestrial source rocks but none is as important as the Cretaceous source rocks.

Oil generating capacity: Cretaceous source rocks

The known original STOIIP of Venezuela was, at least, in the order of 1.5 trillion barrels. McDowell (1975) summarized that global basin data indicate maximum trapping in reservoirs of only one quarter to one third of the oil generated. This implies the original generation of at least 4.5–6 trillion barrels in Venezuela. Rich as the Cretaceous La Luna and its equivalents are, they could not have generated this amount of oil in the existing kitchens, whose reported areal sum is around 120 000 km^2. However, using the proposed models of originally widespread Cretaceous source rock deposition, diachronous kitchen development, and relatively recent limitation of the existing basins by transpressional uplifts, it can be modelled that perhaps 400 000 km^2 of source rock entered the oil window (Krause & James 1990). With a net thickness of 70 metres, and a generating capacity of 290×10^6 bbls km^3 (average figures for the La Luna of the Maracaibo Basin), the source rocks over this area could have generated some six trillion barrels. Two or three times this amount could have been generated if the net thickness increased northwards along with the increase in formation thicknesses. In addition, trapping circumstances may have been especially fortuitous in Venezuela, particularly in the Orinoco Oil Belt where a stratigraphic trap formed in the Cretaceous and greatly increased in capacity during the Oligo–Miocene.

Future exploration objectives

By analogy with the Eastern Venezuela Basin, the Barinas–Apure Basin should contain hydrocarbons stratigraphically trapped in Cretaceous and Tertiary sandstones along its southeastern flank. This would be a western to southwestern continuation of the Orinoco Heavy Oil Belt, which is currently thought to terminate against the eastern flank of the El Baul High. The oil is likely to be biodegraded and heavy, but the volumes may be very large.

On the hinge line of the same basin, and on the hinge line of the North Andean Foredeep of the southern Maracaibo Basin, normal faults could form traps analogous to those of the Las Mercedes/Oficina fields of the Eastern Basin and the Sylvestre fields of the Barinas–Apure Basin. Medium to light crudes might be found here. Stratigraphic traps resulting from Tertiary, synsedimentary fault activity may also exist in these areas.

Also in the Barinas–Apure Basin, it should be suspected that en echelon folds exist along the northeastward continuations of the faults that generated the Cano Limon–Guafita and Arauca structures, or along other NE–SW trending dextral faults in the area.

Overthrust plays should exist along both flanks of the Merida Andes and along the eastern flank of the Perija mountains. The objective horizons would be igneous basement and Cretaceous limestones (both fractured), and Tertiary sandstones, where medium crude may be expected. A similar play may exist to the south of the Falcon Basin where gas may be trapped in Tertiary flysch deposits below Cretaceous metamorphics.

The offshore basins appear to offer only biogenic gas or hydrocarbons thermally derived from rather localized Tertiary source rocks.

Conclusions

The hydrocarbon habitat of Venezuela is intimately related to the plate-tectonic history of the country.

The eastern Caribbean Plate formed in situ by sea-floor spreading during the late Jurassic–early Cretaceous, along with the opening of the Central Atlantic; coeval extensional features formed along the northern flank of the South American Plate. During the early–late Cretaceous phase of South Atlantic opening, South America converged with the Caribbean Plate with consequent island arc activity, arc–continent collision and crustal shortening. Thereafter, South America moved in a more westerly direction, and the Caribbean–South American plate boundary became a roughly east–west, dextral transform. The present position of this boundary is indicated by a chain of Tertiary–Recent pull-apart basins that line up from Trinidad to Panama. Most of the faults and folds of northern Venezuela result from dextral shear along the plate boundary.

Northwestern Venezuela is allochthonous, driven northeastward, since the late Eocene, along Mesozoic extensional faults reactivated as dextral wrenches by compression generated by spreading in the Nazca–Cocos Plate complex. The Gulf of Venezuela, the Falcon Basin, and the Maracaibo Basin formed as pull-apart basins when northwestern Venezuela crossed the Caribbean–South America plate boundary. The Maracaibo Basin and the southern Falcon Basin are structurally dominated by sinistral strike-slip faults, antithetic to the plate boundary, that were enhanced by compressional interaction between northwestern Venezuela and the Caribbean Plate.

The vast reserves of oil and gas in the Maracaibo, Eastern Venezuela and Barinas–Apure basins are derived from Upper Cretaceous source rocks that originally covered an area of at least half a million square kilometres. These source rocks reached maturity in kitchens that migrated southwards during the Late Cretaceous–Early Tertiary plate convergence and in Oligocene–Recent kitchens that formed adjacent to the transpressional uplifts of the Merida Andes and the Coastal and Interior Ranges.

The enormous stratigraphic trap of the Orinoco Heavy Oil Belt is formed by the onlap of Cretaceous and Tertiary sandstones onto the Guayana shield. This belt should have a westward extension along the southeastern flank of the Barinas–Apure Basin. All other accumulations occur in structures, with local stratigraphic complement, resulting from the plate interactions mentioned above. They mostly involve Cretaceous and Tertiary sandstones but fractured Cretaceous limestones and basement are also productive locally. Untested plays for the west, analogous to productive plays of the east, include trapping against normal faults in the south of the Maracaibo Basin, and overthrust traps along the flanks of the Perija Mountains and the Merida Andes.

Considerable amounts of additional reserves are likely to be discovered in Venezuela.

I wish to express my gratitude to colleagues in Maraven, S. A., Caracas, for their encouragement and cooperation during the preparation of this paper. Petroleos de Venezuela, S. A., and Maraven, S. A., Caracas, and Shell, London and The Hague, kindly gave permission to publish.

References

ACOSTA DE MORENO, Z. 1988. Exploratory prospect and prospective areas for heavy and extra heavy crudes in the Eastern Venezuela Basin, Venezuela, (excluding the Orinoco Oil Belt): *Fourth UNITAR/UNDP International Conference on Heavy Crude and Tar Sands*, Paper No. 168.

ARNSTEIN, R., BETORET, C., MOLINA, E., MOMPART, L., ORTEGA, J., RUSSOMANO, F. & SANCHEZ, H. 1982. *Geologia petrolera de la Cuenca de Venezuela, XLV Reunion a Nivel de Expertos, Asistencia Reciproca Petrolera Estatal Latino-Americana, Mexico.*

AYMARD, R. C., QUIJADA DE DAAL, J. & CORIAT, M. 1985. Campo Yucal–Placer trampa estratigrafica gigante de gas en la Cuenca Oriental de Venezuela: *VI Congreso Geologico Venezolano, Memorias, Tomo IV*, 2779–2803.

BARTOK, P. E., RENZ, O. & WESTERMANN, G. E. G. 1985. The Siquisique ophiolites, northern Lara State, Venezuela: a discussion on their Middle Jurassic ammonites and tectonic implications: *G.S.A. Bulletin* **96**, 1050–1055.

BEETS, D. J., MARESCH, W. V., TH. KLAVER, G., MOTTANA, A., BOCCHIO, R., BEUNK, F. F. & MONEN, H. P. 1984. Magmatic rock series and high-pressure metamorphism as constraints on the tectonic history of the southern Caribbean in the Caribbean–South America Plate Boundary and Regional Tectonics: *GSA Memoir* **162**, 95–130.

BELL, J. S. 1972. Geotectonic evolution of the southern Caribbean area IN Studies in Earth and Space. Sciences (Hess Volume), *GSA Memoir* **132**, 367–386.

BELLIZIA, A., PIMENTAL DE BELLIZIA, N. & MUNOZ, M. I. 1981. Geology and tectonics of Northern South American. *Republica de Venezuela, Ministerio de Energia y Minas, Direccion General de Geologica, Publication Especial* **9**, Caracas, 182–183.

—— RODRIGUEZ, D. & GRATEROL, M. 1972. Ofiolitas de Siquisique y Rio Tocuyo y sus relaciones con la Falla de Oca: VI Conferencia Geological del Caribe, Margarita, Venezuela, Memorias, 182–183.

BELLIZIA, G. A. & RODRIGUEZ, G. D. 1968. Consideraciones sobre la estratigrafia de los Estados Lara, Yaracuy, Cojedes y Carabobo: Boletin de Geologia, Caracas, **9**, 515–563.

BIJU-DUVAL, B., MASCLE, A., ROSALES, H. and YOUNG, G. 1982. Episutural Oligo–Miocene basins along the north Venezuelan Margin: *In*: Studies in Continental Margin Geology, *AAPG Memoir* **34**, 347–358.

BLASER, R. & WHITE, C. 1984. Source rock and carbonisation study, Maracaibo Basin, Venezuela. *In*: DEMAISON, G. & MURRIS, R. J. (eds) Petroleum Geochemistry and Basin Evaluation, *AAPG Memoir* **35**.

BOUYSSE, Ph., 1988. Opening of the Granada back–arc Basin and Evolution of the Caribbean Plate during the Mesozoic and Early Paleogene. *Tectonophysics*, **149**, 121–143.

——, ANDREIEFF, P., RICHARD, M., BAUBRON, J. C., MASCLE, A., MAURY, R. C. & WESTERCAMP, D. 1985. Aves swell and northern Lesser Antilles Ridge: rock-dredging results from Arcante 3 cruise. *Geodynamique des Caraibes, Symposium, Paris, Editions Technip*, 65–76.

BURKE, K., COOPER, C., DEWEY, J. F., MANN, P. & PINDELL, J. L. 1984. Caribbean tectonics and relative plate motions. *In*: The Caribbean–South American Plate Boundary and Regional Tectonics. *GSA Memoir* **162**, 31–63.

BUSHMANN, J. R. 1959. Geology of the Barquisimeto area, a summary report: Asociacion Venezolano de Geologica, Mineria y Petroleo, Boletin Informativo, **2**, 65–84.

CAMPOS, V., LANDER, R. & DE CABRERA, S. 1985. Evolucion estructural en el noreste de Anzoategui y su relacion con el norte de Monagas: VI Congreso Geologico Venezolano, Caracas, Memoria, Tomo IV, 2397–2414.

CARNEVALLI, J. 1988. Venezuela nor-oriental: exploracion en frente de montana: III Simposio Bolivariano, Exploracion Petrolera de las Cuencas Subandinas, Caracas, Memoria, 68–89.

CASE, J. E. 1974a. Major basins along the continental margin of northern South America. *In*: *Geology of the Continental Margins*, Springer, New York, 733–742.

—— 1974b. Oceanic crust forms basement of eastern Panama. *GSA Bulletin*, **85**, 645–652.

CASSANI, F., VALLEJOS, C., GALLANGO, O. DE TONI, B. & TALUKDAR, S. 1988. Geoquimica organica de los crudos Venezolanos, III Simposio Bolivariano, Exploracion Petrolera de Las Cuencas Subandinas, Caracas, Memoria, 612–637.

CASTRO, M. & MEDEROS, A. 1985. Litoestratigrafia de la Cuenca de Carupano: VI Congreso Geologico Venezolano, Caracas, Memoria, Tomo I, 201–225.

CHACARTEQUI, F. J. 1985. Estudio Sedimentologico en el Grupo Cogollo del Cretaceo Inferior: VI Congreso Geologico Venezolano, Caracas, Memoria, Tomo I, 278–304.

CHIGNE, N. C. 1985. Aspectos relevantes en la exploracion de Apure: VI Congreso Geologico Venezolano, Caracas, Memoria, Tomo V, 2891–1921.

—— & HERNANDEZ, L. 1990. Main aspects of petroleum exploration in the Apure area, southwestern Venezuela, 1985–87. *In*: BROOKS, J. (ed.) *Classic Petroleum Provinces*, Geological Society, London, Special Publication **50**, 55–76.

CHRISTOFFERSON, E. 1976. Colombian Basin magnetism and Caribbean plate tectonics: *GSA Bulletin*, **87**, 1255–1258.

CORVALAN, J. 1981. Plate-Tectonic Map of the Circum-Pacific Region, Southeast Quadrant. *AAPG Circum-Pacific Map project*.

CRAWFORD, F. D., SZELEWSKI, C. E. & ALVEY, G. D. 1985. Geology and exploration of the Takutu Graben of Guayana and Brazil. *Journal of Pet-*

roleum Geology, **8**, 5–36.
DASHWOOD, M. F. & ABBOTTS, I. L. 1990. Aspects of the petroleum geology of the Oriente Basin, Ecuador *In*: BROOKS, J. (ed.) *Classic Petroleum Provinces*, Geological Society, London, Special Publication **50**, 89–118.
DENGO, G. 1953. Geology of the Caracas region, Venezuela. *AAPG Bulletin*, **64**, 7–40.
DEWEY, J. F. & PINDELL, J. L. 1985. Neogene block tectonics of eastern Turkey and northern South America: continental applications of the finite difference method: *Tectonics*, **4**, 71–83.
DONELLY, T. W., MELSON, W., KAY, R. & ROGERS, J. W. 1973. Basalts and dolerites of Late Cretaceous age from the central Caribbean *In*: *Initial Reports of the Deep Sea Drilling Project*, **15**, Washington, US Government Printing Office, 1013–1016.
—— 1975. The geologic evolution of the Caribbean and the Gulf of Mexico — some critical problem areas *In*: *The Ocean Basins and Margins*, **3**, The Gulf of Mexico, and the Caribbean. New York, Plenum 663–689.
DUNCAN, R. A. & HARGREAVES, R. B. 1984. Plate tectonic evolution of the Caribbean in the mantle reference frame *In*: The Caribbean–South American Plate Boundary and Regional Tectonics. *GSA Memoir* **162**, 81–93.
EVA, A. N., BURKE, K., MANN, P. & WADGE, G. 1989. Four-phase tectonostratigraphic development of the southern Caribbean. *Marine and petroleum Geology*, **6**, 9–21.
FEO-CODECIDO, G., SMITH, F. D., ABOUD, N. & DI GIACOMO, E. 1984. Basement and Paleozoic rocks of the Venezuelan Llanos basins, *In*: The Caribbean–South American Plate Boundary and Regional Tectonics. *GSA Memoir* **162**, 175–187.
FIORILLO, G. J. 1985. Exploration and Evaluation of the Orinoco Oil Belt, final results. *Third International Conference on heavy crudes and tar sands (UNITAR), Long Beach, California*, **1**, 332–361.
FOX, P. J. & HEEZEN, B. C. 1975. Geology of the Caribbean crust, *In*: *The Ocean Basins and Margins*, **3**, The Gulf of Mexico and the Caribbean, New York, Plenum, 421–466.
FUNES, D. 1985. Tipos de acumulaciones de petroleo en el area de Hamaca: VI Conferencia Geologica Venezolano, Caracas, Memoria, Tomo V, 3015–3036.
FURRER, M. 1972. Paleontology, stratigraphy, fossil tintinnids in Venezuela, VI Conferencia Geologica del Caribe, Margarita, Venezuela, Memorias, 451–454.
—— & URBANI, F. 1973. Nuevas localidades fosiliferas en cuevas ubicadas en las formaciones las Mercedes y Guarico, Estado Miranda: Boletin de la Sociedad Venezolana de Espeleologia, **4**, 135–139.
GABELA, V. H. 1985. Campo Cano Limon, Llanos Orientales de Colombia: II Simposio Bolivariano; Exploracion Petrolera de Las Cuencas Subandinas, Bogota, Publicacion I.
GAJARDO, E. & NICOLLE, J. L. 1986. Modelo de corteza de la costa oriental del Lago de Maracaibo: III Congreso Venezolano de Geofisica, Caracas, Memoria, 102–111.
——, ——, ZUCCA, J. J. & BANDA, E. 1990. Crustal seismic refraction studies in western Venezuela. *Bulletin Seismological Society of America*.
GALLANGO, O., TALUKDAR, S. & CHIN A LIEN, M. 1985. Caracteristicas de los crudos marinos en la Cuenca de Maracaibo, Venezuela Occidental: VI Conferencia Geologica Venezolano, Caracas, Memoria, Tomo III, 1661–1693.
GHOSH, N., HALL, S. A. & CASEY, J. F. 1984. Sea floor spreading magnetic anomalies in the Venezuelan Basin *In*: *The Caribbean–South American Plate Boundary*, GSA Memoir **162**, 65–80.
GONZALEZ DE JUANA, C., & VIGNALI, M. C. 1972. Rocas metamorficas e igneas en la Peninsula de Macanao, Margarita, Venezuela: VI Conferencia Geologica del Caribe, Margarita, Venezuela, Memoria, 63–68.
——, ITURRALDE DE AROZENA, J. M. & CADILLAT, X. P. 1980. *Geologia de Venezuela y de sus Cuencas Petroleras*. Caracas, Ediciones. Foninves.
GONZALEZ, L., ESCOBAR, M., LOPEZ, C. & PASQUALI, J. 1985. Biodegradacion de crudos del centro-oeste y sur-este de la Cuenca Oriental de Venezuela: VI Conferencia Geologica Venezolano, Caracas, Memoria, Tomo III, 1694–1725.
GONZALEZ SILVA, L. A. & PICARD, X. 1972. Sedimentacion y aloctonia en el frente de montanas de Guarico: VI Conferencia Geologica del Caribe, Margarita, Venezuela, Memorias 83–88.
HARDING, T. P. 1974. Petroleum traps associated with wrench faults. *APG Bulletin*, **58**, 1290–1304.
HAY, W. W. 1985. Paleooceanography of the Venezuelan Basin: First Geological Conference of the Geological Society of Trinidad and Tobago, Proceedings, 302–307.
HEY, R. 1977. Tectonic evolution spreading centre. *GSA Bulletin*, **88**, 1404–1420.
HEDBERG, H. D. 1931. Cretaceous limestones as petroleum source-rocks in north-western Venezuela. *AAPG Bulletin*, **15**, 229–244.
ISEA, A. 1987. Geological synthesis of the Orinoco Oil Belt, Eastern Venezuela. *Journal of Petroleum Geology*, **10**, 135–148.
JAMES, K. H. 1985. Marco Tectonico, estilos estucturales y habitat de hidrocarburos cretacicos, Venezuela: VI Conferencia Geologica Venezolana, Caracas, Memoria, Tomo IV, 2452–2469.
KISER, G. D. 1981. Resultados preliminares de la exploracion costafuera: Petroleo International, **39**, 78–84.
—— 1987. Exploration results, Machete area, Orinoco Oil Belt: Journal of Petroleum Geology, **10**, 149–162.
—— & BASS, I. 1985. La reorientation del Arco de El Baul y su importancia economica: VI Conferencia Geologica Venezolana, Caracas, Memoria, Tomo VIII, 5122–5135.
KRAUSE, H. H. 1971. La Falla de Icotea en el Bloque I de las concesiones de La Compania Shell de Venezuela en el Lago de Maracaibo: III Jornadas

Tecnicas de Petroleo, Colegio de Ingenieros de Venezuela, Maracaibo.

—— & JAMES, K. H. 1990. Hydrocarbon resources of Venezuela, their source rocks and structural habitat: *Circum Pacific Symposium on the Hydrocarbon and Mineral Resources of the Andes, Santiago, Chile, 1985.*

LADD, J. W. 1976. Relative motion of South America with respect to North America and Caribbean tectonics: *GSA Bulletin*, **87**, 969–976.

——, TRUCHAN, M., TALWANI, M., STOPOFFA, P. E., BUHL, P., HOUTZ, R., MAUFFRET, A. & WESTBROOK, G. 1984. Seismic reflection profiles across the southern margin of the Caribbean *In: The Caribbean–South America Plate Boundary and Regional Tectonics, GSA Memoir* **162**, 153–212.

LARSON, R. L. & LADD, J. W. 1973. Evidence for the opening of the South Atlantic in the Early Cretaceous. *Nature*, **246**, 209–212.

LEHNER, P., DOUST, H., BAKKER, G., ALLENBACH, P. and GUENEAU, J. 1984. Caribbean margin of South America, profiles C-1422, C-1412 and C-1413 *In:* BALLY, A. W. (ed.). Seismic expression of structural styles, *AAPG Studies in Geology, series* **15**, 3.4.2.128.

LEW, M. & ALBARRACIN, J. 1988. Generacion de hidrocarburos en La Bahia de El Tablazo: III Simposio Bolivariano, Exploracion Petrolera de Las Cuencas Subandinas, Caracas, Memoria, 694–722.

LIDDLE, R. A., 1931. Cretaceous limestones as petroleum source-rocks in northwestern Venezuela, Discussion: AAPG Bulletin, *15*, 244–246.

LONSDALE, P. & KLITGORD, K. D. 1978. Structure and tectonic history of the eastern Panama Basin: GSA Bulletin, **89**, 981–999.

MACELLARI, C. E. & DE VRIES, T. J. 1987. Late Cretaceous upwelling and anoxic sedimentation in northwestern South America: *Palaeogeography, Palaeoclimatology, Palaeoecology*, **59**, 279–292.

MACSOTAY, O. 1972. Significado cronologico y paleoecologico de los amonites desenrollados de la Formacion Chuspita del Grupo Caracas: IV Congreso Geologico Venezolano, Memoria, Caracas, 1703–1713.

MALFAIT, B. T. & DINKLEMAN, M. G. 1972. Circum-Caribbean tectonic and igneous activity and the evolution of the Caribbean Plate. *GSA Bulletin*, **83**, 251–272.

MARESCH, W. V. 1974. Plate tectonics origin of the Caribbean Mountain System of northern South America: discussion and proposal. *GSA Bulletin*, **85**, 669–682.

MARTINEZ, A. R. 1987. The Orinoco Oil Belt, Venezuela: Journal of Petroleum Geology: **10**, 125–134.

MATTSON, P. H. 1984. Caribbean structural breaks and plate motions *In:* The Caribbean–South America Plate Boundary and Regional Tectonics, *GSA Memoir* **162**, 131–149.

MAZE, W. B. 1984. Jurassic La Quinta Formation in the Sierra de Perija, northwestern Venezuela: Geology and tectonic environment of red beds and volcanic rocks *In: The Caribbean–South American Plate Boundary and Regional Tectonics: GSA Memoir* **162**, 263–282.

MCCOLLOUGH, C. and PADFIELD, E. 1985. Petroleum Geology of Colombia's Llanos Basin: a summary: Oil and Gas Journal, April 1985, 82–86.

MCDOWELL, A. N. 1975. What are the problems in estimating the oil potential of a basin? *The Oil and Gas Journal*, 85–90.

MINSTER, J. & JORDAN, T. 1978. Present day plate motions. *Journal of Geophysical Research*, **83**, 5331–5354.

MOLNAR, P. & SYKES, L. R. 1969. Tectonics of the Caribbean and Middle America regions from focal mechanisms and seismicity: *GSA Bulletin*, **80**, 1639–1684.

MOTICSKA, P. 1975. Sierra de Perija, Excursion No. 2: Complejo volcanico-plutonico de El Torumo-Inciarte. Congreso Latinoamericano Geologico II, Caracas, 1973, Memoria, Boletin Geologico, Publicacion Especial, **7**, 306–311.

——, 1985. Volcanismo mesozoico en el subsuelo de La Faja Petrolifera del Orinoco, Estado Guarico, Venezuela: *VI Congreso Geologico Venezolano, Caracas, Memoria, Tomo III,* 1929–1943.

MUESSIG, K. W. 1984. Structure and Cenozoic tectonics of the Falcon Basin, Venezuela and adjacent areas, *In: The Caribbean–South American Plate Boundary and Regional Tectonics, GSA Memoir* **162**, 217–230.

ODREMAN, O. R. & BENEDETTO, G. 1977. Paleontologia y edad de la Formacion Tinacoa, Sierra de Perija, Estado Zulia, Venezuela. *V Congreso Geologico Venezolano, Memoria, Tomo I*, 15–25.

PALMER, S. E. & RUSSEL, J. A. 1988. The five oil families of the Llanos Basin. *III Simposio Bolivariano, Exploracion Petrolera de Las Cuencas Subandinas, Caracas, Memoria*, 723–753.

PEIRSON, A. L., SALVADOR, A. & STAINFORTH, R. M. 1966. The Guarico Formation of north-central Venezuela: Association Venezolano de Geologia, Minas y Petroleo, Boletin Informativo, **9**, 183–244.

PEREIRA, J. G. 1985. Evolucion tectonica de la Cuenca de Carupano durante el Terciario: VI Congreso Geologica Venezolano, Caracas, Memoria, Tomo IV, 2618–2648.

PENNINGTON, W. D. 1981. Subduction of the Eastern Panama Basin and seismotectonics of northwestern South America. *Journal of Geophysical Research*, **86**, 10753–10770.

PEREZ, DE MEJIA, D. & TARACHE, C. 1985. Sintesis Geologica del Golfo de Paria: VI Congreso Geologico Venezolano, Caracas, Memoria, Tomo V, 3243–3277.

PIERCE, G. R. 1960. Geologia de la Cuenca de Barinas: III Congreso Geologico Venezolano, Caracas, Memoria, Boletin Geologico, Publication Especial, **3**, 214–276.

PINDELL, J. & DEWEY, J. F. 1982, Permo-Triassic reconstruction of western Pangaea and the evolution of the Gulf of Mexico/Carribean region:

Tectonics, **1**, 179–211.

PITMAN, W. C. & TALWANI, M. 1972. Sea floor spreading in the North Atlantic. *GSA Bulletin*, **83**, 619–646.

PRIEM, H. N. A., BEETS, D. J. & VERDURMEN, E. A. Th. 1986. Precambrian rocks in an early Tertiary conglomerate on Bonaire, Netherlands Antilles (southern Caribbean borderland): evidence for a 300 km eastward displacement relative to the South American mainland? *Geologie en Mijnbouw*, **65**, 35–40.

QUIJADA, E. & CALDERA, A. A. 1985. Geologia del Campo Alpuf, Estado Zulia. *VI Congreso Geologico Venezolano, Caracas, Memoria, Tomo V*, 3328–3350.

RABASSO-VIDAL, J. 1985. Sugerencias/analogias: Exploracion de trampas estratigraficas en una cuenca petrolera madura, region de Anaco-Oficina. *VI Congreso Geologico Venezolano, Caracas, Memoria, Tomo V*, 3351–3364.

RENZ, O., LAKEMAN, R. & VAN DER MEULEN, E. 1955. Submarine sliding in western Venezuela. *AAPG Bulletin*, **39**, 2053–2067.

ROBERTO, M., MOMPART, L., PUCHE, E. & SCHERER, F. 1990. *In*: BROOKS, J. (ed.). New oil discoveries in the Ceuta area, SE Lake Maracaibo, Venezuela. *Classic Petroleum Provinces*, Geological Society, London, Special Publication **xx**, xxx–xxx.

ROSS, M. I. & SCOTESE, C. R. 1988. A hierarchical tectonic model of the Gulf of Mexico and Caribbean region. *Tectonophysics*, **155**, 139–168.

SANTOS, A., & FRONTADO, L. Reservoir geology of the Cerro Negro steam injection area, Orinoco Oil Belt, Venezuela: Journal of Petroleum Geology, **10**, 177–194.

SCHUBERT, C. 1979. Late Pleistocene and Holocene faulting in Lake Valencia Basin, north-central Venezuela. *Geology*, **7**, 289–292.

—— 1982. Origin of the Cariaco Basin, southern Caribbean Sea. *Marine Geology*, **47**, 345–360.

—— 1985. Cuencas cenozoicas tardias de Venezuela: VI Congreso Geologico Venezolano, Caracas, Memoria, Tomo IV, 2663–2679.

—— 1986. Neotectonic Aspects of the Southern Caribbean Plate Boundary. First Geological Conference of the Geological Society of Trinidad and Tobago, Transactions, 265–269.

SIBUET, J. C. & MASCLE, J. 1978. Plate kinematic implications of equatorial fracture zone trends. *Journal of Geophysical Research*, **83**, 3401–3421.

SILVA, L. G. & MARQUEZ, P. 1985. Re-evaluation de la extension noreste de Jusepin: VI Congreso Geologico Venezolano, Caracas, memoria, Tomo V, 3057–3084.

SILVER, E. A., CASE, J. E. and MACGILLAVERY, H. J. 1975. Geophysical study of the Venezuelan borderland, *Geological Society of America Bulletin*, **86**, 213–226.

SOULAS, J. P. 1985. Neotectonica del flanco occidental de Los Andes de Venezuela entre 70 30 y 71 00' (Fallas Bocono de Valera, Tuname, Pinango y del Piedemonte): VI Congreso Geologico Venezolano, Caracas, memoria, Tomo IV, 2687–1811.

—— 1987. Proyecto costa oriental del Lago de Maracaibo, actividad Quaternario y caracteristicas sismogenicas del sistema de fallas de Oca-Ancon y de las fallas Lagarto, Urumaco, Rio Seco y Pedregal; afinamiento de las caracteristicas sismogenicas de las fallas Mene Grande y Valera: Informe Final, Funvisis, Caracas, Venezuela, 78p.

SPEED, R. C. 1986. Cenozoic tectonics of the eastern Caribbean and Trinidad: First Geological Conference of the Geological Society of Trinidad and Tobago, 1985, Transactions, 270–280.

STEPHAN, J. F., BECK, C., BELLIZIA, A. & BLANCHET, R. 1980. La chaine caraibe du Pacifique a l'Atlantique: Memoires Bureau Recherches Geologiques et Minieres, **115**, 38–59.

SYKES, L. R., MCCANN, W. R. & KAFKA, A. L. 1982. Motion of the Caribbean Plate during the last seven million years and implication for earlier Cenozoic movement: *Journal of Geophysical Research*, **87**, 10656–10676.

TALUKDAR, S. & DE BOLIVAR, E. H. 1982. Petroleum Geology of the Tuy-Cariaco Basin, Eastern Venezuelan continental shelf: a preliminary appraisal. *Informes* Tecnicos, Intevep, S. A.

—— & DE TONI, B. 1988. Modelaje geoquimico en la cuenca inexplorada de Guarumen, Venezuela. *III Simposio Bolivariano, Exploracion Petrolera de las Cuencas*.

——, GALLANGO, O. & CHIN-A-LIEN, M. 1985a. Generation and migration of hydrocarbons in the Maracaibo Basin, Venezuela. *12th International Meeting on Organic Chemistry*, Munich, Abstracts, 28.

——, —— & RUGGIERO, A. 1985b. Formaciones La Luna y Querecual de Venezuela: rocas madres de petroleo: *VI Congreso Geologico Venezolano, Caracas, Memoria, Tomo VI*, 3606–3642. Subandinas, Caracas, Memorias, 755–781.

VASQUEZ, E. E. & DICKEY, P. A. 1972. Major faulting in northwestern Venezuela and its relation to global tectonics. *VI Conferencia Geologica del Caribe, Margarita, Venezuela, Memorias*, 191–202.

VASQUEZ, H., C. A. 1988. Mapa estructural del tope del Cretaceo, Cuenca Llanos Orientales de Colombia. *III Simposio Bolivariano, Exploracion Petrolera de las Cuencas Subandinas, Caracas, Memorias*, 580–611.

VEGA, A. & DE ROJAS, I. 1987. *Exploration and evaluation of the Zuata area, Orinoco Oil Belt, Venezuela*.

VIERMA, L. 1985a. Caracterizacion geoquimica de kerogeno en rocas carbonaticas de La Formacion La Luna. *VI Congreso Geologico Venezolano, Caracas, memoria, Tomo IV*, 2263–2280.

—— 1985b. Correlacion de crudos y rocas madres en una porcion de la Cuenca de Maracaibo, Venezuela. *VI Congreso Geologico Venezolano, Caracas, Memoria, Tomo IV*, 2263–2280.

VIGNALI, M. 1979. Estratigrafia y estructura de las

cordilleras metamorficas de Venezuela oriental (Peninsula de Araya-Paria e Isla de Margarita): Geos, Escuela de Geologia y Minas, Universidad Central de Venezuela, Caracas, **25**, 19–66.

WADGE, G. & MACDONALD, R. 1985. Cretaceous tholeiites of the northern continental margin of South America: the Sans Souci Formation of Trinidad: *Journal of the Geological Society of London*, **142**, 297–308.

WEHRMANN, M. 1972. Geologia de la region de Guatire-Colonia Tovar. *Congreso Geologico Venezolano, Caracas, Memoria, Tomo IV*, 2093–2219.

YOUNG, G. A. 1988. La tectonica de la Cuenca Barinas–Apure, Venezuela interpretation magnetica detallada. *III Simposio Bolivariano, Exploracion Petrolera en Las Cuencas Subandinas, Caracas, Memoria*, 962–984.

ZAMORA, L. G., GONZALES, L. & LINARES, L. M. 1988. The Orinoco Delta, a future exploratory province for heavy and extra heavy oils: Fourth UNITAR/UNDP *International Conference on Heavy Crude and Tar Sands*, Paper No. 169.

ZEUSS, H. 1985. Exploracion sistematica de trampas estratigraficas en las areas mayores de Oficina y Anaco, Cuenca Oriente de Venezuela. *VI Congreso Geologico Venezolano, Caracas, Memoria, Tomo VI*, 3671–3698.

Geological integration and evaluation of Northern Monagas, Eastern Venezuelan Basin

R. AYMARD[1], L. PIMENTEL[1], P. EITZ[1], P. LOPEZ[1], A. CHAOUCH[1], J. NAVARRO[1], J. MIJARES[1] & J. G. PEREIRA[2]

[1] *Corpoven, S.A.*
[2] *Lagoven, S.A.*

Abstract: Over the last 60 years, the Eastern Venezuelan Basin has been heavily explored with giant oil accumulations being developed in such areas as Oficina, Anaco, Quiriquire and the Orinoco Oil Belt in the states of Anzoategui and Monagas. Since 1986, interdisciplinary studies and seismic interpretations have led to discovery of a trend of deep structures on the northern flank of the basin, some of them beneath old early Tertiary producing fields. The first exploratory well, FUL-1, tested 12 000 BOPD of 28° API crude. The new reservoirs consist of massive sandstones of the Eocene Caratas Formation and Oligocene Naricual and Los Jabillos Formations between depths of 13 000 and 17 000 feet. The main structures placed in a thrust fold belt as a series of anticlines, extending over 100 km across Northern Monagas; currently four wells are being drilled and six more exploratory wells are planned to be drilled in the new future. To date over 1900 MMBBLS of reserves have been discovered, and possible reserves for the area are close to 8.6 billion BBLS.

The data and some of the concepts presented in this paper have originated from the work carried out by Lagoven, S.A. and Corpoven S.A., both affiliates of PDVSA (Petróleos de Venezuela, S.A.) between 1978 and 1987.

The Eastern Venezuela Basin is the second most important oil province of South America, behind the Maracaibo Basin. It has been developed over the last 60 years in the traditional oil fields of Anaco, Oficina, Santa Bárbara, Orocual and Quiriquire, all of them well known in the world's petroleum geologic history (Fig. 1).

During the late 1970s and 1980s, new and modern techniques of acquisition and processing of seismic information, together with the geological and geophysical integration has led to interpretation of deep exploration targets in this area.

The zone of exploration interest has been extended from previous Upper Miocene−Pliocene traditional plays, to the deeper section with high potential reservoir formations of Oligocene−Eocene and Cretaceous in age.

The most important reservoir beds discovered are shallow marine sandstones deposited on a gently sloping platform. These are Oligocene in age, and correspond to the Naricual Formation.

Four oil fields have been discovered in the last four years. These fields are all found in large anticlinal structures faulted by thrusting and separated by strike-slip faults. These structures have been developed in a compressional regime which was acting in Eastern Venezuelan during Miocene times.

This paper presents the results of the largest integrated exploration campaign undertaken in Northern Monagas and the Interior Range since 1978. It includes the latest geological concepts developed and will show the discoveries achieved as a result of these studies.

Exploration history

Oil exploration in the Eastern Venezuelan Basin began as early as 1890, but commercial production was not established until 1913, when the first well Bababui-1 was drilled in the vicinity of Guanoco Lake in the state of Sucre (Fig. 2).

Oil seepages and the interest of searching for new prospective areas in the Interior Mountain Range, especially in the foothills of the Northern Monagas area, led to important discoveries in the Upper Tertiary sequence. Traditional oil fields like Quiriquire (1928) which has produced over 760 million barrels of oil from alluvial fans and, later on, Jusepin (1936) and Orocual (1938), were the result of the first geologic interpretations of the reconnaissance geophysical surveys (magnetics and refraction) which permitted at that time, the delineation of some of the Upper Miocene−Pliocene events in the subsurface.

Later on during the 1940s and 1950s surface geology and the first seismic reflection surveys were accomplished. As a result, some other

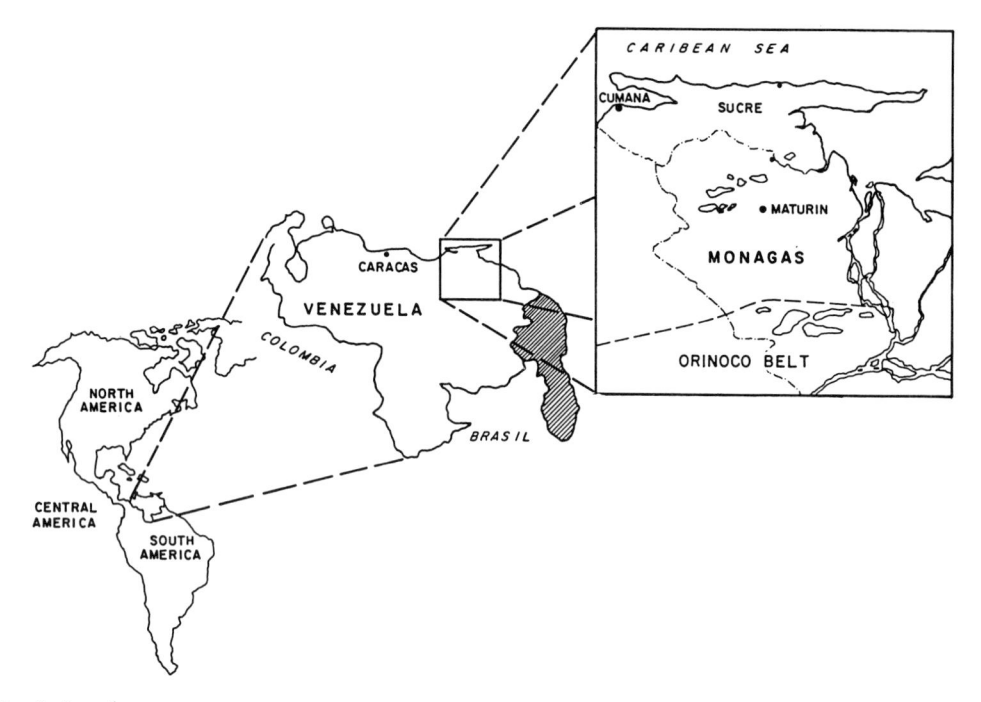

Fig. 1. Location map.

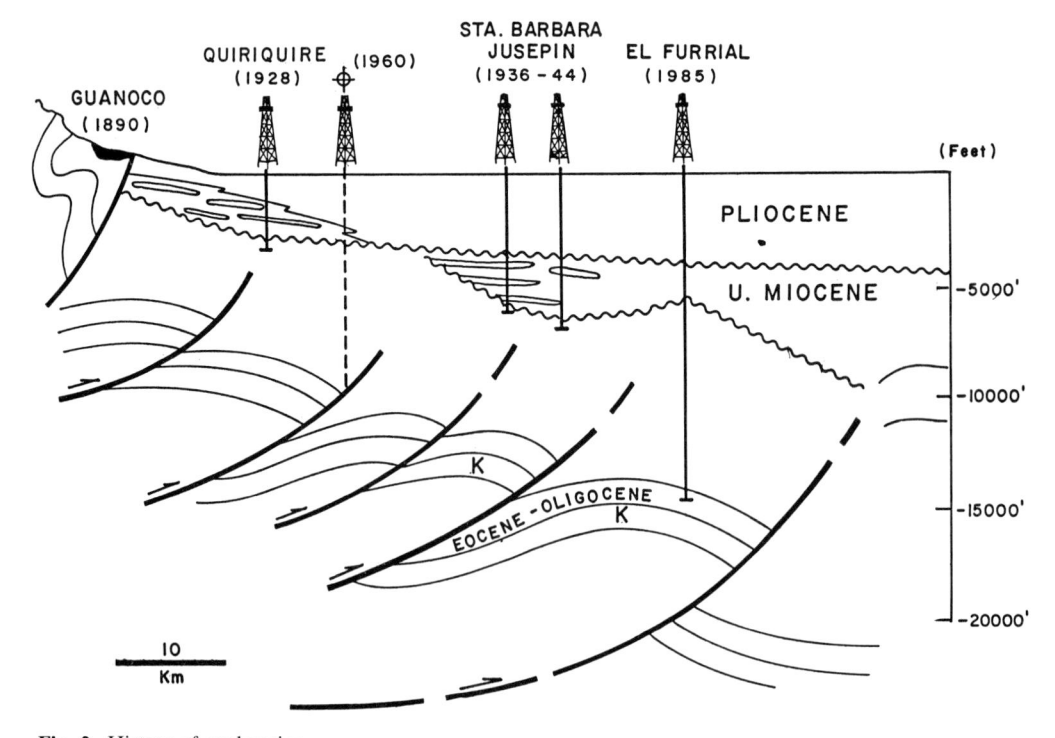

Fig. 2. History of exploration.

fields were discovered: Manresa, Santa Bárbara, Mulata, Muri, later on some deeper structures were evaluated in Quiriquire and Orocual, but without any positive results. Salvador (1958) mentioned that the purpose of most of the wells drilled at the time was to obtain geological information.

During the first period of exploration (1890–1960) shallow targets of the Upper Miocene–Pliocene were evaluated and only seven wells penetrated a sequence over 13 000 feet in depth.

Obviously geologic subsurface knowledge was restricted and the capability and experience in deep drilling was a great constraint at the time.

After 1976 new exploration campaigns were initiated throughout the country, and in 1978 a regional seismic reflection survey of 716 km was carried out. This survey was considered regular to poor in quality as a result of the structural complexities which were not foreseen at that moment.

This information was fundamental in giving the first general indication of the deep structural configuration of the Maturin subsurface area, and it was essential for the following seismic surveys conducted during 1984 and 1987. Over 5000 km of seismic lines (vibroseis and dynamite) have been recorded in this area, together with a gravimetric survey that consisting of 5000 stations.

During 1985, parallel with acquisition of the seismic data, a drilling campaign started with two wells; ORS-52 in the Orocual field and the well FUL-1 in a new area designated as El Furrial. This last well was located a few kilometres South of the traditional field of Jusepin.

Excellent results were obtained in these two wells, considering the production of condensates of the ORS-52 well, and the fact that the El Furrial is a giant field, with more than a billion and a half barrels of oil discovered.

In 1987 the prospect of Musipan and El Carito (Musi-IX and Cari-IX wells) were drilled on adjacent structures to El Furrial, these yielded outstanding results. The El Carito prospect generated another giant discovery with some 2.4 billion barrels of light crude oil.

Regional setting

The Eastern Venezuelan Basin comprises an area of some 165 000 km². It is bounded to the north by the Margarita Island and the Araya–Paria outcrops which form part of the allochtonous metamorphic Caribbean chain (Fig. 3). These rocks are Jurassic–Cretaceous in age and form a series of nappes (Bellizzia 1979).

The main tectonic element present in the North is the El Pilar fault, a right lateral strike-slip fault that has been considered by many authors as the border between the South American and the Caribbean Plates.

Relative motion eastward of the Caribbean plate with respect to South America has been estimated at a rate of 2 to 4 cm/year (Sykes et al. 1982).

To the South the basin is bounded by the Guiana Shield which, is composed of rocks of various lithologies of Pre-Cambrian age. These rocks were an important source of sediments during the passive margin stage (Cretaceous–Early Tertiary).

To the West it is bounded by the El Baul Arch, and to the East towards the Gulf of Paria and Trinidad dips into the Atlantic Ocean.

Two important Geological provinces are identified in the area: The Interior Range in the North and the Maturin Sub-basin towards the foothills and the plains South of the Maturin area (Fig. 4).

Interior Range

A described by Subieta et al. (1988) the Interior Range is represented by a wedge of sediments that were deposited in a continental passive margin during Cretaceous to Early Tertiary times. These sediments were folded and thrusted during the collision of the South American and Caribbean plates, starting in this area during Eocene–Oligocene times.

Main structural features found in the Interior Range are large-scale concentric folds mainly of Cretaceous rocks. These folds have an orientation N 70°–80°E, and some are up to 70 km long by 5 km wide. Depending on competence of the Cretaceous rocks, these anticlines sometimes deform by box folding.

These anticlinal features are accompanied by thrust and overthrust faulting parallel to the axis of the structures, with dip to the north and are believed to reach a detachment zone or a sole thrust in depth.

A geological crossection of the Interior Range based on surface work, estimates a shortening of 40 km that represents 28% (Rossi et al. 1987).

Other important elements are the Urica fault trending NW–SE, considered a vertical right lateral transcurrent fault (Rod 1959) displacing anticlinal structures perpendicular to their axial plane. This fault seems to have been active previous to Cretaceous time.

As cited by Subieta et al. (1988) the San Francisco Fault is a NW–SE right lateral strike-slip fault, described as Rosales (1969). This

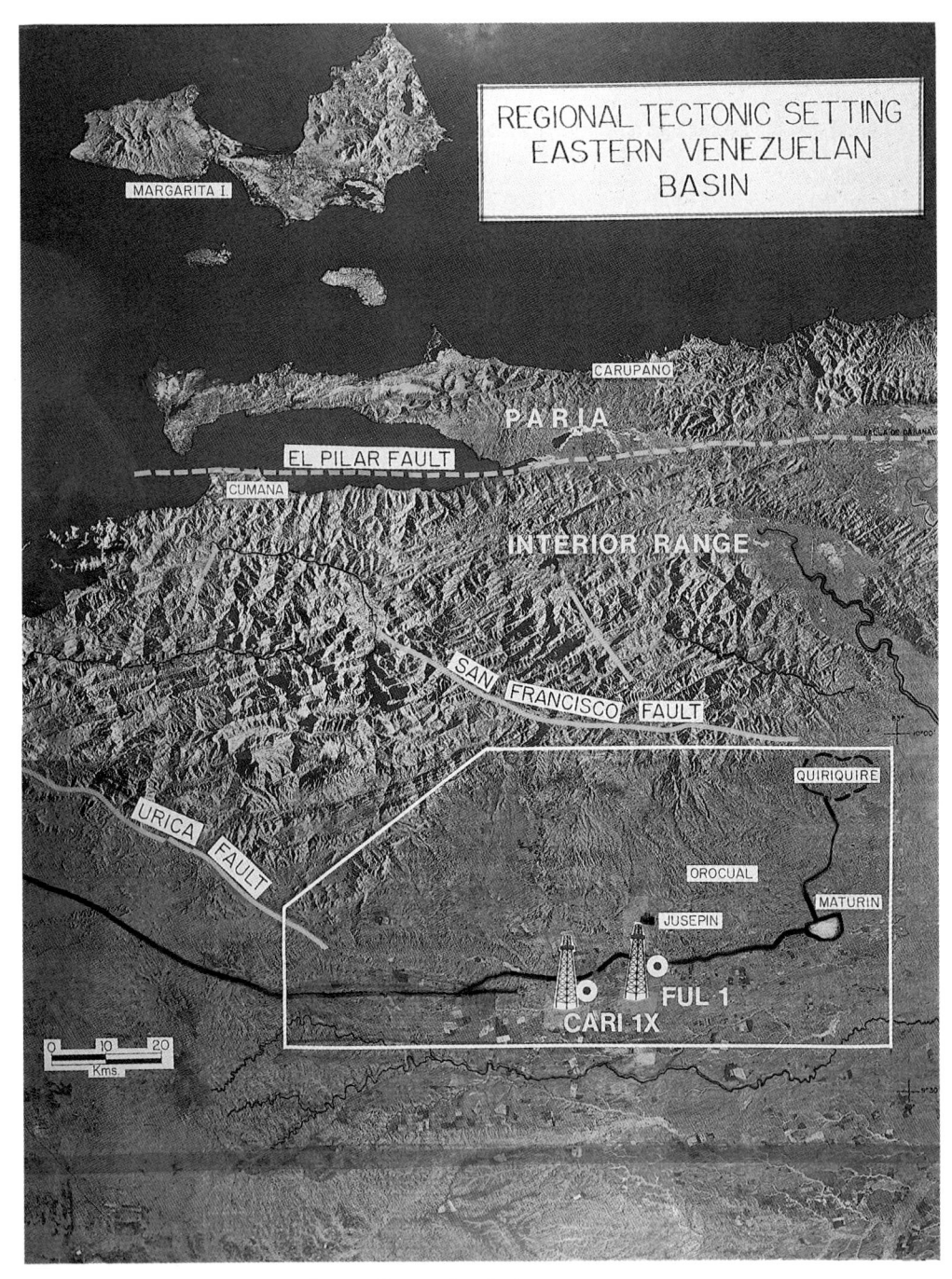

Fig. 3. Regional tectonic setting, Eastern Venezuelan Basin.

fault is vertical in the Northern part, and horizontal towards the South, close to the Quiriquire area, where merges to the Pirital thrust fault.

Rosales estimated a horizontal displacement of the San Francisco Fault of 25 to 35 km based on subsurface on isopach maps of the San Juan

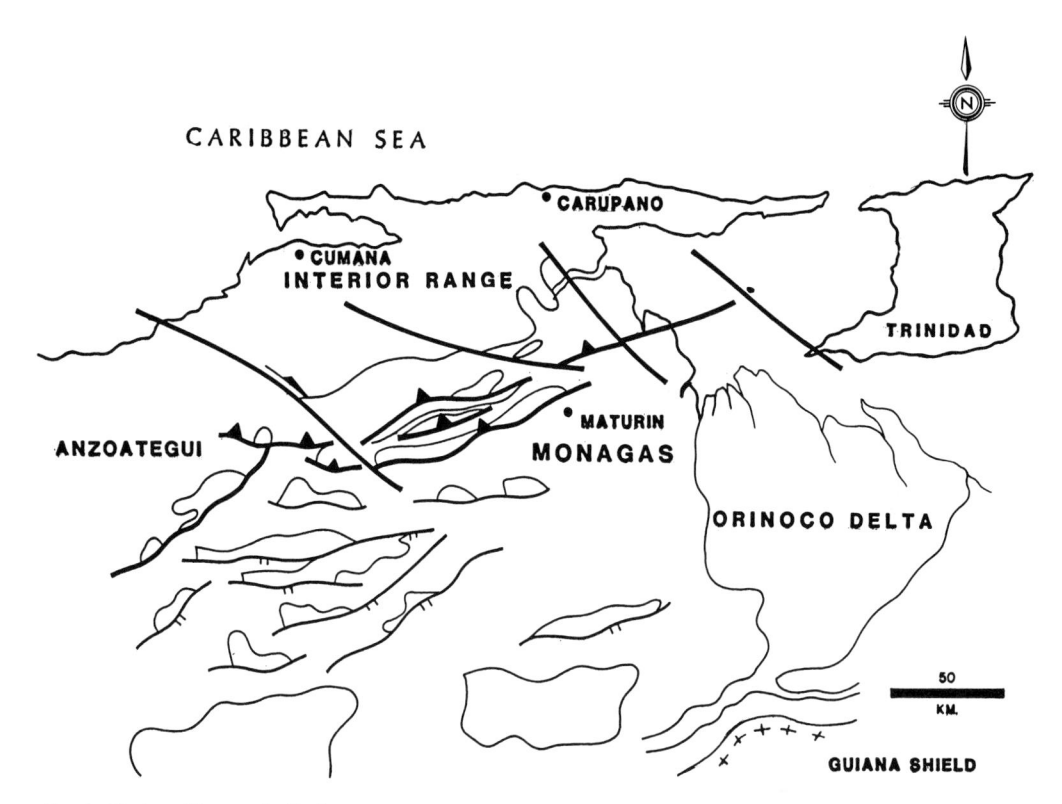

Fig. 4. Eastern Venezuela Basin.

Formation in the Quiriquire Field. The author associates the formation of this fault contemporaneous with folding.

Rossi *et al.* (1987) describe a right lateral horizontal displacement on the San Francisco Fault of 15–20 km on surface work.

Maturin Sub-basin

The Maturin Sub-basin, as a part of the Eastern Venezuelan Basin has a passive southern flank against the Guiana shield which constitutes the basement, and was also an important source of the sediments that where deposited during Cretaceous and Tertiary times (Fig. 4).

This Sub-basin is asymmetric and elongated in shape, trends in a ENE direction, parallel to the Interior Mountain Range 'Serranía del Interior'. More than 30000 feet of sedimentary fill have been deposited, mainly of marine and continental in origin.

The main structural feature in this area is the Pirital overthrust system (Fig. 4) which trends in a N 70°E direction and is considered a low-angle emplaced Cretaceous allocthonous block with a displacement of some 10 km.

By means of the structural configuration, the Maturin Sub-basin can be subdivided into three important provinces; the Compressional Province in the Interior Range, the Frontal Deformation at the foothills, and the Extensional province to the South.

The Tectonic evolution of the Maturin Sub-basin started previous to Cretaceous times, when a predominant extentional regime was active essentially along the Northern borther of South America (Fig. 5).

During the Late Cretaceous and Early Eocene times, a passive continental margin with a stable platform dipping gently to the North was mainly the environment of continuous and uninterrupted sedimentation (mainly deep and shallow marine to the north and continental towards the South) during the early stages of the development of the basin. The distributary province or main source of sediments came from the south.

By Late Eocene to Early Miocene times, uplifting in the northern area started, giving origin to the Interior Mountain Range and the Maturin Sub-basin. During Middle Miocene an oblique transpressive regime produced by the collision of the South American and Caribbean

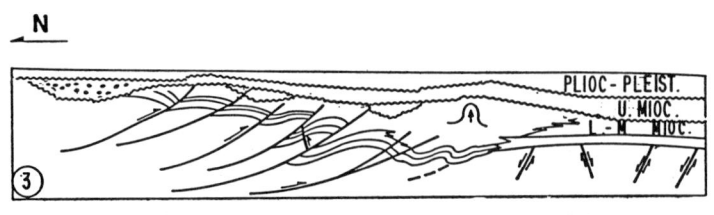

PLIO-PLEIST. DIMINISHED TRANSPRESSURE

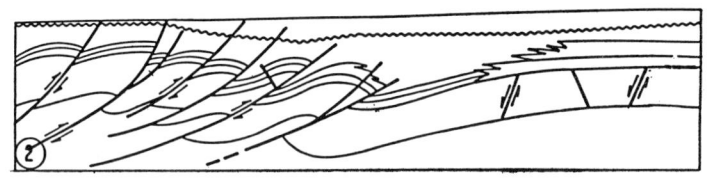

MID.MIOC-FORELAND FOLD AND THRUST BELT

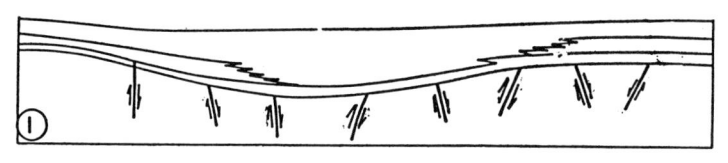

K-EOC-PASSIVE MARGIN

EOC-LOW.MIOC -FORELAND BASIN

Fig. 5. Tectonic evolution of the Maturin Sub-basin.

plates; formed from North to South, the fold and thrust belt and the foreland Maturin Sub-basin parallel to the range. This event is evidenced by a strong unconformity observed in seismic lines and dated by wells as Middle Miocene.

During Pliocene to Pleistocene times, although transpressure seemed to continue, it diminished towards the South. The deformation of the sedimentary cover in the Northern Monagas area shows extentional features, while younger rocks towards the eastern part of Venezuela and Trinidad are still under compression. This means that there has existed an active North to South and West to East migrating compressional front, from Late Eocene to Present.

Mud volcanoes are also observed in the sub-basin, South of Maturin. Some of them are observed on the surface, while on seismic lines they are present in front of the large thrusted anticlinal structures. These features can be followed for many kilometres towards Trinidad and trend parallel to the main anticlines. The upper sequence of La Pica and Las Piedras formations is clearly deformed as a product of this injection of shales.

As explained by Subieta et al. (1987), the origin of diapirs is in great part tectonic. They are related to the high rate of sedimentation with a low initial permeability. The compressive tectonics is responsible for raising the fluid pressure in the pores, and determining the initial flow.

Stratigraphy

The sedimentary rocks found in this area are the result of two major depositional cycles of Cretaceous–Eocene, and post-Eocene ages respectively (Fig. 6).

The Cretaceous–Eocene section was deposited on a stable continental shelf, bordered by the Guyana Shield to the south, with an open sea to the North.

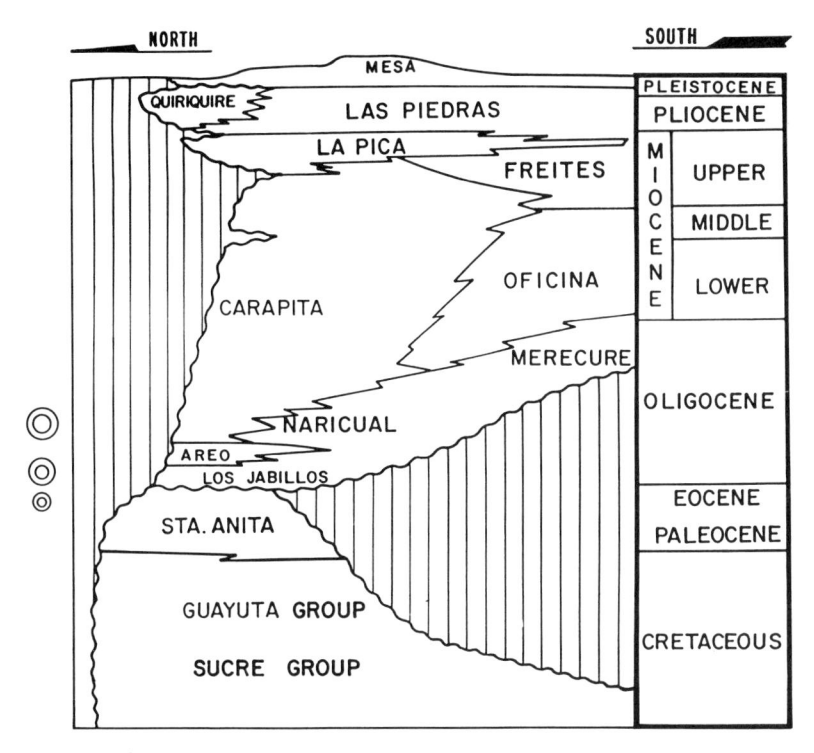

Fig. 6. Stratigraphic column.

The post-Eocene section is characterized by the presence of various regional unconformities caused by successive orogenic events originated in the North. It is this section which contains the most important reservoirs in the Northern Monagas area. These reservoirs of the Naricual and Los Jabillos Formations are overlain by an excellent, thick, shale seal of the Carapita Formation.

The following is a brief description of the formations based on outcrop recognition and information from wells.

Sucre group

The Sucre Group of Lower Cretaceous age includes the Barranquin, El Cantil and Chimana Formations. The Barranquin Formation is composed mainly of sandstones and siltstones. The El Cantil and Chimana are mainly limestones.

Guayuta group

The Guayuta Group of Cenomanian–Campanian age is composed of the Querecual and San Antonio formations which constitute the source rocks for most of the oil in the Northern Monagas area.

Querecual Formation (Upper Albian)

This formation is composed of fine grained calcareous shales, argillaceous limestones and shales and massive black limestones with large calcareous concretions, towards the top presents thin laminated cherts. The Querecual formation is equivalent to the La Luna Formation of Western Venezuela and the Naparima Hill Formation of Trinidad.

San Antonio Formation (Santonian–Campanian)

Composed primarily of sandstones interbedded with shales, silts, and black limestones, similar to those of the Querecual Formation. The sandstones are massive, greyish-white, calcareous and glauconitic. The shales are grey, carbonaceous, and calcareous. The limestones are frequently dolomitic. The environment of deposition is pelagic, with euxinic sea bottom conditions (González de Juana *et al.* 1980). This formation has a diachronic upper contact with the San Juan Formation.

Santa Anita group

The Santa Anita group has an age of Upper Cretaceous to Eocene. It is composed of the San Juan, Vidoño and Caratas Formations.

San Juan Formation (Maestrichtian)

This is the uppermost Cretaceous horizon encountered in the section. It is primarily made up of massive sandstones. These sandstones are fine to coarse grained, occasionally shaly, calcareous, slightly glauconitic, friable, with fine dark shale laminations. Thin dolomitic sandstones are frequent.

Vidoño Formation (Lower Maestrichtian – Lower Eocene)

The lower portion of the Vidoño Formation is made up of interbedded sands and shales. Shaliness increases upwards in the section so that the upper part is primarily composed of shale. These shales are black and calcareous.

Caratas Formation (Lower to Middle Eocene)

The Caratas Formation is made up of interbedded sandstones, silts, and shales. The sandstones are light brown to grey in color, fine to medium grained, calcareous and slightly glauconitic. Thin bedded limestones and anhydrites are present in the upper portion of the Caratas Formation. These are detected by increases in formation density. The environment of deposition is continental to nearshore marine. The contact with the overlying Los Jabilos Formation is unconformable.

Merecure group

This group is composed of the Los Jabillos, Areo and Naricual Formations.

Los Jabillos Formation (Lower Oligocene)

Consists of fluvial sand shale sequences. The sands are shaly, fine to medium grained, with siliceous and authigenic clay cement. Its upper contact is diachronous and transitional with the Areo Formation.

Areo Formation (Middle Oligocene)

The Areo Formation is a regional dark grey shale generally between 100 and 200 feet thick which divides the massive sand shale sequences of the Naricual and Los Jabillos Formations.

Naricual Formation (Upper Oligocene)

The Naricual Formation is made up of massive shallow marine, deltaic, to alluvial sandstones. These sandstones comprise the most important reservoirs in Northern Monagas. These sands are grey to brown, friable, fine to medium grained moderately rounded to subrounded, and well sorted. The massive sandstones contain thin layers of shale and lignite. This formation is generally between 1500 and 1700 feet thick. Its upper contact with the Carapita Formation is concordant.

Carapita Formation (Oligocene–Miocene)

The Carapita Formation is composed predominantly of thick, sub-compacted, overpressured shales; they tend to be silty and calcareous throughout the entire section with traces of pyrite, anhydrite, calcite and chert. Although the Carapita Formation is predominantly a shale sequence, thin sand lenses are sometimes present. These have been interpreted as turbidite sands. The environment of deposition of the Carapita Formation is deep marine to abyssal plain.

La Pica Formation (Upper Miocene)

The lithology is composed primarily of clays and shales with minor amounts of sandstones and silt. Traces of chert, dolomite and anhydrite are common in the upper portion. The presence of siderite in it may indicate be a transition zone between the La Pica and the Las Piedras Formations. The La Pica Formation was deposited in paleolows created by the tectonics which affected this region. There are two intraformational unconformities.

Las Piedras Formation (Pliocene– Pleistocene)

The upper section is composed primilly of grey to red clays. Sands become increasingly more persistent at the base with the appearance of traces of dolomite, and chert. This appears to mark the transition zone between continental and marine deposition. The Las Piedras Formation was deposited in a fluvio-deltaic environment.

Mesa Formation (Pleistocene)

The lithology of the Mesa Formation consists of buff to red shales near the top, grading into coarse grained sands and conglomerates near the base. The resistance to erosion of these conglomerates forms certain topographic features, from which this formation is named.

Geochemical aspects

The results from a regional geochemical evaluation of oils and oil source rock correlation study of the possible Cretaceous and Tertiary source rocks of Northern Monagas, was carried out by Intevep, S.A. (Talukdar et al. 1986).

This study shows that the Upper Cretaceous (Cenomanian–Campanian) Querecual and San Antonio formations of The Guayuta Group are the source rocks responsible for the hydrocarbons present in the Eastern Venezuelan Basin. This source rock was deposited in a passive continental margin and contains a rich marine organic content deposited under highly anoxic conditions (Hunt 1979).

The Guayuta Group has been lithologically described as fine-grained calcareous shales, argillaceous limestones and shales with an average thickness of 2000 feet estimated in subsurface and some 3400 feet measured in outcrops.

Talukdar et al. (1986), gave an estimated that 55% of the total thickness could be considered as a source rock, and has shown values of TOC between 0.25 and 6.6%, and a hydrocarbon yield content of 454 mg Hc/gTOC.

In the study three different organic facies were recognized; a Facies 'A' with 85% amorphous marine organic matter (present in the Interior Mountain range); Facies 'B' with 60% amorphous material and less than 30% of woody material distributed in very thin laminae parallel to the stratification; and Facies 'A + B' located in the Southern part of the sub-basin with low maturity values ($R_0 = 0.5\%$), and a hydrocarbon index of 15.29 mgHc/gTOC.

Oil generation and migration model

Talukdar et al. (1986) proposed a model of oil generation and migration for the basin (Fig. 7). They estimate that a first stage of maturation had occurred prior to the tectonic emplacement (Middle–Miocene) of the Northern Interior Range.

The same characteristics were also observed in the structures along the Maturin Sub-basin in the South; but in the geochemical study they consider that maturity observed in the rocks was not high enough for oil to be expelled from the source rock. They conclude that the generation and migration of oil took place in the foreland basin, during Miocene to Recent times.

The oil found in the Northern flank of the Maturin Sub-basin in the Cretaceous and Tertiary sequences is described as mature, marine, with or without bacterial alteration, sometimes found mixed, apparently from different stages of generation.

The oil seems to have been generated in, and migrated from, the active Upper Miocene to Recent kitchens in the Cretaceous source rocks that are found beneath the main thrust systems.

The updip, southward favorable position of the foreland basin, permitted during the first stages of migration, long distance, lateral migration from the North during Upper Miocene to the present.

In the study Talukdar et al. (1986) proposed that long distance, between 150 and 300 kilometres, of southward lateral migration of oil during Middle Miocene, gave rise to the altered, mature marine oils found in the shallow Miocene stratigraphic reservoirs in the Orinoco belt in the southern flank of the Maturin Sub-basin.

Based on the model proposed and the wide distribution of the Cretaceous source rock in the Monagas area, a great potential exists for in traps especially the structural type, in this area.

Oil occurrence

Oil prospection for deep targets in Northern Monagas is based on broad, well-developed structural closures, which are connected to the upper Cretaceous organic rich, thick sequence of highly potential source rocks, by thrust and overthrust faulting.

The recipient rocks are known to be present in the entire sequence from Upper Cretaceous to Pliocene, these include the San Juan, Caratas, Los Jabillos, Naricual, Carapita and the post tectonic sequences of La Pica and Quiriquire Formations (Fig. 6).

The Cap rock is one of the parameters favorable for trap conception in the area, the presence of a thick transgressive sequence of shales of the Carapita Formation, Miocene in age, constitutes an excellent regional seal.

Oil discoveries

Four fields have been discovered in the area in the last three years. Two of these were discovered in 1985, by the ORS-52 in the Orocual field, and the FUL-1 well in the El Furrial field. Later in, 1987, two more wells were drilled, MUSI-IX and CARI-IX in the Musipan and El Carito structures respectively (Fig. 8). The main characteristics of these fields area are summarized as follows.

Orocual field

The Orocual structure is an elongated anticline with its main axis trending ENE. Like to other structures in the area, it is related to thrust

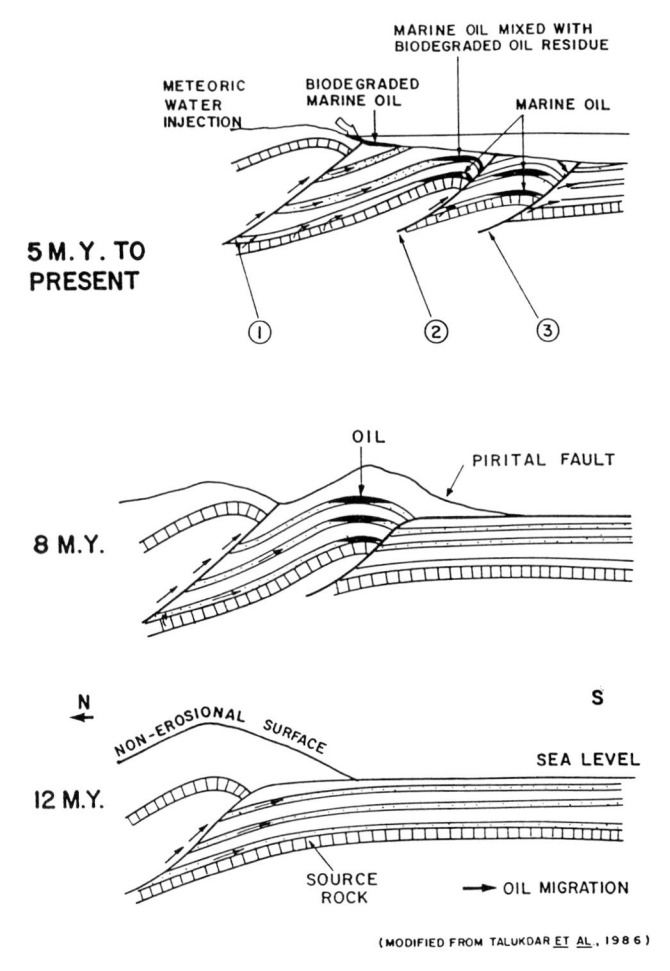

Fig. 7. Oil generation and migration model.

faulting in the South limb, while the other limb dips gently towards the North (Fig. 9).

Five potential reservoirs were tested in this field during 1985 in Cretaceous and Tertiary rocks. These reservoirs are found in the San Juan Formation (Upper Cretaceous) with a total net pay sand of 650 feet, one of them producing up to 2700 barrels of 42° API oil, and 18.2 million cubic feet of gas per day. These beds are found between 13 420 and 14 119 feet which is the total depth. They are marine sandstones with an average porosity of 9%, and water saturation of 25%. Permeability values range between 0.90 and 2.43 md.

Two reservoirs were tested in the Tertiary, both from Oligocene rocks of the Carapita Formation, with a net pay zone of 160 feet. Production was in the order of 1700 barrels per day of 38° API oil, and 11.3 million cubic feet of gas per day.

The two Tertiary reservoirs are turbiditic sandstones which are found at a depth between 11 460 and 12 050 feet, with an average porosity of 13%, and a water saturation estimated at 20%. The field holds total reserves of 50 million barrels of light oil and 230 billion cubic feet of gas.

El Furrial Field

The El Furrial Field is part of a large anticlinal structure truncated in the South by a major thrust fault. This structure extends over 100 km in length as part of one of the blind fold and thrust systems of the Northern Monagas subsurface area. This feature is divided into various

EASTERN VENEZUELAN BASIN 47

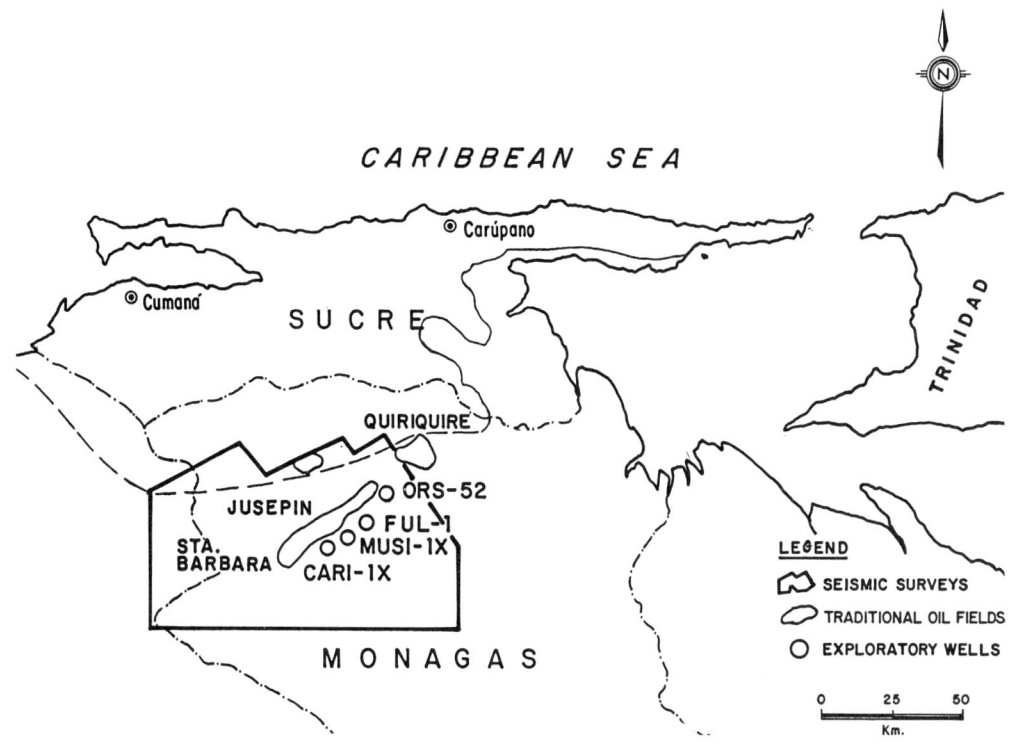

Fig. 8. Exploration activity 1984–1987.

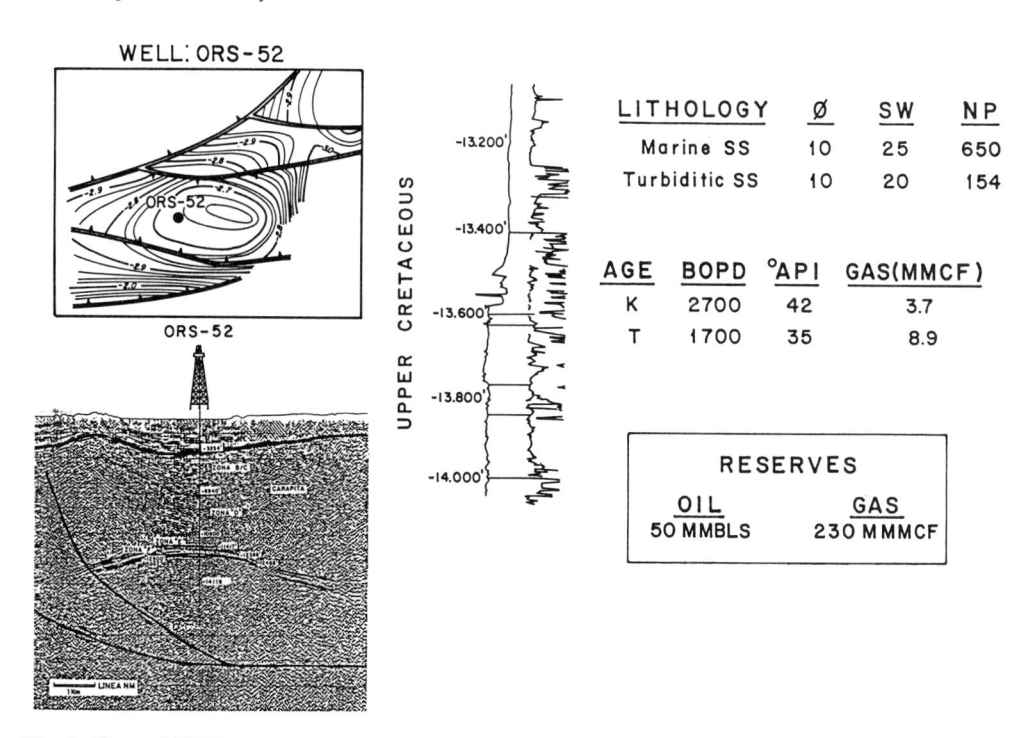

Fig. 9. Orocual Field.

LITHOLOGY	Ø	SW	NP
Marine SS	10	25	650
Turbiditic SS	10	20	154

AGE	BOPD	°API	GAS(MMCF)
K	2700	42	3.7
T	1700	35	8.9

RESERVES

OIL	GAS
50 MMBLS	230 MMMCF

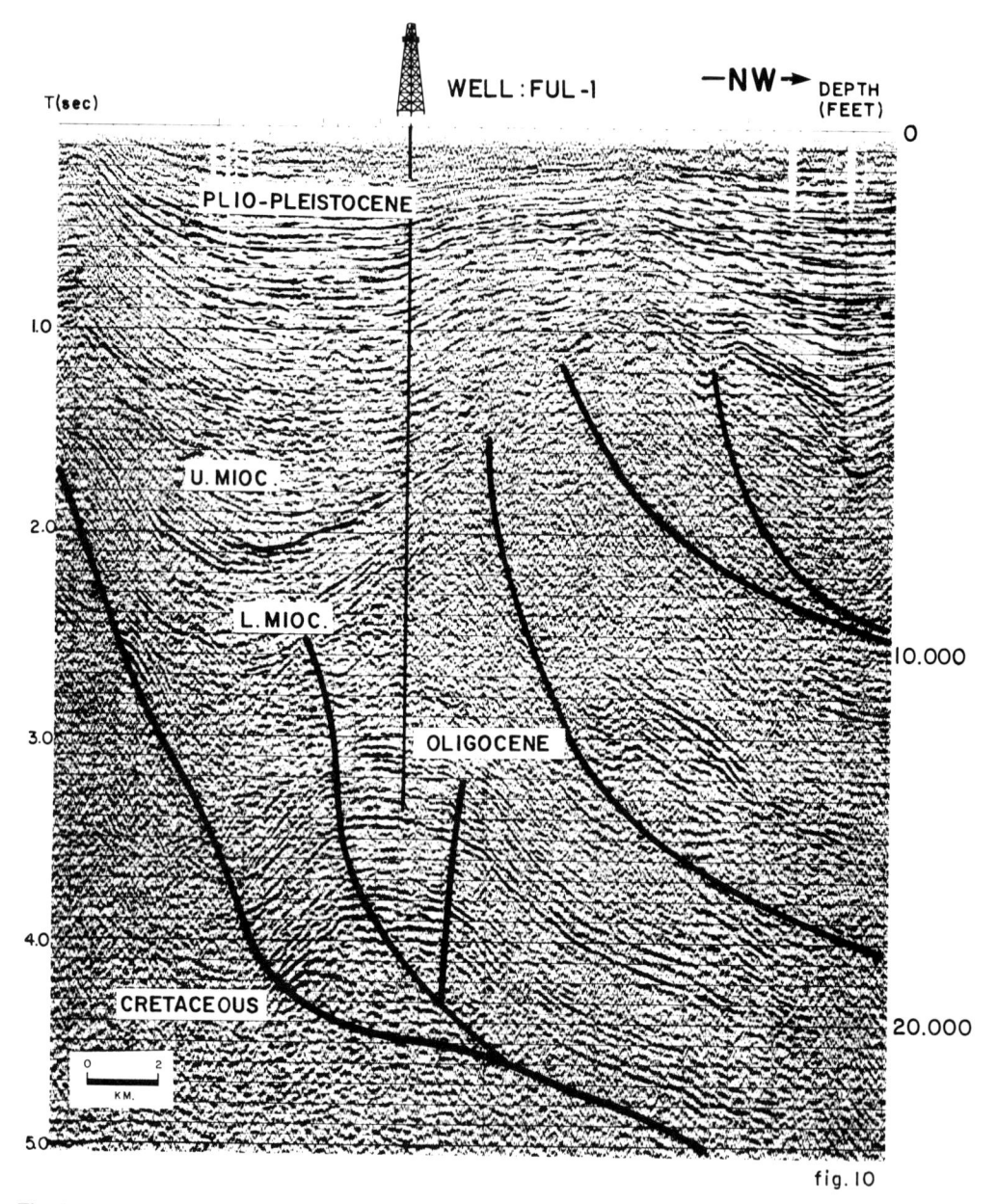

Fig. 10. El Furrial Field.

different blocks by right lateral strike-slip faults which are perpendicular to the axial plane of the anticline.

The El Furrial structural segment has an area of 70 km and a vertical closure of 3000 feet (Fig. 10). The well FUL-1 has a total depth of 14 870 feet, terminating in Oligocene sediments. The most important reservoir rock found in this field correspond to the Oligocene Naricual Formation.

The discovery well, FUL-1 penetrated this reservoir at a depth of 13 000 feet and found a total of 1410 feet of sandstone and a few intercalations of shale. From this total sequence 905 feet are net oil pay. The main reservoir of the upper Naricual Formation is a shallow marine

WELL: FUL-1

LITHOLOGY   **Ø**   **SW**   **NP**

SHALLOW MARINE   15%   24%   900
SS

AGE   **BOPD**   **°API**   **GAS (MMCFG)**
OLIGOCENE   12500   29   12.4

RESERVES OIL

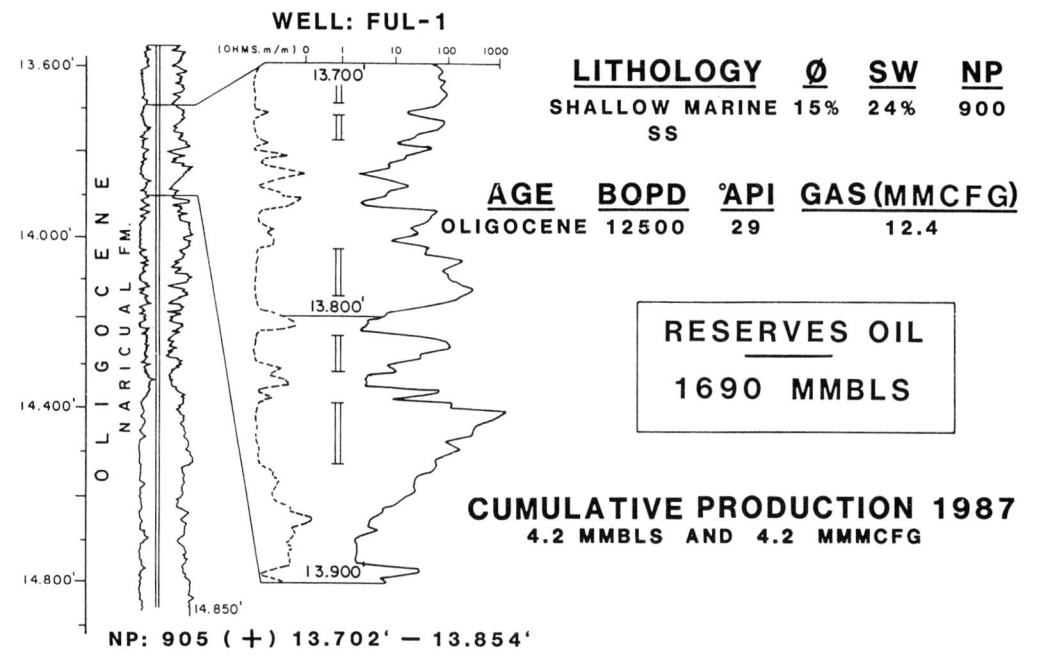

1690 MMBLS

CUMULATIVE PRODUCTION 1987
4.2 MMBLS AND 4.2 MMMCFG

NP: 905 (+) 13.702' − 13.854'

Fig. 11. El Furrial Field.

sandstone, with an average porosity of 15%, low water saturation (24%) and a good permeability (Fig. 11). In general terms, these sandstones are composed of very well sorted quartz grain with some Kaolinite cement. Another well that penetrated the whole prospective Eocene−Oligocene sandstone sequence in the field was FUL-4, which found 1240 feet of net pay sand.

The main reservoir rocks of the Narical Formation are shallow marine sandstones deposited as a transgressive sequence. In general terms, the most common facies found are those of beach and shore face, going inland towards the lagoon, which is separated by long shore bars as represented by Reinson (1979) in a facies model (Fig. 12).

The FUL-1 well was completed on February 1986 in the upper reservoir of the Narical Formation (13690−13842 feet) with a test of 7331 barrels of oil per day on a 1/2-in choke.

Later production of the well gave 12500 barrels of 29° API oil with a choke of 3/4 in. No water whatsoever has been registered to date during the production history of this well. By 1987 it has an accumulated production of 4.2 million barrels of oil during a period of 22 months.

El Carito discovery

A geologic model was established for the Northern Monagas area, based on the integration of geologic and seismic information. This model was used for recording and interpreting the seismic survey of 1986 in the western part of the area.

The geologic interpretation of one seismic line of this survey is shown in Fig. 13. In it, we can observe in the North, the uplifted, deformed and eroded Cretaceous rocks overlying Tertiary sediments. This is a Cretaceous allochthonous block associated with the Pirital thrust fault.

The central portion of the figure shows the location of the El Carito-IX exploratory well. This well is located over a large anticlinal structure that comprises Cretaceous and Tertiary sediments. The structure is limited to the North and South by thrust faults and has an aereal closure of 32 km². The principal oil trapping mechanism is structural with 9000 feet of vertical seal playing a very important role.

Figure 14 shows a typical log of the producing horizons in El Carito-IX well. This well has a total of 1102 feet of net oil sand, with porosity and water saturation between 11−18% and 19−30% respectively. The production tests of

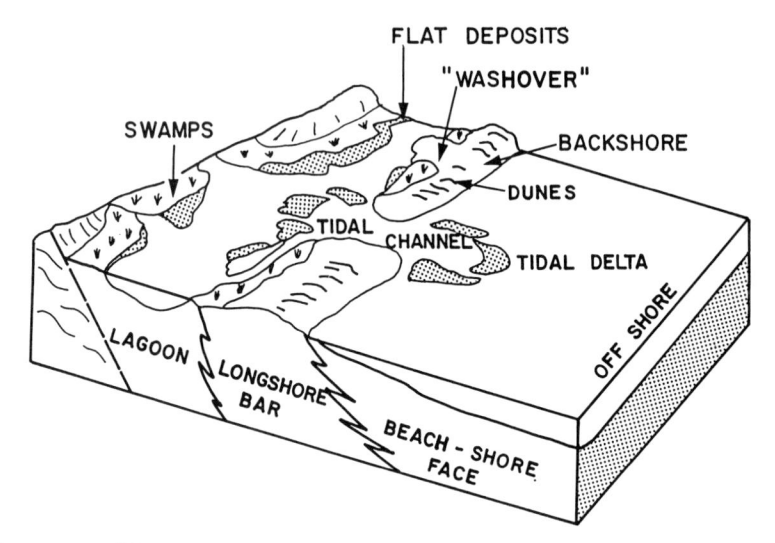

Fig. 12. Sedimentary model.

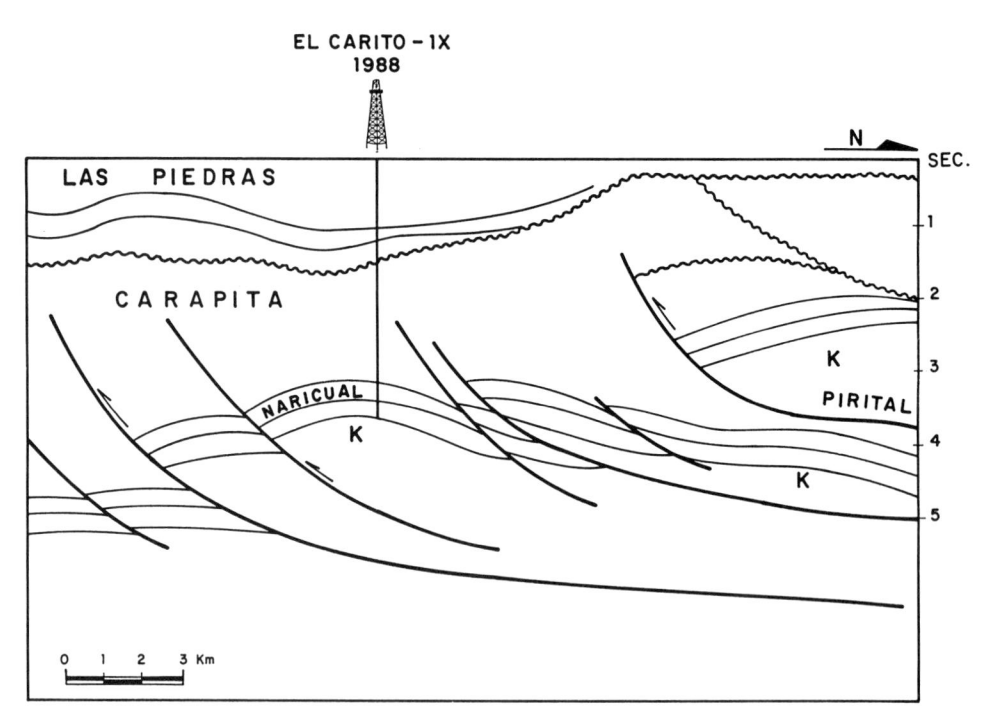

Fig. 13. El Carito discovery: geologic interpretation.

five different intervals of the Narical Formation indicate that each horizon is capable of producing more than 6000 barrels of oil per day with a minimum daily production of 7 and a maximum of 35 million cubic feet of gas.

Reserve estimates for the El Carito structure are in the order of 2.4 billion barrels of light oil (33–36° API) and 2.8 trillion cubic feet of gas. These amounts are an indication of the magnitude of the discovery.

EASTERN VENEZUELAN BASIN

WELL EL CARITO - 1X

	B/D	°API	MMCFG
	8900	34.5	34
	7500	35.7	20
	6800	34	17
	7600	34	15
	6400	33	7

Stratigraphy: OLIGOCENE — CARA., NARICUAL, AREO, L.J.

PETROPHYSICS

ϕ : 11 - 18 %

Sw : 19 - 30 %

NP: 1102'

RESERVES

OIL: 2410 MMBLS

GAS: 2800 MMMCFG

Fig. 14. El Carito discovery 1987–1988.

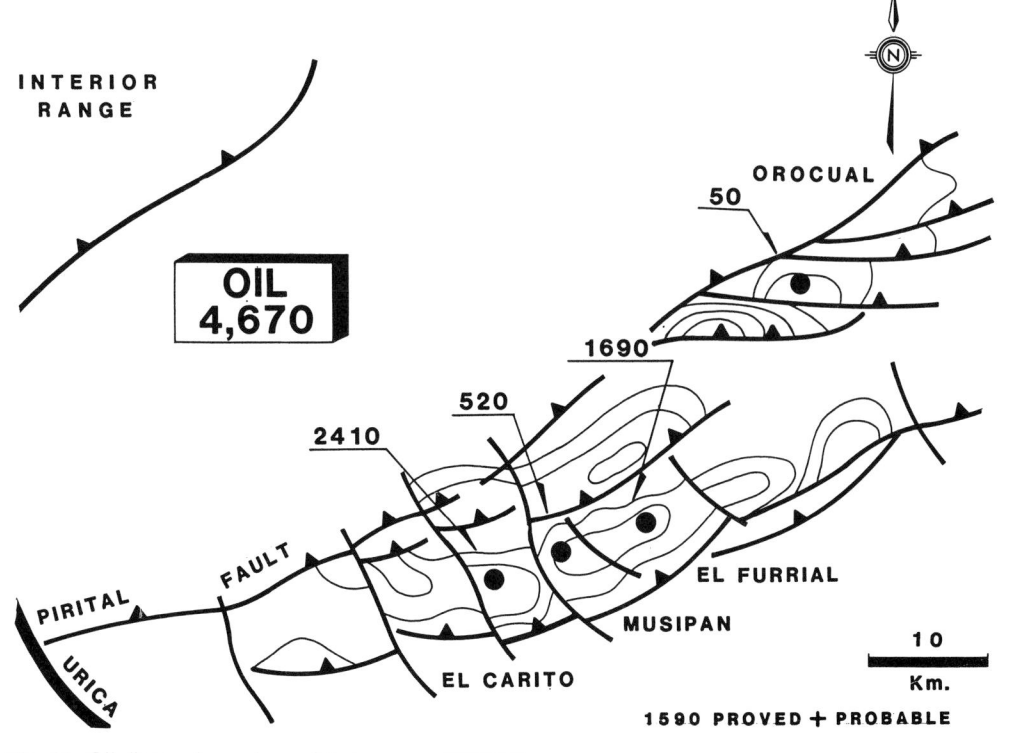

Fig. 15. Oil discoveries and associated reserves MMBLS.

Future exploration and hydrocarbon potential

The present and future exploration of the Northern Monagas Area is influenced mainly by two factor. (i) The discoveries of oil accumulations up to the present time and (ii) the results of the last geophysical and geological integrated interpretations.

Oil discoveries

As we have previously shown, (Fig. 15), three major discoveries, El Carito, Musipan and El Furrial are located in the main east–west over thrust play, whereas the Orocual is located in a nothernmost thrust block.

In the Furrial area and delineation and development drilling is in progress, and so far eight wells have been completed as oil producers in the Narical Formation. The field is producing 60 000 barrels of oil per day and it is planned to increase the production to 200 000 barrels per day in 1990.

In 1987 the delineation drilling began in the Musipan area, with a second producing well and the current drilling of the third well. The delineation drilling in El Carito area will be initiated during this year.

The associated reserves to the discoveries are shown in the Fig. 15. From the total amount of 4670 million barrels of oil, 1590 are proved and probable reserves and 3080 are possible reserves. Present and future drilling in these areas will be mainly focussed to test and evaluate the probable and possible reserves, of fundamental importance in the accelerated development plans for the area.

Result of the last geological-geophysical interpretations

The map in Fig. 16 shows the discoveries and the prospective areas from the last integration studies of the area. There are four major structural plays, where present and future exploratory drilling will take place.

The southernmost thrust fault. Formed by two thrust faults with three associated anticlines each. The most prominent is the Amarilis pros-

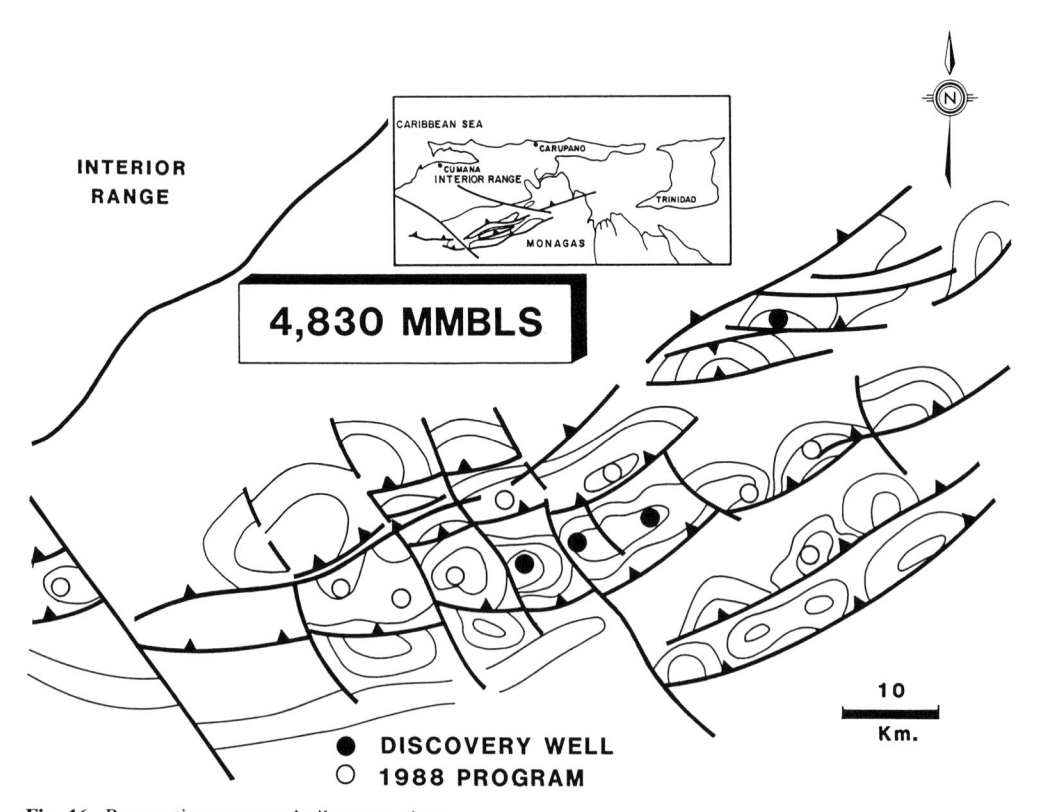

Fig. 16. Prospective areas and oil expectations.

pect and one exploratory well is planned to start by the end of 1988.

The main overthrust fault. This is the most prominent play where the main discoveries have been made. Five exploratory wells are planned to start by mid-1988 in order to test several anticlines with similar characteristics to the ones discovered in El Carito, Musipan and El Furrial.

The Orocual area. This play is located in the Northern part of the area. This is the only area where commercial production have been obtained in the Cretaceous, outpost and delineation drilling will begin this year in a structure west of the discovery well.

The northern thrust and the Pirital Block. This is a new play parallel to the main thrust and located to the North of the discoveries of El Carito and El Furrial. Currently two wells are being drilled and it is planned for 1989 to start a wild cat well North of the Pirital Fault to test Tertiary sediments below the allocthonous Cretaceous block.

The oil expectations for the total prospective areas are 4830 million barrels of oil and for this reason the Northern Monagas area has become the scene of an extensive exploration activity and one of the most promising areas in the western hemisphere.

References

AYMARD, R., PIMENTEL, L., EITZ, P., LOPEZ, P., NAVARRO, J., CHAOUCH, A. & MIJARES J. 1981. Integracion y Evaluación Geológica — Norte de Monagas. Cuenca Oriental de Venezuela. Informe interno Corpoven, S.A., Puerto La Cruz.

FUNES, D. 1986. Proyecto Sismográfico El Tejero — Norte de Monagas. Informe Interno Corpoven, Puerto La Cruz 1986.

BELLIZZIA, A. 1979. "Sistema Montañoso del Caribe, Borde Sur de la Placa Caribe. Es una cordillera Alóctona? *VI Conf. Geol. Caribe, Margarita.*

GONZALEZ DE JUANA, C., ITURRALDE DE AROZENA, J. & PICARD, X. 1980. Geología de Venezuela y sus cuencas petrolíferas.

GUTIERREZ, R. 1986. Nuevas expectativas en la Subcuenca de Maturín, Cuenca Oriental de Venezuela, ARPEL. LX R.A.N.E., Tomo II, Lima Perú.

HUNT, J. M. 1979. *Petroleum Geochemistry and Geology.* Freeman, San Francisco.

PIMENTEL, L., LOPEZ, P. & GONZALEZ, A. 1987. Objectivos Exploratorios Profundos 'Norte de Monagas'. IV Jornadas Técnicas de Corpoven. S.A., Puerto La Cruz.

PIMENTEL, L. 1987. Solicitud del área Musipan — El Carito (Norte de Monagas para la Explotación de los Hidrocarburos). Informe interno Corpoven, S.A., Puerto La Cruz.

REINSON, G. E. 1979. Barrier island systems: *In*: WALKER, R. G. (ed.), *Facies models.* Geological Association of Canada, Reprint Series I, 57—74.

ROD, E. 1959. West End of Serranía del Interior, Eastern Venezuela. *American Association of Petroleum Geology, Bulletin,* **43**, No 4.

ROSALES, H. & CLAXTON, C. D. 1969. Terciaro inferior y Cretáceo del Norte de Monagas. Internal Report Lagoven, S.A. N° 3300.228.11.

ROSALES, H. 1969. La Fall de San Francisco en el Oriente de Venezuela. IV Cong. Geol. Venezolano. Caracas, *Mem. Bol. Geol. Publicación Especial* **5**.

ROSSI, T., STEPHAN, F., BLANCHET, R. & HERNANDEZ, G. 1987. 'Etude Géologique de la Serranía del Interior Oriental (Venezuela). Sur le Transect Cariaco-Maturín. Revue de'l Institut Francais Du Petrole, 1987 N° 1.

SALVADOR, A. 1958. Prospectos Petrolíferos en el Norte de Monagas. Internal report Lagoven, S.A., N° 3300.228.4.

SUBIETA, T. A., CARNEVALI, J. O. & HUNTER, V. 1988. 'Evolución Tectono-estratigrafica de la Serranía del Interior y de la Subcuenca de Maturín. III Simposio Bolivariano, Memoria, Tomo II pg. 548. Caracas — Venezuela.

SYKES, L. R., MC CANN, W. R. & KAFKA, A. L. 1982. 'Motion of Caribbean Plate During Last 7 million years and implications for early Cenozoic Movements. *Journal of Geophysical Research,* **87**.

TALUKDAR, S., GALLANGO, O. & RUGGIERO, A. 1986. Estudio Geoquímico Regional de la Subcuenca de Maturín. INT-01543,86, Intevep, S.A. Caracas.

Main aspects of petroleum exploration in the Apure area of Southwestern Venezuela, 1985−1987

NESTOR CHIGNE & LEROY HERNANDEZ

[1] Corpoven, S.A., Venezuela

Abstract: This paper gives an update of exploration advances made in Apure during the years 1985−1987. It begins with a brief history of exploratory activities in the area, and then reviews present concepts as to its tectonic−sedimentary evolution. New seismic surveys have notably extended structural knowledge and recent geochemical studies have given new insights into the origin and migration of hydrocarbons; these, together with results of all exploratory drilling, allow new ideas as to additional prospects in the area.

As can be seen from Fig. 1, the Barinas/Apure basin is located in the southwestern corner of Venezuela, on the eastern flank of the Merida Andes. This basin belongs to the subandean basin system, aligned parallel to the eastern flank of the Cordillera de Los Andes (Fig. 2).

The Apure region is about 300 km in length east to west and 70 km north to south. The southern limit is the border with Colombia. In the west, the region is flanked by the Merida Andes which are orientated in a southwest−northeast direction. The easternmost limit of the basin has always been the Guayana shield. However, the area of study goes farther to the north into the Barinas area.

The area has been explored for hydrocarbons since 1924 but the steady exploratory effort dates back only to 1980. This last phase resulted in the discovery of the Guafita and La Victoria oil fields, which represent some 400 million barrels of recoverable oil, and an estimated 1000 million barrels of potential reserves.

The results have been obtained from multi-disciplinary studies, including seismic, geochemistry, sedimentology an stratigraphy studies.

Historical review

Geophysics

The first exploration activities in Apure took place in 1924. From then until 1951, the effort was concentrated in the northeastern part of Apure, where gravity and seismic surveys of limited extent were run.

These surveys gave the first structural indications of the Brujas High, in the northwestern part of Apure. A more complete seismic survey, using analogic recording, was initiated in 1958. This survey extended towards the south and southeast of Apure.

In 1976 a sustained exploration effort started in Apure and by 1986 a large volume of seismic and aeromagnetic data had been acquired in the form of detailed, semi-detailed and regional grids, totalling some 9000 km of lines.

Drilling

From 1945 to 1965, nine dry exploratory wells were drilled, all of them located in northern Apure.

In 1982 the exploratory drilling activities were restarted, and by 1985, 25 exploratory wells had been drilled. 11 of these wells found light and/or medium oil and 14 were dry. This effort led to the discovery of the Guafita and La Victoria oil fields.

Regional geology

Tectonic and Sedimentary Aspects

The sedimentary column of Apure includes sediments of Palaeozoic age to Recent. This sequence varies in thicknesses from about 30 000 feet in the deepest part of the Subandean Trough, towards the foothills of the Andes, to some 7500 feet, east of the Guafita field, thinning continuously eastwards towards the Guayana Shield.

The sediments, in general, contain few microfauna and relatively few microflora that makes it difficult to assign an age or establish a reliable geological correlation over the whole area. Nevertheless, recent work (Ortega et al. 1987; Portilla 1988) has made it possible to recorrelate previous work of the Barinas and Apure areas. (Figs 3 & 4).

The integration of sedimentological and bioestratigraphical studies has indicated the presence of five sedimentary cycles in Apure and six cycles in the oil fields of Barinas, to the north (Fig. 5). Each of these cycles commences with a transgression whose maximum advance is represented by a shaly interval followed by a

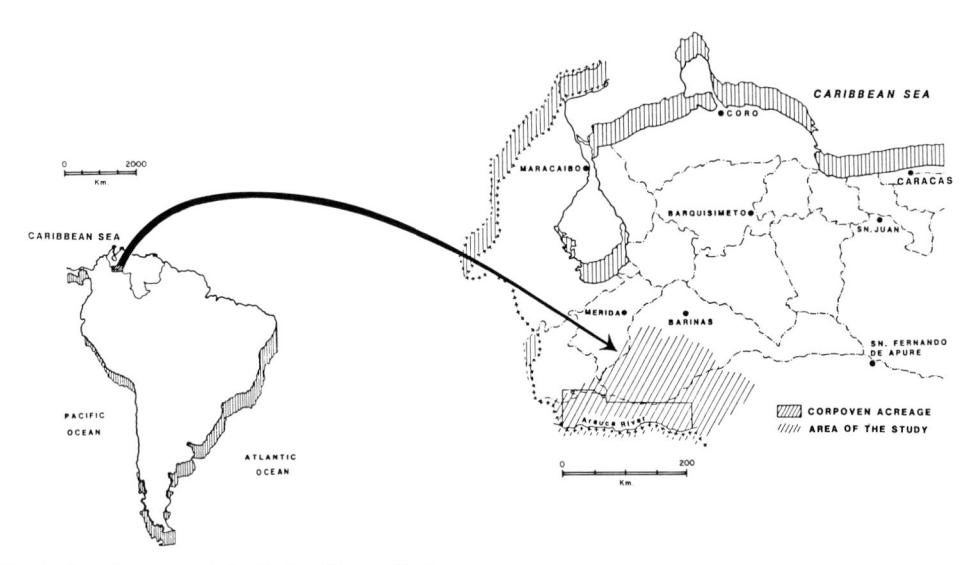

Fig. 1. Location map of the Barinas/Apure Basin.

regression in which increasingly coarser clastics are deposited. Some of these coarse deposits provide a reservoir rock for hydrocarbons in the area. The pre-Cretaceous history of the basin is not very well known. In some places the Cretaceous lies unconformably over oxidised continental sequences, probably of Jurassic age, found to the north of the large Brujas thrust fault (wells LCB-1X and Burgua-3) as well as in wells of the Guafita field and to the south of La Victoria field (well Arauquita-1, in Colombia). These sequences are probably related to 'grabens' of Jurassic age, developed during a Jurassic rifting period associated with the Pangean opening process.

During the Albian to Maestrichtiense, all the area surrounding the Guayana craton, constituted the neritic and coastal provinces where platform deposits passed progressively into deep water deposits towards the west northwest (Fig. 6).

At the end of the Cretaceous, the uplift of the Cordillera Central of Colombia, resulting from a collision between the Pacific Plate and the South American Plate, modified the orientation of the depositional systems, causing them to be rearranged perpendicularly to the preferential directions of the Cretaceous, in such a way that the Tertiary depositional environments became progressively deeper towards the northeast (Fig. 6).

The orogeny at the end of the Cretaceous caused the retreat of the sea to the north (de Juana *et al.* 1980) and the rise of the Arauca, Mérida and El Baúl Arches. The formation of the Arch of Aurucà which started during the late Cretaceous set a limit to the Paleocene sedimentation northwards, towards Barinas. Clastic Paleocene deposits (Barco and Los Cuervos formations) have been identified in the La Ceiba area, of northwestern Apure. It is thought that these clastics are present to the southwest of the Arch of Arauca (Chigne 1985), north of the Arauca Oil Field in Colombia, where Paleocene sediments have been found in the wells drilled there.

During the Middle Eocene, a transgression from the north covered the Barinas/Apure basin, depositing the shales of the Paguey Formation. In the wells of La Ceiba area, in the northwestern part of Apure, very thick packs of deltaic sandstones were penetrated. These were initialy related to the 'Cobre Facies' of the Middle Eocene that had been identified farther to the northeast in the wells Capitanejo-1 and Agualinda-1. However, recent palinological studies indicate an Oligocene age for this section lying discordantly on Paleocene rocks (Fig. 5). Interpretation of seismic data shows the Paleocene and Eocene pinching out in an easterly direction to a limit somewhere close to La Victoria Field.

At the end of the Middle Eocene epirogenic movements caused the retreat of the sea to the north; uplift of the arches is reactivated, permitting peneplanation a deep strong erosion of their crests. During this period, the incipient rising of the Merida Andes started. The

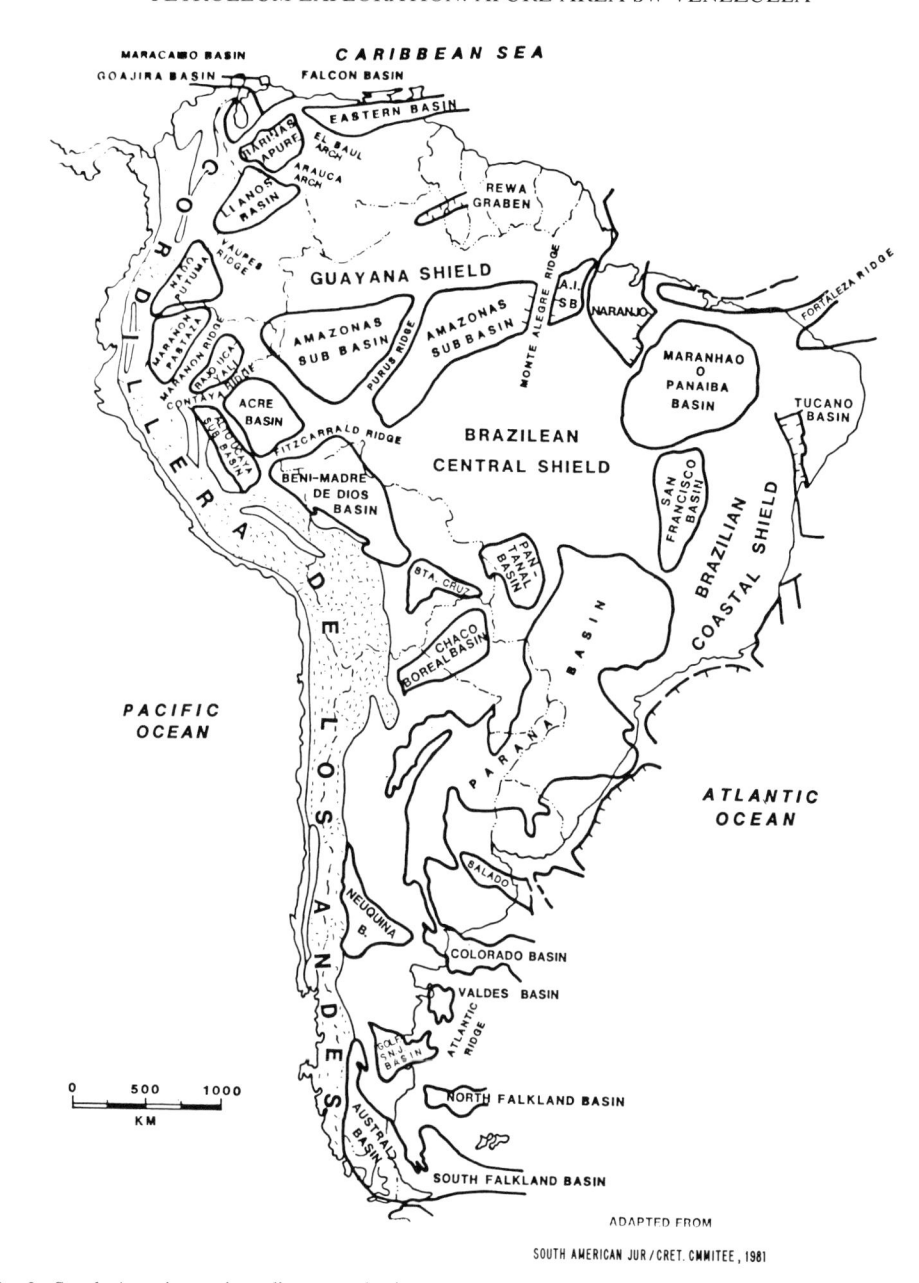

Fig. 2. South America main sedimentary basins.

Eocene formations in Barinas pinched out on the northern flank of the Arauca Arch, as they are not present in the southern part of Apure. Evidence from regional seismic studies supports this configuration (Portilla 1988). In the Llanos Basin of Colombia, Eocene formations have been found in several wells Bogotá-Ruiz (1988) demonstrating the importance of the Arauca Arch as an active element influencing the distribution of the Paleogene sediments in the Llanos and Barina/Apure Basins. The Oligocene marks the beginning of a new sedimentary cycle during which coarse clastics of the Arauca Member of the Guafita Formation were deposited (Ortega

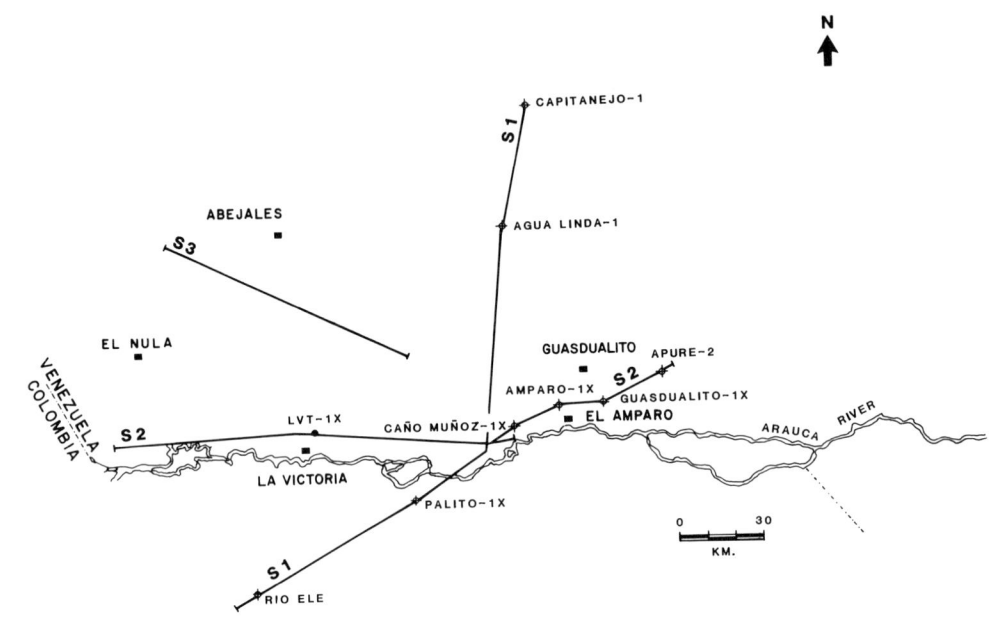

Fig. 3. Index map of cross sections.

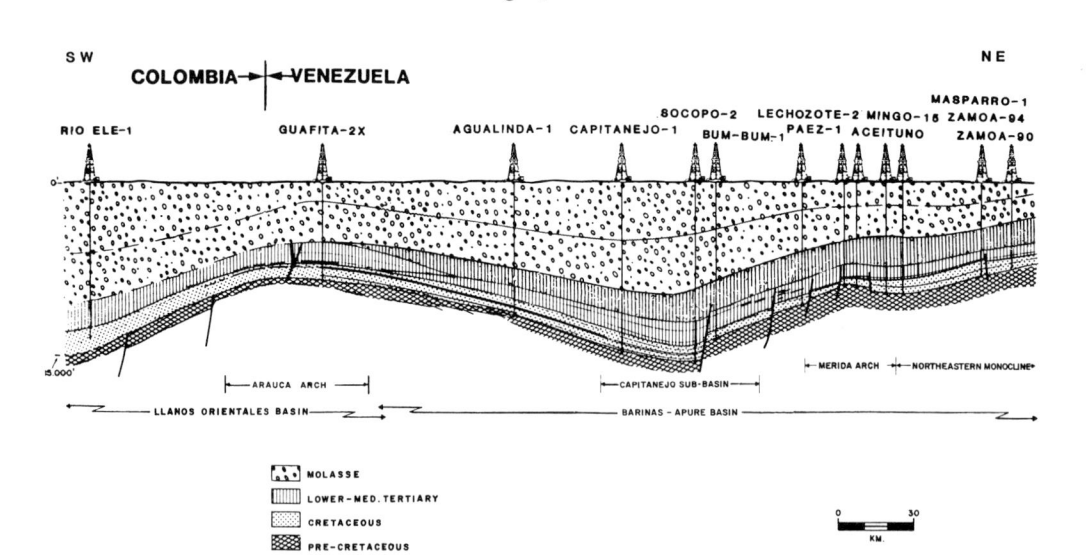

Fig. 4. Apure–Barinas regional geological section.

et al. 1987). These clastics are very good reservoir rocks.

The Oligocene section lies unconformably on the peneplaned Cretaceous. Where in northwestern Apure, the Guafita Formation lies discordantly on Paleocene rocks. In Barinas, a sequence, probably equivalent to the Guafita formation, that belongs to the base of the Parángula formation, overlays rocks of the Middle Eocene cycle, as can be seen in the

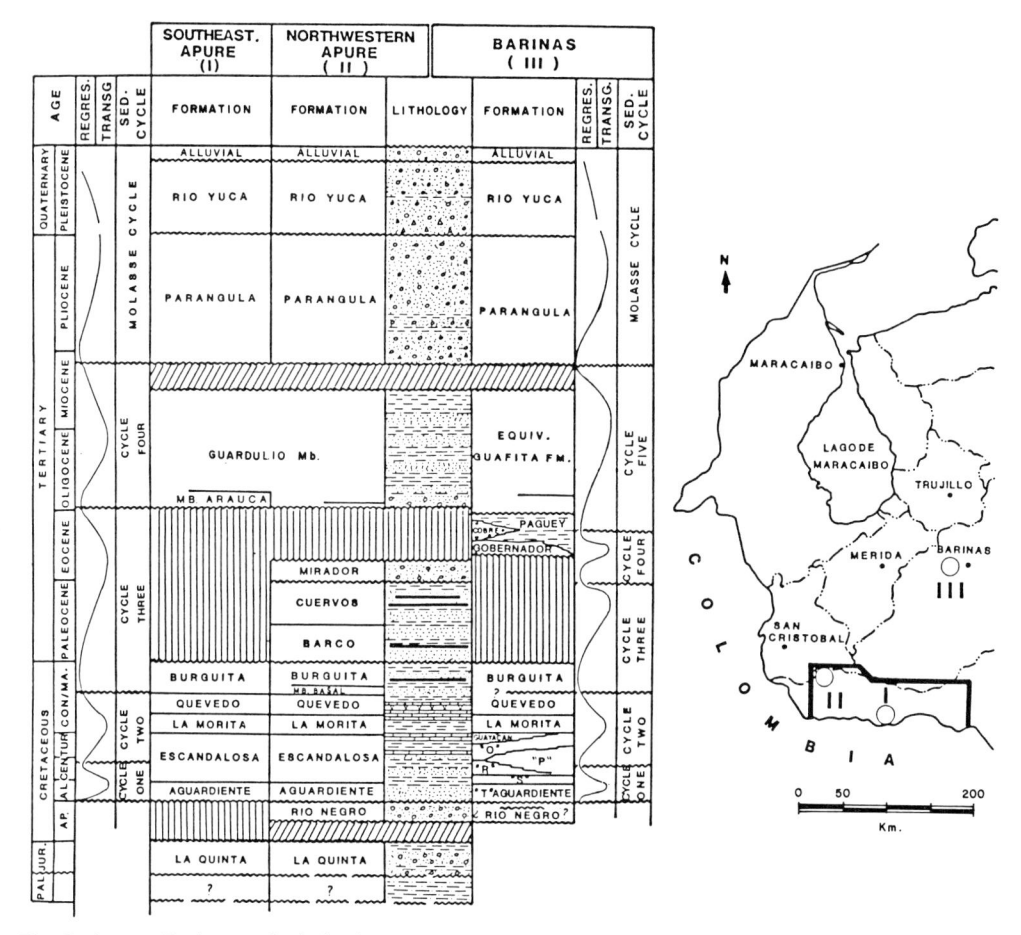

Fig. 5. Apure–Barinas geological columns.

regional seismic correlation (Portilla 1988). This new evidence provides basis for extending the Oligocene–Lower Miocene cycle to Barinas (Figs 3 & 4). In fact, it is possible that the cycle extends even further to the northeast into the Guárico region, where the Oligocene cycle is represented by the La Pascua and Roblecito formations that lies unconformably on Upper Cretaceous rocks. (Chigne 1985; Ortega et al. 1987).

To the west and northwest of the Apure area, on the Merida Andes, Perija and the Maracaibo's Basin, fluvio deltaic and paludal sediments were deposited. These sediments belong to the early facies of the Carbonera Formation that evolved during the lower Miocene to more marine facies of the Leon Formation. The Oligocene–Lower Miocene of Apure was initially assigned to the Carbonera and Leon Formations (Chigne 1985). However, in view of the remoteness of outcrops of these formations from the Oligo–Miocene section in Guafita, they have been designated as Guafita Formation (Ortega et al. 1987).

The Guafita Formation passed into the more marine facies of the Guardulio Member during the Lower Miocene. The Guardulio Member in its turn underlies continental sediments of the molassic cycle, indicating that the regressive phase on top of the Oligo–Lower Miocene cycle had been eroded. This contact is an unconformity that presents the most clear evidence of the riser of the Mérida Andes and the resulting molassic post orogenic sedimentation (Lower Miocene to Recent) which reaches more than 16 000 feet of thickness of the Mérida Andes foothills. It also has large thicknesses on the flanks of the arches (Fig. 4). The younger

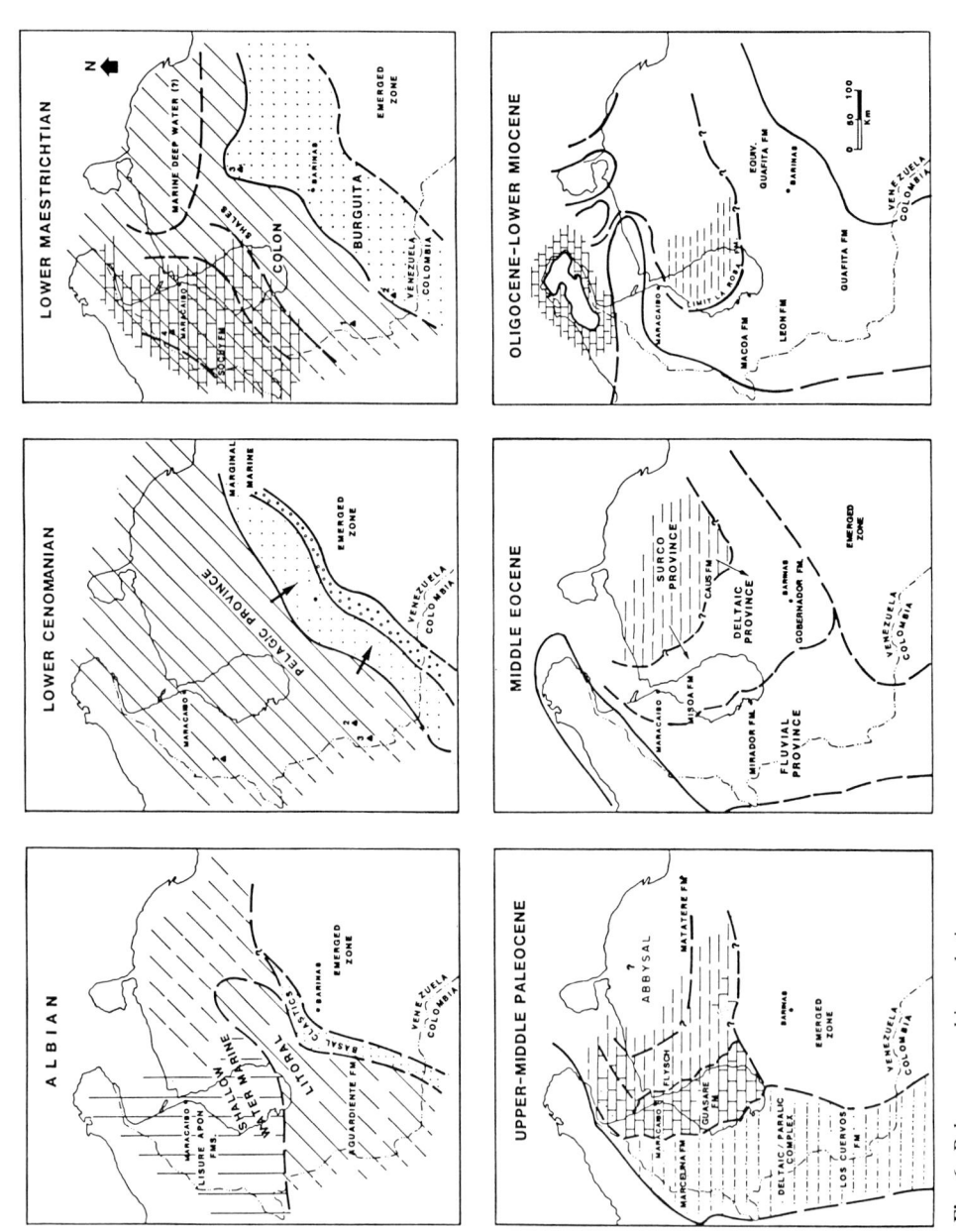

Fig. 6. Paleogeographic evolution.

cycle of the Molasse, probably of Plio-Pleistocene age brings to an end sedimentation in the Barinas/Apure Basin.

Findlay (1988) estimates that more than 20 000 km³ of sediments were eroded from the Mérida Andes; part of this volume was deposited as Molasse on both flanks of the Mérida Andes, and the rest was carried out of the system to be deposited in a marine environment.

Structure

The pre-Cretaceous history is little known, however, the presence of oxidized continental sequences in various parts of the basin could be the result of deposition within troughs formed during a period of jurassic rifting, which may be demonstrated by the interpretations of the magnetic and gravimetric surveys. There is, however, no seismic evidence in support of such possibilities.

The collision between the Pacific and South American plates at the end of the Cretaceous produced uplift in the Cordillera Central of Colombia which is likely to have produced a basin of the Back Arc type. the Guayana Shield was always the eastern limit, and the Cretaceous basin deepened towards the northwest, giving rise to the formation of normal growth faults trending SW–NE with a hade towards the depocenter of the basin.

This Cretaceous normal faulting was reactivated by the tectonic forces that gave rise to the post-Lower Miocene uplift of the Merida Andes, thus producing the present day structure of the Apure Basin as shown in Figs 7–10. The seismic time map on Fig. 7, represents the Base Cretaceous throughout most of the Apure area. The structural contours shows that the basin deepens in a WNW direction with depths varying from 7000 ft in the east to more than 30 000 ft in the west. The fault pattern shown here is very consistent. It shows a high degree of alignment, of possibly transcurrent faults, oriented in a SW–NE direction in subparallelism with the Merida Andes. This is really, the most notable feature of the structural style of the Apura Basin, and suggests a strong relationship with the tectonic stresses that gave rise to the Merida Andes.

These structural trends extend to between 40 and 80 km. They are all formed as a result of reverse faulting of varying lengths, which always retain the same parallel trend. Occasionally, floral structures develop along some of these faults giving further support to the belief that there is a strong component of horizontal movement, possibly dextral, as it should correspond to E–W compressive stresses, reactivating a SW–NE privious normal fault system. On Figs 7 & 10, it can be seen that from the east end of the area to the west, up to La Victoria Field, the dip is quite gentle, and there is just one relevant structure, corresponding to the Guafita–Caño Limon Fault and related oil fields, a possibly dextral transcurrent fault traversing the frontier between Venezuela and Colombia.

Westerly from La Victoria Field the Cre-

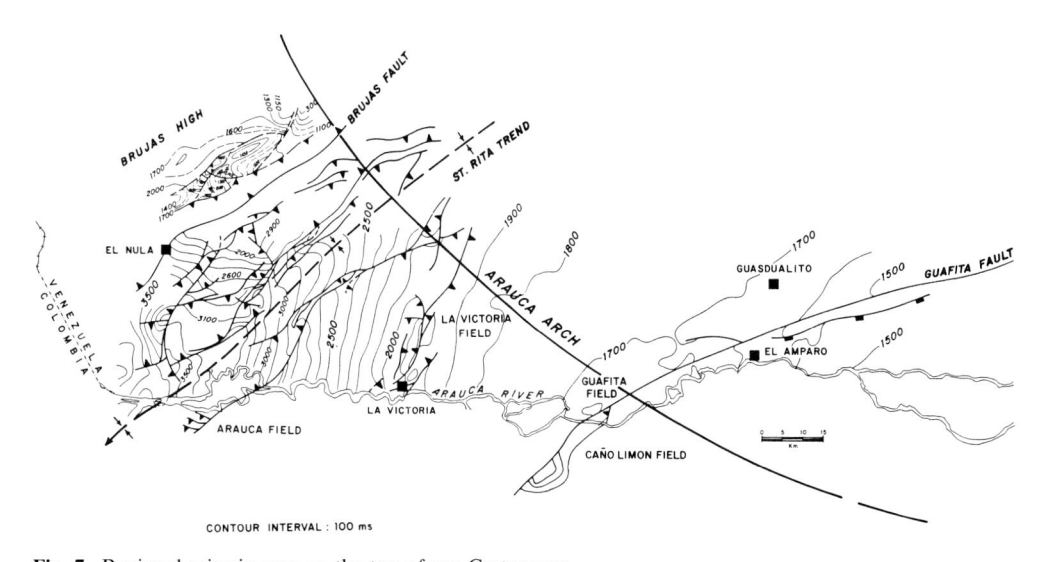

Fig. 7. Regional seismic map on the top of pre-Cretaceous.

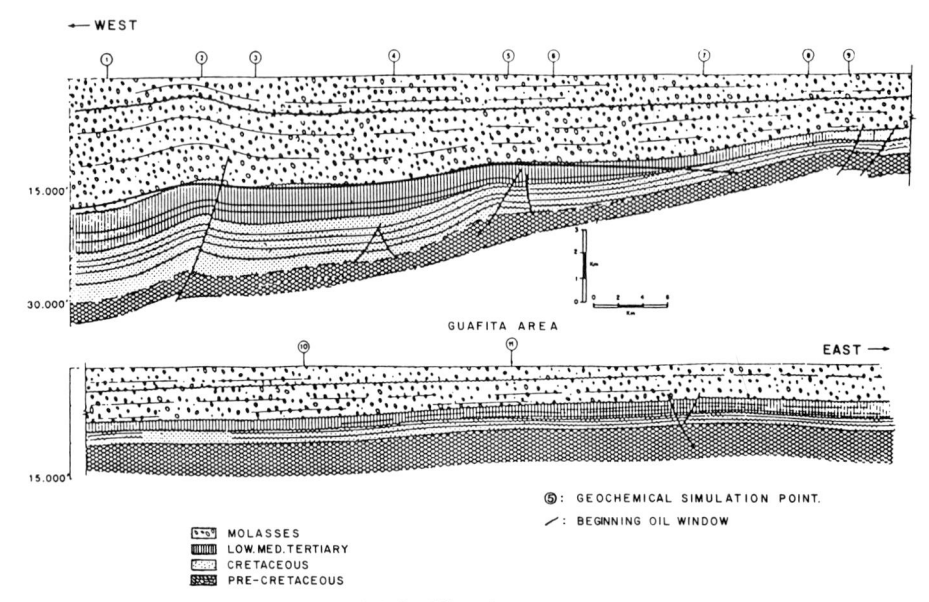

Fig. 8. Apure regional geological section S-4, La Victoria area.

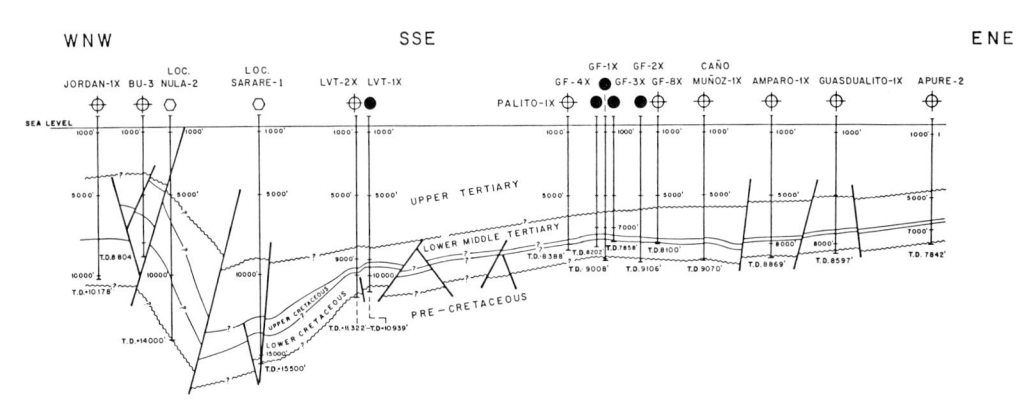

Fig. 9. Apure E–W geological cross section S-2.

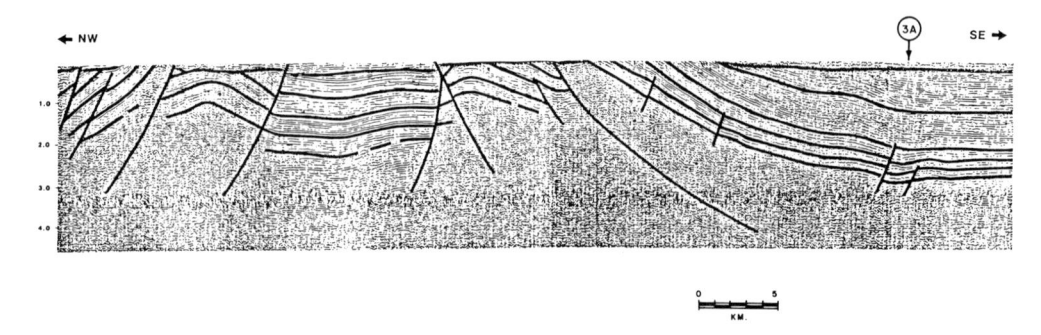

Fig. 10. Apure S-6 structural seismic line crossing Brujas high. 3A: simulation point.

taceous Basin deepens at a much faster rate, reaching depths to the order of 30 000 ft along the axis of the sub-Andean through, which extends in a NE−SW direction. From here towards the northwest, the structures begin to get rapidly shallower by means of a succession of large inverse faults culminating with the Brujas High, where the deepest geological units are brought to the surface, producing outcrops of the Jurassic and the Cretaceous.

Source Rocks

The geochemical analysis carried out on shale samples of the Lower Cretaceous, from wells of the Barinas/Apure region indicate a type-II kerogen content, with predominantly marine organic matter, as been responsible for the hydrocarbon generation in the area.

The areal distribution of the Hydrogen Index (Fig. 11), shows that the marine organic matter is mainly concentrated to the west and northwest of the Apure/Barinas region. This distribution coincide with the more marine facies of the Cretaceous source rocks (Fig. 6). It is also interesting to note that the terrestrial kerogen, has an irregular concentration on top of the Arauca and Merida Arches. This could be interpreted as an evidence of the early formation of these arches. Similar anomalies can be noticed in the distribution of the Total Organic Carbon shown in Fig. 12. The rocks with higher organic content are concentrated in areas of deep water sedimentation, towards the west, northwest and north.

The map of the actual Geothermal Gradient (Fig. 13), shows that the gradient decreases to the south, from the El Baul Arch, where pre-Cretaceous rocks outcrop. It also can be seen that the sub-Andean throughs of Apure are, geothermically could areas, a characteristic which plays and important role on the hydrocarbon generation in Apure.

On the regional geological section shown in Fig. 18, eleven points, (1 to 11) are shown; where geochemical mathematical simulations were done, using the equations of Tissot & Espatalie (1975).

The results of some critical points, are presented in Tables 1, 2 and 3 and Fig. 14. Point 1, located in the deepest part of the Apure area, the oil generation ($R_0 = 0.62$), began 11.2 million years ago (Middle Miocene), when the source rock was at a depth of 5 000 metres. The potential of hydrocarbon generation is 4.2 kg/ton. of rock.

The peak of oil generation ($R_0 = 1\%$), took place 9.6 million years ago (Upper Miocene), when the overburden was 20 200 feet thick, generating approximately 8 kg HC/ton. of rock.

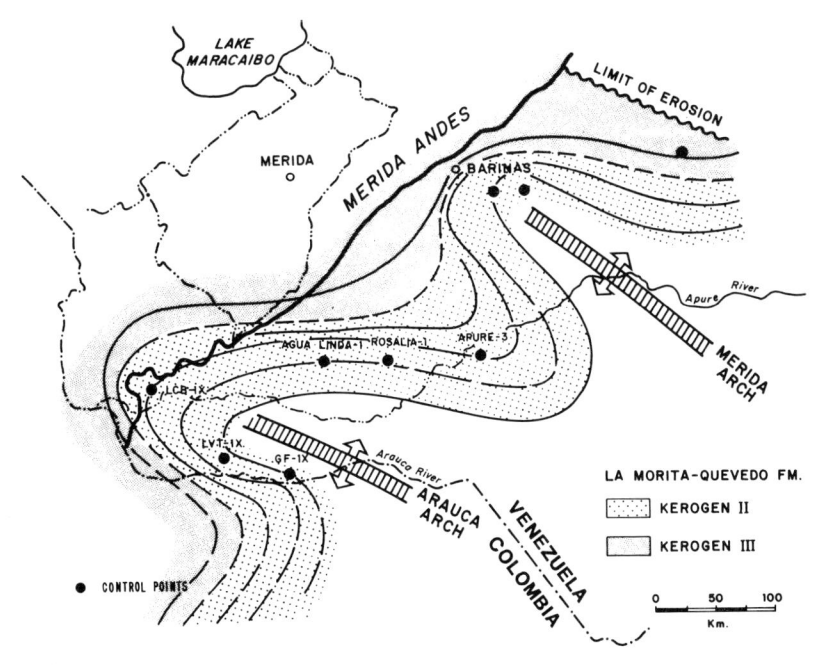

Fig. 11. Distribution of source rocks and hydrogen index (mg Hc/g C org).

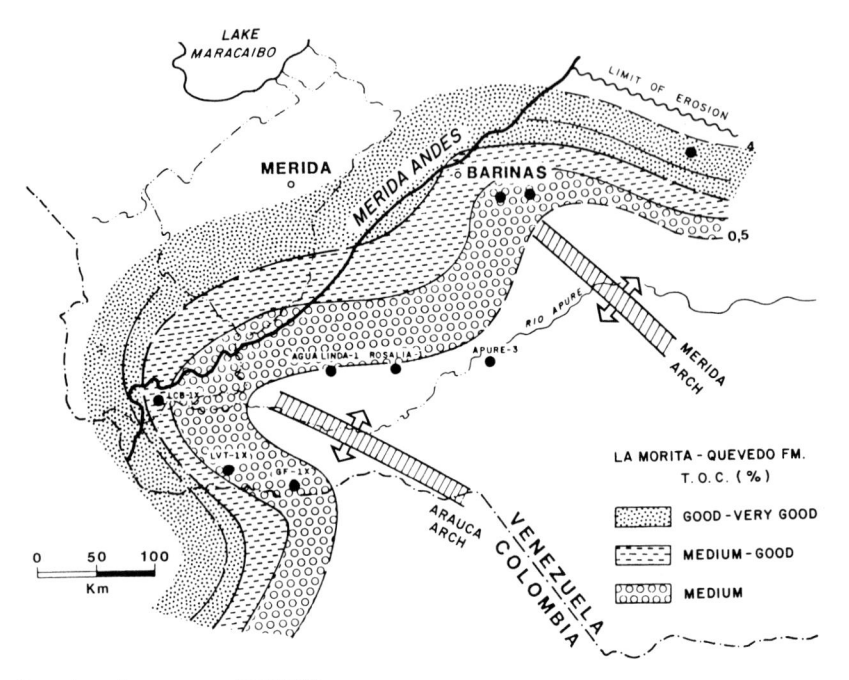

Fig. 12. Organic carbon content (TOC %).

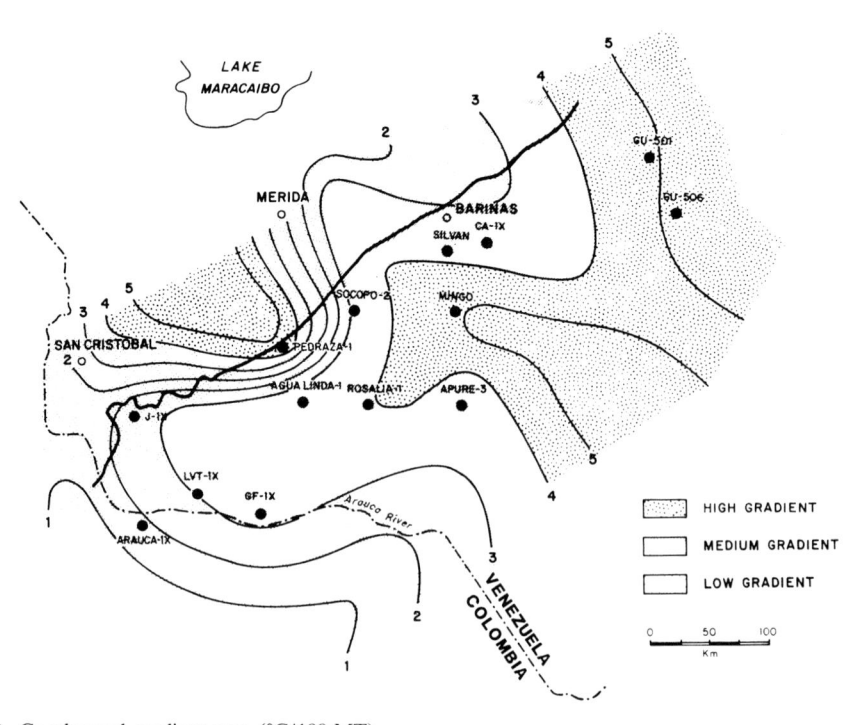

Fig. 13. Geothermal gradient map (°C/100 MT).

PETROLEUM EXPLORATION: APURE AREA SW VENEZUELA 65

Table 1. *Simulation of source rock maturation and hydrocarbon generation base of La Morita member*

Depth (m)	Temperature (°C)	Age (Ma)	Hidroc. (K/Ton)	Transf. (%)	Oil (K/Ton)	Gas (K/Ton)	Vit.Ref. (R_0)
			Section S−2. Point # 1				
0	27	88.5	1.225	0.000	0.000	1.225	0.20
670	39	68.0	1.360	0.006	0.135	1.225	0.21
560	37	65.0	1.383	0.006	0.158	1.225	0.21
1000	45	54.0	1.472	0.010	0.247	1.225	0.21
1300	50	38.0	1.619	0.016	0.394	1.225	0.22
1626	56	32.0	1.678	0.019	0.453	1.225	0.22
1952	62	26.0	1.748	0.022	0.523	1.225	0.23
2280	68	20.0	1.833	0.025	0.608	1.225	0.24
2100	65	17.0	1.872	0.027	0.647	1.225	0.25
2535	73	16.2	1.884	0.027	0.659	1.225	0.25
2970	80	15.4	1.907	0.028	0.682	1.225	0.26
3405	88	14.6	1.953	0.030	0.728	1.225	0.27
3840	96	13.8	2.055	0.034	0.829	1.226	0.30
4275	104	13.0	2.284	0.044	1.056	1.228	0.37
4710	112	12.2	2.800	0.065	1.558	1.241	0.51
5145	120	11.4	3.953	0.112	2.624	1.329	0.60
4810	114	11.0	4.472	0.133	3.086	1.386	0.64
4480	108	10.6	4.661	0.141	3.253	1.408	0.66
5520	126	10.2	5.281	0.167	3.769	1.512	0.70
6015	135	9.8	8.179	0.286	5.538	2.641	0.86
6160	138	9.6	9.955	0.359	5.940	4.015	1.00
6305	140	9.3	11.498	0.422	4.723	6.775	1.16
6450	143	9.0	11.992	0.443	2.655	9.338	1.26
6400	142	7.1	12.471	0.462	0.025	12.445	1.41
6350	141	5.3	12.776	0.475	0.006	12.770	1.48
6640	147	4.2	13.055	0.486	0.009	13.046	1.53
6930	152	3.2	13.690	0.512	0.001	13.689	1.58
7720	166	2.1	17.013	0.649	0.000	17.013	1.65
7510	162	1.1	19.318	0.744	0.000	19.318	1.66
7800	167	0.0	20.396	0.788	0.000	20.396	1.67

Entrance data
Geothermal gradient: 1.80°C/100MT
TOC: 3.5%
Kerogen type: II
Surface temperature: 27°C

Table 2. *Simulation of source rock maturation and hydrocarbon generation base of La Morita member*

Depth (m)	Temperature (°C)	Age (Ma)	Hidroc. (K/Ton)	Transf. (%)	Oil (K/Ton)	Gas (K/Ton)	Vit.Ref. (R_0)
			Section S−2. Point # 7				
0	27	88.5	0.875	0.000	0.000	0.875	0.20
88	29	83.0	0.896	0.001	0.021	0.875	0.21
175	32	78.0	0.917	0.002	0.042	0.875	0.21
264	34	63.0	0.983	0.006	0.108	0.875	0.21
350	36	60.0	0.996	0.007	0.121	0.875	0.21
310	35	58.0	1.006	0.008	0.131	0.875	0.21
270	34	48.0	1.046	0.010	0.171	0.875	0.21
230	33	38.0	1.081	0.012	0.206	0.875	0.21
620	44	32.0	1.106	0.013	0.231	0.875	0.21
1010	54	26.0	1.146	0.016	0.271	0.875	0.22
1400	65	20.0	1.207	0.019	0.332	0.875	0.22
1000	54	18.5	1.220	0.020	0.345	0.875	0.23
600	43	17.0	1.227	0.020	0.352	0.875	0.23

Table 2. *(Continued)*

Depth (m)	Temperature (°C)	Age (Ma)	Section S−2. Point # 7 Hidroc. (K/Ton)	Transf. (%)	Oil (K/Ton)	Gas (K/Ton)	Vit.Ref. (R_0)
1150	58	15.0	1.238	0.021	0.363	0.875	0.23
1700	73	13.0	1.261	0.022	0.386	0.875	0.23
2250	88	11.0	1.332	0.026	0.456	0.875	0.26
2800	103	9.0	1.605	0.042	0.728	0.877	0.37
2750	101	7.0	2.010	0.065	1.125	0.885	0.52
2700	100	5.0	2.265	0.080	1.372	0.893	0.56
2900	105	4.0	2.406	0.088	1.507	0,899	0.59
3100	111	3.0	2.674	0.104	1.756	0,918	0.62
3350	117	2.5	2.980	0.121	2.031	0.949	0.65
4000	135	0.0	7.909	0.405	3.448	4.461	1.08

Entrance data
Geothermal Gradient: 2.7°C/100Mt
TOC: 2.5%
Kerogen type: II
Surface temperature: 27°C

Table 3. *Simulation of source rock maturation and hydrocarbon generation base of La Morita member*

Depth (m)	Temperature (°C)	Age (Ma)	Section S−3. Point # 3A Hidroc. (K/Ton)	Transf. (%)	Oil (K/Ton)	Gas (K/Ton)	Vit.Ref. (R_0)
0	27	89.0	0.270	0.000	0.000	0.270	0.20
150	30	85.0	0.270	0.000	0.000	0.270	0.21
300	34	82.0	0.271	0.000	0.001	0.270	0.21
595	40	75.0	0.272	0.000	0.002	0.270	0.21
850	46	68.0	0.274	0.001	0.004	0.270	0.21
710	43	65.0	0.275	0.001	0.005	0.270	0.21
880	46	61.0	0.276	0.001	0.006	0.270	0.21
1050	50	57.0	0.278	0.002	0.008	0.270	0.21
1051	50	50.0	0.281	0.002	0.011	0.270	0.22
1165	53	44.0	0.284	0.003	0.014	0.270	0.22
1280	55	38.0	0.288	0.004	0.018	0.270	0.22
1590	62	32.0	0.294	0.005	0.024	0.270	0.23
1900	69	26.0	0.307	0.008	0.037	0.270	0.24
2210	76	20.0	0.332	0.013	0.062	0.270	0.27
2370	79	17.0	0.353	0.018	0.083	0.270	0.29
2747	87	15.0	0.381	0.024	0.111	0.270	0.32
3125	96	13.0	0.447	0.038	0.177	0.270	0.39
3503	104	11.0	0.590	0.068	0.318	0.272	0.52
3880	112	9.0	0.841	0.121	0.556	0.285	0.61
3765	110	7.0	1.043	0.164	0.731	0.311	0.69
3650	107	5.0	1.135	0.183	0.805	0.330	0.72
4025	116	2.5	1.253	0.208	0.877	0.375	0.77
4400	124	0.0	1.554	0.272	0.967	0.587	0.88

Entrance data
Geothermal gradient: 2.7°C/100MT
TOC: 1.5%
Kerogen type: III
Surface Temperature: 27°C

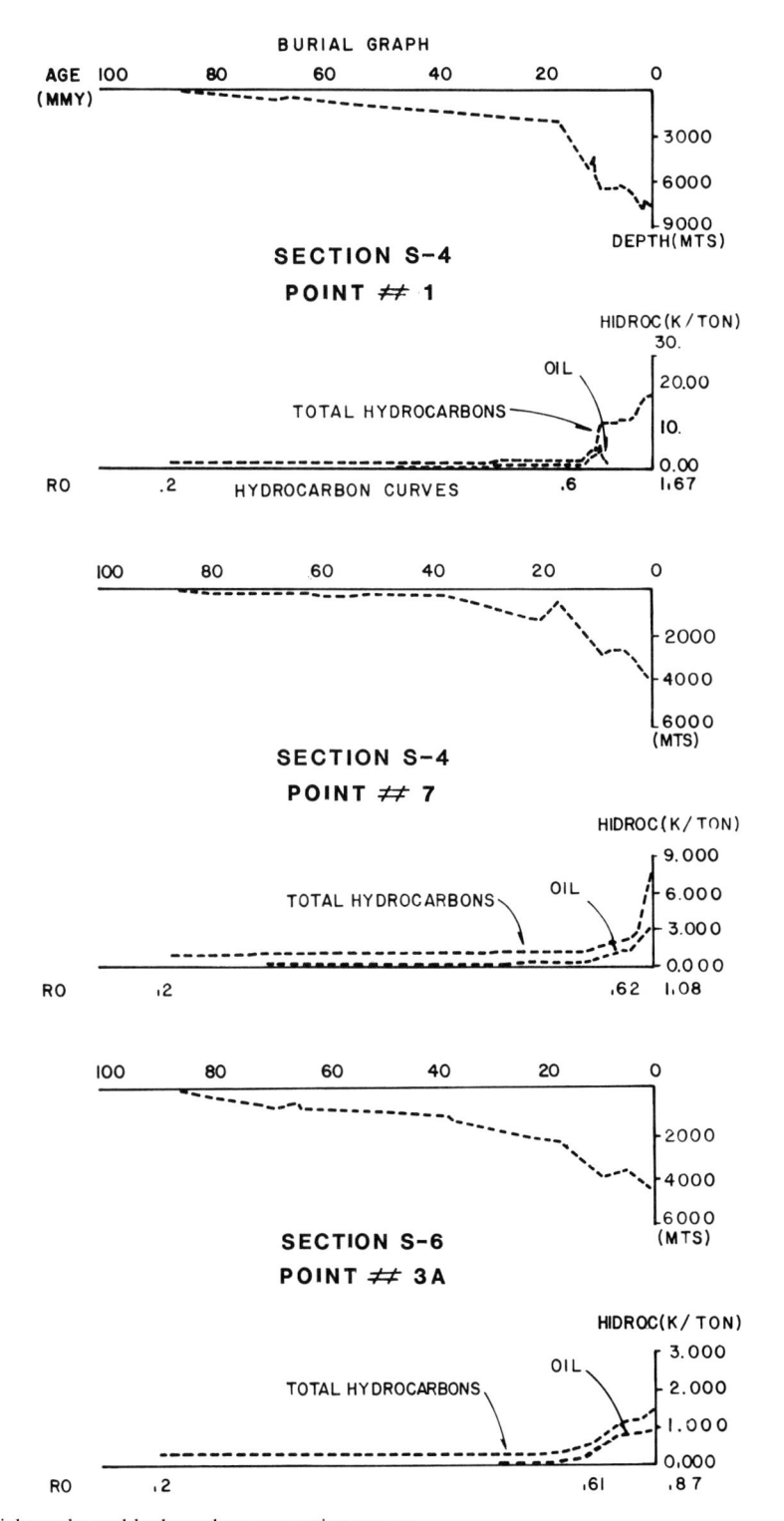

Fig. 14. Burial graphs and hydrocarbon generation curves.

Since these times the gas generation has progressively increased. The peak of gas generation ($R_0 = 1.3\%$) was reached 8.7 million years ago. At the present time the source rock is at a depth of 7800 ft, in the gas window.

It is convenient to mention that, this deep area, simulated on point 1, represents approximately one half of the kitchen of hydrocarbon. The rest of the kitchen, is actually generating as much gas as oil.

On point 7 located some 9 km to the west of La Victoria Field, the simulation shows that we are close to the limit of the sub-andean kitchen. The oil generation ($R_0 = 0.62\%$) began 3 million years ago (Pliocene). At the present time, the source rock has just reached the peak of oil generation ($R_0 = 1\%$), at a depth of 13 120 ft. Up to the present time, the amount of oil generated is 7.9 kg/ton of rock.

The simulation done on a depression to the north of the Arauca Arch, southeast of the Brujas High, point 3A on Fig. 10, where the sedimentary section is 16 000 ft thick, shows that in this area, only a small amount of hydrocarbon has been generated, 1.5 kg/ton of rock. It is considered that in this area, the kerogen is of type III.

The geochemical simulation results presented in this paper are totally in agreement with the results obtained with less complex methods (Chigne 1985). In the Guafita and La Victoria fields the potential source rock is inmature ($R_0 = 0.45\%$), suggesting that the oil accumulated in these fields came from deeper parts of the Subandean Basin.

Fig. 15 shows the thermal Evolution Map, were two main sub-andean kitchens (R_0 0.6%) were delineates.

From the results of the mathematical simulations, it has been calculated that the amount of hydrocarbons generated is 65 000 million barrels of oil. It is assumed that from this volume about 10% will have been expulsed. This means that from the Middle Miocene to the present, some 6500 million barrels of oil will have migrated to higher parts of the area. Considering that 3200 million barrels have been discovered, in the Caño Limon–Guafita trend and La Victoria and Arauca fields, and that 300 million barrels have been lost during the migration processes there may still be some 3000 million barrels of oil to be discovered, from which 1000 million barrels could be recovered in the structures of the southern flank of the Arauca Arch. In the hydrocarbon kitchen located on the northern flank of Arauca Arch (simulation

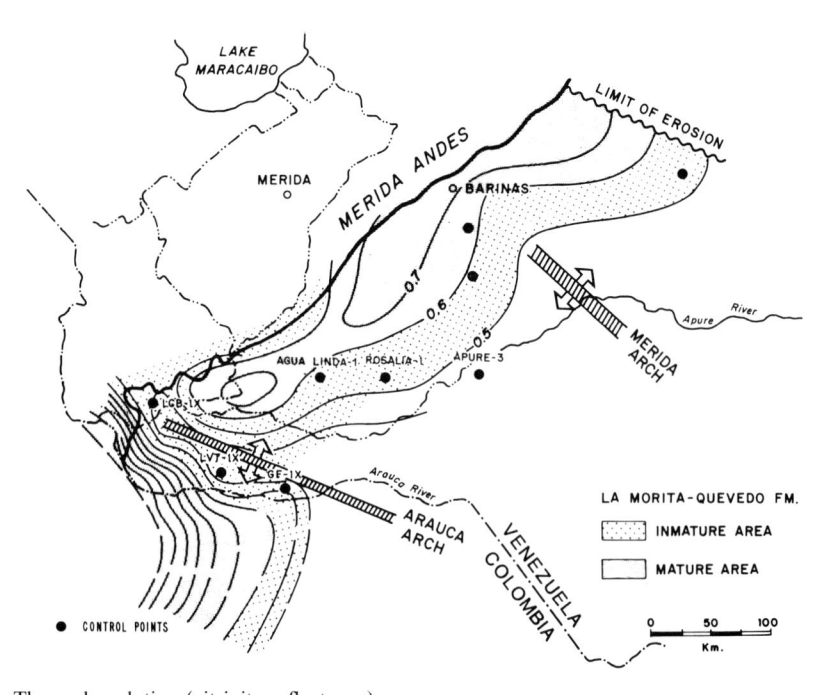

Fig. 15. Thermal evolution (vitrinite reflectance).

point 3A, Fig. 10), the volume of hydrocarbon generated and expulsed is only about 80 million barrels.

Even though all the oil found in the Barinas/Apure area is from marine source, the crudes of the Barinas Fields are more mature than those from Apure (Cassani et al. 1988). This difference can be explained, if the source rocks of Barinas were deposited in a more anoxic environment.

Regarding the origin, migration and trapping of hydrocarbons in the Barinas and the Apure/North of the Llanos Orientales Basin of Colombia, other hypotheses have been proposed. Findlay (1988) suggested that the oil migrated long distances from the Maracaibo and Magdalena Basins respectively, during early to Middle Miocene, before the rise of the Cordillera Oriental of Colombia and the Merida Andes. Such migration routes means distances between 250 and 300 km for the light oil found in La Victoria and the Cano Limon–Guafita trend. For the oil in the Barinas Fields migration distances were between 150 to 175 km. In both cases the possibilities is very low for such long distance migration of light oil.

Findly (1988) also suggested that the structures of pre-Andean age are the most prospective. However, the excellent structures of the La Ceiba area, of Lower Miocene age when drilled were dry, in spite of being closer to the Maracaibo and Magdalena Basins, than La Victoria and Guafita–Caño Limon Fields.

The lack of oil accumulations on the La Ceiba area was due to the fact that the Brujas High, an old structure, acted as a barrier to the migrating oil, coming from the Sub-Andean kitchen on the south.

Moreover the first stage of the uplift of the Merida Andes, was by the end of Middle Eocene, acting as a barrier to any oil coming from the Maracaibo and Magdalena Basins. This considerations plus the estimated amounts of generated oil support the hypothesis that the oil in the Apure/Barinas Basin came from the Sub-andean Kitchen.

Exploratory drilling

Guafita oil field

The Guafita Field is located in the southeastern sector of Apure, as shown in Fig. 7. It is part of the Caño Limon–Guafita trend, which extends from the Llanos Orientales Basin in Colombia through the Apure area, reaching the well Apure-2, some 100 km in length. This structural alignment is part of the wrench fault system that crosses the Apure area in a NE–SW direction, and is parallel with the Merida Andes and Bocono fault. Corpoven discovered the Guafita oil field with the drilling of the Well Guafita-IX in 1984, one year after the discovery of the Caño Limon Field in Colombia. This well proved light oil in the lower part of the Oligocene Basal Sands, on the northern block of the field. In 1985 Well Guafita-5X discovered the southern block of the field and proved oil in the entire Oligocene Basal Sands section, and also in the upper part of the Navay group, of Cretaceous age.

The sedimentary section in Guafita region is 9500 ft thick. The Tertiary section is 8000 ft thick, and lies unconformably on the Cretaceous. This sedimentary sequence lies on sericitic schists of unknown age.

The Guafita Field is a northeast-trending anticline, split into two blocks by the Caño Limon–Guafita strike slip fault zone, which has approximately 400 ft of vertical displacement at the Cretaceous level (Fig. 16). This faulted anticline extends across the colombian border, where it also forms the trap for the Caño Limon field. In the southern block, the Tertiary Basal Sands of the Guafita Formation are 600 ft thick, while in the northern block, are less than 200 feet thick. This is probably due to a growing fault effect during the deltaic sedimentation of the Oligocene. Fig. 17 shows a typical N–S seismic section crossing the field.

The Oligocene Basal Sands, where deposited in a lower deltaic plain, which was part of a constructive deltaic system. Within this section several active distributary channels have been identified together with some abandoned channels, crevasse plays, interdistributary bays and tidal flats environments (Fig. 18).

The Basal Sands are of medium of coarse grain size, poorly sorted and consolidated. The porosity is primary-intergranular with an average values of 26% at overburden conditions. The permeability varies from 1000 to 9000 mD.

The reservoirs produce oil with very low GOR (10 scf/stb). The formation water is fresh, with chloride content which varies from 500 to 13000 ppm. The hydrocarbon production from reservoirs is by a water drive mechanism. The crude oils from the Guafita Field have an average gravity of 29° API, with very low content of sulphur (0.62%), vanadium (21 ppm) and nickel (64 ppm). Table 4 summarizes some reservoir characteristics of the Guafita formation and Navay group.

The average production of the field, up until August 1987, was 40 000 BOPD. The crude is transport by a 20-inch pipeline to the Barinas

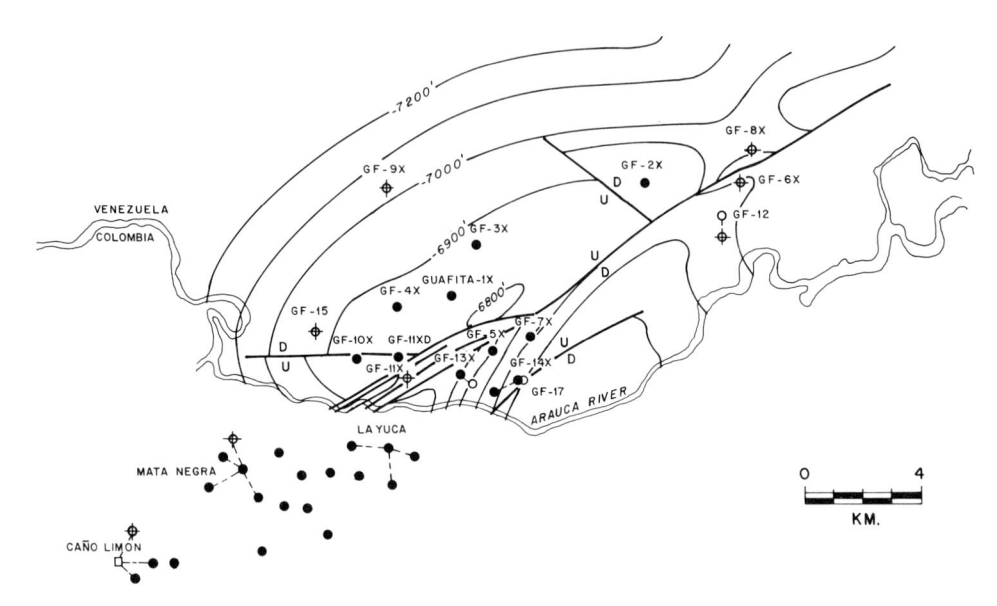

Fig. 16. Guafita Field: structural map on top of Guafita 'B' sand.

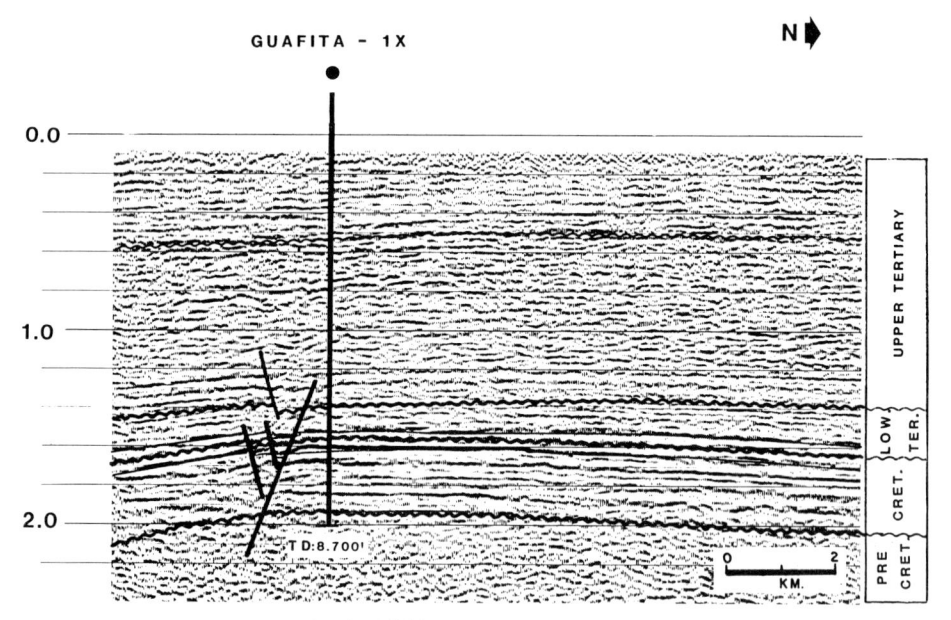

Fig. 17. Structural seismic section: Guafita Field.

Refinery. The original oil-in-place of Guafita field is 1000 million barrels.

La Victoria Field

The La Victoria Field is located 40 km to the east of Guafita field and 6 km to the north of Arauca River. The anticlinal structure of La Victoria was discovered in September 1984, by the well LVT-1X, which tested light oil in the Tertiary Basal Sands and Upper Cretaceous rocks. A further, five more wells were drilled, of which LVT-4X and LVT-6 found oil. Well LVT-4X discovered an additional Cretaceous

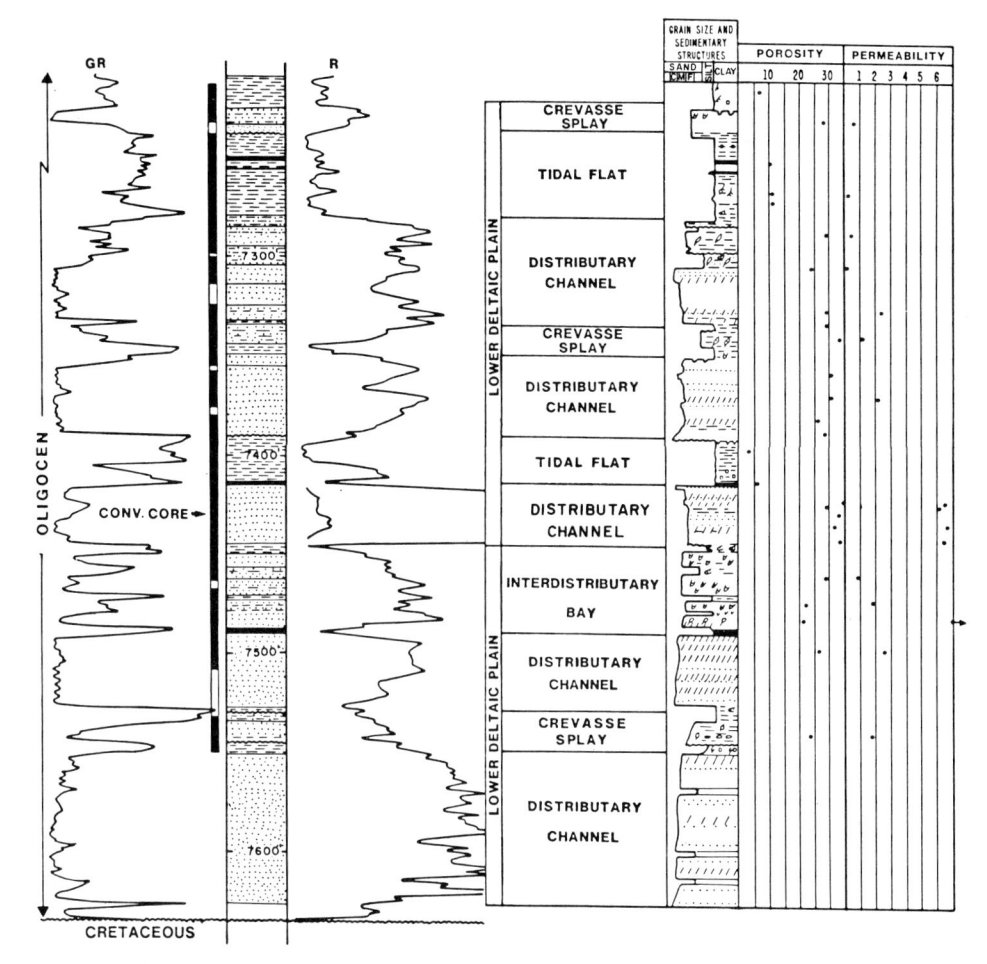

Fig. 18. Well Guafita -2X: sedimentology and petrophysics of basal sands of Guafita Formation.

Table 4

Well	Formation	ϕ	NOS	BOPD	W&S (%)	Gravity (°API)	TP	Choke (")
North block								
GF-1X	Guafita	25	42	1264	1.5	29.7	360	1/4
GF-2X	Guafita	27	30	1046	0.6	28.8	350	1/4
GF-4X	Guafita	26	36	1122	1.2	29.0	394	1/4
South block								
GF-5X	Guafita	29	36	4985	0.1	30.4	110	3/4
	Guafita	30	20	3322	0.2	30.4	80	3/4
GF-7X	Guafita	28	12	3381	0.1	30.0	55	3/4
GF-13X	Navay	27	22	3388	0.2	29.4	40	3/4
	Navay	28	11	1113	0.1	29.4	60	3/8

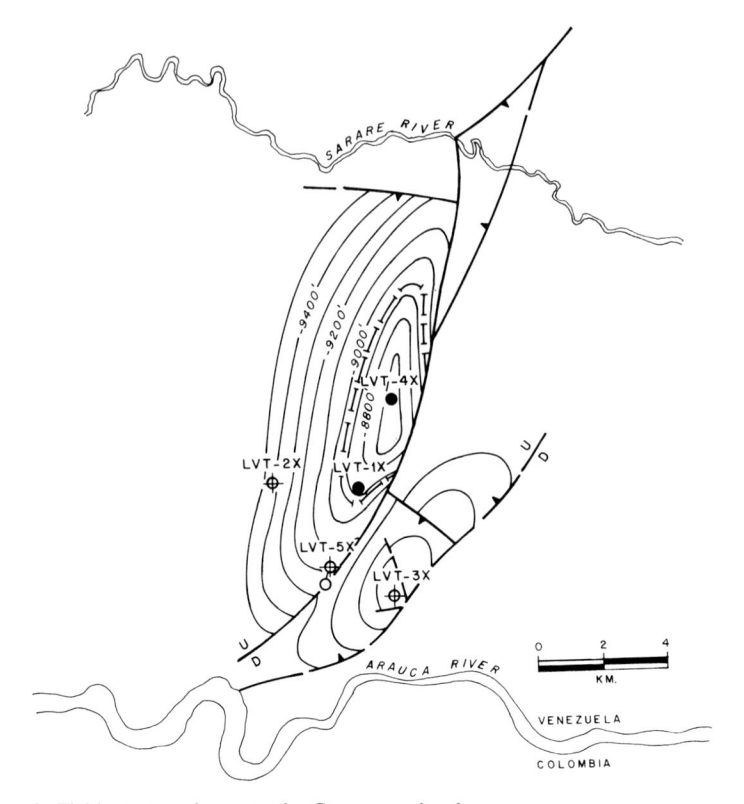

Fig. 19. La Victoria Field: structural map at the Cretaceous level.

reservoir that confirmed this most prolific of the field. Well LVT-3X tested non-commercial oil from the Basal Sands, on a small anticline located 3 km southeast of the main structure (Fig. 19).

The stratigraphic column is about 10 700 ft thick, of which 9100 ft corresponds to the Tertiary sequence. The Cretaceous section has a similar thickness to the Guafita Field, but in this field constitutes the main reservoirs. The lithology consists of calcareous and glauconitic sandstones interbeded with thin beds of limestone.

The La Victoria structure is mapped at the reservoir level as two adjacent, slightly offset, northeast trending, faulted anticlines. Both anticlines have east boundary faults that parallel their anticlinal axes (Fig. 20). The La Victoria Field is located on the northwest anticline and its east boundary fault dies out to the southwest.

The La Victoria anticline is huge and is not filled to spill. The oil is trapped at the crest with maximum oil column of about 275 ft. This suggests the oil has probably escaped via the east boundary wrench fault and the required structural closure is due to east dip into the fault. The vertical closure by dip into the wrench fault on the east side of the structure is about the same as the length of the oil legs (about 275 ft) of the reservoirs. This may indicate the wrench fault is not presently sealing and provides the oil a path of migration and escape to shallower (Bristow 1987) (Fig. 20).

The porosity of the Basal sands averages 17% and varies from 13 to 27% in the Cretaceous reservoirs. The permeabilities varies between 669 to 3095 mD. The crude oils from the La Victoria field are light (33° API) with low content of sulphur (0.36%), vanadium (6.57 ppm) and nickel (27 ppm). The La Victoria field is still not on production. The oil-in-place is estimated at 400 million barrels, but important new prospects to the north of the main structure could considerably increase the reserves of the area.

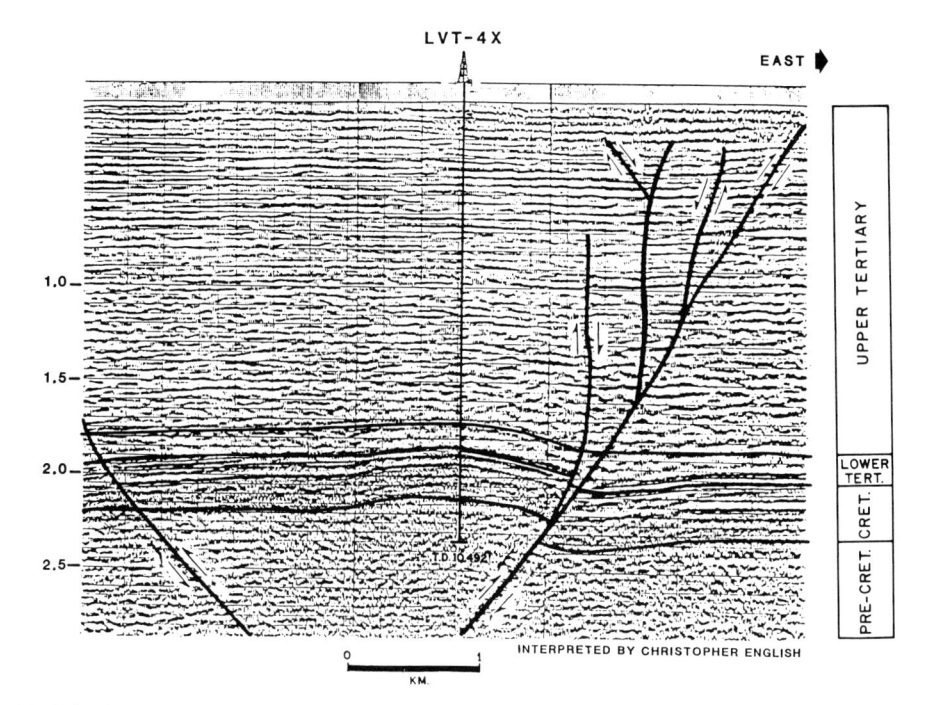

Fig. 20. Seismic line through La Victoria Field.

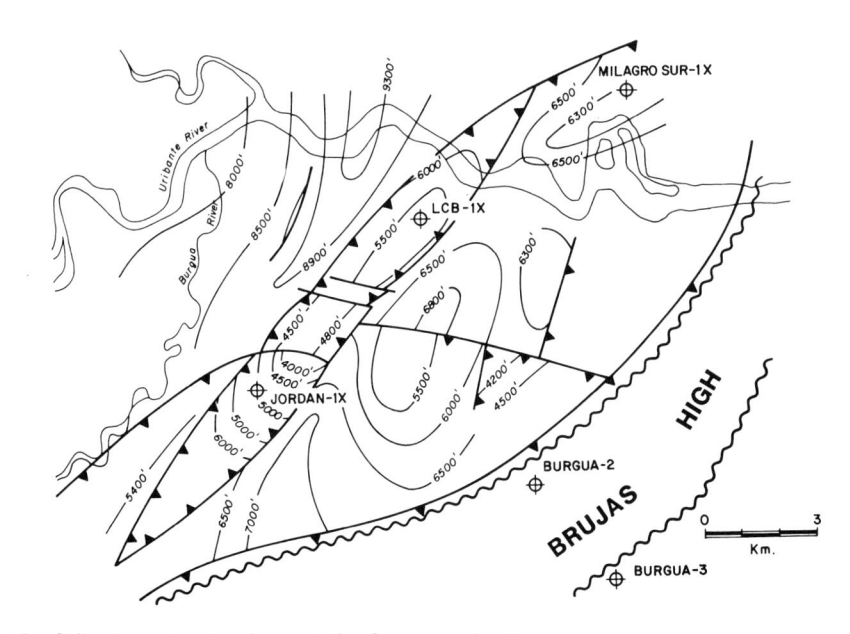

Fig. 21. La Ceiba Area: structural map at the Cretaceous level.

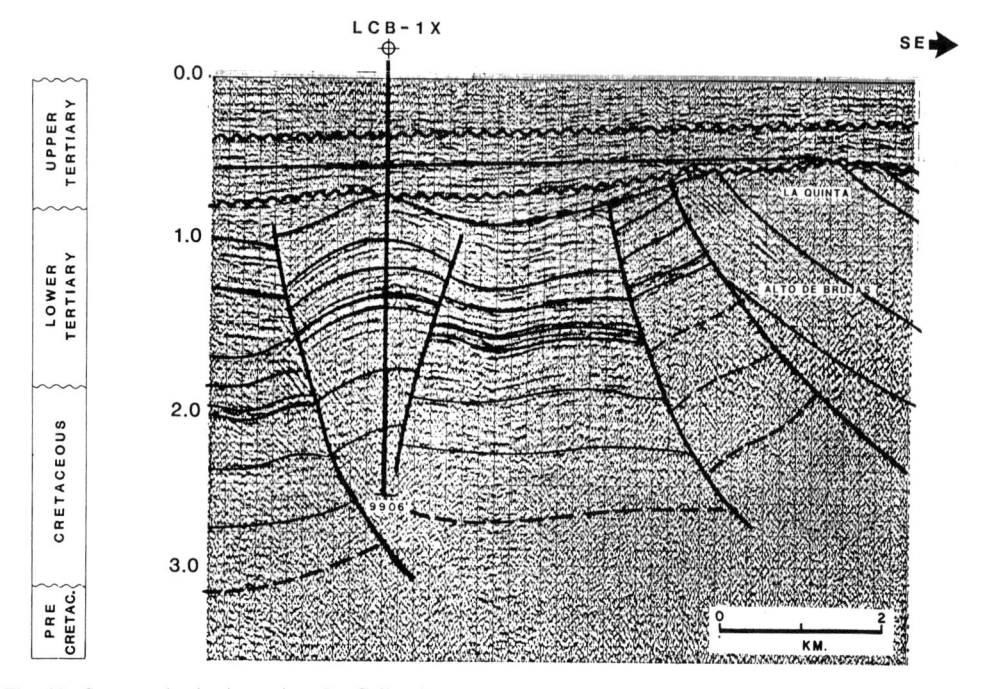

Fig. 22. Structural seismic section, La Ceiba Area.

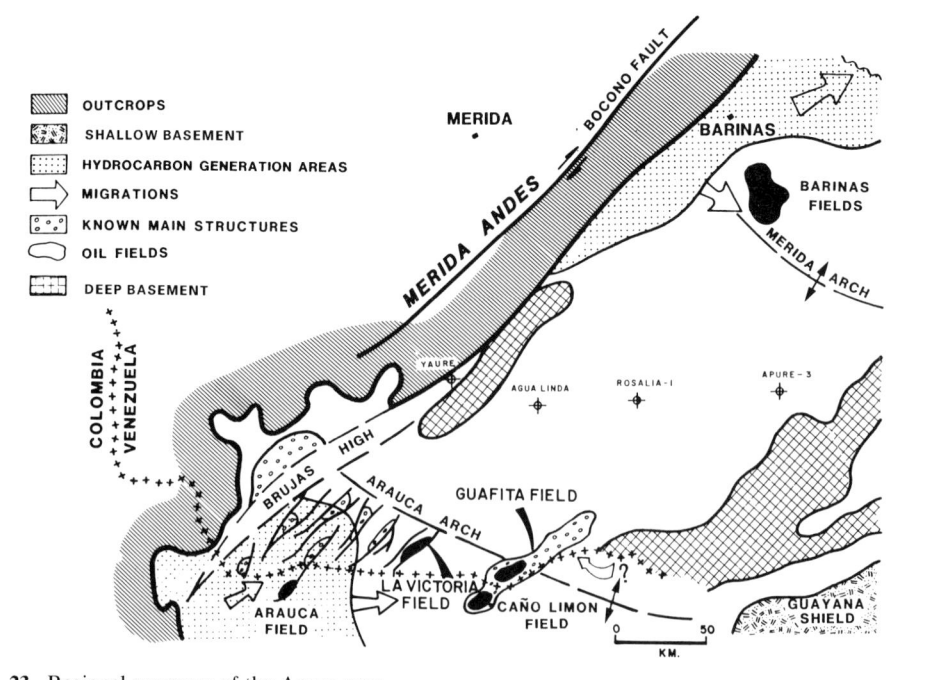

Fig. 23. Regional summary of the Apure area.

La Ceiba area

This area is located to the northwest of the Apure area. Its southern limit is a large inverse fault, known as the Brujas Fault (see Fig. 10). The structures developed in this area are the result of compressive forces acting in a NW–SE direction, probably as late as Upper Oligocene. The undisturbed post-Andean Molass rests unconformably on deeply folded and faulted sedimentary sections.

The structures are of an Oligocene–Lower Miocene age, contemporanous with the pre-Andean structures of other parts of the Basin. Large inverse faults of high angle, with throws sometimes in excess of 3000 ft, define separate blocks (Figs 21 & 22).

In 1983 Corpoven drilled three exploratory wells in the area. The main targets were sandy sections Paleogene and Cretaceous age. The three wells encountered very thick sandy sections at both intervals but discovered only non-commercial hydrocarbon accummulations at Cretaceous level in Well Jordan-1X.

Conclusions

The huge seismic surveys/carried out in the last years,/have identified a wide range of structural and stratigraphic traps.

The evaluation and identification of oil kitchens/along the sub-andean belt/have been studied using various quantitative methods.

New data and recent regional studies are encouraging for new commercial oil discoveries to be made South of the Arauca Arch./Three billion barrels/of new reserves/could be proved up by future drilling based upon current understanding of the area.

References

BOGOTA-RUIZ, J. 1988. Contribucion al Conocimiento Estratigrafico de la Cuenca de los Llanos (Colombia). III Simposio Bolivariano, Exploracion Petrolera en las Cuencas Subandinas, **1**, 308–350.

BRISTOW, J. D., 1987. *The la Victoria Field, Apure State, Venezuela*. Corpoven, S.A.

CASSANI, F., GALLARDO, O., TALUKDAR, S. & VALLEJOS, C. 1988. Geoquimica Organica de los Crudos Venezolanos. III Simposio Bolivariano, Exploracion Petrolera en las Cuencas Subandinas; **2**, 612–637.

CHIGNE, N. A. 1985. Aspectos relevantes en la Exploracion de Apure, Venezuela: II Simposio Bolivariano, Exploracion Petrolera en las Cuencas Subandinas; **1**, 1–48.

FINDLAY, A. L. 1988. Cortes geologicos regionales en los Andes Venezolanos y sus implicaciones petroleras y estructurales: III Simposio Bolivariano, Exploracion Petrolera en las Cuencas Subandinas, **1**, 401–429.

GONZALEZ DE JUANA, C., ITURRALDE, J. & PICARD, X. 1980. Geologia de Venezuela y de sus Cuencas Petroliferas. 1st edn, Foninves, Caracas.

ORTEGA, J. F., VAN ERVE, A. & DE MONROY, Z 1987 Formacion Guafita: Nueva unidad litoestratigrafica del terciario en el subsuelo de la Cuenca Barinas–Apure, Venezuela Suroccidental. *Sociedad Venezolana de Geologos Boletin*, **31**, 9–35.

PORTILLA, A. 1988. Relaciones estructurales y estratigraficas en la Cuenca Barinas–Apure, Venezuela. Corpoven, S.A.

TISSOT, B. & ESPITALIE, J. 1975. L' evolution thermique de la matiere organique des sediments: Applications d'une Simulation Mathematique. *Revue Institut Francois Petroles* **30**, 743–777.

New oil discoveries in the Ceuta area, SE Lake Maracaibo, Venezuela

M. ROBERTO, L. MOMPART, E. PUCHE & F. SCHERER

Corpoven, S.A., Venezuela

Abstract: The Ceuta area forms the southeast extension of a major oil accumulation in the Maracaibo Basin, at present Venezuela's main oil producing area. The Ceuta Field, with current reserves of 600 MMBBls of light to medium gravity oil, is a major wrench fault structure, producing from Miocene and Eocene sandstones. Several major transcurrent faults are recognized in the Maracaibo Basin, each defining major producing oil fields like the Ceuta Field. These structures, roughly parallel to each other, are oriented in a North–South direction.

The principal plays in the Maracaibo Basin consist of Miocene and Eocene sandstones and, at greater depths, fractured Cretaceous carbonate reservoirs and fractured basement, all charged from the rich marine La Luna source rocks of Upper Cretaceous age.

The discovery of the Ceuta Field was the result of an intensive drilling campaign executed in the 1970s, based on conventional seismic techniques. At that time the exploration of the area was oriented towards the main North–South trending structures. The new exploration, undertaken during the last four years, was based on the structural modelling of the area in terms of the known transcurrent faulting. This allowed the recognition, in areas between the major trends, of three new types of traps related to different styles of faulting, such as normal, compressive and antithetic transcurrent.

The successful drilling of these structures has led to the discovery of an additional 1100 MMBBls of oil outside the existing Ceuta Field, confirmed by an intensive drilling appraisal campaign.

Today's total proven oil reserves of Venezuela are 58 000 MMBBls. At present, over 80% of Venezuela's total daily crude production, as well as the major share of light oil, still comes from the classic petroleum province of the Maracaibo Basin. Although this prolific basin has been producing since 1882, its remaining 9250 MMBBls of light to medium oil reserves still constitute over half of the current light to medium oil reserves of Venezuela.

Apart from the fact that to date major parts of the 50 000 km² Maracaibo Basin have still been relatively little explored, it is most interesting to note that substantial amounts of new oil reserves have been discovered during the last few years in recently defined new traps within the classic producing areas (Fig. 1). It must be stressed that part of this success was only possible thanks to the application of new seismic techniques that allowed the integration of land and lake data, in addition to the benefits of an overall increased seismic resolution.

The present paper focuses on the southeasterly extension of a main producing area in the Lake Maracaibo, the Ceuta Field, where in the 1980s substantial new oil reserves have been found. A recent intensive appraisal campaign indicates that the recoverable reserves of these new discoveries are in the range of 1100 MMBBls of light oil, adding to the 650 MMBBls located in the existing fields of the area. Total expectations of currently defined prospects add another 500 MMBBls.

Apart from a brief description of the geological evolution, plate tectonic setting and hydrocarbon habitat, this paper concentrates on the structural model and trap types found in this area. On the basis of present reserves and appraisal results as compared to trapping possibilities, the paper gives a brief outlook of future exploration. Finally, the potential reasons for a continuing rate of success are highlighted.

Geological evolution and plate tectonic setting of the Lake Maracaibo Basin

Paleozoic sediments which have mostly been metamorphosed during a Silurian–Devonian orogeny constitute the economic basement of the Maracaibo Basin (Fig. 2). During the Jurassic–Early Cretaceous, a series of NNE–SSW oriented grabens developed, probably as a result of the regional tectonic events related to the opening of the Central Atlantic (see James 1990). These grabens were filled with red beds and fluvial deposits (La Quinta and Rio Negro Formations).

The subsequent Cretaceous transgression across the newly formed continental margins and gradual subsidence led to the build-up of an

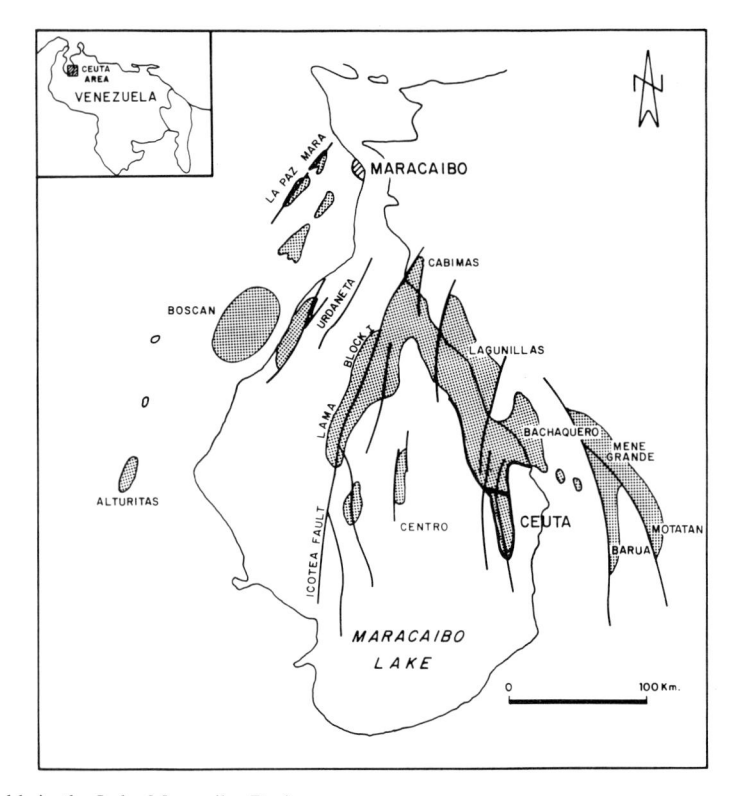

Fig. 1. Oil fields in the Lake Maracaibo Basin.

Fig. 2. Stratigraphy in the Ceuta area.

extensive carbonate platform (Cogollo Group). During Upper Cretaceous time, the organic-rich limestones of the La Luna Formation were deposited in restricted environments, and later were covered by the thick open marine shales of the Colon Formation.

After the deposition of the overlying Palaeocene neritic limestones (Guasare Fm), a cycle of clastic sedimentation began in the basin, along with major changes in the patterns of sediment transport. It culminated during Eocene times, with the progradation in a NE direction of a deltaic sequence consisting of largely regressive clastics (Misoa Fm, B Sands), which in turn were capped by the holomarine transgressive Pauji shales. At the end of the Middle Eocene, erosion of substantial thicknesses of Eocene deposits occurred as the consequence of the regional tilting of the basin to the southwest, creating a regional unconformity.

After an early Miocene transgression (La Rosa shales), regressive fluvial coastal sediments were deposited (Lagunillas Fm) with thick channel sands (Bachaquero member). Finally, the Plio-Pleistocene uplift of the Venezuelan Andes was accompanied by the development of a pronounced foredeep in the southeast of the Maracaibo Basin.

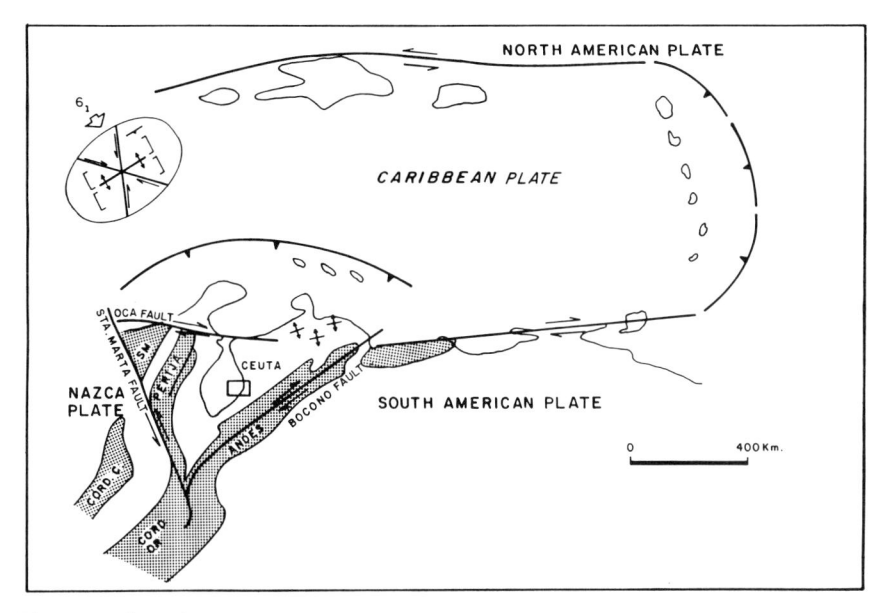

Fig. 3. Plate tectonic setting.

Three major fault systems, the Oca, Bocono and Santa Marta systems (Fig. 3), bound the Maracaibo Block (Kellog 1984). These systems result from a combination of northeasterly directed compression exerted by the Nazca Plate against northwestern South America, and the dextral relative displacement between the Caribbean and South American Plates (see James 1990). Transpression along the Bocono Fault and the sub-parallel Perija Fault has resulted in the isolation of the Maracaibo Basin by two mountain ranges, the Venezuelan or Merida Andes and the Perija Mountains. The basin deepens regionally towards a foredeep adjacent to the Merida Andes.

The Maracaibo Basin is structurally dominated by a southward opening fan of roughly north–south trending sinstral wrench faults (Fig. 1). Transpression along these faults has produced a series of associated highs. These provide traps for hydrocarbons migrating northwards up the regional monocline of the present day basin and they have been the traditional sites of exploration drilling. The main Ceuta Field is developed on one such high.

Other structures recognised in the basin are roughly east–west trending dextral faults and northwest trending normal faults. All of these features may be related to the basin's plate-tectonic setting, described in detail by James (1990), which has resulted in NW–SE compression and NE–SW extension in the basin. Outcrop (Perija Mountains) and core data (Maracaibo Basin) indicate two main compressional phases, the first during the Late Eocene and the second in the Miocene.

Hydrocarbon habitat of the Ceuta area

The principal plays in the Ceuta area (see Fig. 4) are Miocene and Eocene sandstones sealed by interbedded shales, charged from organic-rich marine La Luna source rocks of Upper Cretaceous age (Fig. 2). Cretaceous fractured carbonate reservoirs charged from the same La Luna source rocks are another target at considerably greater depth. To date this objective, which is a prolific producer in the western part of the Maracaibo Basin, has as yet not been tested by the drill to the east of the Ceuta Field.

While geochemical analyses of oil samples from various Ceuta wells indicate a marine La Luna type source rock, further to the west, Eocene source rocks may have contributed some oil, as suggested by the land plant influence noted in oil samples from that area (Talukdar et al. 1985).

The API gravities of the crudes in the Ceuta area vary from values as low as 20° API in Barua, to 35° API in the traditional Ceuta Field. The gravity distribution does not seem to show a clear pattern, in spite of its regional decrease towards the east. However, a vertical gradient seems to be common in the traditional areas,

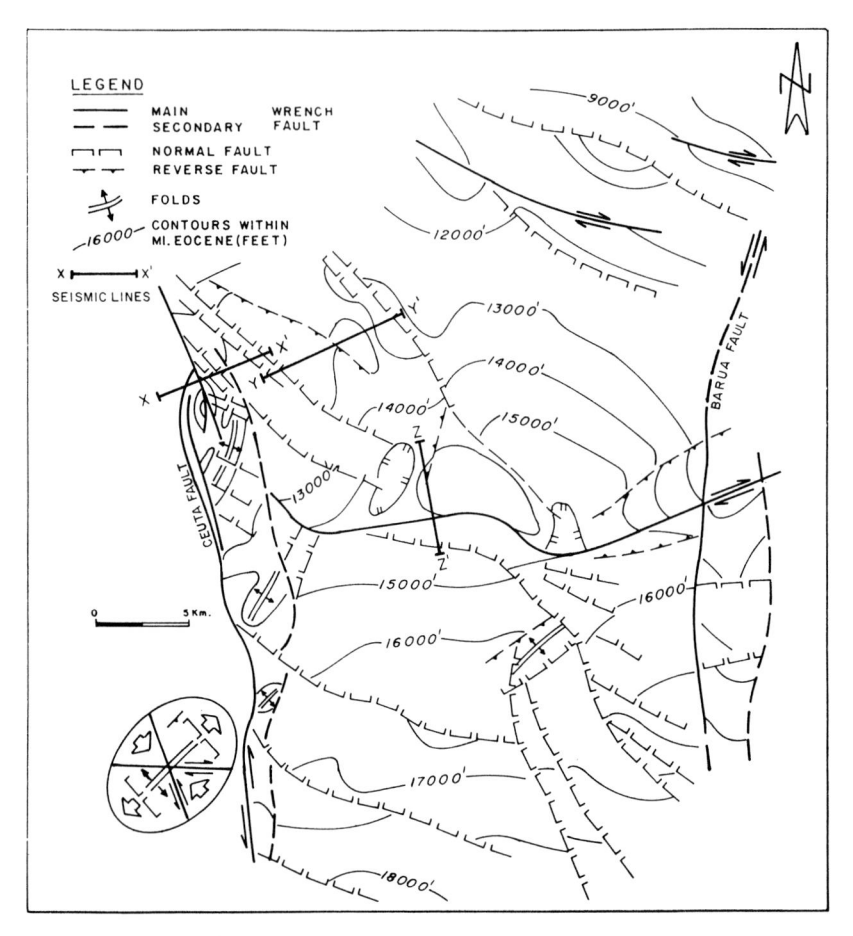

Fig. 4. Tectonic setting of the Ceuta area.

with gravities decreasing towards the upper part of the Eocene section.

The most important reservoir units are Miocene and Middle Eocene regressive sandstones. The Miocene sandstones have the best reservoir properties and the Bachaquero Member of the Lagunillas Formation, consisting of thick channel sands, is the best producer.

The Eocene shows considerable thickness variations across the Ceuta area (Figs 5–7). Most of the deeper parts of the fully developed Eocene sequence have not yet been penetrated by any wells. Where drilled, the deep interval consisted of a mainly transgressive sequence with relatively thin sandstone reservoirs (Misoa C). The thickest sands occur within the overlying, predominantly regressive Misoa B sequence, in particular the lower Misoa B beds, a distributary mouth bar complex.

The overlying holomarine Pauji Formation contains transgressive sands at its base, which are the producing reservoir in the Barua field to the east.

Structural model and new discoveries

The recent exploratory campaign was based on the structural modelling of the area in terms of the known transcurrent faulting. The presence of wrenching has been previously demonstrated in other areas of the Lake Maracaibo (Krause 1971), but this is the first formal effort to interpret observed structures in function of a global model (Fig. 4). Fault geometries and deformational patterns in the Ceuta area appear to fit experimentally derived structural models (Naylor et al. 1986), and indicate a NW–SE oriented compressional stress field in accordance with the regionally observed field.

After exploration had drilled the most prominent high relief structures along the main left lateral wrench fault systems, the search went on for some subtle structural features. These structural elements and the hydrocarbon traps associated with them, as well as the traditionally explored high relief structures linked to the main left lateral strike slip faults are briefly described below.

Structural framework

In the Maracaibo Basin the Eocene section thickens markedly to the northeast of a NW–SE trending hinge-line that runs roughly through the area of study (Van Veen 1972). For example, the area east of the Ceuta Fault contains 7000 ft more of Eocene section than the area to the west. Low-angle (30–40°) basement faulting along this flexure zone produced overburden deformation indicative of varying ratios of dip-to-strike-slip (Naylor et al. 1986).

The deformational history of the Ceuta area is therefore seen to have comprised at least three major phases.
(1) An early Eocene extensional or divergent wrenching phase that resulted in syndepositional faulting and sediment thickness variations.
(2) A late Eocene compressional or convergent wrench phase that produced the flower structure uplift of the Ceuta High. The northern part of the Maracaibo Basin was uplifted at this time (Zambrano et al. 1971) and a regional unconformity developed at the top of the Eocene section.
(3) A Mio-Pliocene phase of transpression along the Bocono and Perija faults, resulting in the rise of the Merida Andes and the Perija Mountains, caused further inversion along the northern part of the Ceuta High trend so that local unconformities truncate major structures at La Puerta Formation level.

Left lateral strike slip

In the Ceuta area, two principal N–S oriented left lateral strike slip zones are present: the first is associated with the Ceuta High described below, the second delimits the Barua Field to the east.

The Ceuta High (Fig. 5) is bounded to the west by a duplex left lateral strike slip zone. The eastern boundary consists of parallel adjustment wrench faults.

The underlying basement fault appears to be

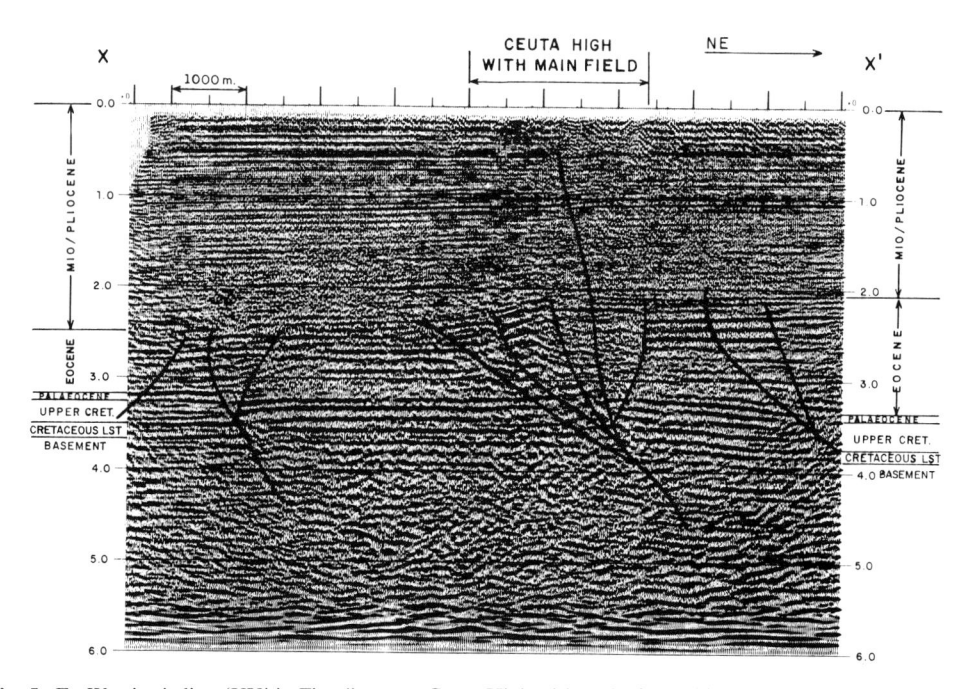

Fig. 5. E–W seismic line (XX' in Fig. 4) across Ceuta High with main field which contains oil accumulations both at Miocene (close Eocene unconformity) and at Middle Eocene levels. Note late Tertiary inversion of the Cueta High.

a 45° reverse fault, but further south the same fault shows a normal throw at basement level. It thus behaves as a 'scissor' fault. Overburden deformation suggests strike-slip and dip-slip displacements to have been about equal. Two NE–SW oriented en echelon folds and a narrow shear – fault bounded pop-up anticline with extremely steep flanks (Fig. 5) constitute the traditional Ceuta High which is cut by four NW–SE oriented extensional faults.

Late Eocene–Oligocene uplift resulted in severe erosion, removing the Pauji Formation over much of the high and cutting progressively deeper westwards into the Misoa Formation. Deposition in Miocene times was largely controlled by intermittent folding and synsedimentary faulting. Miocene stratigraphy and reservoir development is therefore complicated by changes in thickness and lithofacies due to on-laps and off-laps onto the rising anticlines. This results in the pinch-out of some units onto the highs, and thicker lower Miocene units in the lows. The eastern boundary fault was inactive during Miocene times. However, reverse movements along the western boundary fault have resulted in the thickest Miocene development in the downthrown block to the west of the N–S trending boundary fault. This indicates inversion during middle Miocene–Pliocene times. Overpressures in this area might be caused by this process.

Most faults do not reach the younger horizons. Faults with several hundred feet of throw at the Misoa Formation level may still be sealing in the Lower Miocene units, but by top Bachaquero (upper Miocene) level the vertical displacement may amount to only a few tens of feet, thus becoming unimportant for trapping.

Right lateral strike-slip

The largest right lateral (antithetic) strike-slip fault in the Ceuta area appears to be a 30–45° northward hading normal basement fault (Fig. 6). The fault had a very important dip-slip component and was active during early Eocene time as indicated by a greatly increased Eocene thickness on the downthrown side of the fault. This Palaeogene sub-basin is thought to be a branch of the Trujillo Flysch trough, which is known to the northeast of the study area and corresponds to the foredeep that formed in the front of the Caribbean ranges (Gonzalez de Juana et al. 1980). The Trujillo Formation is a shaly sequence including basinal turbidites, but locally coarse clastic aprons may have developed along the subaerially exposed scarp of the active fault. These characteristics had been

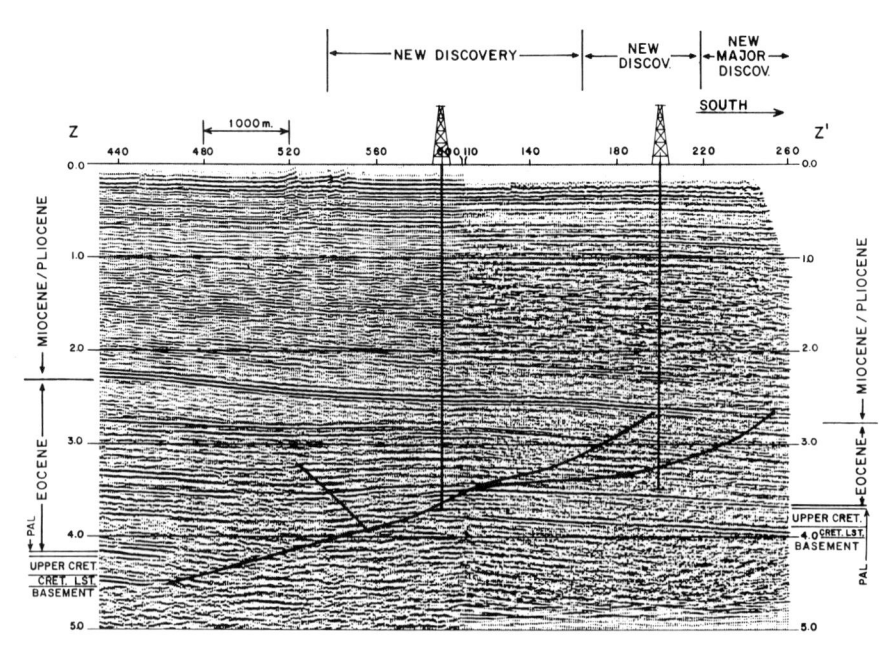

Fig. 6. N–S seismic line across three recently discovered accumulations. Objective level intra-Eocene (ZZ' in Fig. 4). Note right-lateral strike-slip fault with huge normal vertical throw and Riedel-type secondary fault.

partly recognized in earlier seismic interpretations of the area (Araujo 1986), and were confirmed by recent drilling.

On Fig. 4, the fault trace shows two linear segments with a N–S accident offset that may result from dextral offset. The lateral offset and rotation have created a local pull-apart low. The difference in attitude of the associated Riedel-type faults (normal in the west, reverse in the east) is probably related to differences in the strike-slip to dip-slip ratio: along the western segment of the fault, the amounts of strike-slip and dip-slip are probably similar, while in the east the strike-slip is much more important, leading to true Riedel shears. Recent drilling had confirmed that these Riedel shears give rise to structurally high oil-bearing fault blocks (Figs 4 & 6).

It is interesting to note that in the west, the extensive flank behind this fault is largely undisturbed, while in the east it shows considerable faulting.

The above described fault system cuts across the regional monocline that rises northwards from the foredeep adjacent to the Merida Andes, and so forms a trap for migrating hydrocarbons. The recent discoveries suggest that this is a giant trap, containing volumes greater than those of the traditional Ceuta and Barua fields. This fault, with its north hading plane, truncates the monocline of the southern Moracaibo Basin and therefore provides an ideal trap for oil migrating northwards from the Andeau foredeep. The newly discovered oil, exceeding the amount found in the main fields, suggests the presence of a giant hydrocarbon trap.

Folds and reverse faults

Compressional features perpendicular to the main stress field can be mapped by seismic techniques. These are in essence SW–NE orientated folds, generally bounded by reverse faults. These features form important structural trapping elements, as shown by the drilling of the largest one found in the area, with potential reserves in excess of 100 MMBBls (Fig. 6, Shot Point 600).

Detailed studies of the Ceuta High indicate that small-scale folds will generally show synsedimentary growth, and will occur in an en echelon mode, with their axes offset by NW–SE normal faults. This setting will produce a complex distribution of the fluids in the reservoirs, and therefore the recognition and delineation of these folds will be a key factor in the appraisal and development of the area.

The present day seismic coverage is not adequate to map these features in detail, and therefore only the largest ones are shown on the maps. Extensive 3D coverage is at present underway.

For the same reasons, it is not possible to determine the true nature of other features such as the NNW–SSE oriented faults south of ZZ' in Fig. 4. It is possible that they belong to a left-lateral strike-slip system.

Stress – relief normal faults

The preferentially NW–SE oriented normal faults (see Fig. 7) show considerable variation in the amount of throw: some have throws in excess of 1000 ft. Potential hydrocarbon traps of considerable size exist both on the downthrown as well as the upthrown side of major normal faults, if adequate NW closure can be mapped. Such closure may be provided by SW–NE oriented synclines, by a roughly N–S oriented left-lateral strike-slip fault, or occasionally even by a favorable change in the strike of the major fault. The chance for this kind of up-dip closure clearly improves in the vicinity of a left-lateral strike-slip system, where such secondary structural elements are more likely to occur.

The extensional down-to-the-basin faults, with a W to NW direction, are very difficult to differentiate in this area from the antithetic wrench faults because of their similar orientations. Furthermore, it is very likely that these normal faults were overtaken by the later compressional phases, and therefore will show deformational features very similar to other right lateral wrench faults.

Reserves

The traditional Ceuta Field carries ultimate recoverable reserves of 600 MMBBls of light oil, of which 340 MMBBls, are contained in Miocene reservoirs, and around 260 MMBBls are found in the Eocene.

As shown in Fig. 8, the reserves discovered within the last few years actually exceed those previously found in the high relief structures. To date, the newly discovered reserves are all within the Eocene, where folding and faulting is more pronounced than in the overlying section. The oil in place of these new discoveries is estimated to 3065 MMBBls (Dec 1987), with recoverable reserves of 1100 MMBBls, according to the results of the recent intensive appraisal campaign.

Development plans are underway, in order to raise over a period of five years, the production

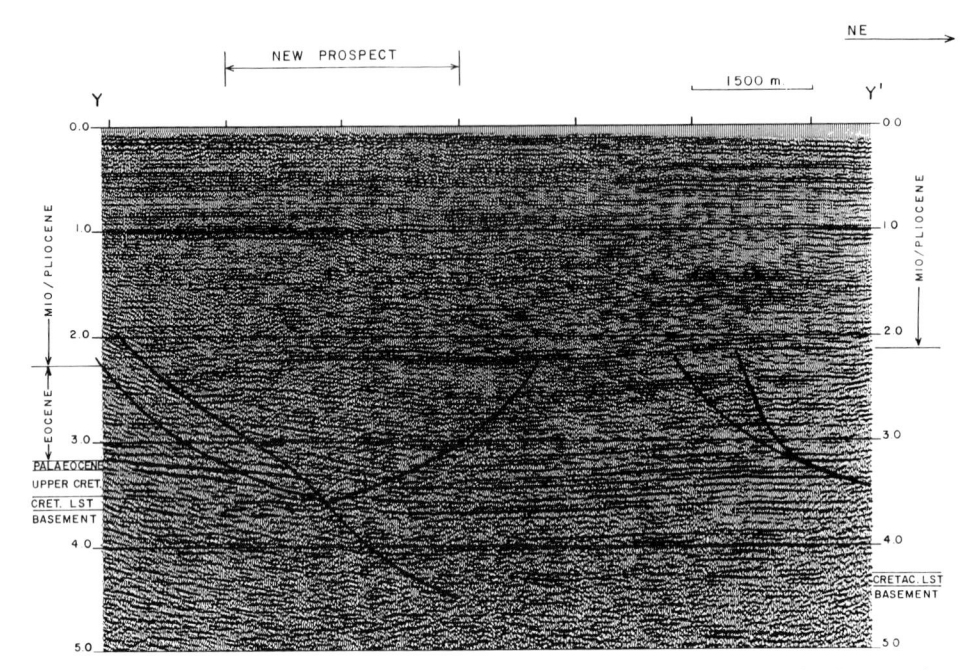

Fig. 7. E−W Seismic line across prospect on down-thrown side of major normal fault (YY' in Fig. 4).

potential of the area from 80 000 BBls/D to a total of 200 000 BBls/D.

Currently defined prospects (Fig. 9) have an additional total expectation of approximately 500 MMBBls of light oil. With a considerable exploration effort ongoing, this figure has a good chance to increase, though much depends on the effectiveness of some of the untested trap types further discussed in a later section.

Appraisal results

First discoveries

The discovery well, VLG-3693 (1X in Fig. 10), a 3 km outstep east of the Ceuta Field, was drilled in 1981. The initial production in the Lower B sands (Misoa Fm) was 3164 BBls/D, with a GOR of 1223 CF/B. In 1982, the exploratory well VLG-3707X (2X in Fig. 10) was drilled, and tested 4500 BBls/D. In both cases, the oil was light, with gravities around 36 °API.

Further to the east, the exploratory well TOM-1X (3X in Fig. 10), located on land, was drilled in 1985. The well discovered oil in Upper and Lower B sands, with an initial production of 2000 BBls/D.

To the west of the Ceuta Field boundary fault, a thick interval of Eocene sediments is missing, and the C sands turn out to be the most productive reservoirs. The first well drilled in this area produced an average of 3000 BBls/D during five years and has accumulated more than 5 MMBBls of oil. An appraisal well drilled 4 km to the south tested 3000 BBls/D.

Appraisal results

Since well 2X, eleven new wells have been drilled and completed, with a cumulative oil production of 14 MMBBls at the end of May 1987. Based on production and pressure data, seismic reinterpretations and geological well data (Figs 11 & 11a), the following reservoir definition was established:

Lower B sands. This is the main producing interval in the area to the southeast of the Ceuta Field, where fault closure exists to the north, east and west. The OWC is as yet unknown. Thicknesses of this interval are in the range of 800 feet in average, slightly increasing in a northeasterly direction (Figs 11 & 11a).

The reservoir is heterogeneous. Several sealing shales can be recognized that separate sands with as yet unknown directional permeability. Well data suggest the presence of coalescing channel sand bodies with widths of 1800 feet (Fig. 12). However, pressure data indicate the presence of three main reservoirs within the Lower B sands, separated by inter-

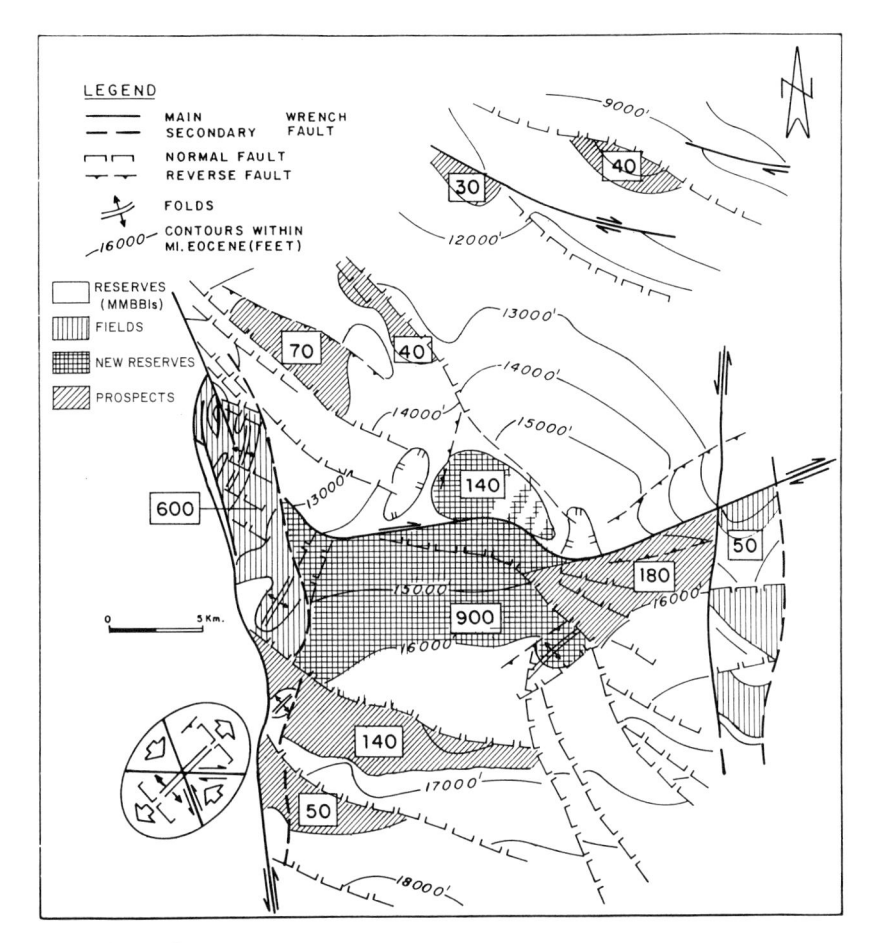

Fig. 8. Reserves in the Cueta area.

bedded shales and correlatable over a distance of 8 km. Pressure differences between these separate reservoirs are as large as 1000 psi. Thus, the Lower B interval appears to correspond to a multilayered reservoir model.

Upper C sands. The interval has been in production only to the west of the Ceuta Field, where the structure is a south-dipping gentle monocline cut by NE–SW trending normal faults which may constitute the main trapping mechanism for the accumulation. Logs define the reservoir units as thin individual sands, 15 to 30 ft thick. No OWC has been detected so far.

Depth of burial is probably an important factor for the porosity and permeability of this interval. To the east of the field, subsidence has been greater than to the west. This is reflected in the poor production of most of the wells that tested the Upper C level in the eastern area, where porosities vary between 9 and 11%.

Future exploration

Given the southerly regional dip, the most promising trapping opportunities are offered by the roughly W–E oriented right-lateral strike-slip faults. An extensive drilling campaign is planned for the future to explore the flank behind the eastern part of the main right lateral fault (Fig. 8). Right lateral strike slip faults also appear to occur in the NE area of Ceuta, which to date has been little explored and is at present poorly covered by seismic.

Further, untested potential hydrocarbon traps are also formed by the NW–SE oriented normal faults which provide up-dip closure against the regional southerly dip. As mentioned earlier, such closure is most likely to be present close to a left-lateral strike-slip system. Another important factor determining the effectiveness of this type of trap is the amount of throw along these normal faults. Although large throws are

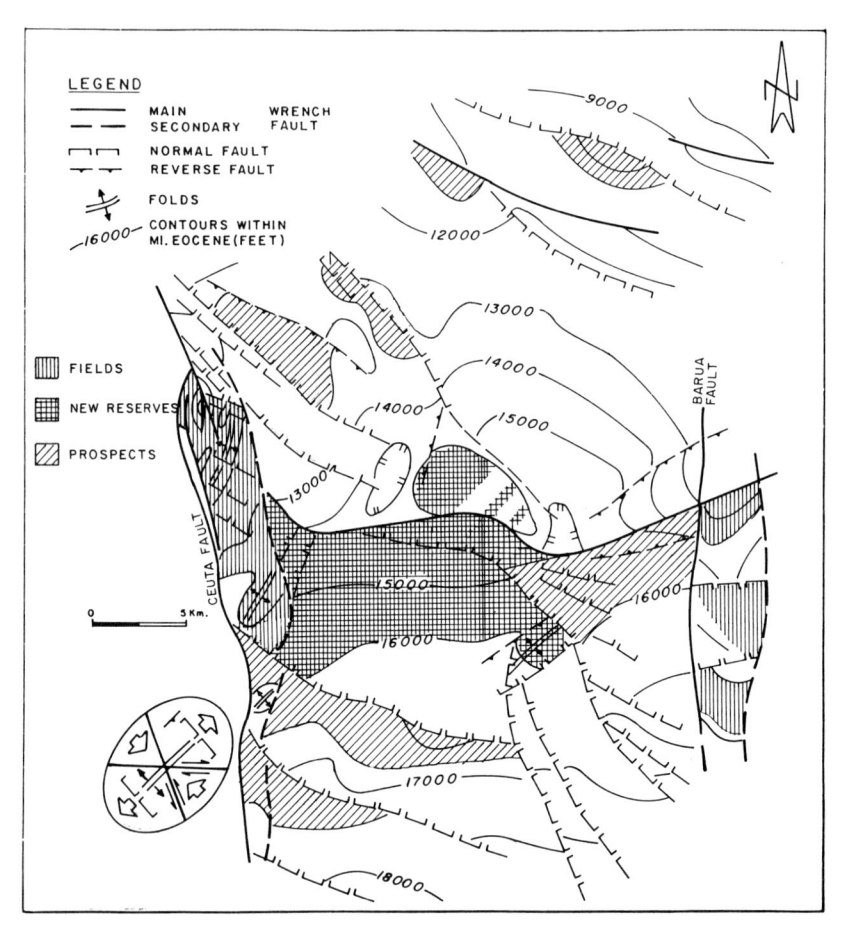

Fig. 9. New discoveries and prospects in the Cueta area.

obviously preferable, faults with as little as 200–300 ft might be sealing. The first of these types of traps against normal faults will be tested later this year.

Reverse-fault bounded folds are another potential type of trap, of which new seismic investigations are likely still to find more. A possible zone of interest is the NNW–SSE oriented normal faults, which might form part of a left-lateral strike-slip system as mentioned earlier. Furthermore, lateral velocity changes appear locally to increase the amount of seismic closure, in particular across areas with rapid thickness variations, such as the zone of wedging of the marine Pauji Formation or the Eocene sub-basin north of the main right-lateral wrench fault.

Finally, there also exists the possibility of some stratigraphic trapping, especially on the downthrown side of major faults, where coarse clastic aprons and thick sand development may shale out further away from the faults. It should also be noted that the oldest section of the Eocene has not yet been penetrated by the drill, as the few deep wells drilled in the area have all been positioned on local highs with an incomplete stratigraphic sequence.

In summary, given the seismic and well data at hand indicate that the probability of discovering between 500 MMBBls to 1000 MMBBls of mainly light oil within the next five years in the Ceuta area is high.

Conclusions

In the Ceuta area, the presence of the main ingredients necessary for oil accumulation has been demonstrated: the source rock occurs as the prolific La Luna organic-rich limestone, currently in an advanced stage of the oil window

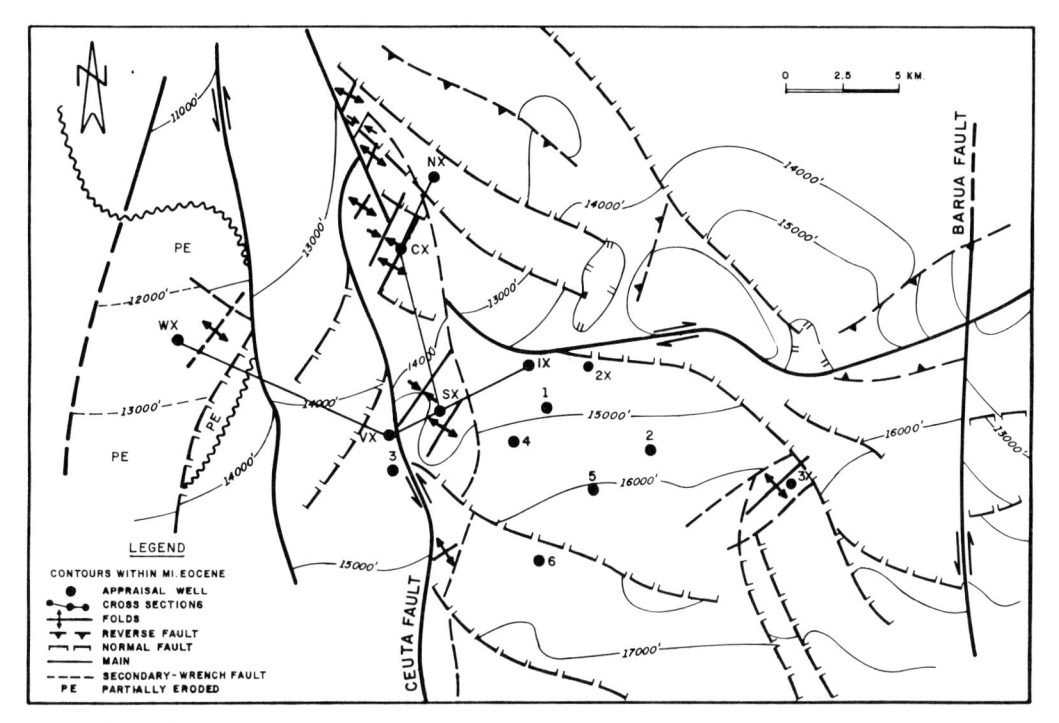

Fig. 10. Cueta area appraisal program.

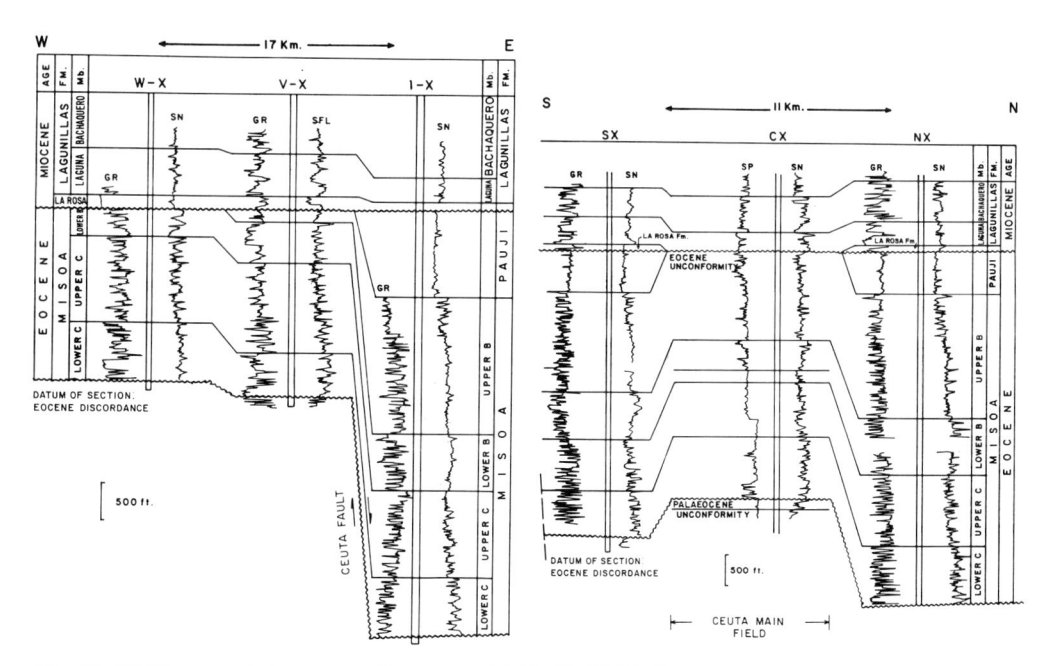

Fig. 11. Well log correlation accross Cueta area: (a) W–E, (b) N–S.

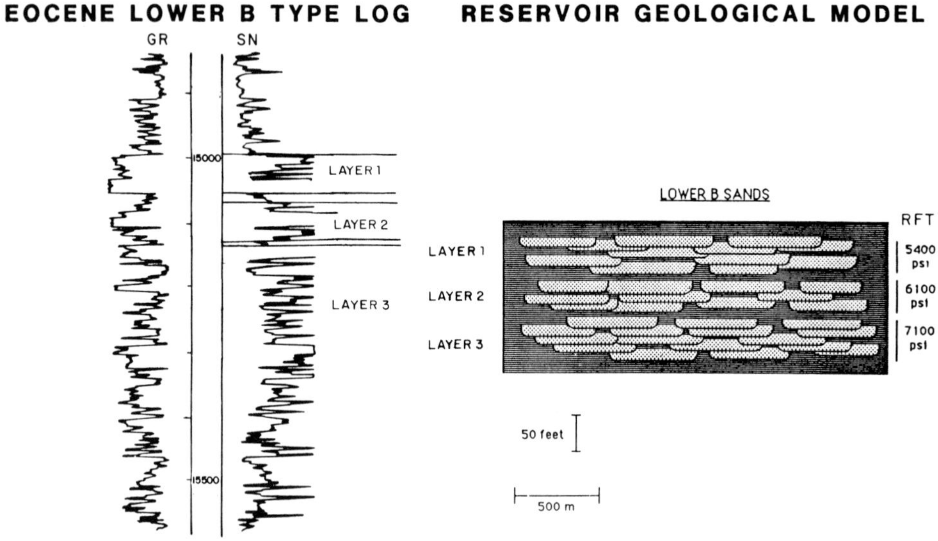

EOCENE LOWER B TYPE LOG **RESERVOIR GEOLOGICAL MODEL**

Fig. 12. Cueta field reservoir model.

over much of the Ceuta area. Maturity was reached only relatively recently, after the Plio-Pleistocene formation of the Andean foredeep in front of the rising Andes. This same foredeep is responsible for the development of a very broad monoclinal flank which provides an extensive oil source drainage area.

Within this setting, the Eocene deltaic Misoa Formation provides ample reservoir capacity. The regional stress pattern of compression along a NW–SE axis generated both folding and faulting that cut across the broad regional monocline, thereby forming a variety of trapping possibilities.

The key to the continuing success is the recognition of the structural features which, according to the model, may control the accumulation. The ability to discover these ever more subtle traps, will depend not only on the seismic data quality, but also on the careful calibration with all existing well information (VSP, well shoots). Depth convertion becomes critical, and seismic stratigraphy a necessary complementary tool.

Finally, continuing success will largely depend on the capacity of the interpreter to produce an imaginative search for further trapping possibilities, both structural and stratigraphic.

References

ARAUJO, M. 1986. Vertical Seismic Profile in Well VLG-3715X and its stratigraphic application. *III Congreso Venezolano de Geofisica, Memorias*.

GONZALEZ DE JUANA, C. et al. 1980. *Geologia de Venezuela y de sus Cuencas Petroliferas*. Ediciones Foninves. Caracas.

JAMES, K. H. 1990. The Venezuelan hydocarbon habitat. *In*: BROOKS, J. (ed) *Classic Petroleum Provinces*. Geological Society, London, Special Publication **50**, 9–35.

KELLOG, J. N. 1984. Cenozoic History of the Sierra de Perija, Venezuela–Colombia and adjacent Basins. *Geological Society of America Memoir* **162**, 239–281.

KRAUSE, H. 1971. La Falla de Icotea en el Bloque I de las concesiones de la CSV en el Lago de Maracaibo. *III Jornadas Tecnicas del Petroleo*, Maracaibo, Oct. 1971.

NAYLOR, M. A. & SIJPESTEIJN, C. H. K. 1986. Fault geometries in basement induced wrench faulting under different initial stress. *Journal of Structural Geology*, **8**.

TALUKDAR, S., GALLANGO, O. & CHIN-A-LIEN, M. 1985. Generation and migration of hydrocarbons in the Maracaibo Basin, Venezuela. *12th International Meeting on Organic Chemistry, Munich, Abstracts*, p. 28.

VAN VEEN, F. R. 1972. Ambientes Sedimentarios de la Formacion Mirador y Misoa del Eoceno Inferior del Lago de Maracaibo. *IV Congreso Geologico Venezolano, Memoria, Tomo II*.

ZAMBRANO, E., VASQUEZ, E., DUVAL, B., LATREILLE, M. & COFINIERES, B. 1971. Sintesis Paleogeografica y Petrolera del Occidente de Venezuela. IV Congreso Geologico Venezolano, Memorias *Bol. Geol., Publicacion Especial*, **5**, Vol. I.

Aspects of the petroleum geology of the Oriente Basin, Ecuador

M. F. DASHWOOD & I. L. ABBOTTS

Clyde Petroleum plc, Coddington Court, Ledbury, Herefordshire, HR8 1JL, UK

Abstract: The Oriente Basin of Ecuador is one of the most productive of the South American Sub-Andean Basins. Cumulative production of oil to the end of 1986 was over one billion barrels, and current production stands at approximately 300 000 barrels per day. At least 20 oil fields have been discovered to date, including five giant fields, and exploration is currently in the early stages of a new and already successful phase in which ten new operators are scheduled to drill over 40 wells between 1985 and 1992.

The Oriente Basin contains a sedimentary fill of Palaeozoic to Recent age. Major commercial interest is confined to the Cretaceous depositional cycle and all of the significant production comes from fluvio-deltaic and marine sandstones of the Hollin and Napo Formations. Most of the productive structures are low-relief, north−south orientated, anticlines of two distinct types: foot wall anticlines associated with normal faults, (Type 1) or hanging wall anticlines associated with reverse faults, (Type 2).

Evidence points to the importance of pre-Miocene structural growth, comprising either early-mid Cretaceous rejuvenation of pre-Cretaceous 'basement' faults in the case of Type-1 structures and 'Early Andean' (latest Cretaceous to Oligocene) compression in the case of Type-2 structures. Few commercial discoveries have been made associated with 'Late Andean' (Miocene−Pliocene) compressional structures, essentially because these structures appear to post-date the main phase of primary oil generation and migration.

The origin of the oils is problematical because the potential source rocks, the marine claystones and limestones of the Cretaceous Napo Formation, are generally immature or marginally mature within the confines of the present-day Oriente Basin. Oil analyses indicate a single family of oils. Available evidence, combined from structural and geochemical data, supports a major phase of Early Andean age oil generation and migration.

There is a considerable variation of oil type from 37° API paraffinic oils with a GOR of 250−300 to altered 10° API oils. Marked variations exist not only between oilfields but also between reservoirs in the same well. The observed trends can in most cases be accounted for systematically in terms of: early primary oil generation, migration and entrapment; and subsequent structural evolution locally involving the processes of bio-degration, water washing and/or flushing, and oil remigration.

Exploration in the Oriente Basin was initiated in 1921 although it was not until 1937 that the first well was drilled by Shell (Tschopp 1953). In 1964, the Texaco/Gulf group acquired a large concession following the discovery of their Orito Field in the Putamayo Basin of Colombia. The first oil field was subsequently discoverd in the Oriente by this consortium at Lago Agrio in 1967. An active phase of exploration followed with operators including Anglo-Ecuadorian Oilfields, Amoco, Sun and Cayman. By 1975 the only remaining concessions were held by Cayman (subsequently taken over by City Investing Co.) and the prolifically successful Texaco concession (in which the Ecuadorian State oil company CEPE had assumed an interest). Texaco has discovered over ten oilfields including Sacha, Shushufindi, Auca, Cononaco and Bermejo (Canfield *et al.* 1982b). CEPE assumed control of all open acreage and by 1985 had discovered five significant oilfields including the giant Libertador Field (Lozada *et al.* 1985, Almeida 1986).

In 1985 Ecuador introduced a new scheme of service contracts and since then have offered acreage throughout the Central and Southern Oriente (Rosania 1985, Anon 1987). Concessions have been acquired by ten consortia with commitments to drill over 35 wildcats by 1992. Significant discoveries have already been made by BP and Conoco and these are under appraisal.

Tectonic setting

The Oriente Basin occupies an area of approximately 100 000 km² and comprises only a small part of the Sub-Andean system of foreland basins, which extends over 6400 km from Venezuela to Argentina (Urien & Zambrano 1985). All of these basins developed during Tertiary times between the Pre-Cambrian

(Brazilian and Guyana) basement shields to the east and the active 'Andean' magmatic arcs and thrust/foldbelts to the west. Plate tectonic geometry and development are beyond the scope of this paper but simplistically the arcs and their complementary back arc basins are the ultimate results of the complex, staged, eastward subduction of the oceanic Nazca (Pacific) Plate beneath the continental South American Plate (see, e.g., Henderson 1979; Henderson & Evans 1980; Lonsdale 1978; Lebrat et al. 1986). The Oriente Basin comprises that part of the back-arc system lying in Ecuador and north eastern Peru between the east−west basement arches of the Vaupes Swell in southern Colombia and the Contoya Arch in northern Peru (Fig. 1).

The basin is strongly asymmetric with a steeper west flank characterized by structural

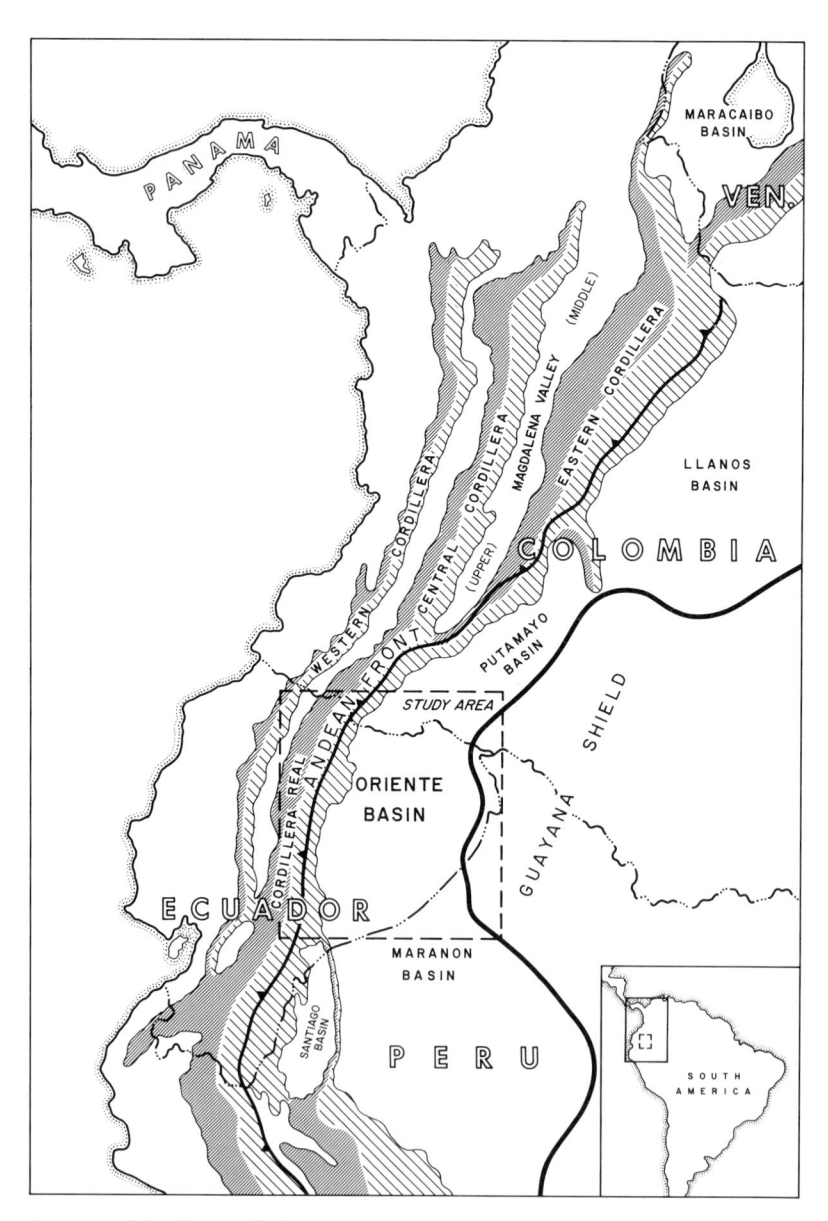

Fig. 1. Location map of the Oriente Basin.

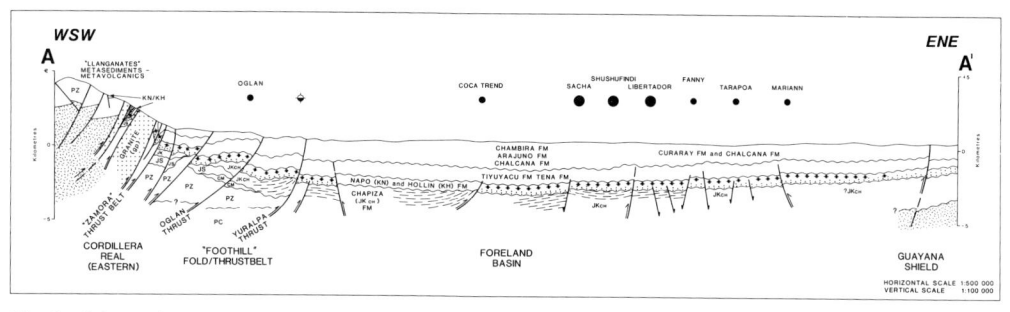

Fig. 2. Schematic cross section of the Cordillera Real and Oriente Basin.

dips of 5–10°, and a more gently dipping east flank, with dips of less than 2° (Fig. 2). The basin axis in Ecuador plunges from north to south to a depocentre in northernmost Peru which contains in excess of 5000 m of Tertiary–Cretaceous sediments (Fig. 3).

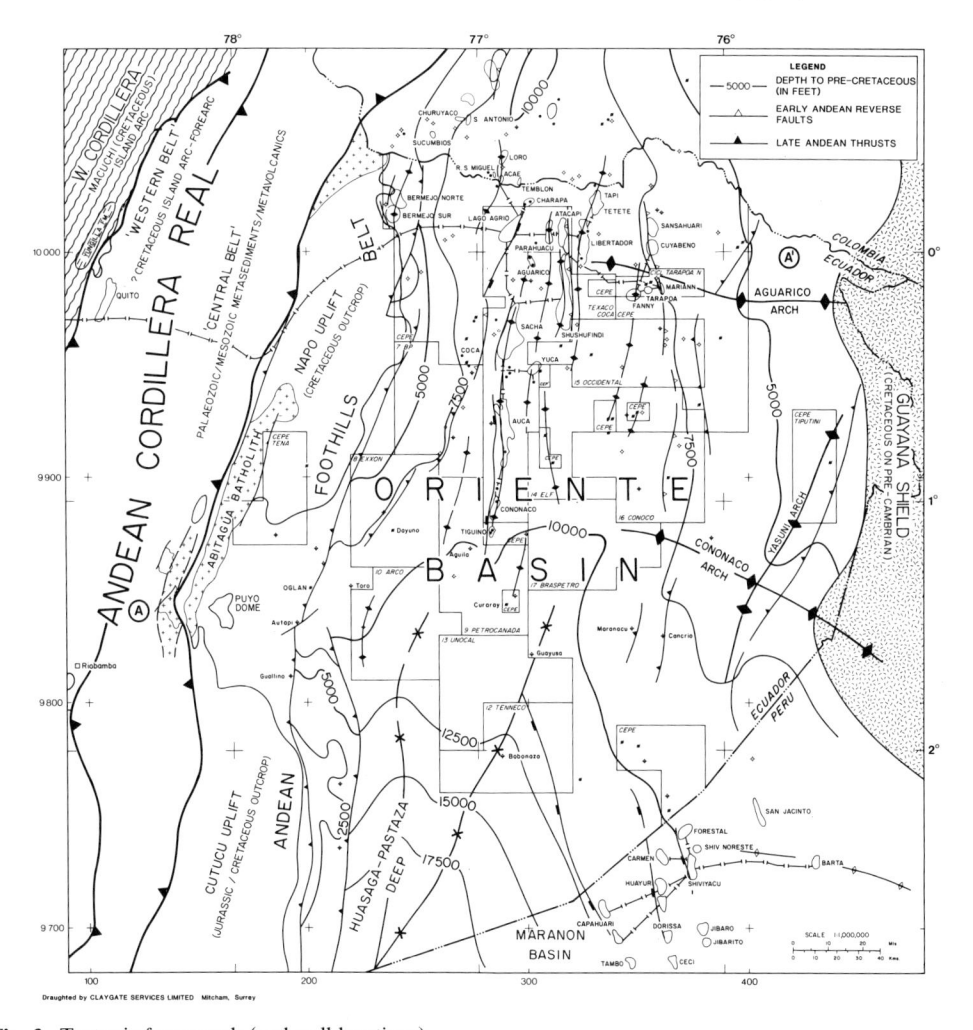

Fig. 3. Tectonic framework (and well locations).

Basin evolution

The sedimentary sequence of the Oriente basin can be divided, for the practical purpose of a petroleum geology synthesis, into three parts: a pre-Cretaceous 'floor', the Cretaceous 'sedimentary cycle', and a post-Cretaceous 'molasse cover' (Fig. 4).

Pre-Cretaceous floor

Four pre-Cretaceous formations have traditionally been recognized and, although recent work (e.g. Rivadenira 1986) suggests this may ultimately be seen to be an oversimplification, the scheme has been followed here. The oldest sedimentary formation drilled to date is of Upper Silurian–Lower Devonian age and comprises deformed and mildly metamorphosed limestones, slates, slaty shales and sandstones of the Pumbuiza Formation. This is overlain by up to 750 m of thinly-bedded, carbonates and shales of the Macuma Formation. These represent the Pennsylvanian to earliest Permian establishment of a shallow marine, carbonate shelf over an extensive area.

A considerable post-Permian hiatus is evident in the subsurface record largely owing to regional pre-rift uplift and erosion during a Mid-Jurassic tectonic event. The Triassic–Lower Jurassic Santiago Formation has been recognized in Ecuador only in the Santiago river area in the Cutucu Uplift of the Andean Foothill Belt and possibly in the Sacha Profundo-1 well (Rivadenira 1986). The Santiago Formation comprises transgressive marine thinly bedded carbonates and black bituminous shales, overlain by a regressive sandstone–siltstone sequence. There is evidence of contemporaneous submarine volcanism to the west. The Santiago transgression came from the south west, from Peru, but appears to have been of limited extent and confined to the extreme western area of the present-day basin. Total formation thickness may be in excess of 250 m.

Following the 'Mid-Jurassic' tectonic event, regional crustal collapse and extension occurred and thick sequences of continental clastics of the Upper Jurassic (to earliest Cretaceous) Chapiza Formation were laid down. The Chapiza Formation is areally extensive west of 70°80', although seismic data indicate that differential subsidence occurred in rift grabens, controlled by north–south basement faults, where up to 1500 feet of sediments accumulated. Graben fill was composed of conglomerates, sandstones and shales interbedded with minor

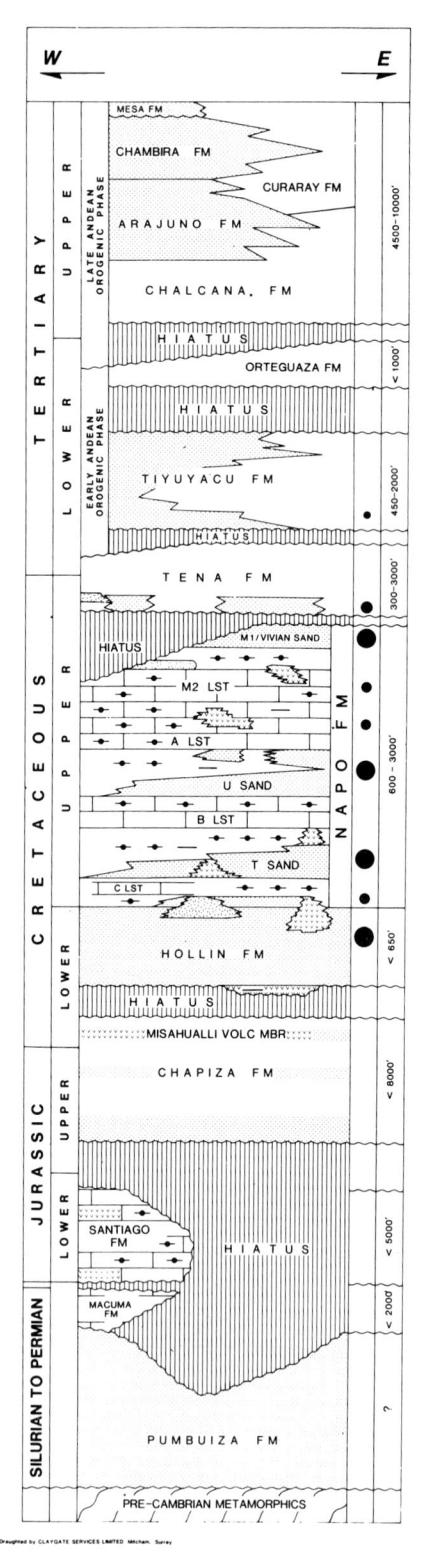

Fig. 4. Generalized stratigraphic column (Oriente Basin).

evaporites. Crustal extension was sufficient to be accompanied in the later stages in the north by subaerial alkaline volcanism. Ultimately, in the Lower Cretaceous, crustal breakthrough occurred, but to the west of the modern basin where a 'marginal sea' spreading centre was established (Lebrat et al. 1986). Regional (?thermal) uplift of the Oriente Basin area occurred accompanied by block faulting, utilizing the pre-existing N–S basement faults, and by extensive peneplanation.

The Cretaceous 'sedimentary cycle'

The Cretaceous is subdivided into three major formations; the basal, Hollin Formation followed successively by the Napo and Tena Formations. Major commercial interest in the basin is confined to this Cretaceous depositional cycle and all of the significant production has come from fluvio-deltaic and marine Cretaceous sandstones.

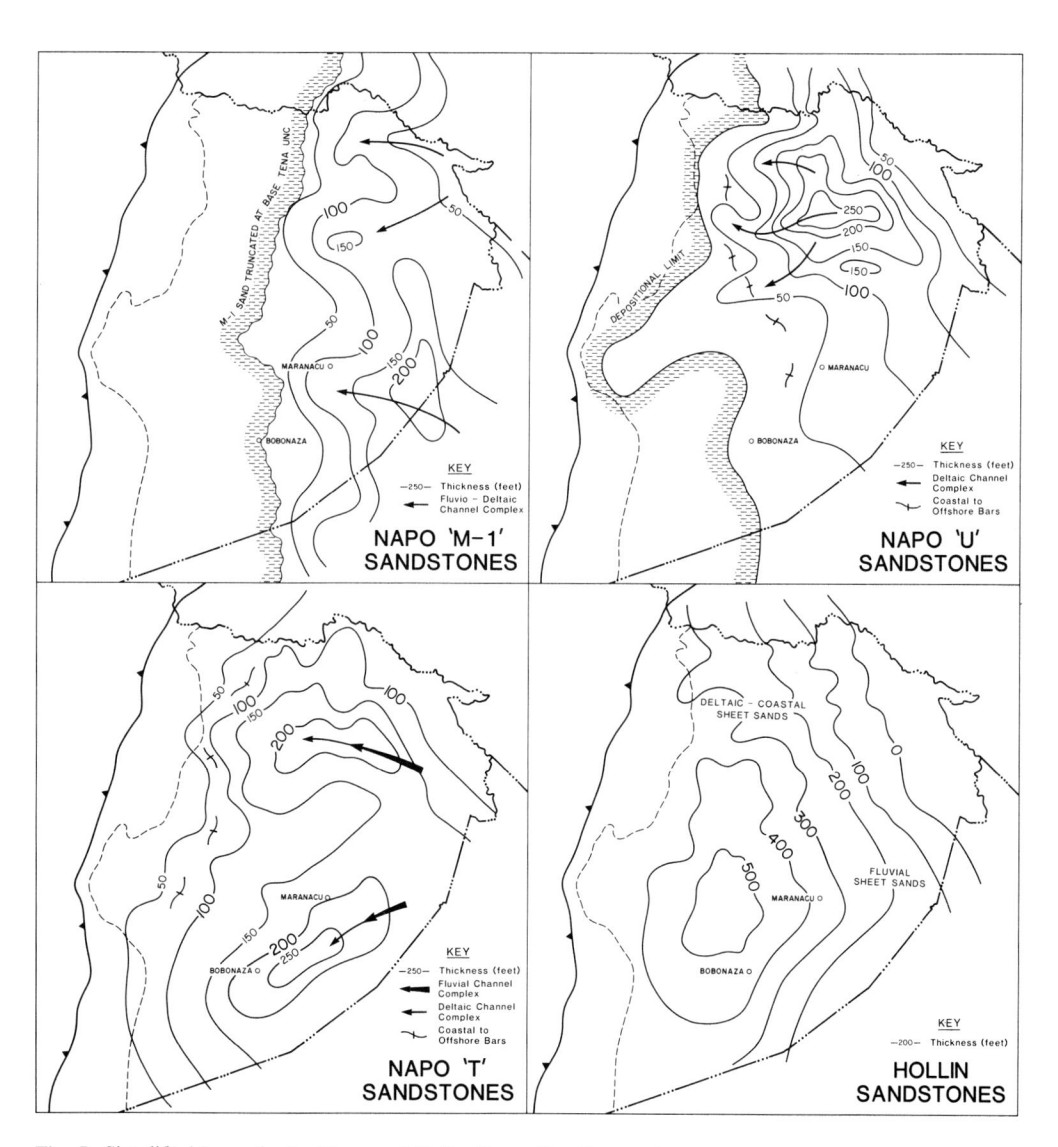

Fig. 5. Simplified Isopachs for Napo and Hollin Formation Reservoirs.

Hollin Formation (Apto–Albian)

The Hollin overlies the Chapiza with marked angular unconformity. It comprises a homogenous basal blanket sandstone which is present in foothill outcrops, in the subsurface throughout the Oriente and, indeed, throughout western Brazil, southern Colombia and northeastern Peru (Campbell 1970; Urien & Zambrano 1985; De Righi & Bloomer 1975). Its maximum thickness in the southwestern Ecuadorian Oriente is approximately 150 m and it thins gadually northestwards onto the Guyana Shield before becoming indistinguishable from overlying Napo sandstones (Fig. 5).

The Hollin comprises a white, thickly bedded to massive, quartz sandstone with only minor, but laterally persistent, carbonaceous claystone and coal interbeds. The widespread Hollin deposition was the result of rapid marine transgression from the western marginal sea, over the peneplaned, block faulted, 'pre-Cretaceous' substructure followed by generally regressive conditions depositing sandstones in fluvial, braided stream to littoral environments (Kummert & Casal 1986). The Hollin sandstone forms the principal oil reservoir in the Oriente Basin. Typical porosity and permeability ranges are 12–25% and 20–2000 millidarcies.

Napo Formation (Albian–Lower Campanian)

The Napo Formation conformably overlies the Hollin Formation. It comprises a sequence of marine claystones, limestones and sandstones (Fig. 4). The claystones and limestones appear to be the main source rocks in the basin while several of the Napo sandstones (the 'T', 'U' and less so the 'M2' and 'M1') are significant oil reservoirs. The maximum Napo Formation thickness exceeds 600 m. Essentially the formation was deposited on a stable marine shelf, on which the claystone-limestone units represent transgressive periods and the sandstones regressive episodes involving westward progradation of littoral to fluvio-deltaic facies belts. The sandstones thicken and progressively coalesce toward the east (Fig. 5).

The Napo 'T' sandstone member comprises a sequence of stacked sandstones with interbedded siltstones and carbonaceous claystones. Gross member thickness varies up to 80 m but it can generally be divided into a lower ('Main') unit of more permeable quartz sandstones and an upper unit of less permeable argillaceous and glauconitic sandstones. Sandstone grain size varies considerably although individual bed sorting is generally good. The general facies association suggests the sands represent tidal channel and barrier to offshore bar deposits within a nearshore marine (and ?estuarine) environment (Alvarado et al. 1982). Coastal swamp conditions are recognized in the extreme east and more offshore marine conditions are represented in the west, in the Foothill area. Typical porosity and permeability ranges are 10–25% and 100–6000 millidarcies respectively.

The Napo 'U' sandstone member is similar, in many respects, to the 'T' with a similar subdivision, particularly apparent towards the west, into a lower (or Main) Sandstone and an upper, more argillaceous sandstone. A similar estuarine channel environment is interpreted (Rosania et al. 1985). Gross sandstone member thickness reaches a maximum of only 150 feet, in the east, and thins rapidly westwards to a zero-edge at approximately 77°W. Porosities and permeabilities are similar to those seen in the 'T' sandstone. The Napo 'M2' sandstone is restricted to the east of the basin and represents a less significant regressive episode. The Nape 'Ml' sandstone is a coarse grained, thickly-bedded to massive, quartz arenite, again divided into a lower main sand and a thinner upper sand (Lozada & Endara 1982). Gross thickness varies up to 120 feet in the east of the basin with porosities and permeabilities in excess of 20% and 1000 millidarcies respectively, (Reed & Wheatley 1983).

Tena Formation (Maastrichtian–? Palaeocene)

The Tena Formation 'red beds' conformably overlie the Napo Formation in the east of the basin but progressively truncate the Upper Napo towards the west, cutting out the 'M1' sandstone, the Upper Napo claystones and limestones and the uppermost Middle Napo claystones. This intra-Upper Cretaceous unconformity represents the earliest of the 'Early Andean' compressional episodes and heralds regional uplift and the establishment of continental Tena deposition. The Tena Formation comprises varicoloured, predominantly red, continental and paralic claystones and siltstones with a local basal sandstone member which varies up to 10 m in thickness and is locally a minor oil-productive reservoir. The formation has an observed maximum thickness of over 750 m in the west of the Oriente and may have been thicker further westwards (Fig. 6).

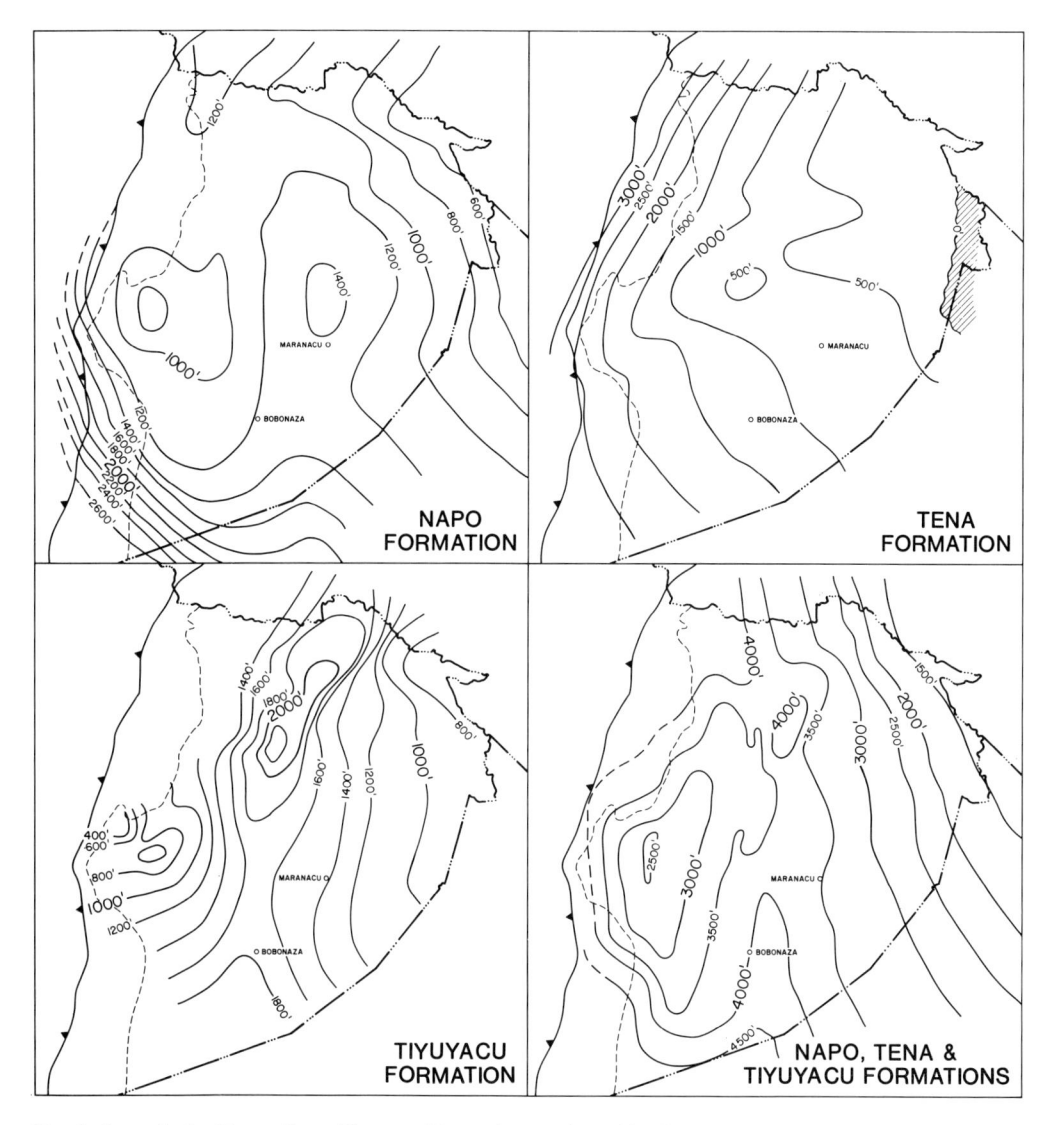

Fig. 6. Isopachs for Napo, Tena, Tiyuyacu Formations and combined Isopach.

Post-Cretaceous cover

The Cenozoic reaches thicknesses of 4000–5000 m in the extreme south of the basin and thins to less than 1500 m in the north. It is made up of several molasse sequences all derived from the west, from the spasmodically active and uplifted Andean Cordillera. The Tiyuyacu Formation (Eocene) overlies the Tena Formation with only a minor degree of angular unconformity but the stratigraphic hiatus is marked by the abrupt appearance of conglomerates, in beds up to 100 m in thickness, and coarse sandstones interbedded with claystones and siltstones comparable with those of the underlying Tena Formation. Total Formation thickness ranges up to 600 m. The overlying Orteguaza (Oligocene) comprises up to 300 m of blue grey claystones and occasional glauconitic sandstones. Sediment fill is completed by up to 3000 m of Miocene to Recent continental red beds of the diachronous Chalcana, Arajuno, Chambira and Mesa Formations.

It is noteworthy that unlike the Putamayo, Middle and Upper Magdalena and Llanos basins of Colombia and the Maracaibo and Barinas basins of Venezuela, no significant oil has been discovered in the post Cretaceous section.

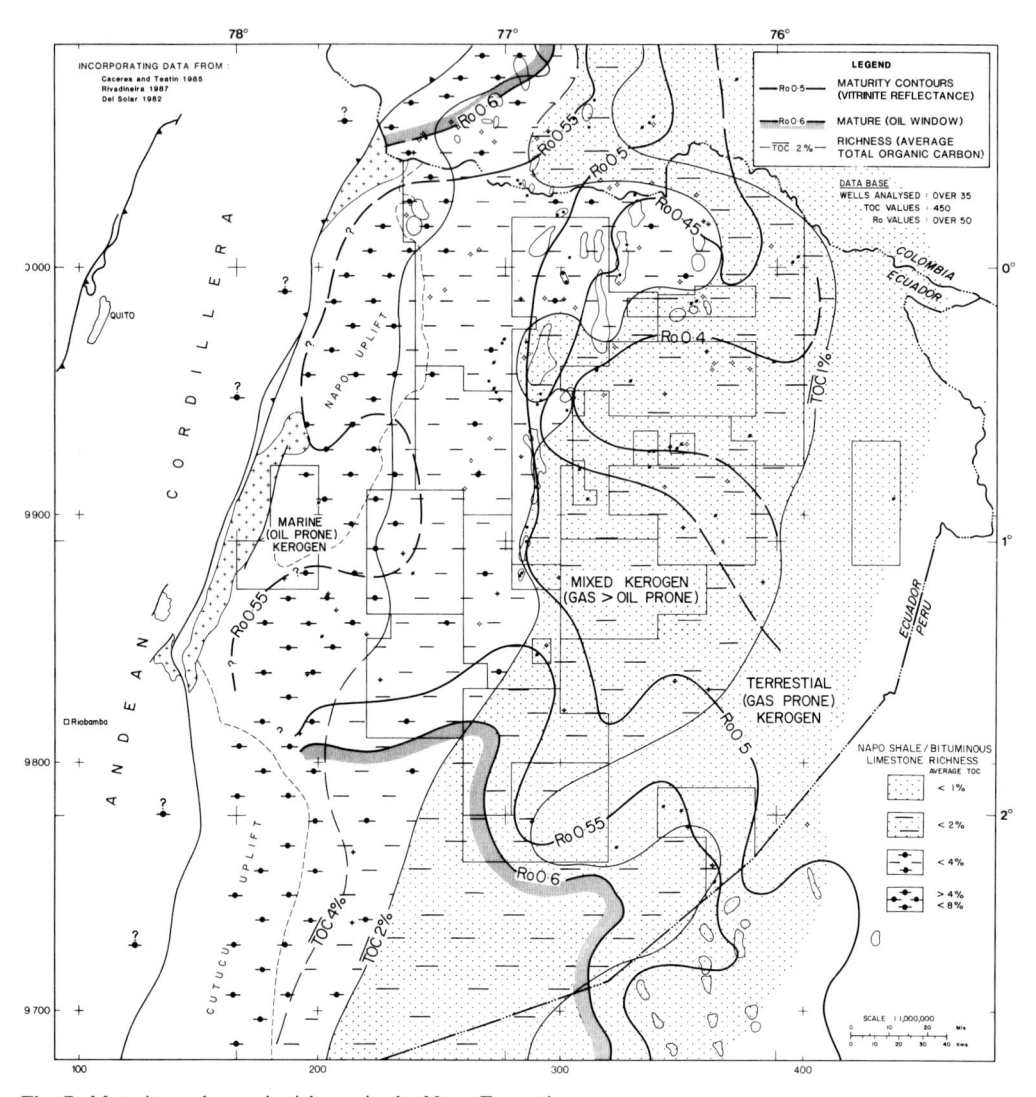

Fig. 7. Maturity and organic richness in the Napo Formation.

Structural development

A regional tectonic elements map (Fig. 3) and cross section (Fig. 2) illustrate the major structural zones of (i) the main magmatic arc (the Andean Cordillera Real); (ii) the back-arc fold-thrust belt with Pre-Tertiary outcrop (the Sub-Andean Zone); (iii) the relatively undeformed Tertiary foreland basin (the Oriente Basin); and (iv) the stable Precambrian basement craton (the Guyana Shield). The main structural trends strike in a roughly north–south direction in all of the zones.

The basin itself does not have a simple synclinal axis; low-relief anticlinal folds distort the axis, particularly in the shallower northern part, and produce an axial region, approximately 50 km wide, separating the gently dipping eastern flank from the steeper west flank. Intra-basin faulting and folding decreases in density and amplitude toward the southern part of the basin. Basin width is much reduced in the north, approaching the cross basin Vaupes Arch, and the increased structural deformation is the result of Jurassic–Cretaceous extension and subsequent Tertiary compression being taken up by a more limited rock volume (Jenks 1956). To the east, the basin fill thins onto the basement shield, whereas on the west flank, approaching the pre-Tertiary outcrops of the Sub-Andean Zone there is little evidence of regional Cretaceous–Lower Tertiary sedimentary thinning (Campbell 1970), other than onto an outer shelf-edge 'swell' probably underlain by granitic intrusions at depth (Figs 2, 5 & 6). This zone of bold relief may have acted as a restriction to marine circulation and allowed anoxicity on the shelf to produce the organic-rich Napo source rocks, during periods of low clastic influx from the east (see Fig. 7).

In the Cretaceous the proto-Oriente basin deepened to the west, into a subsequently consumed oceanic marginal sea. The sedimentary sequence, with the minor exception of some of the regressive sandstones, is exclusively marine and deposited on a stable shelf with sedimentary supply from the east. Pre-Cretaceous rifting involved significant extension on large-scale north–south normal faults but extensive peneplanation must have occurred prior to the late Lower Cretaceous deposition as evidenced by the angular 'Base-Hollin' unconformity and consistent regional Cretaceous depositional thicknesses (Fig. 5). Continued but minor syn-sedimentary movement occurred throughout the mid–early Upper Cretaceous on most of the pre-Cretaceous north–south, high-angle normal faults and produced an early family of basement-drape structures (Structure Type 1). These normal faults can rarely be seen to cut any higher than the lowermost Napo on seismic records.

The first indications of compression, caused by Campanian consumption of the marginal oceanic sea (Lebrat *et al*. 1986) and Macuchi arc–South American plate collision, are seen in the pre-Maastrichtian uplift of the 'Napo Uplift' and progressive westward truncation of the Napo sediment fill. Subduction of the Nazca Plate, with associated collision and compression, resulted in the Palaeocene to Oligocene uplift and unroofing of the Western Cordillera, a major supply for the first time of molasse from the west, and reverse movement on some of the early north–south faults, particularly on the west flank of the rapidly subsiding Oriente Basin. A family of 'early Andean' compressional structures, hanging-wall anticlines over reverse faults, with throws of generally less than 100 m, was produced (Structure Type 2). These faults cut the uppermost Tena, and occasionally the uppermost Tiyuyacu, but apparently no higher in the section. Continued subduction, collision and compression culminated in late Miocene–Pliocene uplift, unroofing of the Cordillera Real and the rejuvenated eastward supply of molasse. Major reverse faulting and thrusting, frequently along the old north–south pre-Cretaceous normal faults, was concentrated in the Sub-Andean zone but is also seen on the eastern flank of the foreland basin (Fig. 2). The result is a family of generally easterly verging folds with occasional back thrust structures. These hanging-wall anticlines, over reverse faults and thrusts which have significant throws ($\geq$1000 m) and frequently extend to the surface, constitute the third important structural style present in the Oriente (Structure Type 3).

The commercial hydrocarbon accumulations of the Oriente Basin are almost exclusively contained in the 'old' (pre-Miocene) structures (Type 1 and Type 2). Examples of Type-1 oilfields are Sacha, Auca, Cononaco and Tiguino; examples of Type-2 oilfields are Coca, Mariann and Shushufindi (Fig. 2). Only one commercial discovery in Ecuador, the Bermejo Oilfield, has been made in a 'young' Type-3 structure. Tests of this structural type have generally only produced water and/or heavy biodegraded oils. Almost all of the Type-1 and Type-2 structures are low relief (less than 60 m), though frequently have a large areal extent. Type-3 structures can be extremely large in area and have significant vertical relief (a few hundred metres), and lie predominantly along the accessible western basin flank. Thus many Type-3 structures were drilled in the earlier exploration phases in the Oriente (Anglo Equadorian Oilfields Ltd 1973).

Petroleum geochemistry

Database

Geochemical studies have been carried out for 15 wells in the Oriente Basin including source rock and oil analyses. The integration of traded data, consultants reports and published data has made it possible for these studies to be reviewed in terms of their regional geological context.

Geochemical analyses for Napo shales from over 30 Oriente wells have been assessed with regard to source richness and maturity. Similar analyses by Rivadaneira (1986) complement this database for the Oriente and published studies by Del Solar (1982) in the Maranon Basin and Caceres & Teatin (1985) in the Putamayo Basin present further data providing a regional database of over 100 wells (Fig. 7).

Oil samples from eight oil wells together with oil shows from seven further wells comprise the database of oils. Published data from the Maranon Basin (Del Solar 1982; Illich *et al*.

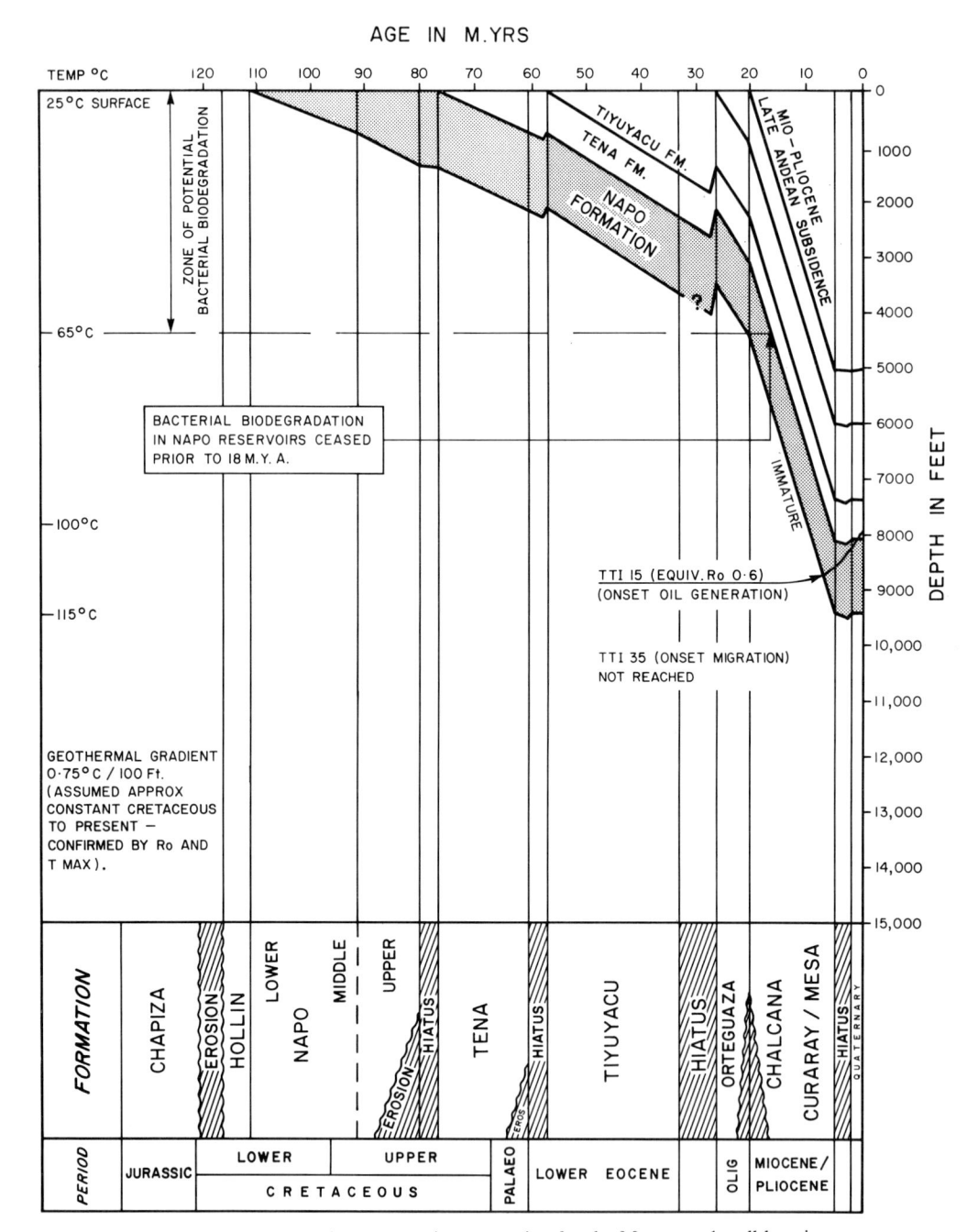

Fig. 8. Lopatin-style time–depth (temperature) construction for the Maranacu-1 well location.

1977; Sofer et al. 1986) provide further data to allow a regional evaluation to be made. It is important to realize that the Putamayo, Oriente and Maranon are one continuous geological basin (Figs 1 & 7).

Source rocks

The Napo Formation is equivalent to the La Luna Formation of Venezuela and the Middle Magdalena Basin of Colombia, to the Gacheta Formation of the Llanos Basin and the Villeta

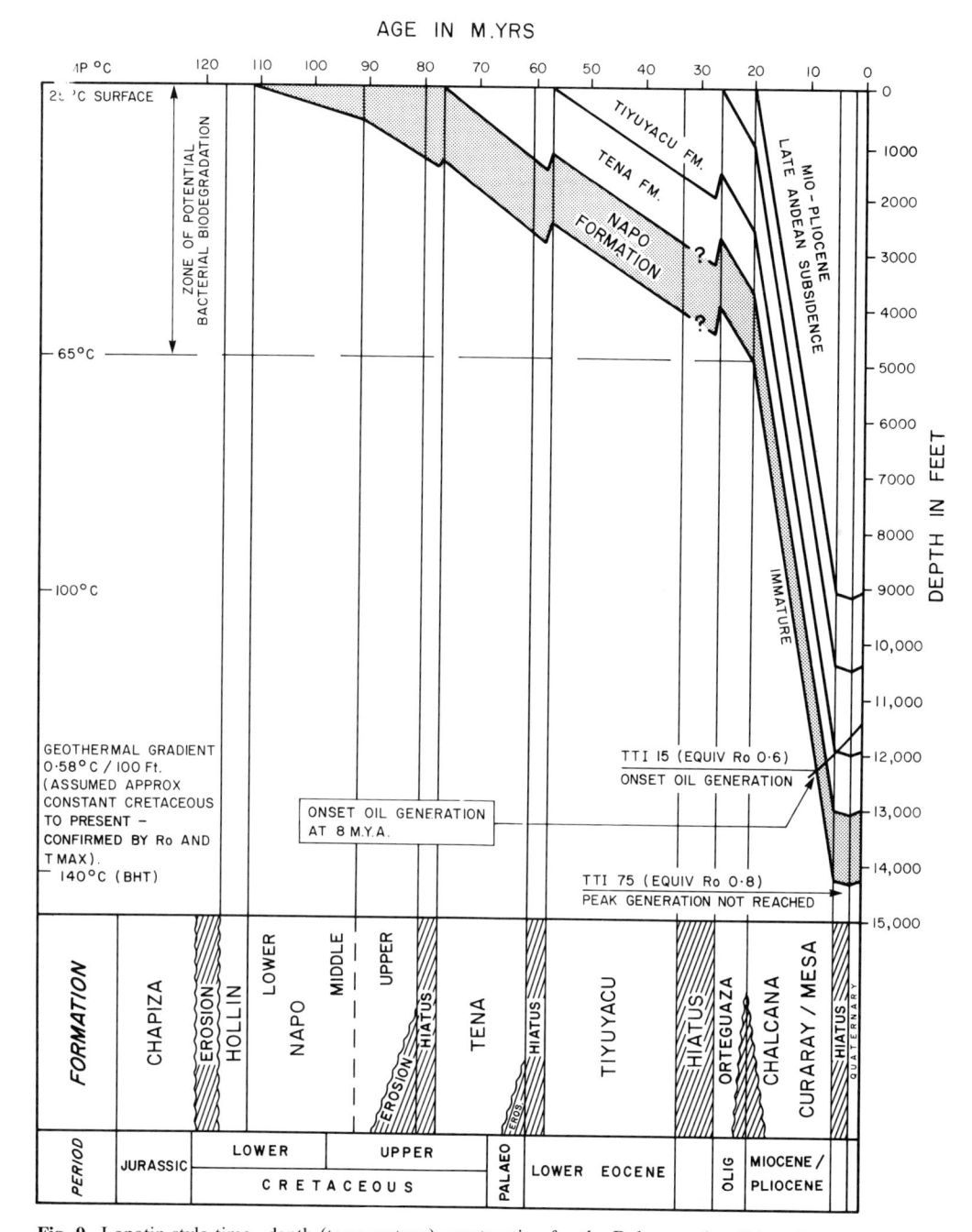

Fig. 9. Lopatin-style time–depth (temperature) construction for the Bobonaza-1 well location.

Formation of the Upper Magdalena Basin (both in Colombia) and to the Chonta Formation of the Maranon Basin (northern Peru) (see Fig. 1). In each of these productive basins the primary oil source rocks have been identified as being these prolific Mid to Late Cretaceous argillaceous sediments (Del Solar 1982; Russomana & Velarde 1982; Zumberge 1986; James 1990; Roberto et al. 1990). The Cretaceous Napo shales are generally accepted as the main source rock for the Oriente oils (Canfield et al. 1982a, 1985; Lozada et al. 1985; Rivadeneira 1986).

In the Oriente the Early Jurassic Santiago Formation may provide a secondary source (Rivadeneira 1986) but this formation is restricted in distribution and has not been definitely encountered in any well drilled in the basin. However, since most wells bottom in the younger Chapiza Formation, the distribution of Santiago sediments is not adequately controlled.

The Santiago Formation outcrops in southwest Ecuador in the Cutucu Uplift and to the south in Peru and may be as much as 1500 m thick in the Santiago Basin (Canfield et al. 1982a).

Richness. Samples of the Napo Formation from over 30 wells indicate that the organic material is of a more terrestrial origin in the east (Type-II Kerogen) and only in a few wells do more oil prone sapropelic and algal kerogen types (Type I) occur, these being generally in the western part of the basin (Fig. 7). The same distribution of kerogens is seen in the Maranon (Del Solar 1982).

Average TOCS from the Napo Formation shales range from less than 1% in the east to rich sediments with over 4.7%, and locally over 10%, in the west. These data suggests that rich Napo source rocks were deposited primarily in the west of the present Oriente Basin. It is proposed, and later further substantiated, that Napo equivalent source rocks were also deposited in an area which is presently the site of the (eastern) Cordillera Real, (see also Del Solar 1982; Sofer et al. 1986). Although the geological model of Feininger (1975) is not accepted, his original concept of a western source for the Oriente oils appears to be valid.

Maturity. Thermal maturity was assessed in over 15 wells using vitrinite reflectance (R_0), Thermal alteration index (TAI) pyrolysis 'T_{max}' and other indices. These, together with published data from over 50 wells, provide a regional distribution of maturity for the Cretaceous Napo Formation (Fig. 7).

The Cretaceous in the Oriente Basin is generally immature (R_0 <0.4%) to early mature (R_0 <0.55%) and only in the Huasaga–Bobonoza Trough in the SW of the Oriente and in the Mocoa area of the Putamayo (Caceres & Teatin 1985) have the Napo sediments reached the onset of oil generation (R_0 >0.6%) (Del Solar 1982; Caceres & Teatin 1985).

No well has encountered a fully mature Napo section in the Oriente, excluding 2–3 anomalous wells each associated with local igneous intrusions (Espin 1980).

In the southwest Maranon of Peru the Cretaceous reached the oil window during the latest Tertiary (Del Solar 1982).

Burial history. The burial history of the Oriente Basin is essentially one of Cretaceous post-rifting basin sag followed by Tertiary subsidence intimately associated with compression and thrusting in the west. Pulsatory compression resulted in thrust emergence and regional uplift, followed by rapid erosion of the thrust sheet into the foredeep. Cretaceous Napo depocentres are indicated on isopach reconstructions in the extreme southwest and probably also northwest of the present basin (Fig. 6). A regional high is apparent in the area of Oglan–Yuralpa and may represent the position of a former shelf break into a deeper water basin to the west (Figs 6 & 7).

The Tena depocentre lay to the west of the present basin where in excess of 1000 m of continental sediments were deposited in the foredeep in front of the emerging thrust sheets (Fig. 6). The Tiyayacu depocentre migrated eastwards in response to tectonic loading in front of the eastwardly advancing thrust front. The lithologies in the Tiyayacu conglomerates may provide indications of the provenance of these sediments and require further study.

The combined Napo–Tena–Tiyayacu section is thickest in the southwest and northwest of the present day basin and was probably thicker still west of the present Foothills Belt (Fig. 6).

Hence source richness, maturity and burial history indicate that areas to the southwest and northwest of the present day basin were the likely source kitchens for the oil in the various Hollin and Napo oilfields, (see also Del Solar 1982; Rivadeneira 1986).

The timing of oil generation is discussed in the next section but it is apparent that any oil generated to the west of the present-day Oriente Basin was only able to migrate eastwards prior to the tectonic disruption of migration conduits (Figs 2 & 3).

Oil generation and migration

Timing of oil generation appears to be a critical factor in evaluating prospectivity of structures in the Oriente Basin. Almost all productive structures show evidence of Cretaceous to Oligocene (Napo–Orteguaza) growth (Canfield et al. 1982b, 1985) and hence it appears that 'palaeostructure' is a critical factor.

Modified Lopatin Time–Temperature Index (TTI) constructions (Waples 1980) have been made for several basinal wells. All show significant Miocene–Pliocene 'Late Andean' subsidence, e.g. Maranacu-1 contains 1500 m of post-Oligocene molasse and Bobonaza-1 contains over 3000m (Figs 8 & 9). Evaluation of the subsidence history and maturation profiles of the Oriente wells assumed an average geothermal gradient equivalent to the present. The assumption appears justified because the 'Lopatin TTI Maturities' correlate with those measured by vitrinite reflectance, Pyrolysis and TAI.

At the Bobonaza-1 well, which is the deepest well in the Ecuadorian Oriente, the onset of oil generation (TTI 15) was calculated at approximately 8 Ma BP for the Base Napo (Fig. 9) and peak generation (TTI 75) has not yet been reached. Vitrinite reflectance values of 0.5–0.57% from the Napo interval confirm the calculation (Fig. 7). By extrapolation to the deepest part of the present-day basin, in the Huasaga–Pastaza Deep (Fig. 3), onset of oil generation was at approximately 8 to 11 Ma BP (see also Schmerber et al. 1986; Del Solar 1982).

These analyses indicate oil generation and migration did not commence in the present-day Oriente Basin until Early Miocene times at the earliest.

Analyses of all of the oils sampled from the Napo and Hollin Formations show them to be biodegraded to some degree. This widespread alteration is confirmed by results from oils in the neighbouring Maranon Basin Oilfields (Illich 1977; Del Solar 1982; Sofer et al. 1986). These biodegraded oils are found in reservoirs currently too hot for bacterial activity (reservoir temperatures >65°C). Indeed, as Del Soler (1982) has shown, examples can be found where reservoir temperatures exceeded this limit for biodegradation as long ago as 20 Ma BP. An example of this is seen in the Maranacu-1 well (Fig. 8) where the subsidence (and thermal) history would dictate cessation of biodegradation by 18 Ma BP.

This example clearly suggests that the oil in this well was trapped prior to the onset of oil generation in the deepest parts of the present-day Oriente (11 Ma BP), and that the oil must have been generated during the early Tertiary to the west of the present basin limits. Del Solar (1982) and Sofer et al. (1986) concluded that an initial main phase of oil generation began (and was possibly completed) as early as the late Cretaceous. However the necessary subsidence rate, heating rate and time period required for expulsion and migration of oil in this model would be exceptional. It is suggested that the additional subsidence and time provided by Tena and Tiyayacu deposition would have provided the necessary conditions for full maturity of the Napo Formation and peak oil generation prior to the end Orteguaza (Oligocene).

Water-washing of the reservoirs by fresh waters from outcrop would probably have been possible periodically during Tena and Tiyuyacu times associated with pulsatory 'Early Andean' thrusting and uplift.

Oriente oils

Analyses

Oils were sampled from five separate reservoirs from eight widely spaced localities across the basin, varying from the Fanny Field in the northeast, Oglan A-1 in the west, Tiguino-1 in the basin centre and Maranacu in the southeast (Fig. 3). Over twenty oil shows from reservoir cores were also analysed. Oils are identified on each of the analyses presented (Figs 8, 9 & 10) and the locations are named on Fig. 3.

Standard analytical techniques were employed, including Gas Chromatography – Mass Spectroscopy (GC–MS) Whole oil, GC–MS Paraffin–Napthene and GC–MS Aromatic chromatography, biomarker distribution and carbon isotope analyses, to compare these oils and to correlate them with potential source rocks.

Oil to Oil correlation

From the analyses conducted (e.g. Fig. 11) it is apparent that the oils sampled, which have widely varying API gravities (10–31° API), are members of the same genetic family of oils (Fig. 11). GC–MS C15+ analyses (Fig. 12) indicate the main differences in these oils reflect alteration processes (Illich et al. 1977, Zumberge 1980).

Oil composition

It is apparent that the oils from the Napo and Hollin reservoirs are of one genetic family and thus the following discussion will concentrate on the variations within that family.

The oils are displayed on a standard plot of Paraffin–Napthene–Aromatics and NSO compounds in Fig. 10. This plot indicates that the oils are not only of similar composition but also relate closely to those oils analysed from the Peruvian Maranon Basin by Del Solar (1982).

The oils all show depletion in normal paraffins and enrichment in aromatics and asphaltenes.

Two main groups of oils are indicated reflecting their different alteration histories. The Mariann, Fanny and Tiguino oils (23°, 22°, 31°

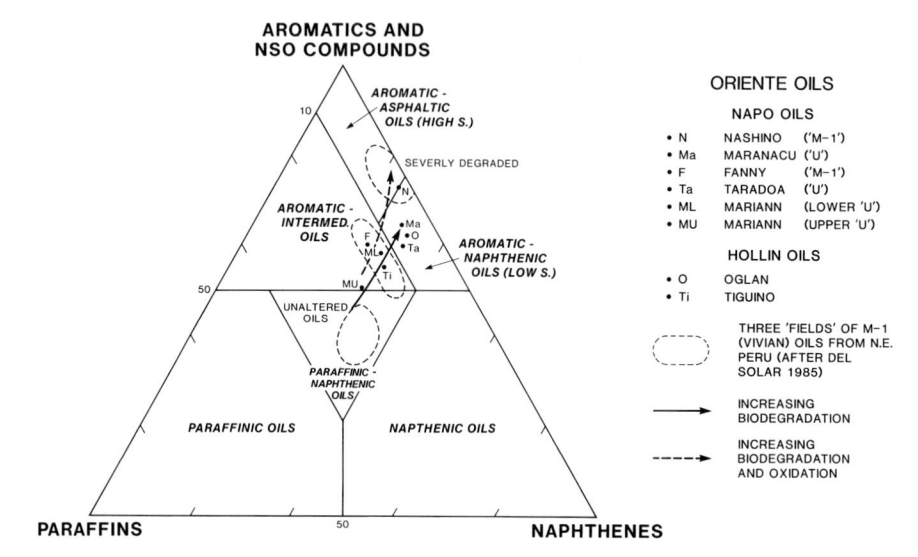

Fig. 10. Oil composition of eight Oriente oils.

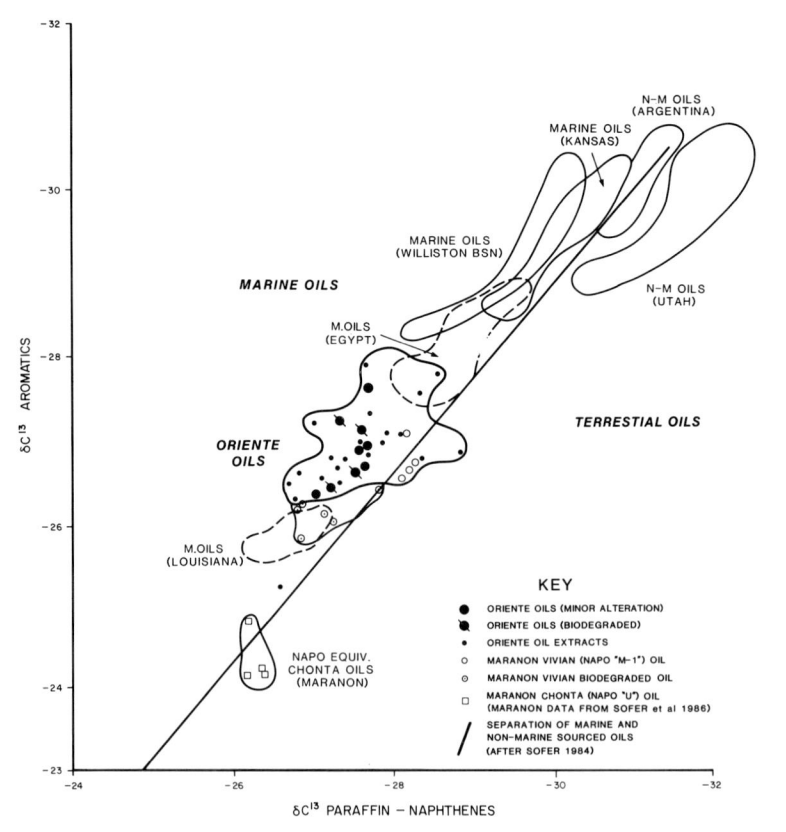

Fig. 11. Stable carbon isotopes for paraffin–napthene and aromatic fractions of Oriente oils compared with selected groups of marine and non-marine oils.

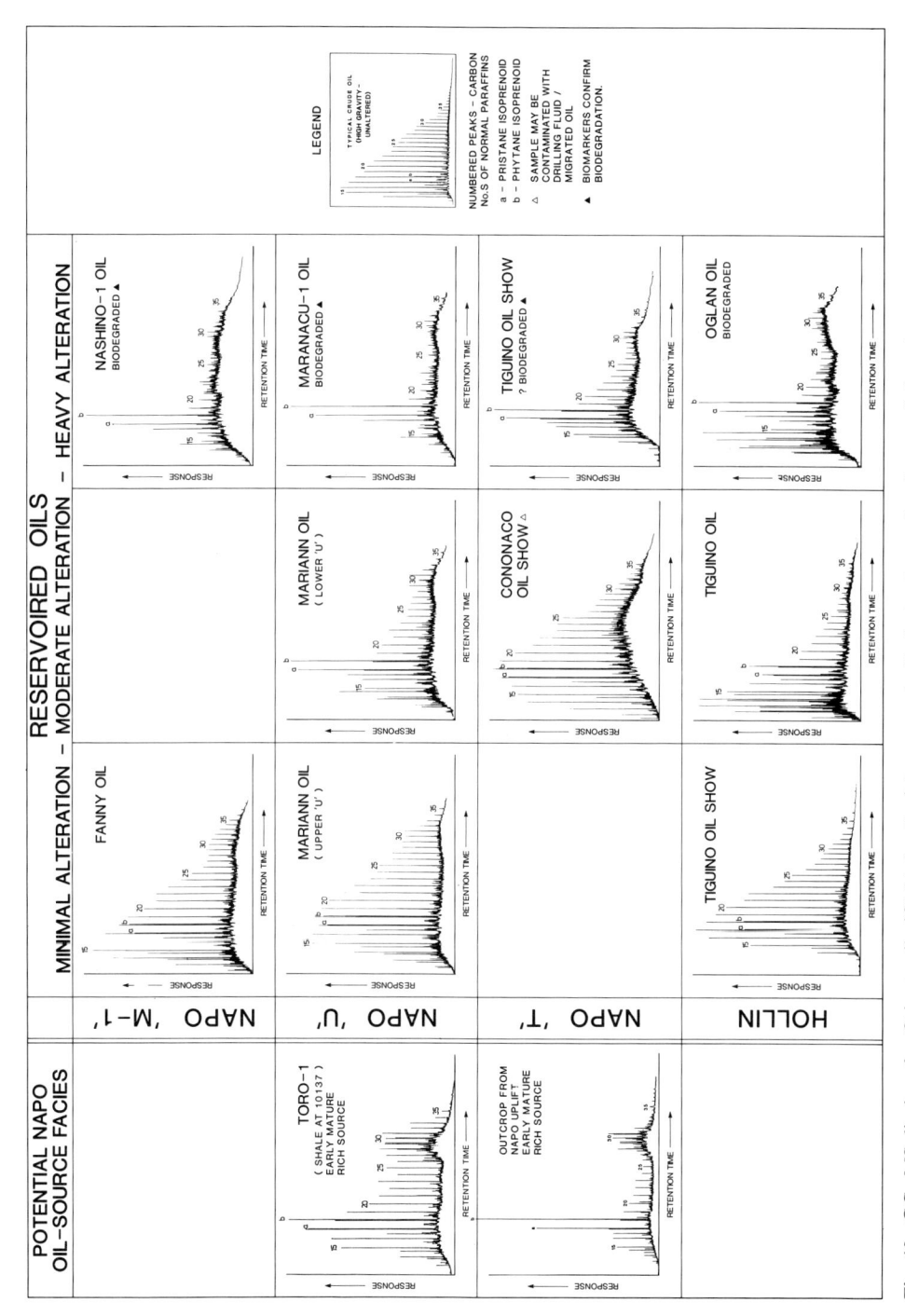

Fig. 12. GC—MS displays for Oriente oils (C15+ fraction) by reservoir displaying increasing degree of alteration.

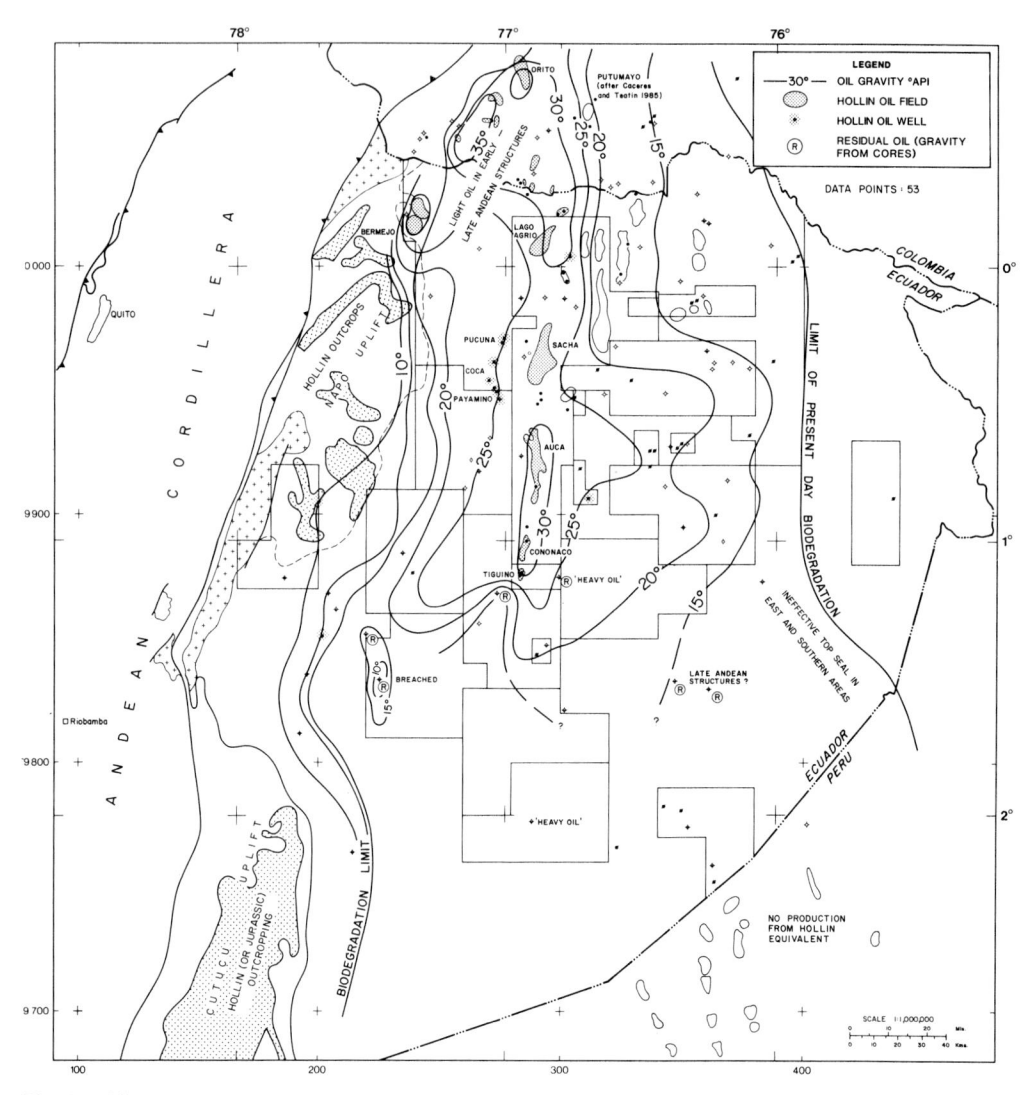

Fig. 13. Oil gravity distribution in the Hollin Sandstone Formation.

API respectively) are least altered; the Tarapoa, Oglan and Maranacu oils (21°, 13° and 10° API respectively) are enriched in aromatic-napthenic compounds due to paraffin depletion and have moderate sulphur. The Nashino oil is rich in asphaltic compounds and has a high sulphur content.

Loss of normal paraffins, enrichment in asphaltenes and sulphur and the presence in the biomarker distribution of demethylated hopanes indicate that each of these oils has been biodegraded to some extent (Sofer et al. 1986). No 'fresh' Oriente oil was sampled although only Nashino shows extensive biodegradation and oxidation.

Oil alteration

The GC-MS C_{15+} displays seen in Fig. 12 shows that a wide distribution of oil quality may be found in each of the major producing reservoirs. The Hollin and each of the three Napo reservoirs ('T', 'U' and 'M-1' sandstones) contain a spectrum of minimally altered (full range of paraffin peaks) to heavily altered oils (few paraffins and increased baseline hump of

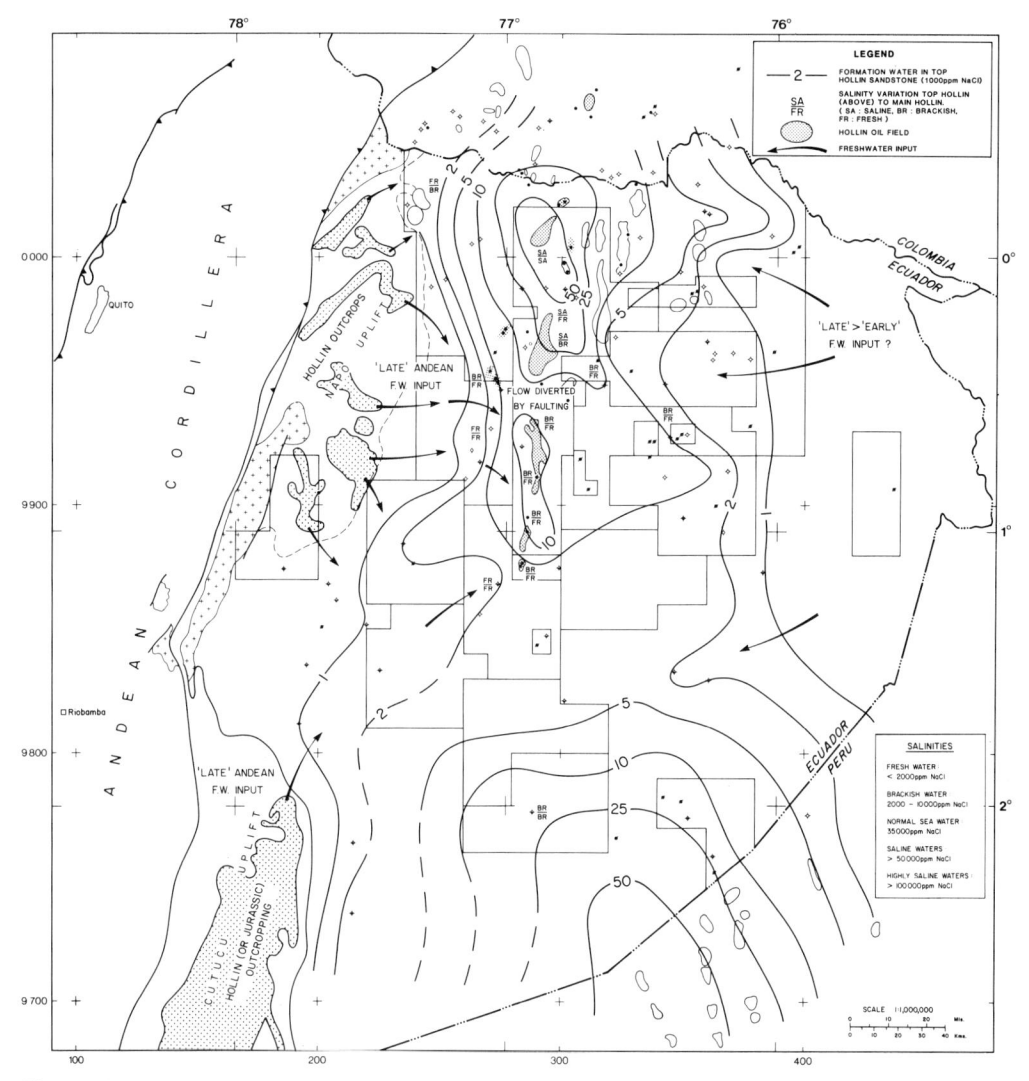

Fig. 14. Water salinity distribution in the Hollin Sandstone Formation.

napthenic compounds). Hence the alteration processes are not reservoir specific.

Carbon isotopes

Isotope analyses support the correlation of the analysed oils. These analyses also indicate that the oils are closely related to the oils produced from the Vivian Formation (Napo 'M-1' equivalent) of the Maranon Basin (Sofer et al. 1986). Oils from the Napo 'U' equivalent (Chonta Formation) in the Maranon appear to be less closely related. The Oriente Basin (and Vivian Formation) oils appear to have strong marine genetic affinities while the data from the Chonta (Sofer et al. 1986) suggest an increased terrestrial influence, probably reflecting a more recent maturation and local migration from mixed kerogen source rocks further east in the Maranon Basin.

Napo shale extracts also display isotopic values generally indicative of terrestrially influenced source rocks (Sofer 1984) and it is unlikely that these shales could have sourced a significant portion of the Oriente oils.

Oriente oil source facies

The integrated analyses of oils and oil shows in this study indicates that the source rocks for

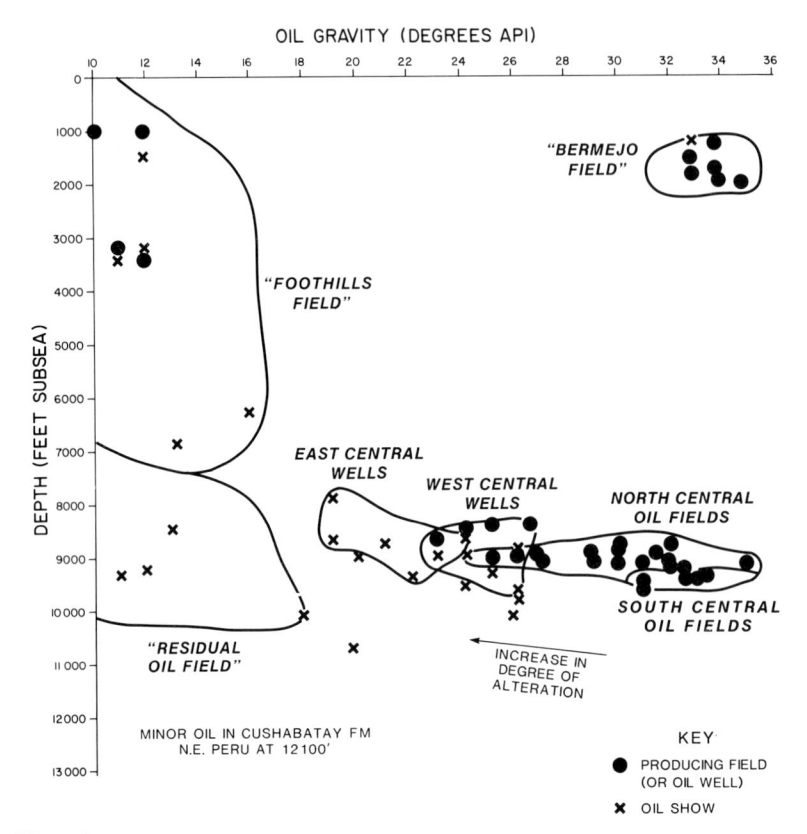

Fig. 15. Oil gravity against depth for the Hollin Sandstone Formation.

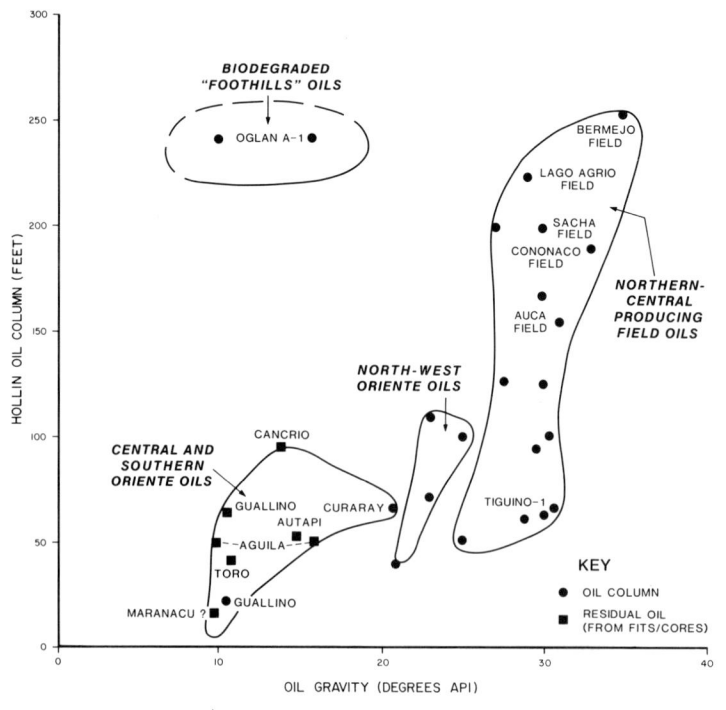

Fig. 16. Oil gravity against oil column for the Hollin Sandstone Formation.

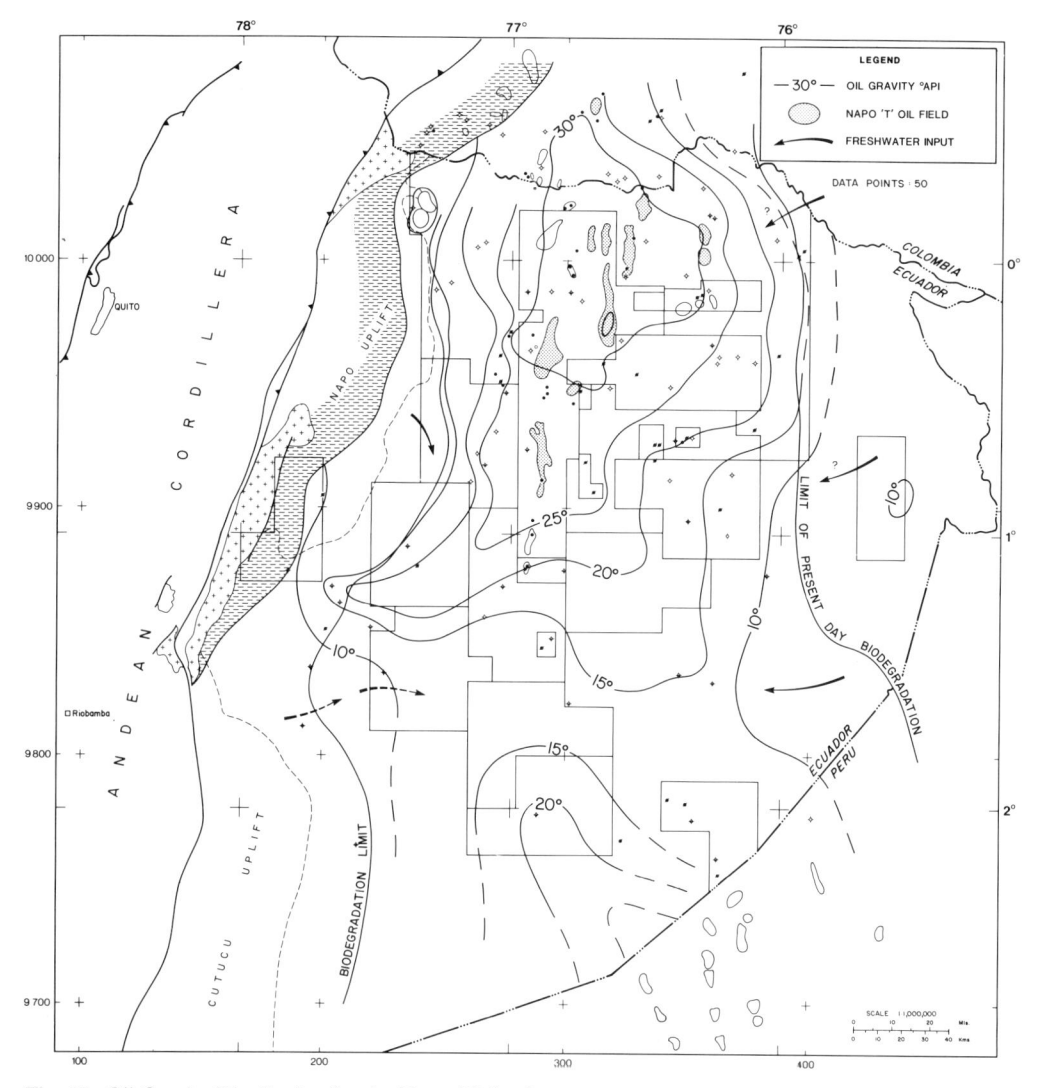

Fig. 17. Oil Gravity Distribution for the Napo 'T' Sand.

these oils were of a common facies type deposited in a predominantly marine environment. The source rocks were only moderately mature when generation occurred. Source rock analysis for the Napo Shales in the Oriente indicate they could not have sourced the oils. The source facies most likely to have sourced the Oriente oils was sampled in the Lower Napo of the Toro-1 well (Western Oriente – Fig. 7) and in outcrop in the Napo Uplift (Fig. 3). Both contained richly organic marine sapropelic kerogens. However, despite these shales being rich and oil prone, they are also immature ($R_0 = 0.55$) and have not generated significant hydrocarbons.

This evidence further supports the hypothesis for a western oil source, where an extensive 'Toro type' Napo shale facies was present during the late Cretaceous. This model requires migration distances of over 300 km but the Hollin sandstone, in particular, provides the Oriente with an excellent migration conduit. The Hollin sandstone is present in the westernmost outcrops of the foothills and is generally oil and tar saturated in these outcrops.

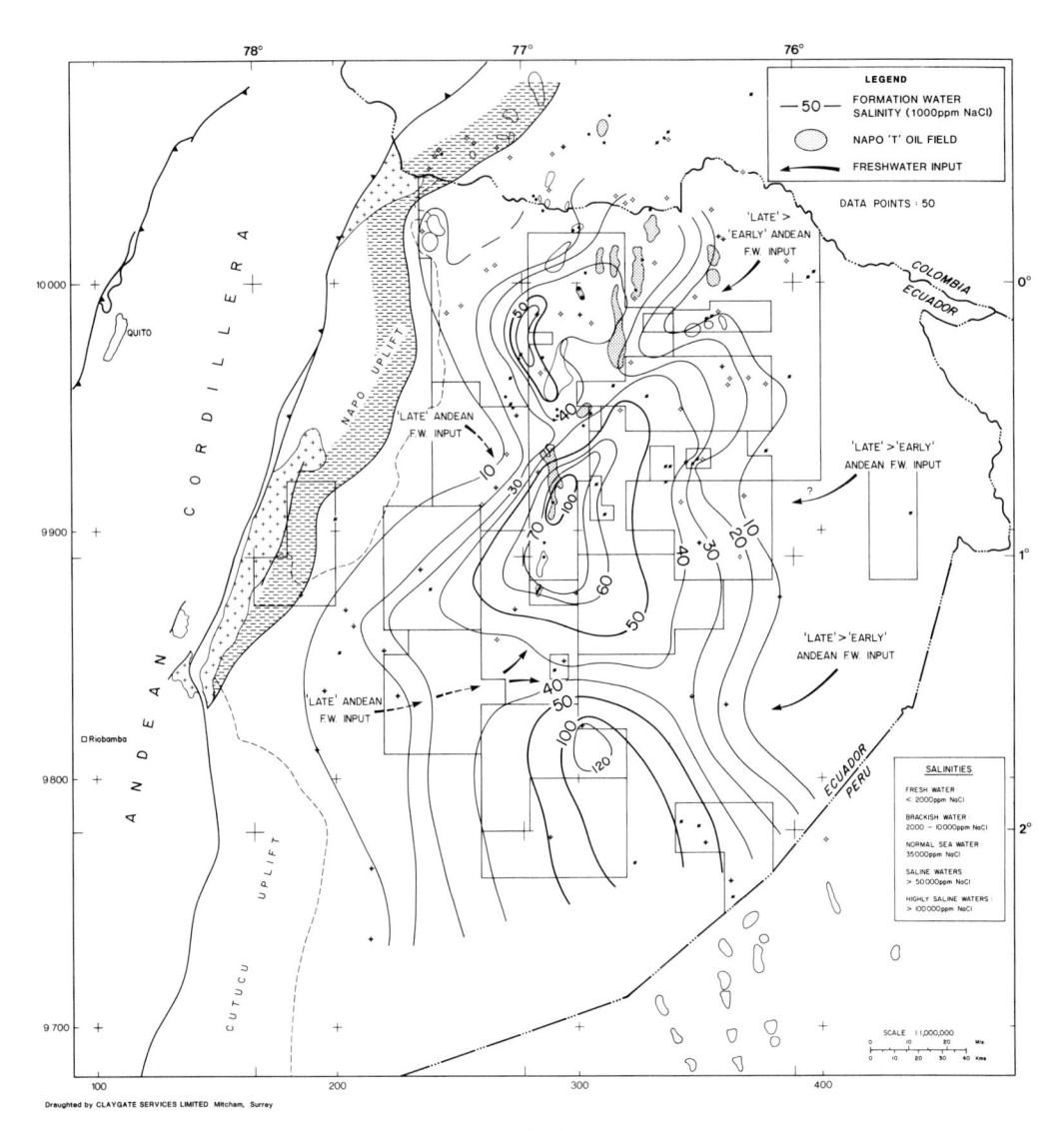

Fig. 18. Water Salinity Distribution for the Napo 'T' Sand.

A model for oil gravity distribution in the Oriente Basin

A model has been developed to explain in a systematic fashion the variation in oil gravity and water salinities in the four main reservoirs, the Hollin, the Napo 'T', the Napo 'U' and the Napo 'M-1'.

One major factor contributing to this model is derived from the preceding geochemical studies, that is that oil generation and migration began prior to the Late Andean (pre-Miocene) orogeny. This oil was generated to the west of the present-day Oriente.

A suite of maps displaying oil gravity and water salinity distribution for each of the four main reservoirs are seen in Fig 13–24. In constructing, and viewing, the interpreted contour maps it is important to appreciate the various processes that have effected fluid distribution. These include the following:

 (i) 'Early Andean' (pre-Miocene) oil migration from the west.
 (ii) 'Early Andean' fresh water influx causing

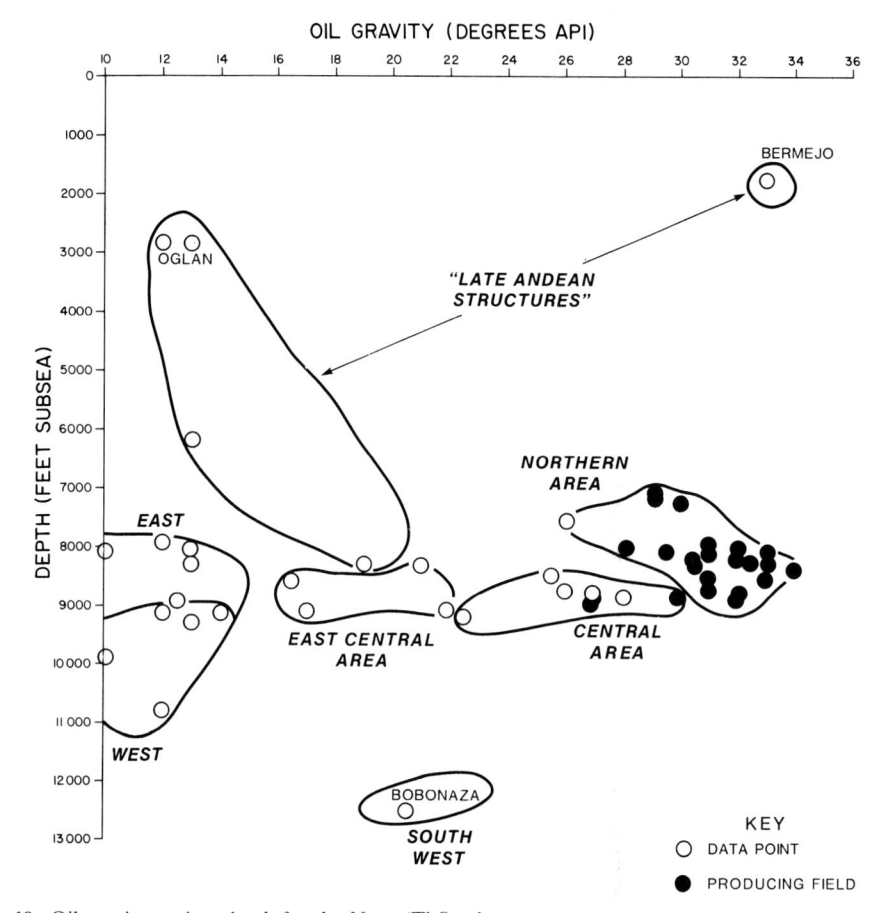

Fig. 19. Oil gravity against depth for the Napo 'T' Sand.

water-washing and biodegradation of trapped and migrating oils.
(iii) 'Late Andean' fresh water influx causing water-washing and biodegradation in shallower reservoirs. The limit for present-day bacterial biodegradation is marked on Figs 13, 17, 20 & 22.
(iv) 'Late Andean' basin subsidence in the southwest which has locally caused oil re-migration (and minor 'late' oil generation).
(v) 'Late Andean' structuring which post-dated the main 'Early Andean' phase of oil generation but locally involved breaching of oil-bearing traps.
(vi) Locally active present-day water drive in some reservoirs.
(vii) On a local scale, increased structural relief and/or an element of stratigraphic trapping can help to preserve unaltered oils (see Fig. 16).

The following section includes observations on the distribution of fluids in each of the main reservoirs as displayed in Figs 13–24.

Hollin Formation

The Hollin sandstone reservoir contains high gravity oil in the north–central, central (axial) and northwestern areas (Fig. 13). These areas generally contain structures with significant prospect structural relief, Early Andean palaeo-structure and faulted western flanks.

Heavy oil in the west reflects Late Andean to present-day water influx from western outcrops causing biodegradation in the shallow reservoirs and water-washing and possibly physical flushing in the deeper reservoirs where hydraulic continuity is not broken by faults.

Local heavy oil on the eastern basin flank probably reflects residual oil in an area of poor sealing capacity because the Lower Napo is predominantly sandy.

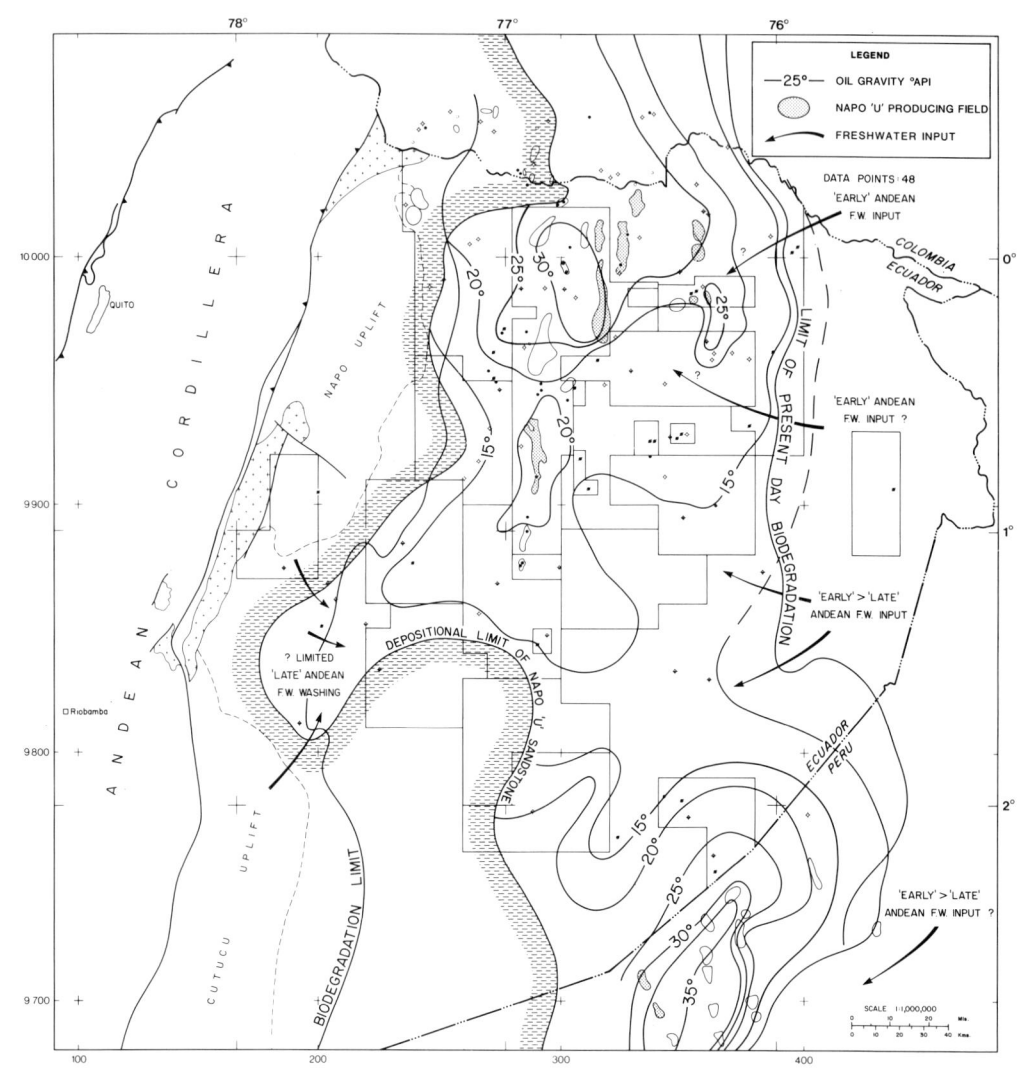

Fig. 20. Oil Gravity Distribution for the Napo 'U' Sand.

Water salinities (Fig. 14) indicate Early Andean fresh water input (possibly from the east) pre-dating or contemporaneous with oil migration, since fresh to brackish connate waters are reported within light oil columns (Canfield et al. 1982b, 1985). Oil locally re-migrated following western uplift and may have been biodegraded in transit (for example at Oglan where a 60 metre column of biodegraded oil is present).

Locally oil is trapped in young, Late Andean, structures, for example at Bermejo, where it is unclear whether there was palaeostructure or whether in situ maturation induced by local volcanicity has produced an anomalously light oil accumulation at a shallow depth. In general, present-day burial depth is not a control on oil gravity (Fig. 13).

Most productive structures occur between 2500 and 3000 m (subsea) with oil gravities between 23° and 35° API. The 'Foothills structures', with the exception of Bermajo, are mainly 'Late Andean' breached structures containing biodegraded oil columns.

A field of 'Residual Oils' (Figs 15 & 16) is thought to reflect low relief structures that have lost their lighter hydrocarbon fraction by structural tilting, water-washing or possibly

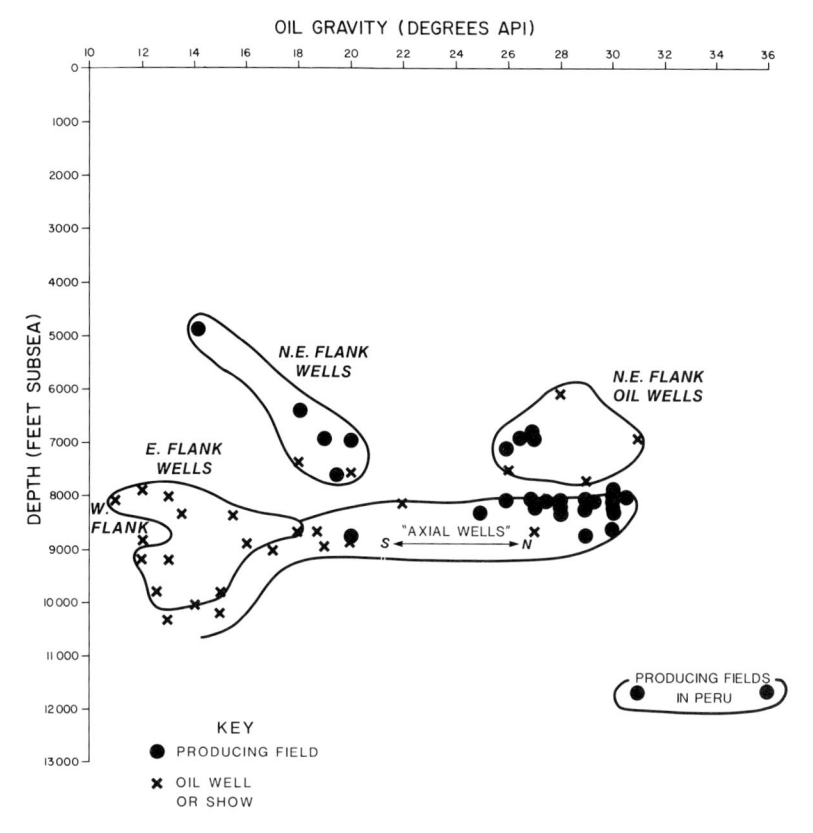

Fig. 21. Oil gravity against depth for the Napo 'U' Sand.

some physical flushing. Even some of the deepest traps such as Tiguino (below 3300 m) may have suffered from biodegradation (Figs 12 & 15). However, the structural relief of a trap does appear to be an important factor (Fig. 16). Protection from the degrading activities of the aquifer may be an important factor in some taller oil columns.

Napo 'T' sand

The Napo 'T' sand oil gravity distribution (Fig. 17) is broadly similar to the Hollin but with some important differences. Light oil is found in the 'T' sand further to the east than for the Hollin.

Low oil gravities are often associated with fresh waters (Fig. 18), suggesting an early phase of water influx, although not all oils associated with brackish waters are altered. A 'Late Andean' input of freshwater would not necessarily have caused biodegradation due to high reservoir temperatures. Oil gravity is clearly not depth dependant (Fig. 19) and there is little evidence of in-reservoir oil maturation, some of the deepest oils found here having suffered most significant alteration.

Napo 'U' sand

The Napo 'U' sand contains light oil in the north-central Oriente and in the north-eastern Peruvian Maranon (Fig. 20). Medium gravity oil (15–25° API) is found throughout much of the central and western parts of the Oriente. Heavy, biodegraded oils are occasionally seen in the Napo 'U', e.g. Maranacu-1 (Fig. 12), but the discontinuous sandstone distribution in the west has probably reduced the degrading effects of flowing water. The oils found in the discontinuous 'U' sands are the most difficult to reconcile with long-range migration from the west. However no evidence for a local sourcing for these sands has yet emerged. Migration of oil across faults juxtaposing the Hollin or 'T' sand against the 'U' sand may provide adequate conduits. Oils found on the sand-rich eastern flank of the basin seem to have suffered signifi-

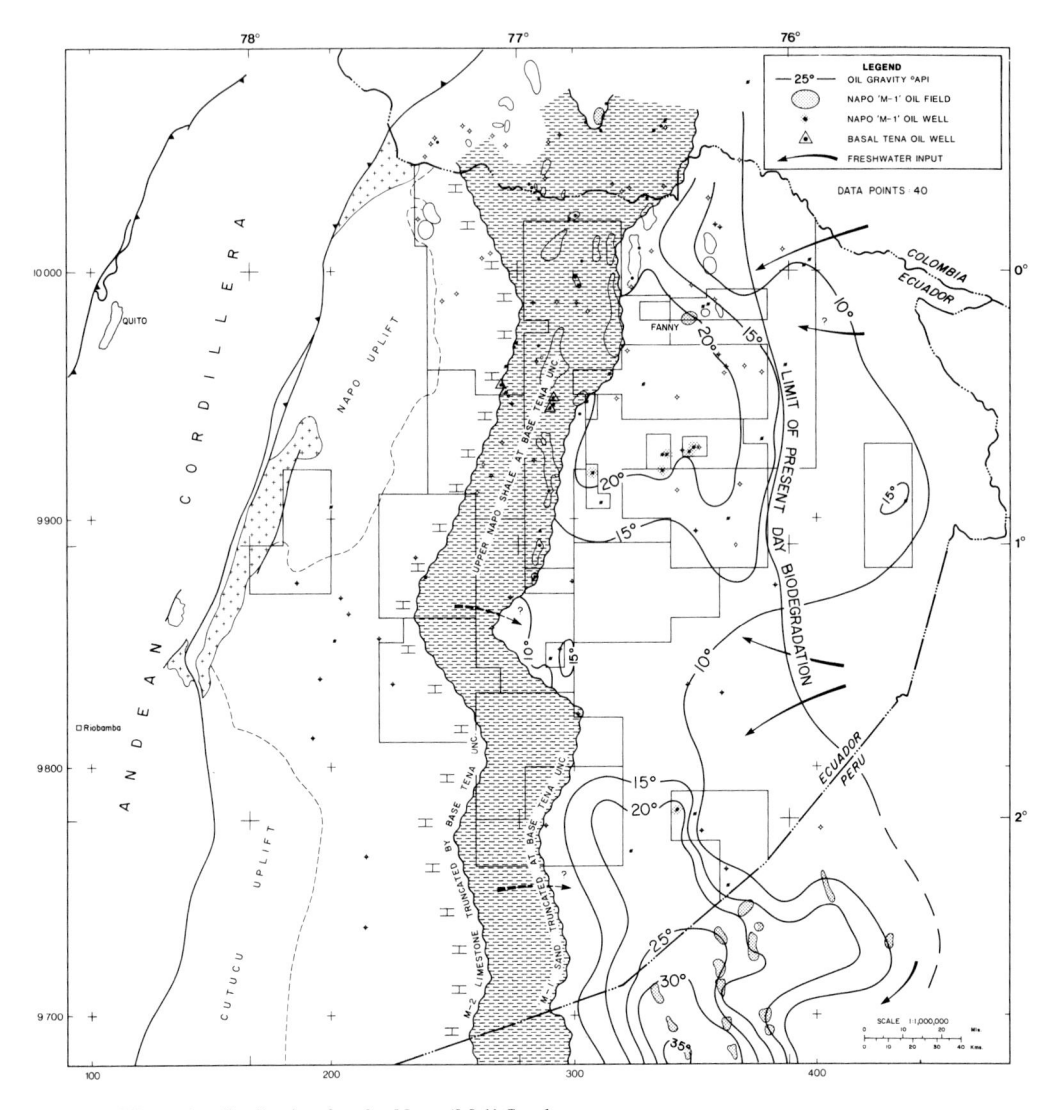

Fig. 22. Oil gravity distribution for the Napo 'M-1' Sand.

cantly from degradation, due to ease of access for the flowing aquifer.

Napo 'M-1' sand

The Napo 'M-1' reservoir contains medium gravity oil in the northeastern and light oil in the southeastern areas of the basin (Fig. 22). Lighter oils are generally associated with more saline formation water. The truncation of the 'M-1' sandstone during 'Early Andean' times may have resulted in an influx of fresh water (Del Solar 1982) but this appears to have been selective and of less importance than the biodegrading effects of fresh waters encroaching from the east along sand-rich fairways probably during both Early and Late Andean times, (Figs 23 & 25).

Notably, areas of significant regional structural elevation appear to have avoided the worst effects of alteration by freshwaters (Fig. 22). Oil gravities show no simple depth relationship (Fig. 23) but do reflect structural and regional position relative to the zones of fresh water influx.

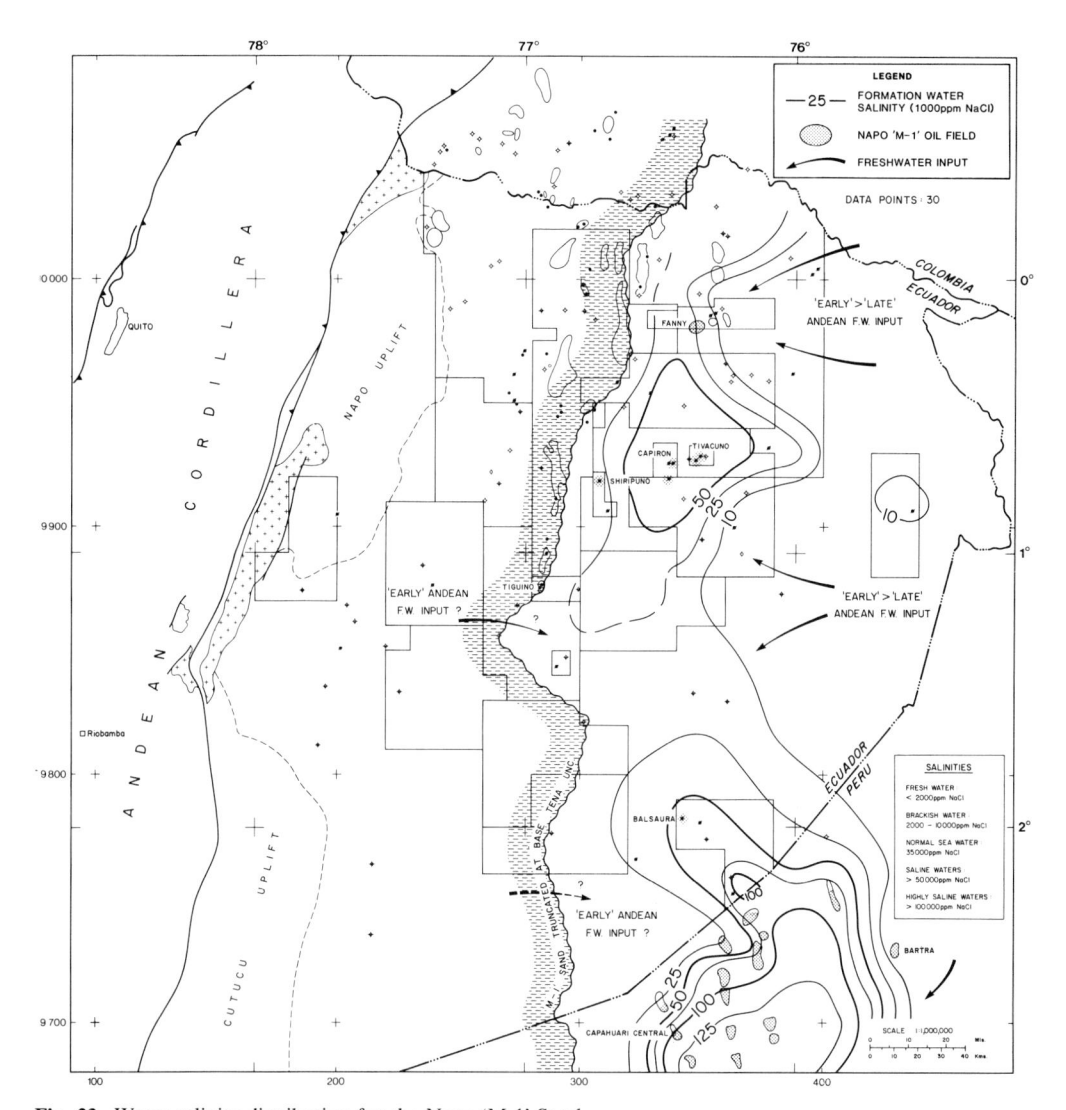

Fig. 23. Water salinity distribution for the Napo 'M-1' Sand.

Summary oil gravity distribution model

A summary figure (Fig. 25) is presented to illustrate the various factors involved in the proposed oil gravity model. The critical factors are as follows:
 (i) Cretaceous source rocks were deposited in a marine basin to the west of the present-day Oriente Basin.
 (ii) During the late Cretaceous and early Tertiary the onset of 'Early Andean' compression and thrusting occurred. Thick clastics shed from the west provided tectonic loading in the foredeep and caused significant Cretaceous oil generation.
 (iii) Fresh water influx, particularly from the east, caused early biodegradation of oil in the shallow reservoirs.
 (iv) During the Late Tertiary continued 'Late Andean' compression thrust out the former Cretaceous marginal basin terminating oil generation (although oil generation continued in the Maranon Basin to the south).
 (v) Up to the present day there has been continued intermittent uplift and differential basin subsidence and further oil alteration on the basin flanks.

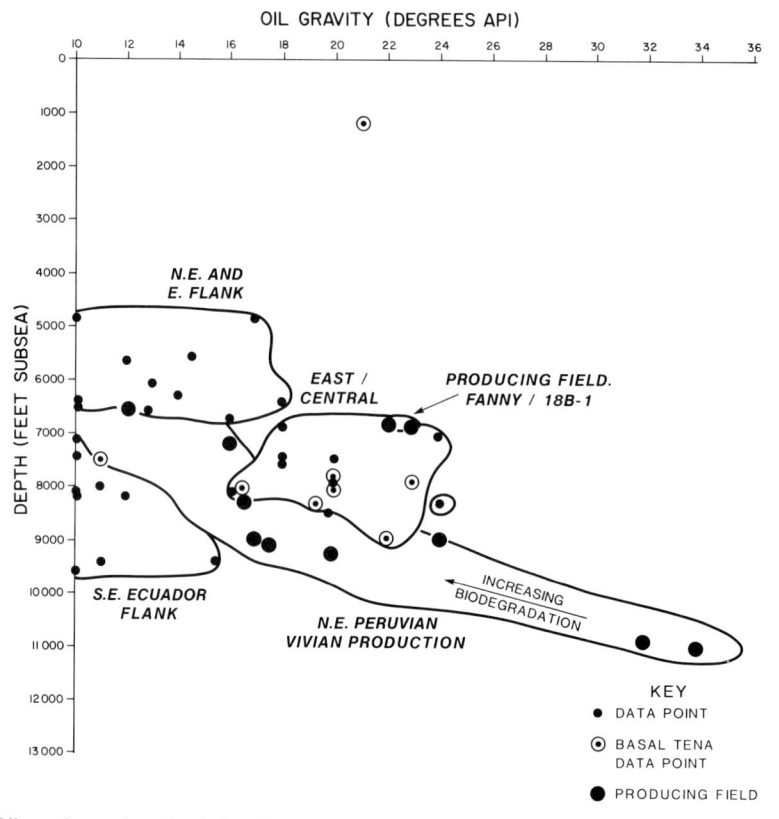

Fig. 24. Oil gravity against depth for the Napo 'M-1' Sand.

Conclusions

The Oriente Basin of Ecuador has many features in common with the other sub-Andean basins of South America, in particular those of northern South America. The structural history of the Early to Late Cretaceous, produced large, albeit low relief structures while the intensive 'Late Andean' Miocene–Pliocene orogeny produced large, high relief structures. Of the two families only the older prove oil productive in the Oriente.

The Cretaceous reservoir and Cretaceous oil source rock pairing is recognized throughout most of the sub-Andean basins, although the Oriente is exceptional in having no significant Palaeocene to Oligocene trapped oil (as seen in the Upper and Middle Magdalena and Llanos basins of Colombia and the Maracaibo and Barinas basins of Venezuela).

The Oriente oils appear to have been generated from a common Cretaceous source rock from areas west of the present basin which have since been 'thrust out' or metamorphosed in the Andean Cordillera. This 'difficult' hypothesis, originally proposed by Feininger (1975), is now invoked for the Maranon (Del Solar 1982; Sofer et al. 1986), the Putamayo (Caceres & Teatin 1985), the Llanos and certain of the basins of Venezuela (James 1990).

Geochemical evidence convincingly argues that the oil was generated prior to the 'Late Andean' orogeny and was not generated from source rocks local to the traps. Early (probably pre-Oligocene) oil generation, migration and entrapment is strongly indicated. The wide variation of oil gravity and composition reflects the alteration processes affecting the various reservoirs in different parts of the basin from as early as the end Cretaceous. The variation can be accounted for systematically. The intensive drilling activity of the coming five years (1988–1993) and active fieldwork programmes in the Cordillera Real should provide valuable new data in refining and testing the models proposed in this paper.

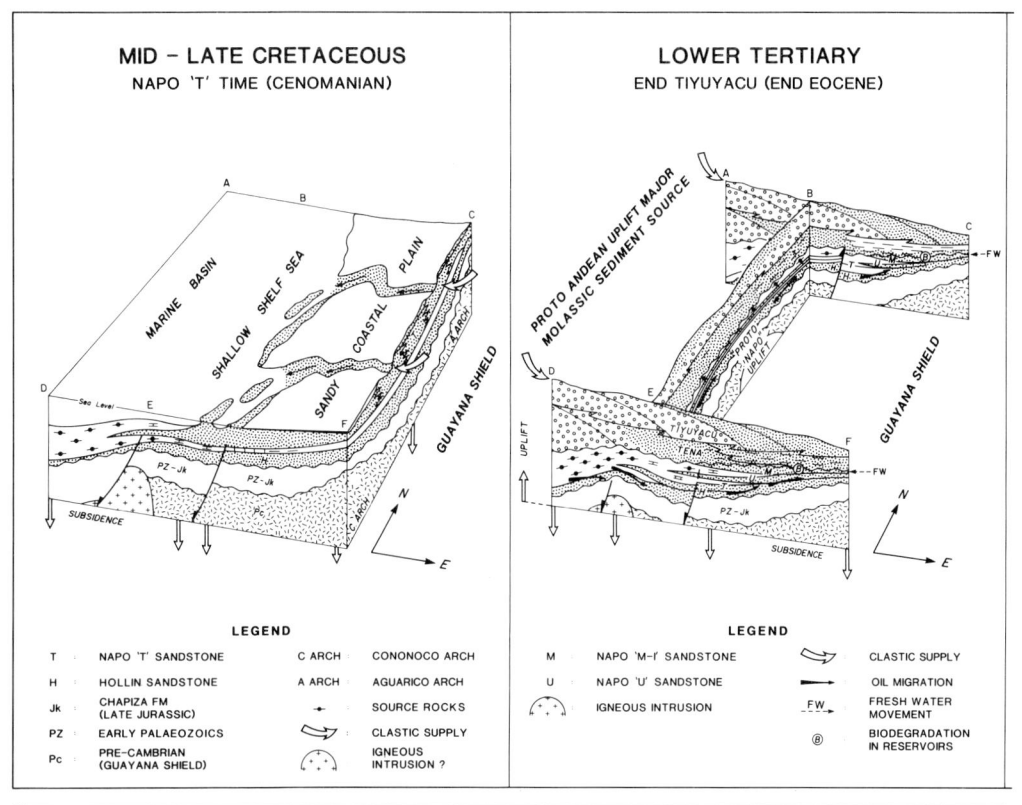

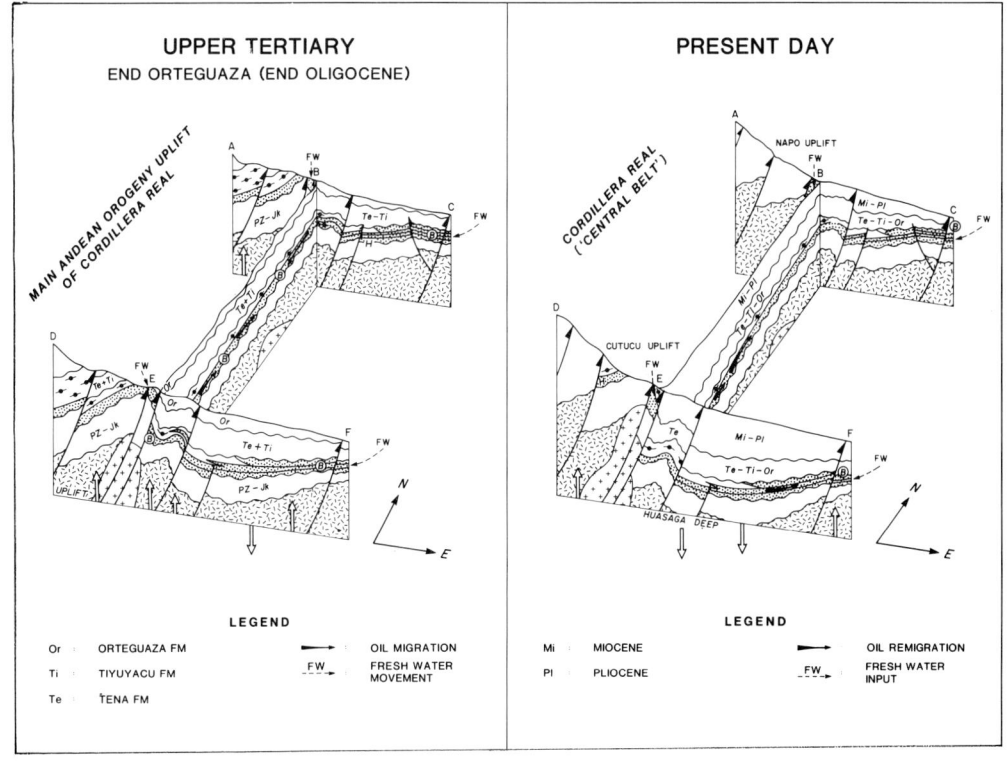

Fig. 25. Summary model for the oil gravity distribution in the Oriente Basin, Ecuador.

The authors thank both the Chairman and Directors of Clyde Expro plc and C.E.P.E. (Corporacion Estatal Petrolera Ecuatoriana) for approval to publish this work. We thank the Anglo-Ecuadorian Oilfields Company for their permission to include certain confidential data in this paper. We also thank Geochem Labs (UK) for geochemical analyses and valuable discussions. Concepts presented in this paper have evolved during regional evaluations, research of Anglo-Ecuadorian Oilfields Co. data, discussions with partners, Clyde Expro colleagues and C.E.P.E. staff. This cooperation is gratefully acknowledged. However, the views expressed in this paper are entirely the responsibility of the authors.

References

ALMEIDA, J. P. 1986. Estudio de litofacies y del Contacto Agua-Petroleo de la Arenisca 'T' del Campo Libertador. *In: IV Congreso Ecuatoriano de Geologia, Minas y Petroleo*. Colegio de Ingenieros Geologos de Minas y Petroleos de Pinchincha, Quito.

ALVARADO, G., BONILLA, G., DE ROJAS, G. & DE ARROYO, R. Z. 1982. Analisis Sedimentologico de la Zona Arenisca Napo 'T' en la Cuencadel Napo. *In: Simposio Bolivariano, Exploracion Petrolera en las Cuencas Subandinas de Venezuela, Colombia, Ecuador y Peru*. Asociacion colombiana de Geologos y Geofisicos del Petroleo, Bogota.

ANGLO ECUADORIAN OILFIELDS LTD. 1969–1973. *Ecuador Oriente geological notes 1–24*. Unpublished internal company reports, Quito.

ANON 1987. *Informe de Ejecucion de los contratos de Prestacion de Servicios Para la Exporacion y Exploitacion de Hidrocarburos al mes de Junio de 1987*. Corporacion Estatal Petrolera Ecuatoriana (CEPE), Quito.

CACERES, H. & TEATIN, P. 1985. Cuenca del Putamayo Provincio Petrolera Meridional de Colombia. *In: II Simposio Bolivanano. Exploracion Petrolera en las Cuencas Subandinas*. Asociacion colombiana de Geologos y Geofisicos del Petroleo, Bogota.

CAMPBELL, C. J. 1970. *Guidebook to the Puerto Napo area, eastern Ecuador, with notes on the regional geology of the Oriente Basin*. Ecuadorian Geological and Geophysical Society, Quito.

CANFIELD, R. W., BONILLA, G. & ROBBINS, R. K. 1982a. Sacha Oil Field of Ecuadorian Oriente. *Bulletin of the American Association of Petroleum Geologists*, **61** 1076–1090.

——, DIAZ, R. N. & MONTENEGRO, J. 1985. El Campo coronanoc del Oriente Ecuatorianto. *In: II Simposio Bolivariano. Exploracion Petrolera en las Cuencas Subandinas*. Asociacion colombiana de Geologos y Geofisicos del Petroleo, Bogota.

——, ROSANIA, G. E. & SAN MARTIN, H. 1982b. Geologia de los campos gigantes del Oriente Ecuatoriano. *In: Simposio Bolivariano. Exploracion Petrolera en las Cuencas Subandinas de Venezuela, Colombia, Ecuador y Peru*. Asociacion colombiana de Geologos y Geofisicos del Petroleo, Bogota.

DE RIGHI, M. R. & BLOOMER, G. 1975. Oil and Gas developments in the Upper Amazon Basin. *In: Proceedings of the Ninth World Petroleum Congress, Tokyo*, **3**, 181–192.

DEL SOLAR, C. 1982. Ocurrencia de hidrocarburos en la formacion Vivian, Cuenca Marañon, Nor-Oriente Peruano. *In: Simposio Bolivariano. Exploracion Petrolera en las Cuencas Subandinas de Venezuela, Colombia, Ecuador y Peru*. Asociacion colombiana de Geologos y Geofisicos del Petroleo, Bogota.

ESPIN, P. 1980. Deteccion de Cuerpos igneos en el Cretacio del Oriente Ecuatoriano. *In: II Congresso Ecuatoriano de Ingenieros*, Geologos de Minas y Petroleos, Quito.

FEININGER, T. 1975. Origin of petroleum in the Oriente of Ecuador. *Bulletin of the American Association of Petroleum Geologists*, **59**, 1166–1175.

HENDERSON, W. G. 1979. Cretaceous to Eocene Volcanic Arc Activity in the Andes of Northern Ecuador. *Journal of the Geological Society of London*. **136**, 367–378.

—— & EVANS, C. D. R. 1980. Ecuadorian Subduction System: Discussion of Lonsdale, 1978. *Bulletin of the American Association of Petroleum Geologists*, **64**, 280–281.

ILLICH, H. A., HANEY, F. F. & JACKSON, T. Y. 1977. Hydrocarbon geochemistry of oils from the Maranon Basin, Peru. *Bulletin of the American Association of Petroleum Geologists*, **61**, 2103–2114.

JAMES, K. H. 1990. The Venezuelan hydrocarbon habitat. *In*: BROOKS, J. (ed.) *Classic Petroleum Provinces*, Geological Society, London, Special Publication **50**, 9–36.

JENKS, W. F. (ed.) 1956. *Handbook of South American Geology*. The Geological Society of America Memoir 65.

KUMMERT, P. & CASAL, C. 1986. Granulometria de Areniscas Cementada con silica, aplicacion a la Determinacion de los Ambientes de sedimentacion en la Formacion Hollin del Campo Bermejo–Sur. *In: IV Congreso Ecuatoriano de Geologia, Minas y Petroleo*. Colegio de Ingenieros Geologos de Minas y Petroleos de Pichnicha, Quito.

LEBRAT, M., MEGARD, F. & DUPUY, C. 1986. Pre-Orogenic Volcanic Assemblages and Position of the Suture Between Oceanic Terrains and the South American continent in Ecuador. *Zbl. Geol. Palaont. Tesl 1, 1985* **9/10** 1207–1214.

LONSDALE, P. 1978. Ecuadorian subduction System. *Bulletin of the American Association of Petroleum Geology*. **62**, 2454–2474.

LOZADA, F. T. & ENDARA, P. 1982. Estudio de litofacies de la zona Arenisca 'M-1', Formación Napo, en la cuenca Oriental. *In: Simposio Bolivariano. Exploracion petrolera en las cuencas Sub-Andinas de Venezuela, Colombia, Ecuador y Peru*. Asociacion colombiana de Geologos y Geofisicos del Petróleo, Bogota.

——, —— & CORDERO, C. 1985. exploracion y desarallo en la Campo del Libertador. *In*: *II Simposio Bolivanano. Exploracion Petrolera en las Cuencas Subandinas*. Asociacion colombiana de Geologos y Geofisicos del Petroleo, Bogota.

REED, R. N. & WHEATLEY, M. J. 1983. Oil and Water Production in a Reservoir with a Significant Capillary Transition Zone. *In*: *Proceedings of Annual Conference of the Society of Petroleum Engineers of AIME, San Francisco*.

RIVADENEIRA, M. 1986. Evaluacion Geochemica de Rocas Madres de la Cuenca Amazonica Ecuatoriana. *In*: *IV Congresso Ecuatoriano de Geologia, Minas y Petroleo*. Colegio de Ingenieros Geologos de Minas y Petroleos de Pichincha, Quito.

ROBERTO, M., PUCHE, E. and MOMPART, L. 1990. New Discoveries in the Ceute area, S.E. Lake Maracaibo, Venezuela. *In*: BROOKS, J. (ed.) *Classic Petroleum Provinces*, Geological Society, London, Special Publication, **50**, 77–88.

ROSANIA, G. 1985. Los Nuevos contratos Petroleos. *In*: *II Simposio Bolivanano. Exploracion Petrolera en las Cuencas Subandinas*. Asociacion colombiana de Geologos y Geofisicos del Petroleo, Bogota.

——, JARAMILLO, P. & OLMEDO, M. 1985. Estudio de Ingenieria de Yacimientos Campo Auca Yacimientos Napo 'U' y 'T'. *In*: *Simposio Bolivariano. Exploracion Petrolera en las Cuencas Subandinas*. Asociacion colombiana de Geologos y Geofisicos del Petroleo, Bogota.

RUSSOMANA, F. & VELARDE, H. 1982. Petroleum Geology of the Barinas–Apure Basin, Venezuela. *In*: *Simposio Bolivariano. Exploración petrolera en las cuencas Sub-Andinas de Venezuela, Colombia, Ecuador y Peru*. Asociacion colombiana de Geologos y Geofisicos del Petróleo, Bogota.

SCHMERBER, G., CHENET, P. Y. & FAUCHER, B. 1986. Ejemplos Practicos del uso de Modelos de geologia Matematica para la Exploracion de Varias Cuencas Sedimentarias. Beicip/I.F.P. Report.

SOFER, Z. 1984. Stable Carbon Isotope Compositions of Crude Oils: Application to Source Depositional Environments and Petroleum Alteration. *Bulletin of the American Association of Petroleum Geologists*, **68**, 31–49.

——, ZUMBERGE, J. E. & LAY, V. 1986. Stable Carbon Isotopes and Biomarkers as Tools in Understanding Genetic Relationship, Maturation, Biodegradation and Migration of Crude Oils in the Northern Peruvian Oriente (Maranon) Basin. Advances in Organic Geochemistry, 1985. *In*: *Organic Geochemistry*, **10**, 377–389.

TSCHOPP, H. J. 1953. Oil explorations in the Oriente of Ecuador 1938–1950, *Bulletin of the American Association of Petroleum Geologists*, **37**, 2203–2347.

URIEN, C. M. and ZAMBRANO, J. J. 1985. The Sub-Andean Basins from Venezuela to the Malvinas Plateau. *In*: *II Simposio Bolivariano, Exploracion Petrolera en las Cuencas Subandinas*. Asociacion colombiana de Geologos y Geofisicos del Petroleo, Bogota.

WAPLES, D. W. 1980. Time and Temperature in Petroleum Formation: Application of Lopatin's Method to Petroleum Exploration. *Bulletin of the American Association of Petroleum Geologists*, **64**, 916–926.

ZUMBERGE, J. E. 1980. Oil to Oil and Oil-source rock correlations of Bacterially Degraded Oils and Cretaceous Outcrops from Colombia, South America. *In*: *Abstracts from 26th International Geological Congress*, Paris, 806.

—— 1986. Source Rocks of the La Luna Formation (Upper Cretaceous) in the Middle Magdalena Valley, Colombia. *In*: (ed.) PALACAS, J. G. *Petroleum Geochemistry and Source Rock Potential of Carbonate Rocks*. AAPG Studies in Geology **18**, 127–133.

Petroleum geology of the Campos Basin, offshore Brazil

W. U. MOHRIAK[1,3], M. R. MELLO[2,3], J. F. DEWEY[1], & J. R. MAXWELL[2]

[1] *University of Oxford, UK*
[2] *University of Bristol, UK*
[3] *Petroleo Brasileiro S. A.*

Abstract: The Campos Basin, offshore Brazil, is the most prolific basin in the western South Atlantic, with more than thirty hydrocarbon accumulations currently accounting for about 60% of Brazilian oil production.

Intensive drilling and seismic, gravity and magnetic data have contributed to the recognition of four tectono-stratigraphic units related to the rifting and break-up of Pangea. The lowest sequence consists of Neocomian clastics deposited on basalt dated at 120–130 Ma, and reflects the fault-controlled subsidence associated with the stretching that preceded the emplacement of oceanic crust. The Aptian proto-oceanic stage is characterized by a sequence of evaporitic rocks that have undergone intense diapiric activity in deep water. An open-marine environment begins with a thick sequence of Albian/Cenomanian limestones, locally with clastic input, which grades upwards and basinwards into deep water marls and shales. This section is structurally associated with detached listric normal faults that sole out on the Aptian evaporites. Finally, the marine Upper Cretaceous to Recent clastic section is characterized by a more quiescent phase of thermal subsidence, with some residual halokinetic activity that increases in intensity towards deeper waters.

The hydrocarbon accumulations are distributed throughout the stratigraphic column of the basin from Neocomian to Miocene. The reservoirs range from fractured basalts and porous bioclastic limestone (coquinas) in the Lagoa Feia Formation, to limestones and sandstones in the Macaé Formation, and sandstones in the Campos Formation.

Detailed geochemical analyses undertaken on cutting, core and oil samples show that almost all the hydrocarbon accumulations discovered to date are sourced mainly from lacustrine calcareous black shales deposited in a closed Upper Neocomian lake system, having saline to hypersaline waters of alkaline affinities. The extreme anoxic conditions in this lacustrine environment resulted in the deposition of fine, well laminated organic-rich (TOC up to 9%) calcareous black shales, with high-quality organic matter composed almost entirely of low-sulphur type-I kerogen, originating from lipid-rich algal and bacterially-derived material. The excellent hydrocarbon source potential of these sediments, combined with the appropriate thermal history, produced the necessary conditions to yield low-density oil (around 30° API) characterized by a low sulphur content (around 0.30%), and significant quantities of alkanes (up to 70%). Diagnostic features in the biological markers from this depositional environment include: low concentration of steranes, presence of β-carotane, gammacerane and 28,30-bisnorhopane, very high concentrations of hopanes and high relative abundances of tricyclic terpanes up to C_{34}.

The distribution of the petroleum resources in the Brazilian continental margin is extremely unequal: several large basins are barren of significant hydrocarbon accumulations, while the relatively small Campos Basin has been proved to be the most prolific oil-producing basin offshore Brazil. This basin has an area of about 100 000 km² extending down to a water depth of 2000 m (Fig. 1). The exploration activity has led to the discovery of several hydrocarbon accumulations, including the recently discovered deep water giant fields. At present, the Campos Basin is responsible for more than 60% of Brazilian oil production, and this proportion is bound to increase in the near future.

The tectono-stratigraphic evolution and the subsidence history of the Campos Basin

The tectonic evolution of the Campos Basin is associated with the Mesozoic rifting in the South Atlantic that resulted in the break-up of Pangea and the development of several basins whose similarities in the tectono-stratigraphic pattern of evolution point towards a common mechanism of basin formation.

Earlier models for the origin of the Campos Basin assumed domal uplift, erosion, and subsequent subsidence (Estrella 1972; Asmus & Ponte 1973; Asmus 1975; Asmus 1982), following the lines of Sleep (1971). However, several

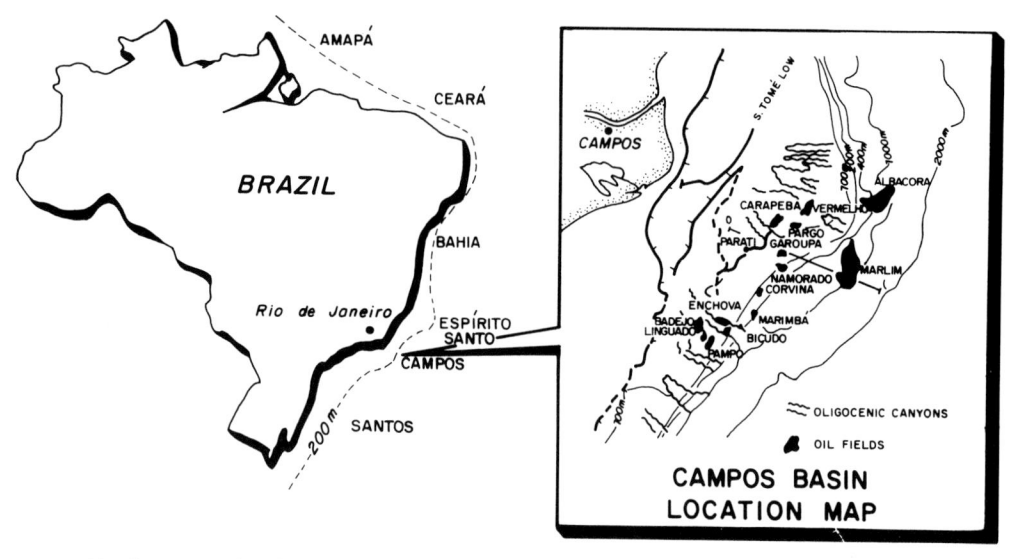

Fig. 1. The Campos Basin, the most prolific hydrocarbon province in Brazil, and the location of several oil fields. Some structural features and submarine canyons are also shown.

backstripped wells show that the subsidence history of the basin can be attributed to an initial rifting and subsequent thermal recovery of the lithosphere, although there are some important deviations from the homogeneous, simple stretching model (McKenzie 1978). The widespread basaltic magmatism contemporaneous with the rifting is highly indicative of a thermal anomaly in the mantle (Furlong & Fontain 1986), and the onlapping of Tertiary sediments on the western margin of the basin suggests flexural control on the subsidence (Mohriak et al. 1987), or distributed, depth-dependent stretching mechanisms (Rowley & Sahagian 1986).

The stratigraphic evolution of the Campos Basin (Schaller 1973 — see Fig. 2) shows four tectono-stratigraphic sequences associated with the following stages of development: rift, proto-oceanic, and oceanic (Asmus & Ponte 1973; Campos et al. 1974; Asmus 1975; Porto & Asmus 1976; Asmus & Guazelli 1981; Asmus 1982).

The Neocomian rift-stage lacustrine deposits are associated with basement-involved, block rotated faulting in a rapidly subsiding crust, and widespread mafic volcanism whose isotopic age shows a mean at 130 Ma BP (Amaral et al. 1966; Cordani et al. 1972).

The Aptian proto-oceanic stage is characterized by the deposition of evaporitic rocks, which are associated with the first sea-water inflows through the Walvis Ridge (Leyden et al. 1976). These two stages constitute the Lagoa Feia Formation.

The lower part of the oceanic stage is characterized by an Albian/Cenomanian section of calcarenites and calcilutites (Macaé Formation), often with clastic intercalations and dolomitization near the base of the sequence. This section is structurally associated with halokinetic features and listric detached normal faulting soling out on salt (Figueiredo & Mohriak 1984). Upwards and basinwards, the section grades into marls and shales (informally called the 'Bota' section).

The marine Upper Cretaceous to Recent clastic section (Campos and Emborê Formations) corresponds to a period of general tectonic quiescence and continued subsidence, although some diastrophic structures occur near the northwestern border of the basin (Lobo et al. 1983; Mohriak 1984). Residual salt movements are still present and increase in intensity in deep water (Lobo & Ferradaes 1983).

The main structural elements in the basin are synthetic and antithetic normal faults formed during the rifting phase, grabens and horsts formed by the rift-phase faulting, hinge lines associated with flexure of the basement, homoclinal structures, listric normal faults detaching on the Aptian salt, and roll-over structures associated with these faults (Ojeda 1982). In the northwest area, basement-involved faulting of

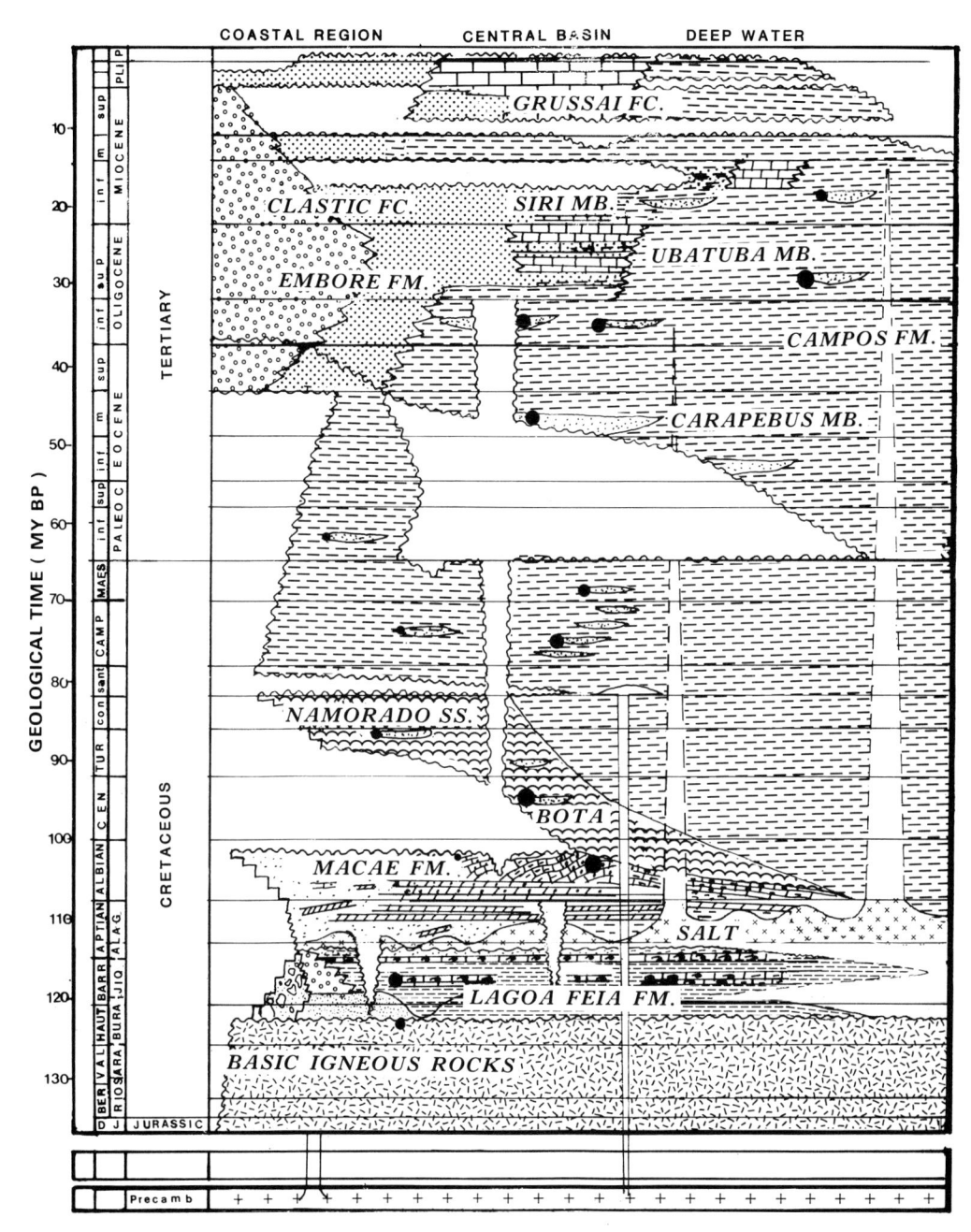

Fig. 2. Lithostratigraphic column of the Campos Basin showing the distribution of producing horizons. ●, oil accumulation. Modified from Beltrami *et al.* (1982) and Schaller (1973).

Tertiary age is present (Lobo *et al.* 1983; Mohriak 1984). Compactional faults are also observed towards the depocentre of the basin.

Figure 3 shows the regional dip seismic line 203-RL-76 in two parts, extending from the northwest region to water depths of about 2 km. The computer-generated geochronostratigraphic section shown in Fig. 4 is based on the shallow (6 s two-way travel time) reflection seismic line, linear interpolation of borehole

Fig. 2. Regional dip seismic line 203-RL-76 showing the block-rotated half-grabens in the Neocomian (Lagoa Feia Fm.) and the listric faults associated with salt

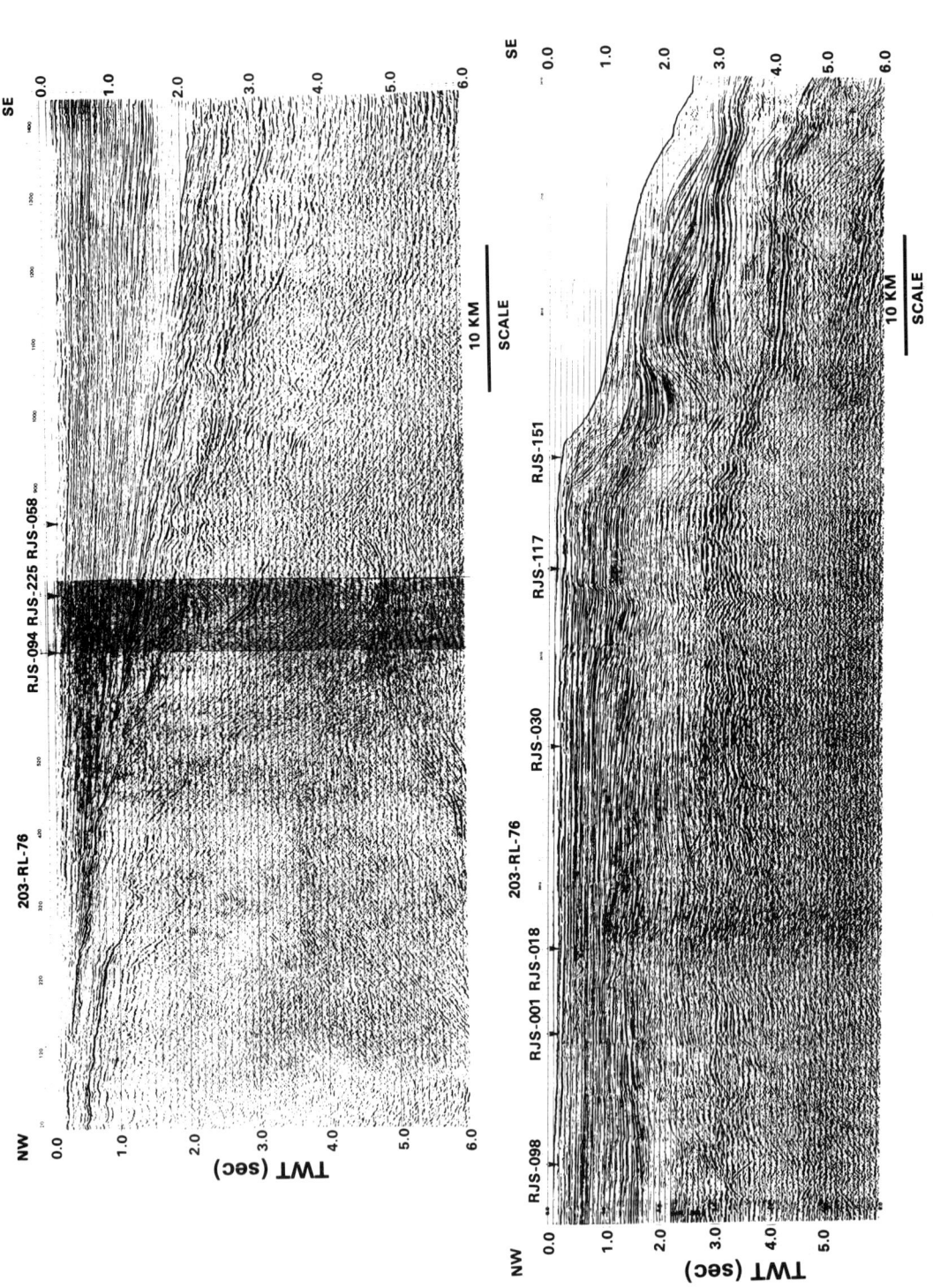

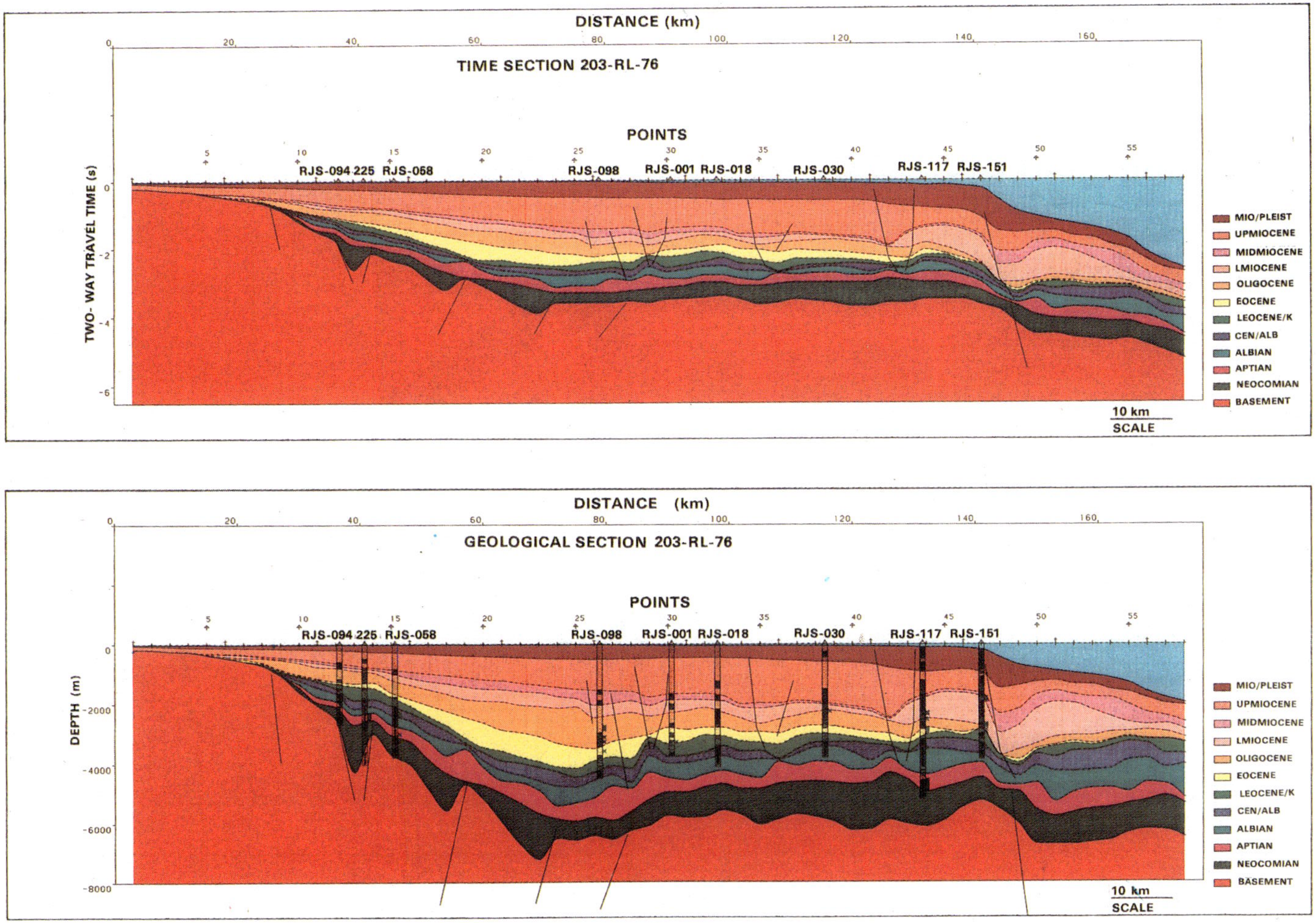

Fig. 4. Interpretation of regional section 203-RL-76 and geochronostratigraphic section based on interpolation on borehole palaeontological dating. Basement-involved, antithetic normal faulting is characteristic of the northwestern part of the basin. Huge roll-overs, controlled by listric faults detaching on the Aptian salt, are observed towards the shelf edge. The Palaeogene sequence tends to condense rapidly towards deep waters.

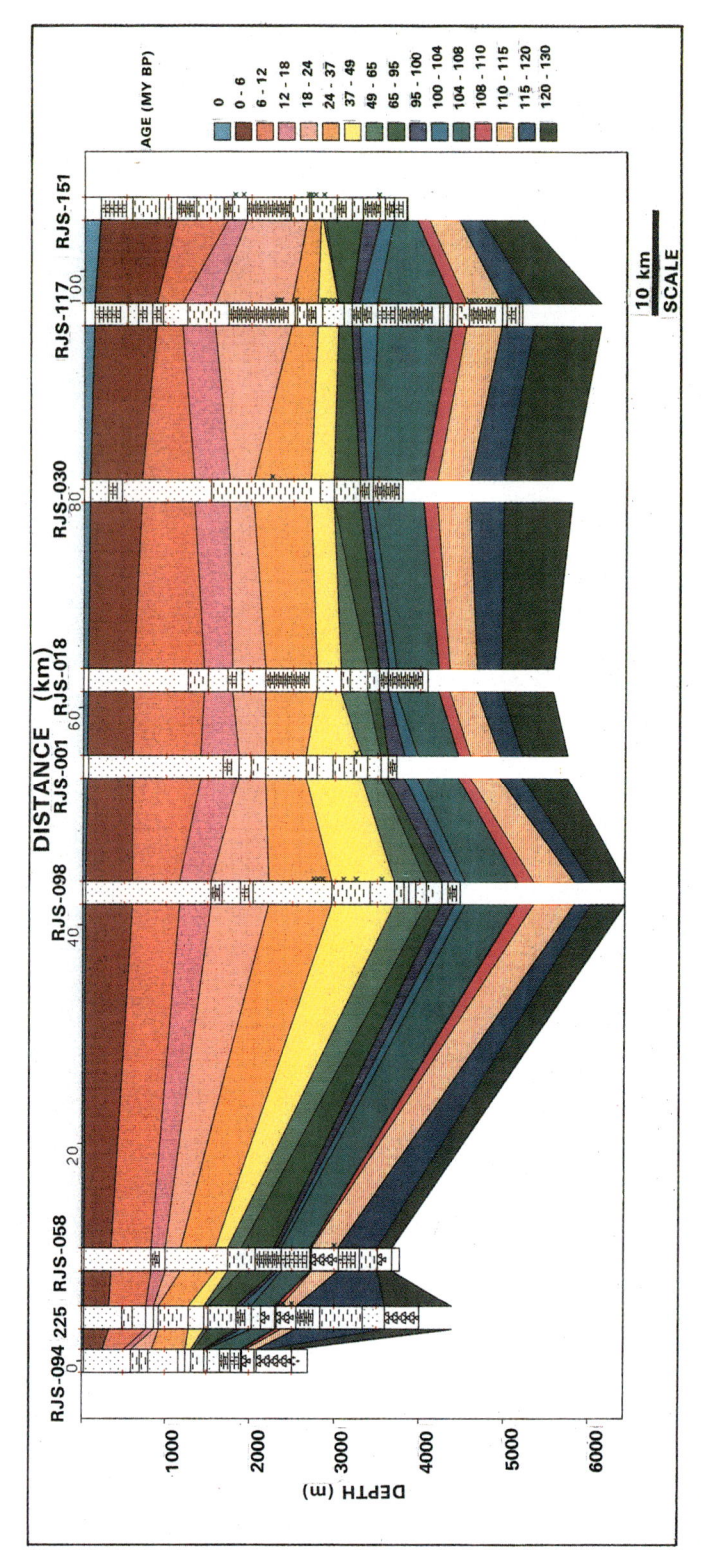

Fig. 5. Synthetic seismic section for the regional line 203-RL-76 based on interpolation of borehole palaeontological dating. Observe condensation of the Eocene strata towards borehole RJS-151.

Table 1. *General geohistory of the Campos Basin*

Early Cretaceous (Neocomian)	lithospheric stretching and crustal thinning; thermal anomaly in the mantle; widespread outpouring of basaltic rocks; synthetic and antithetic faulting in the upper crust; half-grabens filled with fluvio-lacustrine rocks; regional unconformity after the deposition of coquina; renewed tectonic activity resulting in conglomerates.
Aptian	marine water incursions from the southern ocean; deposition of evaporitic rocks.
Albian	shallow-water carbonate platform; high subsidence and sedimentation rates; accumulation of thick section of carbonates; section regionally ruptured by listric normal faults; listric faults detaching on the Aptian evaporites.
Late Cretaceous to Early Tertiary	fast subsidence with low clastic input; deepening of the depositional environment; sediment starvation, bypassing and erosional events; submarine currents and regional unconformities; salt mobilization throughout the period; growth faults and local salt dissolution synclines; deposition of turbiditic sandstones.
Middle Tertiary to Recent	widespread turbiditic sandstone deposition; intense deep water halokinetic activity; submarine canyons; regional offlap and deltaic progradation; shallower marine facies on deep water deposits; onlap of the sediments onto the western margin; localized tectonism in the northwestern area.

velocities, and detailed paleontological dating for industry boreholes along the line. A huge roll-over structure is present near the present-day shelf edge, characterizing a region of marked instability. Paleontological dating obtained from the boreholes allowed the definition of absolute ages for the depositional sequences. Fig. 5 shows a synthetic seismic section (Hinte 1978, 1982) based on the interpolation of borehole paleontological dating along a refined grid of seismic lines. The approximate absolute ages to different depositional sequences were assigned by using Petrobrás geochronostratigraphic charts (Beurlen 1981).

Table 1 summarizes the geological evolution of the Campos Basin.

The theoretical subsidence curves predicted by the simple stretching model (McKenzie 1978) are compared with the subsidence curve obtained by backstripping the borehole RJS-117 (Fig. 6). This borehole drilled more than 5000 m of sediments without reaching the volcanic igneous rocks assumed as the basement, which is probably found around 6000–6500 m, according to the seismic interpretation. The backstripping method (Watts & Ryan 1976; Steckler & Watts 1978) was employed to isolate the tectonic component of the basement subsidence curve through time, after correction for sedimentary loading, sea-level fluctuations, paleobathymetry and compaction. Absolute age estimates are based on paleontological dating and Petrobrás' geochronostratigraphic charts (Beurlen 1982), and paleobathymetry values are estimated from extrapolation of paleoecological studies in the northwestern area of the basin (Koutsoukos 1984) and depositional facies interpretation. Figure 7 shows the estimated paleobathymetry and the undecompacted sedimentation rate through time. There is a remarkable correlation between the low sedimentation rates and the increase in water depth in the Late Cretaceous.

The Airy model is employed as a first approximation to the lithosphere response to loading, because the well is located far away from the hinge line of the basin. The decompaction of the sedimentary column was obtained by least-

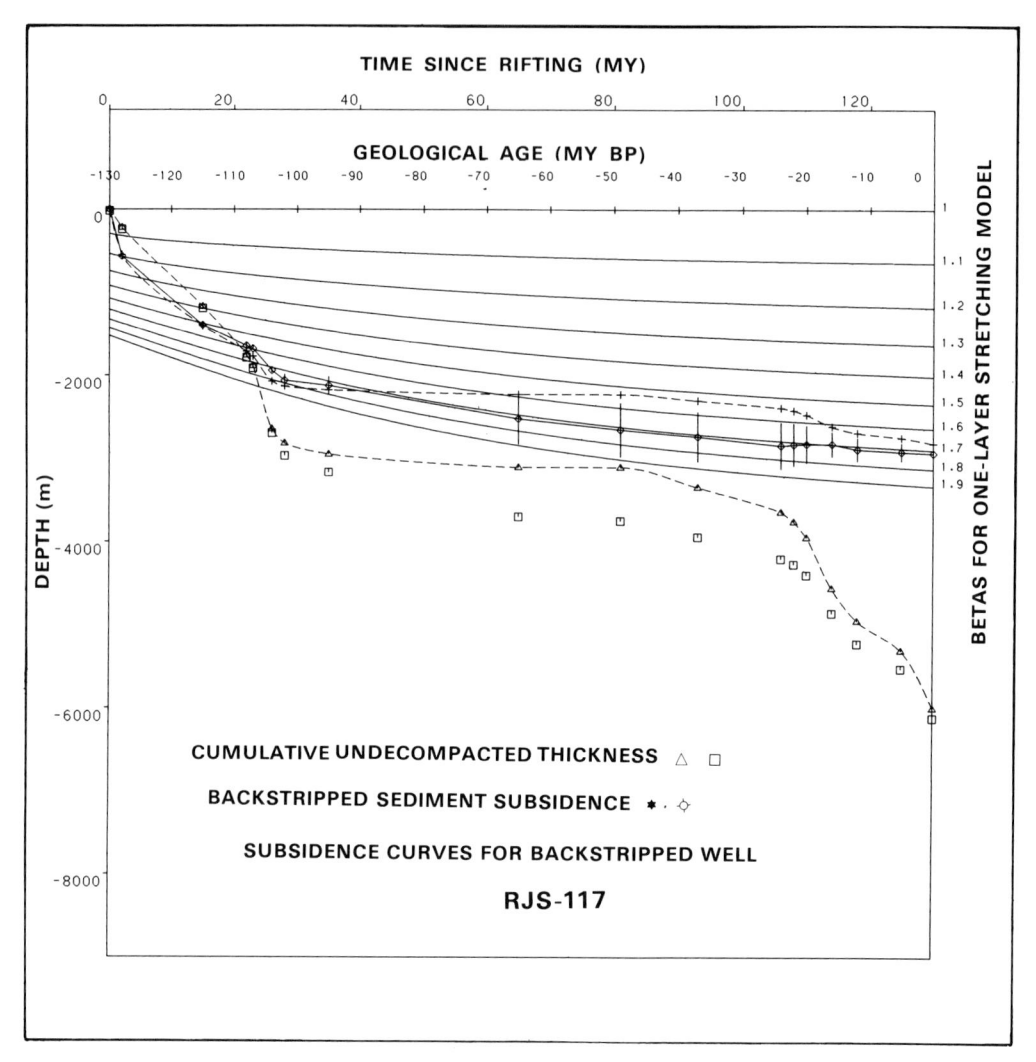

Fig. 6. Geohistory curve for the borehole RJS-117 cross-plotted with the one-layer stretching model subsidence curve. Broken curves correspond to subsidence curves with no correction for palaeobathymetry and eustatic sea-level fluctuation. Vertical bar in the backstripped subsidence curve indicates range of palaeobathymetry estimates.

square fitting an exponential function of porosity variation with depth (Sclater & Christie 1980). The backstripped subsidence of the well, cross-plotted with the simple stretching model subsidence curves (Fig. 6), indicates a rather small value for the extensional parameter β (about 1.7), suggesting a relatively small thinning for the lithosphere and a rather low heat flow for this basin. This is in agreement with calculations of the present-day geothermal gradient as obtained from temperature measurements in logging operations (Roos & Pantoja 1978). The basin shows geothermal gradients varying from about 18 to 30°C km^{-1}, and a mean gradient of about 20°C km^{-1}, with a tendency to increase towards the northern part of the basin, where gravity modelling and deep seismic reflectors are suggestive of rapid lithospheric stretching and crustal thinning (Mohriak & Dewey 1987). This northwestern area of the Campos Basin is also characterized by very deep troughs (half-grabens) formed during the rifting phase (see Figs 3 & 4), the common occurrence of gas accumulations, and the only field in the basin with production of condensates. Geochemical evidence based on vitrinite reflectance and bio-

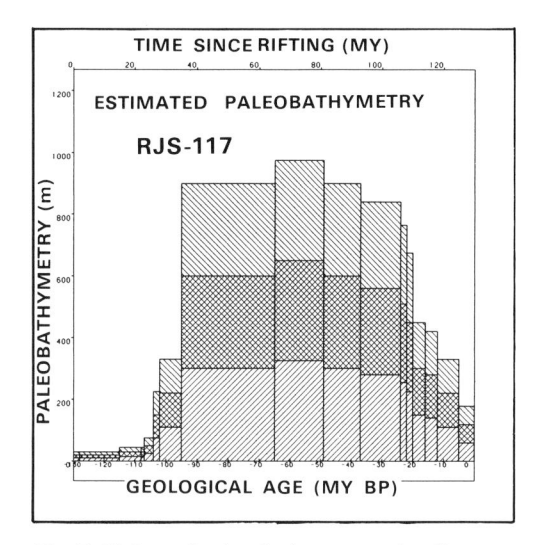

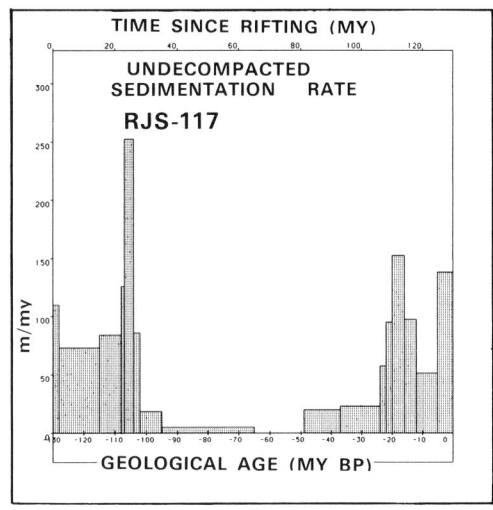

Fig. 7. Estimated palaeobathymetry and sediment accumulation rate for the borehole RJS-117. In the palaeobathymetry diagram, the top of the criss-cross area corresponds to a mean value estimated by palaeoecological studies and by smoothing of the subsidence curve. A range proportional to 50% of the absolute estimate is plotted as slanted lines in this diagram and as error bars in Fig. 6. This implies that the higher the palaeobathymetry, the larger will be the error bar in the estimate.

logical marker data (Figueiredo *et al.* 1983) also indicate low to moderate temperatures throughout the basin history and lends support to the above assumptions.

Oil exploration in the Campos Basin

Several fields have been found in the Campos Basin since the pioneer well 1-RJS-9A discovered the first hydrocarbon accumulation in Albian limestones in 1974. The exploration activity has led subsequently to several other oil finds, and at present, hydrocarbon accumulations have been found throughout the lithostratigraphic column except for the Emborê Formation. Reservoirs range in age from the Neocomian to the Miocene, and include different facies, such as sandstones, limestones, coquinas, and basalts.

The main hydrocarbons accumulations in the Campos Basin are shown in Table 2.

The lithostratigraphic column of the basin with the oil producing horizons is presented in Fig. 2.

Figure 8 shows a composite log depicting an idealized lithostratigraphic column with typical dipmeter, gamma ray and resistivity readings in the different depositional sequences and formations. A summary with most of the hydrocarbon accumulations discovered in the basin since the beginning of exploration, is also included.

The first stratigraphic borehole in the basin was drilled in the onshore region (CST-1), in 1958. Petrobrás started the active exploration by seismic acquisition in the offshore area in 1968 (Campos 1970), which led to the drilling of a wildcat in 1971 (RJS-1). The first hydrocarbon accumulation in the basin (Garoupa field) was found in Albian limestones by the well RJS-9A in December, 1974. Oil-bearing Cretaceous sandstones were found by the well RJS-12 (Pargo area), and by April 1975, a commercial accumulation was found in Eocene sandstones in a structure north of the Garoupa field. The Namorado oil field (Upper Albian/Cenomanian sandstones) was discovered in May, and by November 1975, the discovery of the Badejo field (RJS-13) expanded the stratigraphic range of the reservoirs to the Neocomian coquinas. In May 1976, Eocene sandstones with oil saturation were found in the Enchova area. In April 1977, the Bonito and Cherne fields were discovered, and by July, the hydrocarbon accumulations in the Neocomian coquinas and in the Albian limestones of the Pampo Field were discovered. Unfortunately, the Pampo Eocene sandstones, which also show hydrocarbon saturation, bear heavy oils in the range 13–15° API. Thick deposits of Cretaceous sandstones were found to be oil-bearing in the Marimbá oil field (RJS-216) in May 1982. The Albacora (Albian/Cenomanian sandstone) giant field was discovered by the well RJS-297, in November 1984. By late 1984, the well RJS-305 was drilled in

Table 2. *Main hydrocarbon accumulations in the Campos Basin*

Miocene/Oligocene sandstones	Enchova Oeste, Bonito, Moréia, Marlim, Albacora
Eocene sandstones	Enchova, Bonito, Bicudo, Cherne, Bagre, Viola, Vermelho, Parati, Anequim, Garoupinha
Lower Eocene/Paleocene/Upper Cretaceous sandstones	Pirauna, Marimbá, Corvina, Malhado, Pargo, Carapeba
Cenomanian/Albian sandstones	Namorado, Bagre, RJS-150, RJS-211, Albacora
Albian limestones	Pampo, Enchova, Bonito, Bicudo, Garoupa
Neocomian coquinas	Pampo, Badejo, Linguado, Trilha
Neocomian fractured basalt	Badejo, Linguado

430 m water depth and confirmed the hydrocarbon accumulation that had been drilled by the well RJS-297 and also found Oligocene sands saturated with oil. The Oligocene sands of the giant Marlim oil field (water depth 800–1600 m) were discovered by the well RJS-219 in February 1985.

Petroleum production in the basin started in 1977, with well 3-EN-1-RJS producing through an early production floating system. By April 1979, Petrobrás developed the first sub-sea completion with wet Christmas tree in well RJS-38 at 189 m water depth. In December 1982, oil production started through the float production system in the Bonito Field through a wet sub-sea template. By August 1983, the first oil production through a fixed steel platform in the Campos Basin was initiated in the Namorado-1 well. By December 1983, the Pirauna–Marimbá floating system started operation at 243 m water depth, a world record at the time.

The success rate for drilling in the Campos Basin has been excellent: about 40% of the wells drilled until 1986 were hydrocarbon bearing, and for the deep water area, the success rate is about 80%. The exploration in the deep waters of the basin has proceeded at a very fast pace, and deep-water completions have established successive world records. By April 1985, Petrobrás executed record-breaking deep water operations with the diverless completion of the well RJS-284 at 383 m water depth, and in January 1987, the world record was broken again with the diverless completion of the well RJS-294 at 411 m water depth. Deep water exploration and production are still advancing towards more difficult areas, pushing world records deeper and deeper. In July 1987, the well RJS-367 tested Miocene/Oligocene reservoirs at a water depth of 1565 m.

Petroleum habitat in the Campos Basin

The subsidence history of the boreholes is known with good accuracy, given the detailed paleontological dating and the good correlation with stratigraphic markers in the electrical logs and seismic sections. The calculation of the maturation index was performed by integrating a continuous exponential function of thermal exposure with time (Royden *et al.* 1980). The integrated time–temperature index (TTI) is empirically associated with vitrinite reflectance. The start of oil generation is assumed to correspond to a vitrinite reflectance of 0.6%, or a TTI index of 15, while the gas window corresponds to a vitrinite reflectance of 1.0%, or a TTI value of 160 (Waples 1980; Heroux *et al.* 1979). The results obtained by this technique, decompacting the sediments and employing an optimization of the geothermal gradient and subsidence history, assuming an initial value of 30°C km^{-1} decaying linearly to the present-day gradient of 20°C km^{-1}, are shown in Fig. 9(A). An even more optimistic initial geothermal gradient of about 40°C km^{-1} would imply in oil generation for the Neocomian sediments soon after deposition. A more pessimistic geothermal and subsidence history, with gradients varying

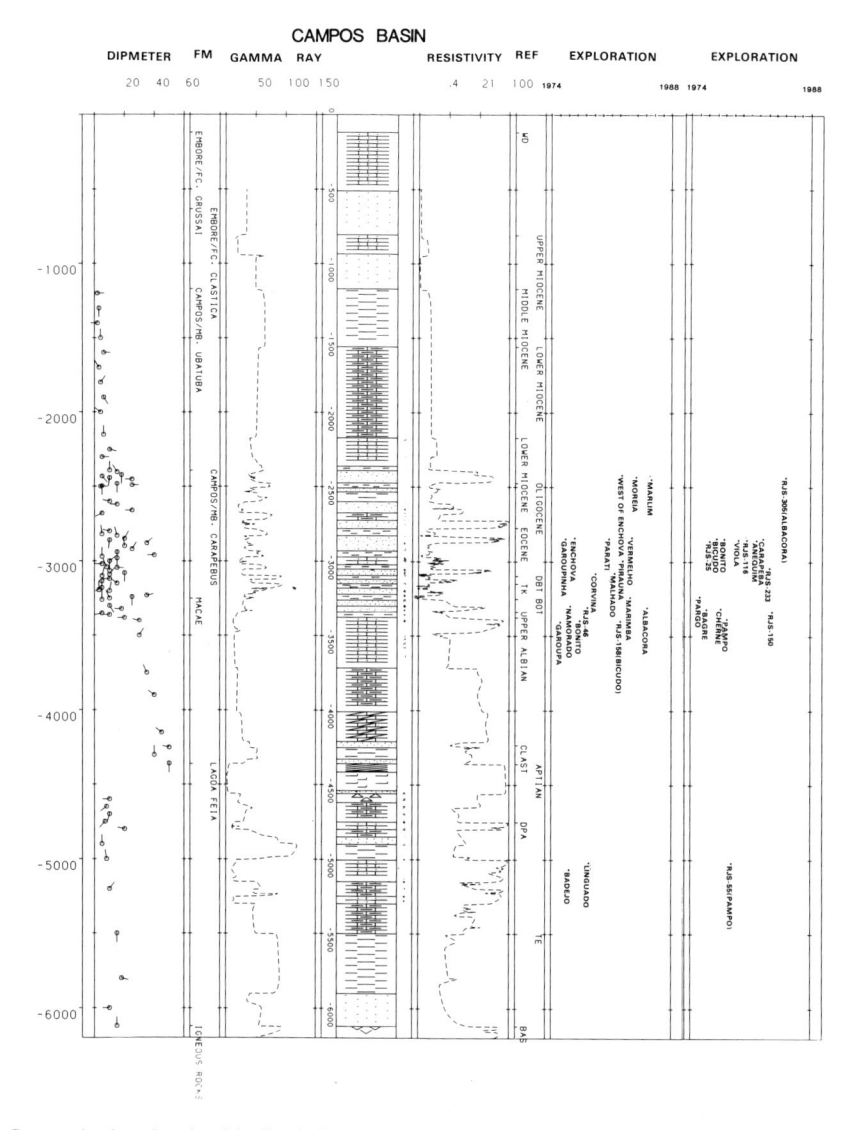

Fig. 8. Composite log showing idealized dipmeter, gamma-ray and resistivity readings in the stratigraphic column with main hydrocarbon accumulations and the results of the exploration effort. In the exploration column, the horizontal axis corresponds to the year of discovery, and the vertical axis corresponds to the stratigraphic level of the producing reservoir.

from 30 to 20°C km^{-1} and no decompaction, is shown in Fig. 9(B). The burial history plots show the dominance of the rifting episode, during which a large proportion of the sedimentary column of the basin was deposited very rapidly with a higher geothermal gradient, resulting in the onset of oil generation in the Albian for the lowermost stratigraphic layers (Neocomian), employing the optimistic geothermal gradient with decompaction of the sedimentary layers. Alternatively, the use of the same geothermal gradient and present-day thicknesses results in maturation occurring very late in the basin history, and the gas window not being reached.

There was limited sediment accumulation in the basin from about 100 Ma to about 50 Ma. The rock units corresponding to the Upper Cretaceous and Lower Tertiary are either absent or represented by very small thickness,

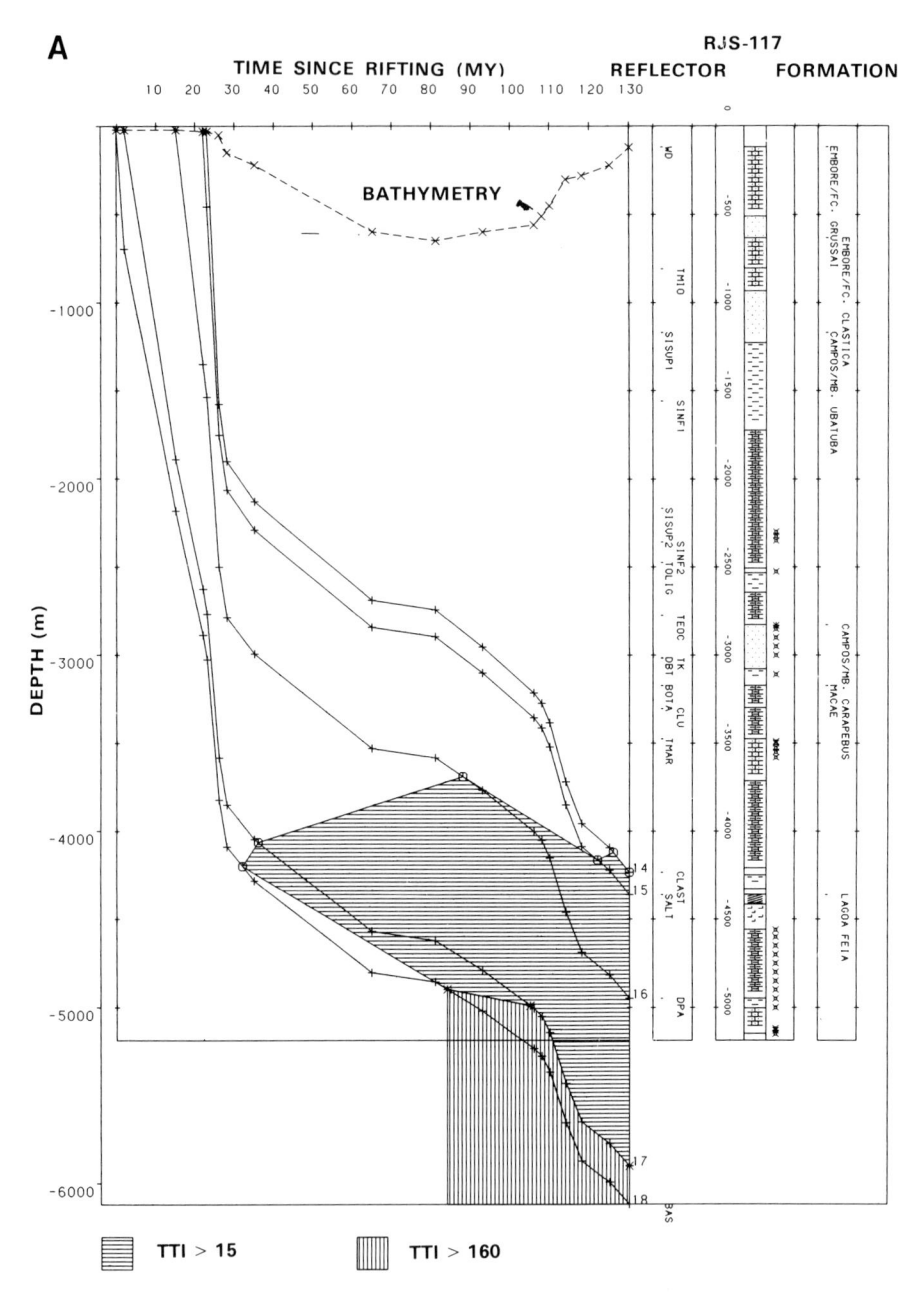

Fig. 9. Subsidence history against maturation for the borehole RJS-117 (A: geothermal gradients varying from 30 to 20°C km^{-1} and sediment decompaction; B: initial geothermal gradient 30°C km^{-1}, with no decompaction). Reflector 14 corresponds to a clastic sequence near the base of the Macaé Fm., reflector 15 corresponds to the top of the Aptian evaporites, reflector 16 is associated with the Pre-Aptian unconformity, reflector 17 is associated with a talc-stevensite section below the coquinas, and reflector 18 corresponds to the top of the Lagoa Feia basalts. Oil shows observed during drilling are also indicated to the right of the lithologic column, from 2300 m to total depth.

PETROLEUM GEOLOGY OF THE CAMPOS BASIN

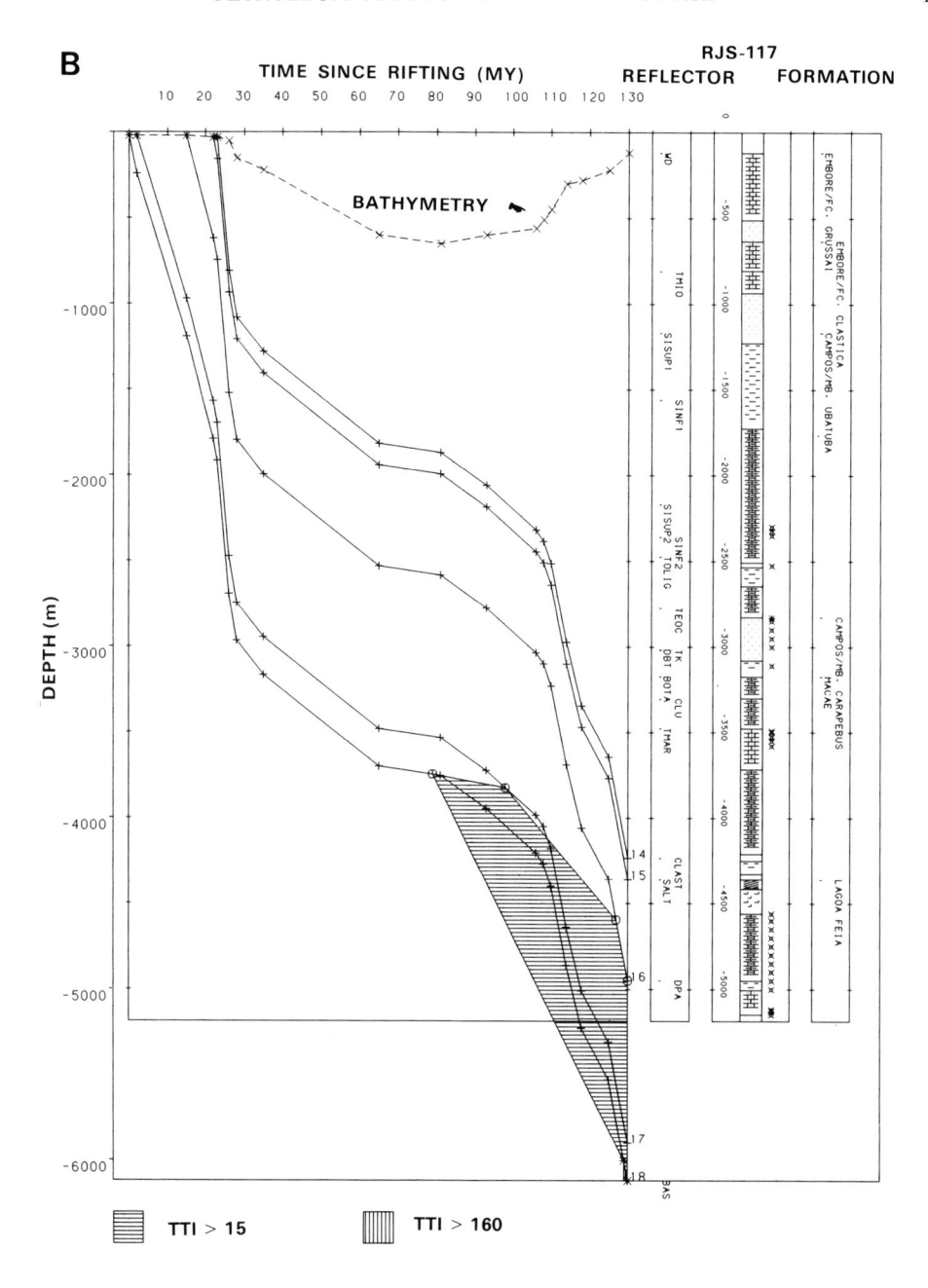

suggesting erosional or non-depositional events. This results in an almost stationary temperature value for all the sedimentary laryers accumulated in the previous phase, and consequently, the thermal maturation of the sediments did not progress significantly during this period. The modelling suggests that no sedimentary layer younger than about 100 Ma is capable of generating significant quantities of hydrocarbons.

Detailed geochemical analysis on samples from the basin, including source rock characterization, oil–oil and oil–source rock correlations for a suite of nearly 20 oils and forty five selected source rock samples has been done. The study has confirmed that almost all the hydrocarbon accumulations discovered to date in the Campos basin are sourced mainly from lacustrine calcareous black shales (Pereira 1982; Pereira *et*

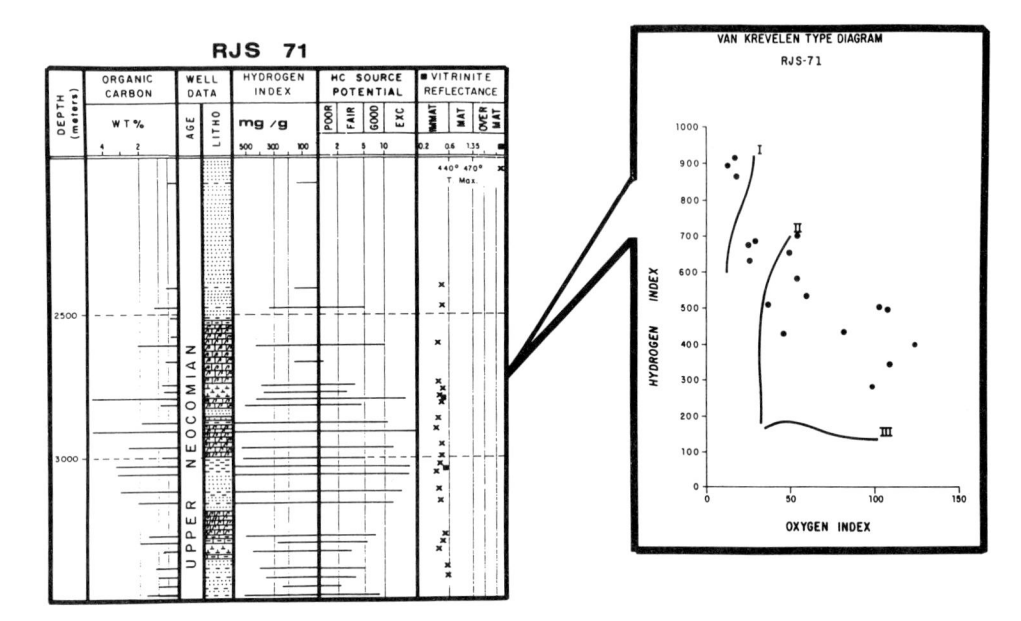

Fig. 10. Geochemical log of a typical well from the Campos Basin (RJS-71), showing the stratigraphic position of the lacustrine saline source rock from the Late Neocomian. The Van Krevelen diagram shows the hydrogen index (S_2/TOC) against oxygen index (S_3/TOC).

al. 1984; Figueiredo et al. 1983; Meister 1984; Mello et al. 1984, 1988a,b), deposited in a closed and shallow Upper Neocomian lake system with saline to hypersaline waters of alkaline affinities (Castro & Azambuja 1980; Bertani & Carozzi 1985; Mello et al. 1986).

Specifically, two source rock systems are present in the Campos basin: Lower Neocomian black shales and marls deposited in a lacustrine environment ranging from brackish to hypersaline water, and an Upper Neocomian system comprising mainly calcareous black shales and marls deposited in lacustrine saline water environment of alkaline affinities (Fig. 10). The younger sedimentary successions, in general, are not considered as source rocks, and are only discussed briefly as follows.

(i) The Aptian sequence consists mainly of thin layers of marine black shales generally intercalated with thicker sections of evaporites. Although these black shales contain good hydrocarbon source potential in some areas of the basin, they are thin and discontinuous. Given the low maturity in all boreholes analyzed, they are not considered as significant oil contributors to the hydrocarbon accumulations in the Campos basin (e.g. Pereira 1982; Pereira et al. 1984).

(ii) The Albo–Cenomanian sequence is mainly composed of marls and calcareous mudstones possessing low organic carbon content (generally less than 1.0%). The predominance of type-III kerogen reflects the relatively shallow, oxygenated, marine shelf environment of deposition for the sediments of this sequence. In some parts of the basin, a deepening of the environment of deposition is recorded by a succession of Upper Albian/Cenomanian thin layers of marls with high organic carbon contents (up to 3%), with good hydrocarbon source potential, made up mainly of type-II kerogen. Nevertheless, the lack of sufficient thermal maturity indicates that they have not significantly contributed to the oil accumulations in the basin (Pereira et al. 1984).

(iii) The Upper Cretaceous to Tertiary sedimentary succession is composed mainly of shales and calcareous mudstones deposited in open marine neritic to bathyal conditions (Koutsoukos 1984). The low organic carbon content, together with the low values of hydrogen index and hydrocarbon source potential of these sediments, suggest that highly oxygenated conditions prevailed during their deposition in most of the basin (Mello et al. 1984; Pereira et al. 1984). Evidence of local anoxic conditions has been detected in the Santonian–Coniacian sedimentary succession, with the presence of thin calcareous mudstone layers possessing moderate organic carbon content, mainly com-

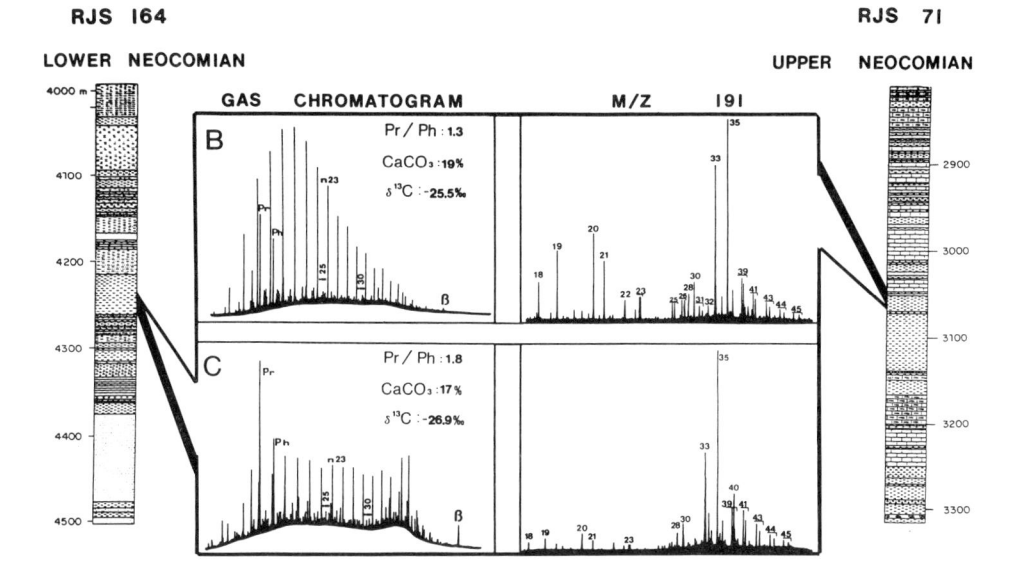

Fig. 11. Lithological logs of the Lower (RJS-164) and Upper (RJS-71) Neocomian succession, depicting the stratigraphic position of the organic-rich samples RJS-164 (4260 m) and RJS-71 (3060 m), for which gas chromatograms of total alkanes, bulk and elemental parameters, and partial m/z 191 mass chromatograms are shown.

prising type-II kerogen (Mello et al. 1988b) However, these sediments remain generally immature in the basin, due to a combination of low geothermal gradients with shallow burial, and are not source rocks (Mello et al. 1984; Pereira et al. 1984; Mello et al. 1988b).

Figure 10 shows a typical geochemical log (well RJS-71) with the stratigraphic position of one important Upper Neocomian organic-rich horizon, chosen as a specific example of the source rock system that has been identified in the Campos basin. This horizon mainly comprises well laminated calcareous ($CaCO_3$ from 5–59%) black shales, very rich in organic matter (TOC up to 9%), with low sulphur content (< 0.3%). The hydrogen and oxygen indices (up to 970 and 70 mg HC g^{-1} of TOC and CO_2, respectively), and organic petrology data identify the organic matter as being mainly Type-I kerogen composed of lipid-rich material. The excellent hydrocarbon source potential of these sediments (S$_2$ from Rock–Eval pyrolysis up to 38 Kg HC ton^{-1} of rock), combined with appropriate maturation conditions indicates good source rock characteristics within this sedimentary succession (Figueiredo et al. 1983; Meister 1984; Mello et al. 1984; Pereira et al. 1984; Figueiredo et al. 1983).

Figure 11 illustrates two typical lithological logs, showing the stratigraphic position of two source rock horizons (Lower Neocomian in RJS-164 and Upper Neocomian in RJS-71). Also shown are gas chromatograms and m/z 191 chromatograms of the alkane fraction of two samples from these horizons. As can be observed, there are some significant differences in the bulk geochemical data and biological marker distribution between the two. The most marked are the presence of higher concentrations of β-carotane (peak β), higher relative abundances of higher molecular weight n-alkanes, higher abundance of gammacerane (peak 40), higher pristane/phytane ratio (around 1.8), lower abundances of tricyclic terpanes (peaks 18 to 26), and lighter carbon isotopic values ($\delta^{13}C \sim -26.9‰$) for the whole extract in the sample from the Lower Neocomian. The absence of C_{30} steranes and dinosterane isomers (see Mello et al. 1988a,b for details) in both samples, lends support to the nonmarine character of these source rocks (such compounds are considered to be diagnostic features of marine organic matter; Moldowan et al. 1985; Summons et al. 1987; Goodwin et al. 1988), and suggest that the Lower Neocomian source rocks in the analyzed samples were deposited in a lacustrine environment with higher salinity and higher plant input than the Upper Neocomian ones (cf. Mello et al. 1988a,b). Taken together, all these features

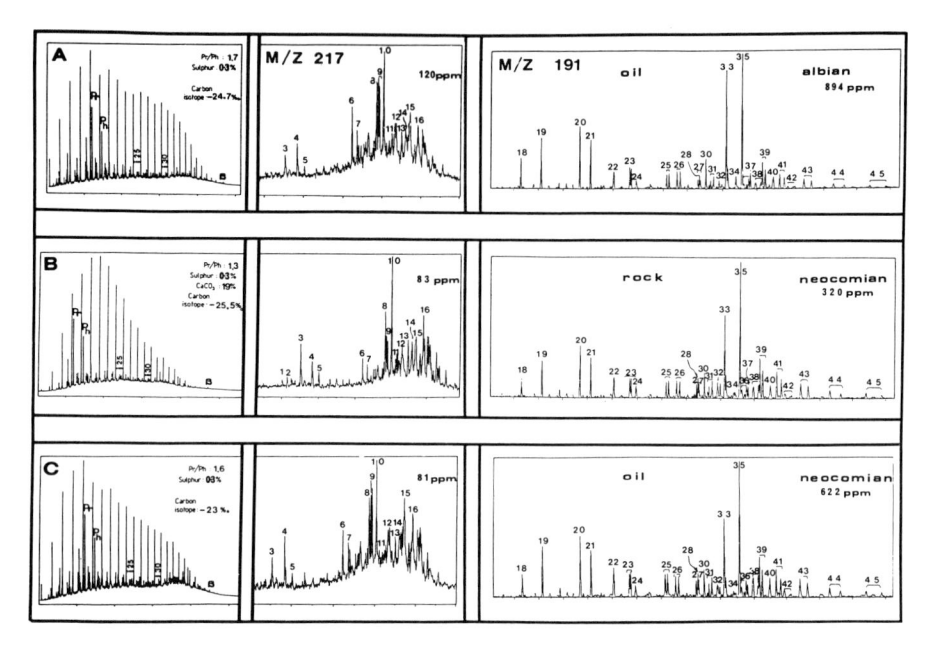

Fig. 12. Oil-source rock correlation using gas chromatograms of total alkanes, bulk and elemental parameters, and partial m/z 217 and m/z 191 mass chromatograms and absolute concentration of $C_{30}\alpha\beta$ hopane and $C_{27}\alpha\alpha\alpha$ 20S + R steranes for sample 3020 m of the well RJS-71 (B) against typical oil samples pooled in different reservoirs in the Campos Basin: Albo–Cenomanian, from the well RJS-305 (A) and Neocomian, from the well RJS-139 (C).

suggest that the Neocomian succession was deposited in a lacustrine saline environment. The differences between the Upper and Lower Neocomian samples in Fig. 11 indicate variations within the depositional environment by way of an enhanced higher plant input and higher salinity for the Lower Neocomian organic-rich sedimentary succession of well RJS-164.

Figure 12 shows gas chromatograms and m/z 217 (steranes) and m/z 191 (terpanes) chromatograms of the alkane fraction of two oils pooled in different reservoirs, along with the same chromatograms for the Upper Neocomian rock sample from the well RJS-71 (Figs 10 & 11, see Appendix). Although there are differences in the n-alkane distributions (see Fig. 13), the similarities in the bulk data and biological marker distributions and concentrations for these samples, together with other geochemical characteristics of a range of sediments from the Upper Neocomian (Mello et al. 1988a), indicate that this sedimentary succession is the major contributor to the hydrocarbon accumulations that have been discovered so far in the Campos basin. This is supported by the fact that the Upper Neocomian source rocks are thicker and have better hydrocarbon source potential than the Lower Neocomian ones.

Despite the overall similarities in the bulk geochemical and biological marker features in the Upper Neocomian source rocks, differences in the molecular properties do exist for samples occurring at different horizons in the succession (see, for example, Upper and Lower Neocomian chromatograms in Fig. 11). Examples from three organic-rich sediments, with similar vitrinite reflectance values, are shown in Fig. 13. As can be observed, there are differences in the n-alkane distribution, β-carotane concentration (peak β), gammacerane abundance (peak 40), Pr/Ph ratio, carbonate content and the tricyclic terpane relative abundances (peaks 18–26; cf. Mello et al. 1988a,b for details). Since these samples share similar maturity, the observed differences suggest fluctuations in the depositional environment of the Upper Neocomian. Indeed, sedimentological and mineralogical studies (e.g. Castro & Azambuja 1980; Bertani & Carozzi 1985), and carbon and oxygen isotopic data (Takaki & Rodrigues 1984) lend support to the presence of such contractions and expansions of the lake system during the deposition of the Upper Neocomian organic-rich sediments. The block diagram in Fig. 14 shows a schematic illustration of the main sedi-

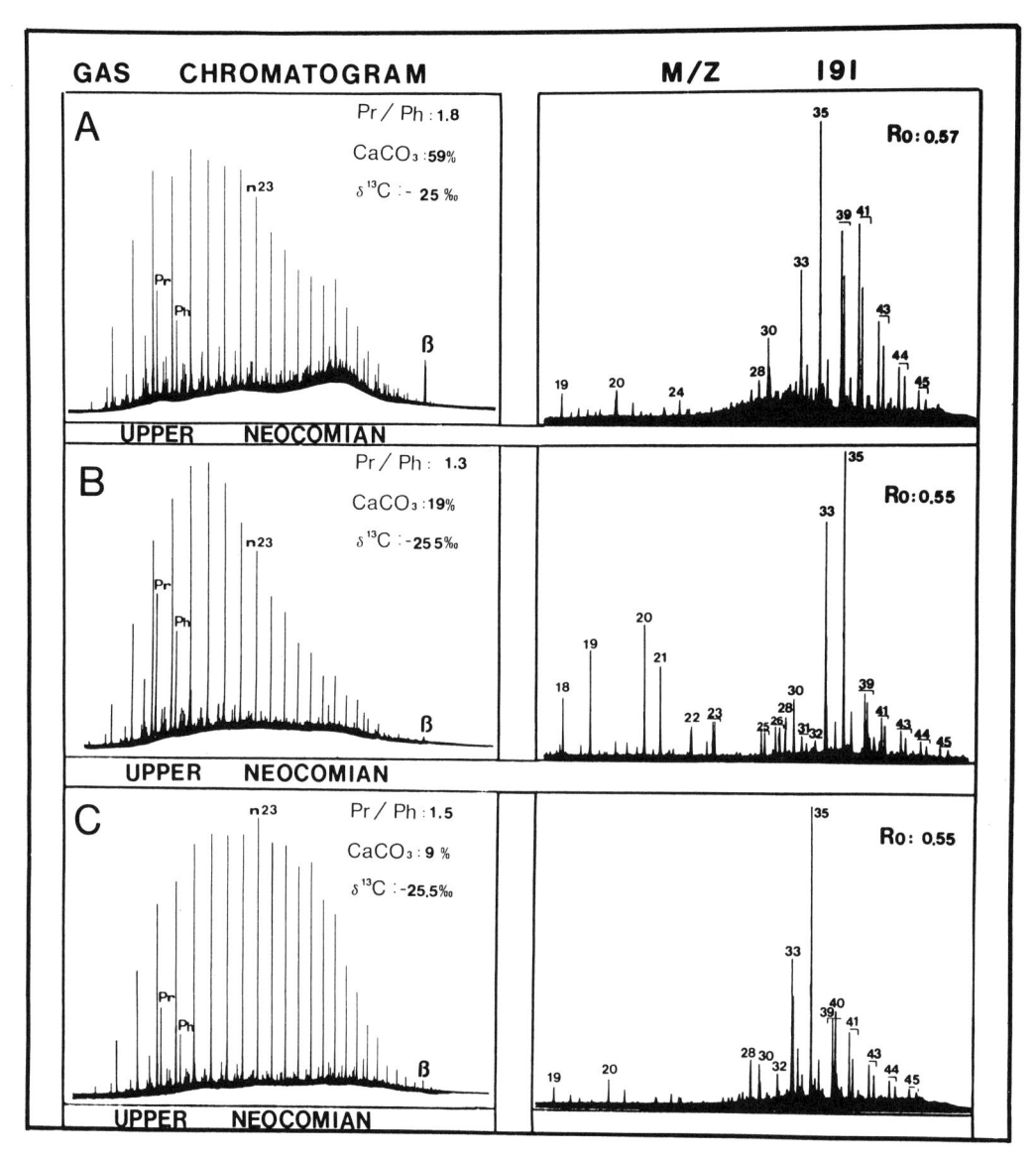

Fig. 13. Gas chromatograms of total alkanes, bulk and elemental parameters, and partial m/z 191 mass chromatograms and vitrinite reflectance (% Ro) for sample RJS-71 (B, see also Figs 10 & 11), and two other source rocks from different horizons (A & C) in the Upper Neocomian sedimentary succession of the Campos Basin.

mentological facies of a shallow saline lake system of alkaline affinity, the proposed palaeoenvironment of deposition for the Campos Basin during the Late Neocomian rift stage.

Present-day lacustrine systems such as that generally occur in areas of high evaporation (semi-arid/moist climates). The high amount of nutrients available in the saline waters, generally associated with perenial alkaline springs, enhances the development of well adapted, limited species that, without competition, show prolific productivity, resulting in a high input of algal and bacterial organic matter within the lake. The differences in salinity between an upper aerobic, less saline layer and a lower anaerobic, very saline (with higher density) and alkaline layer enhance the water column stability, leading to stratification and permanent bottom water

LACUSTRINE SALINE WATER ENVIRONMENT

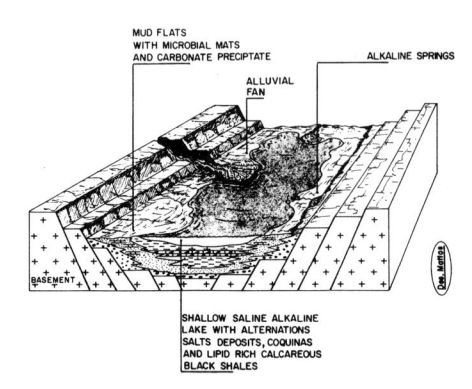

Fig. 14. Idealized block diagram showing lacustrine facies distribution in the Neocomian Lagoa Feia Formation (modified from Williams *et al.* (1986)).

anoxia. These conditions, although enhancing anaerobic bacterial activity, are lethal for microfauna and benthic organisms. Low sulphate concentrations (alkaline character), associated with extreme anoxic conditions in the bottom waters, enhance the degree of organic matter preservation, resulting in the deposition of well laminated, organic-rich calcareous black shales (Demaison & Moore, 1980; Kelts 1988; Decker 1988). Modern analogues of this lacustrine system appear to be lakes Nakuru, Magadi and Bogoria in the Eastern rift system (Eugster 1986; Vincens *et al.* 1986; Degens & Michaelis 1988; Talbot 1988).

Few comparable examples of ancient saline lake systems have been reported in the literature. The best comparisons with the Campos Basin analogue appear to be the well studied Eocene Green River Formation in Uinta Basin, USA (e.g., Tissot *et al.* 1978; Dean & Fouch 1983), the Chaidamu basin in China (Powell 1986), and the Officer basin in Australia (McKirdy *et al.* 1986).

Pathway models of hydrocarbon migration and accumulation

This section will briefly describe the petroleum geology of the main accumulations in the basin in terms of regional features. The detailed description of each field is beyond the scope of this paper, so that only general models will be presented and discussed. It is noteworthy that although the source rocks occur only in the Neocomian sedimentary succession, the greatest share of the hydrocarbon reserves is associated with accumulations in the Albian and post-Albian reservoirs (Barros 1980). The production from the accumulations in the Neocomian coquinas and basalts, which are stratigraphically closer to the source, is relatively meagre, if compared with other basins where the majority of the oil accumulations are stratigraphically close to the source rock. In the Norwegian sector of the North Sea, for example, the Upper Jurassic Kimmeridge Clay Formation is responsible for the majority of the oil generated, while Jurassic sands are responsible for 70% of all recoverable petroleum reserves (Faerseth *et al.* 1986).

The hydrocarbon plays in the Campos basin can be genetically classified by their structural style of trapping and by the age of the reservoir rocks (Marroquim *et al.* 1984). There is a strong link between the trapping mechanism and the tectonic evolution of the basin. The generalized reservoir facies distribution is shown in Fig. 15, and the hydrocarbon plays are shown in Fig. 16. Figure 17 shows a schematic cross section in the basin with the stratigraphic interval and trapping style of the main accumulations. Several fields show multi-storeyed trapping styles, with an oil column composed of different stratigraphic horizons contributing to the oil production.

The Neocomian reservoirs are mainly associated with structural highs in the southern part of the basin, where the coquina lenses are characteristically affected by normal faults with small offsets and the stratigraphic control is provided by the pinch out of the reservoir in the direction of the structural highs and by facies changes towards deeper parts of the basin. In the Badejo area, the basalts are also oil-producing due to open fractures and vesicles caused by dissolution of the calcite that originally filled amygdales (Pimentel and Gomes 1982). These igneous rocks usually have negligible permeability, but, where affected by microfractures, they may have developed brecciated zones with interconnected porosity.

The limestone reservoirs in the Bonito, Bicudo and Pampo fields also show a strong structural control associated with the listric normal faults that created roll-overs and draping on salt that has been partially mobilized. These roll-overs and local highs also provide stratigraphic control by favouring the accumulation of a more porous facies for the limestones.

Isopach maps for the Albo−Cenomanian and the Upper Cretaceous turbidites clearly show the channelled distribution of the reservoirs

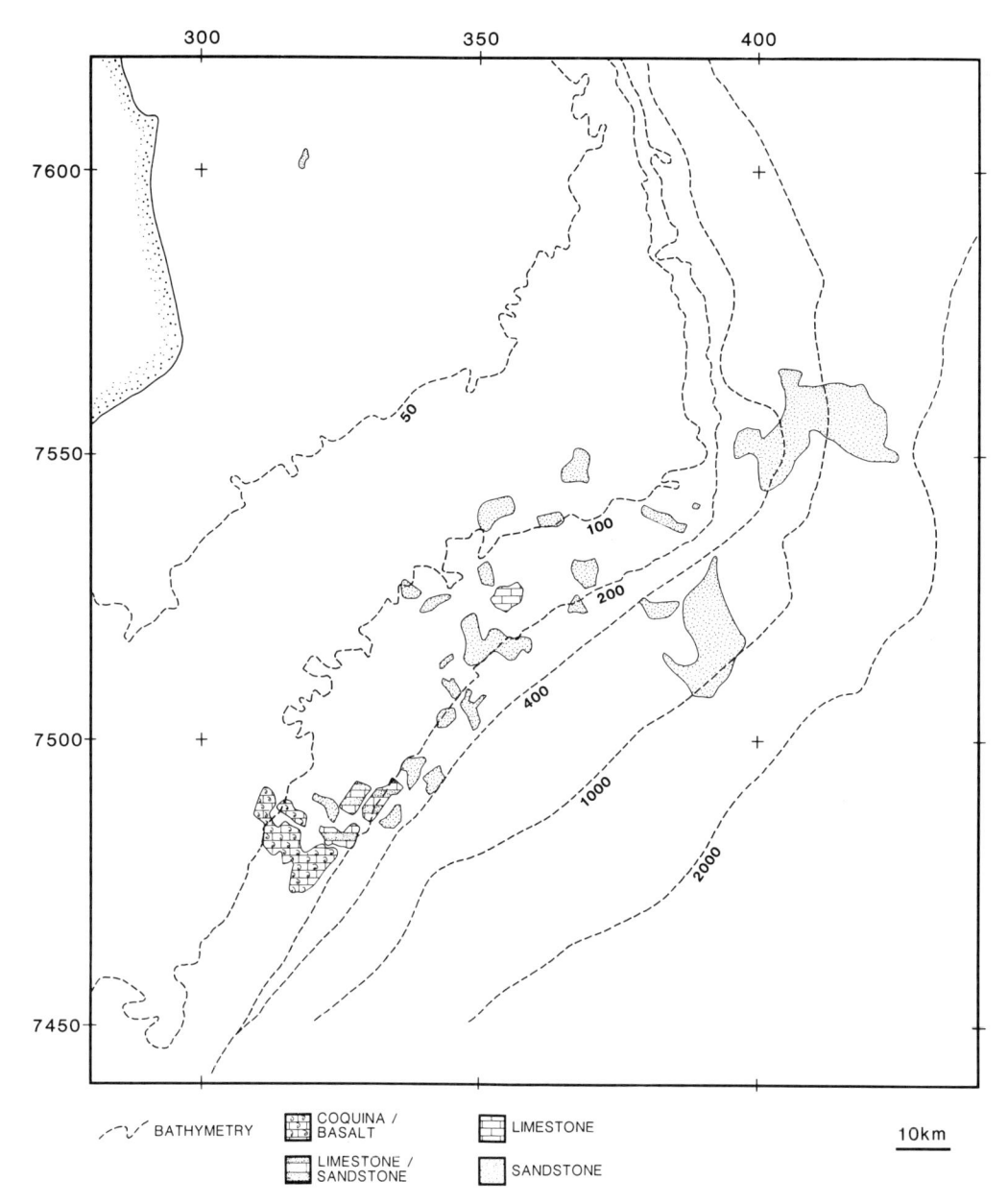

Fig. 15. Generalized map showing reservoir types for the main hydrocarbon accumulations in the Campos Basin. See Fig. 16 for identification of oil fields.

associated with local lows caused by salt movement (Figueiredo & Mohriak 1984).

The Lower Tertiary and Eocene sandstone turbidites are more widespread in the basin and usually blanket the basal Tertiary unconformity. However, they also show local control by draping residual salt highs. The stratigraphic control is provided by pinch-outs towards local highs and the paleo-slope and compactional effects associated with draping residual salt domes and less compacted lithologies such as porous calcarenites. The Oligo–Miocene sandstones are associated with turbiditic deposition in two different situations: locally, they are present in confined troughs — the Enchova Canyon is one example — and areally, they occur as sheets in

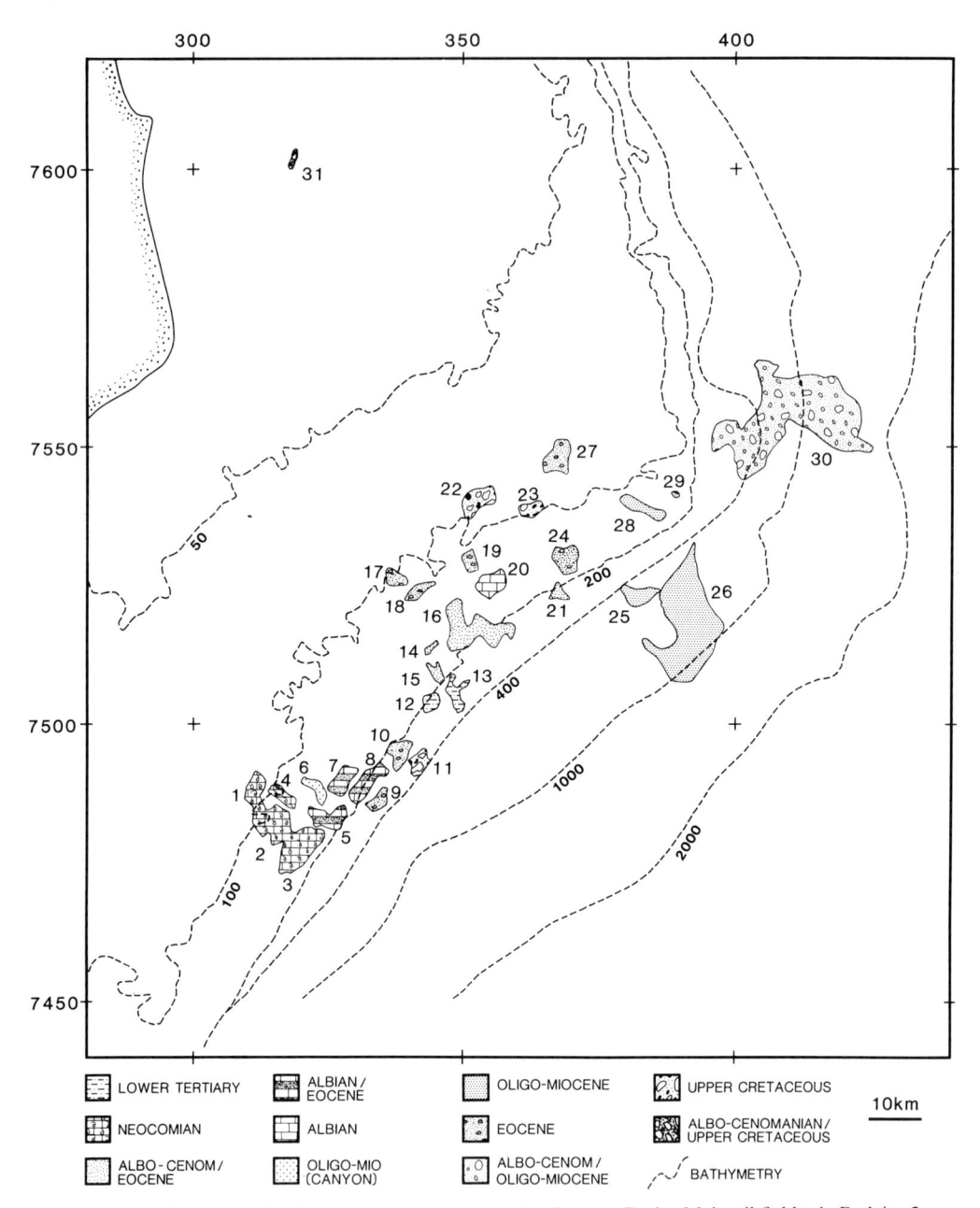

Fig. 16. Generalized map showing hydrocarbon plays in the Campos Basin. Main oil fields: 1. Badejo; 2. Linguado; 3. Pampo; 4. Trilha; 5. Bicudo; 6. Enchova Oeste; 7. Enchova; 8. Bonito; 9. RJS-116; 10. Pirauna; 11. Marimbá; 12. Corvina; 13. Malhado; 14. RJS-046; 15. Bagre/Cherne; 16. Namorado; 17. Parati; 18. Anequim; 19. Garoupinha; 20. Garoupa; 21. RJS-211; 22. Carapeba; 23. Pargo; 24. Viola; 25. RJS-377; 26. Marlim; 27. Vermelho; 28. Moréia; 29. RJS-251; 30. Albacora; 31. RJS-150.

the northeast area of the basin, constituting the reservoir for the Marlim giant field and also occurring in the Albacora giant field. They are associated with roll-overs near the shelf-edge, which is a region of marked instability in the northeast area of the basin, with pinch-outs towards residual highs and on the unconformity surface of the paleo-slope.

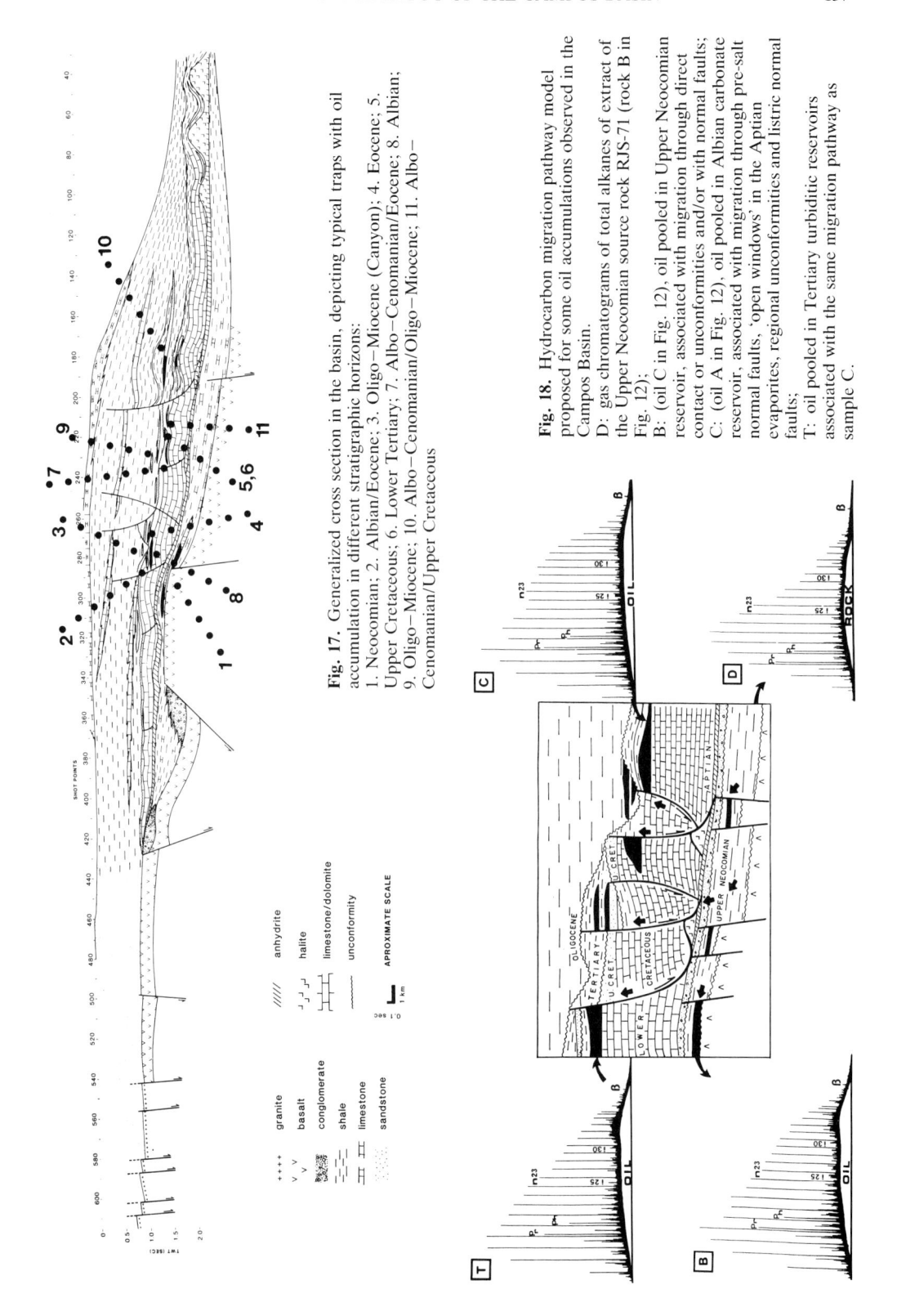

PETROLEUM GEOLOGY OF THE CAMPOS BASIN 137

Fig. 17. Generalized cross section in the basin, depicting typical traps with oil accumulation in different stratigraphic horizons:
1. Neocomian; 2. Albian/Eocene; 3. Oligo–Miocene (Canyon); 4. Eocene; 5. Upper Cretaceous; 6. Lower Tertiary; 7. Albo–Cenomanian/Eocene; 8. Albian; 9. Oligo–Miocene; 10. Albo–Cenomanian/Oligo–Miocene; 11. Albo–Cenomanian/Upper Cretaceous

Fig. 18. Hydrocarbon migration pathway model proposed for some oil accumulations observed in the Campos Basin.
D: gas chromatograms of total alkanes of extract of the Upper Neocomian source rock RJS-71 (rock B in Fig. 12);
B: (oil C in Fig. 12), oil pooled in Upper Neocomian reservoir, associated with migration through direct contact or unconformities and/or with normal faults;
C: (oil A in Fig. 12), oil pooled in Albian carbonate reservoir, associated with migration through pre-salt normal faults, 'open windows' in the Aptian evaporites, regional unconformities and listric normal faults;
T: oil pooled in Tertiary turbiditic reservoirs associated with the same migration pathway as sample C.

Figure 18 shows an idealized diagram of a hydrocarbon pathway model of migration and entrapment, which shows how hydrocarbons migrated from the pre-salt source rocks (e.g. sample D in Fig. 18, B in Fig. 12) to the pre-salt, Upper Neocomian (e.g., sample B in Fig. 18, sample B in Fig. 11) and younger and shallower reservoirs (Albian, e.g. sample C in

Table 3.

Pr–	2, 6, 10, 14–tetramethylpentadecane (pristane).
Ph–	2, 6, 10, 14–tetramethylhexadecane (phytane).
i–C_{25}	2, 6, 10, 14, 18–pentamethyleicosane (regular).
i–C_{25}	2, 6, 10, 15, 19–pentamethyleicosane (irregular).
i–C_{30}	squalane.
β	β-carotane.
1–	13β (H), 17α (H)-diapregnane (C_{21})
2–	5α (H), 14β (H), 17α (H)-pregnane (C_{21})
3–	5α (H), 14β (H) 17β (H) + 5α (H), 14α (H), 17α (H)-pregnane (C_{21}).
4–	4-methyl-5α, 14β (H), 17β (H) + 4-methyl-5α (H), 14α (H), 17α (H) homopregnane (C_{22})
5–	5α (H), 14β (H), 17β (H) + 5α (H), 14α (H), 17α (H) – homopregnane (C_{22})
6–	13β (H), 17α (H)-diacholestane, 20S (C_{27}-diasterane).
7–	13β (H), 17α (H)-diacholestane, 20 R (C_{27}-diasterane).
8–	5α (H), 14α (H), 17α (H), 20S (C_{27}-cholestane).
9–	5α (H), 14β (H), 17β (H), 20R (C_{27}-cholestane).
10–	5α (H), 14α (H), 17α (H), 20R (C_{27}-cholestane).
11–	5α (H), 14α (H), 17α (H), 20S (C_{28}-methylcholestane).
12–	5α (H), 14β (H), 17β (H), 20R + 20S (C_{28}-methylcholestane).
13–	5α (H), 14α (H), 17α (H), 20R (C_{28}-methylcholestane).
14–	5α (H), 14α (H), 17α (H), 20S (C_{29}-ethylcholestane).
15–	5α (H), 14β (H), 17β (H), 20R + 20S (C_{29}-ethylcholestane).
16–	5α (H), 14α (H), 17α (H), 20R (C_{29}-ethylcholestane).
17–	C_{19} tricyclic terpane
18–	C_{20} tricyclic terpane
19–	C_{21} tricyclic terpane
20–	C_{23} tricyclic terpane
21–	C_{24} tricyclic terpane
22–	C_{25} tricyclic terpane
23–	C_{26} tricyclic terpanes
24–	C_{24} tetracyclic (Des-E)
Te–	C_{24} tetracyclic (Des-A)
25–	C_{28} tricyclic terpanes
26–	C_{29} tricyclic terpanes
27–	C_{25} tetracyclic
28–	C_{27} 18α (H)-trisnorneohopane (Ts).
29–	C_{30} tricyclic terpanes
T–	C_{27}, 25, 28, 30-trisnorhopane
30–	C_{27} 17α (H)-trisnorhopane (Tm).
31–	C_{31} tricyclic terpanes
32–	17α (H), 18α (H), 21β (H)-28, 30-bisnorhopane (C_{28}).
N–	25-norhopane (C_{29})
33–	C_{29} 17α (H), 21β (H)-norhopane.
34–	C_{29} 17β (H), 21α (H)-norhopane.
35–	C_{30} 17α (H), 21β (H)-hopane.
36–	C_{33} tricyclic terpanes
37–	C_{30} 17β (H), 21α (H)-hopane.
38–	C_{34} tricyclic terpanes
39–	C_{31} 17α (H), 21β (H)-homohopane (22S + 22R).
40–	C_{30} gammacerane.
41–	C_{32} 17α (H), 21β (H)-bishomohopane (22S + 22R).
42–	C_{35} tricyclic terpanes
43–	C_{33} 17α (H), 21β (H)-trishomohopane (22S + 22R).
44–	C_{34} 17α (H), 21β (H)-tetrakishomohopane (22S + 22R).
45–	C_{35} 17α (H), 21β (H)-pentakishomohopane (22S + 22R).

Figs 18 and 12) and Tertiary (e.g. sample T in Fig. 18). The important hydrocarbon accumulations found in the fractured basement, conglomerates and coquina reservoirs of the pre-salt stage are associated with migration through direct contact or through unconformities associated with normal faults. The oils pooled in the marine sequence (Albian, Upper Cretaceous and Tertiary) migrated through a system associated with pre-salt normal faults, 'open windows' in the evaporitic layers (probably caused by halokinetic tectonism), listric faults and regional unconformities (Estrella et al. 1984, Figueiredo et al. 1983). The most important oil accumulations in the Campos Basin are associated with deep water fans distributed in the stratigraphic column from the Late Cretaceous to the Late Tertiary. They occur as widespread sheets and are also enclosed in submarine canyons. The coalescent turbiditic sand bodies of the Middle Eocene probably acted as a 'hydrocarbon collector' system allowing the migration of the oil arising from the growth fault systems or unconformities.

Conclusions

The Campos Basin shows an almost continuous subsidence from the Early Neocomian to the Recent. The basin evolution is characterized by three tectono-stratigraphic environments: rift, with lacustrine sedimentation, transitional, with evaporitic and carbonatic deposits, and drift, with marine sedimentation. The geodynamic model for the basin involves stretching of continental crust, with an initial phase of fault-controlled subsidence, followed by a subsequent phase of thermal subsidence. The oil accumulations are distributed throughout the stratigraphic column, from the Early Neocomian to the Miocene. The majority of the fields are associated with turbiditic sands deposited in a deep marine environment. The source rocks for these accumulation are Neocomian calcareous black shales.

We gratefully acknowledge Petrobrás support for this study. We thank the board of directors and the Department of Exploration Superintendent, Dr. Milton R. Franke, for permission to publish this paper. Petrobrás Research Center provided all the elemental and bulk analysis, and NERC GC−MS facilities (GR3/2951 and GR3/3758) have been used. We are grateful for Mrs. A. P. Gowar's and Miss L. Dias's advise during analytical work. We thank Mr. N. C. de Azambuja Filho and E. Koutsoukos for their helpful comments, and Dr. H. Reading for the reading of the first draft of this paper.

Appendix. Experimental and analytical procedures

All the samples were submitted to bulk and elemental analysis according to procedures described previously (Mello et al. 1988a,b). The aliphatic hydrocarbons were analysed by gas chromatography (GC) employing a HP 5880 chromatograph equipped with a 30 m, 0.25 mm i.d. fused silica DB-1 column. A temperature program of $50-300°C$ at $6°C$ min^{-1} was used. GC−MS analyses were carried out using two different systems: (i) Finningan 4000 spectrometer coupled to a Carlo Erba 5160 gas chromatograph equipped with on-column injector, and fitted with a 60 m DB-1701 column. A temperature program of $50-310°C$ at $5°C$ min^{-1} was used. Data were acquired and processed using an Incos 2300 data system; (ii) HP 5970 mass selective detector (MSD) coupled to a HP 5880 gas chromatograph fitted with a fused silica 25 m, 0.25 mm DB-1 column. A temperature program of $50-300°C$ at $6°C$ min^{-1} was used. Quantitation of biological markers (ppm of extract or oil) was obtained according to the procedures described previously (Mello et al. 1988a,b). Stable isotope analyses for carbon on whole oil and extracts were undertaken using a vaccum combustion line linked to a high resolution Varian MAT-230 instrument. The data are presented in delta-notation relative to Pee Dee Belemnite (PDB).

References

AMARAL, G. et al. 1966. Potassium argon dates of basaltic rocks from southern Brazil. *Geochimica et Cosmochimica Acta*, **30**, 159−189.
—— & PORTO, R. 1972. Classificação das bacias sedimentares brasileiras segundo a tectônica de placas. *Anais do XXVI Congresso Brasileiro de Geologia*, **2**, 67−90.
ASMUS, H. E. 1975. Controle estrutural da deposição mesozóica nas bacias da margem continental brasileira. *Revista Brasileira de Geociências*, **5**, 160−175.
—— 1982. Significado geotectônico das feições estruturais das bacias marginais brasileiras e áreas adjacentes. *Anais do XXXII Congresso Brasileiro de Geologia*, **4**, 1547−1557.
—— & GUAZELLI, W. 1981. Descrição sumária das estruturas da margem continental brasileira e das áreas oceânicas e continentais adjacentes. *Projeto Remac*, **9**, 187−269.
—— & PONTE, F. C. 1973. The Brazilian marginal basins *In*: NAIR, A. E. & STEHLI, F. G. (eds) *The ocean basins and margins: vol 1, The South Atlantic*, Plenum, New York, 87−132.
—— & —— 1980. Diferenças nos estágios iniciais da evolução da margem continental brasileira:

possíveis causas e implicações. *Anais do XXXI Congresso Brasileiro de Geologia*, **1**, 225–239.

BARROS, M. C. 1980. Geologia e recursos petrolíferos da Bacia de Campos. *Anais do XXXI Congresso Brasileiro de Geologia*, **1**, 254–265.

BERTANI, R. T. & CAROZZI, A. V. 1985. Lagoa Feia Formation (Lower Cretaceous) Campos Basin offshore Brazil: rift-valley stage carbonate reservoirs — I and II. *Journal of Petroleum Geology*, **8**, 37–58, 199–220.

BEURLEN, G. 1981. *Correlação das unidades geocronológicas e biostratigráficas com os ciclos globais de variação do nível do mar*. Petrobrás, Tabela Interna, Depex/Labor/Sepale.

CAMPOS, C. W. M. 1970. Exploração de petróleo na plataforma continental brasileira. *Boletim Técnico Petrobrás*, **13**(3/4), 95–114.

——, PONTE, F. C. & MIURA, K. 1974. Geology of the Brazilian continental margin *In*: BURK, C. A. & DRAKE, C. L. (eds) *The geology of continental margins*. Springer-Verlag, New York, 447–461.

CASTRO, J. & AZAMBUJA, N. C. 1980. Facies, análise estratigráfica e reservatórios da Fm. Lagoa Feia, Cretáceo Inferior da bacia de Campos. *Petrobrás Internal Report*.

CHEN CHANGMING et al. 1984. Depositional models of Tertiary rift basins, Eastern China, and their application to petroleum prediction. *Sedimentary Geology*, **40**, 73–88.

CORDANI, U. G., et al. 1972. Idades Potássio-Argônio da região leste brasileira. *Anais do XXVI Congresso Brasileiro de Geologia*, 71–75.

DEAN, W. & FOUCH, T. D. 1983. Lacustrine environment *In*: SCHOLLE, P. A. et al. (eds) *Carbonate depositional environments*. AAPG Memoir **33**, 97–130.

DEGENS, E. T. & MICHAELIS, W. 1988. Diagenesis of lacustrine organic matter: Black Sea, East African lakes, Messel *In*: FLEET, A. J., KELTS, K. & TALBOT, M. (eds) *Lacustrine petroleum source rocks*, Geological Society, London, Special Publication **40**.

DEDEKKER, P. 1988. Large Australian lakes during the last 20 million years: sites for petroleum *In*: FLEET, A. J., KELTS, K. & TALBOT, M. (eds) *Lacustrine petroleum source rocks*, Geological Society, London, Special Publication, **40**, 45–58.

DEMAISON, G. J. & MOORE, G. T. 1980. Anoxic environments and oil source bed genesis. *AAPG Bulletin*, **64**, 1179–1209.

ESTRELLA, G. O. 1972. O estágio 'rift' nas bacias marginais do leste brasileiro. *Anais do XXVI Congresso Brasileiro de Geologia*, **3**, 29–34.

—— et al. 1984. The Espírito Santo Basin — Brazil: source rock characterization and petroleum habitat *In*: DEMAISON, G. & MOORE, T. (eds) *Petroleum geochemistry and basin evaluation*. *AAPG Memoir* **35**, 253–271.

EUGSTER, H. P. 1986. Lake Magadi, Kenya: a model for rift valley hydrochemistry and sedimentation? *In*: FROSTICK, L. E., RENANT, R. W., REID, I. & TIERCELIN, J. J. (eds) *Sedimentation in the African Rifts*. Geological Society, London, Special Publication, **25**, 177–191.

FAERSETH, R. B., OPPEBOEN, K. A. & SAEBOE, A. 1986. Trapping styles and associated hydrocarbon potential in the Norwegian North Sea *In*: HALBOUTY, M. T. (ed.) *Future Petroleum Provinces of the World*. AAPG Memoir **40**, 585–597.

FIGUEIREDO, A. M. F. et al. 1983. *Fatores que controlam a ocorrência de hidrocarbonetos na Bacia de Campos (com ênfase nos arenitos turbidíticos)*. Petrobrás Internal Report, Depex/Dirsul, Rio de Janeiro.

—— & MOHRIAK, W. U. 1984. A tectônica salífera e as acumulações de petróleo da Bacia de Campos. *Anais do XXIII Congresso Brasileiro de Geologia*, 1380–1394.

FURLONG, K. P. & FOUNTAIN, D. M. 1986. Lithospheric evolution with underplating: thermal considerations and seismic-petrologic consequences. *Journal of Geophysical Research*, **91**, 8285–8294.

GOODWIN, N. S., MANN, A. L. & PATIENCE, R. L. 1988. Structure and significance of C_{30}-4-Methyl Steranes in lacustrine shales and oils. *Organic Geochemistry* **12**, 495–506.

HEROUX, Y., CHAGNON, A. & BERTRAND, R. 1979. Compilation and correlation of major thermal maturation indicators. *AAPG Bulletin*, **63**, 986–996.

HINTE, J. E. 1978. Geohistory analysis — application of micropaleontology in exploration geology. *AAPG Bulletin*, **62**, 201–222.

—— 1982. Synthetic seismic sections from biostratigraphy *In*: WATKINS, J. S. & DRAKE, C. L. (eds) *Studies in continental margin geology*. AAPG Memoir **34**, 675–685.

KELTS, K. 1988. Environments of deposition of lacustrine petroleum source rocks: an introduction *In*: FLEET, A. J., KELTS, K. & TALBOLT, M. (eds) *Lacustrine petroleum source rocks* Geological Society, London, Special Publication, **40**, 3–26.

KOUTSOUKOS, E. A. M. 1984. Evolução paleoecológica do Albiano ao Maestrichtiano na área noroeste da bacia de Campos, Brasil, com base em foraminíferos. *Anais do XXXIII Congresso Brasileiro de Geologia*, **2**, 685–698.

LEYDEN, R. et al. 1976. South Atlantic diapiric structures. *AAPG Bulletin*, **60**, 196–212.

LOBO, A. P. et al. 1983. *Arcabouço tectônico da Bacia de Campos e áreas adjacentes*: Petrobrás Internal Report, Depex, Rio de Janeiro.

—— & FERRADAES. 1983. *Reconhecimento preliminar do talude e sopé continentais da Bacia de Campos*. Petrobrás Internal Report, Depex, Rio de Janeiro.

MARROQUIM, M., TIGRE, C. A., LUCHESI, C. F. 1984. Bacia de Campos: resultados e perspectivas. *Anais do XXXIII Congresso Brasileiro de Geologia*, 1366–1379.

McKENZIE, D. P. 1978. Some remarks on the development of sedimentary basins: *Earth and Planetary Science Letters*, **40**, 25–32.

McKIRDY, D. M. et al. 1986. Botryococcane in a new class of Australian non-marine crude oils. *Nature*, **320**, 57–59.

MEISTER, E. 1984. *A geologia histórica do petróleo na Bacia de Campos*. Petrobrás Internal Report, Depex.
MELLO, M. R., ESTRELLA, G. O. & GAGLIANONE, P. C. 1984. Hydrocarbon source potential in Brazilian marginal basins. AAPG abstract, Annual Convention, San Antonio, Texas.
—— et al. 1988a. Geochemical and biological marker assessment of depositional environment using Brazilian offshore oils. *Marine and Petroleum Geology* **5**, 205–223.
—— et al. 1988b. Organic geochemical characterization of depositional palaeoenvironments of source rocks and oils in Brazilian marginal basins *In*: MATTAVELLI, L. & NOVELLI, L. (eds) *Advances in Organic Geochemistry 1987*, 31–45.
MOHRIAK, W. U. 1984. *Geologia da borda oeste da Bacia de Campos*. Petrobrás Internal Report, Depex, Dirsul/Secasu.
—— & DEWEY, J. F. D. 1987. Deep seismic reflectors in the Campos Basin, offshore Brazil. *Geophysical Journal of the Royal Astronomical Society*, **89**, 133–140.
——, KARNER, G. D. & DEWEY, J. F. 1987. Subsidence history and tectonic evolution of the Campos Basin, offshore Brazil. abstract, AAPG, **71**, 594.
MOLDOWAN, J. M., SEIFERT, W. K. & GALLEGOS, E. J. 1985. Relationship between petroleum composition and depositional environment of petroleum source rocks. *AAPG Bulletin*, **69**, 1255–1268.
OJEDA, H. A. O. 1982. Structural framework, stratigraphy, and evolution of Brazilian marginal basins. *AAPG Bulletin*, **66**, 732–749.
PEREIRA, M. J. 1982. *Tempo e temperatura na formação do petróleo: aplicação do método de Lopatin à Bacia de Campos*. Relatório Interno Petrobrás, Depex.
——, TRINDADE, L. A. F. & GAGLIANONE, P. C. 1984. Origem e evolução das acumulacões de hidrocarbonetos na Bacia de Campos. *Anais do XXXIII Congresso Brasileiro de Geologia*, **10**, 4763–4777.
PIMENTEL, A. M. P. & GOMES, R. M. R. 1982. As rochas ígneas básicas como reservatório do Campo de Badejo, Bacia de Campos. *Anais do XXXII Congresso Brasileiro de Geologia*, **5**, 2383–2391.
PORTO, R. & ASMUS, H. E. 1976. The Brazilian marginal basins — current state of knowledge. *Anais da Academia Brasileira de Ciências*, **48**(sup), 215–240.
POWELL, T. G. 1986. Petroleum geochemistry and depositional settings of lacustrine source rocks. *Marine and Petroleum Geology*, **3**, 200–219.
ROOS, S. & PANTOJA, J. L. 1978. *Estudo Geotérmico da Bacia de Campos*. Petrobrás Internal Report, Depex.
ROWLEY, D. B. & SAHAGIAN, D. 1986. Depth-dependent stretching: a different approach. *Geology*, **14**, 32–35.
ROYDEN, L., SCLATER, J. G. & HERZEN, R. P. 1980. Continental margin subsidence and heat flow: important parameters in the formation of petroleum hydrocarbons. *AAPG Bulletin*, **64**, 173–187.
SCHALLER, H. 1973. Estratigrafia da Bacia de Campos. *Anais do XXVII Congresso Brasileiro de Geologia*, **3**, 247–258.
SCLATER, J. G. & CHRISTIE, P. A. F. 1980. Continental stretching: an explanation of the post–mid-Cretaceous subsidence of the Central North Sea. *Journal of Geophysical Research*, **85**, 3711–3739.
SLEEP, N. H. 1971. Thermal effects of the formation of Atlantic continental margins by continental break-up. *Geophysical Journal of the Royal Astronomical Society*, **24**, 325–350.
STECKLER, M. S. & WATTS, A. B. 1978. Subsidence of the Atlantic-type continental margin off New York. *Earth and Planetary Science Letters*, **41**, 1–13.
SUMMONS, R. E., VOLKMAN, J. K. & BOREHAM, C. J. 1987. Dinosterane and other steroidal hydrocarbon of dinoflagellate origin in sediments and petroleum. *Geochimica et Cosmochimica Acta*, **51**, 3075–3082.
TAKAKI, T. & RODRIGUES, R. 1984. Isótopos estáveis do carbono e oxigênio dos calcários como indicadores paleo-ambientais — Bacia de Campos, Santos e Espírito Santo. *Anais do XXXIII Congresso Brasileiro de Geologia*, **10**, 4750–4762.
TALBOT, M. R. 1988. The origins of lacustrine oil source rocks: evidence from the lakes of tropical Africa *In*: FLEET, A. J., KELTS, K. & TALBOT, M. (eds) *Lacustrine petroleum source rocks*, Geological Society, London, Special Publication, **40**, 29–44.
TISSOT, B., DERRO, G. & HOOD, A. 1978. Geochemical study of the Uinta Basin: formation of petroleum from the Green River Formation. *Geochemica and Cosmochimica Acta*, **42**, 1469–1465.
VINCENS, A., Casanova, J. & TIERCELIN, J. J. 1986. Palaeolimnology of lake Bogoria (Kenya) during the 4500 BP high lacustrine phase *In*: FROSTICK, L. E., RENAUT, R. W., REID, I. & TIERCELIN, J-J (eds) *Sedimentation in the African Rifts*, Geological Society, London, Special Publication, **25**, 323–333.
WAPLES, D. W. 1980. Time and temperature in petroleum formation: application of Lopatin's method to petroleum exploration. *AAPG Bulletin*, **64**, 916–926.
WATTS, A. B. & RYAN, W. B. F. 1976. Flexure of the lithosphere and continental margin basins. *Tectonophysics*, **36**, 25–44.
WILLIAMS, M. A. J., ASSEFA, J. & ADAMSON, P. A. 1986. Depositional context of Plio–Pleistocene Hominid-bearing formations in the Middle Awash Valley, Southern Afar Rift, Ethiopia *In*: FROSTICK, L. E., RENAUT, R. W., REID, I. & TIERCELIN, J-J (eds) *Sedimentation in the African Rifts*. Geological Society, London, Special Publication, **25**, 241–251.

Geological evolution and hydrocarbon habitat of the 'Arctic Alaska microplate'

RICHARD J. HUBBARD[1], STEVEN P. EDRICH[2], & R. PETER RATTEY[3]

Present addresses: [1] *BP Exploration − Frontier & International, London, EC2Y 9BU, UK*
[2] *BP Exploration − Western Hemisphere, 9401 South West Freeway, Suite 1200, Houston, Texas 77074, USA*
[3] *BP Exploration − Europe, 301 St. Vincent St, Glasgow G2 5DD, UK*

Abstract: Depositional sequence mapping has been used to analyse the late Devonian to Recent geologic evolution and hydrocarbon habitat of north Alaska and northwest Canada. Eight depositional megasequences have been identified, each of which records a discrete, major phase of basin evolution. The three oldest megasequences are named the Ellesmerian and reflect deposition on a subsiding fold belt terrane. We name the subsequent two megasequences of early Jurassic to Aptian age, the Beaufortian. They record a 100 Ma period of extension during which a Jurassic failed rift episode was followed by onset of the successful rift episode in the Hauterivian. This extension led to the opening of the oceanic Canada Basin. The final three megasequences record geographically distinct pulses of Brookian orogenesis.

The major proven hydrocarbon habitat occurs on the Barrow Arch of north Alaska. This is a volumetrically large, but greatly restricted, hydrocarbon province, which developed as a result of constructive interference between Beaufortian rift and Brookian orogenic tectonics. Two other, relatively minor, hydrocarbon provinces have also been discovered. They are the Mackenzie Delta and Kugmallit Trough provinces of northwest Canada, which developed in passively subsided basins, located just beyond the influence of Brookian orogenic uplift.

The origin of Arctic North America, from Alaska to the Mackenzie Delta, is one of complex geology in which multiple phases of basin development and structural deformation have taken place (Hea *et al.* 1980; Dutro 1981; Norris & Yorath 1981; Mull 1982; Grantz & May 1984a; McWhae 1986). Two of North America's major petroleum provinces occur in the area. Of these, the most important is the Barrow Arch of north Alaska (Rickwood 1970; Jones & Spears 1976) with estimated reserves of 20 billion barrels of oil. More than 95% of these reserves are located in the supergiant Prudhoe Bay Field and satellite fields on the Kuparuk−Prudhoe High (Edrich 1986). By comparison, the Mackenzie Delta province is small and the larger fields are similar in size to the smaller fields of north Alaska (Procter *et al.* 1983; Lane & Jackson 1980). The largest field in the Delta is thought to be the 1985 Amauligak discovery, where the most recent reserves estimate is 700−800 million barrels of oil (OGJ), 2/3/86). Proven reserves for the Canadian province are now in the range of 1−2 billion barrels.

The development of the prolific petroleum habitats owes much to the complex history of basin development and subsequent deformation. Only a fragmentary record of this complex history is preserved today as the remnants of formerly more extensive sedimentary basins, which collectively comprise the major depocentres of the present day Arctic region. In this paper, we offer an overview of basin development from late Devonian to Recent time and discuss the hydrocarbon habitat in north Alaska and northwest Canada in the context of this interpretation.

Database

The study area is large and extends from the Brooks Range in the south to the oceanic Canada Basin in the north, and from the Mackenzie Delta in the east to the Chukchi Sea in the west, see Figs 1, 4 & 5. The open file geological and geophysical literature for the onshore and offshore areas is voluminous and encompasses studies by academia as well as government agencies, most notably the United States Geological Survey (USGS) and the Geological Survey of Canada (GSC). In addition, several papers by industry discuss individual oil fields and provide summary interpretations of the much more comprehensive database available to industry. The latter database is large and includes some 150 000 km of

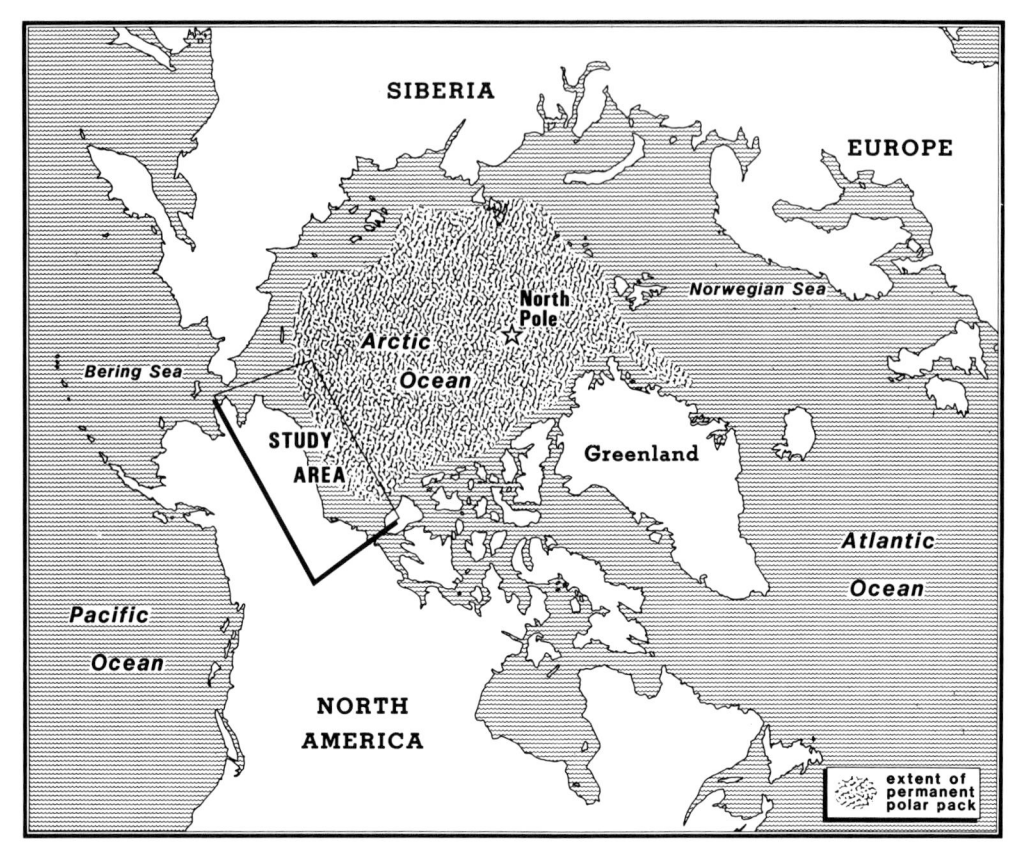

Fig. 1. Arctic Region and study area location map.

seismic data (shown on the inset in Fig. 4) and more than 400 exploration and appraisal wells. This paper is based on a full integration of the subsurface data from industry with published work on Alaska and northwest Canada and proprietary field mapping. In a paper that attempts to present an overview by exploiting this large open file and proprietary database, it is clearly impossible to provide a comprehensive bibliography. Key references based on published work by academia and industry are given throughout.

Stratigraphic analysis and framework

The pattern of basin development in Arctic Alaska and northern Canada is the direct consequence of the plate tectonic evolution of the region from late Devonian time to the present (Tailleur 1973). In the broad sense, the plate tectonic history can be viewed as the consequence of subduction-related processes from late Paleozoic to mid-Jurassic time and to the interaction between collision and sea floor spreading from mid-Jurassic time onward (Smith 1986). However, many aspects of the plate tectonic history are poorly constrained by data, and thus are speculative at best. For example, the age and opening history of the Canada Basin, a key constraint on Arctic margin history, are not known directly because identifiable oceanic magnetic anomalies are absent (Vogt et al. 1979, 1982). In addition, structural and stratigraphic analysis in the allochthons of the Brooks Range has been attempted by a number of workers (Tailleur et al. 1966; Snellson & Tailleur 1968; Tailleur & Brosge 1970; Martin 1970; Mayfield et al. 1978, 1982, 1983a, 1983b; Churkin et al. 1979; Mull 1979; Curtis et al. 1982, 1983; Ellersieck et al. 1979, 1982, 1983); but there is still not enough data. Nonetheless, key constraints can be imposed by rigorous structural and stratigraphic analysis of the geological and geophysical data for the onshore and offshore area. This approach, as discussed below, has allowed analysis of discrete phases of basin initiation and development, which can be used to constrain plate boundary types and in turn, to make plate reconstructions of the Arctic region that allow us to view Alaska in the broader context. This iterative process provides a most likely or best fit solution rather than the desirable exact solution, which cannot be achieved with the incomplete database, especially in the deep offshore area, underlying the polar ice pack.

The summarized stratigraphy for north Alaska is shown in Fig. 2 and in the chronostratigraphic charts of Figs 9, 11, 14, 21 & 23. Our analysis of basin development is based on the depositional megasequences illustrated in Fig. 2. Each megasequence is the sedimentary response to each discrete, major phase of basin evolution, with the architecture and shape a direct result of the tectonic processes (Hubbard et al. 1985a). Major periods of change in basin geometry, for example, the transition from active rifting to passive subsidence, typically results in the development of major regional unconformities, which define the upper and lower boundaries of the megasequences. In the Alaskan area, stratigraphic analysis is made difficult by contemporaneous thrusting and extension on the same small, continental plate, with the consequent overlap of very different basin geometries. Use of this method of analysis has, however, helped us to view each stratigraphic unit in its correct basinal context.

Lerand (1973) was amongst the first to subdivide Beaufort Sea stratigraphy into tectonic sequences, which were defined mainly according to sediment provenance and major lithologies but also closely correlated to tectonics. Lerand's sequences corresponded to complete cycles of basin development, i.e. from pre-rift, through rift, to post-rift, and thus incorporated the interrelated megasequences recognized in this study. However, our analysis, based on newer data, allows refinement by recognition of more phases of basin development and thus requires substantial modification of Lerand's original innovative scheme.

The basement of the study area corresponds to Lerand's (1973) Franklinian of late Cambrian to Devonian age (see Fig. 2). This sequence includes a variable assemblage of rock types. Deformed and weakly metamorphosed Ordovician to Devonian shales (Argillite) are common in the subsurface. At outcrop in the Brooks Range, Cambrian to early Devonian strata include quartzites, graywackes, platform carbonates, radiolarian cherts, graptolitic shales and mafic to intermediate volcanics and volcanogenic sediments (Moore et al. 1986; Dillon et al. 1980; Dutro et al. 1972). This association of strata was progressively deformed into an accretionary assemblage, and the Ellesmerian orogeny was the late Devonian climax to these accretionary events.

Two belts of granitic igneous intrusives with late Devonian (390–360 Ma) radiometric ages intrude this accretionary complex (Sable 1977; Dillon & Tilton 1986). One granitic belt runs along the southern side of the Brooks Range and the other is seen in the Romanzof and British Mountains of the eastern Brooks Range and in the subsurface of north Alaska in the East Teshekpuk #1 well (Bird 1982).

Megasequence boundaries

The boundaries between the eight megasequences overlying Franklinian basement form our regional mapping horizons. The ages of these regional unconformities are shown in the schematic stratigraphic column, Fig. 2, and their typical subsurface configuration over the Barrow Arch is shown in the line drawing of Fig. 3. The absolute age assignment for each regional unconformity, or its correlative disconformity, is taken from the geologic time scale of Harland et al. (1982). The regional unconformities and megasequences are briefly described below.

Megasequence 1

The oldest megasequence overlying Franklinian basement comprises a thick wedge (more than 3 km) of the Devonian sediments, the Kanayut–Hunt Fork, which represents a synorogenic clastic wedge derived from the Ellesmerian fold belt (Nilsen 1981; Moore & Nilsen 1984). This unit is only known from allochthons in the Brooks Range. None of the North Slope exploration wells have been drilled deeply enough to penetrate age-equivalent strata. Hence, the base boundary age (late Devonian, 375 Ma) is assigned by reference to published outcrop data (e.g., Nilsen 1981) and the top boundary age is constrained by subsurface biostratigraphic data from the overlying megasequence.

Megasequence 2

The succeeding megasequence is a dominantly transgressive unit with a pronounced basal angular unconformity and developed due to major infill of Ellesmerian fold belt topography. In the subsurface, the basal unconformity is known from well control to be as old as earliest Mississippian (335 Ma) in depocentre areas and as young as mid to late Mississippian over the

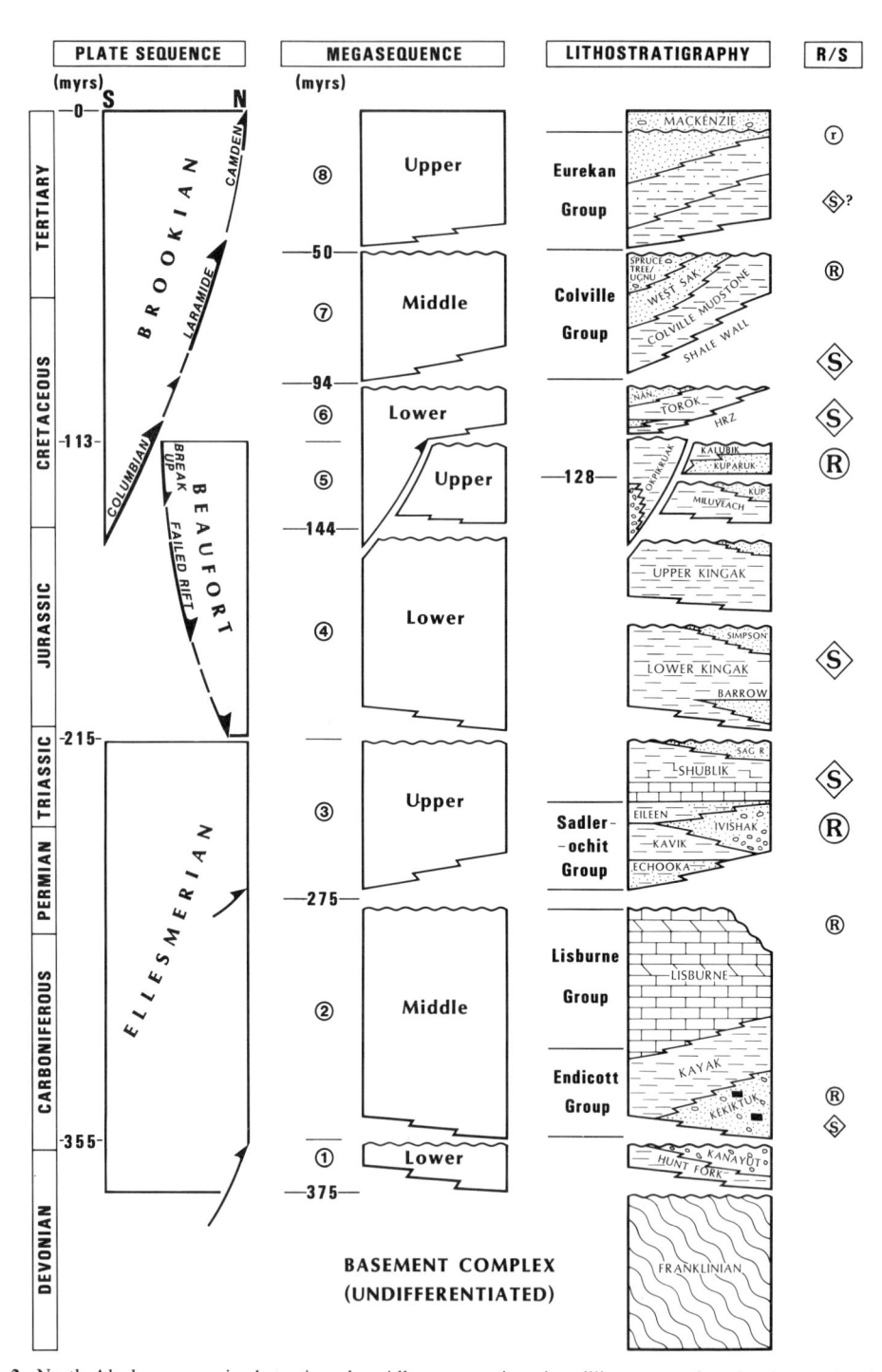

Fig. 2. North Alaska summarized stratigraphy. All ages are given in million years after the time scale of Harland *et al.* (1982). Megasequences are named and numbered 1 through 8, as indicated by the numbers in circles to the left of the central column. The HRZ and Shale Wall lithostratigraphic units are hemipelagic deposits at the base of the Torok and Colville Mudstone terrigenous deposits respectively. The 'S' symbol within a diamond denotes a hydrocarbon source rock and the 'R' symbol within a circle denotes a hydrocarbon reservoir rock.

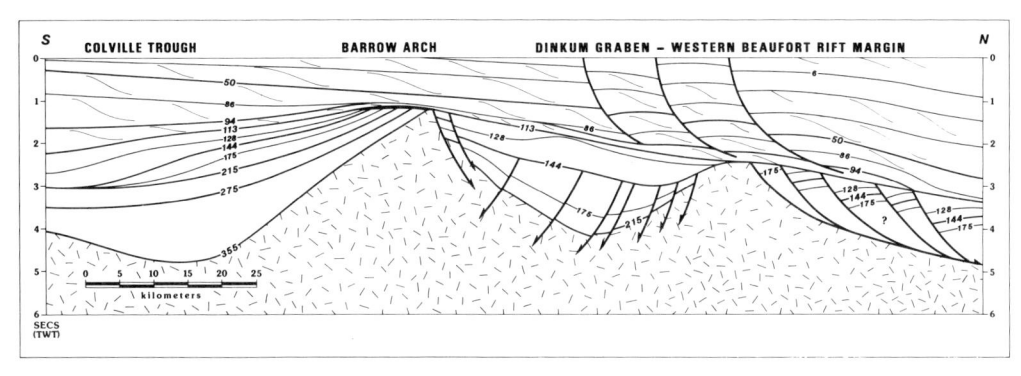

Fig. 3. Schematic cross section from the Colville Trough to the Dinkum Graben and Western Beaufort Rift Margin showing the configuration of regional unconformities and depositional megasequences. The ages of the regional unconformities are in million years after the time scale of Harland et al. (1982).

basement highs. The megasequence contains sediments derived from the north, which include the Endicott Group and Lisburne Group of north Alaskan terminology. (Tailleur et al. 1967; Bowsher & Dutro 1957). Megasequence deposition was terminated in the early Permian (275 Ma) by a period of regional uplift and erosion, which probably affected the entire North American Arctic region.

Megasequence 3

The third megasequence oversteps the early Permian (275 Ma) regional unconformity and contains two transgressive–regressive cycles of mixed clastic–carbonate deposition derived from the north. These cycles are represented by the Sadlerochit Group and the Shublik and Sag River Formations (Leffingwell, 1919; Alaska Geological Society, 1970–71; Detterman et al. 1975). The upper megasequence boundary is a regional seismic marker which, in the North Slope area, coincides with the top of the Sag River sandstone.

Megasequence 4

The Sag River regression was followed by a period of relative still stand, with little clastic input, which persisted during Rhaetian and Hettangian time. The resulting disconformity (215 Ma) formed the basal boundary to Megasequence 4. The boundary is observed on seismic data as a downlap surface (onto the Sag River) on the North Slope, but is buried too deeply for positive identification across part of the continental shelf. Within the Dinkum Graben area (Grantz & May 1983), onlap and localized coeval faulting indicate the precursor to more widespread extensional tectonics.

The megasequences consists largely of the shales of the Jurassic Kingak Formation (Leffingwell 1919; Detterman et al. 1975), but is divided by a Bathonian (175 Ma) regional unconformity into a lower and upper sequence, each topped by a thin regressive sandstone unit. The lower sequence was deposited in basins that formed in response to initial extension, at the onset of the rifting episode, while the upper sequence was deposited during the main phase of the failed-rift episode, with major faults downthrowing to the south.

Megasequence 5

A widespread disconformity surface at the base of the Cretaceous (144 Ma) is marked by a prominent biostratigraphic time gap. Early Tithonian forms at the top of the Jurassic Kingak Formation are typically overlain by Valanginian and younger assemblages at the base of the Cretaceous Miluveach Formation (Carman & Hardwick 1983).

The megasequence contains two sequences separated by a prominent intra-Hauterivian angular unconformity over the crest of the Barrow Arch. Sands of the Kuparuk River Formation (Alaska Geological Society, 1970–71; Carman & Hardwick 1983; Masterson & Paris 1986) are developed immediately above and below this event. The lower sequence developed in response to the onset of a second period of extension and the upper sequence represents the main phase of a syn-rift episode, during which basin polarity reversed and major faults downthrew to the north.

Megasequence 6

The unit includes a series of sequences, which young from south to north. These are possibly as old as middle Jurassic (Bajocian) but mainly of late Jurassic (Tithonian) to early Cretaceous in the south and early Cretaceous to earliest late Cretaceous (early Cenomanian) in the north. They record the northward progradation of the earliest phases of accretionary tectonics in Arctic Alaska and were associated with the early to mid-Cretaceous Columbian orogeny of Cordilleran geology (Roeder & Mull 1978; Douglas 1972). One of the major lithostratigraphic units within the megasequence is the Okpikruak Formation (Gyrc et al. 1951). Extensive published literature is available for this megasequence, much of which is the result of many years of fieldwork in the National Petroleum Reserve—Alaska (NPRA) (Tailleur et al. 1966; Martin 1970; Mayfield et al. 1983a, b; Molenaar 1981a; Ahlbrandt 1979).

Over the Barrow Arch, the base boundary to this megasequence is a widespread disconformity surface, which is recognized by distal downlap of hemipelagic Brookian sediments derived from the south. Precise age dating is hampered by uniform distal shale deposition across the boundary (Pebble Shale or Kalubik Formation below and Highly Radioactive Zone (HRZ) above; Molenaar 1981a, 1983; Carman & Hardwick 1983) and by the poorly age diagnostic impoverished marine faunas and floras. An approximate base Albian (113 Ma) age is assigned.

The upper part of the megasequence, overlying the 113 Ma event, was deposited during the early stages of thermal subsidence of the Beaufortian rift system. A considerable thickness of Brookian orogenic sediments was deposited in the south and west of the study area (Torok Formation and Nanushuk Group, Molenaar, 1981b, 1983; Reiser et al. 1971; Ahlbrandt, 1979) but only a distal veneer of shales covered the crest of the Barrow Arch and the eastern Beaufort Sea rift margin.

Megasequence 7

This megasequence is a strogly regressive, prograding orogenic wedge resting on a regional downlap surface of mid-Cenomanian (94 Ma) age. The basal boundary developed as an erosional unconformity within broad channels and submarine canyons, which formed as the megasequence prograded onto the passively subsiding Beaufort Sea margin.

Sediments were derived exclusively from the Brookian orogen and include the prograding late Cretaceous to Tertiary sequence of the Colville Group and associated formations (Gryc et al. 1951; Detterman et al. 1975; Lyle et al. 1980; Molenaar 1981a, 1983). The Brookian mountain front had shifted a considerable distance to the northeast since the deposition of the previous megasequences and (Laramide) overthrusting of the rift margin began in northeast Alaska/northwest Canada.

Megasequence 8

A mid-Eocene (50 Ma) angular unconformity, which can be mapped throughout the eastern Beaufort Sea and Mackenzie Delta (McWhae 1986; Dietrich & Dixon 1984), forms the base boundary to this megasequence. The base boundary developed as a regional unconformity during a period of orogenic restructuring toward the end of Laramide orogenesis in the Brooks Range.

The eastern Brooks Range thrust front moved further to the northeast across the passive rift margin and active, thrust-controlled depocentres developed on the continental margin during deposition of the megasequence (Grantz et al. 1986). Megasequence geometry differs greatly according to position relative to the mountain front. It forms a thin, passive wedge in the western Beaufort Sea, whereas in northeast Alaska/northwest Canada, it occupies depocentres controlled by thrust faults. In the extreme east it forms the thick, prograding wedge of the Mackenzie Delta. In addition, the megasequence has been strongly modified by late Miocene (6 Ma) to Recent tectonism in northeast Alaska (Grantz et al. 1986), which we have named the Camden orogeny, after its prominent development in Camden Bay.

Plate sequences

The eight megasequences can be grouped into three genetic associations of megasequences, where each association forms a complete plate tectonic cycle. We refer to these megasequence associations as plate sequences because, as will be shown in our interpretation, each was deposited on a separate tectonic plate, see Fig. 2.

Megasequences One, Two and Three comprise sediments derived from the north and were deposited into basins that were located on and between Ellesmerian orogenic belts. Hence, we refer to these units as the Lower, Middle and Upper Ellesmerian megasequences, which collectively form the Ellesmerian plate sequence, (355—215 Ma). Megasequences Four and Five comprise sediments deposited during the failed rift and successful rift episodes, which opened the Canada Basin and formed the Beaufort Sea. We refer to these units as the Lower and Upper Beaufortian megasequences (respectively) and together, they form the

Beaufortian plate sequence, (215–113 Ma). Megasequences Six, Seven and Eight comprise orogenic sediments, which were derived from the Brooks Range to the south. The Brookian orogen formed in response to accretionary tectonics in south–central Alaska and we refer to the sedimentary response as the Lower, Middle and Upper Brookian megasequences. The three Brookian megasequences form the Brookian plate sequence (145–0 Ma).

This gross threefold stratigraphic subdivision into the Ellesmerian, Beaufortian and Brookian plate sequences differs from previously published interpretations (i.e., Lerand 1973; Grantz et al. 1979; Ehm & Haimla 1982) by the explicit recognition of a break-up unit, the Beaufortian plate sequence. This unit was recognized by combining our very detailed database in the Kuparuk River Field with an extensive offshore seismic grid. It represents a rifting episode and forms the transition between Ellesmerian and Brookian geology (Grantz & May 1983; Carman & Hardwick 1983).

The structural configuration of the three plate sequences across the Barrow Arch is shown in the seismic line of Fig. 4 and in the line drawing of Fig. 3. The Barrow Arch was created by uplift during the Beaufortian rift episode. In this area, the Ellesmerian plate sequence subcrops to both the Beaufortian and Brookian plate sequences.

Beaufortian basin geometry was controlled by south-throwing faults in the lower megasequence (Fig. 3) and by north-throwing faults in the upper megasequence (Figs 4 & 3). The strongly regressive nature of the overlying, southern derived Brookian orogenic sediments is well shown by the seismic display in Fig. 4. Illustrative seismic lines have also been published by Kirschner et al. (1983) in northwest Alaska; by Grantz & May (1983), McWhae (1986) and Hubbard, et al. (1985a, b) in the Beaufort Sea; and by the latter authors and Willumsen & Cote (1982) in the Mackenzie Delta.

Sequences and lithostratigraphic formations

In addition to grouping the eight depositional megasequences into three larger megasequence associations, they have also been broken down into 23 component depositional sequences, which are useful for mapping specific parts of the study area. Much of this detailed stratigraphy is beyond the scope of the present overview and is only described where necessary. The summarized stratigraphic column of Fig. 2 does, however, include a generalized lithostratigraphy to enable the reader familiar with north Alaskan geology to correlate the most widely used formation names with our depositional megasequences. The column identifies 31 formations and five lithostratigraphic groups. Additional formations are referred to in the text, where necessary, and all formations are referenced to their original definition, where appropriate. For an introduction to north Alaskan and northwest Canadian lithostratigraphic nomenclature, see Brosge & Tailleur (1971); Detterman et al. (1975); Lyle et al. (1980); Molenaar (1983); Mountjoy (1967a, b); Young et al. (1976); Dixon (1982a); and Bird (1982), for a rock unit report of 228 wells drilled on the North Slope of Alaska.

Total sediment thickness

The post-Franklinian total sediment thickness for the area to the north of the Brooks Range is shown in Fig. 5. The map illustrates the combined thickness of the eight depositional megasequences. In Alaska, the major depocentres of the Hanna Trough, in the Chukchi Sea, and its onshore continuation in the Colville Trough are separated from the depocentre on the Beaufort Sea continental margin by the Barrow Arch. Each depocentre contains in excess of 10 km of sediments. A fourth major depocentre is represented by the Canadian Mackenzie Delta, which contains more than 12 km of sediment.

In the following sections of this paper, the initiation and development of each megasequence is described in terms of plate tectonic setting, the development of basin architecture and the consequences in terms of resource potential and hydrocarbon habitat. An interpretation is presented for the Ellesmerian, Beaufortian and Brookian megasequences, which is then tied together in a final section describing the overall hydrocarbon habitat in the study area.

Ellesmerian plate sequence

Initiation of Ellesmerian basin development

The pre-Ellesmerian plate tectonic setting of Arctic North America is highly speculative because of the fragmentary, and often disrupted, stratigraphic evidence that is available. In this paper we have utilized Smith's (1986) pre-Triassic reconstructions of the Arctic, which showed a collage of terranes (the 'Barents microplate assemblage') considered to result from mid-late Paleozoic collisional events. Evidence for the presence of an Ellesmerian

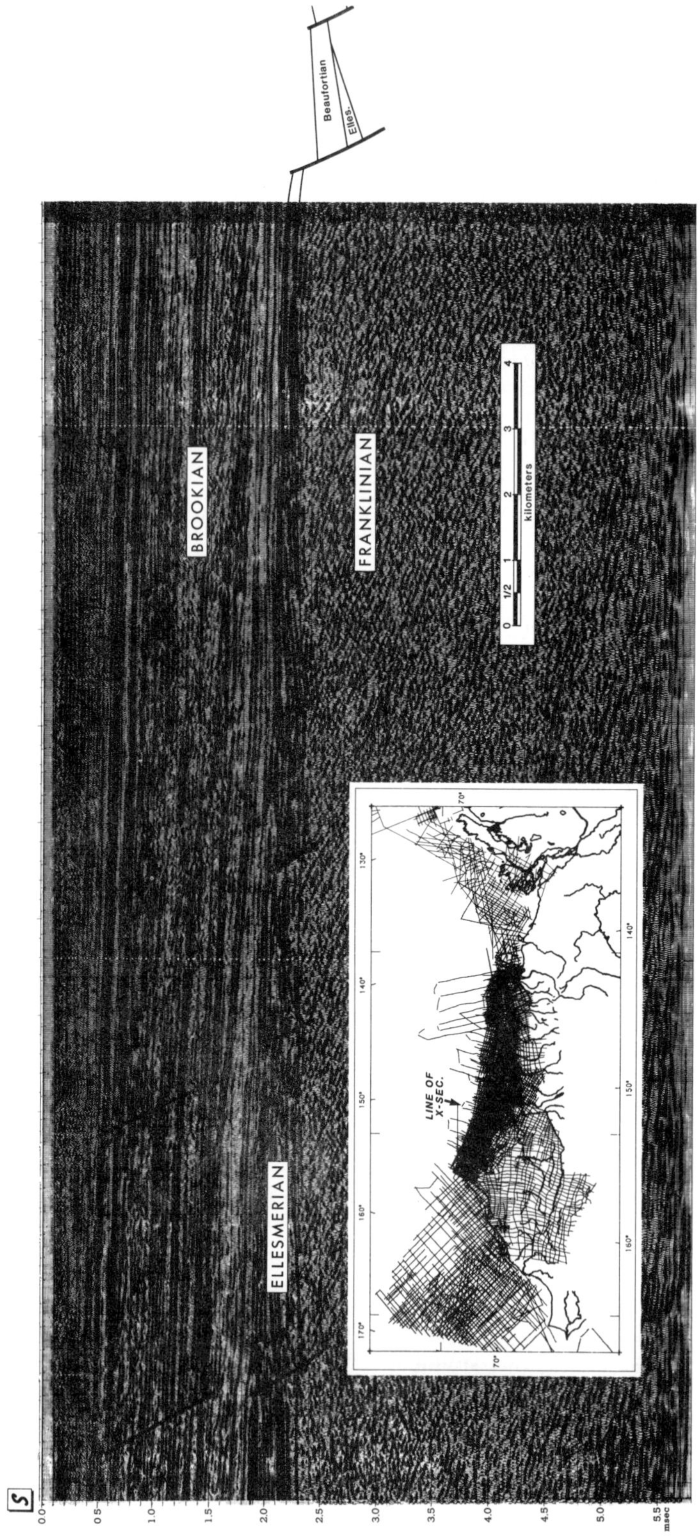
Beaufortian
Elles.
BROOKIAN
FRANKLINIAN
ELLESMERIAN
0 1/2 1 2 3 4
kilometers
LINE OF X-SEC.
S
0.0
0.5
1.0
1.5
2.0
2.5
3.0
3.5
4.0
4.5
5.0
5.5 msec

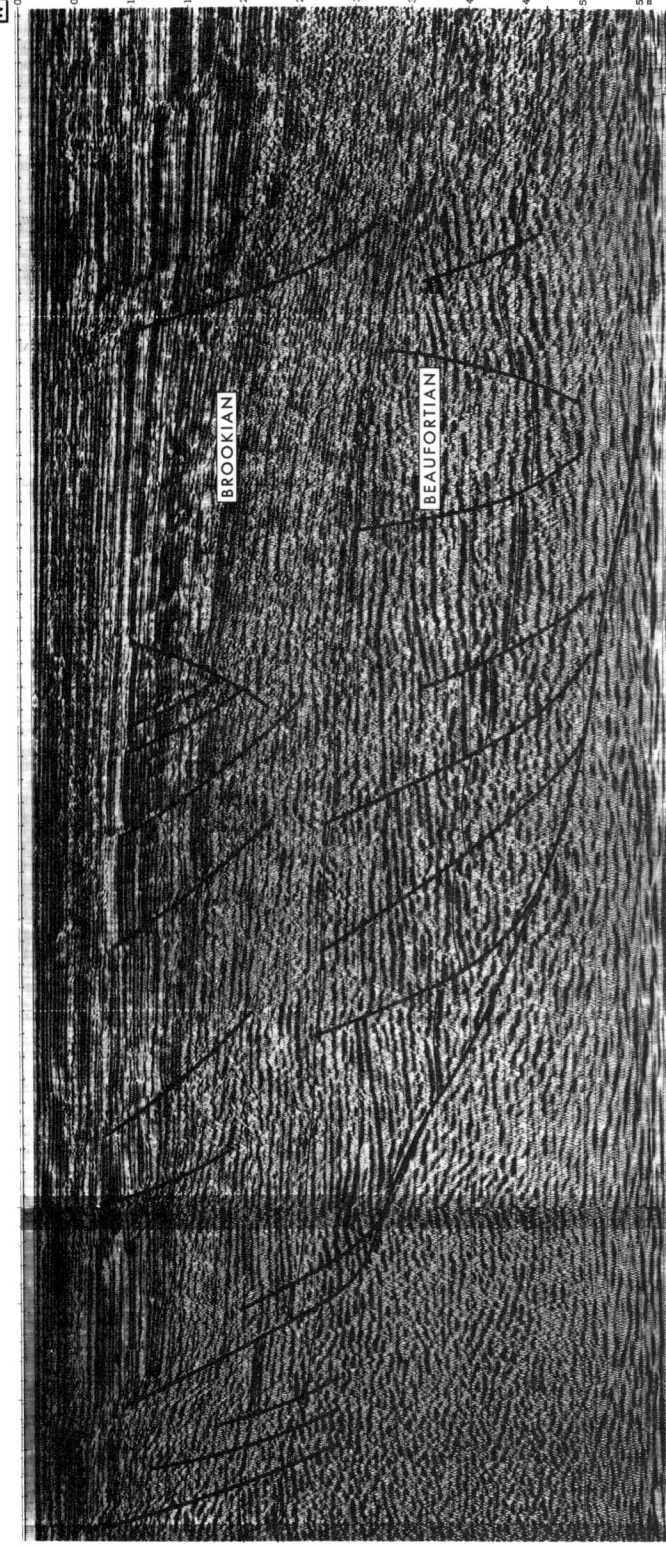

Fig. 4. Composite seismic line across the Barrow Arch showing the configuration of the Ellesmerian, Beaufortian and Brookian. The approximate line of cross section is shown on the insect map. It runs through the Ellesmerian subcrop zone in Harrison Bay (to the NW of the Kuparuk River Field) and onto the rifted margin of the western Beaufort Sea. The inset map shows the grid of seismic reflection data used in this study. It totals some 150000 km and extends from the Brooks Range in the south to the Beaufort Sea in the north and from the Chukchi Sea in the west to the Mackenzie Delta area in the east.

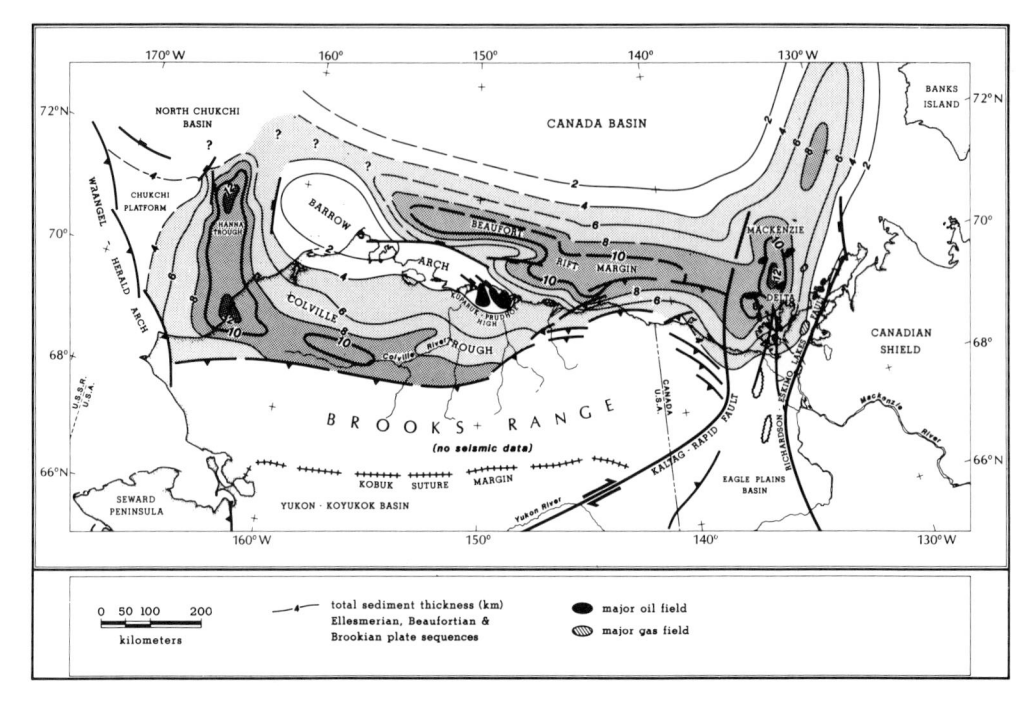

Fig. 5. Study area, total sediment thickness. The map illustrates the combined thickness of the eight post-Franklinian megasequences. It was constructed using seismic isochrons converted to depth.

fold belt terrane throughout the northern Alaska region is suggested by North Slope subsurface data and by outcrop data in the Brooks range, northern Canada and northern Greenland, see Fig. 6.

Beneath the North Slope, we interpret thrust faulting on seismic profiles, which extends to the offshore continuation of the Barrow Arch near Point Barrow (Grantz, pers. comm.). The late Devonian granites, seen in the Teshekput #1 well and in the eastern Brooks Range, are interpreted as post-orogenic intrusives. Seismic data shows that the North Slope thrust faulting has a northern vergence. This is schematically shown in Fig. 7.

To the south and west in Arctic Alaska, Brookian collisional tectonics have destroyed the evidence that is required to determine the presence of Ellesmerian (late Devonian) folding and thrusting. An extensive belt of metaplutonic and metavolcanic rocks has, however, been preserved along the southern side of the Brooks Range (Dutro et al. 1972; Sable 1977; Dillon & Tilton 1986).

From this fragmentary evidence, we suggest the former presence of two Ellesmerian orogenic fold-belts in Arctic Alaska; a northern belt (of northerly vergence) underlying the present North slope and a southern belt (of unknown thrust-fault vergence) currently located on the southern flank of the Brooks Range. From simplified palinspastic reconstruction, we estimate that prior to Brookian compression, the southern Brooks Range plutonics were in excess of 1000 km to the south of the coeval northern plutonic belt.

Ellesmerian fold belts are also known from northern Canada and Greenland (Balkwill et al. 1983; Kerr 1982). The Pearya fold belt is mapped in northern Ellesmere Island and north Greenland, where northwest-directed thrusts have been recorded (Christie 1979). The southern limb of the Ellesmerian orogeny is mapped as the south-verging Parry Islands fold belt, which wraps around the southern flank of the Sverdrup Basin (Embry 1985) before swinging north and striking out to the Arctic Ocean through Prince Patrick Island. Bally (1976) and Eisbacher (1983) suggested that north Alaska originally lay along the projection of the Prince Patrick trend but, a more elegant reconstruction is made by restoring the north-verging Ellesmerian structures under the North Slope of north Alaska to match the structural sense of the remnant north facing limb of the Pearya fold belt (Smith 1986). Using this preferred

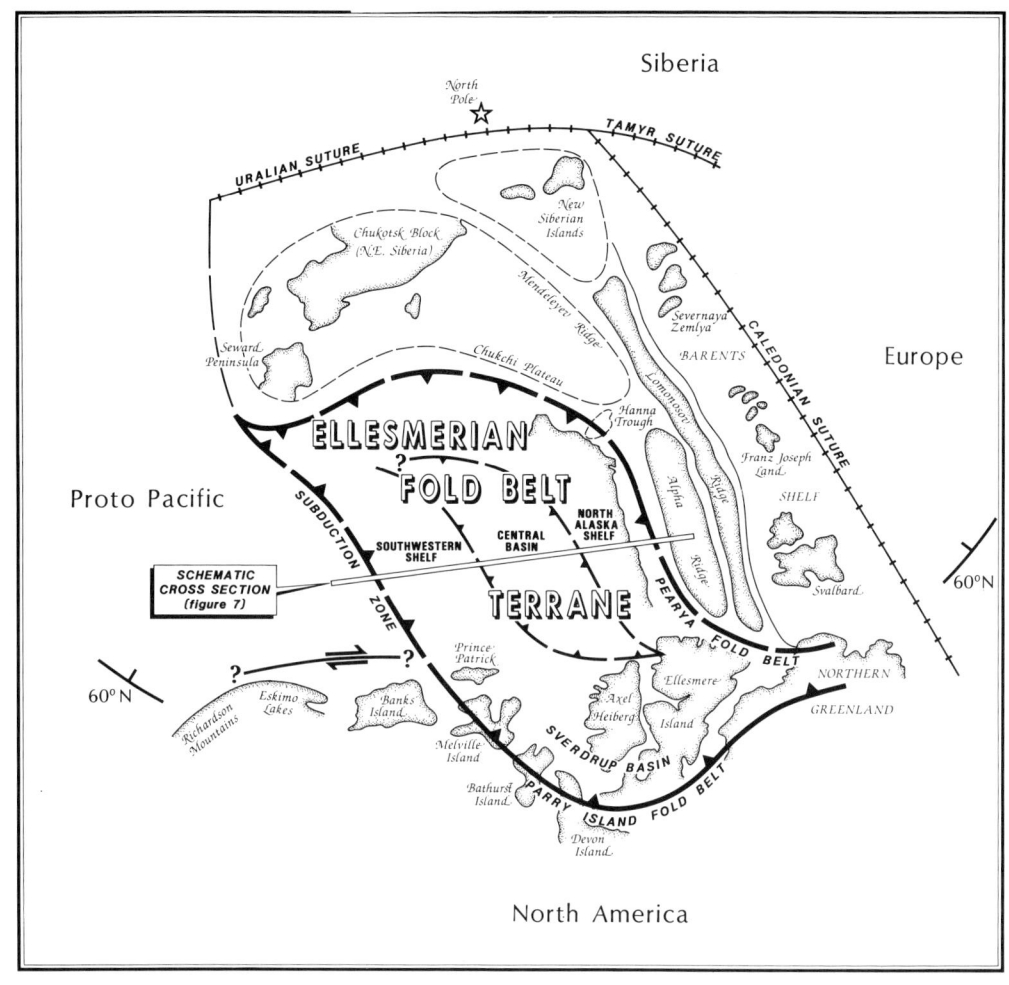

Fig. 6. Plate tectonic setting, Ellesmerian plate sequence, (modified after Figs 8 & 9 in Smith 1986). The Ellesmerian fold belt terrane is thought to have formed as a result of late Devonian (Ellesmerian) orogeny. Forsyth et al. (1989) and Jackson et al. (1985) have shown that the Alpha Ridge is a younger volcanic feature, which probably did not exist at the time modelled by this reconstruction. In this case, the North Alaska Shelf would have been juxtaposed with the Lomonosov Ridge/Barents Shelf, as shown in the regional schematic cross section of Fig. 7.

reconstruction for north Alaska, the southern Brooks Range granites may represent the westerly extension of the south-verging Parry Islands fold belt, see Figs 6 & 7.

Basin development and depositional systems

Collectively, the Lower, Middle and Upper Ellesmerian megasequences record the transition from active, syn-orogenic deposition to passive, post-orogenic subsidence, accompanied by transgression of the entire syn-orogenic basinal area and its peneplaned fold-belt margins. The isopach map for the preserved part of the Ellesmerian plate sequence in north Alaska is shown in Fig. 8. A northern depositional edge is mapped along the northwest continental shelf, but erosional edges are seen elsewhere, suggesting formerly more extensive deposition to the south and west. The isopach map shows a series of north–northwest oriented depocentres. The Umiat Basin attains thickness in excess of 6 km, while several areas within the Hanna Trough (located in the present day Chukchi Sea) have thicknesses greater than 4 km.

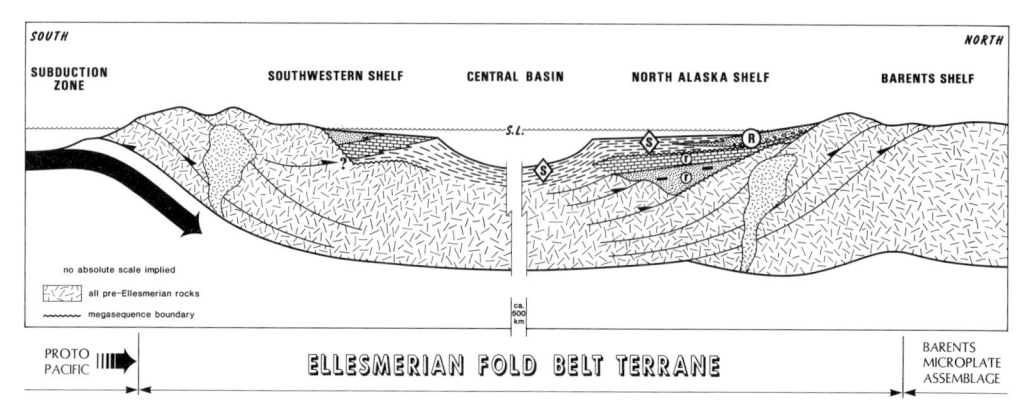

Fig. 7. Regional schematic cross section, Ellesmerian plate sequence. The line of cross section is shown on Fig. 6. Note that approximately 500 km of the Central Basin is not shown. Granitic batholithic intrusions under the North Alaska and Southwestern shelves are shown by the fine stipple. Other lithology symbols are as follows: Coarse clastic deposits, dotted. Limestone-dominated deposits, open bricks. Fine-grained clastics, dashed. The 'S' symbol in a diamond denotes a hydrocarbon source rock and the 'R' symbol in a circle denotes a hydrocarbon reservoir rock.

The Lower Ellesmerian megasequence is only known from outcrop in the Brooks Range and is interpreted as the initial syn-orogenic infill to a subsiding basinal area between the southern and northern Ellesmerian fold belts. The initial transgressive facies (Beaucoup Formation; Dutro et al; 1979) is overlain by progradation of the Kanayut delta system (Tailleur et al. 1967; Nilsen 1981; Moore & Nilsen 1984). This system includes prodelta and delta slope facies (Hunt Fork Shale; Chapman et al. 1964), marine delta top facies (Noatak Sandstone; Smith 1913; Brosge et al. 1979) and nonmarine delta plain, braid plain and meandering stream facies Kanayut Conglomerate; Bowsher & Dutro 1957). On the basis of palinspastic reconstruction and facies patterns observed in outcrop in Brooks Range allochthons, Nilsen (1981) and Moore & Nilsen (1984), interpreted a sediment provenance area for these syn-orogenic clastics to the north and north−east, in the area of the Ellesmerian fold belt underlying the present day North Slope.

Middle Ellesmerian megasequence

The Middle Ellesmerian megasequence is early Carboniferous to early Permian in age. It had a duration of 80 Ma (355−275 Ma) and is a dominantly transgressive unit, which represents the major infill to the Ellesmerian fold belt. Deposition was terminated by regional uplift in the early Permian. A chronostratigraphic chart, shown in Fig. 9, gives the summarized basin stratigraphy.

In north Alaska, the tectonic framework consisted of a major fold belt, surrounded on at least three sides (to the north, south and east) by Ellesmerian orogenic highlands, see Fig. 7. Initial deposition was confined to fault bounded basins, but later sedimentation built out to the south to form a stable marine shelf.

The overall north−south oriented depocentres shown in the isopach map (Fig. 8) are largely filled with nonmarine Middle Ellesmerian clastics of the Endicott Group (Tailleur et al. 1967). Braided to meandering fluvial clastics, interbedded with swamp shales and coals of the Kekiktuk Formation (Brosge et al. 1962; Bird & Jordan 1977; Melvin 1985) are the dominant facies. The tectonic controls on the basin geometry are complex. We interpret the regional control as compressive, but identify local strike-slip and possibly extensional basins. We interpret basement highs as topographic culminations in the underlying Ellesmerian thrust pile while intervening normal fault-bound basins are remnant foredeeps and pull-aparts. The Ikpikpuk Basin (Fig. 10b) is an example of a small foredeep, while the much deeper Umiat Basin (Fig. 10a) is bounded by strike/dip-slip faults, interpreted as lateral structures to regional northwest verging thrust faults.

The tectonic setting of extensional depocentres, such as the Umiat Basin, within a compressive fold belt terrane appears paradoxical. Indeed, Churkin et al. (1979, 1980) postulated that the basinal area to the south of north Alaska did not develop within a fold belt and suggested that Ellesmerian deposits in that area overlie

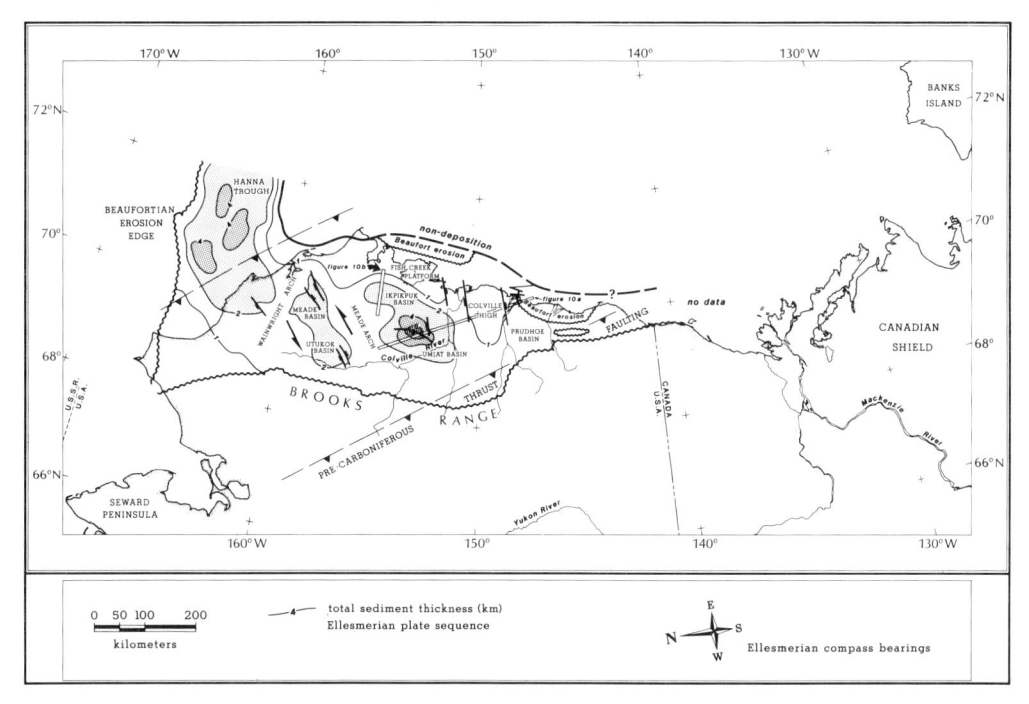

Fig. 8. Sediment thickness, Ellesmerian plate sequence. The map illustrates the combined thickness of the three Ellesmerian megasequences. It was constructed using seismic isochrons converted to depth.

oceanic crust. A variety of geological and geophysical evidence (Crane 1980; Dutro 1980; Mayfield 1980; Metz 1980; Mull 1980; Nelson 1980), appears to negate this possibility. Royden et al. (1983) have analysed a possible analogue to the Ellesmerian basins of north Alaska in the Neogene Pannonian basin system of Europe. Here, extensional basins are connected to each other and to areas of coeval shortening in the Carpathian thrust belt by a system of strike-slip faults. These basins developed immediately behind the orogenic Carpathian arc, which formed due to continental collision between Europe and a series of smaller continental fragments.

Subsequent Middle Ellesmerian facies in north Alaska reflect the gradual transgression of the fold belt and cessation of coarse clastic deposition. Marginal to shallow marine deposition (Kayak Shale; Bird & Jordan 1977; Bowsher & Dutro 1957; and Itkilyariak Formation; Mull & Mangus 1972) is transitional to thick, overlying platform carbonates (Lisburne Group; Schrader 1902; Bowsher & Dutro 1957; Brosge et al. 1962; Bamber & Waterhouse 1971). Wood & Armstrong (1975) suggested a clastic-poor carbonate platform model for the Lisburne Group. They outlined nine facies passing from a low relief shoreline zone, basin-ward through supratidal and lagoonal environments, to restricted, open platform conditions to the south. Subsurface correlation of these facies is possible using the foraminiferal zonation of Armstrong et al. (1970) and Armstrong & Mamet (1970). The Lisburne shelf edge is not seen on seismic data and hence, the nature of the break-of-slope from platform to basin (ramp or rimmed-shelf) is not known. There is no evidence of reefal buildup in the subsurface north of the Brooks Range (Armstrong & Mamet 1977, 1978), although Armstrong and Bird (1976) have interpreted a zone of shoaling to the south of the Umiat Basin, which may indicate proximity to the paleo-shelf edge.

The area to the south of north Alaska was characterized by deposition of basinal deposits throughout the time interval of the Middle Ellesmerian megasequence (Mull et al. 1982). These sediments include thin, black shales, radiolarian cherts and minor basalts and andesites of the Kuna and Siksikpuk Formations (Patton 1957: Mull et al. 1982: Ellersieck et al., 1983). These are preserved in the allochthons of the Brooks Range. The overall structural configuration of this basinal area is speculative, but Mayfield et al. (1983a) concluded that these rocks were deposited in an ensialic bain with

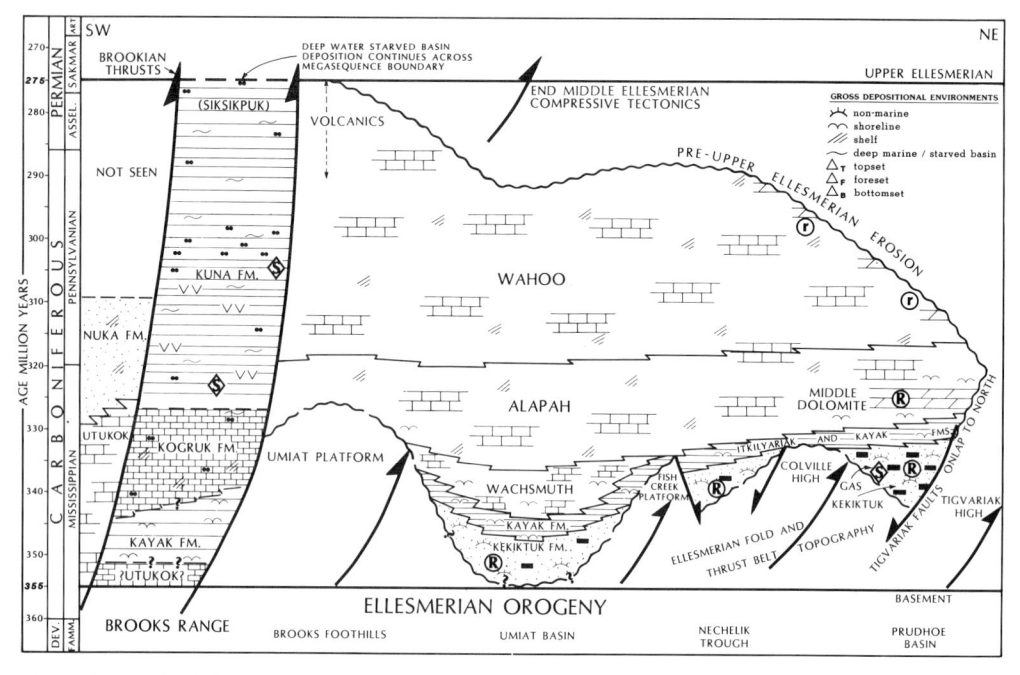

Fig. 9. Summarized chronostratigraphy, Middle Ellesmerian megasequence. Lithology symbols as follows: Coarse-grained clastics, dotted. Coals, solid bricks. Fine-grained clastics, ruled lines. Limestone-dominated deposits, open bricks. Cherts, two small solid circles. Volcanics, 'V' symbols. Hydrocarbon source rocks are denoted by the 'S' symbol in a diamond and the 'R' symbol within a circle denotes a hydrocarbon reservoir rock.

both northern and southern margins. We have sketched this basinal area as the Central Basin on the plate tectonic reconstruction of Fig. 6 and in the regional cartoon cross section of Fig. 9.

The nature of the southern margin to the Central Basin is even less well documented. Mayfield et al. (1983a) estimated that Brookian orogenesis has displaced the remaining fragmentary record of this margin, now only seen in the highest Brooks Range allochthons, by 700–800 km. The presence of a Carboniferous high-relief, southerly granitic provenance area (Nukaland; Mayfield et al. 1983), is, however, suggested by the presence of thick coarse grained arkosic sandstones of the Nuka Formation. Carbonate deposits from the southern basin area, (Utukok, Kogruk and Tupik Formations; Sable & Dutro, 1961; Curtis et al. 1983) are also preserved. As shown on the chronostratigraphic chart of Figure 9, these sediments are broadly coeval with the Lisburne Group of the northern basin margin.

Uplift and partial erosion of the Middle Ellesmerian megasequence occurred in north Alaska during the early Permian. The 275 Ma regional unconformity, which forms the top boundary to the megasequence, developed as a result of this uplift, which has been correlated to the closure of the Uralian Ocean and suture of Siberia to the 'Barents microplate assemblage' and Pangea (Smith 1986).

Middle Ellesmerian hydrocarbon source rock facies developed in two main depositional environments; firstly, the gas prone coal measures of the Kekiktuk Formation, which were deposited in the early, synorogenic fault-bounded basins of north Alaska, and secondly; the oil-prone, basinal deposits of the Kuna Formation, which slowly accumulated in the Central Basin area to the south of north Alaska.

Reservoir development occurs at two levels. Firstly, the fluvial lithofacies of the Kekiktuk Formation, which contains compositionally mature quartzose sands and is productive in the Endicott Field, immediately north of Prudhoe Bay (Woidneck et al. 1986; Melvin 1985). Secondly, in the supratidal facies of the Lisburne Group carbonate platform. Dolomitized limestones of the middle and upper dolomite units (Alapah Formation; Bowsher & Dutro, 1957; Armstrong & Mamet 1970) developed vuggy

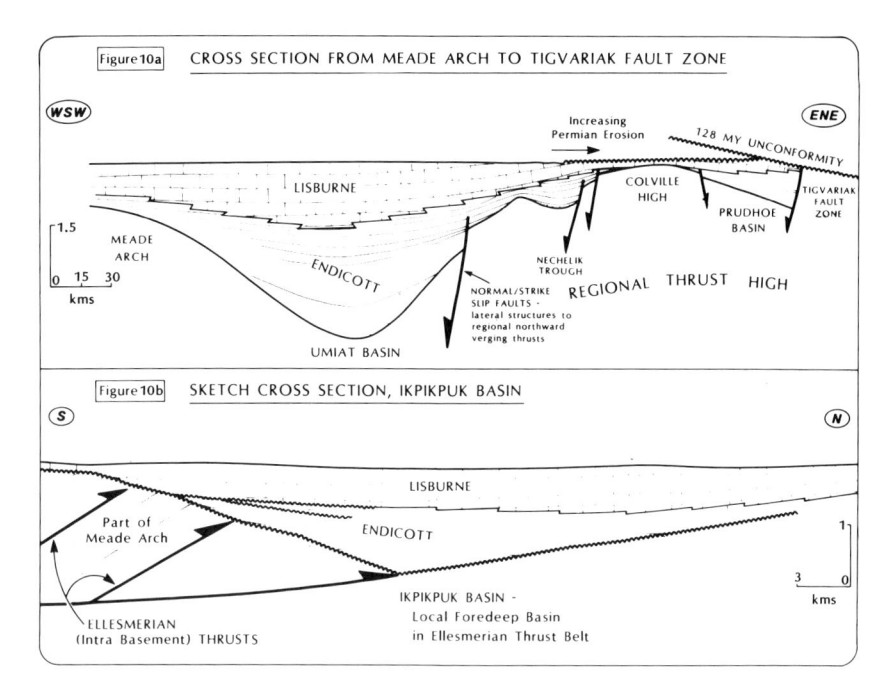

Fig. 10. Basin cross sections: A, Umiat Basin; B. Ikpikpuk Basin. The lines of cross section are shown on Fig. 8. The Endicott and Lisburne Groups are shown on Fig. 2 and include the Kekiktuk/Kayak/Itkilyariak Formations and Wachsmith/Alapah/Wahoo Formations respectively, shown in more detail on Fig. 9.

and inter-crystalline porosity and produce in the Lisburne pool of the Prudhoe Bay Field (Okland *et al*. 1986).

Upper Ellesmerian megasequence

The Upper Ellesmerian megasequence is late-early Permian to late Triassic in age. It had a duration of 60 Ma (275–215 Ma) and comprises a double transgressive–regressive cycle, which was terminated by a period of hiatus and non-deposition at the end of the Triassic. A chronostratigraphic chart summarizing basin stratigraphy is shown in Fig. 11.

In north Alaska, the first depositional cycle began with transgressive deposition across the basal (275 Ma) regional disconformity surface of shallow marine carbonates, cherts, silts and highly glauconitic sands known as the Echooka Formation and marine shales of the Kavik Formation (Jones & Speers 1976; Echooka Member, Keller *et al*. 1961). This was followed by a short-lived regressive phase, lasting perhaps not more than 4 Ma, in which a coalescing alluvial fan-delta system prograded southward onto the shelf from a rejuvenated northern land area. These coarse clastics are the Ivishak Formation of the Sadlerochit Group (Leffingwell 1919; Keller *et al*. 1961; Alaska Geological Society, 1970–1971; Detterman *et al*. 1975; Jones & Spears, 1976; Melvin & Knight 1985), which extend up to some 100 km south of the present day coastline, see Fig. 12a. The size of the alluvial fan systems varies along strike, indicating a high-relief sediment provenance in the east but, substantially lower relief in the west. Evidence for the tectonic processes which led to the reactivated uplift of the provenance area are not seen in north Alaska, but Smith (1986), noted that early Triassic, coarse-grained clastic deposition, related to Uralian orogenesis, is a widespread occurrence throughout the circum-Arctic region.

Subsidence, following the cessation of uplift, led to the second marine transgression of the Upper Ellesmerian megasequence. Alluvial fan deposits were reworked into the informally named Eileen Formation before open-marine conditions were fully re-established. During this period, a facies association known as the Shublik Formation (Leffingwell 1919; Detterman 1970; Tortelot & Tailleur 1971) was deposited. Glauconitic, shoreface sands, silts and bioclastic limestones pass laterally to phosphatic, black

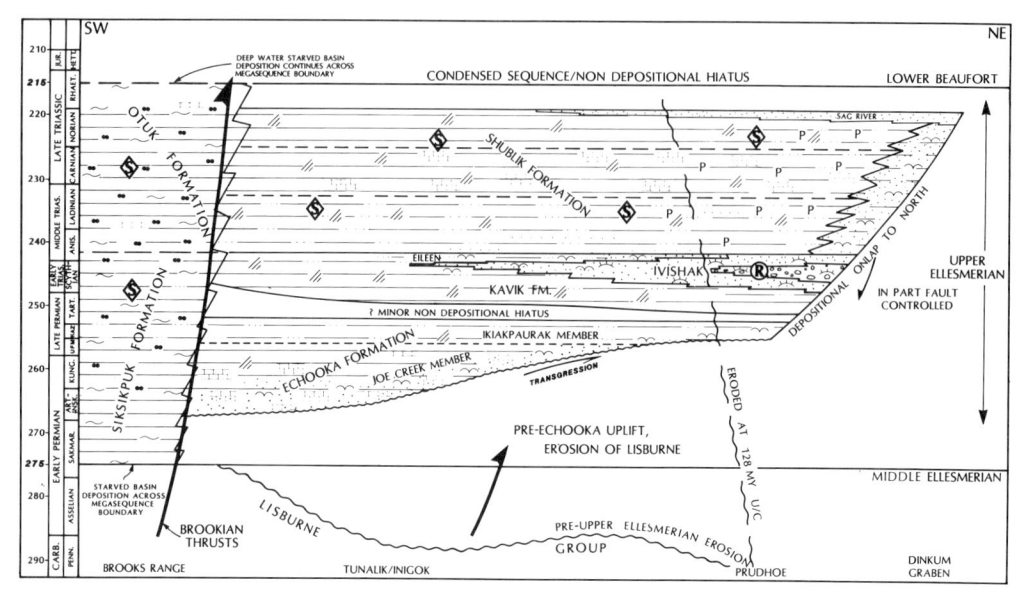

Fig. 11. Summarized chronostratigraphy, Upper Ellesmerian megasequence. See the legend of Fig. 9 for an explanation of gross depositional environment symbols. Lithology symbols as follows: Coarse-grained clastics, dotted. Silts, dotted ruled lines. Fine-grained clastics, solid ruled lines. Limestone-dominated deposits, open bricks. Phosphatic deposits, 'P'. Cherts, two small solid circles. The 'S' symbol in a diamond denotes a hydrocarbon source rock and the 'R' symbol in a circle denotes a hydrocarbon reservoir rock.

shales and grey limestones (rich in Monotis and Halobia bivalves), in turn passing to shelfal, black, organic-rich and calcareous shales. Deposition of these shelfal, phosphatic and organic-rich shales may indicate regional upwelling of oceanic water from the south onto the north Alaskan shelf (Parrish 1986), see Fig. 12b.

A minor, regressive episode in the late Triassic marked the end of the Upper Ellesmerian deposition. Fine-grained glauconitic, bioturbated sands of the Sag River Formation (Alaska Geological Society, 1970–1971; Barnes 1985) extend a short distance south onto the north Alaska Ellesmerian shelf.

Basinal deposits accumulated to the south of the north Alaska shelf throughout deposition of the Upper Ellesmerian megasequence. The age equivalent basinal deposits to the first transgressive–regressive cycle on the shelf are the cherts, shales and silts to the Siksikpuk Formation (Ellersieck et al. 1983). Basinal equivalents to the second shelfal cycle are represented entirely by black, hemipelagic shales and varicoloured, radiolarian cherts, which suggest a complete termination of clastic transport to the basinal area. These rocks are known as the Otuk Formation (Mull et al. 1982), which are partially preserved in allochthons of the west and central Brooks Range.

Hydrocarbon source rock facies are widespread in the Upper Ellesmerian megasequence. The phosphatic and black shale facies of the Shublik Formation, see Fig. 12b, have been considered as the major source rock for much of the oil seen today on the Kuparuk–Prudhoe High (Seifert et al. 1979; Magoon & Claypool 1981). The basinal deposits also preserved very large quantities of marine organic material with good source rock potential, some of which have lithified as oil shale (Chapman et al. 1964).

The Upper Ellesmerian Ivishak Formation is the major reservoir unit on the North Slope of Alaska and reservoirs the supergiant Prudhoe Bay Field, (Jones & Speers 1976; Jasmison et al. 1980). Proximal, coarsening-up, fan-delta sands and conglomerates have high ratios of net reservoir sand to gross sand and, where the formation was uplifted in the early Cretaceous and subareally exposed over the crest of the Barrow Arch, primary porosity is enhanced by diagenetic dissolution of unstable detrital chert and carbonate cement (Melvin & Knight 1985). Reservoir characteristics deteriorate rapidly with depositional changes (decreasing grain size) away from the major sediment provenance and with distance away from the diagenetic porosity enhancement at the subcrop zone. The

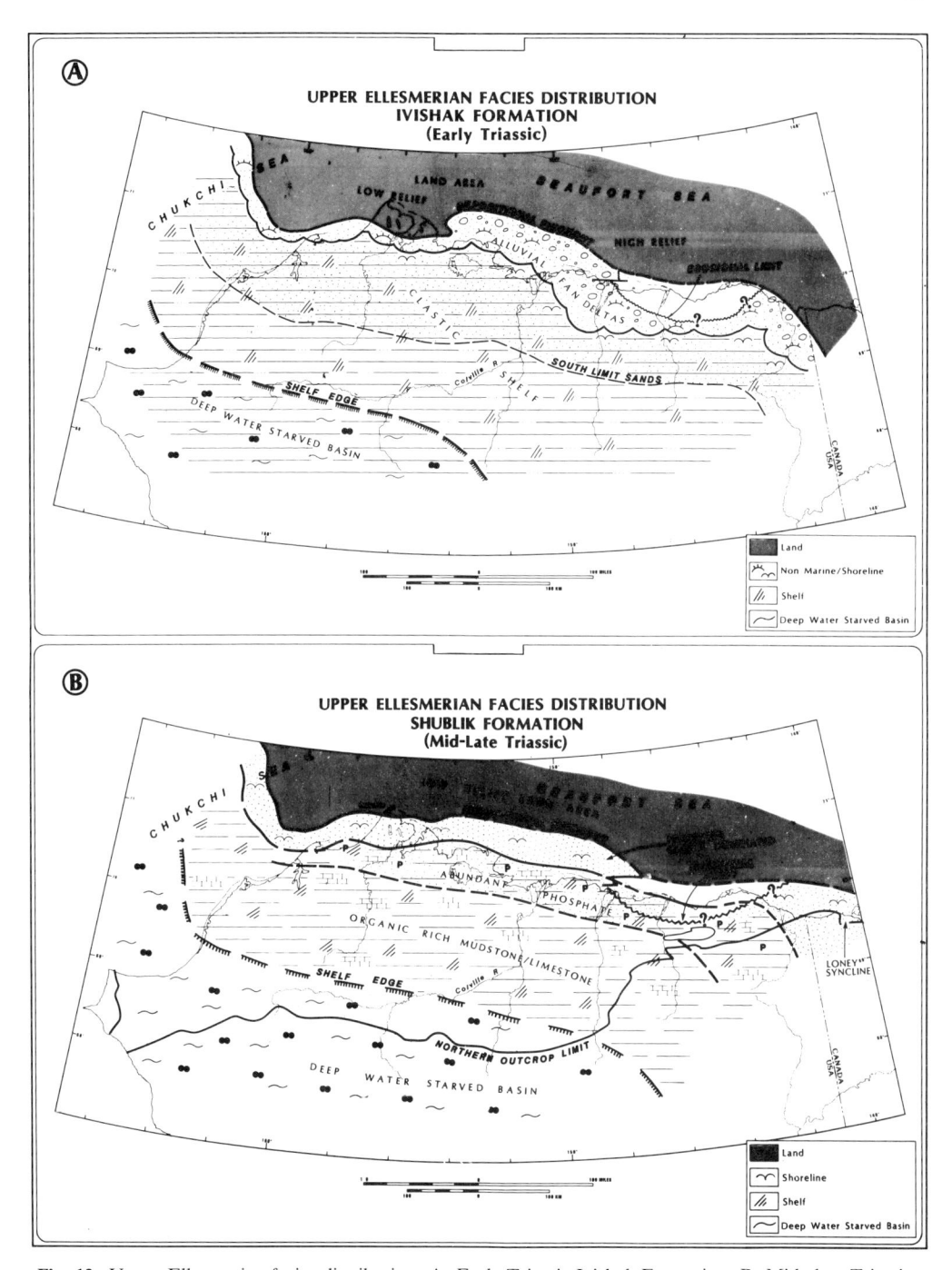

Fig. 12. Upper Ellesmerian facies distribution: A, Early Triassic Ivishak Formation; B, Mid–late Triassic Shublik Formation. Lithology symbols as follows: Conglomerates, open, irregular circles. Sands, dotted. Silts, dotted ruled lines. Fine-grained clastics, ruled lines. Limestone-dominated deposits, open bricks. Phosphatic deposits, 'P' Cherts, two small solid circles.

other, thin clastic units in the megasequence form secondary reservoir targets.

Ellesmerian resource potential

In summary, the Ellesmerian of north Alaska contains a wealth of hydrocarbon source rocks and reservoir rocks. In the basinal area to the south, oil-prone source rocks were deposited at very slow rates over some 150 Ma (early Carboniferous to late Triassic). No carrier beds were, however, juxtaposed with those source rocks.

The major source rock/carrier bed system is the Shublik/Ivishak couplet. During the Ellesmerian, the depth of burial was insufficient to trigger hydrocarbon generation and the stratigraphy and dominantly passive, post-orogenic subsidence was not conducive to hydrocarbon trap development. Hence, at the end of the Ellesmerian (late Triassic), good quality source rocks and reservoir rocks were preserved in thermally immature and largely unstructured sedimentary basins.

Beaufortian plate sequence

Transition from Ellesmerian to Brookian tectonic regimes

The Devonian to Recent tectonic history of north Alaska has been dominated by orogenic processes. For 160 Ma from the time of the late Devonian Ellesmerian orogeny, north Alaska was located at or near the northern margin of the Ellesmerian orogenic basin. A second major orogenic regime was imposed by the Brookian Orogeny, which has continued for some 150 Ma, from the mid to late Jurassic to the present day. Brookian orogenesis caused a complete reversal of north Alaskan basin polarity from that seen in the Ellesmerian. Brookian sediments have been derived from the south.

This transition from northern to southern orogenic provenance areas is one of the most remarkable features of north Alaskan geology and has long been documented (Leffingwell 1919; Mertie 1930; Eardley 1948; Payne & Gryc 1951; Payne *et al.* 1952). Onshore outcrop geology does not, however, reveal the major extensional tectonic episode which is the key to understanding the nature and timing of this reversal in basin polarity. The evidence lies under the North Slope and on the continental shelf, where a Beaufort Sea rifting episode has been described (Tailleur 1969, 1973; Rickwood 1970; Grantz *et al.* 1979; Grantz & May 1983).

We present a new interpretation documenting the significance of this episode, based primarily on our interpretation of North Slope and Beaufort Sea seismic data, but also drawing on the geology of the Mackenzie Delta area.

The Beaufortian plate sequence records the full extent of the rifting episode which opened the oceanic Canada Basin and initiated the reversal of north Alaska basin geometry. Extensional processes persisted from the early Jurassic to early Cretaceous with a duration of approximately 100 Ma. Although the Beaufortian has not been explicitly recognized in the literature, the major elements of our interpretation have been described by earlier authors (i.e., Grantz & May 1983; Tailleur 1973; Bee *et al.* 1984; McWhae 1986). The Beaufortian plate sequence consists of two associated depositional megasequences, which record Jurassic and early Cretaceous rifting episodes. Rifting processes removed the northern land area, which had shed sediment to the south throughout Ellesmerian time.

The plate sequence isopach, illustrating the combined thickness of the Lower and Upper Beaufortian megasequences, is shown in Fig. 13. A depositional edge is mapped to the west while the former southern extension of the plate sequence is now overridden by the Brooks Range. A depositional edge is also seen onlapping the Canadian Shield in the east, while to the north, Beaufortian sediments extend into the oceanic Canada Basin. Beaufortian depocentres display two major structural trends; an east–west trend on the north Alaska continental margin and a north–south trend in northwest Canada. The Dinkum Graben, located to the north of Prudhoe Bay, is the major east–west depocentre in north Alaska and contains sediment thickness in excess of 3 km. In northwest Canada, the major north–south depocentres are the Blow Trough and Kugmallit Trough. These depocentres are located in the Eskimo Lakes area and contain in excess of 4 km and 3 km of sediment respectively.

Lower Beaufortian megasequence

A chronostratigraphic chart, giving summarized basin stratigraphies for the Beaufortian plate sequence, is shown in Fig. 14. The Lower Beaufortian megasequence is of early to late Jurassic age with a duration of 70 Ma (215–145 Ma). It is recognized both in north Alaska and northwest Canada. A two-fold subdivision is made, with Sequence 1 of Hettangian to Bajocian age and Sequence 2 of Bathonian to Tithonian age.

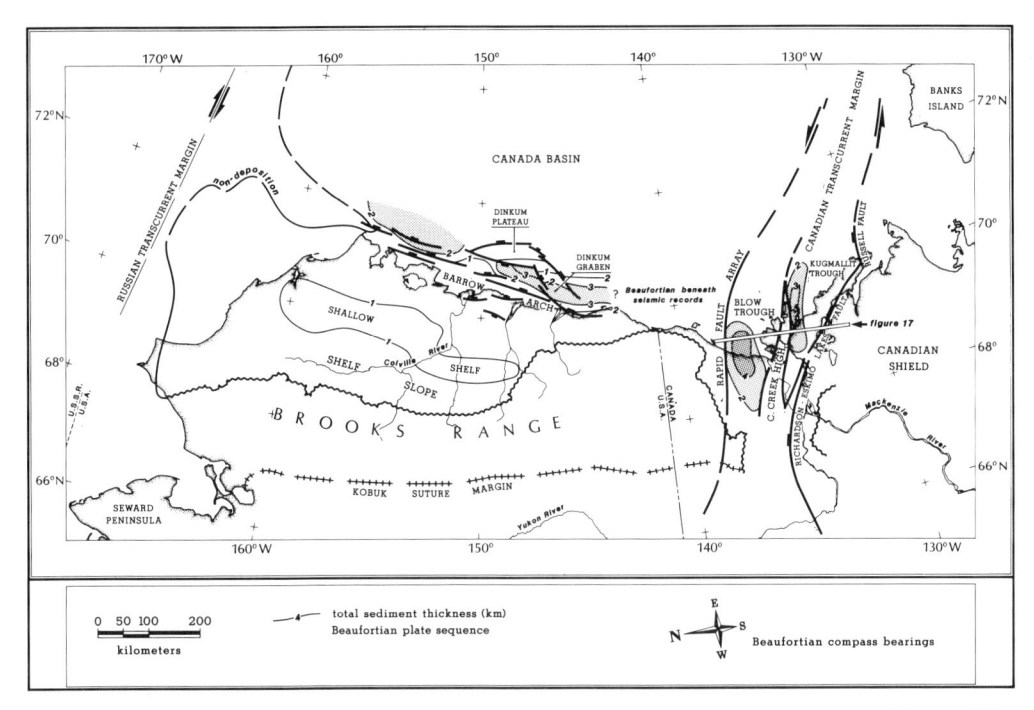

Fig. 13. Sediment thickness, Beaufortian plate sequence. The map illustrates the combined thickness of the two Beaufortian megasequences. It was constructed using seismic isochrons converted to depth.

In north Alaska, Lower Beaufortian basin geometry displays two principal elements; a south-facing marine shelf (Kirschner et al. 1983), and, in marked contrast to earlier Ellesmerian basin geometry, an extensional basin system. (See Fig. 16a.) **Sequence 1** is entirely marine and consists largely of silts and shales of the Lower Kingak Formation (Detterman et al. 1975). Two transgressive-regressive subsequences, the Barrow and Simpson subsequences, are recognized, both of which show fine-grained clastic progradation to the southeast across the marine shelf. Condensed/starved basin deposition occurred to the south of the shelf as represented by the cherts of the Otuk Formation (Mull et al. 1982; Mayfield et al. 1983a, b).

Sequence 2 of the Lower Beaufortian megasequence overlies an early Bathonian (175 Ma) regional disconformity, which is correlatable throughout north Alaska, the Yukon–Mackenzie area of northwest Canada and the Sverdrup Basin of the Canadian Arctic Islands (Embry, 1985). It is equivalent to the upper part of the Kingak Formation in north Alaska (Detterman et al. 1975). Non-deposition during the latest Jurassic/earliest Cretaceous (145 Ma) led to the development of a regional disconformity surface at the top of the Lower Beaufortian megasequence.

The major feature of the Lower Beaufortian Sequence 2 time interval was the development of a series of active, extensional half-grabens along the Arctic margin of north Alaska. The structural configuration of the 175 Ma regional unconformity records the initiation of major graben formation, see Fig. 15. The half-graben system, which includes the Dinkum Graben, was controlled by south-throwing faults. We interpret in excess of 2 km of late Jurassic (Sequence 2) sediments in the axis of this system, and seismic profiles show that the provenance area still lay to the north of north Alaska.

The Lower Beaufortian megasequence of north Alaska contains a major hydrocarbon source rock unit. This is informally termed the Low Velocity Unit (LVU), due to its character on sonic well-logs. It is an organic-rich shale unit within the Lower Kingak Formation. Its distribution is largely restricted to the anoxic environment which developed beyond the break-of-slope of the south-facing Lower Beaufortian shelf. Hemipelagic, starved-basin deposits beyond the slope (Otuk Formation)

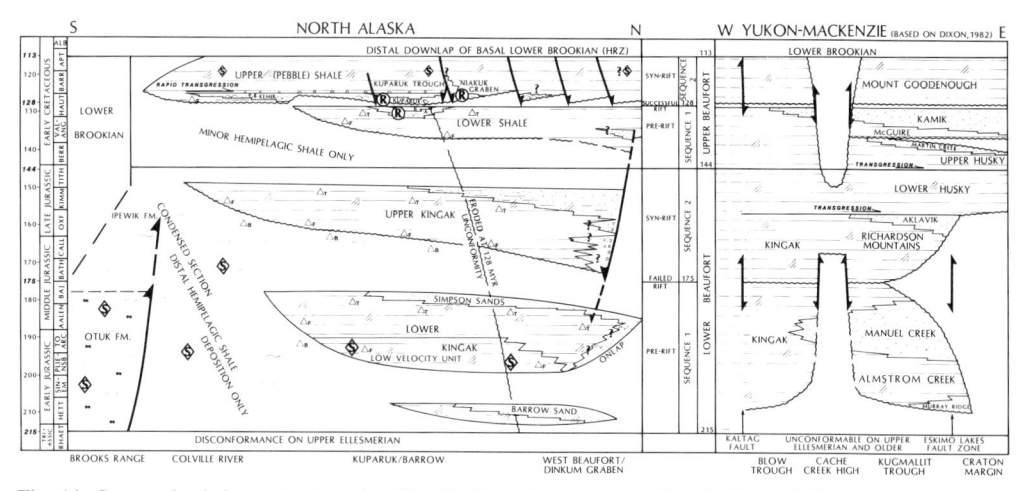

Fig. 14. Summarized chronostratigraphy, Beaufortian plate sequence. See the legend of Fig. 9 for an explanation of gross depositional environment symbols. Lithology symbols as follows: Coarse-grained clastics, dotted. Fine-grained clastics, ruled lines. Cherts, two small solid circles. The 'S' symbol in a diamond denotes a hydrocarbon source rock and the 'R' symbol in a circle denotes a hydrocarbon reservoir rock.

also have significant oil-prone source potential (Bodnar 1984) see Fig. 14 & 16a.

Reservoir sands are poorly developed in north Alaska. The Barrow and Simpson shelf-sands are typically very fine to fine-grained, argillaceous and extremely bioturbated with low permeability. The Barrow Sand provides the reservoir for a minor gas accumulation at Point Barrow.

In northwest Canada, the basal boundary to the Lower Beaufortian megasequence has been recognized where Jurassic strata overstep to rest directly on basement (i.e., Miall, 1973). Sediments in this area are marginal to the North American craton and face a westerly deepening ocean where facies patterns have been described in detail (Young et al. 1976; Dixon 1982a, b). Sequence 1 equivalent deposits are the Hettangian to Bajocian shales and sands of the Murray Ridge, Almstrom Creek and Manuel Creek Formations, interpreted as sediments derived from the Canadian craton and reworked into offshore sand bars (lower Bug Creek Group; Poulton 1982). Sequence 2 consists of coarsening-up cycles, interpreted as marine shelf with barrier islands, offshore sandbars and shoals, of the Richardson Mountains and Aklavik Formations (upper Bug Creek Group), and lower member of the Husky Formation (Young et al. 1976). Facies coarsen and become more proximal to the southeast and pass distally to Kingak-equivalent shales further west. Limited hydrocarbon source rock facies have been described within these shelfal deposits (Husky Formation; Langus 1980).

Upper Beaufortian megasequence

The Upper Beaufortian megasequence is early to mid-Cretaceous in age with a duration of approximately 32 Ma (145−113 Ma). It is shown in the chronostratigraphic chart of Fig. 14. It overlies a basal disconformity surface representing a significant time gap, from Tithonian to Berriasian, and was deposited during a period of extension, which marked the successful rifting episode in the Canada Basin. Two sequences are identified. Sequence 1 (Valanginian to mid-Hauterivian) was deposited during the transition from failed Jurassic rifting to the onset of the successful Neocomian rift. Sequence 2 (late Hauterivian to late Aptian) was deposited during the successful rifting episode, which opened the Canada Basin and removed the northern sediment provenance area that had been present throughout Ellesmerian and early Beaufortian time.

In north Alaska, rift margin uplift during the Valanginian formed the east−west Barrow Arch and rapidly modified the Lower Beaufortian tectonic framework. Upper Beaufortian Sequence 1 is absent over the crest of this east−west trending high, which became a local sediment provenance. Erosion of sub-areally exposed Ellesmerian strata led to sediment accumulation in local troughs along this trend, one of which is the Kuparuk Trough, (Carman and Hardwick, 1983; Materson and Paris, 1986).

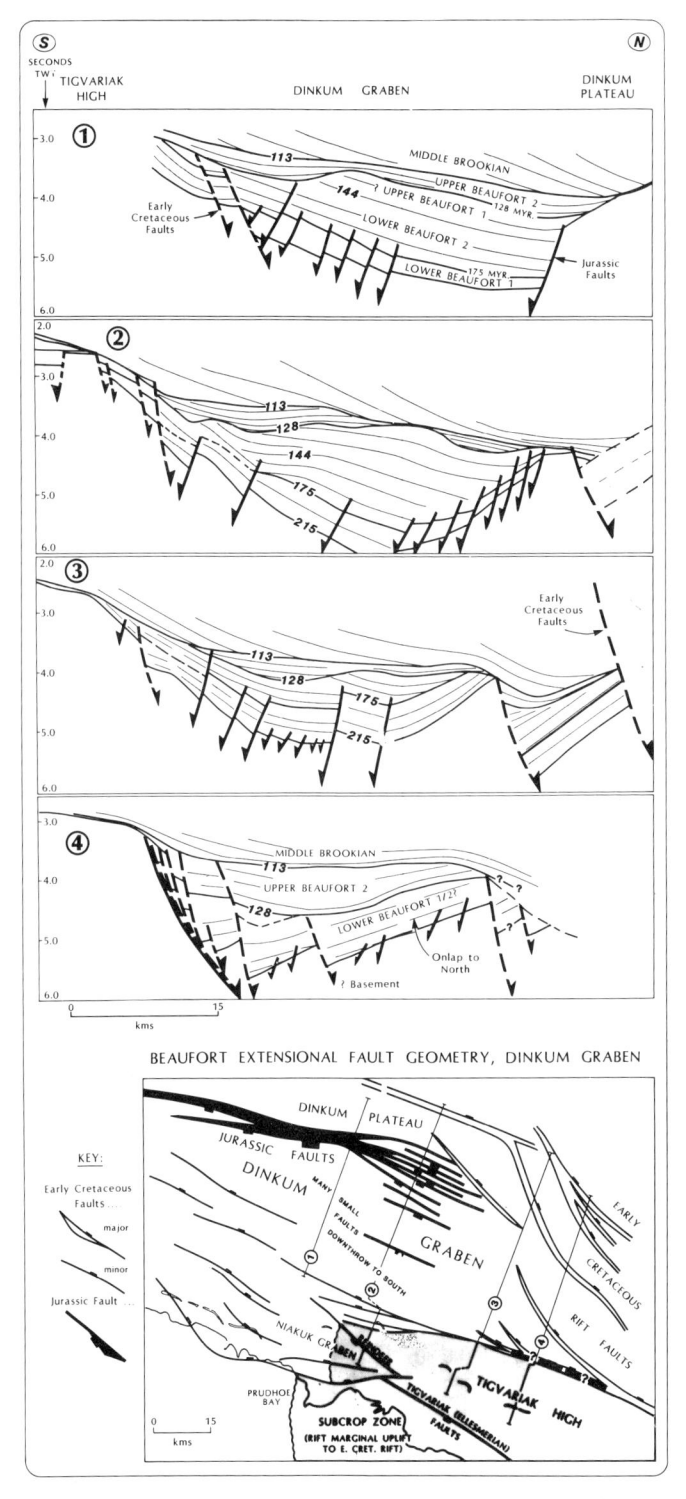

Fig. 15. Basin cross sections, Dinkum Graben. The four lines of cross section are shown on the map. The ages of the regional unconformities are shown in million years after the time scale of Harland *et al.* (1982).

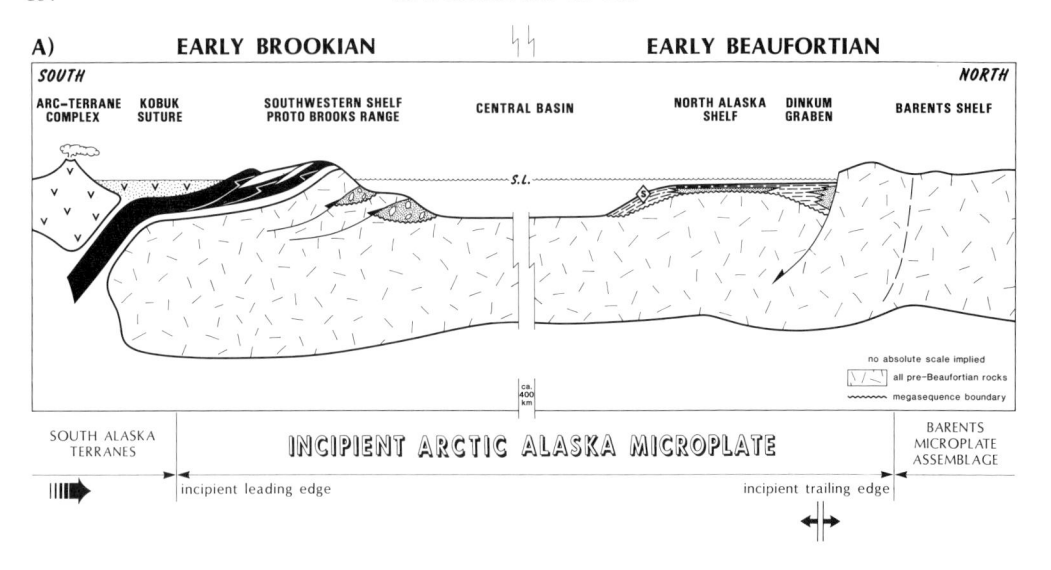

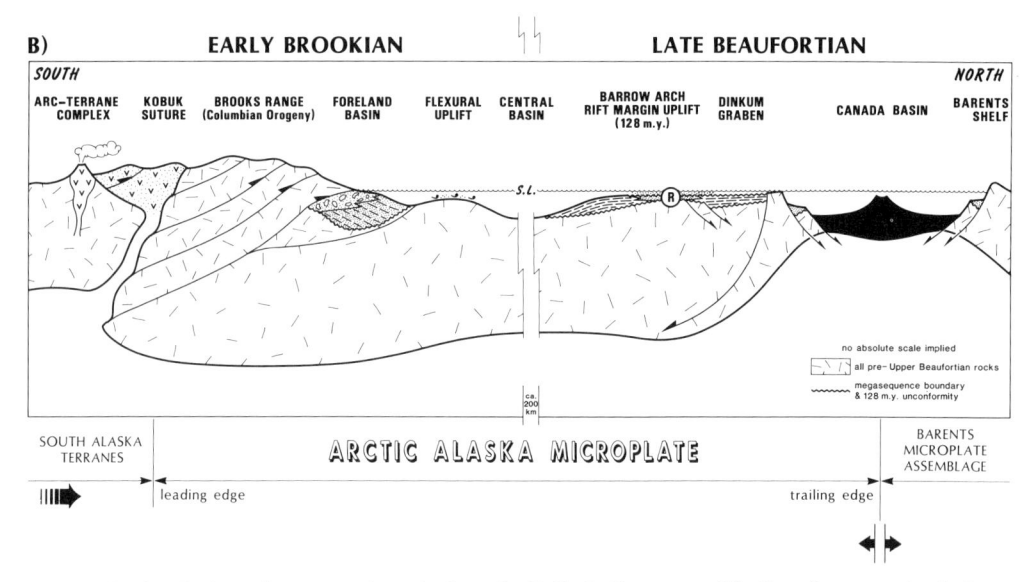

Fig. 16. Regional schematic cross sections: A, Jurassic, B, Early Cretaceous. The line of cross section is shown on Fig. 18. Note that approximately 400 km of the Central Basin is not shown in cross section A and 200 km is not shown in cross section B, indicating northward movement of the Brooks Range during this time interval. Solid black denotes oceanic crust. Lithology symbols as follows: Volcanics, 'V'. Conglomerates, irregular, open circles. Sands, dotted. Fine-grained clastics, dashed lines. Shell hash coquina, shell symbols. The 'S' symbol in a diamond denotes a hydrocarbon source rock and the 'R' symbol in a circle denotes a hydrocarbon reservoir rock.

The fine-grained, Kuparuk 'A' sands were captured in the Kuparuk trough and form an important component of the reservoir in the Kuparuk River Field.

We infer Upper Beaufortian Sequence 1 sediments infill remnant topography in the failed rift half-graben system (Fig. 15 although the base Upper Beaufortian boundary cannot be identified on seismic data in the Dinkum Graben area.

A southerly, prograding mud-prone shelf persisted to the south off the developing rift system, (Miluveach Formation; Carman & Hardwick, 1983). It was, however, of more limited extent than the Lower Beaufortian shelf and the sediment which was transported beyond the shelf edge, became interbedded with coeval, distal Brookian sediments derived from the south (Molenaar, 1981a, 1983).

In northwest Canada, early deposition of the Upper Beaufortian Sequence 1 consisted of Berriasian, dominantly nonmarine sandstones (Martin Creek Formation, Dixon, 1982a; Parsons Sandstone; Cote et al. 1975), which form the reservoir in the Kugpik and Parsons Lake Fields, see Fig. 17. Non-marine deposition was followed by marine transgression, which flooded the developing transtensional basins and a thick shale and sand succession accumulated (Dixon 1982a, b).

Deposition of Upper Beaufortian Sequence 1 terminated at the acme of rift margin uplift. Locally a very important angular unconformity developed during the mid-Hauterivian (128 Ma) along the crest of the rift margin uplift (Barrow Arch) in north Alaska and in northwest Canada. In north Alaska, deposition of the Upper Beaufortian Sequence 2 was strongly controlled by rift tectonics. The east–west rift margin uplift acted as a major provenance for sediments shed into the developing rift system, within which fault-blocks show progressive down-faulting to the north, see Fig. 16b. This fault pattern marks a reversal in basin polarity from the down-to-the-south faults of the Jurassic Dinkum Graben (Fig. 15).

The pattern of lithofacies distribution in the Upper Beaufortian Sequence 2 of north Alaska was controlled by the relative elevation between the areas of rift-margin uplift and the areas of fault-driven subsidence. Both regressive and transgressive facies patterns developed. Sediment type was locally controlled by the nature of the strata being eroded from the rift margin uplift. These strata included the entire suite of Ellesmerian and some of the Franklinian lithofacies. Initially, deposition both to the north and south was entirely regressive. Sediment was transported away from the uplifts and accumulated in the topographic lows on the 128 Ma unconformity surface; either just above sea level, e.g. the fluvial Put River Sandstone (Jamison et al. 1980); or just below sea level, e.g. the Kuparuk 'C' Sands (Carman & Hardwick 1983; Masterson & Paris 1986), the Walapka Sandstone (Hayward & Legg 1983), and the Kemik Sandstone (Mull et al. 1986). As the elevation of the rift margin uplifts became reduced, marine transgression inundated the highs.

Sediments were reworked into transgressive sheets that migrated across the eroded subcrop zone. Only the highest standing uplifts remained above sea level and coarse clastic input was largely cut off. Deposition became dominated by the informally named Pebble Shale (Robinson 1959; Molenaar 1983) or Kalubik Formation (Carman & Hardwick 1983). This unit is a black shale containing very poorly sorted (but well rounded) sand, to pebble or cobble sized clasts of quartz and chert.

In northwest Canada, the intra-Hauterivian unconformity is also well developed at the base of Upper Beaufortian Sequence 2, which consists of shales and siltstones of the Mount Goodenough Formation (Dixon 1982a). Sub-

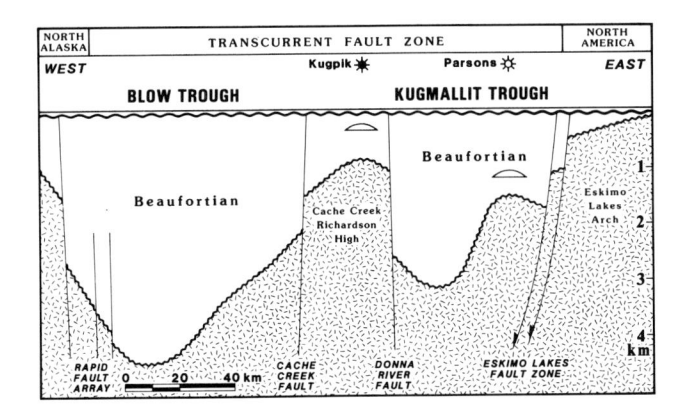

Fig. 17. Basin cross sections, Blow Trough and Kugmallit Trough. The line of cross section is shown on Fig. 13. Kugpik is a small oil field on the Cache Creek High and Parsons Lake is a gas field on the east flank of the Kugmallit Trough.

stantial relative vertical movements were controlled by wrench faulting such as the Cache Creek High and extension in the Blow Trough and Kugmallit Trough (Hea *et al.* 1980), Fig. 17. Transtensional tectonics exerted major control of sediment thickness, but the area appears to have remained regionally below sea level during the late Hauterivian and Barremian. Major regressive clastic wedges are restricted to the age-equivalents of Sequence 1 (Kamik formation; Dixon 1982a, b).

Beaufortian resource potential

The Lower Beaufortian of north Alaska contains major source rock facies on the southern prograding slope of the north Alaska shelf. The principal source rock is the Lower Kingak Low Velocity Unit, which extends basinward into thin, organic-rich basinal shales (Tailleur 1964). The presence of source facies within the rift margin depocentres is not known because the area lies almost entirely beyond the limit of well control. Depositional environments that may have been conducive to the development of source rock facies are, however, predicted such as early restricted marine circulation in the failed rift system (Kingak Formation) and during the initial marine transgression across the successful rift-system (Pebble Shale or Kalubik Formation).

Reservoir sands are thin and have a patchy distribution. The best quality reservoir sands in north Alaska are quartz arenites eroded from subcropping Ellesmerian strata along the crest of the rift margin uplift (Barrow Arch). These were redeposited as well sorted, shallow-marine, glauconitic quartzose sands (variably cemented by siderite) immediately below and above the 128 Ma unconformity i.e., the Kuparuk Formation. The best quality sand in northwest Canada is the dominantly nonmarine, quartzose Parsons Sandstone, which developed near the base of the Upper Beaufortian Sequence 1. This unit forms the reservoir in the Parsons Gas Field (Cote *et al.* 1975).

Beaufortian time is most notable for the development of the Ellesmerian subcrop zone along the crest of the rift margin uplift. This generated the initial subcrop component for the present-day combined structural/stratigraphic traps on the Barrow Arch. The subcrop zone incorporated Ellesmerian strata with favourable resource potential in addition to the Lower Beaufortian LVU source rock interval and the thin Beaufortian reservoir sands. The area was, however, regionally uplifted and regional hydrocarbon generation had not yet begun.

Constraints for the opening of the Canada Basin

Our structural and stratigraphic analysis of north Alaska and northwest Canada provides some constraints for models that describe the opening of the oceanic Canada Basin. Most positively, we have been able to use sequence analysis to document the chronology of the Beaufortian rifting episode. Limited, initial extension occurred from the early Jurassic to Bajocian (Lower Beaufortian, Sequence 1). Failed rifting, with faults downthrowing to the south, prevailed throughout the remainder of Jurassic time (Lower Beaufortian, Sequence 2). A period of restructuring and reversal of basin polarity occurred from Valanginian to mid-Hauterivian times, (Upper Beaufortian, Sequence 1) and the successful rift onset event is marked by the intra-Hauterivian (128 Ma) unconformity. The successful rift phase, with major down-to-the-north extensional faults, lasted some 15 Ma. The 113 Ma boundary, at the base of the Brookian, is interpreted as the termination of the syn-rift phase and the onset of drift in the Canada Basin.

Smith's (1986) paleoreconstructions suggested that the north Alaska margin was juxtaposed with the Lomonosov Ridge prior to and during this rifting episode, see Fig. 18. (The Alpha Ridge is a younger feature with Icelandic type crust that formed after the Canada Basin had opened (Jackson *et al.* 1985; Forsyth 1986)). Hence, in Smith's reconstruction, the northern sediment provenance for the Ellesmerian of north Alaska lay in the vicinity of the Lomonosov Ridge and the Barents Shelf of present day Arctic Europe and Siberia.

Smith's (1986) model necessarily negated the rotational opening of the Canada Basin, which has been preferred by the majority of authors (e.g., Carey 1958; Tailleur 1969, 1973; Rickwood 1970; Grantz *et al.* 1979; Newman *et al.* 1979; Sweeney 1983; Harland *et al.* 1984). Our observations in northwest Canada are, however, inconsistent with the compressional pivotal area which is required by the rotational model. We observe Beaufortian extension and strike-slip tectonics in the Blow Trough and Kugmallit Trough rather than compression. Evidence in favour of a rotational opening is taken, by some workers, not from the implied pivotal area of northwest Canada, but from unpublished paleomagnetic data (Grantz pers. comm.). Those data discussed by Hillhouse & Gromme (1983), however, offered no conclusive support.

As proposed by Bally (1976) and Dutro

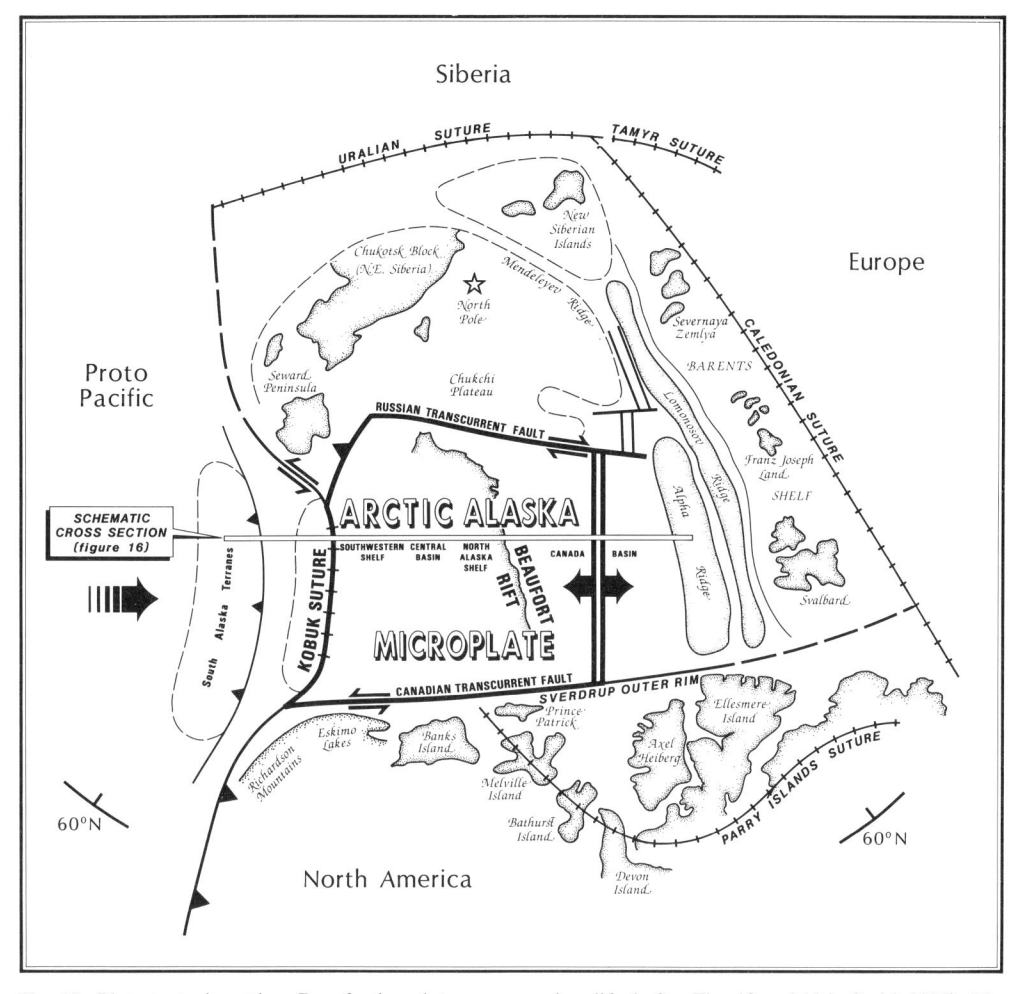

Fig. 18. Plate tectonic setting, Beaufortian plate sequence (modified after Figs 10 and 11 in Smith 1986). The four boundaries to the Arctic Alaska Microplate are: (i) the spreading centre in the Canada Basin, (ii) the Kobuk Suture of the southern Brooks Range, (iii) the Canadian transcurrent fault margin, and (iv) the Russian transcurrent fault margin.

(1981), Smith's 1986 model for the opening of the Canada Basin also required substantial left-lateral transcurrent fault motion to accommodate the opening of the oceanic Canada Basin and to move Arctic Alaska past Arctic Canada. In this scenario, strong evidence for early to mid-Cretaceous strike-slip faulting should be present along the outer rim of the Sverdrup Basin (Meneley 1975; Kerr 1981) and in northwest Canada.

Our present interpretation supports the onset of synchronous (Hauterivian, 128 Ma) faulting in north Alaska and northwest Canada. In addition, the Blow Trough and Kugmallit Trough (Fig. 17) display the anatomy of extensional pull apart basins, which are more likely to have developed on a transtensional strike-slip fault margin rather than within a compressional pivotal zone. Strike-slip faulting as old as Hauterivian has been recognized from field mapping (Balkwill et al. Young et al. 1976) but right-lateral, rather than left-lateral motion has been assumed. There is, however, a strong, right-lateral Brookian overprint in northwest Canada that has been identified by many workers (e.g., Jones 1982). Additional field work is required to document the alternative solutions for left-lateral and right-lateral movement during the Beaufortian.

As a working hypothesis, we support the concept of left-lateral transport of Arctic Alaska as a separate microplate past Arctic Canada. The reconstruction shown in Fig. 18 requires left-lateral movement in excess of 1000 km between the Hauterivian and the Senonian, which was the time of cessation of sea floor spreading in the Canada Basin (Embry 1985; Lawyer & Baggeroer 1983).

The Kobuk suture zone along the southern edge of the Brooks Range (Figs 13, 16 & 18) represented the former leading edge of the 'Arctic Alaska Microplate'. The trailing edge was the Beaufort rift margin. Lateral, or transcurrent fault margins, formed along the outer edge of the Sverdrup rim in Arctic Canada and probably also along the prominent Northwind Escarpment of the Chukchi Plateau (Smith 1986).

Brookian plate sequence

Initiation of Brookian orogenesis

The onset of Brookian orogenesis occurred as a result of collision between 'Pacific realm' exotic terranes and Arctic Alaska in the mid Jurassic (Richter & Jones, 1973), see *Fig. 16a*. Oceanic crust and island arc material were obducted onto Arctic Alaska in the mid to late Jurassic (Patton & Box 1985; Boak *et al.* 1985; Roeder & Mull 1978; Patton *et al.* 1977). The obducted oceanic material has been mapped as the Angayucham Terrane (Fritts 1971; Dillon *et al.* 1981), and the tectonic contact with autochthonous Arctic Alaska has been termed the Kobuk suture zone by Mull (1982, 1985).

Compressional tectonic processes in southern Arctic Alaska were heightened after the Hauterivian (128 Ma) rift-onset event in the Canada Basin (Hitzman *et al.* 1982), see Fig. 16b. The timing of maximum crustal shortening and emplacement of the major allochthons, 145–115 Ma, is broadly synchronous with the Columbian orogeny in the Canadian Rocky Mountains (Douglas 1972). The large number of K–Ar radiometric dates from schists and granitic plutons suggest that the majority of crustal shortening in the west and central Brooks Range had occurred by the late Albian (100 Ma) (Mull 1983; Turner *et al.* 1979).

Basin development and depositional system

The reversal of basin polarity in north Alaska and northwest Canada has been documented by analysis of the Beaufortian plate sequence. The Brookian plate sequence represents a complete change in structural and stratigraphic style and, by definition, it includes all those sediments that were shed across north Alaska from the developing Brooks Range orogenic belt in the south. The oldest, clearly attributable Brookian sediments are of mid to late Jurassic age. These, and Neocomian Brookian strata, are coeval with Beaufortian strata but were deposited hundreds of kilometers to the south of the present Brooks Range. As a result of severe crustal shortening, they are now partially preserved in the allochthons of the Brooks Range (Martin 1970; Mull 1982, 1985; Mayfield *et al.* 1983).

Brookian geological evolution demonstrates a progressive migration of the thrust front, and associated foredeeps, to the north and east, to the point today where active overthrusting of the Canada Basin passive margin is occurring in the eastern Beaufort Sea. Enormous quantities of orogenic sediments have infilled the foredeeps and prograded across the Beaufortian rifted continental margin. Fluvial, deltaic, shelf, slope and basinal depositional environments are widely developed within the Brookian plate sequence (Ahlbrandt 1979; Molenaar, 1983; Young *et al.* 1976).

Three, discrete pluses of Brookian orogenesis are recorded by the three megasequences. The distribution of each orogenic pluse is geographically distinct and has resulted in regional changes to the structural framework of the study area, see Fig. 19. The Lower Brookian megasequence records the southernmost, early to mid-Cretaceous (Columbian) mountain front; the Middle Brookian megasequence records the late Cretaceous to Paleogene (Laramide) mountain front and the Upper Brookian megasequence records the northernmost, Neogene mountain front. Grantz & Mull (1978) have described thrust folds beneath the coastal plain of ANWR and Grantz *et al.* (1986) detailed the offshore extension of the Neogene fold belt in the eastern Beaufort Sea. We have named this third pulse of Brookian orogenesis the Camden Orogeny, due to its strong development in Camden Bay, which is located in northeast Alaska.

The isopach map for the Brookian plate sequence is shown in Fig. 20. By reference to the total sediment thickness isopach map (Fig. 5), it can be seen that the Brookian plate sequence represents the largest component of total sediment accumulation in the study area. Major foreland basins have developed in front of the Brooks Range and contain more than 8–10 km of sediment. More than 8 km of Brookian sediment have also accumulated on the passively

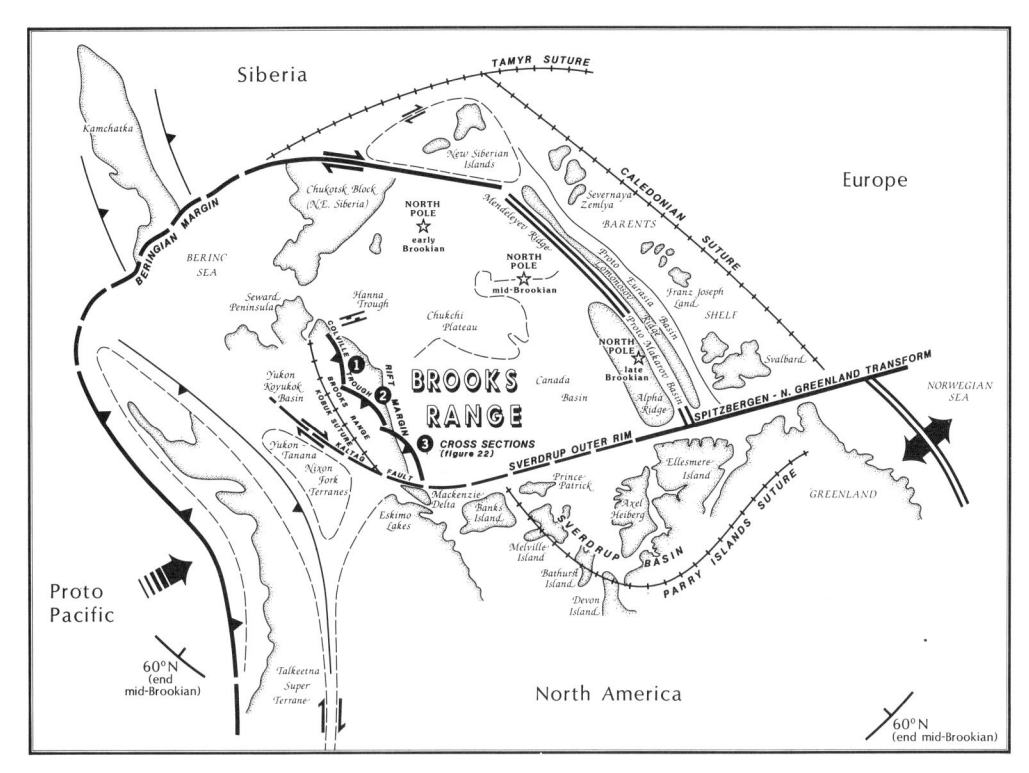

Fig. 19. Plate tectonic setting, Brookian plate sequence (modified after Figs 12 and 13 in Smith 1986). Brooks Range mountain fronts are: (1) Columbian, (2) Laramide and (3) Camden, as shown in the schematic cross sections of Fig. 22.

subsided Beaufortian rifted margin. The locus of deposition in northwest Canada is the Mackenzie Delta which overlies the subsided, Beaufortian transtensional basins.

Lower Brookian megasequence

The Lower Brookian megasequence is possibly as old as middle Jurassic (Bajocian) but mainly Tithonian to early Cenomanian in age. It had a duration of some 55 Ma (150–95 Ma). The basal boundary is a tectonic discontinuity and/or a significant unconformity, which truncates strata of Mississippian to early Cretaceous age. Lower Brookian deposition comprises three regressive sequences. Sequence 1, Bajocian, dominantly late Jurassic to Hauterivian (150–125 Ma) is poorly known from isolated outcrops in the Brooks Range. The relationship of Sequence 1 to Sequence 2, early Albian (113–108 Ma), is largely obscured by tectonism. Lower Brookian deposition in the subsurface of north Alaska and the Beaufort Sea is mainly represented by Sequence 3, early/mid Albian to early Cenomanian (108–94 Ma). A chronostratigraphic chart is shown in Fig. 21.

The first evidence of Sequence 1 deposition is a southerly derived tuffaceous turbidite section containing a Bajocian ammonite fauna in the central Brooks Range (Jones & Grantz 1964). The major lithostratigraphic component is the Opikruak Formation (Gryc *et al.* 1951), which has been dated as Tithonian and younger, and is known from more widespread allochthonous outcrops in the west and central Brooks Range. Equivalent strata have been referred to as the Kongakut Formation by Detterman *et al.* (1975) in the east Brooks Range.

The proximal Opikruak Formation is interpreted as an olistostrome deposit. This was deposited within the active southern thrust front and associated with early accretionary tectonics. These chaotic units are thought to have accumulated in allochthonous basins, which were carried piggy-back (Ori & Friend 1984) within the developing thrust belt, see Fig. 16a. The olistostromes pass distally into submarine fan deposits (Patton & Tailleur 1964). Subsequent

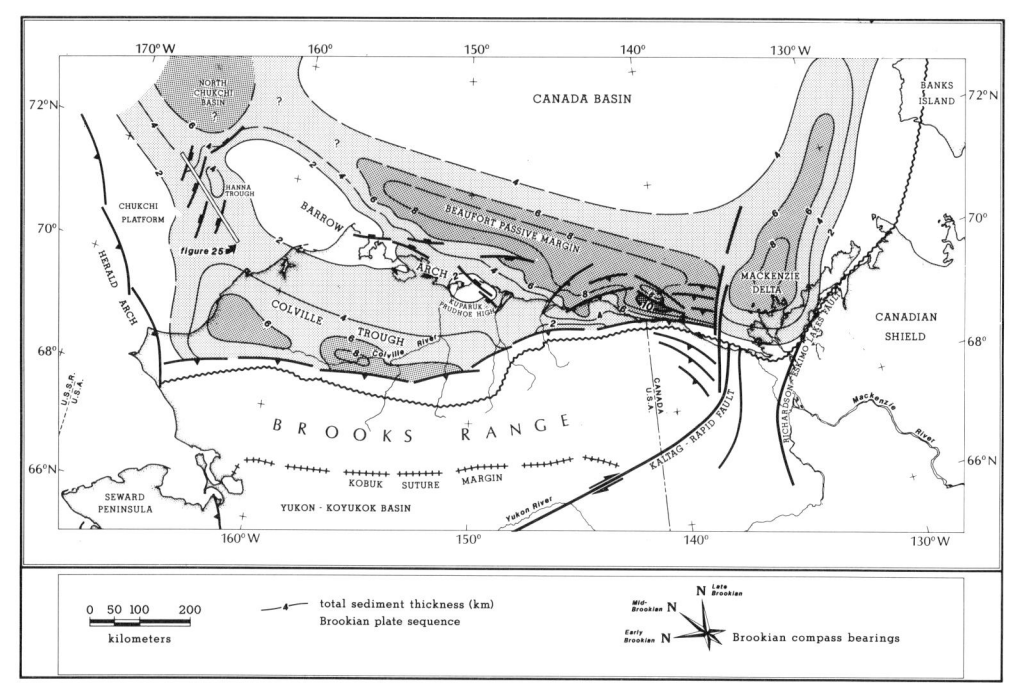

Fig. 20. Sediment thickness, Brookian plate sequence. The map illustrates the combined thickness of the three Brookian megasequences. It was constructed using seismic isochrons converted to depth.

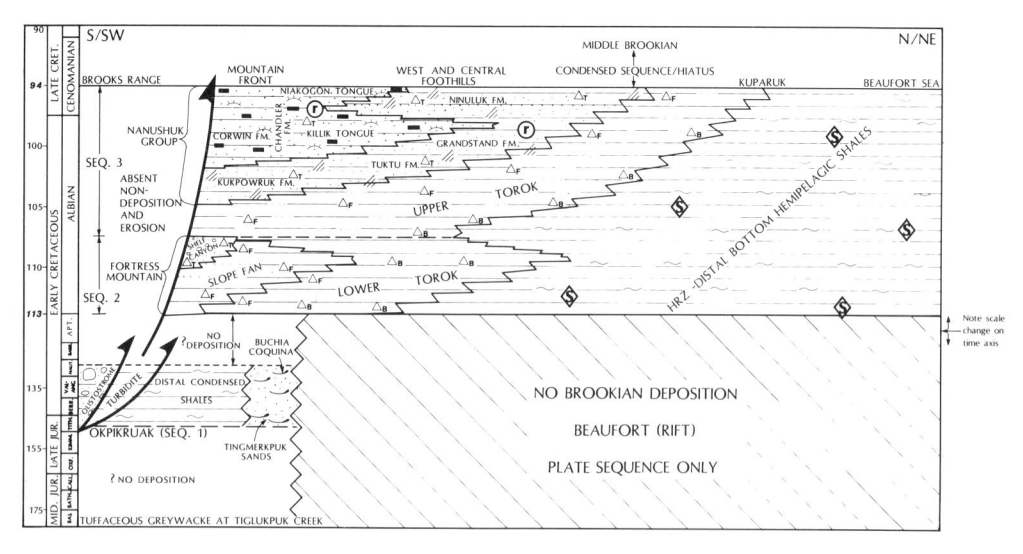

Fig. 21. Summarized chronostratigraphy, Lower Brookian megasequence. The interdigitation of coeval Brookian and Beaufortian sediments is not shown. See the legend of Fig. 9 for an explanation of gross depositional environment symbols. Lithology symbols as follows: Coarse-grained clastics, dotted. Coals, solid bricks. Fine-grained clastics, ruled lines. The 'S' symbol within a diamond denotes a hydrocarbon source rock and the 'R' symbol in a circle denotes a hydrocarbon reservoir rock.

early Cretaceous (Columbian) thrusting carried Sequence 1 units several hundred kilometres to the north, (Mayfield et al. 1983).

The Neocomian 'Buchia Coquina' and shallow-marine Tingmerkpuk Sand (Jones & Grantz 1964, Crane & Wiggins 1976) provide further evidence to help reconstruct Lower Brookian basin geometry. These deposits rest on a structurally elevated east–west trend, which lies to the north of the deep marine Okpikruak deposits. We interpret this east-west shoaling trend as a flexural bulge, which developed around the outer rim of the early Brookian foreland basin, see Fig. 16b.

Much of Hauterivian to Aptian time appears to be unrepresented in the biostratigraphic record, although Crane (1986) has suggested that Hauterivian olistostromes are present. Lower Brookian Sequence 2 generally lies with tectonic discontinuity on the deformed Sequence 1 and its distribution indicates a major northward shift of the mountain front in Barremian to late Aptian time, compare Figs 16 and 22a. Sequence 2 represents a 'flysch' fill to the Albian foreland basin, which was rapidly succeeded by the 'molasse' sediments of Sequence 3.

Sequence 2 comprises a thick classtic wedge, shed into the Colville Trough (Fortress Mountain Formation; Patton 1956; Patton & Tailleur 1964; Molenaar et al. 1981, 1983). Coal bearing, nonmarine and shallow marine delta topsets, which rapidly pass laterally into conglomerate-filled slope-feeder channels and a complex system of overlapping submarine fans, have been mapped in outcrop. The axis of the foreland basin was partly filled by middle to outer fan sediments, becoming progressively thinner bedded and passing into basin plain deposits (Torok Formation; Gryc et al. 1951; Patton 1956; – Lower Torok Formation; Molenaar, 1981a, 1983). The distalmost Sequence 2 sediments onlapped the southfacing shelf of the Beaufortian plate sequence (Kirschner et al. 1983) see Fig. 22a, and pass further northward into a thin section of distal, hemipelagic, radioactive shales, which blanketed the subsiding Beaufortian passive margin. This unit was named the HRZ (Highly Radioactive Zone) by Carman & Hardwick (1983) and has also been referred to as the Pebble Shale Gamma Ray Zone (e.g. Bird 1982).

Sequence 3, the Albian to early Cenomanian Nanushuk Group (Ahlbrandt 1979; Mull 1979; Molenaar 1981b, 1983), represents the most extensive Lower Brookian infill to the Colville Trough. The sequence is a delta/shelf/slope prograding system, in which sediment was transported from west to east along the axis of the foreland basin. Two major deltaic systems, the Corwin Delta and the Umiat Delta, have been recognized.

The early to mid Albian Corwin Delta (Chapman & Sable, 1960) was a mud-prone river-dominated system, in which proximal lithic sands and thick coals accumulated. The provenance area lay to the west, in the Herald Arch thrust complex of the Chukchi Sea (Grantz et al. 1982; Grantz & May 1984b) see Fig. 20, and in the western Brooks Range. The mid-Albian to early Cenomanian Umiat Delta (Brosge & Whittington 1966) is mapped in the central Brooks Range foothills. It has been interpreted as a wave-dominated delta system, in which sands were reworked in beach and shoreface environments.

Seismic mapping of the Lower Brookian in north Alaska shows a well developed topset (Nanushuk) to the foreset (Upper Torok Formation; Molenaar 1981b, 1983) to bottomset (HRZ) relationships, see Fig. 4. A shelf-edge extends from north of Point Barrow, east to Harrison Bay, and from there onshore to intersect the present mountain front southeast of Prudhoe Bay. Multiple phases of submarine erosion are recognized in Torok slope facies and down-to-the-basin slope failure has been described in Harrison Bay (Kerr 1985).

Lower Brookian hydrocarbon source rock facies are well developed in the bottomset depositional environment. The distal, starved-basin setting of the northern and eastern flanks of the Colville Trough and the subsided Beaufortian passive margin accumulated HRZ facies sediments, which are very organic-rich and oil-prone (see the Pebble Shale Gamma Ray Zone analyses published by Molenaar 1983). Traced south and west under the orogenic clastic wedges, these bottom-set shales become thin and less oil-prone. Bottomset shales are absent by submarine erosion over large parts of the western Beaufort Sea.

Lower Brookian reservoir development is poor. The orogenic provenance areas shed lithic and argillaceous material, which was not conducive to good porosity development or permeability retention at depth (Huffman et al., 1985; Bartsch–Winkler, 1979). The wave-reworked, delta front or shoreface, chert arenites and phyllarenites of the Grandstand Formation in the Umiat Delta represent the best known Lower Brookian reservoir sands. They form the reservoir in the non-commercial Umiat Field (Collins 1958; Molenaar 1982), which was discovered during the first phase of

A) EARLY BROOKIAN

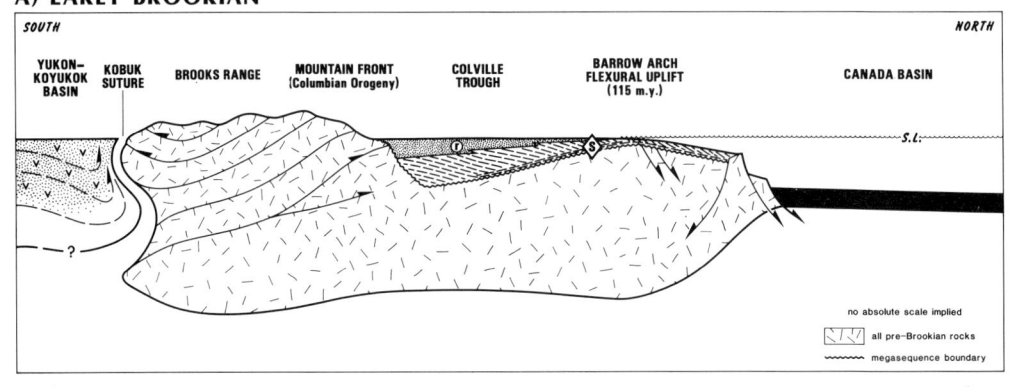

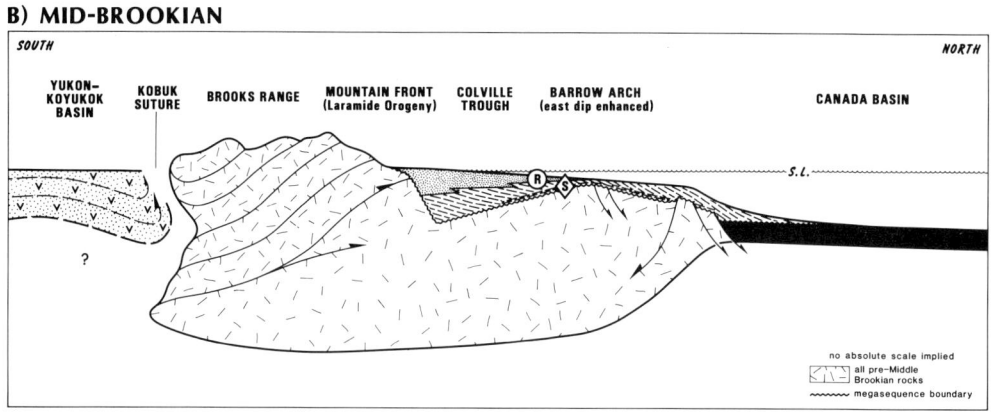

B) MID-BROOKIAN

C) LATE BROOKIAN

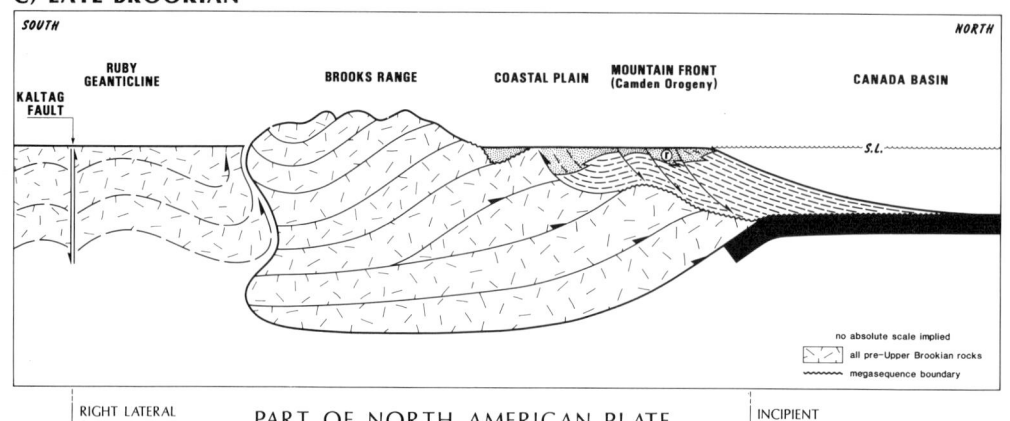

hydrocarbon exploration in the Brooks Range foothills in the post-World War II years.

Middle Brookian megasequence

The Middle Brookian Megasequence is mid-Cenomanian to early Eocene in age. It had a duration of approximately 44 Ma (94–50 Ma) and records a further major change in the tectonic framework and depocentre location, which is part of the overall progression of Brookian geology. Overthrusting affected the Beaufortian passive margin and orogenic depocentres migrated to northeast Alaska/northwest Canada. The nature of the Middle Brookian basins, bounding unconformities and lithofacies, vary greatly according to their position relative to the mountain front. Two principal depositional systems have been recognized in north Alaska, as shown on the chronostratigraphic basin summary chart of Fig. 23.

The Colville Trough depocentre shifted to the east of the Lower Brookian thick and was infilled by a discrete depositional system, the Colville Group (Gryc et al., 1951, Detterman et al., 1963). The base boundary is an erosional unconformity in proximity to the active mountain front, which becomes a downlap surface in the axis of the foreland basin, see Fig. 22b. Middle Brookian sediments also prograded beyond the foreland basin, and were transported north, across the Barrow Arch onto the passive margin. The base boundary in this area shows vigorous submarine erosion.

The Colville Group comprises a prograding clastic wedge, passing from nonmarine through shelf-topset to prograding slope, bottomset and basinal depositional environments. Four principal lithostratigraphic units have been defined. Basinal, bottomset strata are organic-rich and contain abundant bentonites. This is the Shale Wall Member of the Detterman et al. (1975), Palmer et al. (1979) and Lyle et al. (1980). The bottomset to slope deposits are shales, silts and sandstones, which were primarily deposited by sheet and channelized turbidites. Molenaar (1981a, 1983) referred to these deposits as the 'shales of the Colville Group'. Marine shelf deposits prograded over the slope environment, and sheet sand packages at the slope/topset transition include fine grained, coarsening-up units such as the informally defined West Sak Sandstone (Jamison et al. 1980; Werner 1986). Topset strata also comprise coarsening-up deltaics, overlain by braided to meandering, fluvial nonmarine strata, which include the conglomerates, sands, shales and coals of the Ugnu and Spruce Tree Formations (Werner 1986).

A second, discrete Middle Brookian depositional system extends from northeast Alaska to northwest Canada. This is the Moose Channel delta system (Mountjoy 1967a; Young et al. 1967), which infilled allochthonous basins that were carried piggy-back in the thrust front of the eastern Brooks Range. These deposits record the climax of the late Cretaceous to early Eocene Laramide orogeny. A seismic line through the mountain front was published by Hubbard, Pape & Roberts (1985b, Fig. 11b). Syndepositional tectonics have also been discussed, by McWhae (1986), Grantz & Mull (1978) and Grantz et al. (1986).

Outcrops close to the western limit of the Moose Channel system have been mapped in the Arctic National Wildlife Refuge (ANWR) by Buckingham (1986), where a thick (approximately 3 km) section of lithic, fluviatile sands, conglomerates and shales is preserved. This depositional system is best known in northwest Canada where distal bottomset strata, the Upper Boundary Creek Formation (Young 1975; Brideaux & Myhr 1976) are age-equivalent to the Shale Wall Member. The prograding marine deposits of the Tent Island Formation (Young 1975) are age-equivalent to the shales of the Colville Group. The major difference between the Colville & Moose Channel systems is in the nature of the topset strata. The Moose Channel consists of lithic and immature sediments, which were deposited in very close proximity to the developing (Laramide) orogenic highlands as a series of coalescing alluvial fans and delta lobes.

A third, discrete Middle Brookian depositional system infilled a structural low, which had developed between the uplifted Kaltag Fault margin at the eastern end of the Brooks Range and the Eskimo Lakes Arch of northwest

---

Fig. 22. Regional schematic cross sections: A, Early Brookian; B, Mid-Brookian; C, Late Brookian. The three lines of cross section are drawn through the three Brookian mountain fronts (Columbian, Laramide and Camden) shown on Fig. 19. Sediment thicknesses are diagramatic. In particular, there is more late Brookian sediment in the Canada Basin than shown. Solid black denotes oceanic crust. Lithology symbols as follows: Volcanics, 'V'. Coarse-grained clastics, dotted. Fine-grained clastics, dashed lines. The 'S' symbol in a diamond denotes a hydrocarbon source rock and the 'R' symbol in a circle denotes a hydrocarbon reservoir rock.

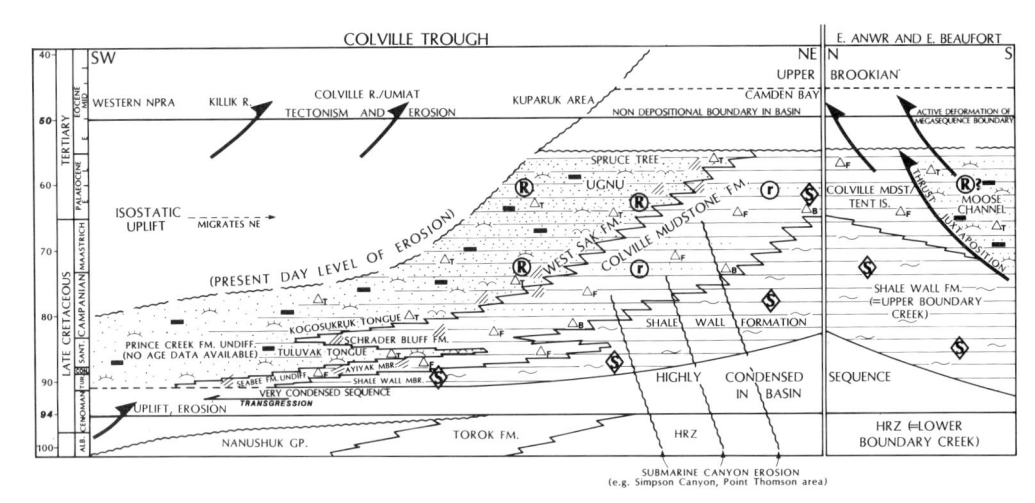

Fig. 23. Summarized chronostratigraphy, Middle Brookian megasequence. See the legend of Fig. 9 for an explanation of gross depositional environment symbols. Lithology symbols as follows: Coarse-grained clastics, dotted. Coals, solid bricks. Fine-grained clastics, ruled lines. The 'S' symbol in a diamond denotes a hydrocarbon source rock and the 'R' symbol in a circle denotes a hydrocarbon reservoir rock.

Canada. Here, a mid-Paleocene transgressive marine unit (Ministicoog Member; Young 1975) is overlain by a series of three, late Paleocene to mid-Eocene deltaic progradations. This is the Reindeer Formation of Mountjoy (1967a), which has been described in more detail by Bowerman & Coffman (1975); Dixon (1981); Nentwich & Yole (1982). A river dominated delta developed at the head of the proto-Mackenzie River that was draining the Canadian craton to the south and southwest. Petrologically, it is a fairly mature system and contains porous litharenites.

Middle Brookian hydrocarbon source rock facies are widespread at the base of the megasequence. The initial transgressive-marine and starved-basin deposits of the Shale Wall Member and Upper Boundary Creek Formation are oil-prone and very organic-rich (Molenaar 1983; Snowdon 1980a). Detterman (1963) has described the Shale Wall in the foothills of the central Brooks Range as a low grade oil shale and Yorath et al. (1975) described the Upper Boundary Creek on the Anderson Plain of northwest Canada as a sulphurous, organic-rich unit that undergoes self-combustion (Smoking Hills Formation).

Significant reservoir potential is developed in the Colville Group of the Middle Brookian megasequence. The marine shelf sheet-sands of the West Sak Formation and the fluvial−deltaic sands of the Ugnu Formation, which were deposited in front of the Middle Brookian deltas over the crest of the Barrow Arch, contain up to 40 billion barrels of oil-in-place (Werner 1986). The West Sak Formation consists of very fine grained, unconsolidated sands with interbedded silt and mud; while the shallowly buried Ugnu Formation consists of fine to medium grained unconsolidated sands, which contain heavy ($8°-12°$ API), biodegraded oil.

The abundance of labile, lithic grains in the Moose Channel system downgrades reservoir potential. Thick, gross sand sections contain only thin net sand intervals after compaction in the subsurface. The Reindeer Delta system contains improved reservoir development, due to the more mature clast mineralogy, and provides the reservoir for the Taglu gas field (Bowerman & Coffman 1975; Hawkins & Hatfield 1975) and for some of the small, nearby oil fields (e.g. Adgo, Garry, Niglintak).

Upper Brookian megasequence

The Upper Brookian megasequence is mid-Eocene to Recent in age. It has a duration of approximately 50 Ma and records the third, major phase in the progressive development of the Brooks Range. Basin geometries in northeast Alaska reflect overthrusting of the Beaufortian passive margin and the continued crustal shortening in the active thrust-front. (See position 3 of the mountain front in Fig. 19 and Fig. 22c). The base boundary is a strong, angular unconformity in the eastern Beaufort

Sea, (McWhae 1986), where end-Middle Brookian (Laramide) deformation was most intense. Further west, the boundary is a depositional downlap surface and, in part, a submarine erosion surface.

Four principal basin elements are recorded by the Upper Brookian megasequence in the Beaufort Sea. The westernmost element is a prograding passive wedge, which overlies the rifted margin of the west/central Beaufort Sea. The second element is the Camden Bay depocentre, which lies further east and represents a structural depression on the thermally contracted passive margin. The Camden Bay depocentre remained sediment starved until late Brookian time, see Fig. 24a. Five Upper Brookian sequences prograded into this depocentre and are now deformed by uplift, reverse faulting and normal faulting (Grantz et al. 1986). We interpret deep-seated thrust duplexes to have generated the uplift known as the Camden Anticline, see Fig. 24b, over which Upper Brookian sediments (cut by syn-depositional normal faults) have been passively deformed by folding and post-depositional normal faulting. The climax to this (Camden) orogenic restructuring is marked by a very prominent, late Miocene (6 Ma) angular unconformity, which records considerable erosion along the crests of the shallow anticlines draped over the deep thrust duplexes.

The third element of Upper Brookian basin geometry is seen in the eastern Beaufort Sea. Here, basin shape was controlled by underlying Middle Brookian (Laramide) topographic relief, see Fig. 24c. Two sequences prograde into this area and seismic data show topset, foreset and chaotic, mounded-bottomset depositional environments.

The fourth basin element is the Mackenzie Delta depocentre, which developed over the axis of the subsided Beaufortian transtensional basins. Here, distal mudstones at the base of the Upper Brookian megasequence pass up to a series of delta cycles (Hubbard et al. 1985b; Willumsen & Cote 1982). Two major, delta-lobe progradations have been mapped (the Pullen and Akpak cycles of Willumsen & Cote 1982), and sands are present in fluvio–deltaic and base-of-slope settings. Most of the coarser clastics appear to have been trapped in fluvio–deltaic topsets, where deposition was strongly controlled by syn-depositional, listric-normal, delta-front growth-faulting.

To the west of the Beaufort Sea, a further element of late Brookian basin development is present. Here, in the Chukchi Sea, the Upper Brookian is characterized by extensional tectonics, see Fig. 25. The axis of subsidence lies to the west of the Hanna Trough, between the Barrow Arch and the Chukchi Platform, see Fig. 20. Grantz et al. (1982) and Grantz & May (1984b) have described major, north–south fault trends and they defined depositional systems from seismic data. These include a major, downslope fan at the base of the Upper Brookian megasequence, see Fig. 25.

There is no public well data in north Alaska to confirm the presence of absence or Upper Brookian hydrocarbon source rock facies. The bottomset environment where, by analogy to the Lower and Middle Brookian, organic-rich facies may be present, lies largely beyond the limit of well control in the Chukchi Sea, Camden Bay and in the east Beaufort Sea. Source rock potential in age-equivalent strata of the Mackenzie Delta is also poorly known from the literature. It is probable, however, that Eocene (and possibly Oligocene) bottomset shales have generated some of the hydrocarbons seen in that area. In addition, Snowdon (1980b) has suggested that resinite in delta-front deposits may provide oil-prone rock potential.

Upper Brookian reservoirs are largely undocumented in north Alaska. In the Mackenzie Delta, however. Upper Brookian sands represent the most important reservoir development. Hydrocarbon occurrence in Upper Brookian topset environments typically consist of interbedded oil, gas and water bearing sands. In the delta front environment, channel and bar sands reservoir major oil accumulations (Amauligak, Tarsiut, Issungnuk). In the pro-delta environment, mounded, turbidite sands also form important reservoir units (Kopanor, Koakoak).

Brookian resource potential

In summary, the distal mudstones of the Lower and Middle Brookian megasequences developed a concentration of organic material and they are often rich petroleum source rocks (e.g. HRZ, Molenaar 1983; Boundary Creek, Snowdon 1980a). Traced to more proximal basin positions, under and within the prograding clastic wedges, the organic-rich shales thin and become less oil prone. In north Alaska, the distal shales of the Upper Brookian megasequence lie almost entirely beyond the limit of well control and hence, their source rock potential is unknown. In northwest Canada, equivalent strata in the Mackenzie Delta are thought to have source rock potential, but published source rock data is sparse.

In general, Brookian reservoir systems are

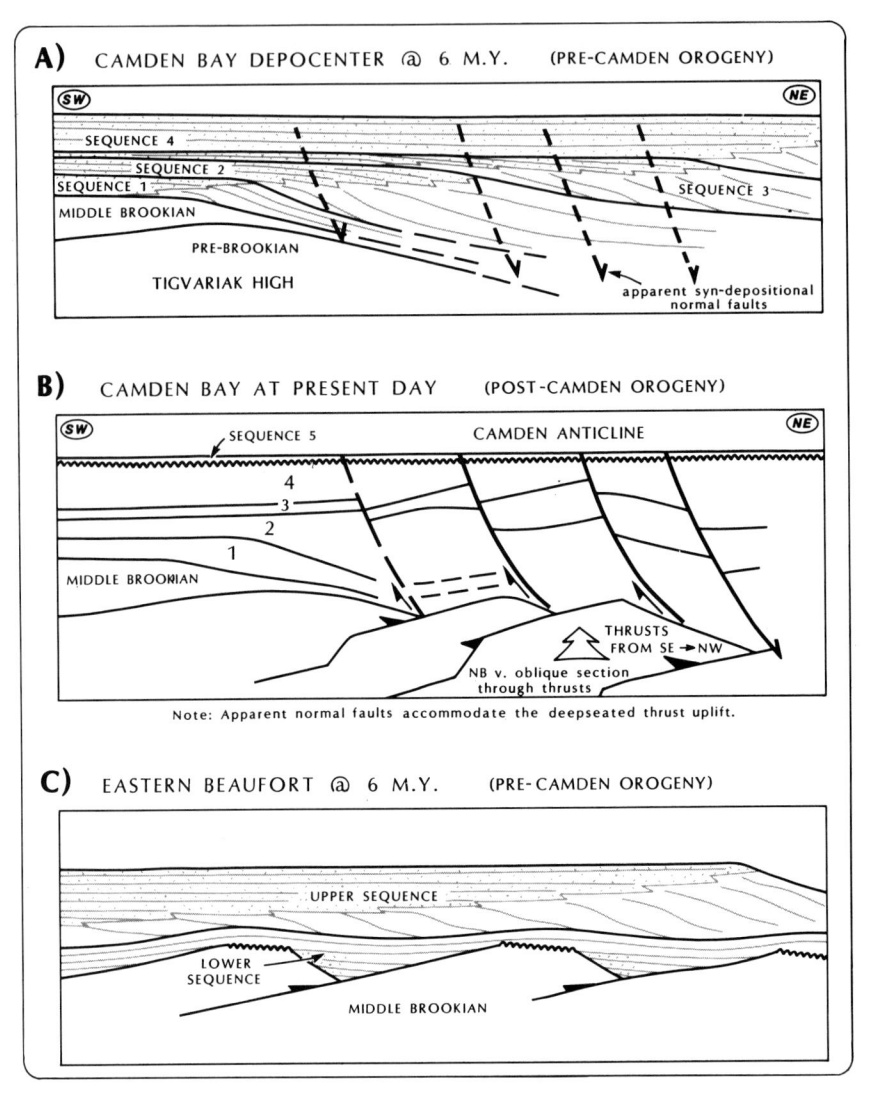

Fig. 24. Basin cross sections A, B, Camden Bay; C, Eastern Beaufort. The line of the cross sections is through mountain front 3 (Camden), shown on Fig. 19. The Upper Brookian megasequence has been divided into Sequences 1 through 5 in Camden Bay and into a Lower and Upper Sequence in the eastern Beaufort Sea.

poorly developed due to two main factors. Firstly, the immature lithofacies have caused a rapid deterioration of reservoir quality during burial and secondly, sand bodies in the prograding depositional environments are often localized and lenticular. In addition, the overall depositional style of the Brookian prograding clastic wedges has often resulted in a wide stratigraphic separation between topset/forest reservoir sandstones and bottomset source rocks. In consequence, good hydrocarbon carrier beds are seldom present between the two elements of Brookian source rock and reservoir rock development.

Possible constraints for incipient plate boundaries

Analysis of the Brookian plate sequence identifies the location of possible incipient plate boundaries.

Firstly, crustal shortening is most severe in

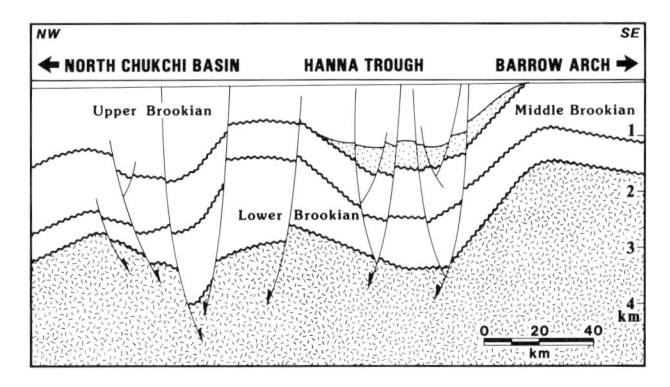

Fig. 25. Basin cross section, Hanna Trough, Chukchi Sea. The line of cross section is shown on Fig. 20. The base of the Upper Brookian in the Hanna Trough is a down-slope fan, shown by the dotted pattern.

northeast Alaska, where Rattey (1986) estimated that the Beaufortian 'Arctic Alaska Microplate' has been shortened by as much as a factor of three. Late Miocene (Camden) overthrusting has transported continental curst to the edge of the early Cretaceous oceanic Canada Basin. Hence, we speculate that a subduction zone may be developing, as shown on Fig. 22c.

Secondly, right-lateral movement on the Kaltag Fault to the south and east of the Brooks Range has been documented by several authors for mid and late Brookian time (e.g., Patton and Hoare 1968; Young et al. 1976; Jones 1982). This fault cuts north to the Rapid Fault Array (Yorath & Norris 1975), where it forms the lateral boundary to the British Mountain thrust belt and from there extends offshore to the west of the Mackenzie Delta and Banks Island, where it is a major boundary fault, (McWhae 1986) utilizing the former Beaufortian transcurrent fault zone, see Fig. 20. Hence, the Kaltag may represent a right-lateral, transcurrent fault-margin to an incipient Alaskan microplate.

Thirdly, extension in the Chukchi Sea represents a Brookian rifting episode. Major, north–south trending faults extend north into the North Chukchi Basin, which is assumed by Grantz & May (1984) to be floored by oceanic crust.

In summary, evidence is observed to indicate the possible development of three, incipient (post-Brookian) microplate boundaries. Compression in the eastern Beaufort Sea may represent the transition from a trailing edge rifted-margin to a leading edge subduction zone. In this event, north Alaska will be transported north as the oceanic Canada Basin is consumed, and right-lateral movement past Arctic Canada will be accommodated by the Kaltag Fault zone.

Extensional tectonics in the Chukchi Sea indicates the progressive development of a north–south trending rift zone to the west of north Alaska.

Hydrocarbon habitat

Our evaluation of the megasequences has shown that enormous hydrocarbon potential existed at the end of Beaufortian time (Aptian), but two factors were required for a truely giant, or world-scale, hydrocarbon province to develop. These were: (i) additional source rock burial to trigger regional hydrocarbon generation, and (ii) more widespread trap development, especially in the largely unstructured Middle and Upper Ellesmerian basins.

As an order of magnitude, we estimate that approximately 10 trillion barrels of oil were generated on the 'Arctic Alaska Microplate' in Brookian time. Concurrent crustal shortening was, however, very severe and many of the oil generating kitchens were rapidly uplifted and incorporated into the orogen. As a result, many of the hydrocarbon habitats have been 'lost'. In this section, we briefly describe the history of oil generation and the Brookian development of hydrocarbon habitat in the study area, see Fig. 26.

Early Brookian

By early Brookian time (late Jurassic), the depth of burial was sufficient to have formed an oil kitchen in the depocentre of the Beaufortian rifted margin, see Fig. 26a.

To the south, the Brooks Range mountain front advanced as far north as the present head of the Colville River. This destroyed almost the

a) EARLY BROOKIAN OIL GENERATION

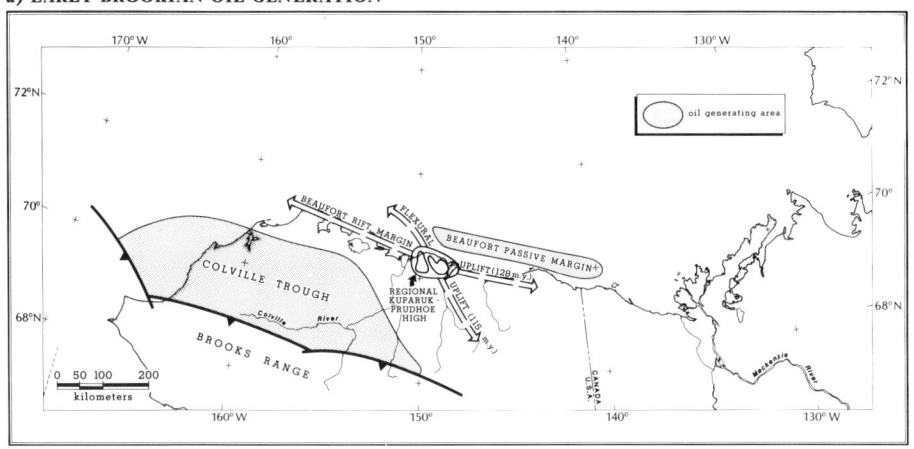

b) MID-BROOKIAN OIL GENERATION

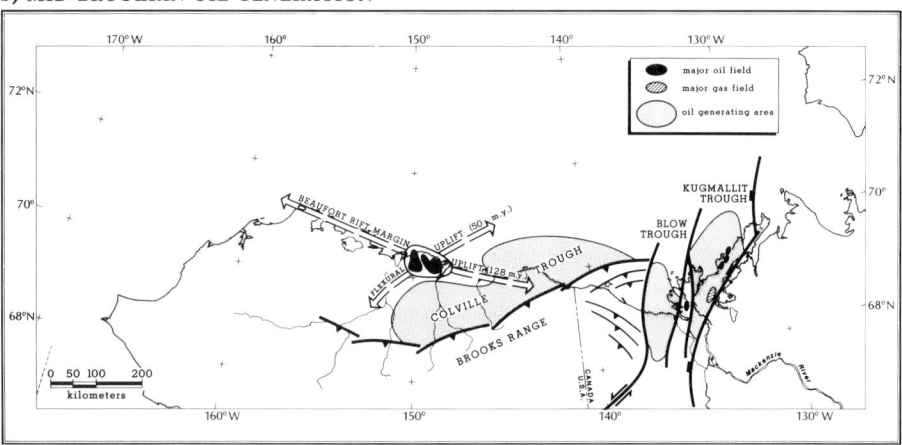

c) LATE BROOKIAN OIL GENERATION

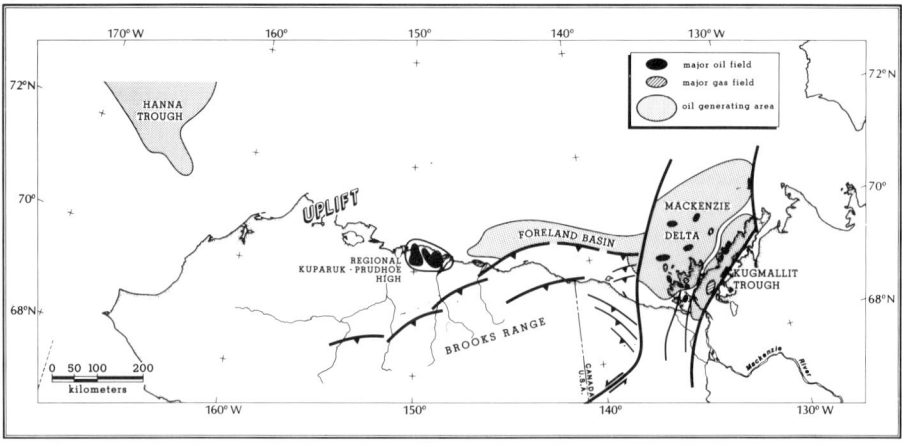

entire Ellesmerian and Beaufortian central basin, which contained the greatest concentration of source rocks in the study area. These source rocks were buried by the advancing foreland basin and then were very rapidly uplifted, incorporated into the early Brookian (Columbian) orogen and largely eroded.

By the late Aptian (115 Ma), the foreland basin (Colville Trough) occupied a position immediately to the south of the Beaufortian rift shoulder, see Fig. 22b. The flexural bulge, migrating in front of the foreland basin, cross cut the rift shoulder and structural elevation was enhanced at the point of constructive interference between the two high trends, see Figs 22b & 26a. The regional anticlinal closure that formed at the point of constructive interference is referred to as the Kuparuk/Prudhoe High. Here, the subcrop geometry, beneath the intra-Hauterivian (128 Ma) unconformity, developed four-way dip closure and combined structural–stratigraphic traps were formed.

The major depocentre of the early Brookian Colville Trough formed a large oil kitchen, as shown on Fig. 26a. The principal oil migration paths moved north, away from the mountain front and continued unhindered to the uplifted shoulder of the western Beaufortian rift margin (western Barrow Arch), well to the west of the developing Kuparuk/Prudhoe High. Early Brookian hydrocarbon traps had not developed on the western Barrow Arch and much of the oil is believed to have escaped through surface seepage.

Middle Brookian

Middle Brookian time (Cenomanian to Eocene) saw a considerable change in the pattern of oil generation, controlled by the shift in the position of the orogenic front and foreland basin, see Fig. 26b.

The Beaufortian passive margin was partly buried by the prograding Middle Brookian clastic wedge, which pushed the oil generating area on the margin into the thermal gas window.

The central and eastern Brooks Range (Laramide) mountain front advanced to the northeast during deposition of the Middle Brookian megasequence, when active foreland basin subsidence was centred in central and northeast Alaska. The advancing orogenic front uplifted the final remnant of the Ellesmerian and Beaufortian central basin, which had been briefly buried in the early Brookian.

As the axis of the foreland basin moved to the northeast through mid-Brookian times, so also did the axis of the flexural bulge migrating in front of the active depocentre (compare Figs 26a & 26b). This enhanced the northeasterly dip of the regional Kuparuk/Prudhoe high and increased the size of the structural–stratigraphic traps.

Large quantities of hydrocarbons were generated in the mid Brookian Colville Trough and these hydrocarbons migrated toward the rift shoulder of the Beaufortian margin. The focus for oil migration was the regional Kuparuk/Prudhoe High. Middle Brookian deposition buried the anticline, and we interpret that every effective structural–stratigraphic trap within the regional high was filled to spill in mid-Brookian times. The major accumulations of this age are the Prudhoe Bay, Endicott, Lisburne and Kuparuk River Fields.

During the mid-Brookian, the former oil generating area of the western early Brookian foreland basin lay well to the west of the area of active thrust loading and foredeep subsidence. It entered an inactive or passive phase, expressed by uplift, which cooled the source rocks and the oil generation process was 'switched off'. Very large volumes of oil were undergoing secondary migration when generation was arrested, as evidenced by the high residual oil saturations still seen in the strata of this basin in NPRA.

In northwest Canada, Middle Brookian burial pushed the source rocks of the Blow Trough and Kugmallit Trough into the oil window.

Fig. 26. Hydrocarbon habitat: A, Early Brookian; B, Mid-Brookian; C, Late Brookian. The major hydrocarbon fields are, Kuparuk River/West Sak/Ugnu on the west side of the Kuparuk-Prudhoe High and Prudhoe Bay/Lisburne/Endicott on the east side of the Kuparuk-Prudhoe High. Point Thomson is the gas field to the east. The oil field between the Blow Trough and Kugmallit Trough (on the Cache Creek High) is Kugpik. The Kugmallit Trough contains the Parsons Lake gas field and the Atkinson, Mayogiak and other small oil fields. The significant fields in the onshore portion of the Mackenzie Delta are the Taglu gas field and the Adgo, Garry and Niglintak oil fields. In the offshore, the Netserk gas field is located in the west and the Ukalerk/Tingmiark gas fields in the east. The larger oil fields in the delta front are (from west to east) Tarsiut, Issungnak and Amauligak and in the distal pro-delta, Kopanoar and Koakoak.

Independent, minor oil provinces began to develop in these two basins.

Late Brookian

The late Brookian (Camden) mountain front advanced entirely into the offshore area of northeast Alaska and northwest Canada, see Fig. 26c. The actively subsiding foredeep basins lie to the north of the limit of Ellesmerian source rock development and overlie the subsided Beaufortian margin. Here, the present day oil migration paths are moving toward the outer shelf and slope.

Today, uplift is occurring everywhere to the south and west of the active late Brookian mountain front. The initial effect of this uplift on the regional Kuparuk/Prudhoe High was tilting in the Oligocene (Bird 1985). Spillage of oil from the mid-Brookian accumulations occurred and leakage up fault zones charged the shallow West Sak and Ugnu Formation reservoirs overlying the Kuparuk River and Prudhoe Bay Fields, (Carman & Hardwick, 1983). If shortening of Arctic Alaska continues, the Barrow Arch oil province will eventually be uplifted to surface.

In Canada, mid to late Brookian reactivation on the Kaltag−Rapid Fault Array resulted in severe structural deformation, which pushed the oil generating area in the Blow Trough into the thermal gas window. The Kugmallit Trough was largely unaffected by structural deformation and the minor, mid-Brookian hydrocarbon province was preserved. This includes the Kugpik Field, see Fig. 17, Parsons Lake and the other small fields on the Tuktoyaktuk Peninsula.

The significant oil generating area in the Mackenzie Delta developed in late Brookian time. The kitchen lies just beyond the area of Brookian orogenic uplift and it developed as an independent, passively subsided hydrocarbon province. Hydrocarbon traps are syndepositional and developed as a result of two main processes: (i) down-to-the-basin growth faulting in the delta front environment (e.g., Amauligak, Issungnak), and (ii) drape of deep water sands over and around mud diapirs in the pro-delta setting (e.g., Kopanoar, Koakoak). The main exploration risk in this hydrocarbon province is in locating the thin reservoir sands within the thick, prograding clastic wedges.

Another possible, independent oil province is the unexplored Hanna Trough in the northern Chukchi Sea. Late Brookian subsidence may have pushed this area into the oil generating window. Frontier exploration potential exists at Ellesmerian, Beaufortian, and Brookian stratigraphic levels.

Conclusions

In conclusion, we have presented a stratigraphic analysis of eight depositional megasequences, which track the geological evolution of north Alaska and northwest Canada from the late Devonian to Recent. During the Ellesmerian, basin development on a fold belt terrane is postulated. These basins accumulated a wealth of hydrocarbon source rocks and reservoir rocks. Then, as the Pangea supercontinent broke up in the Mesozoic, a period of extensional tectonics led to the opening of the Canada Basin. The rifting episode is recorded by two Beaufortian megasequences (early Jurassic to Aptian), which provide evidence for a Jurassic failed rift episode and onset of the successful rift episode in the Hauterivian. During Beaufortian time, hydrocarbon potential was enhanced by the initial development of the subcrop traps on the uplifted shoulder of the rift margin. In addition, further hydrocarbon source rock, reservoir rock and trap development occurred. Thus, by the Aptian, the scene was set for a world-scale hydrocarbon province to develop.

Brookian orogenesis, however, has meant that this enormous potential has not been fully realized. As the 'Arctic Alaska Microplate' was subjected to collisional events and crustal shortening, much of the hydrocarbon potential was destroyed. Many of the oil generating areas were either uplifted into the orogen and the oil 'lost', or they were rapidly buried too deeply and pushed into the thermal gas window. The pattern of hydrocarbon kitchens and structural trends conspired to produce a single, regional focus for liquid hydrocarbons migrating in the subsurface.

Hence, when the overall hydrocarbon habitat is examined, we see one major (but areally restricted) hydrocarbon province, the regional Kuparuk/Prudhoe High, which resulted from constructive interference between Beaufortian and Brookian tectonic elements. Away from this focus, the combination of rift tectonics and orogenesis have been less favourable. The Blow Trough, for example, in northwest Canada, contains hydrocarbon habitats that have been completely destroyed by excessive subsidence and tectonism.

The only other hydrocarbon provinces which have been discovered to date occur in passively subsided basins, lying just beyond the influence of Brookian orogenic uplift. These provinces are the Mackenzie Delta and Kugmallit Trough of northwest Canada, and, for future exploration, possibly the Hanna Trough of the Chukchi Sea.

Permission to publish this paper has been granted by BP Exploration and BP Alaska Exploration. D. G. Roberts, A. Grantz, R. J. Bailey and S. A. Graham provided critical readings of the manuscript. The authors wish to thank the many BP geologists and geophysicists who have been involved with this work. R. J. Hubbard also wishes to thank David Roberts, Terry Adams, Tina, Philip and Andrew for their support throughout. Special thanks are due to Debbie Linton for the drafting and Sandra Seiki and Cynthia Turner for the word processing.

This article was first published in *Marine and Petroleum Geology* **4** 2–34 and appears by permission of the publishers, Butterworth & Co. (Publishers) Ltd.

References

AHLBRANDT, T. S. (ed.) 1979. *Preliminary geological, pertrological and palaentological results of the study of Nanushuk Group rocks, North Slope Alaska*: USGS Circ. **794**.

ALASKA GEOLOGICAL SOCIETY 1970–1971. *West to east stratigraphic correlation section, Point Barrow to Ignek Valley, Arctic North Slope, Alaska*: North Slope Stratigraphic Committee, Anchorage, Alaska.

ARMSTRONG, A. K. & BIRD, K. J. 1976. Facies and environments of deposition of Carboniferous rocks, Arctic Alaska *In*: (MILLER, T. P. ed.) *Recent and ancient depositional environments of Alaska symposium*, Alaska Geol. Soc. P. A1–A16

ARMSTRONG, A. K., B. L. 1970. Biostratigraphy and dolomite porosity trends of the Lisburne Group, *In*: ADKINSON, W. L. & BROSGE, M. M. (eds) *Geological seminar on the North Slope of Alaska*, Proceedings, Los Angeles AAPG Pacific Section P. N1–N16

—— & —— 1977. *Carboniferous microfacies, microfossils, and corals, Lisburne Group, Arctic Alaska*: USGS Prof. Paper 849.

ARMSTRONG, A. K. and Mamet, B. L. 1978. Microfacies of the Carboniferous Lisburne Group, Endicott Mountains, Arctic Alaska, *In*: (eds STELCK, C. R. and CHATTERTON, B. D. E.), *Western and Arctic Canadian biostratigraphy*, Geol. Soc. of Canada Special Paper 18.

ARMSTRONG, A. K., MAMET, B. L. and DUTRO, J. T. 1970. Foraminiferal zonation and carbonate facies of Carboniferous (Mississippian and Pennsylvanian) Lisburne Group, central and eastern Brooks Range, Arctic Alaska *AAPG Bull.* **54** (5), 687–698.

BALKWILL, H. R., COOK, G. G., DETTERMAN, R. L., EMBRY, A. F., HAKANSSON, E., MIALL, A. D., POULTON, T. P., and YOUNG, F. G. 1983. Arctic North America and Northern Greenland, *In*: (MOULLADE, M. and NAIRN, A. E. eds) *The Phanerozic geology of the world, II: The Mesozoic, B*, p. 1–31.

BALLY, A. W. 1976. Canada's passive continental margin — a review *Mar. Geophys. Res.* **2**, 327–340.

BAMBER, E. W. and WATERHOUSE, J. B. 1971. Carboniferous and Permian stratigraphy and paleontology, northern Yukon Territory, Canada *Bull. Can. Pet. Geol.* **19** (1), 29–250.

BARNES, D. A. 1985. Sag River Formation, Prudhoe Bay, Alaska: Depositional environment and diagenesis: Abstract *AAPG Bull.* **69**, (4) 656.

BARTSCH–WINKLER, S. 1979. Textural and mineralogical study of some surface and subsurface sandstones from the Nanushuk Group, western North Slope, Alaska, *In*: (AHLBRANDT, T. S. ed.) *Preliminary geologic, petrologic and paleontologic results of the study of Nanushuk Group rocks, North Slope, Alaska* U.S. Geological Survey Circular **794**, p. 61–76.

BEE, M., JOHNSON, S. H. and CHIBURIS, E. F. 1984. Marine seismic refraction study between Cape Simpson and Prudhoe Bay, Alaska *J. Geophys. Res.* **89** (B8), 6941–6960.

BIRD, K. J. 1982. *Rock unit report of 228 wells drilled on the North Slope, Alaska* U.S. Geological Survey Open-File Report 82–278, 106 p.

BIRD, K. J. 1985. North Slope Oil and Gas: The Barrow Arch Paradox *Abstract, AAPG Bull.* **69**, (4), 656.

BIRD, K. J. and JORDAN C. F. 1977. Lisburne Group (Mississippian and Pennsylvanian), potential major hydrocarbon objective of Arctic Alaska *AAPG Bull.* **61**, (9), 1493–1512.

BOAK, J. L., TURNER, D. L., WALLACE, W. K. and MOORE, T. E. 1985. Potassium–Argon ages of allochthonous mafic and ultramafic complexes and their metamorphic aureoles, Western Brooks Range, Alaska *Abstract, AAPG Bull.* **69**, (4) 656.

BODNAR, D. A. 1984. *Stratigraphy, age, depositional environments, and hydrocarbon source rock evaluation of the Otuk Formation, north–central Brooks Range, Alaska* Fairbanks, University of Alaska, M. S. thesis, 232 p.

BOWERMAN, J. N. and COFFMAN, R. F. 1975. The Geology of the Taglu Gas Field in the Beaufort Basin, NWT *In*: (YORATH, C. J., PARKER, E. R. and GLASS, D. J. eds) *Canada Society Petroleum Geologists, Memoir* **4**, p. 649–662.

BOWSHER, A. L. and DUTRO, Jr. J. T. 1957. *The Paleozoic sections in the Shainin Lake area, central Brooks Range, Alaska* USGS Prof. Paper 303A, 39 p.

BRIDEAUX W. W. and MYHR, D. W. 1976. *Lithostratigraphy and dinoflagellate cyst succession in the Gulf–Mobil Parsons N–10 well* Geol. Surv. Canada. Paper 76–1B p. 235–249.

BROSGE, W. P. and WHITTINGTON, C. L. 1966. *Geology of the Umiat — Maybe Creek Region, Alaska* USGS, Prof. Paper 303-H p. 501–638.

BROSGE, W. P. and TAILLEUR, I. L. 1971. Northern

Alaska petroleum province, *In*: (CRAM, I. H. ed.) *Future petroleum provinces of the United States — their geology and potential*, AAPG memoir **15**, 68–99.

BROSGE, W. P., DUTRO, J. T., MANGUS, M. D. and REISER, H. N. 1962. Paleozoic sequence in eastern Brooks Range, Alaska *AAPG Bull.* **46**, (12), 2174–2198.

BROSGE, W. P., REISER, H. N., DUTRO, J. T. and DETTERMAN, R. L. 1979. *Bedrock geologic map of the Philip Smith Mountains quadrangle, Alaska* USGS Miscellaneous Field Studies Map MF-879 B, Scale 1:250 000, two sheets.

BUCKINGHAM, M. L. 1986. Fluvio–Deltaic Sedimentation Patterns of the Upper Cretaceous to Lower Tertiary Sabbath Creek Formation, Arctic National Wildlife Refuge (ANWR), Northeastern Alaska, *In*: (TAILLEUR, I. L. and WEIMER, P. eds) *Alaskan North Slope Geology*, SEPM–Pacific Section/Alaska Geol. Soc. (in prep.).

CAREY, S. W. 1958. A tectonic approach to continental drift, *In*: (CAREY, S. W. ed.) *Continental Drift. A symposium — Hobart, University of Tasmania*, p. 177–355.

CARMAN, G. J. and HARDWICK, P. 1983. Geology and regional setting off the Kuparuk Oilfield, Alaska *AAPG Bull.* **67**, 1014–1031.

CHAPMAN, R. M. and SABLE, E. G. 1960. *Geology of the Utukok–Corwin region northwestern Alaska* USGS Prof. Paper 303-C, p. 47–174.

CHAPMAN, R. M. DETTERMAN, R. L. and MANGUS, M. D. 1964. *Geology of the Killik–Etivluk Rivers Region, Alaska* USGS Prof. Paper 303-F, p. 325–407.

CHRISTIE, R. L. 1979. The Franklinian geosyncline in the Canadian Arctic and its relationship to Svalbard *Skr. norsk Polarinst.* **167**, 263–314.

CHURKIN, Jr., M., NOKLEBERG, W. J. and HUIE, C. 1979. Collision-deformed Paleozoic continental margin, western Brooks Range, Alaska *Geology* **7**, (8), 379–383.

CHURKIN, Jr., M., CARTER C. and TREXLER Jr., J. 1980. Collision-deformed Paleozoic continental margin of Alaska — a foundation for microplate accretion, *Geol. Soc. Am. Bull., Part 1* **91**, 648–654.

COLLINS, F. R. 1958. *Test wells, Umiat Area, Alaska, with micropaleontologic study of the Umiat Field, northern Alaska*, by H. R. Bergquist. USGS, Prof. Paper 305-B, p. 71–206.

COTE, R. P., LERAND, M. M. and RECTOR, R. J. 1975. Geology of the Lower Cretaceous Parsons Lake gas field, MacKenzie Delta, NWT, *In*: (YORATH, C. J., PARKER, E. R. and GLASS, D. J. eds) *Can. Soc. Pet. Geol. Memoir* **4**, 613–632.

CRANE, R. C. 1980. Comments and replies on Churkin, M. Jr., Nokleberg, W. J. and Huie, C., 'Collision-deformed Paleozoic continental margin, western Brooks Range, Alaska' *Geology*, **8**, (8), 354.

CRANE, R. C. 1986. Cretaceous Olistostrome Model, Brooks Range Alaska, *In*: (TAILLEUR I. L. and WEIMER, P. eds) *Alaskan North Slope Geology*, (Geol. Soc. (in prep.).

CRANE, R. C. and WIGGINS, V. D. 1976. Ipewik Formation, significant Jurassic–Neocomian map unit in northern Brooks Range foldbelt. Abs.: *AAPG Bull.* **60**, 2177.

CURTIS, S. M., ELLERSIECK, I. F., MAYFIELD, C. F. and TAILLEUR, I. L. 1982. *Reconnaissance geologic map of southwestern Misheguk Mountain Quadrangles, Alaska* USGS Open File Report 82–611, Scale 1:63 360.

CURTIS, S. M., ELLERSIEK, I. F. MAYFIELD, C. F. and TAILLEUR, I. L. 1983. *Explanation to accompany reconnaissance geologic map of the DeLong mountains A1, B1, and part of C1 quadrangles, Alaska* USGS Open File Report, OF 83–185, 53 p., Scale 1:63, 360.

DETTERMAN, R. L. 1970. Sedimentary history of the Sadlerochit and Shublik Formations in northeastern Alaska, *In*: (ADKINSON, W. L., and BROSGE, M. M. eds) *Proceedings of the Geological Seminar on the North Slope of Alaska* Pacific Sec. AAPG, p. 0–1 to 0–13.

DETTERMAN, R. L. BICKEL, R. S. and GRYC, G. 1963. *Geology of the Chandler River region, Alaska* U.S. Geological Survey Professional Paper 303-E, p. 223–324.

DETTERMAN, R. L., REISER, H. N., BROSGE, W. P. and DUTRO, Jr., J. T. 1975. *Post-Carboniferous stratigraphy, northeastern Alaska* U.S. Geol. Survey Prof. Paper 886, 46 p.

DIETRICH, J. R. and DIXON, J. 1984. Regional stratigraphy of Upper Cretaceous to Holocene sediments of the Beaufort–Mackenzie and Banks basins (abs.) *Can. Soc. Pet. Geol. Res.*, **12**, (8), 1–2.

DILLON, J. T. and TILTON, G. R. 1986. Devonian magmatism in the Brooks Range, Alaska: Abstract *In*: (TAILLEUR, I. L. and WEIMER, P. eds) *Alaskan North Slope Geology*, SEPM–Pacific Section/Alaska Geol. Soc. (in prep.).

DILLON, J. T., HAMILTON, W. B. and LUECK, L. L. 1981. *Geologic map of the Wiseman A-3 Quadrangle* Alaska Division of Geological and Geophysical Surveys Open-file Report 1 119, scale 1:63 360, 1 sheet.

DILLON, J. T., PESSEL, G. H., CHEN, J. A. and VEACH, N. C. 1980. Middle Palaeozoic Magmatism and Orogenesis in the Brooks Range, Alaska *Geology* **8**, 338–343.

DIXON, J. 1981. *Sedimentology of the Eocene Taglu Delta, Beaufort–MacKenzie Basin: Example of a river-dominant delta* Geol. Surv. Canada Paper 80–11.

DIXON, J. 1982a. Jurassic and Lower Cretaceous subsurface stratigraphy of the MacKenzie Delta — Tuktoyaktuk Peninsula, N.W.T. *Geol. Surv. Can. Bull.* **349**, 52 p.

DIXON, J. 1982b. Upper Oxfordian to Albian geology, Mackenzie Delta, Arctic Canada, *In*: (EMBRY, A. F. and BALKWILL, H. R. eds) *Can. Soc. Pet. Memoir* **8**, p. 29–42.

DOUGLAS, R. J. W. 1972. *Geology and economic minerals of Canada* Geol. Surv. Canada Economic Geology Report No. 1.

DUTRO, Jr., J. T. 1980. Comments and replies on CHURKIN, M., Jr., NOKLEBERG W. J. and HUIE, C., 'Collision-deformed Paleozoic continental margin, western Brooks Range, Alaska' *Geology* **8**, (8), 355−356.

DUTRO, Jr., J. T. 1981. Geology of Alaska bordering the Arctic Ocean, *In*: (NAIRN, A. E. M., CHURKIN, M. Jr. and F. G. STEHLI eds) *The ocean basins and margins: The Arctic Ocean*, Plenum Press, vol. 5, ch. 2, p. 21−36.

DUTRO, Jr., J. T., BROSGE, W. P. and REISER, H. N. 1972. The significance of recently discovered Cambrian fossils and the reinterpretation of the Neruokpuk Formation, northeast Alaska *AAPG Bull.* **56**, 806−815.

DUTRO, Jr., J. T., BROSGE, W. P., DETTERMAN, R. L. and REISER, H. N. 1979. Beaucoup Formation, a new Upper Devonian stratigraphic unit in the central Brooks Range northern Alaska, *In*: (SOHL, N. F. and WRIGHT W. B. eds) *Changes in stratigraphic nomenclature by the U.S. Geological Survey*, 1978 USGS Bull. 1482-A, p. A63−A69.

EARDLEY, A. J. 1948. Ancient Arctica, *J. Geol.* **56**, (5), 409−436.

EDRICH, S. P. 1986. Geologic setting of North Slope oil fields, Alaska: Abstract, *In*: (TAILLEUR I. L. and WEIMER, P. eds) *Alaskan North Slope Geology*, SEPM−Pacific Section/Alaska Geol. Soc. (in prep).

EHM, A. and HAIMLA, N. 1982. *Geology and hydrocarbon potential of the Beaufort Sea and environs* Petroconsultants nonexclusive report.

EISBACHER, G. H. 1983. Devonian−Mississippian sinstral transcurrent faulting along the cratonic margin of western North America, *Geology* **11**, 7−10.

ELLERSIECK, I. F., CURTIS, S. M., MAYFIELD C. F. and TAILLEUR, I. L. 1982. *Reconnaissance geologic map of south−central Misheguk Mountain quadrangle, Alaska* USGS Open File Report 82−612, Scale 1:63 360.

ELLERSIECK, I. F., CURTIS, S. M., MAYFIELD, C. F. and TAILLEUR, I. L. 1983. *Reconnaissance geologic map of the DeLong Mountains, A3, B3, and part of C2 quadrangles, Alaska* USGS Open File Report 83−184, 53, p., Scale 1:63,360.

ELLERSIECK, I. F., MAYFIELD, C. F., TAILLEUR, I. L. and CURTIS, S. M. 1979. Thrust sequences in the Misheguk Mountain quadrangle Brooks Range, Alaska *In*: (JOHNSON, K. M. and WILLIAMS, J. R. eds) *USGS Circular* 804-B p. B-8.

EMBRY, A. F. 1985. Mesozoic stratigraphy of Canadian Arctic archipelago and implications for opening of Amerasian Basin: *AAPG Bull.* **69**, (2), 253.

FORSYTH, D. A., ASUDEH, I., GREEN, A. G. and JACKSON, H. R. Crustal structure of the northern Alpha Ridge *Nature* London, (in press).

FRITTS, C. E., EAKINS, G. R. and GARLAND, R. E. 1971. *Geology and Geochemistry Near Walker Lake, Southern Survey Pass Quadrangle, Arctic Alaska* Alaska Division of Geological and Geophysical Surveys Annual Report, p. 19−27.

GRANTZ, A. and MAY, S. D. 1983. Rifting history and structural development of the continental margin north of Alaska, *In*: (WATKINS J. S. and DRAKE, C. L. eds) *Studies in continental margin geology* AAPG Memoir **34**, p. 77−100.

GRANTZ, A. and MAY, S. D. 1984a. Sedimentary Basins and Geologic Structure of the Continental Margin North of Alaska. *27th International Geological Congress, Arctic Geology, Colloquium* Pub. 'Nauka' Moscow, 04 v. 4 p. 125−142.

GRANTZ A. and MAY, S. D. 1984b. *Summary geologic report for Barrow Arch outer continental shelf (OCS) planning area, Chukchi Sea, Alaska* USGS Open File Report 84−395, 40 p.

GRANTZ, A. and MULL, C. G. 1978. *Preliminary analysis of the petroleum potential of the Arctic National Wildlife Refuge, Alaska* USGS Open File Report 78, 20 p.

GRANTZ, A. DINTER, D. A. and CULOTTA, R. C. 1986. Offshore Structure, Chapter III-D-5 ANWR Report (in prep.).

GRANTZ, A., EITTREIM, S., DINTER, D. A. and EITREIM, S. 1979. Geology and tectonic development of the continental margin north of Alaska *Tectonophysics* **59**, 263−91.

GRANTZ, A., DINTER D. A., HILL, E. R., HUNTER, R. E., MAY, S. D., MCMULLIN, R. H. and PHILLIPS, R. L. 1982. *Geologic framework, hydrocarbon potential and environmental conditions for exploration and development of proposed oil and gas lease sale 85 in the central and northern Chukchi Sea* USGS Open File Report 82−1053, 84 p.

GRYC, G., PATTON, Jr., W. W. and PAYNE, T. G. 1951. Present Cretaceous stratigraphic nomenclature of northern Alaska *Washington Acad. Sci. J.* **41**, (5), 159−167.

HARLAND, W. B., COX, A. V., LLEWELLYN, P. G., PICKTON, C. A. G., SMITH, A. G. AND WALTERS, R. 1982. *A geologic time scale*, Cambridge University Press.

HARLAND, W. B., GASKELL, B. A., HEAFFORD, A. P., LIND, E. K. and PERKINS, P. J. 1984. Outline of Arctic post-Silurian continental displacements, *In*: (SPENCER A. M. ed.) *Petroleum Geology of the North European Margin* Graham and Trotman, for Norwegian Petroleum Society, p. 137−148.

HAWKINS, T. J. and HATFIELD, W. G. 1975. The regional setting of the Taglu Field, *In*: (YORATH, C. J. PARKER, E. R. and GLASS, D. J. eds) *Can. Soc. Pet. Geol., Memoir* **4**, pp. 633−647.

HAYWARD, H. and LEGG, G. 1983. *Geological Report, Walakpa Test well No. 1, Husky Oil NPR Operations, Inc.*, (BROCKWAY R. G. ed.) US Geological Survey, Office of the National Petroleum Reserve in Alaska, Dept. of Interior.

HEA, J. P., ARCURI, J., CAMPBELL, G. R., FRASER, I., FUGLEM, M. O., O'BERTOS, J. J., SMITH, D. R. and ZAYAT M. 1980. Post Ellesmerian basins of Arctic Canada: Their depocentres, rates of sedimentation and petroleum potential, *In*: (MIALL ed.) *Can. Soc. Pet. Geol. Memoir* **6**, 447−488.

HILLHOUSE, J. W. and GROMME, C. S. 1983. Paleomagnetic studies and the hypothetical rotation of Arctic Alaska, *J. Alaska Geol. Soc.* **2**, 27−39.

HITZMAN, M. W., SMITH, T. E. and PROFFETT, J. M. 1982. *Bedrock geology of the Ambler District, southwestern Brooks Range, Alaska* Alaska Division of Geological and Geophysical Surveys, Geologic Report #75, Scale 1:250 000, two plates.

HUBBARD, R. J. PAPE, J. and ROBERTS D. G. 1985a. Depositional sequence mapping as a technique to establish tectonic and stratigraphic framework and evaluate hydrocarbon potential on a passive continental margin, *In*: (BERG, O. R. and WOOLVERTON, D. eds) *Seismic stratigraphy II* AAPG Memoir **39**, p. 79–91.

HUBBARD, R. J., PAPER J. and ROBERTS, D. G. 1985b. Depositional sequence mapping to illustrate the evolution of a passive continental margin, *In*: (BERG, O. R. and WOOLVERTON, D. eds) *Seismic stratigraphy II* AAPG Memoir **39**, p. 93–115.

HUFFMAN, A. C., AHLBRANDT, T. S., PASTERNACK, I., STRICKER, G. D. and FOX, J. E. 1985. Depositional and sedimentologic factors affecting the reservoir potential of the Cretaceous Nanushuk Group, central North Slope, Alaska *In*: (HUFFMAN, A. C. ed.) *Geology of the Nanushuk Group and related rocks, central North Slope, Alaska* U.S. Geological Survey Bulletin 1614.

JACKSON, H. R., MUNDIE, P. J. and BLASCO, S. M. (eds) 1985. Initial geological report on Cesar – The Canadian expedition to study the Alpha Ridge, Arctic Ocean *Geol. Surv. Canada Paper* 84–22.

JAMISON, H. C., BROCKETT, C. D. and McINTOSH, R. A. 1980. Prudhoe Bay: A ten-year perspective, *In*: (HALBOUTY, M. T. ed.) *Giant oil and gas fields of the decade 1968–1978* AAPG Memoir **30** p. 289–314.

JONES, D. L. and GRANTZ, A. 1964. Stratigraphic and structural significance of Cretaceous fossils from Tiglukpuk Formation, northern Alaska *AAPG Bull.* **48**, (9), 1462–1474.

JONES, H. P. and SPEERS, R. G. 1976. Permo-Triassic reservoirs of Prudhoe Bay Field, North Slope, Alaska, *In*: (BRAUNSTEIN, J. ed.) *North American oil and gas fields* AAPG Memoir **24**, p. 23–50.

JONES, P. B. 1982. Mesozoic rifting in the western Ocean Basin and its relationship to Pacific sea floor spreading, *In*: (EMBRY, A. F. and BALKWILL, H. R. eds) *Can. Soc. Pet. Geol., Memoir* **8**, p. 83–100.

KELLER, A. S., MORRIS, R. H. and Detterman, R. L. 1961. *Geology of the Shaviovik and Sagavanirktok River region, Alaska, Part 3* U.S. Geol. Survey Prof. Paper 303-D, p. 169–222.

KERR, J. Wm. 1981. Evolution of the Canadian Arctic Islands; a transition between the Atlantic and Arctic oceans, *In*: (NAIRN, A. E. M., CHURKIN, Jr., M. and STEHLI, F. G. eds) *The Ocean Basins and Margins Vol. 5, The Arctic Ocean* Plenum Press, p. 105–199.

KERR, J. Wm. 1982. Evolution of sedimentary basins in the Canadian Arctic, *In*: (KENT, P. *et al*. eds) *The evolution of sedimentary basins*. Phil. Trans. Roy. Soc. Lond. A305, p. 193–205.

KERR, R. S. 1985. Seismic expression of catastrophic slope failure: Lower Cretaceous Torok Formation, North Slope of Alaska: Abstract, *AAPG Bull.* **69**, (2), 273.

KIRSCHNER, C. E., GRYC, G. and MOLENAAR, C. M. 1983. Regional seismic lines in the National Petroleum Reserve in Alaska, *In*: (BALLY, A. W. ed.) *Seismic expression of structural styles – a picture and work atlas* AAPG Studies in Geology 15, v. 1, p. 1.2.5–1 to 1.2.5–14.

LANE, F. H. and JACKSON, K. S. 1980. Controls on occurrence of oil and gas in the Beaufort-Mackenzie Basin, *Facts and principles of world petroleum occurrence* (MAILL, A. D. ed.) Can. Soc. Pet. Geol. Memoir 6, p. 489–508.

LANGHUS, B. G. 1980. Generation and migration of hydrocarbons in the Parsons Lake Area, NWT, Canada *In*: (MAILL A. D. ed.) Facts and principles of world petroleum occurrence Can. Soc. Pet. Geol., Memoir 6, p. 523–534.

LAWYER, L. A. and A. BAGGEROER 1983. A note on the age of the Canada Basin, *J. Alaska Geol. Soc.* **2**, 57–66.

LEFFINGWELL, E. de K. 1919. *The Canning River region, northern Alaska* U.S. Geol. Survey Prof. Paper 109, 251 p.

LERAND, M. 1973. Beaufort Sea, *In*: (McCROSSAN, R. G. ed.) *Future petroleum provinces of Canada – their geology and potential* Can. Soc. Pet. Geol., Mem. 1, p. 315–386.

LYLE, W. M., PALMER, Jr., I. F., BOLM, J. G. and MAXEY L. R. 1980. *Post–Early Triassic formations of northeastern Alaska and their petroleum reservoir and source-rock potential* Alaska Division of Geology and Geophysical Surveys Geological Report 76, 100 p.

MAGOON, L. B. and CLAYPOOL, G. E. 1981. Two oil types on North Slope of Alaska – Implications for exploration, *AAPG Bull.* **65**, 644–652.

MARTIN, A. J. 1970. Structure and tectonic history of the Western Brooks Range, De Long Mountains and Lisburne Hills, Northern Alaska, *Geol. Soc. Am. Bull.* **81**, 3605–3622.

MASTERSON, W. D. and PARIS C. E. 1986. Depositional history and reservoir description of Kuparuk River Formation, North Slope, *In*: (TAILLEUR, I. L. and WEIMER, P. eds) *Alaskan North Slope Geology* SEPM-Pacific Section-Alaska Geol. Soc. (in prep.).

MAYFIELD, C. F. 1980. Comments and replies on CHURKIN, M. Jr., NOKLEBERG, W. J. and HUIE, C. 'Collison-deformed Paleozoic continental margin, western Brooks Range, Alaska, *Geology* **8**, (8), 357–359.

MAYFIELD, C. F., TAILLEUR, I. L. and ELLERSIECK, Inyo 1983a. *Stratigraphy, structure and palinspastic synthesis of the western Brooks Range, northwestern Alaska*: U.S. Geological Survey Open-file Report OF83–779, 58 p., scale 1:1 000 000, 5 plates.

MAYFIELD, C. F., CURTIS, S. M., ELLERSIECK, I. F. and TAILLEUR I. L. 1982. *Reconnaissance Geologic Map of the Southeastern Part of the Mishequk Mountain Quadrangle, Alaska* USGS Open File, Report 82–613.

MAYFIELD, C. F., CURTIS, S. M. and ELLERSIECK, I. F. 1983b. *Reconnaissance geologic map of the De Long Mountains A3, B3 and parts of A4 and B4 Quadrangles, Alaska* USGS Open File Report 83–183, 59 p., Scale 1:63 360.

MAYFIELD, C. F., TAILLEUR, I. L., MULL, C. G. and SABLE, E. G. 1978. *Bedrock geologic map of the southern half of the NPRA, Alaska* USGS Open File Report 78–70B, Scale 1:500 000 two sheets.

McWHAE, J. R. 1986. Tectonic History of Northern Alaska, Canadian Arctic and Spitsbergen Regions Since Early Cretaceous, *AAPGG Bull.* **70**, (4), 430–450.

MELVIN, J. 1985. Sedimentological evolution of Mississippian Kekiktuk Formation, Sagavanirkok Delta Area, North Slope, Alaska: Abstract, *AAPG Bull.* **69**, (4), 669.

MELVIN, J. and KNIGHT A. S. 1985. Lithofacies diagenesis and porosity of the Ivishak Formation, Prudhoe Bay Area, Alaska, *In*: (McDONALD, D. A. and SURDAM, R. C. eds) *Applications in exploration and production* AAPG, Memoir 37, Part 3, p. 347–365.

MENELEY, R. A., Henao, D. and MERRITT, R. K. 1975. The northwest margin of the Sverdrup Basin, *In*: (YORATH, C. J., PARKER, E. R. and GLASS, D. J. eds) *Can. Soc. Pet. Geol.*, Memoir 4, 531–44.

MERTIE, Jr., J. B. 1930. Mountain Building in Alaska, *Am. J. Sci.* **20**, 101–124.

METZ, P. A. 1980. Comments and replies on Churkin, M., Jr., Nokleberg, W. J. and Huie, C., Collision-deformed Paleozoic continental margin, western Brooks Range, Alaska' *Geology* **8**, (8), 360.

MIALL, A. D. 1973. Regional Geology of Northern Yukon, *Bull. Can. Soc. Petr. Geol.* **21**, (1), 81–116.

MOLENAAR, C. M. 1981a. Depositional history and seismic stratigraphy of Lower Cretaceous rocks, NPRA and adjacent areas, USGS Open File Report 81–1084, 42 p.

MOLENAAR, C. M. 1981b. Depositional history of the Nanushuk Group and related strata, *In*: (ALBERT, N. R. D. and HUDSON, T. eds) *The United States Geological Survey in Alaska: Accomplishments during 1979* US Geological Survey Circular 88823-B, p. B4-B6.

MOLENAAR, C. M. 1982. *Umiat Field, An oil accumulation in a thrust-faulted anticline, North Slope of Alaska*, Rocky Mountain Assoc. Geol. p. 537–548.

MOLENAAR, C. M. 1983. Depositional relations of Cretaceous and lower Tertiary rocks, northeastern Alaska *AAPG Bull.* **67** (7), 1066–1080.

MOLENAAR, C. M., EGBERT, R. M. and KRYSTINIK, L. F. 1981. *Depositional facies, petrography and reservoir potential of the Fortress Mountain Formation (Early Cretaceous), Central North Slope, Alaska* USGS Open File Report 81–967, 32 p.

MOORE, T. E. and NILSEN, T. H. 1984. Regional variations in the fluvial Upper Devonian and Lower Mississippian Kanayut conglomerate, Brooks Range, Alaska, *In*: (NIELSEN,T. H. ed.) *Fluvial Sedimentation and Related Tectonic Framework, Western North America, Sedimentary Geology* **38**, 465–497.

MOORE, T. E., BROSGE, W. P., CHURKIN, M., Jr. and WALLACE, W. K. 1986. Pre-Mississippian Accreted Terranes of northeastern Brooks Range, Alaska: Abstract, *In*: (TAILLEUR, I. L. and WEIMER, P. eds) *Alaskan North Slope Geology* SEPM–Pacific Section/Alaska Geol. Soc. (in prep.).

MOUNTJOY, E. W. 1967a. *Upper Cretaceous and Tertiary Stratigraphy. Northern Yukon Territory and Northwestern District of MacKenzie*, Geol. Surv. Can. Paper 66–16.

MOUNTJOY, E. W. 1967b. *Triassic Stratigraphy of Northern Yukon Territory*, Geol. Surv. Can. Paper 66–19.

MULL, C. G. 1979. Nanushuk Group deposition and the Late Mesozoic structural evolution of the central and western Brooks Range and Arctic Slope, *In*: (ALBRANT, T. ed.) *Preliminary geologic, petrologic, and paleontologic results of the Nanushuk Group rocks, North Slope, Alaska* USGS Circular 794, p. 1–46.

MULL, C. G. 1980. Comments and replies on Churkin, M., Jr., Nokleberg, W. J. and Huie, C. 'Collision-deformed Paleozoic continental margin, western Brooks Range, Alaska' *Geology* **8**, (8), 361–362.

MULL, C. G., 1982. Tectonic evolution and structural style of the Brooks Range, Alaska: An illustrated summary, *In*: (POWERS, R. B. ed.) *Geologic studies of the Cordilleran Thrust Belt* **1**, p. 1–46.

MULL, C. G. 1985. *Guidebook to bedrock geology along the Dalton Highway, Yukon River to Prudhoe Bay, Alaska* Alaska Division of Geological and Geophysical Survey Guidebook #7, (Preliminary Draft).

MULL, C. G. and MANGUS, M. D. 1972. Itkilliariak Formation: New Mississippian formation of the Endicott Group, Arctic Slope of Alaska *AAPG Bull.* **56**, (8), 1364–1369.

MULL, C. G., PARIS, C. E. and ADAMS, K. E. 1986. Kemik Sandstone, Arctic National Wildlife Refuge (ANWR), Alaska, *In*: (TAILLEUR, I. L. and WEIMER, P. eds.) *Alaskan North Slope Geology*, SEPM–Pacific Section/Alaska Geol. Soc. in prep.

MULL, C. G., Tailleur, I. L., MAYFIELD, C. F., ELLERSIECK, INYO and CURTIS, S. 1982. New Upper Paleozoic and Lower Mesozoic stratigraphic units, central and western Brooks Range, Alaska *AAPG Bull.* **66**, (3), 348–362.

NELSON, S. W. 1980. Comments and replies on CHURKIN, M., Jr., Nokleberg, W. J. and Huie, C., 'Collision-deformed Paleozoic continental margin, western Brooks Range, Alaska' *Geology* **8**, (8), 364–365.

NENTWICH, F. W. and YOLE, R. W. 1982. Sedimentary petrology and stratigraphic analysis of the subsurface Reindeer Formation (Early Tertiary), Mackenzie Delta – Beaufort Sea Area, Canada, *In*: (EMBRY, A. F. and BALKWILL, H. R. eds) *Can. Soc. Pet. Geol.* Memoir **8**, p.

55–82.
NEWMAN, G. W., MULL, C. G. and WATKINS, N. D. 1979. Northern Alaska paleomanetism, plate rotation and tectonics, *In:* (Sissan, A. ed.) *The relationship of plate tectonics to Alaskan Geology and Resources* Anchorage Alaska Geol. Soc. p. C1–C7.

NILSEN, T. H. 1981 Upper Devonian and Mississippian red beds, Brooks Range, Alaska, *In:* (MIALL, A. D. ed.) *Sedimentation and tectonics in alluvial basins* Geol. Assoc. Can. Special Paper 23, p. 187–219.

NORRIS, D. K. and YORATH, C. J. 1981. The North American Plate from the Arctic Archipelago to the Romanzof Mountains, *In:* (NAIRN, A. E. M., CHURKIN, Jr., M. and STEHLI, F. G. eds) *The Ocean Basins and Margins* Plenum Press, vol. 5, Ch 3, p. 37–104.

OKLAND, L. E., CHANCEY, D. K., CVITASH, J. G., KLINGER, M. E., MAJEWSKI, O. P., OWENS, R. T., SHAFER, D. C. and SMITH, M. T. 1986. Facies Analysis and Correlation in the Lisburne Development Area, Prudhoe Bay, Alaska *In:* (TAILLEUR, I. L. and WEIMER, P. eds) *Alaskan North Slope Geology*, SEPM–Pacific Section/ laska Geol. Soc. (in prep.).

Oil and Gas Journal 1986. *Amauligak tests brighten Beaufort Outlook*, Feb. 3rd, p. 32.

ORI, G. G. and FRIEND, P. F. 1984. Sedimentary basins formed and carried piggyback on active thrust sheets *Geology* **12**, 475–478.

PALMER, Jr., I. F., BOLM, J. G., MAXEY, L. R. and LYLE, W. M. 1979. *Petroleum source rock and reservoir quality data from outcrop samples, onshore North Slope of Alaska east of prudhoe Bay* US Geological Survey Open-File Report 79–1634, 54 p.

PARRISH, J. T. 1986. Lithology Geochemistry, and Depositional Environment of the Shublik Formation (Triassic), Northern Alaska, *In:* (TAILLEUR, I. L. and WEIMER, P. eds) *Alaskan North Slope Geology*, SEPM–Pacific Section/ Alaska Geol. Soc. (in prep.).

PATTON, Jr., W. W. 1956. New and redefined formations of Early Cretaceous age, *In:* (GRYC, G. et al. eds) Mesozoic sequence in Colville River region, northern Alaska *AAPG Bull.* **40** (2), 219–223.

PATTON, Jr., W. W. 1957. *A new upper Paleozoic formation, central Brooks Range, Alaska* US Geological Survey Professional Paper 303-B, p. 41–43.

PATTON, Jr., W. W. and BOX, S. E. 1985. Tectonic setting and history of the Yukon – Koyukuk Basin, Alaska, Abst. AAPG Annual Sym. p. 1101.

PATTON, Jr., W. W. and HOARE, J. M. 1968. *The Kaltag Fault, West–Central Alaska* USGS Prof. Paper 600-d, p. D147–153.

PATTON, Jr., W. W. and TAILLEUR, I. L. 1964. *Geology of the Killik-Itkillik region, Alaska, Part 3: Areal Geology* USGS Prof. Paper 302-G, p. 409–500.

PATTON, Jr., W. W., TAILLEUR, I. L., BROSGE, W. P. and LANPHERE, M. A. 1977. Preliminary report on the ophiolites of northern and western Alaska, *In:* (COLEMAN, R. G. and IRWIN, W. P. eds) *North American ophiolites: State of Oregon,* Dept. of Geol. and Mineral Industries Bulletin, vol 95, p. 51–57.

PAYNE, T. G. and GRYC, G. 1951. Alaska Petroleum in Relation to Geologic History, Abstract – *AAPG Bull.* **62**, (12), Part 2, 1157.

PAYNE, T. G. et al. 1952. *Geology of the Arctic Slope of Alaska* USGS Oil and Gas Investigations Map OM-126.

POULTON, T. P. 1982. Paleogeographic and tectonic implications of the Lower and Middle Jurassic facies patterns in northern Yukon Territory and adjacent Northwest Territories, *In:* (EMBRY, A. F. and BALKWILL, H. R. eds) *Arctic geology and geophysics* Can. Soc. Pet. Geol. Memoir **8**, p. 13–27.

PROCTOR, R. M. TAYLOR G. C. and WADE, J. A. 1983. *Oil and natural gas resources of Canada, 1983* Geological Survey of Canada, Paper 83–31.

RATTEY, R. P. 1986. Northeastern Brooks Range, Alaska: New evidence for complex thin-skinned thrusting: Abstract, *In:* (TAILLEUR, I. L. and WEIMER, P. eds) *Alaskan North Slope Geology,* SEPM–Pacific Section/Alaska Geol. Soc. (in prep.)

REISER, H. N., BROSGE, W. P., DUTRO, Jr., J. T. and DETTERMAN, R. L. 1971. *Preliminary geologic map, Mt. Michelson quadrangle, Alaska* US Geological Survey Open-File Report 71–237, scale 1:200000.

RICHTER, D. H. and JONES, D. L. 1973. Structure and Stratigraphy of Eastern Alaska Range, Alaska, *In:* (PITCHER, M. G. ed.) *AAPG Memoir* **19**, p. 408–420.

RICKWOOD, F. K. 1970. The Prudhoe Bay Field, *In:* (ADKINSON, W. L. and BROSGE, M. M. eds) *Proceedings of the geological seminar on the North Slope of Alaska*, Pacific Section AAPG, p. L1–L11.

ROBINSON, F. M. 1959. *Test Wells, Simpson Area, Alaska*, US Geological Survey Prof. Paper 305-J, p. 523–568.

ROEDER, D. and MULL, C. G. 1978. TECTONICS OF BROOKS RANGE OPHIOLITES, ALASKA, *AAPG Bull.* **62** (9), 1696–1713.

ROYDEN, L., HORVATH, F. and RUMPLES, J. 1983. Evolution of the Pannonian Basin System Tectonics, v. 2, No. 1, Part One, p. 63–90. Part Two, Subsidence and Thermal History, p. 91–137.

SABLE, E. G. 1977. *Geology of the western Romanzof Mountains, Brooks Range, northeastern Alaska* Geol. Surv. Prof. Paper 897.

SABLE, E. G. and DUTRO, J. T. 1961. New Devonian and Mississippian Formations in DeLong Mountains, northern Alaska *AAPG Bull.* **45**, (5), 585–593.

SCHRADER, F. C. 1902. Geologic section of the Rocky Mountains in northern Alaska *Geol. Soc. Am. Bull.* **13**, 233–252.

SEIFERT, W. K., MOLDOWAN, J. M. and JONES, R. W. 1979. Application of biological marker chemistry to petroleum exploration, *In: 10th World Petroleum Congress, Bucharest*, v. 2, Special Paper, Heyden and Sons Ltd., p. 425–440.

SMITH, P. S. 1913. The Noatak–Kobuk region, Alaska *USGS Bull.* **536**, 160.

SMITH, D. G. 1986 Late Paleozoic to Cenozoic reconstructions of the Arctic, *In:* (TAILLEUR, I. L. and WEIMER, P. eds) *Alaskan North Slope Geology*, SEPM–Pacific Section/Alaska Geol. Soc. (in prep.).

SNELLSON, S. and TAILLEUR, I. L. 1968. Large-scale thrusting and migrating Cretaceous foredeeps in the western Brooks Range and adjacent regions of northwestern Alaska [abs.] *AAPG Bull.* **52**, (3), 567.

SNOWDON, L. R. 1980a. Petroleum source potential of the Boundary Creek Formation, Beaufort–Mackenzie Basin *Can. Soc. Pet. Geol. Bull.* **28**, no. 1.

SNOWDON, L. R. 1980b. Resinite – a potential petroleum source in the Upper Cretaceous/Tertiary of the Beaufort–Mackenzie Basin *Can. Soc. Pet. Geol. Mem.* **6**, 509–522.

SWEENEY, J. F. 1983. Evidence for the origin of the Canada Basin margin by rifting in Early Cretaceous time. The origin of the Arctic Ocean (Canada Basin) Proc. of 1981 mini-symposium. *J. Alaska Geol. Soc.* **2**, 17–23.

TAILLEUR, I. L. 1964. *Rich oil shale from Northern Alaska* USGS Geol. Surv. prof. Paper 475-D, Art. 148, p. 131–133.

TAILLEUR, I. L. 1969. Rifting speculation on the geology of Alaska's North Slope *Oil Gas. J.* **67**, (39), 128–130.

TAILLEUR, I. L. 1973. Probable rift origin of Canada Basin, Arctic Ocean, *In: Proceedings of the Second International Symposium on Arctic Geology*, AAPG Memoir # 19, p. 526–535.

TAILLEUR, I. L. and BROSGE, W. P. 1970. Tectonic history of northern Alaska, *In:* (ADKINSON, W. L. and BROSGE, M. M.) *Proceedings of the geological seminar on the North Slope of Alaska*, AAPG Pacific Section, Menlo Park, p. E1–E19.

TAILLEUR, I. L., BROSGE, W. P. and REISER, H. N. 1967. Palinspastic analysis of Devonian rocks in northwestern Alaska, *In:* (OSWALD, D. H. ed.) *International Symposium on the Devonian System: Calgary, Alberta, Canada*, Alberta Society of Petroleum Geologists, v. 2, p. 1345–1361.

TAILLEUR, I. L., KENT, Jr., B. H. and REISER, H. N. 1966. *Outcrop/geologic maps of the Nuka–Etivluk region, northern Alaska* USGS Open File Report 66–128, Scale 1:63 360, seven sheets.

TOURTELOT, H. A. and Tailleur, I. L. 1971. *The Shublilk Formation and adjacent strata in N.E. Alaska: Description, Minor elements, Depositional environments and Diagenesis* USGS Open File Report 71–284 (462), 62 p.

TURNER, D. L., FORBES, R. B. and DILLON, J. T. 1979. K-Ar geochronology of the southwestern Brooks Range, Alaska *Can. J. Earth Sci.* **16**, (9), 1789–1804.

VOGT, P. R., TAYLOR, P. T. KOVACS, L. C. and JOHNSON, G. L. 1979. Detailed aeromagnetic investigation of the Arctic Basin, *J. Geophys. Res.* **84**, 1071–1089.

VOGT, P. R., TAYLOR, P. T., KOVACS, L. C. and JOHNSON, G. L. 1982. The Canada Basin – Areomagnetic constraints on structure and evolution, *In:* (JOHNSON, G. L. and SWEENEY, J. F. eds) Structure of the Arctic, *Technophysics* **89**, (1–3), 295–336.

WERNER, M. R. 1986. West Sak and Ugnu Sands: Low gravity oil zones of the Kuparuk River area, North Slope, Alaska, *In:* (TAILLEUR, I. L. and WEIMER, P.) *Alaskan North Slope Geology*, SEPM–Pacific Section/Alaska Geol. Soc. (in prep.).

WILLUMSEN, P. S. and COTE, R. P. 1982. Tertiary sedimentation in the Southern Beaufort Sea, Canada, *In:* (EMBRYO, A. F. and BALKWILL, H. R. eds) *Can. Soc. Pet. Geol. Mem.* **8**, 43–54.

WOIDNECK, K., Behrman, P., SOULE, C. and WU, J. 1986. Reservoir description of the Endicott Field, North Slope, Alaska *In:* (TAILLEUR, I. L. and WEIMER, P.) *Alaskan North Slope Geology*, SEPM–Pacific Section/Alaska Geol. Soc. (in prep.).

WOOD, G. V. and Armstrong, A. K. 1975. *Diagenesis and stratigraphy of the Lisburne Group limestones of the Sadlerochit Mountains and adjacent areas, northeastern Alaska* US Geological Survey Professional Paper 857, 47 p.

YORATH, C. J. and NORRIS, D. K. 1975. The tectonic development of the southern Beaufort Sea and its relationship to the origin of the Arctic Ocean Basin, *In:* (YORATH, C. J., PARKER, E. R. and GLASS, D. J.) *Canada's continental margins and offshore petroleum potential* Can. Soc. Pet. Geol. Memoir **4**, p. 589–612.

YORATH, C. J., Balkwill, H. R. and KLASSEN, R. W. 1975. *Franklin Bay and Molloch Hill map areas, District of Mackenzie* Geol. Surv. Canada, Paper 74–36.

YOUNG, F. G. 1975. Upper Cretaceous stratigraphy, Yukon coastal plain and northwest Mackenzie Delta *Geol. Surv. Can. Bull.* **249**.

YOUNG, F. G., MYHR, D. W. and YORATH, C. J. 1976. *Geology of the Beaufort–Mackenzie basin* Geol. Surv. Can., Paper 76–11, p. 1–65.

Hydrocarbon generation and migration in the Western Canada sedimentary basin

S. CREANEY & J. ALLAN

Esso Resources Canada Limited, 237–4th Avenue S.W., Calgary, Alberta, T2P OH6, Canada

Abstract: A total of ten marine, petroleum source rocks ranging in age from Middle Devonian to Late Cretaceous exist within the Western Canada Basin. Principal hydrocarbon reserves occur in the Lower Cretaceous Mannville Group and subcropping Paleozoic carbonates, where 270×10^9 m^3 (1.7 trillion barrels) of heavy oil form one of the largest deposits of oil on Earth. Biomarker analyses of these heavy oils, many conventional oils and most source rocks in the basin have been combined with basic geology to establish the source and migration history of the Western Canada Basin. Several pre-Cretaceous source rocks (principally the Lower Jurassic Nordegg Formation, the Mississippian Exshaw/Bakken Formation, the Triassic Doig/Phosphate Formation and the Upper Devonian Duvernay Formation) have contributed to the Mannville accumulations. The Joli Fou shale, overlying the Mannville, is a regional seal for the Mannville and older section. Oils present in the overlying Viking, Cardium, and Belly River sands are all of a single family, sourced from the Colorado Shale section and are quite distinct from the oils in the Mannville and older section.

Meteoric water incursion at the eastern limb of the basin permitted biodegradation, which converted the conventional Mannville crude oil to heavy oil. The Western Canada Basin also contains an estimated 5×10^{12} m^3 (180 TCF) of gas. Gas sources are the marine, oil-prone source rocks where overmature, as well as Lower Cretaceous Mannville coals and coaly shales. Hydrogen sulphide is contributed from overmature Paleozoic evaporites/carbonates with an estimated 36% of the total in-place gas reserves being classified as sour.

To facilitate the discussion that follows, a general location map is shown in Fig. 1, and a general stratigraphic summary for the Western Canada Basin is provided in Fig. 2.

The Western Canada Basin contains approximately 1.75 trillion barrels of crude oil (initial-in-place), of which 2.4% (41 billion barrels) is considered conventional. Thus, the Western Canada Basin is (from a source rock point of view) one of the most prolific hydrocarbon provinces on Earth. Exploration has been extremely intense in western Canada for over 70 years, with considerable success occurring prior to the routine study of source rock potential. For this reason, the presence of source rocks has largely been 'assumed' (obviously correctly) and little rigorous study has been reported.

In 1892, J. B. Tyrrell described 'possible oil shales' along the Manitoba escarpment, and McInnes (1913) generated a 7 gallon/ton oil yield from samples of these shales. Ells (1923) described these rocks as having been deposited in a 'muddy sea', and considered them to be age equivalent to the Niobrara Formation of the United States. These descriptions represent the first attention paid to some of the more significant petroleum source rocks in the Western Canada Basin.

As early as 1954, William Gussow discussed hydrocarbon migration in the Upper Devonian Rimbey–Meadowbrook reef chain of Alberta, and speculated that this oil may have contributed to the Athabasca Cretaceous deposit. Gussow (1955) also suggested a Jurassic source for these heavy oil deposits. So began a series of many research studies which dealt almost exclusively with oil and/or gas samples, and yet sought to understand the source of hydrocarbons in Western Canada.

Deroo *et al.* (1977) were the first authors to publish source rock data combined with reservoired hydrocarbon data. Using a variety of analyses which were then 'state-of-the-art', Deroo *et al.* (1977) divided 105 oil samples into three genetic groups.

Group 1: Belly River, Cardium, and Viking reservoirs.

Group 2: Lower Mannville, Jurassic, and Mississippian reservoirs.

Group 3: Nisku, Leduc, and Beaverhill Lake reservoirs.

Approximately 800 rock samples were screened for source potential using Total

Fig. 1. The Western Canada Sedimentary Basin.

Organic Carbon (TOC) and organic extract data. The vast majority of these samples were not from source facies, and so presented a somewhat 'blurred' picture of the properties of Western Canada's source rocks.

Du Rouchet (1984), while discussing migration mechanisms for the tar sand deposits, provided evidence for a Triassic source for these deposits based on aromatic hydrocarbons and benzothiophene distributions. However, the evidence again relied on oil–oil correlation (Triassic-reservoired oil at Inga to Mannville oil), with no source rock data.

Stoakes & Creaney (1985) in a detailed organic geochemical, sedimentological and stratigraphic study of the Upper Devonian Woodbend Group shales showed that the Duvernay/Majeau Lake basinal, organic-rich laminites were the source of the hydrocarbons in the Leduc reefs of west–central Alberta and could, on geologic grounds, have contributed to the heavy oils in the Mannville.

Macauley (1984a, b) and Macauley et al. (1985) provided the first stratigraphically constrained geochemical data on organic-rich rocks within the marine Colorado Shale section. These authors analyzed ~450 samples from 29 wells and potash shafts across Manitoba, Saskatchewan, and eastern Alberta. Although all samples were from the immature portion of the Western Canada Basin they, for the first time, showed that the First White Speckled Shale (Boyne Formation) and the Second White Speckled Shale (Favel Formation) were organically enriched (TOC contents to 11%) and dominated by marine, type-II (oil-prone) organic matter (Rock–Eval pyrolysis hydrogen indices to 600).

Following an earlier description of massive gas deposits in low porosity Mesozoic clastics (Masters 1979), Welte et al. (1984) described the origin of these gases. They considered them to have been generated from Jurassic–Lower Cretaceous coals and coaly shales at advanced levels of maturity. TOC, reflected light organic petrography, and Rock–Eval pyrolysis data were used to identify the gas-prone nature of this part of the section. Moshier & Waples (1985) provided a mass balance calculation, based on the low average TOC of intra-Mannville shales and generally low hydrogen indices, which showed this section to be incapable of generating the required 1.7 trillion barrels of oil in the tar belt. However, these authors did not suggest an alternative source.

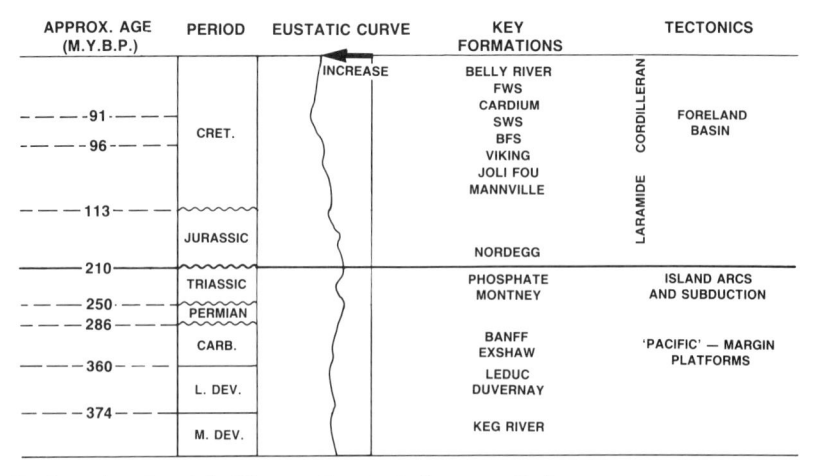

Fig. 2. Generalized stratigraphy of the Western Canada Sedimentary Basin.

The present study will attempt to provide this additional information and incorporate it into an analysis of this prolific hydrocarbon basin.

Methodology

The principal resource behind this study is an extensive collection of crude oils, source rock cores, gas analyses and geophysical well logs from the Western Canada Basin. To date, over 300 oil samples, over 200 source rock cores and over 400 gas analyses have been interpreted in the basin model. The classification of the hydrocarbons into families is complex and based on numerous factors including age and location of reservoir, reservoir type, oil gravity, sulphur content, oil geochemistry and source rock geochemistry.

Gas chromatography (GC) provides the initial geochemical screen of oils whereby n-paraffin and isoprenoid distributions can be examined, maturities approximately assessed, presence or absence and extent of biodegradation determined, and so on. The final analysis is the evaluation of naphthenic and aromatic biomarker distributions using computerized gas chromatography–mass spectrometry (C-GC-MS).

Biomarker analysis provides the main evidence for oil-source rock correlations, and all chemical correlations reported here are based exclusively on examination of core samples. In like manner to the oils, presumed source rocks are extracted (dichloromethane as solvent), and GC and C-GC-MS analyses are performed routinely on unfractionated extract or separated saturated hydrocarbon and aromatic hydrocarbon fractions. Prior to this step, however, Rock–Eval pyrolysis provides base data on organic richness (TOC), organic matter type via hydrogen indices (HI), and maturity via temperature of maximum rate of evolution of kerogen pyrolysis products (T_{max}).

Natural gas analyses provide information on compositions from which maturities can be determined. In combination with wetness indices and H_2S content (if any), these data complement the oil analyses in compiling evidence for the likely hydrocarbon migration pathways which have been active. The key control on this modelling is detailed knowledge of the geological structure of the basin and the timing of significant events which impact hydrocarbon generation, migration and entrapment.

Hydrocarbon systems of the Western Canada basin

The Devonian

The Elk Point Group (Middle Devonian). A single commercially-effective source rock occurs in the Elk Point Group of the Middle Devonian. The source facies occurs in basinal areas of the Lower Keg River and Muskeg Formations (Fig. 3) and is, in part, time-equivalent to reef growth in the Rainbow/Zama/Shekilie sub-basins.

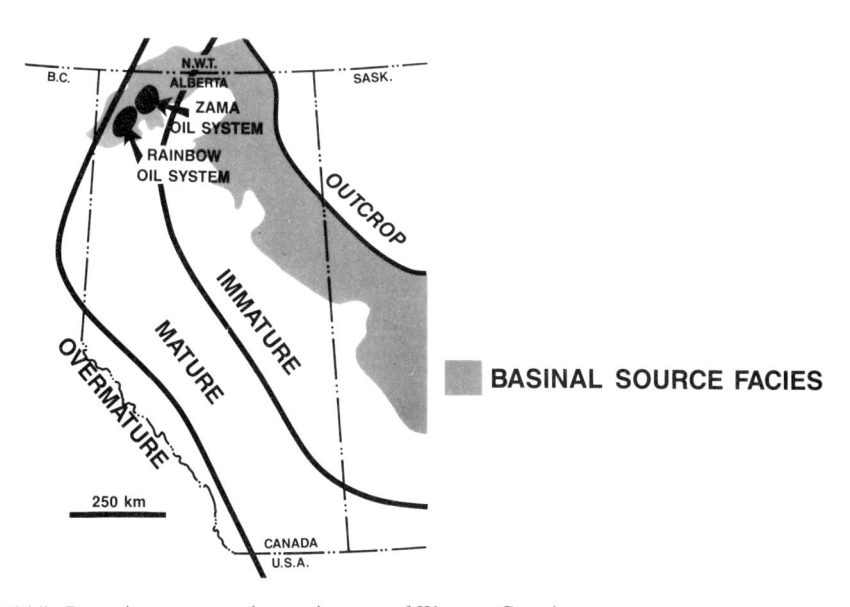

Fig. 3. Middle Devonian source and maturity map of Western Canada.

Powell (1984) described this unit from its occurrence around the Pine Point ore deposit in the southern Northwest Territories. The source sediment was deposited under anoxic, hypersaline conditions with low siliciclastic influx which allowed significant amounts of sulphur to become incorporated into the kerogen. This feature enables heavy bitumen to be generated at low levels of maturity (LOM 6−7, R_0 = 0.40−0.45%), as is found at Pine Point. TOCs up to 15.5% and hydrogen indices to 438 (ie: type II) have been measured within this facies.

The source facies, described above, is immature over much of its present occurrence, only attaining full maturity in the northwest in the Rainbow/Zama/Shekilie sub-basins (Fig. 3) and in southern Saskatchewan. The occurrence of a basinally-restricted source facies between a series of pinnacles which are capped by shale and/or evaporites provides an ideal generation/migration/entrapment situation. Local sourcing in these pinnacles results in a systematic progression of oil properties from low gravity, lower maturity oils at the eastern end of the Zama sub-basin to high gravity oils, condensates, and gases as the onset of overmaturity is approached to the west. No evidence exists to date that any Elk Point-sourced hydrocarbons have migrated out of the Middle Devonian strata and this, therefore, represents a closed system.

The Beaverhill Lake and Woodbend Groups (Late Devonian). A single source rock occurs in the Beaverhill Lake and Woodbend Groups of the Upper Devonian. Stoakes & Creaney (1985) have provided detailed descriptions of the Duvernay Formation. This source is time-equivalent to Leduc-age reef buildups and occurs throughout the basin between these buildups (Fig. 4). TOCs range from 2−17%, with hydrogen indices in the 500−600 range (typical marine, type-II kerogen). The inorganic matrix is a variable mixture of carbonate and clay minerals. Again, this source is immature over much of its occurrence, with a band of progressive 'oil window' maturities and ultimately overmaturity in the deep, western part of the basin adjacent to the Disturbed Belt.

Stoakes & Creaney (1985) have provided a detailed description of hydrocarbon migration in the Woodbend Group, principally along the Rimbey−Meadowbrook reef chain. Oils generated from mature Duvernay source facies have migrated locally into porous and permeable dolomitized Cooking Lake Platform margins and Leduc-reef buildups. This oil then migrated updip along the platform margin, some accessing overlying pinnacle reefs such as Leduc, Redwater, Acheson, etc., with the possibility of further updip migration until it could leak to the overlying Grosmont Formation and then to subcrop beneath the Mannville.

In the area around the Bashaw complex, oil from mature Duvernay has migrated into adjacent Leduc pinnacles, but the proximity of the overlying Camrose Member and Nisku Formation (platform facies) has allowed significant leakage to the Nisku. This oil subsequently

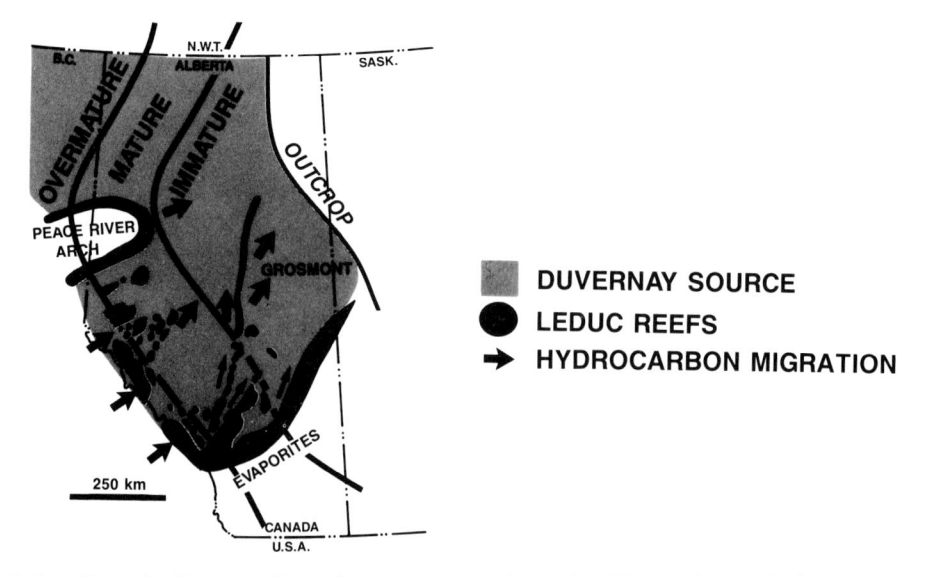

Fig. 4. Late Devonian Duvernay Formation occurrence and maturity, Western Canada Basin.

accumulated updip in Nisku drape structures over underlying Leduc buildups, as well as in stratigraphic traps. Excess oil migrated to Nisku subcrop beneath the Mannville and thence into the Mannville in the area of Bellshill Lake to Lloydminster.

Northwest of the Rimbey–Meadowbrook trend, the Nisku shale basin is developed, effectively top sealing the Duvernay system in that direction. Thus, Duvernay oil filled Leduc buildups, spilled into underlying Woodbend platforms, and migrated updip to the platform margin. The nature of platform and reef development in western Alberta resulted in 'stacking' of Leduc platforms and reefs directly on Beaverhill Lake platforms. This local stacking geometry allowed Duvernay-sourced oil to spill from Woodbend platforms into stratigraphically older Beaverhill Lake platforms. This process continued, with oil ultimately spilling into, and filling, Swan Hills buildups at Judy Creek, Carson, etc., and the Swan Hills platform margin at Deer Mountain. The remaining excess oil spilled into Middle Devonian Gilwood sands, and was prevented from further updip migration only by the stratigraphic pinch-out of these sands at Nipisi and Mitsue.

A similar process of stratigraphically down, but structurally updip, migration occurred on the Peace River Arch, resulting in peak maturity (API = 40°) Duvernay-sourced oil in Keg River carbonates as far east as Senex. The Keg River shelf margin on the edge of the La Crete Basin provided the ultimate trap for Duvernay-sourced oil in the Peace River Arch area.

Overmaturity in the deep western portion of the basin has produced significant quantities of gas, often H_2S-rich.

The Winterburn Group (Late Devonian). Chevron exploration staff (1979) described basic geochemical properties from the Cynthia shale in the Nisku shale basin. This shale is generally thin, with TOC values up to 2.0%, and contains typical type-II marine organic matter. No information exists at this time on the distribution of this source outside the area of the Nisku pinnacle reefs of West Pembina.

As mentioned earlier, the Nisku platform sediments to the southeast of the Nisku shale basin received much of their hydrocarbons by leakage from underlying Leduc reefs. Only at the southern margin of the Nisku shale basin did Nisku pinnacles receive locally-sourced oil and gas from the Cynthia Member. This local sourcing is reflected in the progressive increase in API gravity from east to west along the pinnacle belt, ultimately ending in condensate and gas at the western edge where overmaturity is reached. Some sourcing into the platform margin and updip migration may have occurred.

Late Devonian–Carboniferous. A single, prolific source rock occurs in this section, namely the Mississippian Exshaw/Bakken shale. This shale was deposited immediately following a major drowning event which terminated the carbonate reef systems of the Devonian. The Bakken Shale is a major hydrocarbon source in the Williston Basin (Webster 1984) and has been proposed by Leenheer (1984) as the principle source for the massive heavy oil accumulations of Western Canada.

The source is thin (<10 m), but often extremely organic-rich (TOCs ≤20%) and oil prone with hydrogen indices to 600. These high values suggest a type-I organic matter and *Tasmanites* can be frequently observed in the shales. Maturity distribution is essentially as shown for the Duvernay in Fig. 4.

Exshaw-sourced hydrocarbons have accessed both the underlying Wabamun and the overlying Banff Formations. In the Peace River Arch area, the Wabamun is a low porosity, low permeability ramp carbonate and reservoir is developed only along fault plane dolomitization at fields such as Tangent. To the south (further up the Wabamun shelf), the porosity and permeability improves, providing a significant updip migration conduit. Thus, in the southern portion of Alberta, Exshaw-sourced hydrocarbons can migrate into the Mannville via the top of the subcropping Wabamun.

A similar situation exists for the overlying Banff, with Exshaw-sourced oils trapped in the Banff at fields such as Twining, but with significant volumes expected to have leaked to the Mannville at the Banff subcrop.

The Permo-Triassic

A number of organic-rich sediments of Triassic age occur within the basin, although with limited present-day extent. The most prolific source is the basal radioactive phosphatic facies of the Doig Formation (referred to here, informally, as the Phosphate Zone), with measured TOCs to 10% and hydrogen indices to 400 where mature. Organic deposition of 'Phosphate type' continued through the overlying Doig and Lower Halfway Formations (distal facies), but did not attain the prolific source richness of the Phosphate Zone itself. The Phosphate Zone is underlain by the thick (up to 300 m) Montney Shale which has a different organic geochemical signature to the Phosphate Zone and later shales. The Montney has known TOCs to 3%,

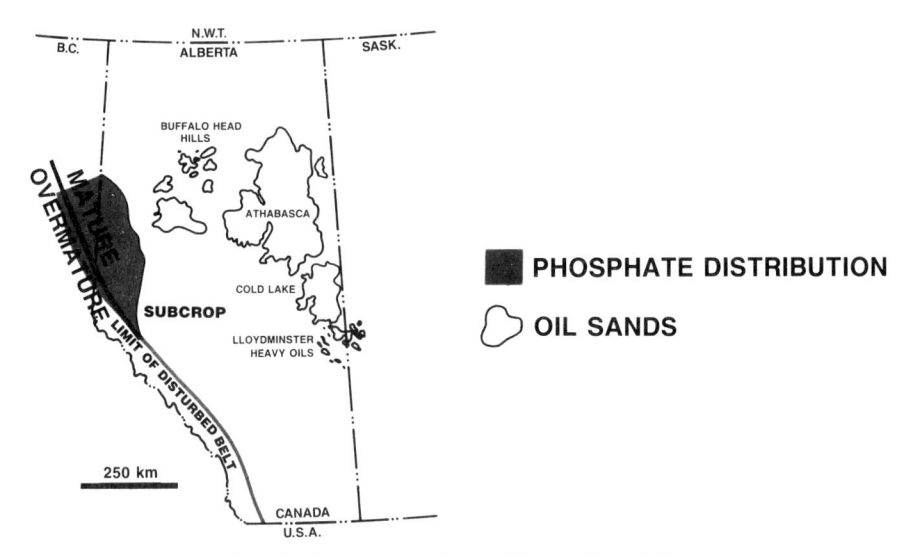

Fig. 5. Distribution and maturity of Triassic sources in the Western Canada Basin.

with hydrogen indices to 500, but orgainc-rich intervals in the Montney are localized rather than regional.

The Phosphate Zone ranges from mature to overmature, with the onset of maturity line approximately coinciding with its erosionally-truncated eastern limit (Fig. 5). The Montney subcrops the Jurassic–Cretaceous section further east than the Phosphate and, consequently, has an extensive immature interval east of the mature zone.

Most oils and gases reservoired in the Triassic have been generated from the Phosphate Zone. Oils currently reservoired in the Halfway, Doig, and Charlie Lake show biomarker signatures which correlate them to the Phosphate. Du Rouchet (1984) considered this section to be the principal source of hydrocarbons for the massive tar sands accumulations of eastern Alberta. However, over much of its occurrence, the Triassic section is isolated from the Cretaceous Mannville by the shaly Jurassic section, and access for Triassic oils to the Mannville is expected to have been restricted.

Passage downward through the thick Montney Shale is known to have occurred locally, as Phosphate oil has been identified in Upper Belloy (Permian) reservoir. However, these are only limited occurrences, and the Montney acts, for the most part, as a bottom seal for Phosphate Zone oil. The Montney itself has supplied oil to the Upper Belloy, as evidenced by matching stains to a Montney source, and this provides a probable conduit for Triassic-age oil to be leaked to the Mannville. Interestingly, the Peace River heavy oil deposit within the Mannville straddles the Montney subcrop.

The Jurassic

An extremely rich, marine, type II, organic-rich source rock exists within the Jurassic section of western Canada. The Lower Jurassic (Sinemurian) Nordegg Formation has TOCs up to 27% and hydrogen indices up to 600 (when immature). The Nordegg, which is often less than 20 m thick, was deposited following a major marine transgression in Early Jurassic time.

The characteristic gamma log response of the Nordegg (Fig. 6) in northwest Alberta reveals a doublet character, suggesting two cycles of relative oceanic deepening with concomitant organic enrichment. In west–central Alberta, the Nordegg thins and is represented by a single gamma high and, further south, finally undergoes a facies change to more proximal chert-rich sediments. The more basinal facies of the Nordegg contains a sulphur-rich kerogen following accumulation of organic detritus in a strongly reducing, sulphur-rich and iron-poor environment. Thus, the Nordegg generates migratable oils at very low maturity levels (LOM 6–7, $R_0 = 0.40-0.45\%$).

Figure 7 is a present-day maturity map of the Nordegg. This source is within the oil window over much of its occurrence, only attaining overmaturity in the far west of the basin. This extensive area of rich, mature source rock makes the Nordegg a major contributor to the hydrocarbons of the Western Canada Basin.

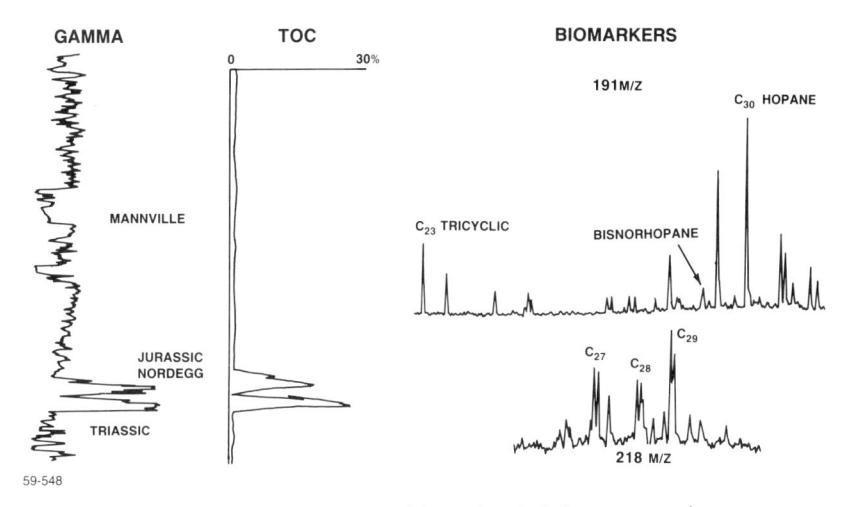

Fig. 6. Log profile of the Jurassic Nordegg source with geochemical data annotated.

The Lower Jurassic Nordegg source rock occupies a unique position in the stratigraphy of the Western Canada Basin. The Jurassic section overlies the Triassic and older formations unconformably and, thus, has the opportunity to source and seal oil directly into older, subcropping sections. Such oils, therefore, have properties which reflect the maturity and organic composition of the overlying Nordegg. Thus, toward the eastern part of the Nordegg occurrence, its maturity is low and subcropping oils are often low gravity and have high sulphur contents.

The total thickness of Jurassic sediments is commonly small. However, the Fernie shales overlie the Nordegg and effectively seal off much of the directly overlying Cretaceous sediments from significant amounts of Nordegg oil. The consequence of this is that Nordegg oil is channelled updip in the sandier facies of the middle Nordegg itself (Fig. 6) to Mannville subcrop and thence has made its way to the eastern limb of the basin.

The Mannville Group (Early Cretaceous)

The Lower Cretaceous Mannville Group is a sand-prone deltaic section containing predomi-

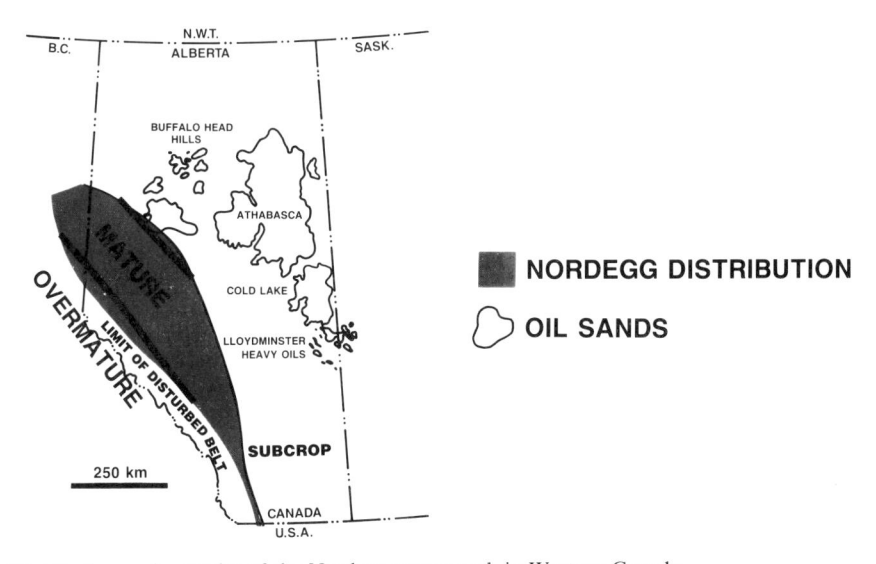

Fig. 7. Distribution and maturity of the Nordegg source rock in Western Canada.

nantly terrigenous organic detritus. TOC values in shales are often <2%, with very low hydrogen indices (Moshier & Waples 1985). However, this section also contains some very significant coal deposits. Welte et al. (1984) have described basic geochemical properties of this coaly section.

The predominant hydrocarbon maturation product of the Mannville Group is gas because of the type III/IV kerogen in the sediments. However, the presence of locally-restricted, delta plain shales containing concentrated amounts of liptinites cannot be discounted in this sedimentary environment. Hence, localized oil or condensate potential may exist, but regional geochemical studies preclude the presence of Mannville-age sources as major contributors to the oils of the basin.

Welte et al. (1984) have shown that the approximately 17 TCF of proved and probable natural gas reservoired in low porosity Lower Cretaceous sands of Western Canada's Deep Basin was largely generated from local, high maturity coals and migrated into adjacent low porosity and permeability sands. Masters (1979) and Welte et al. (1984) have suggested that these gases are hydrodynamically trapped since the sands are water-filled updip from the 'tight gas sands'.

Oils from several pre-Mannville sources have accessed the Mannville at various locations across the basin. Figure 8 is a summary map of sand occurrence in the Mannville. The subcrops of these sources are annotated on Fig. 8. Structural dip of the Mannville is generally up to the northeast, and this is the expected general direction of hydrocarbon migration within the Mannville.

The Post-Mannville Section (Late Cretaceous)

The Colorado Shale Group of Late Cretaceous age is a thick, marine shale and siltstone sequence which contains several oil source horizons. In the southern part of the basin, two principal effective source zones are the Second White Speckled Shale (Fig. 9) and the Fish Scale Zone. These condensed sections lie close together on the eastern limb of the basin (distal facies) and become increasingly separated stratigraphically toward the Alberta Syncline (more proximal facies). Both contain marine, type-II organic matter, with TOCs to 7% and hydrogen indices to 450, where immature. Between the condensed sections are other effective marine, though less rich, source intervals where TOCs are in the 2-3% range and hydrogen indices range up to 300.

In the central part of the basin, a younger condensed section (First White Speckled Shale) becomes mature and hence an effective source rock. This is usually less prominent on natural gamma logs than the older condensed sections, but has source quality characteristics similar to the older Second White Speckled Shale.

The Colorado Shales are mature only in the most westerly part of their occurrence, adjacent to the Disturbed Belt (Fig. 10). Peak maturity of the Second White Speckled Shale and the

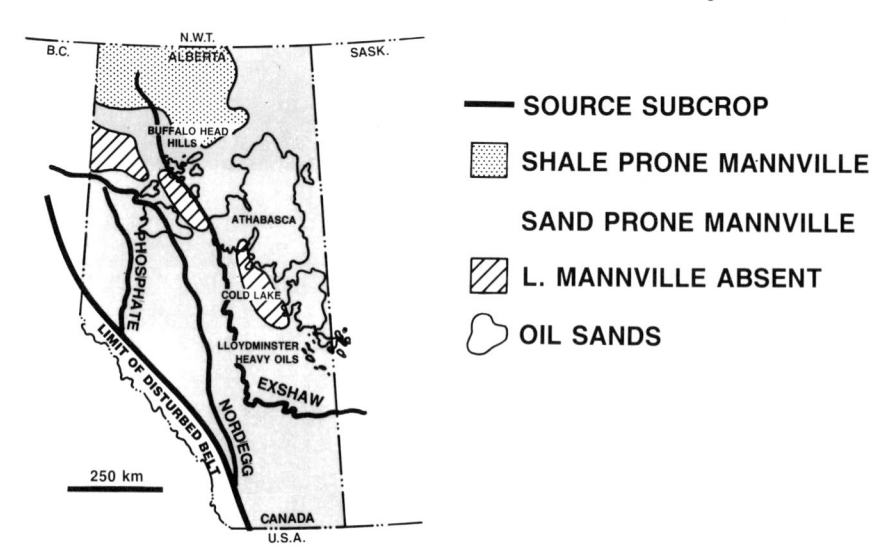

Fig. 8. Approximate Mannville facies distribution and pre-Mannville source rock subcrop.

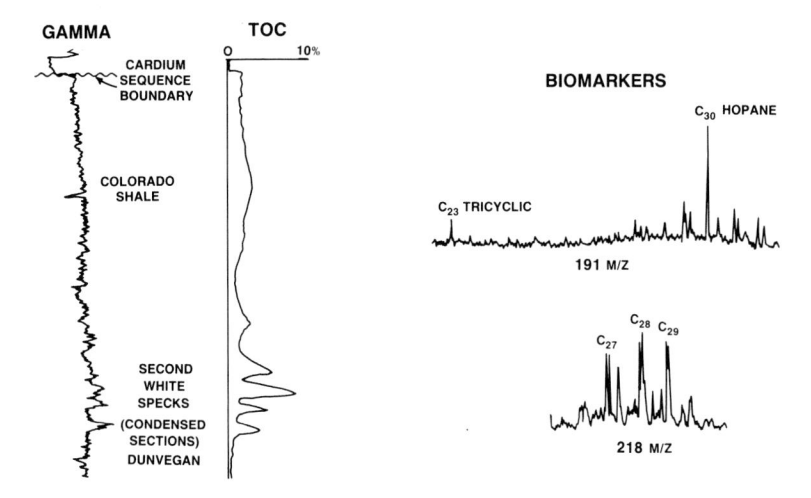

Fig. 9. Log signature and measured TOC data through the Colorado Shales.

Fish Scale Zone lies close to the Disturbed Belt in the southern part of the basin. Passing northward, deeper maximum burial causes these two source zones to trend toward overmaturity and the First White Speckled Shale then becomes mature along the basinal axis. In central Alberta, the maturity contours swing east–west and run into the Disturbed Belt. Thus, there is a broad area from central Alberta north to outcrop where the Colorado Shale Group is immature.

The Colorado Shale outcrop runs east from Alberta into central and southern Saskatchewan. Very low maturities are present throughout Saskatchewan, and the condensed sections have been locally classified as oil shales (Macauley, 1984a, b; Macauley et al. 1985).

Oil derived from the Colorado Shales is principally found in Viking, Cardium and Belly River sands. There is no currently known occurrence of Late Cretaceous-sourced oil in reservoirs older than Viking age and, concomitantly, there is no occurrence of pre-Late Cretaceous oil in Viking or younger sands.

Variable secondary migration pathways have been effective in the dispersal of Colorado Shale oils. Some Cardium nearshore sands are well sealed within shale sequences and clearly have

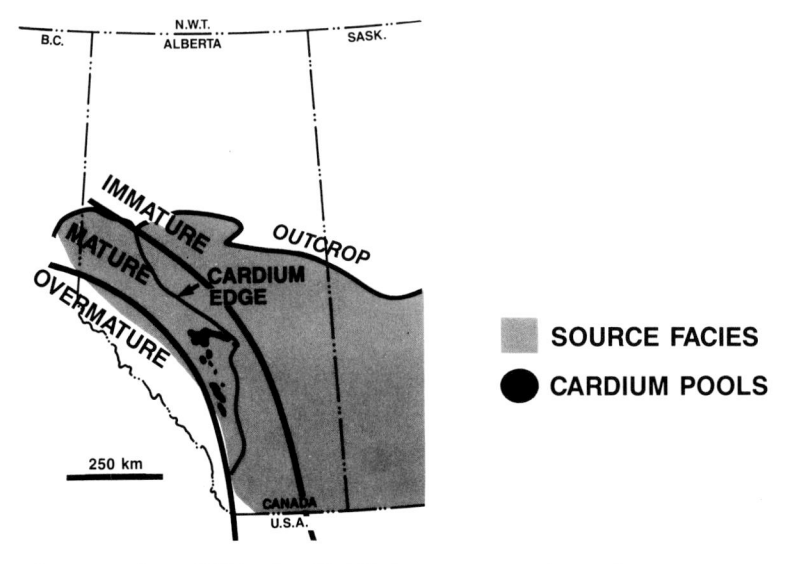

Fig. 10. Late Cretaceous Second White Speckled Shale occurrence and maturity.

been charged with oil from locally adjacent sources. In these cases, the oil is not necessarily derived from the condensed sections and the maturities of pooled oils are similar to the maturity levels of the reservoir rocks.

Secondary migration distances of many tens of kilometres updip have been effective for other Cardium sands and Belly River sands, where peak maturity oil is found in low maturity reservoirs. Connectivity via faults associated with the Disturbed Belt allows oil into the reservoir horizons. Continuous permeability updip to stratigraphic traps (pinch-outs, shoreline sands, etc.) has led to significant accumulations, the most notable of which is the 7 billion BBL giant Pembina field.

The Viking represents a third migration case. This is a formation which underlies a shale sequence beneath the Fish Scale Zone and was charged with oil of peak maturity close to the Disturbed Belt. Continuous permeability in this formation has facilitated easterly updip migration of several hundred kilometres to distant traps located as far away as the Dodsland and Eureka fields in Saskatchewan. This is the greatest secondary migration distance identified to date in the Western Canada Basin.

General hydrocarbon migration synthesis

Migration Pathways

Figure 11 is a pair of schematic cross-sections of the Western Canada Sedimentary Basin, marked with the principal migration conduits for oil and gas derived from the source rocks described in this paper.

Support for the overall hydrocarbon migration dynamics described above is obtained from analysis of basic reserves data. Figure 12 shows conventional oil initial-in-place reserves by geological period for western Canada. Sixty-two per cent of the conventional crude oil occurs in the more or less 'closed system' of the Late Cretaceous and Devonian periods. In this context, the term 'closed system' is used to describe a system which is sourced internally, and there are structural/stratigraphic restrictions on the ability of these hydrocarbons to leak out of the system. In contrast, the relatively open systems of the Jurassic and Mississippian, which contain the prolific Nordegg and Exshaw source rocks, contain only 8.6% of the total conventional reserves.

Migration timing

Conventionally, burial history curves constructed for the Western Canada Basin lead to the conclusion that the development of full maturity for any source rock did not occur until Late Cretaceous time or later. The event responsible for this is the crustal downwarping (Laramide orogeny) associated with foreland basin development. Previous geochemical studies (e.g. Deroo et al. 1977) have supported this dating of peak oil generation. An example of maturity development profiles in the southern part of the basin, currently an axial location, is shown in Fig. 13. This modelling shows an approx. 20 million year time range over which oils from Duvernay Shale to Colorado Shale may have been generated. With the exception of Late Cretaceous-age oils, the vast majority of these liquids migrated into Grosmont and Mannville sediments and were subsequently biodegraded. The maturity model predicts a 15 milion year time gap between peak maturity of the Duvernay Formation and that of the Nordegg Shale, the youngest and oldest source rocks, respectively, to contribute oils to the heavy oil deposits. Thus, it is reasonable to suppose that oils arrived at the accumulation sites at Wabasca, Athabasca, Cold Lake and Grosmont over roughly the same span of time. Biodegradation may have occurred after all the oils were in place, or it may have been an active process during the time over which the oils accumulated. If the former case is correct, the degree of biodegradation might be expected to be relatively uniform, while, if the latter case is correct, then the earliest pooled oil might well show a greater degree of biodegradation than the latest pooled oil simply because the microorganisms will have had an additional 15 million years to feed on the hydrocarbons.

Figure 14 shows m/z 191 fragmentograms of pentacyclic triterpanes in a Mannville and Grosmont heavy oil. The Mannville example is typical of almost all heavy oils examined from both Mannville and Grosmont. The Grosmont example in Fig. 14 is seen only occasionally. Using the scheme of relative alteration proposed by Alexander et al. (1983), these two examples show widely differing degrees of biodegradation although the two sampling sites are (geographically) very close together. A possible conclusion is that the Grosmont sample has been biodegraded over a greater time period than the Mannville sample. Geological considerations suggest that the Duvernay Formation could be the source rock which would first charge oil to the Grosmont. Geochemical typing suggests that the Mannville sample was derived from the Nordegg Shale, or that Nordegg oil has provided the latest overprint to earlier oil (e.g. Exshaw source) already in place. The hypothesis is,

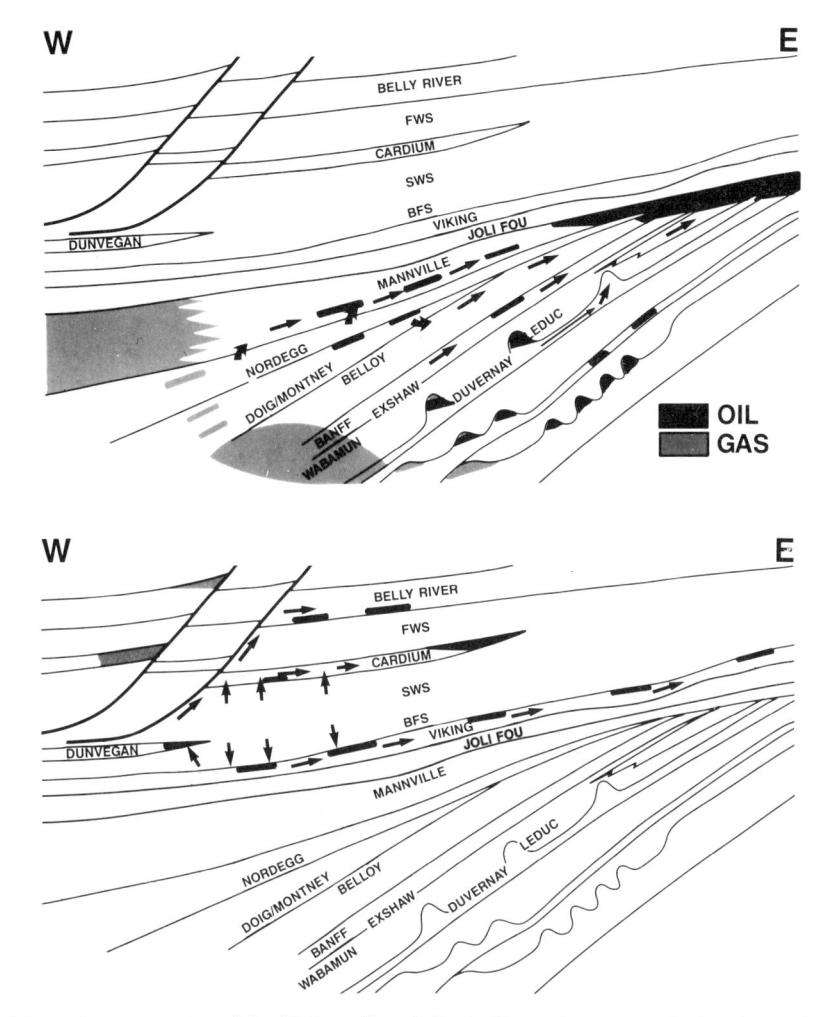

Fig. 11. Schematic cross section of the Western Canada Basin illustrating proposed migration pathways in the pre-Joli Fou section (top) and the post-Joli Fou section (bottom).

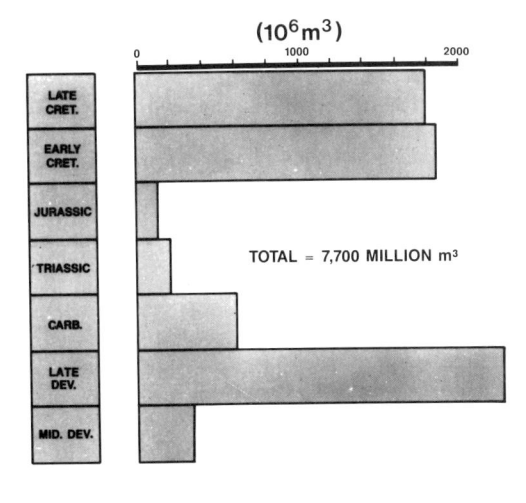

therefore, proposed that biodegradation had commenced in oil accumulating in the Grosmont before Nordegg oil arrived at Athabasca and, possibly, as early as the arrival time of Duvernay oil in the Grosmont. A fuller discussion of this is reserved for a separate publication.

Liquid hydrocarbon mass balance

Having established the presence of source rocks which can reasonably be expected to have provided hydrocarbon 'charge' to the heavy oil

Fig. 12. Conventional oil reserves by geological period for the Western Canada Basin (data source ERCB).

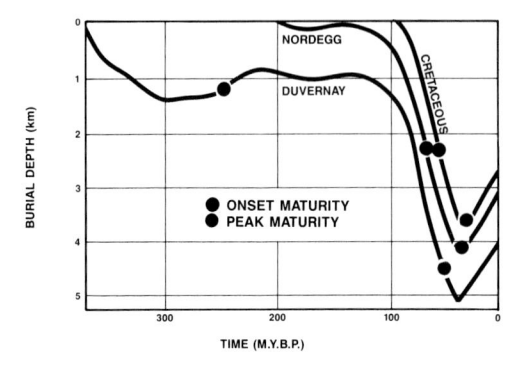

Fig. 13. Burial history curves for Duvernay, Nordegg and Colorado Shale source rocks.

accumulations of western Canada, then the calculation of their total potential yield is straightforward. The general calculation is as follows:

Source × Thickness × Drainage Area × Fractional TOC
× Fractional Liquid Yield = Hydrocarbon Charge

The (conservative) maximum net liquid yield factor used for this calculation is 0.28 for rocks which have passed through peak maturity (LOM 10.5, $R_0 = 0.82\%$), and no liquid yield prior to LOM 7 ($R_0 = 0.45\%$). In this calculation, the generation of gas and its ultimate fate in the subsurface have been ignored.

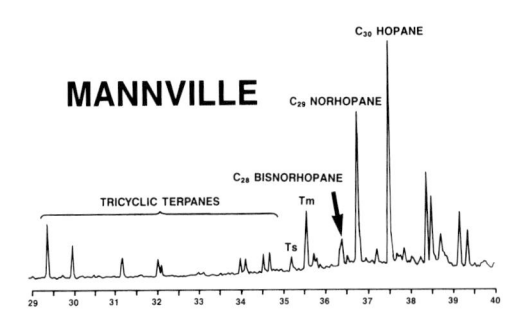

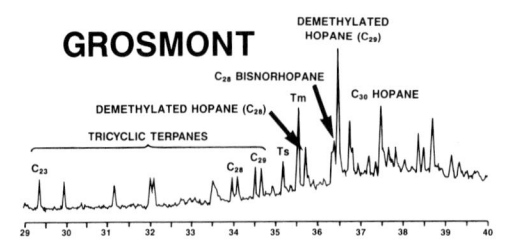

Fig. 14. Terpane fragmentograms (191 m/z) of Mannville and Grosmont heavy oil samples.

Drainage area is geologically constrained and was kept constant, as is shown in Fig. 15. Figure 16 displays a sensitivity to total drainage area (i.e. the sum of drainage areas of four source rocks). The 100% mark on the abscissa represents the total drainage area (155 000 square miles) used for the analysis. At 0%, the drainage area = 0 and, therefore, yield = 0. Variables in the analysis are source thickness and TOC. Using geophysical logs and measured core data, ranges of values for the two variables were assessed from which the low, most likely, and high combined liquid yields are computed. The known heavy oil-in-place reserve is also marked on Fig. 16.

This analysis shows that while the reserves are enormous, they are not unrealistic for the overall geological setting which has been described in this paper.

Conclusions

Geochemical data and geological modelling have led to the following conclusions.

Ten marine, organic-rich (TOCs = 2–20%), kerogen type I/II source facies (HIs up to 600) have been documented in the Western Canada Basin:

Middle Devonian–Lower Keg River;
Late Devonian–Duvernay Formation;
Late Devonian–Cynthia Member;
Early Mississippian–Exshaw/Bakken Formation;
Triassic–Montney Formation;
Triassic–Phosphate Zone;
Early Jurassic–Nordegg Formation;
Late Cretaceous–Fish Scales Zone;
Late Cretaceous–Second White Speckled Shale;
Late Cretaceous–First White Speckled Shale.

Two sections exist which contain coal seams and extensive deltaic, kerogen type III/IV facies having gas source potential. They are the Late Cretaceous–Tertiary Edmonton Group and the Early Cretaceous Mannville Group. Only the latter reaches sufficiently high maturity levels to have generated large volumes of methane, mostly in the Deep Basin.

All source rocks show a progressive increase in thermal maturity from an easterly zone of immaturity, through to sufficient maturity for oil generation, and ultimately to gas-generating overmaturity in western Alberta and northeastern British Columbia.

The oils in Late Cretaceous reservoirs (Cardium, Viking and Belly River) are a distinct family sourced from Late Cretaceous sources

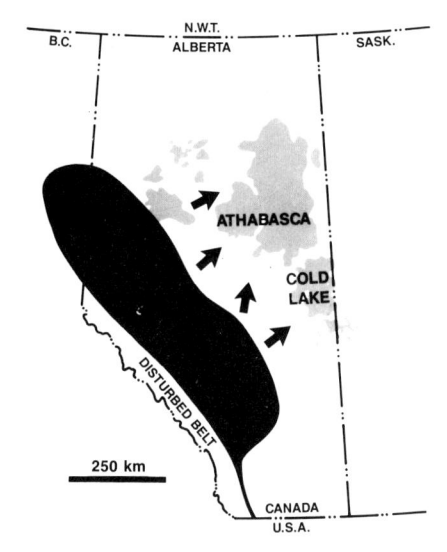

Fig. 15. Source rock drainage areas of sources that have contributed to the heavy oils of Western Canada.

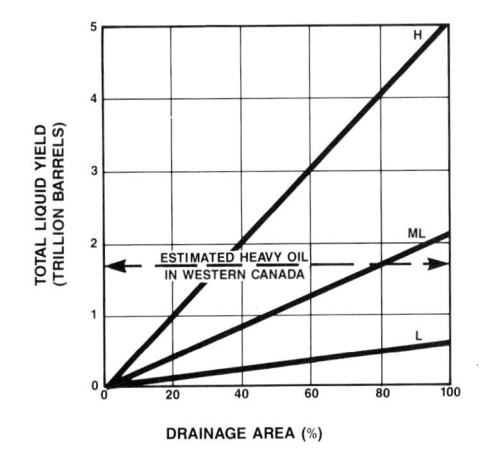

Fig. 16. Yield versus drainage area for the heavy oils of Western Canada.

(Fish Scales Zone, Second White Speckled Shale, undifferentiated Colorado Group Shales and First White Speckled Shale). These oils have migrated up to several hundred kilometres from their source area through continuous sands to updip stratigraphic pinch-outs. This system can be regarded as 'closed' in that the source is contained within the Late Cretaceous and no oils can be documented as having migrated out of the Belly River–Viking section.

The Devonian system of western Canada is also relatively 'closed'. Hydrocarbons in the Middle Devonian have been locally sourced from the Lower Keg River and equivalents into adjacent pinnacle reefs with no remnant evidence of significant leakage out of the Middle Devonian.

The Duvernay Formation has sourced prolific quantities of oil in the Late Devonian. This oil has migrated stratigraphically downward, but structurally updip, in the Peace River Arch and Swan Hills areas of west–central Alberta. This migration of Duvernay oil into older Devonian strata terminates in a series of stratigraphic pinch-out traps in basal Devonian sands and carbonate platform margins, with no leakage from the Devonian system.

A leak point of the Devonian 'closed' system is in the northeastern end of the Rimbey–Meadowbrook–Leduc trend, where Duvernay-sourced oil is stratigraphically trapped in Leduc-age reefs, but has also escaped via overlying Grosmont and Nisku Formations into the Early Cretaceous Mannville system.

Nisku-age pinnacles in the West Pembina age reefs, but may have also escaped via overlying Grosmont and Nisku Formations into the Early Cretaceous Mannville system.
River hydrocarbon system to the north.

The Mannville Group sands of the Early Cretaceous have received at least 1.7 trillion barrels of oil from subcropping and 'open' (i.e. lower frequency of internal stratigraphic trapping mechanisms) rocks of the following ages (in order of relative importance):

Nordegg Formation (Early Jurassic)
Exshaw Formation (Early Mississippian)
Phosphate Zone (Triassic)
Duvernay (Late Devonian).

Biomarker data confirm the relative dominance (because of increased yield, access ef-Nordegg source and Mississippian exshaw.
ficiency and latest arrival) of the Early Jurassic

The high porosity, permeability sands of the Mannville, which acted as a regional hydrocarbon 'drain' for subcropping formations, are also a major aquifer for incoming fresh meteoric water, which allowed bacteria to biodegrade these hydrocarbons to 'heavy oil' and tar.

The relatively thin, clay-rich Joli Fou Shale, which separates the Mannville and Viking

Formations, has acted as the major regional seal for the Mannville and older section. No oils have been documented to have crossed the Joli Fou in either stratigraphic direction.

Toward the western edge of the basin, high levels of maturity and overmaturity have promoted significant gas generation from all of the sources described above. In the carbonate section below the Late Triassic, this includes significant H_2S generation with irregular co-reservoiring of this with hydrocarbon gases in Late Devonian–Triassic reservoirs. Under current market conditions, the sulphur content of many pooled gases adds significant economic value to the resources.

The authors would like to thank Esso Resources Canada Limited for fostering the research required to produce this paper and permission to publish.

Further, we would like to thank the members of the western Canada exploration teams at Esso Resources Canada Limited (past and present) whose detailed knowledge of this basin provided critical control on the ideas presented herein.

D. L. Layton, S. D. Triffo, R. B. Scott, P. M. D. Riggins and L. J. Hasiuk are thanked for their assistance.

Without the core facility maintained by the Energy Resources Conservation Board in Calgary, this paper would not have been possible.

References

ALEXANDER, R., KAGI, R. I., WOODHOUSE, G. W. & VOLKMAN, J. K. 1983. The Geochemistry of some Biodegraded Australian Oils. *APEA Journal*, **23**, 53–63.

CHEVRON EXPLORATION STAFF 1979. The Geology, Geophysics and Significance of the Nisku Reef Discoveries West Pembina Area, Alberta, Canada. *Bulletin of Canadian Petroleum Geology*, **27**, 326–359.

DEROO, G., POWELL, T. G., TISSOT, B. & MCCROSSAN, R. G. 1977. The Origin and Migration of Petroleum in the Western Canadian Sedimentary Basin, Alberta: A Geochemical and Thermal Maturation Study. *Geological Survey of Canada, Bulletin* **262**.

DU ROUCHET, J. 1984. Migration in Fracture Networks — An Alternative Interpretation of the Supply of the 'Giant' Tar Accumulations in Alberta, Canada. Parts 1 and 2, *Journal of Petroleum Geology*, **7**, (4); **8**, (1).

ELLS, S. C. 1923. *Cretaceous Shales of Manitoba and Saskatchewan as a Possible Source of Crude Petroleum*. Canada Department of Mines Summary Report of Investigations, 1921, Report 588, 34–41.

GUSSOW, W. C. 1954. Differential Entrapment of Oil and Gas: A Fundamental Principle. *American Association of Petroleum Geologists Bulletin*, **38**, 816–853.

—— 1955. Discussion of 'In-Situ Origin of McMurray Oil'. *American Association of Petroleum Geologists Bulletin*, **39**, 1625–1631.

LEENHEER, M. J. 1984. Mississippian Bakken and Equivalent Formations as Source Rocks in the Western Canada Basin. *Organic Geochemistry*, **6**, 521–533.

MACAULEY, G. 1984a. *Cretaceous Oil Shale Potential of the Prairie Provinces, Canada*. Geological Survey of Canada Open File Report OF-977.

—— 1984b. Cretaceous Oil Shale Potential in Saskatchewan. *In: Oil and Gas in Saskatchewan 1984*, 255–269.

—— SNOWDON, L. R. & BALL, F. D. 1985. *Geochemistry and Geological Factors Governing Exploitation of Selected Canadian Oil Shale Deposits*. Geological Survey of Canada, paper 85–13.

MCINNES, W. 1913. *The Basins of Nelson and Churchill Rivers*. Geological Survey of Canada, Memoir **30**, 68, 127, 133–134.

MASTERS, J. A. 1979. Deep Basin Gas Trap Western Canada. *American Association of Petroleum Geologists Bulletin*, **63**, 152–181.

MOSHIER, S. O. & WAPLES, D. W. 1985. Quantitative Evaluation of Lower Cretaceous Mannville Group as Source Rock for Alberta's Oil Sands. *American Association of Petroleum Geologists Bulletin*, **69**, 161–172.

POWELL, T. G. 1984. Some Aspects of the Hydrocarbon Geochemistry of a Middle Devonian Barrier–Reef Complex, Western Canada. *In*: PALACAS, J. G. (ed). *Petroleum Geochemistry and Source Rock Potential of Carbonate Rocks*. AAPG Studies in Geology **18**.

STOAKES, F. A. & CREANEY, S. 1985. *Controls on the Accumulation and Subsequent Maturation and Migration History of a Carbonate Source Rock*. SEPM Core Workshop Proceedings; Golden, Colorado; August 1985.

TYRRELL, J. B. 1892. North–Western Manitoba with Portions of the Districts of Assiniboia and Saskatchewan. Geological Survey of Canada Annual Report, 1890–1891, **5**, part 1, report E.

WEBSTER, R. L. 1984. Petroleum Source Rocks and Stratigraphy of the Bakken Formation in North Dakota. *In*: WOODWARD, J., MEISSNER, F. F. & CLAYTON, J. L. (eds). Hydrocarbon Source Rocks of the Greater Rocky Mountain Region. Rocky Mountain Association of Geologists.

WELTE, D. H., SCHAEFER, R. G., STOESSINGER, W. & RADKE, M. 1984. Gas Generation and Migration in the Deep Basin of Western Canada. *In*: MASTERS, J. A. (ed.) *Elmworth — Case Study of a Deep Basin Gas Field*. American Association of Petroleum Geologists Bulletin, Memoir **38**, 35–47.

Structural styles of growth faults in the U.S. Gulf Coast Basin

J. A. LOPEZ

Amoco Production Company, New Orleans, Louisiana, USA

Abstract: Structure and sediments related to shelf margin growth faults are an important exploration objective in the Cenozoic in the U.S. Gulf Coast basin. The syndepositional nature of these fault systems has been recognized for decades with well control and seismic. Recent seismic of higher quality and deeper penetration has begun to reveal more clearly the variety of structural styles of the 'typical' Gulf Coast shelf margin growth fault.

Generally five different shelf margin growth fault types are recognized: (1) listric faults which detach above undeformed older section, (2) listric faults which detach on mobile, overpressured shales above older undeformed sediments, (3) faults which sole into autochthonous salt withdrawal basins, (4) faults which detach along allochthonous salt sill intrusions, and (5) non-detached faults caused by differential compaction of thick shale intervals or basement faulting. The first four are the dominant structural types and their different structural styles reflect various modes of accommodation of fault offset within the sedimentary section. The fifth structural type, although not as commonly recognized, may be obscured by other structural complexities.

These five fault types have different structural and stratigraphic characteristics which create different trap-source-reservoir relationships for each type. By understanding and identifying the distribution of these structural types one may better understand the complex production trends established in the Gulf Coast basin and more accurately assess potential production on prospective traps.

The U.S. Gulf Coast basin lies on the northern rim of a small oceanic basin which is the Gulf of Mexico. Although the Gulf Coast basin is one of the most highly explored basins in the world, the complexity of the basin suggests that major hydrocarbon reserves have still eluded the drill bit, and will only be discovered as the basin is better understood. The most prolific producing portion of the Gulf Coast basin is underlaid by Jurassic Louann Salt. Within this province reservoirs range in age from Jurassic to Pleistocene, multiple source rocks are present, and diverse structural and stratigraphic traps have been recognized. Of particular economic importance are the Cenozoic growth faults south of the Comanchean (lower Cretaceous) carbonate shelf margin (Fig. 1). From the Early Cenozoic to the present, growth faults have expanded clastic intervals which have prograded 300 km into the Gulf over the basinal Mesozoic section. The syndepositional nature of these faults has been recognized for decades (Ocamb 1961) and the associated 'rollover' anticline is a common hydrocarbon trap. Additionally these fault-expanded intervals are commonly involved in numerous other traps such as those associated with salt domes. This paper describes the structural styles of these faults that typically expand the Cenozoic in the Gulf Coast basin.

Shelf margin growth faults

The faults described herein are referred to as shelf margin growth faults and have the following general characteristics.
 (i) The fault surfaces are slightly to strongly concave upward on seismic or on true scale depth cross sections.
 (i) The fault movement is syndepositional, i.e. sedimentation is coeval with fault movement and results in greater sediment accumulation on the downthrown block.
 (iii) Outer shelf or upper slope sediments have the greatest amount of expansion. Winker (1983) has also emphasized shelf margin sedimentation associated with the active expansion phase of these faults and refers to these systems as unstable progradational clastic shelf margins.
 (iv) The faults which expand any particular interval coalesce into a regional fault system which is located near the shelf margin for that age interval.
 (v) The strike of these faults is roughly subparallel to the paleo-shelf margin, but may dip either toward the paleo-shelf or paleo-slope.

The character of these shelf margin growth faults suggests that they were probably initiated

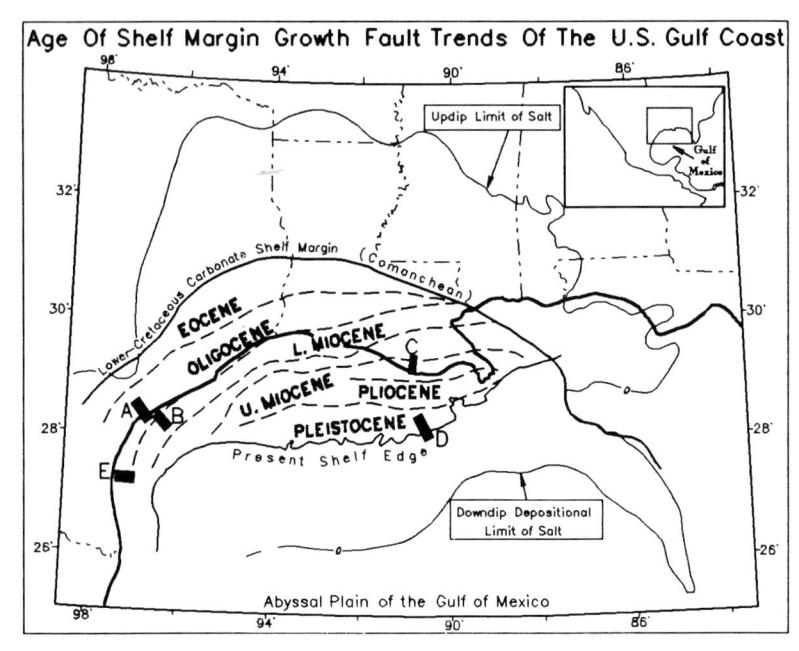

Fig. 1. General ages of sediment expansion on Cenozoic Shelf Margin growth faults of the U.S. Gulf Coast and location of examples used to illustrate the structural styles.

by differential sedimentary loading, and that further movement is primarily dependent on additional sedimentation across the downthrown fault block (as opposed to some external cause like basement rifting). The sediments which continue the fault movement are not necessarily relatively dense sandstones deposited over less dense shale. Jackson & Talbot (1986) have demonstrated that a differential loading condition can also be created by variable accumulation of sediment of the same lithology. As a consequence, thick shale or any other lithology may be expanded across shelf margin growth faults. It is not fortuitous, however, that these growth faults expand shelf margin sediments. In the downthrown fault block, sea floor subsidence responds to sedimentation, and it is this continued subsidence which geomorphically defines and perpetuates the shelf slope break. As Winker & Edwards (1983) point out, these faults separate the unstable slope conditions from the stable shelf. It is the updip portion of the unstable slope which will be differentially loaded by sediments transported across the stable shelf. Therefore the subsidence and sediment accumulation occurs at the shelf–slope interface. When conditions on a shelf margin growth fault do not permit adequate subsidence, the expansive phase ceases and the shelf margin progrades. If conditions permit, the next shelf margin growth fault develops. The inactive growth fault then becomes part of the stable shelf, but may have continued minor movement due to differential compaction. Winker & Edwards (1983) recognized this fundamental distinction between the unstable slope and a stable shelf. This paper emphasizes the varying geological settings for this instability.

The difference in sediment accumulation across the shelf margin growth faults requires differential subsidence of the sea floor across the fault. The peak rates of sea floor subsidence on these faults have been estimated to be as high as 5.5 km Ma^{-1}, and can result in fault throws of thousands of metres. The throw across these faults must be accommodated by either basement movement or by deformation within the sediments above basement. Since the vast majority of these faults probably do not offset basement, the accommodation is usually achieved within the sedimentary section. A useful way to categorize the various types of shelf margin growth faults is by defining the accommodation mechanisms. Significantly, each structural type may have a different source/reservoir/trap relationship. As will be seen, the occurrence of these various structural types is mainly dependent on the presence and thickness of the Jurassic Louann salt. Therefore the distribution of salt will be discussed briefly.

Distribution of the Louann Salt in the Gulf Coast Basin

Pindell (1985) gives a thorough synthesis of the late Paleozoic reconstruction of the southern U.S. continental margin and the subsequent Triassic/Jurassic opening of the Gulf of Mexico. The Louann Salt is interpreted to have been deposited during the late-rift stage (Callovian) and therefore deposited on rifted-transitional crust. The central, abyssal Gulf of Mexico basement is interpreted as the subsequently developed oceanic crust (drift phase) which split the Louann salt basin into a northern and southern half by Tithonian time. Thermal subsidence of the transitional and oceanic crust subsequently created the deep Gulf basin floor and allowed accumulation of 3000–6000 m of upper Jurassic and Cretaceous sediment over salt or oceanic basement.

It is within these salt basins that burial of the salt created the numerous salt structures, many of which have been drilled in both the U.S. and Mexico (Martin 1980, 1984; Halbouty 1979). Salvador (1987) has studied the present distribution of these various salt features and has made an estimate of the original undeformed salt thickness. Fig. 2 is an isopach map of the Louann Salt using both Amoco proprietary data and Salvador's data. The variability in salt thickness is probably due to the basement relief at the time of deposition (Salvador 1987). Regionally the salt had its greatest original thickness (>4000 m) in offshore Louisiana, and appears to thin (<500 m) westward toward Texas. In general, if the salt is less than 500 metres thick in the Gulf Coast, little deformation is found associated with it. However, eastward in offshore Louisiana where the greatest thickness of salt is estimated; the subsequent salt deformation has greatly influenced the development of the shelf margin growth faults.

Potential hydrocarbon sources for tertiary sandstone reservoirs

Although there continues to be much discussion on which units source the Gulf Coast Tertiary sandstones, one fact seems to be clearly supported by the vast majority of the geochemical data from the Tertiary. In general, the typical sandstone reservoirs associated with the shelf margin growth faults could not have been charged by adjacent shales whose organic richness and thermal maturity are typically too low for oil or gas generation. Two alternatives exist:

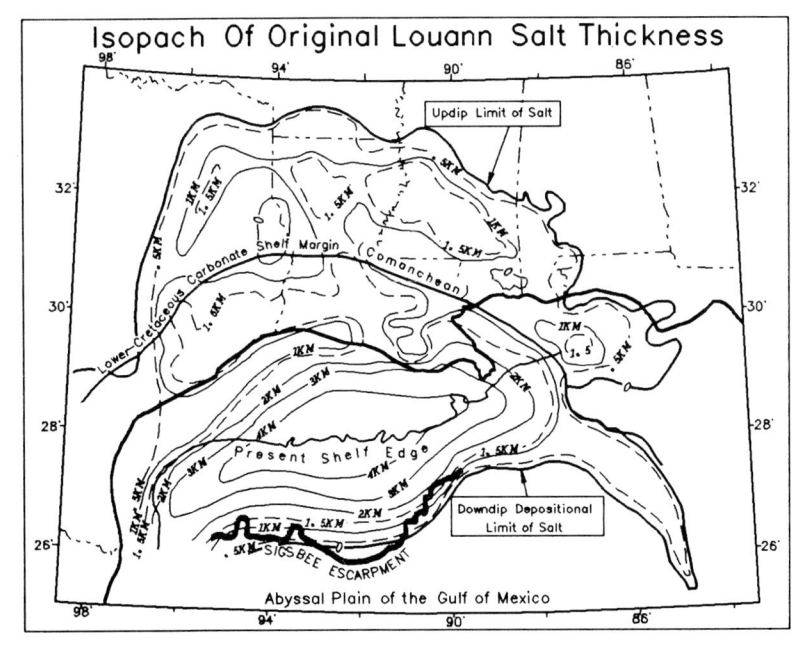

Fig. 2. Isopach of original Louann salt thickness of the U.S. Gulf Coast (redrawn from Salvador 1987). Original salt thickness is estimated from present distribution of salt structures. Salt thickness varies dramatically over the area of Tertiary shelf margin growth faults (Fig. 1).

(1) distal Tertiary units may be organically rich due to local restricted circulation conditions in salt withdrawal basins (Dow 1984) and (2) a non-deltaic basinwide source(s) exists within the older basinal section (Eocene to Jurassic). If the former situation occurs, it could still be argued that the structural complexity in the Tertiary basin precludes sourcing the reservoir from age equivalent source units. For example, if a source unit is deposited in a middle or lower slope environment, the age equivalent reservoir sandstone deposited along a shelf margin growth fault would be 100–250 km updip. Once these units were buried by progradation of subsequent shelf margin sedimentation, the reservoir would be several growth fault systems separated from its age equivalent source. Therefore it is unlikely to be sourced by it. A more plausible thesis is for shelf margin sandstones to be charged by older distal units buried deeply below, or alternatively by some regional non-deltaic source horizon which may be present even deeper in the section. In either case, significant vertical migration would be required. As such, the ability of the deeper section to allow vertical migration by fracturing, faulting or other migration pathways and its relationship to the shelf margin growth fault surfaces may be the primary control of the charging of hanging wall reservoirs.

Description of structural styles

The following section includes a general description and example of each fault type. (Fig. 3 is a summary). A seismic section and a true scale depth section serve to illustrate the interpreted fault geometry. A chronostratigraphic chart of the hanging wall of the growth fault is also used in some examples. On these chronostratigraphic charts, accumulation rates are shown to provide a continuous and more precise depiction of subsidence history than the traditional expansion index which is a relative value and therefore variable. An expansion index is a ratio of a unit thickness downthrown to the same unit upthrown (Thorsen 1963). A weakness of the technique used here is that its accuracy is dependent on the accuracy of the absolute ages assigned to the units to calculate the accumulation rates. Even if the absolute ages are incorrect, relative changes in depositional rates seen in a given interval are still valid; and therefore contouring of the values gives a continuous and more complete depiction of the subsidence history than a simple expansion index.

Most of the structural types described here have been described in the literature previously. There is a tendency, however, to consider one structural type or general growth fault model as representative of the whole Gulf Coast basin. This is clearly not the case. Fundamental differences exist in the causes, geometry and associated production. This paper attempts to outline these differences. Of particular importance to the explorationist are the possible trap–source–reservoir relationships. Fig. 3 illustrates the five basic structural types and their general characteristics.

Simple stratigraphic detachment

The essential feature of this fault type is that the hanging wall has broken away on a listric fault which detaches along a stratigraphic horizon above relatively underformed units. This type of fault is most common in areas of thin salt as in the Oligocene trend of the Texas coastal plain, but is also seen in interdomal areas of south Louisiana in the Lower Miocene and Oligocene. This type of fault has been described in association with the Mexican Ridges in the western Gulf of Mexico by Buffler (1979). The stratigraphic detachment levels seen in the Gulf Coast are variable, but in general, older expansion systems — like the Oligocene or Eocene — detach at older stratigraphic horizons. Older shelf margin growth faults are more likely to detach deeper and therefore nearer to a hydrocarbon source, whereas younger systems may be more removed from potential sources and are less likely to be charged.

The listric shape of these faults and the rigid footwall require that the vertical component of translation is accommodated by rotation to horizontal motion, and suggest regional downdip translation (Cloos 1968). Crans et al. (1980) have described the experimental mechanics of this type of system and conclude that either a downdip toe thrust (reverse fault) or lateral compaction of the hanging wall must accommodate for horizontal translation. Generally neither of these compressional phenomena are recognized in the distal shelf margin growth faults. In both cases described by Crans et al. (1980) the section is shortened horizontally within the hanging wall, yet often what is observed downdip within the hanging walls of the Gulf Coast is simply additional normal faulting. Some authors argue that distal compressional zones are not recognized because of subsequent progradation and extensional 'overprinting' by younger depositional systems (Winker 1983). This may occur in some instances. However, the stratigraphic detachment shelf margin

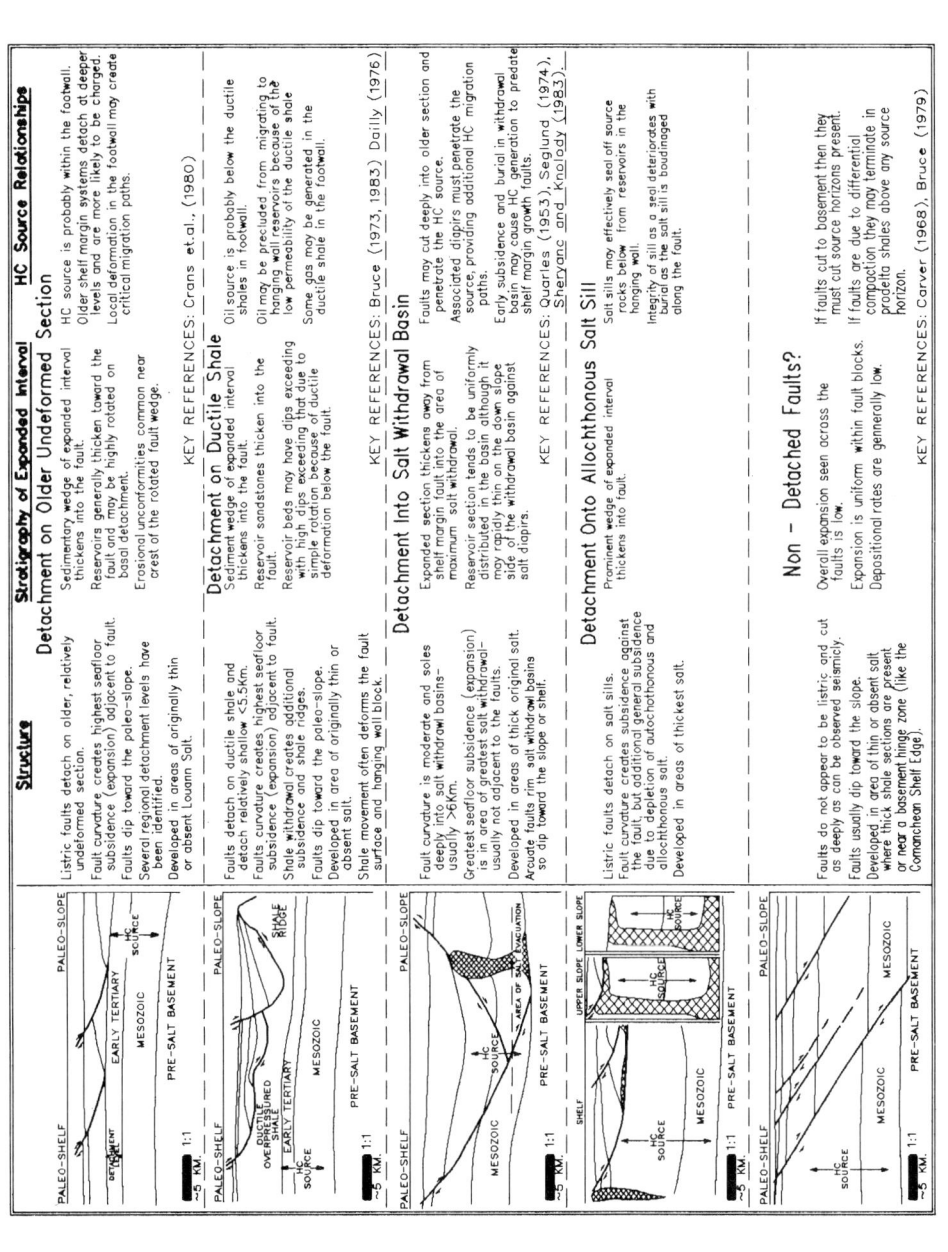

Fig. 3. General characteristics of the five structural types of shelf margin growth faults in the U.S. Gulf Coast Basin.

growth fault often changes downdip to other shelf margin growth fault types which may not require a distal compressional zone. This suggests that the horizontal translation in one system is balanced by other types of shelf margin fault systems downdip.

Example A: The example of simple stratigraphic detachment is within the Oligocene trend of the Texas coastal plain (Fig. 1). A migrated dip line (Fig. 4) shows several shelf margin growth faults which segment the expanded interval into several rotated fault blocks. Reflections within the hanging walls diverge toward each of the expanding faults. A regional detachment level is suggested by the listric projection of the faults which appear to sole on relatively flat and continuous reflections and also, by the structural discordancy between the hanging wall and deeper reflections. This relationship is particularly clear at the southeast end of the line. Fig. 5 is a true scale cross-section of Example A. Well control indicates much of the expanded interval is shale. Regional seismic lines indicate that the detachment level is at the Wilcox Formation (Lower Eocene) but could be at the Jackson Formation (Upper Eocene). The high-amplitude reflection approximately one second below the detachment is interpreted as the top of Cretaceous. Throughout the Gulf of Mexico the top of Cretaceous is recognized as a high-amplitude event which usually separates the slower velocity Cenozoic from higher velocity Mesozoic carbonate units. Faults which appear to offset the Cretaceous are interpreted as basement faults related to the isostatic loading as the basin filled. The depth to basement is estimated from regional refraction data (Locker & Chatteriere 1984) and the general thickness of the Mesozoic section seen in the abyssal Gulf of Mexico.

Effective hydrocarbon sources have been demonstrated in the Wilcox and several units in the Cretaceous in the eastern Gulf Coast (Sassen 1988). If the Wilcox is an effective source here there is a good chance it has charged hanging wall structures. If on the other hand the only effective hydrocarbon sources are deeper than the detachment level, then vertical migration and charging of hanging wall reservoirs may be prevented.

Detachment on ductile shale

The essential feature of this type of shelf margin growth fault is the presence of ductile-overpressured shale below the fault. This type

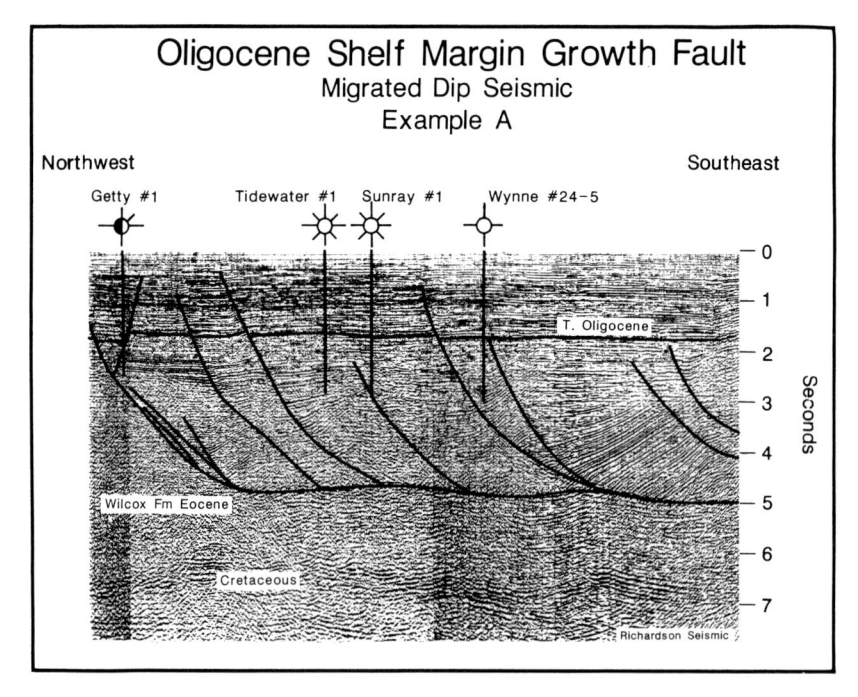

Fig. 4. Migrated seismic line of Oligocene shelf margin growth faults from coastal Texas illustrating simple stratigraphic detachment (Example A). Sediments below the detachment area are relatively undeformed.

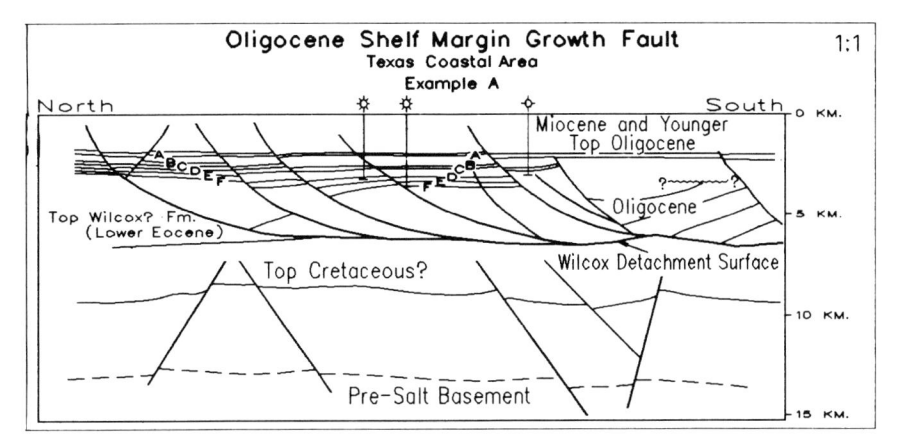

Fig. 5. Cross-section of Oligocene shelf margin growth fault from coastal Texas seismic in Figure 4. Detachment at the Wilcox level may preclude vertical migration from deeper hydrocarbon sources.

of faulting has been described in the Gulf Coast by Bruce (1973, 1983) and Dailly (1976). This fault type is developed downdip of major Tertiary expansion trends and where salt is thin or absent. The Middle and Lower Miocene shelf margin growth faults in central and eastern offshore Texas are good examples.

Although both Bruce (1973) and Dailly (1976) recognized the importance of sedimentation in the activation of these faults, they emphasized the differential loading condition caused by relatively denser sandstone being deposited over less dense, viscous shale. Jackson & Talbot (1986) have shown that a differential loading condition can also be created by unequal sediment accumulation over a ductile substratum of salt. If the mechanical behavior of the undercompacted shale is similar to salt, shale or any sediment type may be expanded across these faults. Therefore, expansion across a shelf margin growth fault does not imply the presence of any specific lithology. The expanded shales of the upper Miocene in offshore Texas serve as a good example.

Although the mechanical behaviour of the overpressured shale may be analogous to the salt, the initial relative position of these units to the shelf-margin growth faults in the Gulf Coast is generally different (salt sills may be in a similar structural position as ductile shale). The ductile, overpressured shale associated with these faults is typically mid-slope sediment of the interval slightly older than the expanded section on the shelf margin growth fault. Whereas the salt is Jurassic and must intrude through the entire Upper Mesozoic section to reach the Tertiary. The initial physical proximity of ductile shale to the shelf margin growth faults has two major influences on the faults subsequent development.

First, the ductile shale provides a shallow, low-friction detachment surface upon which the listric faults develop. Therefore, these faults tend to be more concave than the other fault types. Seafloor subsidence on the downthrown block is principally accommodated by translating the vertical motion to horizontal motion. Second, withdrawal of ductile shale creates additional seafloor subsidence. Shales are often forced downdip into shale ridges (Bruce 1973). Also the ductile shale often warps the detachment surface itself.

Although the ductile shales below the shelf margin growth faults of this type may have been buried deep enough to have generated hydrocarbons, they are typically organically lean with only marginal gas sourcing potential. Fields associated with this fault style are generally gas productive and some gas contribution may come from the ductile shale zones. However, it seems more likely that a deeper source is present in the section below the ductile shale zone. If the hydrocarbon source is in this deeper section hydrocarbons must migrate through the ductile shale to reach reservoirs in the hanging wall. This may preclude oil migration and may also explain why these structures are predominately gas productive. Alternative explanations are that oil generation predated the traps or that only gas prone organic facies are present in these trends.

Example B: This example of a detachment on ductile shale shows two lower Miocene faults from offshore Texas (Fig. 6). Rollover anticlines are well developed due to the strong upward concavity of the faults. The

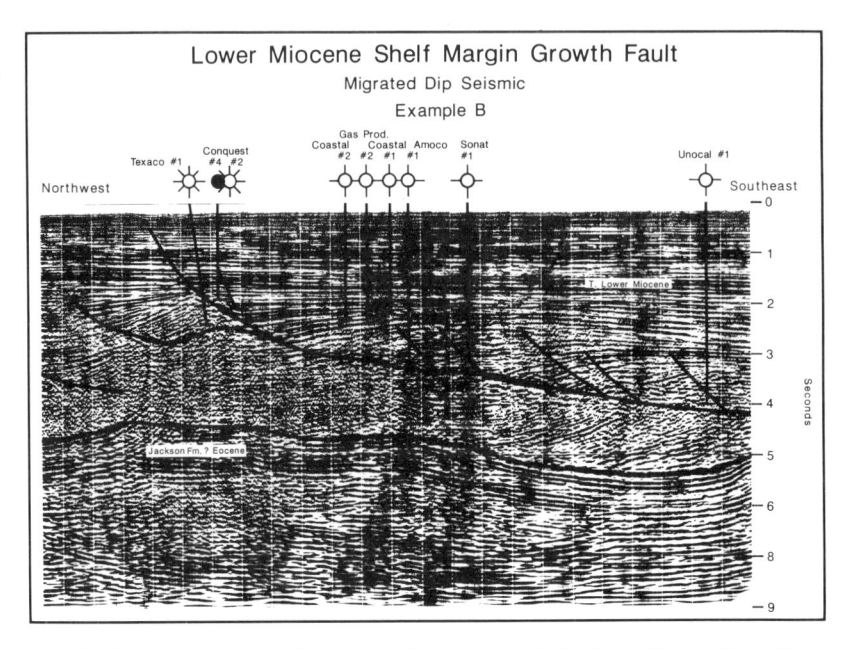

Fig. 6. Migrated seismic line of Lower Miocene shelf margin growth fault of offshore Texas illustrating detachment on ductile shale (Example B). The chaotic zone below the detachment is ductile overpressured shale.

fault surfaces detach upon a zone of generally chaotic reflections interpreted as ductile shale. The low-amplitude flat reflections are interpreted to be multiples. Some zones of coherent dipping reflections may indicate sub-detachment structures. Below the chaotic zone at 4.5–6.5 s is a series of high amplitude continuous reflections which are only gently folded. Fig. 7 is a true-scale depth cross section along the seismic line. Within the hanging wall sandstones rapidly disappear southeastward (basinward) away from the expanded wedge. Benthonic foraminifera assemblages indicate the expanded interval was deposited in an outer shelf to upper slope environment indicative of a paleo-shelf/slope break. The Mesozoic section and basement are estimated from regional refraction data (Locker & Chatteriere 1984), and the estimated thickness of the Mesozoic in the abyssal Gulf. The shallow detachment level of the faults make it unlikely that they penetrate any hydrocarbon source which might be present in the lower Tertiary or older section. Production in this trend is predominately dry gas.

Fig. 8 is a chronostratigraphic chart of the hanging wall structure developed from the cross section and shows contoured accumulation rates for the units identified. Seafloor subsidence rates are highest immediately downthrown to the fault where the listric geometry translates the vertical motion to horizontal. Secondary subsidence in the interval B–C can be seen away from the fault and is probably due to evacuation of shale into nearby shale ridges. The very slow rate of deposition in the youngest interval reflects the end of the growth phase of the fault and the progradation of shelf deposition across the faults.

Salt withdrawal faults

The principal feature of these faults is that the seafloor subsidence which allows expansion of section across the faults is accommodated by salt evacuation from the withdrawal basin into diapiric salt structures. Quarles (1953) originally described this fault type along the Texas Gulf Coast. Fault concavity is less than with other fault types because the faults sole more deeply into the withdrawal basin and may actually terminate into the mother layer of salt. The faults rim the salt withdrawal basin and are most common where the original salt thickness was probably greater than 2000 m. This type of fault is referred to by Seglund (1974) as a 'collapse-fault'. He describes six 'collapse-fault systems' in South Louisiana. Similar faults have been described in the Gulf Coast by Skeryanc & Kolodny (1983), Larberg (1983), and Spindler (1977). This fault style is currently active in

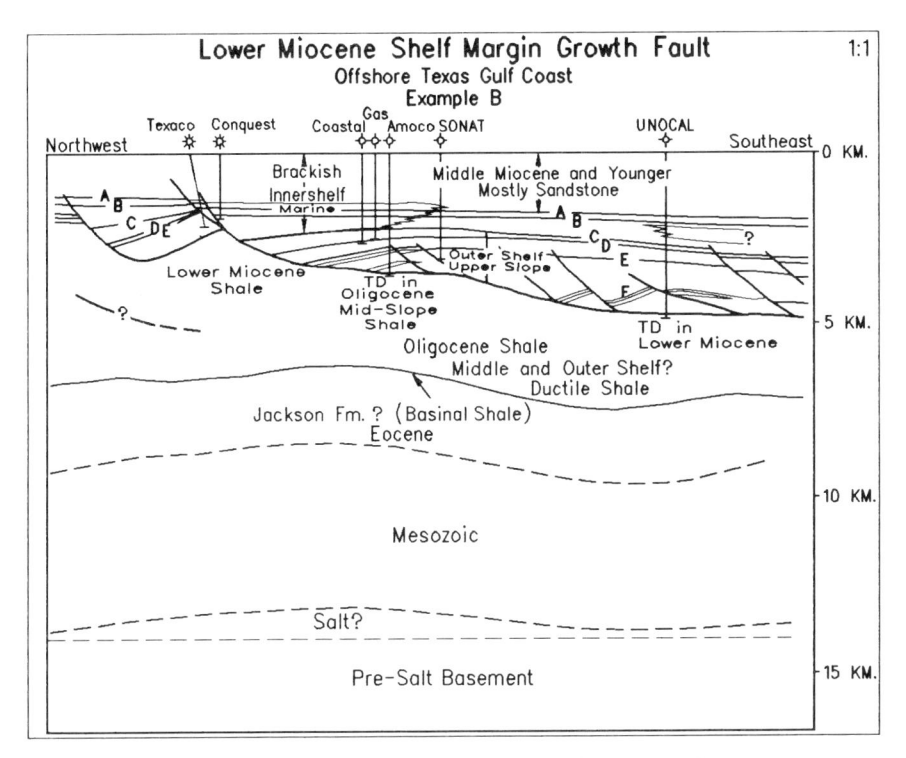

Fig. 7. Cross-Section of Lower Miocene shelf margin growth fault offshore Texas from seismic in Figure 6. The thick ductile shale zone may inhibit vertical oil migration and may explain the predominance of gas production in this trend.

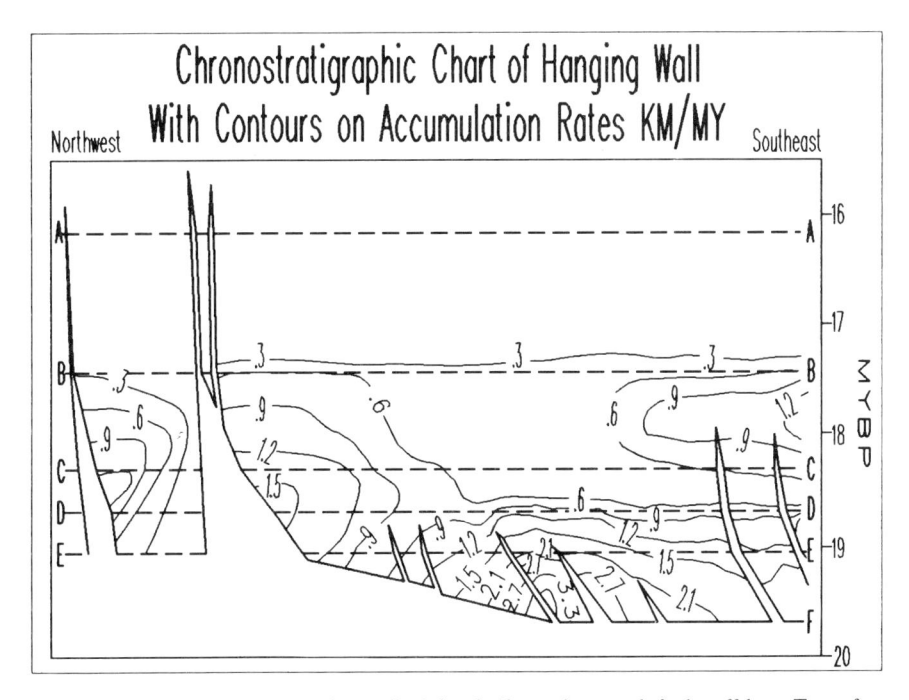

Fig. 8. Chronostratigraphic chart of hanging wall of the shelf margin growth faults offshore Texas for cross-section in Figure 7. Accumulation rates are contoured in km/my. The greatest accumulation rates are found adjacent to the shelf margin growth faults.

many of the 'intra-slope basins' on the present continental slope described by Martin (1980, 1984).

The instability of differential sediment accumulation over the ductile salt substratum causes the seafloor to subside as rapidly as sediments may accumulate. The faults are located where the subsidence is not equal and the sediments are consequently faulted. The mechanics of this type of fault are principally controlled by the mechanical properties of salt.

Example C: The Timbalier Trough is one of the largest salt withdrawal basins in southeast Louisiana. Major oil accumulations are found around salt diapirs and major gas reserves are trapped on growth fault structures. The principal productive interval is the Upper Miocene section expanded across shelf margin faults located north and south of the trough. As shown on Fig. 9 the expanded interval has little rotation and consequently no 'rollover anticline' is present. Instead, reflections diverge and dip away from the bounding shelf margin growth faults and toward the area of maximum salt withdrawal. Two large diapirs are located at the southern margin of the basin at this location and numerous other domes are found along the northern and southern edges of the trough. On the northern side of the basin, deep reflections below the faults are found to be downwarped into the trough. This inflection and the presence of numerous salt diapirs rimming the basin are evidence of salt withdrawal in the trough. The cross section shown in Fig. 10 illustrates a more complete structural interpretation. In the centre of the basin sandstone and shale deposited in an outer-shelf to upper-slope setting have subsided to over 7000 m. The sand content is fairly uniform except on the distal side of the trough where sandstones rapidly thin toward the salt diapirs. Fig. 11 is a chronostratigraphic section across the trough illustrating that the maximum rate of sea-floor subsidence occurred away from the fault and was positioned centrally between the bounding faults. The slow rates of deposition in the youngest intervals represent the end of the growth phase of the trough and the progradation of sand rich shelf sediments over the trough.

The depth to basement is from regional refraction for the Gulf Coast from refraction surveys (Locker & Chatteriere 1984). The Mesozoic thickness here is estimated from the equivalent section in the abyssal Gulf of Mexico.

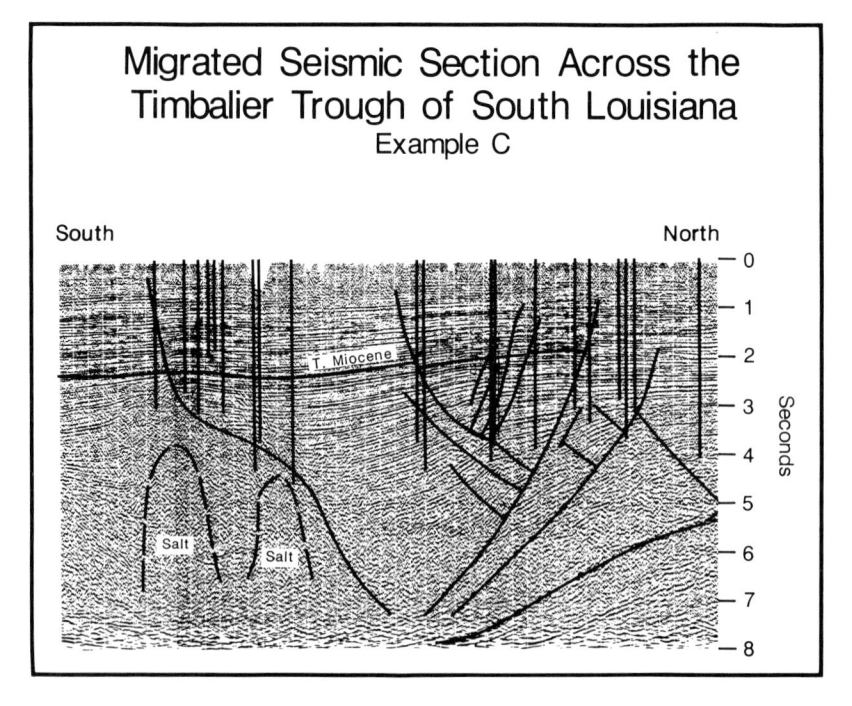

Fig. 9. Migrated seismic line across the Timbalier Trough of south Louisiana illustrating subsidence within a salt withdrawal basin (Example C). Reflections diverge away from the bounding shelf margin growth faults and toward the basin center.

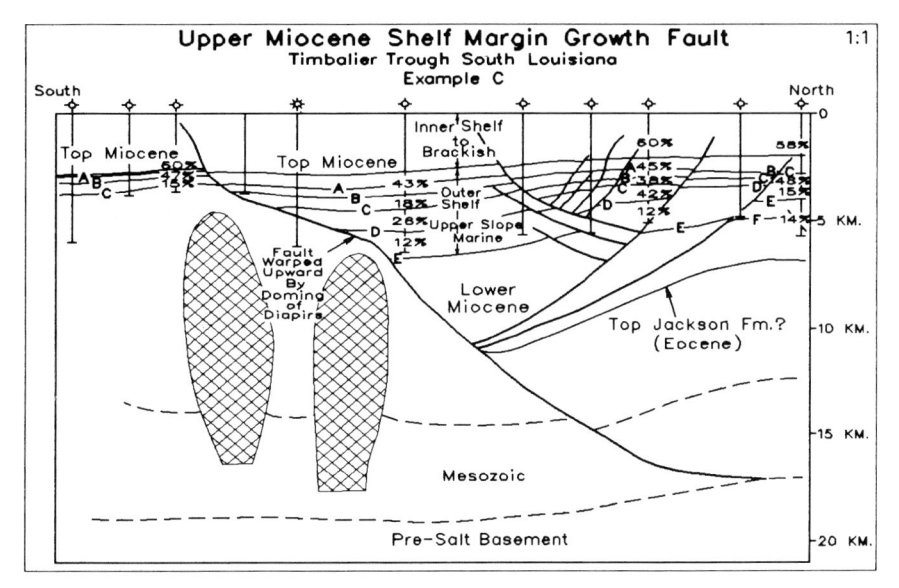

Fig. 10. Cross-section across the Timbalier Trough of south Louisiana from seismic on Figure 9. Faults may sole deeply into the withdrawal basin, but hydrocarbon generation in these deep basins may pre-date the shelf margin expansion. Sandstone percents are shown for representative wells.

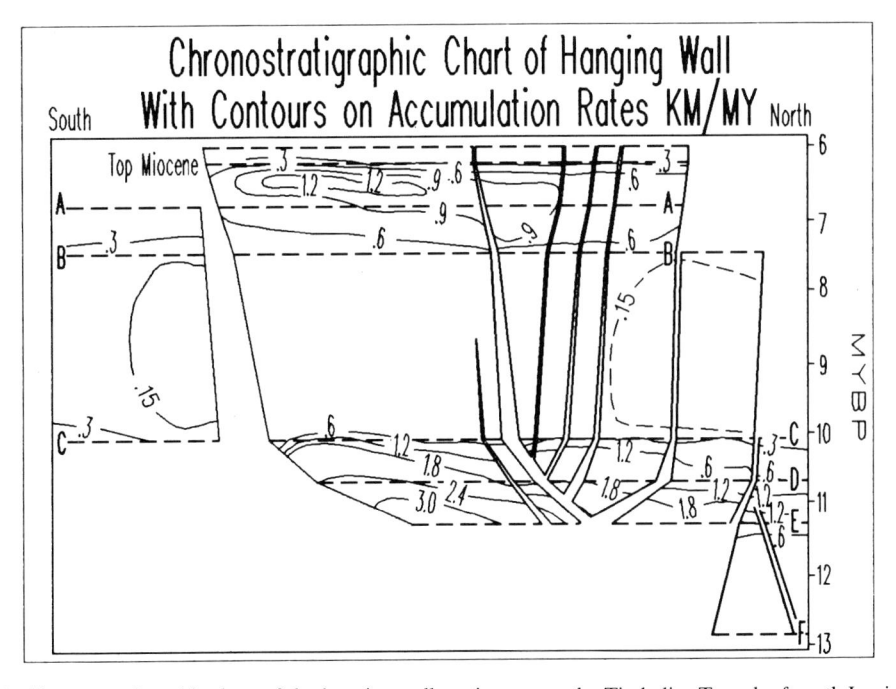

Fig. 11. Chronostratigraphic chart of the hanging wall section across the Timbalier Trough of south Louisiana for cross-section in Figure 10. Accumulation rates are contoured in km/my. The greatest accumulation rates are away from the shelf margin growth faults.

Faults project down into the trough and may sole out above the Mesozoic or at the top of pre-salt basement where the salt has evacuated. Because these faults cut more deeply than the other fault types they are more likely to provide a conduit for migration from whatever source horizons are present. Early burial in these deep basins, however, may cause hydrocarbon generation to predate the major expansion phase of these faults.

Salt-sill detachment faults

The Sigsbee escarpment, located at the base of the continental slope (Fig. 2), has been recognized as a leading front of a shallow lateral salt flow for many years (Amery 1969; Buffler 1983). More recently it has become evident that salt 'sills' are present elsewhere on the slope and shelf (Wang 1988). It is now recognized that many of the older shelf margin growth faults located on the present shelf developed over salt sills emplaced on paleo-slopes and that sills often acted as a detachment surface for the shelf margin growth faults (Fig. 3). The sills are initially emplaced as smooth continuous bodies up to 4 km thick. As they are progressively buried and later become zones of detachment, shearing, differential-loading and buoyancy apparently remobilize the salt in the sills. The result is that the once continuous salt sill is 'boudinaged' along the fault surface. Secondary diapirism from the allochthonous sills may also develop. The mechanical behavior of fault detachment on these salt sills is analogous to that of detachment on ductile shale.

Obviously this type of shelf margin growth fault can only develop in areas where salt sills were initially emplaced. This probably occurs in areas where the original salt layer is sufficiently thick to allow diapirism near the seafloor without depleting the basal layer of salt. In the Gulf Coast this process is actively occurring on the present lower continental slope of offshore Louisiana and Texas. On the Louisiana shelf remnants of salt sills have been recognized increasingly deeper in the Tertiary. The extent of salt sill emplacement in the Gulf Coast is not known, but salt sill emplacement has probably been a major process in Miocene and younger sediments.

The presence of a salt sill along the basal portion of a shelf margin growth fault has obvious implications for charging of expanded hanging wall reservoirs by deeper hydrocarbon sources. The effectiveness of the salt sill as a barrier to migration is dependent on how continuous the salt is at the time of generation and migration. Because this structural type is also found with the shelf margin faults which sole into the salt withdrawal basins, detailed mapping of these structures is necessary to recognize how they may have influenced local production trends. In areas where the extensive salt sill detachment faults have been recognized, the established production is relatively minor and tends to be gas prone.

Example D: This example is from the Plio-Pleistocene trend of offshore Louisiana near the present shelf/slope break. Fig. 12 is a migrated dip-line over several faults. Reflections in the Pleistocene and Pliocene diverge toward the expanding faults. The faults are listric and sole into a southeast dipping zone of discontinuous, high-amplitude reflections. These high-amplitude zones are interpreted to be salt lenses. The Amoco #1 well drilled salt at TD. The well did not drill through the salt but sub-salt sediments are indicated by numerous reflections below the detachment zone. Elsewhere in offshore Louisiana other wells have drilled through similar salt detachments and found Cenozoic age sediments below the salt. Since the salt is almost certainly Jurassic Louann salt, these salt lenses are interpreted to be allochthonous salt bodies which have subsequently been buried by prograding shelf margin growth faults. The high-amplitude events at 3.75–4.25 s and 4.25–4.75 s are interpreted to be the top and base of the basinal Mesozoic carbonates.

Fig. 13 shows a more complete structural and stratigraphic interpretation and illustrates the thick outer-shelf/upper-slope sediments expanded on the shelf margin growth faults. The chronostratigraphic chart (Fig. 14) shows that high rates of deposition are still occurring, allowing for continued expansion of outer shelf deposition. The cross section shows a thick Tertiary section below the salt detachment zone. Any hydrocarbon source below the detachment must find a window in the salt. This may preclude significant migration of oil or gas into hanging wall structures.

Non-detached faults

This fault 'type' is probably not a single structural style, but a group of less well understood faults which are observed to cut deeply into the basin. Depending on velocities used to convert seismic sections to depth they may exhibit slight concavity, and they do not appear associated with mobile shale or salt. Faults which fall into this category are either caused by differential compaction as Bruce (1973) and Carver (1968) have described or by basement faulting. Both

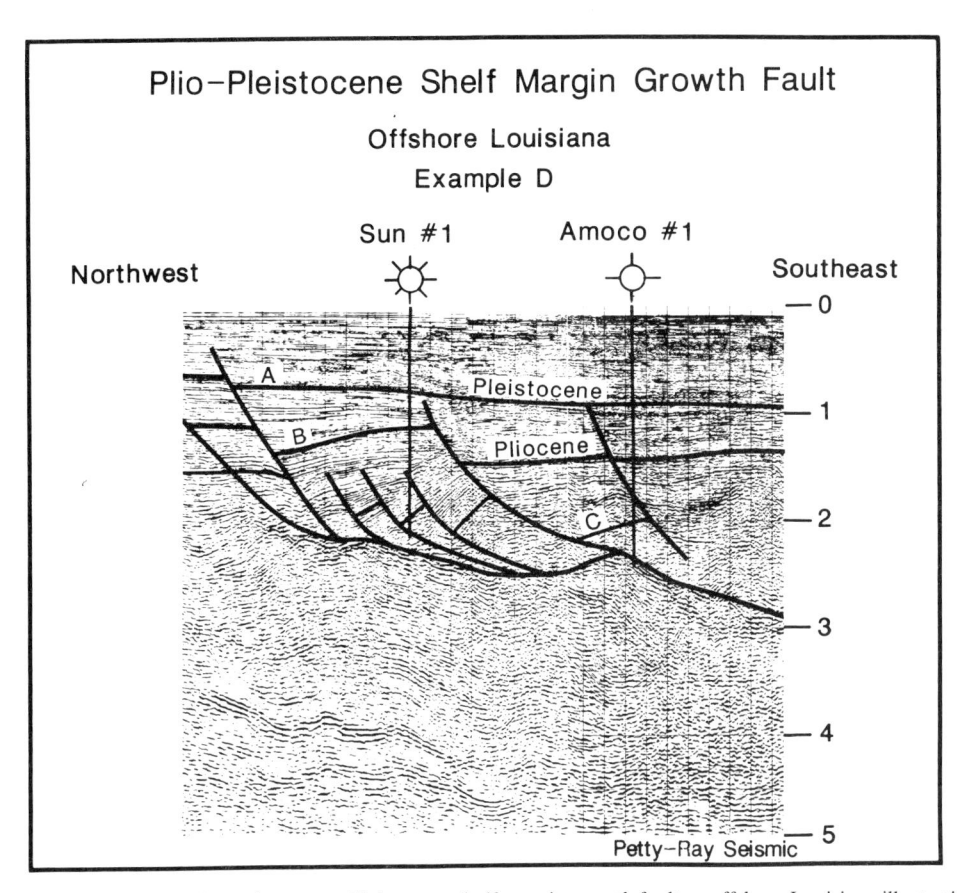

Fig. 12. Migrated seismic section across Pleistocene shelf margin growth fault — offshore Louisiana illustrating detachment on allochthonous salt sills (Example D). Discontinuous high amplitude events along the fault are probably salt lenses.

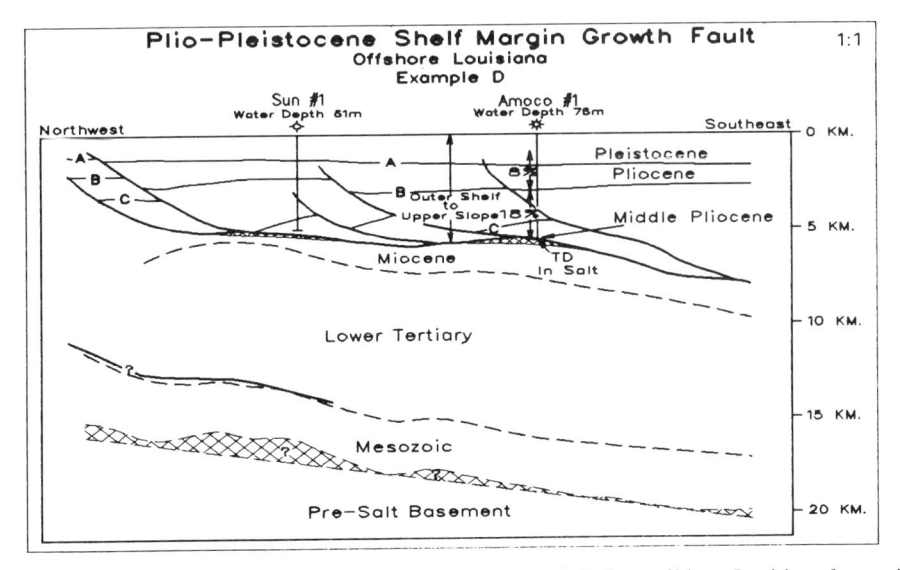

Fig. 13. Cross-section across a Plio-Pleistocene shelf margin growth fault — offshore Louisiana from seismic on Figure 12. Allochthonous salt sills may prevent vertical migration to hanging wall structures. Sandstone percent is shown for the Amoco well.

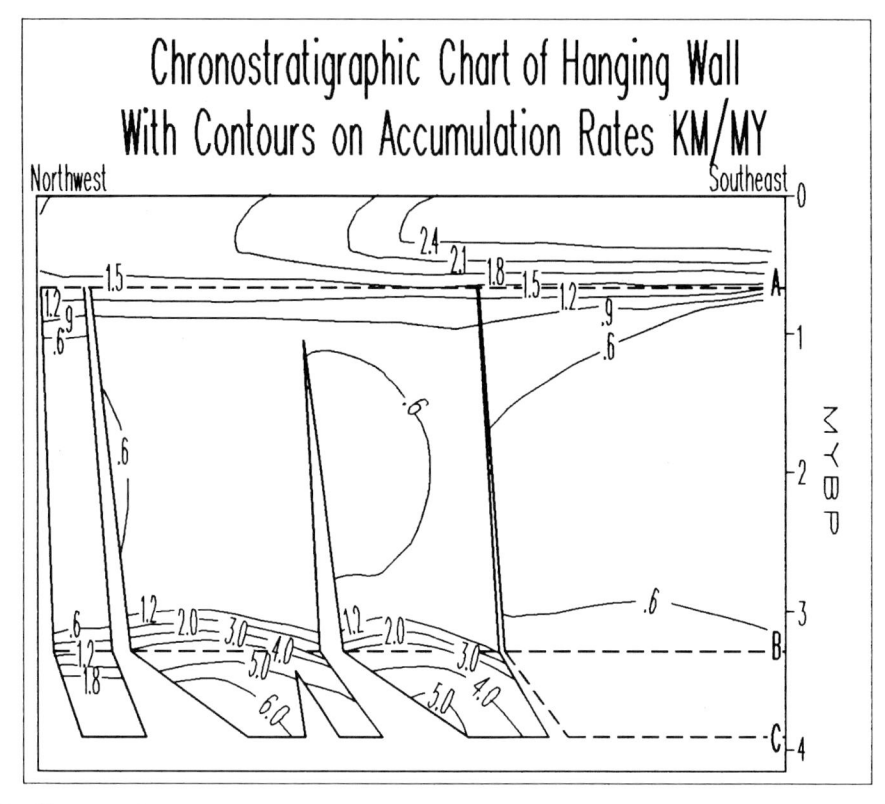

Fig. 14. Chronostratigraphic chart across the hanging wall from the cross-section in Figure 13. High accumulation rates still occur and expand outer shelf sedimentation.

mechanisms may be operative. Faults of this type are found south of the Comanchean shelf margin which is probably a major basement hinge zone in the Gulf Coast. Here, faults are probably related to basement reactivation along the basement hinge. This fault type is also recognized in the southern part of offshore Texas where it seems more likely the faults are related to differential compaction of thick shale intervals present. These faults are recognized where salt is generally thin or absent but may be a more widespread structural type which is simply not recognized because of structural complexities found in most other settings. The presence of these faults in other trends might play a significant role in migration since they do cut deeply into the basin.

Example E: The seismic line across the Lower Miocene and Upper Oligocene of offshore Texas show that these sediments have relatively small offsets with little or no rotation of these units even deep on the section (Fig. 15). The faults appear to have little concavity on the seismic or the cross section (Fig. 16). The apparent small offset seen on these faults may be simply explained by differential compaction in the thick Tertiary section, but without better seismic or well control this relationship cannot be demonstrated.

Relationship of structural types to each other

The relationship of the fault types to each other is important in understanding the overall dynamics of the basin. In particular, the possible role of downdip compressional zones which accommodate updip extension is a process which has been invoked as integral to the basin, but as yet such compressional zones are not commonly recognized within the basin. A nearly continuous fold and thrust belt is found near the base of the present slope along the clastic margin of the Gulf of Mexico (Blickwede & Queffelec 1988); however, the total compressional shortening found in this belt is small (<10%) and cannot account for the cumulative extension found updip in the Cenozoic trends. Until it is shown

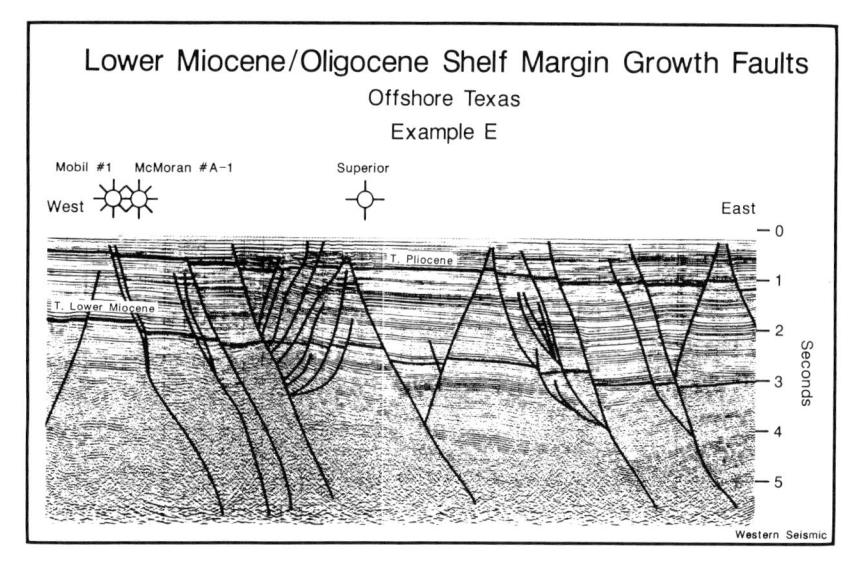

Fig. 15. Migrated seismic line of Lower Miocene/Upper Oligocene faults offshore Texas illustrating non-detached faults (Example E). Faults appear planar and downthrown reflections have little apparent rotation.

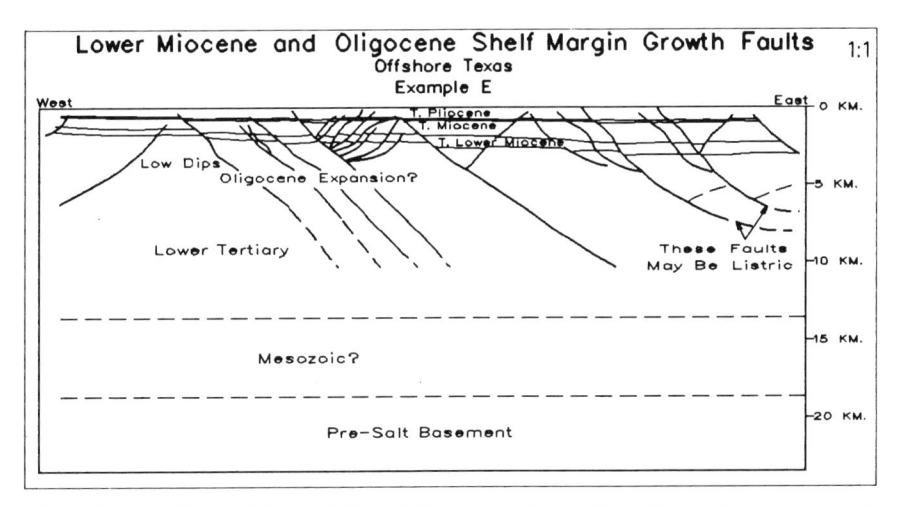

Fig. 16. Cross-Section of Lower Miocene/Upper Oligocene faults — offshore Texas from seismic on Figure 15. These planar faults are probably due to differential compaction, but basement faulting cannot be ruled out.

that these compressional zones are more common than currently recognized it is difficult to account for the total extension found in the Tertiary. As previously mentioned, some insight may be gained by looking internally at the basin and considering the relationship of the different fault types to each other.

Two of the shelf margin growth fault types may be considered structurally self-contained in that the vertical motion on the faults is accommodated within that system and clearly do not require compressional zones. These are the non-detached faults and the autochthonous salt withdrawal basin faults. The other types have, in part, their vertical motion translated to horizontal motion on listric faults. Because of this mechanism it is argued that these faults require downdip compressional zones (Crans et al. 1980). The simple stratigraphic fault type which has a rigid footwall is the best analogy to this model and would seem to require a compressional zone. However, regional seismic lines suggest that the horizontal translation on this type — like the Oligocene trend of Texas — is translated downdip to the lower Miocene trend offshore where the faults detach on ductile shale. Within this type of fault system shortening may occur by means which do not record the shortening within the hanging wall sediments. For example, units which are expanded and translated on this type of fault will sometimes onlap on to the shale ridge downdip (Fig. 17). As the ridge rises it may become narrower and allow the onlapping section on either side to move toward the ridge centre, such that the horizontal translation is accommodated by lateral collapse of the shale ridge. Any folding due to diapiric rise of shale or salt will also allow for lateral shortening. Similar processes were suggested by Dailley (1976). If these processes are insufficient, the horizontal translation may be carried further downdip to the present continental slope where large salt structures are developed. Within this area, a more significant process in the overall accommodation of updip extension may be occurring. The large salt massifs present may be collapsing into narrower salt ridges or diapirs and may account for a significant amount of shortening of the overall Cenozoic basin with little evidence of the shortening in the sediments between the salt massifs (Fig. 17).

The processes described here may explain why compressional zones are not more commonly recognized in the Gulf Coast basin. However, the basin is highly complex and still poorly imaged on seismic in the deeper sections — so new levels of complexities will undoubtedly be revealed as technology improves and drilling continues.

Conclusions

The typical Gulf Coast growth fault is more than just a syndepositional fault. These are faults driven by sedimentation because sea-floor subsidence is allowed to keep pace with sediment accumulation in various geologic settings in the Gulf Coast. It is the subsidence related to these faults which geomorphically defines the shelf/slope break. Therefore these faults expand sediments associated with the shelf margin and are referred to as shelf margin growth faults.

Shelf margin growth faults are found in at least five different geological settings in the Gulf Coast. They are: (1) simple stratigraphic detachment; (2) detachment on ductile shale; (3) detachment into autochthonous salt basins; (4) detachment on allochthonous salt sills; and

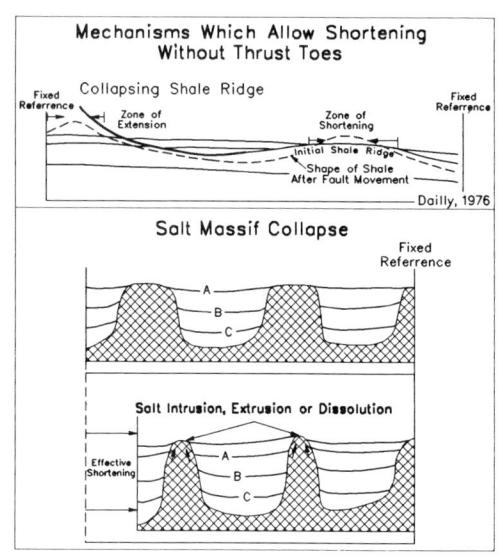

Fig. 17. Diagrams illustrating shale and salt ridges collapse may account for downdip shortening. Both salt and shale ridges often have onlapping adjacent sediments. Narrowing of these ridges allows for shortening of the section and may balance lateral downdip translation from updip listric normal faults. Dailly (1976) has suggested a similar mechanism for shale ridges associated with growth faults.

(5) non-detached fault due to differential compaction or basement faulting.

These five types have different structural and stratigraphic characteristics, the most important of which may be the different structural relationships to deep source units in the Gulf Coast.

References

AMERY, G. B. 1969. Structure of the Sigsbee Scarp, Gulf of Mexico. *Bulletin of the American Association of Petroleum Geologists*, **53**, 2480–2482.

BLICKWEDE, J. F. & QUEFFELEC, T. A. 1988. Perdido Foldbelt: a New Deep-Water Frontier in Western Gulf of Mexico. *Bulletin of the American Association of Petroleum Geologists*, **72**, 163.

BRUCE, C. H. 1973. Pressured shale and related sediment deformation: mechanism for development of regional contemporaneous faults. *Bulletin of the American Association of Petroleum Geologists*, **57**, 878–886.

—— 1983. Shale Tectonics, Texas Coastal Area Growth Faults. *In*: BALLY, A. W. (eds) *Seismic Expressions of Structural Styles*. Studies in Geology American Association of Petroleum Geologists **15**, 2.3.1–1.

BUFFLER, F. T. 1983. Structure of the Sigsbee Scarp, Gulf of Mexico. *In*: BALLY, A. W. (eds) *Seismic*

Expression of Structural Styles. Studies in Geology American Association of Petroleum Geologists, **15**, 2.3.2–50.

——, SHAUB, F. J., WATKINS, J. S. & WORZEL, J. L. 1979. Anatomy of the Mexican Ridges, southwestern Gulf of Mexico. *In*: WATKINS, J. S. *et al.* (eds) *Geological and Geophysical Investigations of Continental Margins*, Memoir American Association of Petroleum Geologists, **29**, 319–327.

CARVER, R. E. 1968. Differential Compaction as a cause of regional contemporaneous faults. *Bulletin of the American Association of Petroleum Geologists*, **52**, 414–419.

CLOOS, E. 1968. Experimental Analysis of Gulf Coast Fracture Patterns. *Bulletin of the American Association of Petroleum Geologists*, **52**, 420.

CRANS, W., MANDL, G., & HAREMBOURE, J. 1980. On the Theory of Growth Faulting a Geomechanical Delta Model Based on Gravity Sliding. *Journal of Petroleum Geology*, **2**, 165–307.

DAILLY, G. C. 1976. A possible mechanism relating progradation, growth faulting, clay diapirism and overthrusting in a regressive sequence of sediments. *Bulletin of Canadian Petroleum Geology*, 24–1, 92–116.

Dow, W. G. 1984. Oil Source beds and oil prospect definition in the upper Tertiary of the Gulf Coast. *Transactions of the Gulf Coast Association Geology Society*, **XXXIV**, 329–339.

HALBOUTY, M. T. 1979. *Salt domes, Gulf Region, United States and Mexico*, 2nd edn. Gulf, Houston, Tx.

JACKSON, M. P. A. & TALBOT, C. J. 1986. External Shapes, strain rates, and dynamics of salt structures. *Bulletin of the Geological Society of America*, **97**, 305–323.

LARBERG, G. M., BYRD. 1983. Contra-regional faulting: Salt withdrawal compensation, offshore Louisiana, Gulf of Mexico. *In*: BALLY, A. W. (edn) *Seismic Expression of Structural Styles*. Studies in Geology American Association of Petroleum Geologists, **15**, 2.3.2–42.

LOCKER, S. D. & CHATTERIERE, S. K. 1984. *Seismic Velocity structure. In*: BUFFLER, R. T. & PILGER, R. H. (eds) *Gulf of Mexico Atlas*. Ocean Margin Drilling Program. Atlas series 6, 4.

MARTIN, R. G. 1980. *Distribution of salt structures in the Gulf of Mexico; map and descriptive text*. U.S. Geological Survey map MF-1213.

—— 1984. Diapiric trends in the deep-water Gulf basin: Characteristics of Gulf Basin Deep-Water sediment sand. Their Exploration Potential. *Proceedings SEPM Research Conference*, **6**, 60–62.

OCAMB, R. D. 1961. *Growth Faults of South Louisiana*. Transactions, Gulf Coast Association of Geological Societies **XI**, 139–176.

PINDELL, J. 1985. Alleghenian reconstruction and subsequent evolution of the Gulf of Mexico, Bahamas and proto-Carribean. *Bulletin Tectonics*, **4**, 1–39.

QUARLES, M. Jr. 1953. Salt-ridge hypothesis on origin of Texas Gulf Coast type of faulting. *Bulletin of the American Association of Petroleum Geologists*, **52**, 399–413.

SALVADOR, A. 1987. Late Triassic–Jurassic Paleogeography and Origin of Gulf of Mexico. *Bulletin of the American Association of Petroleum Geologists*, **71**, 419–451.

SASSEN, R. 1988. Origin of Crude Oil in Eastern Gulf Coast: Upper Jurassic, Upper Cretaceous, and Lower Tertiary Source Rocks. *Bulletin of the American Association of Petroleum Geologists*, **72**, 243.

SEGLUND, J. A. 1974. Collapse-Fault Systems of Louisiana Gulf Coast. *Bulletin of the American Association of Petroleum Geologists*, **58**, 1289.

SKERYANC, A. J. & KOLODNY, C. R. 1983. Syndepositional structures on Mobile Substratums. *In*: BALLY, A. W. (ed.) *Seismic Expression of Structural Styles*. Studies in Geology. American Association of Petroleum Geologists **15**, 2.3.1–41.

SPINDLER, W. M. 1977. Structure and stratigraphy of a small Plio-Pleistocene depocenter, Louisiana Continental shelf. *Transactions of the Gulf Coast Association of Geological Societies*, **XXVII**, 180–196.

THORSEN, D. C. 1963. Age of growth faulting southeast Louisiana. *Transactions of the Gulf Coast Association of Geological Societies*, **XIII**, 103–110.

WANG, Y. F. 1988. Salt Tongues in Northern Gulf of Mexico. *Bulletin of the American Association of Petroleum Geologists*, **72**, 156.

WINKER, C. D. & EDWARDS, M. B. 1983. Unstable Prograding Clastic Shelf Margins. *In*: STANLEY, D. J. & MOORE, G. T. (eds) *The Shelfbreak: Critical Interface on Continental Margins*. Special Publication of the Society of Economic Paleontologists and Mineralogists, **33**, 139–157.

The Northern Gulf Coast Basin: a classic petroleum province

T. G. FAILS

Independent Petroleum Geologist and Consultant, 1777 Larimer Street, Denver, CO 80202, USA

Abstract: The US portion of the Gulf Coast Basin is that country's greatest petroleum province, having produced about 31% of the oil and about 48% of the gas. Jurassic evaporites, deposited in basins opened by Mesozoic drift, underlie much of an area which received marine, mainly clastic, upper Mesozoic and Cenozoic deposits. Three small tensional basins were filled by late Eocene time. A large paralic basin remains an area of active deposition. Petroleum is produced from Mesozoic carbonates and clastics in the tensional basins, often from salt-related features, and from Cenozoic clastics in the paralic basin, where salt domes are common. Eocene fault-trapped production dominates in salt-free south Texas. Oligocene production, often from diapiric structures, is most important in coastal Texas and southwestern Louisiana. Miocene and younger production, usually salt-related, dominates in onshore and offshore south Louisiana. Favourable factors include large volume deposition and rapid burial, early and continuing structural development, lack of sand cementation and excellent source characteristics.

The United States portion of the greater Gulf Coast Basin is the most important producing province in that country. Through 1979, 45 billion barrels of oil and more than 347 TCF of gas had been produced or were in proven reserves (Rollins 1982). Compared to total produced and proven US reserves, including Alaska, the Gulf Coast accounted for 31% of the oil and 48% of the gas. The Gulf Coast portion of Texas alone produced 13.2 bbls and over 120 TCF of gas in the period 1934–1986 (C. Garrett, Texas BEG, pers. comm. 1988). Production is now slowly declining, but no other US producing province contributes a greater share. The Gulf Coast Basin is one of only three US areas with ongoing large-scale exploration programs.

Prominence of the Gulf Coast Basin as a producing province did not come early or easily. A well drilled for water at Corsicana, Texas in 1896 (37 years after the Drake well) encountered oil; a large, shallow field was developed. Few geologists or oilmen of the day considered the Gulf Coast to be prospective for petroleum; a notion quickly and quietly abandoned after the Spindletop discovery in southeast Texas, which blew in early in 1901 at an estimated rate of 100 000 barrels per day. Jennings Field, in south Louisiana, was discovered the same year. Both fields are on salt domes, as are most large Gulf Coast fields. Gulf Coast exploration and production expanded as rapidly as the technology of the times permitted. Drilling of wells in shallow waters from barges started early. Gravity measurement permitted detection of salt domes lacking surface expression, while refraction and reflection seismic defined configuration of salt diapirs and the structures above and around them. Micropalaeontologic control systems and electric logs became invaluable tools for geologists dealing with the thick, monotonous, unconsolidated strata. One offshore field had been discovered and developed during the late 1930s. After World War II, exploration and production operations were aggressively extended offshore, especially off Louisiana. The Gulf of Mexico off Louisiana, Texas and Alabama is still the site of major exploration and development, some in waters several thousand feet deep. Onshore operations, including exploration, continue at a slackening pace, but can still be highly profitable.

Geological setting

The petroleum geology of the greater Gulf Coast Basin is simple in principle, given the non-compressional, gravity-dominant nature of the area since Jurassic times, but often extremely complex and challenging in detail, due to the effects on ongoing deposition of contemporaneous diapirism and growth faulting. Most hydrocarbons occur on diapiric structures. The rapidly deposited regressive strata, particularly post-Oligocene, may vary markedly over relatively short distances. Faults with throws of more than 400 m are common, and throws of 1250 m or more are occasionally encountered.

The northern Gulf Coast Basin is bounded on the cratonic side by a prominent peripheral graben system (Fig. 1). While upper Cretaceous

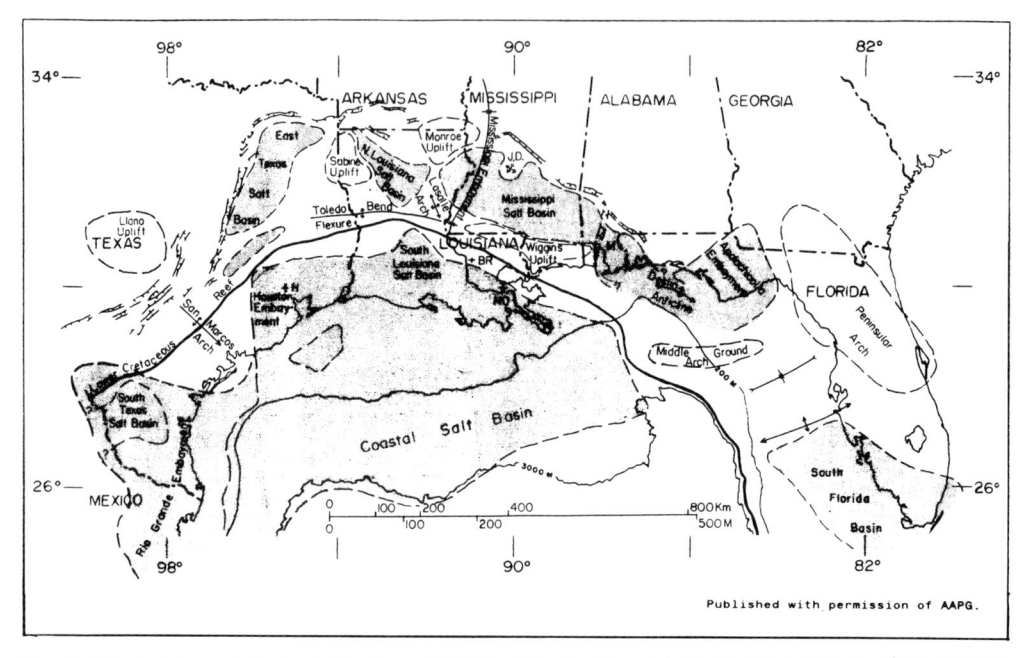

Fig. 1. Principal tectonic features, Northern Gulf Coast Basin. The peripheral graben system, where present, is considered to be the cratonic-flank boundary, although some Gulf Coast depositional sequences extend across it. Basin areas, usually but not always underlain by salt, are stippled. The Lower Cretaceous shelf-edge barrier-reef complex separates the shelf/basin area (north) from the paralic basin (south). After Murray (1961, 1966).

and lower Tertiary strata were widely deposited landward from the peripheral graben system, the greater Gulf Coast Basin is limited to the basinward side. Two areas with distinct structural styles and stratigraphic sequences comprise the greater Gulf Coast basin. An inner 'shelf' area (hereafter 'shelf/basin area') of non-compressional uplifts and basins containing Jurassic, Cretaceous and Eocene strata is underlain by transitional−continental crust. Basinward lies a major, passive-margin, paralic basin (hereafter 'paralic basin') containing a thick regressive sequence of mainly Cenozoic clastic strata. The paralic basin, underlain by thinned transitional−continental crust, continues along the Mexican coast to Yucatan. The Comanchian (later lower Cretaceous) carbonate shelf edge complex marks a convenient line of division between these two areas which are both characterized by the following features.
(1) Non-compressive, extensional tectonics. Subsidence, salt diapirism, lateral displacement of salt and clay, gravity sliding and normal faulting dominate.
(2) Rapid alluvial, transitional and marine deposition.
(3) Salt diapirism in most basin areas (Fig. 2).
(4) Prolific hydrocarbon production.

Development of the greater Gulf Coast Basin appears to be related to the early Mesozoic opening of the Atlantic. Triassic rift basins in the southeastern US and eastern Mexico may have counterparts under the present Basin area. Salt deposition (about 90% halite) was widespread during much of the Jurassic (Pindell 1985; Salvador 1987). All Gulf Coast Basin salt (Louann formation) appears to be of the same age. Opening of the Gulf Coast Basin is often viewed as a consequence of southeasterly drift and rotation of the Yucatan block, although other scenarios have been proposed (Pilger 1981). Continuing drift after salt deposition caused the existing division into two Jurassic salt basins (Fig. 3) separated by a deep about 300 km wide lacking salt and underlain by oceanic crust (Buffler & Sawyer 1985; Humphris 1979; Pindell 1985; Salvador 1987). Four smaller evaporite basins occur in the shelf/basin area (Figs 1 & 2), either as embayments off the central evaporite basin or as isolated basins separated from it by uplands, and contain numerous salt diapirs. Salt occurs outside of these basins as well, but is thinner, sometimes discontinuous and usually undeformed. Three are elongate, with relatively straight boundaries. While the presence of classic, fault-bounded

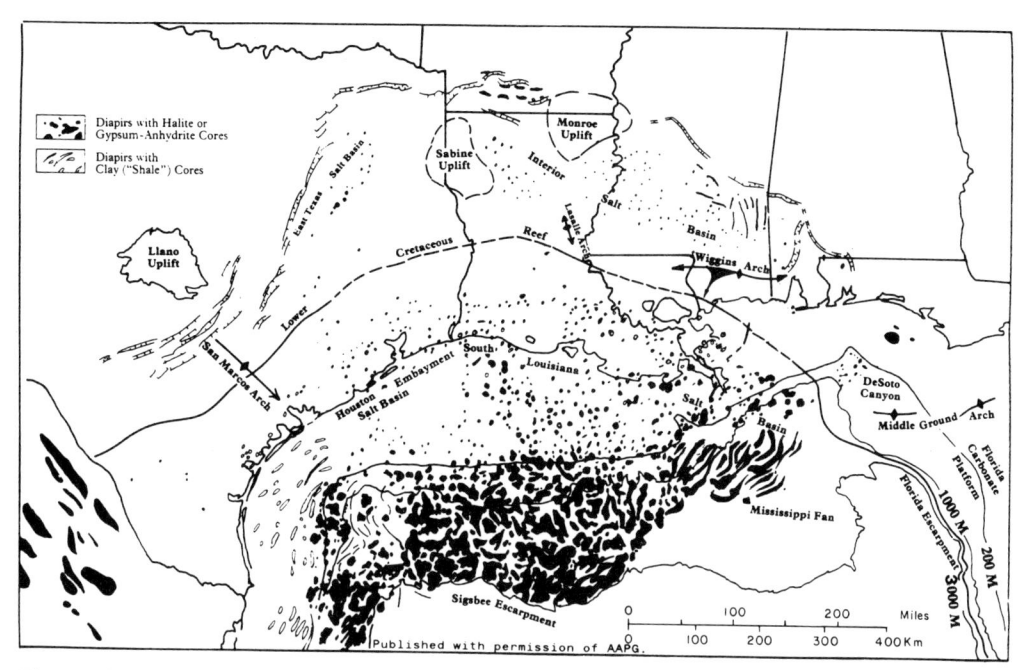

Fig. 2. Diapir distribution, Northern Gulf Coast Basin. The Coastal Salt Basin (the Houston Embayment and South Louisiana Salt Basins) extends south beneath the continental slope to the Sigsbee Escarpment and under the Mississippi Fan. Offshore, most salt diapirs are outlined at approximately 1.0 s (2000 m). Onshore, only salt diapirs which crest within about 2000 m of the surface are shown. The more numerous deeply buried salt diapirs are not shown in either area, for the sake of clarity. Clay diapirs, onshore and offshore, are shown to the degree known. Published with permission of AAPG. After Martin (1980) and Murray (1961, 1966).

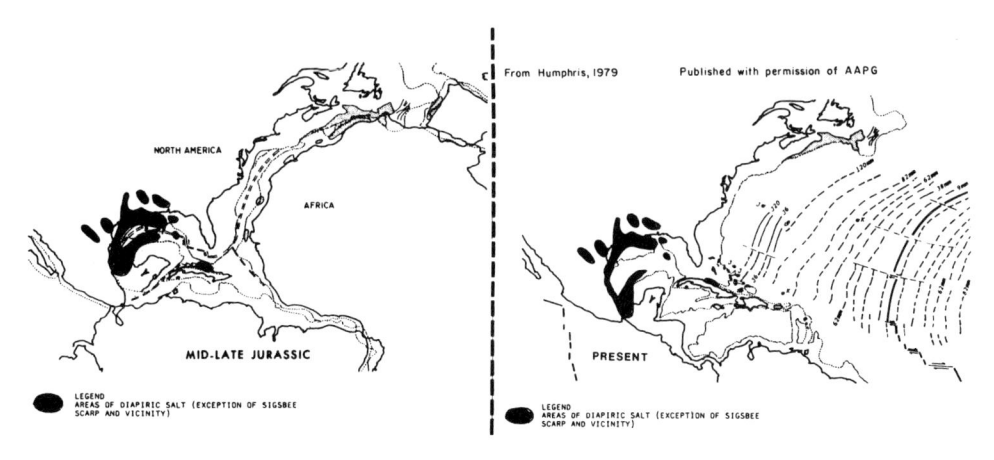

Fig. 3. Present configuration and Jurassic reconstruction, Greater Gulf of Mexico Basin. Areas of salt diapirs are stippled, with exception of laterally displaced salt in area north of Sigsbee Escarpment. Note drift and rotation of Yucatan Block, broadening and deepening the post-Jurassic Greater Gulf of Mexico Basin. From Humphris (1979), Figs 12 & 13. Published with permission of AAPG.

grabens cannot be proved, these basins appear to have developed in response to tension resulting from drift-associated stretching and adjustment in an area where vertical crustal-block movement was more important.

Shelf/basin area structural features

Most of the principal structural features of the shelf/basin area play an important role in hydrocarbon production, and are described below. Mesozoic and lower Tertiary sediments are both source and reservoir rocks. A mixed clastic—carbonate sequence with generally similar characteristics is present throughout the main producing areas in Texas, Louisiana, Mississippi and Alabama. Minor oil production exists in the South Florida Basin, where carbonate rocks predominate. Three major transgressive—regressive cycles characterize the Jurassic through Eocene stratigraphic sequence (Fig. 4). The regressive sequences pass downdip from redbeds through transitional and marine sandstones and shales to outer shelf marine shales, sometimes with minor carbonate beds. Updip sandstones are usually unimportant in the marine—transgressive carbonate and shale sequences. Evaporites sometimes occur with the carbonates, suggesting lack of clastic influx was sometimes more important than water depth during 'transgressions'.

The principal structural features of the shelf/basin area are described below and illustrated in Figs 1 & 2. Murray (1961) and Anderson (1979) are important references.

Rio Grande Embayment

A triangular basin shared with Mexico, the Embayment broadens and plunges southeast toward the paralic basin. They probably merge; basement depth at the coast is estimated to be about 10 000 m. Fill is mainly Cretaceous and lower Tertiary clastics, with some Cretaceous carbonates. Limited Jurassic strata occur, as do a small number of salt domes. Compressive structures are present in the Mexican portion. Hydrocarbons are produced mainly from fault traps and folded or gravity-slide structures in the lower Tertiary and Cretaceous.

San Marcos Arch

A broad, gentle, southeast-plunging nose off the Llano Dome (a persistent cratonic uplift in central Texas), the San Marcos Arch area is unaffected by salt diapirism. All pre-Miocene strata thin crossing the arch. Most hydrocarbons are trapped in low-relief, normally-faulted structures. Small-scale upper Cretaceous volcanism was widespread. Clay diapirism is important in development of structures productive from Oligocene and Miocene clastics in the paralic basin portion.

East Texas, North Louisiana and Mississippi Salt Basins

Three elongate basins with near-parallel sides, the East Texas, North Louisiana and Mississippi alt Basins were the sites of thick salt deposition. (The latter two are sometimes collectively called the 'Interior Salt Basin'.) Salt diapirs and pillows form hydrocarbon traps in overlying or flanking Mesozoic strata; 'turtle' structures, truncation traps, stratigraphic traps and faulted anticlines are also productive. The development, form and stratigraphy of all three basins are similar, as they developed at the same time under similar conditions as a continuous unit, being segmented into three smaller, deepening basins during middle and upper Cretaceous and early Tertiary times. Although grabens are suggested, prominent boundary—fault systems occur on just one flank of only the Texas and Mississippi Basins. Petroleum has been produced since 1905, from Jurassic, Cretaceous and Eocene strata and exploration remains at a relatively high level in Texas, Mississippi and Alabama.

Sabine and Monroe Uplifts, Jackson Dome

Three somewhat similar uplifts affect the most productive portion of the shelf/basin area: the Sabine Uplift separates the East Texas and North Louisiana Salt Basins, the Monroe Uplift lies between the North Louisiana and Mississippi Salt Basins, and the much smaller Jackson Dome intrudes into the north flank of the latter basin. The Sabine and Monroe appear to have resulted from intermittent late Palaeozoic and Mesozoic basement uplift, especially at the end of the lower Cretaceous; volcanism strongly affected the latter as well. Both are relatively flat-topped, with thinned but largely complete sections and truncation traps are common. The high-relief Jackson Dome appears to have formed mainly in response to upper Jurassic and Cretaceous igneous

---

Fig. 4. Stratigraphic Section, Northern Gulf of Mexico Basin, excluding Florida. Mesozoic section is that characteristic of the shelf/basin, and Cenozoic of the paralic basin. After Figs 2 & 3, Lafayette and New Orleans Geological Societies (1968) and Fig. 2A, Shreveport Geological Society (1968).

THE NORTHERN GULF COAST BASIN

SYSTEM	SERIES	POSSIBLE EUROPEAN STAGE EQUIVALENT	GROUP OR FORMATION AS USED BY INDUSTRY (TEXAS)	GROUP OR FORMATION AS USED BY INDUSTRY (LA.-MISS.)	LITHOLOGY TEXAS	LITHOLOGY LA.-MISS.
QUATERNARY	RECENT				alluvial, transitional	
	PLEISTOCENE			PLEISTOCENE		
TERTIARY	PLIOCENE	ASTIAN / PLAISANCIAN	PLIOCENE	PLIOCENE	marine sands, shales & clays	
	MIOCENE UPPER	PONTIAN / SARMATIAN	UPPER MIOCENE	"FLEMING"		
	MIOCENE MID.	TORTONIAN / HELVETIAN	MIDDLE MIOCENE			
	MIOCENE LOWER	BURDIGALIAN / AQUITANIAN	LOWER MIOCENE	ANAHUAC (U/M/L)	marine shales & sands; marine shales & sands; marls, limestone & sh	
	OLIGOCENE?	CHATTIAN / RUPELIAN	FRIO (U/M/L)		alluvial, transitional & marine sands, silts, shales & clays	marine shales & clays
	EOCENE	TONGRIAN	VICKSBURG	VICKSBURG		
		LUDIAN / BARTONIAN	JACKSON (YAZOO) / MOODY'S BRANCH / YEGUA	COCKFIELD	transitional & marine sands & shales	transitional & marine sands & shales
		AUVERSIAN / LUTETIAN	COOK MOUNTAIN / SPARTA (Q/R)	CANE RIVER	marine shales, occasional sands	
		CUISIAN / YPRESIAN	WILCOX	WILCOX	alluvial, transitional & marine sands & shales	
	PALEOCENE	LANDINIAN / DANIAN	PORTER'S CREEK / CLAYTON		marine shales & clays shales, marls, chalks	

SYSTEM	SERIES	POSSIBLE EUROPEAN STAGE EQUIVALENT	GROUP OR FORMATION AS USED BY INDUSTRY (TEXAS)	GROUP OR FORMATION AS USED BY INDUSTRY (LA.-MISS.)	LITHOLOGY TEXAS	LITHOLOGY LA.-MISS.
CRETACEOUS	UPPER (GULFIAN)	MAASTRICHTIAN	NAVARRO	"SELMA"	sand & shale; marl, sand & shale; chalk	marl, chalk & shale; mar & chalk; chalk & limestone
		CAMPANIAN	TAYLOR			
		SANTONIAN / CONIACIAN	AUSTIN			
		TURONIAN	EAGLE FORD	EUTAW	alluvial, transitional & marine sandstone & shale	
		CENOMANIAN	WOODBINE	TUSCALOOSA		
	LOWER (COMANCHEAN)	ALBIAN	DANTZLER / WASHITA- FREDERICKSBURG / EDWARDS / GLEN ROSE OF TEXAS / PALUXY	MOORINGSPORT (RUSK) / RODESSA / PEARSALL	limestone, dolomite & shale; alluvial, transitional occasional marine clastics; alluvial, transitional & marine sandstone, shale, minor limestone; anhydrite, sh, limestone	limestone
		APTIAN		JAMES	limestone, shale, sandstone	
	(COAHUILAN)	NEOCOMIAN	SLIGO / TRAVIS PEAK	HOSSTON	shale & limestone "sandwich"; limestone & shale	
JURASSIC	UPPER AND MIDDLE	PORTLANDIAN	COTTON VALLEY		alluvial, transitional & shallow marine sandstone & shale	
		KIMMERIDGIAN	HAYNESVILLE		anhydrite, shale & limestone	
		OXFORDIAN	SMACKOVER		limestone & dolomite	
		CALLOVIAN / BAJOCIAN / AALENIAN-	NORPHLET		aeolian & alluvial ss.	
	LOWER	PLIENSBACHIAN / SINEM.-HETT.	LOUANN SALT / WERNER / EAGLE MILLS		90% halite; anhydrite?; redbed clastics?; OFTEN MISSING	
TRIASSIC	UPPER		MOREHOUSE		limestone, shale & sandstone	
PENN. ?						

activity and volcanism. The crest and flanks of the Sabine are major producing areas, but the Monroe Uplift and Jackson Dome are relatively unimportant in comparison.

Wiggins Uplift

An irregular, low-relief uplift, the Wiggins is poorly known. Located between the Mississippi Salt Basin and the paralic basin, the Wiggins received only late Jurassic sediments (no salt) plus a thin Cretaceous section. It was probably emergent Eocene through Miocene and has been relatively high since. Most of the Wiggins is non-productive. It may be representative of a broad, structurally featureless area to the west, crossing central Louisiana south of the Sabine Uplift and extending into East Texas between the East Texas and Coastal Salt Basins, which, together with the Wiggins, is termed the Toledo Bend Flexure by Anderson (1979).

Apalachicola Embayment

This southwest-plunging triangular basin is lightly explored, with only limited data available. Relatively shallow, the onshore section is composed mainly of Cretaceous and upper Jurassic redbeds, shales and carbonates. Salt is present offshore and under a small onshore area.

Peninsular Arch

East of the Apalachicola Embayment, the proportion of carbonate rocks increases in Mesozoic and Cenozoic strata. The Tertiary and Cretaceous of the Peninsular Arch are predominantly limestone and dolomite, with varying lesser amounts of clastics and evaporites. This broad, low-relief uplift underlies most of the Florida peninsula, flanking the Apalachicola Embayment and the South Florida Basin. Relatively high through the lower Cretaceous, it has gradually subsided since. The Peninsular Arch and satellite Ocala Arch have been lightly explored, as no commercial accumulations have been discovered, but are not considered to be particularly prospective.

South Florida Basin

This carbonate shelf basin of upper Mesozoic and Cenozoic age comprises the southern third of the Florida Peninsula plus part of the Florida Shelf offshore to the west. An area of southerly-thickening, structurally undisturbed shelf carbonates, with subordinate shales and anhydrites, the South Florida Basin has been sporadically explored since World War II. Fourteen small stratigraphic oil fields have been discovered in lower Cretaceous bioclastic mounds.

Paralic Basin

Basinward from the shelf/basin area of the greater Gulf Coast Basin lies the Cenozoic-filled, paralic basin. The Comanchian carbonate bank edge provides a convenient line of division with geologic significance between the two. It formed the lower Cretaceous continental shelf edge along the basinward flank of the Wiggins Uplift, Toledo Bend Flexure and San Marcos Arch through Mississippi, Louisiana and Texas, and parallels or overlies the transition from thick to thinned transitional−continental crust across this area. To the east, off Alabama and Florida, the Comanchian bank edge directly underlies the present Florida Escarpment, which forms the edge of the contemporary continental shelf west of the Florida peninsula (see Figs 1 & 2). The nature of the paralic basin in both structural and stratigraphic senses is very different from the small 'shelf' basins described above. It rims a far larger basin, the post-drift, subsided Gulf of Mexico low, and is sediment-filled to varying degrees along the flanks. Deposition has been continuous since the late Jurassic or earlier. Continental shelves began to develop on the basinward side of the Comanchian bank edge during earliest upper Cretaceous time.

The paralic basin underlying the present coastal plains and continental shelves and slopes of northeastern Mexico, Texas and Louisiana contains a great volume of rapidly-deposited, regressive, clastic, Cenozoic strata which steadily prograded into the abyss as sedimentation rates consistently exceeded the rate of subsidence. The structural nature of the paralic basin is essentially the same across Texas and Louisiana, as well as the area offshore from Mississippi, Alabama and Florida ('Mafla' area) beyond the shelf edge, although subsidence rates have varied greatly. Thick sedimentary fill is derived largely from the landward side except off Mafla, where it is thin and largely of pelagic origin. Unlike the shelf basins and most well known petroleum-producing basins, the paralic basin lacks a structural limit or flank on the basinward side. The resulting lack of lateral constraint has greatly affected development of lateral displacements and diapiric structures during most of the Cenozoic.

Sedimentary fill within the paralic basin varies significantly. Areas of thick strata (sometimes exceeding 15 000 m) and rapid basinward progradation (up to 500 km) are overwhelmingly clastic-dominant. Periods of limited clastic influx appear as marine transgressions with shale and minor carbonate deposition. A major late Cretaceous depocenter developed in north-eastern Mexico—south Texas. During the Tertiary, major clastic depocentre locations migrated northeasterly across coastal Texas and then east across south Louisiana in response to continuing uplift of the Cordilleran system, reaching a maximum eastern extent in south-eastern Louisiana during the upper Miocene. Section thickening across growth faults is common.

In the eastern Gulf off Mafla, where the rivers are small, with limited loads, relatively small volumes of slowly deposited, mainly carbonate muds are the dominant lithotype.

Depositional rate and volume appear to have increased progressively through time in the clastic-dominant areas. Sedimentation rate estimates for coastal Louisiana and Texas increase from about 5.0 cm/1000 years for the Cretaceous—late Jurassic to about 46.0 for the Oligocene, 25.0 (Texas) to 62.5 (Louisiana) for the Miocene and exceed 240.0 cm/1000 years off Louisiana during the Pleistocene (Lafayette and New Orleans Geological Societies 1968). Steady basinward progradation occurred as well. Sufficient evaporite thicknesses to permit diapirism may not occur in some areas, but salt domes are common elsewhere. While good production exists in some non-salt areas, approximately 59% of all hydrocarbons in the Texas—Louisiana Gulf Coast Tertiary basin are associated with salt diapirs (Woodbury et al. 1980).

Gulf Coast Basin salt diapirism

Salt movement or diapirism is rare in most areas outside of the salt basins shown on Figs 1 & 2. Within the salt basins, salt movement and diapirism are ubiquitous and salt-movement-related structures have attracted and trapped vast quantities of petroleum.

In all salt basins, the petroleum explorationist's primary concern is with the salt diapir-objective section relationship. Most wells drilled in a typically regressive Gulf Coast sequence, especially in the Tertiary, will first penetrate a non-objective, predominantly alluvial sand sequence. Deeper, the objective section, usually several thousands of metres in thickness, follows, deposited in deltaic and continental shelf environments, and composed of interbedded sands and shales. Below is a thick, pre-objective sequence of outer shelf and continental slope shales and clays which are usually abnormally pressured. This may be a source of hydrocarbons where buried sufficiently deep.

In all Gulf Coast Salt Basins, diapiric salt may penetrate all or part of the objective section, or may arch overlying objective strata without penetrating them. Due to the thickness of the pre-objective shale section in the Coastal Salt Basin, most salt dome structures there appear to result from salt diapirism, including many where the objective section is not pierced by salt. In the East Texas and Interior Salt Basins, however, non-diapiric salt pillows produced by lateral salt flow alone underlie some salt-cored structures; similar salt pillows are common in other salt basins, but have not been recognized in the Coastal. Turtle structures form important producing structures in the Coastal, East Texas and Interior Salt Basins.

Hydrocarbon accumulation patterns vary with respect to the various types of salt-cored structures. As exploration and exploitation operations are influenced as well, pragmatic terms describing salt domes have developed (Fig. 5).

(1) Penetrant, piercement or shallow salt domes: salt penetrates the entire objective section, and crests at, or close to, the surface. Most accumulations occur flanking the salt.

(2) Semi-penetrant, piercement, intermediate or deep salt domes: salt penetrates only a lower portion of the objective section, with the overlying strata arched above the diapir. Accumulations occur above the diapir and on its flanks.

(3) Non-penetrant, deep or deep-seated salt domes: salt does not penetrate the objective section, which is arched or anticlinally folded above it. Accumulations usually occur above the buried diapir.

Clay diapirs form the cores of some non-penetrant structures in the Coastal Salt Basin and in some south Texas areas lacking salt diapirism.

Salt diapirism occurs in somewhat different forms in the Interior and East Texas Basins, compared with the Coastal Salt Basin. The effect on hydrocarbon occurrence is important. In the East Texas and Interior Salt Basins, most hydrocarbons are found on semi-penetrant and non-penetrant salt domes, above pillows, and on turtle structures, with very little on penetrants. In the Coastal Salt Basin, penetrant salt domes are the most-productive on an equivalent barrels per uplift-volume basis. Semipenetrants, less-common than penetrants, are highly productive as well. Non-penetrant structures are the most common numerically, but on an equivalent-barrels per uplift-volume basis are

DIAPIRS CLASSIFIED BY RELATIONSHIP TO OBJECTIVE SECTION

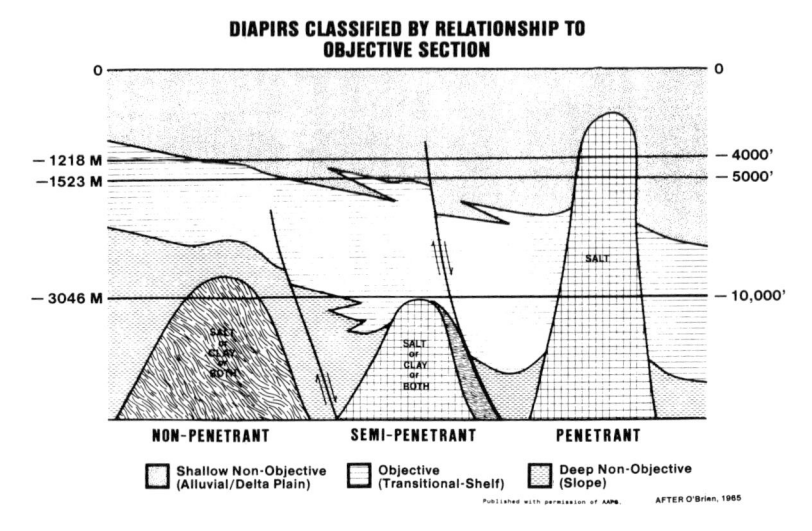

Fig. 5. Diapirs classified by relationship to objective section. Irrelevancy of depth-of-burial-based classifications is obvious. From Fails *et al.* (1990), Figure 1–10. Published with permission of AAPG.

the least productive class of Coastal Basin salt domes.

Rapid deposition of great sedimentary volumes, thick unconsolidated marine shales, extensive clay diapirism, widespread lateral salt displacement and low salt-to-sediment volume ratios characterize the Coastal Salt Basin. The East Texas and Interior Salt Basins are in total contrast. Differences in salt diapirism style and in hydrocarbon accumulation patterns between the two areas probably arise from these contrasts.

Diapiric structures contain most Coastal Salt Basin reserves. The structural type versus hydrocarbon distribution data plotted in Fig. 6 are based on a sample of 690 south Louisiana fields productive from Oligocene or younger strata (Spillers 1965). Thirty-nine per cent are diapiric structures, which contain 74.6% of the total hydrocarbons. Of the fields with original reserves exceeding 100 million barrels, 82.5% are on diapiric structures. Although this study is over twenty years old, the current proportions are about the same and the importance of Coastal Salt Basin salt domes in petroleum productivity continues.

Fig. 6. Structural type versus hydrocarbon accumulation, south Louisiana Oligocene, Miocene and Pliocene strata. Gas has been converted to barrels-equivalent hydrocarbon at 6000 cubic feet per barrel. Data from Spillers (1965). From Fails *et al.* (1990), Figures 1–7. Published with permission of AAPG.

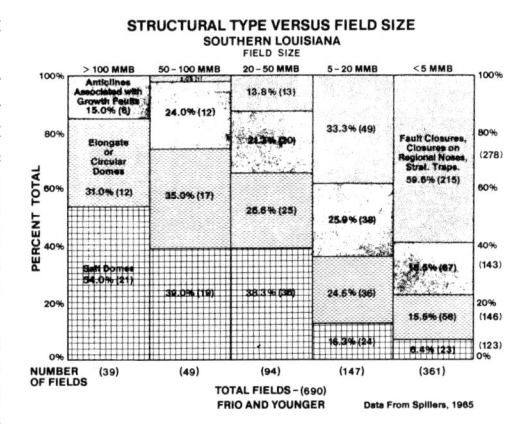

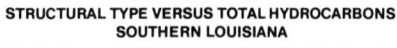

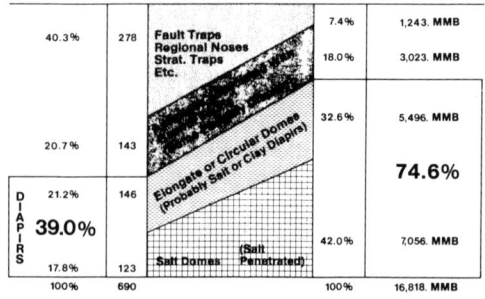

Significant papers on Gulf Coast Salt Diapirism include Atwater & Forman (1959), Braunstein (1965, 1967), Braunstein et al. (1973), Campbell & Sheller (1964), Davis & Lambert (1963), Denham (1962), Fails et al. (1990), Franscogna (1957), Frey & Grimes (1970), Galloway et al. (1983), Halbouty (1967, 1968), Harrison et al. (1970), Hughes (1968), Humphris (1979), Johnson & Bredeson (1971), Layfayette & New Orleans Geological Societies (1968), McCormick & Kline (1983), Murray (1961, 1966), Pilger (1981), Pindell (1985), Steward (1981) and Woodbury et al. (1980).

Upper Jurassic production

Widespread desert deposits (Norphlet) unconformably overlie both salt and pre-salt formations. A marine transgression (Smackover) followed. Evaporites and transitional/marine clastics and limestones (Haynesville) were laid down as regression commenced. The post-drift Gulf of Mexico basin cooled during latest Jurassic, and major subsidence occurred, with the East Texas and Interior Salt Basin areas being eventually filled by a clastic wedge (Cotton Valley) thickening into the basins. Salt movement and diapirism started during Norphlet time, and increased in response to rapid Cotton Valley deposition. With the 'shelf' basinal areas temporarily filled, a broad continental shelf platform with a carbonate barrier shelf edge existed by earliest Cretaceous times.

Three Jurassic stratigraphic units are important petroleum producers (Figs 4 & 7) in the shelf/basin area. The *Norphlet formation*, usually redbed clastics less than 30 m thick, directly overlies the Louann Salt in much of the shelf/basin area and is normally non-objective. Where the Norphlet thickens, the Denkman facies, up to 500 m of porous dune sands, occurs. Several major deep gas fields productive from the Denkman facies have been discovered since 1978 in the Mobile Bay area (Mary Ann field, Fig. 7). Most production is below 6700 m, but several TCF appear to be present. Several smaller, shallower Norphlet fields have been discovered onshore in Alabama and west Florida as well. All Norphlet fields are on domal or fault-line structures resulting from salt movement.

The Smackover formation

This is productive over a broad area (Fig. 7) where the upper portion occurs in two facies. A

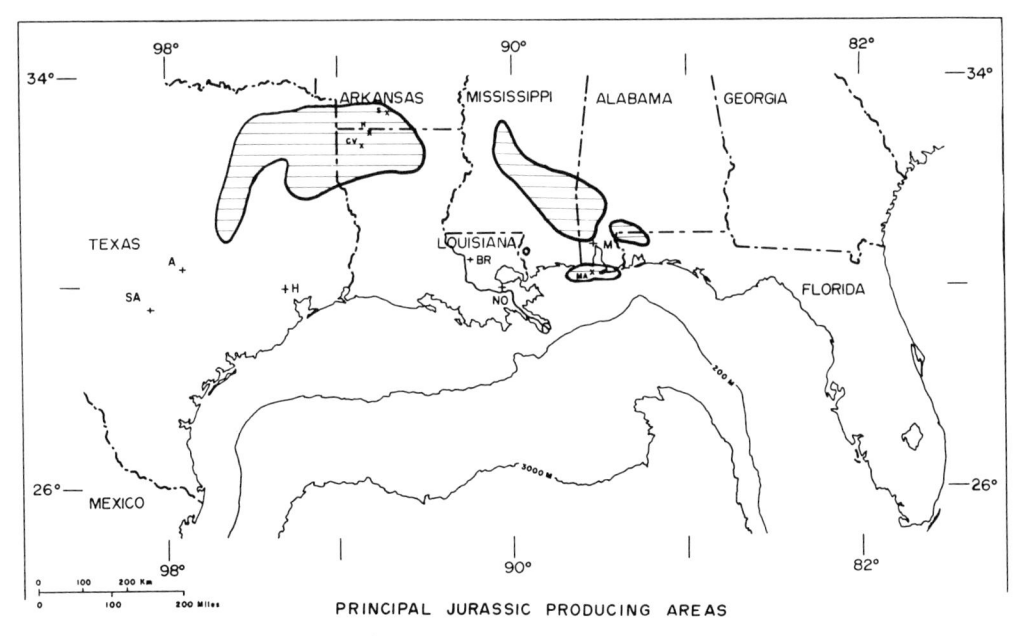

Fig. 7. Principal Upper Jurassic Producing Areas. Mary Ann Field was the first discovery in the Norphlet deep-gas area at the mouth of Mobile Bay, offshore Alabama. Locations of the giant Cotton Valley (2.5+ TCFG) and Smackover (500+ MMBO) fields are shown as well.

non-objective, tight, clastic shelf facies is present in northeast Louisiana and west–central Mississippi. In the remainder of the area, the upper Smackover is a shallow-marine, carbonate ramp facies, with oolite shoal, leached Omoldic, and detrital limestones deposited on bathymetric highs. Important oil reservoirs occur in this facies. Salt domes and pillows, low-relief anticlines and tilted fault-block traps probably resulting from salt movement account for nearly all Smackover fields, which are often relatively small (10–30 MM bbls oil), but are often prolific producers on a per-well basis. A lower Smackover facies is everywhere composed of a transgressive, low energy, sub-tidal carbonate mudstone sequence. It is non-productive but an excellent source rock.

The Cotton Valley group

This group, uppermost Jurassic and continuing into the early lower Cretaceous, is a very thick, regressive, southerly-thickening clastic wedge. It is an important source and reservoir sequence for natural gas. Two facies, the Schuler and Bossier formations, occur. The Schuler is composed of sandstone, conglomerate and shale deposited in alluvial, transitional and shallow marine environments; deltaic and lagoonal sands are productive. Up to 1000 m thick, the Schuler overlies and is laterally transitional with up to 850 m of Bossier marine shale. An important gas producer in northwest Louisiana, the Schuler is also productive in northeast Texas, south Arkansas, east–central Mississippi and, recently, southeast Mississippi. Anticlinal and domal structures resulting from salt movement, semi-penetrant and non-penetrant salt domes, and fault-block structures are all productive, often together with underlying Smackover and/ or higher lower Cretaceous objectives. Exploration for stratigraphic traps has been relatively successful in north Louisiana during the past decade.

Significant papers on Gulf Coast Jurassic production include Anderson (1979), Beebe (1968b), Davis & Lambert (1963), Hughes (1968), Montgomery (1987 a,b), Moore (1983), Murray (1961), Newkirk (1971), Nicholls et al. (1968), Pearcy & Ray (1986), Rainwater (1968), Salvador (1987), Sheveport Geological Society (1968) and Ventress (1984).

Lower Cretaceous Production

Regression continued into the lower Cretaceous; an unconformity often occurs between the earliest Cretaceous Cotton Valley and overlying Coahuilan strata (Fig. 4). Lower Cretaceous deposition took place on a broad, low-relief continental shelf limited by a carbonate bank-edge complex, mainly reefal, along the basinward shelf edge, which occupied the basin/shelf area of the greater Gulf Coast Basin. Three transgressive–regressive cycles occurred during the lower Cretaceous. A widespread erosional unconformity occurs at the top of the lower Cretaceous, between the lower (Albian) and upper (Cenomanian) Cretaceous sequences.

Source areas adjacent to the Gulf Coast were more subdued during the lower Cretaceous than during Cotton Valley time. A broad coastal plain and transitional–shallow marine area of alluvial, deltaic, tidal flat/lagoonal and shallow shelf environments graded seaward into a restricted carbonate shelf environment landward of the shelf-edge barrier. Strata deposited seaward of the barrier are undrilled. Subsidence was the dominant structural influence on the shelf, modified by subtle basement uplifts and by salt flow and diapirism. Differing rates of subsidence, variations in sediment supply and sea level fluctuations determined the type, thickness and reservoir–source characteristics of the strata deposited. The lower Cretaceous is thus the Gulf Coast sequence most varied in stratigraphic nature, vertically and laterally. Three distinct sub-provinces are discussed in some detail, as considerable exploratory potential still exists.

The San Marcos Arch–South Texas–Rio Grande embayment sub-province

Although a thick, shallow marine lower Cretaceous sequence composed mainly of carbonate rocks is present, only one formation, the Edwards limestone (Fredericksburg Group), is an important source of hydrocarbons (see Fig. 8). The Buda, Glen Rose and James limestones also produce in a large number of small fields, but only the James in the Rio Grande Embayment has much potential. Sligo/Pettet and Hosston/Travis Peak clastics are prospective in parts of this sub-province, but at great depths.

Both the Edwards and its barrier reef equivalent, the Stuart City limestone, are productive. The Edwards is an important producer of gas-condensate, but only small gas fields have been found in the rudistid and coral reef facies, a portion of the lower Cretaceous shelf-edge barrier. Organic carbonate muds, dolomites, shell mounds, patch reefs and shales comprise the Edwards, deposited during a transgression on a broad, slowly-subsiding, medium- to high-energy back-reef shelf. Edwards fields are productive in a trend over 350 km long, down-

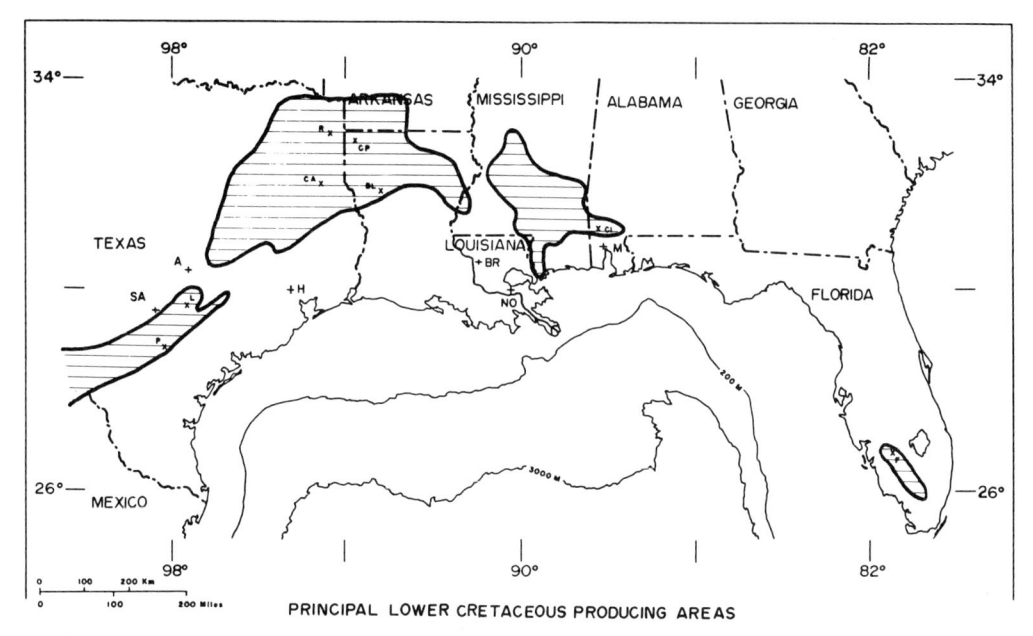

Fig. 8. Principal Lower Cretaceous Producing Areas. The Edwards–Stuart City trend in south Texas is productive mainly of gas-condensate; Person and Luling are among the larger fields. The remaining areas are productive of oil, gas and gas-condensate. Among the larger fields are Carthage (4+ TCFG, mostly Sligo fm.) Rogers (1988); Rodessa (200+ MM eq. bbls, mostly Rodessa fm.) and Caddo–Pine Island (350+ MMBO, multiple fms.). Fourteen stratigraphic-trap oil fields occur in the South Florida Basin (Sunniland fm.); produceable reserves total only about 100 MMBO.

thrown to three major growth fault zones down-thrown toward the basin southwest of and crossing the San Marcos Arch. Major field locations are shown in Fig. 8. Most occur in tilted fault blocks upthrown to up-to-the basin (up to the southeast) faults. Dolomitized tidal flat and restricted platform deposits in the upper Edwards Person facies are the most important reservoirs. Field size ranges up to 150 MM bbls oil or 300 BCFG. Gas or gas-condensate is most common; the oils have high gas–oil ratios. The Edwards also has excellent source-rock characteristics. Potential for deep but costly exploration exists.

Significant papers on Lower Cretaceous production include Beebe (1968a), Galloway et al. (1983), Keahey (1968), Murray (1961) Perkins & Martin (1985), Rainwater (1968, 1971), Roberts (1982) and Winker (1982).

The East Texas and Interior Salt Basins sub-province

This area includes the East Texas and Interior Salt Basins and the uplifts and shelf areas in between (Figs 1, 2, 4 & 8). Lower Cretaceous strata of importance to petroleum explorationists are the products of three transgressive–regressive cycles.

The Hosston/Travis Peak, Sligo/Pettet, Rodessa and Paluxy formations are major lower Cretaceous hydrocarbon producers in this sub-province; the James limestone and the Dantzler sandstones (Washita group) are productive locally. Fredericksburg group rocks are unimportant as producers here. Marine shales and dark carbonates throughout the Coahuilan and lower Comanchian are source rocks. Areas of lower Cretaceous production are shown in Fig. 8.

Hosston or Travis Peak Formation

Hosston/Travis Peak deposition occurred in a band extending from northeast Mexico to central Florida, as the top Cotton Valley unconformity was transgressed. Transitional clastics grade downdip through shallow marine sandstones and shales into marine shales except in south Texas where marine carbonates occur. Up to 1300 m thick, the Hosston/Travis Peak is an important objective for gas-condensate exploration from northeast Texas to southwest Alabama. Deltaic, strand-line and marine sandstones are productive in thick sections with numerous relatively thin and sometimes tight sandstone beds from structures related to salt movement, tilted fault blocks and faulted anticlines; stratigraphic traps occur. As with the Cotton Valley, less is known of the Hosston than of shallower objectives due to depth and lack of drilling. It is productive or potentially productive clear across the sub-province (northeast Texas to southwest Alabama) and probably possesses the greatest exploration potential of any lower Cretaceous formation.

Sligo or Pettet Formation and James Limestone of Pearsall Formation

These units were deposited during a transgression in areas lacking clastic influx. Oolitic, fossiliferous shelf limestones which become reefal basinward comprise the Sligo/Pettet in northeast Texas; shales and limestones occur in north Louisiana and grade into an arenaceous facies in Mississippi. Porous carbonates produce in northeast Texas and northwest Louisiana, while Sligo clastics are productive in the East Texas and Mississippi Salt Basins in settings similar to those of the Hosston.

Marine shale became dominant during the transgressive peak. A shale–limestone–shale 'sandwich' comprises the Pearsall; the James limestone is productive in northeast Texas and north Louisiana. Pearsall strata are usually minor producers outside of Texas.

Rodessa Formation and Paluxy Formation

Both formations are regressive-phase, basinward-thickening, progradational clastic wedges, separated by the Ferry Lake Anhydrite. Both grade into marine shales and shelf carbonate downdip.

The Rodessa, up to 300 m thick, is a major petroleum producer across the area. Deltaic sandstones form the major reservoirs east of the Sabine Uplift, while on and west of the Sabine, interbedded oolitic, bioclastic limestones and calcareous sandstones are productive. Salt domes, salt pillows, turtles and tilted fault blocks form the productive structures for both the Rodessa and the shallower Paluxy.

Up to 500 m thick, the Paluxy was deposited across the sub-province in coastal plain, transitional and shallow marine environments with several large deltaic depocentres. Paluxy deltaic sandstones, sometimes wave-modified into barrier complexes, form important oil and gas reservoirs, usually on structures similar to those productive from the Rodessa, although salt involvement appears to be less important. Both the Rodessa and Paluxy formations have exploratory potential in some areas.

Significant papers on the lower Cretaceous include Anderson (1979), Beebe (1968b), Davis & Lambert (1963), Denham (1962), Franscogna (1957), Galloway et al. (1983), Hughes (1968), Montgomery (1987a), Murray (1961, 1966), Nichols et al. (1968), Pearcy & Ray (1986), Perkins & Martin (1985), Rainwater (1968, 1971), Rogers (1968), Sheveport Geological Society (1968) and Woodbury et al. (1980).

The South Florida Basin sub-province

A complete lower Cretaceous sedimentary section is present in the South Florida Basin (see Figs 1 & 4); only the Sunniland formation is productive. The structurally undisturbed basin covers about 130 000 km^2, about 40% onshore, and contains a shallow marine, carbonate platform or shelf bounded on the southwest by the lower Cretaceous shelf-edge complex. Approximately 2700–3300 m of limestones, dolomites and anhydrites were deposited there during the lower Cretaceous in shallow, marine shelf environments. They overlie about 1000 m of possibly prospective dolomites, evaporites, limestones and clastics of upper Jurassic age (Wood River formation).

Lower Cretaceous Pre-Sunniland Strata

The South Florida Basin pre-Sunniland strata are considered to have some prospectivity, but are not well known due to low well density.
Bone Island Formation (lower Coahuilan). About 700 meters in thickness, the Bone Island is largely limestone and is overlain by anhydrite. *Pumpkin Bay Formation (upper Coahuilan).* A 30–500 m thick limestone, anhydrite and dolomite sequence, the Pumpkin Bay contains a 100 m thick zone with porosities up to 30% and total organic carbon values exceeding 3% toward the top. Oil shows are common. *Lehigh Acres Formation (lowermost Comanchean).* About 200 m of limestones, dolomites, anhydrites and a thin shale member. The Beacon Dolomite is most prospective, due to attractive reservoir characteristics and an overlying anhydrite.

Sunniland Formation

An anhydrite-carbonate-anhydrite 'sandwich' nearly 350 m thick contains the Sunniland. Anhydrites envelope the 100 m thick Sunniland carbonates, possibly even updip. Sunniland reservoirs occur in the upper portion as limestone–dolomite complexes originally deposited as bioclastic debris mounds and patch reefs within a shallow lagoon and tidal flat environment. Porosities range from 10 to 30% and permeabilities are usually in the 50 to 200 mD range. Argillaceous, organic limestones composing the lower Sunniland are believed to be the source of the 24°–26° API gravity crudes. The productive-zone facies grades into chalky limestones updip and into carbonate mudstones and anhydrites downship.

Fourteen fields are productive in the 'Sunniland Trend', which extends northwest–southeast for about 130 km in a belt 15–25 km. wide, and contains about 100 million barrels of recoverable oil. All fields are stratigraphic and depend upon the vertical relief penetration of the bioclastic debris mounds and patch reefs into overlying lime muds and anhydrite for closure. Leaching and dolomitization have improved permeability; original porosities, also enhanced, were good. Most fields discovered to date are small. The 'Felda complex', deposited in a somewhat higher energy environment, is the largest, with an ultimate of about 60 million barrels. Sunniland field, about 20 million barrels, is next largest. Exploration continues despite the problems associated with the Sunniland and other south Florida plays: drilling depths of 3700–5300 m, seismic quality, an unexciting wildcat success ratio and stringent environmental constraints.

Post-Sunniland Strata

Lower Cretaceous limestones, dolomites and anhydrites up to 1300 m thick are present, some of which may be prospective at greater depth offshore. Some 2600 m of younger limestones, chalks and occasional dolomites and anhydrites also occur, but are not considered to be prospective.

Significant papers on the South Florida Basin include Applegate & Lloyd (1985), Applegate & Pontigo (1984), Halley & Schmocker (1983), Mitchell-Tapping (1984, 1985, 1986) Montgomery (1987b), Murray (1961), Pearcy & Ray (1986), Rainwater (1986, 1971) and Tyler & Erwin (1976).

Upper Cretaceous production

Upper Cretaceous (Gulfian) strata in the greater Gulf Coast Basin (Fig. 4) were deposited in one great marine transgressive–regressive cycle, and range from 100 to 3000 m in thickness. The lower–upper Cretaceous boundary is marked by an angular unconformity representing removal of up to 800 m of strata over a large area in northeast Texas, south Arkansas, north Louisiana and much of Mississippi. Continuous deposition occurred only in the Coastal Salt Basin and isolated shelf/basin area locales. Sands (Woodbine, Tuscaloosa) deposited at the base of the transgressive sequence are among the Gulf Coast's most prolific producers. With continuing transgression and diminished sand influx, thinner, often tight sandstones interbedded with shales (Eagle Ford, Tokio, Eutaw) were laid down. Overlying thick sequences of marine shales, chalky limestones and chalks (Austin, Taylor, Selma) deposited during the peak of the great upper Cretaceous marine transgression comprise most of the section. Reef limestones are rare. As regression commenced, sandstone deposition (Navarro) returned, although marine shale, chalky limestone and chalk deposition (Gas Rock, Selma) persisted in some localities. Deposition continued uninterrupted into the Paleocene in most areas. See Fig. 9 for areas productive from the Gulfian.

Volcanism was common during the Gulfian, especially on the San Marcos Arch, Monroe Uplift and Jackson Dome. Locally important gas-productive bioherms (Gas Rock) developed on the crests of the Monroe and Jackson Uplifts as the Cretaceous ended. Differential subsidence resulting in more linear, graben-like configurations occurred in the North Louisiana and Mississippi Salt Basins.

Woodbine/Tuscaloosa group

This basal transgressive clastic sequence deposited above the widespread mid-Cretaceous unconformity is present from the San Marcos Arch to northern Florida. Termed Woodbine west of the Sabine Uplift and Tuscaloosa to the east, it is missing only on the Sabine and Monroe Uplifts and in the Rio Grande Embayment. An ashy, grey sandstone at the base is the single most important productive unit in the shelf/basin area. Sandstones with good reservoir qualities were deposited in alluvial, deltaic, interdeltaic and shallow marine environments as the sea transgressed. Grey marine shale, probably a source rock, is present everywhere above, and downdip from, the basal sandstone. As deposition usually occurred in coastal plain and deltaic environments, local and semi-regional unconformities are common and the largest fields are truncation traps. Many Woodbine/Tuscaloosa fields are stratigraphic traps, so exploratory potential remains high.

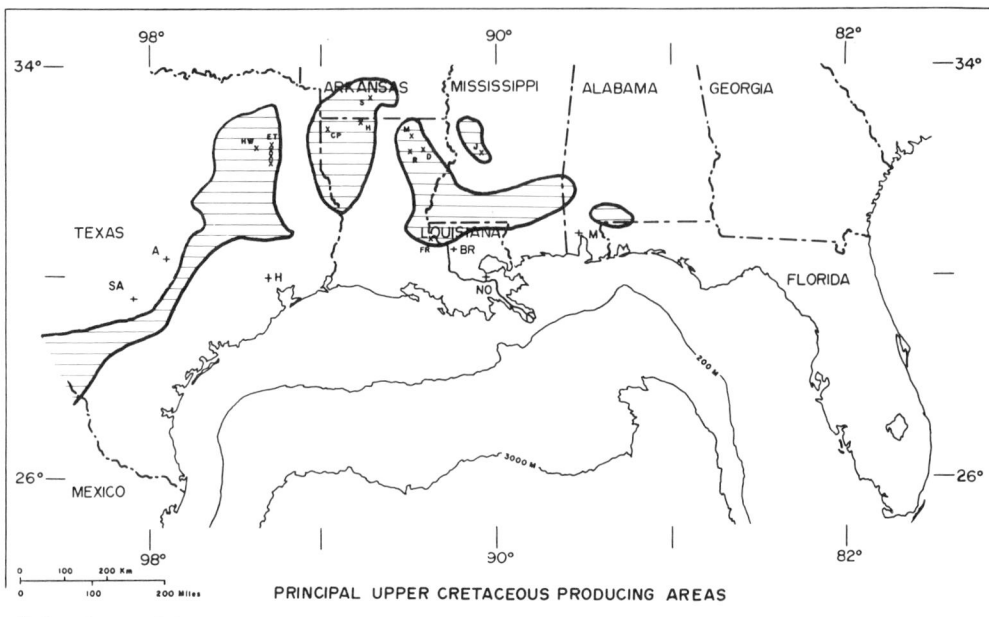

CP = Caddo-Pine Island, D = Delhi-W.Delhi, ET = East Texas, FR = False River
H = Haynesville, HW = Hawkins, J = Jackson, M = Monroe, R = Richland, S = Smackover

After: AAPG Mem.15 (1971), Rainwater (1968), Murray, et al (1985), Petroleum Frontiers v. 4, n. 2 (1987)

Fig. 9. Principal Upper Cretaceous Producing Areas. The Woodbine/Tuscaloosa sandstone is the principal reservoir sequence in all areas except in south Texas, where thin Navarro and Taylor sands and fractured Austin chalks produce oil; on the Sabine Uplift south flank (fractured oil-productive Taylor chalks); and on the crests of the Monroe (7+ TCFG) and Jackson uplifts, where Navarro bioherms are gas-productive. Woodbine/Tuscaloosa is generally oil productive, except at Richland and in the paralic-basin False River trend. Important fields include East Texas (5+ BBO), Hawkins (850+ MMBO) and Delhi (200+ MMBO), Powell (1972).

Tuscaloosa is also the oldest productive unit within the paralic Coastal Salt Basin. Sands crossed the Comanchean carbonate shelf-edge barrier in a 300 km long area in east–central Louisiana and sandstones with gross thicknesses of up to 200 m were deposited in a southerly thickening wedge on the previously-existing fore-reef continental slope. Several deep (5200 to 7000 m) major gas-condensate fields (False River trend) characterized by high abnormal pressures were discovered during the late 1970s. Due to declining gas prices, disappointing producing characteristics and the limited area within which reservoir quality sandstones are present, this once promising play had ended by the mid-1980s.

Woodbine/Tuscaloosa fields include truncation traps, stratigraphic traps, tilted fault block structures and structures resulting from salt movement. The East Texas field, with an ultimate reserve approaching 6 billion barrels, is a classic truncation trap on the west flank of the Sabine Uplift. Delhi is similar, with smaller reserves, on the south flank of the Monroe Uplift. Woodbine/Tuscaloosa fields on salt-related structures vary widely in size. Large-reserve structures formed above deep, non-penetrant salt diapirs or salt pillows. Smaller fields occur in salt-related tilted-fault-block structures.

Post-Woodbine/Tuscaloosa groups

While only the Woodbine/Tuscaloosa Group is a major producer of petroleum, marine shales of the Gulfian form one of the most important Gulf Coast source sequences. Sandstones in the Eagle Ford Group are occasionally productive in northeast Texas and northwest Louisiana and in the Mississippi Salt Basin (locally termed 'lower Eutaw'). Fractured chalks/chalky limestones (Austin Group) produce oil in usually uneconomic quantities on the flanks and crest of the San Marcos Arch. Similar fractured chalks in the overlying Taylor Group are productive on the Sabine Uplift south flank. Taylor sandstones are locally important producers in the Rio Grande Embayment and northeast Texas. The uppermost Navarro Group contains locally productive sandstones in the northeast Texas –

northwest Louisiana – south Arkansas area and in the Rio Grande Embayment, as well as the gas-productive bioherms on the Monroe Uplift and Jackson Dome. Shales and marls are the dominant lithotypes, however, in all post-Woodbine/Tuscaloosa groups except the Austin west of central Mississippi.

Significant papers on Upper Cretaceous production include Anderson (1979), Beebe (1968b), Davis & Lambert (1963), Denham (1962), Franscogna (1957), Galloway *et al.* (1983), Holcomb (1971), Montgomery (1987a), Murray (1961), Nichols *et al.* (1968), Perkins & Martin (1985), Powell (1972), Rainwater (1968), Roberts (1982), Sheveport Geological Society (1968), Steward (1981) and Woodbury *et al.* (1980).

Palaeocene

The Paleocene Midway Group (Danian) overlies the Gulfian; the contact is unconformable in some areas but gradational in most. Chalks, marls, shales and occasional sandstones comprise a relatively thin Clayton formation, which is conformably overlain by the Porters Creek formation. A dark marine shale, the Porters Creek ranges in thickness from 100 to 900 m. The lower portion of the Wilcox group, described below with the Eocene, was deposited in late Paleocene time as the first of several clastic sequences associated with pulses of the Larimian orogeny. The Clayton and Porters Creek are unproductive except for a few small Texas fields, and do not have exploratory potential.

Eocene production

Major changes of source areas and depositional regimes occurred in the greater Gulf Coast Basin during the Eocene. Uplift of the Cordilleran system in northern Mexico and the Rocky Mountain area supplied a vast influx of clastic materials to the Gulf Basin, especially in northeast Mexico and south Texas, with lesser thicknesses deposited to the northeast and east. Approximate Eocene thicknesses are 7000 m in south Texas, 3000 m in southwest Louisiana, 1100 m in southeast Mississippi, and range from 800 to 1200 m in the Interior Salt Basin area. Overall a great regressive to transgressive cycle, most of the Eocene exhibits an illusionary cyclic transgressive–regressive nature; the cycles are more a function of the volume of land-derived materials available than of sea-level fluctuation. The Eocene is unique among Gulf Coast series in that objective sections were deposited in both the shelf/basin and paralic basin areas. The shelf basins were filled to grade during the Eocene as differential vertical movement in the shelf/basin area diminished.

Wilcox group

Although Eocene strata are exposed at the surface over a large area, numerous shallow Eocene fields produce oil from alluvial, deltaic and transitional sands in southwest Mississippi, central Louisiana and south Texas in shelf/basin areas landward from the lower Cretaceous shelf-edge (Fig. 10), usually from stratigraphic traps. Average field size is small in this 'updip Wilcox' trend, a very mature area with declining activity.

Thick deltaic, interdeltaic and shallow marine sands occur over a large paralic basin area immediately basinward from the Comanchean shelf-edge. Numerous large Eocene fields, mainly gas productive, occur in the Rio Grande and Houston Embayments and in the area in between. Continental and brackish, often swampy, delta-plain environments were dominant; lignites are common. These strata are interbedded with inner neritic shales. The 'downdip Wilcox' thickens rapidly basinward across numerous down-to-basin growth faults, and becomes more marine. Massive blanket sands occur within the marine shale section, but permeabilities are often low and erratic. Downdip Wilcox thicknesses exceed 2800 m in the Rio Grande Embayment but thin crossing Louisiana from 2000 to 700 m close to the Wiggins Uplift.

Claiborne group

Three productive sand intervals occur in Claiborne group strata overlying the Wilcox group (see Fig. 4).

Queen City and Reklaw sands. These are locally productive across coastal Texas, within the predominantly marine shale Cane River formation.

Sparta Formation. Sparta sands were deposited in deltaic/inter-deltaic-marsh environments and are interbedded with shales, silts and lignites. Abrupt lateral facies changes are common; the sands grade downdip into marine shales. A moderate oil producer in Mississippi, Louisiana and the Rio Grande Embayment, but poorly developed with minor productivity in between. One hundred to 500 m thick, thinner east of the Mississippi Embayment.

Cockfield or Yegua Formation. At the top of the Claiborne group, the Cockfield (Louisiana and east) or Yegua (Texas) is unconformably overlain by Jackson group marine shales. Very similar in nature and origin to the Sparta and similarly grades downdip into marine shales.

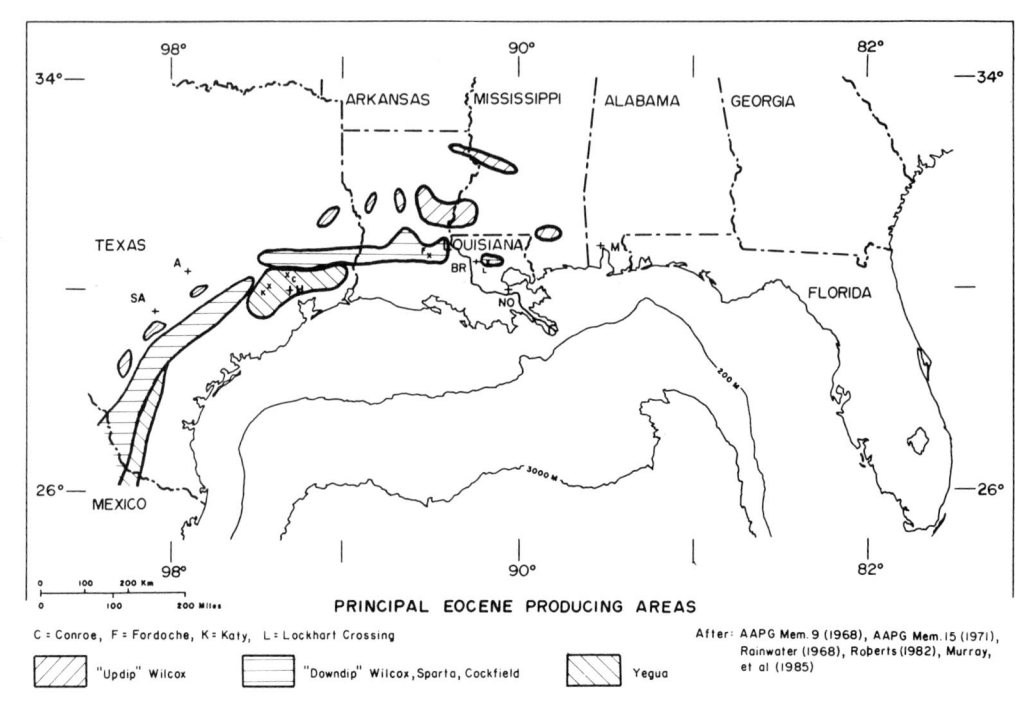

Fig. 10. Principal Eocene Producing Areas. Numerous small stratigraphic-trap oil fields occur in the 'Updip' Wilcox, Williams (1969). 'Downdip' Wilcox is a major gas-condensate producer in Texas, but productive quality, usually of oil, declines in Louisiana where Sparta and Cockfield production may be locally important. Yegua (Cockfield equivalent) sands are prolific producers' on salt domes in the Houston Embayment, where Katy (6+ TCFG) and Conroe (700+ MMBO) are major fields.

Two hundred to 700 m thick, thinner east of the Mississippi Embayment.

Cockfield/Yegua and Sparta sands are usually productive on the same structures as the underlying downdip Wilcox (Fig. 10). An exception exists in the Rio Grande Embayment where Yegua and Jackson sands are productive from multiple stratigraphic traps deposited in strandline environments. 'Downdip Wilcox' is an important producer of oil and especially gas across coastal Texas, and the overlying Yegua sands are important oil producers in the Houston Embayment; Yegua porosities and permeabilities are better than those of the Wilcox. Both penetrant and non-penetrant salt domes are the sites of major Yegua/Wilcox fields there. Katy, with 6 TCF of gas above a deep salt diapir, is the largest Eocene field in the area. In coastal Texas areas unaffected by salt diapirism (Figs 1 & 2), numerous gas-condensate fields of all sizes occur on faulted anticlines along down-to-basin regional growth faults; upthrown blocks, downthrown blocks, or both, may be productive. Some stratigraphic trapping occurs.

In southwest Louisiana, Eocene productive structures are similar to those in the Houston Embayment but usually trap smaller accumulations. Both the Wilcox and Claiborne thin east across south Louisiana. While the Sparta is sometimes productive here it does not compensate for the usually disappointing Wilcox, often non-marine and with low permeability. Some large non-penetrant salt dome fields in south Louisiana (Opelousas, Tepetate) produce from both Cockfield and Sparta, with the underlying Wilcox non-productive or sub-commercial. The Cockfield is productive in many more fields than is the Sparta, but is more likely to produce gas-condensate and is often thin as well. It is usually considered to be a secondary objective. However, as good quality 'deep' Wilcox oil fields (Fordoche, Lockhart Crossing) are occasionally discovered, high-risk 'deep' Wilcox exploration will continue.

Wilcox recovery factors are in the 300 to 400 barrel per acre-foot range for oil and 800 to 1000 MCF per acre-foot for gas. In the Yegua/Cockfield, recovery factors are higher, averaging 400 to 600 barrels or 1000 to 1500 MCF per acre-foot, respectively. The transitional and marine clays and shales of the Eocene are probably major hydrocarbon source rocks in areas where sufficient subsidence has occurred.

Significant papers on Eocene production include Beebe (1968b), Braunstein (1965), Campbell & Sheller (1964), Curtis & Picou (1978), Davis & Lambert (1963), Denham (1962), Franscogna (1957), Galloway et al. (1983), Harrison et al. (1970), Lafayette & New Orleans Geological Societies (1968), Lofton & Adams (1971), Murray (1961), Perkins & Martin (1985), Rainwater (1968), Roberts (1982), Sheveport Geological Society (1968), Williams (1969) and Winker (1982).

Oligocene production

Oligocene and Miocene are used herein in the conventional Gulf Coast sense (Fig. 4). Correlation of the Oligocene−Miocene boundary with the European type locality is not clear. The upper Oligocene Anahuac formation is believed to be roughly equivalent to the Aquitanian and Burdigalian, the mid-Oligocene Frio formation is tentatively correlated with the Rupelian and Chattian, and the lowermost Miocene with the Helvetian.

Oligocene strata are of economic importance only in the Tertiary paralic basin, with production limited to Texas and Louisiana (Fig. 11). As with the Eocene, the Oligocene section is thicker and more sandy along the Texas coast than in Louisiana, in both the Rio Grande (5300 m) and Houston (4000 m) Embayments, as well as across the San Marcos Arch. Thinning east across south Louisiana, the Oligocene is generally non-productive and composed of shale, marl and limestone east of the Mississippi River.

Two groups, the Vicksburg (lower Oligocene) and the Catahoula comprise the Gulf Coast Oligocene; the Frio (mid-Oligocene) and Anahuac (upper Oligocene) formations make up the latter. The Vicksburg/Frio represent a transgressive−regressive cycle, while the maximum Gulf Coast Tertiary marine transgression occurred during the Anahuac. Frio sands are highly productive, but the largely shale Vicksburg and Anahuac strata are objective only locally.

Frio Formation

The Frio contains the most important and prolific Gulf Basin objectives in coastal Texas. This productive trend extends into south Louisiana as well, but there the Miocene is the single most prolific Gulf Coast sequence. Deltaic, delta fringe, strand plain and shallow marine sands interbedded with marine shales were deposited in tectonically-active settings in southeast Texas and southwest Louisiana characterized by contemporaneously-developing diapiric structures and growth faults. Growth-fault-associated thickening and structural development contemporary with deposition occurred in non-diapiric areas of south Texas as well.

In Texas, the Frio thickens rapidly to 2000 + m across the Vicksburg and Frio growth fault systems (or 'flexures'), both downthrown toward the basin. Two-fold to six-fold section expansion is not unusual. Large-scale shelf-edge slumping and gravity sliding apparently initiated both systems. In Louisiana, the fault zones are not as well defined, with a larger number of faults being spread-out over a broader area, but produce sections up to 3300 m thick. Frio strata were deposited in deltaic environments in the Rio Grande and Houston Embayment areas, in laterally-shifting deltaic and delta-fringe settings with extensive inner and mid-shelf bar development in south−central Louisiana, and in non-deltaic strandplain environments in the areas in-between. The south−central Louisiana Frio is thicker but less sandy than in Texas; it thins rapidly farther east to a hundred metres and becomes non-productive and calcareous. Numerous large fields occur downthrown to major faults: in deltaic strata in the two Texas embayments, in the nearshore and marine bars in south−central Louisiana (deltaic strata are largely upthrown and less productive), and in stacked barrier sequences up to 700 m thick in the strandplain areas.

Anahuac Formation

Dark marine shales with minor local sands, sometimes productive, characterize the Anahuac in coastal Texas and southwest Louisiana; sand content is typically 15% or less. With diminished clastic influx, carbonate content of the Anahuac is the greatest of any Gulf Coast Cenozoic unit: the 'Heterostegina lime', which represents the transgressive culmination, persists west into the Houston Embayment. 'Het' reefs, usually non-productive, occur on the flanks of several penetrant salt domes in south−central Louisiana. Carbonate content increases to the east, as the section thins.

Where Anahuac or Vicksburg production exists, it usually occurs on structures productive from the Frio as well, or from similar structures (Fig. 11). Both Anahuac and Vicksburg production are more important in Texas than in Louisiana. In Louisiana, Anahuac production is

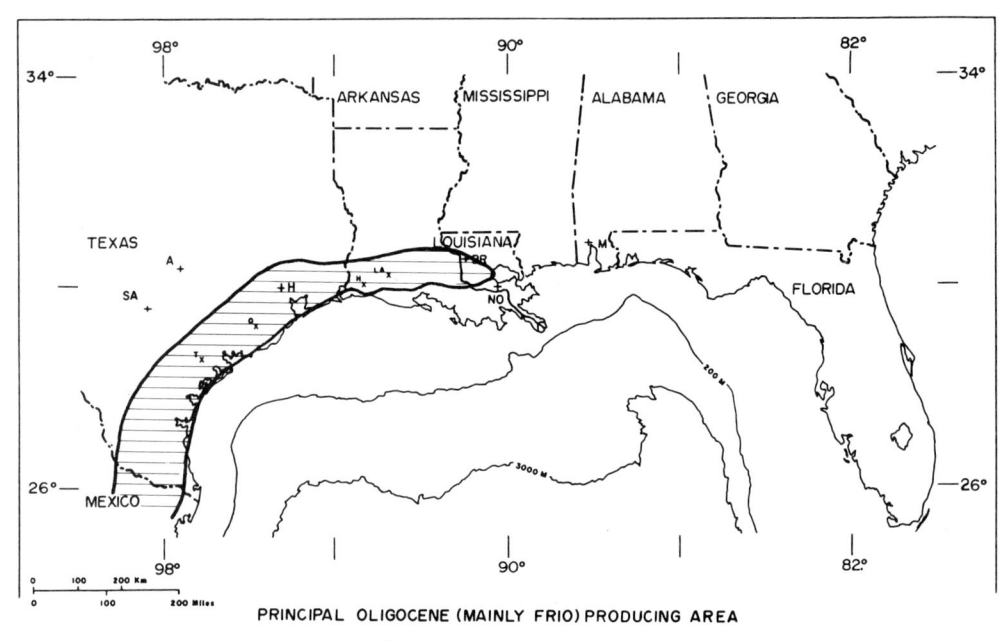

Fig. 11. Oligocene Producing Area. Locally important production is obtained from Vicksburg and Anahuac sands in Texas, but the Frio formation accounts for most Oligocene petroleum, both oil and gas-condensate, in the paralic basin. Major fields include Tom O'Connor (500+ MMBO), Mills (1970); Old Ocean (2+ TCFG), Halbouty (1968); Hackberry (200+ MMBO) and Lake Arthur (1+ TCFG).

limited to the updip production of the Frio productive area.

Most productive Oligocene structures formed contemporaneously with deposition in response to growth faulting and to movement of diapiric materials: salt and/or plastic clay. Oil and gas-condensate are produced from penetrant, semi-penetrant and especially non-penetrant diapiric structures and (in Texas) possible turtle structures. Most occur downthrown to down-to-basin growth faults; the structures are complex and heavily faulted as well. Exploratory potential therefore still exists. Productivity is high: estimated ultimate recovery from Frio salt dome fields in the Houston Embayment alone approaches 4 billion barrels. Smaller growth-fault-associated rollover anticlines and tilted fault block structures are important as well; both often result from salt or plastic clay movement.

Southwest of the Houston Embayment, most Oligocene production is obtained from large, simple, low-relief anticlines downthrown to the Vicksburg and Frio 'flexures'. Strandline sands thicken across the faults, are thickest close to the fault plane, and rollover anticlines develop. Movement of plastic clays may also play a role. Stacked Frio and Vicksburg barrier sands produce objective sections exceeding 1000 m in thickness. Some fields are large, e.g. Tom O'Connor with over 500 million barrels and 1 TCF, and Old Ocean with over 2 TCF (Fig. 11). Over 3 billion barrels have been produced from this trend. Frio exploratory potential exists downdip.

Significant papers on Oligocene production include Atwater & Forman (1959), Bornhauser (1960), Braunstein (1965, 1967), Campbell & Sheller (1964), Curtis & Picou (1978), Denham (1962), Galloway et al. (1982, 1983), Halbouty (1967, 1968), Harrison et al. (1970), Lafayette & New Orleans Geological Societies (1968), Martin (1969), McCormick & Kline (1983), Mills (1970), Murray (1961, 1966), New Orleans Geological Society Study Group (1962), Paine (1966, 1971), Perkins & Martin (1985), Rainwater (1968), Roberts (1982), Spillers (1965), Stanley (1970), Tipsword et al. (1971) and Winker (1982, 1984).

Miocene production

Renewed uplift of the Cordilleran System plus easterly tilting during the Miocene provided a vast, continuing supply of derital materials to

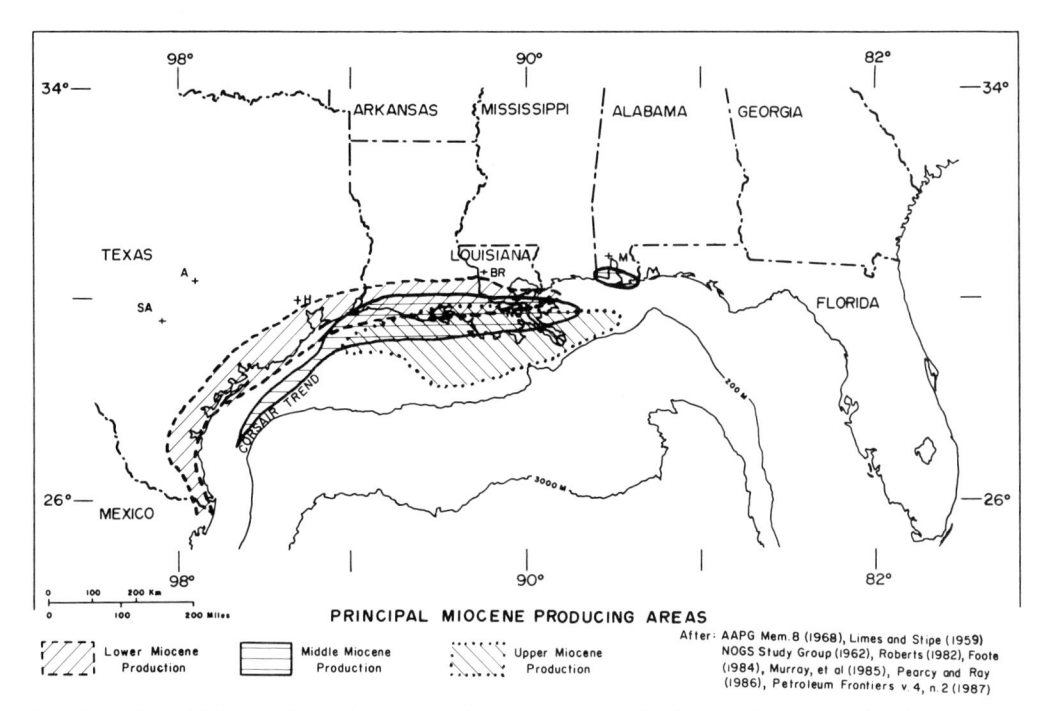

Fig. 12. Principal Miocene Producing Areas. Miocene strata contain the most important oil and gas-condensate production in Louisiana, and are locally important, usually for gas-condensate, in Texas. Shallow mid-Miocene dry gas production occurs in southwest Alabama.

the Gulf Coast Basin. Major river systems which migrated progressively farther east of the Eocene and Oligocene systems may have occupied the Mississippi Embayment by late Miocene times, where one remains today. Most Miocene strata of interest to petroleum explorationists were deposited in south and offshore Louisiana where the Miocene is the single most important producing unit (Fig. 12).

Miocene subsidence rates and associated growth fault activity were great in terms of both the number of faults and the amount of section expansion (occasionally up to 1 to 10) across individual faults. Nevertheless, subsidence was unable to accommodate the sedimentary influx. Rapid progradation into the basin occurred, with the shoreline migrating south approximately 100 km in south–central Louisiana. Gross Miocene progradation was to the southeast (Fig. 13), shifting from southwest to extreme southeast Louisiana. Maximum Miocene thicknesses must thus be estimated from many wells, as no lower, middle or upper Miocene 'type section' exists. The lower Miocene, best developed and thickest in southwest Louisiana (Fig. 13), is estimated to have a maximum aggregate thickness of about 5000 m, and to thin to about 2000 m in southeast Louisiana. Maximum middle Miocene thicknesses are estimated to range from 4000–6000 m, and the upper Miocene maximum may be 6000 to 6600 m in southeast Louisiana.

Lower, middle and upper Miocene as used herein are established on a pragmatic basis on the Gulf Coast and do not correlate with the classic Miocene type section. No groups or formations as such exist; division of the Miocene is based entirely on palaeontologic faunas or assemblages. 'Fleming formation' was once used on the Gulf Coast for the Miocene or Miocene plus Pliocene, but is mainly of historic significance. Reliable E-log correlations can be established locally, sometimes for tens of kilometres along strike but for considerably less in dip directions. Semi-regional and regional correlations are thus dependent upon paleontologic data. Semi-regional seismic correlation of stratigraphic markers is sometimes possible.

A well drilled at any location in the Louisiana Miocene or Pliocene will penetrate three depositional facies. Starting in predominantly sandy strata deposited in alluvial or delta plain environments, it then penetrates progressively older, interbedded sand and shale facies de-

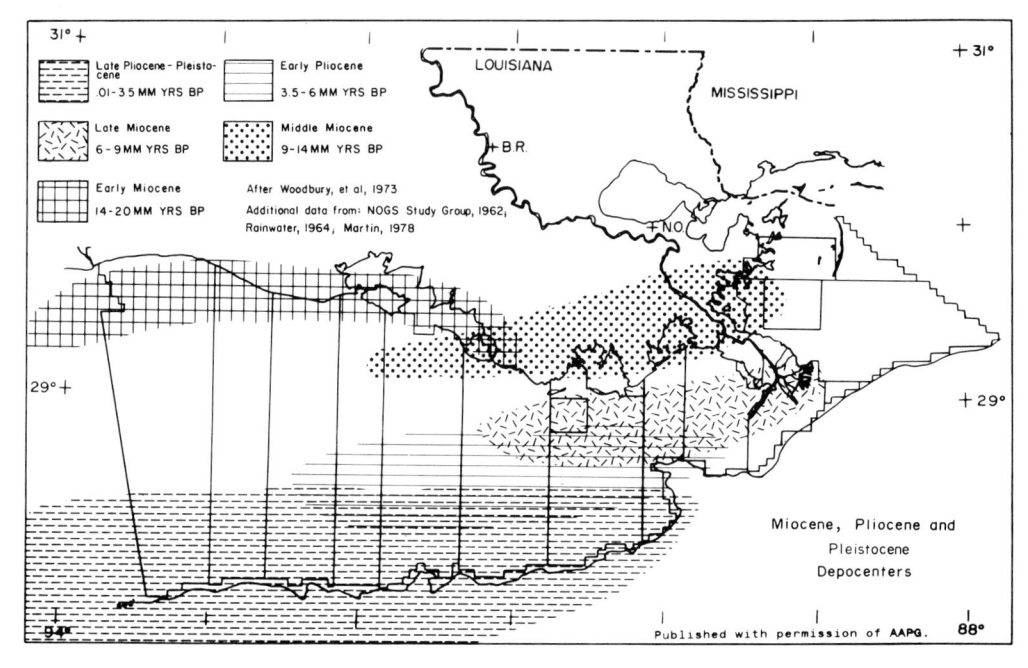

Fig. 13. Main Miocene, Pliocene and Pleistocene Depocentres, South Louisiana. Thinner objective strata occur outside of these main depocentre areas. Substantial southeasterly progradation occurred during the Miocene. After Woodbury et al. (1973), with revisions.

posited in delta-front, delta-fringe and, to a lesser degree, inter-deltaic environments in brackish and inner–middle neritic waters on the continental shelf. Still deeper a predominantly shale facies consisting of outer shelf shales containing occasional shelf sand-bars or sheets will be penetrated. Each of those three facies will usually be 1000 m or more in thickness. Deeper in the shale facies, continental slope clays and shales occur beneath the outer shelf sequence; occasional submarine canyon-fill and slope fan-delta sands occur as well. Slope strata are typically abnormally pressured; the sands may be good producers in some areas but lack commercially produceable hydrocarbons in others, depending on the nature of the abnormal pressure regime. The configuration, distribution and sedimentological nature of any individual depositional unit will be affected by important syndepositional influences: thickening across growth faults and diapiric and lateral displacement movement of plastic materials, both salt and high water content clays. Contemporaneous sedimentation, growth faulting, differential subsidence and diapiric movement interact in the formation of structures, traps and potential reservoirs and the subsequent entrapment and retention of hydrocarbons to transform a monotonously repetitive, uncorrelatable, unstable sedimentary mass into one of the world's great producing sequences.

Along the Texas coast, the Miocene is thinner, less deltaic and less affected by growth faulting and diapirism than in Louisiana. It is less productive as well. In the Mississippi, Alabama and west Florida coastal area, thin clastic Miocene strata were deposited in an area characterized by small rivers, slow deposition and limited subsidence. Nearly 30 shallow dry gas fields productive from 'middle' Miocene delta-destructive marine bars have been discovered onshore in southwest Alabama and in Mobile Bay since 1979. The fields are relatively small, but the wells and the 'bright-spot' exploration method employed are inexpensive. This play may eventually involve a much larger area.

Most Louisiana Miocene accumulations occur in deltaic/inner and middle-shelf interbedded sands (50% to 15%) and shales. Marine shales seal unconsolidated, well sorted sands with porosities usually ranging from 28% to 35% and permeabilities typically exceeding 1000 mD. Individual sand bodies may range from 3 to more than 40 m in thickness. Deeper middle and outer shelf sand-bars and sheets, usually less than 15 m thick, within a predominantly shale section are often productive as well, although porosities and permeabilities may be lower than with deltaic/inner and middle-shelf deposits.

Miocene accumulations are strongly structure oriented. Combination reservoirs are very common, but structure is nearly always more important than is stratigraphy. Stratigraphic pinchouts on the flanks of diapiric structures are common and often trap large accumulations. True stratigraphic traps not associated with diapiric structures (i.e. those on monoclines, regional noses, etc) are common as well but unimportant economically, as accumulations are typically small or absent. Structures resulting from movement of plastic materials form most Louisiana fields productive from the Miocene. Salt domes are common and form the most important fields, some of which will ultimately produce nearly 1 billion barrels. Clay-diapiric structures are much less common, but of importance locally. Numerous smaller fields produce from rollover anticlines downthrown to regional, down-to-basin growth faults; many probably formed in response to movement of plastic materials.

Lower Miocene production occurs along the southeast Texas coast, onshore and offshore, from structures similar to those productive in southwest Louisiana (Fig. 12). Additionally, the 'Corsair Trend', under the Texas continental shelf area downdip from the San Marcos Arch, has become an important gas producing area, with 2 TCF gas possibly available. Abnormally pressured mid-Miocene deltaic sands occur in rollover anticlines downthrown to a major, downthrown-to-basin listric growth fault system. Continued exploration in the Corsair trend will be price dependent, as costs are high in this play.

Significant papers on Miocene production include Atwater & Forman (1959), Braunstein (1965, 1967), Braunstein et al. (1973), Campbell & Sheller (1964), Curtis & Picou (1978), Foote (1984), Frey & Grimes (1970), Galloway et al. (1983), Halbouty (1967), Harrison et al. (1970), Johnson & Bredeson (1971), Lafayette & New Orleans Geological Societies (1968), Limes & Stipe (1959), McCormick & Kline (1983), Montgomery (1987a), Murray (1961, 1966), New Orleans Geological Society Study Group (1962), Pearcy & Ray (1986), Perkins (1961), Perkins & Martin (1985), Rainwater (1964, 1968), Roberts (1982), Shaub et al. (1984), Shinn (1971), Sloan (1971), Spillers (1965), Studlick et al. (1988), Thorsen (1964), Tipsword et al. (1971), Winker (1984) and Woodbury et al. (1973, 1980).

Pliocene production

Marine Pliocene strata produce natural gas and oil offshore Louisiana, usually from fields productive from either the Miocene or Pleistocene as well (Fig. 14). The Pliocene is much thinner than the Miocene, ranging in thickness from about 1400 m off southeast Texas to more than 2600 m off Louisiana. The strong basinward progradation of the Miocene did not continue during the Pliocene, as the main depocentre shifted laterally southwest from that of the upper Miocene (Fig. 13). This may be the result of both deeper waters occurring over the area of the present continental shelf than were present in late Miocene times due to a long period of Pliocene marine transgression, and the presence of uplifted ridges in the shelf area which resulted from load-induced large-scale lateral displacement of Miocene sedimentary materials toward the basin. These ridges may have limited Pliocene basinward sediment transport, with the main Pliocene depocentre ponded behind the ridges (Shaub et al. 1984; Reed 1988).

Pliocene strata were deposited in laterally shifting, delta-dominant systems, predominantly in continental shelf environments and are similar to the Miocene. Pliocene production is strongly structure-oriented and is found on the same types of structures (often the same structures) as Miocene or Pleistocene production.

Significant papers on Pliocene production include Braunstein (1965, 1967), Braunstein et al. (1973), Foote (1984), Frey & Grimes (1970), Lafayette & New Orleans Geological Societies (1968), Martin (1969), McCormick & Kline (1983), Pearcy & Ray (1986), Perkins & Martin (1985), Rainwater (1968), Roberts (1982), Sacks (1973), Studlick et al. (1988), Shinn (1971), Winker (1982, 1984) and Woodbury et al. (1973).

Pleistocene production

Natural gas and oil are produced in commercial quantities from marine Pleistocene strata offshore Louisiana and southeast Texas; the Pliocene and Pleistocene are of about equal importance as producing units, but Pleistocene potential is greater. Much of the current exploration activity in the Gulf is in areas with Pleistocene–Pliocene objectives.

Both distribution and deposition of Pleistocene statra were profoundly influenced by glacio-eustatic fluctuations, some of which exceeded 200 m. Exposure of the late-Pliocene continental shelf during the Nebraskan glacial period lead to removal of varying thicknesses of upper Pliocene strata. Precise definition of the Pliocene–Pleistocene boundary remains in dispute. Channels and canyons crossed the exposed continental shelf during glacial periods, with deltas restricted to a narrow band along the

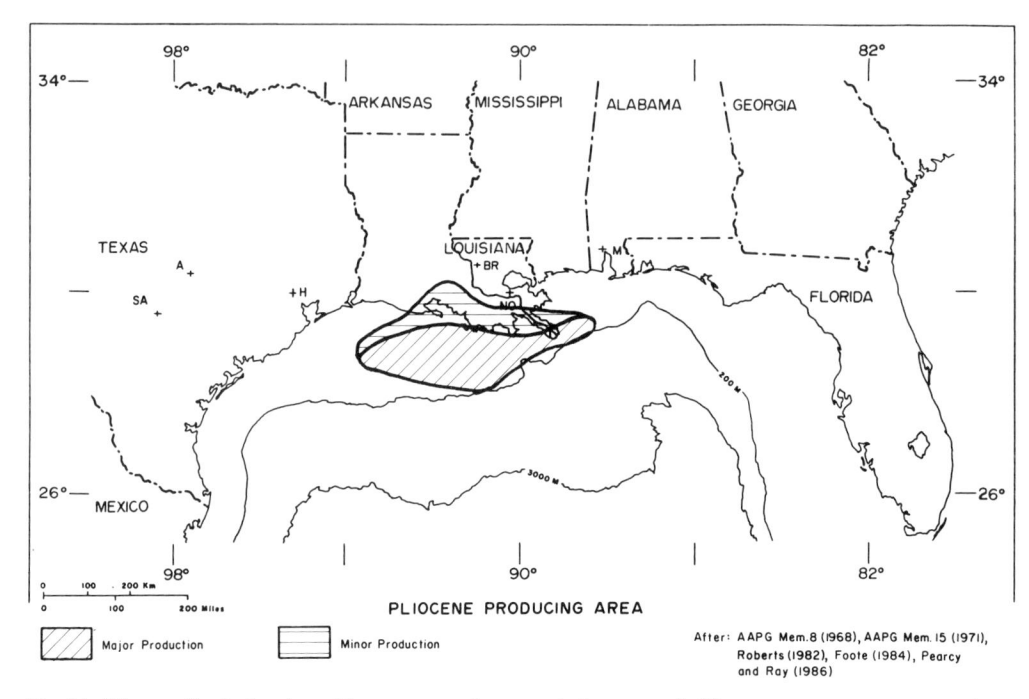

Fig. 14. Pliocene Producing Area. Numerous small accumulations occur in Pliocene strata on structures where the Miocene is the principal producer. Important Pliocene and Pleistocene production often occurs in the same fields on the middle and outer shelf.

shelf break. Turbidite, submarine canyon and fan delta deposition on the continental slope and abyssal plain were important during low sea-level periods. Deltaic sedimentation was restricted to inner shelf environments during interglacial, high sea-level periods, when relatively deep marine transgressions submerged the entire shelf area. Part of each up-dip section eroded during the subsequent glacial period. Thus, while hydrocarbons are found in Pleistocene sands across the present inner and middle shelf and onshore in southeastern Louisiana (Fig. 15), the bulk of Pleistocene reserves occur under the present middle and outer shelf and upper slope. Given the nature of Pleistocene deposition, thickness is widely variable, ranging from less than 700 m onshore to possibly more than 6700 m in salt-withdrawal basins within the main depocentre under the present outer shelf–upper slope. The main producing areas on the middle and outer-shelf are off southeast Texas and southwest Louisiana. Well developed sands occur there in both glacial and interglacial period sediments and natural gas is dominant. In the outer shelf/upper slope producing area, termed the 'Flexure Trend', most discoveries to 1988 have been made in an area offshore Louisiana where oil is the principal product. As inter-glacial strata on the shelf in this area are mainly shale, most of the sands were deposited during glacial intervals in local upper slope depocentres formed in response to salt and plastic clay movement. Turbidite and mass-wasting deposits are common. Pliocene sands are productive in this 'Flexure Trend' setting as well.

Pleistocene production on the middle- and outer-shelf occurs in the same types of structural settings as Miocene and Pliocene strata, and generally in fields with Miocene and/or Pliocene production as well. Most formed in response to movement and diapirism of salt and plastic clay, to growth faulting, or to both. In the 'Flexure Trend', an area of intense, episodic salt movement, the salt structures are much larger than whose under the shelf, forming ridges, massifs and walls, as well as individual salt domes. Laterally extruded salt tongues also occur. Very little has been published on the 'Flexure Trend', where fields are being developed in waters from 200 to 600 m deep; drilling depths range from 2700 to 6000 m. As with the 'Corsair Trend', continuing exploration and development will be price dependent.

Significant papers on Pleistocene production include Braunstein *et al.* (1973), Foote (1984),

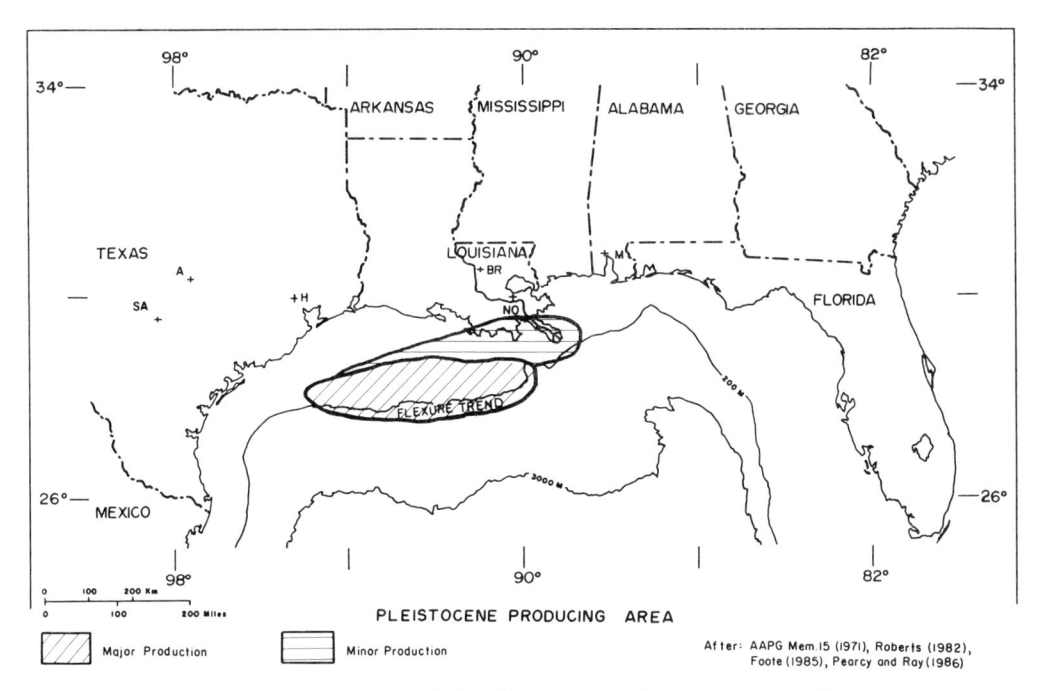

Fig. 15. Pleistocene Producing Area. As with the Pliocene, minor Pleistocene production occurs on many structures mainly productive from older strata. Gas-condensate production dominates on the shelf off southwest Louisiana and southeast Texas, while the shelf-break, 'Flexure Trend' area is oil-dominant.

Humphris (1979), Lafayette & New Orleans Geological Societies (1968), Pearcy & Ray (1986), Powell & Woodbury (1971), Reed (1988), Roberts (1982), Sacks (1973), Shaub et al. (1984), Winker (1982, 1984) and Woodbury et al. (1973).

Future exploration possibilities

Much of the greater Gulf Coast Basin appears to have reached exploratory maturity. Yet, activity has continued at a relatively high rate, albeit reduced during the mid-1980s slump. Many experienced Gulf Coast explorationists are confident that exploration and development will continue and will be rewarding in the future. This is due to a combination of prolific production and complex geology and to the continuing extension of operations into deeper waters. As both represent high-cost petroleum, the level of future activity will be price dependent to a large extent, although strategic considerations may play a role as well.

Development drilling and secondary/tertiary recovery operations will continue at a high level in many producing areas. Exploration is expected to continue or to increase as discussed below.

(1) Outer shelf-upper slope new field exploration in the Pleistocene and Pliocene will continue, will be extended along strike and will expand into deeper waters. As Miocene objective sands may be present over a much larger area than originally believed, new Miocene exploration may be undertaken as well, especially beneath salt tongues. Salt tongues also occur in older Coastal Basin strata.

(2) The offshore Rio Grande Embayment area is lightly explored. Jurassic, Cretaceous and Miocene strata are prospective in the lightly explored Mafla offshore, partly in waters of more than 200 m. Exploration is expected to continue on the shelf west of peninsular Florida, probably at a slow rate.

(3) If a better cost-return relationship develops, exploration for smaller fields (<20 million barrels equivalent) will commence in inner and middle continental shelf areas.

(4) Miocene and Oligocene producing areas onshore in Louisiana and Texas may see increasing field-extension exploration. Mature and semi-mature fields on extremely geologically complex diapiric structures have continued to be explored, with a recent consistent 1 in 3 success ratio. Deep exploration of the downdip Frio and lower and middle Miocene (plus

the Yegua and Wilcox in Texas) for outer shelf and upper slope sands will continue and may increase. Deep Wilcox new field and field extension exploration may continue in south Louisiana. Stratigraphic trap exploration in all trends should increase, seeking to repeat the widespread success in the Tuscaloosa/Woodbine.

(5) In the shelf/basin area, successful exploration of shelf-facies Tuscaloosa should continue and may expand. Most accumulations are stratigraphic, but structures sometimes occur. Lower Cretaceous and Cotton Valley exploration in the East Texas and Interior Salt Basins and adjacent areas offer promise as deep drilling in many areas is surprisingly limited. Exploration for lower Cretaceous and Smackover reefs (barrier and pinnacle) has been disappointing but on a relatively small scale except in south Texas. The Smackover play is relatively mature in presently productive areas but has exploratory potential in Texas, Alabama and west Florida. Prior to the Mobile Bay deep gas discoveries, the Denkman facies of the Norphlet (a marine-reworked dune field facies) was at best a secondary objective. It is now the primary objective of a fast-developing, albeit very deep, play offshore the Alabama coast. Additional areas with similar potential are suggested by palaeogeology and the play could broaden.

(6) Exploration of Palaeozoic strata beneath the salt has occurred on a small scale, without success. Marine upper Palaeozoic rocks are present in some areas and systematic exploration will eventually be undertaken.

(7) A number of horizontally-drilled wells have been completed recently in fractured Austin chalk (south Texas). If this play proves to be economic, it will expand and may involve other Mesozoic fractured carbonates and tight sandstone reservoirs.

Conclusions

Early Mesozoic continental drift, crustal stretching and subsidence formed the Gulf of Mexico small-ocean basin, as well as several smaller basins on part of the surrounding shelf, in which evaporites were deposited until the Oxfordian. Strata of upper Jurassic and lower Cretaceous ages were deposited in and around the small shelf basins in a continental shelf environment and are composed of interbedded sandstones and shales with significant cyclic alternations to carbonates and shales. Salt movement and diapirism commenced before the end of the Jurassic and were important in formation of many petroleum-producing structures, especially in the East Texas and Interior Salt Basins. Additional petroleum traps are fault-related or stratigraphic. Broadscale uplift and erosion marked the end of the lower Cretaceous. The Woodbine–Tuscaloosa sandstones, near the base of the upper Cretaceous transgressive sequence, are among the most important Gulf Coast producing formations. Thick shales, marls, chalks and occasional sandstones comprise the remaining upper Cretaceous. Little is known of the Mesozoic strata in the paralic basin which rings most of the Gulf of Mexico small-ocean basin.

Large-volume deposition shifted to the paralic basin during the Eocene. This was the last period of significant deposition in most of the shelf and basin area landward from the paralic basin. Tertiary and Quaternary uplift of the Cordilleran system provided a vast, continuous supply of detritus to the northwestern portion of the paralic basin, increasing in volume through time. Major depocentres shifted progressively northeasterly from the Rio Grande Embayment (Eocene) to southeastern Louisiana (upper Miocene) and have remained in the Louisiana continental shelf area since then. Rapidly prograding clastic strata, deposited in deltaic, inter-deltaic, shelf and slope environments dominate. These were strongly affected by contemporaneous diapirism and growth faulting, and numerous productive structures were formed.

A total resource (produced, proved, undiscovered) of about 60 billion barrels and 600 TCF appears to be present in an area about the size of onshore France plus the Low Countries! Why such prolificacy in a passive margin basin? Important factors appear to be as follows.

(i) Rapid, large volume deposition in rapidly-subsiding depocentres.
(ii) Rapid burial of potential source rocks with good organic content before extensive oxidation had occurred, in a 'warm' area.
(iii) Numerous large growth faults connect source and reservoir strata.
(iv) Tectonic activity in the greater Gulf Coast Basin area has been limited to gravity-influenced extension since the Jurassic. Gravitational instability in the form of salt and plastic clay movement, diapirism and lateral displacement, gravity sliding, differential subsidence, and growth faulting has occurred on a broad scale forming numerous productive structures.
(v) Individual structures formed in response to contemporaneous deposition, diapirism, and growth faulting. Traps on

penetrants, semi-penetrants and pillows were formed early, with most being preserved.
(vi) Reservoir sands usually are clean and well sorted, unconsolidated and un-cemented, and have high porosities and permeabilities.
(vii) High gas–oil ratios and active water drives are common, especially in the paralic basin.
(viii) While Gulf Coast productive structures are generally relatively small in area, they often form early, have multiple, high quality reservoirs, are prolific for their size and are numerous.

The discussions of geologic setting and salt diapirism are drawn in part from 'Diapirs and Diapiric Structures of South Louisiana', an AAPG Methods in Exploration Series publication now in preparation, as are Figs 2, 5 & 6. Figure 3 is from AAPG Bulletin vol. 63, no. 5. Permission by AAPG to publish these materials is gratefully acknowledged. A. Berthe is thanked for her assistance.

References[†]

ANDERSON, E. G. 1979. *Basic Mesozoic Study in Louisiana – The Northern Coastal Region and the Gulf Basin Province.* Louisiana Geological Survey, Folio Series, **3**.

APPLEGATE, A. V. & LLOYD, J. M. 1985. *Summary of Florida Petroleum Production and Exploration Onshore and Offshore, Through 1984.* Florida Geological Survey, Information Circular, **101**.

—— & PONTIGO, F. A. 1984. *Stratigraphy and Oil Potential of the Lower Cretaceous Sunniland Formation in South Florida.* Florida Geological Survey, Report of Investigations, **89**.

ATWATER, G. I. & FORMAN, M. J. 1959. Nature of Growth of Southern Louisiana Salt Domes and Its Effect on Petroleum Accumulation. *Bulletin of the American Association of Petroleum Geologists*, **43**, 2592–2622.

BEEBE, B. W. 1968a. Deep Edwards Trend of South Texas. *In*: BEEBE, B. W. & CURTIS, B. F. (eds) *Natural Gases of North America.* American Association of Petroleum Geologists, Memoir **9**, Vol. 1, 961–975.

—— 1968b. Natural Gas in Post-Paleozoic Rocks of Mississippi. *In*: BEEBE, B. W. & CURTIS, B. F. (eds). *Natural Gases of North America.* American Association of Petroleum Geologists, Memoir **9**, Vol. 1, 1176–1226.

BORNHAUSER, M. 1960. Depositional and Structural History of Northwest Hartburg Field, Newton County, Texas. *Bulletin of the American Association of Petroleum Geologists* **44**, 458–470.

BRAUNSTEIN, J. (ed.) 1965. *Oil and Gas Fields of Southeast Louisiana,* Vol. 1. New Orleans Geological Society.

—— (ed.) 1967. *Oil and Gas Fields of Southeast Louisiana,* Vol. 2. New Orleans Geological Society

——, HARTMAN, J. A., KANE, B. L. & VAN AMRINGE, J. H. (eds) 1973. *Offshore Louisiana Oil and Gas Fields.* New Orleans and Lafayette Geological Societies Vol. 1.

BUFFLER, R. T. & SAWYER, D. S. 1985. Distribution of Crust and Early History, Gulf of Mexico Basin. *Gulf Coast Association of Geological Societies Transactions*, **35**, 333–344.

CAMPBELL, J. C. & SHELLER, J. W. 1964. *Typical Oil and Gas Fields of Southwestern Louisiana,* Vol. 1 with Supplement. Lafayette Geological Society

CURTIS, D. M. & PICOU, E. B. 1978. Gulf Coast Cenozoic: A Model for the Application of Stratigraphic Concepts to Exploration on Passive Margins. *Gulf Coast Association of Geological Societies Transactions, Vol. 28*, Part 1, 103–120.

DAVIS, D. C. & LAMBERT, E. H. 1963. *Mesozoic–Paleozoic Producing Areas of Mississippi and Alabama,* Vol. 2 with Supplement. Mississippi Geological Society

DENHAM, R. L. (ed.) 1962. *Typical Oil and Gas Fields of Southeast Texas,* Horston Geological Society

FAILS, T. G., O'BRIEN, G. D., HARTMAN, J. A., GREENE, E. F., & GILREATH, J. A. 1990 *Diapirs and Diapiric Structures of South Louisiana.* American Association of Petroleum Geologists Methods in Geology Series.

FOOTE, R. Q. (ed.) 1984. *Summary Report on the Regional Geology Petroleum Potential, Environmental Consideration for Development, and Estimates of Undiscovered Recoverable Oil and Gas Resources of the United States Continental Margin in the Area of Proposed Oil and Gas Lease Sales Nos. 81 and 84.* US Geological Survey Open-File Report 84–339.

FRANSCOGNA, X. M. 1957. *Mesozoic–Paleozoic Producing Areas of Mississippi and Alabama,* Vol. 1. Mississippi Geological Society

FREY, M. G. & GRIMES, W. H. 1970. Bay Marchand – Timbalier Bay – Caillou Island Salt Complex, Louisiana. *In*: HALBOUTY, M. T. (ed.) *Geology of Giant Petroleum Fields.* American Association of Petroleum Geologists, Memoir **14**, 277–291.

GALLOWAY, W. E., HOBDAY, D. K. & MAGARA, K. 1982. Frio Formation of the Texas Gulf Coast Basin – Depositional Systems, Structural Framework, and Hydrocarbon Origin, Migration,

---

[†] In addition to these references, the Geological Surveys of Alabama, Arkansas, Florida, Louisiana and Mississippi, as well as the Texas Bureau of Economic Geology and the U.S. Geological Survey and Minerals Management Services are sources of numerous publications of fair-to-excellent quality dealing with the petroleum geology of these respective states. The Transactions of the Gulf Coast Association of Geological Societies, published annually since 1950, are the single most important collections of exploration oriented Gulf Coast reference materials. Local geological society publications are often useful and of high quality.

Distribution and Exploration Potential. *Texas Bureau of Economic Geology Reports of Inv.*, **122**.

GALLOWAY, W. E., EWING, T. E., GARRETT, C. M., TYLER, N. & BEBOUT, D. G. 1983. *Atlas of Major Texas Oil Reservoirs*. Texas Bureau of Economic Geology

HALBOUTY, M. T. 1967. *Salt Domes — Gulf Region, United States and Mexico*. Gulf Publishing Co.

—— 1968. Old Ocean Field, Brazoria and Matagorda Counties, Texas. *In*: BEEBE, B. W. & CURTIS, B. F. (eds). *Natural Gases of North America*. American Association of Petroleum Geologists, Memoir **9**, Vol. 1, 295–305.

HALLEY, R. B. & SCHMOCKER, J. W. 1983. High Porosity Cenozoic Carbonate Rocks of South Florida: Progressive Loss of Porosity with Depth. *Bulletin of the American Association of Petroleum Geologists*, **67**, 191–200.

HARRISON, F. W., JONES, R. K. & SEARLES, L. C. 1970. *Typical Oil and Gas Fields of Southwestern Louisiana*, Vol. 2. Lafayette Geological Society

HOLCOMB, C. W. 1971. Hydrocarbon Potential of Gulf Series of Western Gulf Basin. *In*: CRAM, I. H. (ed). *Future Petroleum Provinces of the United States — Their Geology and Potential*. American Association of Petroleum Geologists, Memoir **15**, Vol. 2, 887–900.

HUGHES, D. J. 1968. Salt Tectonics as Related to Several Smackover Fields Along the Northeast Rim of the Gulf of Mexico Basin. *Gulf Coast Association Geological Societies Transactions*, **18**, 320–330.

HUMPHRIS, C. C. 1979. Salt Movement on Continental Slope, Northern Gulf of Mexico. *Bulletin of the American Association of Petroleum Geologists*, **63**, 782–798.

JOHNSON, H. A. & BREDESON, D. H. 1971. Structural Development of Some Shallow Salt Domes in Louisiana Miocene Productive Belt. *Bulletin of the American Association of Petroleum Geologists*, **55**, 204–226.

KEAHEY, R. A. 1968. Fashing Field, Atascosa — Karnes Counties, Texas. *In*: BEEBE, B. W. & CURTIS, B. F. (eds). *Natural Gases of North America*. American Association of Petroleum Geologists, Memoir **9**, Vol. 1, 976–981.

LAFAYETTE AND NEW ORLEANS GEOLOGICAL SOCIETIES 1968. Geology of Natural Gas in South Louisiana. *In*: BEEBE, B. W. & CURTIS, B. F. (eds) *Natural Gases of North America*. American Association of Petroleum Geologists, Memoir **9**, Vol. 1, 376–581.

LIMES, L. L. & STIPE, J. C. 1959. Occurrence of Miocene Oil in South Louisiana. *Gulf Coast Association of Geological Societies Transactions*, Vol. **9**, 77–90.

LOFTON, C. L. & ADAMS, W. M. 1971. Possible Future Petroleum Provinces of Eocene and Paleocene, Western Gulf Basin. *In*: CRAM, I. H. (ed.). *Future Petroleum Provinces of the United States — Their Geology and Potential*. American Association of Petroleum Geologists, Memoir **15**, Vol. 2, 855–886.

MARTIN, G. B. 1969. The Subsurface Frio of South Texas; Stratigraphy and Depositional Environments as Related to the Occurrence of Hydrocarbons. *Gulf Coast Association of Geological Societies Transactions*, **19**, 489–501.

MARTIN, R. G. 1980. *Distribution of Salt structures in the Gulf of Mexico*. U.S. Geological Survey Map MF-1213.

MCCORMICK, L. L. & KLINE, R. S. (eds) 1983. *Oil and Gas Fields of Southeast Louisiana*, Vol. 3 with Supplement. New Orleans Geol. Soc.

MILLS, H. G. 1970. Geology of Tom O'Connor Field, Refugio County, Texas. *In*: HALBOUTY, M. T. (ed.) *Geology of Giant Petroleum Fields*. American Association of Petroleum Geologists, Memoir **14**, 292–300.

MITCHELL-TAPPING, H. J. 1984. Petrology and Depositional Environment of the Sunniland Producing Fields of South Florida. *Gulf Coast Association of Geological Societies Transactions*, **34**, 157–173.

—— 1985. Petrology of the Sunniland, Forty Mile Bend, and Bear Island Fields of South Florida. *Gulf Coast Association of Geological Societies Transactions*, **35**, 233–242.

—— 1986. Exploration Petrology of the Sunoco Felda Trend of South Florida. *Gulf Coast Association of Geological Societies Transactions*, **36**, 241–256.

MONTGOMERY, S. L. (ed.) 1987. Exploring the Eastern Gulf: The Case for Expansion. *Petroleum Frontiers*, **4**, No. 2. Petroleum Information Corp.

—— 1987. Success and Sensibility in South Florida. *Petroleum Frontiers*, **4**, No. 3. Petroleum Information Corp.

MOORE, T. 1983. Cotton Valley Depositional Systems of Mississippi. *Gulf Coast Association of Geological Society Transactions*, **33**, 163–167.

MURRAY, G. E. 1961. *Geology of the Atlantic and Gulf Coastal Plain Environment of North America*. Harper and Brothers.

—— 1966. Salt Structures of Gulf of Mexico Basin — A Review. *Bulletin of the American Association of Petroleum Geological*, **50**, 439–478.

——, RAHMAN, A. V. & YARBOROUGH, H. 1985. Introduction to the Habitat of Petroleum, Northern Gulf (of Mexico) Coastal Province. *In*: PERKINS, B. F. & MARTIN, G. B. (eds) *Habitat of Oil and Gas in the Gulf Coast*. Proceedings, Fourth Annual Research Conference, Gulf Coast Section, Society of Economic Paleontologists and Mineralogists, 1–24.

NEW ORLEANS GEOLOGICAL SOCIETY STUDY GROUP 1962. The Potential Down-Dip Limits of Production from the Oligocene and Miocene of Southeastern Louisiana. *Gulf Coast Association of Geological Societies Transactions*, **12**, 63–87.

NEWKIRK, T. F. 1971. Possible Future Petroleum Potential of Jurassic, Western Gulf Basin. *In*: CRAM, I. H. (ed.) *Future Petroleum Provinces of the United States — Their Geology and Potential*. American Association of Petroleum Geologists, Memoir **15**, Vol. 2, 927–953.

NICHOLS, P. H., PETERSON, G. E. & WUESTNER, C. E. 1968. Summary of Subsurface Geology of Northeast Texas. *In*: BEEBE, B. W. & CURTIS, B. F. (eds) *Natural Gases of North America*. American Association of Petroleum Geologists, Memoir 9, Vol. 1, 982–1004.

PAINE, W. R. 1966. Stratigraphy and Sedimentation of Subsurface Hackberry Wedge and Associated Beds of Southwestern Louisiana. *Gulf Coast Association of Geological Societies Transactions*, **16**, 261–273.

—— 1971. Petrology and Sedimentation of the Hackberry Sequence of Southwest Louisiana. *Gulf Coast Association of Geological Societies Transactions*, **21**, 37–55.

PEARCY, J. R. & RAY, P. K. 1986. The Production Trends of the Gulf of Mexico: Exploration and Development. *Gulf Coast Association of Geological Societies Transactions*, **36**, 263–268.

PERKINS, H. 1961. Fault Closure-Type Fields, Southeast Louisiana. *Gulf Coast Association of Geological Societies Transactions*, **11**, 177–196.

PERKINS, B. F. & MARTIN, G. B. (eds). 1985. *Habitat of Oil and Gas in the Gulf Coast*, Proceedings, Fourth Annual Research Conference, Gulf Coast Section, Society of Economic Paleontologists & Mineralogists.

PILGER, R. H. 1981. The Opening of the Gulf of Mexico: Implications for the Tectonic Evolution of the Northern Gulf Coast. *Gulf Coast Association of Geological Societies Transactions*, **31**, 377–382.

PINDELL, J. L. 1985. Alleghenian Reconstruction and Subsequent Evolution of the Gulf of Mexico, Bahamas, and Proto-Caribbean. *Tectonics*, Vol. 4, 1–39.

POWELL, J. B. 1972. Exploration History of Delhi Field, Norteastern Louisiana. *In*: KING, R. E. (ed.) *Stratigraphic Oil and Gas Fields*. American Association of Petroleum Geologists, Memoir **16**, 548–558.

POWELL, L. C. & WOODBURY, H. O. 1971. Possible Future Petroleum Potential of Pleistocene, Western Gulf Basin. *In*: CRAM, I. H. (ed.), *Future Petroleum Provinces of the United States – Their Geology and Potential*. American Association of Petroleum Geologists, Memoir **15**, Vol. 2, 813–823.

RAINWATER, E. H. 1964. Regional Stratigraphy of the Gulf Coast Miocene. *Gulf Coast Association of Geological Societies Transactions*, Vol. 14, 81–124.

—— 1968. Geological History and Oil and Gas Potential of the Central Gulf Coast. *Gulf Coast Association of Geological Societies Transactions*, Vol. 18, 124–165.

—— 1971. Possible Future Petroleum Potential of Lower Cretaceous, Western Gulf Basin. *In*: CRAM, I. H. (ed.) *Future Petroleum Provinces of the United States – Their Geology and Potential*. American Association of Petroleum Geologists, Memoir **15**, Vol. 2, 901–926.

REED, C. J. 1988. Correlation of Cenozoic Sediments on the Gulf of Mexico OCS. *Oil and Gas Journal*, **86**, 54–59.

ROBERTS, W. H. 1982. Gulf Coast Magic. *Gulf Coast Association of Geological Societies Transactions*, **32**, 205–216.

ROGERS, R. E. 1968. Carthage Field, Panola County, Texas. *In*: BEEBE, B. W. & CURTIS, B. F. (eds) *Natural Gases of North America*. American Association of Petroleum Geologists, Memoir 9, Vol. 1, 1020–1057.

ROLLINS, T. W. 1982. Future Energy Invulnerability. *Gulf Coast Association of Geological Societies Transactions*, **32**, 3–15.

SACHS, J. B. 1973. Pleistocene – Pliocene Stratigraphy of the Louisiana Continental Shelf. *In*: BRAUNSTEIN, J. *et al*. (eds) *Offshore Louisiana Oil and Gas Fields*. New Orleans and Lafayette Geological Societies, **1**, 1–11.

SALVADOR, A. 1987. Late Triassic–Jurassic Paleogeography and Origin of Gulf of Mexico Basin. *Bulletin of the American Association of Petroleum Geologists*, **71**, 419–451.

SHAUB, F. J., BUFFLER, R. T. & PARSONS, J. G. 1984. Seismic Stratigraphy of Shelf and Slope, Northeastern Gulf of Mexico. *Bulletin of the American Association of Petroleum Geologists*, **68**, 1790–1802.

SHREVEPORT GEOLOGICAL SOCIETY 1968. Stratigraphy and Selected Gas-Field Studies of North Louisiana. *In*: BEEBE, B. W. & CURTIS, B. F. (eds) *Natural Gases of North America*. American Association of Petroleum Geologists, Memoir **9**, Vol. 1, 1099–1175.

SHINN, A. D. 1971. Possible Future Petroleum Potential of Upper Miocene and Pliocene, Western Gulf Basin. *In*: CRAM, I. H. (ed.) *Future Petroleum Provinces of the United States – Their Geology and Potential*. American Association of Petroleum Geological, Memoir 15, Vol. 2, 824–835.

SLOAN, B. J. 1971. Recent Developments in the Miocene Planulina Gas Trend of South Louisiana. *Gulf Coast Association of Geological Societies Transactions*, **21**, 199–210.

SPILLERS, J. P. 1965. Distribution of Hydrocarbons in South Louisiana by Types of Traps and Trends – Frio and Younger Sediments. *Gulf Coast Association of Geological Society Transactions*, **15**, 37–40.

STANLEY, T. B. 1970. Vicksburg Fault Zone, Texas. *In*: HALBOUTY, M. T. (ed.) *Geology of Giant Petroleum Fields*. American Association of Petroleum Geologists, Memoir **14**, 301–308.

STEWARD, D. B. (ed.) 1981. *Tuscaloosa Trend of South Louisiana*. New Orleans Geological Society.

STUDLICK, J. R. J., BRYANT, J. G., HARTMAN, J. A. & SHEW, R. D. (eds) 1988. *Offshore Louisiana Oil & Gas Fields*. New Orleans Geological Society, Vol. 2.

THORSEN, C. E. 1964. Miocene Lithofacies in Southeast Louisiana. *Gulf Coast Association of Geological Societies Transactions*, **14**, 193–201.

TIPSWORD, H. L., FOWLER, W. A. & SORRELL, B. J. 1971. Possible Future Petroleum Potential of Lower Miocene–Oligocene, Western Gulf Coast Basin. *In*: CRAM, I. H. (ed.) *Future Petroleum*

Provinces of the United States — Their Geology and Potential. American Association of Petroleum Geologists, Memoir **15**, Vol. 2, 836–854.

TYLER, A. N. & ERWIN, W. L. 1976. Sunoco–Felda Field, Hendry and Collier Counties, Florida. *In*: BRAUNSTEIN, J. (ed.) *North American Oil and Gas Fields*. American Association of Petroleum Geologists, Memoir **24**, 287–299.

VENTRESS, W. (ed.) 1984. *The Jurassic of the Gulf Rim*. Proceedings, Third Annual Research Conference, Gulf Coast Section, Society Economic Paleontologists & Mineral.

WILLIAMS, C. H. (ed.) 1969. *Wilcox Fields of Southwest Mississippi*. Mississippi Geological Society

WINKER, C. D. 1982. Cenozoic Shelf Margins, Northwestern Gulf of Mexico. *Gulf Coast Association of Geological Societies Transactions*, **32**, 427–448.

—— 1984. Clastic Shelf Margins of the Post-Comanchean Gulf of Mexico: Implications for Deep-Water Sedimentation. *In*: PERKINS, B. F. & MARTIN, G. B. (eds) 1984. Habitat of Oil and Gas in the Gulf Coast. Proceedings, Forth Annual Research Conference, Gulf Coast Section, Society Economic Paleontologists & Mineral., 109–120.

WOODBURY, H. O., MURRAY, I. B. & OSBORNE, R. E. 1980. Diapirs and Their Relation to Hydrocarbon Accumulation. *In*: MIALL, A. D. (ed.) *Facts and Principles of World Petroleum Occurrence*. Canad. Society Petroleum Geologists, Memoir **6**, 119–142.

——, ——, PICKFORD, P. J. & AKERS, W. H. 1973. Pliocene and Pleistocene Depocentres, Outer Continental Shelf, Louisiana and Texas. *Bulletin of the American Association of Petroleum Geologists*, **57**, 2428–2439.

Yates and other Guadalupian (Kazanian) oil fields, U. S. Permian Basin

D. H. CRAIG

PO Box 661, Littleton, CO 80160−0661, USA

Abstract: More than 150 oil and gas fields in west Texas and southeast New Mexico produce from dolomites of Late Permian (Guadalupian [Kazanian]) age. A majority of these fields are situated on platforms or shelves and produce from gentle anticlines or stratigraphic traps sealed beneath a thick sequence of Late Permian evaporites. Many of the productive anticlinal structures are elongate parallel to the strike of depositional facies, are asymmetrical normal to facies strike, and have flank dips of no more than 6°. They appear to be related primarily to differential compaction over and around bars of skeletal grainstone and packstone. Where the trapping is stratigraphic, it is due to the presence of tight mudstones and wackestones and to secondary cementation by anhydrite and gypsum. The larger of the fields produce from San Andres−Grayburg shelf and shelf margin dolomites. Cumulative production from these fields amounts to more than 12 billion bbl (1.9×10^9 m^3) of oil, which is approximately two-thirds of the oil produced from Palaeozoic rocks in the Permian Basin. Eighteen of the fields have produced in the range from 100 million to 1.7 billion bbl ($16-271 \times 10^6$ m^3). Among these large fields is Yates which, since its discovery in October 1926, has produced almost 1.2 billion bbl (192×10^6 m^3) out of an estimated original oil-in-place of 4 billion bbl (638×10^6 m^3). Flow potentials of 5000 to 20 000 bbl (800 to 3200 m^3) per day were not unusual for early Yates wells.

The exceptional storage and flow characteristics of the Yates reservoir can be explained in terms of the combined effects of several geologic factors: (1) a vast system of well interconnected pores, including a network of fractures and small caves; (2) oil storage lithologies dominated by porous and permeable bioclastic dolograinstones and dolopackstones; (3) a thick, upper seal of anhydrite and compact dolomite; (4) virtual freedom from the anhydrite cements that occlude much porosity in other fields which are stratigraphic analogues of Yates; (5) unusual structural prominence, which favourably affected diagenetic development of the reservoir and made the field a focus for large volumes of migrating primary and secondary oil; (6) early reservoir pressures considerably above the minimum required to cause wells to flow to the surface, probably related to pressures in a tributary regional aquifer.

This paper has two principal objectives. The first is to identify the major geologic controls which were responsible for the many large reservoirs of oil and gas found in shelf carbonates of Late Permian (Guadalupian [Kazanian]) age in the Permian Basin of west Texas and southeast New Mexico. The second is to describe the geologic attributes of Yates, one of the larger fields producing from these shelf rocks. Reservoir and production characteristics of the 18 largest Permian Basin fields producing from Guadalupian carbonate rocks, including the Yates Field, are summarized in Tables 1 & 2.

It will be shown that most of the large Guadalupian fields occupy the shelf areas of the Late Permian. They produce from shelf and shelf margin carbonates (mostly dolomites) of the San Andres and Grayburg formations which underlie a thick regional seal of anhydrite and compact silty dolomite.

Most of the reservoir structures are gentle anticlines with only 400 to 500 ft (120 to 150 m) of relief and flank dips of no more than 5 or 6°. Those located on the Central Basin platform tend to be asymmetrical and result largely from differential compaction over and around San Andres−Grayburg bars of shallow subtidal skeletal or oolitic grainstone. Dips over these bars appear to be consistently higher to the east or northeast, the presumed directions from which prevailing winds and waves approached during the Late Permian.

It also will be shown that the special characteristics of the Yates Field are the outcome of a succession of geologic events which conserved, and even enhanced, storage and flow capacity of the dolomites which make up the principal stratigraphic units of the reservoir. Other events produced the reservoir structure, effectively sealed the trap, provided the migrating oil needed to build a reserve of significant size, and established a regional aquifer with sufficient potential energy to produce much of the oil by natural flow.

From BROOKS, J. (ed.), 1990, *Classic Petroleum Provinces*, Geological Society Special Publication No 50, pp 249−263.

Table 1. *Characteristics of 18 major San Andres–Grayburg Fields, U.S. Permian basin*

No.	Field	Discovery Year[2]	Approximate Depth[2] (ft)	Temperature[1] (°F)	Oil Gravity[2] (°API)	Cumulative Oil[5] (MMBO)	Original Oil Column[1] (ft)
1	Wasson	1936	5000	106	31	1712	450
2	Slaughter-Levelland	1936; 1945	4900	–	30	1514	–
3	Yates	1926	1500	82	31	1172	450
4	Goldsmith	1935	4300	95	36	756	300
5	Seminole	1936	5200	108	37	525	262
6	S. Cowden-Foster	1932	5400	119	35	501	800
7	N. Cowden	1930	4400	114	35	488	800
8	McElroy	1926	2900	86	32	466	400
9	Vacuum	1929	4600	101	38	432	–
10	Hobbs	1928	4050	–	34	297	250
11	Sand Hills	1944	3200	86	33	248	500
12	Midland Farms	1944	4800	102	46	242	–
13	Means	1934	4400	100	29	229	230
14	Dune	1938	3300	88	34	183	800
15	Eunice-Monument	1929	3600	–	32	175	–
16	Maljamar	1926	3600	–	37	145	–
17	Jordan	1937	3700	88	35	129	300
18	Waddell	1927	3500	88	34	101	300

Principal Sources:
[1] Galloway et al. (1983)
[2] Mason Map Service Inc & International Oil Scouts Assoc. (1987)
[3] Bebout & Harris (1986)
[4] Longacre (1980)
[5] *Oil & Gas Journal* (30 January 1989)

Table 2.

No.	Synchronous Differential Compaction	Later Vertical Bending	Later Faulting	Unconformity	Stratigraphic Seal(s)	Remarks
1	++	– –	– –	+	+	
2	– –	– –	– –	– –	++	Complex of Stratigraphic Traps.
3	◊ +	++	++	++	+	
4	++	– –	– –	– –	– –	
5	++	– –	– –	– –	– –	
6	++	– –	– –	– –	– –	
7	++ (1)	– –	– –	– –	– –	
8	++ (1)(3)	– –	++ (4)	+	+	
9	++ (3)	++	– –	+	+	
10	++	++	– –	++	– –	
11	◊ ++	– –	– –	– –	– –	
12	– –	– –	– –	– –	– –	Apparent Basin Slope Carbonate
13	◊ ++	– –	+	– –	– –	Faulted at Basinward Margin?
14	◊ ++	– –	– –	– –	– –	
15	++	++	– –	– –	– –	
16	++	– –	– –	– –	++	
17	– –	– –	– –	– –	– –	
18	++	– –	– –	– –	– –	

Reservoir Strata Deformed By

\+ Minor } Inferred Importance
++ Major
◊ Anticlinal Axis Overlies Asymmetrical Ridge of Skeletal or Oolitic Grainstones

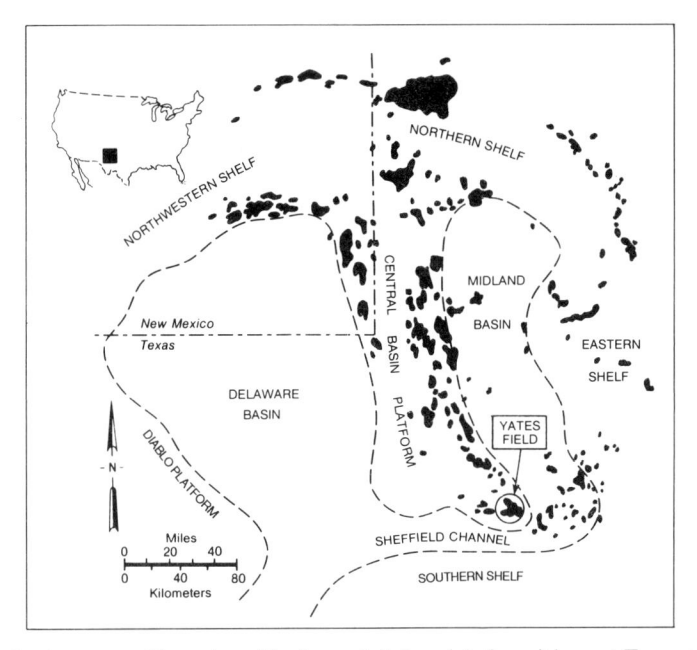

Fig. 1. Regional paleogeographic setting of basins and shelves (platforms) in west Texas and southeast New Mexico during the Late Permian. Black areas are oil and gas fields, including the Yates Field, which produce largely from carbonates of Guadalupian (Kazanian) age and are sealed above by evaporites of later Guadalupian age. (Modified from Shirley (1987). Reprinted with permission of the American Association of Petroleum Geologists.)

Late Permian paleogeography & lithofacies

The shallow Late Permian paleogeography of the Permian Basin (Fig. 1) and its deeper structural controls were produced by tectonic partitioning of a simpler, roughly circular early and middle Paleozoic cratonic depression, the Tobosa Basin (Galley 1958). Tobosa Basin stratigraphy is summarized in Fig. 2. Structural modification of the Tobosa Basin began during the Pennsylvanian (Late Carboniferous) Period with block faulting and ended during the Early Permian (Wolfcampian [Asselian–Sakmarian)] with thrust and wrench faulting (Hills 1984). The final 20 Ma of the Permian, which included the Guadalupian (Kazanian) Age, was a time of relative tectonic quiescence.

Regional patterns of sedimentation in the Permian Basin during Late Permian time were controlled by three major paleogeographic elements, the Delaware and Midland basins and the Central Basin platform (Fig. 1). These were framed by broad, ramp-like shelves or platforms which bordered the entire Permian Basin. Throughout Guadalupian time, the basins were dominated by siliciclastic sedimentation. The shallow marine environments of the Central Basin platform and other shelf areas controlled the areal distribution, the kinds, and the diagenesis of the carbonate sediments which now as dolomites comprise the reservoirs of the largest Permian Basin fields.

The two groups of regional lithofacies associated with Guadalupian sedimentation and oil production in the Permian Basin are shown in Fig. 2. Early and middle Guadalupian carbonates (Fig. 2a) are the principal stratigraphic units in the major oil and gas reservoirs of the Central Basin platform and other, adjacent, shelf areas. The late Guadalupian calcium sulfate evaporites of Fig. 2b, hundreds of feet thick across the Central Basin platform, provide the impervious seal overlying these same reservoirs.

There are more than 150 fields which produce, or have produced, from one or more of four contiguous stratigraphic units of early to middle Guadalupian age. The units from oldest to youngest, typified by those in the Yates Field section (Fig. 3), are the San Andres Formation (dolomite, minor limestone and clay shale), the Grayburg Formation (dolomite and siltstone), the Queen Formation (sandstone, siltstone, dolomite), and a sandstone facies of the lower Seven Rivers Formation (predominantly anhydrite and dolomite). In many San Andres–

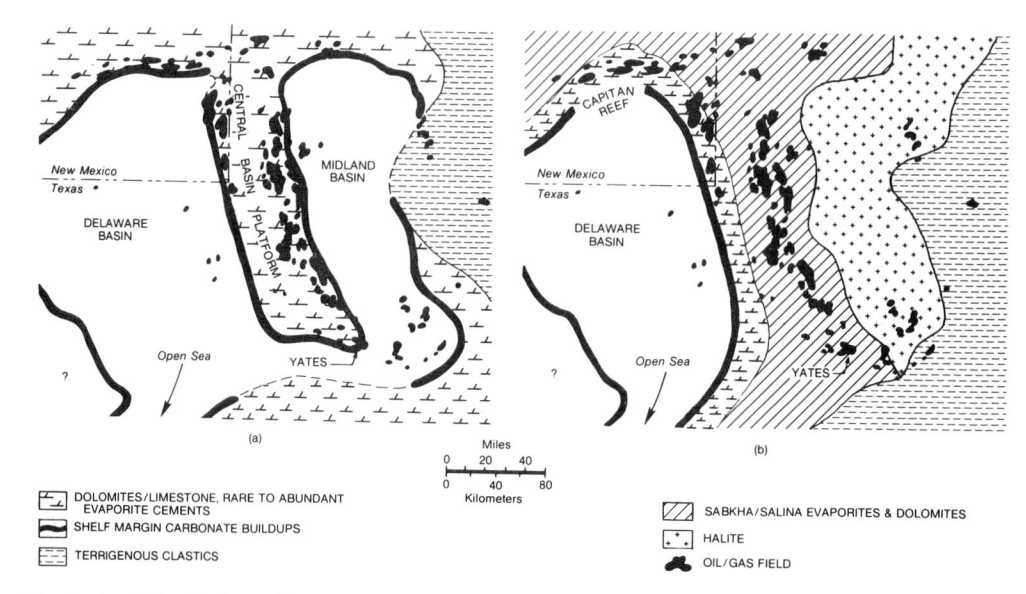

Fig. 2. Areal distribution of San Andres–Grayburg shelf and shelf margin carbonates (a) and of slightly younger Seven Rivers evaporites (b) in the Permian Basin. The shallow water carbonates contain the principal reservoirs, and the overlying complex of sabkha and salina anhydrite and gypsum forms an effective seal on the reservoirs.

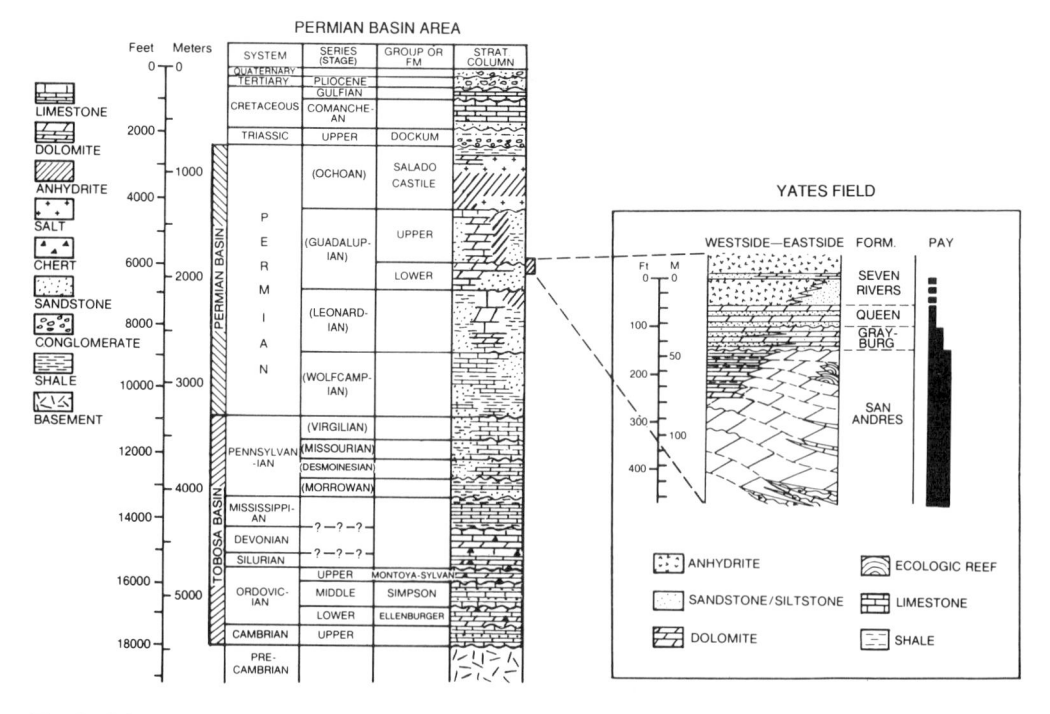

Fig. 3. Columnar sections showing the lithologies and formations that comprise the stratigraphic section regionally in the Permian Basin (left) and locally in the reservoir of the Yates Field (right). Virtually all of the series/stages shown in the regional column are, or have been, productive of oil or gas in the central area of the basin.

Grayburg fields, the Queen and lower Seven Rivers do not produce oil or are only marginally productive. The Yates Field produces from all four of the units, although as implied by the bar diagram of pay in Fig. 3, the best porosity and permeability in the reservoir are found in the San Andres and Grayburg dolomites. The two carbonate units are separated by a significant unconformity at Yates which has been recognized in other reservoirs, principally on the Central Basin platform (Tables 1 & 2).

On the platforms the four formations record a gradual upward change from dolomite (originally limestone) to silty dolomite and siltstone to evaporites. This sequence is interpreted as a transition from open marine to restricted marine conditions. Accompanying the changes in depositional and diagenetic environments was an increase in the amount of mud-supported carbonate and quartzose silt and sand. The platforms were broadened by basinward progradation of the carbonates at their margins. Accompanying this broadening was a gradual impairment of sea water circulation in the lagoons and across the mud flats of the platform interiors, such that salinities increased, and precipitation of calcium sulphates and sodium chloride took place beneath supratidal surfaces and in saline lakes. A result was widesrpead plugging of pores in adjacent carbonate sediments or the partial replacement of calcium carbonate by calcium sulphate.

In areas of shallow water like the Central Basin platform, thick sequences of shelf and shelf margin carbonates were deposited as limestone during Middle and Late Permian time. Subtidal bars composed of skeletal grains and fragments (now dolograinstone and dolopackstone) were deposited in areas of higher wave and current energy, chiefly at the platform margins. Shelfward and generally west of the bars were lagoons and tidal flats containing sediments dominated by lime mud and deposited under conditions of lower energy. These close associations of higher energy and lower energy carbonates in the San Andres–Grayburg are common across the eastern two-thirds of the Central Basin platform, and they have been recognized as shoaling upward or prograding sequences in wells with adequate log and core control (Garber & Harris 1986). Facies transitions from platform peritidal muddy carbonates to platform margin skeletal sand shoals to basinal slope muddy carbonates within the area of a single field have been documented for various San Andres–Grayburg platform reservoirs, as in Galloway *et al.* (1983), Garber & Harris (1986), Ruppel (1986), Craig (1988), and Fig. 4 in this paper. The relationship of these depositional facies to water depth and patterns

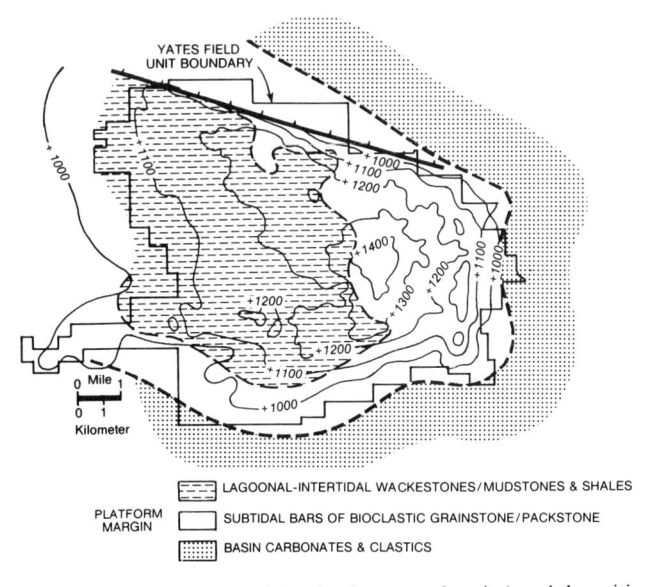

Fig. 4. Structure on the top of the San Andres dolomite (an unconformity) and depositional lithofacies that comprise the upper San Andres in the Yates Field area. Structure contours are in feet above sea level. As in some other large fields on the Central Basin platform, the San Andres carbonates in this reservoir contain both shelf and shelf margin facies in a prograding sequence.

of water movement over carbonate platform margins has been identified with low to moderate levels of wave and current energy (Wilson 1974, 1975). Except for effects of carbonate progradation at the platform margins and gradual constriction of the Midland basin by sediment filling and dessication, the areas occupied by shelves and basins in the Permian Basin remained fairly constant from Middle to Late Permian time.

Late Permian structure

San Andres–Grayburg reservoirs on the Permian platforms of west Texas and southeast New Mexico are thought to have developed structurally by a variety of processes. These include drape over deeply buried fault scarps, anticlines, and hills; differential compaction of lime mudstones and grainstones under conditions of shallow burial; normal and lateral (?) faulting; anticlinal folding related to fault movement; regional jointing; and local tectonic fracturing. Some of the reservoirs show clear evidence of several, or even all, of these structural processes. The Yates Field is one example. However, most of the reservoir structures appear to be best explained in terms of either drape over deep structural highs or differential compaction over and around bars of skeletal grainstone deposited at the platform margin. Hills (1984), Bebout (1986) and Ward et al. (1986) support the view that development of field structures during the Late Permian was largely by drape folding controlled by deep pre-Permian structural and palaeotopographic features. The author believes that differential compaction during and following deposition of shelf margin skeletal grainstone bars and associated lagoonal and intertidal mudstones was the dominant structure-forming process among San Andres–Grayburg oil reservoirs. The process appears to have been most common among the structures located on the Central Basin platform.

At the stratigraphic level of the San Andres–Grayburg on the Central Basin platform, field anticlines typically have only 400 to 600 ft (120 to 180 m) of relief over distances of 1 to 2 miles (2 to 3 km); that is, they have relatively low flank dips of 2 to 6°. Much of the relief is probably due to folding by differential compaction which took place during deposition of the San Andres–Grayburg carbonate sediments and continued, although with gradually diminishing effect, throughout deposition of the overlying Seven Rivers evaporites. This kind of superficial drape appears to have occurred as a response to the contrasting compactibilities of the mud-rich and grain-rich carbonate rocks which are associated with the shoaling upward prograding sequences recognized in many San Andres–Grayburg reservoirs on the platform. Galloway et al. (1983) called upon differential compaction as an explanation for a few San Andres–Grayburg structures on the Central Basin platform, but it may apply to many more.

The shape of San Andres grainstone bars in the Yates Field reservoir (they are recurved to the southwest) and the greater abundance of coralline algae in the northeast-facing portions of the bars suggest that prevailing winds blew across the Permian Basin seas from the northeast during the Late Permian. As a result of these winds, higher flank dips tended to be on the eastern and basinward, paleo-windward, side of the structures; that is, on the side occupied by bars of relatively incompactible skeletal grainstone. The highest points on the compaction structures overlie the bars. Lower dips to the west, shelfward and paleo-leeward of the bars, coincide with highly compactible muddy lagoonal and tidal flat deposits. Asymmetry of dip on the structures and the relation of structural apexes to underlying skeletal sands point to a clear genetic link between the structures and the lithofacies. For those fields that happen to lie at the present eastern edge of the Central Basin platform, asymmetry has been further accentuated by the steeper paleotopographic slope which marks the zone of facies transition from the shelf margin to the Midland basin (Galloway et al. 1983).

The relatively consistent compass directions of the asymmetry in San Andres–Grayburg structures is a characteristic which would seem most unlikely if the irregularities of fault direction and throw and the shapes and orientations of anticlines or hills at depth were controlling the shallow structure. Furthermore, the axes of elongation of individual field structures and the extension of these axes to form trends of several fields have compass bearings that are at least 30° from the trends of Late Pennsylvanian cross faults identified by Hills (1984) on the Central Basin platform. The trends of the fields probably represent lines of depositional strike at points of arrested progradation during the eastward outbuilding of the platform. It seems unlikely that faults or other structural features at depth and with radically different trends can be the main controls on the location and orientation of individual shallow fields, or on the alignment of groups of these fields along trends of probable depositional strike.

Faults in relation to San Andres–Grayburg

reservoir structure have been reported for only two fields on the edge of the Central Basin platform. The faults are associated with apparent basinward dips of from 6 to 22°. The first example is from the McElroy Field (Table 1, number 8) and is based on interpretation of a seismic section through the field. The section (Harris & Walker 1988) shows one major and several minor faults beneath the shallow structure, but these do not appear to extend to Permian depths or involve the rocks of the San Andres–Grayburg reservoir. On the other hand, Longacre (1980, 1986) interprets at least one major fault displacing the reservoir strata in McElroy. The second example is from the Yates Field (Fig. 4 and Table 1, number 3). Identification of a fault that limits oil production along the north flank of Yates is based on wireline logs and production data. The characteristics of the fault are described in a subsequent section of this paper.

The rarity of faults reported for San Andres–Grayburg structures may testify to the fact that direct evidence of their presence and their effects on San Andres–Grayburg carbonates is difficult to get from subsurface seismic and well data. Or they may simply not be present in significant numbers in the Late Permian, as claimed by Hills (1984) and others.

San Andres–Grayburg stratigraphic traps

About two-thirds of the fields with reservoirs in Permian rocks produce from stratigraphic traps or from combined structural–stratigraphic traps (Ward *et al.* 1986). It is uncertain whether this estimate would apply to San Andres–Grayburg reservoirs alone. Nevertheless, it appears that the abundance and erratic distribution of calcium sulphate cements in and around these reservoirs, over large areas of the Central Basin platform, has provided innumerable barriers to horizontal and vertical movement of oil. Some of these barriers have the size and spatial configuration needed to produce giant stratigraphic traps, as at Slaughter–Levelland Field (Table 1, number 2). Many more are irregular zones of local impermeability in what are otherwise essentially structural traps.

The apparent systematic distribution of major calcium sulphate cementation in the San Andres–Grayburg carbonates on the Central Basin and other platforms, and the likely role of the cements as controls on forming some traps, are shown in Figs 5 & 6. The map and the cross section demonstrate the areal expansion of calcium sulphate cementation from its local beginnings in the interior of the Central Basin platform during deposition of San Andres car-

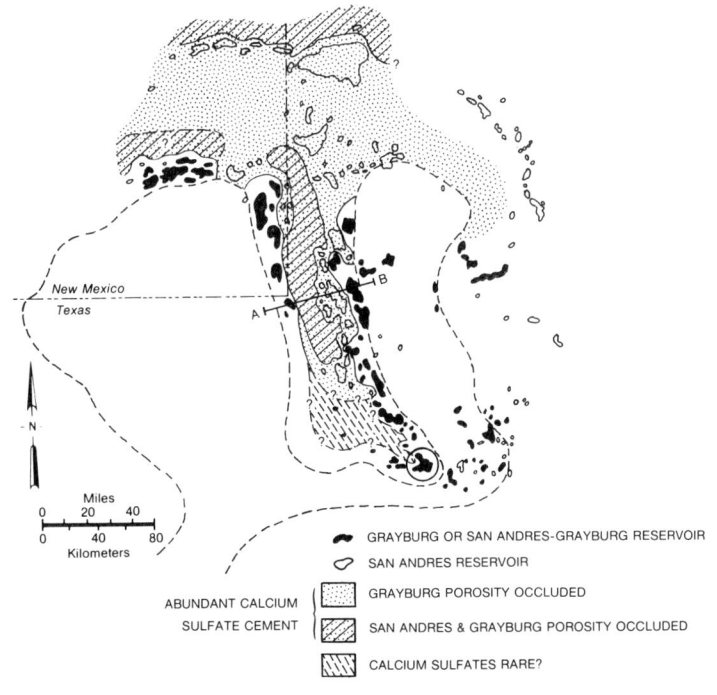

Fig. 5. San Andres and Grayburg dolomite reservoirs on the Central Basin platform and other shelf areas in relation to inferred major occlusion of porosity by calcium sulphate cements.

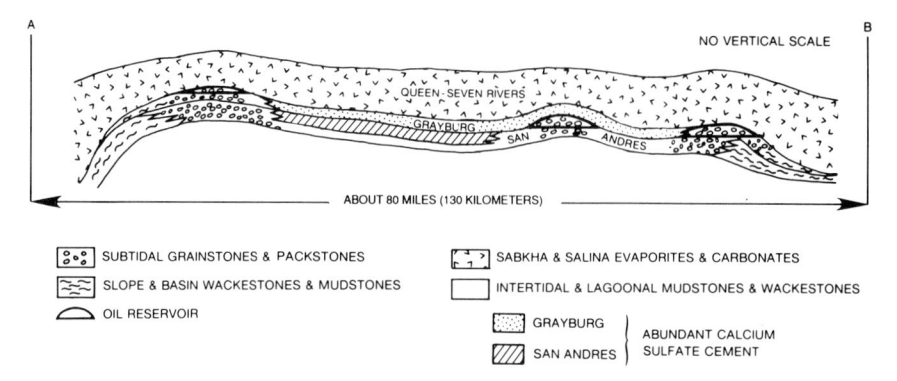

Fig. 6. Generalized structural-stratigraphic section along line A-B of Fig. 5, illustrating the inferred control of calcium sulphate cementation on areal and stratigraphic distribution of San Andres–Grayburg traps on the central Basin platform.

bonates to a more nearly regional distribution across the Central Basin and other platforms during deposition of the Grayburg. Presumably a dry climate and progressive restriction of the Permian Basin sea behind a partial barrier to open circulation (the late Guadalupian Capitan reef complex which rimmed the Delaware basin; Fig. 2b) were major controls on the amount and distribution of sulphate cement. Also important, especially to the formation of reservoirs on the Central Basin platform, was the effect of widening the shelves by basinward progradation of carbonates and thus expanding the areas of restricted marine circulation and hypersalinity in the shelf interiors.

Source and migration of San Andres–Grayburg oil

J E Adams (1930) was the first geologist to speculate in print concerning the source of oil in the Yates Field and, by implication, in other fields along the Central Basin platform. He looked chiefly to the organic carbonates and clastics basinward and downdip from the shelf margin carbonates at Yates as the most probable sources of Yates oil. There is no apparent need to revise this hypothesis except perhaps to add a second possible source of primary oil. That source may have been euxinic carbonates deposited in saline lakes and lagoons (Warren 1986), which were located in the interior of the Central Basin platform and were coeval with the San Andres and Grayburg carbonates of the platform margins (Figs 5 & 6).

Variations in the densities of San Andres–Grayburg oils in 54 reservoirs across the Central Basin platform, expressed in terms of API gravity, are mapped in Fig. 7. Included among

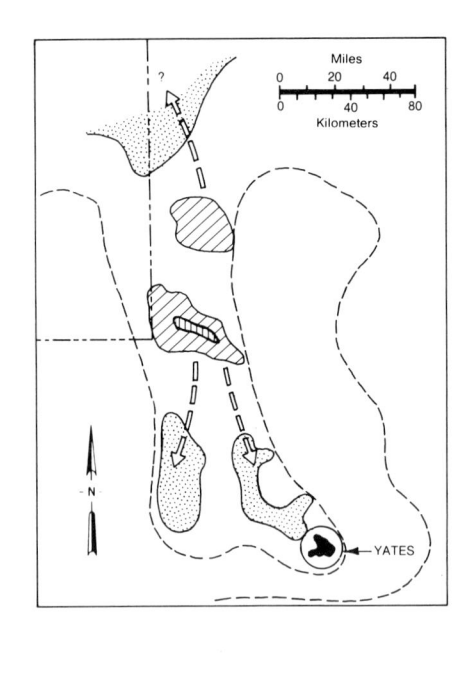

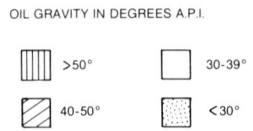

Fig. 7. Areal variation of oil gravity in 54 San Andres–Grayburg reservoirs on the Central Basin platform. Arrows indicate probable directions of secondary oil migration. Data from Mason Map Service, Inc. & International Oil Scouts Association (1987).

the 54 gravities are those from the top 18 producing fields listed in Tables 1 & 2. The range of all values is from 26 to 53° API, with the highest occurring at the approximate mid-point of the platform and the lowest at the platform's southern tip. Between these two points, a distance of about 90 miles (145 km), there are gradients of oil gravity, reservoir temperature, and reservoir depth, all decreasing southward to the area of the Yates Field. The changes in temperature (3.3°F/1000 ft of depth, 5.9°C/1000 m) and gravity (1.7° API/1000 ft of depth, 0.029 g/cm^{-3}/1000 m) bear a direct and linear relationship to reservoir depth (or to reciprocal reservoir elevation). The temperature range is 80 to 90°F (27 to 32°C), and the gravity (density) range is 30 to 35° API (0.876 to 0.850 g/cm^{-3}) at reservoir depths of 1500 to 4000 ft (450 to 1220 m). Present depths of burial in the southern part of the Central Basin platform are not greatly different from what they probably have been during the past 250 Ma, and it is therefore unlikely that significantly higher temperatures have been imposed on the San Andres–Grayburg reservoirs. It follows from this assumption that the systematic change in gravity of the oil must be a consequence of physical segregation rather than thermochemical cracking, since the latter does not take place at temperatures much below 212°F (100°C) (H. Dembicki Jr. pers. comm.). In fact, the direct relationship between oil gravity and reservoir depth (or reciprocal elevation) can be interpreted as one of the effects of differential migration and entrapment of oil and gas described by Trumbell (1916) and Gussow (1954). That effect is to lower the gas saturation and API gravity of remigrating oil as it moves by spillage from the bottoms of over-filled reservoirs to ever higher reservoirs up a regional dip.

Spillage from reservoirs north of Yates may have been due to an increase in the northward structural dip of the Central Basin platform (Fig. 8). Or it may have resulted from a relatively modest increase in geothermal heat along the platform, causing expansion of the oil and gas in reservoirs and expulsion of some of these fluids from traps which were filled to the spill point. Present evidence is too scanty to allow one to choose between these hypotheses.

Discovery and development of San Andres–Grayburg oil

The first successful well drilled in the Permian Basin was completed in 1921 in Mitchell County, Texas, in San Andres dolomite, in a field named Westbrook, located in an area that would later be called the Eastern shelf of the Midland basin (Fig. 1). A second discovery well, drilled in 1924, and completed in both San Andres and Grayburg carbonates, opened up the Big Lake

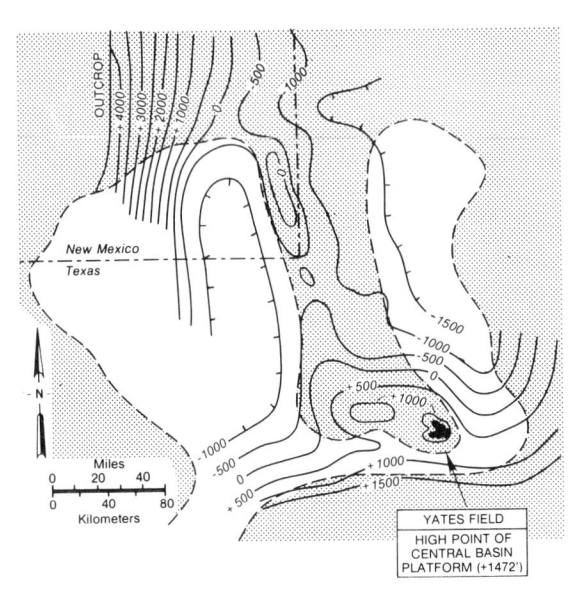

Fig. 8. Regional structure on top of San Andres–Grayburg shelf and shelf margin carbonates and coeval basin carbonates and clastics. Where the regional San Andres–Grayburg aquifer exists (Fig. 10), these contours describe approximately the shape of its upper surface. Structural datum is mean sea level.

Field 20 miles (32 km) northeast of Yates. In 1925, oil production from San Andres–Grayburg carbonates on the Central Basin platform was established in the McCamey Field, 20 miles (32 km) NNW of Yates. These discoveries, including the one at Yates and most of the exploratory wells drilled before the mid-1930s, were located on subtle surface structures detected by mapping with plane table and alidade. By 1935 enough wells had been drilled to provide a fair amount of subsurface control, especially along the Central Basin platform. Drilling prospects came to be chosen from linear trends of structures, from alignments of producing wells, and from maps showing the thickness of shallow Late Permian (Ochoan [Tatarian]) redbeds and salt. Areas in which the Ochoan units were thin or absent frequently were found to overlie structural or paleotopographic highs in the deeper subsurface. In fact, such isopach anomalies were the first indications of the presence of the Central Basin platform (King 1959; Harris & Walker, 1988).

Yates Field

The Yates Field is located in easternmost Pecos County, Texas, at about 30°52′ N latitude and 101°55′ W longitude. The discovery well was drilled in October 1926 at the apex of a broad but low-relief surface anticline which was mapped on thin-bedded Early Cretaceous (Aptian to Albian) limestones. Unitized area of the field is 26 000 acres (10 522 ha), and productive closure on the trap at the stratigraphic level of the San Andres–Grayburg is approximately 400 ft (120 m).

A large and well integrated network of pores combined with artesian reservoir pressures resulted in a spectrum of well potentials in the early years of Yates production that are remarkable for a carbonate reservoir with a relatively small productive area and closure. During open flow tests of 315 wells conducted in June 1929, 203 were potentialed for more than 10 000 bbl (1596 m^3) per day, and 26 for more than 80 000 bbl (12 768 m^3) per day. In the fall of 1929, well 4930 was drilled and potentialed for more than 200 000 bbl (31 900 m^3) per day (Fig. 9). This well actually was gauged open flow at 4833 bbl (771 m^3) of oil in 34 min through 10.75 in (27 cm) casing and three flow lines, one with an internal diameter of 6 in (15 cm) and two with internal diameters of 4 in (10 cm). As further evidence of the untapped excess capacity present at that time in the history of the field, the well maintained a back-pressure of 165 psi (1038 kPa) during peak flow (Oil and Gas Journal 1976).

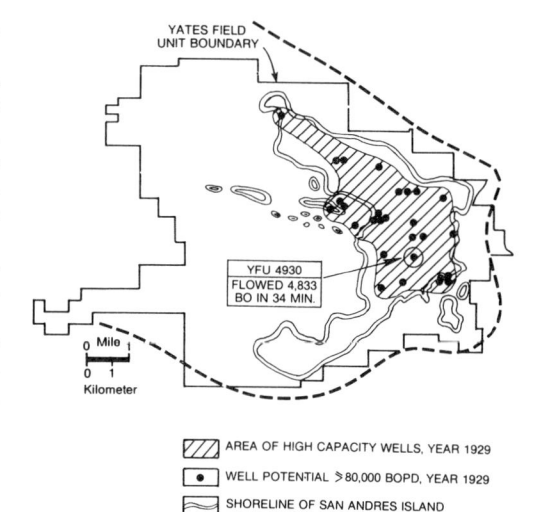

Fig. 9. Yates Field area showing the relationship of high well productivities to an island cluster that formed during emergence of upper San Andres carbonates above sea level, before deposition of the Grayburg Formation. A karst system, including open fractures and thousands of small caves, developed beneath the limestone islands.

The unusual storage and flow characteristics of the Yates Field reservoir resulted from a convergence of the following favorable geologic factors:

Carbonate pore systems

The reservoir has a vast network of interconnected pores developed chiefly in the dolomites of the San Andres Formation, which is the oldest and thickest of the reservoir stratigraphic units and contains about 80% of the reservoir pore volume. These pores range in size from fine intercrystal and intergrain voids to solution-enlarged fractures and small caves. The size of intergrain pores is controlled primarily by the differing depositional and diagenetic facies which characterize the two halves of the field, west and east (Figs 3 & 4). Finer pores are found in the mud-rich lagoonal and intertidal carbonates which, with discontinuous thin clay shales, comprise much of the upper San Andres reservoir in the western part of the field. Coarser intergrain pores occur in the sand- and gravel-sized bioclastic grainstones and packstones which predominate in the eastern part of the field. These grainy San Andres rocks were deposited at the margin of the Central Basin platform as an eastward prograding complex of shallow subtidal bars (Fig. 6). The grain-rich eastside dolomites have contributed the bulk of historical oil production from Yates and have matrix porosities that average between 15 and 20%. These same rocks have horizontal matrix permeabilities that lie typically in the range from 50 to 100 mD.

Larger voids (open fractures and caves) in the San Andres dolomites are products of regional jointing, local tectonic folding and fracturing, and karstic solution and brecciation. The karst was related to Late Permian emergence of the San Andres as a cluster of limestone islands in the Yates Field area (Adams 1930; Craig 1988) and in at least one other field in the southern part of the Central Basin platform (Taylor–Link West San Andres Field: Kerans & Parsley 1986). More than 300 caves have been detected by wells drilled in Yates, and thousands probably exist in between wells, especially on the east side of the field. Most of the caves were produced by dissolution associated with the activity of dynamic lenses of fresh water beneath the islands (Craig 1988). They range in height from one to 21 ft (0.3 to 6.4 m) and have a modal height of 2 ft (0.6 m). Coincidence of high capacity wells with the northern two-thirds of the principal island (Fig. 9) indicates a probable relationship of these wells to elements of the cave and fracture network.

Regional & local structure

A second factor contributing to the unusual characteristics of the Yates reservoir is its location at the structural apex of the 150 mile (240 km) long Central Basin platform (Fig. 8). The difference in structural elevation between the reservoir high point at Yates and the tops of reservoirs in San Andres–Grayburg fields 100 mile (160 km) north of Yates is about 2500 ft (760 m).

At Yates the effects of San Andres depositional facies and shallow paleotopography are seen in persistent patterns of compactional thickening and thinning over the field which extend, though subtly, even to the shallowest beds of the Seven Rivers Anhydrite, about 600 ft (183 m) above the top of the San Andres. Local non-compactional structure on the reservoir is a broad asymmetrical dome with a prominent anticlinal axis extending to the northwest (Fig. 4). The northern limit of production coincides with a normal fault which is downthrown 190 ft (58 m) to the north and trends approximately N 73° W. This trend closely parallels the trends of Precambrian and younger faults reported by Hills (1970, 1984) for the southern part of the Central Basin platform. Along the plane of the fault, evaporites of the Seven Rivers seal lie against oil- and water-charged dolomites of the San Andres–Grayburg reservoir and aquifer. Structural warping and faulting at reservoir level in the Yates Field area at drilled depths of less than 2500 ft (762 m) (the effects of which may extend into Cretaceous strata at ground level) possibly are related to Late Permian (Ochoan) reactivation of Pennsylvanian (Late Carboniferous) faults about 5000 ft (1520 m) below the reservoir.

Absence of evaporite cements

Anhydrite and gypsum occlude much of the porosity in many San Andres–Grayburg reservoirs on the Central Basin platform north of Yates (Longacre 1986; Bebout 1986); whereas, calcium sulphate cements are virtually absent from the reservoir rocks of the Yates field itself and probably from a large area across the southern part of the Central Basin platform. These cements may be largely missing from the western and northwestern edge of the platform as well (Fig. 5). It is probable that the two major subaquifers in the San Andres–Grayburg regional aquifer, the southern and the northwestern (see Fig. 10 and discussion on the aquifer below), are without significant amounts of these cements. At present these seems to be no way of determining whether the paucity of

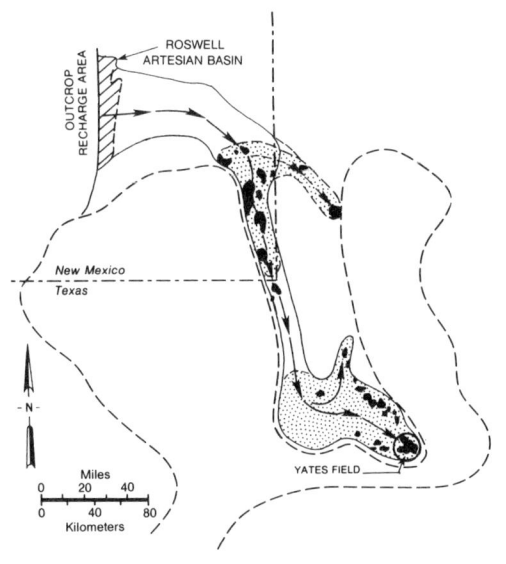

Fig. 10. Outline of regional aquifer that controlled original pressures in San Andres–Grayburg oil fields in the southern and northwestern areas of the Central Basin platform. Arrows indicate likely directions of hydrodynamic flow. Stipples define two areas of pressure drawdown caused by high rates of oil and water production from large fields.

sulphates is related to non-deposition of the cements or to their dissolution, or to some combination of the two. Strongly positive structure on the top of the San Andres–Grayburg (Fig. 8) coincides with the area now occupied by the regional aquifer (Fig. 10), which suggests a relationship between structure and aquifer. The author favors the view that the San Andres–Grayburg carbonates deposited in relatively shallow water over the highs in the area of the aquifer were laid down under conditions of open marine circulation and near normal salinity, so that precipitation of anhydrite and gypsum was absent or rare.

Oil source & volume

Development of the oil column in the Yates Field presumably began by primary migration from the basin areas that adjoin the field on three sides (Adams 1930). This earliest oil may have been augmented by contributions from evaporitic carbonates (Warren 1986) which were deposited in the area of restricted marine circulation in the interior of the Central Basin platform (Fig. 5). Later the oil column may have been increased still further by secondary migration from over-filled Guadalupian reservoirs north of the field due to northward tilting of the platform or to increased temperatures in some of the reservoirs (Fig. 7).

The hypothesis of secondary oil migration into the Yates reservoir is supported by two lines of evidence from the field itself. The first has to do with interpretation of the fact that at the time of discovery Yates oil had a relatively low gas saturation. This fact may be considered as evidence in support of the interpretation stated in the paragraph above that a significant amount of Yates oil migrated into the reservoir by spillage from reservoirs to the north filled to capacity (Gussow 1954; see Fig. 7 of this paper). A second line of evidence comes from a study of secondary calcites which are found at different paleo levels in the San Andres dolomite in the Yates reservoir (D. H. Mruk, pers. comm.). These calcites are of different ages and are interpreted as the metabolic products of sulphate-reducing bacteria that lived on the hydrocarbons at the oil–water contact of the filling reservoir. Carbon and oxygen isotope analyses of the calcites suggest that the earlier calcite represents migration of primary oil which moved into the trap by compactional and hydrodynamic gradients from environments of high fluid temperature in the depths of the basins adjoining the reservoir. The later calcite appears to be a product of lower fluid temperatures and somewhat more intense (prolonged?) biodegradation in the reservoir. These are conditions that fit waters of lower temperature and salinity (probably diluted by meteoric water) which are characteristic of the southern part of the Central Basin platform.

The times of primary and secondary oil migration into the Yates structure are unknown, apart from the fact that primary migration can not have occurred before deposition and lithification of the lower part of the Seven Rivers Formation, the reservoir seal. Secondary migration could have occurred as recently as Late Cretaceous or Paleocene (Maastrichtian to Thanetian?) in conjunction with tectonic or thermal events accompanying the Laramide orogeny.

Upper seal

The Seven Rivers Formation, a 350 ft (107 m) thick unit consisting mostly of saline lake (salina) and salt flat (sabkha) anhydrites and dolomites (Spencer & Warren 1986), is the efficient upper seal that completes the anticlinal trap of the Yates Field. The presence of this virtually impervious regional unit over the San Andres–Grayburg shelf and shelf margin carbonates accounts for the large number and size of major reservoirs on the central Basin platform (Fig. 3b).

Adams & Rhodes (1960) hypothesized that precipitation of calcium sulphates in the hypersaline lagoons and sabkhas of the Upper Guadalupian generated high-magnesium brines which seeped downward and basinward through the limestones of the San Andres–Grayburg, converting them to dolomites. Their 'seepage refluxion' model is supported by evidence from petrographic and carbon–oxygen isotope studies of San Andres dolomites from the Yates reservoir (D. H. Mruk, pers. comm.) and from seven San Andres reservoirs in the northern half of the Central Basin platform (Leary & Vogt 1986). Samples of dolomites from these eight reservoirs are characterized by $\delta 13$ C with values in the range from +1 to +7‰ and $\delta 18$O values from +2 to +6‰. The data suggest that formation of the dolomite took place in the presence of sea water with greatly elevated salinity, probably due to evaporation.

Reservoir–aquifer dynamics

Bottom-hole pressures at Yates at discovery were significantly above hydrostatic pressure for the burial depth of the reservoir, so that most of the wells drilled during the early years flowed naturally and strongly. In a study by the writer in the late 1960s evidence was presented that original pressure in the field was, at least in part, controlled by a regional aquifer in the San Andres–Grayburg carbonates. Primary stratigraphic and areal controls on the aquifer facies are related to deposition over shoals above the structural-palaeotopographic highs which occupied the southern toe and western margin of the Central Basin platform. The most active part of the aquifer (Fig. 10) was first identified by mapping areal variations in water density, chemistry, and pressures in the San Andres–Grayburg carbonates throughout the Permian Basin.

Recharge of the aquifer appears to have taken place at an outcrop some 275 streamline miles (440 km) northwest of the Yates Field and at an elevation 2000 ft (610 m) above the present elevation of the Yates Field reservoir (Fig. 10). The average paleohydrodynamic gradient is estimated to have been less than 1 psi/mile (4.3 kPa/km), an amount which on a regional scale approximates hydrostatic conditions. Paleoflow from northwest to southeast through the aquifer probably was slow and presumably was maintained by leakage at various points along the path of flow as depth and net hydrostatic pressure increased away from the recharge area. Loss of water from the aquifer may have taken place upward along faults to shallower aquifers with lower pressures. Over geologic time the inflow of meteoric water was sufficient to establish and maintain a column of brackish water in the aquifer with an average salinity of 10 000 to 15 000 mg of total dissolved solids per litre. These physical conditions prevailed until the mid 1920s. Then oil was discovered in Yates and in other fields trapped by structural and stratigraphic closures along the roof of the aquifer, and withdrawal of large volumes of oil (and water) began.

The regional aquifer is not a highly transmissive system, and recharge has not kept pace with oil and water production from the fields within it. Evidence for this is that, since the 1920s, pressure sinks have developed in the areas of highest oil and water withdrawal. The two largest sinks are shown in Fig. 10. Both now are bounded by relatively higher pressure and are separated from each other along the western edge of the Central Basin platform by an area of even more severe pressure drawdown, which presumably has the lowest permeability.

Comparative quality, San Andres–Grayburg reservoirs

The Yates Field has produced almost 1.2 billion bbl (192×10^6 m^3) of oil out of an estimated original oil-in-place of 4 billion bbl (638×10^6 m^3). Cumulative total oil production and cumulative oil per acre for the top eighteen San Andres–Grayburg reservoirs in the Permian Basin are shown graphically in Fig. 11. The

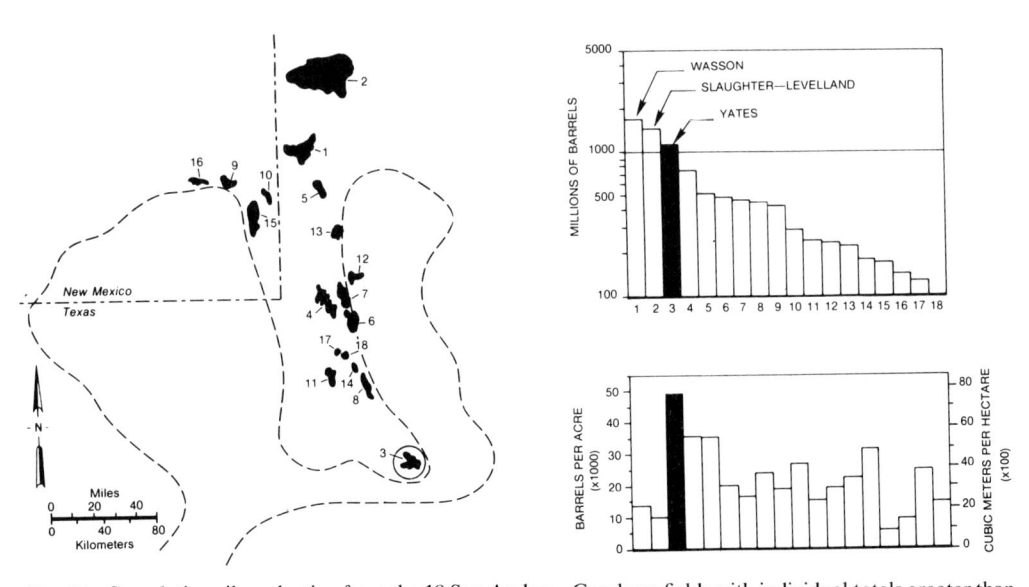

Fig. 11. Cumulative oil production from the 18 San Andres–Grayburg fields with individual totals greater than 100 million bbls (16×10^6 m^3). The numbers on the map and bar graph are keyed to the fields listed in Table 1. Statistics on cumulative production taken from *Oil and Gas Journal*, issue of 30 January 1989.

grand total for these fields is 9.315 billion bbl (1.5×10^9 m^3), to which the top three fields, including Yates, have contributed 4.398 billion bbl (702×10^6 m^3) or 47%.

It is clear from the data shown in Fig. 11 that Yates is not preeminent in cumulative oil production, but that it stands well above its companions in what it has yielded per unit of field area: that is, 49000 bbl/acre (19300 m^3/ha). Although not as rigorous as indices expressed in terms of oil volume per unit volume of reservoir rock (numbers not available for Yates or other San Andres–Grayburg fields), this statistic is nevertheless a fair index for comparing reservoir quality among fields with similar geology.

Acknowledgments. The author thanks Marathon Oil Company for release of information on the Yates Field and for permission to publish this paper. He is especially grateful to P. W. Choquette, M. J. Heymans, and D. H. Mruk who reviewed the paper and were most helpful with suggestions for its improvement; and to H. Dembicki, Jr for a critical discussion of evidence for remigration of some San Andres–Grayburg oil. T. J. Melton drafted the illustrations.

References

ADAMS, J. E. 1930. Origin of oil and its reservoir in Yates pool, Pecos County, Texas. *American Association of Petroleum Geologists Bulletin* **14**, 705–717.

—— & RHODES, M. E. 1960. Dolomitization by seepage refluxion. *American Association of Petroleum Geologists Bulletin* **44**, 1912–1920.

BEBOUT, D. G. 1986. Facies control of porosity in the Grayburg Formation: Dune Field, Crane County, Texas. *In*: BEBOUT, D. G. & HARRIS, P. M. (eds) *Hydrocarbon Reservoir Studies, San Andres/Grayburg Formations, Permian Basin* Research Conference Proceedings, PBS-SEPM Pub. No. **86–26**, 107–11.

CRAIG, D. H. 1988. Caves and other features of Permian karst in San Andres dolomite, Yates Field reservoir, west Texas. *In*: JAMES, N. P. & CHOQUETTE, P. W. (eds) *Paleokarst*, Springer, New York, 342–363.

GALLEY, J. E. 1958. Oil and geology in the Permian Basin of Texas and New Mexico. *In*: WEEKS, L. G. (ed.) *Habitat of Oil*, The American Association of Petroleum Geologists, Tulsa, Oklahoma, 395–446.

GALLOWAY, W. E., EWING, E. T., GARRETT, C. M., TYLER, N. & BEBOUT, D. G. 1983. *Atlas of Major Texas Oil Reservoirs*, Bureau of Economic Geology, The University of Texas at Austin, 139.

GARBER, R. A. & HARRIS, P. M. 1986. Depositional facies of Grayburg–San Andres dolomite reservoirs: Central Basin platform, Permian Basin. *In*: BEBOUT, D. G. & HARRIS, P. M. (eds) *Hydrocarbon Reservoir Studies, San Andres/Grayburg Formations, Permian Basin* Research Conference Proceedings, PBS-SEPM Pub. No. **86–26**, 61–66.

GUSSOW, W. C. 1954. Differential entrapment of oil and gas: a fundamental principle. *American Association of Petroleum Geologists Bulletin* **38**, 816–853.

HARRIS, P. M. & Walker, S. D. 1988. McElroy field, Central Basin platform, U.S. Permian basin. *In*: BEAUMONT, E. A. & Foster, N. H. (eds) *Treatise of Petroleum Geology, Atlas of Oil and Gas Fields, No. 1*, American Association of Petroleum Geologists, Tulsa, Oklahoma, 32.

HILLS, J. M. 1970. Late Paleozoic structural directions in southern Permian Basin, west Texas and southeastern New Mexico. *American Association of Petroleum Geologists Bulletin* **54**, 1809–1827.

—— 1984. Sedimentation, tectonism, and hydrocarbon generation in Delaware basin, west Texas and southeastern New Mexico. *American Association of Petroleum Geologists Bulletin* **68**, 250–267.

KERANS, C. & PARSLEY, M. J. 1986. Depositional facies and porosity evolution in a karst-modified San Andres reservoir: Taylor–Link West San Andres, Pecos County, Texas. *In*: BEBOUT, D. G. & HARRIS, P. M. (eds) *Hydrocarbon Reservoir Studies, San Andres/Grayburg Formations, Permian Basin* Research Conference Proceedings, PBS-SEPM Pub. No. **86–26**, 133–134.

KING, P. B. 1959. *The Evolution of North America* Princeton University Press, Princeton, New Jersey, 190.

LEARY, D. A. & VOGT, J. N. 1986. Diagenesis of the Permian (Guadalupian) San Andres Formation, Central Basin platform, west Texas. *In*: BEBOUT, D. G. & HARRIS, P. M. (eds) *Hydrocarbon Reservoir Studies, San Andres/Grayburg Formations, Permian Basin* Research Conference Proceedings, PBS-SEPM Pub. No. **86–26**, 67–68.

LONGACRE, S. A. 1980. Dolomite reservoirs from Permian biomicrites. *In*: HALLEY, R. B. & LOUCKS, R. G. (eds) *Carbonate Reservoir Rocks*, Notes for SEPM Core Workshop No. 1, Denver, Colorado, June 8 & 12, 1980, Society of Economic Palaeontologists & Mineralogists, 105–117.

—— 1986. The Grayburg Formation at North McElroy: A Cinderella reservoir. *In*: BEBOUT, D. G. & HARRIS, P. M. (eds) *Hydrocarbon Reservoir Studies*, San Andres/Grayburg Formations, Permian Basin Research Conference Proceedings, PBS-SEPM Pub. No. **86–26**, 123–126.

MASON MAP SERVICE, Inc. & International Oil Scouts Association 1987. *International Oil and Gas Development, Review of 1984–1985, Part II (Production)*, 55–56, 999.

OIL AND GAS JOURNAL. 1976 (12 April). Marathon unveils details of Yates unit, 30–31.

RUPPEL, S. C. 1986. San Andres facies and porosity distribution: Emma Field, Andrews County, Texas. *In*: BEBOUT, D. G. & HARRIS, P. M. (eds) *Hydrocarbon Reservoir Studies, San Andres/Grayburg Formations, Permian Basin* Research

Conference Proceedings, PBS-SEPM Pub. No. **86**–26, 99–103.
SHIRLEY, K. 1987. Colorful history, odd geology — Yates field celebrates 60 years. *AAPG Explorer*, **8**, 4–5.
SPENCER, A. W. & WARREN, J. K. 1986. Depositional styles in the Queen and Seven Rivers formations: Yates Fields, Pecos County, Texas. *In*: BEBOUT, D. G. & HARRIS, P. M. (eds) *Hydrocarbon Reservoir Studies, San Andres/Grayburg Formations, Permian Basin*; Research Conference Proceedings, PBS-SEPM Pub. No. **86**–26, 135–37.
TRUMBELL, L. W. 1916. The effect of structure upon migration and separation of hydrocarbons. *Bulletin of the Science Service*, **1**, 17–27.

WARD, R. F., KENDALL, G. St. C., & HARRIS, P. M. 1986. Upper Permian (Guadalupian) facies and their association with hydrocarbons: Permian Basin, west Texas and New Mexico. *American Association of Petroleum Geologists Bulletin* **70**, 239–262.
WARREN, J. K. 1986. Perspectives: shallow-water evaporitic environments and their source rock potential. *Journal of Sedimentary Petrology* **56**, 442–454.
WILSON, J. L. 1974. Characteristics of carbonate platform margins. *American Association of Petroleum Geologists Bulletin* **58**, 810–824.
—— 1975. *Carbonate Facies in Geologic History*: Springer, New York, 471.

Geochemistry of carbonate source rocks and crude oils in Jurassic salt basins of the Gulf Coast

R. SASSEN

Basin Research Institute and Department of Geology and Geophysics, Louisiana State University, Baton Rouge, LA 70803, USA

Abstract: The lower member of the Jurassic Smackover Formation is a source rock for crude oil across the northern Gulf Rim from east Texas to Florida. Type-I crude oils occur in Smackover reservoirs of east Texas, north Louisiana, south Arkansas, and Mississippi, and were generated by algal-derived kerogen preserved in an anoxic environment. Type-II crude oils in Smackover reservoirs of Alabama and Florida reflect an origin from kerogen with a terrestrial component preserved in a less anoxic environment. Thermal maturity differences across the Gulf Rim caused generation and destruction of crude oil to occur at much shallower depths in east Texas, north Louisiana, and south Arkansas than in Mississippi, Alabama and Florida. Jurassic through Cretaceous timing of migration and faulting permitted vertical migration of Type-I Smackover crude oils to many Upper Jurassic, Lower Cretaceous and Upper Cretaceous reservoirs along the Gulf Rim. Vertical migration of Type-II crude oils was less significant, especially in Alabama, because of evaporite barriers to migration. The high contents of hydrogen sulphide and carbon dioxide in deep, hot Smackover reservoirs are related to thermochemical sulphate reduction that has driven early destruction of methane. Compositionally distinct crude oils from Upper Cretaceous and Lower Tertiary siliciclastic source rocks are found in reservoirs south of the Gulf Rim.

Crude oil, gas condensate and methane associated with non-hydrocarbon gases such as hydrogen sulphide have been produced from carbonate reservoirs of the upper member of the Jurassic Smackover Formation in salt basins across much of the northern Gulf Rim. The high-sulphur crude oils of the Upper Smackover have distinctive geochemical compositions dissimilar to other crude oils south of the Gulf Rim that are derived from Upper Cretaceous and Lower Tertiary siliciclastic source rocks (Koons *et al.* 1974; Sassen *et al.* 1988; Sassen 1989). Smackover crude oils display compositional characteristics indicating an origin from a carbonate source rock deposited in an anoxic environment. These characteristics include carbon isotopic compositions (Sofer 1988), generally low pristane/phytane ratios, even-carbon predominance of n-alkanes (Sassen & Moore 1988), abundant aromatic sulphur biomarkers (Hughes 1984) and hydrocarbon biomarkers described by Sofer (1988).

Geochemical studies provide evidence that the lower member of the Smackover Formation, deposited in an anoxic and perhaps hypersaline carbonate environment, is the main source rock for Upper Smackover hydrocarbons (e.g., Oehler 1984; Sassen *et al.* 1987a, Wade *et al.* 1987; Sassen 1988; Sassen & Moore 1988). Moreover, crude oil is found in numerous reservoir rocks, ranging in age from Upper Jurassic through Upper Cretaceous, that often overlie known Smackover production along the northern Gulf Rim. Research suggests that much of the crude oil in these overlying reservoirs has been emplaced by vertical migration from the Smackover source rock (Sassen *et al.* 1987b; Sofer 1988; Sassen 1989).

Clearly, geochemical research in this area has reached a juncture where a regional synthesis is possible. The primary objective of the present paper is to review previous research on the Smackover source rock and hydrocarbons in the context of new results that provide insight to the areal and stratigraphic distribution of crude oil across the northern Gulf Rim.

This paper was also presented at the Ninth Annual Research Conference, Gulf Coast Section Society of Economic Paleontologists and Mineralogists Foundation, New Orleans, Louisiana, U.S.A., December 4–7, 1988, and is reprinted with permission of the Gulf Coast Section Society of Economic Paleontologists and Mineralogists Foundation.

Geological setting

The main structural features affecting the Mesozoic sequences of the northern Gulf rim, including the upper Jurassic Smackover, are a series of interior salt basins (East Texas, North

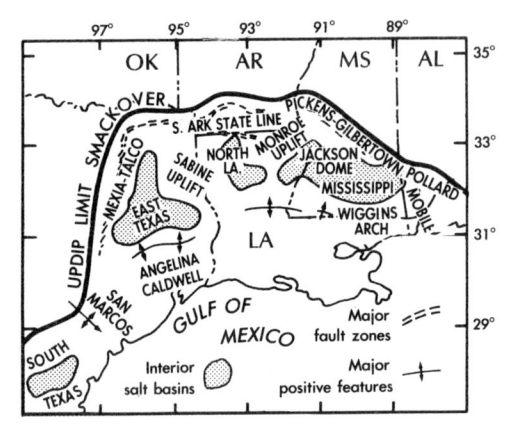

Fig. 1. Generalized structural setting of the northern Gulf (modified from Sassen & Moore, 1988).

Louisiana and Mississippi salt basins) extending from south Texas to Alabama (Fig. 1). These basins formed during the rifting stage of the formation of the Gulf of Mexico by extensional crustal thinning and attenuation (Buffler & Sawyer 1985). They are bounded on their Gulf of Mexico flanks by marginal positive features held up by blocks of continental crust including the Angelina–Caldwell Flexure to the west and the Wiggins Arch to the east (Fig. 1).

During the first major marine incursion of the Gulf region in the Middle Jurassic, the interior salt basins were floored with thick evaporite sequences (Werner/Louann formations), and today are marked by numerous high-rise salt structures. Rapid burial of sediments along the northern Gulf Rim occurred, during the Jurassic and Lower Cretaceous, with little additional burial thereafter (Nunn & Sassen 1986; Driskill et al. 1988).

A major tensional graben system occurs along the updip limits of Upper Jurassic marine deposits from Texas to Florida, and is known from west to east as the Mexia–Talco, south Arkansas State Line, and Pickens–Gilbertown fault zones (Fig. 1). This system is thought to be related to tensional forces caused by gliding of Upper Jurassic sediments toward the adjacent salt basin as the sequences were tilted toward the basin during subsidence. It is important to note that activity on these faults, which are potential conduits for vertical migration of crude oil, commenced during the Upper Jurassic and extended into the Cretaceous (Seni & Jackson 1983).

The Smackover source rock

Source potential of the Smackover Formation is mainly restricted to the laminated lime mudstone facies of the Lower Smackover, whereas the Upper Smackover serves mainly as a reservoir rock (Oehler 1984; Sassen et al. 1987a; Sassen & Moore 1988). The Smackover source facies extends from east Texas to Florida (Sassen et al. 1987a). There is evidence based on programmed pyrolysis that localized source potential for crude oil could exist in the Upper Smackover of the Manila Embayment in Alabama (Wade et al. 1987).

The Lower Smackover source rock was deposited during the Upper Jurassic in an anoxic and perhaps hypersaline carbonate environment that favored preservation of algal-derived kerogen (Sassen et al. 1987a). Total organic carbon contents (TOC) are strongly related to carbonate facies (Sassen et al. 1987a) (Fig. 2). Quite organic rich laminated lime mudstones are present, especially in the vicinity of large oil fields in Florida (Sassen et al. 1987a). However, the mean TOC of the entire lower Smackover member (including source and nonsource facies) is only about 0.5% (Oehler 1984; Sassen et al. 1987a; Sassen & Moore 1988).

Much of the Smackover Formation is deeply buried and has experienced advanced levels of thermal maturity (Nunn & Sassen, 1986; Driskill et al. 1988), and TOC contents were higher in the geological past prior to time of crude oil generation and migration. Expulsion of crude oil from the source rock was influenced by depositional and diagenetic features characteristic of carbonate rocks (Fig. 3). The concentration of source kerogen and siliciclastics along laminations of depositional origin could facili-

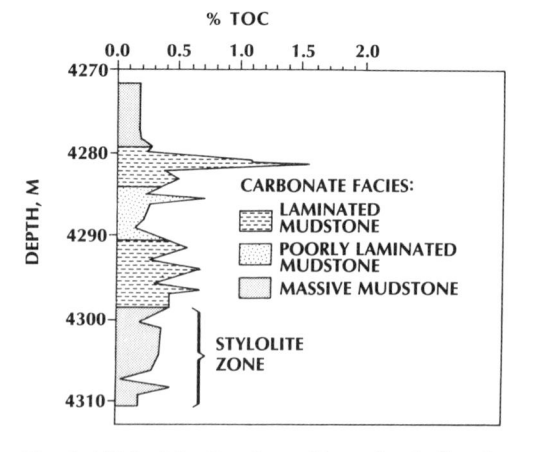

Fig. 2. TOC of the Smackover Formation is directly related to carbonate facies. The highest TOC contents occur in laminated lime mudstones, whereas massive mudstones have the lowest TOC contents. Pressure solution along stylolite zones can increase TOC (after Sassen et al. 1987).

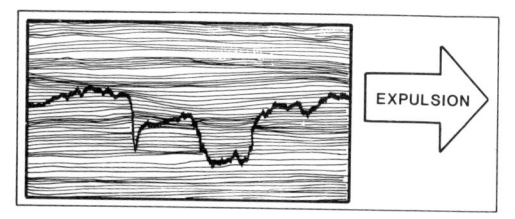

Fig. 3. Crude oil expulsion from the laminated lime mudstone source facies is unusually efficient because source kerogen and siliciclastics are concentrated along laminations of depositional origin and along pressure solution features such as stylolites.

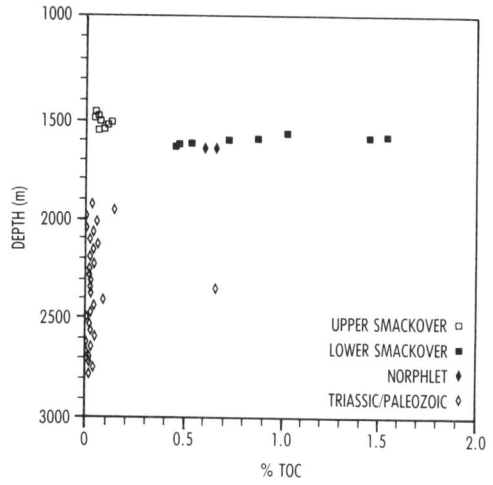

Fig. 4. The high TOC of the oil-prone lower member of the Smackover Formation in Arkansas stands in contrast to TOC of underlying Triassic and Paleozoic rocks. Some high TOC values were noted in Norphlet shales here, but programmed pyrolysis suggests the presence of gas-prone kerogen.

tate expulsion from the Smackover source rock (Oehler 1984; Sassen et al. 1987a). Moreover, expulsion efficiency is thought to be further increased by pressure solution of carbonate rock (Sassen et al. 1987a). Pressure solution redistributed source kerogen and siliciclastics preserved during deposition, forming thin but exceptionally organic-rich layers (mean TOC = 10.3%) along stylolites and solution seams that also served as conduits for fluid movement (Sassen et al. 1987a).

Visual kerogen assessment suggests that algal-derived amorphous kerogen dominates in the Smackover source rock across much of the northern Gulf Rim to include the East Texas and Mississippi salt basins (Sassen et al. 1987a; Thompson-Rizer 1987). Programmed pyrolysis of thermally immature Lower Smackover source rock from shallow depths (1561–1585 m) in Bradley County, Arkansas, provides excellent evidence of hydrogen-rich, unoxidized, algal-derived kerogen prone to generation of crude oil (Sassen & Moore 1988). Exceptions exist, especially where calcareous shales and sandstones are present near the Smackover updip limit or in the vicinity of some basement highs. For example, terrestrially-derived kerogen can be dominant along the updip limit of Smackover deposition nearest the paleoshoreline in Alabama (Wade et al. 1987).

Although localized exceptions are probable, rocks other than those of the Smackover Formation generally lack meaningful source potential. TOC values of the Upper Jurassic Haynesville and Norphlet formations are generally too low to signify meaningful hydrocarbon source potential (Sassen et al. 1987a). The source potential of the Triassic Eagle Mills Formation and Paleozoic rocks that underlie the Jurassic Trend have not been well studied. Measurements of TOC provide some insight into the organic richness of these older rocks in Bradley County, Arkansas (Fig. 4). As can be seen, the pre-Jurassic rocks are generally characterized by low TOC values that contrast with the organic-rich lower member of the Smackover Formation.

Crude oil subfamilies and regional variation in the Smackover source rock

Analytical results on 26 crude oils and condensates from Upper Smackover reservoirs from across the northern Gulf Rim from east Texas to Florida are presented by Sassen (1989). Although these crude oils and condensates are all derived from the Smackover source rock and represent a single family, their compositions vary regionally and correlate with differences in the Smackover source rock. Carbon isotopic compositions suggest that two main subfamilies of Upper Smackover crude oils are present (Fig. 5). Type-I Smackover crude oils are most widely distributed, and are found from the East Texas Salt Basin east to the Mississippi Salt Basin, whereas Type-II Smackover crude oils are less widely distributed and are found in Alabama and Florida (Fig. 6).

Many Type-I Smackover crude oils from east Texas to Mississippi display pristane/phytane ratios <1 (mean = 0.86), whereas many Type-II Smackover crude oils from Alabama and Florida display pristane/phytane ratios >1 (mean = 1.14), even at low levels of thermal maturity (Sassen 1988). The mean carbon pref-

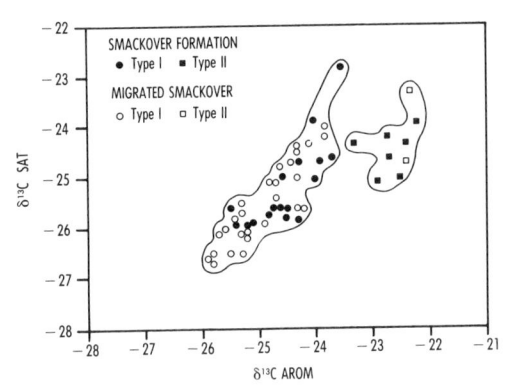

Fig. 5. Carbon isotopic compositions of crude oils in Smackover reservoirs suggest two subfamilies. Type-I Smackover crude oils occur from east Texas to Mississippi, and Type-II crude oils occur in Alabama and Florida. Similarity of carbon isotopic compositions of crude oils from Upper Jurassic through Upper Cretaceous reservoirs suggests that vertical migration of Type-I crude oils was widespread (modified from Sassen 1989).

erence index (CPI) of all 26 crude oils from Smackover reservoirs is 0.94, with no significant differences between subfamilies (Sassen 1989) (Fig. 7). Even-carbon predominance is often observed in carbonate-sourced crude oils (Palacas 1984).

The Type-I crude oils appear to be sourced by the algal-derived kerogen with little or no terrestrial contribution that is characteristic of the Smackover source rock from the East Texas Salt Basin east to the Mississippi Salt Basin (Sassen *et al.* 1987a). Moreover, preservation of source kerogen occurred in an anoxic environment of deposition (Sassen *et al.* 1987) that gave rise to low pristane/phytane ratios (often <1). The occurrence of distinctive aromatic sulfur biomarkers in these crude oils (Hughes 1984) provides compelling evidence of an anoxic source rock depositional environment.

The higher pristane/phytane ratios of Type-II crude oils could relate to the presence of terrestrial kerogen in the Smackover of Alabama and parts of Florida (Sofer 1988). Increased terrestrial kerogen in the Smackover Formation in Alabama is suggested by programmed pyrolysis of Smackover rock samples (Wade *et al.* 1987), and by differences in crude oil carbon isotopic compositions (Sofer 1988). Although aromatic sulfur biomarkers also occur in Alabama crude oils (Hughes 1984), the Smackover environment of deposition in Alabama appears to have been somewhat less anoxic (Mancini *et al.* 1986; Sofer 1988).

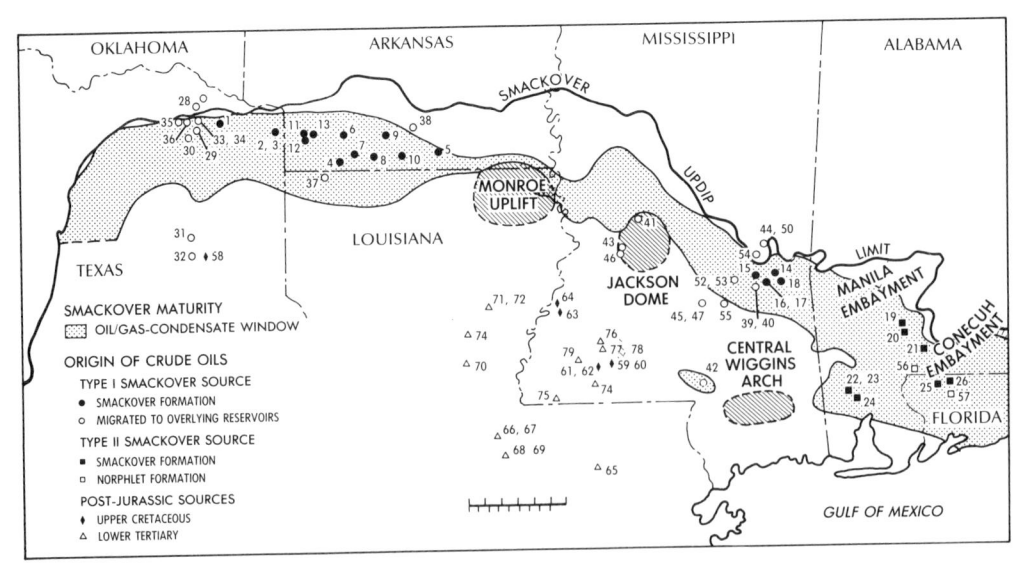

Fig. 6. Thermal maturity map of the Smackover Formation from east Texas to Florida expanded from Moore & Sassen (1988). Map shows Smackover Formation within the thermal maturity window for generation of crude oil (equivalent to about 0.55% vitrinite reflectance) and preservation of meaningful gas-condensate (equivalent to about 1.3% vitrinite reflectance). Map shows locations and inferred source of 79 crude oils (identified in Table 1).

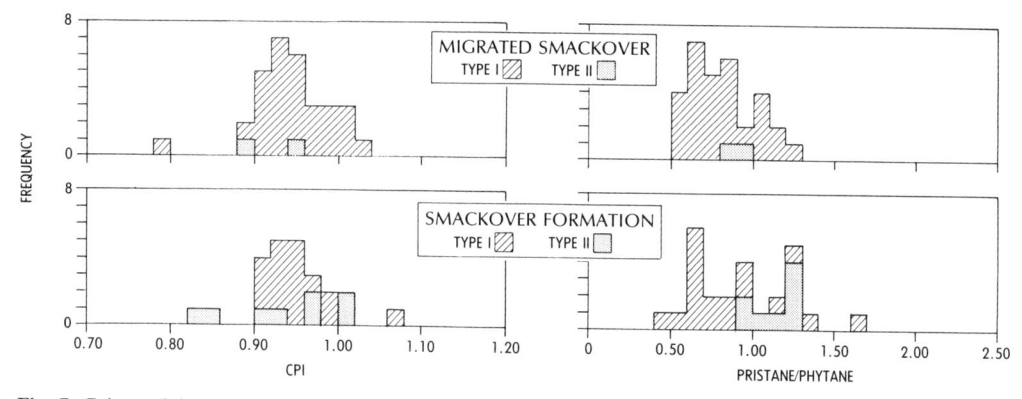

Fig. 7. Pristane/phytane ratios of Type-I and Type-II crude oils in Smackover reservoirs differ. There is no clear difference in CPI between groups. Similarity of pristane/phytane ratios in crude oils from Upper Jurassic through Upper Cretaceous reservoirs suggests that vertical migration of Type-I crude oils was widespread (modified from Sassen 1989).

Thermal maturity and phases of crude oil generation and destruction

The relationship between thermal maturity and generation and destruction of crude oil in the Smackover Formation has been studied previously (Sassen 1988; Sassen & Moore 1988). Generation of high-sulphur bitumens can occur in some carbonate rocks at low levels of thermal maturity (Tannenbaum & Aizenshtat 1985). The concept of early generation from high-sulphur kerogen in carbonate source rocks contributes to understanding the origin of rare immature Smackover crude oils.

The composition of heavy crude oil (10–12° API) from shallow reservoirs of North New London Field (1844 m) in Union County, Arkansas, is unusual when compared to other Smackover crude oils (Sassen 1989, Table 1). The C_{15+} fraction comprises 87.7% of the total crude oil, indicating that early migration from the source rock could have occurred prior to significant thermal cracking. The low saturate/aromatic ratio of the North New London crude oil (0.16) further implies thermal immaturity. The chromatographic signature of the crude oil displays characteristics that are not attributable to the effects of biodegradation or water washing (Fig. 8).

The onset of generation and migration of medium-gravity crude oil by the Smackover Formation appears to occur at a level of thermal maturity equivalent to a vitrinite reflectance of 0.55% or perhaps lower (Sassen 1988; Sassen & Moore 1988). The chromatographic signature of a medium-gravity Type-II crude oil is shown in Fig. 8. The effects of maturation on Smackover crude oils after emplacement in reservoirs result in progressive cracking of heavier molecules to yield lighter molecules. C_{15+} fractions decrease in percentage as heavier components are cracked, and C_{15+} compositions are progressively altered. Saturate/aromatic ratios of Smackover crude oils increase markedly during maturation (0.16–12.04) (Sassen 1989). The carbon isotopic compositions of saturated and aromatic hydrocarbons become progressively heavier with increasing maturation (Sassen 1988; Sofer 1988). The even-carbon predominance of n-alkanes becomes less obvious with maturation (Sassen & Moore 1988), and pristane/phytane ratios increase with maturation (Sofer 1988). The chromatographic signature of a high-gravity Type-I condensate is shown in Fig. 8.

Rapid thermal cracking of crude oil in Smackover reservoirs appears to commence at a level of thermal maturity equivalent to about 1% vitrinite reflectance reflectance (R_0) (Sassen & Moore 1988). This phase of thermal cracking is accompanied by widespread precipitation of solid bitumens in reservoir rocks (Sassen 1988), and by the onset of rapid thermochemical sulfate reduction and increased abundance of hydrogen sulphide (Orr 1977; Sassen 1988). The maximum thermal maturity for preservation of meaningful volumes of condensate is equivalent to about $R_0 = 1.3\%$ or higher (Sassen & Moore 1988).

Thermal maturity differences across the Smackover trend

There are significant differences in depths at which crude oil is generated and destroyed across the Smackover Trend. As suggested by Fig. 9, which relates API gravity and depth, the

Table 1. List of crude oils, with key geochemical characteristics, featured in Fig. 6. Crude oils include the two subfamilies found in Smackover reservoirs (A), migrated Smackover sourced crude oils in Jurassic and Cretaceous reservoirs (B), Upper Cretaceous sourced crude oils in Tuscaloosa and Woodbine reservoirs (C), and Lower Tertiary sourced crude oils in Wilcox reservoirs (D). Additional analytical data on these families of crude oils are presented by Sassen (1989), and by Sassen et al. (1988).

Location	Formation	Average Depth (m)	County/parish	State	Field	PR/PHY	C_{15+} SAT CPI	$\partial^{13}C$ PDB SAT	AROM
A									
1	Smackover	2,191	Red River	TX	Megan	0.58	0.91	−24.7	−24.3
2	Smackover	2,297	Bowie	TX	West Texarkana	1.64	0.95	−24.7	−23.9
3	Smackover	2,313	Bowie	TX	West Texarkana	0.64	0.94	−24.6	−23.7
4	Smackover	3,377	Columbia	AR	Keoun Creek	0.88	0.94	−25.6	−24.6
5	Smackover	1,844	Union	AR	North New London	0.44	0.90	n. d.	n. d.
6	Smackover	2,182	Lafayette	AR	Buckner	0.65	0.98	−25.0	−24.6
7	Smackover	2,704	Columbia	AR	Dorcheat-Macedonia	0.95	1.07	−25.6	−24.5
8	Smackover	2,568	Columbia	AR	Pine Tree	0.88	0.93	−26.0	−25.2
9	Smackover	2,176	Union	AR	North Mt. Holly	0.73	0.96	−25.9	−25.1
10	Smackover	2,295	Union	AR	East Schuler	0.92	0.95	−25.9	−25.4
11	Smackover	2,588	Miller	AR	North Fouke	1.16	0.99	−25.0	−24.0
12	Smackover	2,736	Miller	AR	Fouke	1.38	n. d.	−22.8	−23.5
13	Smackover	2,548	Miller	AR	Cypress Lake West	1.26	n. d.	−23.9	−24.0
14	Smackover	3,962	Clarke	MS	Pachuta Creek	0.65	0.93	−25.6	−24.7
15	Smackover	4,440	Jasper	MS	South Paulding	0.77	0.90	−25.8	−24.3
16	Smackover	4,237	Clarke	MS	West Nancy	0.68	0.93	−25.7	−24.8
17	Smackover	4,200	Clarke	MS	West Nancy	0.65	0.92	−25.8	−24.5
18	Smackover	4,145	Clarke	MS	East Nancy	0.61	0.94	−25.6	−25.5
19	Smackover	3,895	Monroe	AL	Not Named	1.01	0.83	−25.1	−22.9
20	Smackover	3,963	Monroe	AL	Lovett's Creek	0.96	0.84	−25.0	−22.5
21	Smackover	4,115	Conecuh	AL	Barnett	0.97	0.91	−24.6	−22.7
22	Smackover	5,639	Mobile	AL	Chunchula	1.22	1.00	−24.3	−23.3
23	Smackover	5,600	Mobile	AL	Chunchula	1.29	1.00	−24.2	−22.7
24	Smackover	5,600	Mobile	AL	Hatter's Pond	1.27	0.96	−23.9	−22.2
25	Smackover	4,724	Santa Rosa	FL	Jay	1.22	0.93	−24.6	−22.7
26	Smackover	4,816	Santa Rosa	FL	Black Jack Creek	1.18	0.96	−24.3	−22.4
B									
27	Walnut	545	Red River	TX	Buzbee	0.59	0.95	−24.8	−24.6
28	Walnut	701	Red River	TX	I & L	0.60	0.92	−24.4	−24.3
29	Woodbine	1067	Titus	TX	Trix Liz	0.64	0.92	−26.0	−25.6
30	Rodessa	1311	Titus	TX	Talco Graben	0.79	0.92	−25.0	−24.3
31	Pettet (Sligo)	2575	Smith	TX	Moon Rock	0.86	0.91	−26.2	−25.2
32	Travis Peak	2469	Smith	TX	Chapel Hill	0.82	1.00	−25.9	−24.9
33	Cotton Valley	1661	Red River	TX	Clarksville	0.52	0.95	−24.7	−24.4
34	Cotton Valley	1676	Red River	TX	Clarksville	0.83	0.79	−24.2	−23.8
	Cotton Valley	1615	Red River	TX	Shelton	0.90	0.90	−24.3	−24.1

SOURCE ROCKS OF THE GULF COAST

37	Haynesville	3241	Bossier	LA	Shelton	0.63	0.88	−24.0	−23.8
38	Nacatoch	549	Bossier	LA	Arkana	1.00	0.96	−25.4	−24.7
39	Eutaw	1506	Union	AR	Smackover	0.63	0.93	−25.7	−25.3
40	Eutaw	1489	Jasper	MS	Heidelberg	0.59	0.94	−25.8	−25.4
41	Selma Gas-Rock	1334	Jasper	MS	Heidelberg	0.64	0.93	−26.6	−25.4
42	Lower Tuscaloosa	2743	Madison	MS	Flora	0.76	0.93	−26.1	−25.9
43	Paluxy	2840	Lamar	MS	Baxterville	0.79	0.97	−26.7	−25.7
44	Paluxy	1829	Hinds	MS	Bolton	0.99	1.01	−25.1	−25.8
45	Paluxy	3830	Clarke	MS	Enterprise	0.64	0.90	−26.1	−24.8
46	Mooringsport	3206	Smith	MS	Not Named	0.72	0.93	−26.5	−25.3
47	Mooringsport	3962	Hinds	MS	S. Bolton	1.07	0.99	−26.1	−25.8
48	Rodessa	3353	Smith	MS	Not Named	0.71	1.00	−26.5	−25.2
49	Rodessa	3316	Hinds	MS	Bolton	1.14	0.98	−26.2	−25.5
50	Sligo	2297	Hinds	MS	S. Bolton	1.09	0.98	−25.5	−25.2
51	Hosston	3701	Clarke	MS	Enterprise	0.56	0.90	−26.5	−25.3
52	Cotton Valley	4420	Hinds	MS	Bolton	1.23	1.02	−25.6	−24.2
53	Cotton Valley	4420	Jasper	MS	Bay Springs	0.83	0.96	−25.6	−24.3
54	Cotton Valley	3479	Jasper	MS	Bay Springs	0.84	0.94	−24.5	−24.3
55	Haynesville	4200	Jasper	MS	Missionary	0.63	0.91	−25.1	−23.7
56	Norphlet	4633	Jones	MS	S. Summerland	0.83	0.95	−23.3	−22.3
57	Norphlet	4663	Escambia	AL	Flomaton	1.15	0.95	−24.7	−22.4
			Santa Rosa	FL	Mt. Carmel	1.02	0.89		
C									
58	Woodbine	2469	Smith	TX	East Texas	1.75	1.00	−28.8	−27.1
59	Upper Tuscaloosa	3243	Amite	MS	Smithdale	2.20	1.06	−28.0	−26.3
60	Lower Tuscaloosa	3365	Amite	MS	Smithdale	1.71	1.07	−28.0	−26.3
61	Lower Tuscaloosa	3383	Amite	MS	W. Smithdale	2.02	1.03	−28.3	−26.4
62	Lower Tuscaloosa	3353	Amite	MS	W. Smithdale	1.92	1.10	−28.1	−26.5
63	Lower Tuscaloosa	2957	Jefferson	MS	Fayette	1.69	1.05	−27.6	−26.6
64	Lower Tuscaloosa	2865	Jefferson	MS	Rodney	1.86	1.06		−25.9
D									
65	Wilcox	3048	Livingston	LA	Livingston*	1.95	1.06	−26.4	−26.2
66	Wilcox	4050	St. Landry	LA	Elba	1.89	1.07	−27.0	−26.1
67	Wilcox	4074	St. Landry	LA	Elba	1.85	1.06	−27.0	−26.3
68	Wilcox	3962	Pointe Coupee	LA	Fordoche	2.14	1.04	−26.5	−25.8
69	Wilcox	3962	Pointe Coupee	LA	Fordoche	2.23	1.11	−26.4	−25.4
70	Wilcox	1798	Rapides	LA	Big Island	1.79	1.05	−26.9	−26.4
71	Wilcox	1463	Catahoula	LA	Rawson	2.04	1.05	−26.2	−25.6
72	Wilcox	1494	Catahoula	LA	Rawson	1.97	1.06	−26.4	−25.4
73	Wilcox	1219	La Salle	LA	Nebo	2.06	1.12	−26.8	−25.9
74	Wilcox	1699	Amite	MS	Beechwood*	1.66	1.07	−26.7	−26.5
75	Wilcox	2682	Wilkinson	MS	Turnbull	1.73	1.06	−26.4	−25.9
76	Wilcox	1201	Franklin	MS	Willis Branch*	1.80	1.05	−26.8	−26.3
77	Wilcox	1301	Franklin	MS	Hominy Ridge*	1.52	1.08	−26.5	−26.5
78	Wilcox	1256	Franklin	MS	Hominy Ridge*	n. d.	n. d.	−26.5	−26.3
79	Wilcox	1417	Franklin	MS	Clear Springs*	1.81	1.07	−26.9	−26.2

* biodegraded or water-washed
nd: not determined

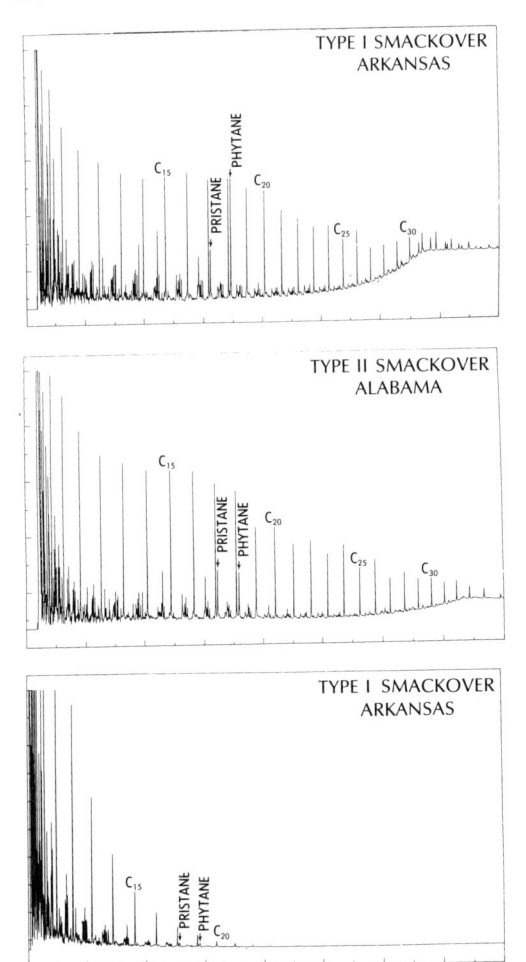

Fig. 8. The chromatographic signature of saturated hydrocarbons isolated from a low-gravity crude oil from North New London Field in Arkansas is consistent with early generation and migration rather than water washing or biodegradation (top). The chromatographic signatures of medium-gravity Smackover crude oil (middle) and of high-gravity condensate (bottom) are also shown.

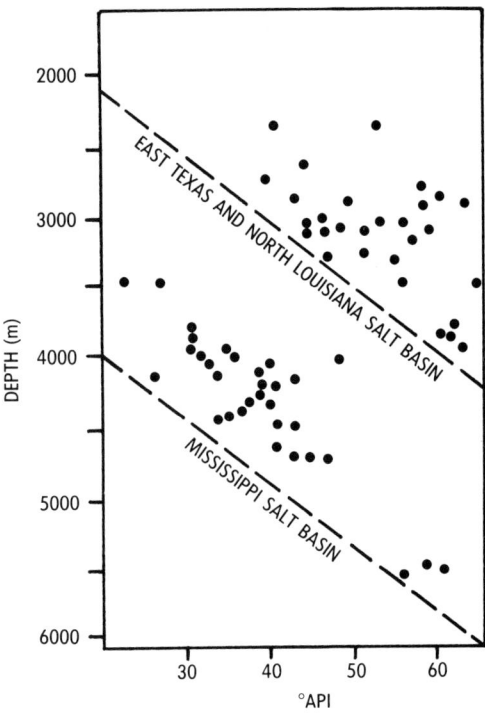

Fig. 9. Relationship between depth and API gravity of crude oils in Smackover reservoirs from the East Texas, North Louisiana, and Mississippi salt basins. Mississippi Salt Basin data includes the deep Chunchula and Hatters Pond gas-condensate fields in Alabama. Differences between the basins relate to different thermal maturity profiles across the Smackover Trend.

maximum depths at which crude oil and condensate are preserved in Smackover reservoirs are shallower in the East Texas and North Louisiana Salt basins than to the east in the Mississippi Salt Basin. This observation implies that thermal maturities are highest, at a given depth, west of the Monroe Uplift in northeast Louisiana. Moreover, the relationship between API gravity and depth appears more chaotic in the eastern half of the trend, especially in the East Texas Salt Basin (Fig. 9).

Results of programmed pyrolysis on samples of the Smackover Formation from east Texas, Louisiana, and Arkansas and east of the Monroe Uplift in Mississippi, Alabama, and Florida help explain observed differences in maximum depths for preservation of crude oil and condensate (Fig. 10). Pyrolysis T_{max} values suggest that thermal maturity is generally higher, at a given depth, in the west than in the east.

Thermal maturity mapping of the Smackover Formation

As emphasized by Demaison (1984), mapping the areal extent of thermally mature source rocks for crude oil can provide insight to the geographic distribution of hydrocarbon accumulations across productive basins. This approach can also find application in the search for overlooked hydrocarbons in areas with a long drilling history. Sassen & Moore (1988) presented a thermal maturity map of the

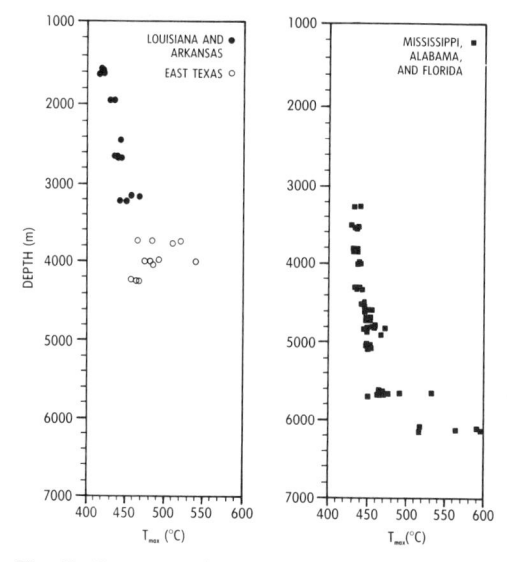

Fig. 10. Programmed pyrolysis results on organic-rich samples of the Smackover Formation suggest great differences in thermal maturity across the Smackover Trend. Generation and destruction of crude oil occurred at much shallower depths in east Texas, Louisiana, and Arkansas than in Mississippi, Alabama, and Florida.

Smackover Formation in eastern Mississippi, Alabama, and Florida that contributes to explaining the known geographic distribution of crude oil, gas condensate and methane.
. A thermal maturity map of the Smackover Formation from east Texas to Florida is shown in Fig. 6. This preliminary map was produced using generalized depths for the onset of crude oil generation (equivalent to about 0.55% vitrinite reflectance) and for preservation of meaningful volumes of gas-condensate (equivalent to about $R_0 = 1.3\%$). West of the Monroe Uplift, available thermal maturity data suggests that crude oil generation commences at about 1800 m, and thermal destruction of condensate approaches completion at roughly 4100 m. East of the Monroe uplift, crude oil generation commences at about 3300 m and thermal destruction of condensate approaches completion at about 5800 m.

There are, of course, limitations to this simple interpretation because of localized differences in burial and thermal history. In the Wiggins Arch area, for example, thermal maturities have been measured equivalent to the wide $R_0 = 1.5-3\%$ range at similar depths (Sassen et al. 1987b). Moreover, anomalously high thermal maturities appear to occur in the vicinity of the Monroe Uplift in Louisiana and the Jackson Dome in Mississippi. This thermal maturity map nevertheless serves as a base that can be improved through additional thermal maturity measurements and through use of calculated thermal maturity models.

Migration of Smackover crude oils to overlying reservoirs

Results of geochemical analyses for 31 crude oils from Upper Jurassic, Lower Cretaceous, and Upper Cretaceous reservoir rocks from across the northern Gulf Rim suggest an origin by vertical migration from the Smackover source rock (Sassen 1989). Carbon isotopic compositions of crude oils in Upper Jurassic to Upper Cretaceous reservoirs (i.e., Haynesville to Nacatoch formations) of the northern Gulf Rim from the East Texas Salt Basin to the Mississippi Salt Basin are consistent with vertical migration of Type-I Smackover crude oils (Fig. 5). Mean pristane/phytane ratios (0.81) and CPI values (0.94) of migrated crude oils and Type-I Smackover crude oils also suggest a genetic relationship (Fig. 7). Sofer (1988) presented hydrocarbon biomarker data on a more limited sample suite that also suggest vertical migration of Type-I Smackover crude oils in the Mississippi Salt Basin.

Vertical migration of Type-I Smackover crude oils to overlying reservoirs was widespread along the northern Gulf Rim from east Texas to Mississippi. Calculated thermal maturity histories for the Mississippi Salt Basin (Nunn & Sassen 1986) and for Alabama and Florida (Driskill et al. 1988) are consistent with early to late Cretaceous crude oil generation. Thermal maturity results imply that migration generally occurred earlier west of the Monroe Uplift. Petrographic evidence suggests that migration in east Texas commenced prior to the end of the Jurassic (Moore 1984). Faulting that commenced along the northern Gulf Rim during the Jurassic and extended to the Cretaceous (Seni & Jackson 1983) probably provided migration conduits. Some migrated Smackover crude oils are found in reservoirs that overlie thermally overmature Smackover source rocks in the East Texas and Mississippi salt basins (Fig. 6). Once migration occurs from deep, hot source rocks to shallower and cooler reservoirs, source rock and migrated crude oils will mature at greatly different rates (Nunn & Sassen 1986; Driskill et al. 1988). Some short-range lateral migration also occurred, as shown by Smackover crude oils trapped in Cretaceous reservoirs north of the updip limit of the Smackover in east Texas and in Mississippi (Sassen 1989) (Figure 6).

The presence of Type-I Smackover crude oils in reservoirs as young as Upper Cretaceous suggests that crude oil migration largely ended before the Tertiary across the northern Gulf Rim (Sassen 1988). The relative paucity of Type-II Smackover crude oils in overlying reservoirs is consistent with geologic evidence of Haynesville evaporites that often served as impermeable barriers to migration, especially in Alabama (Sassen & Moore 1988). Crude oils and condensates in the underlying Norphlet Formation in Alabama and Florida are derived from Smackover source rocks by downward migration.

Younger source rocks south of the study area

Geologically younger source rocks, including the Upper Cretaceous Tuscaloosa and Eagleford formations and the Lower Tertiary Wilcox Formation, have given rise to crude oils in reservoir rocks south of the study area. Locations of some of these crude oils from post-Jurassic source rocks are shown in Fig. 6. The Upper Cretaceous Tuscaloosa Formation appears to be a source rock for crude oils within the Tuscaloosa Formation of southwest Mississippi (Koons et al. 1974; Evans 1987). The similarity of Tuscaloosa crude oils to those from Woodbine reservoirs of East Texas Field, presumably derived from the Upper Cretaceous Eagleford Formation, has been noted (Sassen 1988). Lower Tertiary source rocks appear to be present that have given rise to crude oils of the Wilcox Trend in Louisiana and Mississippi (Evans 1987; Sassen et al. 1988; Walters & Dusang 1988). It should be emphasized that the compositions of crude oils from Upper Cretaceous and Lower Tertiary siliciclastic source rocks are grossly dissimilar to crude oils from the Smackover carbonate source rock (Sassen et al. 1988).

Early methane destruction and the origin of non-hydrocarbon gases

Hydrogen sulphide and carbon dioxide are economically undesirable constituents of gas in deep, hot reservoirs of the Smackover Formation (Parker 1974; Orr 1977). Dolomitized Lower Smackover reservoirs of Black Creek Field in Perry and Stone counties, Mississippi, provide an excellent opportunity to study the final stages of hydrocarbon destruction and the generation of these non-hydrocarbon gases. Reservoir temperatures at a depth of about 6100 m in Black Creek Field approximate 200°C. Major reservoir gases are hydrogen sulphide (77.8%) and carbon dioxide (20.2%), whereas methane (1.5%) and nitrogen (0.3%) are minor components.

Abundant solid bitumens rich in elemental sulphur in Lower Smackover reservoir rocks show that Black Creek Field is the thermally degraded residue of a former crude oil accumulation (Sassen et al. 1987b; Sassen 1988; Fig. 11 of this paper). Calculated thermal maturity models for the deep Mississippi Salt Basin suggest that crude oil was emplaced in Black Creek Field during the early Cretaceous and was cracked to form methane during the early Tertiary (Nunn & Sassen 1986). Methane destruction now approaches completion.

Precise measurement of thermal maturity in the Smackover Formation at Black Creek Field is difficult because of the rarity of vitrinite. Measurements based on Thermal Alteration Index (TAI) suggest that thermal maturity is equivalent to $R_0 = 1.5-1.7\%$ range (Sassen et al. 1987a). The calculated thermal maturity models of Nunn & Sassen (1986) suggest some-

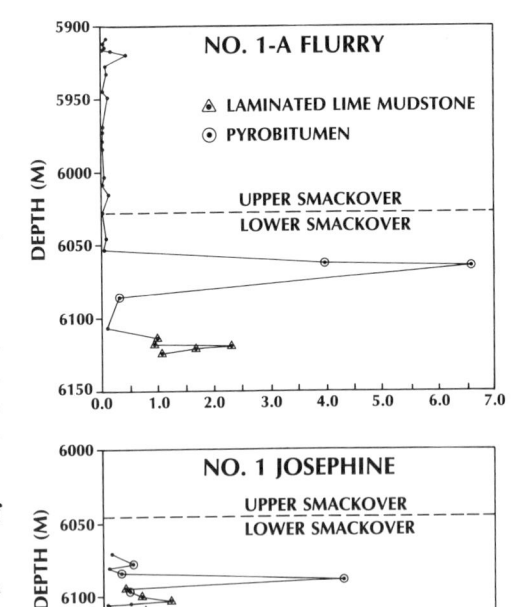

Fig. 11. High TOC contents in dolomitized Lower Smackover reservoir rocks in Black Creek Field reflect abundant solid bitumens (pyrobitumen) resulting from thermal destruction of a former crude oil accumulation (after Sassen et al. 1987).

what higher levels of thermal maturity than measurements of TAI. Sassen & Moore (1988) report some pyrolysis T_{max} values (>500°C) that could be interpreted to suggest higher levels of thermal maturity equivalent to $R_0 = 2\%$.

The presence of abundant hydrogen sulphide and carbon dioxide indicate that intense thermochemical sulphate reduction and oxidation of methane has occurred in Black Creek Field (Sassen & Moore 1988). Abundant elemental sulphur and some late-forming carbonate cements characterized by isotopically-light carbonate carbon provide further evidence of this process (Sassen et al. 1987).

Clearly, methane is a very stable molecule that can persist at much higher levels of thermal maturity than measured in Black Creek Field. Methane is known to persist in anthracite coals and in inert quartz-dominated reservoir rocks at levels of thermal maturity equivalent to $R_0 > 4\%$. Thermochemical sulphate reduction appears to have driven early destruction of methane in deep sulphate-rich carbonate reservoirs of Black Creek Field (Sassen et al. 1987a; Sassen & Moore 1988). Siebert (1985) has suggested specific reactions involving thermochemical sulphate reduction that contribute to explaining early destruction of methane in deep Smackover reservoirs.

In contrast to Smackover reservoirs of Black Creek Field, large volumes of methane with modest contents of hydrogen sulphide are preserved in deeper sandstone reservoirs of the Norphlet Formation offshore Alabama (Mancini et al. 1987). Thermal maturity of the offshore Alabama gas fields is estimated to be equivalent to $R_0 = 1.8-2.1\%$ range (Thompson-Rizer 1987; Sassen & Moore 1988). Methane preservation in the deep Norphlet reservoirs could be related to decreased availability of sulphate to drive the hydrocarbon destruction process, and to removal of hydrogen sulphide by formation of pyrite and other metal sulphides (Sassen & Moore 1988).

The sporadic occurrence of nitrogen gas in reservoirs across the Smackover Trend from east Texas to Florida has been more difficult to explain than the origin of hydrogen sulphide and carbon dioxide. There appears to be an empirical relationship between the occurrence of nitrogen and minimum levels of thermal maturity that correspond to thermal destruction of crude oil. The origin of nitrogen gas could be related to late thermal alteration of aromatic nitrogen compounds in solid bitumens precipitated during crude oil destruction. Siebert (1985) suggested that as gases such as methane, carbon dioxide, and hydrogen sulphide are converted to minerals at quite advanced maturities, relatively inert nitrogen becomes an important residual gas.

However, migration of nitrogen from underlying high-rank Paleozoic basement rocks to Smackover reservoirs could also occur. For example, the geographic distribution of nitrogen occurrences in Smackover reservoirs of east Texas shows some relationship to the deep Mexia-Talco fault zone. Nitrogen also occurs in the vicinity of the Wiggins Arch in Mississippi where Paleozoic basement rocks are present.

Late carbonate diagenesis

Although initial migration of hydrocarbons into carbonate reservoir rocks at shallow depths appears to slow or halt diagenesis, during deeper burial thermochemical sulphate reduction and hydrocarbon oxidation appear to drive diagenesis in deep Smackover reservoir rocks. It has been suggested that many late diagenetic features such as late carbonate cements, saddle dolomites, elemental sulphur, metal sulphides (pyrite, sphalerite, and galena), and late porosity generation can be related to thermochemical sulphate reduction and hydrocarbon oxidation (Moore & Druckman 1981; Sassen & Moore 1988). A similar relation between thermochemical sulphate reduction, hydrocarbon oxidation, and late carbonate diagenesis in Canada is documented by Krouse et al. 1988).

Conclusions

The lower member of the Upper Jurassic Smackover is a significant source rock for crude oil across the northern Gulf Rim from east Texas to Florida. Regional differences in source kerogen and environment of deposition resulted in two crude oil subfamilies in upper Smackover reservoirs. Type-I Smackover crude oils are found from east Texas to Mississippi, whereas Type-II Smackover crude oils are found in Alabama and Florida.

Some early generation of low-gravity crude oil occurs in the Smackover Formation. The main phase of crude oil generation, however, commences at a thermal maturity equivalent to a vitrinite reflectance of about 0.55% or lower. Rapid thermal cracking of crude oil and thermochemical sulphate reduction begin at a level of maturity equivalent to about $R_0 = 1\%$, and the last hydrocarbon liquids are destroyed at a level of maturity equivalent to 1.3% vitrinite reflectance or higher. Higher thermal maturities in east Texas, Louisiana, and Arkansas caused generation and thermal destruction of crude oil at shallower depths than in Mississippi,

Alabama, and Florida. An interpreted thermal maturity map of the Smackover Formation roughly defines a belt of Smackover source rock from east Texas to Florida within the thermal maturity window for generation of crude oil and preservation of gas-condensate.

A Jurassic to Cretaceous timing of crude oil migration and faulting favored vertical migration of Type-I Smackover crude oils that charged many overlying Upper Jurassic and Cretaceous reservoirs from east Texas to Mississippi. Some lateral migration of Type-I Smackover crude oils to Cretaceous reservoirs north of the Smackover updip limit occurred in east Texas and in Mississippi. Some Smackover crude oils are preserved in reservoirs even though the underlying source rock is now thermally overmature. Vertical migration of Type-II Smackover crude oils was limited, especially in Alabama, by Haynesville evaporites that served as impermeable barriers to migration.

The high contents of hydrogen sulphide and carbon dioxide in deep, hot Smackover reservoirs are related to thermochemical sulphate reduction and oxidation of hydrocarbons. Thermochemical sulphate reduction has driven destruction of methane at lower levels of thermal maturity than would occur in coals or in inert, quartzose reservoirs. Preservation of methane with lower hydrogen sulphide contents in deep sandstone reservoirs of the Norphlet Formation, offshore Alabama, is related to lesser availability of sulphate and to hydrogen sulphide removal by metal ions. Nitrogen gas in deep Smackover reservoirs appears to be related to late thermal breakdown of nitrogen compounds in reservoir solid bitumens, although some migration from Paleozoic basement rocks cannot be excluded. Thermochemical sulphate reduction and hydrocarbon oxidation have strongly influenced carbonate reservoir diagenesis.

Acknowledgements. This work was supported by the Basin Research Institute of Louisiana State University and by a number of oil companies. The assistance of E. W. Chinn and C. H. Moore, Basin Research Institute, and M. L. Eggert, Louisiana Geological Survey, is greatly appreciated.

References

BUFFLER, R. T., & SAWYER, D. S. 1985. Distribution of crust and early history, Gulf of Mexico basin: *Transactions of the Gulf Coast Association of Geological Societies*, **35**, 333–344.

DEMAISON, G. 1984. The generative basin concept. *In*: DEMAISON, G. & MURRIS, R. J. (eds) *Petroleum Geochemistry and Basin Evaluation. American Association of Petroleum Geologists Memoirs* **35**, 1–14.

DRISKILL, B. W., NUNN, J. A., SASSEN, R. & PILGER, R. H. 1988. Tectonic subsidence, crustal thinning, and petroleum generation in the Jurassic Trend of Mississippi, Alabama, and Florida. *Transactions of the Gulf Coast Association of Geological Societies.* **38**, 257–265.

EVANS, R. 1987. Pathways of migration of oil and gas in the south Mississippi Salt Basin. *Transactions of the Gulf Coast Association of Geological Societies* **37**, 75–86.

HUGHES, W. B. 1984. Use of thiophenic organosulfur compounds in characterizing crude oils derived from carbonate versus siliciclastic sources. *In*: PALACAS, J. G. (ed.) *Petroleum Geochemistry and Source Rock Potential of Carbonate Rocks: American Association of Petroleum Geologists Studies in Geology*, **18**, 181–196.

KOONS, C. B., BOND, J. G. & PEIRCE, F. L. 1974. Effects of depositional environment and post-depositional history on chemical composition of Lower Tuscaloosa oils. *Bulletin of the American Association of Petroleum Geologists*, **58**, 1272–1280.

KROUSE, H. R., VIAU, C. A., ELIUK, L. S., UEDA, A. & HALAS, S. 1988. Chemical and isotopic evidence of thermochemical sulphate reduction by light hydrocarbon gases in deep carbonate reservoirs. *Nature*, **333**, 415–419.

MANCINI, E. A., MINK, R. M. & BEARDEN, B. L. 1986. Integrated geological, geophysical, and geochemical interpretation of Upper Jurassic petroleum trends in the eastern Gulf of Mexico. *Transactions of the Gulf Coast Association of Geological Societies*, **36**, 219–226.

——, ——, ——, & HAMILTON, P. 1987. Recoverable natural gas reserves from the Jurassic Norphlet Formation, Alabama coastal waters area. *Transactions of the Gulf Coast Association of Geological Societies*, **37**, 153–160.

MOORE, C. H. 1984. The Upper Smackover of the Gulf Rim: Depositional systems, diagenesis, porosity evolution and hydrocarbon production. *In*: VENTRESS, W. P. S., BEBOUT, D. G., PERKINS, B. F. & MOORE, C. H. (eds) *The Jurassic of the Gulf Rim: SEPM Gulf Coast Section Third Annual Research Conference Proceedings* 283–307.

—— and DRUCKMAN, Y. 1981. Burial diagenesis and porosity evolution, Upper Jurassic Smackover, Arkansas and Louisiana. *Bulletin of the American Association of Petroleum Geologists*, **65**, 597–682.

NUNN, J. A. & SASSEN, R. 1986. The framework of hydrocarbon generation and migration, Gulf of Mexico continental slope. *Transactions of the Gulf Coast Association of Geological Societies*, **36**, 257–262.

OEHLER, J. H. 1984. Carbonate source rocks in the Jurassic Smackover trend of Mississippi, Alabama, and Florida. *In*: PALCAS, J. G. (ed.) *Petroleum Geochemistry and Source Rock*

Potential of Carbonate Rocks. American Association Petroleum Geologists Studies in Geology, **18**, 63–70.

ORR, W. L. 1974. Changes in sulfur content and isotopic ratios of sulfur during petroleum maturation — Study of Big Horn Basin Paleozoic oils. *Bulletin of the American Association Petroleum Geological*, **58**, 2295–2318.

—— 1977. Geologic and geochemical controls on the distribution of hydrogen sulphate in natural gas. *In*: CAMPOS R. & GONI J. (eds) *Advances in Organic Geochemistry*, **7**, 572–597.

PALACAS, J. G. 1984. Carbonate rocks as sources of petroleum: Geological and chemical characteristics. *Proceedings 11th World Petroleum Congress*, **2**: London, 1983, 31–43.

PARKER, C. A. 1974. Geopressures and secondary porosity in the deep Jurassic of Mississippi. *Transaction of the Gulf Coast Association of Geological Societies*, **24**, 69–80.

SASSEN, R. 1988. Geochemical and carbon isotopic studies of crude oil destruction, bitumen precipitation, and sulphate reduction in the deep Smackover Formation. *Organic Geochemistry*, **12**.

—— 1989. Migration of crude oil from the Smackover source rock to Jurassic and Cretaceous reservoirs of the northern Gulf Rim. *Organic Geochemistry*, **14**, 51–60.

—— & MOORE, C. H. 1988. Framework of hydrocarbon generation and destruction in eastern Smackover Trend. *Bulletin of the American Association of Petroleum Geologists*, **72**, 649–663.

——, ——, & MEENDSEN, F. C. 1987a. Distribution of hydrocarbon source potential in the Jurassic Smackover Formation. *Organic Geochemistry*, **11**, 379–383.

——, ——, NUNN, J. A., MEENDSEN, F. C. & HEYDARI, E. 1987b. Geochemical studies of crude oil generation, migration and destruction in the Mississippi Salt Basin. *Transactions of the Gulf Coast Association of Geological Societies*, **37**, 217–224.

——, TYE, R. S., CHINN, E. W. & LEMOINE, R. C. 1988. Origin of crude oil in the Wilcox Trend of Louisiana and Mississippi: Evidence of long-range migration. *Transactions of the Gulf Coast Association of Geological Societies*. **38**, 27–34.

SENI, S. J. & JACKSON M. P. A. 1983. Evolution of salt structures, East Texas diapir province, Part 2: Patterns and rates of halokinesis: *Bulletin of the American Association of Petroleum Geologists*. **67**, 1245–1274.

SIEBERT, R. M. 1985. The origin of hydrogen sulfide, elemental sulfur, carbon dioxide, and nitrogen in reservoirs. *SEPM Gulf Coast Section Sixth Annual Research Conference*, 30–31.

SOFER, Z. 1988. Biomarkers and carbon isotopes of oils in the Jurassic Smackover Trend of the Gulf Coast states, U.S.A.: *Organic Geochemistry*, **12**, 421–432.

TANNENBAUM, E. & AIZENSHTAT, Z. 1985. Formation of immature asphalt from organic-rich carbonate rocks-I: Geochemical correlation. *Organic Geochemistry*, **8**, 181–192.

THOMPSON-RIZER, C. L. 1987. Some optical characteristics of solid bitumen in visual kerogen preparations: *Organic Geochemistry*, **11**, 385–392.

WADE, W. J., SASSEN, R. & CHINN, E. W. 1987. Stratigraphy and source potential of the Smackover Formation in the northern Manila Embayment, southeast Alabama. *Transactions of the Gulf Coast Association Geological Societies*, **37**, 277–286.

WALTERS, C. C. & DUSANG, D. D. 1988. Source and thermal maturity history of oils from Lockhart Crossing, Livingston Parish, Louisiana. *Transactions of the Gulf Coast Association of Geological Societies*, **38**, 37–44.

Origin of the Infracambrian Salt Basins of the Middle East

MOUJAHED I. HUSSEINI & SADAD I. HUSSEINI

P.O. Box 1916, Dammam 31441, Saudi Arabia

Abstract: During the Infra and Lower Cambrain, the Arabian Gulf and Zagros Mountains (Hormuz Formation), South Oman (Ara Formation), Kerman in Central Iran (Ravar Formation), and the Salt Range Province in Pakistan (Salt Range Formation) were all sites of salt precipitation in subsiding rift basins along the Middle Eastern edge of Gondwanaland. Also from 600 to 540 Ma, a rift system extended westwards from the Sinai Peninsula and North Egypt across NE Africa, while a second rift system (about 600 to perhaps 510 Ma) extended northwards from the Sinai Peninsula across the Jordan Valley and Dead Sea into SE Turkey.

The development of these rifts was contemporaneous with the left-lateral Najd strike-slip fault system which dislocated the Arabian Peninsula by 300 km. The Najd system may be interpreted as a transform fault system which accommodated rifting between NE Egypt, the Jordan Valley, North Pakistan, Kerman and South Oman. The Najd system is parallel to the Zagros Line, and the latter is also the eastern depositional limit of the Infracambrian Hormuz Formation. The Infracambrian Zagros Line may be interpreted as a right-lateral, strike-slip fault system which caused Infracambrian rifting in the Arabian Gulf and Zagros Mountains.

Palaeontological and radiometric age constraints, together with stratigraphic considerations, suggest that the syn-rift evaporites were first precipitated during a transgression which commenced in the Latest Precambrian (600 Ma) and again during a Lower Cambrian regression. The regression terminated a period dominated by marine carbonates and evaporites and initiated the deposition of progradational alluvial to marginal marine sediments. Finally, in the late Lower Cambrian and Middle Cambrian, transgressive marine deposits swept across the Middle East.

During the Infracambrian (from about 600 to 540 Ma) many areas in the Middle East, particularly the Arabian Gulf and Zagros Mountains, South Oman, Central Iran and North Pakistan, were great basins where massive evaporites including rocksalt were precipitated (Fig. 1). These evaporitic formations include the Hormuz Formation of the Arabian Gulf and Zagros Mountains, the Ara Formation of South Oman, the Ravar Formation of Kerman in Central Iran, and the Salt Range Formation of North Pakistan. These massive evaporites (greater than 1000 m in the Arabian Gulf; Kent 1970) not only form the salt core of many of the hydrocarbon-rich structures of the Middle East, but also surround major basement-controlled, hydrocarbon-bearing structures of Infracambrian origin.

The Infracambrian system, besides providing an important structural contribution to the giant oil fields of the Middle East, deposited sediments which are intrinsically of economic importance. For example, in South Oman, the Ara evaporites are interbedded and laterally pass into the algae-rich source rocks of the Huqf Group (Gorin et al. 1982). More than seven billion barrels STOIIP are geochemically attributed to this group and these hydrocarbons are trapped in Early and Late Palaeozoic sandstone reservoirs by Permian shales (Al-Marjeby & Nash 1986). Yet the origin of these basins, the depositional environment of their evaporitic formations and, in general, the tectono-depositional landscape which dominated the Middle East during this critical early phase is poorly understood.

In this paper, key Infracambrian and Cambrian outcrops are described with a view to illustrate a tectonic and depositional model of the Infracambrian salt basins of the Middle East. Other outcrops, which are not discussed here, do not contradict the proposed extensional model; but rather would only render this paper more complex by introducing incomplete sections with local formation names without clarifying the regional geology. Wolfart (1981, 1983) provides a complete enumeration and description of the Cambrian rocks of the Middle East and his study will not be repeated here.

The Infracambrian Stage

During the Precambrian (about 900 to 620 Ma), the Nubian–Arabian shield was fused together by a complex process of accretionary collisional tectonics which terminated with the formation

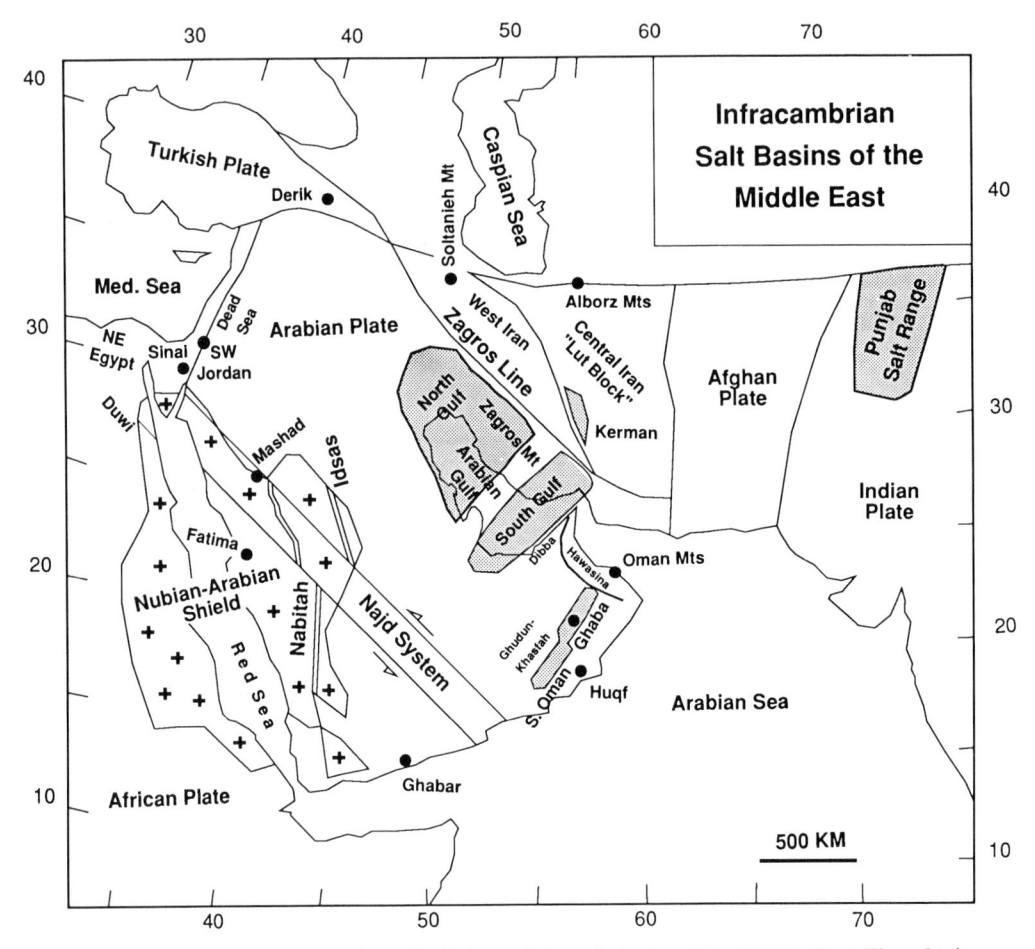

Fig. 1. The shaded areas represent the major Infracambrian salt basins of the Middle East. These basins include the area of the Arabian Gulf and Zagros Mountains, South Oman and Ghaba, Kerman and the Punjab Salt Range.

of the Nabitah (680–640 Ma) and Idsas (640–620 Ma) thrust belts (Al-Shanti & Mitchell 1976; Schmidt et al. 1979; Stoesser & Camp 1985; Johnson et al. 1987). These accretionary processes closed ancient seas (whose remnants are represented by ophiolitic suites which are intercalated in the Nabitah and Idsas belts shown in Fig. 1) and merged the terranes of the Middle East with the NE African edge of Gondwanaland (by 640 Ma). These Late Precambrian orogenies undoubtedly raised the Nubian–Arabian shield area to a mountainous stature.

While the final Idsas orogeny (640–620 Ma) was uplifting the western part of the Arabian plate, the eastern part of this plate and the Iran platform were folded and thrust-faulted along a NS grain (Idsas-trending), and wrenched by conjugate NW–SE (Najd-trending) and NE–SW fault systems. The regional extent of these collisional processes into eastern Saudi Arabia is evident from steeply-dipping, grey and black shales and slates (Table 1), encountered by deep boreholes, which were dated by radiometric techniques at 671 to 604 Ma (Kiilsgaard & Greenwood 1976, pers. comm.) and 636 to 605 Ma (Bass 1981, pers. comm.). In Iran, Stocklin (1968) noted similar lithology and structure in the Precambrian Morad Series and Kahar Formation (Table 1). The Morad Series are intricately folded slates and sandstone with Algonkian fauna which are cut by the volcanics of the Rizu Formation (760–595 ± 120), while the Kahar Formation (up to 1600 m) consists mostly of green–grey shales which are cut by the post-tectonic Precambrian Doran Granite.

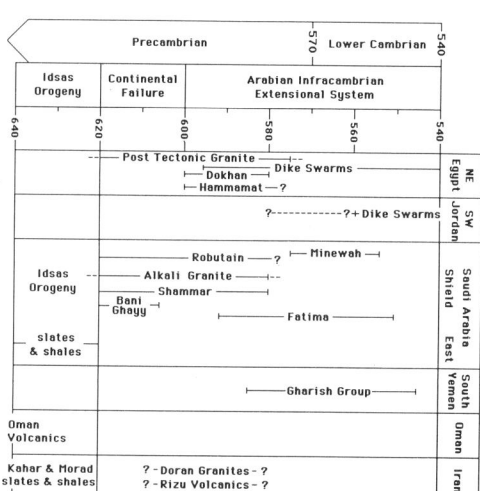

Table 1. Radiometric Ages of Igneous Rocks.

	Precambrian		Lower Cambrian	
		570		540
Idsas Orogeny	Continental Failure	Arabian Infracambrian Extensional System		
640 – 620 – 600 – 580 – 560 – 540				
	Post Tectonic Granite — Dokhan — Dike Swarms — Hammamat —?			NE Egypt
	?————————?+Dike Swarms			SW Jordan
Idsas Orogeny	Robutain —? — Minewah — Alkali Granite —?— Shammar — Bani Ghayy — Fatima			Saudi Arabia Shield
slates & shales	Gharish Group			South Yemen East
Oman Volcanics				Oman
Kahar & Morad slates & shales	?- Doran Granites -? ?- Rizu Volcanics -?			Iran

By about 620 Ma, the structural evolution of the Middle East appears to have taken a noticeable change in course away from collisional to extensional tectonics. Firstly, the Arabian shield appears to have failed under tension as is evident by the intrusion of post-tectonic, alkali granites (Table 1; generally 620–580 Ma; Stoesser 1986) and the extrusion of the coeval Shammar rhyolites (particularly 600 Ma; Schmidt *et al.* 1979). Secondly, NS-trending, extensive and deep, fault-bounded basins developed between 620 to 608 Ma, in the Central Arabian shield. These extensional grabens were filled with the coarse clastics and volcanics of the Bani Ghayy Group (Agar 1985). Similarly in Iran (Doran Granite), Jordan and Egypt, post-tectonic plutonic activity dominated the period from about 620 to 580 Ma (Table 1; Stocklin 1968; Berberian & King 1981; Stern & Hedge 1985).

Berberian & King (1981) associated the Infracambrian stage with tectonic extension in Iran, the Arabian Gulf and the Arabian shield, while Stern (1985) concluded that a similar environment simultaneously existed in NE Egypt and the Sinai Peninsula. Husseini (1988) proposed joining the extensional system of North Egypt across the Arabian Peninsula into South Oman and Pakistan by interpreting the extensive NW–SE-trending Najd strike-slip system, which moved between about 600 and 540 Ma, as a transform fault system (Fig. 2; based on Morel & Irving 1978; Gorin *et al.* 1982; Husseini 1988; also Lower Cambrian latitudes after Smith *et al.* 1981).

The interpretation of the Najd system as a transcurrent fault system was also suggested by A. F. Howland in a discussion of a paper presented by J. M. Moore on this subject at a meeting of the Geological Society of London in 1979. This interpretation is in contrast to the model proposed by Schmidt *et al.* (1979) which compared the Najd system to similar fault systems of Central Asia which are driven by the collision between the Indian and Asian plates. The collisional model of the Najd system may be appropriate for the very early phase of Najd-trending deformation when the Arabian shield was fused together along the Idsas and Nabitah sutures.

The period from about 600 Ma to 540 Ma was dominated by a major phase of intra-continental rifting and wrenching; informally referred to as the 'Infracambrian Arabian extensional system' (Fig. 2; note that in the text present-day po-

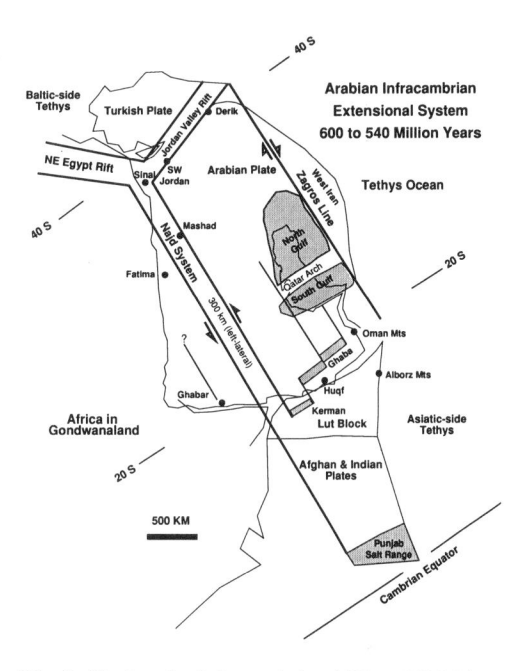

Fig. 2. During the Infracambrian (600 to 540 Ma), a triple junction was centered near the Sinai Peninsula with one rift branch extending along NE Egypt, a second rift branch extending along the Jordan Valley and Dead Sea to Derik in SE Turkey, and a third wrench-rift branch extending along the Najd left-lateral, strike-slip, fault system. Radiometric age determinations (Table 1) show that these three branches were active simultaneously. The salt basins of the Middle East were rift basins which were accommodated by the Najd and Zagros transcurrent systems as well as other Najd-parallel faults. The western extent of the Infracambrian evaporites is poorly constrained.

larities, such as east, north, etc., are used for simplicity even though the Cambrian polarities were oriented in the opposite sense). This extensional system not only included the development of great rift basins, but was also accompanied by a structural style of wrenching, mild folding, and normally-faulted horst and graben systems.

Regional correlation

In this paper the rocks which were deposited while the Infracambrian Arabian extensional system was active are interpreted as syn-rift sediments. These sediments reflect deposition in basinal areas which were mostly continental near the western part of the Arabian plate, Africa and Turkey (Gondwana-side) and increasingly marine towards the eastern rift basins of the Arabian Gulf, South Oman, Kerman and the Punjab Salt Range in Pakistan (Tethys-side). Initially, the paper focuses on the evidence for a triple junction centered around the Sinai Peninsula and NW Saudi Arabia as sketched in Fig. 2. This triple junction may have consisted of one rift branch which extended from Sinai westwards across NE Egypt; a second rift branch which extended from the Jordan Valley (SW Jordan), across the Dead Sea into SE Turkey (Derik); and a third branch, the Najd rift-wrench system, which cut across the Arabian Peninsula.

The paper then describes key outcrops, in the western and southern Arabian Peninsula (Mashad, Wadi Fatima and Wadi Ghabar) which bridge the stratigraphic correlation from the Sinai triple junction into South Oman (Oman Mountains and Ghaba-Huqf), Central Iran (Kerman), the Arabian Gulf (Gulf) and Pakistan (Punjab Salt Range). In Fig. 3, the stratigraphic correlation is presented in a manner which represents increasingly marine conditions to the right where the Alborz Mountains is chosen as the most complete marine-type section of the syn-rift group.

As shown in Fig. 3 (and idealized in Fig. 4)

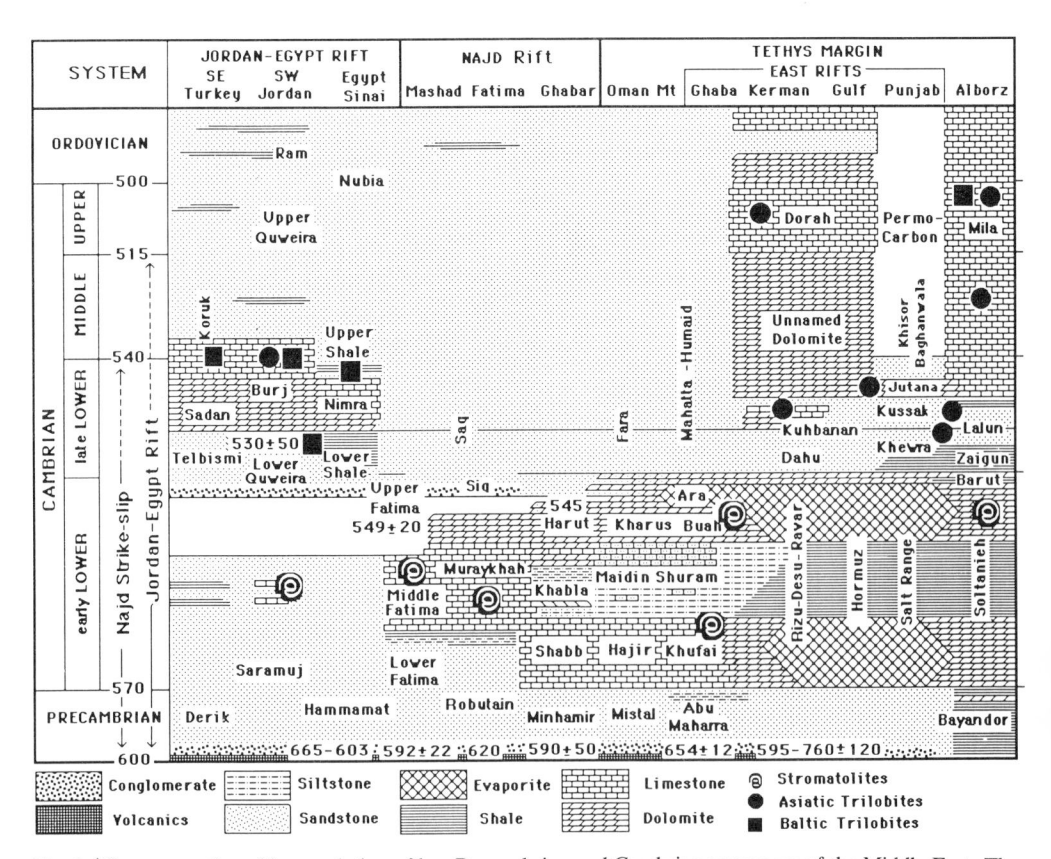

Fig. 3. Chrono-stratigraphic correlation of key Precambrian and Cambrian sequences of the Middle East. The outcrop sites are arranged to emphasize paleogeographic locations as shown in Fig. 2.

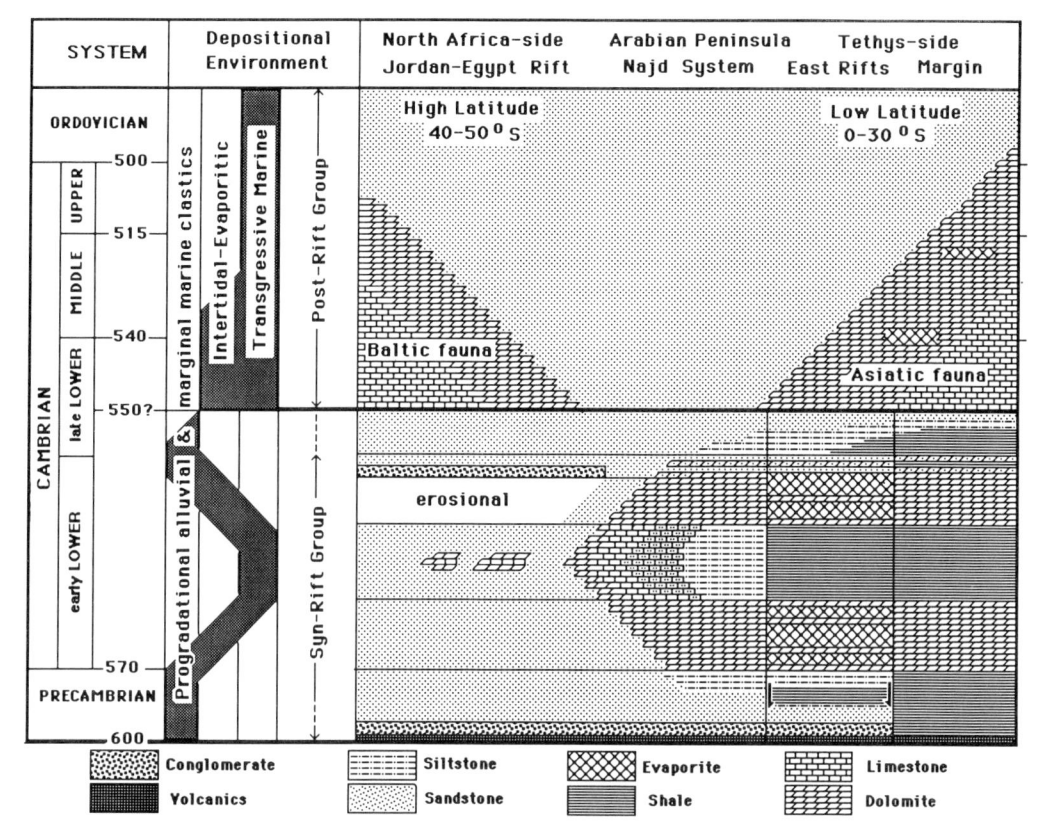

Fig. 4. Idealized regional interpretation of the Infracambrian and Cambrian depositional system for a conceptual traverse from the Sinai triple junction, across the Arabian Peninsula to the eastern Tethys rifts (South Oman, Kerman, Arabian Gulf and the Salt Range) and the Tethys Ocean margin.

the syn-rift rocks overlie the basement and include, at the base, fluviatile conglomerates and clastics which grade upwards and seawards (Tethys-side) into transgressive marginal marine clastics and eventually to marine carbonates. The transgressive sequence is followed by regressive shallow-marine carbonates and evaporites or an erosional hiatus landward (Gondwana-side). The post-rift sediments are separated from the syn-rift sediments by massive, progradational alluvial and marginal marine transgressive clastics. Finally, the postrift clastics grade to shallow-marine carbonates which dominate the Middle Cambrian.

NE Egypt and Sinai Peninsula Rift Branch

In NE Egypt, the basement includes granites and grano-diorites which range in age from 670 to 540 Ma, but mostly 620 to 575 Ma; the Dokhan rhyolites and andesites dated at 600–580 Ma; and E–W to NE–SW-trending, dike swarms dated at 595–540 Ma (Stern & Hedge 1985). Further east in the NW Arabian shield, the rift-related Minewah Volcanics are dated at about 575 Ma and an alkali–feldspar syenite is dated at 553 ± 4 Ma (Clark 1985).

Stern (1985) concluded that the NE Desert of Egypt was a rift basin which terminated in the Sinai Peninsula. Clark (1985) also concluded that the NW edge of the Arabian Peninsula underwent rifting after 625 Ma. Piper (1983), based on palaeomagnetic reconstructions of the Cambrian, extended this rift system westwards across North Africa and interpreted the Najd system and the areas of the Jordan Valley as two more rift branches.

The oldest sediments in NE Egypt, the Hammamat Formation (up to 5 km thick), consists of fluviatile clastics dated at about 600 Ma, (Massey 1984; Stern & Hedge 1985). Further east, in the Timna area of the Sinai Peninsula, cross-bedded fluviatile sandstones (0–80 m; Bentor 1960, in Wolfart 1983) rest on the basement complex. These fluviatile sandstones, which may correspond to the Hammamat Formation, are overlain by the late Lower Cambrian Lower Shale (12–13 m) which consists of a basal conglomerate, red sandstones and shale, which in turn pass to the limestones and dolomites of the late Lower Cambrian Nimra Formation (23 m). The Nimra Formation is overlain by the Zebra Sandstone (16 m) and the Upper Shale (3 m) of very early Middle Cambrian age. Finally, the Upper Shale is overlain by the Nubian Sandstone which includes sediments ranging from the Cambrian through the Cretaceous. The top part of the Lower Shale, the Nimra and the Upper Shale all yield Baltic trilobite fauna.

Therefore, in NE Egypt, Sinai and NW Arabia, the igneous rocks suggest a complex NE–SW to E–W trending, extensional environment between about 620 to 540 Ma. The disconformity between the syn-rift Hammamat Formation (600 Ma) and the much younger, late Lower Cambrian Lower Shale Formation of Timna suggest that an erosional phase separates the syn-rift from the post-rift groups.

Jordan Rift Branch

In the Wadi Araba area of the Jordan Valley, acid to basic N–S striking dikes, which cut older plutonic rocks, are peneplained by younger Cambrian clastics (Bender 1982). Also elongated N–S trending rhyolithic to basic andesitic and quartzporphyric intrusive rocks are prevalent. These igneous rocks are dated between 560 ± 6 and 510 ± 10 Ma. Bender interpreted the Jordan Valley geosuture as an Infracambrian rift basin. The N–S-trending Jordan Valley rift may have joined the NE–SW trending North Egypt rift branch in the Sinai Peninsula and NW Saudi Arabia.

In SW Jordan the Saramuj Formation consists of conglomerates, sandstones, thin beds of stromatolitic limestones which are intruded by acid and basic dikes (Mitchell 1955, in Gorin *et al.* 1982). In other outcrops (Bender 1963), the basal conglomerate (up to 50 m thick) of the Lower Quweira Formation (Siq Sandstone in NW Arabia; Powers *et al.* 1966) overlies the crystalline basement and includes boulders with a diameter of one metre, followed conformably by the Bedded Arkose Sandstone of the top part of the Lower Quweira Formation (up to 200 m in the area of the Dead Sea). The Lower Quweira is overlain by the Burj Formation (124 m in Suweilah-1 near the Dead Sea) which consists of a lower portion of sandy shale and an upper carbonate which yielded both Baltic and Asiatic trilobites fauna and inarticulate brachiopods of late Lower to early Middle Cambrian age.

Towards NW Saudi Arabia, the Burj Formation passes laterally and eventually is overlain by the marine White Fine Grained Sandstone which corresponds to the lower part of the Upper Quweira Formation. The top part of the Upper Quweira (Massive Brownish Weathered Sandstone) consists of massive, cross-bedded medium to coarse-grained sandstones which grade into interbedded silty shale (394 m in Suweilah-1). At Wadi Ram in SW Jordan, the Upper Quweira Formation is overlain by the Ram Sandstone Formation (246 m, Quennel 1951; also Bedded, Brownish Weathered Sandstone) which contains Lower Ordovician Cruziana.

The volcanic dikes together with the clastics and stromatolitic limestones of the Saramuj Formation, may correspond to the early syn-rift sediments. The Lower Quweira basal conglomerates may correspond to the final syn-rift regression, while the top part of the Lower Quweira, Burj and Upper Quweira Formations may comprise the post-rift group which spans the late Lower Cambrian to the Upper Cambrian.

Extent of the Jordan Valley Rift into Derik, SE Turkey

In the Derik area of SE Turkey, the Derik Formation consists of andesites, porphyries, spilites, volcanic breccia and tuffs which are interbedded with reddish sandstone and shale (> 500 m; Kellog & Tayyar, in Wolfart 1981). The Derik Formation is overlain with a slight disconformity by the Telbismi Formation (580 m; Dean 1975, in Wolfart 1983) which consists of alternating conglomerates containing volcanic pebbles, thin-bedded sandstone, shale and sandy limestone. The Telbismi Formation passes gradually into the dolomites of the Sadan Formation (260 m). The Dolomite Formation, in turn, passes transitionally into the limestone unit (30 m) of the Koruk (also Sosink) Formation (260 m). The Sadan Formation, limestone unit which includes Middle Cambrian Baltic trilobite fauna, the Koruk Formation consists of interbedded limestone, shale and siltstone (200 m) followed by 560 m of cross-bedded, shallow-marine sandstone which are overlain by the Middle Ordovician Bedinan Formation.

The Derik area lies on strike with the Jordan Valley geosuture and the igneous rocks of the syn-rift Derik Formation support the possibility of extending the Jordan Valley rift branch into SE Turkey. The Sadan Formation is palaeontologically dated as early Middle Cambrian, and therefore the underlying Telbismi may include both the syn-rift regressive sequence and the post-rift transgressive sequence.

Najd Wrench-Rift Branch, Saudi Arabia

The exposed Najd fault system consists of a series of NW–SE-trending, left-lateral fault zones which are 1100 km long and cut an area about 300 km wide in the Arabian shield (Brown & Jackson 1960). Agar (1985) concluded that the early Najd system may have been dextral between 640 and 600 Ma, while the subsequent, left-lateral movement is assigned to the period between 600 and 540 Ma. This age assignment is generally consistent with many radiometric estimates of igneous rocks which are associated with the Najd system (e.g. Hadley 1972; Fleck et al. 1976; Schmidt et al. 1979).

The Najd System consists of four primary shear zones which dislocated the older N–S Nabitah ophiolitic suture by 165, 125, 22 and 32 km resulting in a total of 300 to 350 km of left-lateral movement (Schmidt et al. 1979). These fault zones may be extrapolated into NW–SE-trending faults into NW Arabia, SW Jordan and NE Egypt (e.g. Duwi shear zone in Egypt) where they terminate against the rift branches of North Egypt and the Jordan Valley. The synchronicity of the left-lateral phase of the Najd system with extension in North Egypt and the Jordan Valley, together with the convergence of these fault systems into an area centred in the Sinai Peninsula supports the interpretation of an Infracambrian triple junction in this area. Coincidently these three branches also subtend angles of about 120° which further suggests the geometry of a triple junction.

Moore (1979) described the early phase of Najd movement as ductile and this early phase was accompanied by greenschist regional metamorphism. However, the subsequent displacement was brittle and resulted in the development of 'pull-apart' basins which received the syn-rift J'Balah group sediments (Delfour 1970). In the Mashad area in the northern part of the Arabian shield, Hadley (1972) identified three formations in the J'Balah group. The basal Robutain Formation lies uncomformably on granite which is dated at younger than 620 Ma and consists of poorly sorted, cobbles and boulders (up to 1 m in diameter) which grade into red–brown graywackes composed of volcanic rocks and further upwards into siltstone, and shale. The Badayi Basalt Formation cuts the Robutain Formation. Above the Badayi Formation the Muraykhah Formation consists of a lower limestone member (135 m) and an upper dolomite member (150 m) which yielded Lower Cambrian stromatolites such as *Conophyton* (Cloud et al. 1979). The J'Balah Group is unconformably overlain by the relatively flat-lying, Cambro–Ordovician Saq Sandstone (Delfour 1970) or the older fluviatile Siq Formation (Hadley 1972).

The Siq Sandstone (Powers et al. 1966; also Yatib Formation of Vaslet 1988) in Central and NW Saudi Arabia consists of a basal conglomerate of angular and rounded boulders and pebble in a matrix of arkosic sand followed by massive sandstone. The Siq Sandstone is correlated to the Lower Quweira Formation of SW Jordan (Powers et al. 1966) and is overlain by the Saq Sandstone. The Saq Sandstone has been divided into a lower fluvial–deltaic portion which is commonly crossbedded sandstones or siltstones, and an upper part of shallow marine sandstones (Delfour et al. 1982; Vaslet 1988). Near its top, the Saq Sandstone includes the Cruziana Series of Lower Ordovician age (Helal 1964).

Therefore in the areas which were affected by the Najd system, the J'balah sediments (Robutain and Muraykhah Formations) and volcanics (Badayi Basalt Formation) represent the early syn-rift sediments. The Siq Sandstone corresponds to the Lower Cambrian regression, and the Saq Sandstone represents the post-rift progradational alluvial to shallow marine transgression.

Wadi Fatima, Western Saudi Arabia

On the western edge of the Arabian shield, at Wadi Fatima, the syn-rift Fatima Series has been bio-stratigraphically correlated to the J'Balah Group (Cloud et al. 1979; Basahel et al. 1984). The Lower Fatima Formation (10–150 m) consists of a basal conglomerate, pyroclastics and sandstones, which grades into siltstones and shales. The Middle Fatima Formation (90–130 m) consists of massive dolomitized carbonates which are intercalated with reddish clastics, while the Upper Fatima Formation consists of reddish siltstone, pyroclastics and lavas (500 m). The Fatima Series are associated with igneous rocks which range in age from 592 ± 22 Ma to 549 ± 20 Ma (Gettings & Stoesser 1981) and the Middle Fatima carbonates include stromatolites *Conophyton*, archaeocyathids and ichonofossils which Basahel et al. (1984) assign to the Lower Cambrian.

Wadi Ghabar, South Yemen

In Hadramawt, South Yemen, the oldest sedimentary rocks are the Ghabar Group. Beydoun (1966) describes this group at two nearby localities: Wadi Al Ghabar and Wadi Al-Minhamir. This group is strongly folded by tectonic movements but shows no metamorphism. The Ghabar Group rests on lavas, is overlain by Cretaceous sediments, and consists of four formations. The basal formation is the Minhamir Formation (95–143 m) which consists of water deposited lithic and crystal tuffs, tuffaceous mudstones with pebbles of lava, dull red tuffaceous limy conglomeratic sandstones. The Minhamir Formation is intruded by silicic and mafic dikes.

The Shabb Formation (13–42 m) overlies the Minhamir Formation conformably and consists of limestone that are partly sandy and includes chert bands and some gypsum. Conformably overlying the Shabb Formation, the Khabla Formation at Wadi Ghabar (210 m) consists of dolomites and calcareous sandstones, siltstones, and some gypsum. At Wadi Minhamir the Khabla Formation (142 m) consists mostly of tuffaceous or arkosic sandstone. The Harut Formation overlies the Khabla Formation and consists of 31 m of sandy dolomites, massive calcareous quartzites with shaly beds of impure dolomite and fairly pure limestone.

In the nearby Wadi Garish and Jabal Madbi area, the Gharish Group has been compared to the Ghabar Group (Beydoun 1966). The Gharish Group consists of tuffaceous conglomeratic sands and shales and thin bedded limestones which are metamorphosed to the albite–epidote–amphibolie facies. Samples of pegmatite from the Gharish Group have been dated by potassium-argon methods at 590 ± 50 Ma (York, in Beydoun 1966) and 545 Ma (Schurmann, in Beydoun 1966).

Both Wadi Ghabar and Wadi Minhamir appear to be basins which strike parallel to the Najd system. The preservation of the Ghabar sediments in apparent Najd-trending basins and their age consistency with Najd movements suggests that these sediments are also syn-rift sediments. The folding and metamorphism of these syn-rift sediments at Wadi Gharish may be largely due to folding by Najd wrench deformation and thermal metamorphism. This interpretation leads to the conclusion that the Najd system not only produced localized pull-apart basins, but also deformed both the shallow crust and the syn-rift sediments. Therefore the syn-rift and post rift sediments may be conformable in some basins and greatly disconformable in others. Furthermore, a clear separation of deformation due to the collisional Nabitah–Idsas phase (640–630 Ma) from Najd-related folding (younger than 600 Ma) may only become apparent from radiometric age determinations, or metamorphic and sedimentary lithological considerations.

Huqf and Oman Mountains, South Oman

In the Huqf–Ghaba area, South Oman, the oldest sedimentary rocks are the Huqf Group. As in South Yemen, the Huqf Group also consists of four formations: the Abu Mahara, Khufai, Shuram and Buah Formations. These four formations may be correlated to the Mistal, Hajir, Maidin and Kharus Formations in the Oman Mountains (Tshopp 1967). A core hole drilled in the Huqf area bottomed beneath the basal Abu Mahara Formation in volcanic rock dated at 654 ± 12 Ma (Gorin *et al.* 1982) while the topmost Buah Formation is unconformably overlain by the Mahatta-Humaid Formation of the Cambro–Ordovician Haima Group (Morton 1959; Beydoun 1960; *Glennie 1977*, Glennie *et al.* 1974).

In outcrop the Abu Maharra Formation (> 600 m) rests on metamorphic basement and consists of quartz-cemented sandstones and argillaceous siltstones to silty shales. Near the top the siltstones are increasingly dolomitic and alternate with silty dolomites. In the Oman Mountains the stratigraphically equivalent Mistal Formation consists of conglomerates which have been compared to glacial tillites. These conglomerates resemble other basal syn-rift sediments in the Middle East and a glacigenic origin appears unlikely as South Oman was nearly at 20° south of the equator at that time.

The Abu Maharra passes into the silty dolomites of the Khufai Formation (240–340 m) which includes well bedded dolomites near the top. The Khufai Formation is overlain disconformably by the Shuram Formation (230–330 m) which consists of red–brown shaly siltstones which pass upwards to siliciclastics, lime grainstones and mudstones. The Shuram Formation is overlain by the Buah Formation (0–340 m) which consists of thinly bedded limestones and dolomites in the lower part which pass to thickly bedded dolomites with chert nodules. The top part of the Buah includes crystalline dolomite which alternate with sandstones up to 20 m in thickness.

The clastics in the top part of the Buah Formation are overlain by the Mahatta-Humaid Formation (550 m) which consists near the base of poorly sorted, cross bedded sandstones which pass into sandstone and siltstone. In the Ghaba well, Lower Ordovician graptolite shales overlie the Mahatta-Humaid, and therefore this latter formation is of Cambro−Ordovician age. In the Oman Mountains the Fara Sandstone Formation corresponds to the Mahatta-Humaid Formation. In South Oman the Huqf Group and its equivalent in the Oman Mountains are syn-rift sediments while the Mahatta-Humaid and Fara Formations are representative of the post-rift transgressive sequences.

Ghaba and South Oman Graben System

The Ghaba and South Oman graben system trends NE−SW and is about 100 km wide and 600 km long (Gorin et al. 1982). The Ghaba graben is abruptly terminated to the east by the Hawasina fault. The South Oman graben is bounded to the north by the normal, NE−SW-trending, Ghudun-Khasfah fault and is offset from the Ghaba basin by a possible Najd-trending fault. The Ghudun-Khasfah fault has a normal throw of about 2 km (Al-Marjeby & Nash 1986).

In the Ghaba graben, the Ara Formation outcrops in six surface salt plugs. Wells drilled in southern Oman show that the Ara Formation consists of a lower part of halite with subordinate anhydrite, shales, siltstones and thin carbonates and an upper part of dolomites, limestones with subordinate anhydrite, shale, siltstone, and sandstone intercalations. Gorin et al. (1982) correlate the Ara Formation laterally to part of the Buah Formation.

Kerman Graben, Central Iran

In the Kerman area of Central Iran, the Rizu Formation consists of siliceous shale and dolomites (12 m); 100 m of alternating sandstone and dolomite; 500 m of sedimentary−volcanic deposits of sandstone, dolomites, conglomeratic layers and red−brown tuffs (Huckriede et al. 1962; in Wolfart 1983). The Rizu Volcanics are capped by red−brown lavas and radiometric dating of a syn-sedimentary ore deposits yielded an age range of 760 to 595 ± 120 Ma. The Rizu Formation is the extrusive equivalent of the coeval post-tectonic, alkali-rich Doran Granite which is correlated to the alkali-rich granites of Saudi Arabia (mostly 620−570 Ma) and the granites of NE Egypt (mostly 620−575 Ma).

North of Kerman the Rizu Formation is believed to pass laterally to the Desu Formation which consists of dolomites, gypsum and red sandstones with volcanic rocks (up to 300 m), and in the Kerman Graben, the dolomitic facies pass into the salt facies of the Ravar Formation. The Desu-Rizu Formations are overlain by the Dahu Formation (0 to 1000 m) which consists of red−violet−brown sandstones. In turn the Kuhbanan Formation overlies the Dahu Formation and consists of a lower sandstone (55 m), a dolomite member (80 m); a limestone member (40 m) which yielded middle to late Lower Cambrian Asiatic trilobite fauna; and an upper part of fine grained, red, massive sandstone with dolomitic intercalations (130 m). The Kuhbanan Formation is overlain by an unnamed dolomite (450 m) which is followed by the Dorah Formation (145 m) which yielded Upper Cambrian Asiatic trilobite fauna.

Therefore in the Kerman graben area the Rizu, Desu and Ravar Formations may be assigned to the syn-rift group while the Dahu marks the onset of the post-rift group in the late Lower Cambrian.

Gulf Basins, Arabian Gulf and Zagros Mountains of Iran

The Infracambrian tectonic nature of the Zagros and Dibba faults is not known; however, several striking structural and depositional patterns suggest that these faults played an important role in the Infracambrian extensional system. Firstly the Zagros and Dibba faults define the northeastern and southeastern limits of the Hormuz evaporites. This suggests that both of these faults had a normal component of movement. Secondly, the Zagros fault is parallel to the Najd system, while the Dibba fault is parallel to the South Oman graben system and perpendicular to the Najd and Zagros systems. These geometries suggest that the Zagros line may have been a right-lateral, strike-slip system which complemented the Najd movement. Such an interpretation for the Zagros line would be consistent with the opening of the Arabian Gulf and Zagros Mountains rift basins along lines parallel with the Dibba fault.

The salt basins of the Arabian Gulf are numerous: the salt of the South Gulf basin probably extends into parts of the Rub Al-Khali in SE Saudi Arabia and is separated from the North Gulf salt basins by a NE−SW basement high, which extends from the Qatar Arch to the Zagros Mountains. The North Gulf basin extends northwards into Iraq, Kuwait and eastwards into Eastern Saudi Arabia, Bahrain and Qatar. The lack of Infracambrian salt along the NE−SW-trending Qatar-Zagros arch, which separates the two Gulf basins suggests that the Arabian Gulf includes NW−SE-trending structural elements.

In the Zagros Mountains, the syn-rift Hormuz Formation (> 1000 m) is composed of rocksalt, gypsum, fetid limestone, cherty dolomite, red sandstone and shale and volcanic rocks which are probably syn-sedimentary (Kent 1970). Kent deduced that the saline units may be two or more in number and overlain and underlain by over 1000 m of non-evaporitic beds. In the Zagros Mountains, the Hormuz Formation outcrops as salt plugs and salt glaciers (Wolfart 1981). King (1937) found late Lower Cambrian Asiatic trilobite fauna in an unnammed limestone unit near Isfahan in the Zagros Mountains. Therefore Stocklin (1968) and Wolfart (1981) both suggest that this formation spans the Precambrian–Cambrian boundary and extends into the Lower Cambrian.

Punjab, North Pakistan Rift Basin

In North Pakistan, the Salt Range Province is a large fault-controlled Infracambrian salt basin which is over 250 km wide and may extend under the Hindu–Kush fold belts of the Himalayan suture. The Salt Range Formation consists of the Punjab halite and gypsum and associated carbonates and clastics and this formation overlies the Precambrian metamorphic and crystalline basement (Shah 1977; Wolfart 1983). The Salt Range Formation is overlain by the Khewra Sandstone Formation (185 m; also Purple Sandstone Series). This formation include Asiatic trilobite prints which are middle Lower to late Lower Cambrian.

The Khewra Formation is overlain by the late Lower Cambrian Kussak Sandstone Formation (50 m), the late Lower Cambrian Jutana dolomite (50–80 m), the Khisor Formation consisting of gypsum of unknown thickness, and the Baghanwala Formation (40–116 m). The final Baghanwala Formation, which consists of alternating red shales and sandstones with salt pseudomorphs, is overlain by the early Permian Tobra Formation.

Alborz Mountains, NW Iran

The areas of the Alborz and Soltanieh Mountains may have been closest to the margin of the Tethys Ocean in Infracambrian time. In Fig. 2, the Alborz outcrop is tentatively placed along the eastern Tethys margin of the Central Iran Lut Block; however, it is quite possible that certain parts of the Soltanieh and Alborz Mountains were actually attached to either west or Central Iran. This confusion arises from geological complexities caused by subsequent collisions along the boundary between Central and West Iran. Despite this, the Alborz type section is sufficiently prevalent in other parts of west and Central Iran as to serve as a representative section for the margin of the Tethys Ocean.

In the Alborz Mountains, the Doran Granite intrudes the Kahar Formation, and this granite, in turn, is truncated and transgressively overlain by the Bayandor Formation (500 m). The Bayandor Formation consists of sandstones and shales with thin dolomitic intercalations (Stocklin 1968, 1972; Ruttner et al. 1968). The Soltanieh Formation overlies the Bayandor Formation and it consists of three members: the Lower Dolomite (120 m), the Chapoghlu Shale (247 m), and the Upper Dolomite (790 m). The Soltanieh Formation, which is considered to be broadly correlative to the Hormuz Formation of the Arabian Gulf and Zagros Mountains, and to the Rizu, Desu and Ravar Formations of Central Iran, yields only stromatolites, mostly *Collenia*.

Within the Soltanieh Formation, the Chapoghlu shale, which consists of dark, green–grey, slaty-shale is interbedded with thin limestone and calcareous shale and may be interpreted as the maximum transgressive phase between the two overlying and underlying shallow-marine dolomites of the Soltanieh Formation and the two evaporitic cycles of the Hormuz salt. In the Central Alborz and in the Soltanieh Mountains, the Lower Dolomite is missing and the Chopaghlu shale rests directly on the Precambrian Kahar Formation. The lack of deposition of the Bayandor Formation and Lower Dolomite of the Soltanieh Formation, suggest that these areas were elevated and transgressed only when the relative sea level was at a maximum.

The Soltanieh Formation is overlain by the Barut Formation (453 m) which consists of argillaceous, silty and fine sandy shales containing more than fifty intercalations of dolomite and limestone. Above the final stromatolitic limestone, the Barut Formation passes gradually into the Zaigun Formation (714 m) which consists of silty to fine, cross-laminated red shales. Next, the Zaigun Formation is overlain by the sandy quartzitic facies of the Lalun Sandstone Formation. Near the top of the Lalun Formation, Gansser & Huber (1962) and Allenbach (1966) identified *Cruziana*-like tracks which are middle to late Lower Cambrian. The Lalun Sandstone passes into the Middle Cambrian to Ordovician Mila Formation (485 m) which consists mostly of limestone.

The syn-rift group which consists of the Bayandor and Soltanieh Formations appears to be entirely marine. These syn-rift sediments pass to the post-rift group very gradually and the Barut Formation clearly marks the termination of carbonate conditions associated with the Soltanieh Upper Dolomite in favour of progradational continental alluvial system of the Zaigun ShaleLalun Sandstone. This interpretation leads to the assignment of the Zaigun, Lalun and Mila Formations to the post-rift group.

Discussion

In Fig. 4 an idealized schematic interpretation of the regional depositional system, described in Fig. 3, is presented. A significant time-stratigraphic line defines the lower boundary of the syn-rift rocks. This boundary is demarcated by the onset of the Najd rift system, around 600 Ma, and may be characterized by rift-related volcanic rocks (also dated from about 620 to 540 Ma) and Precambrian continental conglomerates and coarse clastics which grade to marginal marine clastics towards the Tethys-side. The siltstones and silty-shales of the Abu Maharra and Bayandor Formations clearly represent marginal marine conditions. The lower boundary includes the Precambrian Derik, Saramuj, Hammamat, Robutain, Lower Fatima, Minhamir, Abu Maharra, Mistal, Lower Rizu and Bayandor Formations.

The upper siltones and shales of these Precambrian rocks represent a marine transgression which eventually introduced thick carbonates and evaporites into the Middle East and these carbonates may be an appropriate lower boundary for the Cambrian system. In areas of the Arabian Peninsula, which subsided during the Infracambrian extensional phase due to the Najd rift-wrench system, the maximum phase of the Lower Cambrian transgression deposited the Fatima carbonates as far as the Red Sea coast, the Muraykhah carbonates in the Central Arabian shield, the Khabla silty carbonates in South Yemen. Further towards the Tethys Ocean, deeper marine conditions resulted in the deposition of siltstones and limestones of the Maidin and Shuram Formations in Oman, and finally, at Alborz Mountains along the Tethys margin, the calcareous shales of the Chapoghlu member of the Soltanieh Formation.

The passage from continental and marginal marine conditions in the Precambrian to transgressive marine in the Lower Cambrian must have involved an intertidal−evaporitic phase which would correspond to a Lower Cambrian (Infracambrian) evaporite. The transgression was followed by a regression of the Tethys Sea away from the Central Arabian Peninsula and a second intertidal−evaporitic phase accompanied this regression. Towards the Tethys Ocean margin, this transgressive−regressive model would clearly result in two carbonate−evaporite units separated by deeper-marine calcareous clastics. This pattern may be clearly seen in the Soltanieh Formations. Of the two evaporitic phases, the second appears to be the most important and this is reflected by the much greater thickness of the Upper Dolomite member of the Soltanieh Formation (710 m) relative to its Lower Dolomite member (120 m). Also in Oman, the Ara Formation is generally correlated to the second evaporitic phase of the Buah regressive carbonates.

The dolomite members of the Tethys-side outcrops (e.g. Alborz) indicate that during the two evaporitic phases, relatively frequent and short-lived sea level variations occurred. During periods of supratidal conditions the gently-subsiding, rift basins and shelfs were carbonate platforms while during periods of subtidal conditions, the shelfs suffered an hiatus while the subsiding basins accumulated evaporites consisting mostly of halite.

The upper boundary of the syn-rift group is less clear to identify as it involves a regression as defined by a sharp erosional unconformity landwards and a very thick facies change in marine areas. This upper boundary may be characterized by the onset of an epi-continental prograding alluvial to marginal marine clastic system which gradually terminated the prevailing syn-rift carbonate system. In the areas which were continental in the earliest rift phase (600−570 Ma), this Lower Cambrian regression resulted in erosion and fluvial conditions for a second time. In these areas two similar sets of fining-upwards, clastic sequences occur, and these may both lie on the crystalline−metamorphic basement. From both bio-stratigraphic correlations and lithological considerations, this unconformity probably occurs between the Telbismi and Derik Formation in SE Turkey, Lower Quweira and the Saramuj Formations in SW Jordan, Lower Shale conglomerate and the Hammamat Formations in Sinai and the Saq and Siq Sandstones in Saudi Arabia.

The regression which closed the syn-rift phase is dramatically expressed to the Tethys margin areas. In Oman, the alternation of dolomites with sandstones near the top of the Buah Formation indicates a sharp influx of clastics. In Iran, the transitional Barut with its 700 m of alternating dolomite and limestone with silty to fine shales, clearly shows the arrival of clastics which eventually terminated the carbonate environment and induced a restricted water environment typified by the Barut shale.

Therefore, the basal conglomerates at the

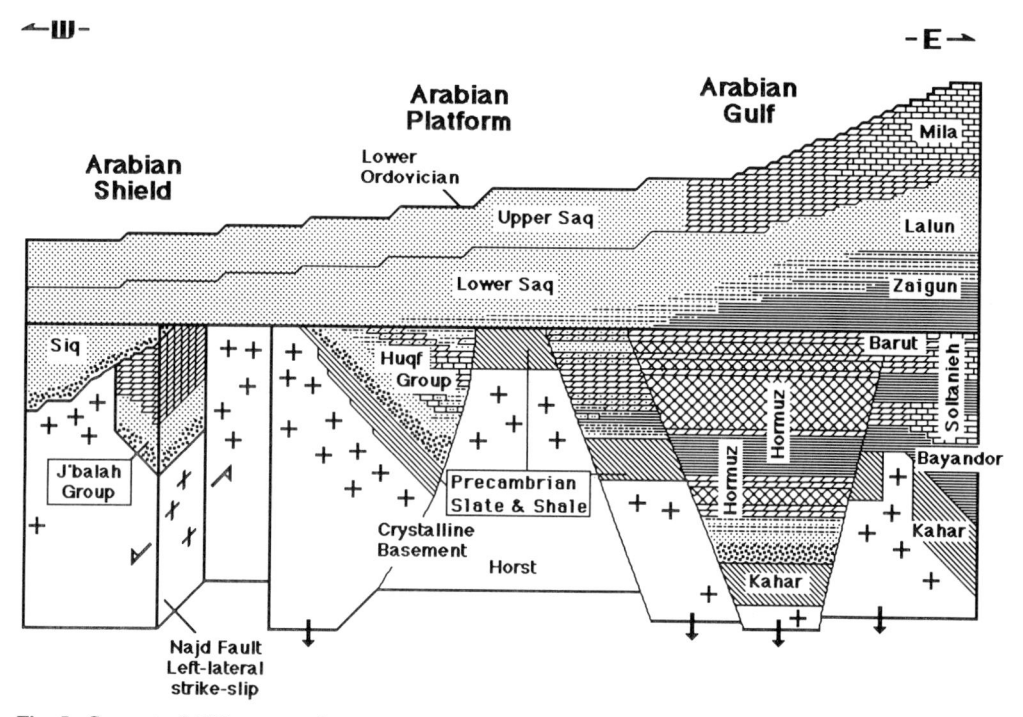

Fig. 5. Conceptual EW cross section through the Arabian plate datumed on the base of the post-rift Saq Sandstone and Zaigun Shale.

onset of the Lower Cambrian regression in the continental areas could well correlate to the base of the facies change from carbonates to clastics in the Tethys margin. This interpretation provides a natural, lithological and stratigraphic correlation between the Lower Saq, Mahatta Humaid, Fara and Lalun Sandstones. This correlation is further supported by the return of a very broad, bio-stratigraphically controlled, marine transgressive system in the late Lower Cambrian, with Asiatic fauna on the eastern Tethys-side, and Baltic fauna on the northern Tethys-side. During this final Cambrian transgression, the Arabian Peninsula was in a deltaic to shallow marine environment, and the non-fossiliferous sediments of the Saq Sandstone probably separated a low-latitude Asiatic Tethys in Iran and Pakistan (equatorial to 30° South) from a high latitude Baltic Tethys in Jordan and Turkey (40° to 50° South).

In Fig. 5, the interplay of the regional sequences of Fig. 4 are conceptually perturbed by wrench faults and normally-faulted, horst and grabens. The complexity of the resulting tectono-depositional system describes the numerous permutations of sediments which may be encountered beneath the Cambro–Ordovician transgressive sequences typified by the Saq Sandstone.

Conclusion

The evolution of the Arabian plate during the Late Precambrian indicates that accretionary collisional tectonics gradually evolved into intracontinental extensional tectonics. This process occurred over a time span of about 40 million years (640 to 600 Ma). The mountainous areas in the western part of the Arabian Peninsula, which were formed during the collisional stage, were largely eroded by the end of the Cambrian (500 Ma). Between 600 and about 540 Ma, the Middle East was dominated by wrenching, extension and subsidence. Since sealevel was at a low stand in the Lower Cambrian (Vail et al. 1977) the carbonate transgression between 570 and about 550 may be attributed to tectonically induced regional subsidence rather than a global sea-level rise.

The Infracambrian subsidence of the Middle East introduced two evaporitic phases resulting in salt precipitation in rift basins. These shallow marine to evaporitic facies may extend into many areas of the Middle East. In fact the basement-controlled structures may have the least representative, if any syn-rift section. The Infracambrian syn-rift sediments are ideal source rocks which may source structures as well as stratigraphic traps along the flanks of major basement-controlled structures. The absence of Infracambrian carbonates and salt in Jordan Valley and NE Africa may be due to the dominance of clastic deposition in the interior of the Gondwanaland continent and cooler conditions which prevailed at these higher latitudes.

References

AGAR, R. A. 1985. The Najd fault system revisited: a two-way strike-slip orogen. Directorate General of Mineral Resources, Saudi Arabia DGMR-OF05-18.

ALLENBACH, P. 1966. Geologie und Petrographie des Damavand und seiner Umgebung (Zentral Elburz), Iran. *Mitteilungen aus dem Geologischen Institute der Eidgenossischen Technischen Hochschule und der Universität Zürich*, **63**.

BASAHEL, A. N., Bahafzallah, A., OMARA, S. & JUX, U. 1984. Early Cambrian carbonate platform of the Arabian Shield. *Neues Jahrbuch fur Geologie und Palaontologie, Monatshefte*, **2** 113-128.

BENDER, F. 1963. Jordanie (Extreme Sud de la Jordanie), *Lexique Stratigraphique International*, 3, (Asie).

—— 1982. On the evolution of the Wadi Araba — Jordan Rift. *Jahrbuch der Geologischen Bundesanstalt Wien*, **45**, 3-20.

BENTOR, Y. K. 1960. *Lexique Stratigraphique International*, 3, (Asie).

BERBERIAN, M. & KING, G. C. 1981. Towards a paleogeography and tectonic evolution of Iran. *Canadian Journal of Earth Science*, **18**, 210-265.

BEYDOUN, Z. R. 1960. Synopsis of the geology of east Aden Protectorate. 21st International Geological Congress, Copenhagen, 131-149.

—— 1966. *Geology of the Arabian Peninsula, Eastern Aden Protectorate and part of Dhufar*. United States Geological Survey Professional Paper 560-H.

BROWN, G. F. & JACKSON, R. O. 1960. *The Arabian Shield*. 21st International Geological Congress, Copenhagen, 69-77.

CLARK, M. D. 1985. Late Proterozoic crustal evolution of the Midyan region, northwestern Saudi Arabia. *Geology*, **13**, 611-615.

CLOUD, P., AWRAMIK, S. M., MORRISON, K. & HADLEY, D. G. 1979. *Earliest Phanerozoic or latest Proterozoic fossils from the Arabian shield, Kingdom of Saudi Arabia*. United States Geological Survey, Saudi Arabia Proj. Rep. 260.

DEAN, W. T. 1975. Cambrian and Ordovician correlation and trilobite distribution in Turkey. In: (ed) J. BERGSTORM. *Evolution and morphology of the Trilobita, Trilobitoidea and Merostomata, fossils and Strata*, **2**, 353-373.

DELFOUR, J. 1970. Le groupe de J'Balah. Une nouvelle unite du Bouclier Arabe, *Bulletin Bureau Recherches Geologique et Mineraux*, (2nd series), **4**, 19-32.

—— DHELLEMMES, R., ELSASS, P., VASLET, D., BROSSE, J., LE NINDRE, Y. & DOTTIN, O. 1982. *Geologic Map of the Ad Dawadimi quadrangle, sheet 24G, Kingdom of Saudi Arabia*. Directorate General of Mineral Resources, Saudi Arabia.

FLECK, R. J., COLEMAN, R. G., CORNWALL, H. R., GREENWOOD, W. R., HADLEY, D. G., PRINZ, W. C., RATTE, V. C. & SCHMIDT, D. L. 1976. Geochronology of the Arabian shield, western Saudi Arabia: KAR results. *Bulletin of the Geological Society of America*, **87**, 9-21.

GANSSER, A. & HUBER, H. 1962. Geological observations in the Central Elburz, Iran. *Schweizerische Mineralogische und Petrographische Mitteilungen*, **42**, 593-630.

GETTINGS, M. E. & STOESSER, D. B. 1981. A tabulation of radiometric age determinations for the Kingdom of Saudi Arabia, **20**, Directorate General of Mineral Resources, Saudi Arabia.

GLENNIE, K. W. 1977. Outline of the geology of Oman. *Mémoire de la Société Géologique de France* (h. ser.), no. **8**, 25-31.

—— BOEUF, M. G., HUGHES CLARKE, M. H., MOODY-STUART, M., PILAAR, W. F. & RIENHART, B. M. 1974. Geology of the Oman Mountains. *Transactions of the Royal Dutch Geological Mining Society*, **31**, 1-423.

GORIN, G. E., RACZ, L. G. & WALTER, M. R. 1982. Late Precambrian-Cambrian sediments of Huqf group, Sultanate of Oman. *Bulletin of the American Association of Petroleum Geologists*, **66**, 2609-2627.

HADLEY, D. G. 1972. *The taphrogeosynclinal J'Balah group in the Mahad area, northwestern Hijaz, Kingdom of Saudi Arabia*. Report 151, Directorate General of Mineral Resources, Saudi Arabia.

HELAL, A. H. 1964. On the occurrence of Lower Paleozoic rocks in Tabuk Area, Saudi Arabia. *Neues Jahrbuch fur Geologie und Palaontologie, Monatshefte*, **7**, 391-414.

HUCKRIEDE, R., KURSTEN, M. & VENZLAFF, H. 1962. Zur geologie des Gebietes zwischen Kerman und Sagand (Iran). *Geologisches Jahrbuch, Reihe*, B, **51**.

HUSSEINI, M. I. 1988. The Arabian Infracambrian extensional system. *Tectonophysics*. **148**, 93-103.

JOHNSON, P. R., SCHEIBNER, E. & SMITH, A. 1987. Basement fragments, accreted tectono-stratigraphic terranes, and overlap sequences: elements in the tectonic evolution of the Arabian Shield. *Geodynamics Series, American Geophysical Union*, **17**, 323-343.

KENT, P. E. 1970. The salt plugs of the Persian Gulf region. *Transactions of the Leicester Literary & Philosophical Society*, **64**, 55-88.

KING, W. B. R. 1937. Cambrian trilobites from Iran (Persia). *Memoir, Geological Survey of India*, **22**.

MARJEBY, A. Al- & NASH, D. 1986. A summary of the geology and oil habitat of the Eastern Flank

Hydrocarbon Province of South Oman. *Marine and Petroleum Geology*, **3**, 306–314.

MASSEY, K. W. 1984. *Rubidium-strontium geochronology and petrography of the Hammamat Formation in the northeastern desert of Egypt*. M.S.c. Thesis, University of Texas at Dallas.

MITCHELL, R. C. 1955. Note sur le Precambrien de Transjordanie. *Compte-Rendu de la Société Géologique de France*, 262–263.

MOORE, J. M. 1979. Tectonics of the Najd transcurrent fault system, Saudi Arabia. *Journal of the Geological Society of London*, **136**, 441–454.

MOREL, P. & IRVING, E. 1978. Tentative paleocontinental maps for the early Phanerozoic and Protorozoic. *Journal of Geology*, **86**, 535–561.

MORTON, D. 1959. The geology of Oman. *Proceedings of the 5th World Petroleum Congress*, **1**, 277–294.

PIPER, J. D. A. 1983. Dynamics of the continental crust in Protorozoic times. *Memoir of the Geological Society of America*, **161**, 11–34.

POWERS, R. W., RAMIREZ, L. F., REDMOND, C. D. & ELBERG Jr., E. L. 1966. Geology of the Arabian Peninsula. Sedimentary geology of Saudi Arabia. United States Geological Survey Professional Paper 560D.

QUENNEL, A. M. 1951. The geology and mineral resources of (former) Transjordan. *Colonial Geology & Mining Resources*, **2, H. 2**, 85–115.

RUTTNER, A., NABAVI, M. H. & HAJIAN, J. 1968. Geology of the Shirghest area (Tabas area, East Iran). Report, *Geological Survey of Iran*, **4**.

SCHMIDT, D. L., HADLEY, D. G. & STOESSER, D. B. 1979. Late Proterozoic crustal history of the Arabian Shield, Southern Najd Province, Kingdom of Saudi Arabia, *In*: *Evolution and Mineralization of the Arabian–Nubian Shield*. Institute of Applied Geology, Jeddah, **2**, 41–58.

SHAH, S. M. IBRAHIM, 1977. Stratigraphy of Pakistan. *Memoirs Pakistan Geological Survey*, **12**.

SHANTI, A. M. Al- & MITCHELL, A. H. G. 1976. Late Precambrian subduction and collision in the Al Amar–Idsas region, Arabian Shield, Kingdom of Saudi Arabia. *Tectonophysics*, **30**, 41–47.

SMITH, A. G., HURLEY, A. M. & BRIDEN, J. C. 1981. *Phanerozoic paleocontinental world maps*. Cambridge University Press, Cambridge.

STERN, R. J. 1985. The Najd fault system, Saudi Arabia and Egypt: a late Precambrian rift-related transform system? *Tectonics*, **4**, 497–511.

—— & HEDGE, C. E. 1985. Geochronological and isotope constraints on late Precambrian crustal evolution in the eastern desert of Egypt. *American Journal of Science*, **185**, 97–127.

STOCKLIN, J. 1968. Structural history and tectonics of Iran: A Review. *Bulletin of the American Association of Petroleum Geologists*, **52**, 1229–1258.

—— 1972. *In*: *Lexique Stratigraphique International*, **3**, Asie.

STOESSER D. B. 1986. Distribution and tectonic setting of plutonic rocks of the Arabian Shield. *Journal of African Earth Sciences*, **4**, 21–46.

—— & CAMP, V. E. 1985. Pan-African microplate accretion of the Arabian Shield. *Bulletin of the Geological Society of America*, **96**, 817–826.

TSHOPP, R. H. 1967. The general geology of Oman. *Proceedings of the 7th world Petroleum Congress*, **2**, 231–242.

VAIL, P. R. MITCHUM Jr., R. M. & THOMPSON III, S. 1977. Global cycles of relative changes of sea level, *In*: PAYTON C. E. (ed.), Seismic Stratigraphy applications to hydrocarbon exploration. *Memoir of American Association of Petroleum Geologists*, **26**, 83–98.

VASLET, D. 1988. *Lithostratigraphie des Series Paleozoiques* (Ante Permien Superieur). PhD Thesis, University of Paris.

WOLFART, R. 1981. Lower Paleozoic rocks of the Middle East, *In*: HOLLAND, C. H. (ed) Lower Paleozoic rocks of the Middle East, Eastern and Southern Africa, and Antartica. Wiley, Chichester, 6–130.

—— 1983. *The Cambrian System in the Near and Middle East*. **15**, International Union Geological Sciences.

The Middle East Basin: a summary overview

R. STONELEY

Department of Geology, Imperial College, Prince Consort Rd, London SW7 2BP, UK

Abstract: In a review of classic petroleum provinces, it is instructive to consider the most prolific of them all and once again to ask ourselves why it is that the Middle East is so rich in oil and gas. The question has been addressed many times (e.g. see the discussion by Beydoun) but, as new data become available, it will continue to be asked and new insights may be gained by fresh attempts to provide the answer.

The part of the region considered in this brief overview surrounds the Arabian/Persian Gulf, and does not extend either northwest far into Iraq or southeast into Oman, where much of the oil is Palaeozoic in age. Neither is the abundant gas in the Permian Khuff Formation included, since little information about it is available — and even less on its source. We are therefore concerned with the Mesozoic and Tertiary accumulations in a limited part of the basin.

The story begins in the Permian with a marine transgression that covered most of the region (Koop & Stoneley 1982). It was probably related to the onset of rifting within a continental block which then embraced not only the whole of Afro-Arabia and southwest Iran but also present-day central Iran. Rifting led, in the late Triassic, to the separation of central Iran from the Arabian continent and to the formation of a new ocean, the Southern Tethys (or Neotethys), along the line of the later Zagros suture (Stöcklin 1974). Thereafter, the entire region of eastern Arabia, the Gulf and southwest Iran evolved as an extensive continental shelf at the passive southern margin of Southern Tethys. It was not until this ocean finally closed again early in the Neogene, reuniting the Arabian shelf with central Iran, that the situation changed and the resulting Zagros folding began to affect its outer part.

The Mesozoic–early Tertiary continental shelf covered an area 2000–3000 km wide and at least twice as long (Murris 1980), lying between Southern Tethys to the north and the basement outcrops of the Arabian Shield to the south and west. The shelf included a number of variably persistent sub-basins; they were at least in part tectonically controlled but may have been enhanced by carbonate build-up in shallower water areas surrounding the relatively sediment starved depressions. The surrounding highs, or perhaps areas of slower subsidence, were governed at least from the middle Cretaceous by gentle vertical movements in the basement, and were aligned approximately north–south. Towards the end of the Cretaceous, the outer part of the shelf (present area of the Zagros Mountains) was overwhelmed by northwest to southeast trending features, parallel to the edge of the shelf: they probably reflect events related to the closure of southern Tethys, including the late Cretaceous emplacement of allochthonous oceanic sediments and ophiolites (e.g. Stoneley 1990). The sediments of this vast shelf sequence are predominantly carbonates interspersed with argillaceous and evaporite intervals, and the sub-basins were often anoxic.

The whole picture is one of essential stability with continued steady, although somewhat differentiated, subsidence of unusual persistence; unconformities are extremely gentle, and some formations can be followed over great distances. Only occasional influxes of sand derived from the Arabian Shield in the middle Cretaceous and in the Miocene disturbed this tranquility. It was eventually brought to an end by the closure of Southern Tethys, which was complete by the Miocene. Once the subsequent Zagros folding had commenced, the Gulf region again became more nearly separated from the open ocean. Thick Miocene evaporites accumulated over the earlier carbonates, and were in turn covered by a great thickness of debris eroded from the rising mountains.

Much of the region is underlain by pre-rift, highest Pre-Cambrian or Lower Cambrian salt which has given rise to numerous salt-domes and plugs: their growth dates most commonly from the beginning of the Cretaceous but, in a few instances, had commenced as early as the Triassic. This incompetent salt also provides a medium for the detachment underlying the late Tertiary to Holocene concentric folds of the

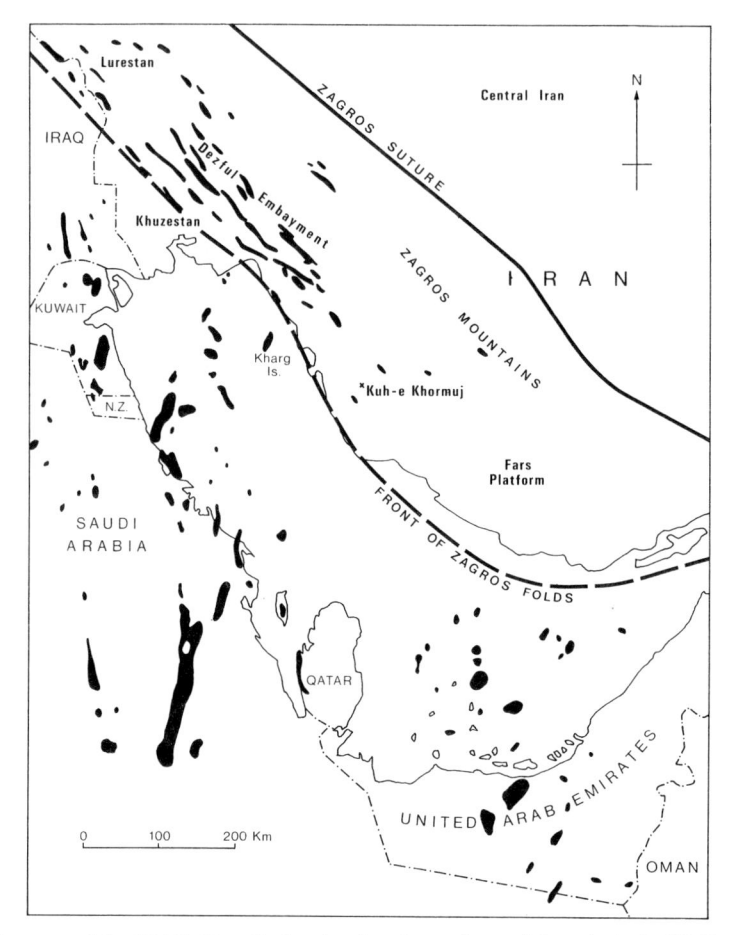

Fig. 1. Locality map of the Middle East Basin, showing the outlines of the principal oilfields.

Zagros, the present front of which more or less coincides with the northern side of the Gulf (see Fig. 1).

In this broad setting, let us consider the controls on the occurrence of petroleum.

Source rocks

Throughout the Jurassic, Cretaceous and Palaeogene, the Arabian shelf was the scene of high organic productivity. This is shown by the tremendous volume of skeletal debris in the predominantly carbonate succession, as well as by high organic contents in the fine-grained sediments where they accumulated under anoxic conditions. Such conditions prevailed in many of the intra-shelf depressions referred to above, producing potential source rocks of somewhat limited lateral extent at various levels (see Fig. 2).

The reasons for this high productivity must be climatic and oceanographic. Possibly the margin of Southern Tethys to the north was the site of oceanic upwelling, although Irving et al. (1974) suggested that the organic material might have been concentrated in a narrowing oceanway by prevailing southeasterly trade winds. As far as the local anoxicity is concerned, basins within a continental shelf provide one of the classic environments for oxygen deficiency below a thermo/halocline (Demaison & Moore 1980).

The organic matter in the potential source rocks of the Mesozoic of the region is almost entirely Type II marine sapropel. Organic contents commonly exceed 10% and may reach more than 15% (Stoneley 1987).

Effective sources have been established in lime mudstones of the Middle and Upper Jurassic, which accumulated in temporary intra-

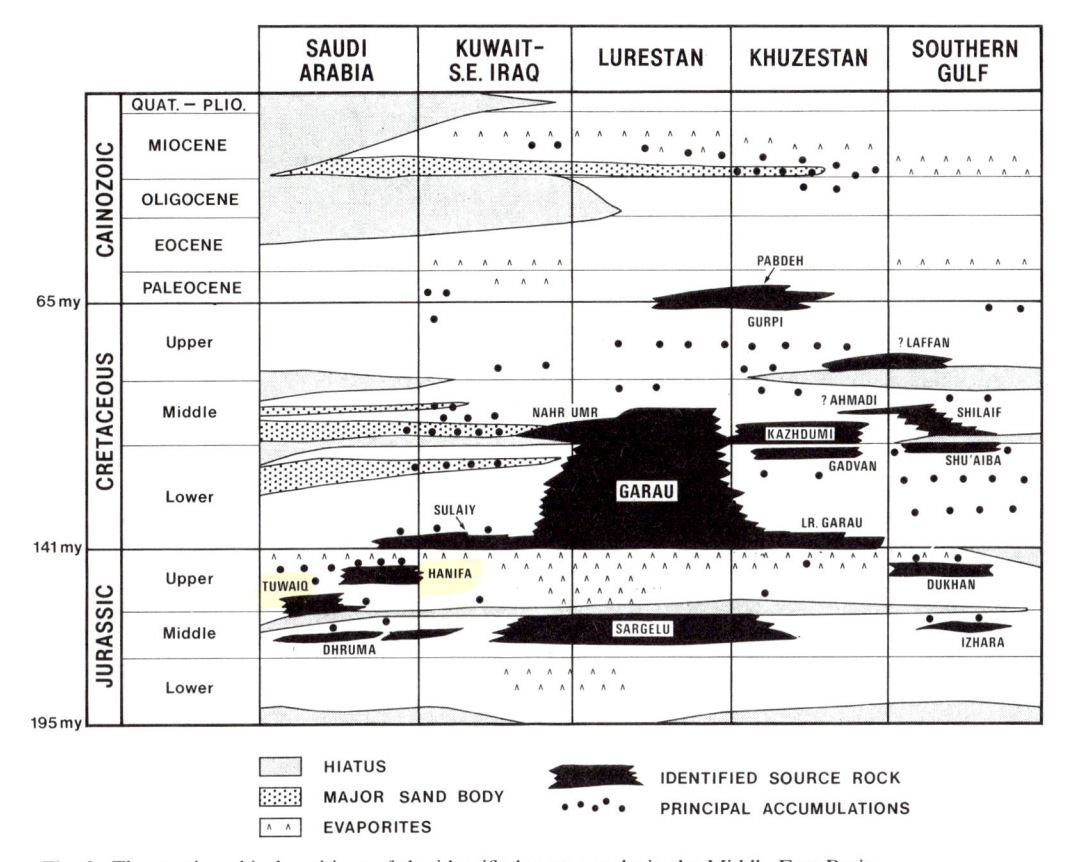

Fig. 2. The stratigraphical positions of the identified source rocks in the Middle East Basin.

shelf depressions. Two of the most important overlapped each other sequentially in the Saudi Arabia–Qatar region, the Tuwaiq Mountain and Hanifa (Dukhan) Formations: they are believed to have provided the source for much of the oil in this part of the region (Ayres *et al.* 1982), although important contributions may have come from deeper in the Jurassic.

A more extensive and deeper water basin persisted from middle Jurassic to middle Cretaceous times around the head of the Gulf (Sargelu–Garau), where shales and argillaceous limestones are of source rock quality. Shales also filled temporary anoxic basins in the Gulf and in Iran during the Cretaceous. The Albian Kazhdumi Formation (Ala 1982) characterizes one of them, over the area of the later Dezful Embayment, and is the source of most of the oil in Iran and Kuwait: its lateral representatives to the north and east are grey oxic mudstones.

Such is the distribution of the various subbasins in time and space, that most parts of the Gulf region are underlain by one or more source rock formation of high quality. The Fars Platform of southern Iran may be the main exception.

Maturity

The Jurassic and Lower Cretaceous source rocks are currently within the oil generation window over much of the region: it is apparently only over the north–south high of the Qatar Arch that they are immature. The Middle Cretaceous is marginally mature over most of the Gulf; in southwest Iran, however, the Kazhdumi Formation has reached peak generation and even the uppermost Cretaceous has locally yielded oil.

In the deeper parts of the Sargelu basin, in southern Iraq and western Iran, the Middle Jurassic is believed to have reached the gas generation zone and to have been the source of sour gas encountered in the Upper Jurassic Najmah Formation.

Reservoirs

Carbonates provide the majority of the reservoirs of the region. It is where these deteriorate somewhat in quality, in Kuwait and neighbouring areas, that they are largely replaced by clean sands in the Middle Cretaceous and in the Miocene.

Primary porosity is extensively preserved in grainstones and packstones of the Middle and Upper Jurassic in Saudi Arabia. It is particularly well developed in the oolitic grainstones of the uppermost Jurassic Arab Zones: some reduction in the water zones beneath oil accumulations suggests porosity preservation until the arrival of the oil, which in turn suggests that there was no movement of the formation waters for periods of some 100 million years. Packstones and grainstones, together with localized rudist reefs, form the reservoirs of the Cretaceous in the southern Gulf and United Arab Emirates. Even in the Cretaceous sands at the head of the Gulf, primary porosity is unusually well preserved.

Limestones in Iran, including the renowned Mid-Tertiary Asmari Formation and the Upper Cretaceous Bangestan Group, on the other hand, are considerably more cemented and rock porosities are low. It the extensive fracturing due to the Zagros folding that makes them prolific producers.

As with the source rocks, reservoirs of high quality but very different character are developed at one or more level in the Mesozoic–Tertiary sequence in most parts of the region.

Seals

Excellent seals are provided by evaporites, particularly in the uppermost Jurassic, where they both cap and are interbedded with the high porosity reservoirs. The extensive evaporites of the Miocene Gach Saran Formation similarly provide a near perfect seal to the Asmari reservoir of Iran and eastern Iraq.

Elsewhere in the succession, shales and occasionally tight limestones act as effective seals to reservoirs containing high oil and gas columns.

Traps

In the Gulf itself and on the Arabian side, regions which have not been affected by the late Tertiary Zagros folding, large and gentle traps are provided by non-compressional anticlines. They owe their origin to one of two causes: drape and compaction over nearly north–south, fault bounded basement highs (e.g. Ghawar); and to doming by the infra-Cambrian salt (e.g. Kuwait, Umm Shaiff). The very large size of such structures gives rise to enormous accumulations.

In Iran and eastern Iraq, on the other hand, most of the accumulations are in the large compressional anticlines of the Zagros foldbelt. These are essentially concentric in form and most of them involve all of the prospective sequence (e.g. Coleman-Sadd 1978).

A recently reported accumulation in a thin Albian sand north of Qatar is apparently hydrodynamically trapped (Wells 1988).

Migration paths

In general, it seems that migration distances have been relatively short, and that most accumulations were derived from sources that were nearby areally and stratigraphically.

The Arabian oils probably migrated from the more basinal areas to the adjacent highs, predominantly laterally. Cross-stratal migration, however, will have to be postulated if it is established that the Middle Jurassic contributed to the oils in the Arab reservoirs. There is also difficulty in explaining the distributions of the oils in the various, evaporite-separated reservoirs of the Arab zones: it is common to postulate off-structure fault channels.

In the northern Gulf, Kuwait and southern Iraq, the Middle Cretaceous sands are believed to have been fed laterally from the contemporaneous rich source rocks of the Kazhdumi Formation to the east (Ayres et al. 1982; Ibrahim 1983).

Only in Iran, and a few areas such as the Neutral Zone, has significant vertical migration taken place. The oils in the prolific Asmari Formation and in the lower, Upper Cretaceous Bangestan limestone reservoir have been matched positively with extracts from the Albian Kazhdumi source (Ala et al. 1980), and the migration clearly took place through fractures created by the Zagros folding: in a few instances the Asmari and Upper Cretaceous reservoirs are still in effective communication. Volumetric considerations suggest that the oils in individual fields were derived from present source drainage areas.

Timing of generation, migration and trap formation

Questions of relative timing were discussed in Stoneley (1987): in summary, the position is as follows (see Fig. 3).

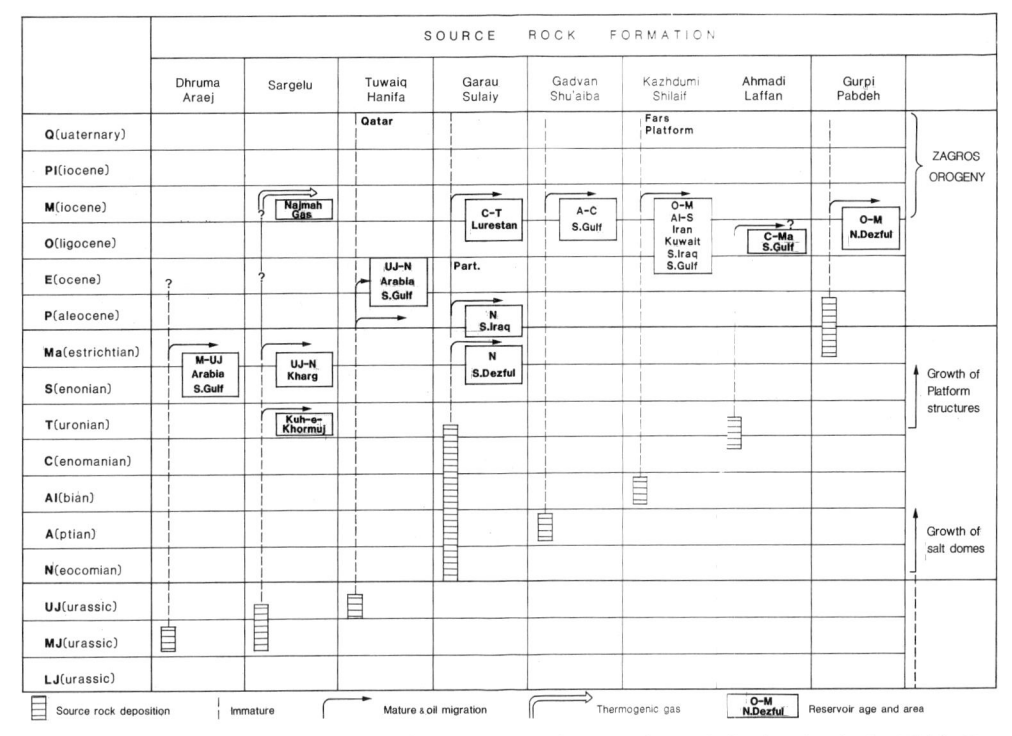

Fig. 3. Diagram showing tentatively the timing of source rock maturation and oil migration in the Middle East Basin.

Structural traps are of three kinds. Basement controlled drape–compaction anticlines in Arabia and the Gulf date from the Turonian (Steineke et al. 1958; Ibrahim et al. 1981), although the structure at Kharg Island was formed by early Cretaceous faulting. The majority of the features due to doming of the Hormuz salt, for example in Kuwait and the southern Gulf, originated early in the Cretaceous (Al-Refai 1967; Kent 1979); however, at least two of those on the Iranian side of the Gulf were growing by the Triassic. The folding of the inner Zagros commenced very early in the Miocene, and there is evidence that the oil-bearing anticlines in the foothills belt date from this time.

Direct evidence that oil was migrating during the late Cretaceous is confined to a fossil seepage in the Turonian, adjacent to a salt plug at Kuh-e Khormuj near the southeastern edge of the Dezful Embayment in Iran (Kent 1979). Other estimates are based on the calculation of TTI values (e.g. Ibrahim et al. 1981) and suggest that the first generation from the Mesozoic was in the Sargelu basin in the Turonian: this is consistent with the evidence from Kuh-e Khormuj. Lower Jurassic sources in Arabia may have been mature by the latest Cretaceous, but peak generation from the prolific Upper Jurassic source formations is believed to have been reached in the Paleocene to early Eocene. In Iran, the Albian Kazhdumi Formation entered the oil window locally in the late Eocene (Ala 1982), but it was not widely mature until the Miocene.

It thus seems that, in all regions where Mesozoic source rocks have yielded oil, nearby structural traps were available and ready to receive it, but in some cases only just!

Preservation

Little has happened to disturb the accumulations in the Gulf and in Arabia since their original formation: there have been no problems of preservation. Apparent partial redistribution of a little of the oil in the Gulf, by hydrodynamic flow from the Arabian side (Wells 1988), is likely to have been a Neogene to Recent phenomenon.

In Iran, however, the accumulations in the Zagros anticlines had to survive the culmination of the folding. This was achieved through the integrity of the overlying evaporite seal. In a few cases, erosion appears to have breached this seal, and some accumulations may have been lost. There is also the possibility that some oil may have undergone partial remigration to neighbouring anticlines, but this is unlikely to have been extensive; the few cases where structures have been found barren of oil were probably due to such remigration, made possible by tilting or by hydrodynamic flow during the foundering of the Dezful Embayment.

In instances where gas has been found at deep levels in the Jurassic, it is possible that any oil once present has now been destroyed by deeper burial.

Conclusions

Each of the factors described above has, in the past, been postulated as the reason why the Middle East is so rich in oil. It now appears that it is due to a combination of all of them: if any one of them had failed, then the story would have been very different. Not only are all of the requirements satisfied, but each of them is developed to an unusually favourable extent.

Moreover, favourable juxtaposition of favourable factors occurs in different ways in different parts of the basin. Thus in Arabia, we find rich Jurassic source rocks that reached maturity when suitable large traps had already been formed; the reservoirs have abnormally well preserved primary porosity in the carbonates and, where these deteriorate, then clean sands appear. In Iran, on the other hand, younger source rocks became mature when the enormous compressional anticlines were developing; otherwise rather poor reservoirs were enhanced by fracturing, which also permitted migration to higher levels, and the oil was retained by a near perfect evaporite seal.

Its geological development has indeed favoured the region!

References

ALA, M. A. 1982. Chronology of Trap Formation and Migration of Hydrocarbons in the Zagros Sector of Southwest Iran. *Bulletin of the American Association of Petroleum Geologists*, **66**, 1535–41.
——, KINGHORN, R. R. F. & RAHMAN, M. 1980. Organic Geochemistry and Source Rock Characterisation of the Zagros Petroleum Province, Southwest Iran. *Journal of Petroleum Geology*, **3**, 61–89.
AL-REFAI, B. H. 1967. The Stratigraphy and Sedimentation of Jurassic and Lower Cretaceous of Kuwait. *6th Arab Petroleum Congress*, **47** (B-3), Baghdad.
AYRES, M. G., BILAL, M., JONES, R. W., SLENTZ, L. W., TARTIR, M. & WILSON, A. O. 1982. Hydro-carbon Habitat in Main Producing Areas, Saudi Arabia. *Bulletin American Association Petroleum Geologists*, **66**, 1–9.
BEYDOUN, Z. R. 1988. *The Middle East: Regional Geology and Petroleum Resources*. Scientific Press, Beaconsfield.
COLEMAN-SADD, S. P. 1978. Fold Development in Zagros Simply Folded Belt. *Bulletin American Association Petroleum Geologists*, **62**, 984–1003.
DEMAISON, G. J. & MOORE, G. T. 1980. Environments and Oil Source Bed Genesis. *Bulletin American Association Petroleum Geologists*, **64**, 1179–1209.
IBRAHIM, M. W. 1983. Petroleum Geology of Southern Iraq. *Bulletin American Association Petroleum Geologists*, **67**, 97–130.
——, KHAN, M. S. & KHATIB, H. 1981. Structural Evolution of Harmaliyah Oil Field, Saudi Arabia. *Bulletin American Association Petroleum Geologists*, **65**, 2403–16.
IRVING, E., NORTH, F. K. & COUILLARD, R. 1974. Oil, Climate and Tectonics. *Canadian Journal Earth Science*, **11**, 1–17.
KENT, P. E. 1979. The Emergent Hormuz Salt Plugs of Southern Iran. *Journal of Petroleum Geology*, **2**, 117–44.
KOOP, W. J. & STONELEY, R. 1982. Subsidence History of the Middle East Zagros Basin, Permian to Recent. *Philosophical Transactions of the Royal Society of London A*, **305**, 149–68.
MURRIS, R. J. 1980. Middle East: Stratigraphic Evolution and Oil Habitat. *Bulletin American Association Petroleum Geologists*, **64**, 579–618.
STEINEKE, M., BRANKAMP, R. A. & SANDER, N. J. 1958. Stratigraphic Relations of Arabian Jurassic Oil. *In*: WEEKS, L. G. (ed.) *Habitat of Oil*, American Association Petroleum Geologists, 1294–1329.
STÖCKLIN, J. 1974. Possible Ancient Continental Margins in Iran. *In*: BURK, C. A. & DRAKE, C. L. (eds) *The Geology of Continental Margins*, Springer, New York, 889–903.
STONELEY, R. 1987. A Review of Petroleum Source Rocks in Parts of the Middle East. *In*: BROOKS, J. & FLEET, A. J. (eds) *Marine Petroleum Source Rocks*, Geological Society London Special Publication **26**, 263–9.
—— 1990. The Arabian Continental Margin in Iran during the Late Cretaceous. *In*: ROBERTSON, A. H. F., SEARLE, M. P. & RIES, A. C. (eds) *Geology and Tectonics of the Oman Region*. Geological Society London Special Publication **49**, 787–796.
WELLS, P. R. A. 1988. Hydrodynamic Trapping in the Cretaceous Nahr Umr Lower Sand of the North Area, Offshore Qatar. *Journal of Petroleum Technology*, **40**, 357–61.

Geology and reservoir characteristics of Lower Cretaceous Kharaib Formation in Zakum Field, Abu Dhabi, United Arab Emirates

A. S. ALSHARHAN

Geology Department, United Arab Emirates University, P.O. Box: 15551, Al Ain, United Arab Emirates

Abstract: The Zakum oil field is an east–west plunging anticline, located in offshore Abu Dhabi, UAE. The field was discovered in 1963 by seismic survey, and came on production in 1967 at an initial rate of 50 000 bopd from the Lower Cretaceous carbonate sediments. The Zakum structure is a broad asymmetric anticlinal feature characterized by gentle dips throughout; the steepest dips occur in the north and northeast where the average dip is about 2.5°. The structure is about 45 km long and 28 km wide, and has a vertical closure in excess of 1200 ft.

At Zakum field the Lower Cretaceous Thamama Group is a thick shallow-marine carbonate sequence with an average thickness of about 2300 ft, and is divided into four formations: Shuaiba, Kharaib, Lekhwair and Habshan (in descending order). An informal reservoir-oriented classification was set up for this group. It was subdivided into six zones given a numerical notation from top to bottom (Zones I to VI).

The Kharaib Formation represent shallowing upward cycles formed in very shallow epicontinental seas. The energy level grades from low-energy subtidal to high-energy setting at or above wave base. It is characterized by dense-argillaceous pyritic bioclastic lime mudstones/wackestones, peloidal-ooidal packstones/grainstones, dolomitic limestones and thin black shale laminae.

The reservoir quality of the formation is controlled by vertical and lateral distribution of porosity, permeability and stylolites. The porosities ranges from 5 to 28%, and permeabilities from less than 1 to 100 Md. The influence of initial environment of deposition, diagenesis and petrophysical properties on the reservoir characteristics is evaluated.

On March 1953 the ruler of Abu Dhabi granted a concession to D'Arcy Exploration Co. for the exclusive right to search for oil in the offshore areas of Abu Dhabi. The company began with a topographical and geological investigation of the sea bed, combined with a reconnaissance gravity survey. In May 1954, Abu Dhabi Marine Area Ltd (ADMA — a joint Anglo–French Company) was formed, to operate the concession, and extended the exploration using an extensive seismic survey from December 1954 to April 1955. These surveys identified several structural features and determined the location for the test well Zakum-1, which was spudded on 3rd April 1963.

The Zakum field is one of the largest offshore fields in the Arabian Gulf and lies under an average water depth of 20 metres. The field is located approximately 80 km NW of Abu Dhabi city (Fig. 1) in the UAE. After the successful discovery of Zakum-1 which proved oil in the Lower Cretaceous carbonates (Thamama Group); extensive drilling and development programmes were applied to attain production from the field. The field first produced commercial oil in October 1967 at about 50 000 bopd; further development led to increased production of 304 000 bopd in 1973. Closure of the high gas wells in February 1975 decreased production to about 95 668 bopd but this was again increased to 250 000 bopd in 1979. In 1977 a development plan was introduced by the Zakum Development Co (ZADCO) for developing the Thamama reservoirs to optimize the recovery from each reservoir. The Lower Cretaceous reservoir was divided into two: the upper reservoir or upper Zakum (Shuaiba and Kharaib Formations-operated by ZADCO) and the lower reservoir or lower Zakum (Lekhwair and Habshan Formations-operated by ADMA–OPCO).

This paper describes the lithology and depositional setting of the Lower Cretaceous (Kharaib Formation), with emphasis on the reservoir zonations, petrophysical characteristics and diagenetic history and its effects on the reservoir quality.

Structure

The Zakum structure appears as a broad, gentle low-amplitude east–west trending asymmetric anticlinal feature with a gentle but distinct plunge to the west. The field is about 45 km

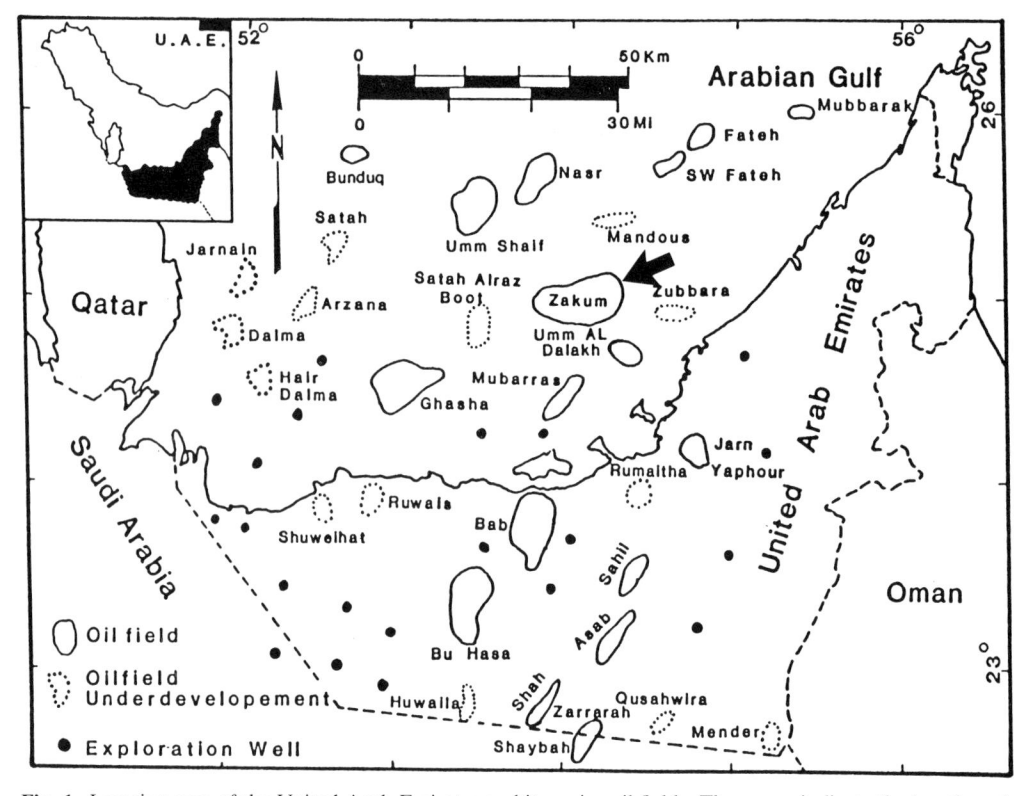

Fig. 1. Location map of the United Arab Emirates and its major oil fields. The arrow indicate the location of Zakum field.

in length (E–W), and at its widest (N–S) measures 28 km. Closure of the structure at top Kharaib Formation level is in excess of 1200 ft. The field is characterized by gentle dips throughout with the steepest dips occurring in the north and northeast, where the average dip is about 2–5°. Elsewhere within the structure, dips are usually less than 2°, whilst in the western part of the field they are flatter and dips can be less than 1°. Recent down flank drilling by ADMA–OPCO and ZADCO has clarified this structural picture and the wells drilled in the east shown that the structure is relatively steeper than previously indicated, and the wells drilled on the western flank have confirmed that the plunge of the structure is due west and not WSW as previously reported. A structural contour map at the top of Zone II (Fig. 2) illustrates the extent of the field. No major faults have so far been identified or mapped in the Jurassic or Cretaceous section. One small fault occurs in the ESE part of the field, and cuts out about 10 ft in the lower part of the Lekhwair Formation. At the Khuff (Upper Permian) level seismic interpretations indicates the presence of faults in the northern and eastern part of the field.

General stratigraphy

The Lower Cretaceous is a thick carbonate sequence deposited over a period of about 30 Ma and extends in age from Berriasian to

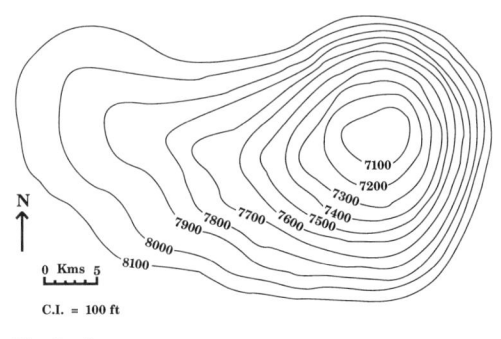

Fig. 2. Structure contour map of top Thamama Zone II, Zakum field.

mid–late Aptian. The Thamama Group in the Zakum field ranges in thickness from 2250–2500 ft and can be divided into four formations of ascending order Habshan, Lekhwair, Kharaib and Shuaiba (Fig. 3).

The Habshan Formation in the western part of the Zakum field rests unconformably on the top of the Hith anhydrite and towards the east, the anhydrite of the Hith pass laterally to pellety intraclasts grainstones with minor oolites and are called the Hith equivalent or Asab Formation. The early phase of the Habshan sedimentation is characterized by oolitic-peloidal grainstones of high-energy beach ridge setting, interbedded with lagoonal and supratidal dolomites and anhydritic dolomites, graded upward to lime mudstones and bioclastic packstones and subordinate sucrosic dolomites deposited in low-energy lagoonal-intertidal setting.

The Lekhwair Formation represents the development of normal unrestricted marine conditions characterized by a number of cyclic sequences. The formation is dominated by band of argillaceous lime mudstones and shales at the base, graded upward to argillaceous pyritic lime mudstones, bioclastic-intraclastic packstone/wackestones, peloidal grainstones and dolomitic limestones. These were laid down in open subtidal to intertidal setting. This cyclic deposition resulted from a series of minor transgressions and regressions which were probably initiated by minor tectonic pulses.

The Kharaib Formation represents shallowing upward cycles formed in very shallow epicontinental seas. The energy level grades from low-energy subtidal to high-energy settings, at or above wave base. It is characterized by dense-argillaceous pyritic bioclastic lime mudstones/wackestones, peloidal-ooidal packstones/grainstones, dolomitic limestones and thin black shale laminae.

The Shuaiba Formation shows an extensive diversification in lithology, due to minor changes in eustatic sea level, which produced a major change in the lithofacies distribution across the field, these range from a deep water basinal to a shallow marine setting. The lower part of the Shuaiba Formation is characterized by peloidal-bioclastic packstones/wackestones minor grainstones, and argillaceous-shaly lime mudstones and wackestones. The upper part dominated by argillaceous globigirinal limestones graded upwards to bioclastic dolomitic limestones and shales interbedded with ooliticbioclasts-intraclasts limestones, representing a shallow shelf to high energy shoal deposits.

The contact between the Nahr Umr shale (Albian) and the Shuaiba limestone (Aptian) is unconformable and matches the regional unconformity which affected the area. The contact characterized by abrupt changes from bioclastic limestones to greenish-grey shales with some pyrite, quartz grains, phosphate and glauconite with very rare echinoderm fragments and ostracods.

AGE		GROUP	FORMATION	RESERVOIR ZONATION ONSHORE	OFFSHORE		
Albian		WASIA	Nahr Umr				
LOWER CRETACEOUS	Aptian	THAMAMA	Shuaiba	Shuaiba buildup/ Bab Member	Zone I	Bab Member	Subzone I C
							Subzone I B
				Zone A			Subzone I A
	Barremian		Kharaib	Zone B	Zone II		
				Zone C	Zone III		Subzone III A
	Hauterivian		Lekhwair	Zone D			Subzone III B
				Zone E			Subzone III C
				Zone F	Zone IV (Zakum Member)		
				Zone G			
				Zone H			
	Berriasian-Valanginian		Habshan	Upper	Zone V		
				Middle			
				Lower	Zone VI		
Upper Jurassic	Tithonian	SILA	Hith/ Asab				

Fig. 3. Classification of the Thamama Group showing the reservoir zonation.

Reservoir definitions and classifications

The Lower Cretaceous reservoir in the Zakum field is subdivided into six zones which are designated from top to bottom as Zone I to Zone VI and each zone is divided to sub-zones. This classification is based on gamma ray/neutron log characteristics; lithology and stylolitic intervals. Based on this reservoir classification Zone I is equivalent to the Shuaiba Formation; Zone II and subzone IIIA are equivalent to the Kharaib Formation; subzone IIIB, C and Zone IV are equivalent to the Lekhwair Formation, and Zones V and VI are equivalent to the Habshan Formation.

In this study the proposed classification in-

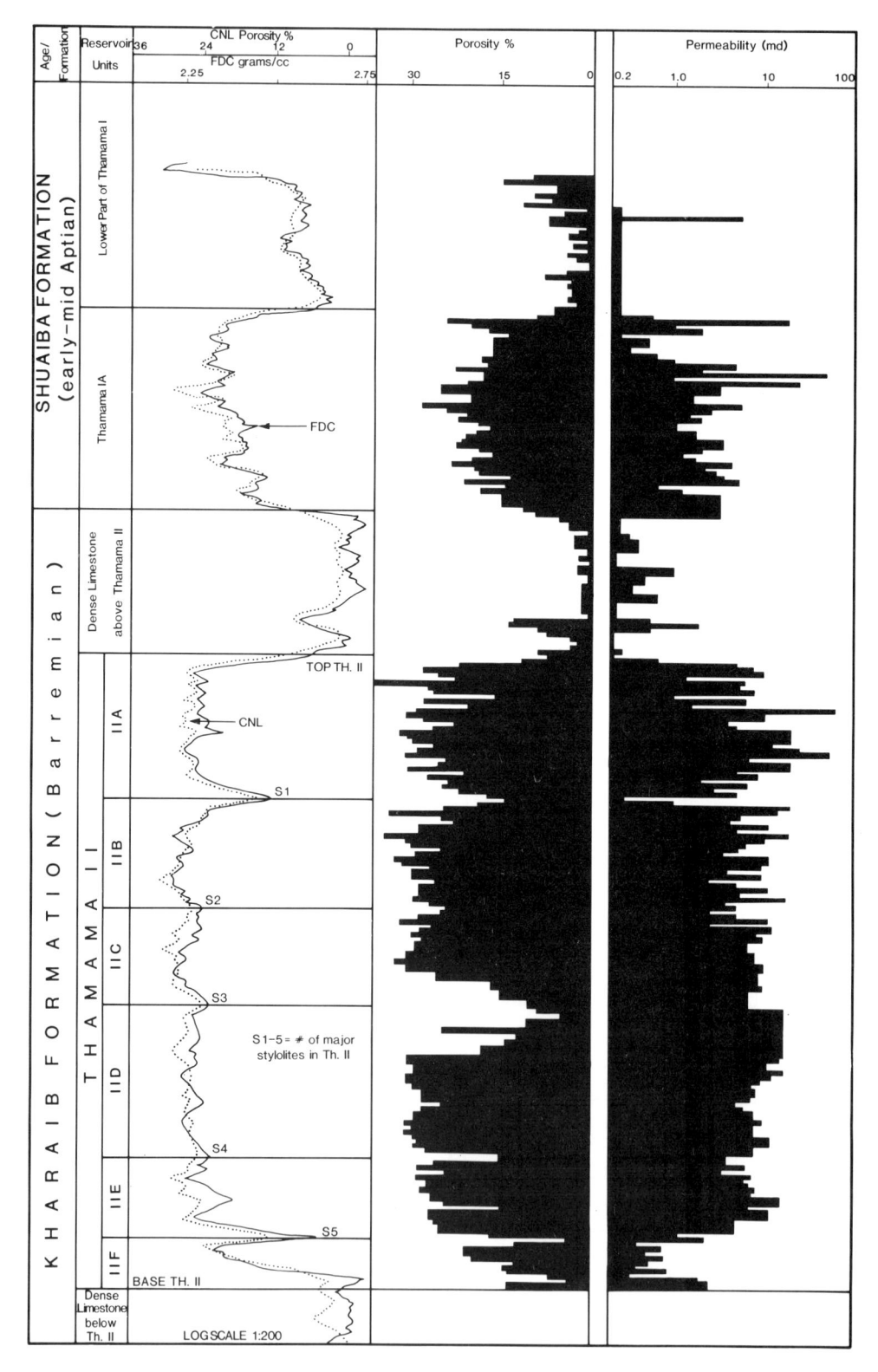

Fig. 4. Shows the reservoir classification of Zone II and the relationship between logs, stylolites, porosity and permeability.

volves the subdivision of Zone II (upper part of the Kharaib Formation) into six distinct subzones which are given alphabetical notation from top to bottom as: IIA−F. The basis of this classification is shown in Fig. 4. Zone II can be subdivided into a number of subzones but such does not imply that they cannot be in fluid communication with each other. The subzones are definitely in free communication with each other in the field. Essentially the classification is based on a number of marker horizons which are variably stylolitic and can be easily correlated throughout the field.

The proposed classification in Zone III is divided into three distinct subzones, which are given an alphabetical notation from top to bottom as: IIIA−C. In subzone IIIA (lower part of the Kharaib Formation) some microstylolites occur in the uppermost and lowermost part of the section which have reduced the porosity and permeability within the section (Fig. 5). Subzone IIIA confirms the general trend of lower permeability in the western part of the field, where the reservoir section exhibits a uniformly low permeability. Based on petrophysical characteristics subzone IIIB can be divided into three units IIIB1−B3. Horizontal permeability in subzone IIIB1 is locally variable but overall is quite good. Subzone IIIB2 and IIIB3 are characterized by very low permeabilities throughout the units. Subzone IIIC shows general decrease in both porosity and permeability within the lower parts of the reservoir.

Petrographical description

Dense limestone above Zone II (see Fig. 6)

This dense limestone section is composed of two different lithologies (Fig. 7). The upper part consists of wackestone with interbedded packstone with abundant orbitolinoids, echinoderms, pelecypods and peloid and lithoclast sediment particles. Slightly recrystallized sediments and some syntaxial overgrowth on echinoderm fragments are also observed. Quartz, shale, pyrite and glauconite are also present together with abundant pyritized and blackened grains. The lower part of the section is composed of interbedded peloidal wackestone and packstone with local grainstone containing miliolids, textularids, Dasycladacean algae, pelecypods, ostracods and echinoderms. Porosity is of vuggy and moldic pore types.

Zone II (Fig. 6)

This zone consists of six units (A−F) from top to bottom as follows:

Unit A consists of peloidal packstone with local grainstones containing *Bacinella aggregatum* algae, echinoderms, miliolids and pseudocyclamina foraminifers. The porosity is in vuggy, moldic, matrix and interparticle pores. It is followed by interbedded recrystallized wackestone and packstone with abundant peloids, algal clasts, echinoderms, rudistid fragments, pelecypods, orbitolinoids and miliolids. Most of the constituents have undergone micritization. Particle recrystallization is brought about by alteration of original components to a weakly developed calcite mosaic. Porosity is mainly vuggy, moldic, framework, interparticle intercrystalline and fractured (Fig. 8).

Unit B consists of recrystallized packstones (which locally grade into wackestones) interbedded with grainstone (which locally becomes boundstones). There is an alternation of peloids and lithoclasts which are associated with echinoderms and rudistid fragments. These rocks are all more or less bound by *Bacinella aggregatum* and *Lithocodium irregularis* algae, Dasycladacean algae and encrusting foraminifers. Neomorphic micrite enlargement creates products resembling cements. The porosity is vuggy, moldic, interparticle and intercrystalline.

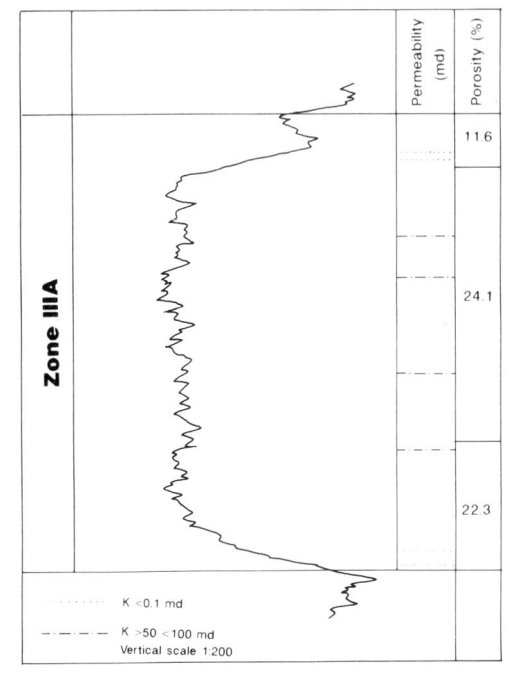

Fig. 5. Zone IIIA shows log character and average porosity−permeability measurements.

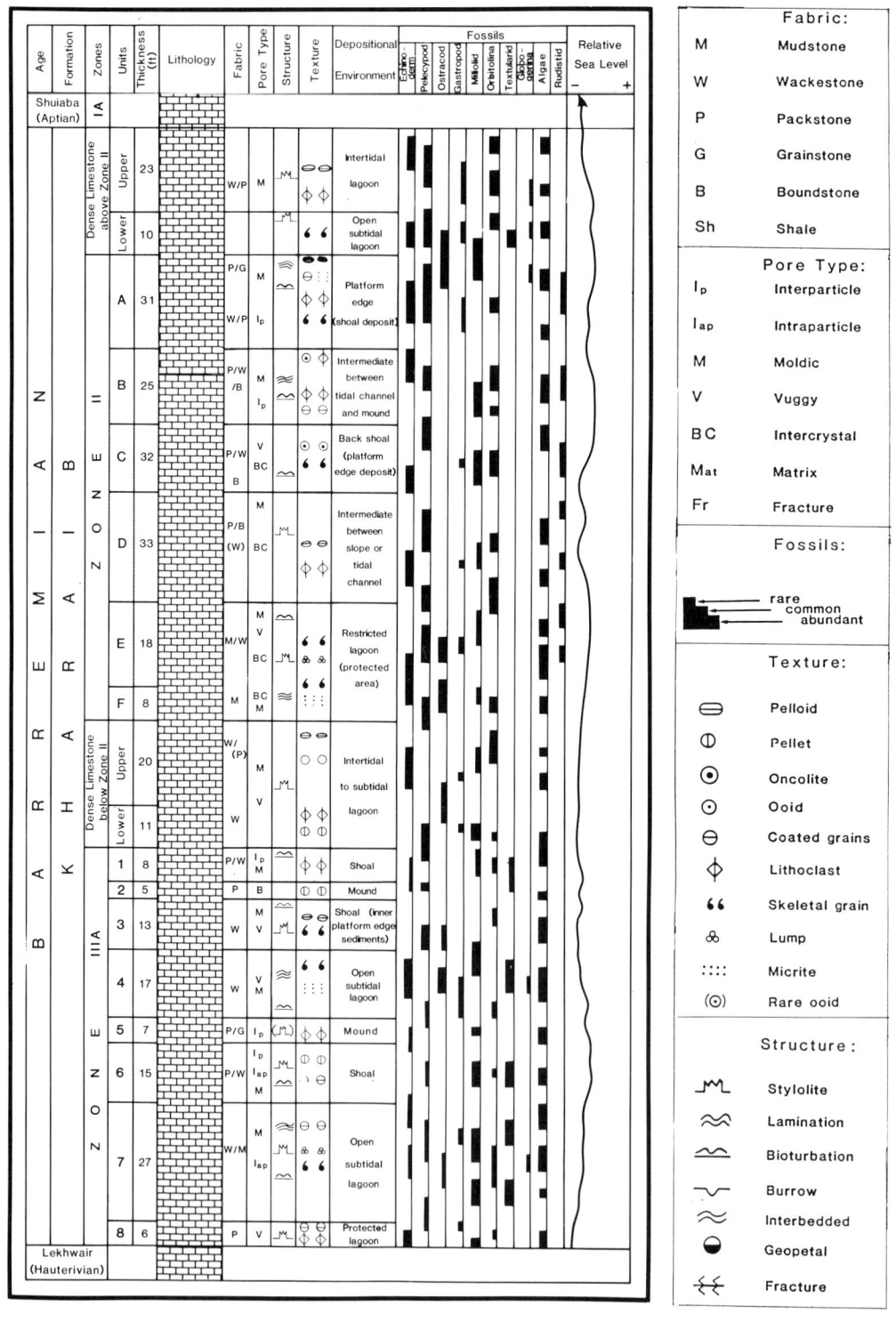

Fig. 6. Composite log shows lithology, pore types, environment of deposition, fossil occurrences and relative sea level in the Kharaib Formation, Zakum field.

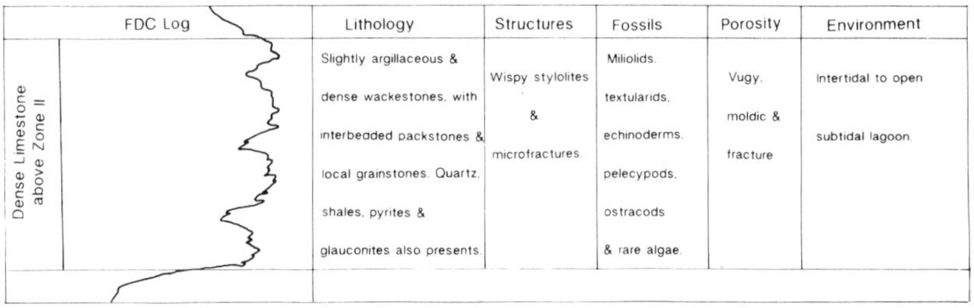

ZAKUM OIL FIELD, UAE

Fig. 7. Log character, general lithology, faunas, type of porosity and depositional setting in the dense limestone above Zone II.

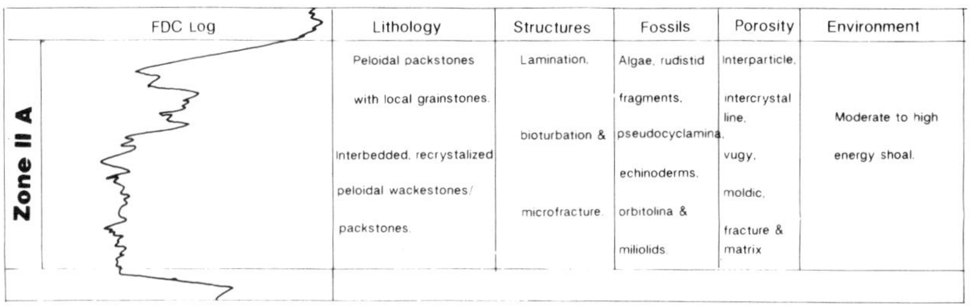

Fig. 8. Log character, general lithology, faunas, type of porosity and depositional setting in Zone II (Unit A).

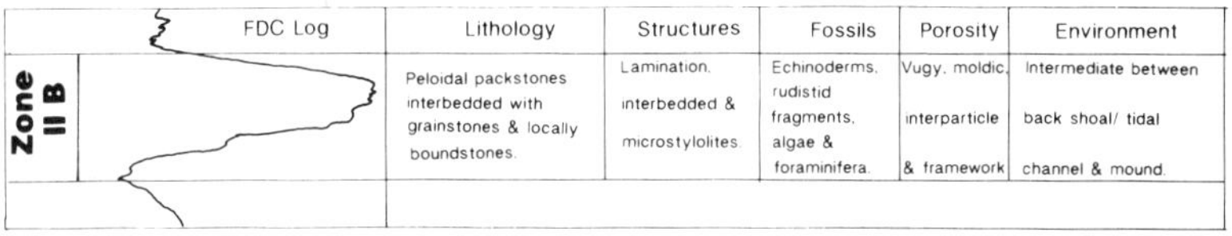

Fig. 9. Log character, general lithology, faunas, type of porosity and depositional setting in Zone II (Unit B).

Locally framework pores appear to occur associated with the algal binding (Fig. 9).

Unit C consists of two parts, the upper one is interbedded packstone and wackestone which is more or less recrystallized, with scattered rhombic dolomite and abundant algae. The peloids are abundant and associated with lithoclasts and bioclasts. The main bioclasts are *Bacinella aggregatum*, *Lithocodium irregularis* and Dasycladacean algae, echinoderms, pelecypods. Some miliolids and orbitolinoids occur. Porosity is in vuggy, moldic, interparticles and matrix pore types. Framework pore space occurs within the binding algae.

The lower part of this unit consists of packstone interbedded with boundstone, and some wackestone. There are also scattered rhombic dolomites within the unit. Alternations of peloids and lithoclasts are associated with echinoderms, pelecypods, all of which are more or less bound by rudistid, algae and orbitolina. The porosity is in vuggy, interparticle, moldic and locally framework pore type (Fig. 10).

In Unit D the upper part consists of packstone with interbedded boundstones which are slightly dolomitized. Binding organisms include rudistid, algae and encrusting foraminifers. The pore types are mainly vuggy, matrix and locally framework types.

The lower part is composed of dolomitized wackestone to packstone with abundant peloids, lithoclasts and bioclasts, mainly of pelecypods, echinoderms, codiacean green algae, Dasycladacean and orbitolina. The porosity is mainly of vuggy, matrix, moldic and intercrystalline types (Fig. 11).

Unit E consists of interbedded wackestone and mudstone which is highly dolomitized with local patches of packstones and streaks of shales. The dominant bioclasts are ostracods,

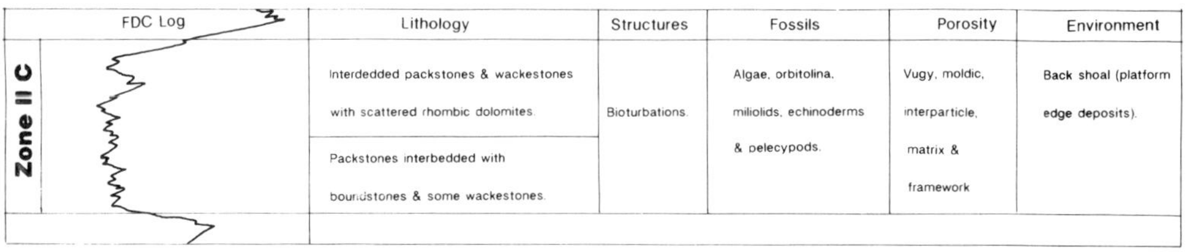

Fig. 10. Log character, general lithology, faunas, type of porosity and depositional setting in Zone II (Unit C).

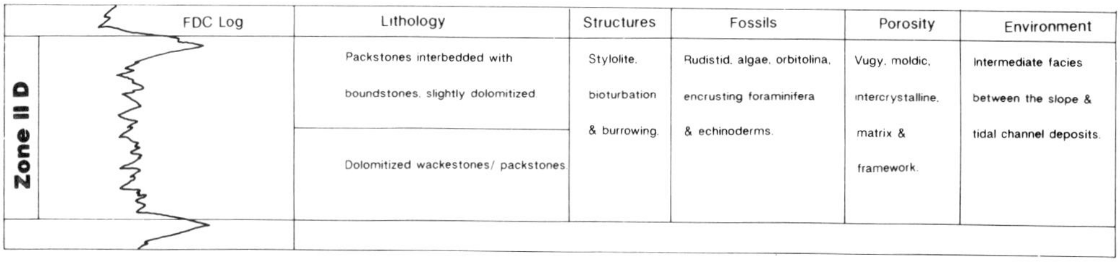

Fig. 11. Log character, general lithology, faunas, type of porosity and depositional setting in Zone II (Unit D).

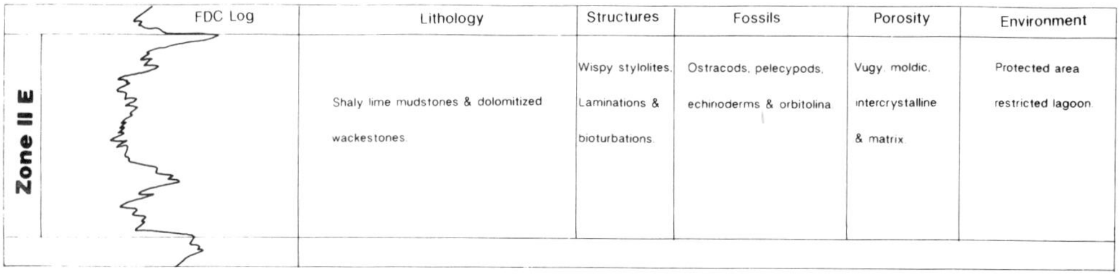

Fig. 12. Log character, general lithology, faunas, type of porosity and depositional setting in Zone II (Unit E).

pelecypods, echinoderms, *Bacinella aggregatum*, Dasycladacean algal and orbitolina. The pore types are mainly vuggy, moldic, matrix and intercrystalline types (Fig. 2).

Unit F is composed of dolomitized mudstone, with local wackestone and abundant orbitolina and some ostracods, pelecypods, echinoderms with some Dasycladacean algae. The porosity is mainly vuggy and moldic and locally intercrystalline types (Fig. 13).

Dense limestone below Zone II (Fig. 6)

The upper part of this section (Fig. 14) is composed of wackestone with abundant orbitolina and echinoderms which are slightly recrystallized in the matrix, and some are syntaxial overgrowths. Slightly argillaceous beds with quartz are present with abundant pyritized grains. Abundant microstylolites and microfractures occur between the constituents. The porosity is of vuggy and moldic types.

The lower section (Fig. 14) consists of wackestone to packstone, which are partly recrystallized with abundant pelecypods, miliolids, textularids, orbitolinoids and Dasycladacean algae. The low porosity occurs in solution vugs and moldic pores.

Zone IIIA: (Fig. 6)

Eight lithological units have been recognized in this zone and confirmed from the core description (Fig. 15). These are as follows, from top to bottom.

Unit 1 is composed of peloidal packstone to wackestone (Pelbiomicrite), locally recrystallized and bioturbated. The major organisms include algae (*Bacinella aggregatum* and *Lithocodium irregularis*), pelecypods, echinoderms,

Zone II F

FDC Log	Lithology	Structures	Fossils	Porosity	Environment
	Dolomitized lime mudstones to locally wackestones	Lamination, bioturbation, stylolite & fracture	Orbitolina, pelecypods, ostracods & echinoderms.	Vugy, moldic & intercrystalline	Protected area of restricted lagoon.

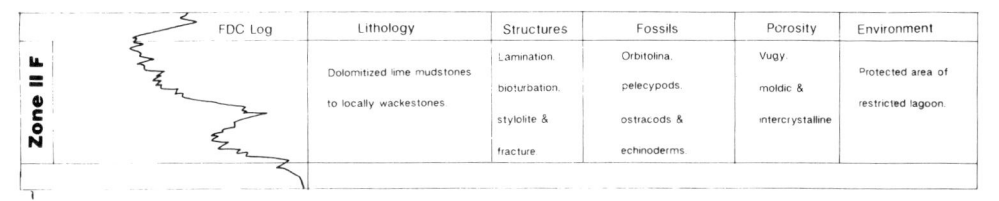

Fig. 13. Log character, general lithology, faunas, type of porosity and depositional setting in Zone II (Unit F).

Dense Limestone below Zone II

FDC Log	Lithology	Structures	Fossils	Porosity	Environment
	Slightly argillaceous wackestones with abundant quartz & pyritized grains.	Microfractures & microstylolites	Orbitolina, textularids, echinoderms, miliolids & pelecypods.	Vugy & moldic	Intertidal to open subtidal lagoon.
	Slightly argillaceous wackestones to packstones.				

Fig. 14. Log character, general lithology, faunas, type of porosity and depositional setting in the dense limestone below Zone II.

Zone IIIA

	Sonic Log	Lithology	Structures	Fossils	Porosity	Environment
Unit 1		Peloidal packstones/ wackestones.	Bioturbations.	Algae, textularids, miliolids, echinoderms & pelecypods.	Vugy, moldic & interparticle	Back shoal.
Unit 2		Peloidal packstones/ grainstones.	Laminations.	As above.	Interparticles.	Fore shoal.
Unit 3		Dolomitized wackestones	Bioturbations & microstylolites.	Algae, foraminiferas & pelecypods.	Vugy, moldic, interparticle, fenestral	Back shoal.
Unit 4		Dolomitized lime mudstones/ wackestones	Bioturbations, laminations & stylolites.	Algae, Choffatella sp., textularids, miliolids & echinoderms.	Intercrystalline.	Open subtidal lagoon.
Unit 5		Packstones/ grainstones.	Microstylolites.	Algae, rare foraminifera, pelecypods & echinoderms.	Interparticles.	Fore shoal.
Unit 6		Wackestones/ packstones.	Bioturbations & stylolite seams.	Textularids, miliolids, Trocholina sp., algae, pelecypods & echinoderms.	Vugy, moldic interparticle & intraparticles.	Back shoal
Unit 7		Lime mudstones/ wackestones.	Bioturbations, Laminations & stylolites.	Algae, foraminifera, ostracods, pelecypods & sponge spicules.	Vugy, moldic & interparticle	Open subtidal lagoon.
Unit 8		Peloidal packstones.	Seam of stylolites.	Choffatella sp., pelecypods & echinoderms.	Vugy.	restricted lagoon.

Fig. 15. Log character, general lithology, faunas, type of porosity and depositional setting in Zone IIIA.

textularids and miliolids. The porosity is mainly vuggy, moldic and interparticle pore types.

Unit 2 consists mainly of packstone (Intrapeloomicrite), dominated by aggregates (Lumps), peloids and ooclasts. There are a few codiacean green algae, pelecypods, echinoderms, miliolids and textularids. The pore type is mainly; interparticle.

Unit 3 is composed of wackestone (pelmicrite), and is highly dolomitized and locally bioturbated, with some stylolite seams, pelecypods, foraminifers and codiacean green algae. The porosity is dominated by vuggy and moldic types, with few interparticles and fenestral types.

Unit 4 consists of wackestone (Biomicrite), which is bioturbated and dolomitized, with stylolites present in several intervals. The dominant components are dasyclacean algae, some pelecypods, echinoderms, gastropods, *Choffatella* sp., textularids, miliolids and sponge spicules. The pyrite and quartz are also present in the unit. The porosity occurs within the dolomitized intervals.

Unit 5 is composed of grainstone to packstone (biopelspar micrite). The unit is dominated by peloids bioclasts and few lithoclasts. Pelecypods, echinoderms and foraminifer are rare, and there are some codiacean and dasycladacean algae. The components show compaction throughout the unit, with the presence of stylolite intervals. The porosity is mainly interparticle.

Unit 6 consists of packstones to wackestones (recrystallized biopelmicrite), which are bioturbated, and contain stylolite seams. Scattered quartz, pyrite and dolomite are present throughout the unit. Foraminifera include textularids, miliolids and *Trocholina* sp. Pelecypods, echinoderms with some dasycladacean and codiacean green algae are also present. The porosity is mainly of the vuggy and moldic types with some interparticle and intraparticle forms.

Unit 7 composed of wackestone to mudstone (Biopelmicrite, biomicrited), which are bioturbated and stylolitized and contain scattered quartz, pyrite and dolomite throughout the unit. There are a few pelecypods, ostracods, echinoderms, sponge spicules and dasycladacean algae present. The foraminifers include *Choffatella*, *Pesudocyclamina*, miliolids and textularids. The porosity includes vuggy, moldic and intraparticle forms.

Unit 8 consists mainly of packstones (recrystallized biopelintramicrite). Quartz, pyrite and glauconite are also present. The stylolites are apparently superimposed on shaly seams. Dominant components are echinoderms, pelecypods, *Choffatella* sp. and dasycladacean algae. The porosity is mainly of vuggy forms and is associated with stylolites.

Interpretation of depositional setting

Shelf sediments commonly consists of depositional cycles in which repetitive sequences of rock types are spread over vast areas (Wilson 1975). The cyclicity in the strata described here can be seen in some detail in the wells examined in this study.

It is considered that the dense limestone above and below Zone II was deposited in an intertidal to open subtidal lagoonal depositional setting (Alsharhan 1985). The rock is texturally variable (mudstone to packstone, but contains considerable amounts of lime mud, predominantly as a wackestone). Most of the sediments consist of a mixture of peloidal mud with varying amounts of skeletal grains. These are highly burrowed with abundant miliolids, textularids, dasycladacean algae, scattered pelecypods, echinoderms and sponge spicules suggesting deposition in the open circulation of an intertidal to subtidal quiet shallow lagoonal environment.

Five depositional settings can be recognized from the six lithological units described from Zone II. *Unit A* represents shoal deposits, which consist of a bioturbated to brecciated grain supported limestone containing considerable amounts of lime mud which grade into peloidal packstone to grainstone and/or into boundstone strata. These beds accumulated as a shoal facies rich in interclasts, algae, rudists, and encrusting foraminifers. Most of the grains show abundant boring with the resulting micritization often destroying the texture and the whole primary structure. Rudistids appear to be present but are not extremely common. They are commonly disoriented in micrite matrix.

Unit B represents an intermediate facies from a back shoal or slope and channel deposit between mounds. It consists of wackestones/packstones which include fine and coarse debris, composed of both reworked bioclasts and cemented lithoclasts. They appear to be mixed with the mud mound facies and contain varying amounts of material, derived from adjacent algal mounds of codiacean green algae.

Unit C represents platform edge deposition of bioturbated grain supported limestones, but contains considerable amounts of limemud with algae and peloidal sediment. Most of the compound grains are composed of lithified algae and skeletal material which also occur as discrete grains too, suggesting the intraclasts were locally formed.

The dominant sedimentary structures are associated with bioturbation, suggesting that these structures were produced by alternating low and medium wave or that the current energy pulses were destroyed by organisms. The biota are common to the platform edge, situated just behind the seaward shoal in very shallow water, which was probably exposed to moderate wave energy.

Unit D represents an intermediate facies between the slope and tidal channel deposits. It is composed of mainly bioturbated wackestone with some laminations and burrows. These sediments are predominantly rich in skeletal debris from organisms which inhabited the lagoon. The peloidal carbonate mud material was probably washed in from the adjacent marine environment and also includes sediments deposited in the tidal channels of a medium energy environment, which is characterized by peloids, lithoclasts, codiacean green algae and orbitolina.

Units E and F were deposited in a protected area of restricted lagoon. They consist of dark and slightly argillaceous laminated and bioturbated wackestones grading occasionally into grain supported packstones. Small amounts of terrigenous material are present as disseminated grains of quartz which may be locally concentrated in thin discrete layers. Seams of argillaceous carbonate occur in structures and burrows in conjunction with horsetail and laminated shales.

Four depositional settings can be recognized from the eight lithological units described for Zone IIIA. The lower unit (*Unit 8*) has the typical characteristics of a protected restricted lagoon of an inner shelf. The predominance of packstone indicates that the sediment movement was probably an intertidal restricted lagoon.

Units 4 and 7 contain abundant miliolids, ostraods, sponge spicules and dasycladacean algae in a wackestone to mudstone matrix, implying a quiescent environment, unaffected by wave energy, probably of an open subtidal lagoon.

Units 1, 3 and 6 consist of packstones with codiacean green algae and an absence of pyrite and blackened grains, containing rare quartz suggesting a backshoal environment characteristic of the platform edge.

Units 2 and 5 consist of peloidal grainstones to packstones suggesting deposition in a foreshoal setting which was subjected to wave movement.

Diagenetic history

Different diagenetic events and their response to both time and diagenetic setting in the Zakum field are shown in Fig. 16. The magnitude of these diagenetic events are minor in the intervals studied in the field.

Marine phreatic

Micritization. Micritization is one of the first stages of diagenesis seen in these sediments and it occurred during deposition. Micritization of grains by boring algae and fungi result in the formation of micrite envelope similar to those described by Kendall & Skipwith (1969) and by Bathurst (1975). This processes has been observed throughout the section, and goes on

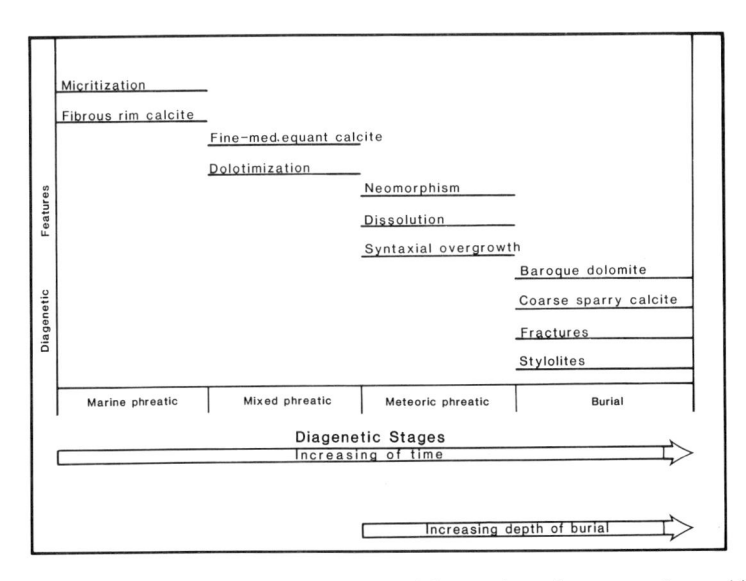

Fig. 16. Different diagenetic events through both time and diagenetic environments observed in the Kharaib Formation, Zakum field.

centripetally within the grain and may completely convert it to micrite (Fig. 17a). These encrusting algae and boring organisms micritize the outer portions and occasionally the inner portions of bioclastics; this has been observed in Recent sediment of Abu Dhabi by Kendall & Skipwith (1969).

Fibrous calcite crust. Isopachous crusts of fibrous calcite are common and are considered to be very early post-depositional submarine cement (Becher & Moore 1976) of a phreatic environment. Folk (1974) reported that the high Mg/Ca ratios found in normal marine water is responsible for these fibrous habits of marine calcite cement. A thin rim of calcite crystals (normal to grain surface) surrounding grains (Fig. 17a) observed in some thin sections are followed by equant calcite or coarse sparry calcite which represents different stages of cementation (as discussed below).

Mixed phreatic

Equant calcite crust. Fine to medium equant calcite crusts occur only as a pore fill cement on the surface of allochems (Fig. 17a). Continued cementation results in the crystals coarsening toward pore centers and commonly produces an interlocking mosaic that fills all pore spaces. It may represent cementation in a meteoric vadose environment. The fine rhombohedral calcite cement is interpreted by Folk (1974) as the product of rapid precipitation from water with low Mg/Ca ratios as found in meteoric water. The coarsening of calcite crystals towards pore centers is thought to be a distinctive characteristic of early cementation in a fresh water environment (Loucks 1977).

Dolomitization. The dolomites of these rocks commonly occurs as dolomite rhombs and are a minor constituent seen in thin section (Fig. 17b). In most of the intervals studied in the Zakum field dolomite usually averages about 15–20%, but may increase locally up to 80% in a few intervals where it is accompanied by significant intercrystalline porosity (Fig. 17b). No important changes in reservoir characteristics occur where the rhombs are scattered in a micritic matrix.

Meteoric phreatic

Neomorphism. The dissolution and its reprecipitation of the aragonite of shells (mainly molluscs) as a mosaic of sparry calcite mainly occurs in Zone II and to a lesser degree in IIIA (Fig. 17c). In some intervals a micritic coating is observed associated with recrystallization of some grains such as peloids into microcrystalline calcite. Relics of the original grains can be sometimes observed in outline.

Carbonate minerals which compose limestones described in this Formation proceed by micrite enlargement, a process by which tiny crystals of limestone (only a few microns in diameter) enlarge into crystals measuring tens or even hundreds of microns, similar to cement in crystal size. These new crystals are known as pseudospar (Folk 1965) or neospar (Flugel 1982).

Dissolution. Evidence of extensive dissolution is found in most of the sections studied in the Zakum Field. It has a broad range of intensity and generally increases the pore system. Dissolution forms solution vugs and moldic voids (Fig. 17d) which are locally interconnected by the enlargement of the basic interparticle porosity. This has resulted in a vuggy pore system interconnected by a solution enlarged interparticle network, to high permeability anomalies.

Syntaxial overgrowth. Syntaxial overgrowths on echinoderm and/or crinoid fragments (Fig. 17e) have been described mainly in Zone II and to a lesser degree in Zone IIIA. Syntaxial calcite growing in optical continuity with echinoderm fragments generally surrounds the grains and are elongate along the C-axis similar to that described by Evamy & Shearman (1969). These syntaxial overgrowth are developed in meteoric phreatic environment, and Longman (1980) suggested that although syntaxial overgrowths form most commonly in fresh water phreatic environment, they may also form, but rarely, in vadose and deeper subsurface diagenetic environments.

Burial Phase

Baroque dolomite. This type of dolomite can be seen in most parts of Zone II and IIIA, and is

Fig. 17. (a) Peloidal bioclastic packstone/grainstones. These sediments have a high percentage of micritized grains (M) formed by algae or fungi, followed by isopachous crust of calcited cement (I) and fine to medium crystalline equant calcite (E). Note that the pores of some grains are filled by baroque dolomite (BD). (b) Fine to medium rhombic dolomite form intercrystalline porosity. (c) The entire grains comprise an irregular mosaic of crystals of varying sizes and shapes which is a product of neomorphism. (d) Peloidal-bioclastic wackestone-packstones showing dissolution of the grains form solution vugs (V). (e) Syntaxial overgrowth (S) around the echinoderm plate (E). (f) Coarse crystalline baroque dolomite (BD), anhedral to subhedral crystals with curved crystal faces and cleavage. (g) Coarse crystalline calcite are commonly euhedraland partly fill the voids. (h). Microstylolites (S) stained with oil and associated with dolomites (D).

a

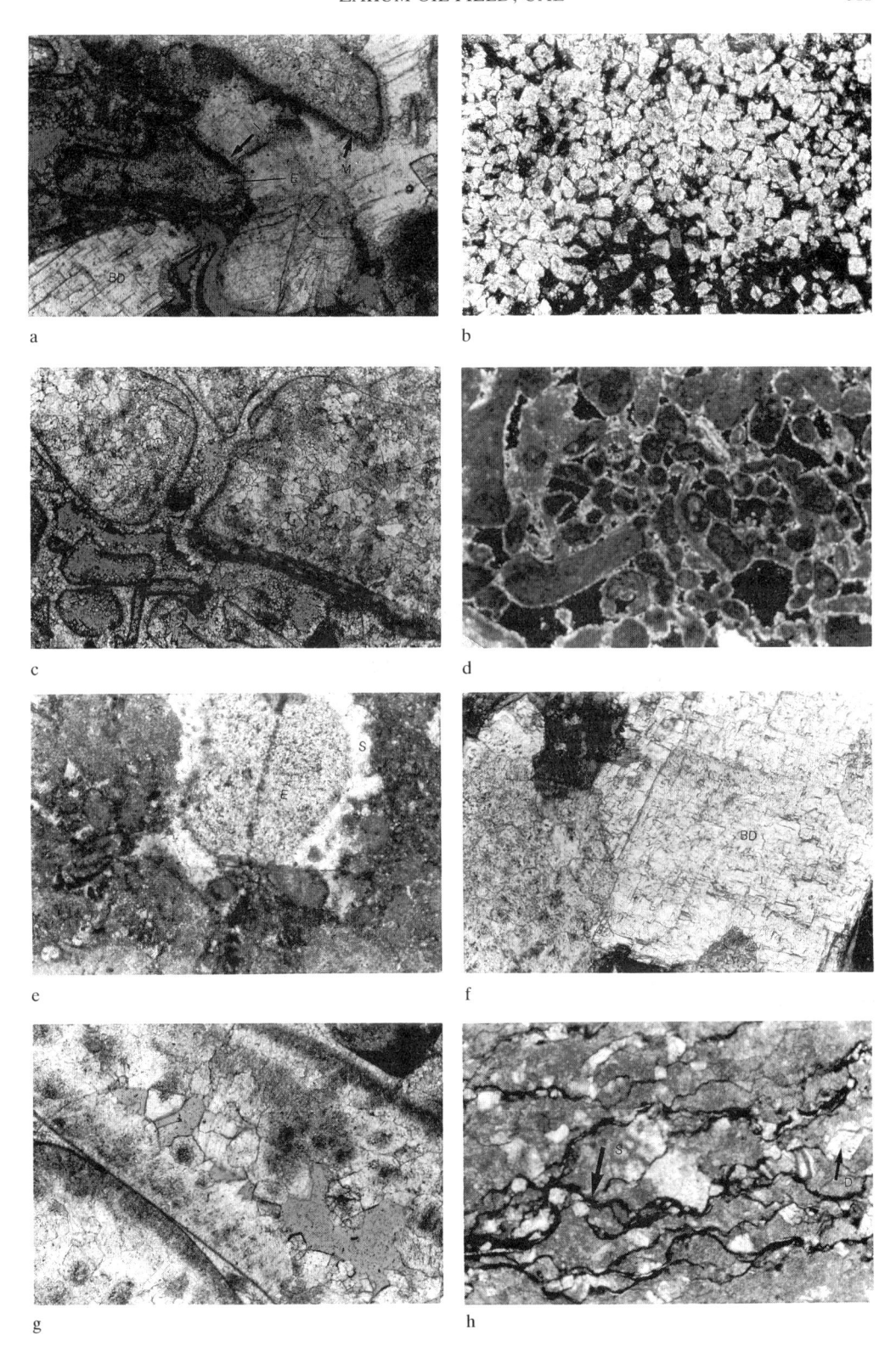

b

c

d

e

f

g

h

interpreted as late stage diagenesis based on isotope analysis ($\delta\ O^{18}$ ranges from -4 to -9%) (Alsharhan & Williams 1987). The dolomite, characterized by coarse crystalline curved crystal faces, undulose extinction and generally clouded appearance, (Fig. 17f) sometimes replaces the calcite. The name 'baroque' was first coined by Folk & Assereto (1974); beforehand the material was termed 'white saddle' dolomite.

Folk & Assereto (1974) suggested that the baroque dolomite was precipitation from mixed hypersaline-meteoric water as an early postdepositional cement. Loucks (1977) described the same kind of dolomite in south Texas and postulated that it formed from mixing of subsurface brine and diagenetic waters from clays. Radke & Mathis (1980) believe the baroque (saddle) dolomite precipitated under mesogenetic or burial conditions at temperatures of 60° to 150°C.

Coarse sparry calcite. Sparry calcite cement consists of clean blocky crystals and its formation is the most significant of possible diagenetic events in terms of porosity reduction. In thin sections described from the Kharaib Formation of the Zakum field (Fig. 17g) the coarse sparry calcite commonly overlie the finer crystals of the crust precipitation developed in early meteoric phreatic environment similar to the interpretation of Becher & Moore (1979). They reported that the meteoric phreatic environment encompasses those portions of the subsurface below the water table which contains water that is totally or partially of meteoric origin.

Stylolitization and Fracturing. The seams stylolites observed in the section studied usually consist of relatively insoluble residues as layers of clay, organic matter, quartz, dolomite and pyrite. The organic matter in stylolites is composed of aggregates of small particles, probably primary marine amorphous organic matter. A secondary asphaltic or bituminous substance fills up the spaces between the crystals. Some micro-stylolites described from Zone II are filled with fine-grain rhombs of dolomite (Fig. 17h). This could have formed from Mg^{2+} released during pressure solution as described by Wanless (1979). He suggested that permeability controlled the fluid migration, the composition of concentrate, and the differential pressure which provided preferential conditions for dolomitization. The residue of the stylolite seams consists of relatively insoluble material. Dolomite associated with stylolitization is local and rare. Some intervals characterized by the predominance of laminations. These laminations are of the linear or undulate pattern and consist mainly of organic-rich substance.

Microfractures are associated with stylolites and are predominant in the lime mudstone and wackestone intervals and resemble calcite veins. Some are plugged while others are open or partially filled with calcite. Fracture frequency appears to be larger than that of stylolites in the studied intervals. Most of the fractures are small in dimensions (maximum length measured from the cores is around 15 cm with a width on the order of a few millimetres). Their occurrence is predominantly in mud-supported sediments and these are filled with calcite or dolomite, which can be clearly identified in thin sections.

In the Kharaib reservoir, the stylolites are associated with small scale fractures. These fractures are usually less than 2 cm in length and are often vertical or subvertical, but a few horizontal fractures cutting across the vertical ones have been observed and some of them partially filled in with calcite. Commonly the fractures are on the upper or lower parts of the stylolites but do not cut across them indicating perhaps that they post-date stylolite formation.

It is difficult to assess the potential contribution of these fractures to the production of hydrocarbons. It is likely that the very small scale fractures (cm) would not enhance the permeability of the reservoir significantly. The contribution of the larger fractures to increased permeability are expected to be significant, provided they are open under reservoir conditions.

In the Zakum field stylolites seen in cores are poorly developed in the central part of the field. The degree of stylolitization was found to decrease from the flank towards the tope of the structure, and are responsible for the overall reduction in thickness on the east and west flanks. Stylolite development towards the northeast of the field is diminished probably due to either an anomaly in the oil–water contact or high hydrocarbon saturations. The highest intensity and the greatest amplitude of stylolites were found in the mud-supported sediment intervals. The intensity and thickness of stylolitic intervals are not uniform throughout the field.

Five densely stylolitic intervals separating the different reservoirs within Zone II in Zakum field were defined, from wire-log data (see Fig. 18) and core descriptions. The majority of stylolites in this zone are characterized by their low amplitude. Few intervals are stylolitized and do not show any response on either electrical or poro-perm logs. It is interesting to note that the cores described in this study, show laminations occur to some extent in the intervals

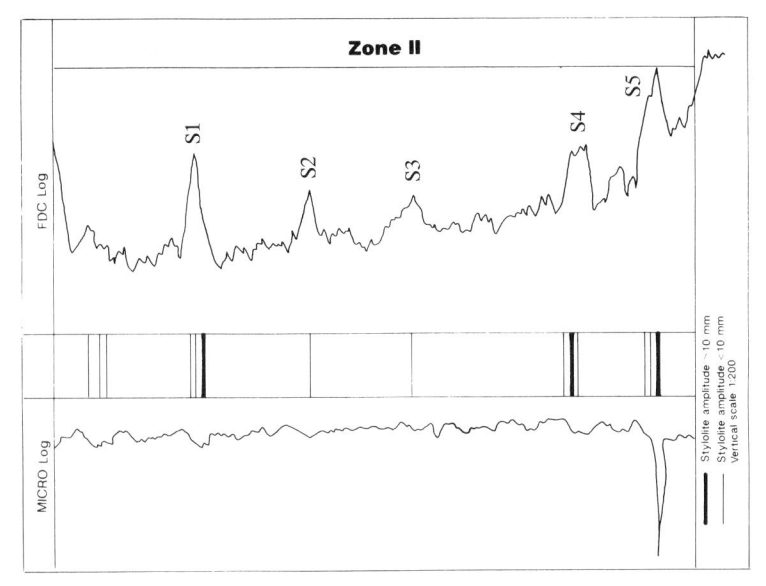

Fig. 18. The relationship between FDC Log and stylolites and their amplitudes (S1–S5).

that are stylolitized. In fact, stylolite development could be enhanced or governed by the presence of such laminations.

Reservoir characteristics

The Kharaib reservoir contains a high quality, low sulphur crude oil (35° API). The oil is under-saturated by about 600–1200 psi. The initial pressure was 3825 psi at reservoir datum 7550 ft subsea. A summary of general information on the field, and the basic reservoir data for the Kharaib is given in Tables 1 & 2.

The porosity in Zone II ranges from 7.3% to 31.8%, and their is a general decrease in porosity from the upper to the lower part of the Zone. The horizontal permeability varies from 0.05–62 Md, and the permeability values also show decrease in value down through the zone.

There is a general decrease in porosity from the upper part of Zone II downward although, of course, in the vicinity of the major stylolite zones (S1 and S5 see Fig. 18), there is consistently a reduction of porosity.

The variation in porosity in the Thamama reservoir is based upon original work of Dunnington (1967) on the stylolite development in the Thamama Zone B (Zone II) of the Bab

Table 1. *Basic information on the Zakum field*

Hydrocarbon type:	Oil
Reservoir type:	Peloidal-bioclastic packstone-grainstone and wackestone.
Formation/Age:	Thamama Group (from older: Habshan, Lekhwair, Kharaib and Shuaiba Formations). Lower Cretaceous (Berriasian-mid Aptian).
Geologic setting:	Northwestern part of the Rub Al Khali basin (Arabian platform).
Reservoir characteristics:	Porosity and water saturation values are exceptionally uniform within each individual porous interval, while permeability shows considerable variations.
Trapping mechanism:	The Thamama Group comprises of a multilayer limestone series of porous beds (reservoirs), separated by impermeable horizons of dense argillaceous beds (seals).
Source:	Diyab Formation (Upper Jurassic), probably minor source comes from the late Lower Cretaceous (Bab Member).
Time of maturation:	Early Tertiary.
Time of trap formation:	Late Cretaceous to Early Tertiary
Time of migration:	Early Tertiary.

Table 2. *Basic reservoir data of the Kharaib Formation in the Zakum field, Abu Dhabi*

Total thickness:	290–320 ft
Pay thickness:	210–190 ft
Porosity:	5–28%
Permeability:	<1 to >100 Md
Oil gravity:	35° API
Initial reservoir pressure:	3825 psig
Initial reservoir temperature:	215°F
Bubble point pressure:	1000–1200 psig
Total producing gas/oil ratio:	250–350 standard ft/STB
Formation volume factor:	1–22 RB/STB
Sulphur content:	1.8%
Water saturation:	11%
Hydrogen sulphide:	0.5 wt%

field (onshore Abu Dhabi). In this field it was demonstrated (Dunnington 1967) that with increasing depth of burial there is (i) a conspicuous increase in degree of stylolitization of a given stylolitic interval and consequently greater reduction in porosity; (ii) an increasing incidence of stylolitic development. However, in the Zakum field such developments are not observed, or at the very most are seen to occur in a minor way. It appears that the vertical and lateral variations in porosity are mainly due to the diagenetic conditions. Towards the end of the sedimentation of Zone II, and the concurrent uplift of a deep-seated salt plug, the Zakum area became subaerially exposed. The consequent subaerial diagenetic model appears to be mainly the cause of the vertical and lateral distribution of porosity. Porosity was further modified to some extent after burial by the development of stylolites.

Permeability, like porosity, also tends to vary vertically. There is a general decrease in permeability down through the zone, but this variation in permeability is much more pronounced than changes in porosity. The permeability is much reduced in the vicinity of stylolite zones 1 and 5 (Fig. 4). The high-permeability intervals or streaks (K > 100 Md) are usually related to grain supported limestone with excellent interparticle and intraparticle porosities in addition to vuggy porosity. Figure 19 shows the relationship between porosity and permeability in Zones II and IIIA. The relatively high mean permeability occurs in the central part of the field. Away from the central area and in particular to the west of the field, there is a gradual decrease in permeability, and on the far flanks of the Zakum structure the mean permeability < 5 Md. This

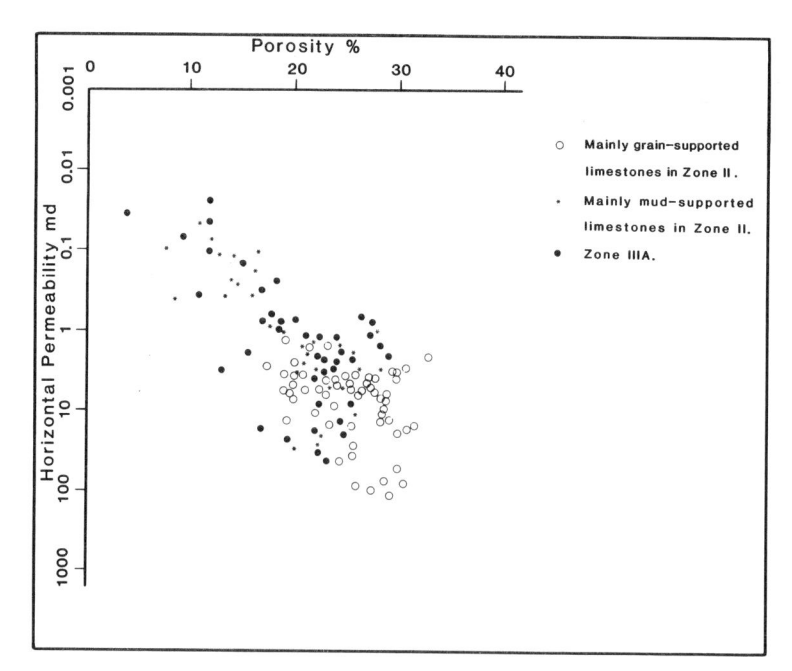

Fig. 19. The relationship between porosity and permeability measurements in the Kharaib Formation, Zakum field.

decrease in permeability is directly related to the gradual loss of decrease in permeability is directly related to the gradual loss of interparticle porosity in the structurally higher parts of the field.

A relationship between limestone texture and porosity/permeability has been established and the grain supported limestone texture tends to have higher permeabilities than the mud supported limestone textures (Fig. 19). Plotting individual porosity core plug values against the horizontal permeability values produces a large scatter of points and no meaningful relationship is apparent for Zones II and IIIA. However, plotting the mean porosity versus the mean permeability for each of the cored wells it is possible to obtain a general idea of these readings in the field (Fig. 19).

The porosity in Zone IIIA ranges from 5% to 32.8% (average of 22.9%). The permeability ranges from 0.04 to 86 Md, and is uniformly low regardless of texture.

The result of differing diagenetic histories of the zones is reflected in variations in amount and kind of porosity within the formation. Concentration of calcite spar, void filling calcite and/or dolomite, compaction, fracturing and pressure solution has caused porosity destruction. Porosity distribution in the reservoir is complex and contains primary and secondary pore types: primary interparticle (Fig. 20a) and intraparticle porosity (Fig. 20b); leached moldic and vuggy porosity (Fig. 17d); intercrystalline porosity within dolomite (Fig. 17b) and fracture porosity. These different types could have been modified by solution and cementation. Permeability has probably been enhanced to some extent by both pre- and post-lithification, fracturing and solution.

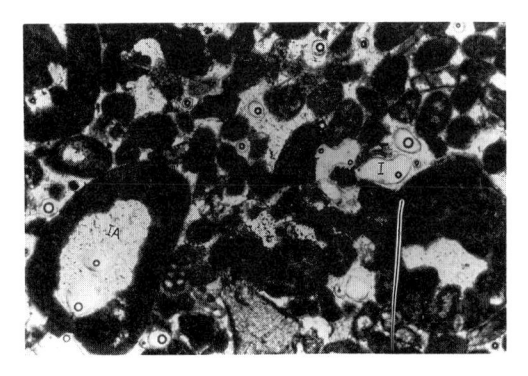

Fig. 20. Interparticle porosity (I) between the pellets and ooids and intraparticle porosity (IA) within the pelloids and foraminiferas.

Conclusion

The Kharaib Formation in Zakum field is a shelf sediment, in which repeating sequences of rock types are spread over a large area. This transgressive–regressive sequence is made up of restricted marine overlain by a sequence of open marine shelf deposits.

Different diagenetic histories explain the variations in amount and kind of porosity within the formation. Concentration by calcite spar, void filling calcite and/or dolomite, compaction, fracturing and pressure solution have all contributed to porosity destruction in the formation. Porosity in the Zakum reservoir is complex and contains primary and secondary pore types. These different types may have been modified by solution and cementation. Permeability has probably been enhanced to some extent by both pre- and post- lithification fracturing and solution.

The porosity in Zone II ranges from 7.3% to 31.8%. There is a general decrease in porosity from the upper to the lower part of the zone. The horizontal permeability varies from 0.05–62 Md and also permeability intervals are usually related to grain-supported limestone with excellent primary porosity in addition to vuggy porosity. In Zone IIIA the porosity ranges from 5% to 32% (average of 22.9%). The permeability ranges from 0.04 to 86 Md and is uniformly low regardless of textures.

I thank the geologists (past and present) of ADMA–OPCO and ZADCO for their assistance and fruitful discussion on different aspects of Zakum reservoir. Part of this paper is extracted from my PhD thesis carried out at University of South Carolina, USA. Professor C. G. St. Kendall supervised this thesis, and his advice and encouragement is greatly appreciated.

References

ALSHARHAN, A. S. 1985. Petrography, sedimentology, diagenesis and reservoir characteristics of the Shuaiba and Kharaib Formations (Barremian–mid Aptian) carbonate sediments of Abu Dhabi, United Arab Emirates. PhD thesis, University of South Carolina, USA.

—— & WILLIAMS, D. F. 1987. Petrography and stable isotope composition of baroque dolomite from the Shuaiba Formation (Lower Cretaceous), Abu Dhabi, United Arab Emirates. *Journal of African Earth Sciences*, **6**, 881–890.

BATHURST, R. G. C. 1975. Carbonate sediments and their diagenesis, *Developments in sedimentology* **12**, Elsevier, Amsterdam.

BECHER, J. W. & MOORE, C. H. 1979. The Walker Creek field: a Smackover diagenetic trap. *In*: MOOR, C. H. (ed.) *Geology of carbitate po-*

rosity. American Association of Petroleum Geology continuing education, course notes series **11**, A24−A46.

DUNNINGTON, H. V. 1967. Aspects of diagenesis and shape change in stylolitic limestone reservoirs. *7th World Petroleum Congress, Mexico*, **2**, 339−352.

EVAMY, B. D. & SHEARMAN, D. J. 1969. Early stages in development of overgrowth on echinoderm fragments in limestones, *Sedimentology*, **12**, 317−322.

FLUGEL, E. 1982. *Microfacies analysis of limestones*. Springer Verlag; Berlin.

FOLK, R. L. 1965. *Some aspects of recrystallization in ancient limestones*. Society of Economic Paleontologists and Mineralogists, special publication **13**, 14−48.

—— 1974. The natural history of crystalline calcium carbonate: effect of magnesium content and salinity. *Journal of Sedimentry Petrology*, **58**, 40−53.

—— & ASSERETO, R. 1974. Giant aragonite rays and barouque white dolomite in tepee-fillings, Triassic of Lombardy, Italy. *Society of Economic Paleontologists and Mineralogists Annual meeting, San Antonio*. Abstract with Program, 34−35.

KENDALL, C. G. ST. C. & SKIPWITH, Sir P. A. d'E. 1969. Holocene shallow water carbonates and evaporites sediments of the Khor al bazam, Abu Dhabi, SW Persian Gulf. *American Association of Petroleum Geologists Bulletin*, **53**, 841−870.

LONGMAN, M. W. 1980. Carbonate diagenetic textures from nearshore diagenetic environment. *American Association of Petroleum Geologists Bulletin*, **64**, 461−487.

LOUCKS, R. G. 1977. *Porosity development and distribution in shallow water carbonate complexes − subsurfaces Pearsal Formation (Lower Cretaceous), South Texas*. Texas University Bureau of Economic Geology, Report of Investigation **89**, 97−126.

RADKE, B. M. & MATHIS, R. L. 1980. On the formation and occurrence of saddle dolomite. *Journal of Sedimentary Petrology*, **50**, 1149−1168.

WANLESS, H. R. 1979. Limestone response to stress: pressure solution and dolomitization. *Journal of Sedimentary Petrology*, **49**, 437−462.

WILSON, J. L. 1975. Carbonate facies in geologic history. Springer, Berlin.

Geochemistry of crude oils in Oman

P. J. GRANTHAM[1], G. W. M. LIJMBACH & J. POSTHUMA

Koninklijke/Shell Exploratie en Produktie Laboratorium, Postbus 60, 2280 AB Rijswijk (Z. H.), Netherlands
[1]*Present address: NAM B. V., Postbus 28000, 9400 MM Assen, The Netherlands*

Abstract: The petroleum geochemistry of Oman provides a picture of considerable variety since crude oils and their source rocks are found both throughout the country and throughout the stratigraphic column from the Infra-Cambrian to the Tertiary. The oils can be geochemically classified into five groups. Three of them can be related to good oil source rocks found in the Precambrian Huqf Group, the Silurian Safiq and the Cretaceous Natih Formations. Another group of oils probably originates from the Upper Jurassic Diyab Formation whilst the fifth group of crudes (name 'Q') cannot be correlated to a known source rock but is inferred to have originated from an unpenetrated Huqf level.

The 'Huqf oils' are those that have been correlated to known Infra-Cambrian Huqf source rocks and are characterized by a strong C_{29} sterane predominance and very light carbon isotope values (δC_{13}) of around -36.0‰. In contrast, the 'Q' crudes, derived from the unknown source are characterized by a C_{27} sterane predominance and carbon isotope ratios of around -30.5‰. Both the Huqf and 'Q' crudes also contain a series of characteristic compounds referred to as the 'X' compounds (all isomers of methyl and dimethyl alkanes). Oils thought to originate from Silurian Safiq source rocks have a weak C_{29} sterane predominance, a significant content of rearranged steranes and carbon isotope ratios of approx. -30.5‰. The oils correlatable to the Jurassic Diyab Formation have a similar sterane distribution but heavier carbon isotope values of around -26.5‰. Finally, the crude oils from the mid-Cretaceous Natih Formation source rocks are characterized by steranes with an equivalent distribution of C_{27}, C_{28} and C_{29} isomers and carbon isotope values of around -26.9‰.

Oil exploration in the Sultanate of Oman began in 1937. The first well was spudded in 1955 by Dhofar Cities Service Petroleum Co. and in 1956 with Marmul-1 they discovered heavy oil, which was then uneconomic to extract. In 1960, exploration in the northern areas by Petroleum Development Oman (PDO) came under the direction of the Shell Group and in 1962 oil in commercial quantities was discovered at Yibal. This was followed by the discovery of the Natih Field in 1963 and the Fahud Field in 1964. In 1969, PDO also took over exploration in Dhofar and reopened interest in the heavy oils around Marmul. Through the 1970s a number of other oil fields in both the north and south were discovered and progressively brought into production. Presently some 50 fields have been developed and a total production of over 500 000 bbls per day has been established. A pipeline network leads to the terminal West of Muscat (see Fig. 1).

Oils and potential source rocks in Oman have been the subject of geochemical study in the Shell Exploration and Production Research Laboratory (KSEPL) since the 1970s. Potential oil source rocks have been identified in several levels of the stratigraphy ranging from Proterozoic to mid-Tertiary (Fig. 2). Oils have been accumulated throughout the stratigraphic column, but mostly in Cretaceous, Jurassic and Permian reservoirs due to the positioning of the best seals (see Fig. 2). Except in the south, most Oman crudes are generally low-sulphur, low-wax light oils. Standard geochemical techniques (Lijmbach 1975; Lijmbach *et al.* 1983) show that all Oman crude oils appear to have been generated from marine structureless or sapropelic organic matter with no contribution from land-plant material (see Figs 3 & 4). The most significant variations are in biomarker distributions and carbon isotope values which are distinct enough to ensure a high degree of certainty in the grouping of the crude oils described here. Five chemically distinct types of crude oil can be recognized. Three of them are linked to source rocks in Proterozoic (Huqf), Silurian (Safiq) and Mid-Cretaceous (Natih) levels, whilst the fourth is probably derived from an Upper Jurassic (Diyab) source. The fifth group has characteristics and occurrences that argue strongly for a distinct, as yet unidentified, source.

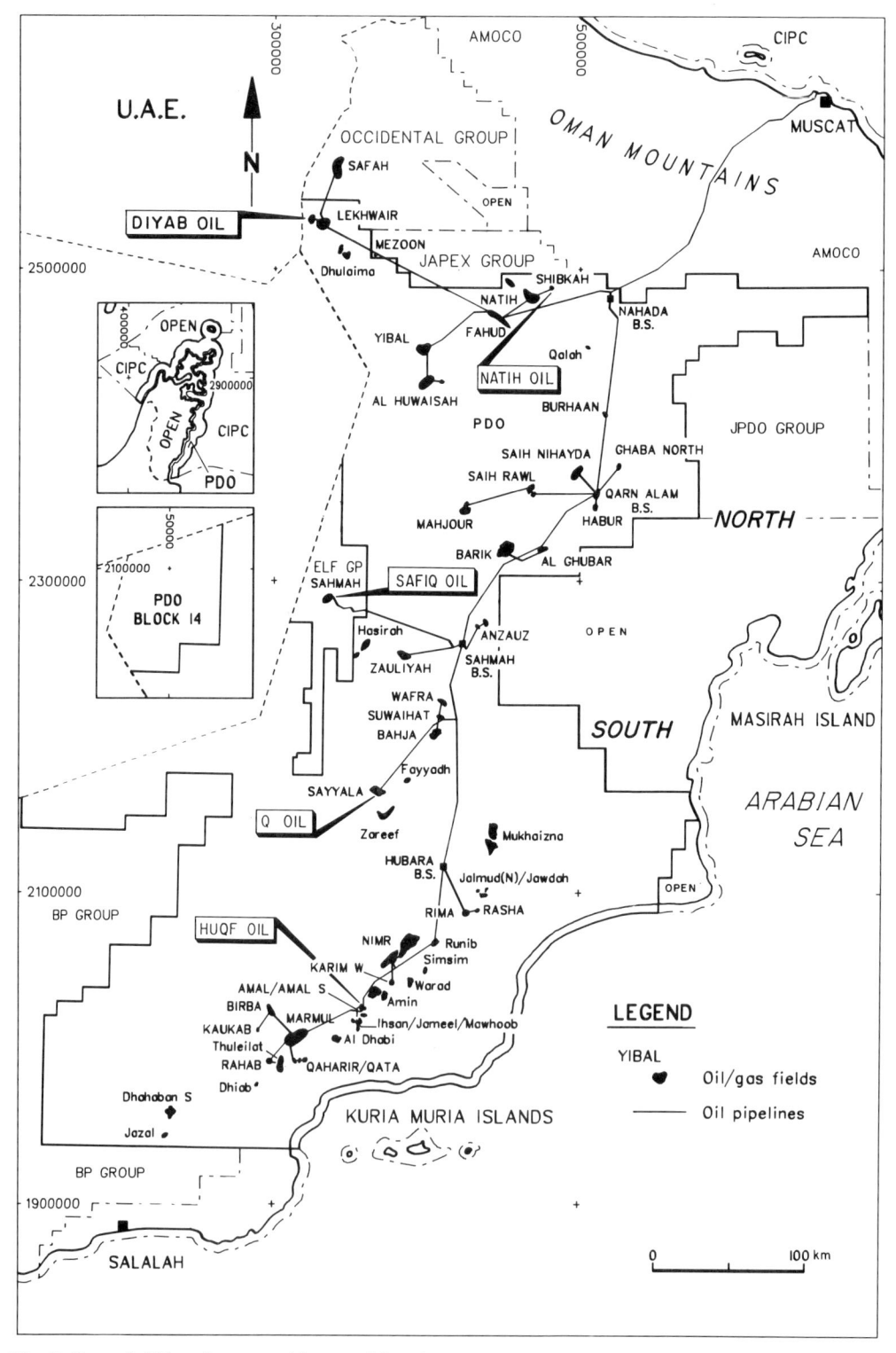

Fig. 1. Oman field location map with type oil locations.

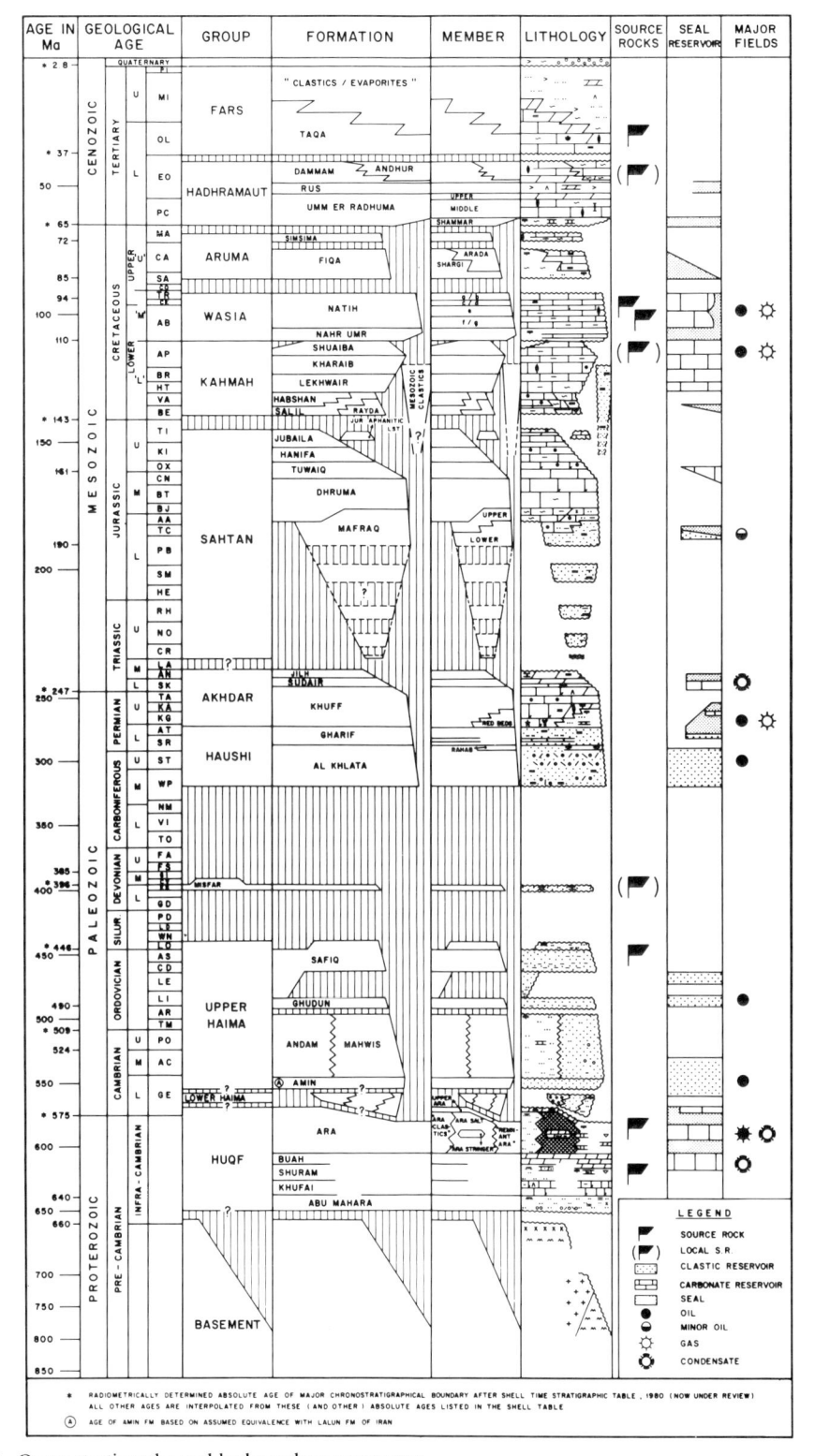

Fig. 2. Oman stratigraphy and hydrocarbon occurrence.

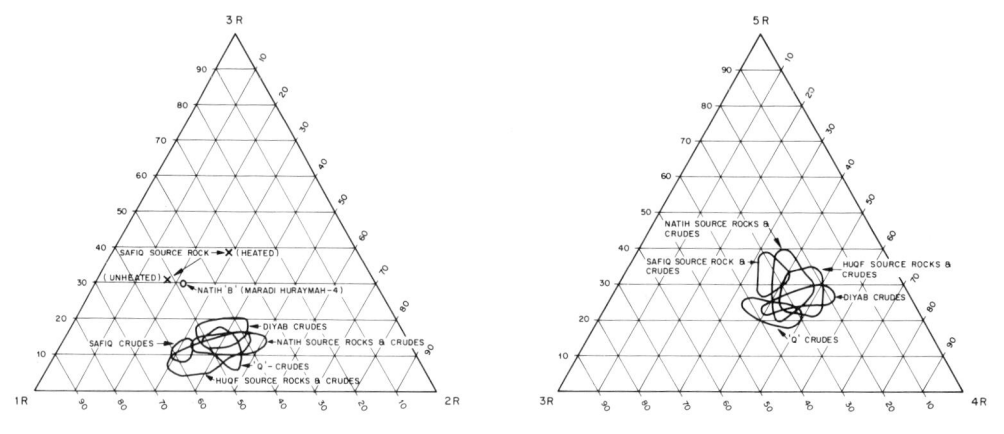

Fig. 3. C15 and C30 ring distributions of Oman crude oils and source rock extracts.

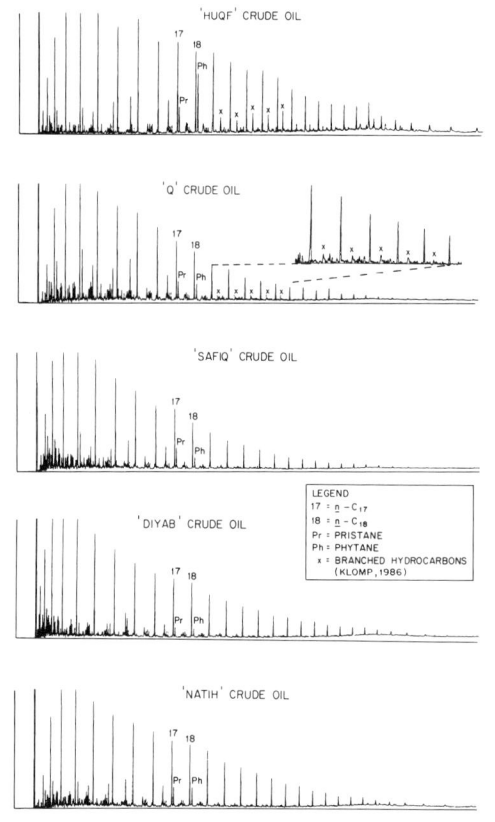

Fig. 4. Typical gas chromatograms for the saturated hydrocarbons of the five main groups of Oman crude oils.

The stratigraphy (Fig. 2) in the oil productive areas of Oman has recently been described by Hughes Clarke (1988). For the petroleum geology of Oman see the synopsis in Grantham et al. (1988). The most significant feature of Oman's hdyrocarbon habitat is that sedimentary rocks of such diverse ages can all lie within the hydrocarbon window. Hence, organic-rich horizons forming potential source rocks in any of these age intervals can have generated and may still be generating oil and gas.

Crude oil characterization and oil/source rock correlations

Tables 1–5 present the geochemical data of specific examples of the five groups of Oman crude oils and their source rocks. Representative gas chromatograms of the saturated fraction, sterane fragmentograms and sterane carbon number distributions are given in Figs 4–6.

'Huqf oils' (typified by oil from Amal South-1)

These oils occur in the eastern flank province of South Oman (Aley & Nash 1984; de la Grandville 1982). They are generally heavy and sulphur-rich and occur in Palaeozoic clastic reservoirs lying above the Huqf sediments (see Fig. 2).
  The main distinguishing geochemical features of these crudes are as follows:
(i) Pristane/phytane ratios of generally less than 1.
(i) The presence of the 'X' compounds (Fig. 4; Klomp 1986).
(iii) A strong predominance of C_{29} steranes (Figs 5 & 6). C_{29} sterane predominances have also been reported for Siberian crude oils of preCambrian origin (see Arefev et

'HUQF' CRUDE OIL

'Q' CRUDE OIL

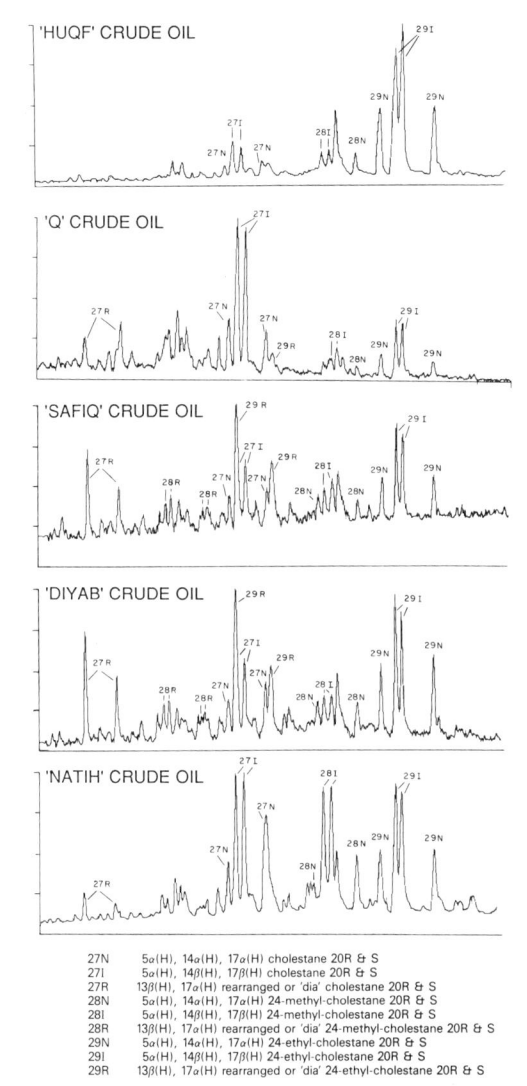

'SAFIQ' CRUDE OIL

'DIYAB' CRUDE OIL

'NATIH' CRUDE OIL

27N	5α(H), 14α(H), 17α(H) cholestane 20R & S
27I	5α(H), 14β(H), 17β(H) cholestane 20R & S
27R	13β(H), 17α(H) rearranged or 'dia' cholestane 20R & S
28N	5α(H), 14α(H), 17α(H) 24-methyl-cholestane 20R & S
28I	5α(H), 14β(H), 17β(H) 24-methyl-cholestane 20R & S
28R	13β(H), 17α(H) rearranged or 'dia' 24-methyl-cholestane 20R & S
29N	5α(H), 14α(H), 17α(H) 24-ethyl-cholestane 20R & S
29I	5α(H), 14β(H), 17β(H) 24-ethyl-cholestane 20R & S
29R	13β(H), 17α(H) rearranged or 'dia' 24-ethyl-cholestane 20R & S

Fig. 5. Typical sterane fragmentograms (m/z 217 + 218 + 259) for the five main groups of Oman crude oils.

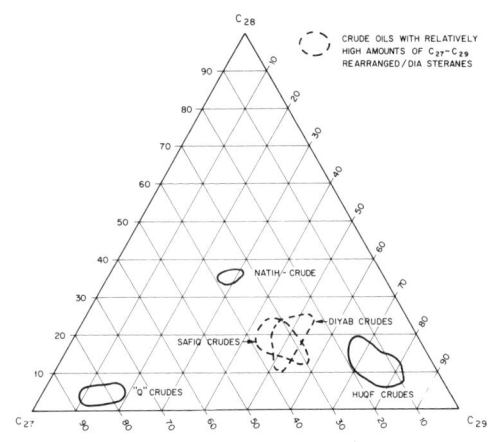

Fig. 6. Sterane carbon number distribution for the five main groups of Oman crude oils.

al. 1980; Moldowan et al. 1985; Fowler and Douglas 1987). Fowler and Douglas also report the presence of the 'X' compounds in their Siberian crudes. Measured concentrations of total steranes in Huqf crude oils approach 600 ppm. Similar concentrations of steranes are also found in the Huqf source rock.

(iv) The absence or very low relative contents (approx. 5% or less) of rearranged steranes (Fig. 5; Grantham 1986).
(v) Highly negative carbon isotope values of around −36‰. These carbon isotope values are amongst the most negative crude oil values known.

Good source rocks for oil occur in south Oman in several formations of the Huqf group (see Fig. 7 and Meyer & Nederlof 1984). They are marly to shaly and contain type-II organic matter, most probably originating from bacteria and cyano-bacteria. The oil/source rock correlation work was carried out with source rock samples from wells throughout the Huqf sediments.

The crude oils correlate very well with extracts of Huqf source rocks (see Table 1). The 'X'-branched hydrocarbons are always present in Huqf-derived crude oils but not always in all source rock extracts. The variation in carbon isotope ratios of source rock extracts and kerogen is somewhat larger than that in the oils attributed to them. However, this phenomenon is frequently observed in other oil/source rock pairs.

Interestingly, the same geochemical features are found in a tar from well Karampur-1 in Pakistan (see Table 1). The Pakistan and south Oman salt basins were in close proximity and possibly continuous during the infra-Cambrian (Gorin et al. 1982).

The 'Q' crude oils (typified by oil from Sayyala-1)

In Central Oman a group of light oils is found in Upper Palaeozoic reservoirs (see Table 2) which are characterized by the following.
  (i) The presence of the 'X' compounds (as in the Huqf oils).
  (ii) Pristane/phytane ratios greater than 1.
  (iii) Strong predominances of C_{27} steranes and

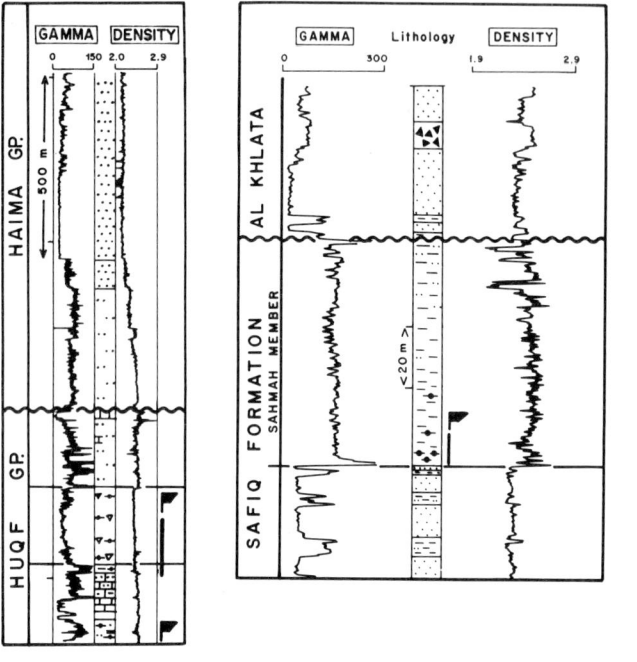

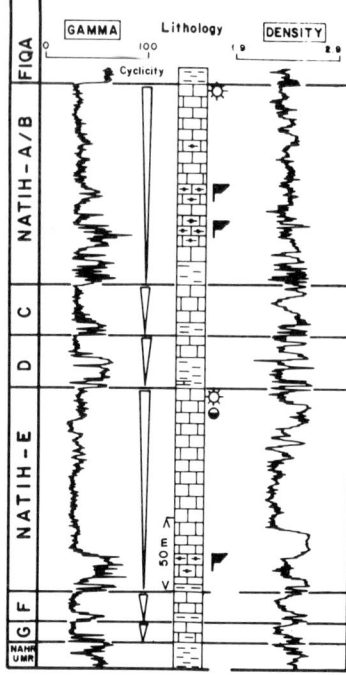

Fig. 7. Oman source rocks.

relatively high concentrations of tricyclic terpanes (see Figs 5 & 6 and Grantham 1986). The unusual sterane distribution of these oils is discussed in more detail by Grantham & Wakefield (1988). Sterane concentrations in the 'Q' crudes are roughly equivalent to those found in the Huqf crudes.

(iv) Moderate levels of rearranged steranes (~ 20%, see Fig. 5).

(v) Carbon isotope ratios of around −30.5‰.

None of the known source rocks can have sourced this unusual oil. The 'Q' oils contain the 'X'-branched hydrocarbons but, relative to the n-alkanes, in lower concentrations than in the Huqf crude oils (see Fig. 4). So far, within Oman, the 'X'-branched hydrocarbons have been found only in the Huqf and 'Q' crudes, although the same compounds have been identified in East Siberian crude oils (Fowler & Douglas 1987) and tentatively in the worlds oldest known crude oil from the Proterozoic of Australia (Jackson et al. 1986). If these compounds prove to be indicative for the pre-Cambrian, then a source for the 'Q' crudes also in the Huqf group is likely, a suggestion supported by geological reasoning. The oil occurs where Huqf levels are deep and unpenetrated by the drill and the oils are light and well matured, suggesting a deep source. Hence, an undrilled Proterozoic Huqf level might be the source for this distinctive oil type found in reservoirs in upper Palaezoic clastics.

Safiq oils (typified by oil from the Sahmah-1 well)

In southwest Oman a distinct crude oil type is found (see Table 3), in Upper Palaeozoic clastics, with the following geochemical features.

(i) Low sulphur contents and medium API gravities.

(ii) Pristane/phytane ratios of 1.6–1.7 (Table 3).

(iii) Sterane carbon number distributions which show a reduced predominance of C_{29} steranes compared to Huqf crudes (approximately 50%, Fig. 5).

(iv) Relatively high concentrations of rearranged-steranes (30–39%, Fig. 5).

(v) Carbon isotope values occur in the range −29.9 to −30.5‰.

Most geochemical parameters support the correlation of the oil from Elf Oman's Sahmah

Table 1. Geochemical data: Precambrian Huqf source rocks and Huqf-derived crude oils

Crude oil	Reservoir	API grav.	% Lt. fract.	% S	Gross comp* Sats	Arom	Het	Isoprenoids Pr/Ph	Pr/17	Ph/18	C15 dist 1R	2R	3R	C30 dist 3R	4R	5R	C29 VRE	Sterane type Iso	Nor	Rear	Ster carb. no. C27	C28	C29	Carbon isotope ‰
Amal S-1	Gharif	24.4	4.7	2.3	33	37	12	0.6	0.3	0.7	61	30	9	20	48	32	0.73	66	34	0	14	12	74	−35.9
Nuham-1	A1 Kh.	10.0	0.0	2.6	22	32	11	0.6	0.6	1.3	48	39	13	21	46	33	0.69	60	40	0	17	14	69	−37.2
Rima-7	A1 Kh.	32.4	7.0	1.3	38	26	5	0.9	0.3	0.5	62	30	8	26	52	22	0.85	65	35	0	18	14	68	−36.2
Runib-1	Natih	25.0	0.3	1.4	42	50	8	0.8	1.2	1.7	53	36	11	27	46	27	0.80	56	38	6	14	13	73	−36.5
Pakistan Karampur-1 (Tar)		–	–	5.1				0.4	0.5	1.6	58	26	16	18	47	35	0.72	52	48	0	14	14	72	−37.0

Source rock extracts	% extract	% C	Gross comp Sats	Arom	Het	Isoprenoids Pr/Ph	Pr/17	Ph/18	C15 dist C30 dist 1R	2R	3R	3R	4R	5R	C29 VRE	Sterane type Iso	Nor	Rear	Ster carb. no. C27	C28	C29	Carbon isotope ‰
Amal−9	3.0	6.2	28	49	23	0.6	0.6	1.2	61	33	5	22	50	28	0.67	68	32	0	20	20	60	−34.7
Nimr−4	0.4	2.6	18	47	35	1.0	0.3	0.3	57	34	9	31	44	26	0.92	66	34	0	23	10	67	−35.4
Runib−1	0.2	2.2	19	48	33	0.8	0.6	0.9	–	–	–	24	45	31	0.80	66	34	0	23	13	64	−35.2
Runib−1 heated	0.2	4.9				1.0	0.7	0.8	25	47	28	20	32	47	0.73	60	40	0	13	37	50	−34.4
									44	35	21	32	39	39								

% ext = % ethyl acetate extract of source rocks: % C = % organic carbon:
% S = % organic sulphur:
API grav. = API gravity: % Lt. fract. = % light fraction boiling below 120°C
Gross comp. = Gross composition of sample; Sat = % saturates: Aro = % aromatics:
Het = % heterocompounds; *In cases where the gross composition does not sum to 100%, the difference is the percentage material lost due to evaporation
Isoprenoids
Pr = pristane; Ph = phytane; 17 = n-C_{17}; 18 = n-C_{18};
C15 dist. & C30 dist. = C_{15} and C_{30} ring distributions (Lijmbach et al. 1983);
1R = 1-ring saturates; 2R = 2-ring saturates; 3R = 3-ring saturates;
4R = 4-ring saturates; 5R = 5-ring saturates; C29 VRE = vitrinite reflectance equivalent based on the C_{29} ring distributions (Lijmbach et al. 1983);
Sterane type = stereochemical composition of C_{27}-C_{29} steranes;
Iso = 5α (H), 14β(H), 17β(H) steranes;
Nor = 5α(H), 14α (H), 17α(H) steranes;
Rear = 13β(H), 17α(H) rearranged or 'dia' steranes;
Ster. carb. no. = sterane carbon number distribution

Table 2. *Geochemical data: 'Q' crude oils*

Crude oil	Reservoir	API grav.	% Lt. fract.	% S	Gross comp* Sats	Arom	Het	Isoprenoids Pr/Ph	Pr/17	Ph/18	C15 dist 1R	2R	3R	C30 dist 3R	4R	5R	C29 VRE	Sterane type Iso	Nor	Rear	Ster carb. no. C27	C28	C29	Carbon isotope ‰
Sayyala-1	Gharif	50.1	17.9	0.3	76	5	19	1.2	0.7	0.7	46	37	16	40	36	24	1.04	60	18	22	84	3	13	−30.3
Bahja-1	Gharif	48.7	14.3	0.1	49	7	2	1.1	0.5	0.5	48	43	9	37	46	17	1.00	65	16	20	89	0	11	−30.4
Bahja-1	Gharif	42.5	9.7	0.2	63	10	3	1.0	0.5	0.6	56	35	9	45	37	18	1.04	66	18	16	72	11	17	−30.9
Zareef-1	Gharif	46.3	13.5	0.2	50	8	3	1.1	0.8	0.8	53	40	7	41	38	21	1.00	67	16	18	75	5	20	−31.0
Fayyadh-1	Gharif	27.3#	0.9#	0.4	70	17	4	—#	—#	—#	39	46	15	38	44	19	0.92	66	18	16	72	6	22	−30.6

Parameters affected by bacterial degradation.
See Table 1 for an explanation of abbreviations.

Table 3. *Geochemical data: Silurian Safic Fm. Source rocks and Sahmah field oils*

Crude oil	Reservoir	API grav.	% Lt. fract.	% S	Gross comp* Sats	Arom	Het	Isoprenoids Pr/Ph	Pr/17	Ph/18	C15 dist 1R	2R	3R	C30 dist 3R	4R	5R	C29 VRE	Sterane type Iso	Nor	Rear	Ster carb. no. C27	C28	C29	Carbon isotope ‰
Sahmah-1	Gharif	39.6	9	0.1	65	32	3	1.7	0.5	0.4	56	32	12	35	37	28	0.96	40	24	36	32	24	44	−30.5
Sahmah-1	Gharif	36.5	6	0.4	64	33	3	1.7	0.6	0.4	56	3	13	32	38	30	0.92	49	18	33	30	15	57	−30.6
Shamah-1	Gharif	36.4	6	0.3	56	41	3	1.7	0.5	0.4	60	29	11	35	36	29	0.96	44	17	39	38	18	44	−30.6

Source rock extracts	% extract	% C	Gross comp* Sats	Arom	Het	Isoprenoids Pr/Ph	Pr/17	Ph/18	C15 dist 1R	2R	3R	C30 dist 3R	4R	5R	C29 VRE	Sterane type Iso	Nor	Rear	Ster carb. no. C27	C28	C29	Carbon isotope ‰
Hasirah-2	0.09	1.0	29	39	30	0.9	0.9	0.6	51	18	31	16	48	36	0.84	30	37	33	37	18	45	−30.4
U. Safiq Heated sample	0.8	0.5	18	29	51	0.9	0.4	0.3	32	29	39	31	32	37	1.08							−30.7

See Table 1 for an explanation of abbreviations.

field with the extracts of combined sidewall samples of the Silurian Safiq Fm. from well Hasirah-2 (see Table 3). This Formation has a high organic content in its basal shale which constitutes a source rock (Fig. 7, Meyer & Nederlof 1984).

The correlation of oils and extracts is not perfect; there are differences in the C_{15} distributions (Fig. 3). Hydrous pyrolysis of the not fully mature Hasirah-2 sample also resulted in an extract similar to the oils but not identical (see Table 3). However, the most diagnostic parameters (steranes and carbon isotopes) correlate very well. We therefore conclude that the Sahmah oils most probably originate from the Silurian Safiq source rock.

The oil character (naphthenic, low sulphur) indicates a source environment in a marine setting but with terrigenous clastic influx. This is in keeping with the setting of the global transgressive cycles of late Ordovician and early Silurian age covering the northern and eastern shelves of the Afro-Arabian cratons.

Diyab oils (typified by oil from Lekhwair-8)

In the northwest of Oman a characteristic type of crude oil is found in the east Lekhwair field (Fig. 1). This oil is produced from Barremian–Aptian (Shuaiba–Kharaib) reservoirs whilst two non-commercial accumulations are found in Upper Jurassic Tuwaiq Mountain Fm. reservoirs.

This oil type has the following characteristics (see Table 4).
  (i) Low sulphur and medium API gravities.
 (ii) Pristane/phytane ratios above 1 (except in the less mature Tuwaiq crudes).
(iii) C_{29} steranes which are just predominant within total steranes (approximately 50%, Fig. 5, Table 4).
(iv) Relatively high concentrations of rearranged steranes (31–39%, Fig. 5).
 (v) Carbon isotope values around $-26.5‰$.

The oils are geochemically identical with the oil found in the Upper Jurassic (Arab zone) and Lower Cretaceous (Thamama) reservoirs in the United Arab Emirates and Qatar (see Frei 1984). There is no good source rock developed in Oman that matches these oils. Studies of the oils and source rocks in Abu Dhabi (Hassan & Azer 1985; Alsharan 1985) conclude that the oil in the Jurassic– Lower Cretaceous reservoirs originates from Upper Jurassic source rocks (Hanifa or Diyab Formations). The east Lekhwair oils are therefore interpreted to have migrated laterally from such a source outside Oman. The occurrence of a non-commercial accumulation of this oil in an Upper Jurassic reservoir (Tuwaiq Mountain Formation) in well Dhulaima-1 close to Lekhwair supports the postulate that the oil is from a Jurassic rather than a Cretaceous source. There is also the possibility that some of the oil may have been supplied by organic-rich Aptian marine sediments (Bab facies of the Shuaiba Formation) but no satisfactory oil/source rock correlations for the Shuaiba could be established.

Natih oils (typified by oil from Shibkah-1)

The Natih Formation forms the reservoir for many of the northern Oman oil fields but also includes two organic-rich intervals forming source rocks (Fig. 7). There is no geochemical difference between these two intervals, but the upper level (Natih 'b') is probably the more important source rock, whereas in Saudi Arabia the lower level (Natih 'e', Safaniya source rock) is the better developed potential source (Newell & Hennington, 1983; Lehner 1984).

The Natih oils have the following characteristics (see Table 5).
  (i) Low API gravity and moderately high sulphur.
 (ii) Pristane/phytane ratios of c. 1.0.
(iii) Rearranged steranes are relatively low (5%, Fig. 5).
(iv) Steranes with an equivalent distribution of the C_{27}, C_{28} and C_{29} isomers (Fig. 6).
 (v) Carbon isotope values of around $-26.9‰$.

The sterane distributions are very similar to those of a worldwide range of crude oils derived from Cretaceous carbonates (Grantham & Wakefield 1988).

Table 5 gives geochemical data on a number of extracts of the Natih source rock. The extracts correlate well with three Natih crude oils from wells Shibkah-1, Suqtan-1 and Natih West-57 (same table).

Conclusions

The petroleum geochemistry of Oman provides a picture of considerable variety. Oil accumulations and source rock occurrences are found geographically spread throughout the country and through the stratigraphic column from the Infra-Cambrian (c. 650 Ma) to the Tertiary. The main levels of source rocks have been identified in the Infra-Cambrian Huqf, the Silurian Safiq, the Upper Jurassic Diyab and the Cretaceous Bab and Natih formations.

Five types of crude oil are recognized, four of which can be correlated to one of the source

Table 4. Geochemical data: crude oils derived from the Divab source rock

Crude oil	Reservoir	API grav.	% Lt. fract.	% S	Gross comp* Sats	Arom	Het	Isoprenoids Pr/Ph	Pr/17	Ph/18	C15 dist 1R	2R	3R	C30 dist 3R	4R	5R	C29 VRE	Sterane type Iso	Nor	Rear	Ster carb. no. C27	C28	C29	Carbon isotope ‰
Lekhwair-8	Shuaiba	34.7	10.3	0.6	55	37	8	1.1	0.3	0.3	48	37	15	37	40	23	1.00	43	22	34	37	10	53	−26.4
Lekhwai-35	Shuaiba	35.8	6.7	0.7	51	42	7	1.1	0.3	0.3	51	35	14	30	44	26	0.92	42	21	37	33	16	51	−26.6
Lekhwai-27	Tuwaiq	29.3	5.7	1.9	—	—	—	0.8	0.3	0.4	50	33	17	20	54	27	0.85	47	22	31	31	19	50	−26.7
Dhulaima-1	Tuwaiq	30.1	7.7	2.3	—	—	—	0.8	0.3	0.4	39	43	18	26	46	28	0.92	51	22	27	29	18	53	−25.5

See Table 1 for an explanation of abbreviations.

Table 5. Geochemical data: Mid-Cretaceous Natih source rocks and Natih-derived crude oils

Crude oil	Reservoir	API grav.	% Lt. fract.	% S	Gross comp* Sats	Arom	Het	Isoprenoids Pr/Ph	Pr/17	Ph/18	C15 dist 1R	2R	3R	C30 dist 3R	4R	5R	C29 VRE	Sterane type Iso	Nor	Rear	Ster carb. no. C27	C28	C29	Carbon isotope ‰
Shibkah-1	Natih	26.4	9.0	1.5	33	27	8	1.0	0.4	0.4	52	36	12	31	40	29	1.00	69	28	3	38	35	27	−26.9
Suqtan-1	Natih	16.2**	16.0	1.7	23	27	14	1.0	0.5	0.6	52	34	14	25	42	33	0.92	68	27	5	35	35	30	−26.6
Natih W-57	Natih	13.2	1.7	1.0	33	48	19	1.0	0.4	0.4	59	31	10	24	48	28	0.92	70	27	3	36	34	30	−27.0

** The Suqtan-1 sample is heavily contaminated with water. This has probably influenced the API gravity and percent light fraction.

Source rock extracts		% extract	% C	Gross comp Sats	Arom	Het	Isoprenoids Pr/Ph	Pr/17	Ph/18	C15 dist 1R	2R	3R	C30 dist 3R	4R	5R	C29 VRE	Sterane type Iso	Nor	Rear	Ster carb. no. C27	C28	C29	Carbon isotope ‰
Jebel Qusaybah CH-3	Natih 'b'	0.3	1.9	39	44	17	0.9	0.5	0.6	49	39	12	35	45	21	0.92	74	26	0	34	38	28	−26.9
	Natih 'b'	0.4	3.1	39	45	16	1.4	0.5	0.5	51	37	13	35	44	21	0.96	73	27	0	32	40	28	−27.1
Maradi Huraymah-4	Natih 'b'	0.5	3.1	23			0.7	0.4	0.5	47	23	30	29	38	33	0.82	67	30	3	36	37	27	−26.6 (kerogen) −26.9
	Natih 'e'	0.6	1.4	31			1.1	0.8	0.9	42	16	27	42	31		0.76	57	15	28	42	32	26	(kerogen)

See Table 1 for an explanation of abbreviations.

Table 6. *Main geochemical characteristics of Oman crude oils*

	API grav	Isoprenoids	'X' compounds	Sterane carbon number	Rearr. sterane content	Carbon isotope ratio ‰
Huqf	25–30	Ph predom over Pr	present	strong C_{29} predominance	low	~ −36.0
'Q'	40–50	Pr predom over Ph	present	strong C_{27} predominance	low	−30.5
Safiq	37–39	ditto	absent	weak C_{29} predominance	high	−30.0
Diyab	~ 35	ditto	absent	weak C_{29} predominance	high	−26.5
Natih	~ 30	ditto	absent	C_{27}, C_{28}, C_{29} equivalent	low	−26.5

Ph = phytane; Pr = pristane

rocks described above. For one type of oil ('Q' oils) the source is as yet unknown but there are implications that it is also from a Huqf Group source rock. Table 6 provides a summary of the significant geochemical features of the five Oman oil groups.

This interpretation of the oil habitat of Oman is the result of the work of many Shell geologists and geochemists who cannot all be mentioned here. Their contributions are appreciated. In particular we are indebted to Dr. M. W. Hughes Clarke and Dr. R. J. Willink for their helpful discussions in integrating the geochemistry into the geological context of Oman. This publication is supported and authorized by Shell International Petroleum Maatschappij N.V., Shell Research B.V., Petroleum Development Oman LLC and the Ministry of Petroleum and Minerals in the Sultanate of Oman. We thank these authorities for their permission to publish this article. Additionally, Elf-Aquitane Oman has kindly agreed that we may include our analyses of oils from their Sahmah-1 field. Members of the Geochemistry Department of KSEPL from 1975 to 1985 are thanked for their analytical assistance.

References

ALEY, A. A. & NASH, D. F. 1984. A summary of the geology and oil habitat of the eastern flank hydrocarbon province of South Oman. *Proceedings of OAPEC Seminar 'Source and Habitat of Petroleum in the Arab Countries'* Kuwait, October, 1984.

ALSHARAN A. S. 1985. Depositional environment, reservoir units evolution and hydrocarbon habitat of Shuaiba Formation, Lower Cretaceous, Abu Dhabi, United Arab Emirates. *American Association of Petroleum Geologists Bulletin.*, 69(6), 899–912.

AREFEV, O. A., ZABRODINA, M. N., MAKUSHINA, V. M. & PETROV, A. A. 1980. Relic tetra- and pentacyclic hydrocarbons in the old oils of the Siberian Platform. *Izv. Akad. Nauk. SSSR, Ser. Geol.*, 3, 135–140.

FOWLER, M. G. & DOUGLAS, A. G. 1987. Saturated hydrocarbon biomarkers in oils of Late Pre-Cambrian age from eastern Siberia. *Organic Geochemistry.*, 11, 201–213.

FREI, H. O. 1984. Mesozoic source rocks and oil accumulations in Qatar. *Proceedings of OAPEC Seminar 'Source and Habitat of Petroleum in the Arab Countries'* Kuwait, October, 1984.

GORIN, G. E., RACZ, L. G. & WALTER, M. R. 1982. Late PreCambrian–Cambrian sediments of Huqf group, Sultanate of Oman. *American Association of Petroleum Geologists Bulletin*, 66(12), 2609–2627.

GRANDVILLE, B. F. de la 1982. Appraisal and development of a structural and stratigraphic trap oil field with reservoirs in glacial and periglacial clastics. *A.A.P.G. Memoir*, 32, 267–286.

GRANTHAM, P. J. 1986. The occurrence of unusual C_{27} and C_{29} sterane predominances in two types of Oman crude oil. *Organic Geochemistry*, 9, 1–10.

——, LIJMBACH, G. W. M., POSTHUMA, J., HUGHES CLARKE, M. W. & WILLINK, R. J. 1988. Origin of crude oils in Oman. *Journal of Petroleum Geology*, 11, 61–80.

—— & WAKEFIELD, L. L. 1988. Variations in the sterane carbon number variations of marine source rock derived crude oils through geological time. *Organic Geochemistry*, 12, 61–73.

HASSAN, T. H. & AZER, S. 1985. The occurrence and origin of oil in offshore Abu Dhabi. *Proceedings of the SPE 1985 Middle East Oil Technical Conference, Bahrain, March 1985*. 143–149. (SP 13696).

HUGHES CLARKE, M. W. 1988. Stratigraphy and rock unit nomenclature in the oil producing area of interior Oman. *Journal of Petroleum Geology*, 11, 5–60.

JACKSON, M. J., POWELL, T. G., SUMMONS, R. E. & SWEET, I. P. 1986. Hydrocarbon shows and petroleum source rocks in sediments as old as 1.7×10^9 years. *Nature*, **322**, 727–729.

KLOMP, U. C. 1986. The chemical structure of a pronounced series of iso-alkanes in South Oman crudes. *Organic Geochemistry*, **10**, 807–814.

LEHNER, P. 1984. Mesozoic source rocks of the Arabian Platform: Palaeogeography and position in sedimentary cycle. *Proceedings of OAPEC Seminar 'Source and Habitat of Petroleum in the Arab Countries' Kuwait, October, 1984.*

LIJMBACH, G. W. M. 1975. On the origin of petroleum. *Proc. 9th World Petroleum Congress, Sect. 2*, 357–369.

LIJMBACH, G. W. M., VAN DER VEEN, F. M. & ENGELHARDT, E. D. 1983. Geochemical characteristics of crude oils and source rocks using Field Ionisation Mass Spectrometry. *In*: BJOROY M. *et al.* (ed.), *Advances in Organic Geochemistry 1981*. John Wiley, Chichester, 788–798.

MEYER, B. L. & NEDERLOF, M. H. 1984. Identification of source rocks on wireline logs by density/resistivity and sonic transit time/resistivity cross-plots. *American Association of Petroleum Geologists Bulletin*, **68**, 121–124.

MOLDOWAN, J. M., SEIFERT, W. K. & GALLEGOS, E. J. 1985. Relationship between petroleum composition and depositional environment of petroleum source rocks. *American Association of Petroleum Geologists Bulletin*, **69**(8), 1255–1268.

NEWELL, K. D. & HENNINGTON, R. D. 1983. Potential petroleum source rock deposition in the Middle Cretaceous Wasia Formation, Rub'Al Khali, Saudi Arabia. *Proceedings of the SPE 1983 Middle East Oil Technical Conference, Bahrain, March 1983.* 151–156.

People's Democratic Republic of Yemen: a future oil province

S. K. PAUL

Sanpra Ltd., Bute House, 97 Longdown Lane South, Epsom, Surrey KT17 4JJ, UK

Abstract: The People's Democratic Republic of Yemen has recently joined the ranks of oil producing countries and it is hoped that in the near future there will be every likelihood that it will become a major oil producing country and 'a Classic Petroleum Province', similar to the surrounding Arabian Gulf region.

The People's Democratic Republic of Yemen (PDRY) covers the southwestern coastal segment of the Arabian Peninsula. It is bounded on the south by the Gulf of Aden, while inland it shares a common border with the Yemen Arab Republic, Saudi Arabia and Oman. Offshore, PDRY includes the Islands of Soccotra, Abd al Kuri, Samha, Darsa and Perem Islands (Fig. 1).

The generally accepted total area of the country amounts to about 337 000 km^2 (130 000 square miles), including the islands of Socotra, Abd al Kuri, Samha, and Darsa in the Arabian Sea off the eastern coast of Africa; and Perim, at the south end of the Red Sea. The continental shelf, along the southern edge of the country is generally narrow, dropping off steeply to depths in excess of 4500 m. The area of the shelf, to a depth of 1000 m totals approximately 78 000 km^2. Of this amount, about 32 000 km^2 (40% of the area) surrounds Socotra and its neighbouring islands.

Geological history

Geologically, PDRY occupies the southern edge of the Arabian Shield (see Fig. 2). In the extreme west, the Red Sea, the Gulf of Aden and the Ethiopian rift valleys meet in the region referred to as the Afar Triangle.

The Arabian Shield was formed in Precambrian time (see Husseini & Husseini 1990). After peneplanation, the shield tilted slightly northeast towards the Tethys trough. As tilting and subsidence continued, shallows seas advanced across the bevelled surface of crystalline rocks, burying it beneath thin sheets of almost flat-lying sediments. This belt of low-dipping, relatively undisturbed beds forms the stable region structural province: the Arabian shelf. The eastern three quarters of the Republic is situated in this shelf province, the northern extension of which becomes the Rub al Khali Basin. Sediments of Palaeozoic, Mesozoic and Cenozoic age occupying this basin in Saudi Arabia, the United Arab Emirates and Oman contain the giant oil and gas fields of the Arabian Gulf region.

In the eastern PDRY, at the southern edge of the Rub al Khali Basin, the total sedimentary column, consisting of alternating sandstone, shales, marls and limestones, reaches a thickness in excess of 3000 m at the edge of the sand desert. Greater thicknesses may be preserved in structurally depressed areas in the east–central part of the country and in the coastal offshore areas.

Sedimentary basins

There are six basinal areas in the PDRY which are prospective for oil and gas (see Fig. 3). Three are mainly or entirely onshore; the Rub al Khali, the Hadramaut–Jeza Trough, and the Shabwa–Balhaf Graben. Three others are mainly offshore; the Aden–Abyan, Sayhut, and Socotra basins.

The *Rub al Khali Basin* contains a relatively thin shelfal sequence of Mesozoic and Lower Tertiary rocks which overlies a northward-thickening pre-Jurassic clastic wedge. The section resembles units known from deep drilling in Saudi Arabia to the north, but estimated to be thinner.

The *Hadramaut–Jeza Trough* is an interior graben paralleling the Gulf of Aden. It contains up to 6 km of sedimentary section of probable mid-Mesozoic to Tertiary age. Its eastern end opens into the offshore area of the Gulf of Qamar.

The *Sayhut Basin* lies offshore in eastern PDRY where AGIP has drilled six exploration wells and have found one hydrocarbon discovery. The basin is characterized by down-to-gulf faulting with numerous fault blocks offering a variety of trap types.

The *Shabwa–Balhaf Graben* is another interior fault trough with its southeastern end

Fig. 1. People's Democratic Republic of Yemen. Regional location map.

open to the Gulf of Aden. The interior portion contains salt domes with associated seeps of live oil and the seaward portion is an area of intersecting fault trends of the NW (Red Sea) trend and the NE (Gulf of Aden) trend.

The *Aden–Abyan Basin*, discovered during a recent seismic reconnaissance survey, is an extension basin with down-to-gulf normal faults forming the dominant structural style.

The *Socotra Basin* lies on a detached continental block. The structure is characterized by three large, complex, north-facing tilted fault blocks.

Exploration has been initiated in all basins, with hydrocarbon shows in the Rub al Khali Basin and the Hadramaut–Jeza trough. Oil discovery has been made in the Sayhut Basin and the Shabwa–Balhaf Graben.

Depth to magnetic basement

These depth to basement contours (Fig. 4), taken from a recent aeromagnetic survey, are supported by drilling results. The location of deep magnetic basement coincides with the axial troughs of the graben in the onshore areas covered by the survey. The down-to-gulf fault patterns are supported by seismic surveys.

In the Rub al Khali, magnetic basement at a depth of greater than 3 km is confirmed by wells which have penetrated more than 2700 m of sedimentary rocks without encountering basement. These exploration wells, however, are some distance south of the northern border of the PDRY.

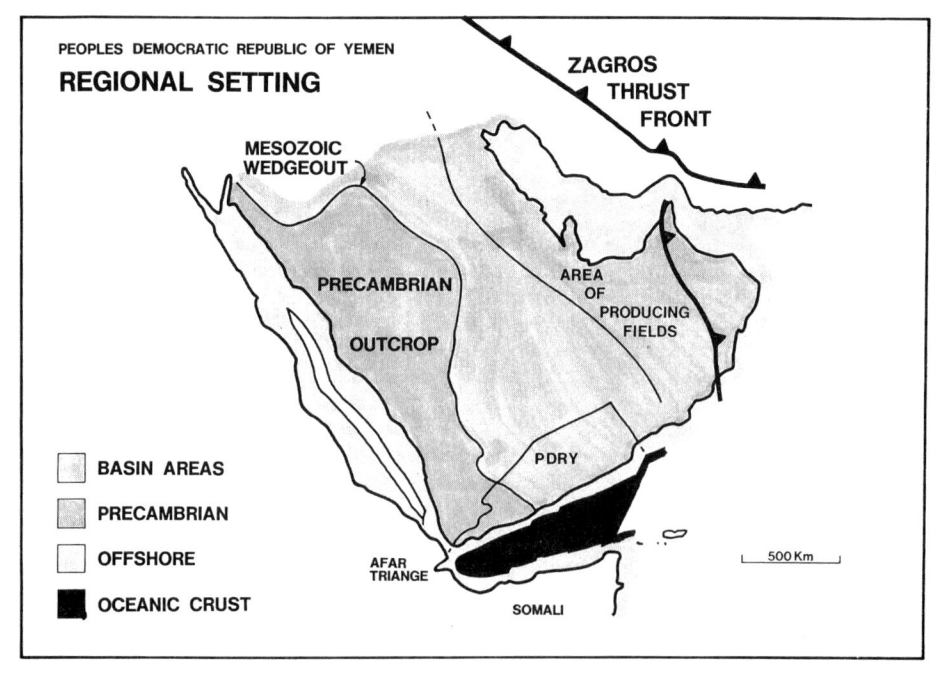

Fig. 2.

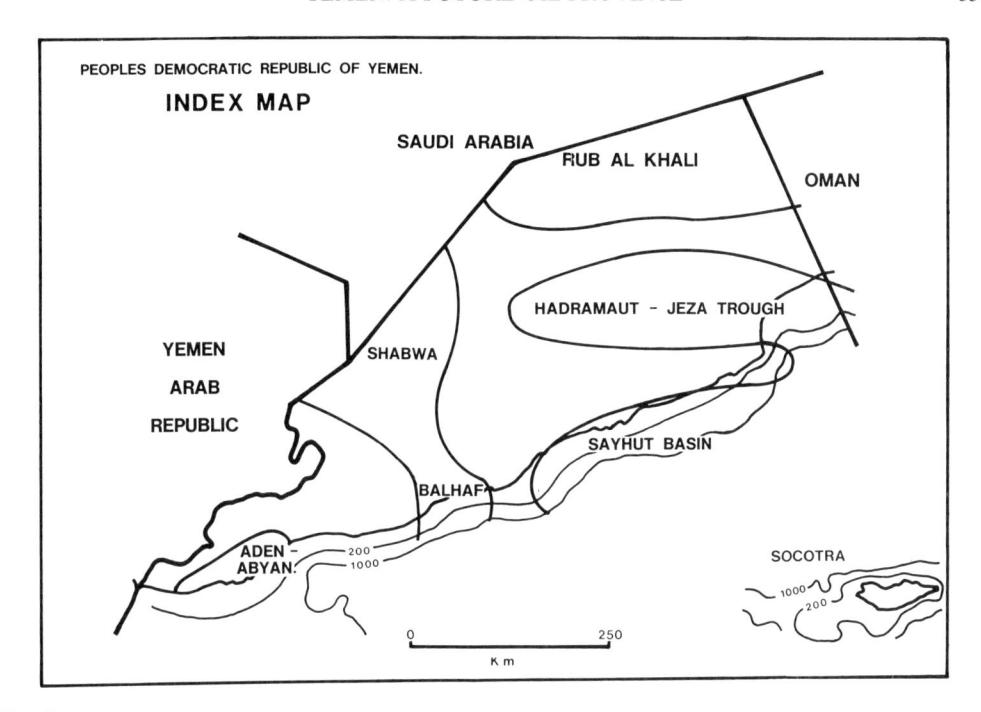

Fig. 3.

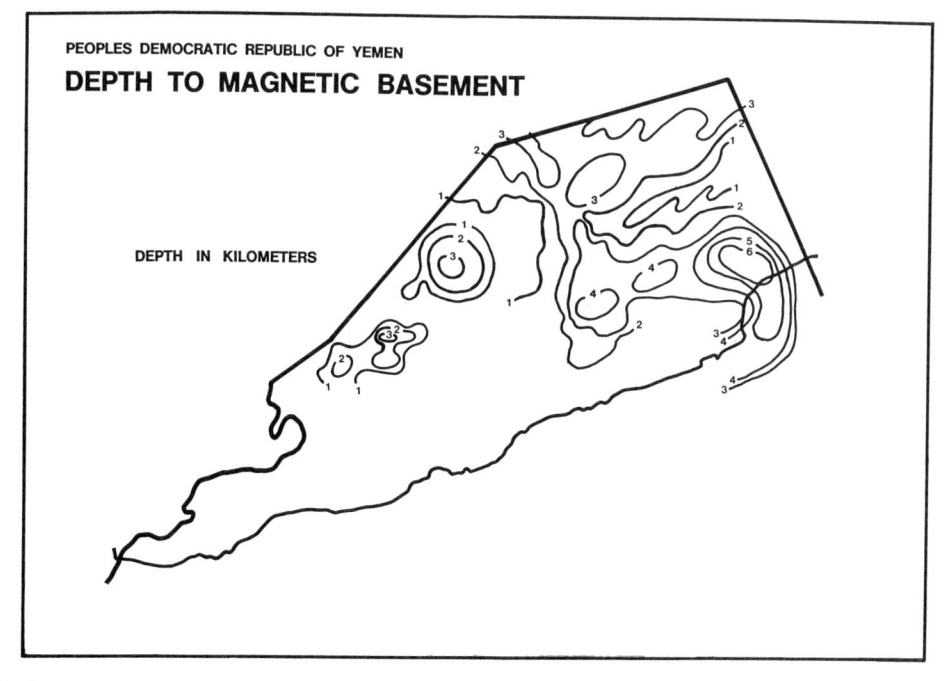

Fig. 4.

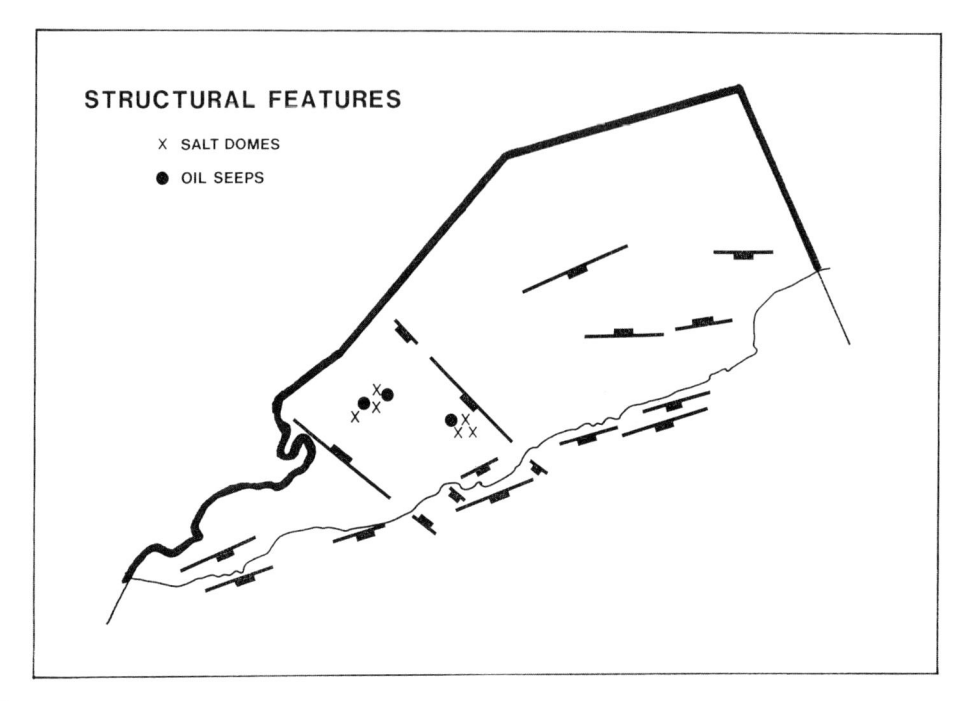

Fig. 5.

Structure features

The fault patterns and salt dome regions shown in Fig. 5 are complementary to the magnetic basement pattern. They indicate a regional structural style dominated by normal faulting with moderate amounts of salt movement. The upper Jurassic salt, which forms pillows rather than piercement structures, is associated with live oil seeps.

Normal faults form two intersecting trends. The NW/SE trend of the Shabwa–Balhaf Graben parallels the direction of the Red Sea rifting, and the ENE-trending Hadramaut–Jeza pattern follows the Gulf of Aden rifting direction. A logical inference from this relationship is that the Shabwa–Balhaf and Hadramaut–Jeza Graben are aborted rifts, similar in origin to the UKCS Viking Graben and the Sirte Embayment of Libya. If these assumptions are valid then their prospectiveness is enhanced.

Rub al Khali to Gulf of Aden section

The interpretive geological cross section (Fig. 6) relates the onshore northeastern region with both the Hadramaut–Jeza Trough and the offshore Sayhut Basin. It also correlates and lies in three of the wells having oil and gas shows. The diagrammatic representation illustrates, the structural style of the basin. Thicknesses shown are consistent with the magnetic basement depth map.

In the Rub al Khali basin, platform Lower Tertiary and Cretaceous beds overlie a wedge of Jurassic and a wedge of pre-Jurassic rocks, both of which thicken northward into Saudi Arabia. The Al Fatk well, drilled in the offshore extension of the Hadramaut–Jeza Trough, shows the graben nature of the trough and the rapidly thickening Cretaceous deposits in its centre.

The offshore area at the southern end of the section displays down-to-gulf normal faulting as well as the semi-restricted nature of the Sayhut Basin.

Columnar Section, Rub al Khali

The diagrammatic columnar section (Fig. 7) summarizes the major lithologies, thicknesses, and known hydrocarbon occurrences of the five unconformity-bounded sequences which make up the sedimentary geology of onshore northeast PDRY. It also tabulates and predicts the critical play parameters of each of these sedimentary packets: reservoir, source and seal.

Reservoir rocks are found in the carbonates and sandstones of the Cretaceous, Jurassic, and

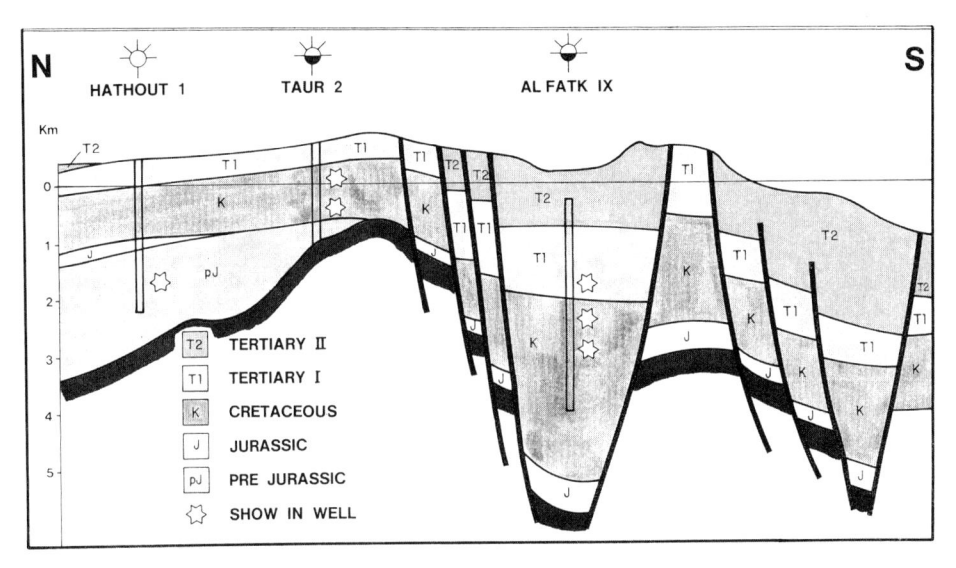

Fig. 6. Schematic cross section Rub'Al Khali to Gulf of Aden.

pre-Jurassic strata. Hydrocarbon shows have been logged in each formation. The presumed source rocks, in the absence of geochemical studies, are the mid-Cretaceous shales, the downdip Jurassic shales, and the Devonian shales overlying the Siluro–Devonian gas shows in Hathout-1. Seals are provided by the thick shale sequences in the Tertiary, Cretaceous, and Devonian, and also the bedded evaporites of Jurassic and Eocene age.

Columnar section offshore

This columnar section (Fig. 8) summarizes current knowledge of the offshore PDRY section and attempts to itemize the critical play

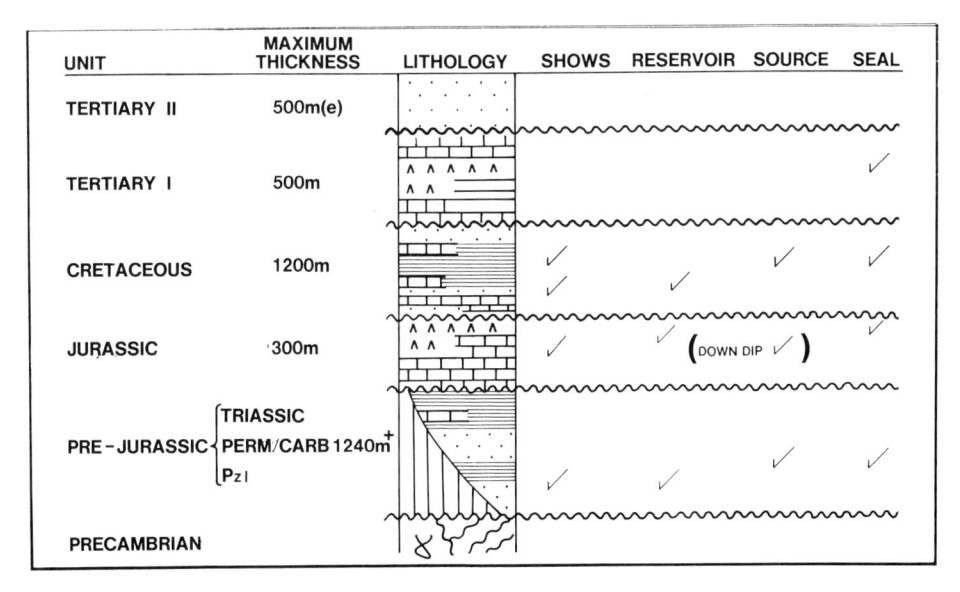

Fig. 7. Columnar section Rub'al Khali.

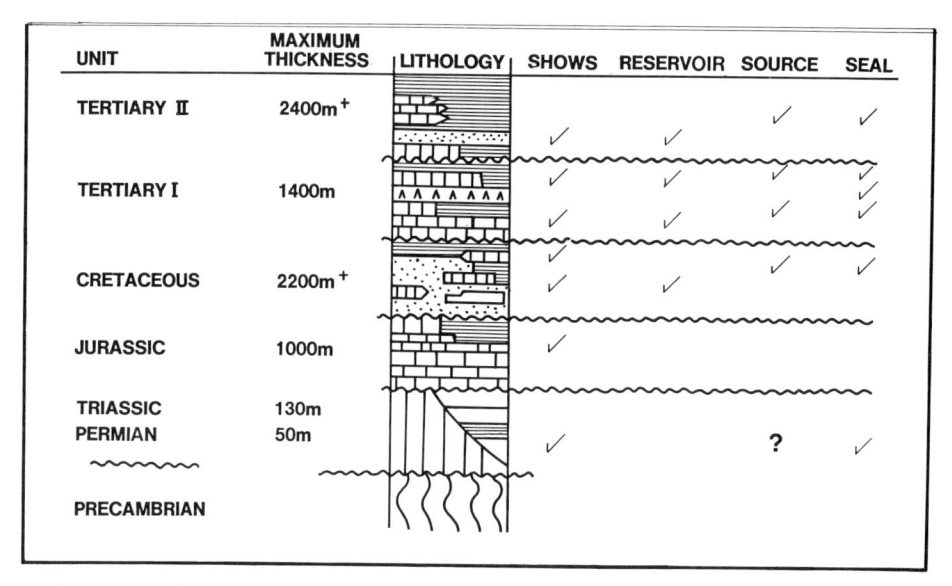

UNIT	MAXIMUM THICKNESS	LITHOLOGY	SHOWS	RESERVOIR	SOURCE	SEAL
TERTIARY II	2400m+		✓	✓	✓	✓
TERTIARY I	1400m		✓ ✓	✓ ✓	✓ ✓	✓ ✓
CRETACEOUS	2200m+		✓	✓	✓	✓
JURASSIC	1000m		✓			
TRIASSIC	130m					
PERMIAN	50m		✓		?	✓
PRECAMBRIAN						

Fig. 8. Columnar section offshore.

parameters. Similar to the onshore there are five unconformity-bounded sedimentary packages which have been identified, together with maximum known thicknesses and hydrocarbon shows that have so far been encountered in the formations.

The offshore composite column (Fig. 8) is both thicker and more marine that the equivalent onshore section, especially in the post-Jurassic section. There are known source rocks and reservoir plays, and abundant seals are present in both the seaward-thickening shale wedges and in the blanket evaporites of Eocene age. Hydrocarbon have been recorded from all of the five sedimentary units, and an important oil discovery in Middle Eocene beds is currently being evaluated. Source rocks are abundant in the seaward-thickening shale wedges, and attractive trapping possibilities exist in the down-to-gulf faults with associated rollovers and drape.

Geophysical studies

Geophysical surveys (see Figs 9–11) have been carried out in all six basins. Approximately 17 000 km of seismic data has been acquired in the country both onshore and offshore up till late 1987. Aeromagnetic surveys were also carried out in all six basins. Bouger gravity surveys were run in Shabwa–Balhaf, Aden–Abyan and Hadramaut–Jeza trough.

Seismic surveys have shown sedimentary thickness of approximately 3 km in the Rub Al Khali Basin. Seismic studies have also indicated thick Cretaceous sections in the Hadramaut–Jeza trough. The Aden–Abyan Basin was discovered by seismic survey. The cross section shows rapidly thickening section to the south with down-to-gulf normal faults. Five wells drilled in the salt exposed Shabwa Basins have been drilled on the basis of these seismic and gravity surveys.

Seismic data quality in all basins is generally good. Strong seismic reflectors can be correlated to stratigraphic units of Eocene (Mishrif Formation), Middle Cretaceous, Lower Cretaceous (Shabwa Formation), Middle and base Jurassic.

Drilling activity and present

Onshore wells

Fifteen onshore wells have been drilled in the PDRY. Of these fifteen wells, oil and gas shows have been encountered in twelve wells. The hydrocarbon shows range from minor gas shows during drillings, to live oil in cores, dead oil in cuttings, and oil recovered on DST evaluation. The principal zones containing the hydrocarbons are the Lower and Middle Cretaceous, but shows of gas were logged in the Pre-Jurassic section of the Hathout-1 well.

The high percentage of hydrocarbon shows is made more remarkable by the fact that many of

YEMEN: A FUTURE OIL PROVINCE 335

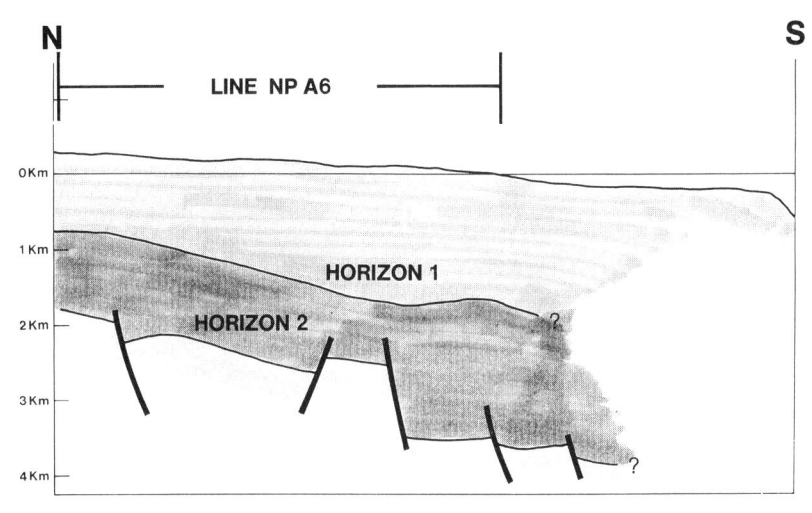

Fig. 9. Aden cross section.

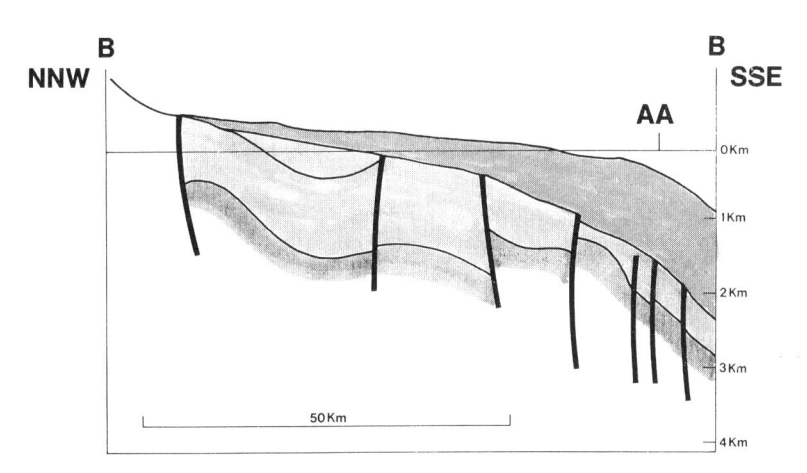

Fig. 10. Balhaf cross section BB.

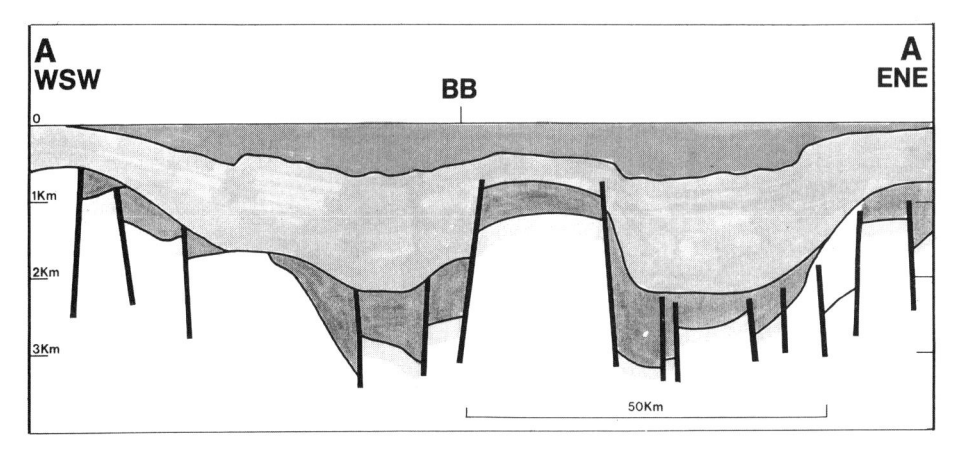

Fig. 11. Balhaf: profile parallel to coast AA.

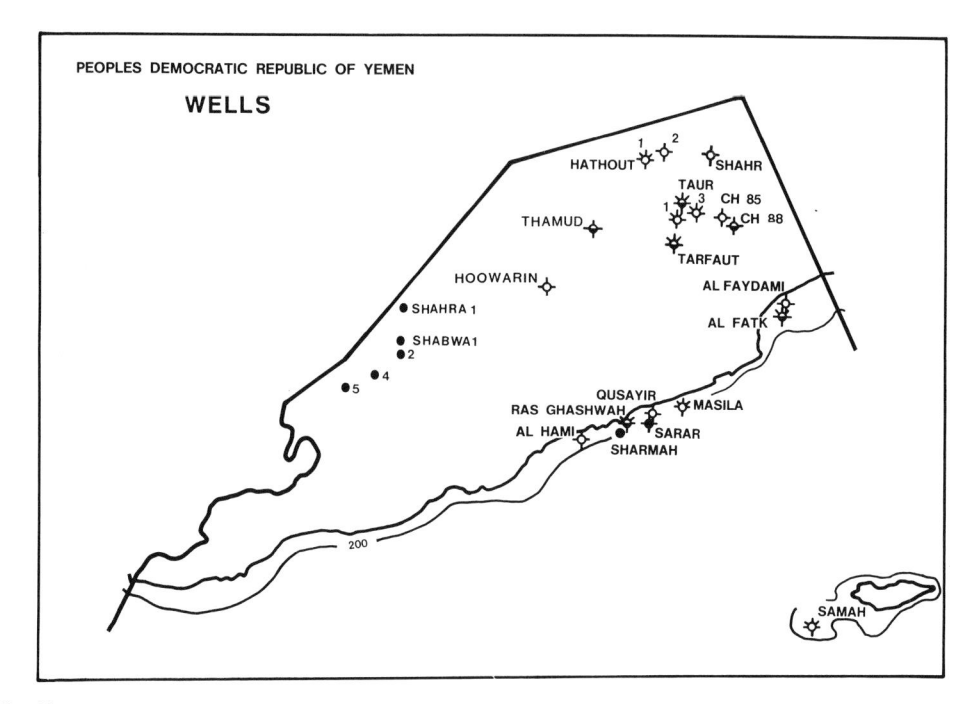

Fig. 12.

Fig. 13.

```
PEOPLES DEMOCRATIC REPUBLIC OF YEMEN

    9 OFFSHORE WELLS DRILLED
             1978 - 1982
         TD 1656 - 4397m

        6 WELLS WITH SHOWS
    1 DISCOVERY - 3000b/d 43.6° OIL

            OLIGOCENE OIL
          MIDDLE EOCENE OIL
          LOWER EOCENE OIL
  UPPER AND MIDDLE CRETACEOUS OIL AND GAS
            JURASSIC GAS
            PERMIAN GAS
```

Fig. 14.

```
PEOPLES DEMOCRATIC REPUBLIC OF YEMEN

  16 ONSHORE WELLS DRILLED 1964 - 1983
         TD 1100 - 2704m
        7 WELLS WITH SHOWS

   MIDDLE CRETACEOUS    OIL AND GAS
   LOWER CRETACEOUS     OIL AND GAS
   PRE JURASSIC         GAS

         5 WELLS PRODCING OIL
```

the wells were located on poorly mapped prospects and some could not be classified as exploration wells, but were core holes drilled for stratigraphic information without being located on a structural prospect.

The best shows were 7 bbl of 32° API oil recovered on DST in the Tarfaut well and the oil bleeding from cores in Taur 2. They were both Middle Cretaceous shows.

Five wells have been drilled in the Shabwa area. All five wells are producing. Currently 3000–10 000 barrels per day of oil is being produced and transported to the refinery in Aden.

Offshore wells

Ten offshore wells have been drilled since 1978. Eight offshore wells were drilled by AGIP. Hydrocarbon shows have been logged in six of these wells, and the Sharmah-IX well (1982) was a discovery with 3000 barrels per day of 42.3° API oil tested from a Middle Eocene play.

Hydrocarbon potential

In discussing the oil and gas prospects of PDRY, it is convenient to divide the country into four stratigraphic–tectonic regions, each having a somewhat different potential for petroleum exploration (Fig. 15). These regions are desig-

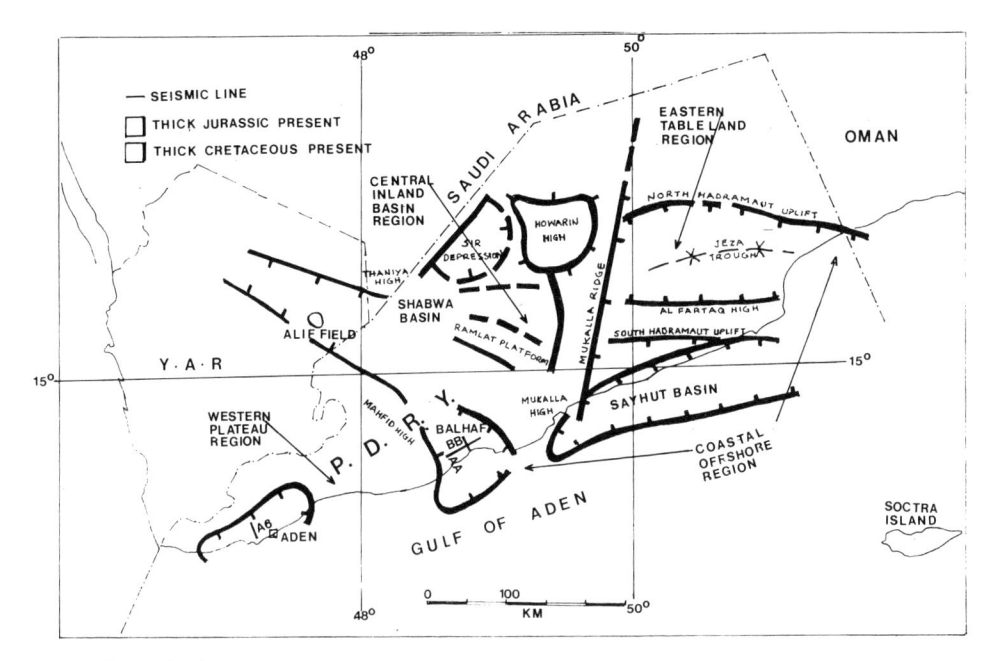

Fig. 15. Tectonic elements.

nated as follows: the Western Plateau Region, the Central Inland Basin Region, the Eastern Tableland Region and the Coastal Offshore Region. The continental shelf region surrounding the island of Socotra and its adjacent islands is considered separately.

The Western Plateau Region

The Western Plateau Region includes the Aden–Abyan basin which extends down to the Gulf of Aden. Seismic work has shown closed structures bounded by faults in the Cretaceous and Jurassic sequence. The sedimentary section thickens rapidly southwards to the sea. More geological and geophysical work is needed to determine the oil potential in this area.

The Central Inland Basin Region

The Central Inland Basin Region consists of the area between the Ataq fault zone to the west, and the Ayad fault system to the east. The western boundary is well defined by the faulted contact of Precambrian basement to the west against Mesozoic and Tertiary sediments to the east, but it is only inferred where the fault enters the Ramlat Sabatayn sand desert. The eastern boundary is more diffuse. It becomes complexly faulted at its southern end, and can only be projected in the Wadi Hadramaut–Ramlat Sabatayn region. The Central Inland Basin Region contains the inland Jurassic evaporite basin with its explosed salt domes, and it extends from near Balhaf village on the coast northwestward into Yemen Arab Republic.

Surface indications of the presence of hydrocarbons are commonly associated with the salt domes and a few isolated seeps have been recorded in the region. These include: (i) in Wadi Ahmad, northwest of Habban village, oil collects in a cave from fractures in Cretaceous Tawilah sandstone, and is also observed impregnating the sandstone; (ii) in Wadi Hajar, east of Mahmada village, oil oozes from fractures in sandstone of the Cretaceous Tawilah series, overlain by alluvium; and (iii) in Wadi Hajar, northwest of Maabir village, methane gas bubbles constantly from the bed of the Wadi.

Eight salt domes are exposed at the surface in the Central Inland Basin Region: three in the Mintaq area at the south end of the basin; three in the Shabwa area in the central part of the basin; one at Ayad, 50 km south of Shabwa; and one at Layadim, 70 km northwest of Ayad. A ninth dome is located at Safir, in Yemen Arab Republic, 60 km north of Layadim. Petroleum occurrences are found at each of these domes, particularly in the Jurassic members of the Sabatain formation. The basal Shabwa salt member commonly contains yellow–brown streaks and spots, described as bituminous and scattered asphalt impregnations. The overlying Layadim shale is generally bituminous, and often contains gilsonite beds. The Ayad−Mqah member of the northern and western domes contains interbedded thin bands of fetid dolomite and bituminous limestone, as well as asphaltic streaks in clastic interbeds. Live oil can be collected from seeps at the edge of the Ayad salt dome.

The Central Inland Basin Region, occupying a large graben structure which extends from northern Yemen Arab Republic to coastal PDRY, is considered a highly prospective area for hydrocarbon exploration.

In the Yemen Arab Republic, the Alif Field (discovered by the Hunt Oil Company) lies in the northern extension of the basin. The Alif-1 well flowed 7800 bopd. It is reported that some 40 wells have now been drilled in this field and production will be about 100 000 bopd by the end of 1988. In the Shabwa/Balhaf graben the geological and geochemical requirements of an oil province appear to be quite favourable. Structurally the traps will be either fault related or possibly salt related. Reservoir rocks are of basal and Upper Jurassic sandstones, Middle and Upper Cretaceous sandstones. Source rocks are Jurassic Middle Shale/limestone unit. Seal is provided by Jurassic Evaporites of Upper Cretaceous sandstones.

The PDRY Ministry of Energy has drilled 5 wells with the help of Technoexport. All 5 wells are supposedly on separate structures and producing good oil. Production is rumoured to be 3000−10 000 bopd.

The Eastern Tableland Region

The Eastern Tableland Region comprises roughly the eastern one half of PDRY, and it contains Beydoun's broad, regional biaxial arch with its associated Wadi Hadramaut−Jeza depression. In the northeast, the gently-dipping south flank of the Rub al Khali basin extends northward from the crest of the North Hadramaut structure disappearing under desert sands. The south boundary of the region is the coastal plain, beyond which the Coastal Offshore Region reaches to the edge of the continental shelf. A relatively thick sedimentary section, averaging 3000 m or more, generally correlatable with the formations containing prolific oil fields in Saudi Arabia and Oman, make this region attractive for the discovery of oil and gas in Cretaceous, Jurassic and older sediments. Tertiary formations are regarded as non-prospective as they are widely exposed over most of the region; the Paleocene limestone is an important regional aquifer. Pre-Jurassic prospects exists in the northeast, where beds of possible Cambrian, Silurian, Devonian, Permian, and Triassic age pinch out along the south edge of the Rub al Khali basin. These sediments may also be present in the depths of the Wadi Hadramaut−Jeza trough.

In the northern part of the Cretaceous and Jurassic, sediments onlap to form stratigraphic traps against the northern Hadramaut uplift and its juxtaposition with the Mukalla ridge. In the Jeza trough the traps are fault related and possibly reefs. Reservoir rocks are Middle and Upper Cretaceous sandstones, possibly Lower Cretaceous reefs. Source rocks are primarily Upper Cretaceous shales but possibly also other Cretaceous shales. Seal is provided by Upper Cretaceous or local shale.

The Ministry of Energy (formerly Petroleum & Minerals Board) in association with oil companies, drilled several waterholes, core holes and deep holes. All of these holes are located close to the crest of the North Hadramaut uplift or on its north flank. Some of these wells encountered strong shows of oil and gas. Tarfaut well flowed 7 bbl of 32° API on DST.

Interpretation of recent seismic survey in this area has shown that most wells were incorrectly located. In the Jeza trough, Braspetro drilled three wells. All were dry. Our study indicates that the central part of Jeza trough is a promising exploration area. In this region the prominent central east−west high trend and a north−south high trend collide.

The Coastal Offshore Region

The Coastal Offshore Region extends from the Bab al Mandab, at the western end of PDRY, along the Gulf of Aden to Ras Darbat at the Oman border. It consists of a very narrow continental shelf; from the coastline to a water depth of 1000 m, the shelf averages only 20−30 km in width. The widest shelf areas are located west of Aden, opposite the mouths of the Wadi Tuban and Wadi Bana drainage systems; in the Sayhut−Ras Sharwayn area, opposite the mouth of Wadi Masila; and at Qamar Bay, opposite the mouth of Wadi Jeza. In these areas, the shelf averages 60−70 km wide. The total area of the continental shelf is approximately 25 000 km^2 (not including Socotra).

Presently, the coastal offshore region of PDRY is the most prospective area for finding significant accumulations of oil and gas, based solely on the presence of a thick sedimentary section containing significant shows of hydrocarbons. Structurally traps will be fault related and porosity related. Reservoir rock is Paleocene–Eocene limestones. Source rock is evaporite shales. Seal is provided by evaporite shales, Anhydrite and local shales.

AGIP drilled eight wells in the licence areas of Sayhut, Mukalla and Qamar. In most wells good hydrocarbon shows of gas and oil were found. However, the most important discovery of oil presence was made in the Sharman IX well. A production test within Oligocene Ghadyah Formation produced oil (40.5° API) at the rate of 300 barrels per day. A second test recovered some heavy oil (10° API) after circulation was reversed. After production casing was run, the well flowed oil (43.6° API) at the rate of 1800 barrels per day through perforations testing the Eocene Habshiya Formation. Immediately after acidization of the interval, production increased to 3000 barrels per day. The well has been abandoned, but it can be re-entered for further evaluation. It is apparent that important hydrocarbon objectives are to be found in the offshore Eocene Habshiya and the overlying Oligocene Ghaydah formations. Other intervals in the Middle and Lower Eocene sediments also show potential for future exploration.

Conclusion

This review has attempted to present possibilities of the presence of hydrocarbons in a very large, as yet unexplored, region of the Middle East. Only 25 exploratory wells have been drilled in an inland area of approximately 337 000 km^2 and in the offshore shelf of approximately 78 000 km^2.

The People's Democratic Republic of Yemen has recently joined the ranks of oil producing countries. In the near future with concerted exploration effort by major oil companies, there is every likelihood that this region will become a major oil producing country.

I am pleased to acknowledge the extensive contribution to this paper that derives from a long association with the Director and his officers of the Petroleum & Minerals Department of the Government of P. D. R. Yemen. I thank F. Lomis and R. W. Murphy, from whose reports material has been used to prepare this paper.

Reference

HUSSEINI, M. I. & HUSSEINI, S. I. 1990. Origin of the Infracambrian Salt Basins of the Middle East. *In*: BROOKS, J. (ed.), *Classic Petroleum Provinces*. Geological Society Special Publication **50**, 279–292.

Sub-salt imaging in the Gulf of Suez

R. ZAKI, H. MCDOWELL, I. THREADGOLD & O. OLDFIELD

Gulf of Suez Petroleum Company (GUPCO), Palestine Street, 4th Sector, New Maadi, Cairo, Egypt

Abstract: Sub-salt imaging on seismic data is a serious geophysical problem affecting exploration in the Gulf of Suez. Here, thick Miocene evaporite, shale and salt sequences with salt tectonics overlie rift structures involving pre-Miocene and Miocene oil-producing formations. This paper discusses the seismic imaging problems associated with salt swells in the Gulf of Suez and presents experimental work by GUPCO that has provided a better understanding and possible solution to the imaging problem. A realistic model for the Gulf was used to generate synthetic shot records using finite difference modelling. Processing of the simulated data in conjunction with actual field data led to improved understanding of the wave propagation behaviour in the subsurface, and to a realization of the validity of the common midpoint stacking technique and post-stack depth migration.

The Gulf of Suez is a Miocene graben, in which clastic sediments (sand and argillaceous matter) as well as non-clastic sediments (limestone and evaporites) were deposited in varying thicknesses and under varying conditions of sedimentation (Mostafa 1974). Hydrocarbon accumulations are present in both Miocene and pre-Miocene reservoir rocks.

The Miocene evaporites, which overlie the Miocene clastics, contain two formations that are the main sources of most seismic problems in the Gulf. There are namely the Zeit and the underlying South Gharib salt formations. The Zeit formation consists of a sequence of relatively thin layers of interbedded shales and anhydrites with occasional salts. The South Gharib formation consists of an extensive salt body with occasional anhydrite streaks.

Seismic studies

The lack of identifiable reflectors below the base of the South Gharib salt is, in general, the major seismic problem in the Gulf of Suez. A second problem, in areas of salt diapirism, is imaging (or the ability to map an underlying structure). The lack of reflector definition beneath the Miocene evaporites decreases the effectiveness of the seismic method as an exploration and/or exploitation tool. These inadequacies stem from three coupled reasons:
(i) Sharp lateral velocity contrasts, due to salt diapirs, cause the ray paths to be severely distorted as energy propagates through the earth. In such a case, common midpoint (CMP) geometry will not represent common reflection point geometry, and the validity of conventional stacking and migration will be in doubt.
(ii) Seismic energy suffers severe attenuation as it is transmitted through the highly cyclic shale and anhydrite sequences of the Zeit formation. This results in time and frequency dependent changes in the seismic wavelet, exhibited as variable time delays in the recorded data.
(iii) Strong reverberating multiples are generated during transmission through the Zeit formation (Tsay 1976, pers. comm.; Moaty 1988, pers. comm.). These multiples have move-out nearly parallel to primary energy and so velocity filtering approaches are often ineffective in trying to attenuate them, as have been conventional deconvolution methods.

This paper documents a study of the first problem, that of ray path distortion caused by salt diapirs. This was addressed by studying, through modelling, the effect of salt swells on seismic reflection data. The objective was to determine the severity of the ray path distortion and its impact on imaging of sub-salt data, when conventional CMP stacking and migration techniques are used.

Another objective was to develop ideas that might lead to new acquisition, processing, and modelling techniques which would result in improved imaging of sub-salt structures in the Gulf of Suez.

Study method

A realistic model, based on the known geological structure in one area of the Gulf, was used for the study. A finite difference (FD) modelling technique was used to generate synthetic shot records. This technique has been utilized and described previously, by Whitmore & Lines

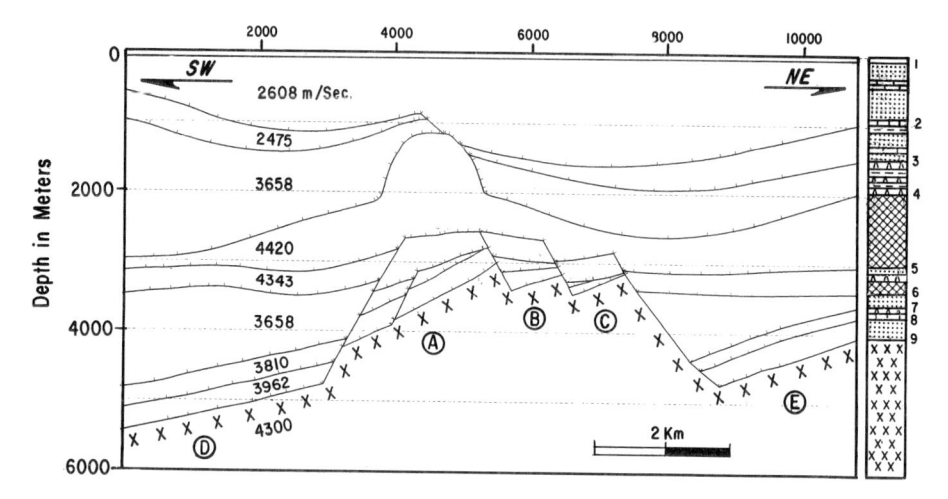

Fig. 1. The input model: 1, sea bottom; 2, post-Zeit; 3, Zeit; 4, S. Gharib; 5, Belayim; 6, Miocene clastics; 7, Nukhul evaporites; 8, Nubia; 9, basement.

(1986). Instantaneous wave field records were generated to help identify reflectors on the records and to study the wave propagation behaviour in the subsurface. The simulated records were sorted into CMP gathers and passed through a standard CMP processing flow to produce a stacked section. Time as well as depth migration were then applied. Normal incidence inverse modelling was used to compare the different outputs with the original model.

The actual seismic line, across the structure which was used to define the model, was processed for comparison. At several locations along the model profile, corresponding synthetic and actual records were compared.

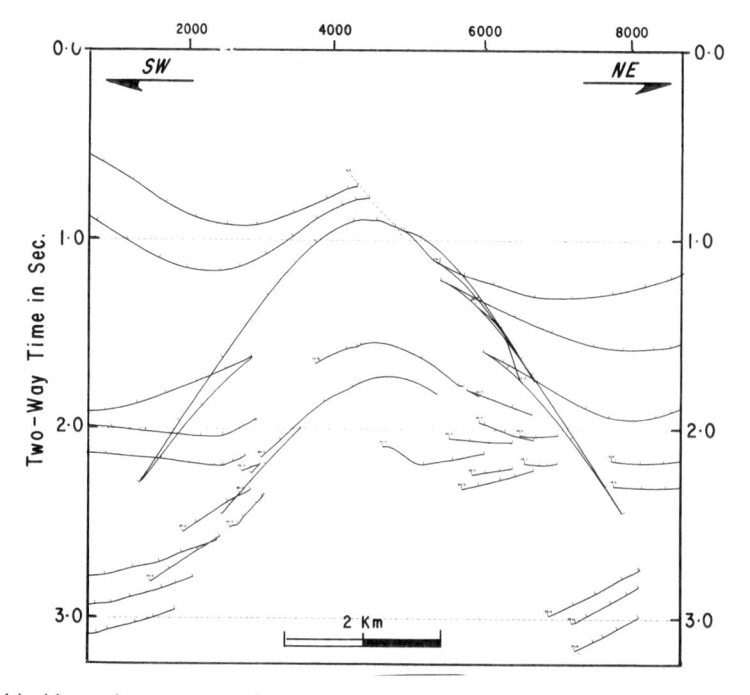

Fig. 2. Normal incidence time response of the input model.

Input model

The input model (Fig. 1) was constructed based on well data over a salt diapir. It represents a typical Gulf structure where salt has swelled on the top of faulted and tilted horst blocks. The individual pre-Miocene blocks are annotated for reference. The normal incidence time response of the input model is shown in Fig. 2. The complications of the time section, due to the refraction effects at the salt wall, raise a question about the validity of the CMP stacking method for this model.

Finite difference modelling

The finite difference modelling system used in this study, solves the two dimensional acoustic wave equation by a numerical finite difference technique. Since the technique obtains a solution to the complete wave equation, the solution automatically includes all diffractions and multiples, as well as any interference phenomena and amplitude effects associated with wave theory (Regone & Whitmore 1981, pers. comm.).

Synthetic shots were generated, at the equivalent of 50 m intervals, using zero offset, single shot, and single receiver geometry. The synthetic spread length was 3000 m with receiver intervals corresponding to 25 m. The top of the model represents sea level, and it was considered a non-reflecting surface to minimize multiples and emphasize concentration on the imaging problem. Wavefront displays were generated every 100 ms at ten shot intervals. These displays were very useful in studying the sub-salt imaging.

Synthetic shot records

Figures 3 and 4 show respectively, one synthetic shot record and the corresponding wavefront displays. Significant wavefront distortion is evident on the two displays, including the following:
  (i) Two reflections from the salt are recorded, one from the salt directly below the shot, and another shallower reflection from the left salt wall to the east of the shot. This shallower reflection dips steeply on the near traces and would probably be attenuated in conventional processing if an F/K filter were used on actual field records.
  (ii) Double values for the base salt reflections from blocks D and A. The base salt reflector of block A is pulled up in time as it passes through the high velocity salt.

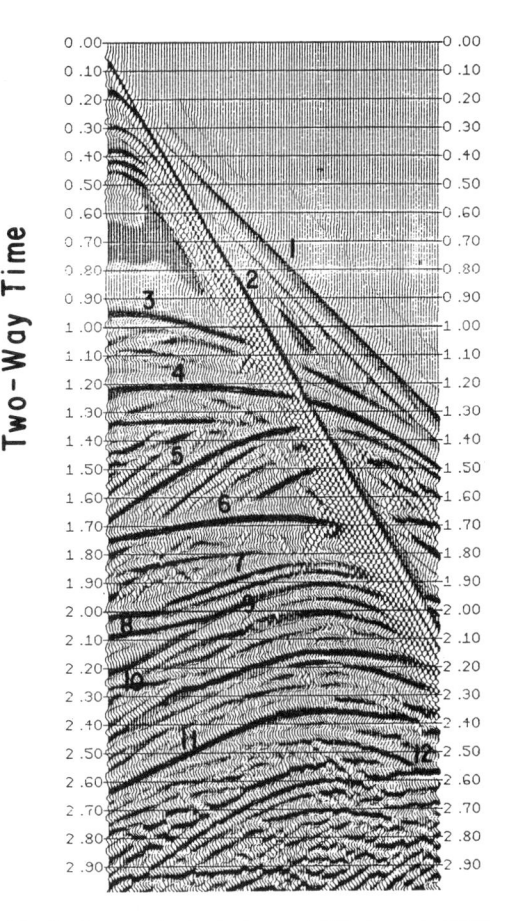

Fig. 3. One-synthetic-shot record: 121 channels, 3000 m cable. The identified reflections represent: 1, sea bottom; 2, direct arrivals; 3, post Zeit; 4, Zeit; 5, salt wall; 6, basinal salt; 7, Miocene clastics of block A; 8, Belayim of block D; 9, multiple; 10, Miocene clastics of block D; 11, basement of block A; 12, basement of block B.

  (iii) A strong basement reflection (no 11 on Figs 3 & 4) is recorded from block A. Another is recorded from block B (no 12) at the far end of the spread.

Processing of synthetic records

The simulated records were sorted into CMP gathers. Velocity analysis was applied every 50 CMP. Normal moveout (NMO) corrections and muting were then applied to these gathers. The primary reflections were identified on the corrected gathers guided by the interpreted shots. Figure 5 shows one CMP gather before and after NMO correction.

A stacked section was produced (Fig. 6).

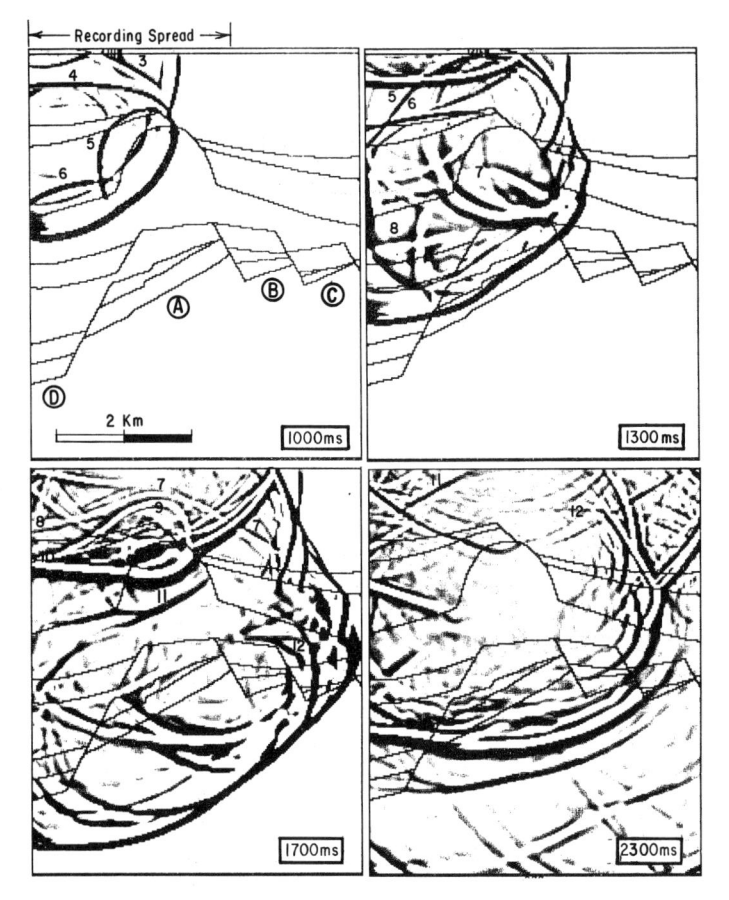

Fig. 4. Some instantaneous wave field displays of the record of Fig. 3. Reflections are annotated as referred to in Fig. 3.

Reflections marked on this stack were based on information from previous steps. While some reverbatory coherent noise trains are evident (these are artifacts of the FD modelling algorithm) several points can be noted concerning primary and multiple reflections.

(i) Shallow horizons are readily defined.
(ii) The almost flat base salt reflector of block A appears as a curved event. This is due to the pull-up effect of the high-velocity salt. The terminations of this event do not represent the actual width of the block. This is because the reflected energy is diffracted once it propagates through the curved corners of the salt swell.
(iii) The pre-Miocene dipping horizons of block A are stacked nearly one kilometre to the west of their actual locations as very steep dipping reflections. These events are almost parallel to the diffractions from the salt wall and fault edges. Quite often such events are removed from actual seismic data by a post-stack dip filter.
(iv) The basement reflection from block B is showing another interesting phenomenon. The reflection was split into two dipping segments, one is 100 ms shallower than the other. This can be predicted from the wave field record analysis, since the reflected wave front propagated parallel to the salt wall, where part of it travels through thick salt with faster velocity than the rest of the energy. The result is two separate events that could be mis-interpreted as a fault.
(v) The basement reflector from block C was easily imaged since it was far from the salt swell.
(vi) Multiples of the Zeit and the post Zeit reflectors are masking the deeper primary reflections in blocks D and E.

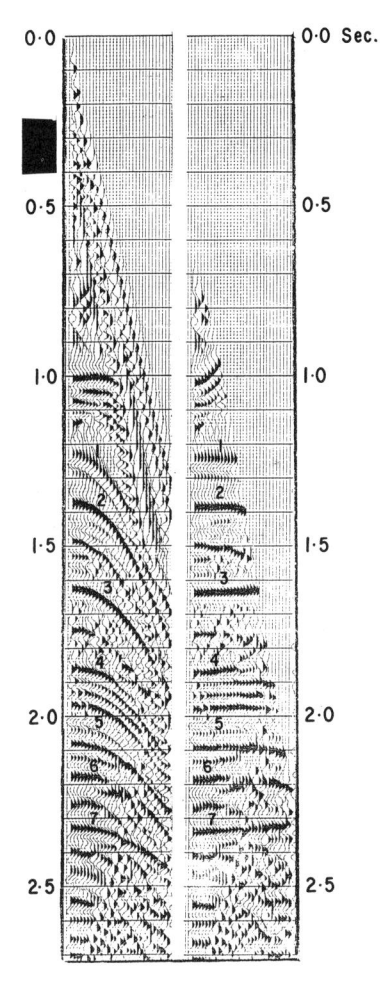

Fig. 5. Synthetic CMP, before and after NMO correction. 1, post Zeit; 2, Zeit; 3, S. Gharib; 4, Belayim; 5, Miocene clastics; 6, basement (B); 7, basement (C).

Model extraction

The comparison between the stacked section (Fig. 6) and the normal incidence time response of the original model (Fig. 2) was outstanding. The two sections were almost coincident even in the areas where the CMP method was expected to fail. This observation is very important as it demonstrates on an experimental basis, the validity of the CMP stacking method in imaging the sub-salt structures in the Gulf of Suez.

The question remains, however, whether the depth model can be derived from the stacked section. Several different techniques were tried in an attempt to resolve this problem. These are discussed below:

Event time to depth migration

The normal procedure followed in normal incidence raytrace modelling of structures associated with salt swells in the Gulf is divided into two steps.
(i) The first is to identify the shallow events including the salt reflections from the top of the swell and from salt in the basin, migrate these events, modify the output depth by connecting the shallow event terminations on each side of the salt bulge to determine the salt walls.
(ii) The second step is to pick the base salt reflector and any deeper reflection pieces that could be primaries and migrate them.

The same procedures were applied to the stack section of Fig. 6. Figure 7 represents reflections picked by an independent interpreter as a blind test. Figure 8 represents the migrated depth conversion of Fig. 7 using correct interval velocities. Here the following can be observed:
(i) Block C was not recognized on the time interpretation.
(ii) The salt swell is dislocated.
(iii) Wide gaps occur between sub-salt blocks.

Consequently, the deeper events, if identified, will not be migrated properly. This method has been proved unreliable, but it will benefit significantly from knowledge of subsurface control.

Time migration

Figure 9 shows the frequency domain time migration of the stacked section using stacking velocities. On this section the following are observable:
(i) Erroneous migration effects on attempts to migrate steep dips.
(ii) Failure to correct pull-up effects.
(iii) Failure to correct refraction problems.

An interpretation based on this migrated section will definitely lead to an incorrect mapping of the structure beneath the salt bulge.

Depth migration

In the depth migration program used, the acoustic wave equation is solved by finite difference techniques. The solution obeys Snell's law at boundaries, and therefore properly handles the refractions involved. Any dip can be handled, as long as the returned signal is received at the surface. Figure 10 shows the output of depth migration. This migration was done using the correct velocity model. From this figure it can be seen that:
(i) All horizons were successfully migrated, with the exception of a small part at the

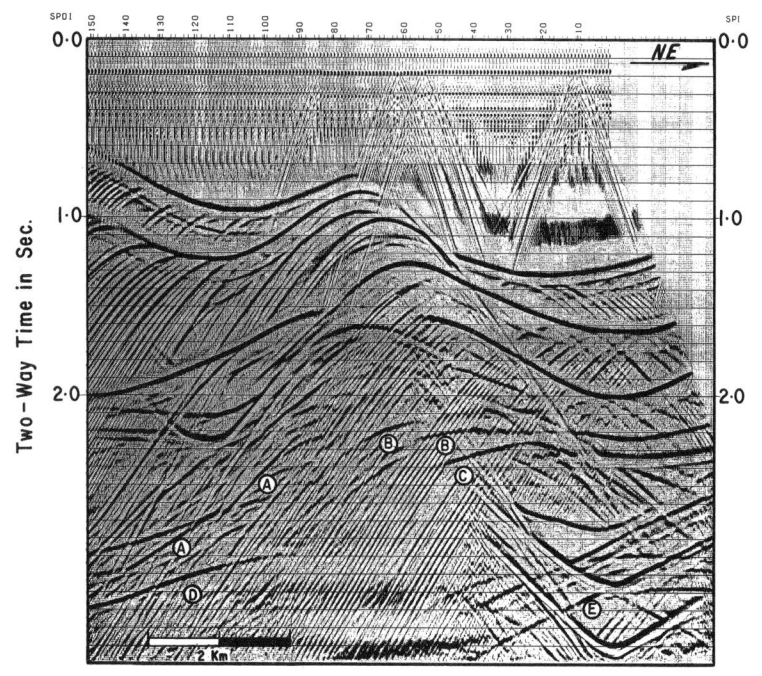

Fig. 6. CMP stack of the synthetic data. Basement reflections from different pre-Miocene blocks are annotated as referred to in Fig. 1.

base of the eastern salt wall.
(ii) Steep salt wall reflections were migrated to their proper positions.
(iii) Pull-up effects of the salt were removed.

(iv) Refraction problems were solved and reflections like that of the basement block B were reconstructed back and properly imaged.

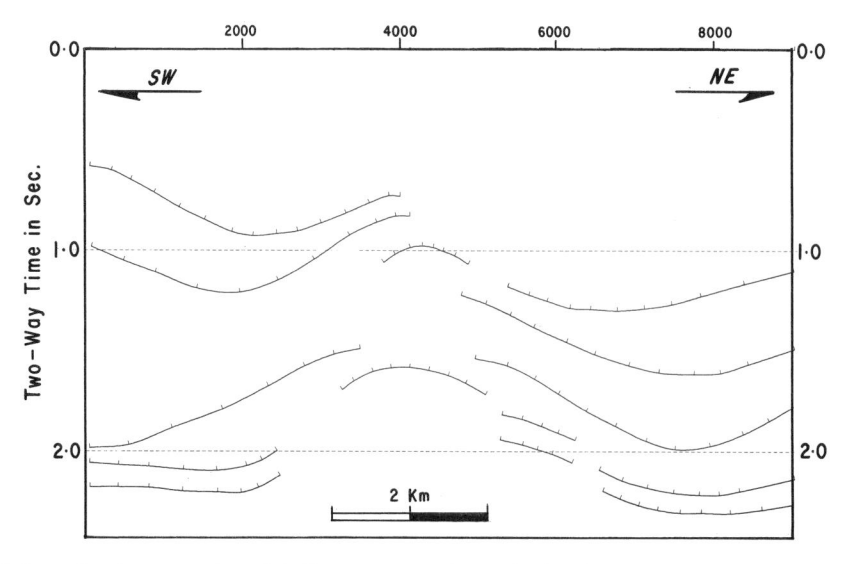

Fig. 7. Independent interpretation of shallow events on the stacked section of Fig. 6.

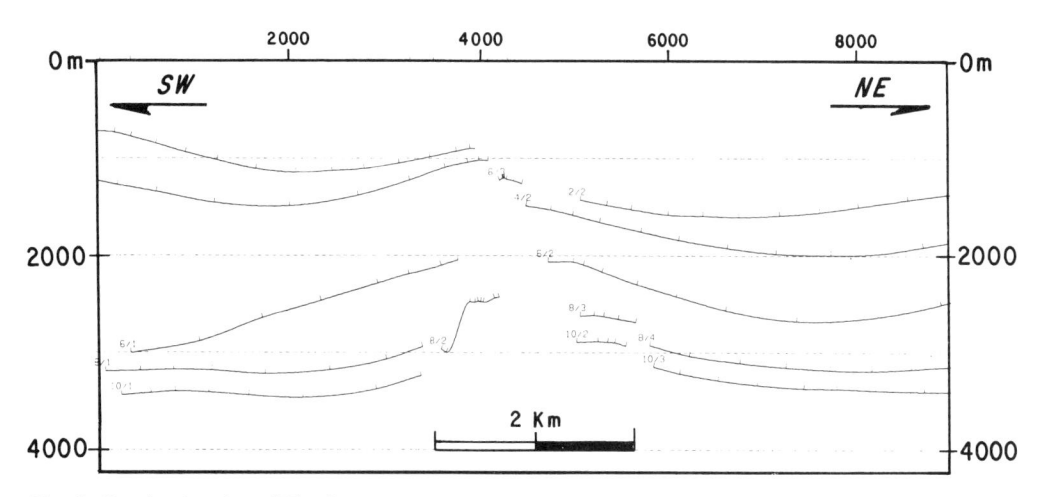

Fig. 8. Depth migration of Fig. 7.

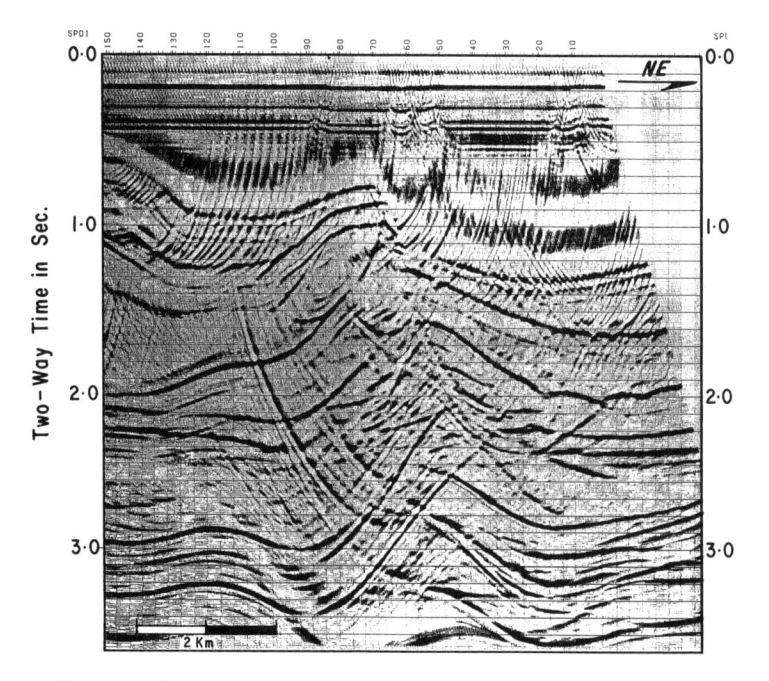

Fig. 9. Post-stack time migration.

(v) Pinchouts are not resolved because of the low cutoff frequency used before migration (20 Hz).
(vi) Multiples remain a problem that need to be solved using other techniques.

Even with multiples present, the section remains interpretable and an experienced interpreter can identify the structure.

In order to find out how effective post-stack depth migration is when the velocities are not precisely known, the same stack was depth migrated using interval velocities determined from the model stacking velocities. The model stacking velocities had been derived from the velocity

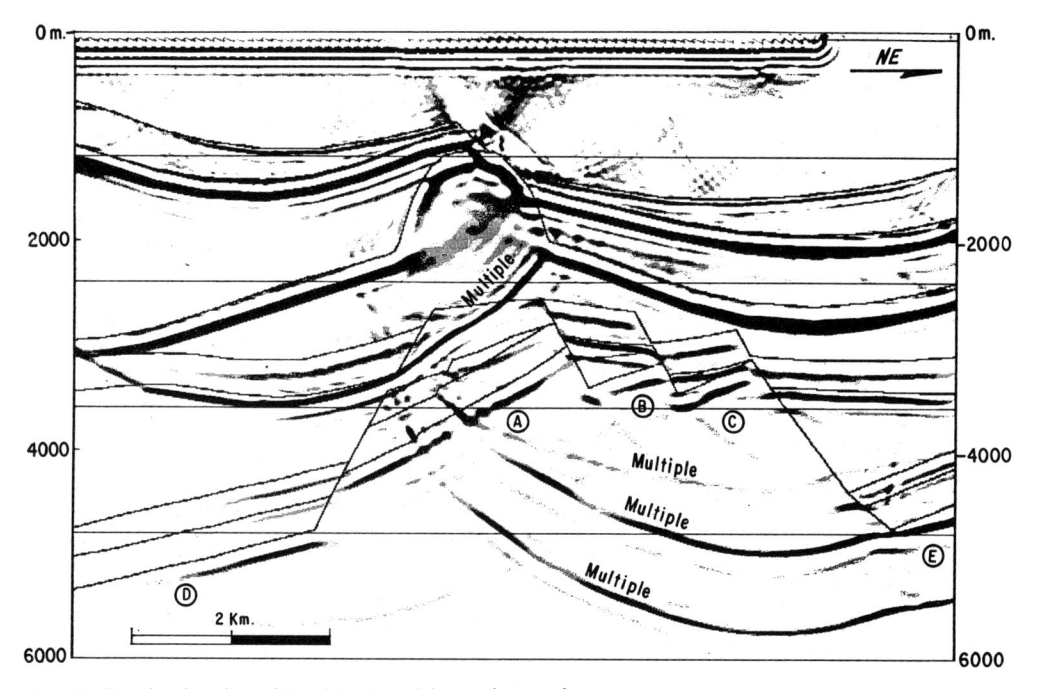

Fig. 10. Depth migration with original model superimposed.

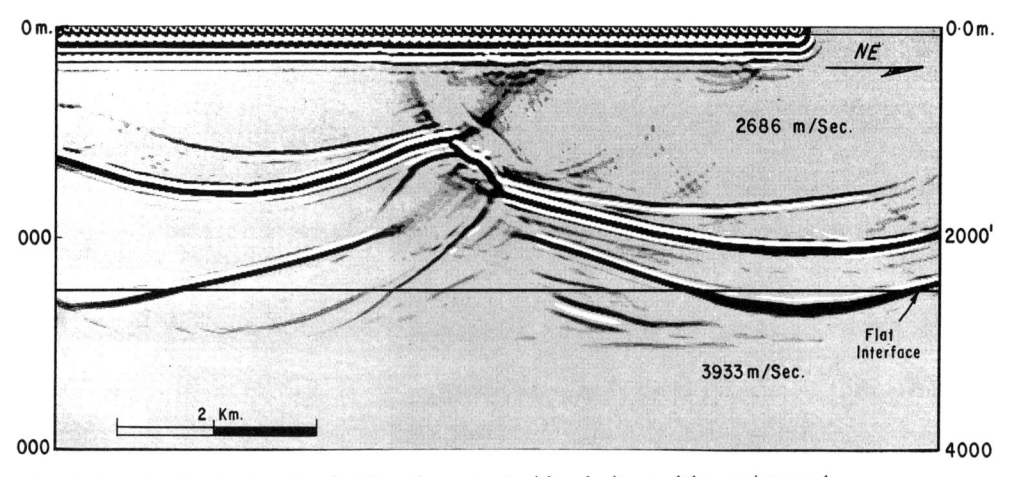

Fig. 11. Iterative depth migration first iteration output with velocity model superimposed.

analysis run during the processing of the synthetic data. Using these velocities the actual depth field to be used with them was iteratively determined.

In the first iteration of depth migration, shallow data down to the top of the salt were migrated using a simple velocity model of two flat layers. The output is shown in Fig. 11. This section was then used to digitize a new velocity model with more details in terms of dipping layers and salt relief. The output is shown in Fig. 12, although it does not represent the exact depth picture, due to slightly incorrect velocities, it shows good imaging results after only a few iterations, and using velocities determined from the data itself. A more precise velocity

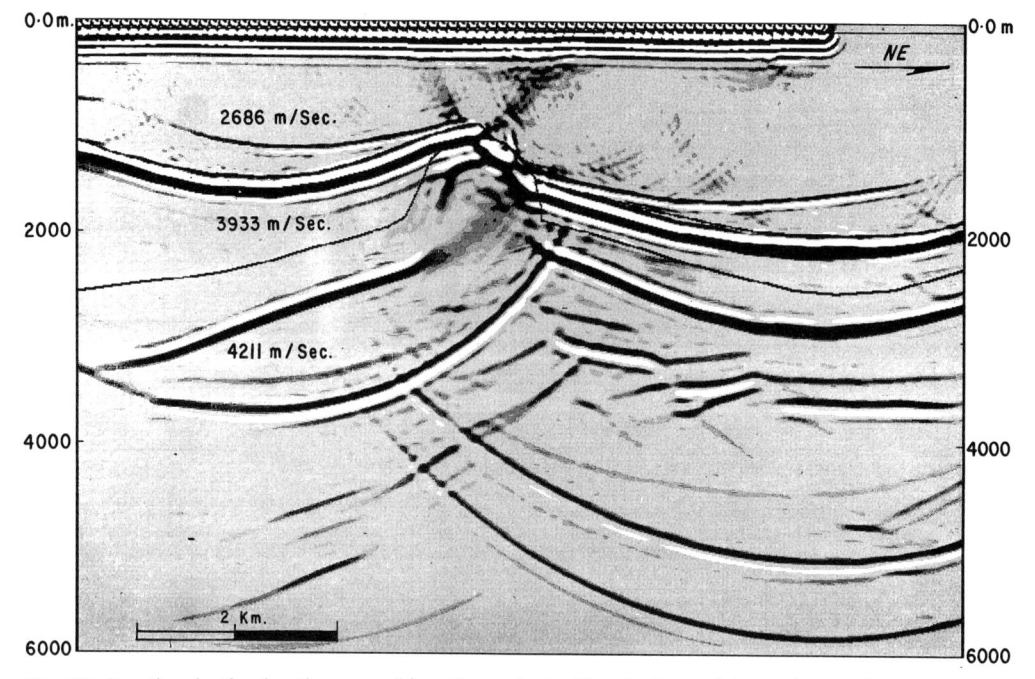

Fig. 12. Iterative depth migration second iteration output with velocity model superimposed.

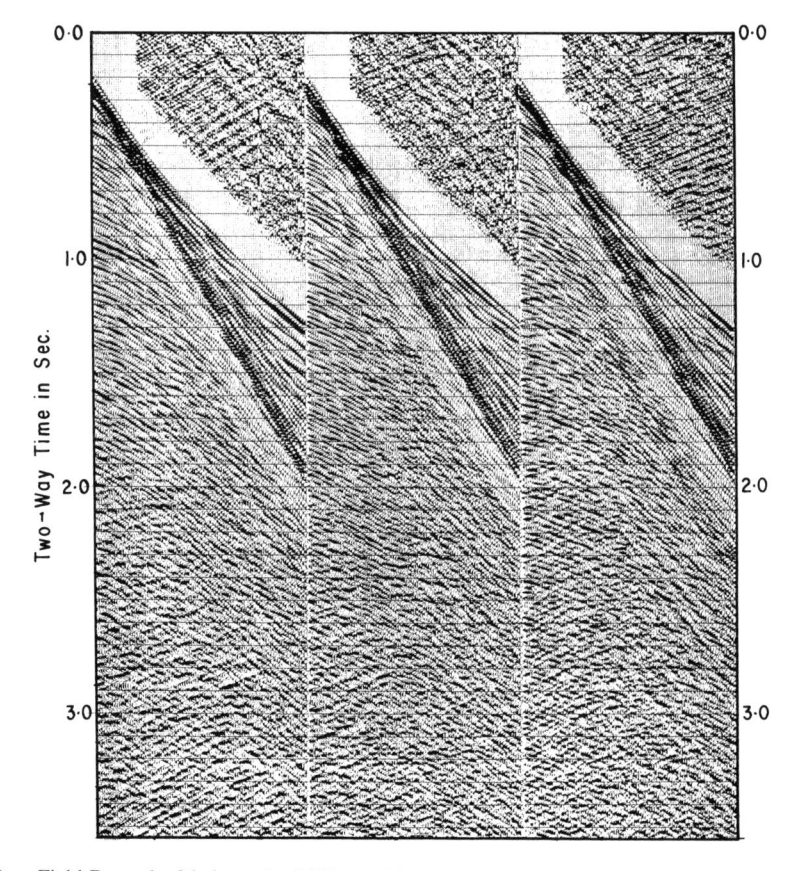

Fig. 13. Raw Field Records: 96 channels, 2400 m cable.

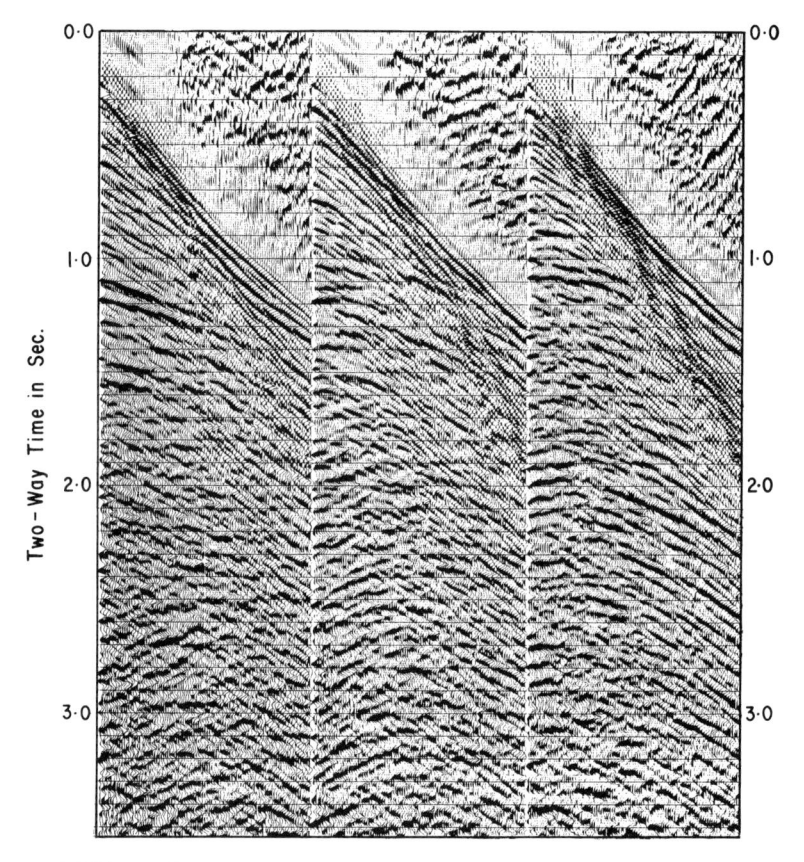

Fig. 14. Processed field records.

field could be calculated using tomography or travel-time inversion techniques (Whitmore & Lines 1986). As with any technique like depth migration that requires a velocity field, the more accurately the field is known the more precise the results will be.

Comparison with real data

With encouraging results on the model data, it remains necessary to find out how this applies to real seismic data recorded in the Gulf of Suez. The actual seismic line along the model profile was reprocessed guided by the previous observations from the modelling process. Figure 13 shows three of the raw field records. Primary reflections are almost invisible here due to the ringing multiples and low signal to noise ratio. These are significant problems which have not been addressed in this paper. Figure 14 shows the same records after F/K filtering followed by deconvolution and bandpass filtering.

The F/K filter was used carefully to remove the direct arrivals only. The records were improved and primaries became more evident. Figure 15 shows a comparison between two coincident, field and synthetic, shot records. The primary reflections on the synthetic records could be defined on the field record with less confidence as time increases. The interference effects due to varying dips and salt pull-up are recognized on the field record, while the pre-Miocene reflections are very weak and can hardly be seen. CMP stacking and subsequent depth migration will not be effective unless the low signal-to-noise problem is solved in this part of the Gulf of Suez.

Conclusions

On model data, the CMP method has been shown to still be a valid technique in spite of large lateral velocity variations associated with salt swells. Iterative depth migration coupled with CMP stacking produced the best final imaged representation of the original model.

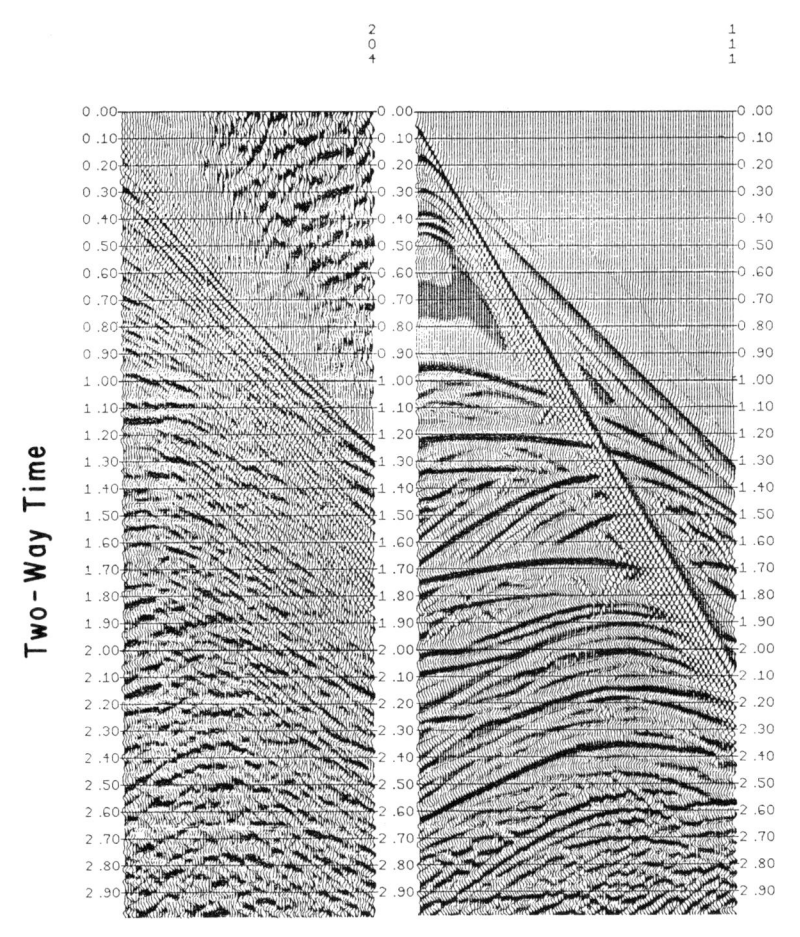

Fig. 15. Two coincident field record (Left). Synthetic record (Right).

Velocities can be derived from the raw data that will generate a reasonably good image of the pre-Miocene data. Time migration fails to correctly migrate data beneath salt swells. Iterative inverse and forward event modelling is an impractical method of mapping below salt swells in the Gulf of Suez.

References

MOSTAFA, A. M. 1974. Review of diapiric salt structures, Gulf of Suez, 4th Exploration Seminar, EGPC, Cairo, Egypt.

WHITMORE, N. D. & LINES, R. L. 1986. Vertical seismic profiling depth migration of a salt dome flank. *Geophysics*, **51**, 1087–1109.

Southern Gulf of Suez, Egypt: structural geology of the B-trend oil fields

H. M. HELMY

International Petroleum Corporation, P.O. Box 9211, Dubai, UAE

Abstract: The Nubia sandstone, the primary reservoir in B-trend oil fields (GS 365, Sidki, East Zeit, and Hilal) was mapped by projecting from the deepest reliable seismic event (base of Miocene evaporites), using the intermediate isopach maps. This was based on Gebel El Zeit stratigraphy and structure. The Gebel El Zeit outcrop lying 10 km to the west was used to describe a geological model analogous to the fields; since a close correlation was found between structural style and Miocene stratigraphy in both the surface and the subsurface data sets. The implementation of this subsurface model results in the enhancing of the oil potential of the B-trend.

The Neogene Suez rift extends north–northwestwards for about 350 km across the once continuous Sinai–North Africa continental crust. The rift is dominated by north–northwest oriented normal faults (named clysmic faults by Hume (1921)) and several cross elements indicated in both subsurface (Sultan & Schutz 1984; Helmy & Zakaria 1986), and surface (Moustafa & Khalil 1987). The present-day Gulf of Suez occupies approximately the axial one third of the rift complex. The stratigraphy and structure of the Suez rift has been described by Sadek (1959), Ghorab (1961), Said (1962), Garfunkel & Bartov (1977) and Beleity (1982) among others.

Moustafa (1976) indicated that the Suez rift consists of three dip provinces, each with a predominant direction of dip as shown in Fig. 1. The northern province as well as the southern one are characterized by southwesterly dipping fault blocks; however the middle province is dominated by northeasterly dipping blocks. He also defined the boundaries between these provinces as east–northeast oriented hinge zones. The Galala–Zenima and the Morgan hinge zones, respectively. Moustafa & Fouda (1986) and Moustafa & El Shaarawy (1987) accurately defined the location of these boundaries and studied the nature of deformation in the Morgan hinge zone.

The three dip provinces of the Suez rift are dissected by Hume's clysmic faults into several north–northwest oriented tilted fault blocks. The major clysmic faults bounding these blocks are shown in Fig. 2. The southernmost dip province has predominant southwest dip and includes several fault blocks which include the B-trend with its associated oil producing fields.

Southern Gulf of Suez

The study area is situated in the southern part of the Gulf of Suez to the south of the Morgan hinge zone (Figs 1 & 2). The structural setting of the southern Gulf is characterized by the existence of north–northwest oriented structural trends. These are the Mellaha trend, the Zeit trend, the B-trend and the Garra trend (Fig. 3). They comprise Miocene and pre-Miocene southwest dipping fault blocks bounded by major down-to-the-northeast clysmic normal faults (Fig. 4), and dissected by cross faults oriented northeast–southwest. To the north of B-trend oil production is mainly from syn-rift sequences while the B-trend is characterized mainly by oil production from pre-rift and syn-rift reservoirs.

The stratigraphy and characteristics of the southern part of the Gulf of Suez with particular emphasis on the area of study in B-trend are shown in Fig. 5, where the stratigraphic section shows the pre-rift, syn-rift and post-rift stratigraphic sequences identified by Garfunkel & Bartov (1977).

The pre-rift sequence in the southern Gulf of Suez comprises the following.
(i) The Precambrian basement consists mainly of igneous (mainly granitic) and metamorphic rocks. In Sidki field the fractured basement acts as an oil reservoir.
(ii) Nubia sandstone overlies the basement. It is a thick sedimentary section of Paleozoic (?) to Lower Cretaceous age composed mainly of non-marine, well sorted, arkosic and kaolinitic sandstones. The Nubia sandstone ranges in thickness from 1600 ft in the north to about 500 ft in the south,

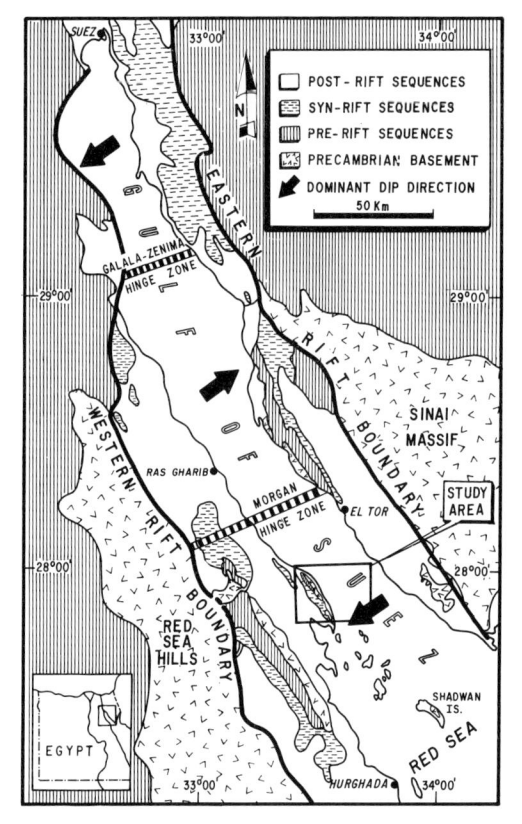

Fig. 1. Gulf of Suez dip provinces and hinge zones. Modified after Moustafa (1976).

and it is considered to be the best oil reservoir in the study area.

(iii) Marine Deposits unconformably overlie the Nubia sandstone with an average thickness of 750 ft, they include good oil source rocks. These marine deposits make up the Wata, Matulla and Sudr formations and are composed of limestones, shales and sandstones of Turonian to Late Senonian in age. An unconformity separates these rocks from highly calcareous marine shales (Palaeocene–Lower Eocene Esna Shale) and cherty limestones (Lower Eocene Thebes Formation). An angular unconformity separates the pre-rift and the syn-rift sequences. The syn-rift rocks are made up of Lower to Middle Miocene clastics and Middle to Upper Miocene evaporites. The Miocene clastics include several hiati (Beleity 1982) which are recognized in the study area by stratigraphic, and electric log correlations. The thickness of these clastics ranges between 1000 ft in structurally high areas to 5000 ft in basinal areas. The Miocene clastics unit is subdivided from base to top into the following.

(i) The Nukhul Formation is composed of slightly calcareous sand and sandstone with streaks of reddish shales at the base and anhydrite at the top which are referred to as Ghara and Shoab Ali members respectively (Khalil & Saoudi 1984). The Shoab Ali member is a reservoir rock in the South Gulf of Suez particularly in East Zeit and Hilal fields.

(ii) The Rudeis Formation unconformably overlies the Nukhul and is mainly composed of highly fossiliferous shales and limestones.

(iii) The Kareem Formation also unconformably overlies the Rudeis and consists predominantly of shales and limestones. Local feldspathic sandstones form a secondary reservoir in Sidki, East Zeit and Hilal oil fields.

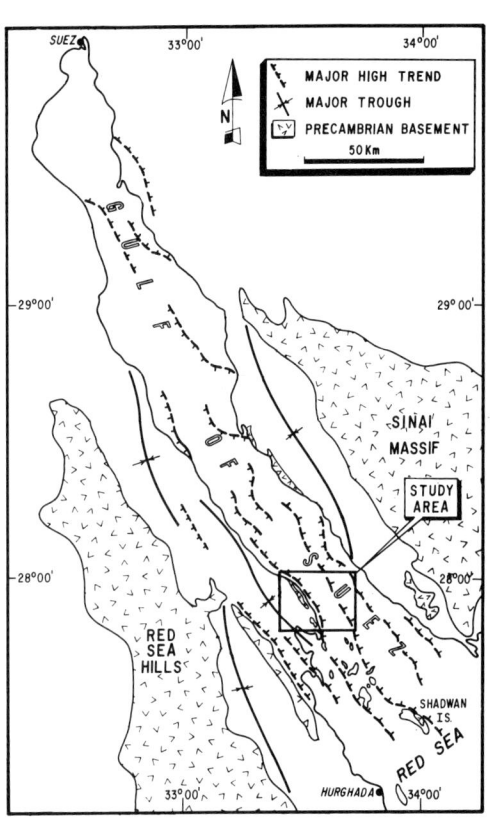

Fig. 2. Gulf of Suez major clysmic structural trends. After Steen & Helmy (1982).

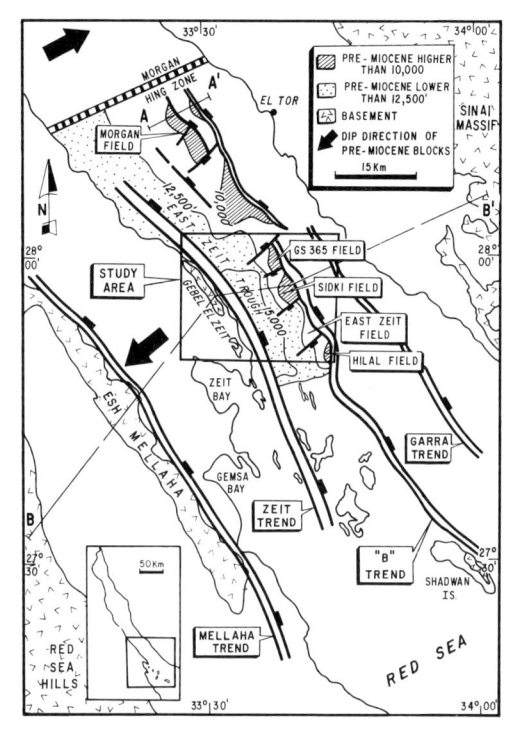

Fig. 3. Southern Gulf of Suez trends and Pre-Miocene structure map, C.I. = 2500 ft.

These sands could result from erosion of the updip basement blocks and deposited as sand tongues. To the north of the study area the Kareem sandstones are the primary reservoirs in the Morgan oil field.

The Miocene evaporites unconformably overlie the Miocene clastics and consist mainly of halite and anhydrite. They are considered to be the ultimate seal in the Suez rift. They range in thickness within the study area between 750 and 8000 ft. The evaporites are locally thickened by diapirism. This phenomenon has been described by several authors, e.g. Moustafa (1974). The Miocene evaporites are subdivided from base to top into the following.

(i) The Belayim Formation is mainly composed of evaporites at the base (Baba member, mainly anhydrite and salt with shale streaks), followed by shales (Sidri member), evaporites (Feiran member), and a marly section (Hammam Faraun member) at the top.

(ii) The South Gharib Formation is mainly a massive section of halite interbedded with anhydrite and shale streaks. Its thickness reaches 5000 ft in diapirs which are charac-

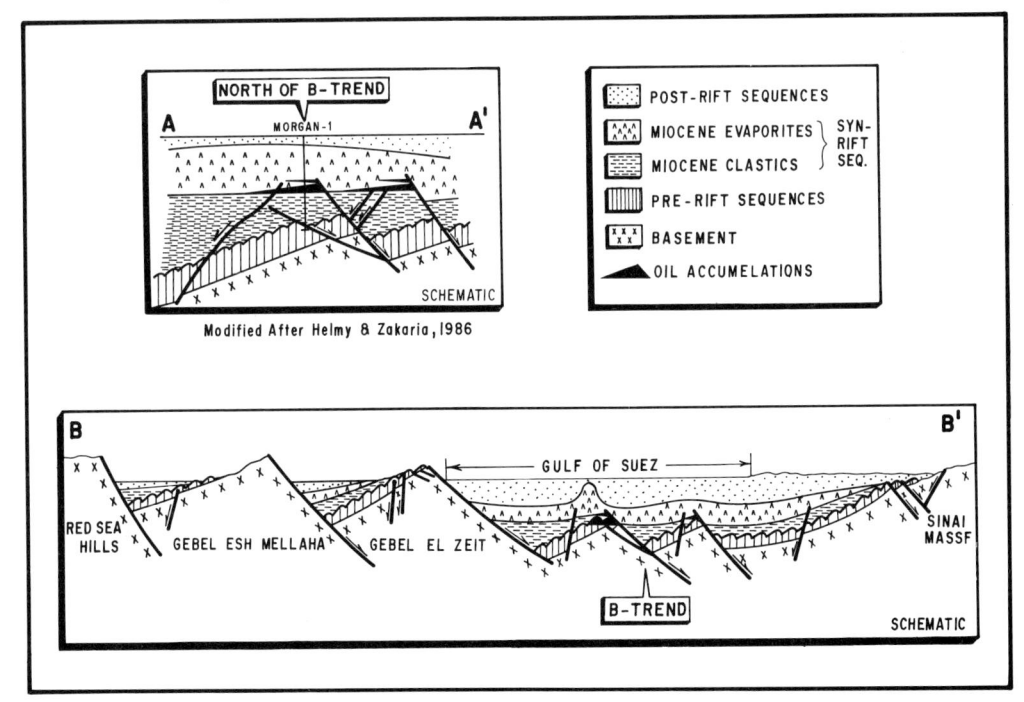

Fig. 4. Structural cross section (A) North of B-trend, (B) across the south Gulf of Suez, for line of section see Fig. 3.

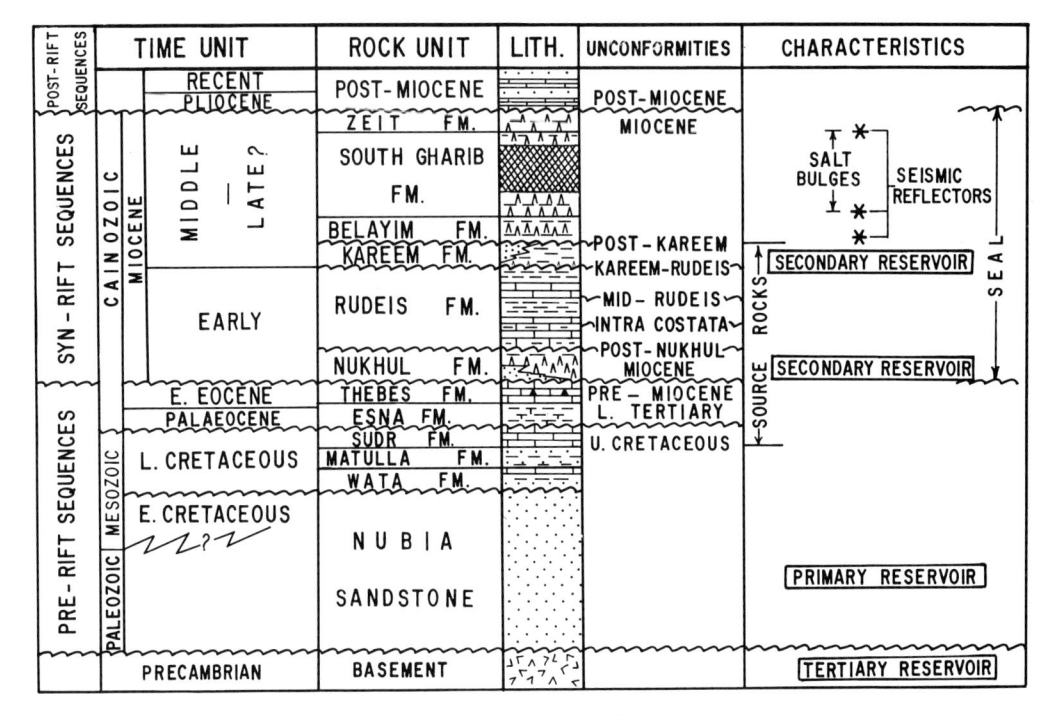

Fig. 5. Southern Gulf of Suez stratigraphic section and chracteristics.

teristic in the southern part of the Suez rift.
(iii) The Zeit Formation is the top part of the Miocene evaporites. Its thickness reaches 2000 ft between South Gharib diapirs, however, it can be absent on their crests.
The post-rift sequence is represented by a widespread post-Miocene deposits consisting of loose sands and limestones.

Gebel El Zeit

Gebel El Zeit (the oil mountain in Arabic) is named after oil seepages in the syn-rift rocks. It is a 30 km long northwest–southeast trending fault block that lies on the western side of the Gulf of Suez. Colleta et al. (1986) show that Gebel El Zeit is a southwest dipping fault block bounded to the east by a major down-to-the-east northwest–southeast clysmic fault (Zeit fault).

The author considers Gebel El Zeit as two parts: the northern one called Gebel Zeit and the southern one is called Gebel Little Zeit.

The stratigraphic section of Gebel El Zeit is preserved in its northern part (Fig. 6A), where pre-Miocene sedimentary section (Lower Eocene to Nubia sandstone) is highly eroded and the leading edge of the block has Precambrian basement outcrop. This is overlain unconformably by onlapping Miocene clastics

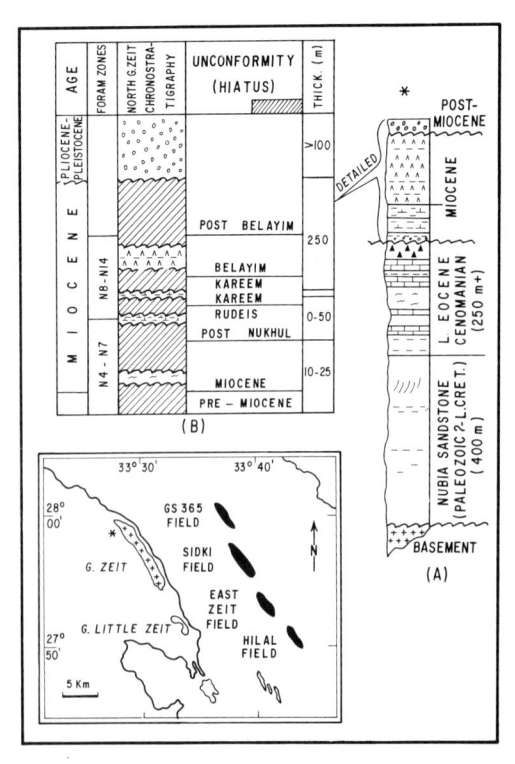

Fig. 6. Gebel el Zeit stratigraphic section (A) type section. (B) Miocene stratigraphy, after Evans & Moxon (1986).

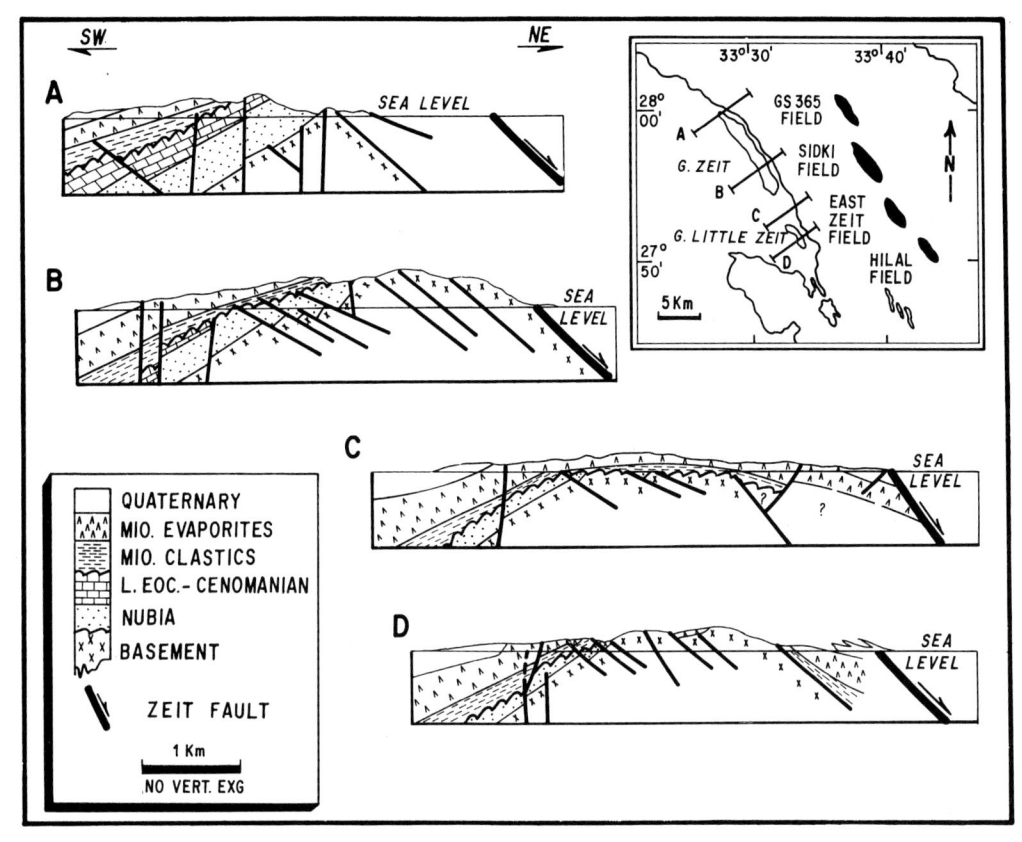

Fig. 7. Structural dip sections across Gebel el Zeit, after Colletta et al. (1986).

and evaporites. The Miocene clastics section in the northern Gebel El Zeit was studied in detail by Evans & Moxon (1986). It shows several stratigraphic hiati (Fig. 6B).

The structural dip cross sections of Gebel El Zeit (Fig. 7) indicate that the only correlatable fault is the Zeit fault which exists at the leading edge of the Zeit block. Other faults exist in these sections and cannot be correlated between them. The strike sections of Gebel El Zeit (Fig. 8) indicate that cross faults are also important. The author interprets a structural saddle caused by cross faults to be responsible for the separation of Gebel Zeit and Gebel Little Zeit.

The stratigraphy and structure of Gebel El Zeit were used in this study as an analog to model the subsurface data obtained from wells of the B-trend within the study area.

B-trend stratigraphic mapping

The B-trend extends from the GS 365 oil field southeastward to Shadwan island at the extreme end of the Gulf of Suez (Fig. 5). Several oil fields exist on this trend. The fields in the area of study (GS 365, Sidki, East Zeit and Hilal) are characterized by oil production mainly from pre-rift sequences. These four oil fields are described in this study, and lie in the northern part of the B-trend.

Electric log correlations of the wells in the four oil fields (Fig. 9), show missing sections within the pre-Miocene sequence interpreted to be due to faulting, while those in the Miocene rocks are interpreted to be due to the unconformities, similar to those described by Beleity (1982) and interpreted by him to have a regional extent. The interpreted unconformities are also supported by paleontological data, and are inferred to be due to non-deposition rather than erosion. However, other missing sections within the Miocene formations are found to be due to minor faulting and are scattered throughout the wells.

The Miocene unconformities detected in the study area from base to top are as follows.

(i) The Miocene/pre-Miocene unconformity

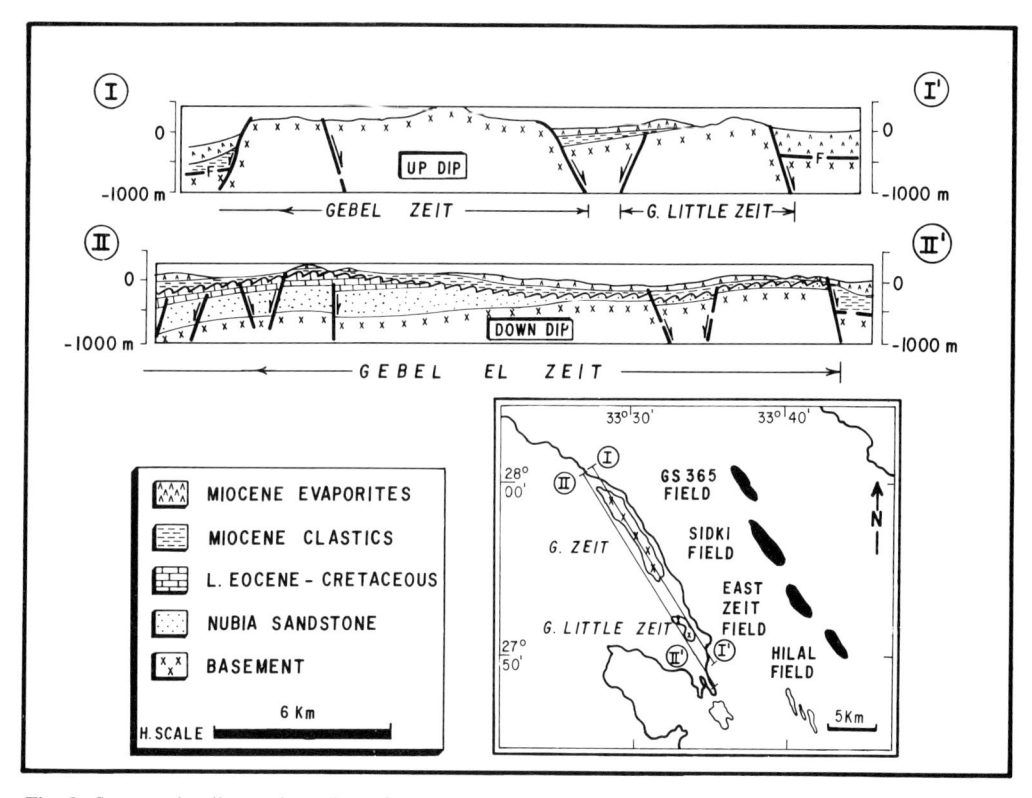

Fig. 8. Structural strike sections along Gebel el Zeit. Modified after Colletta et al. (1986).

is most severe in the GS 365 and Sidki fields, where Miocene rocks rest directly on units older than lower Eocene Thebes formation. This unconformity represents the erosion of the youngest pre-Miocene Formations and non deposition of Nukhul clastics. It has its least effect on East Zeit and Hilal oil fields.

(ii) The post-Nukhul unconformity is clear in GS 365 and Sidki oil fields.

(iii) The Intra Costata unconformity like that at the top of pre-Miocene affects the northern fields (GS 365 and Sidki) more severely with lesser effect in the East Zeit and Hilal fields.

(iv) The mid-Rudeis unconformity is found in all the four fields.

(v) The Kareem-Rudeis unconformity is detected between the top of Miocene clastics and R-1 electric log marker in all four fields.

(vi) The Miocene clastics–Miocene evaporites unconformity is demonstrated in Sidki field by the complete absence of the Belayim formation. However, in the East Zeit and Hilal oil fields this unconformity is represented by the absence of only the lower members of the Belayim and the upper units of the Kareem formations.

The identification of missing stratigraphic section in the study area facilitates the construction of controlled isopach maps. The Lower Eocene to Nubia Isopach map (Fig. 10) shows a thickening in the northern part of the area up to 1000 ft while only 500 ft are encountered in the south. The depositional strike of this pre-rift sequence is more or less northeast–southwest, and is unaffected by the clysmic trends.

The isopach map of the total Miocene clastics represented by the Kareem, Rudeis and Nukhul formations, shows variation in thickness across the structural trends (Fig. 11). The Zeit trend shows that the entire section of these clastics thins eastward from 4800 ft in the Zeit Bay-1 borehole to 450 ft on the surface in Gebel El Zeit (Evans & Moxon 1986). The studied fields show the same phenomenon. The Zeit trend and the B-trend are separated by the East Zeit trough, where the estimated thickness of the Miocene clastics is 5000 ft. This trough is con-

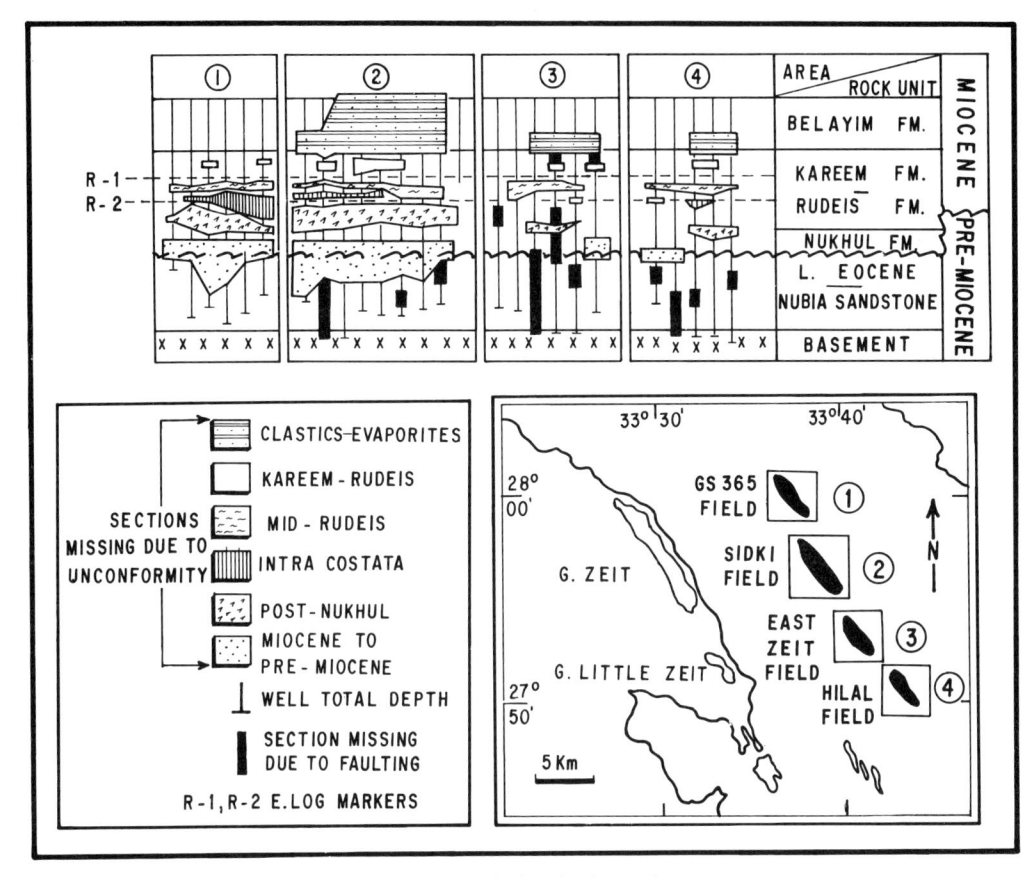

Fig. 9. Schematic presentation of electric log correlations in the study area.

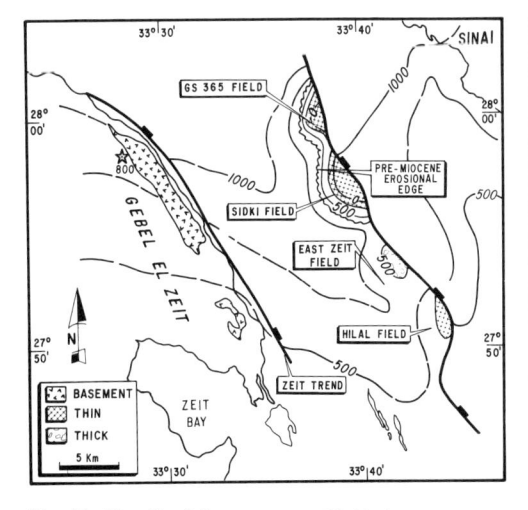

Fig. 10. Top Pre-Miocene to top Nubia isopach map, C.I. = 250 ft.

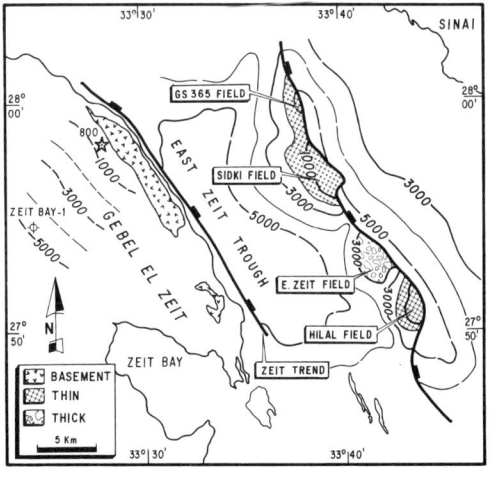

Fig. 11. Total Miocene clastics isopach map, C.I. = 1000 ft.

sidered to be a good oil generating trough from Miocene clastics as well as the pre-rift sequences for the four fields in the study area.

Several conclusions can be deducted from the isopach maps and stratigraphic correlations.
  (i) Three main areas can be identified from the isopach maps discussed above, each area has its own characteristics. The first area to the north includes GS 365 and Sidki fields, and is characterized by partially eroded pre-Miocene section, and thin Miocene clastics. The second area (East Zeit field) which lies in the middle. Here the top of the pre-Miocene section is not eroded and thick Miocene clastics are found. The third, southernmost area, includes Hilal field and is similar to the second, since uneroded pre-Miocene section occurs, but the total Miocene clastics are thinner relative to East Zeit. Cross faults possibly form boundaries between the above mentioned three areas.
  (ii) The erosion of the pre-rift sequence is a result of block faulting and rotation on the clysmic trend, as in the first northern area (GS 365 and Sidki).
  (iii) Block faulting affects the distribution of syn-rift deposits so that the Miocene clastics thin on the leading edges and thicken in downdip and downthrown sides of the blocks. This is similar to the situation identified in the Mellaha trend by Burchette (1986).

B-trend structural mapping

The regional structure of the study area is easily identified geophysically at the base of Miocene evaporites by seismic reflections and at basement level by aeromagnetic data.

The reduced-to-pole magnetic map shows that the area of study and Gebel El Zeit are high magnetic anomalies (Fig. 12). Such interpretation shows that both are controlled by major down to the northeast clysmic faults and are also dissected by cross Gulf of Suez faults.

The seismic reflection data at the contact between the Miocene evaporites and the Miocene clastics (the deepest reliable seismic reflector in the Gulf of Suez) led to the same conclusion as was derived from the magnetic maps. In Fig. 13, the seismic map shows that the bounding fault has an approximate throw at the top of Miocene clastics ranging between 1000 and 2500 ft, and also reveals that the dip range of Miocene clastics is between 0° and 4° SW in the GS 365 field, 11° SW in Sidki field, 12° SW in East Zeit field and 10° NE in Hilal field.

The seismic map at the top of Miocene clastics, the borehole data and isopach maps (total Miocene clastics and Top pre-Miocene to top Nubia) were used to construct several dip and strike structural sections across and along the area of study. A dip section across Sidki field (Fig. 14) as an example, shows that the pre-Miocene block and the Miocene clastics have a

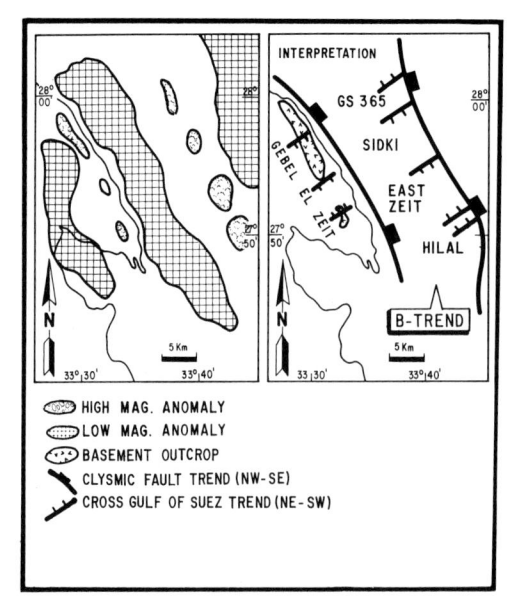

Fig. 12. Magnetic maps of the study area (Hammouda pers. comm.).

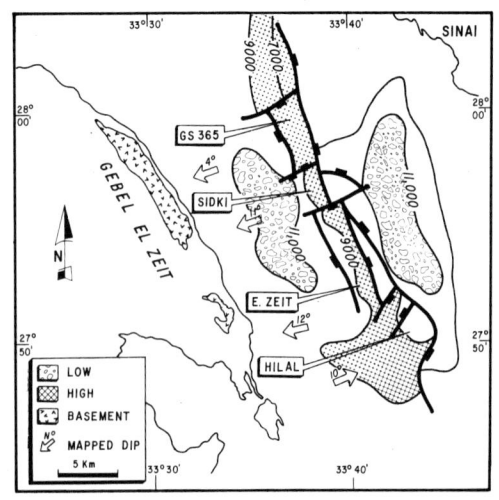

Fig. 13. B-trend top Miocene clastics (base evaporites) seismic map, C.I. = 1000 ft.

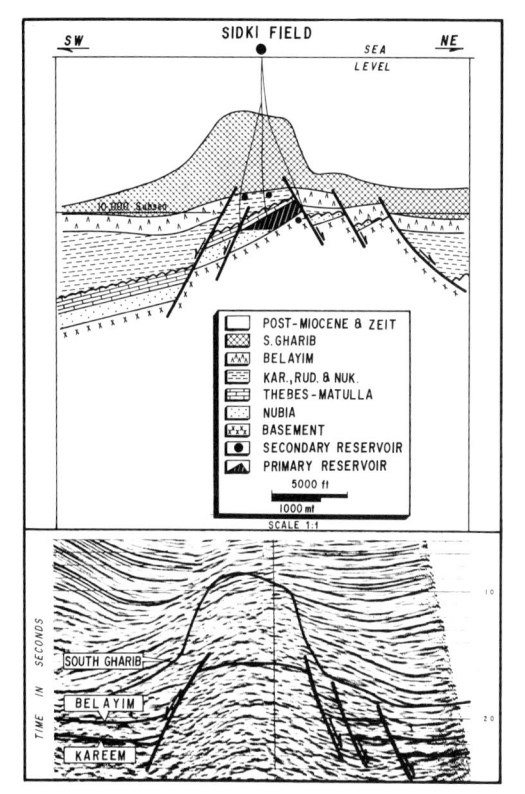

Fig. 14. Structural dip cross section across Sidki field and its corresponding seismic profile.

predominant southwesterly dip, also, the block is bounded to the east by a major clysmic fault (B-trend fault) with an approximate throw of 5000 ft at the pre-Miocene level. Other faults of lesser magnitude of throw exist with either northeast or southwest dipping fault planes.

The relationship between the fields along strike is shown in Fig. 15. The updip strike section along the top of the blocks shows that the maximum oil column is encountered in the Nubia sandstone. The cross faults also play an important role in defining the southern boundaries of the fields. Meanwhile the northern boundaries are controlled by structural plunge of the pre-Miocene blocks and the associated increase in thickness of the Miocene clastics. The downdip strike section also reveals the importance of cross faults and shows the oil accumulation within the Miocene clastics in both Sidki and East Zeit fields.

It is concluded that the study area can be treated as one regional block dipping to the southwest and bounded to the east by a major fault. The strike cross sections along the fields under study reveal that they are also dissected by cross faults trending mainly northeast–southwest.

Gebel El Zeit and B-trend comparison

The study of Gebel El Zeit stratigraphically by Evans & Moxon (1986) and structurally by Colletta et al. (1986), and the data displayed for the B-trend oil fields in the current study show that Gebel El Zeit and the four oil fields are similar in the following respects.

(i) The dimensions of both Gebel El Zeit and the study area of B-trend are similar. Also, cross faults dissect them and the three areas (GS 365 and Sidki, Hilal and East Zeit fields) in the B-trend could be correlated to Gebel Zeit, Gebel Little Zeit and the trough in between.

(ii) Unconformities detected in both are similar in their stratigraphic positions and resulted from non-deposition on the updip part of the structure.

(iii) The only correlatable faults in Gebel El Zeit and the fields are the respective bounding faults.

(iv) The pre-Miocene sequence is eroded in both areas, and maximum erosion occurs at the leading edges of the main bounding faults (Fig. 16).

Nubia structural map

The Nubia sandstone is the primary reservoir rock in the oil fields of the B-trend. An accurate map of this reservoir is therefore of great importance. Unfortunately, the Nubia sandstone cannot be identified on seismic profiles, since the deepest identifiable events is the base of the Miocene evaporites. The map therefore, was generated from serveral cross section across and along the fields area using Gebel El Zeit as a structural model with the aid of isopach and base evaporites seismic maps.

The Nubia structure map of the study area (Fig. 17) shows that the average angle of dip from dipmeters in the fields reaches a maximum of 30° SW in Sidki, and is more gentle in the other three fields, where the maximum is 23° SW. The variation of the oil-water contact across the area is also indicated on this map. With the deepest oil-water contact in East Zeit field (−11 800 ft) and the shallowest in GS 365 at (−10 140 ft). The Nubia map shows that the oil fields on the B-trend are bounded by faults on two sides (B-trend bounding fault on the northeast side and a cross fault to the south side), the northern and western sides are controlled by structural plunge.

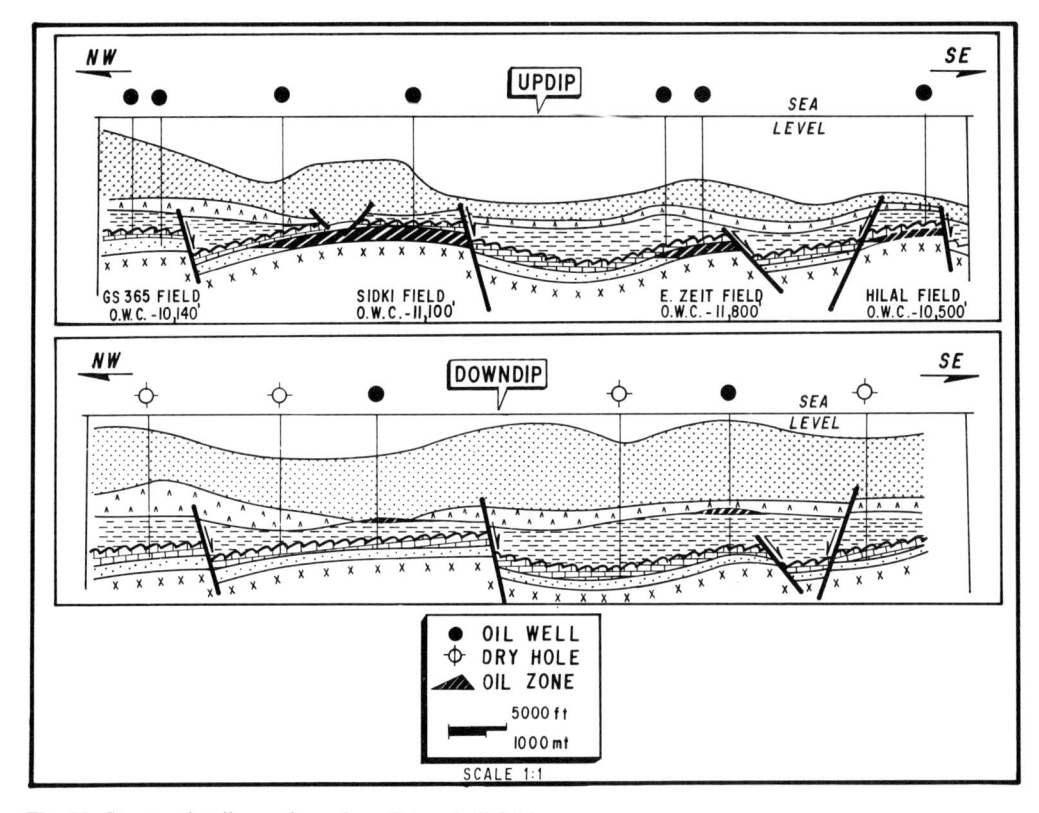

Fig. 15. Structural strike sections along B-trend oil fields.

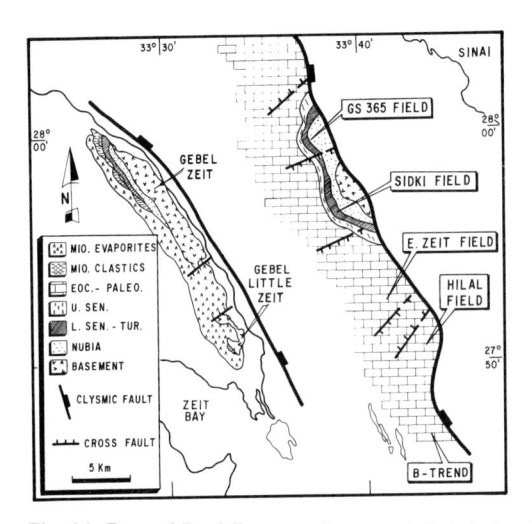

Fig. 16. B-trend Pre-Miocene subcrop and Gebel el Zeit geological map.

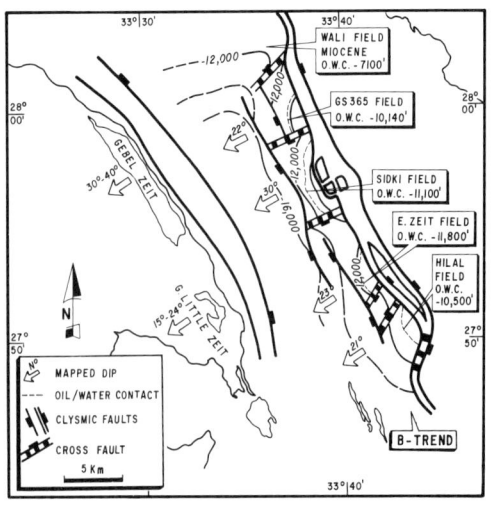

Fig. 17. Area of Study Nubia structure map (the main reservoir in the area) C.I. = 2000 ft.

The lateral seal in most of the fields on the eastern side is the Miocene clastics or the Miocene evaporites. The latter is the ultimate seal for the shallower reservoirs (e.g. Kareem reservoirs in Sidki field as seen on the dip cross section).

Conclusion

Known surface models can be successfully used for modelling of stratigraphic and structural subsurface data in adjacent areas. As a result more accurate maps can be drawn, and the hydrocarbon potential enhanced.

This work was conducted in 1988 when the author was part of the exploration staff with Gulf of Suez Petroleum Company.

The author wishes to express his sincere thanks to the management of Amoco International in Houston, EGPC and Gulf of Suez Petroleum Company (GUPCO) for granting him the permission to publish this paper. Grateful thanks to all GUPCO staff, and in particular to Mr Shawky Abdine and Mr Ahmed Gad for their guidance and assistance in generating this project.

References

BELEITY, A. 1982. The composite standard and definition of paleo-events in Gulf of Suez, 6th EGPC Exploration Seminar, Cairo.
BURCHETTE, T. P. 1986. Mid-Miocene tectonics and sedimentation, Esh El Mellaha range, southwest Gulf of Suez, 8th E.G.P.C. Exploration Seminar, Cairo.
COLLETTA, B., MORETTI, I., CHENET, P. Y., MULLER, C. & GERARD, P. 1986. The structure of Gebel El Zeit area: A field example of tilted block crest in the Gulf of Suez, Institut Francais Du Petrole. Guide book for workshop on tectonics and sed. of Gulf of Suez and Red Sea, Hurgada.
EVANS, A. L. & MOXON, I. W. 1986. Gebel El Zeit chronostratigraphy: Neogene syn-rift sedimentation atop a long-lived paleohigh, 8th E.G.P.C. Exploration Seminar, Cairo.
GARFUNKEL, Z. & BARTOV, Y. 1977. The tectonics of the Suez rift, *Geological Survey of Israel Bulletin* **71**.
GHORAB, M. A. 1961. Abnormal stratigraphic feature in Ras Gharib oil field 3rd Petroleum Conference, Alexandria, Egypt, 10p. E.G.P.C. Exploration Seminar, Cairo.
HELMY, H. & ZAKARIA, A. 1986. El Morgan area geological framework, 8th E.G.P.C. Exploration Seminar, Cairo.
HUME, W. F. 1921. Relations of northern Red Sea and associated Gulf area to the 'rift' theory, *Proceedings Geological Society London*, **77**, 96–101.
KHALIL, B. A. & SAOUDI, A. M. 1984. Distribution and hydrocarbon potential of Nukhul sediments in Gulf of Suez, 7th E.G.P.C. Exploration Seminar, Cairo.
MOUSTAFA, A. M. 1974. Review of diapiric salt structures, Gulf of Suez, 4th E.G.P.C. Exploration Seminar, Cairo.
—— 1976. Block faulting in Gulf of Suez, 5th E.G.P.C. Exploration Seminar, Cairo.
MOUSTAFA, A. R. & FOUDA, H. G. 1986. The Gebel Gharamul Gebel Sufr El Dara wrench zone, Gulf of Suez, Egypt (abs.), 24th Annual Meeting of Geological Society of Egypt, Cairo, 2–3.
—— & EL SHAARAWY, O. A. 1987. Tectonic setting of the northern Gulf of Suez (abs.), 5th Annual Meeting of Egyptian Geophysical Society, Cairo, 14.
—— & KHALIL, M. H. 1987. The Durba–Araba fault, southwest Sinai, Egypt, Egypt. *Journal of Geology*, **31**, 15–32.
SADEK, H. 1959. *The Miocene in the Gulf of Suez region, Egypt*, Geological Survey and Mineral Research Department.
SAID, R. 1962. *The Geology of Egypt*, Elsevier, Amsterdam.
STEEN, G. E. & HELMY, H. 1982. Pre-Miocene evoltion in the Gulf of Suez, Gulf of Suez Petroleum Company, Internal Report (Unpublished).
SULTAN, N. & SCHUTZ, K. 1984. Cross faults in the Gulf of Suez, 7th E.G.P.C. Exploration Seminar Cairo, Egypt.

Petroleum geology of the Niger Delta

H. DOUST

Shell International, Hague

Abstract: The Niger Delta is one of the World's largest Tertiary delta systems and an extremely prolific hydrocarbon province. It is situated on the West African continental margin at the apex of the Gulf of Guinea, which formed the site of a triple junction during continental break-up in the Cretaceous. Throughout its history, the delta has been fed by the Niger, Benue and Cross rivers, which between them drain more than 10^6 km^2 of continental lowland savannah. Its present morphology is that of a wave-dominated delta, with a smoothly seaward-convex coastline traversed by distributary channels. From apex to coast the subaerial portion stretches more than 300 km, covering an area of 75 000 km^2. Below the Gulf of Guinea, two enormous lobes protrude a further 250 km into deeper waters.

The delta sequence comprises an upward-coarsening regressive association of Tertiary clastics up to 12 km thick. It is informally divided into three gross lithofacies: (i) marine claystones and shales of unknown thickness, at the base; (ii) alternations of sandstones, silstones and claystones, in which the sand percentage increases upwards; (iii) alluvial sands, at the top.

Delta structure and stratigraphy are intimately related, the development of each being dependant on the interplay between sediment supply and subsidence rates. The dominant subsurface structures are syn- and post- sedimentary listric normal faults which affect the main delta sequence. They die out upwards into the alluvial sands and sole out at depth near the top of the marine claystones. Major growth-fault trends cross the delta from northwest to southeast, dividing it into a number of structural and stratigraphic belts, called depobelts, which become younger towards the south. The deltaic sequence in each of these depobelts is distinct in age, so that they actually represent successive phases in the delta's history.

Hydrocarbons have been located in all of the depobelts of the Niger Delta, in good quality sandstone reservoirs belonging to the main deltaic sequence (the 'paralic sequence' of common usage). Most of the larger accumulations occur in roll-over anticlines in the hanging-walls of growth faults, where they may be trapped in either dip or fault closures. In the 30 or 50 years since the first discoveries were made, approximately 4×10^9 m^3 of oil and an unestablished, but very substantial quantity of associated and unassociated gas have been discovered. Most fields are small, ranging up to 50×10^6 m^3, though several larger fields contain recoverable reserves in excess of 80×10^6 m^3. The hydrocarbons are found in multiple pay sands with relatively short columns, and adjacent fault blocks usually have independent accumulations.

Niger Delta reservoir geology: historical growth of the sedimentological model and its application to field development

K. J. WEBER

Shell International, Hague

Abstract: In 1953 when the exploration in Nigeria moved towards the Niger Delta, it was realised that the recent delta could be used as a guide to the past. Consequently many studies were carried out between 1953 and 1964, ranging from surveys of the shallow offshore, the coastal zone and the rivers to an extensive shallow coring campaign (1960–62).

The studies of the recent delta were soon augmented by the analysis of cores from several newly discovered fields (1963–69) and also by the description and grain size analysis of numerous side-wall samples. Around 1970 the basic facies types comprising the individual deltaic cycles of the hc-bearing paralic sequence were well established. The facies types occur in a very repetitive pattern of combinations and successions which facilitates recognition and correlation. A fairly reliable system of diagnostic criteria was worked out based on petrophysical logs and side-wall samples and calibrated by the cores and the recent delta.

With the sequencial introduction of new logging tools the diagnostic methods are being refined while additional cored wells continuously enlarge the data base. Detailed seismic analysis at reservoir level is showing considerable promise for sand body delineation. For each facies a representative data base of sand body geometry could be compiled based on detailed reservoir correlations and recent examples. There is a good understanding of typical permeability and grain-size distributions. Knowledge on shale intercalation continuity is particularly useful in the planning of well completions to retard gas or water breakthrough to the well bore. It was shown that permeability can be computed from log derived porosity and grain-size distribution measured on side-wall samples.

The practical value of the reservoir geological research lies in the possibility to correlate and model reservoirs in a reliable and consistent manner. Reservoir and well behaviour can be predicted and where necessary reservoir simulation models can be constructed. Continuous updating of reservoir studies has demonstrated the validity of this approach by checking past predictions and analysis of infill wells. In this way the production planning and reserve estimation is given a solid base.

Petroleum geology of the 'Fossa Bradanica' (foredeep of the Southern Apennine thrust belt)

M. SELLA[1], C. TURCI[1] & A. RIVA[2]

[1] *DESI/GERM, AGIP S.p.A., San Donato Milanese (Milano), Italy*
[2] *SGEL/GEOL, AGIP S.p.A., San Donato Milanese (Milano) Italy*

Abstract: The 'Fossa Bradanica' is located in Southern Italy, East of the Apennine mountain chain. It has undergone a complex geotectonic evoluton: from platform (Mesozoic carbonates) to unstable foreland (Eocene–Miocene) and finally to foredeep of the Apennine thrust belt (Plio–Pleistocene). Petroleum exploration targets in this province are represented by: (a) Gas and oil in carbonates of the pre-Pliocene substratum. In this domain traps are only structural; (b) Gas and subordinately oil in Plio–Pleistocene sands, where structural, stratigraphic and combination traps can be found. The oils produced range in gravity from 3° to 45° API, with prevailing values of 15–20° API. Two families of oils have been recognized: oils generated by a shaly source deposited in oxygenated waters and oils derived from a carbonate source deposited in a euxinic environment. Most of the oils are classified as non-mature or marginally mature oils. Gas is mostly dry: 79% of the gas discovered in Fossa Bradanica being of biogenic, 6% of thermogenic and 15% of mixed origin. Many fields have high CO_2 concentrations (50 to 90% in the carbonates). The CO_2 content decreases, moving upwards in the stratigraphic column.

The area of this study pertains geographically to the 'Fossa Bradanica' *sensu strictu*, located westward of the Murge plateau, and to the Tavoliere delle Puglie, bordered westwards by the Southern Apennine mountain chain and eastwards by the Gargano Peninsula (Fig. 1). The area extends over part of Molise and Puglia and over Basilicata for about 9600 km^2; it is 240 km long (NW–SE) and 40 km wide (SW–NE). In this province, mature to petroleum exploration, are located the major hydrocarbon finds of Southern Italy. The geological framework of the province provided in this paper is based on the work of Crescenti (1975), Ippolito *et al.* (1975), Mostardini & Merlini (1986) and Pieri & Mattavelli (1986).

Geology and petroleum exploration targets

The northernmost portion of the African Plate, called the Apulia Plate, was a site of carbonate platform deposition until the Upper Triassic. Starting from the Liassic, extension faulting generated local basins but in the area of the study platform conditions persisted until the Miocene ('Apulia Platform').

During the Jurassic and Cretaceous, dolomites ('Dolomie di Ugento') and limestones ('Calcari di Cupello') was deposited.

The end of the Cretaceous saw the inception of the Apennine thrust belt formation. Several structural units were carried eastwards along thrust surfaces and the Fossa Bradanica area turned into an unstable foreland (Eocene–Miocene). During the Eocene, shaly limestones with Nummulites ('Calcari a Nummuliti') and calcareous breccias ('Brecce di Lavello') were deposited; some basalt flows were also present. At the end of the Eocene a general regression occurred were and Oligocene sediments are missing all over the area.

Lower and Middle Miocene sediments were again deposited in a carbonate shelf environment, but the terrigenous influence became important ('San Ferdinando' Formation, known in outcrops as 'Calcari a Briozoi'). During the Messinian, large evaporitic basins are formed in this region as in the whole Mediterranean Sea.

By Pliocene time the Fossa Bradanica area had developed into the foredeep of the Southern Apennine thrust belt. The load of the rising mountain chain is in fact at the origin of downwarping in Fossa Bradanica and of uplifting in the Murge plateau (Ricchetti 1980). The extension tectonics is mainly active through NW–SE and NNW–SSE fault alignments. In the West, below the thrust sheet which represents the more external unit of the orogenic belt (the 'Allochthonous'), the carbonates of the Apulia Platform are involved in the Apennine thrust tectonics (Fig. 2). A more recent wrench tectonics phase has affected the area, with horizontal displacements along SW–NE and subordinately WSW–ENE directions, which may have induced local compressions on the carbonate substratum.

From BROOKS, J. (ed.), 1990, *Classic Petroleum Provinces*, Geological Society Special Publication No 50, pp 369–378.

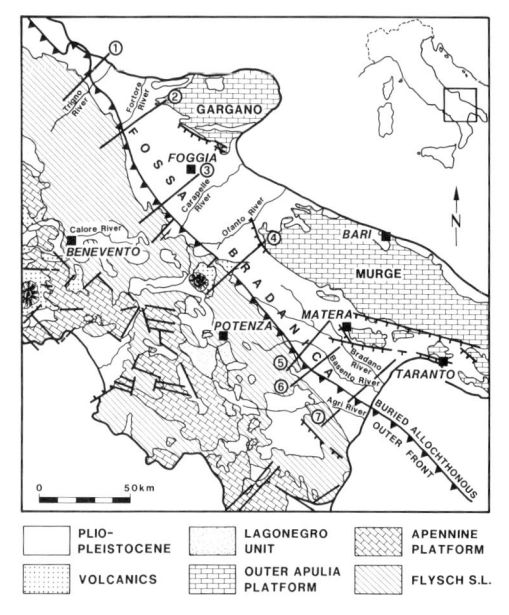

Fig. 1. Geological sketch map with location of geological sections.

During Lower Pliocene the sea flooded the Fossa Bradanica depression and marine conditions persisted until Pleistocene. The Plio–Pleistocene cycle began with shaly and marly sediments (pre-turbiditic phase), followed by sequences of sands and shales deposition (turbiditic phase) (Balduzzi et al. 1982a,b; Casnedi et al. 1981, 1982; Rossi 1986). The Lower Pliocene sea is not continuous from Molise to the present Taranto Bay, but it is interrupted in its middle by a tectonic ridge. This obstruction disappears in the Middle Pliocene and the basin takes the shape of a long and narrow seaway. From the Lower Pliocene to Pleistocene the eastern limit of the basin, as well as its depocentre, moves progressively eastwards, ahead of the advancing Allochthonous front. The evolution of the western border cannot be retraced, because it is obliterated by the Allochthonous eastwards motion.

West of Taranto Bay, a Plio–Pleistocene semi-allochthonous basin develops on top of the Allochthonous sheet (Fig. 2b, Geological section no 7). It is fairly deep (more than 1500 m) and extends from land (about 800 km^2) into the adjacent offshore.

The Plio–Pleistocene clastics of the Fossa Bradanica have been to some extent involved in the Apennine formation process, which has produced gentle folds and minor thrusts, mainly in the proximity of the buried front of the Allochthonous sheet. In the central sector of Fossa Bradanica, in connection with some 'highs' of the pre-Pliocene substratum, a number of structural closures are present in the terrigenous sequence. Their origin relates to synsedimentary tectonics and/or differential compaction.

The petroleum exploration targets in Fossa Bradanica are represented by: (a) carbonates of the Apulia Platform, where oil and gas accumulations are possible in structural traps; (b) sand bodies of the Plio–Pleistocene sequence where gas and subordinately oil accumulation can be expected in structural, stratigraphic and combination traps.

Exploration history and results

A short résumé of the petroleum exploration history in the province is herewith provided (see Fig. 3 for wells and fields location).

The exploration for oil and gas in Fossa Bradanica began in the late 1930s. Three dry holes were drilled by AGIP at Genzano, near Matera, on the basis of the studies carried out by the geologist Migliorini, between 1937 and 1942.

In 1949, AGIP and RPM (Ricerche Petrolifere Meridionali) resumed exploration work in the area by carrying out seismic surveys. In 1952, another dry hole, Gaudiano 1, was drilled by AGIP. In 1957 a new Petroleum Law was issued, which provided incentives for the exploration for oil and gas in Italy. As a result, the years following recorded a sharp increase both in the number of oil companies involved and in the acreage covered by exploration licences. The first significant results in this new exploration phase were obtained by AGIP with the discoveries of the Grottole–Ferrandina gas field and the Pisticci gas and oil field, in 1959 and 1960 respectively. The 1960s were extremely successful years: six gas fields and one gas and oil field were discovered, most of them containing important reserves (Candela, Torrente Tona, Roseto–Montestillo). In the following years exploration continued, but at a decreasing rate of success.

At the end of 1987, 65 fields had been discovered in the Fossa Bradanica petroleum province, including 38 one-well fields. At present 33 fields are in production, out of which only 9 produce from more than 6 wells. In the near future 10 additional fields will be put on stream. Seven fields have been completely depleted, while 9 are being considered for future development and production.

Oil production in Fossa Bradanica is definitely

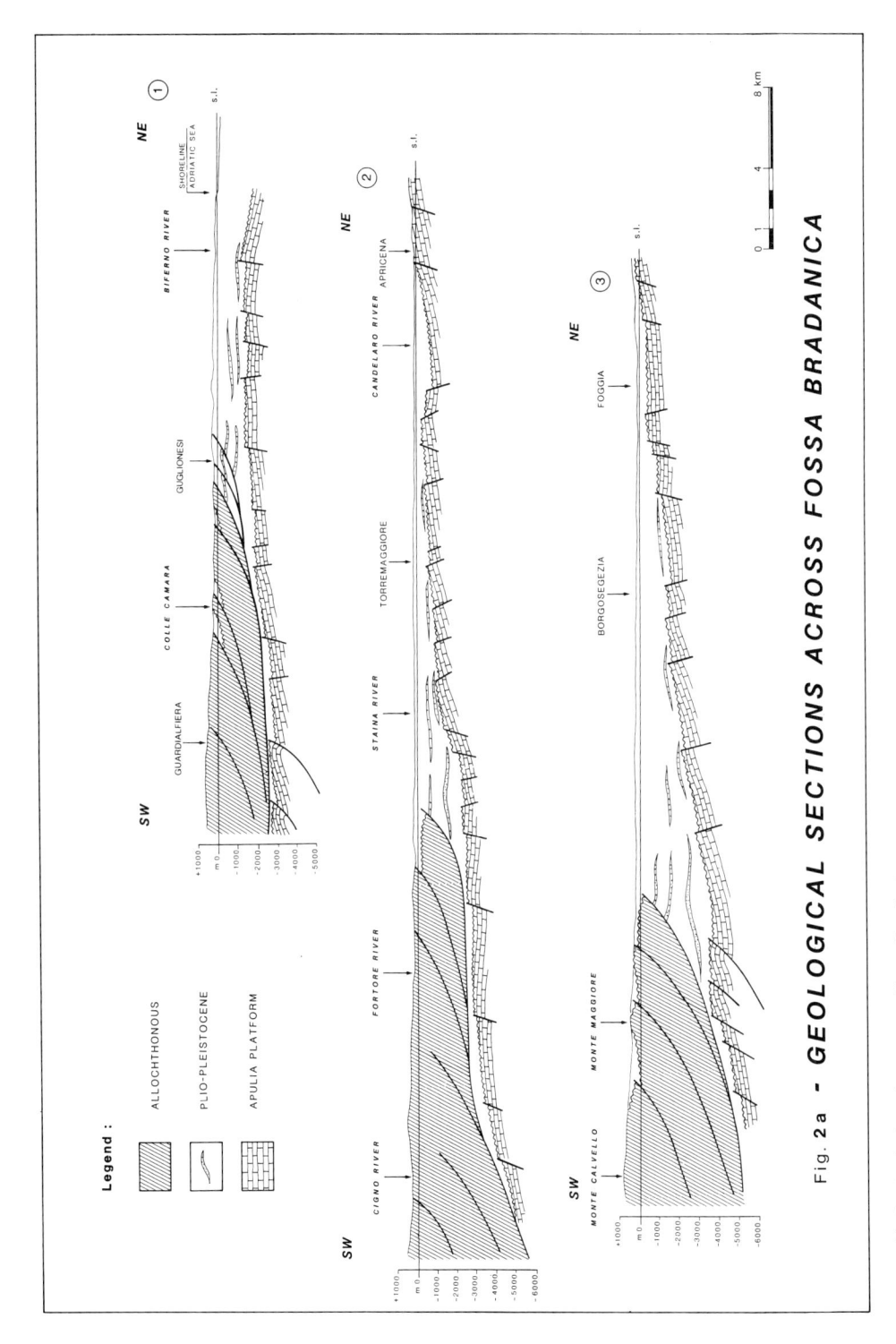

PETROLEUM GEOLOGY OF THE 'FOSSA BRADANICA' 371

Fig. 2a - *GEOLOGICAL SECTIONS ACROSS FOSSA BRADANICA*

Fig. 2. (a) Geological sections across Fossa Bradanica.

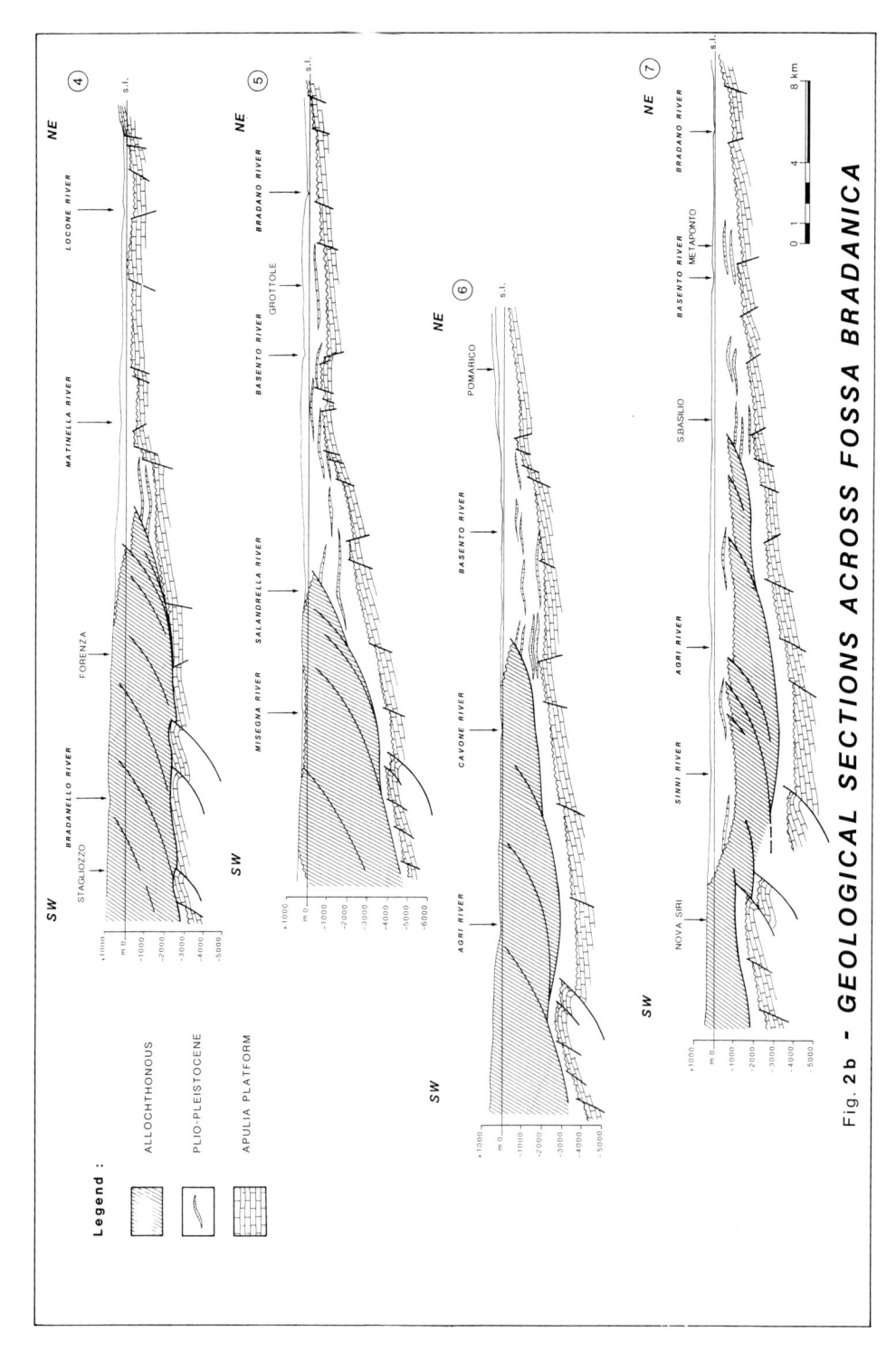

Fig. 2. (b) Geological sections across Fossa Bradanica.

PETROLEUM GEOLOGY OF THE 'FOSSA BRADANICA'

combination traps and to the location of the petroleum accumulation in the stratigraphic column. In some cases, as in Pisticci (Fig. 4), more than one single trapping geometry does exist in the same field.

Pre-Pliocene substratum

Structural traps: thrust features in the carbonate substratum below the Allochthonous thrust sheet (Fig. 5a)

Fields. Two oil fields, Monte Taverna and Strombone; their assessment is still in progress.
This trapping configuration has been found in the more 'external' sector of the Apennine thrust belt, where the carbonate substratum appears to have been involved in the compression tectonics phase.

Structural trap: horsting in the carbonate substratum close to the Allochthonous outer front (Fig. 5b)

Fields. One major oil producing field, Pisticci (Fig. 4) and one gas field, Melanico. (The second major oil field in Fossa Bradanica, Torrente Tona, is producing from Pliocene.)
These features represent in Fossa Bradanica the most 'internal' positive structural trends originated by extension faulting.

Structural trap: horsting in the eastern sector of Fossa Bradanica (Fig. 5c)

Fields. Two gas fields are on stream, the major one is Grottole–Ferrandina. These positive structural trends, usually oriented NW–SE, represent the easternmost limits of prospective acreage in Fossa Bradanica.
In a more 'internal' position, towards the outer front of the Allochthonous sheet, horsting in the substratum is uncommon. Only three small producing gas fields have this configuration.

Structural trap: tilted fault blocks in the central sector of Fossa Bradanica (Fig. 5d)

Fields. One producing gas field, Mezzanelle, and two oil fields, Orsino and Masseria Pepe, currently under evaluation.
This trapping geometry is peculiar to structures located in the central sector of Fossa Bradanica.

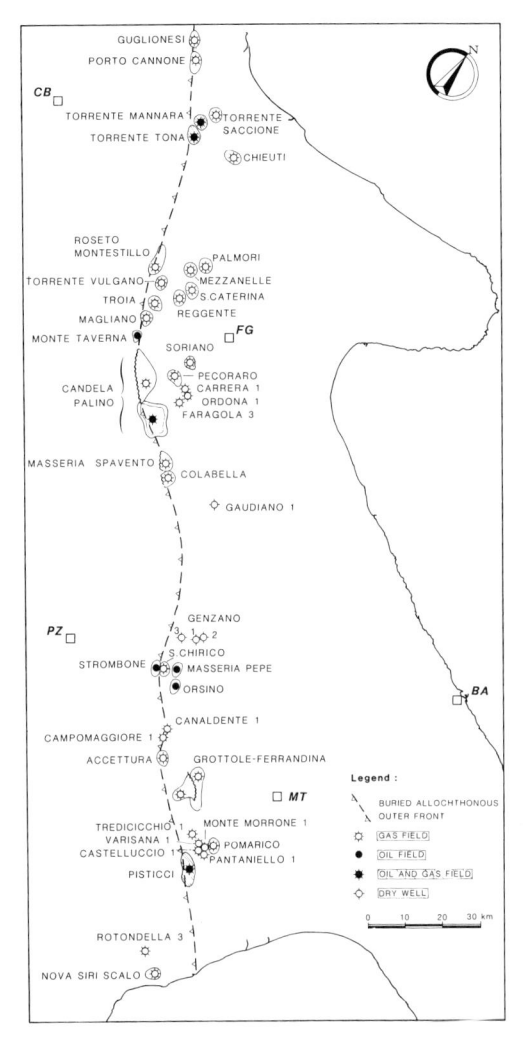

Fig. 3. Gas and oil fields in Fossa Bradanica.

poor; Pisticci (Upper Cretaceous) and Torrente Tona (Middle Pliocene) are the only oil fields in production. A few fields contain sour gas (CO_2 exceeds 2%), which requires special production treatments. Also, a number of gas pools are quite shallow, hence the flowing pressure is low and pumping stations are often needed to tie the wells to the existing gaslines network.

Petroleum trapping configurations in Fossa Bradanica

A variety of petroleum trapping configurations have been recognized in the Fossa Bradanica and have been classified according to the classic subdivision into structural, stratigraphic and

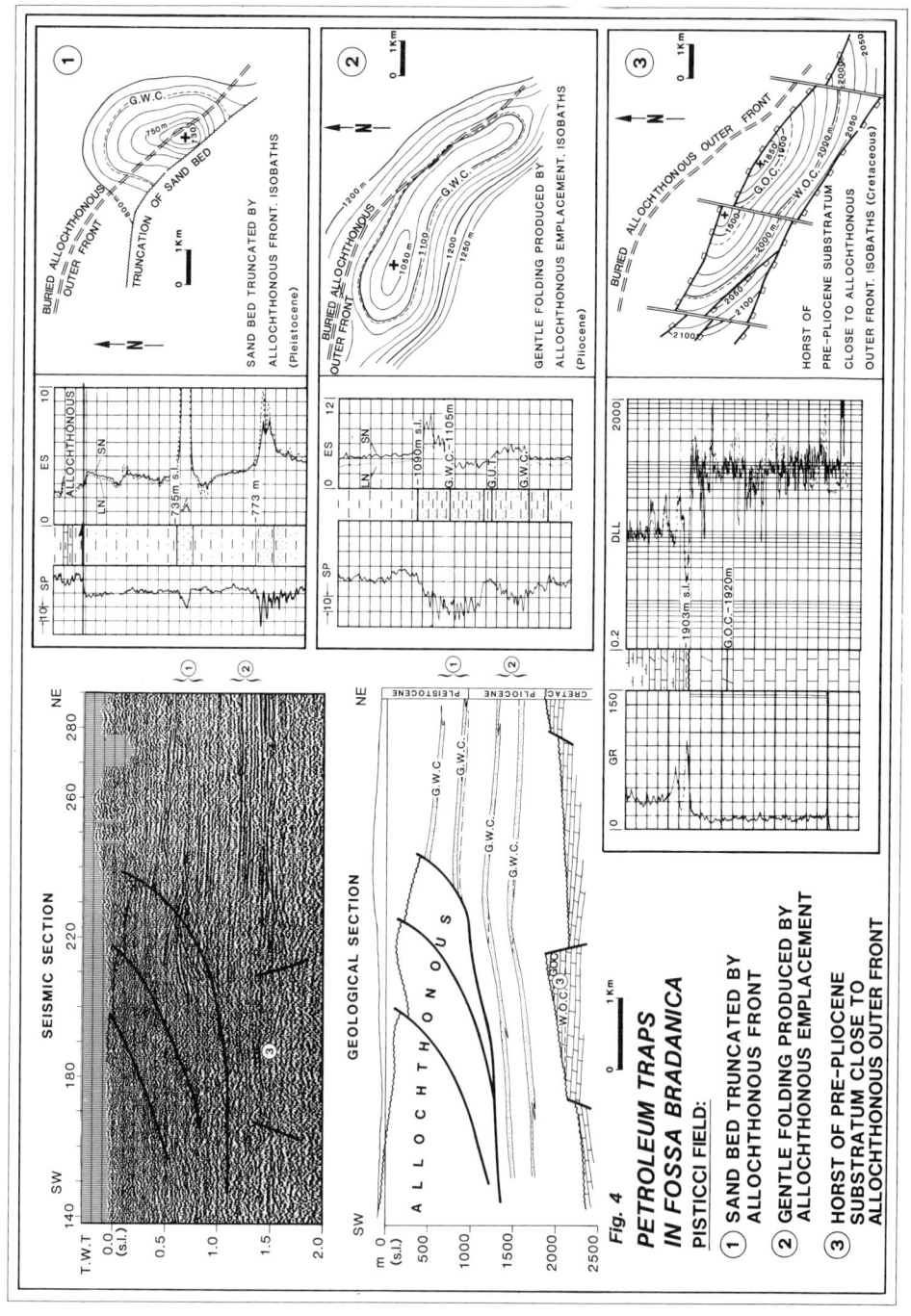

Fig. 4. Petroleum traps in Fossa Bradanica: Pisticci field: (1) sand bed truncated by Allochthonous front; (2) gentle folding produced by Allochthonous emplacement; (3) horst of pre-Pliocene substratum close to Allochthonous outer front.

Stillo, while the third one, of major importance, Torrente Tona, is gas and oil bearing.

Traps of this type are located in the northern sector of Fossa Bradanica, in front of the eastern termination of the Allochthonous thrust sheet.

Structural trap: sand bed truncated by the Allochthonous thrust surface (Fig. 6b)

Fields. Six gas fields in production, the most important being Candela. This trapping geometry is very common all over Fossa Bradanica, but the size of the fields discovered, with the exception of Candela, is limited.

Structural trap: gentle folding produced by Allochthonous emplacement (Fig. 6c)

Fields. Six fields in production, including the majors Candela–Palino and Pisticci (Fig. 4); another one will be soon on stream.

Trapping configurations of this type have been found close to the front of the Allochthonous sheet or just below it. In some cases trap efficiency and volume are enhanced by shaling out of the reservoir bed.

Structural trap: draping of Pliocene beds upon pre-Pliocene high (Fig. 6d)

Fields. Eight gas fields in production.

This type of trap is quite common. It is originated by synsedimentary tectonics and/or differential compaction. It has often the characteristics of a combination trap; the reservoir sand is in fact concordant to the underlying carbonate high, but the closure is controlled to some extent by shaling out of the reservoir.

Stratigraphic trap: pinching out of sand bed (Fig. 6e)

Fields. Two gas fields, Torrente Saccione already in production and Soriano, a recent new discovery.

This trapping geometry is created by a gradual updip shaling out of the sand beds.

Stratigraphic trap: sand onlapping basal marls and shales (Fig. 6f)

Fields. Six gas fields, one already in production, the other to be put on stream soon.

This type of trap is common in the eastern sector of Fossa Bradanica, where Middle Pliocene sand bodies onlap onto impervious basal marls and shales.

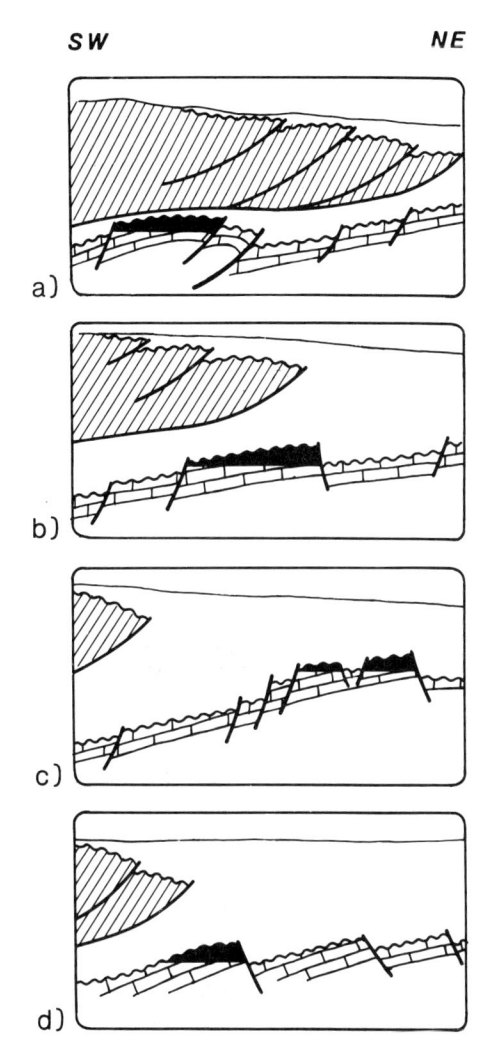

Fig. 5. Models for petroleum traps: Pre-Pliocene substratum. (a) Structural trap: thrust features within the Appennine thrust belt underlying the Allochthonous cover. (b) Structural trap: horst of pre–Pliocene substratum close to Allochthonous outer front. (c) Structural trap: horst of pre-Pliocene substratum in the eastern Fossa Bradanica. (d) Structural trap: tilted fault blocks in the central Fossa Bradanica.

Pliocene

Structural traps: thrust feature in the Pliocene sequence and fault truncated Pliocene sands (Fig. 6a)

Fields. Three fields are producing; two are gas bearing, Porto Cannone and Roseto–Monte

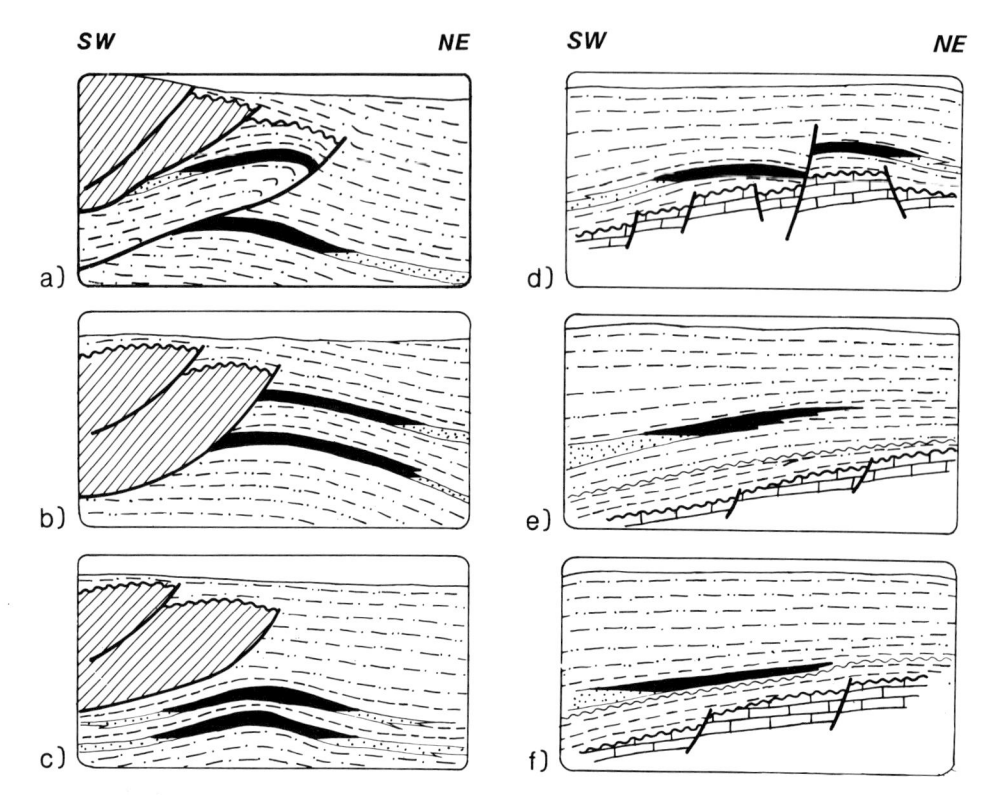

Fig. 6. Models for petroleum traps: Pliocene. (a) Structural traps: thrust in the Pliocene sequence and fault truncated Pliocene sand. (b) Structural trap: sand bed truncation by Allochthonous front (the same model occurs in the Pleistocene). (c) Structural trap: gentle folding produced by Allochthonous emplacement. (d) Structural trap: draping of Pliocene beds upon pre-Pliocene high (the same model occurs in the Pleistocene). (e) Stratigraphic trap: shaling out. (f) Stratigraphic trap: sand onlapping basal marls and shales.

Pleistocene

Structural trap: sand bed truncated by Allochthonous thrust surface (Fig. 6b)

Fields. One field only, Pisticci (Fig. 4).

This trap has already been commented in the previous paragraph as a Pliocene feature. In Southern Fossa Bradanica, traps of this type are present also in the Pleistocene because the Allochthonous sheet is here overthrust upon the Pliocene sequence.

Structural trap: draping of sand beds upon pre-Pliocene high (Fig. 6d)

Fields. One field in the Pleistocene, Grottole–Ferrandina.

This type of trap has been described before because it also occurs as a Pliocene feature.

Stratigraphic trap: turbidity fans (Fig. 7a)

Fields. Eight fields are on stream.

In Southern Fossa Bradanica, at the end of the Pliocene but mainly during the Pleistocene, basin morphology and synsedimentary tectonics have favoured the deposition of turbidites. Turbidity currents originated either from the Allochthonous front outcropping in the West, or from emerging highs within the basin itself. The turbidity fans mentioned here are among the most common deposits in the Southern sector of Fossa Bradanica and tend to form stacked, individually sealed, sand reservoirs.

Semi-allochthonous Pleistocene (in the semi-allochthonous Plio–Pleistocene basin SW of Taranto)

Structural trap: draping upon Allochthonous ridge (Fig. 7b)

Fields. One field only, Nova Siri Scalo. This is a new discovery under evaluation.

The Allochthonous sheet is made up of thrust blocks, which are buried under Pleistocene deposits. Trapping conditions are originated by

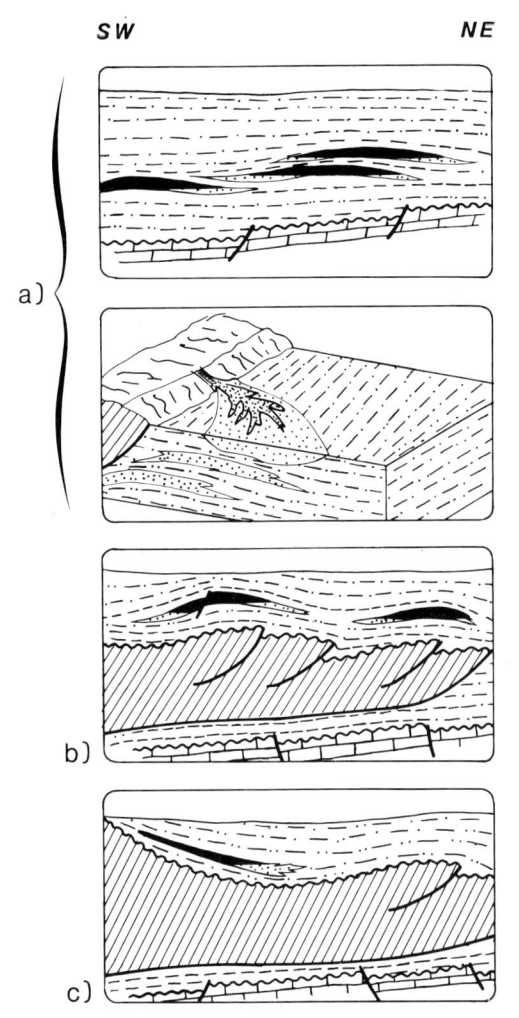

Fig. 7. Models for petroleum traps: Pleistocene. (a) Stratigraphic trap: turbidity fan originating from Allochthonous ridge or from pre-Pliocene substratum high. Semi-Allochthonous Pleistocene: (b) Structural trap: draping upon Allochthonous ridge. (c) Stratigraphic trap: shaling out towards boundary of semi-Allochthonous basin.

gentle draping of Pleistocene sand beds upon the Allochthonous thrust structures.

Stratigraphic trap: shaling out towards the boundary of the basin (Fig. 7c)

Fields. One field only, Rotondella, not on stream.

In the direction of the termination of the Taranto semi-allochthonous basin, where the Allochthonous substratum outcrops, sand beds tend to shale out or to be truncated by new sedimentary cycles, which creates a trapping geometry.

Petrophysical properties of reservoirs in Fossa Bradanica

Plio–Pleistocene reservoirs are represented by the sandy portions of turbidity deposits. As expected in this type of sediments, the petrophysical characteristics of the reservoirs are extremely variable, with porosities ranging between 30% in clean sands and 10–15% in more shaly intervals.

Because the clastics source was mainly in the West, from the Allochthonous outcrops, the sand beds have a tendency to shale out eastwards. In the easternmost sector of Fossa Bradanica producing reservoirs are in fact made up of very thin silt beds.

Carbonate reservoirs have very low average porosity values (1–5%). Economic production can therefore only be achieved in presence of important fracture systems. More so if the hydrocarbon to be produced is oil.

Petroleum geochemistry considerations

Geochemical characteristics of the hydrocarbons

Oils produced in Fossa Bradanica fall within a wide density range, from 3° to 45° API, but their average density is 15–20° API. The high density is explained by early generation and expulsion from the source and, in one case only, by biodegradation (Pisticci).

A high sulphur content has been observed in oils of a number of fields (Pisticci, Torrente Tona, Orsino: 4–8%) and has been related to sources deposited in highly reducing environments.

The gas discovered is mostly dry gas. Nitrogen and carbon dioxide are locally present in high concentration. The presence of these components is related to the volcanism which has been active in the central area of Fossa Bradanica in Pleistocene times. Only in few wells sulphydric gas has been observed.

Origin of oil and gas in Fossa Bradanica

Oils correlation studies point to the presence of two major families.
Group 1: (Candela–Palino and Soriano fields). These oils have been originated from a shaly source deposited in an oxygenated environment. The organic matter is of continental type.

They are all quite mature oils.

Group 2: (Torrente Tona, Torrente Mannara, Orsino, Strombone, Masseria Pepe, Pisticci). These oils have been originated in a reducing carbonatic environment, from an organic matter of marine type, with some continental influence. They are characterized by early generation and expulsion from their source. These oils can therefore be classified as non-mature or only marginally mature oils. Biodegradation is evident in the Pisticci oil only. Within this group minor differences have allowed a further subdivision into four subgroups.

In Italy, 80% of gas reserves are located in foredeep regions, 10% in fold/thrust belts, 10% in foreland domains. In Fossa Bradanica 79% of the discovered gas is of biogenic origin, i.e. generated under the influence of bacterial and/or thermochemical activity, at low temperature and shallow depth. Thermogenic gas, obtained by cracking of the organic matter, at high temperature, is uncommon (6% only). 15% of the gas reserves are considered to be of a mixed origin. The low percentage of thermogenic gas is explained with the low geothermal gradient. Some fields contain gas of one type only, either biogenic or thermogenic (Pieri & Mattavelli 1986). In some cases, as in Candela field, a transition has been observed from thermogenic gas, in the deepest reservoirs, to mixed gas in the intermediate and finally to biogenic gas in the more shallow accumulations.

High concentrations of carbon dioxide

CO_2 can be found both in carbonate and Pliocene reservoirs. The highest concentrations are in the carbonates, where they can be as high as 90%. A high CO_2 content can also be found in the basal levels of the Pliocene sequence. In Pleistocene sands, CO_2 has been found in one well only. In the case of a multipay, CO_2 concentrations has a tendency to decrease upwards, as in Candela−Palino and Torrente Tona fields.

The highest CO_2 concentrations are located in the central sector of Fossa Bradanica and are related to the presence of the Vulture volcanic complex of Pleistocene age. It is accepted that CO_2 is in Fossa Bradanica mainly of volcanic origin.

The presence of CO_2 is a major problem in gas production. Gas with a CO_2 content in excess of 2% is considered a sour gas and must be cleaned upstream. At the moment, only three fields are produced in Fossa Bradanica through gas cleaning facilities.

The concepts expressed in this paper are based on ideas developed over the years by many AGIP geologists and geophysicists. The authors are indebted to AGIP S.p.A. management for their permission to publish this paper. They also wish to thank SELM for authorization to divulge data of the Candela Field. Finally, the authors thank Dr F. Frigoli of AGIP S.p.A. for his helpful assistance and criticism.

References

BALDUZZI, A., CASNEDI, R., CRESCENTI, U. & TONNA, M. 1982. Il Plio−Pleistocene del sottosuolo del bacino pugliese (Avanfossa Appenninica). *Geologica Romana*, 21, 1−28.

——, ——, —— & —— 1982. Il Plio−Pleistocene del sottosuolo del bacino lucano (Avanfossa Appenninica). *Geologica Romana*, 21, 89−III.

CASNEDI, R., CRESCENTI, U., D'AMATO, C., MOSTARDINI, F. & ROSSI, U. 1981. Il Plio−Pleistocene del sottosuolo molisano. *Geologica Romana*, 20, 1−42.

——, ——, & TONNA, M. 1982. Evoluzione dell'avanfossa adriatica meridionale nel Plio−Pleistocene, sulla base dei dati di sottosuolo. *Bollettino della Società Geologica Italiana*, 24, 243−260.

CRESCENTI, U. 1975. Sul substrato pre−pliocenico dell'avanfossa appenninica dalle Marche al Mar Ionio. *Bollettino della Società Geologica Italiana*, 94, 583−634.

IPPOLITO, F., D'ARGENIO, B., PESCATORE, T. & PESCATORE, P. 1975. Structural-stratigraphic units and tectonic framework of Southern Appennines. *In*: SQUYRES, C. (ed.) *Geology of Italy*. Earth Sciences Society of Lybian Arab Republic, 317−328.

MOSTARDINI, F. & MERLINI, S. 1986. Appennino Centro Meridionale. Sezioni geologiche e proposta di modello strutturale. *Memorie della Società Geologica Italiana*, 35, 177−202.

PIERI, M. & MATTAVELLI, L. 1986. Geologic Framework of Italian Petroleum Resources. American Association of Petroleum Geology Bulletin, 70, 103−130.

RICCHETTI, G. 1980. Contributo alla conoscenza strutturale della Fossa Bradanica e delle Murge. *Bollettino della Società Geologica Italiana*, 99, 420−430.

ROSSI, M. E. 1986. Sequence stratigraphy and geometric characteristics of the Apenninic foredeep basin in the subsurface of the Torrente Tona oilfield area (Central Italy). *Memorie della Società Geologica Italiana*, 35, 139−155.

Zechstein reservoirs in The Netherlands

DIETERT VAN DER BAAN

Petroleum Development Oman LLC, P.O. Box 81, Muscat, Sultanate of Oman

Abstract: Thick, shallow-water gypsum banks, formed during Zechstein times at the site of regional palaeohighs along the margins of the Southern Permian Basin, acted as foundations for shallow-water carbonate platforms. The carbonate platform margins, in particular their northern to northeastern windward margins, were characterized by deposition of grainstones dominated by aragonitic ooids, pisoids, intraclasts and bioclasts, while muddy, stromatolitic carbonates and nodular anhydrites accumulated on tidal flats and in lagoons behind the platform margins.

Usually the platforms are surrounded by thick slope deposits consisting of carbonate mudstone and redeposited intervals, but locally the platforms deepened gradually and changed, without intervening slope deposits, into a shallow-marine facies devoid of ooids. The slope aprons grade basinwards into thin successions of finely laminated carbonate.

Porosity creation in Zechstein carbonates was due to meteoric-water leaching and to subsurface leaching by carbon dioxyde-enriched formation waters during burial. Porosity destruction was mainly caused by anhydrite pore plugging and subsurface calcitization. The best reservoir properties are found in the unrestricted platform facies and in parts of the slope facies. Fractures are important in Zechstein carbonates; their presence is a prerequisite for high production rates in gasfield wells.

Commercial quantities of hydrocarbon gas in Dutch Zechstein reservoirs were discovered in 1948 by the Nederlandse Aardolie Maatschappij (NAM) in the East Netherlands. The first gasfield, the Coevorden field, was soon followed by the discovery of other gasfields, all in the same general area (Fig. 1). The gas is sourced from Upper Carboniferous coal beds and contains variable amounts of hydrogen sulphide. In one well 38 wt % hydrogen sulphide has been measured but usually the sulphur content measurements range from 0.006 wt % to 3.8 wt %. The sulfur is extracted from the gas in a recently commissioned desulphurization plant and is marketed as a commodity. Table 1 lists the cumulative reserve and production figures from the East Netherlands fields.

Although the cumulative gas reserves in dutch Zechstein reservoirs constitute but a small fraction of the total Dutch gas reserves (due to the large contribution of the giant Groningen gasfield, whose gas is stored in Upper Rotliegend clastic reservoirs), the Zechstein remains a target of active exploration in the Netherlands and new discoveries continue to be made.

Stratigraphy and geological setting

The Zechstein Group in western Europe is composed of cyclic successions of claystone, carbonate, anhydrite, halite, and potash salt. The number of cycles varies, depending on the relative position within the Zechstein basin. The deepest, central part of the basin, is characterized by up to seven cycles, of which the upper three cycles do not contain carbonate beds; this part is found in the subsurface of northern Germany and southern Denmark (Best 1987; Kulick & Paul 1987). In England, up to five cycles have been recognized (Smith *et al.* 1986), whereas in Poland and in the Netherlands four cycles have been observed.

In the Dutch stratigraphic nomenclature, formation status is assigned to the total of all deposits that formed in a single cycle, and member status to the lithologically distinct units. The Dutch subdivision (Fig. 2) follows the German subdivision.

The cycle subdivision cannot be upheld in the predominantly clastic basin fringe areas. In these areas the lower part of the lateral equivalent of the Zechstein Group is named the Fringe Zechstein Formation. The upper part is lithologically undistinguishable from the overlying Bunter Formation and is included in the latter.

The Dutch Zechstein forms part of the infill of the Southern Permian Basin (Fig. 3). This basin formed as a post-orogenic foreland collapse basin in the aftermath of the Variscan orogeny, probably in response to a decaying thermal anomaly (Ziegler 1982). During deposition of the Upper Rotliegend the basin was the site of a large desert lake, surrounded by

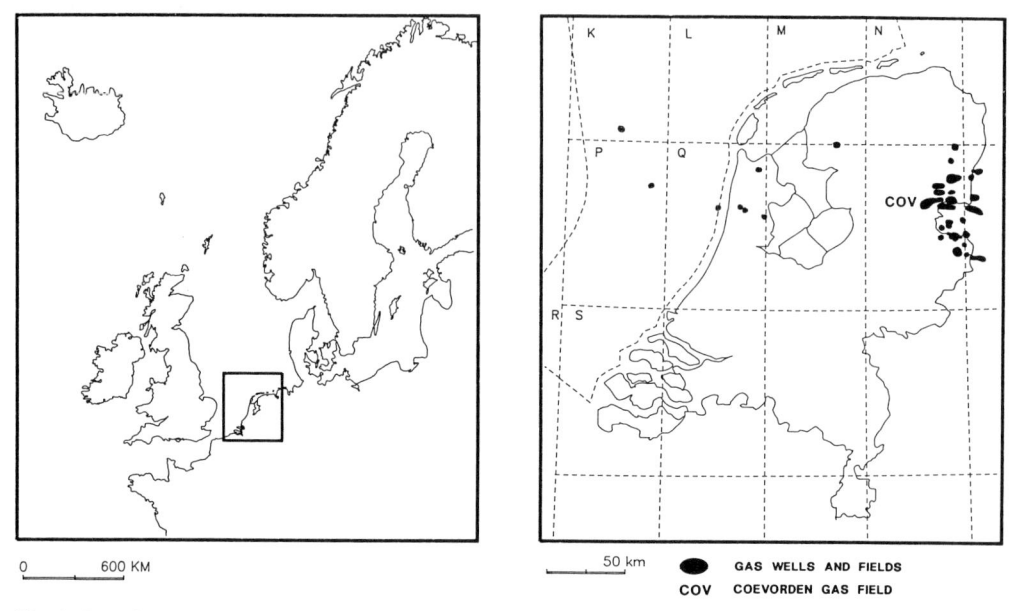

Fig. 1. Location map of gas in Zechstein reservoirs, The Netherlands.

dune fields and wadis (Glennie 1972). This desert lake was situated several hundreds of metres below global sea level. In the Late Permian the basin was very rapidly flooded, probably because of a eustatic sea-level rise due to a Gondwana deglaciation cycle. As a result, deposition switched from continental clastics to marine claystone and later to marine carbonates.

As the global sea-level fell, probably reflecting a Gondwana glaciation cycle, the basin became increasingly isolated from outside

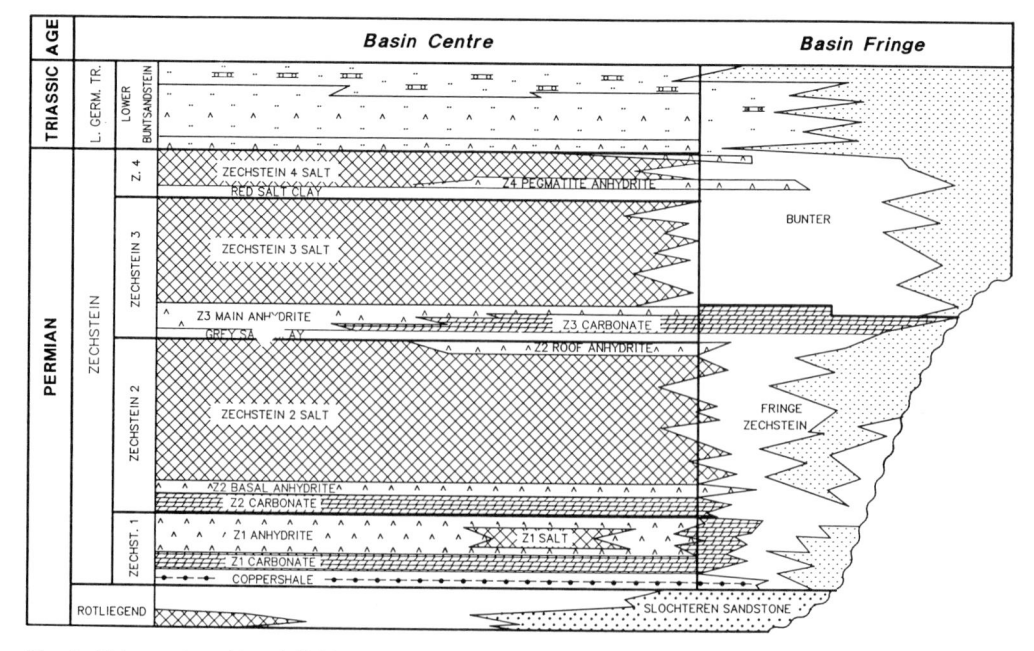

Fig. 2. Lithostratigraphic subdivision of the Zechstein Group in the Netherlands (NAM & RGD 1980).

Table 1. *Gas reserves and production from Zechstein reservoirs, East Netherlands*

Gas in place	80×10^9 m^3
Ultimate recovery	60×10^9 m^3
Already produced	20×10^9 m^3
Daily production	8×10^6 m^3

The figures are approximate values. They refer to the situation at 1.1.1988

marine environments, and deposition switched to the sulphates and rock salts of the first Zechstein cycle.

The overlying Zechstein cycles reflect similar eustatic sea-level movements. During the third Zechstein cycle the sea-level rose particularly high, and the Zechstein basin achieved its maximum marine extension.

The fourth and subsequent Zechstein cycles show a progressive infilling of the basin, associated with a predominance of playa-type halite and potash salt deposits in the basin centre, and fine-clastic redbeds along the wide, flat basin margins (Fig. 4).

General aspects of carbonate/evaporite sedimentation

Of the many factors that contribute towards the development of facies patterns in carbonate/evaporite systems, water depth is probably the

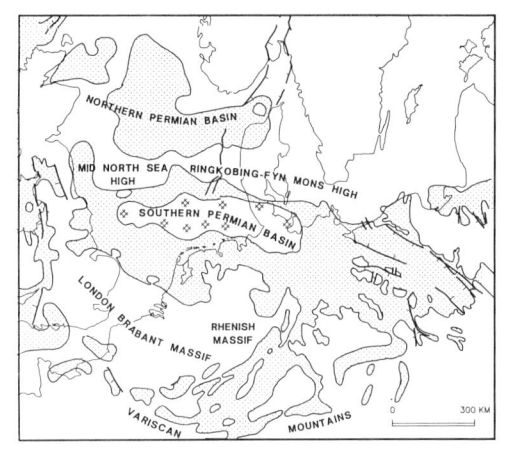

Fig. 3. Upper Rotliegend palaeogeography of northwestern Europe, showing the Southern Permian Basin desert lake (after Ziegler 1982).

most important. Shallow water environments in evaporitic settings are characterized by maximum deposition rates that are orders of magnitude higher than those in deeper water environments (see Table 2).

There are two reasons for this difference. Firstly, evaporation in shallow to very shallow

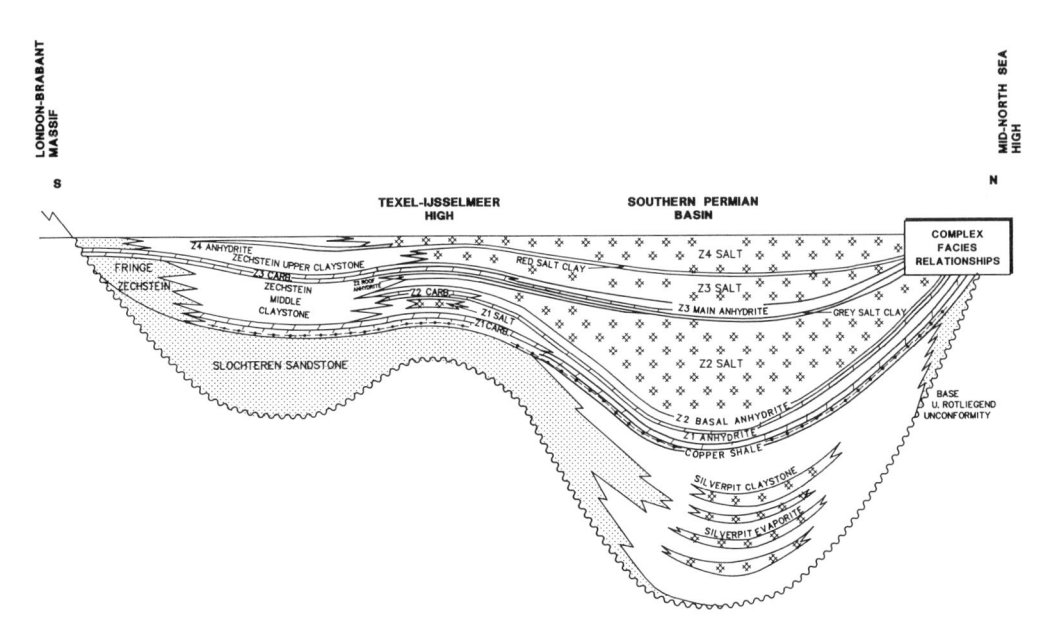

Fig. 4. Conceptual cross section through the Southern Permian Basin. Complex facies relationships in the north are due to intersection of faultblocks with the Mid-North Sea High.

Table 2. *Maximum observed deposition rates of evaporites*

Sediment	Environment	Max. Dep. Rate
Carbonate + Sulphate	sabkha	1 m/1000 y
Gypsum	salina + solar pond	40 m/1000 y
Gypsum	deep basin	0.3 m/1000 y
Halite	playa	100 m/1000 y

Slightly modified after Schreiber & Hsü (1980).

water environments causes relatively fast nucleation and growth of evaporite minerals. Secondly, the tiny gypsum crystals formed by evaporation at the water surface stand a much better chance of growing or becoming preserved in the predominantly oxic environments of shallow-water settings than in the stratified oxic/anoxic environment of deeper water settings.

In the latter, populations of sulphate-reducing bacteria thrive along the chemocline, the oxic/anoxic boundary. Their metabolic activities cause dissolution of the small gypsum crystals as they settle downward and pass through the chemocline. As a result, these crystals cannot reach the basin floor. This applies to conditions of a stable chemocline. The chemocline is destroyed, however, when oxygenated surface waters mix with underlying anoxic waters, for instance during storms or seasonal turnovers. The oxygen kills the populations of sulphate-reducing bacteria and the small gypsum crystals can reach the basin floor and form a thin lamina. Single laminae of gypsum in deeper water environments of evaporite basins reflect, therefore, a weather-related event of relatively short duration, because the chemocline is restored relatively fast. Richter–Bernburg (1960, 1963) concluded that deeper water evaporite laminites represent seasonal deposits comparable to varvites, and based on this interpretation the value for maximum gypsum deposition in deep basinal environments can be calculated (Table 2).

The difference between the evaporite deposition rates in shallow water and deeper water environments plays a major role in shaping the history of evaporite basins, because it is responsible for the development of thick gypsum platforms in shallow water environments along the basin margins and on top of offshore shoals.

In recent coastal salina environments, large selenite crystals are known to grow within the laminated gypsum mush that forms the bottom sediment, and as selenite meadows on the salina floor (Warren 1982, 1985). These rigid masses of fast growing, interlocking selenite crystals can probably develop in any oxic shallow water environment that is perenially saturated with respect to gypsum. Undoubtedly they are able to form gypsum banks that grow in the zone of wave action where their debris will behave just like other detrital sediment and produce deposits as beaches, spits and offshore bars.

In such a way, large sulphate platforms composed of selenite crystals, discoidal and detrital gypsum, and nodular anhydrite, can rapidly form.

Zechstein sedimentation and palaeostructures

During the Zechstein, large sulphate platforms developed along the basin margins. They acted as foundations upon which, during later Zechstein cycles, shallow marine carbonate platforms developed. The carbonate platforms and their adjacent slopes that formed during the second and third Zechstein cycles provide the reservoirs of all Dutch Zechstein gasfields.

The search for these platforms and slopes is greatly facilitated by the fact that their geographic distribution coincides with that of the underlying sulphate platforms, and by the fact that the seismic expression of the latter is usually far better — due to the thickness of their deposits (up to 280 metres) — than that of the overlying carbonate platforms.

The search for good reservoirs in the Zechstein 2 and 3 Carbonate Members is, therefore, in practice often the same as the search for thick platform and slope deposits in the Zechstein 1 Anhydrite Member. The younger Zechstein Anhydrite members (see Fig. 2) cannot be used for this purpose as they are much thinner. The thick platform deposits of the Zechstein 1 Anhydrite Member developed in shallow water environments associated with regional palaeohighs along the margins of the Zechstein Basin. These palaeohighs were produced during the Variscan orogeny, although in many cases they reflect older, probably Caledonian structures.

During the Permian, two structural trends predominated in the northeastern Netherlands: a northwest–southeast trend and a roughly north–south trend. Both are expressed in the shape of an outlier of the Rhenish Massif that extends into the Netherlands (Fig. 3). The three main structural elements of this outlier are shown in Fig. 5: the Ems Low, the East Netherlands High, and the Proto Texel–IJsselmeer High. Their control on the facies development of the Zechstein is evident from the palaeogeographic map shown in Fig. 6.

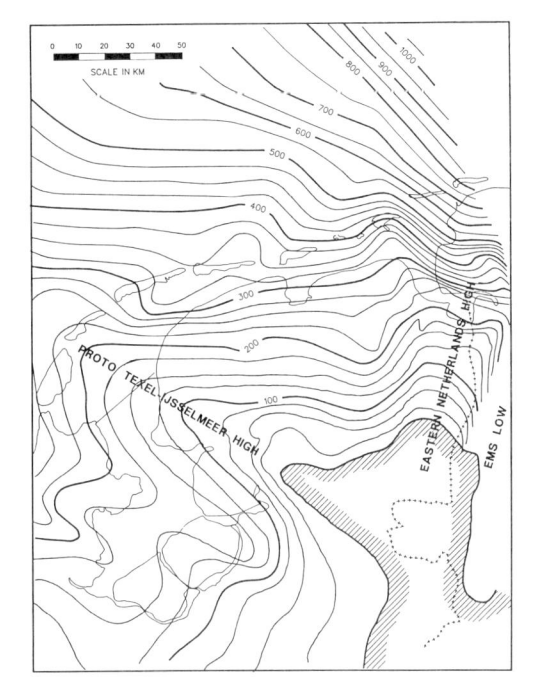

Fig. 5. Total thickness map of Upper Rotliegend deposits, northeastern Netherlands, showing important palaeostructures that controlled Zechstein sedimentation.

The Ems Low is a Variscan graben that probably reflects a much older structure since it lies on the trend of an important Devonian regional structure, situated more to the south. During the Early Permian, the Ems Low became partly filled with thick volcanic deposits, but remained a palaeotopographical low. During Zechstein deposition, the Ems Low was the site of predominantly deeper water sedimentation.

The north-trending, uplifted and slightly tilted faultblock that bounds the Ems Low to the west, and acted as graben 'shoulder', constitutes the East Netherlands High. This faultblock, which is transected by faults of pre-Permian to Early Permian age — including east–west striking normal faults — formed the substrate on which several sulphate and carbonate platforms developed (Fig. 7).

The Proto Texel–IJsselmeer High is a regional northwest–southeast trending swell, bounded and transected by pre-Permian or Early Permian faults. Towards the southeast, the high intersects and merges with the East Netherlands High, producing a triangular-shaped plateau (Fig. 5) that underlies the

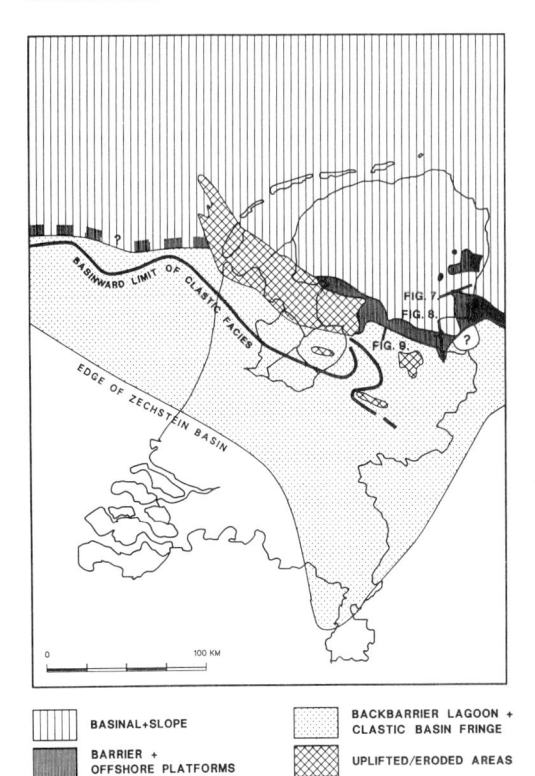

Fig. 6. Simplified palaeogeographic map of the Zechstein 2 Carbonate. German part based on Richter–Bernburg (1985).

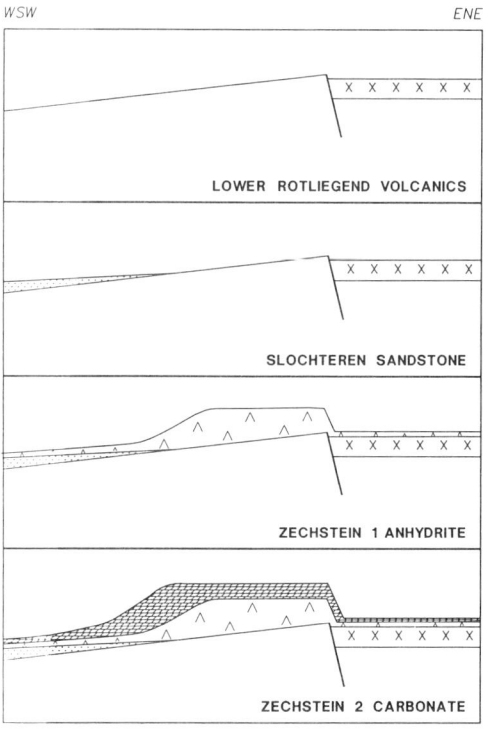

Fig. 7. Diagrammatic cross-sections showing development of shallow water sulphate and carbonate platforms on top of the East Netherlands High. Not to scale. See Fig. 6 for location.

prolific Coevorden gasfield. Towards the northwest, just off the Dutch coast, the high looses its expression. During the Mesozoic part of the Proto Texel–IJsselmeer High was uplifted and much of its cover, including the Zechstein Group, was removed by erosion.

During Zechstein times, an offshore barrier composed of sulphate and carbonate deposits developed on the Proto Texel–IJsselmeer High. This barrier was associated with a wedge of thick slope deposits along its northern margin (Fig. 8). The back-barrier area to the south was characterized by local subsidence due to syndepositional faulting, which resulted in local thickening of sulphate and carbonate successions and in the occurrence of halite lenses (Fig. 9). These salt lenses can be correlated to the thick (more than 200 metres) Zechstein 1 Salt successions of the Lower Rhine Basin, located towards the southeast in Germany (Richter–Bernburg 1985).

Figure 7–9, which are based on good quality seismic sections, show that in The Netherlands very thick (i.e. seismically detectable) anhydrite successions characteristically occur in two settings: in platform settings and in backbarrier settings.

The belts of thick anhydrite along the Zechstein basin margins are known in Germany as 'Anhydritwall' or 'Sulphat Wall' (Richter–Bernburg 1985), and their distribution is, like in The Netherlands, intimately linked to basement ridges (Rockel & Ziegenhardt 1979; Piske & Schretzenmayr 1984; Richter–Bernburg 1986). Carbonate platforms and offshore shoals, similar to those of the East Netherlands, have been described from Poland (Peryt 1977), Denmark (Clark & Tallbacka 1980), Germany (Sannemann et al. 1978; Paul 1980) and England (Smith 1980). In some of these descriptions terms such as 'shelf' and 'reef' have been used rather than 'platform' and 'offshore shoal'.

These carbonate platforms, although different in detail, share common characteristics. The

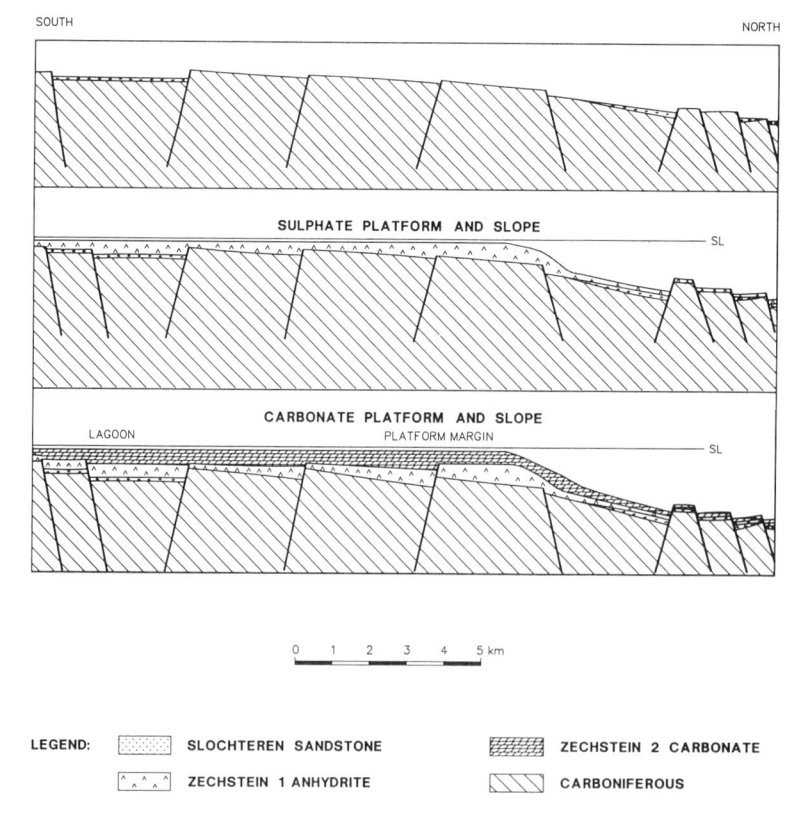

Fig. 8. Sections showing sulphate and carbonate development across the north flank of the Proto Texel–IJsselmeer High. Note thick sulphate development at platform margin and at syndepositionally subsiding intra-platform areas. See Fig. 6 for location.

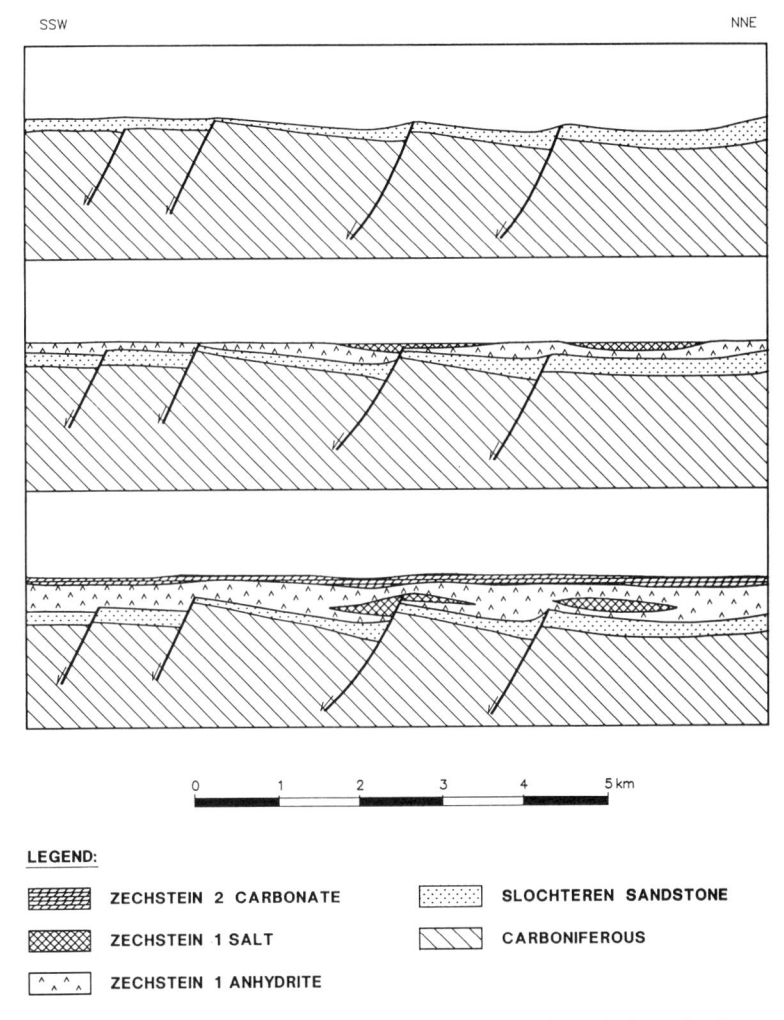

Fig. 9. Sections showing the effects of syndepositional faulting on sulphate platform development in the backbarrier area immediately south of the crest of the Proto Texel–IJsselmeer High. See Fig. 6 for location.

most important of these is their response to the dominant Permian palaeowind (northeast tradewind) direction. Leeward platform margins appear, as a rule, to be characterized by very gentle progradational slopes with dips of less than 1.5 degrees. Windward margins, on the contrary, are often better detectable on seismic sections as their slopes can be steeper; occasionally they are formed by vertical escarpments of more than 200 metres high. Windward margins are furthermore characterised by the occurrence of a belt of very porous carbonates that appears to be absent along leeward platform margins.

Depositional model of Zechstein carbonates

A combination of much core material, many well penetrations and good quality 3D seismic data was used to construct a depositional model for Zechstein carbonates in The Netherlands.

This model is very similar in outline to earlier published models from various parts of Europe (Clark 1980 a, b; Antonowicz & Knieszner 1981; Paul 1987). This similarity is considered an indication that uniform patterns existed, basinwide, during Zechstein carbonate sedimentation.

The model is based on an area in the East Netherlands that is characterized by a barrier (including offshore platforms), slope and basin palaeogeography (Fig. 6), and shows four depositional facies (Fig. 10): basin, slope, shallow marine and platform. In the back-barrier and basin fringe area more to the south, two additional depositional facies can be recognised: deeper lagoon and arid flood-plain/playa. The main characteristics of these six facies are discussed below. The arid floodplain/playa facies (Fig. 11) is characterized by successions of red-brown and grey-green claystones and siltstones, intercallated with streaks of nodular anhydrite. Sheet flood deposits and dessiccation cracks are common (Fig. 12). The facies becomes more sandy towards the west, where a major source of clastic supply existed, just off the dutch coast (Van Adrichem Boogaert & Burgers 1983).

The deeper lagoon facies consists of monotonous, well bedded dolomite mudstones that are devoid of fossils. The facies is developed in the back-barrier area between the shoreline and the carbonate platform barrier (see Fig. 6). During Zechstein times, this area was probably comparable to the present-day Khor al Bazm lagoon behind the Great Pearl Bank barrier in the Arabian Gulf (Purser 1973). The platform facies has been subdivided in a restricted platform facies and an unrestricted platform facies.

The restricted platform facies is composed of sediments that accumulated on supratidal flats and in very shallow lagoons (Fig. 14). They include supratidal pavement breccias (Figs 15, 16), storm layers (Fig. 17), calcrete hardpan crusts (Fig. 17), thrombolites (Fig. 18) and beds of nodular anhydrite. Excellent and reliable recent analogs of these rock types can be found in Shark Bay, western Australia (Logan et al. 1974).

In many Zechstein 2 Carbonate successions in the East Netherlands, this facies records the gradual drowning of a supratidal flat (Fig. 14). This event is correlatable along part of the basin margin and may reflect a basinwide slight sea-level rise.

The unrestricted platform facies (Figs 19-21) contains a variety of rock types, including ooid grainstones, pisoid grainstones, intraclast grainstones, bioclastic grainstones and grapestones. A characteristic feature of this facies is the common occurrence of flat, angular and rounded clasts that range in size from less than a millimetre to several centimetres (Fig. 19). These clasts may consist of the above mentioned rock types, but frequently they are composed of diagenetically altered pelleted fabric (Fig. 20). Similar pelleted clasts are well known from Shark Bay where they develop in the upper intertidal zone during the summer, when long periods of dessication are interspersed with tidal floodings (Logan 1974, p 217). The deposits of the unrestricted platform facies formed in extremely shallow, occasionally subaerially exposed environments, and include beach deposits, reworked supratidal pavement breccias, bar and spit deposits and calcretes. Frequently the rocks show imprints of vadose diagenesis,

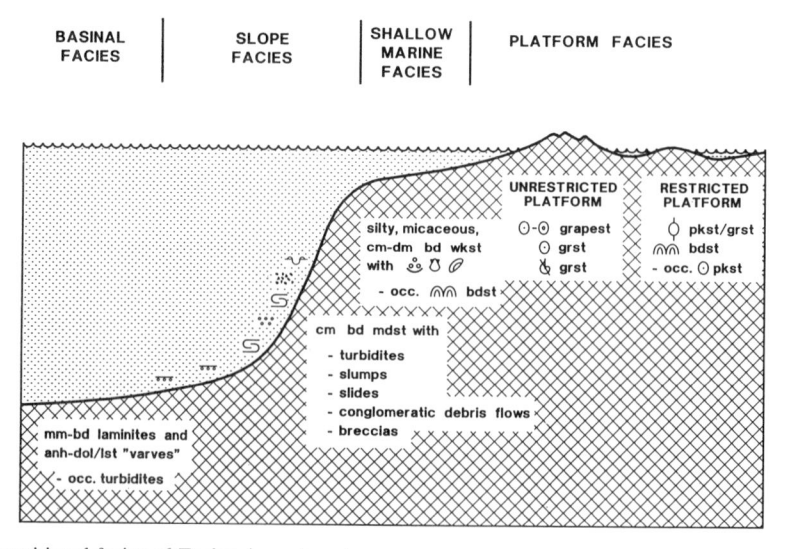

Fig. 10. Depositional facies of Zechstein carbonates.

Fig. 11. Succession of laminated redbrown and greygreen claystones and siltstones with intercalated streaks of nodular anhydrite, formed in an arid floodplain/playa environment. Fringe Zechstein Formation. Sedimentary top to top-left.

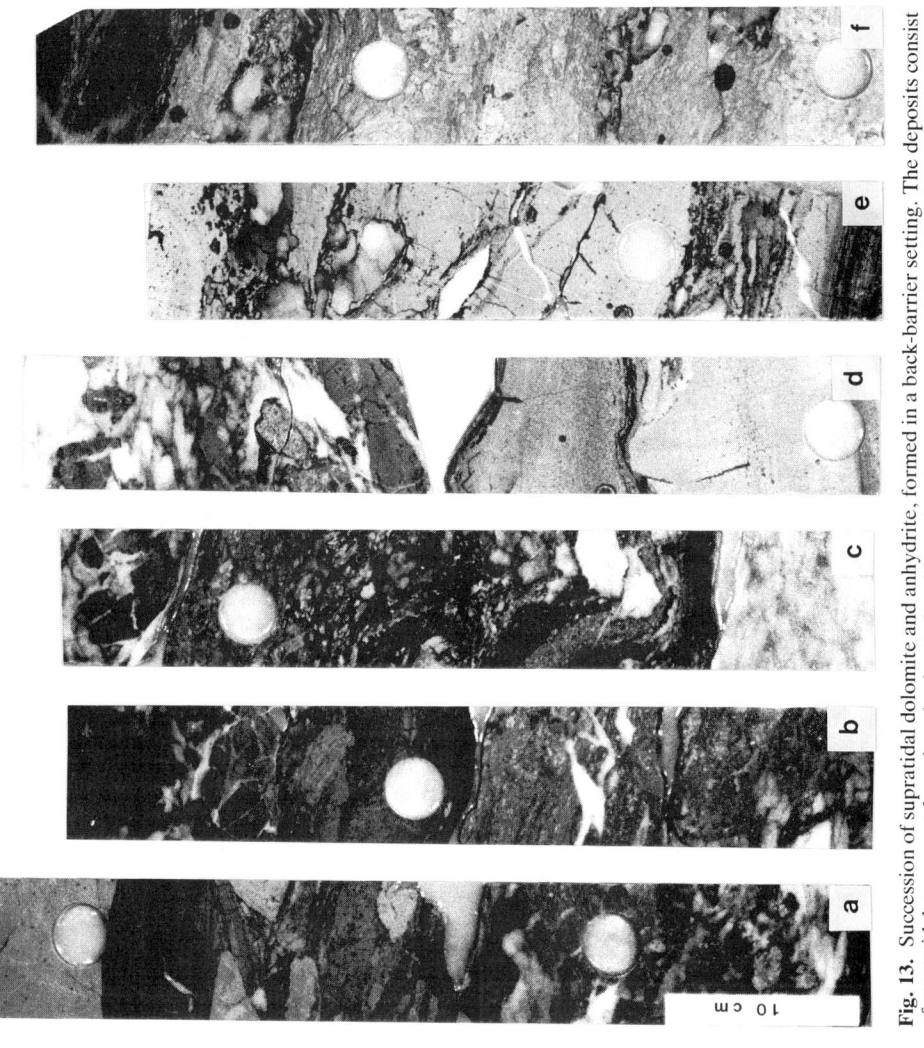

Fig. 12. Detail of Fig. 11, showing dessiccation cracks and light coloured sheet-flood deposit in laminated redbrown claystone.

Fig. 13. Succession of supratidal dolomite and anhydrite, formed in a back-barrier setting. The deposits consist of supratidal pavement breccias, dessiccated and brecciated storm layers, nodular anhydrite, and calcrete hardpans. The light rock at the base of (f) represents the top of the Zechstein 1 Anhydrite.

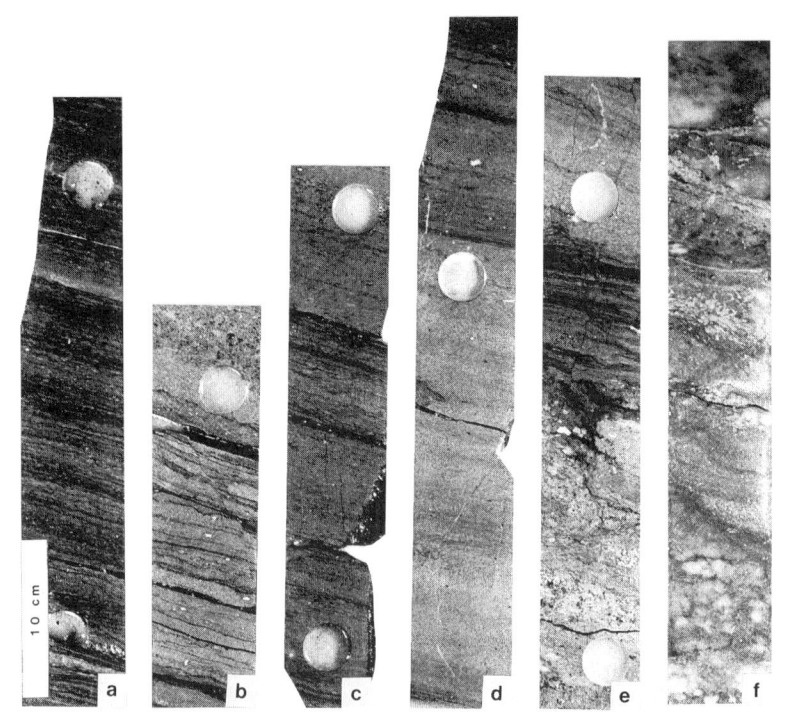

Fig. 14. Example of the transition from tidal flat to shallow lagoonal deposits in a back-barrier setting. The lower part of the interval (f) consists of stromatolitic dolomite interbedded with nodular anhydrite and reflects deposition in hypersaline pools on a poorly drained supratidal flat. The overlying succession consists of thrombolitic dolomite (see Fig. 18) that is sharply overlain, halfway (e), by well bedded, very shallow lagoonal deposits.

including vadose compaction (Peryt 1985) and calcretisation (Fig. 21).

The shallow marine facies consists of laminated secondary limestones and dolomites that are very similar in appearance to deposits from the basinal facies. It differs from the latter in containing fossils (bivalves, ostracods, nodosarid foraminifers) and occasionally cyanobacterial boundstones. The facies is not widely distributed. If present, its deposits usually occupy an intermediate position between unrestricted platform deposits and slope deposits, but they have also been found on top of submerged shoals, from where they grade laterally into basinal deposits. The slope facies is characterized by the occurrence of sedimentary structures indicative of slope instability, such as slumps (Fig. 22), slides, and conglomeratic debris flows (Fig. 23). The latter may contain platform-derived debris but more usually, debris flows are composed of fragments of early cemented slope sediment in a dolomite mudstone matrix. Slope successions are typically composed of intervals of autochtonous, laminated mudstones of bacteriogenic origin (Dawans 1988), that alternate with beds of turbiditic origin (Fig. 24). Lithologically the successions consist of dolomite and secondary limestone.

The basinal facies consists predominantly of secondary limestone and occasionally of dolomite. The rocks are characterized by fine lamination (Fig. 25), which is emphasised by abundant dark organic-rich laminae and less common, white, anhydritic laminae.

Diagenesis and reservoir development

Carbonate/evaporite systems are notorious for their diagenetic complexities and the Zechstein proves no exception in this respect. The diagenetic history of Zechstein carbonates has not only resulted in a variety of rock fabrics and sedimentary structures that are occasionally difficult to interpret, but has also determined their reservoir development.

The outlines of the diagenetic history of Zechstein carbonates are now relatively well

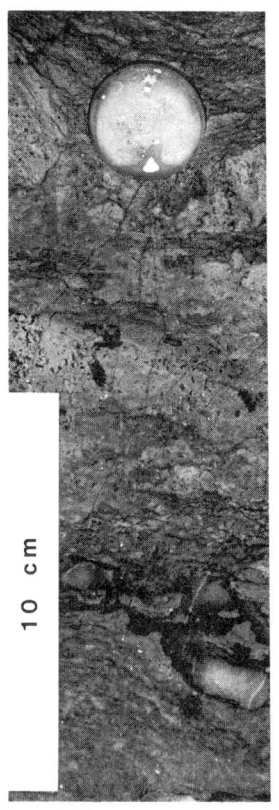

Fig. 15. Supratidal pavement breccia. Detail of Fig. 13.

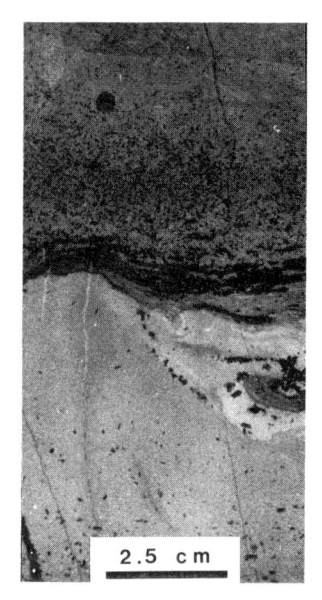

Fig. 17. Laminar calcrete hardpan crust, capping a supratidal storm layer. Note ghosts of spherical structures in overlying interval. These probably represent diagenetically altered caliche glaebules. Detail of Fig. 13.

Fig. 16. Dolomite supratidal pavement breccia, sharply overlain by a storm layer. Note clast imbrication in the top of the breccia, probably reflecting storm reworking. Detail of Fig. 13.

Fig. 18. Thrombolitic dolomite, produced by cyanobacterial pustular mat. These mats are characteristic for the middle to upper intertidal zone of tidal flats in environmentally restricted settings. Detail of Fig. 14.

ZECHSTEIN RESERVOIRS IN THE NETHERLANDS 391

Fig. 19. Ooidal grapestone clasts from the unrestricted platform facies.

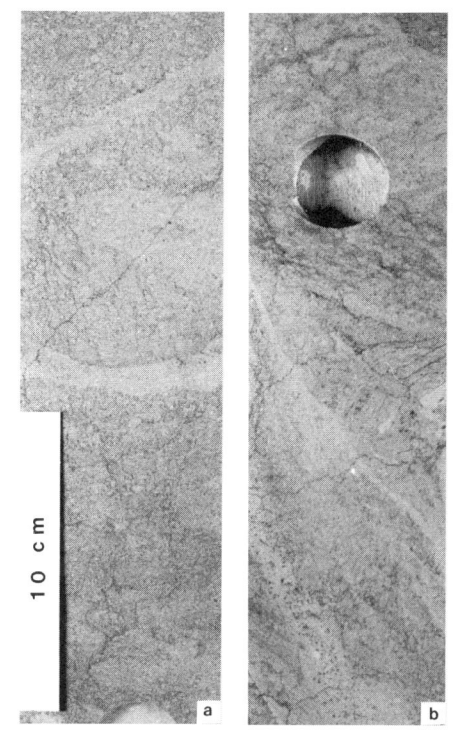

Fig. 21. Characteristic appearance of calcretized dolomite from the unrestricted platform facies. The rock is an ooid/pisoid grapestone breccia. (a) Partly micritized lithoclast and micritized fissure(s). (b) Micritized network of haircracks.

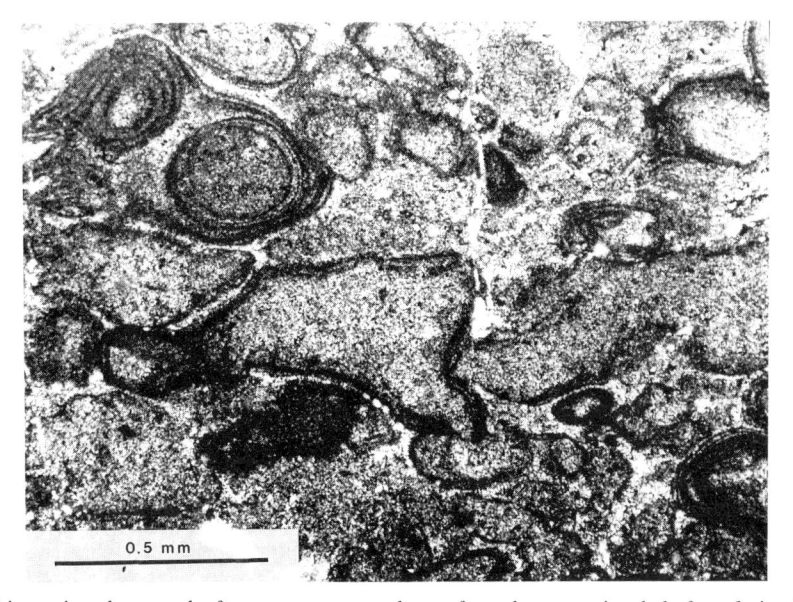

Fig. 20. Thin section photograph of a very common rock type from the unrestricted platform facies. The nature of the particles and of the diagenetic overprint indicates deposition in an extremely shallow, probably occasionally subaerially exposed environment.

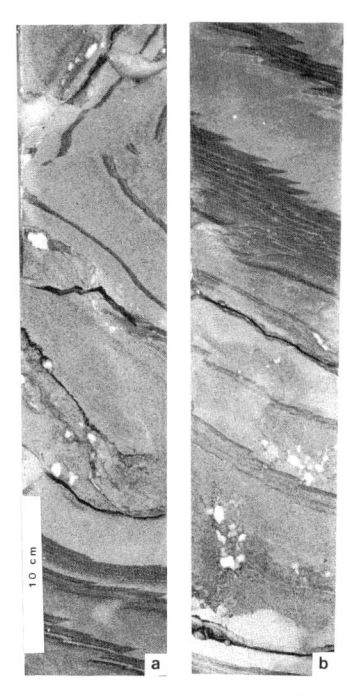

Fig. 22. Features from the slope facies. The two core slabs are continuous; sedimentary top is top-left. (a) Slumped interval of well bedded dolomite mudstone. (b) Microthrusts, immediately underlying the slump.

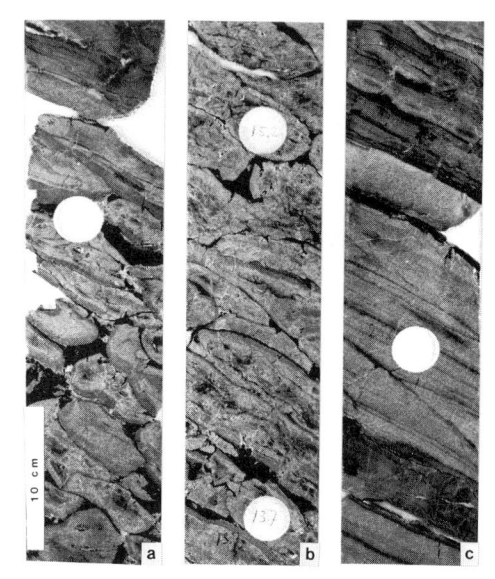

Fig. 23. Conglomeratic debris flow in slope dolomite. The clasts represent broken-up fragments of early cemented crusts of slope mudstone.

understood (Clark 1980; Kaldi 1986; Harwood 1986; Huttel 1987), and this understanding permits general statements about the reservoir properties of the various depositional facies. Accurate reservoir prediction is, however, not always possible. This is partly so because some reservoirs, in particular those in slope successions, are distributed along diagenetic trends that reflect ancient, difficult to predict, fluid migration pathways. Fractures pose another problem. They often contribute substantially to the reservoir properties of Zechstein carbonates and in several East Netherlands gasfields production rates are determined by fracture presence. Fractures are difficult, if not impossible, to predict and may be the reason for reservoirs in otherwise tight carbonates.

In the following, the reservoir development of the various depositional facies will be characterized in terms of the most important porosity creating and destroying diagenetic processes. Subsequently, the facies will be characterized in terms of reservoir potential, neglecting the contribution of fractures. The diagenetic history of almost all Zechstein carbonates started with dolomitization, which occurred in all depositional facies. Cathode luminescence microscopic investigations often show imprints of several dolomitization events. Textural evidence suggests that the first dolomitization event occurred during early diagenesis.

Excluding the dolomitization event(s), the diagenetic history of platform carbonates is fundamentally different from that of the other carbonate facies. The former was dominated by the effects of early, surface related diagenesis, whereas the latter was dominated by late, burial related diagenetic processes. Many platform carbonates, in particular grainsupported carbonates from the unrestricted platform facies, were initially characterized by a network of large, interconnected pores. These depositional pores are occasionally preserved unaltered (Fig. 26A), but more usually they acted as pathways for various diagenetic fluids, that, during early diagenesis, ranged from meteoric waters to hypersaline brines. These fluids caused porosity destruction by dolomite and anhydrite cementation (Fig. 26B), or porosity enhancement through leaching (Fig. 26C) depending on their composition. The leached fabrics were produced by meteoric-water leaching and possibly by mixing zone corrosion (Back *et al.* 1986; Smart *et al.* 1988); they are often closely associated with diagenetic structures indicative of vadose

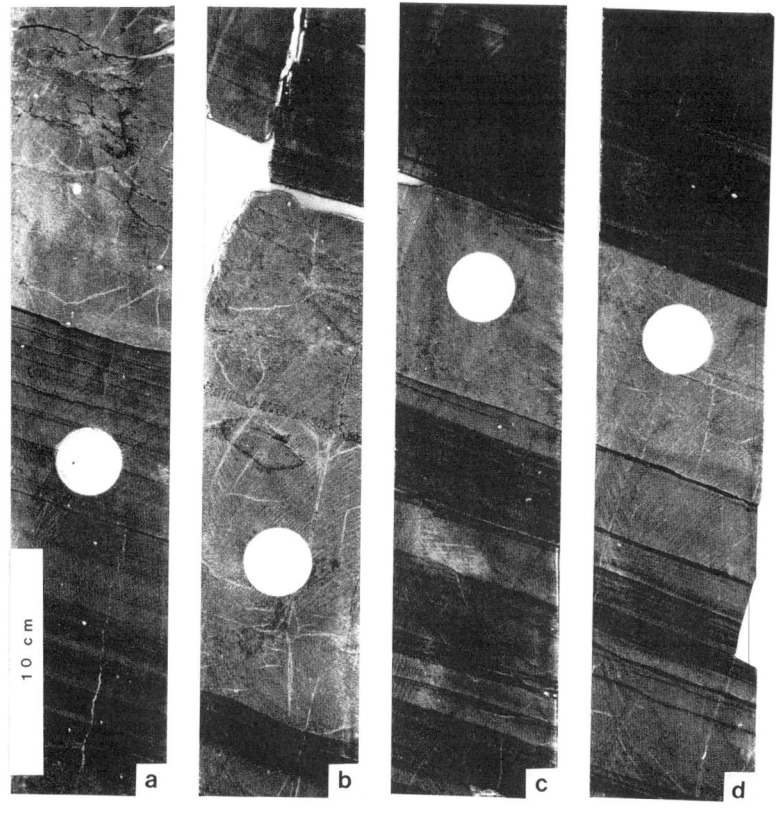

Fig. 24. Tight secondary limestone of the slope facies showing laminated autochthonous sediment and interbedded turbidites.

Fig. 25. Laminated secondary limestone of the basinal facies.

environments (meniscus cements, vadose compaction).

The early diagenetic history of carbonates from the restricted platform facies was frequently characterized by conditions of hypersaline groundwater which resulted in rapid porosity destruction through dolomite cementation and anhydritization. This facies is, as a result, usually characterized by poor reservoir properties.

Early diagenesis produced relatively minor modifications in non-platform carbonates. Probably the most important modification was the formation of dolomite concretions and cemented crusts in the topmost layers of slope and deeper lagoon sediments. These indurated layers were occasionally broken up and transported by debris flows (Fig. 23). Such debris flow deposits formed laterally extensive inhomogeneities in otherwise tight successions; they consequently frequently acted as pathways for migrating formation fluids during deeper burial and were therefore particularly susceptible to diagenetic alteration.

During deeper burial, two closely related diagenetic processes, calcitization and dolomite leaching, caused important changes of the reservoir properties of non-platform carbonates. Both processes modified the initial dolomite rocks (Fig. 27C) and produced a large number of diagenetic carbonate rock types, including leached dolomite (Fig. 27A), tight secondary limestone (Fig. 27D), porous dolomitic limestone (Fig. 27E), and vuggy secondary limestone (Fig. 27F). Burial related calcitization and dolomite leaching depend on the availability of carbon dioxide, which has the effect of rendering carbonate minerals unstable. Small amounts of carbon dioxide cause conditions in which dolomite is unstable, but calcite is stable. The dolomite consequently is calcitized, i.e. replaced by calcite in a texture-preserving process of 'incongruent dissolution' (Clark 1980). This process proceeds relatively rapidly in the presence of an external supply of calcium and sulphate ions. During burial diagenesis these ions were supplied to Zechstein carbonates, largely as a result of temperature dependent gypsum-anhydrite transformations in enclosing gypsum bodies.

Huttel (1987) found, on the basis of fluid inclusion studies, that the temperature of the calcitizing fluid was in the range of 80–100 °C and the salinity about 24%. These values demonstrate conclusively that the calcitization occurred under conditions of deep burial. The source of the carbon dioxide required for the calcitization was traced by Huttel (1987) to organic matter within Zechstein carbonates, based on geochemical and isotope analysis. His measurements show δC^{13} values in the range of +6 to +8‰ for calcitized carbonates (dedolomites) from the slope facies. Such 'heavy carbon' values are characteristic for carbon dioxide that is derived from organic matter by fermentation (Irwin et al. 1977).

In the presence of large amounts of carbon dioxide, dolomite is unstable, but calcite cannot nucleate or grow. Dolomite will consequently dissolve, but calcitization will not occur. Such a phase of dolomite leaching can occur at different times and at different places during burial diagenesis and as a result it will produce a variety of rock fabrics.

In the case that dolomite leaching represents the final phase of a gradual build-up of carbon dioxide supply, most carbonate will already have been transformed, partly or completely, into secondary limestone. Dolomite leaching will then affect the remnant patches of dolomite in these limestones (Fig. 27E), ultimately producing a vuggy secondary limestone (Fig. 27F).

Alternatively, dolomite leaching may affect a rock without a previous phase of calcitization, in which case the precursor dolomite will be transformed into a leached, potentially very porous dolomite (Fig. 27A). This would be the case whenever highly carbon dioxide-charged formation fluids were suddenly introduced in a dolomite rock body, for instance as a result of overpressured formation-water breakthrough.

Another possibility for dolomite leaching occurs when pathways of migrating formation fluids are shifting while the supply of carbon dioxide is increasing with time.

Burial related calcitization and dolomite leaching are not expected to occur in platform carbonates which characteristically do not contain much organic matter. This is corroborated by core investigations which show that the imprints of both processes are limited to carbonates from the basinal, slope, and shallow marine facies. In basinal carbonates, the effects of calcitization predominate and dolomite leaching usually was of minor importance. In slope carbonates, dolomite leaching can be very important and may locally affect up to half of slope intervals. In the absence of a supply of carbon dioxide during burial diagenesis, slope and basinal carbonates develop into tight overdolomite (Fig. 27B).

The considerations outlined above make it clear that the distribution of diagenetically produced rock fabrics in Zechstein carbonates can be extremely complex, in particular in slope carbonates. The diagram (Fig. 28), showing a model of the distribution of the main diagenetic rock types of Zechstein carbonates, should therefore be regarded as a simplified, state of the art summary of our present understanding

---

Fig. 26. Thin section photographs that illustrate characteristic porosity development in unrestricted platform dolomites. (A) Interparticle porosity of depositional origin — here largely plugged by solid hydrocarbons (black) — occurs in many unrestricted platform successions. (B) The main diagenetic porosity-destroying processes in unrestricted platform successions are: anhydrite pore-plugging (shown here), dolomite cementation, and vadose compaction. Photo taken with crossed polars and gypsum plate. (C) Vuggy porosity, as shown in this coated intraclast grainstone, is very common in unrestricted platform deposits. The vuggy intervals are frequently associated with vadose diagenetic imprints and were probably formed during early-diagenetic meteoric water infiltration.

A

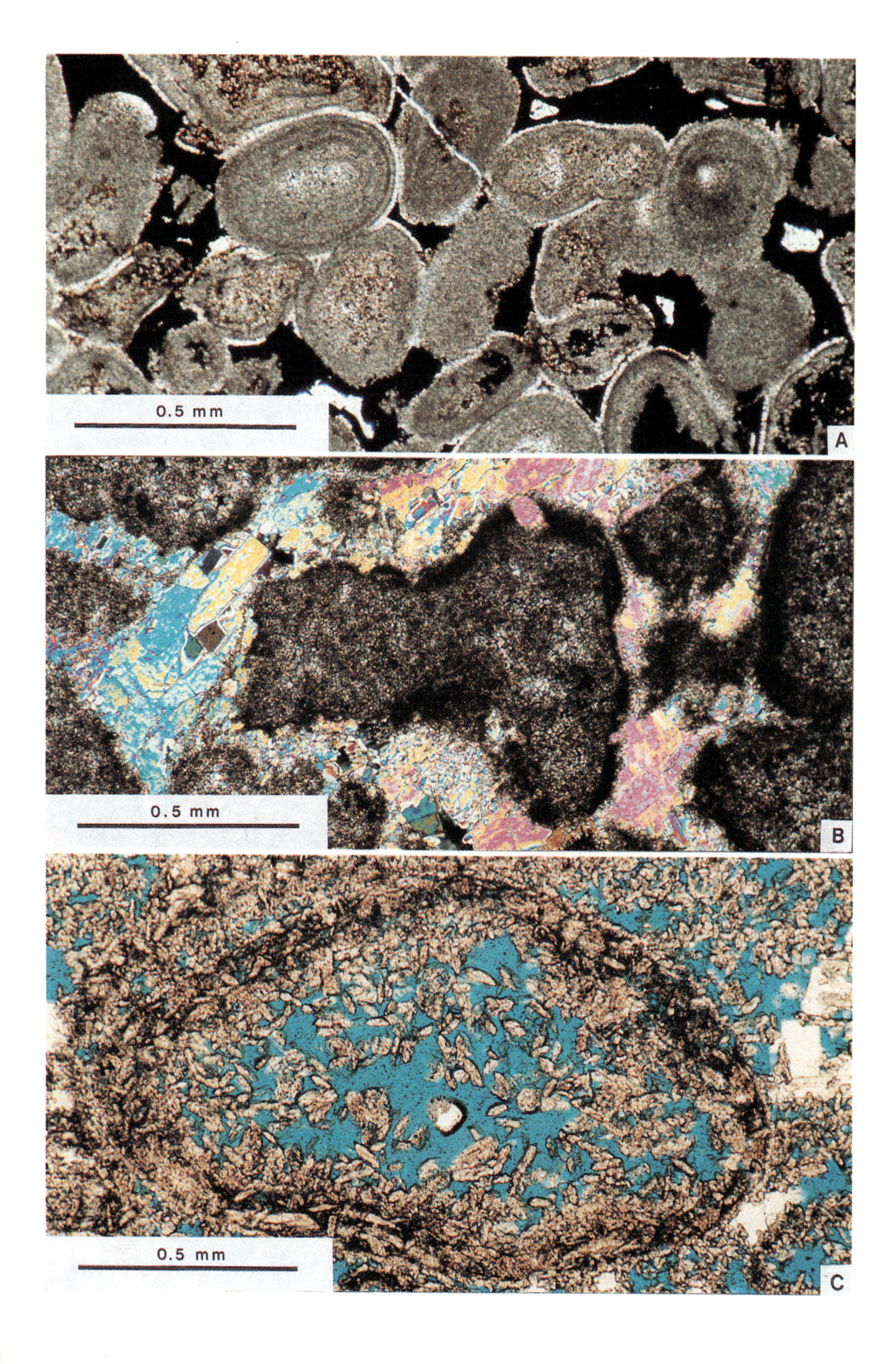

0.5 mm

B

0.5 mm

C

0.5 mm

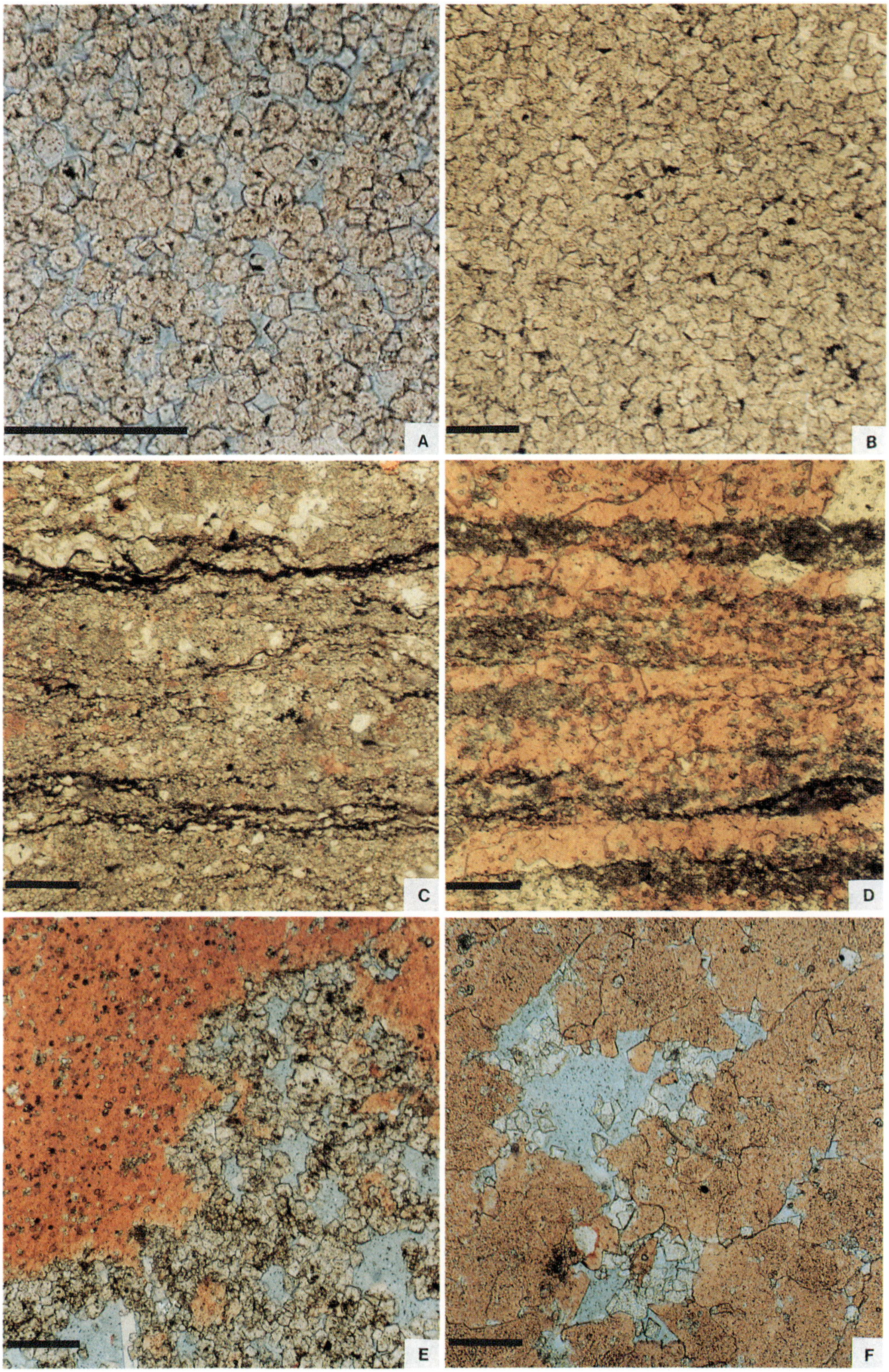
A
B
C
D
E
F

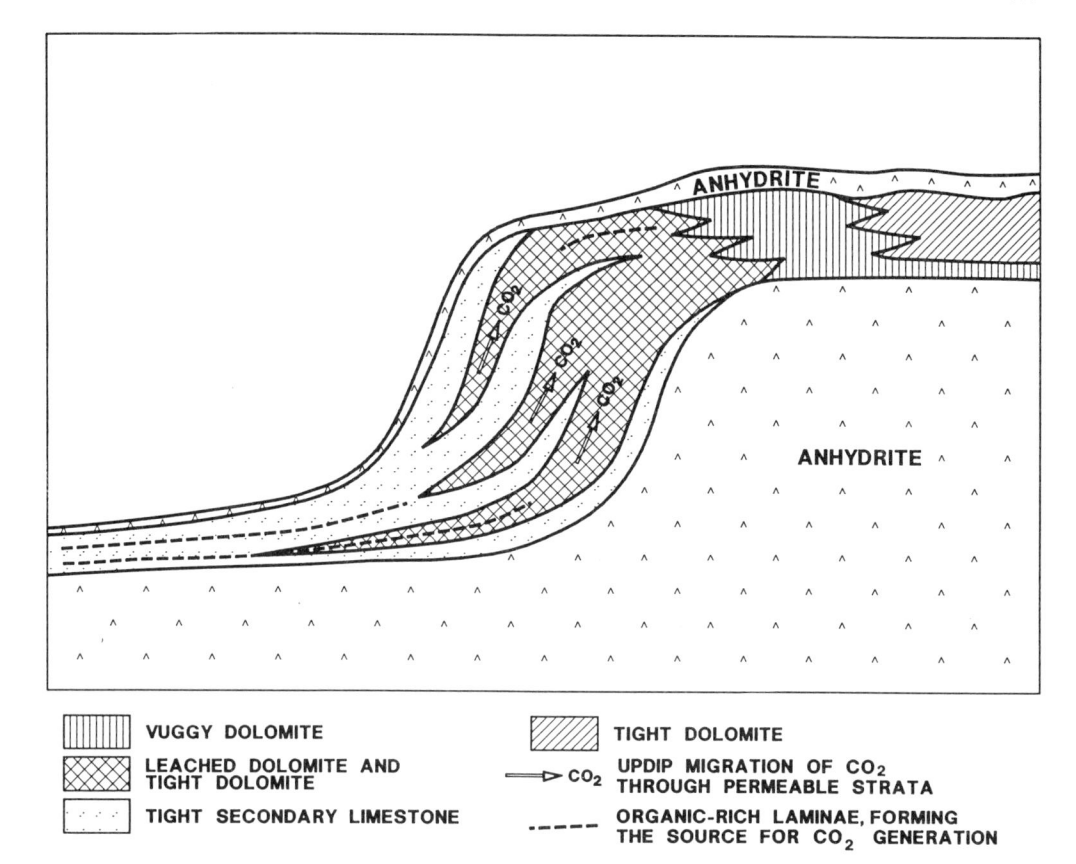

Fig. 28. Simplified distribution of main diagenetic rock types.

Legend:
- VUGGY DOLOMITE
- LEACHED DOLOMITE AND TIGHT DOLOMITE
- TIGHT SECONDARY LIMESTONE
- TIGHT DOLOMITE
- ⟹ CO_2 UPDIP MIGRATION OF CO_2 THROUGH PERMEABLE STRATA
- ----- ORGANIC-RICH LAMINAE, FORMING THE SOURCE FOR CO_2 GENERATION

of the diagenetic history of Zechstein carbonates. Excluded from the diagram are rock types that were produced as a result of Mesozoic uplift and subsequent erosion/subaerial exposure of Zechstein carbonates.

Figure 29 shows representative core-plug porosity and permeability values for most of the Zechstein carbonate depositional facies. The figure illustrates that high porosities are associated with the distribution of ooids (and pisoids, intraclasts, etc) and leached slope deposits (Fig. 29E). The latter are frequently, but not always, associated with redeposited intervals. Figure 29 also illustrates that the matrix permeability of most Zechstein carbonates (excluding leached slope dolomites) is usually very low. This is borne out by the fact that highly productive wells (400 000 m³/day) in East Netherlands gas fields invariably drain from fractured reservoirs whereas non-fractured reservoirs are character-

---

Fig. 27. Thin section photographs showing characteristic diagenetic fabrics and porosity development of the slope facies. Thin sections have been stained with Alizerin Red S: calcite is red, dolomite is buff-grey, anhydrite is white. Porosity is blue. Scale bar equals 0.2 mm. (A) Leached dolomite of probably bacteriogenic origin (note dark centres and subspherical shape of crystals). This type of dolomite fabrics is associated with high core-plug porosity and permeability values (up to 24% and 65 mD, respectively). (B) Tight, impermeable dolomite, texturally similar to (A), but not subjected to leaching. (C) Laminated, silty, micaceous dolomite with organic seams. This nonporous rock type represents the initial, pre-calcitization stage of basinal deposits and of the autochtonous part of slope deposits. (D) Calcitized version of (C). These tight, laminated secondary limestones constitute the most common rock type of the slope and basinal facies. Note patches of anhydrite. (E) Patch of intensely leached dolomite within a tight secondary limestone. The photo illustrates the close diagenetic relationship between subsurface calcitization and dolomite leaching and the importance of these dolomite patches for the reservoir properties of the rock. (F) Vuggy secondary limestone. Compared with (E), this fabric represents the final stage of dolomite leaching in which all dolomite has been removed.

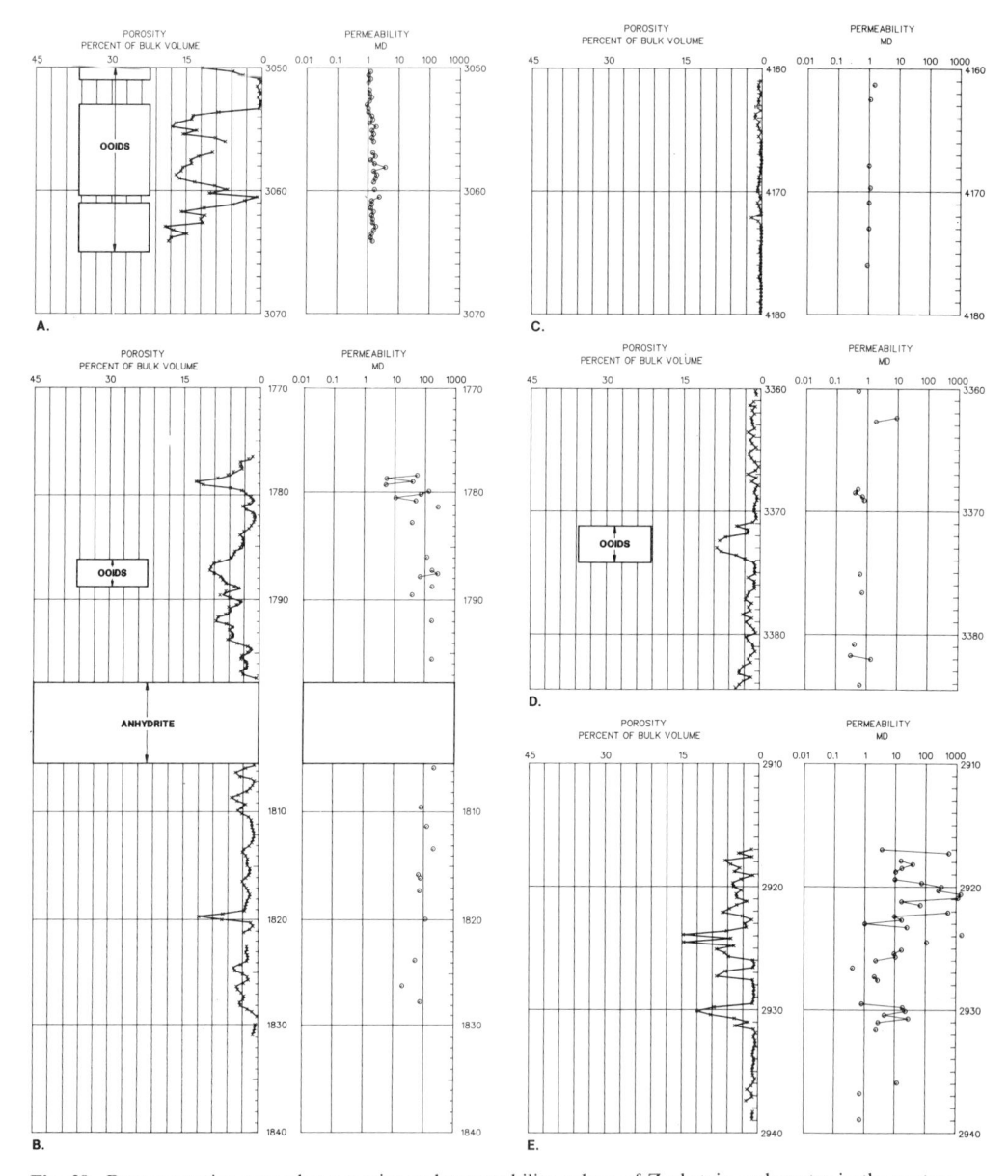

Fig. 29. Representative core-plug porosity and permeability values of Zechstein carbonates in the eastern Netherlands, plotted against depth (in metres). (A) Unrestricted platform dolomites. The tight interval at top represents a succession of inter/supratidal deposits. (B) Restricted platform dolomites, composed mainly of lagoonal and tidal flat mudstones. Note anhydrite interval composed of salina deposits and relatively porous oolite bed. (C) Tight dolomite mudstone from the slope facies. (D) As C, but including a relatively porous interval composed of redeposited, platform derived ooids and flat lithoclasts. (E) Slope succession composed of tight secondary limestone and porous, leached dolomite. The latter consist of slope breccias and conglomeratic debris flows.

ized by much lower production per well (down to 4000 m^3/day). Dramatic productivity increases can be achieved through artificial fracturing of Zechstein reservoirs.

Conclusions

In conclusion it can be stated that diagenetic processes were largely responsible for the reservoir development of Zechstein carbonates.

Carbonates from the platform facies, deposited under shallow marine to sunaerially exposed conditions, were subjected to a diagenetic history that was dominated by the effects of surface related, early diagenetic processes, whereas the diagenetic history of carbonates from the other depositional facies was dominated by the effects of burial related, late diagenetic processes. The reservoir development of platform carbonates as a result follows different, more predictable, patterns compared with the reservoir development of the other facies.

Unrestricted platform carbonates, which were initially composed of aragonitic particles (ooids, pisoids, intraclasts, bioclasts) and characterized by high depositional porosities and permeabilities, consistently developed into porous, frequently vuggy, dolomites characterized by average porosity values in the 10–15% range.

Restricted platform carbonates, initially mainly composed of stromatolitic, muddy carbonates and nodular anhydrites, usually developed into relatively tight dolomites.

Slope carbonates, initially composed of laminated mudstones with intercalations of redeposited, occasionally platform-derived intervals, developed into a variety of diagenetic rock types characterized by very different reservoir properties. Good reservoirs can be expected in slope successions, but their accurate prediction is in many cases not possible due to the diagenetic complexities involved. Basinal carbonates characteristically developed into tight, non-reservoir, secondary limestones, although localized occurrences of leached dolomite cannot be excluded.

Deeper lagoonal carbonates developed into tight non-reservoirs, usually composed of argillaceous dolomite.

The reservoir properties of the shallow marine facies are not yet sufficiently known, but they are probably relatively poor.

Regional prediction of the distribution of the various depositional facies of Zechstein carbonates relies heavily on the ability to recognise the subsurface expression of the ancient Zechstein carbonate platforms. These platforms can be identified on good quality seismic sections, not so much by the seismic expression of the carbonates themselves, as by the seismic expression of the underlying thick sulphate deposits.

Finally, the importance of fractures must be stressed. The productivity of many wells in the East Netherlands gasfields depends not on the reservoir properties of the rock matrix itself, but on the presence and density of fractures in the carbonates.

I wish to express my thanks to colleagues and former colleagues in NAM's exploration and petroleum engineering departments who provided information on many aspects related to the subject of this paper. They include H. J. G. Anemaat, A. Axelrod, H. van Dierendonck, A. J. E. Donzé, J. F. Karlo, L. W. Kuilman, P. Landmesser and I. Young. The paper benefitted from the presentations and discussions during the International Zechstein Symposium 1987 in Hannover, and in particular from an excursion to the classical Zechstein outcrops of the Harz Mountains under the guidance of Prof. Dr. G. Richter–Bernburg. I am indebted to the managements of Nederlandse Aardolie Maatschappij B.V., Shell Internationale Petroleum Maatschappij B.V., and Exxon International for permission to publish this paper.

References

Antonowicz, L. & Knieszner, L. 1981. Reef zones of the Main Dolomite, set out on the basis of paleogeomorphologic analysis and the results of modern seismic techniques. *Proceedings of the International Symposium of the Central European Permian, Warsaw, 1978*, Wydawnictwa Geologiczne, Warsaw. 356–368.

Back, W., Hanshaw, B. B., Herman, J. S. & Van Riel, J. N. 1986. Differential dissolution of a Pleistocene reef in the groundwater mixing zone of coastal Yucatan, Mexico. *Geology*, **14**, 137–140.

Best, G. 1987. Die Grenze Zechstein/Buntsandstein in NW Deutschland nach Bohrlochmessungen. *Abstracts of the International Symposium on Zechstein 1987*, Hannover, 18–19.

Clark, D. N. 1980a. The sedimentology of the Zechstein 2 Carbonate Formation of Eastern Drenthe, The Netherlands. *In*: Füchtbauer, H. & Peryt, T. M. (eds), *The Zechstein Basin with Emphasis on Carbonate Sequences*. Contribution Sedimentation **9**, 131–165.

—— 1980b. The diagenesis of Zechstein carbonate sediments. *In*: Füchtbauer, H. & Peryt, T. M. (eds) *The Zechstein Basin with Emphasis on Carbonate Sequences*. Contribution Sedimentation **9**, 167–203.

—— & Tallbacka, L. 1980. The Zechstein Deposits of southern Denmark. *In*: Füchtbauer, H. & Peryt, T. M. (eds) *The Zechstein Basin with Emphasis on Carbonate Sequences*. Contribution Sedimentation **9**, 205–231.

Dawans, J. M. 1988. *Microbial biodiagenesis and its role in the formation of carbonate rock, evidence from dolomite spherulites*. Abstr. 9th IAS regional meeting, Leuven.

Glennie, K. W. 1972. Permian Rotliegendes of North West Europe interpreted in the light of modern desert sedimentation studies. *Bulletin of the American Association of Petroleum Geologists* **56**, 1048–1071.

Harwood, G. M. 1986. The diagenetic history of Cadeby Formation carbonate (EZ1Ca), Upper Permian, eastern England. *In*: Harwood, G. M.

& SMITH, D. B. (eds), *The English Zechstein and Related Topics*, Geological Society, London, Special Publication **22**, 75–86.

HUTTEL, P. 1987. Das Stassfurth-Karbonat (Ca2) in Südoldenburg. Fazies und Diagenese in Sedimenten am Nordhang der Hunte-Schwelle. *PhD thesis*, University of Göttingen.

IRWIN, H., CURTIS, C. D. & COLEMAN, M. 1977. Isotopic evidence for source of diagenetic carbonates formed during burial of organic-rich sediments. *Nature*, **269**, 209–13.

KALDI, J. 1986. Diagenesis of nearshore carbonate rocks in the Sprotbrough Member of the Cadeby (Magnesian Limestone) Formation (Upper Permian) of eastern England. *In*: HARWOOD, G. M. & SMITH, D. B. (eds), *The English Zechstein and Related Topics*, Geological Society, London, Special Publication **22**, 87–102.

KULICK, J. & PAUL, J. 1987. Zur Stratigraphie und Nomenklatur des Zechsteins. *Glossar. Int. Symp. Zechstein 1987, Hannover, Exkursionf.* **I**, 25–34.

LOGAN, B. W. 1974. Inventory of Diagenesis in Holocene–Recent Carbonate Sediments, Shark Bay, Western Australia. *In*: LOGAN *et al*. 1974. *Evolution and diagenesis of Quaternary carbonate sequences, Shark Bay, Western Australia*. American Association of Petroleum Geologists Memoir **22**, 195–247.

——, READ, J. F., HAGAN, G. M., HOFFMAN, P., BROWN, R. G., WOODS, P. J. & GEBELEIN, C. D. 1974. Evolution and diagenesis of Quaternary carbonate sequences, Shark Bay, Western Australia. *American Association of Petroleum Geologists Memoir* **22**.

MAUSFELD, S. 1987. Der Plattformrand des Stassfurtkarbonates (Z2) südlich von Oldenburg: Sedimentologie, Fazies, und Diagenese. *PhD thesis*, University of Marlburg.

NEDERLANDSE AARDOLIE MAATSCHAPPIJ B. V. & RIJKS GEOLOGISCHE DIENST. 1980. Stratigraphic Nomenclature of The Netherlands. *Verh. Kon. Ned. Geol. Mijnb. Genoot.* **32**.

PAUL, J. 1980. Upper Permian algal stromatolitic reefs, Harz Mountains (F.R. Germany). *In*: FÜCHTBAUER, H. & PERYT, T. M. (eds), *The Zechstein Basin with Emphasis on Carbonate Sequences*. Contribution Sedimentation **9**, 253–268.

—— 1987. Der Zechstein am Harzrand: Querprofil über eine permische Schwelle. *International Symposium Zechstein 1987, Hannover, Exkursionf.* **II**, 193–276.

PERYT, T. M. 1978. Sedimentology and paleoecology of the Zechstein Limestone (Upper Permian) in the Fore-Sudetic area (western Poland). *Sedimentary Geology* **20**, 217–243.

—— 1985. A Permian beach in the Zechstein dolomites of western Poland: influence on reservoirs. *Journal of Petroleum Geology* **8**, 463–474.

PISKE, J. & SCHRETZENMAYR, S. T. 1984. Sedimentations-Zyklen im Stassfurtkarbonate und ihre Modifikation durch azyklische Vorgänge. *Zechstein geology of Wiss.* **12**, 83–100.

PURSER, B. H. (ed) 1973. *The Persian Gulf — Holocene carbonate sedimentation and diagenesis in a shallow epicontinental sea*. Springer, Berlin.

RICHTER–BERNBURG, G. 1960. Zeitmessung geologischer Vorgänge nach Warven-korrelationen im Zechstein. *Geological Rdsch.* **49**, 132–148.

—— 1963. Solar cycle and other climatic periods in varvitic evaporites. *In*: NAIRN, A. E. (ed.), *Problems in Palaeoclimatology*, 510–520. Processing of NATO Conference Newscastle, London.

—— 1985. Zechstein–Anhydrite, Fazies und Genese. *Geology Jb*, A, **85**, 3–82.

—— 1986. Zechstein 1 and 2 Anhydrites: facts and problems of sedimentation. *In*: HARWOOD, G. M. & SMITH, D. B. (eds), *The English Zechstein and Related Topics*, Geological Society, London, Special Publication **22**, 157–163.

ROCKEL, W. & ZIEGENHARDT, W. 1979. Strukturelle Kriterien der Lagunenbildung im tieferen Zechstein im Raum südlich von Berlin. *Zechstein geology of Wiss.*, **7**, 847–60.

SANNEMANN, D., ZIMDARS, J. & PLEIN, E. 1978. Der Basale Zechstein (A2–T1) zwischen Weser und Ems. *Zechstein dt. geology Ges.*, **129**, 33–70.

SCHREIBER, B. C. & HSÜ, K. J. 1980. Evaporites. *Applied Science Development on Petroleum Geology* **2**, 87–138.

SMART, P. L., DAWANS, J. M. & WHITAKER, S. 1988. Carbonate dissolution in a modern mixing zone, south Andros Island, Bahamas. *Nature*.

SMITH, D. B. 1980. The evolution of the English Zechstein basin. *In*: FÜCHTBAUER, H. & PERYT, T. M. (eds), *The Zechstein Basin with Emphasis on Carbonate Sequences*. Contributions of Sedimentology **9**, 7–34.

——, HARWOOD, G. M., PATTISON, J. & PETTIGREW, T. H. 1986. A revised nomenclature for Upper Permian strata in eastern England. *In*: HARWOOD, G. M. & SMITH, D. B. (eds), *The English Zechstein and Related Topics*, Geological Society, London, Special Publication **22**, 9–17.

VAN ADRICHEM, BOOGAERT, H. A. & BURGERS, W. F. J. 1983. The development of the Zechstein in the Netherlands. *Geologie en Mijnbouw*, **62**, 83–92.

WARREN, J. K. 1982. The hydrological setting, occurrence and significance of gypsum in Late Quaternary salt lakes in South Australia. *Sedimentology*, **29**, 609–637.

—— 1985. On the significance of evaporite lamination. *Sixth International Symposium on Salt, 1983*, **1**, 161–170, Salt Institute.

ZIEGLER, P. A. 1982. *Geological Atlas of Western and Central Europe*. Elsevier Shell International Petroleum Maatschappij B.V.

Lower Permian Rotliegend reservoir of the Southern North Sea gas province

K. W. GLENNIE[1] & D. M. J. PROVAN[2]

Shell UK Exploration Production, Shell-Mex House, Strand, London, UK

Abstract: Rotliegend aeolian sandstone of Early Permian age forms the main producing horizon in the Southern North Sea gas province. Currently some 30–40 fields have been identified in a broad zone measuring 450 by 100 km stretching from the English coast, through the UK and Dutch sectors of the North Sea and on to The Netherlands and West Germany. North Sea field sizes range between 10–300 Bcm (0.3–10.5 TCF). Ultimate recoverable reserves for the province are estimated at some 4250 Bcm (150 TCF). The Rotliegend is underlain by Carboniferous coals and carbonaceous shales, which form the source rock, while Late Permian Zechstein evaporites form the seal. The aeolian sandstones occur interbedded with wadi sandstones and sabkha siltstones. Dune sandstones with large-scale cross bedding generally form the best reservoirs, though even here reservoir quality is highly variable. Core porosities range between 5 and 30% and permeabilities from <1 to 3000 mD. Reservoir impairment through the effects of burial diagenesis (compaction, pressure solution, clay mineral transformations and cementation) is evident in many fields. In others, the occurrence of natural fractures serves to enhance production. As a result, individual well productivities show wide variation.

Lower Permian Rotliegend sandstones form the reservoir for gas of Carboniferous origin in some 30 fields that have been developed or are under development in the Southern North Sea gas province (Fig. 1).

Basin-wide studies (e.g. Glennie 1972, 1986a,b; references in Falke 1972, 1976; Ziegler 1980, 1982) and those of a more local nature within Shell/Esso companies in Britain (e.g. van Veen 1975; Glennie et al. 1978; ten Have & Hillier 1986), The Netherlands (Stäuble & Milius 1970; Lutz et al. 1975; van Wijhe et al. 1980; Oele et al. 1981) and West Germany (Plein 1978; Drong et al. 1982; Betz et al. 1987), and by the staff of other oil companies (e.g. Arthur et al. 1986; Butler 1975; Conway 1986; France 1975; Goodchild & Bryant 1986; van Lith 1983; Roos & Smits 1983; Walker & Cooper 1987) have led to a fairly clear understanding of how this important gas province was formed and the factors governing gas emplacement.

The aim of this paper is to provide an overview of the geological evolution of the province, the characteristics of the Rotliegend reservoir and the factors governing reservoir quality. Although concerned mainly with the Southern North Sea basin, occasional forays are made into the adjacent land areas of the Netherlands and West Germany to illustrate specific points on reservoir development.

Structural and stratigraphic evolution

The oldest rocks normally encountered in wells in the province belong to the Carboniferous Coal Measures Group, though Tubb et al. (1986) report Lower Carboniferous in wells in the UK sector. Upper Namurian and Westphalian coals and carbonaceous shales probably form the source rock for all the gas of the province, irrespective of reservoir age (Cornford 1986). The Coal Measures were deposited over a broad east–west trending plain between the northern Caledonian ranges and the flanking Proto-Tethys ocean (Ziegler 1982; see also Fig. 2).

The humid equatorial conditions of coal-swamp deposition ceased during the later Westphalian in response to a passive northward drift, which carried the area of former Coal Measure deposition into the northern hemisphere desert belt, and the growth of the Variscan Mountain Belt, which formed a barrier to moist southerly winds.

Transpressive movements resulted in inversion uplift of the sedimentary fill of some sub-basins (e.g. Cleveland Hills and Sole Pit sub-basins, and the Texel–IJsselmeer area) with erosion locally extending down to the Upper Namurian (Kent 1980; Glennie & Boegner 1981; van Hoorn 1987; van Wijhe 1987; see also Figs 3 & 7).

---

[1] Now consultant, Ballater, Crampian, UK.
[2] Now A/S Norske Shell, Risavika, Norway.

From BROOKS, J. (ed.), 1990, *Classic Petroleum Provinces*, Geological Society Special Publication No 50, pp 399–416.

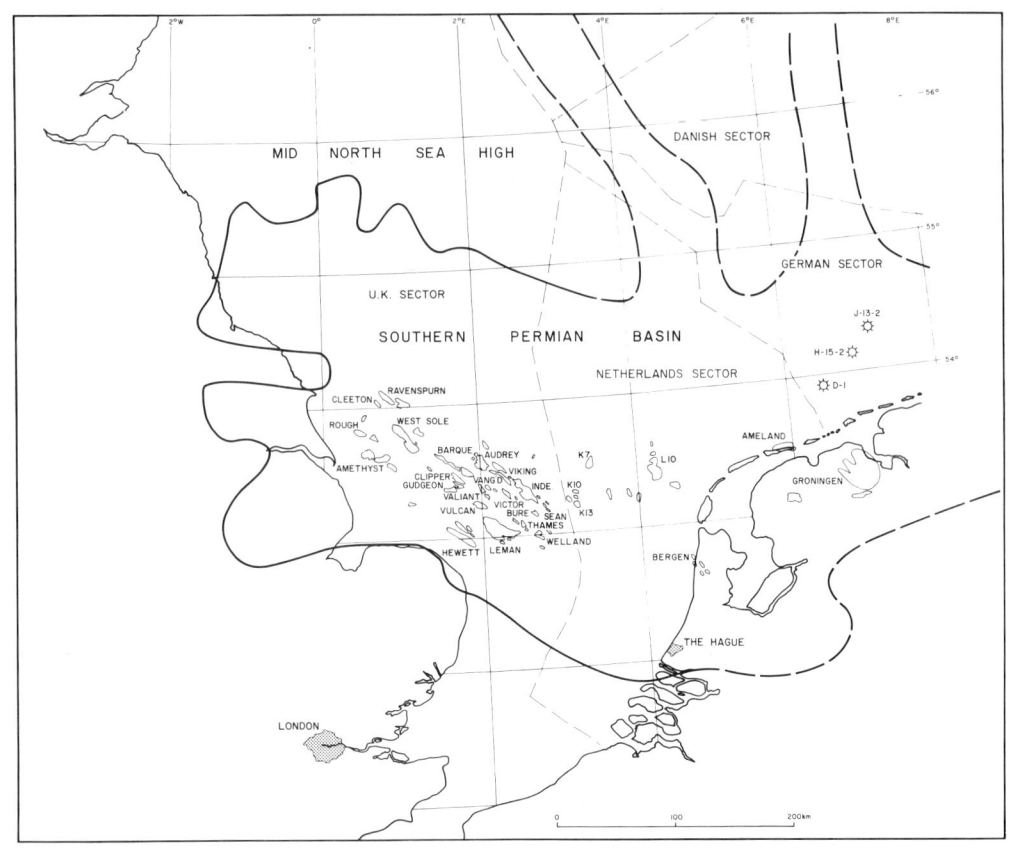

Fig. 1. Southern North Sea Province, showing Rotliegend gas fields.

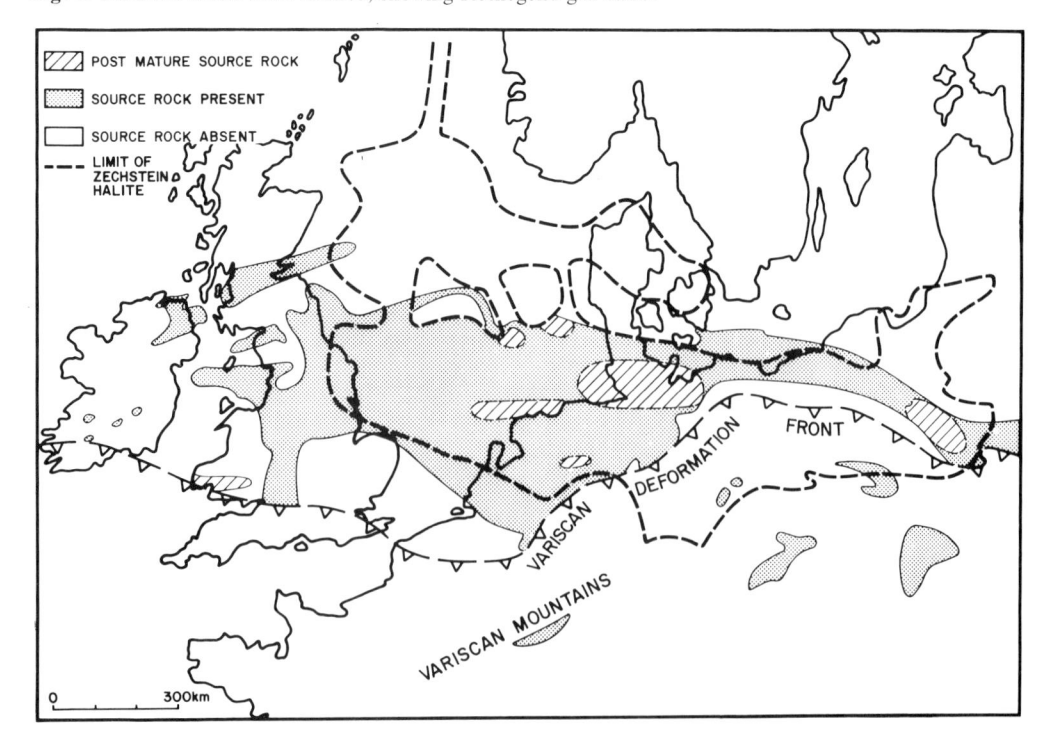

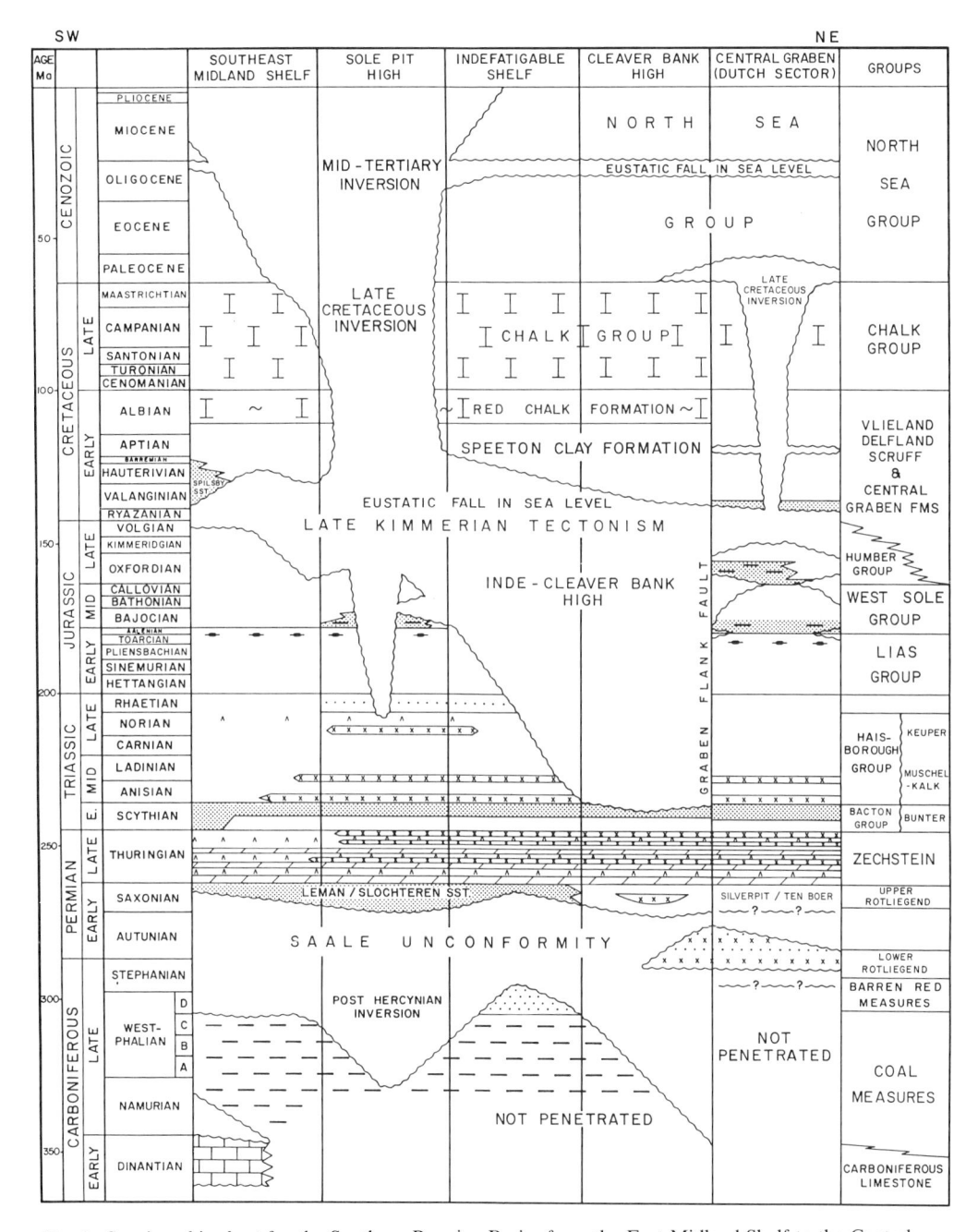

Fig. 3. Stratigraphic chart for the Southern Permian Basin, from the East Midland Shelf to the Central Graben.

Fig. 2. Distribution of preserved Coal Measures in NW Europe: areas of deep burial or known high heat flow, where the coals are probably post-mature are shown hatched (see also Teichmuller 1986). Outline of Zechstein salt seal also shown.

Subsidence of the Variscan northern foreland in the latter part of the Early Permian (Saxonian) initiated development of the Southern Permian Basin. A Northern Permian Basin formed simultaneously north of the east–west trending Mid North Sea–Ringkøbing–Fyn system of highs (Fig. 4). Within the southern basin, erosional products from the surrounding upland areas were carried northwards by ephemeral streams, or wadis. The end point for these streams was a large, probably permanent desert lake. Wadi and lake-margin sediments were deflated and the sand-size fraction redeposited in broad dune systems between the wadi channels. The aeolian dune and wadi sediments now form the reservoir for the Rotliegend gas fields (cf. Figs 1 & 4). In the UK sector they are termed the Leman Sandstone Formation, while in The Netherlands and Germany the equivalent is the Slochteren Formation and Hauptsandstein respectively.

The siltstones of the sabkhas and desert lake now form the Silverpit Formation in the western part of the province and the Ten Boer Formation in the Netherlands (Fig. 3). This sequence attains a thickness of 600 m in Poland (Depowski 1978) and 1500 m at the foot of the Danish peninsula (Ziegler 1982).

Subsidence of the southern basin rapidly outstripped fill, and eventually the shallow desert lake expanded to form a large inland sea — the Zechstein Sea. The supply of marine water was through a narrow northern channel; it would have been reduced or even stopped during any eustatic lowering of sea level. Under these conditions, the salinity of the sea increased to such an extent that halite and other associated evaporites were deposited in thick sequences, with those of Zechstein Cycle 2 (Stassfurt Halite) reaching to well over 1000 m (Taylor 1986). Glacially induced fluctuations in global sea level probably determined the characteristic cyclic nature of Zechstein deposition.

Thus, by the end of the Permian, the Southern Permian Basin was already endowed with the basic requirements for a hydrocarbon province: Carboniferous source rock, Rotliegend reservoir and Zechstein seal. However, it was the post-Permian history of the area that determined which parts of the basin were to become prospective. The distribution of prospective Rotliegend reservoir is outlined in Fig. 5. Derived from Figs 2 & 4, it shows the areas where the Carboniferous source rocks are mature for gas, the Rotliegend is in a reservoir facies, and the overlying Zechstein seal is of sufficient

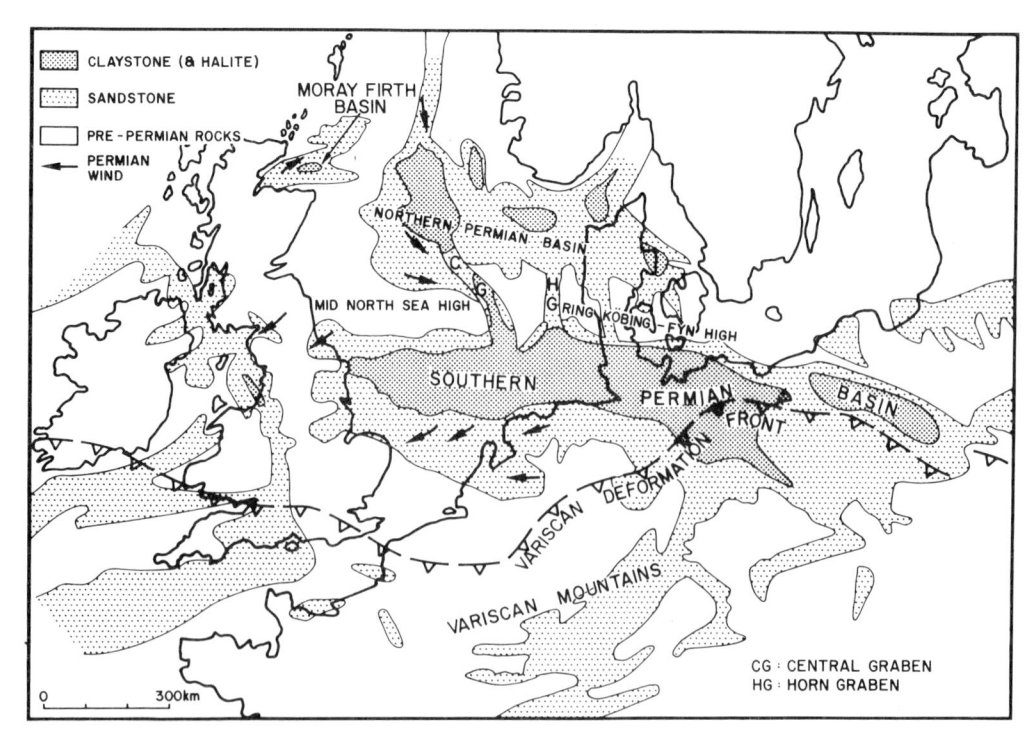

Fig. 4. Distribution of Rotliegend sandstones and desert lake/sabkha claystones in NW Europe. Modified from Ziegler (1982), and Sørensen and Martinsen (1987).

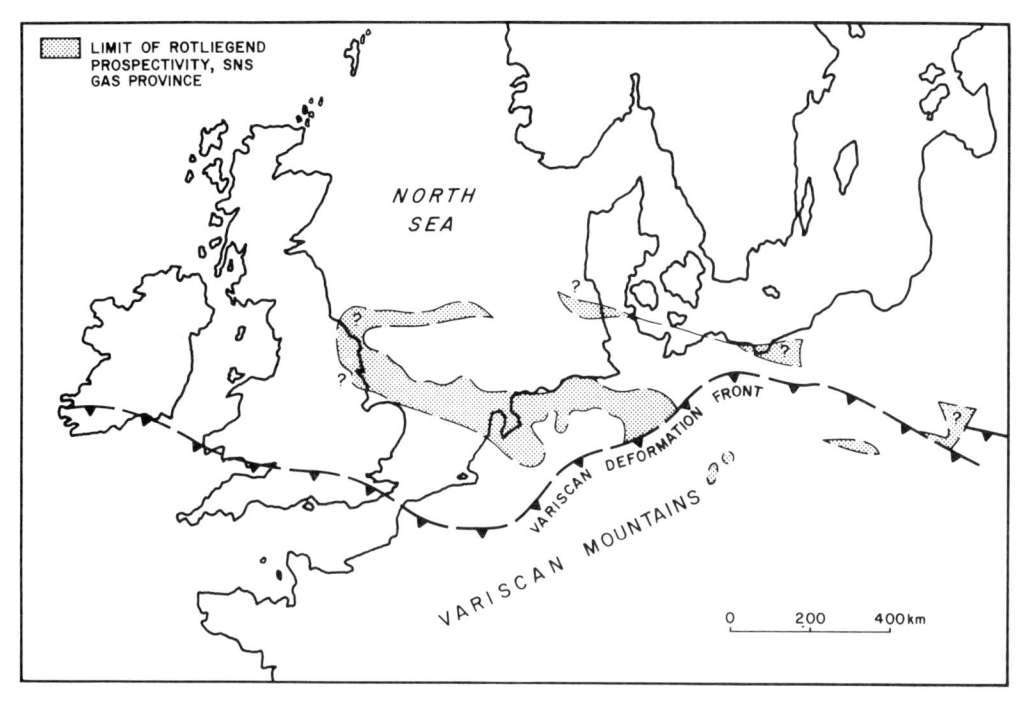

Fig. 5. Areas where the Rotliegend is prospective for gas. Few wells have been drilled in the northern part of the Southern Permian Basin, and the prospectivity in this area is speculative.

thickness to be effective. The belt of proven viability extends from just east of the English coast, through the northern Netherlands to the southern half of the Lower Saxony basin in western Germany (Plein 1978; Betz et al. 1987).

At the end of the Permian, the cyclic evaporation of the Zechstein Sea was succeeded by a long period of continental conditions. The Zechstein sub-basins now became the sites of terrestrial lakes with, initially, the deposition of Bunter Shales. The shales were overlain by fluvial sands transported towards the basin centres. Marine conditions (Muschelkalk) returned to eastern Germany in the Mid-Triassic, whereas in more western areas evaporation of lake-filled basins resulted in precipitation of the Röt, Muschelkalk and Keuper halites.

Over much of the Southern Permian Basin, the Triassic was a period of relative tectonic calm. The Horn and Central grabens, however, together with their extensions into the Southern Permian Basin, began to subside early in the Triassic (Best et al. 1983; Olsen 1983; van Wijhe 1987) in response to active crustal extension. Towards the end of the Triassic the Rhaetian marine transgression spread across the area from the southeast. A global rise in sea level during the Late Triassic–Early Jurassic allowed the accumulation of Liassic shales. Sandstones and associated coals were deposited locally in the Mid Jurassic, to be followed by renewed transgression and deposition of the Kimmeridge Clay.

Sedimentation was largely uninterrupted within the subsiding basins and grabens, and thick sequences of Jurassic strata accumulated. In the Sole Pit and Broad Fourteens sub-basins, this had the effect of burying the Rotliegend reservoir sands and Westphalian source rocks to depths of 3500 m or more, (Cope 1986; Glennie & Boegner 1981; van Hoorn 1987; van Wijhe 1987).

During the Jurassic there was active crustal extension across the North Sea graben systems, reaching a climax with the Late Kimmerian movements during the Late Jurassic–Early Cretaceous (Fig. 3). This tectonism involved reactivation of the post-Variscan wrench system, and helped define the NNW–SSE orientation of most of the UK gas fields.

Under these conditions of tectonic instability, the thick sequences of Zechstein halite were induced to flow diapirically, particularly in the central part of the province. This region is characterized by NW–SE trending salt-induced anticlines and salt piercements (van Hoorn 1987; see also Fig. 6).

The Late Kimmerian earth movements were

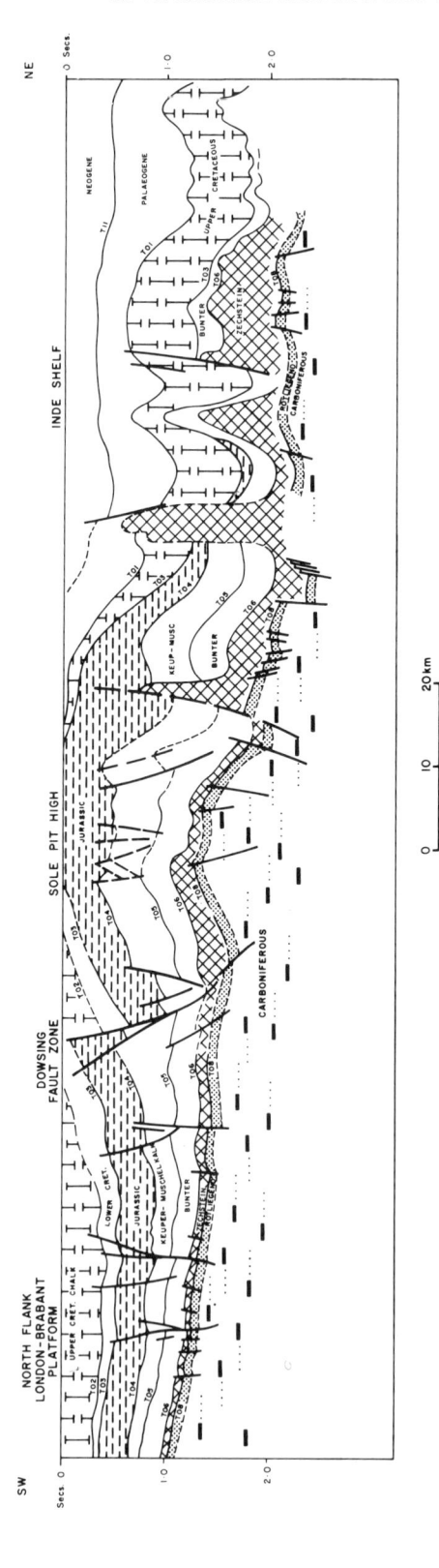

Fig. 6. Seismic line drawing crossing the UK sector of the Southern Permian Basin. Note the thick Jurassic section over the Sole Pit inverted high, and the progressive onlap of the Upper Cretaceous towards the east. For line of Section, see Fig. 7.

also associated with a global lowering of sea level, so that the pattern of uplift-related erosion of the flanks of the graben systems was accentuated during the latest Jurassic–earliest Cretaceous (Fig. 7). In UK waters, the Late Kimmerian uplift and subsequent erosion increases towards the east. Thus in the Sole Pit area, Early Cretaceous sediments rest directly on Lias and locally Keuper sediments, while on the Inde High, truncation extends down almost as far as the Zechstein (Fig. 6).

During the Early Cretaceous, most of the North Sea area was the site of shale deposition, to be followed by a thick deposit of Chalk in the Late Cretaceous. Climatic changes early in the Cenozoic, coupled with uplift of the basin margins, resulted in the accumulation of mainly siliciclastic sediments in the Tertiary sequence.

Transpressive tectonic stresses associated with the Alpine Orogeny affected the gas province in both the Late Cretaceous and the Tertiary. They had the effect of inducing inversion of the sedimentary fill of several sub-basins including the Sole Pit, Broad Fourteens, West and Central Netherlands basins (Figs 6 & 7). Uplift of the Rotliegend reservoir sands by some 1500 m or more created a series of structural closures at Rotliegend level into which gas migrated (Glennie and Boegner 1981; Cope 1986; van Hoorn 1987; Walker and Cooper 1987).

Upper Rotliegend stratigraphy

The Rotliegend desert sediments were deposited in a basin occupying the greater part of the southern North Sea and extending on to the neighbouring land areas of the UK and the Continent (Fig. 4). To the south and southwest lay the London-Brabant Massif, while to the north the uplands of the Mid North Sea High and Ringkøbing–Fyn High separated this basin from the Northern Permian Basin.

Between the foothill regions along the southern margin of the desert and the margins of the desert lake lay a wide plain dominated by aeolian dune systems. Onshore in the UK the dunes are considered to be elongate seifs (Fig. 8), while offshore they are thought to be of the transverse or barchan type (Fig. 9). Evidence from dipmeter logs and oriented cores indicates that the prevailing wind direction was from the east or northeast (van Veen 1975; Glennie 1972).

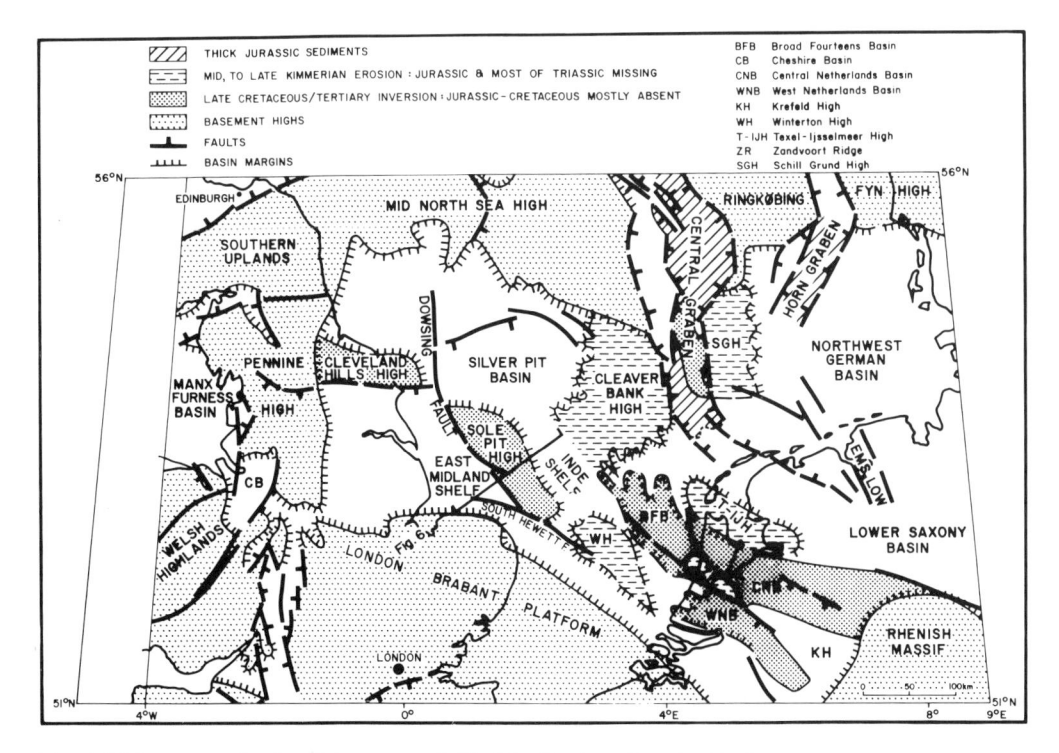

Fig. 7. Main structural units of the western half of the Southern Permian Basin.

Fig. 8. Permian aeolian dune sandstone, Crime Rig quarry, Co. Durham, UK. These dunes are considered to be of the elongate seif type. Height of cliff face approximately 12 m.

Fig. 9. Barchan dunes and extensive interdune sandsheet areas in the Oman desert. Photo courtesy of W. Wiemer.

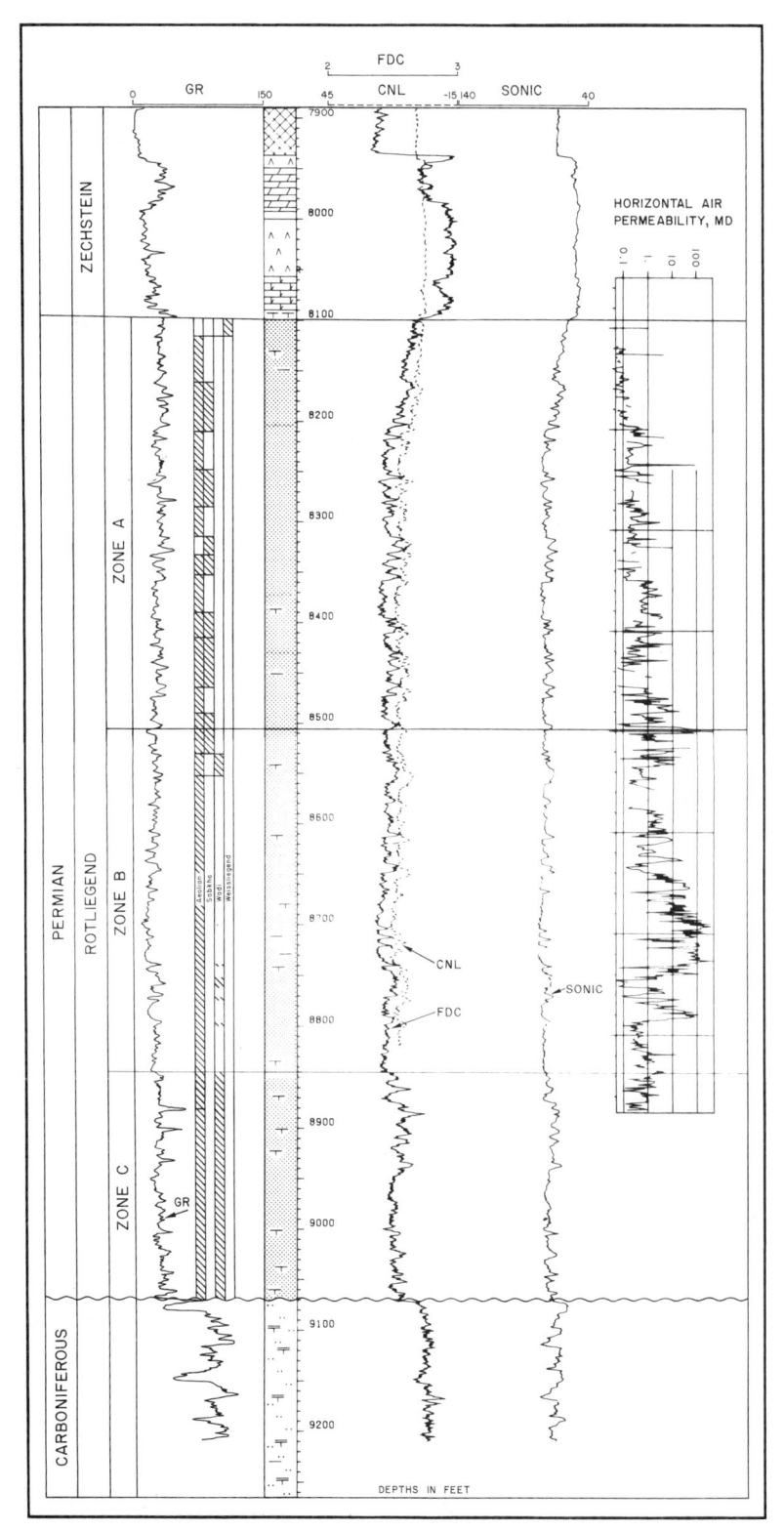

Fig. 10. Log of Well 48/20−1, Sole Pit area, showing Rotliegend reservoir zonation, facies distribution and horizontal air permeability.

In the western part of the province, the Leman Sandstone Formation can be subdivided in up to four or five units depending on the amounts of interbedded wadi, interdune or lake-margin sabkha sequences. In the Indefatigable–Sean area, for example, the Leman Sandstone is relatively thin and uniform, and no subdivision is made (France 1975; ten Have & Hillier 1986). Further west, in the Sole Pit area, three units are generally recognized (Fig. 10). The lowermost unit (C) comprises alternations of wadi and thin dune sands. This is overlain by a middle unit (B) dominated by dune sands, while the upper unit (A) is predominantly dune sands interbedded with thin wadi sandstones and sabkha siltstones. To the north and west, as evident in the West Sole field (Butler 1975), this unit grades into the Silverpit Formation clays and silts.

A fourth sub-unit, homogenized and partially cemented sediments extending over the uppermost 20 to 100 ft of the reservoir, may be recognized in many of the Sole Pit fields. This facies, the Weissliegend, is considered to have formed by internal deformation and partial reworking of the dune sediments during the Zechstein transgression (Glennie & Buller 1983).

In the northern Netherlands, the Upper Rotliegend can be subdivided into five units, of which the second and fifth form the Ameland and Ten Boer claystones (see e.g. van Lith

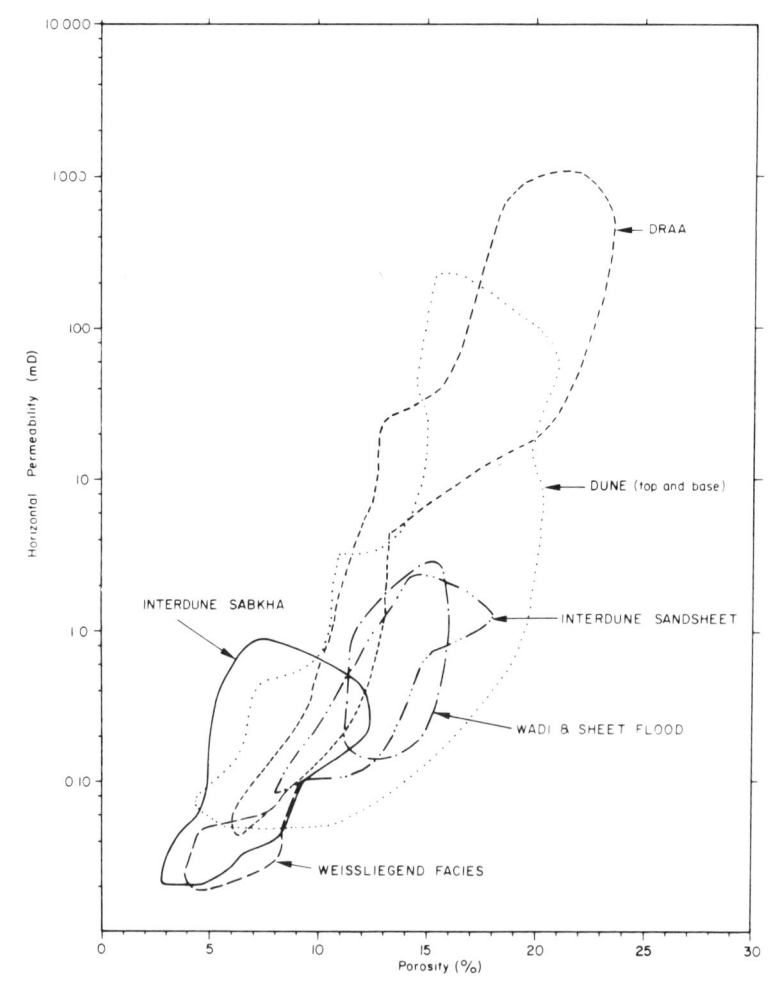

Fig. 11. Permeability/porosity relationships for differing Rotliegend facies, Sole Pit area, UK sector. Note the progressive improvement in reservoir quality from interdune sabkhas through dune to draa sandstones, associated with decreasing clay content and better sorting.

1983), representing phases of expansion of the adjacent desert lake. The coarser Rotliegend clastics in the Groningen field (the Slochteren Formation) were deposited in a relatively narrow strip between the desert lake and the northern margins of the Variscan Mountains. This is illustrated by the rapid increase in thickness from 80 m in the southern part of the field to 240 m in the north (Stäuble & Milius 1970).

Controls on reservoir quality

Primary depositional factors

The reservoir quality of the Rotliegend sediments show wide variation both within and between the individual fields of the province. These variations are the result of a complex interaction on both a regional and local scale between primary depositional factors, post-depositional diagenetic effects, and the tectonic history of the province.

Reservoir sands are of two main types: wadi sands and conglomerates, and aeolian dune sands. The wadi sands and conglomerates of the basin margins form the reservoir rock in at least two fields. In addition to dune sands, the reservoir of the giant Groningen field in the Netherlands comprises fluvial clastics in which the preserved porosity is in the range 15–20% and permeabilities between 0.1 and 3000 mD (Stäuble & Milius 1970). In the Rough field in the western part of the UK sector, the reservoir consists of proximal fluvial sandstones and conglomerates deposited in an alluvial fan environment, interbedded with aeolian dunes at the top of the section, (Goodchild & Bryant 1986). Sand quality deteriorates eastwards across the field as the proximal sands grade into more distal sediments.

Further out in the desert basin, aeolian dune sands generally form the best reservoir rock. The highest porosities and permeabilities are encountered in the 'draa' dunes (Wilson 1972; see also Fig. 11), very thick cross-bedded units of relatively coarse sand deposited as large-scale aeolian bedforms. Generally only the lowest parts of the avalanche-slope sequences are preserved, the upper part being truncated by the horizontal-laminated dune base sands of the next sequence. The foreset beds are normally well sorted, fine to medium grained with laminations dipping up to 25 degrees. Clay content is generally low. Dune-base sandstones show poorer sorting and a higher clay content (ten Have & Hillier 1986).

The shallow water table and damp sabkha conditions of the interdune areas allowed the capture and incorporation of wind-borne clays into the interdune deposits. These areas are now represented by interbedded sheet sandstones and adhesion-ripple siltstones; the higher clay content results in lower porosities and permeabilities than in the dune sandstones (Nagtegaal 1979 and Fig. 11). Moreover, the siltstones are subject to cementation through the evaporation of sulphate and carbonate-rich groundwater.

A similar relationship between reservoir quality and facies is described by Lindquist (1983) from the Jurassic–Triassic Nugget sandstone of Wyoming.

In the Sole Pit fields in the UK sector, the larger dunes are generally restricted to the middle part of the section where the bedding sets attain thicknesses of 10–20 m (Conway 1986). As a consequence, this zone shows the best overall reservoir quality, decreasing upwards and downwards into the more heterogeneous upper and lower zones (Fig. 10).

In contrast, fields in the eastern part of the UK sector, e.g. Leman, Thames–Yare–Bure and the Sean fields (van Veen 1975; Arthur et al. 1986; ten Have & Hillier 1986), the reservoir consists almost entirely of dune sandstones with minor wadi deposits at the base. Although the variation between the different fields is masked to a certain extent by post-depositional effects (see below), reservoir quality in these fields is generally higher than those to the north west, where the section contains a higher proportion of sabkha and wadi deposits.

The primary depositional environment thus had a considerable impact on the resultant matrix properties of the reservoir. Superimposed on these effects are post-depositional diagenetic processes which, in general, have led to a deterioration in reservoir properties.

Diagenetic factors

The primary porosity of aeolian sands varies between 37–41% for the dune base sands and 44–48% for the foreset beds (Beard and Weyl 1973; Hunter 1977). In contrast, the porosity preserved in the Rotliegend sands ranges between 10 and 30% (Table 1). Apart from the influence of pressure solution related to compaction, the majority of the Rotliegend fields have suffered reservoir impairment through a variety of diagenetic factors.

The diagenetic processes may be grouped into early, intermediate and late-stage events. Early diagenesis occurred at shallow burial depth under the influence of ground water movements. Apart from some early cemen-

Table 1. *Reservoir parameters, Rotliegend gas fields, Southern North Sea basin*

Field	On stream	Ultimate reserves (BCM)	Plateau rate (MMCM/d)	No. Platforms
FIELDS ON PRODUCTION:				
UK Sector				
West Sole	1966	54	8	3
Leman	1968	298	46	18
Indefatigable	1971	127	20	7
Viking	1972	84	18	13
Victor	1984	23	7	1
Rough	1975	10	28*	5
Sean	1986	13	17*	3
Thames, Yare, Bure	1986	13	4	2
NETHERLANDS				
Groningen	1959	2720	–	–
Bergen	1972	–	–	–
K8 FA	1977	20	–	–
K14 FA	1977	12	–	–
K15 FA	1977	10	–	–
K15 FB	1977	19	–	–
K/13-E	1979	–	–	1
K/13-F	1978	–	–	1
Ameland	1986	28	–	–
UK FIELDS UNDER DEVELOPMENT:				
Audrey	1988	30	3	1
V-Fields (Vulcan, N&S Valiant, Vanguard)	1988	53	8	8
Villages (Cleeton, S. Ravenspurn)	1988	34	6	4
UK POSSIBLE DEVELOPMENTS:				
Amethyst	1990	8		
Barque/Clipper	1990	25		
North Ravenspurn	1990(?)	34		
Gudgeon	1990	18		
Scram, Wissey, Welland	1990	?12		

* Peak shaving field.
Source: Published data, UK Dept. Energy, *Oil Gas Journal*, June 1984.

tation, it involved the mechanical infiltration of grain-coating clays and the dissolution of unstable ferro-magnesian minerals and the precipitation of the haematite which gives the Rotliegend its characteristic red colour. Some samples also show the development of an early stage of framework-supporting quartz cementation.

With increasing depth of burial and rising temperatures (Fig. 12), clay mineral transformations and feldspar dissolution comes into play.

Authigenic chlorite is common, particularly in samples from the wadi and sabkha facies. Perhaps the most widespread diagenetic effect, however, is the growth of blocky and fibrous forms of illite (Nagtegaal 1979; see Fig. 13).

This phase is thought to have developed during the Cretaceous, when, in the Sole Pit and Broad Fourteens areas, the reservoir was buried to a depth of some 4000 m, i.e. some 1500–2000 m below its present depth (Glennie & Boegner 1981; van Hoorn 1987; Oele et al. 1981; van Wijhe 1987a, b). There is little doubt that illitization is the major factor contributing to the reservoir impairment seen in the Sole Pit and other deeply-buried areas (see also Oele et al. 1981). Depth of burial, and hence the impact of depth-related diagenesis, decreased to the east towards the Inde Shelf. Thus Conway (1986) reports only limited diagenetic impairment in samples from the Victor field reservoir.

No. Devn. Wells	Reservoir Depth (m)	Thickness (m)	Average porosity (%)	Air Permeability (mD)	Major facies	Reference
20	2700	125	15	1	Wadi, Dune	Butler (1975)
160	1800	180–270	14	60	Dune	van Veen (1975)
54	2400	35–130	15	0.1–2000	Dune	France (1975)
–	2800	150	16	60	Dune, Wadi	Gray (1975); Gage (1980)
5	2600	115	16	130–300	Dune, Wadi, Sabkha	Conway et al. (1986)
9	2800	30	15	1–1200	Wadi, Dune,	Goodchild & Bryant (1986)
10	2600	35–70	20	5–650	Dune	ten Have, Hillier (1986)
–	2500	85–210	19	–	Dune	Arthur et al. (1986)
150?	2750	70–240	17	0.1–1000	Dune, Wadi	Stauble & Milius (1970)
22?	2200	200–270	17	–	Dune, Wadi	Van Lith (1983)
–	3300	200–300	10–15	–	Dune, Wadi	Oele et al. (1981)
–	2950	"	"	–	Dune, Wadi	Oele et al. (1981)
–	3300	"	"	–	Dune, Wadi	Oele et al. (1981)
–	3800	"	"	–	Dune, Wadi	Oele et al. (1981)
3	2500	100–200	21	–	Dune	Roos & Smits (1983)
4	2500	100–200	21	–	Dune	Roos & Smits (1983)
15						Oil Gas Journal, June 1984

Late stage diagenetic influences are reflected in widespread quartz, gypsum/anhydrite and dolomite cementation. The gypsum/anhydrite and dolomite cements are most prevalent near the top of the reservoir, and were probably precipitated from percolating brines originating from the overlying Zechstein cap rock.

Not all the post-depositional processes led to reservoir impairment. Evidence from cores from the West Sole field (Butler 1975), and the neighbouring Barque and Clipper fields point to the presence of a natural fracture system in parts of the reservoir (Fig. 14). Most of the fractures are interpreted as shear or dilational shear fractures, the latter having been formed by re-opening of the shear fractures. The timing of fracture development is uncertain, but it seems reasonable to assume that the majority were developed during the Late Cretaceous and Tertiary inversion movements (Fig. 12).

Individual fractures seen in cores may be up to 10 mm wide and several metres long. Many are filled with cement (Fig. 15), but others are open over at least part of their length. By providing a conduit between the well bore and the surrounding matrix, the fractures serve to enhance productivity in those wells which are fortunate enough to encounter them.

Lower Rotliegend reservoirs

The above discussion has concentrated on Upper Rotliegend reservoirs. In the Sohlingen gas field in West Germany, however, Lower Rotliegend aeolian sandstones form a reservoir. The 400 m thick Schneverdingen Sandstone

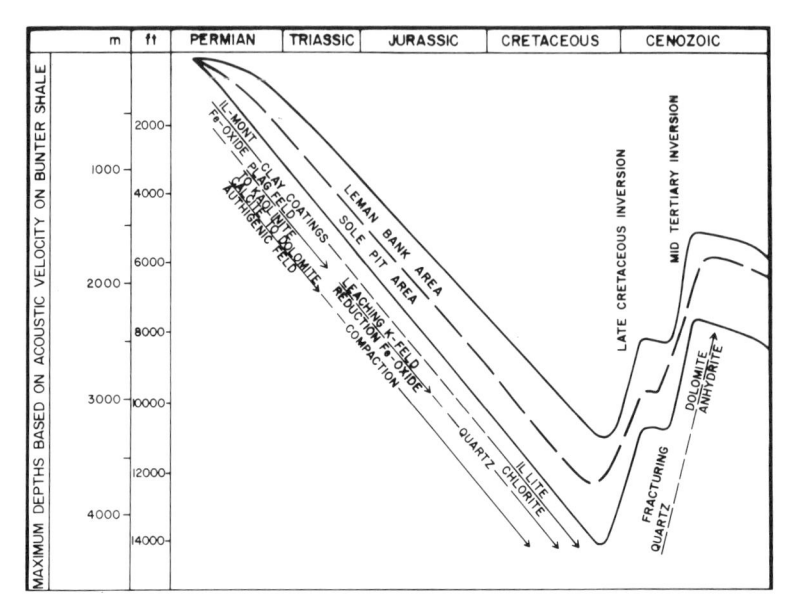

Fig. 12. Burial graph and associated diagenetic sequence for Rotliegend sandstone in the Leman Bank and Sole Pit areas, UK sector. Modified from Glennie *et al.* (1978).

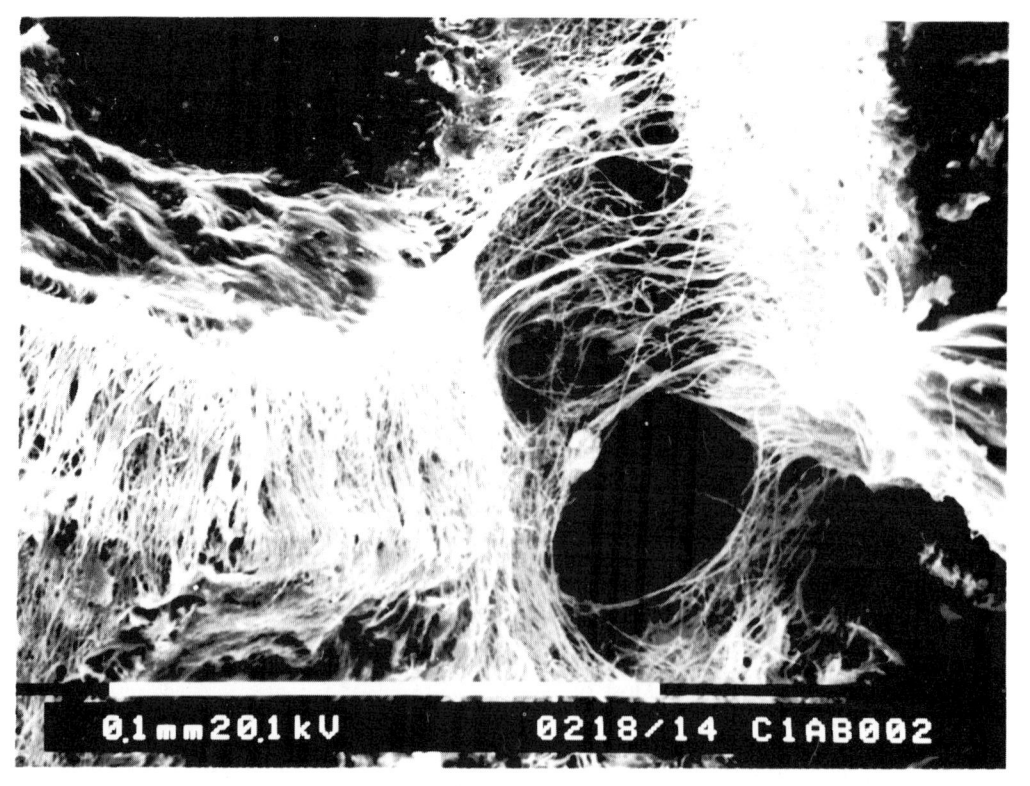

Fig. 13. SEM photograph of fibrous illite bridging pores in aeolian dune sandstone, Sole Pit area, UK sector. Courtesy of Poroperm UK.

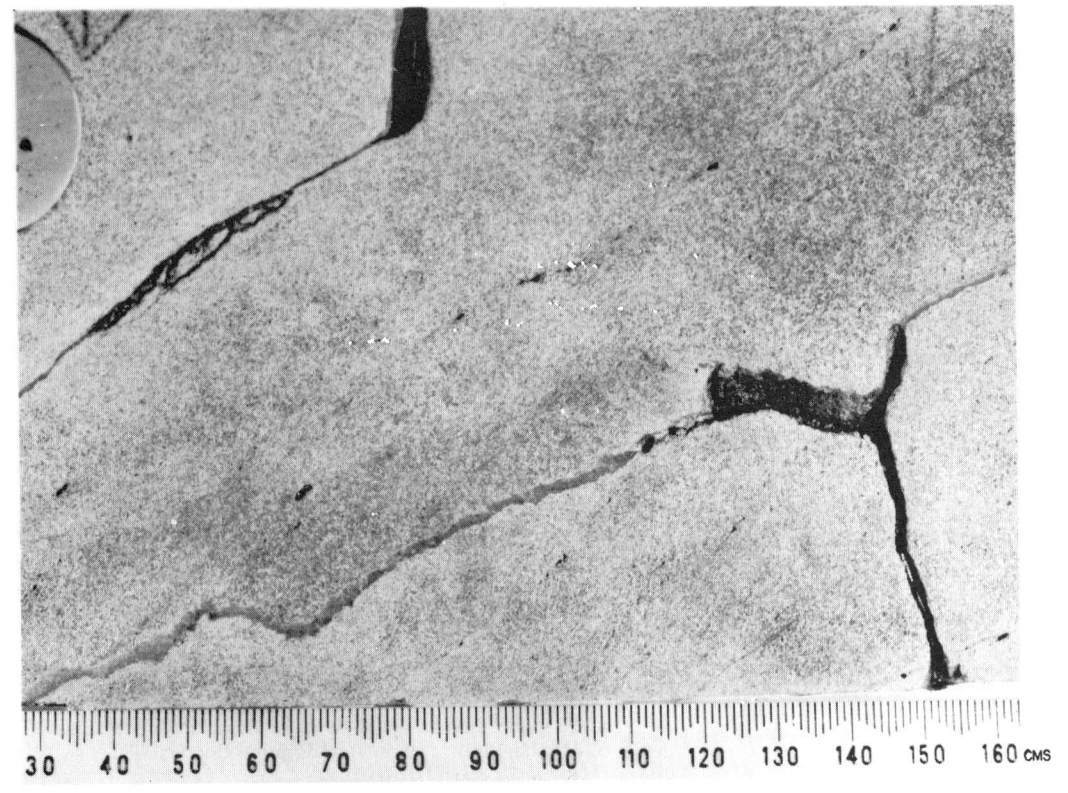

Fig. 14. Core photograph showing partially cemented fractures in wadi sandstone, Sole Pit area, UK sector.

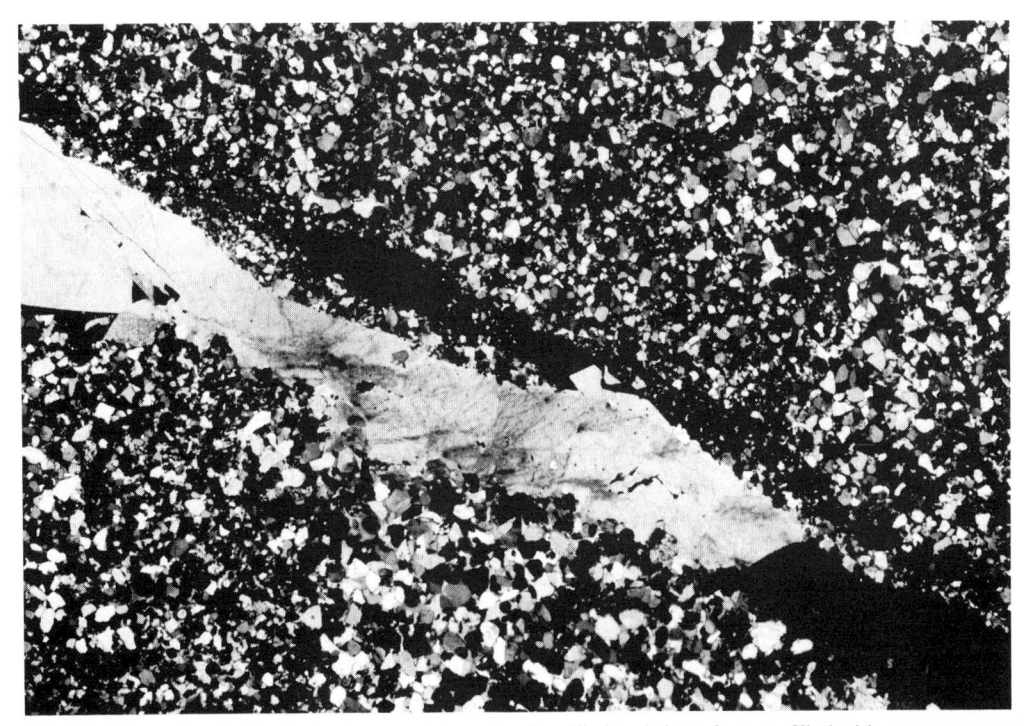

Fig. 15. Thin section (×9) of interdune sandstone showing dilational shear fracture filled with quartz cement developed parallel to shear fracture with cataclasis. Courtesy of KSEPL, Netherlands.

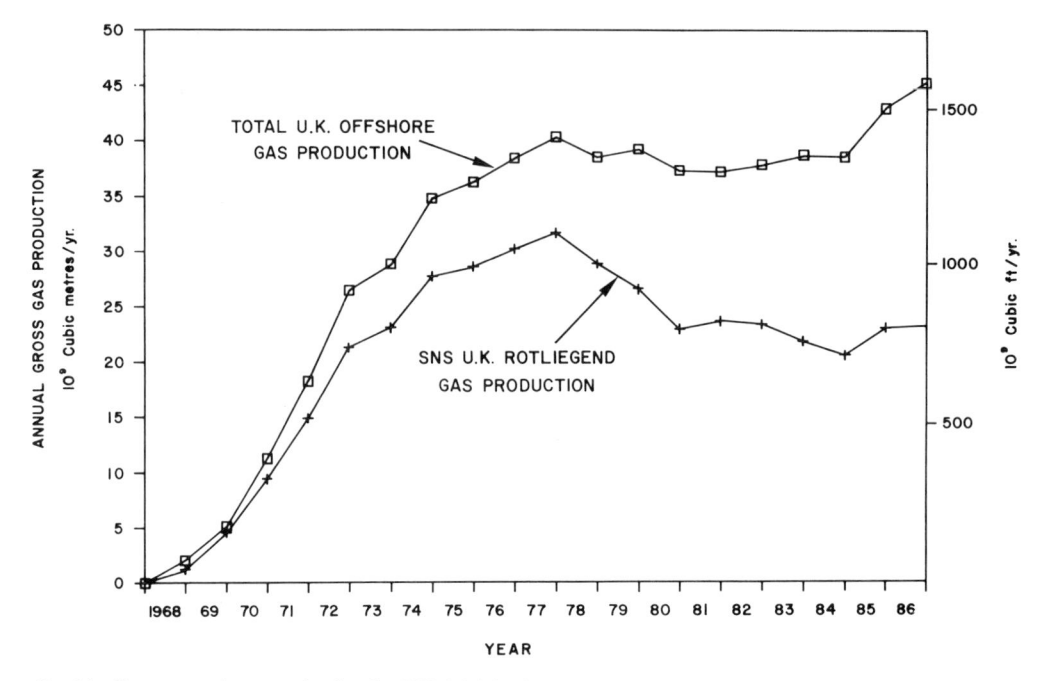

fig. 16. Gross annual gas production for UK fields in the period 1968–1986. Note the significant contribution from the Rotliegend gas fields of the Southern Permian Basin. Figures from Dept. of Energy, UK.

seems to have been deposited in a narrow graben. The sandstones are separated from the overlying and more widespread Upper Rotliegend Hauptsandstein (= Slochteren Formation) by a conglomerate rich in volcanic clasts. Surprisingly, the Schneverdingen Sandstone at a depth of over 5000 m forms a better reservoir (porosity up to 15%, permeability 1–10 mD), than the overlying Hauptsandstein (porosity 1–5%, permeability 0.1 mD). Drong et al. (1982) ascribe the variation in reservoir quality to differences in climate at the time of deposition. It seems just as likely, however, that even with the effects of deep burial diagenesis, the coarser-grained Schneverdingen Sandstone was capable of retaining more of its primary reservoir characteristics than the finer-grained dune sands that overlie it.

Reserves

Published reserves and other salient parameters for Rotliegend gas fields in the Southern North Sea province are shown in Table 1.

Considerable uncertainty is attached to the reserve estimates, particularly for the fields under development or listed as potential developments. The giant Groningen field onshore in The Netherlands dominates the list, with ultimate reserves of 2720 bcm (96 TCF). Offshore, the Leman field in the UK sector has an ultimate recovery of 298 bcm (10.5 TCF), although the majority of the fields have reserves in the range 10–50 bcm (0.35–1.75 TCF).

The current estimate for total reserves in the southern UK Rotliegend fields which are developed, under development, or potential developments, stands at some 850 bcm (30 TCF). It is estimated that a further 160–700 bcm (5.6–24.5 TCF) remain to be discovered in the UK Southern Basin and Irish Sea (Dept. Energy 1987).

The significant contribution of the Rotliegend gas fields to the total UK offshore gas supply (including the Norwegian fields) is illustrated in Fig. 16, which shows the gross annual production for the period 1968–1986. Up until 1973, the Rotliegend fields accounted for some 80% of the supply, but thereafter supply from other parts of the North Sea increased in importance. Nevertheless, after peaking in 1977, the Rotliegend fields have maintained a relatively steady production and in 1986 still accounted for over 50% of the total amount of offshore gas supplied.

Conclusions

The Permian Rotliegend sandstone is the main reservoir of the Southern North Sea gas province. The province itself is the outcome of the sequential deposition of Carboniferous Coal Measures source rock, Rotliegend reservoir and Zechstein cap rock, followed by burial to depths at which the source rock reached maturity.

The Rotliegend reservoir sands were deposited in a desert environment. The best reservoir quality occurs in foreset-bedded sandstones from the avalanche slopes of transverse dunes. Wadi sandstones generally show poorer sorting, a higher clay content and more cementation; consequently they form much poorer reservoirs.

Reservoir quality was impaired by the effects of depth-related diagenesis. Sandstones that were buried deeply in subsiding basins had primary porosities averaging over 40% reduced, in extreme cases, to less than 10%. On the other hand, the development of an open natural fracture system in certain fields has served to enhance productivity.

The authors wish to thank Shell UK Ltd and Esso Petroleum Co Ltd for permission to publish this paper. We acknowledge the contributions by numerous geologists from these and other companies to our understanding of the Rotliegend reservoir. We would also like to thank V. Long for her assistance.

References

ARTHUR, T., PILLING, D., BUSH, D. & MACCHI, L. 1986. Block 49/28 sedimentation, diagenesis and burial history. *In*: BROOKS, J., GOFF, J. & VAN HOORN, B. (eds) *Habitat of Palaeozoic Gas in NW Europe*. Special Publication of the Geological Society, **23**, 251–266.

BEARD, D. C. & WEYL, P. K. 1973. Influence of texture on porosity and permeability of unconsolidated sand. *Bulletin American Association of Petroleum Geologists*, **57**, 349–369.

BEST, G., KOCKEL, F. & SCHONEICH, H. 1983. Geological history of the southern Horn Graben. *In*: KAASSCHIETER, J. P. H. & REIJERS, T. J. A. (eds) *Geologie en Minbouw*, **62**, 25–33.

BETZ, D., FUHRER, F., GREINER, G. & PLEIN, E. 1987. Evolution of the Lower Saxony Basin. *In*: ZIEGLER, P. A. (ed.) *Inversion Tectonics*. Tectonophysics, **137**, 127–170.

BUTLER, J. B. 1975. The West Sole Gas Field. *In*: WOODLAND, A. W. (ed.) *Petroleum Geology and the Continental Shelf of NW Europe*. Heyden, 213–223.

CONWAY, A. M. 1986. Geology and petrophysics of the Victor field. *In*: BROOKS, J., GOFF, J. & VAN HOORN, B. (eds) *Habitat of Palaeozoic Gas in NW Europe*. Geological Society Special Publication, **23**, 85–100.

COPE, M. J. 1986. An interpretation of vitrinite reflectance data from the Southern North Sea Basin. *In*: BROOKS, J., GOFF, J. & VAN HOORN, B. (eds) *Habitat of Palaeozoic Gas in NW Europe*. Geological Society Special Publication, **23**, 237–250.

CORNFORD, C. 1986. Source rocks and hydrocarbons of the North Sea. *In*: GLENNIE, K. W. (ed.) *Introduction to the Petroleum Geology of the North Sea*. Blackwell, 197–236.

DEPOWSKI, S. 1978 (ed.) *Lithofacies–Palaeogeographical Atlas of the Permian Platform areas of Poland*. Institute of Geology, Warsaw.

Dept. of Energy (UK), 1987. *Development of the Oil and Gas Resources of the United Kingdom*, HMSO.

DRONG, H. J., PLEIN, E., SANNEMANN, D., SCHEUPBACH, M. A. & ZIMDARS, J. 1982. Der Schneverdinger Sandstein des Rotliegenden — eine aolische Sedimentfullung alter Graben Strukturen. *Zeit. deutsch. geol. Ges*, **133**, 699–725.

FALKE, H. 1972. (ed.) Rotliegend. Essays on European Lower Permian. Inter. Sediment. Petrograph. Ser, **XV**.

—— 1976. (ed.) The continental Permian in Central, West and South Europe. Nato Advanced Study Institutes Series. Reidel, Dordrecth.

FRANCE, D. S. 1975. The geology of the Indefatigable gas field. *In*: WOODLAND, A. W. (ed.) *Petroleum Geology and the Continental Shelf of NW Europe*. Heyden, 233–241.

GAGE, M. 1980. A review of the Viking gas field. *In*: HALBOUTY, M. T. (ed.) *Giant oil and gas fields of the decade 1968–1978*. American Association of Petroleum Geologists Memoir, **30**, 39–55.

GLENNIE, K. W. 1972. Permian Rotliegendes of North West Europe interpreted in the light of modern sedimentation studies. *Bulletin American Association of Petroleum Geologists*, **56**, 1048–1071.

—— 1986a. Development of NW Europe's Southern Permian gas basin. *In*: BROOKS, J., GOFF, J. & VAN HOORN, B. (eds) *Habitat of Palaeozoic Gas in NW Europe*. Geological Society Special Publication, **23**, 3–22.

—— 1986b. Early Permian – Rotliegend. *In*: GLENNIE, K. W. (ed.) *Introduction to the Petroleum Geology of the North Sea*. Blackwell, 63–86.

——, MUDD, G. C. & NAGTEGAAL, P. J. C. 1978. Depositional environment and diagenesis of Permian Rotliegendes sandstones in Leman Bank and Sole Pit areas of the UK Southern North Sea. *Journal of the Geological Society London*, **135**, 25–34.

—— & BOEGNER, P. 1981. Sole Pit inversion tectonics. *In*: ILLING, L. V. & HOBSON, G. D. (eds) *The Petroleum Geology of the Continental Shelf of NW Europe*. Heyden, 110–210.

—— & BULLER, A. T. 1983. The Permian Weissliegend of NW Europe: the partial deformation of aeolian dune sands caused by the Zechstein transgression. *Sedimentary Geology*, **35**, 43–81.

GOODCHILD, M. W. & BRYANT, P. 1986. The geology

of the Rough field. *In*: BROOKS, J., GOFF, J., & VAN HOORN, B. (eds) *Habitat of Palaeozoic Gas in NW Europe*. Geological Society Special Publication, **23**, 223–235.

HAVE, A. TEN & HILLIER, A. 1986. Reservoir geology of the Sean North and South gas fields, UK Southern North Sea. *In*: BROOKS, J., GOFF, J. & VAN HOORN, B. (eds) *Habitat of Palaeozoic Gas in NW Europe*. Geological Society Special Publication, **23**, 267–273.

HOORN, B. VAN 1987. Structural evolution, timing and tectonic style of the Sole Pit Inversion. *In*: ZIEGLER, P. A. (ed.) *Inversion Tectonics*. Tectonophysics, **137**, 239–284.

HUNTER, R. E. 1977. Basic types of stratification in small aeolian dunes. *Sedimentology*, **24**, 361–387.

KENT, P. E. 1980. Subsidence and uplift in East Yorkshire and Lincolnshire: a double inversion. *Proceedings of the Yorkshire Geological Society*, **42**, 502–524.

LINDQUIST, S. J. 1983. Nugget Formation reservoir characteristics affecting production in the Overthrust Belt of Southwestern Wyoming. *Journal of Petroleum Technology*, **35**, 1355–1365.

LITH, J. G. J. VAN 1983. Gas fields of Bergen Concession, The Netherlands. *In*: KAASSCHIETER, J. P. H. & REIJERS, T. J. A. (eds) Geologie en Mijnbouw, **62**, 63–74.

LUTZ, M., KAASSCHIETER, J. P. H. & WIJHE, H. D. VAN, 1975. Geological factors controlling Rotliegend gas accumulation in the Mid-European Basin. *Proceedings of the 9th World Petroleum Congress*, **2**, 93–97.

NAGTEGAAL, P. J. C. 1979. Relationship of facies and reservoir quality in Rotliegendes desert sandstone, Southern North Sea Region. *Journal of Petroleum Geology*, **2**, 145–158.

OELE, J. A., HOL, A. C. P. J. & TIEMENS, J. 1981. Some Rotliegend gas fields of the K and L blocks, Netherlands offshore (1968–1978) – a case history. *In*: ILLING, L. V. & HOBSON, G. D. (eds) *The Petroleum Geology of the Continental Shelf of NW Europe*. Heyden, 289–300.

OLSEN, J. C. 1983. The structural outline of the Horn Graben. *In*: KAASSCHIETER, J. P. H. & REIJERS, T. J. A. (eds) *Geologie en Mijnbouw*, **62**, 47–50.

PLEIN, E. 1978. Rotliegend-Ablagerungen im norddeutschen Becken. *Zeit. deut. geol. Ges*, **129**, 71–97.

ROOS, B. M. & SMITS, B. J. 1983. Rotliegend and main Buntsandstein gas fields in block K/13 – a case history. *In*: KAASSCHIETER, J. P. H. & REIJERS, T. J. A. (eds) *Geologie en Mijnbouw*, **62**, 75–82.

SORENSEN, S. & MARTINSEN, B. B. 1987. A paleogeographic reconstruction of the Rotliegendes deposits in the northeastern Permian basin. *In*: BROOKS, J. & GLENNIE, K. W. (eds) *Petroleum Geology of North West Europe*. Graham & Trotman, 497–508.

STAUBLE, A. J. & MILIUS, G. 1970. Geology of Groningen gas field. *American Association of Petroleum Geologists Memoir*, **14**, 359–369.

TAYLOR, J. C. M. 1986. Late Permian – Zechstein. *In*: GLENNIE, K. W. (ed.) *Introduction to the Petroleum Geology of the North Sea*. Blackwell, 87–111.

TEICHMULLER, M. 1986. Coalification and natural gas deposits in NW Germany. *In*: BROOKS, J., GOFF, J. and VAN HOORN, B. (eds) *Habitat of Palaeozoic Gas in NW Europe*. Geological Society Special Publication, **23**, 101–112.

TUBB, S. R., SOULBY, A. & LAWRENCE, S. R. 1986. Palaeozoic prospects on the northern flanks of the London-Brabant Massif. *In*: BROOKS, J., GOFF, J. & VAN HOORN, B. (eds) *Habitat of Palaeozoic Gas in NW Europe*. Geological Society Special Publication, **23**, 55–72.

VEEN, F. R. VAN, 1975. Geology of the Leman gas field. *In*: WOODLAND, A. W. (ed.) *Petroleum Geology and the Continental Shelf of NW Europe*. Heyden, 223–233.

WALKER, I. M. & COOPER, W. G. 1987. The structural and stratigraphic evolution of the north–east margin of the Sole Pit Basin. *In*: BROOKS, J. & GLENNIE, K. W. (eds) *Petroleum Geology of North West Europe*. Graham & Trotman, 263–275.

WIJHE, D. H. VAN, 1987a. Structural evolution of inverted basins in the Dutch offshore. *In*: ZIEGLER, P. A. (ed.) *Inversion Tectonics*. Tectonophysics, **137**, 171–219.

—— 1987b. The structural evolution of the Broad Fourteens Basin. *In*: BROOKS, J. & GLENNIE, K. W. (eds) *Petroleum Geology of North West Europe*. Graham & Trotman, 315–335.

—— LUTZ, M. & KAASSCHIETER, J. P. H. 1980. The Rotliegend in the Netherlands and its gas accumulations. *Geologie en Mijnbouw*, **59**, 3–24.

WILSON, I. G. 1972. Aeolian Bedforms – their development and origins. *Sedimentology*, **19**, 173–210.

ZIEGLER, P. A. 1980. Geology and hydrocarbon provinces of the North Sea. *Canadian Society of Petroleum Geologists Memoir*, **6**, 635–706.

—— 1982. *Geological Atlas of Western and Central Europe*. Elsevier, Amsterdam.

A regional assessment of the intra-Carboniferous play of Northern England

A. J. FRASER[1], D. F. NASH, R. P. STEELE & C. C. EBDON

BP Petroleum Development Ltd., Kirklington Rd, Eakring, Newark, Notts NG22 0DA, UK

[1]*Present address: BP Petroleum Development Ltd., 301 St Vincent Street, Glasgow, G25 2DD, UK*

Abstract: Hydrocarbons have been produced from the Carboniferous of the East Midlands since the 1920s. Some 30 discoveries have been made to date and although field sizes are generally small, the relatively low costs have made exploration and development an attractive commercial proposition. Many of the factors which contribute to success in the East Midlands occur in other basins in Northern England but have not produced significant hydrocarbon accumulations. This paper examines the key attributes of the intra-Carboniferous play and through an integrated regional geological framework contrasts the successful East Midlands oil province with the other Carboniferous basins of Northern England.

North of the Variscan front in England, Carboniferous sediments form viable targets for hydrocarbon exploration. The sediments were deposited in a complex network of fault bounded basins which escaped the penetrative deformation and pervasive, low-grade metamorphism that affects much of the Carboniferous further south (see Besly (1988a) for an excellent summary).

Hydrocarbons have been produced from the Carboniferous of Northern England since the 1920s (Fig. 1) (Less & Cox 1937; Lees & Taitt 1945; Falcon & Kent 1960; Kent 1985). Some 30 discoveries have been made to date, the majority in the East Midlands (Fig. 2). Despite field sizes being generally small (typically 1−2 mm bbls recoverable), the relatively low costs have made exploration and development an attractive commercial proposition.

A cursory inspection of the number of tests and hydrocarbon shows in the Carboniferous of Northern England would suggest that plays with Carboniferous targets had been intensely explored: some 280 exploration wells have been drilled to test the Carboniferous, of which almost 60% have encountered shows. On closer inspection, however the number of valid (structure/reservoir-seal present) tests are substantially fewer and for the intra-Carboniferous are concentrated in the East Midlands with as little as a dozen valid tests outside this area, and none of commercial significance. The lack of success outside the East Midlands requires some explanation, particularly given the widespread distribution throughout Northern England of Carboniferous sediments and basins apparently similar to the East Midlands (Fig. 3).

The object of this paper is to attempt to answer the question: 'Is the East Midlands unique as an oil province in Northern England?' To achieve this, data from an extensive inventory of 'successful' intra-Carboniferous tests has been synthesized and used to identify and calibrate the key factors controlling the specific hydrocarbon environments. These factors have in turn been used to assess the remaining 10−15 sub-basins in Northern England where Carboniferous sediments were deposited and where the play may be potentially developed.

Regional geological model

The basis for understanding the distribution of hydrocarbons in the intra-Carboniferous of Northern England is to be found in an interpretation of the complex geological history which the area has undergone in Phanerozoic times (Anderton *et al.* 1979; Andre *et al.* 1986; Leeder 1976, 1982, 1987; Matte 1986; Thorpe *et al.* 1984; Whittaker 1985; Zeigler 1978, 1982, 1986). Four important tectonic phases are recognized, each of which fundamentally influences the processes which shape the hydrocarbon environment. Modern sequence stratigraphic techniques applied to regional seismic lines have been used along with outcrop and well data to provide a consistent interpretational framework, while recent ideas on Caledonian

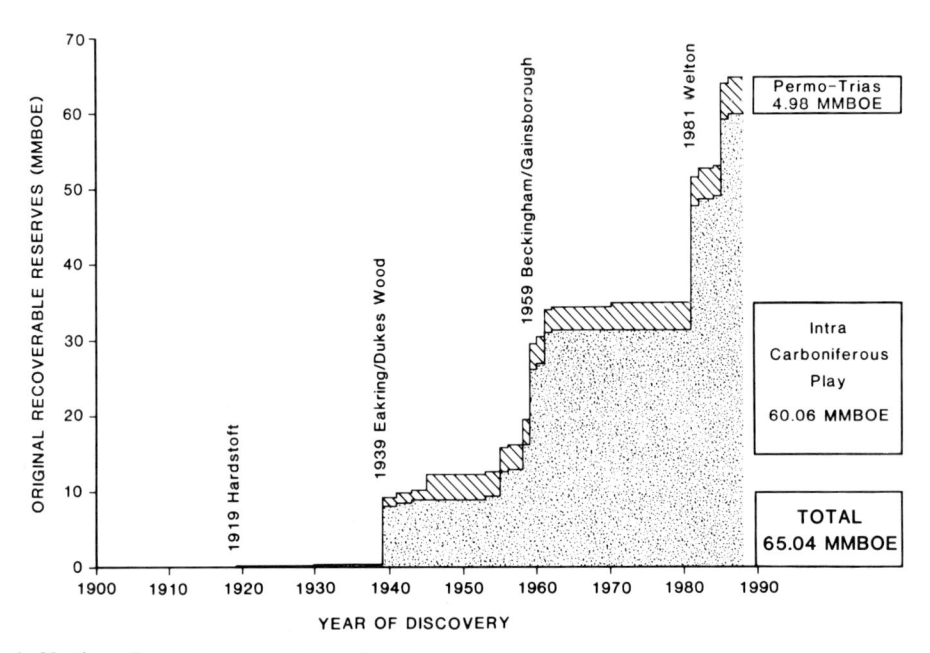

Fig. 1. Northern England — cumulative original recoverable reserves.

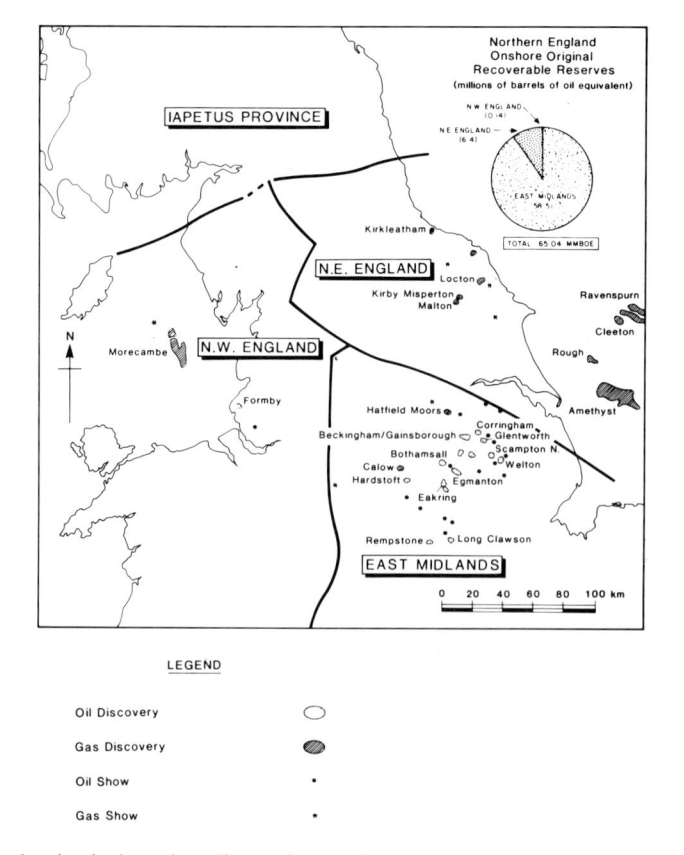

Fig. 2. Study area showing hydrocarbon discoveries and shows.

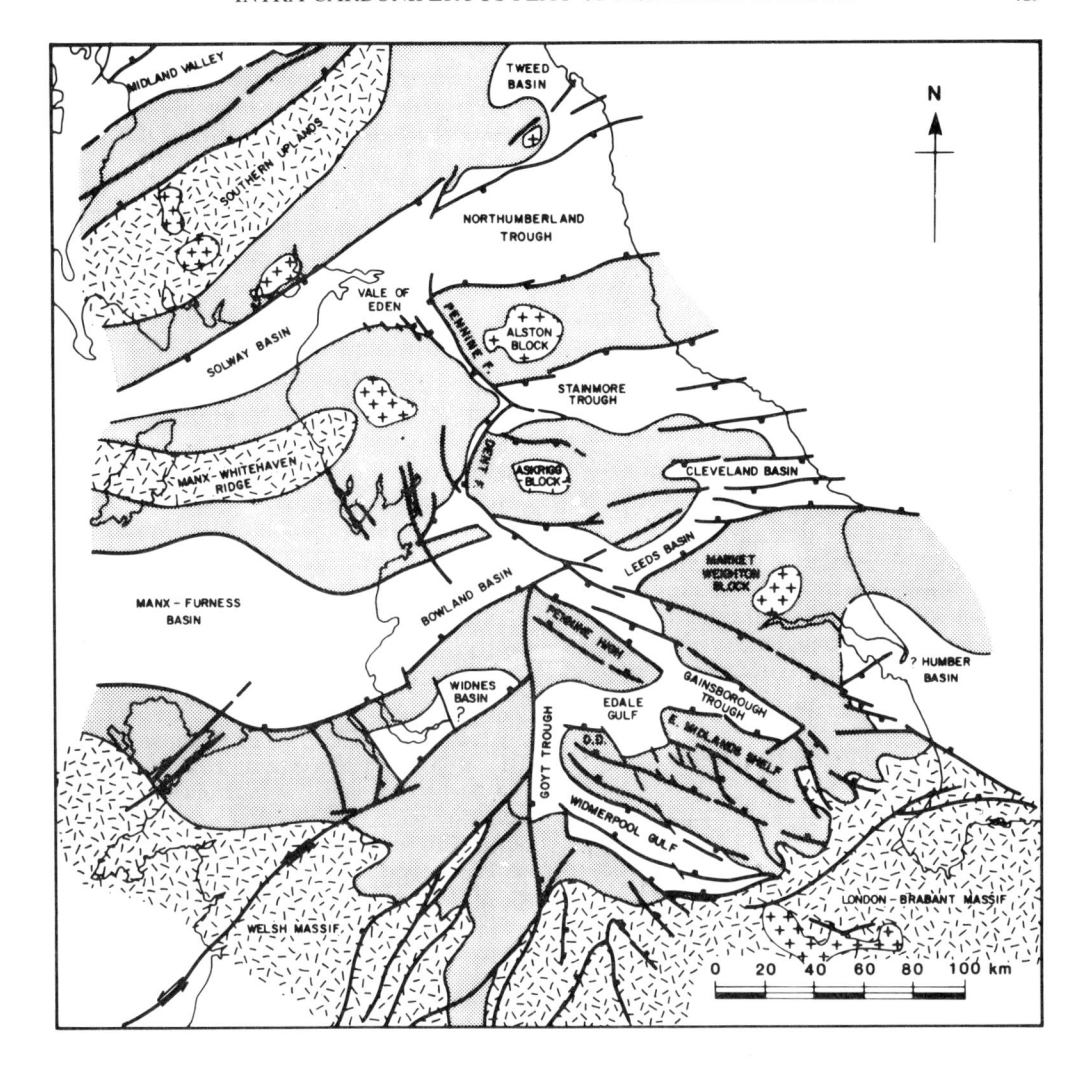

Fig. 3. Early Carboniferous structural elements.

and Variscan tectonics have been incorporated and developed to provide the framework for a regional basin development model in Northern England.

Caledonian structural framework

It is widely recognized that the subsidence and inversion patterns which emerged from late Palaeozoic times onwards were controlled by the underlying structure of the Caledonian basement (Arthurton 1986; Turner 1949; Wills 1973, 1978). Reconstructing the Caledonian structure and understanding its origin is an important pre-requisite for interpreting later crustal events that controlled basin polarity, subsidence history, depositional patterns, uplift etc.

On the basis of Caledonide tectonic trends, sutures or major fault zones and other geological features (e.g. proportion of granite), five provinces are recognized (Fig. 4).

The Central England Province. A triangular, fault-bounded basement area with a thin, weakly deformed covering of Lower Palaeozoic sediments, terminated to the south by the Variscan front. Geodynamically the feature is interpreted as a microcraton indenter that collided and joined with other crustal fragments during the closing of the Iapetus ocean in early Palaeozoic times (Andre *et al.* 1986, Pharaoh *et al.* 1987, Soper & Hutton 1984, Soper *et al.* 1987).

The East Midlands Province. This exhibits a pronounced NW−SE (so-called Charnoid) trend, which can be identified both in the basement areas to the southeast and in the post-Caledonian sedimentary cover to the northwest (Turner 1949; Kent 1968; Le Bas 1972; Evans 1979; Bott 1987; Wills 1973). The same trend is typical of the Caledonian grain as far east as the Tournquist line in mainland Europe (Ziegler 1982). The province is bound to the north by the Craven−South Humberside−Hewitt line, to the east by the Pennine−Dent−Malvern line and to the south by the northeastern margin of the Central England microcraton.

The Northeast England Province. This occurs to the north of the Craven−South Humberside−Hewitt line, and is bounded to the west by the Pennine−Dent line and to the north by the Iapetus suture. The eastern boundary is not clear: it may extend as a distinct block into the Mid North Sea High or be truncated by the precursor of the Dowsing fault zone. Within the province the dominant tectonic trends are east−west and most of the areas which remained positive during subsequent subsidence phases are cored with granites of late Caledonian age (Bott 1966).

The Northwest England Province. On the opposite side of the Pennine−Dent line, the Caledonian grain assumes the more familiar NE−SW trend of the Scottish Caledonides. Basement faults in the Welsh Massif active prior to collision (Coward & Siddans 1979) can be traced into Northwest England where they are truncated by the Pennine−Dent−Malvern line. The boundary of the indenter (Pontesford−Church Stretton fault) forms the southeastern margin, but the northern boundary is poorly constrained: whether the Lake District should be counted as part of the Northwest England province, or whether it represents a separate distinctive block (microcraton) is not clear.

The Iapetus Province. Finally, overlying the site of the Iapetus suture (Dewey 1971, 1982; Soper & Hutton 1984, Beamish & Smythe 1986; Klemperer & Matthews 1987) encompassing the Northumberland/Solway basin, a narrow province is identified where the Caledonian structure trends strongly ENE−WSW.

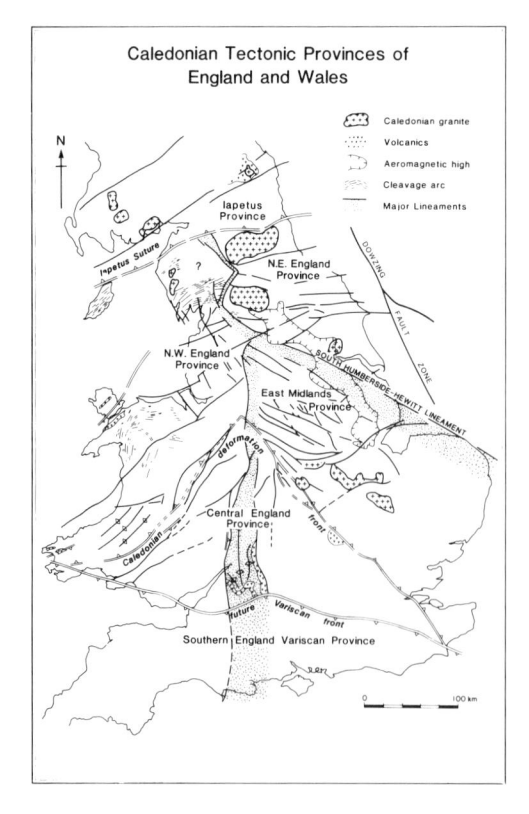

Fig. 4. Caledonian tectonic provinces of England and Wales.

Variscan cycle

Framework and interpretation. The Variscan Cycle extends from post Caledonian consolidation in late Devonian times to the cessation of inversion in the late Carboniferous — early Permian. Rifting during this period produced a complex series of basins by extension along the pre-existing Caledonian framework (Fig. 4).

Traditionally the Carboniferous structure and facies of these basins has been interpreted in terms of a series of steep-sided 'blocks' and 'troughs' with shelf carbonates covering the blocks and basinal mudstones accumulating in the troughs (e.g., Kent 1966). Typically 'reefs' were interpreted as flanking the margins of the blocks (Hudson 1930; Parkinson 1957; Miller & Grayson 1982). More recently tilt block models have become popular to explain the marked asymmetry of the basins revealed by deep drilling and modern multifold seismic (e.g., Grayson & Oldham 1987; Lee 1988; Leeder 1987; Leeder & Gawthorpe 1987; Smith *et al.* 1985). These models typically use straight faults which are associated with folding in the hangingwall. However, this particular geometrical relationship will only occur above curved listric faults.

Seismic data has shown the Carboniferous basins to have a very characteristic form (Figs 5–7). They appear as strongly inverted half-graben with a rollover flexure often associated with antithetic faulting on the shallow side of the basin. This flexure which represents the return to regional in the hangingwall is often associated with the development of rimmed carbonate margins which prograde basin-wards. Similar margins are developed contemporaneously above the footwall scarp of the graben. The hangingwall shows strong inversion with pronounced angular unconformity at the base of the Permian as a result of reactivation of the main boundary faults during the Variscan orogeny (late Carboniferous).

Variscan stratigraphy and palaeofacies interpretation. A chronostratigraphic diagram for the Carboniferous of Northern England has been compiled to illustrate the relationship between conventional Carboniferous stages (George *et al.* 1976; Ramsbottom 1973; Ramsbottom *et al.* 1978) and the sequence stratigraphy developed from seismic, well control and outcrop studies (Fig. 8). The sequence stratigraphic scheme proposed here for the Carboniferous (Fig. 9) is based mainly on the seismo-stratigraphic interpretation of high-quality, modern multifold reflection seismic lines in Northern England and in particular the East Midlands. These have

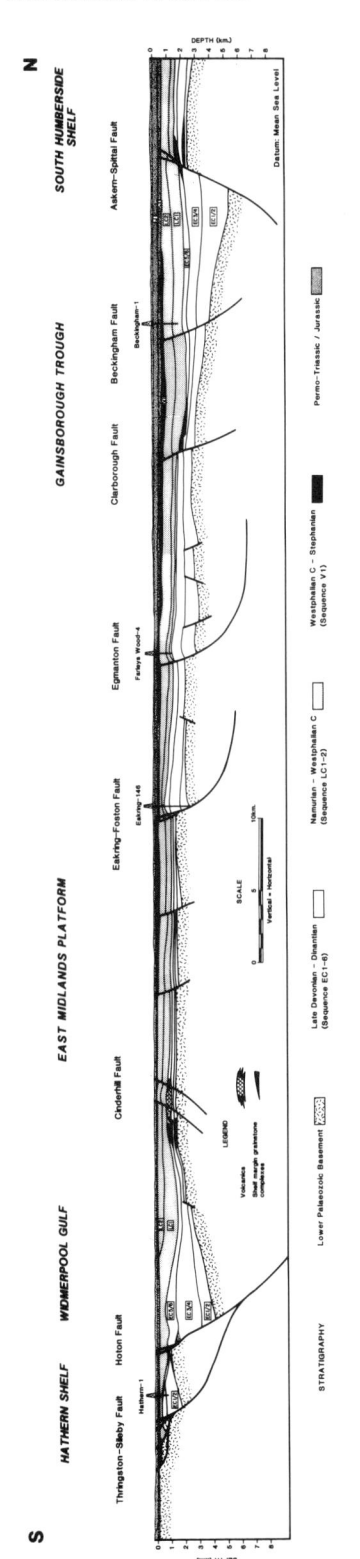

Fig. 5. East Midlands province – Regional cross section.

Table 1. *Sequence stratigraphy of the Variscan cycle in the Central England rift system*

Megasequence		Sequence	Age	Tectonics/Facies Description
Inversion		V1	Late Westphalian C–Stephanian	Strong uplift and erosion of syn-rift depocentres. Accumulation of barren red beds and coarse pebbly sands and fanglomerates in internally drained molasse basins.
Post-rift		LC2	Early Westphalian A–late Westphalian C	Thermal subsidence. Establishment of upper delta plain, coal swamp conditions over much of Northern England. Progradation of Pennine delta southwards across Wales–Brabant massif.
		LC1	Late Brigantian–early Westphalian A	Onset of post-rift thermal subsidence. Drowning of shelf margins and development of intra-shelf basins (see Gutteridge 1987) following marine transgression over inverted topography. Progressive advance of deltaic systems from northeast drowning carbonate platforms.
Syn-rift	III	EC6	Early–mid Brigantian	Basin inversion event resulting in a regressive phase. Carbonate ramp to rimmed shelf development in the East Midlands with margins both aggradational and progradational.
		EC5	Late Asbian–early Brigantian	Reactivation of extensional fault regime with significant back-stepping of fault system in Widmerpool Gulf (Fig. 6). Renewed footwall rotation and erosion and development of boulder beds and slumps basinward of drowned margins.
	II	EC4	Late Holkerian–mid Asbian	Stillstand or regressive phase. Carbonate ramp to rimmed shelf development with margins both aggradational and progradational.
		EC3	Late Chadian–late Holkerian	Rejuvenation of extensional faults causing rotation of fault blocks and significant footwall erosion. Development of boulder beds and slumps (see Gawthorpe 1987) and drowning of carbonate shelf margins.
	I	EC2	Chadian	Stillstand or regressive phase. Carbonate ramp to rimmed shelf development with the growth of Waulsortian-type mud mounds on the upper part of ramp.
		EC1	Late Devonian–Courceyan	Initial development of half graben. Downlapping subaerial fan sequence developing into Old Red Sandstone fluvial plain deposits. Marine transgression marked by progressive onlap of basement by carbonate slope deposits.
Pre-rift		PZ	Pre–late Devonian	Caledonian granite and folded lower Palaeozoic basement complex.

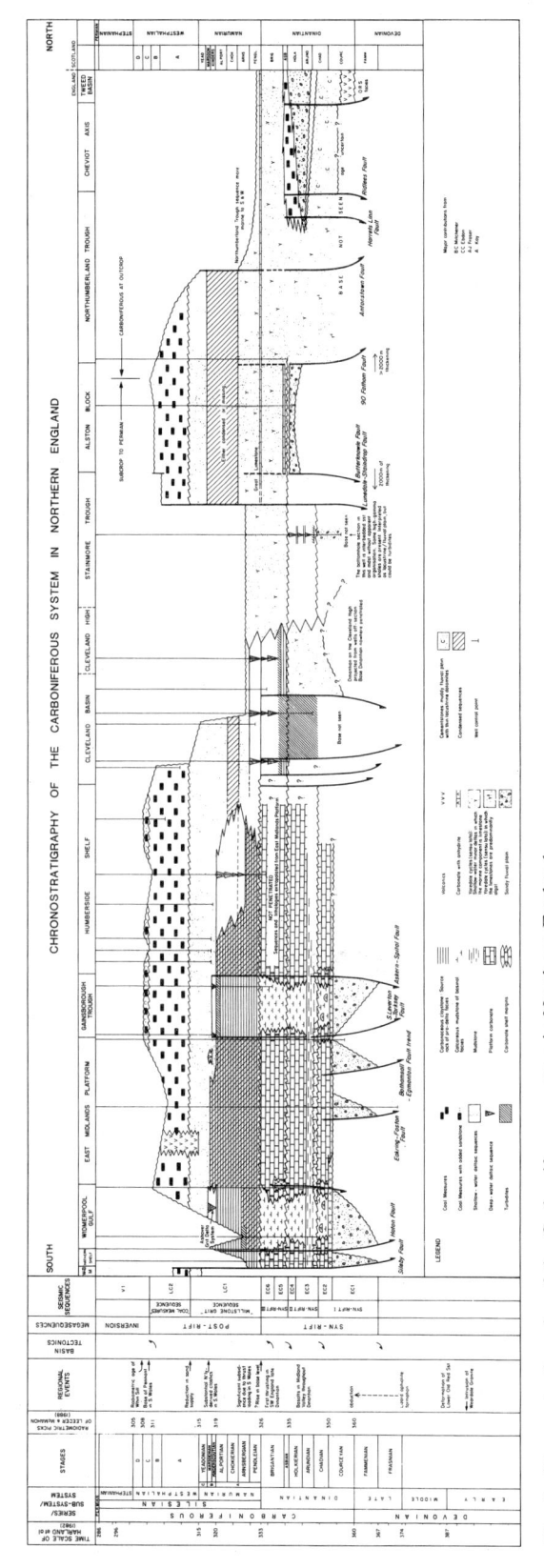

Fig. 8. Chronostratigraphy of the Carboniferous system in Northern England.

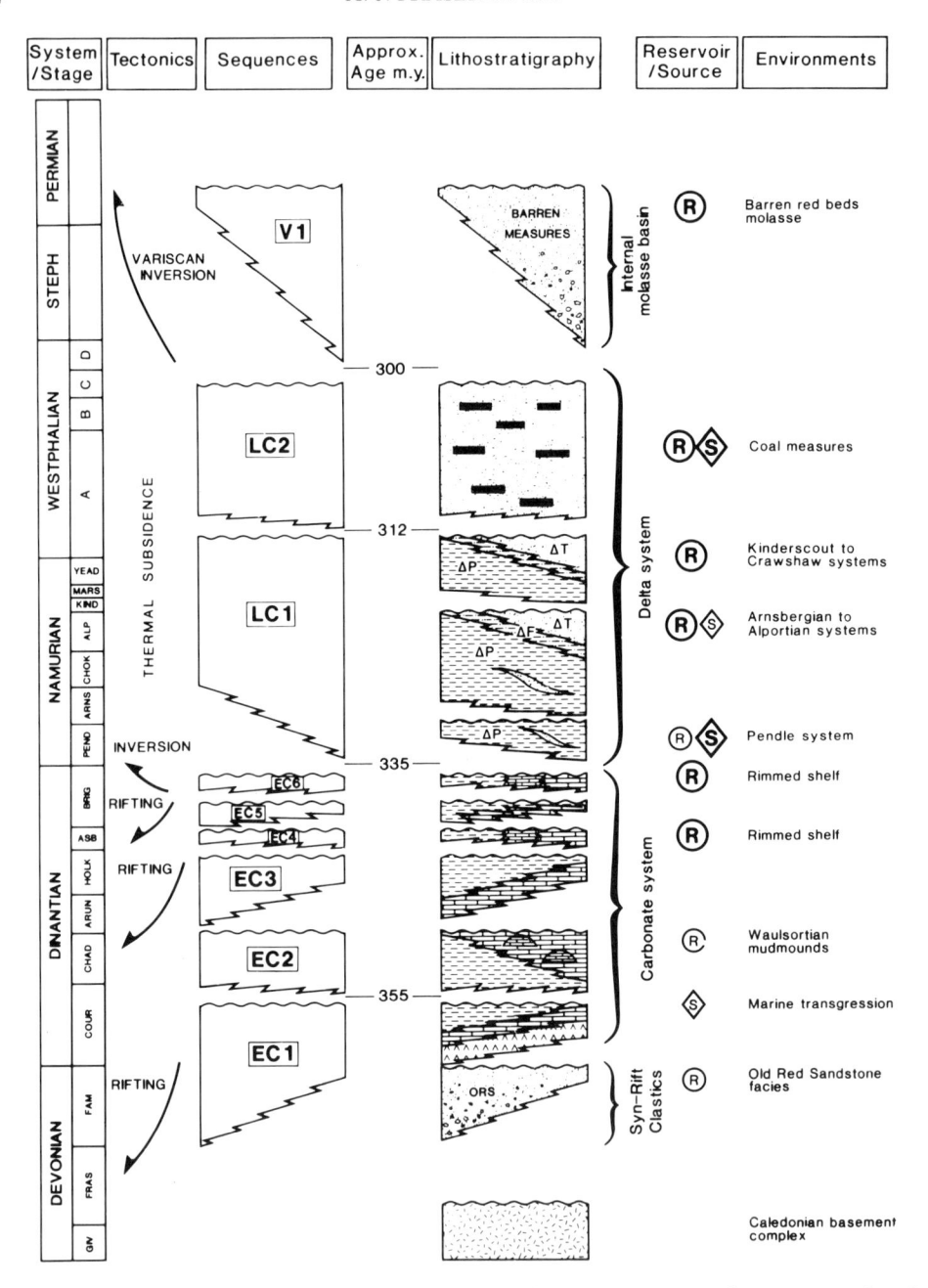

Fig. 9. Summarised stratigraphy of the East Midlands — showing megasequence and sequence stratigraphy.

been rigorously tied to biostratigraphic data from near-by well control and correlated with equivalent sections at outcrop. Table 1 summarizes the sequences presently observed within the Carboniferous of the Bowland Basin, Derbyshire Dome and East Midlands. The sequence stratigraphic approach has been adopted as the basis for compiling a series of twelve palaeofacies maps for the Variscan cycle (Fig. 10). These maps permit an analysis of the evolution of reservoir, seal and source facies by time slice.

The fundamental control on sedimentation in the Northern England Carboniferous basins was

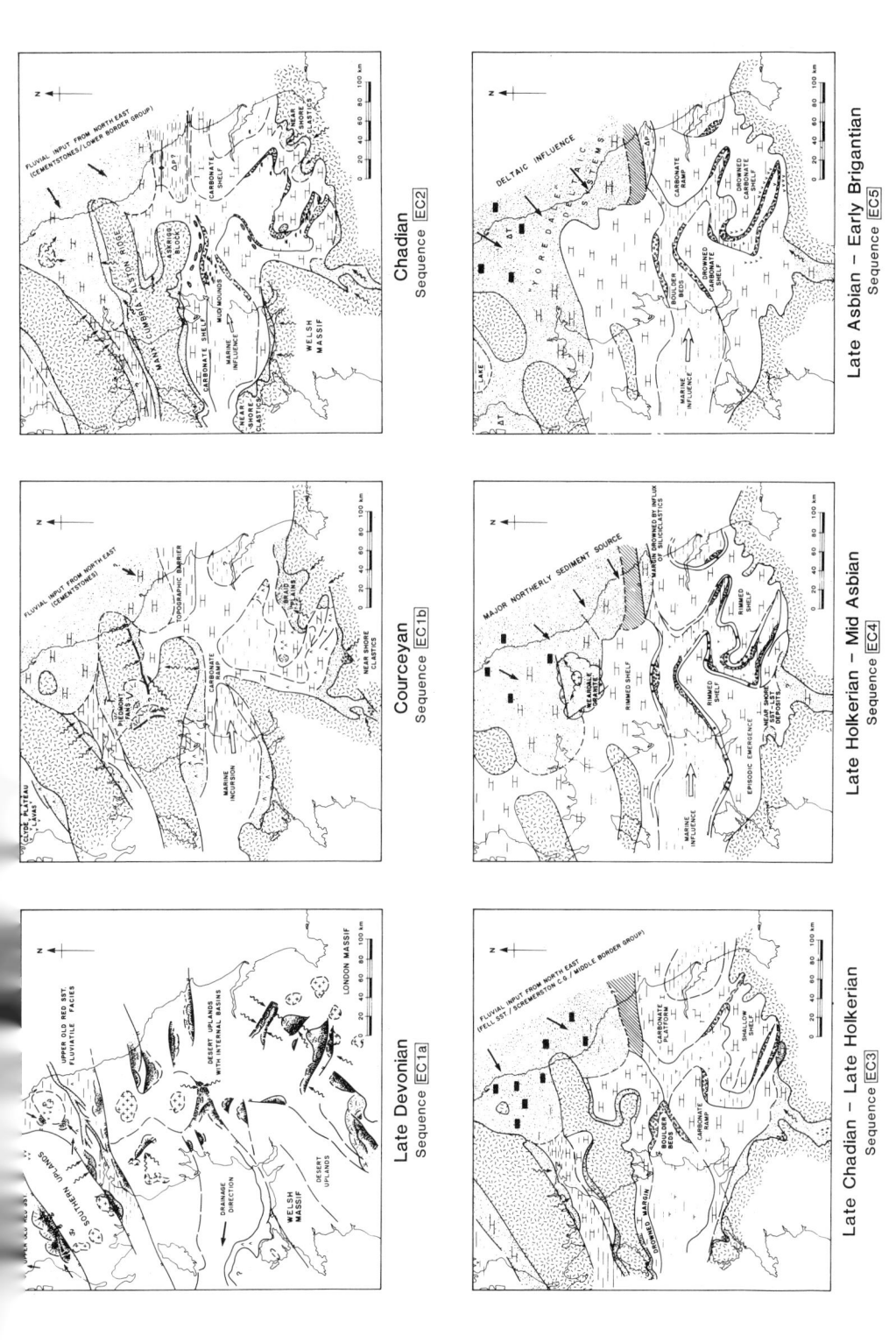

Fig. 10. (a) Late Devonian–Early Brigantian palaeofacies.

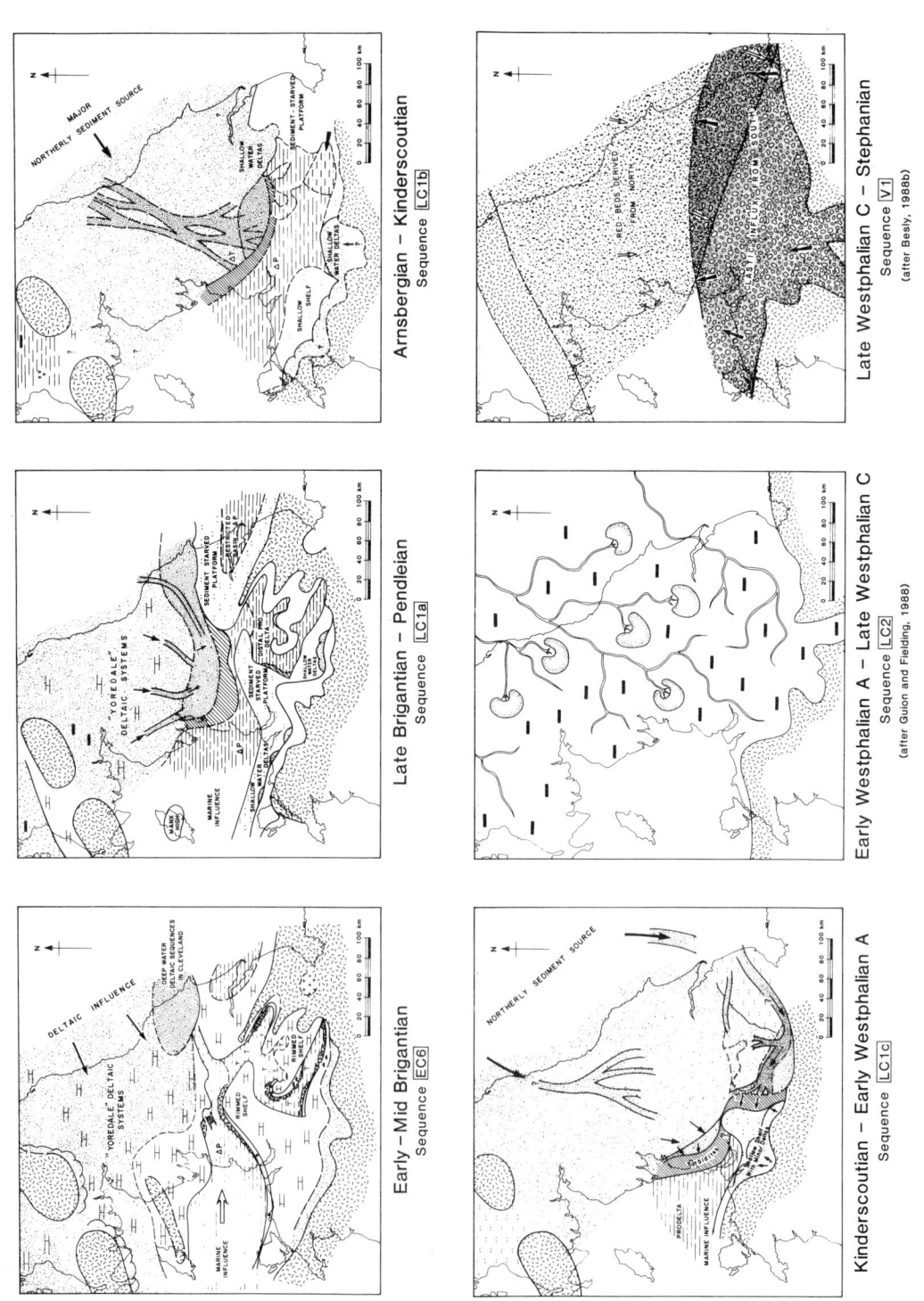

Fig. 10. (b) Early Brigantian–Carboniferous palaeofacies.

the presence of a major clastic source area to the north. Thus the predominant sediment transport direction was from north to south controlled and hindered by the developing rift topography. Whilst clastics were trapped by the more northerly basins, carbonates accumulated in the south. The more southerly basins only received a clastic fill during the post-rift when an inherited sediment-starved topography was rapidly infilled.

Geodynamic interpretation. There is much controversy surrounding the origin of the Northern England extensional basins. Essentially the geodynamic models fall into two categories, with subsidence controlled either by plate subduction and back-arc spreading (Leeder 1982 & 1987) or related to large-scale transtensional shear (Badham & Halls 1975; Dewey 1982). A recent model proposed by Haszeldine (1984, 1988) has suggested an alternative mechanism based on east–west extension related to early rifting in the Atlantic province. The work carried out for the study reported here, favours the 'Leeder model' and interprets Northern England rifting south of the Iapetus suture specifically in terms of domal arching followed by incipient north–south directed back-arc spreading centred on Central England. Some evidence for the presence of a thermally driven dome is provided by the widespread development of continental facies at the onset of rifting (Fig. 10), the semi circular southern rim to the rift system formed by the northern margin of the Wales Brabant Massif (Fig. 3) and the 'bulls eye' pattern of the post-rift thermal sag (Fig. 1).

The sequence stratigraphic analysis has suggested that doming and rifting commenced in late Devonian times and continued to the end of the Dinantian, with continental deposition gradually replaced by more marine conditions from the Courceyan as subsidence progressed. The prolific clastic hinterland to the north (Leeder 1987) of the Central England basins, however, did not prograde markedly southwards in the Dinantian, due probably to the very substantial subsidence of the Iapetus basins (Midland Valley, Northumberland/Solway) keeping pace with sediment supply but, also, to the barrier formed by the granite-cored Alston and Askrigg blocks (Fig. 3). The palaeofacies shown on Fig. 10 illustrate the isolation of the Central England rifts and the predominance of carbonate over clastic sedimentation at this time.

In late Brigantian times this partitioning of sedimentary facies was terminated by tectonic movements, which are reflected in a widespread early-mid Brigantian inversion in Central England, that elevated basin margins and induced intra-Carboniferous erosion surfaces (Strank 1987). The inversion was relatively minor but corresponds, significantly, to the switchover from active rifting to thermally-induced 'sag' subsidence. Temporally the inversion correlates with collision events as recorded by nappe emplacement in the Variscan orogenic zone lying to the south (Sellwood & Thomas 1986). It would seem likely that this event disabled the subduction which controlled back-arc spreading leading to the termination of active rifting in Northern England. This switch in subsidence patterns is particularly well demonstrated by comparison of the restored syn and post-rift isopachs over Northern England. (Figs 11 & 12). The early Carboniferous (syn-rift) isopach shows deposition essentially confined to isolated fault bounded half-graben. In contrast, the post rift isopach displays a pattern of thermally driven, broad, regional subsidence. Over 3.5 km of sediments were deposited during the post-rift phase with subsidence centred in the southwest of the study area. The depocentre migrated southwards with time in response to initial phases of basement uplift and breaching of the southern shoulder of the collapsed dome (Wales-Brabant Massif) during the Westphalian (Guion & Fielding 1988).

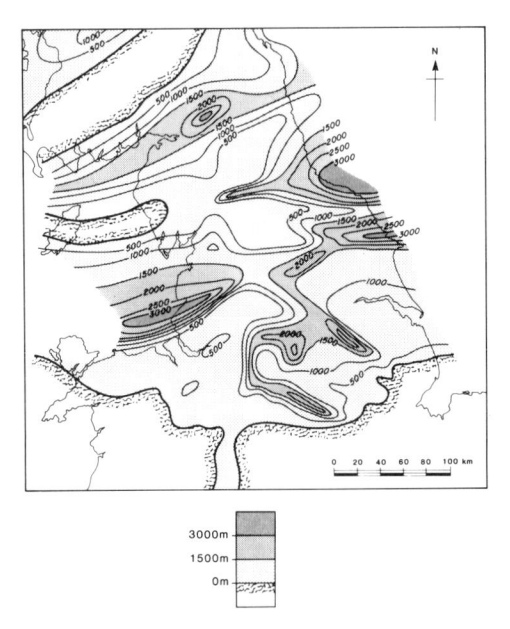

Fig. 11. Restored Chadian–early Brigantian Isopachs. Syn-rift sequences EC2–EC6.

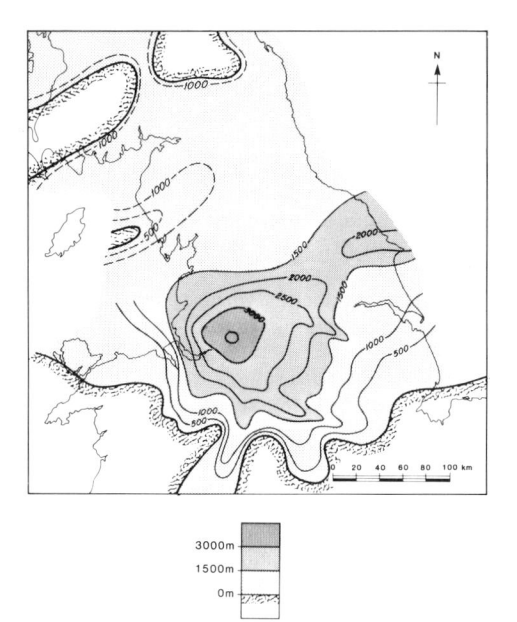

Fig. 12. Restored Namurian–mid-Westphalian C isopachs. Post-rift sequences LC1–LC2.

Late Westphalian, Variscan movements extensively uplifted Northern England (Fig. 13). Faults in Northwest England trending NE–SW and at a high angle to the translation direction of Variscan convergence were massively inverted. Importantly the NW–SE grain in the East Midlands was only mildly inverted with the inversion anticlines often occurring at intersections with NNW–SSE basement lineaments (e.g. Eakring, Hardstoft).

Permo–Trias events

By late Permian times a new tectonic regime had become established over Northwest Europe, with widespread regional subsidence in the area of the North Sea where the east–west trending Southern Permian Basin formed (Ziegler 1978). To the west, incipient rifts, once again reactivating the NE–SW Caledonian trends, formed the Rockall, Celtic Sea and Western Approaches rift systems (Ziegler 1981).

Northern England straddled these two provinces with the subsequent evolution of the hydrocarbon environment in each province profoundly controlled by this differentiation. Whereas the basins to the west were repeatedly extended and tectonically reactivated, the East Midlands and Northeast England provinces were located on a gently dipping rift shoulder with little differential subsidence or thickness variation (Fig. 14).

Later Mesozoic Events

An accurate reconstruction of the Jurassic and Cretaceous history of Northern England is hampered by the widespread erosion of Mesozoic sediments in Cenozoic times.

Fission track analysis (Green 1986) vitrinite reflectance, shale velocity plots and conodont colour data have all been used in an attempt to deduce the likely post-Triassic overburden. Thick Jurassic and Cretaceous sediments have been penetrated offshore in the Southern North Sea Basin. These thin markedly towards the Pennine High which remained a positive feature for much of the remainder of geological time (Fig. 15). In Northwest England, continued subsidence can be inferred from the rift histories derived from the offshore extensions of the basins. However, the profound event, in terms of the hydrocarbon habitat, occurred on the eastern·side of the Pennine High where extension and inversion events related to the evolving Sole Pit Trough (Kent 1975; Glennie & Boegner 1981; Van Hoorn 1987), disrupted the passive subsidence and quiescent setting of the Northeast England province (Fig. 14). This rifting of early Jurassic and early Cretaceous age locally induced exceptional burial and was followed by successive phases of basin inversion in the Cleveland and Stainmore Basins during

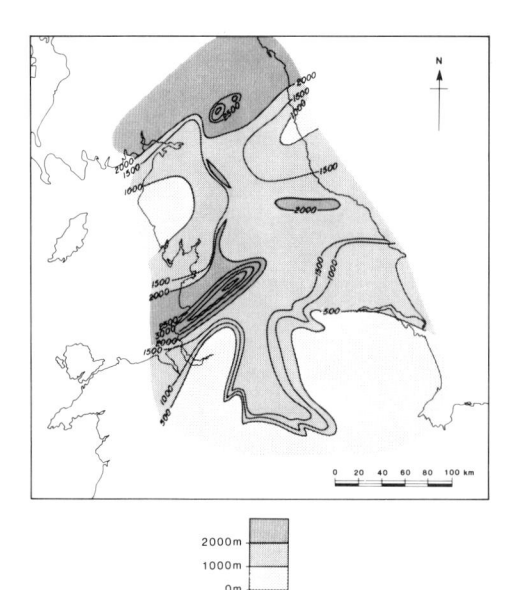

Fig. 13. Estimated Variscan uplift and erosion.

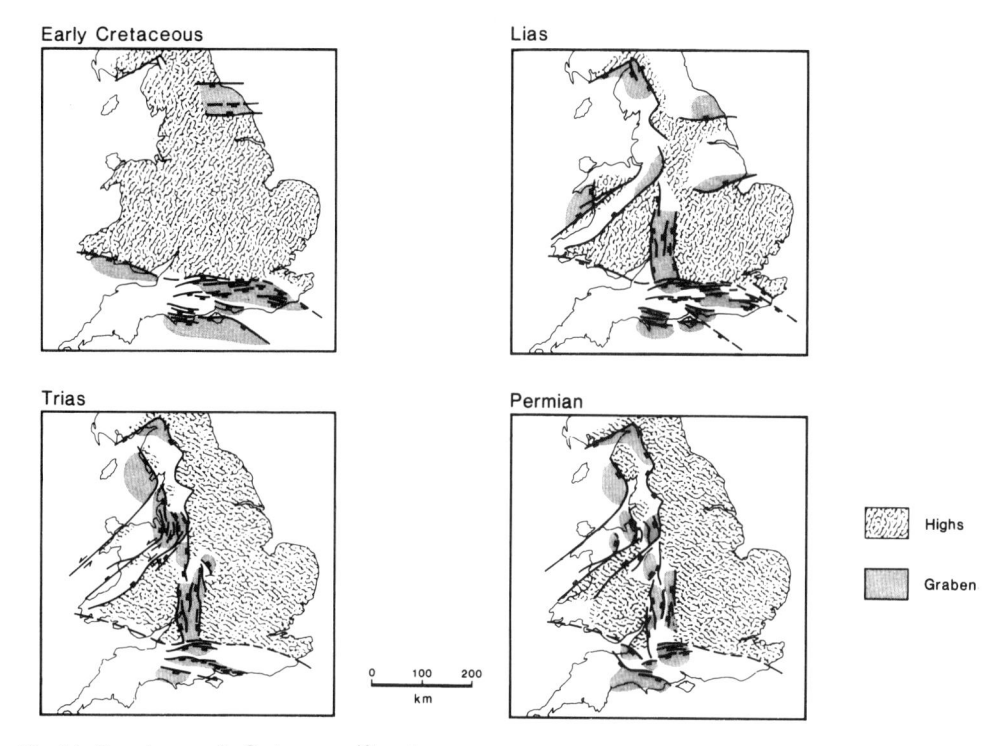

Fig. 14. Permian–early Cretaceous rift systems.

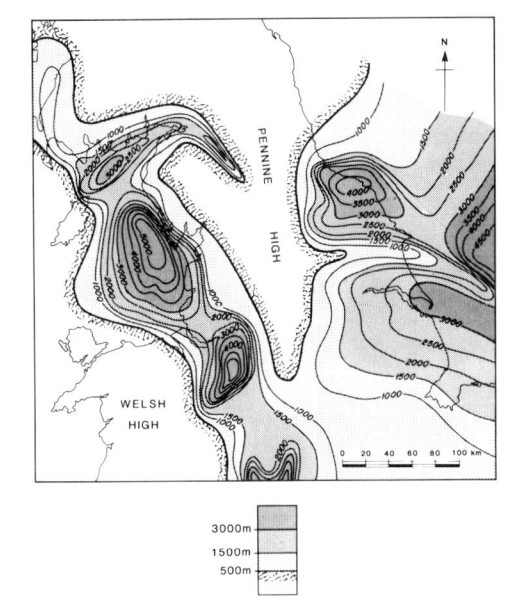

Fig. 15. Restored post-Carboniferous isopachs.

the Cretaceous–Tertiary. In stark contrast, the East Midlands province escaped these movements except for minor activity on the southern bounding faults of the Widmerpool Gulf (Kent 1966).

Cenozoic Events

Recent fission track analyses in Northern England (Green 1986) show that regional uplift began around 60 Ma with up to 3 km of overburden removed during the Palaeogene in the Lake District area. There is virtually no stratigraphic record of the Neogene onshore except for isolated fissure and pocket deposits (Walsh et al. 1972). Elsewhere evidence exists for weathered zones of this age. Walsh et al. (1972) correlate the base Neogene weathering zone and sediment surface with an important high-level erosion platform in the Pennines described by Sissons (1954). This occurs at 550 m above sea level and represents a bench which truncates the geological dip.

Figure 16 is a contour map of proposed Neogene and Quaternary uplift. This is based on the assumption that the base Neogene sediments in the Southern North Sea basin can be

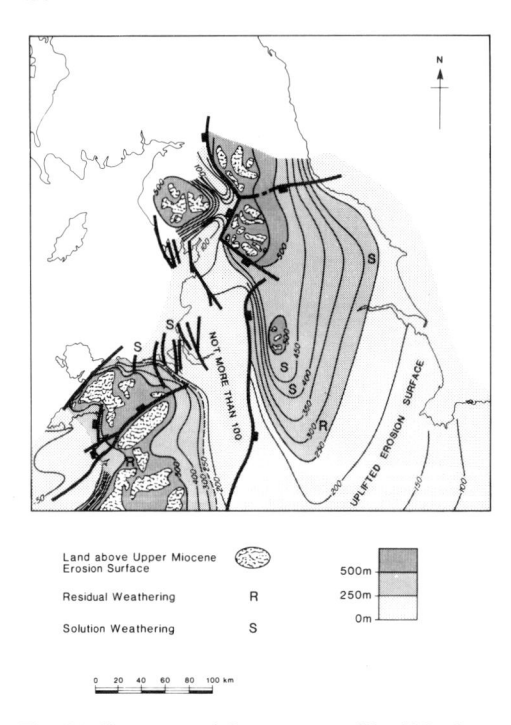

Fig. 16. Neogene and Quaternary uplift of Northern England.

correlated with the base Neogene sediment surface and high level Miocene erosion surfaces observed in Northern England.

A southeasterly tilt of eastern England during Cenozoic times of between 1 and 2° is recorded which has implications for hydrocarbon retention and aquifer characteristics in the region.

Controls on the hydrocarbon environment

Figure 17 shows a conceptual model of the potential intra-Carboniferous targets in Northern England. All have been recognized by BP but only the clastic delta play has been extensively explored. The remaining two major plays, the syn-rift clastics and the carbonate shelf margins will therefore only be discussed briefly.

The importance of the preceding events on the evolution of the hydrocarbon environment of intra-Carboniferous plays will be discussed in the following section. In order to make an assessment of the play system the key factors controlling the occurrence and distribution of hydrocarbons will be examined. These factors are discussed under three headings which address reservoiring, retaining and charging.

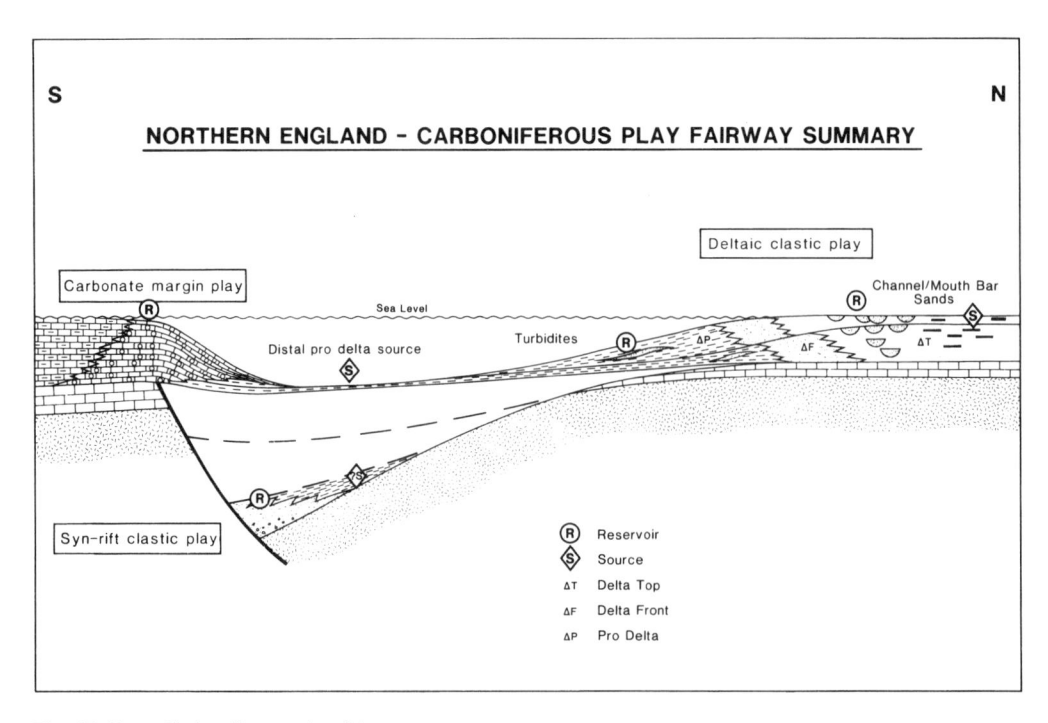

Fig. 17. Intra-Carboniferous play fairway summary.

Reservoiring

Distribution of potential reservoir units. The conceptual location of primary reservoir facies is shown in Fig. 17. For the deltaic clastic play these include turbidite sands in the pro-delta basinal setting, but of major importance are delta-top mouthbar and channel sands. The distribution of the latter in the Carboniferous is extensive both laterally and vertically. As palaeofacies maps show, potential reservoir sands occur in the shallow water Dinantian delta systems in the Iapetus province in the north of the area and in the deeper water Silesian deltas deposited in the Central England rifts. The trunk channels forming the distributary network are commonly medium to coarse grained with a moderate feldspar content. Sands in the LC2 sequence are similarly feldspathic but generally fine to medium grained. Quartz-rich sands, probably sourced from the Wales—Brabant Massif to the south, are also observed in parts of the East Midlands following the mid Brigantian inversion.

The most prospective reservoir facies for the carbonate margin play are the rimmed shelf complexes comprising grainstone shoals, buildups and boulder beds. Peritidal and shallow shelf limestones lying landwards of the margin commonly exhibit a mudstone—wackestone texture with low primary porosity which only produce hydrocarbons when fractured in tight Variscan anticlines (e.g., the Eakring and Hardstoft oilfields). High-energy environments near wavebase may enable the development of grainstone shoals behind the shelf margin towards palaeo-shorelines, but these are very difficult to predict in the subsurface. As a general rule the carbonate reservoirs are restricted to the Chadian, Asbian and early-mid Brigantian (sequences EC2, EC4 and EC6) when rimmed shelves were able to develop in periods of reduced tectonic activity. These facies are areally restricted to basins within the Central England rift system (Fig. 10).

Burial Effects (deltaic clastics). Burial diagenesis is the key post-depositional process controlling poroperm development in the deltaic clastic reservoirs. The major features are a progressive quartz-clay diagenesis that occludes primary porosity and the leaching of feldspars to create secondary porosity. The destruction of primary porosity is the more important in restricting permeability and is the principal limiting factor on reservoir quality.

Figure 18 shows a crossplot of maximum burial depth against core porosities for Namurian channel sands based on over 1.5 km of core data from a number of East Midlands wells. Relationships between log and core porosities,

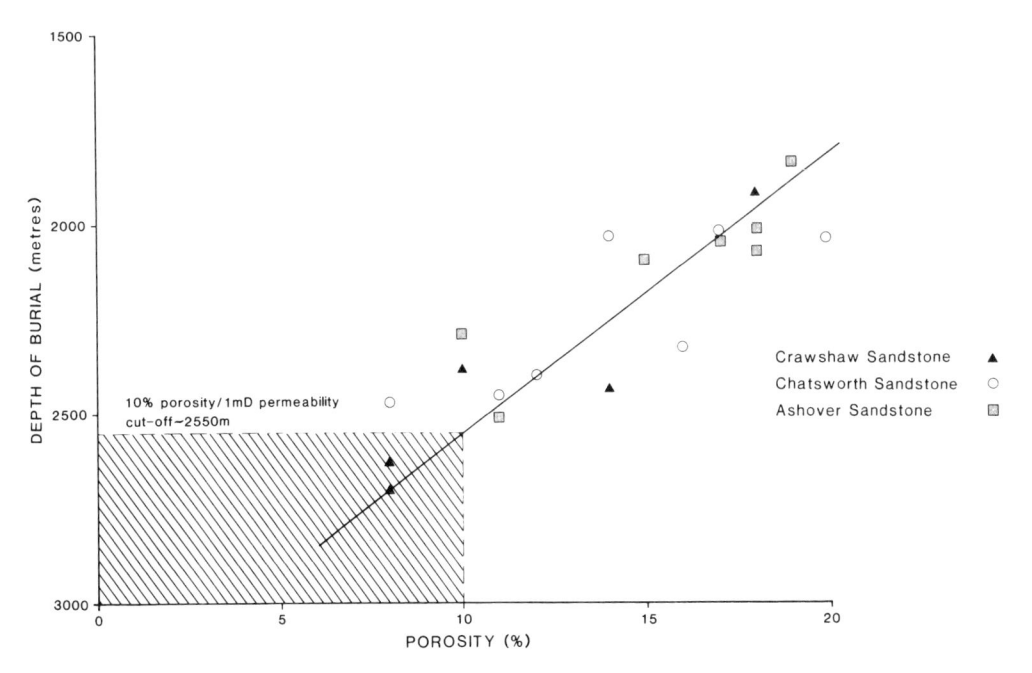

Fig. 18. Porosity against depth plot for Namurian—early Westphalian channel sandstone facies from wells in the East Midlands.

and between porosities and permeabilities permit the prediction of porosity in different sandstone types to within 1 or 2 porosity units. For channel sandstones the porosity cut-off for commercially viable reservoirs (10% = c.1mD) corresponds to a maximum depth of burial of 2550 m ± 200 m.

Retaining

Retaining includes two aspects of the hydrocarbon environment: seal capacity/extent and trap occurrence/integrity.

Seal Capacity and Extent. This aspect of the hydrocarbon system in Northern England is important, particularly for gas, but extremely difficult to define. The two most common reservoir-seal pairs in the Carboniferous delta environment are delta-top sands sealed by marine bands or interdistributary muds and turbidites encased in pro-delta shales. The marine bands of the East Midlands are highly effective regionally distributed seals with negligible sand/silt contents as reflected by their overall gamma-ray log response. The bands are rarely in excess of 1–2 m thick in the East Midlands, but where unfaulted can support oil columns of 40–50 m. Whereas the marine bands are thin but extensive (Calver 1968), the interdistributary shales may locally be relatively thick (10–15 m in the East Midlands fields). However, the latter are likely to be less effective due to their limited extent and relatively high silt content. The sealing mechanism developed by an interdistributary shale may be quite complex and not involve a single stratigraphic layer but an interdigitating network of shales.

Seal capacity for gas has been measured confidently on only one structure — Calow, a dip-closed, inversion anticline in the East Midlands (Fig. 19). Gas columns of 30–40 m are sealed at two separate reservoir levels by delta top shale and argillaceous sandstone where structural closure is in excess of 100 m.

With source rock generation frozen in the early Tertiary following regional uplift, seals are required to have remained effective for up to 60 Ma. As a result significant gas accumulations can only survive in Northern England where overlying evaportie seals are present in the Permo-Triassic.

Trap occurrence and integrity. Trap occurrence is not thought to be a limiting factor on the hydrocarbon environment of Northern England. The Variscan orogeny was pervasive and created an extensive array of variably faulted, inversion anticlines of which the Eakring Field is a typical example (Fig. 7). The amplitude and wavelength of the structures was determined by the relative orientations of the controlling faults to the northwesterly directed compression. Only in Northwest England is the presence of Variscan closures difficult to demonstrate. Here post-trap faulting, associated particularly with Permo-Triassic rifting, has

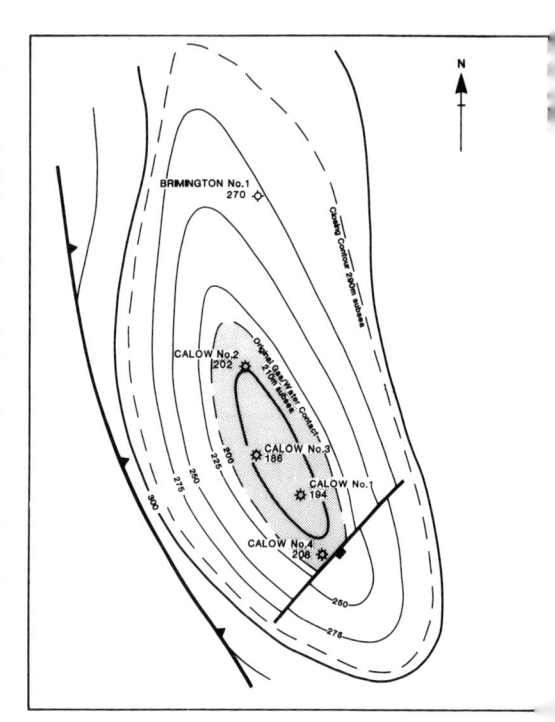

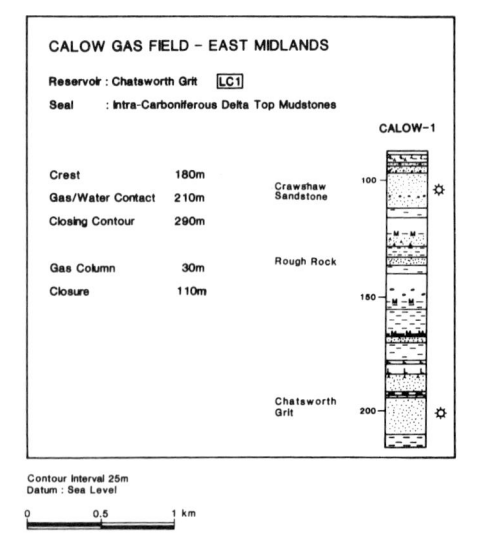

Fig. 19. Calow gas field — depth to Chatsworth Grit reservoir.

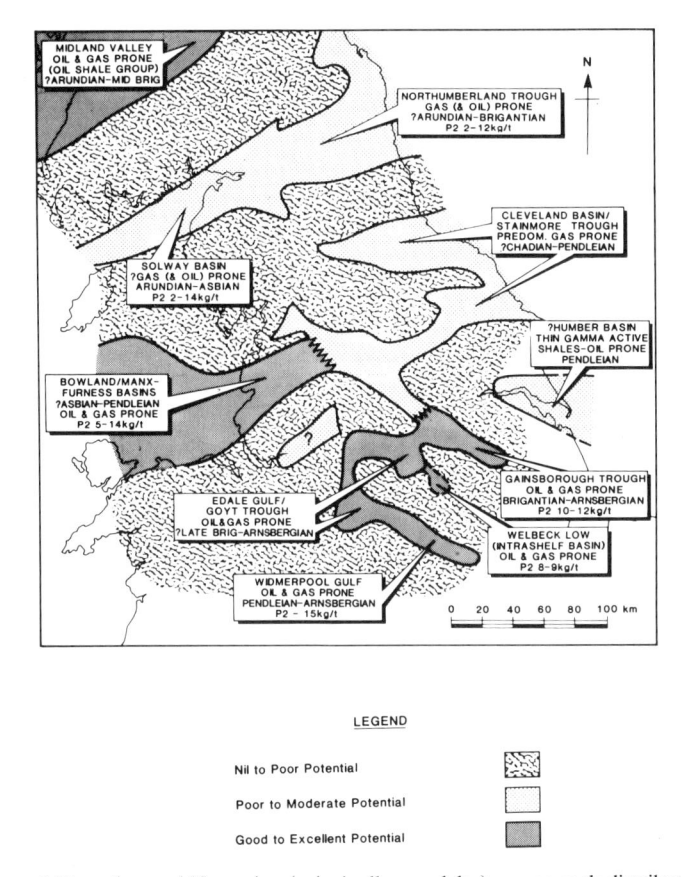

Fig. 20. Summary of Dinantian and Namurian (principally pro-delta) source rock distribution and potential.

fragmented and severely modified earlier Variscan structures. In sharp contrast, traps in the East Midlands province are almost entirely Variscan in origin with only Tertiary tilting significantly disturbing accumulations. The latter is reflected throughout the East Midlands by the prominence of transition zones and residual oil columns, which if reconstructed within a pre-Tertiary structural setting may be found to correspond to structural spill point. The Cleveland Basin (Northeast England Province) was subject to a more profound phase of late Cretaceous–Tertiary trap tilting and inversion which has considerably modified earlier Variscan structures.

Charging

Charging addresses the critical questions of source rock distribution, oil quality and maturity /migration timing.

Source rock distribution. In general terms the source rocks in Northern England young southwards in step with the southward progradation of the clastic delta systems throughout the Carboniferous (Figs 10 & 20). The best quality oil prone source rocks were deposited in the Central England rift basins in the south of the study area. These source rocks are associated with the distal pro-delta of relatively deep water (200–300 m) delta systems and correspond to the Bowland, Edale and Sabden shales. These exhibit gross thicknesses of around 100 m with yields of up to 14 kg tonne^{-1}. Shallower water deltas (50–150 m) which were established in basins such as Northumberland, Solway and Stainmore during the early Carboniferous are characterized by thinner, poorer quality source rock intervals (Figure 20).

Oil–source rock correlations in the East Midlands confirm the pro-delta mudstones as the main source intervals for oils discovered in the province (Fig. 21). However, alternative source facies have been identified, notably in delta top environments, where marine bands, coals and interdistributary mudstones can form

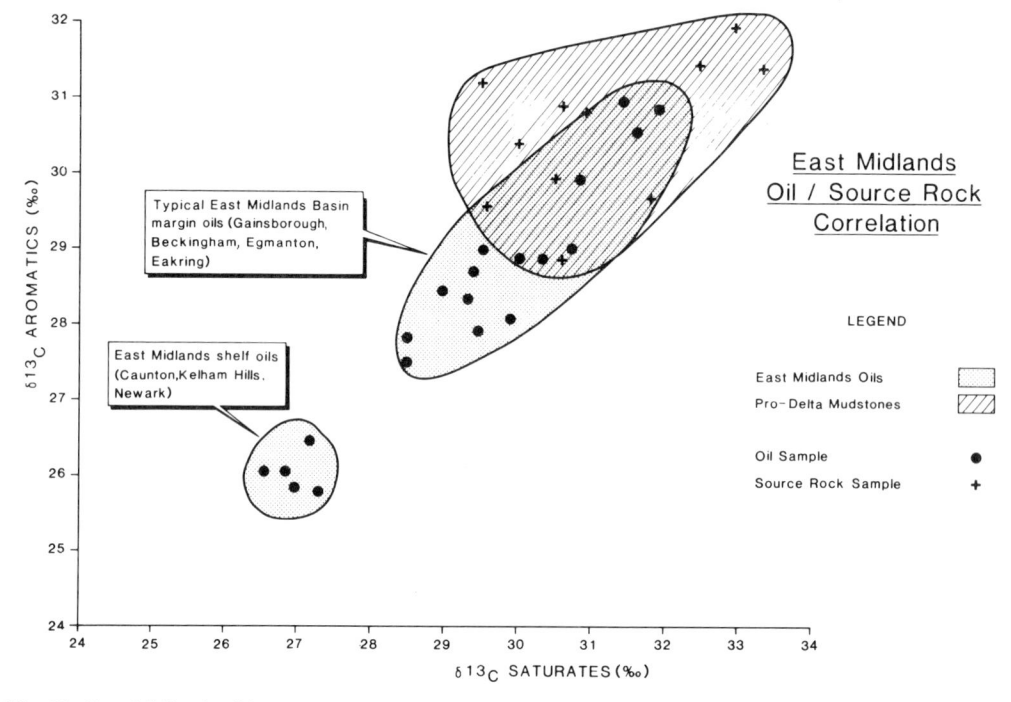

Fig. 21. East Midlands oil/source rock correlation.

local sourcing mechanisms (e.g., Caunton & Kelham Hills Fields, Fig. 21). None of these are regarded as having significantly contributed to the hydrocarbon budget in the East Midlands; however, all are areally extensive and in the case of the marine bands, locally give high yields (7–10 kg/tonne is not unusual). Coals are ubiquitous and rich but wholly gas prone and requiring significant burial.

The optimal development of the pro-delta source rock probably occurs in the Gainsborough Trough of the East Midlands where two basin-confined horizons occur with a third, thinner unit extending beyond the basin margins. The lowermost unit is Pendleian in age and comprises a black, pyritous organic rich shale. These 'gamma active' shales have average TOC contents of 3–4% and hydrocarbon yields of over 15 kg/tonne^{-1}. Gross thickness varies in this lower unit from 80 to 200 m. The middle source unit is Arnsbergian in age and is strictly confined to the Gainsborough, Widmerpool and Welbeck basins (Fig. 20). (The Welbeck basin is unique as a source kitchen in the East Midlands, being initiated as an intrashelf basin following the mid-Brigantian inversion and subsequent thermal subsidence). The main difference when compared with the lower unit is an increased coarse clastic component which reduces the source potential to 5–10 kg/tonne^{-1}. The upper source unit occurs towards the top of the Arnsbergian and continues into the earliest Marsdenian. It extends onto the East Midlands shelf and includes several major reservoir units. Source quality is considerably poorer than the basin-bound units but it forms the basis of a local source mechanism in the platform areas. Net thicknesses are 5–15 m with original yields estimated at 5–7 kg/tonne^{-1}.

Oil Characteristics. The characteristics of the different oils and source rocks are illustrated on saturates against aromatic residue plots for some 20 sample points in the East Midlands (Fig. 20). The oils can be grouped into two main types based on their carbon isotope values. Typical East Midlands crudes reservoired on the margins of the major basinal areas form the largest group. These oils are principally sourced from the distal pro-delta 'gamma-active' shales. An isotopically heavier suite of oils occurring more distally from the basinal areas (Kelham and Caunton fields) reflect increased contributions from marine bands and the interdistributary mudstones. Oil quality is not regarded as a critical factor, except for the occasional occurrence of sour crudes (Welton) and wax contents

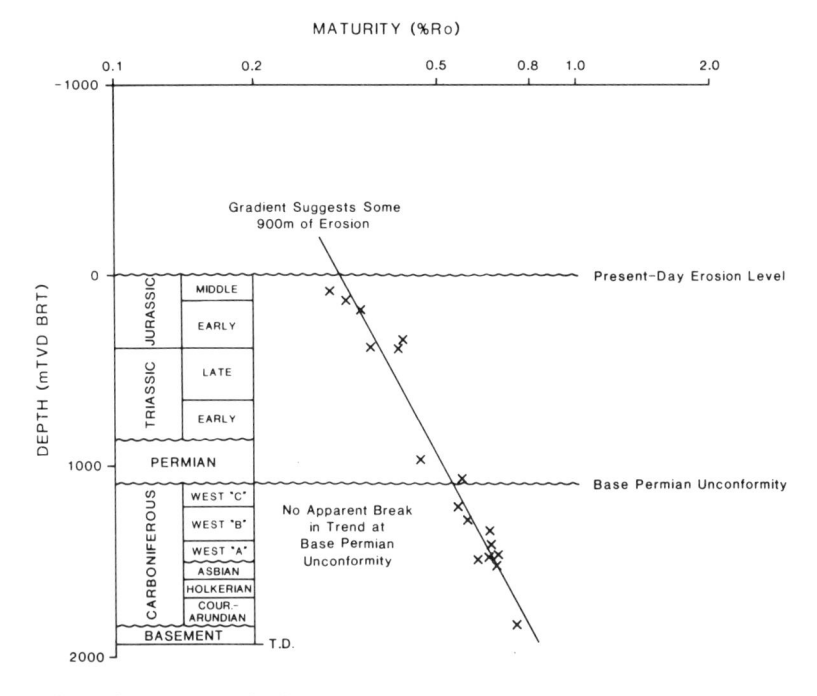

Fig. 22. Maturity against depth profile for the well Bardney-1 in the East Midlands based on vitrinite reflectance data.

which tend to be high (10–20%) due to the contribution from cuticular wax from land plants.

Maturity thresholds. Some 40 key wells with maturity data (i.e. vitrinite and spore colour) from the Central England rift basins have been analysed and compared to interpret maturity thresholds in late Dinantian–Silesian source rock types. Insufficient data exist for the late Devonian–Dinantian sequences in this area to draw any firm conclusions. Over most of the East Midlands maturity/depth trends based on vitrinite reflectance show little variation in gradient between the pre and post Variscan section suggesting that post-Carboniferous burial has overprinted any signature of end-Carboniferous erosion. (Fig. 22).

Approximate figures used for the top and base of the oil window are 1900 m and 3600 m with a maximum at around 2700 m +/− 200 m. The gas generation maximum is assessed at around 3450 m +/− 200 m.

Summary of critical factors

Of the critical factors which can be quantified and are amenable to mapping, the most important for the assessment of the intra-Carboniferous play of Northern England are:
  (i) critical burial depth for preservation of reservoir quality (maximum 2550 m).
 (ii) critical burial depth for the oil window (1900–3600m).
(iii) distribution of good quality basinal pro-delta source rocks
(iv) degree of post-Variscan, trap modification.

Application of these to the principal intra-Carboniferous play (deltaic clastic) allows the potential importance of the various Northern England basins to be assessed and the 'uniqueness' of the East Midlands examined.

Evolution and assessment of the intra-Carboniferous play in Northern England

Evolution (Deltaic Clastics)

Post-rift Carboniferous Events. Two key events critically affect the potential of the intra-Carboniferous play at this time.
  (i) Restricted distribution of basinal shales. The data from the East Midlands demonstrate the volumetric importance of the lower Namurian shales as source rocks and hence the necessity to have available the distal pro-delta shales within an effec-

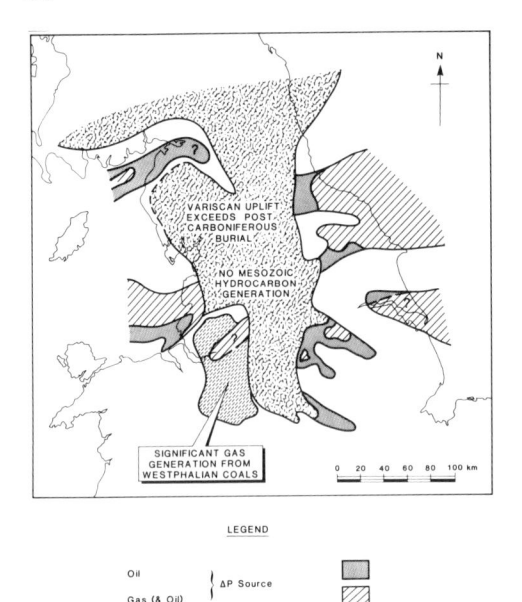

Fig. 23. Summary of Late Mesozoic hydrocarbon generation (post-Variscan trap formation).

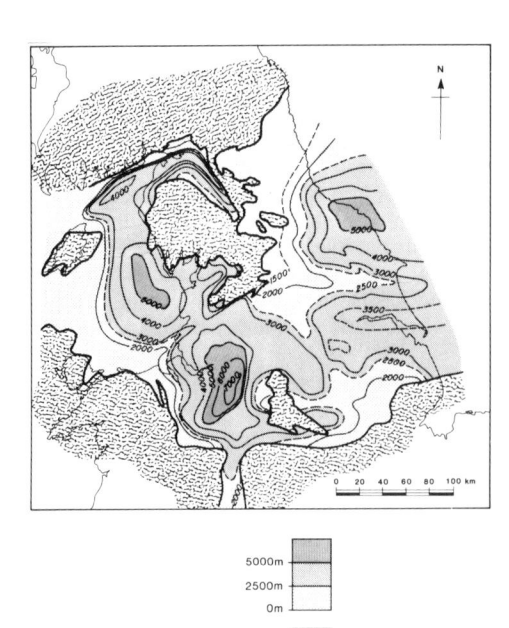

Fig. 24. Estimated maximum palaeo burial depths for the basal Namurian reservoir and source intervals.

tive kitchen. This generally limits the play fairway to within a few kilometres of the early Carboniferous depocentres. In the East Midlands the maximum horizontal migration distance from this source is 12 km; the average is 5 km. As discussed previously source rock quality in the pro-delta varies across Northern England with the Central England rifts displaying the greater potential.

(ii) Location of the Silesian depocentre. As Fig. 12 shows, the basinal shales located in the Northwest England province were probably mature and producing hydrocarbons during the late Carboniferous thereby reducing oil potential in this area prior to Variscan trap formation. Basal Namurian reservoir quality would also have been adversely affected.

End Carboniferous Variscan inversion. The Variscan inversion has been identified as the dominant trap-forming event in Northern England. The uplift associated with the Variscan unfortunately also led to the freezing of oil generation from source rocks throughout the province. The hydrocarbon system was essentially reset at this time necessitating a second, Mesozoic, phase of burial and source rock generation post dating the formation of Variscan traps. Areas where significant post-Carboniferous hydrocarbon generation occurred are highlighted in Fig. 23.

Figure 24 shows the areas where the main deltaic clastic target was removed by end Carboniferous erosion related to Variscan uplift and erosion. This includes the Derbyshire Dome, parts of the Bowland Basin and extends along the Dent–Pennine line into the Northumberland Trough where the removal of the deltaic clastic sequence is only partial.

Permo–Triassic Extension. The distinct burial and tectonic evolution in Permo–Triassic times substantially downgraded the hydrocarbon potential of basins on the western side of the Pennine High. The critical effects on the intra-Carboniferous play are:

(i) In the major Permo–Triassic depocentres (Cheshire Basin and parts of Manx–Furness) burial cut-offs for both reservoir quality and oil generation were exceeded (Fig. 24).

(ii) Pervasive and repeated rifting throughout the period fragmented earlier traps and reactivated trap bounding faults. Examples of trap modifications at this time occur in the Formby–Fylde area (onshore Manx–Furness) and in parts of the Cheshire Basin (Fig. 14).

Gas generation commenced in the major depocentres towards the end of the Triassic, but repeated faulting of marine band seals and trap modification make prospects in basins lying in the Northwest England Province unsuitable for gas retention. With continued gas generation, flushing of oil accumulations becomes an additional risk.

In contrast, basins lying east of the Pennine High, in both geological provinces, underwent gradual burial at this time with renewed oil generation in the basinal areas and little trap modification.

Jurassic–Cretaceous Subsidence and Local Inversion. The later Mesozoic history of the Northwest England province is highly conjectural but it has been shown from fission track analysis that some 1–1.5 km of Jurassic and Cretaceous sediments were deposited. This continued burial left only minor onshore areas in the south of the Formby–Fylde below the oil maximum for base Namurian shales. Gas generation from Upper Carboniferous coals, where preserved following Variscan uplift and erosion, was widespread, particularly in the Cheshire Basin (Fig. 23).

Prior to early Jurassic times the hydrocarbon environments of the intra-Carboniferous basins lying in the Northeast England and East Midland provinces had been subject to a remarkably similar burial and thermal history. Both provinces had been strongly structured in late Variscan times, forming an array of anticlinal and fault-related traps and had been subject to gentle regional subsidence during the Permo–Triassic. The contrast between basins of the Northeast and East Midlands provinces commenced in early Jurassic times and continued into the Tertiary as the Cleveland and Stainmore basins became involved in rifting and inversion related to the Sole Pit fault system (Fig. 14). This had essentially two effects:
(i) Widespread disruption of traps due to localized fault movements and successive phases of basin inversion in the late Cretaceous–Tertiary.
(ii) Localized deep burial, principally as a product of early Cretaceous rifting in northern Cleveland. This event pushed most of the basin into gas generating zone (Fig. 23) with consequent flushing of liquid hydrocarbons updip towards the south.

Cenozoic Uplift and Erosion. From the end of late Cretaceous chalk deposition onwards, the Northern England basins experienced strong uplift and erosion with a consequent freezing of hydrocarbon generation. A regional heating event which has elevated present day geothermal gradients in Northern England (Wheildon & Rollin 1986) may have been associated with Tertiary volcanism. However, these elevated heat flows are unlikely to have significantly re-triggered generation from source rocks, which in most areas were already in an advanced stage of maturity. Regional estimates of uplift and erosion range from 1100–3000 m in the Northwest England province to 1000–2000 m in the East Midlands. The effects of uplift and tilting on existing accumulations is well demonstrated in many of the East Midlands fields, where palaeo-oil columns can be related back to pre-tilt closures. Surface and shallow seeps still give a fair indication of the pre-tilt distribution of subsurface accumulations.

Play assessment

An assessment of the deltaic clastics play in the Northern England basins is illustrated on Fig. 25. From the point of view of the evolution of the hydrocarbon system described above, it is clear that the geological history of the East Midlands province is somewhat unique. The area works as a successful oil environment because of the following.

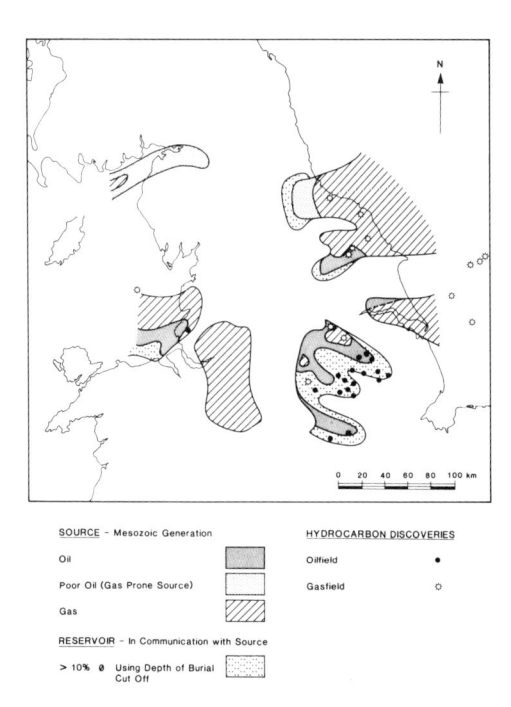

Fig. 25. Clastic delta play assessment.

(i) It includes several isolated early Carboniferous rift basins containing thick basal Namurian source rocks which are encountered at several levels providing source 'back-up'.
(ii) Except in areas to the west of the province, maturation occurred in Mesozoic times after Variscan trap formation.
(iii) There is an abundance of reservoir-seal pairs providing reservoir-seal 'back-up' with burial depths over most of the area above the critical threshold for commercial oil production.
(iv) The area was tectonically quiescent following Variscan trap formation with passive burial accompanied by a mild easterly tilting.

The burial data provided here suggests that the East Midlands (in a fairly restricted fairway surrounding the basinal areas) is the only oil province likely to be developed in the intra-Carboniferous of Northern England. In contrast the intra-Carboniferous gas play is likely to be more extensive, particularly in Northeast England. However, gas potential may be hampered by poor reservoir quality and, particularly in the Northwest England province, by contemporaneous trap reactivation and the increased need for an effective sealing facies.

Conclusions

This paper began with a question concerning the uniqueness of the East Midlands in terms of the intra-Carboniferous plays of Northern England. It would seem that even with a small data set of fully tested targets there exists sufficient evidence from the reconstructed hydrocarbon environment to suggest that the East Midlands is geologically a very unique setting and most probably the only Northern England province where commercial oil accumulations will be found (Fig. 26).

Permission to publish this paper has been granted by the British Petroleum Company plc. We are also grateful to partners, Gas Council (Exploration) and Floyd Oil for permission to publish seismic data from the East Midlands. The authors wish to thank the many BP Eakring geologists and geophysicists who have been involved in this work. In particular we would like to mention Dr A. Kay, Dr J. Hossack, B. Mitchener, Dr A. Strank, C. Einchcomb and N. Taylor who made important contributions to the study.

References

ANDERTON, R., BRIDGES, P. H., LEEDER, M. R. & SELLWOOD, B. W. 1979. *A dynamic stratigraphy of the British Isles. A study in crustal evolution.* Allen & Unwin.

ANDRE, L., HERTOGEN, J. & DEUTSCH, S. 1986. Ordovician–Silurian magmatic provinces in Belgium and the Caledonian orogeny in Middle Europe. *Geology*, 14, 879–882.

ARTHURTON, R. S. 1986. The Ribblesdale fold belt, NW England — a Dinantian–Early Namurian dextral shear zone. *In*: HUTTON, D. H. W. & SANDERSON, D. J. (eds). *Variscan Tectonics of the North Atlantic Region.* Geological Society, London, Special Publication 14, 131–138.

BADHAM, J. P. N. & HALLS, C. 1975. Microplate tectonics, oblique collisions, and the evolution of the Hercynian orogenic system. *Geology*, 3, 373–376.

BEAMISH, D. & SMYTHE, D. K. 1986. Geophysical images of the deep crust: the Iapetus suture. *Journal of the Geological Society*, 143, 489–497.

BESLY, B. M. 1988a. Late Carboniferous sedimentation in northwest Europe: an introduction. *In*: BESLY, B. M. & KELLING, G. (eds). *Sedimentation in a synorogenic basin complex: The Carboniferous of north west Europe.* Blackie, Glasgow, 1–7.

—— 1988b. Palaeogeographic implications of late Westphalian to early Permian red beds, Central England. *In*: BESLY, B. M. & KELLING, G. (eds). *Sedimentation in a synorogenic basin complex: The Carboniferous of north west Europe.* Blackie, Glasgow, 200–221.

BOTT, M. H. P. 1966. Geophysical investigations

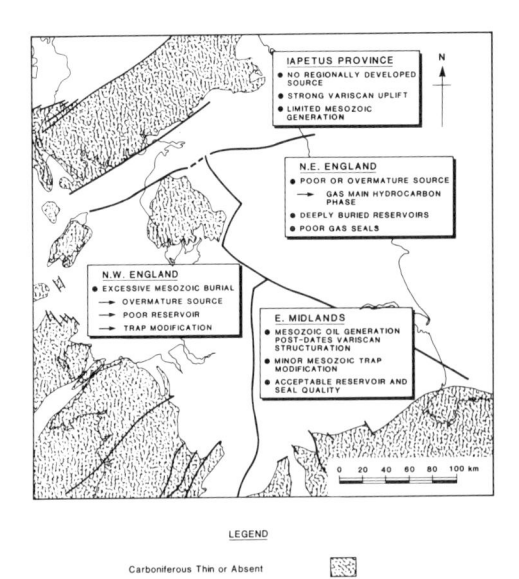

Fig. 26. Intra-Carboniferous play summary.

of the Northern Pennine basement rocks. *Proceedings of the Yorkshire Geological Society*, **36**, 139–168.
—— 1987. Subsidence mechanisms of Carboniferous basins in Northern England. *In*: MILLER, J., ADAMS, A. E. & WRIGHT, V. P. (eds). *European Dinantian Environments*. John Wiley, Chichester, 21–32.
CALVER, M. A. 1968. The distribution of the Westphalian marine faunas in northern England and adjoining areas. *Proceedings of the Yorkshire Geological Society*, **37**, 1–72.
COWARD, M. P. & SIDDANS, A. W. B. 1979. The tectonic evolution of the Welsh Caledonides. *In*: HARRIS, A. L., HOLLAND, C. H. & LEAKE, B. E. (eds). *Caledonides of the British Isles – reviewed*. Geological Society, London, Special Publication **8**, 187–198.
DEWEY, J. F. 1971. A model for the Lower Palaeozoic evolution of the southern margin of the early Caledonides of Scotland and Ireland. *Scottish Journal of Geology*, **7**, 219–240.
—— 1982. Plate tectonics and the evolution of the British Isles. *Journal of the Geological Society of London*, **139**, 371–412.
EVANS, A. M. 1979. The East Midlands Aulacogen of Caledonian age. *Mercian Geologist*, **7**, 31–42.
FALCON, N. L. & KENT, P. E. 1960. *Geological results of petroleum exploration in Britain 1945–1957*. Geological Society, London, Memoir **2**.
GAWTHORPE, R. L. 1987. Tectono-sedimentary evolution of the Bowland Basin northern England, during the Dinantian. *Journal of the Geological Society*, **144**, 59–71.
GEORGE, T. N., JOHNSON, G. A. L., MITCHELL, M., PRENTICE, J. E. RAMSBOTTOM, W. H. C., SEVASTOPULO, G. D. & WILSON, R. B. 1976. A correlation of Dinantian rocks in the British Isles. *Geological Society, London, Special Report*, **7**.
GLENNIE, K. W. & BOEGNER, P. L. F. 1981. Sole Pit Inversion tectonics. *In*: ILLING, L. V. & HOBSON, G. D. (eds). *Petroleum geology of the continental shelf of North–West Europe*. Institute of Petroleum, London 110–120.
GRAYSON, R. F. & OLDHAM, L. 1987. A new structural framework for the northern British Dinantian as a basis for oil, gas and mineral exploration. *In*: MILLER, J., ADAMS, A. E. & WRIGHT, V. P. (eds). *European Dinantian Environments*. John Wiley, Chichester, 33–59.
GREEN, P. F. 1986. On the thermo-tectonic evolution of Northern England: Evidence from fission track analysis. *Geological Magazine*, **123**, 493–506.
GUION, P. D. FIELDING, C. R. 1988. Westphalian A and B sedimentation in the Pennine basin, UK. *In*: BESLY, B. M. & KELLING, G. (eds). *Sedimentation in a synorogenic basin complex: The Carboniferous of north west Europe*. Blackie, Glasgow. 153–177.
GUTTERIDGE, P. 1987. Dinantian sedimentation and basement structure of the Derbyshire Dome. *Geological Journal*, **22**, 25–41.
HASZELDINE, R. S. 1984. Carboniferous North Atlantic palaeogeography: stratigraphic evidence for rifting, not megashear or subduction. *Geological Magazine*, **121**, 443–463.
—— 1988. Crustal lineaments in the British Isles: their relationship to Carboniferous basins. *In*: BESLY, B. M. & KELLING, G. (eds). *Sedimentation in a synorogenic basin complex: The Carboniferous of north west Europe*. Blackie, Glasgow, 53–83.
HUDSON, R. G. S. 1930. The Carboniferous of the Craven Reef Belt: the Namurian unconformity of Scalber, near Settle. *Proceeding of the Geologists' Association*, **41**, 290–322.
KENT, P. E. 1966. The structure of the concealed Carboniferous rocks of NE England. *Proceedings of the Yorkshire Geological Society*, **35**, 323–352.
—— 1968. The buried floor of Eastern England. *In*: SYLVESTER-BRADLEY, P. C. & FORD, T. D. (eds). *The Geology of the East Midlands*. Leicester University Press, 138–148.
—— 1975. The tectonic development of Britain and surrounding seas. *In*: WOODLAND, A. W. (ed.). *Petroleum and the continental shelf of North West Europe*. Vol. 1 Geology. Applied Science Publishers. 3–28.
—— 1985. UK Onshore Oil Exploration, 1930–1964. *Marine and Petroleum Geology*, **2**, 56–64.
KLEMPERER, S. & MATTHEWS, D. 1987. Iapetus suture located beneath the North Sea by BIRPS deep seismic reflection profiling. *Geology*, **15**, 195–198.
LE BAS, M. J. 1972. Caledonian igneous rocks beneath central and eastern England. *Proceedings of the Yorkshire Geological Society*, **39**, 71–86.
LEE, A. G. 1988. Carboniferous basin configuration of Central England, modelled using gravity. *In*: BESLY, B. M. & KELLING, G. (eds). *Sedimentation in a Synorogenic basin complex: The upper Carboniferous of north west Europe*. Blackie, Glasgow, 69–84.
LEEDER, M. R. 1976. Sedimentary facies and origins of basin subsidence along the northern margin of the supposed Hercynian ocean. *Tectonophysics*, **36**, 167–179.
—— 1982. Upper Palaeozoic basins of the British Isles – Caledonian inheritance versus Hercynian plate margin processes. *Journal of the Geological Society, London*, **139**, 479–491.
—— 1987. Plate tectonics and sedimentation in Lower Carboniferous Europe. *In*: ADAMS, A. E., MILLER, J. & WRIGHT, V. P. (eds). *European Dinantian Environments*. John Wiley, Chichester, 1–19.
—— & GAWTHORPE, R. L. 1987. Sedimentary models for extensional tilt block/half graben basins. *In*: COWARD, M. P., DEWEY, J. F. & HANCOCK, P. L. (eds). *Continental extensional tectonics*. Geological Society, London, Special Publication **28**, 139–152.
—— & MCMAHON, A. H. 1988. Upper Carboniferous (Silesian) basin subsidence in northern Britain. *In*: BESLY, B. M. & KELLING, G. (eds). *Sedimentation in a synorogenic basin complex: The Carboniferous of north west Europe*, Blackie, Glasgow, 43–52.

LEES, G. M. & COX, P. T. 1937. The geological results of the present search for oil in Great Britain by the D'Arcy Exploration Company Ltd. *Quarterly Journal of the Geological Society*, **93**, 156–194.

—— & TAITT, A. H. 1945. The geological results of the search for oilfields in Great Britain. *Quarterly Journal of the Geological Society*, **101**, 255–317.

MATTE, P. 1986. Tectonics and plate tectonics model for the Variscan belt of Europe. *Tectonophysics*, **126**, 329–374.

MILLER, J. & GRAYSON, R. F. 1982. The regional context of Waulsortian facies in N. England. In: BOLTON, K., LANE, H. R. & LEMONE, D. V. (eds). *Symposium on the Environmental Setting and Distribution of the Waulsortian facies*. The El Paso Geological Society and University of Texas at El Paso, 17–33.

PARKINSON, D. 1957. Lower Carboniferous reefs of Northern England. *Bulletin of the American Association of Petroleum Geologists*, **41**, 512–537.

PHAROAH, T. C., MERRIMAN, R. J., WEBB, P. C. & BECKINSDALE, R. D. 1987. The concealed Caledonides of eastern England: preliminary results of a multidisciplinary study. *Proceedings of the Yorkshire Geological Society*, **46**, 355–369.

RAMSBOTTOM, W. H. C. 1973. Transgressions and regressions in the Dinantian: a new synthesis of British Dinantian stratigraphy. *Proceedings of the Yorskhire Geological Society*, **39**, 567–607.

——, CALVER, M. A., EAGAR, R. M. C., HODSON, E., HOLLIDAY, D., STUBBLEFIELD, C. J. & WILSON, R. B. 1978. *A correlation of Silesian rocks in the British Isles*. Geological Society, London, Special Report **10**.

SELLWOOD, E. B. & THOMAS, J. M. 1986. Variscan facies and structure in central S.W. England. *Journal of the Geological Society, London*, **143**, 199–207.

SISSONS, J. B. 1954. The erosion surfaces and drainage system of S. W. Yorkshire. *Proceedings of the Yorkshire Geological Society*, **29**, 305–342.

SMITH, K., SMITH, N. J. P. & HOLLIDAY, D. W. 1985. The deep structure of Derbyshire. *Geological Journal*, **20**, 215–225.

SOPER, N. J. & HUTTON, D. H. W. 1984. Late Caledonian sinistral displacements in Britain: implications for a three plate collision model. *Tectonics*, **3**, 781–794.

——, WEBB, B. C. & WOODCOCK, N. H. 1987. Late Caledonian (Acadian) transgression in North West England: timing, geometry and geotectonic significance. *Proceedings of the Yorkshire Geological Society*, **46**, 175–192.

STRANK, A. R. E. 1987. The stratigraphy and structure of Dinantian strata in the East Midlands, UK. In: ADAMS, A. E., MILLER, J. & WRIGHT, V. P. (eds). *European Dinantian Environments*. John Wiley, Chichester, 157–175.

THORPE, R. S., BECKINSDALE, R. D., PATCHELT, P. J., PIPER, J. D. A., DAVIES, G. R. & EVANS, J. A. 1984. Crustal growth and Late Precambrian–Early Palaeozoic plate tectonic evolution of England and Wales. *Journal of the Geological Society*, **141**, 521–536.

TURNER, J. S. 1949. The deeper structure of Central and Northern England. *Proceedings of the Yorkshire Geological Society*, **27**, 280–297.

VAN HOORN, B. 1987. Structural evolution, timing and tectonic style of the Sole Pit inversion. *Tectonophysics*, **137**, 239–284.

WALSH, P. T., BOULTER, M. C., IJTABA, M. & URBAN, D. M. 1972. The preservation of the Neogene Brassington formation of the Southern Pennines and its bearing on the evolution of Upland Britain. *Journal of the Geological Society*, **128**, 519–559.

WHEILDON, J. & ROLLIN, K. E. 1986. Heat flow. In: DOWNING, R. A. & GRAY, D. A. (eds). *Geothermal Energy — The Potential in the United Kingdom*, HMSO, London, 8–20.

WHITTAKER, A. (ed.) 1985. *Atlas of onshore sedimentary basins in England and Wales: Post Carboniferous tectonics and stratigraphy*. Blackie, Glasgow.

WILLS, L. J. 1973. *A palaeogeographical map of the Palaeozoic floor beneath the Permian and Mesozoic formations in England and Wales*. Geological Society, London, Memoir **7**.

—— 1978. *A palaeogeographical map of the Lower Palaeozoic floor below the cover of Upper Devonian, Carboniferous and later formations*. Geological Society, London, Memoir **8**.

ZIEGLER, P. A. 1978. North–Western Europe: tectonics and basin development. *Geologie en Mijnbouw*, **57**, 589–626.

—— 1981. Evolution of sedimentary basins in north–west Europe. In: ILLING, L. V. & HOBSON, G. D. (eds). *Petroleum geology of the Continental Shelf of North–West Europe*, Heyden, London, 3–39.

—— 1982. *Geological Atlas of Western and central Europe*. Shell International Petroleum Maatschappij B. V., The Hague.

—— 1986. Geodynamic model for the Palaeozoic crustal consolidation of Western and Central Europe. *Tectonophysics*, **126**, 303–328.

Hydrocarbon plays in the northern North Sea

R. M. PEGRUM & A. M. SPENCER

Statoil, Forushagen (UND-GE), Postboks 300, 4001 Stavanger, Norway

Abstract: The hydrocarbon finds of the Norwegian and British sectors of the North Sea, north of 56°, can be grouped into six discrete plays. The key to all the plays is the presence of organic-rich Upper Jurassic shales (source rocks) and a rift system of the same age. The rifting provided the structures (the traps) and post-rift cooling caused the subsidence necessary for hydrocarbon generation. The pre-Jurassic Play is of least importance. The Lower–Middle Jurassic Play, with about 40% of the resources, results from pre-rift uplift in the south providing clastic input for a delta system in the north, the thick sandstone reservoirs of which were preserved during the subsequent rifting. In late Jurassic times graben formation by rift collapse was accompanied by erosion of marginal uplifts, resulting in thick sand sequences, which interfinger with graben shales containing the source rocks. This ideal relationship provides the Upper Jurassic Play which contains 30% of the resources. The Lower Cretaceous Play occurs in submarine fan sandstones and is of minor importance. Chalk deposition in a tectonically unstable environment, with subsequent rapid burial beneath Tertiary muds and clays, is responsible for the Chalk Play. Although geographically limited, it is prolific and accounts for nearly 10% of the hydrocarbon resources. Geologically youngest is the Paleogene Play. Uplift of the Orkney–Shetland Platform in early Tertiary times was a consequence of sea-floor spreading in the North Atlantic. Great quantities of sand derived from this uplift spilled eastwards into the northern North Sea, blanketing the western flank of the Tertiary basin and extending axially far south along the Central Graben. These sands have captured vertically migrating hydrocarbons and have allowed extensive lateral migration. Although important volumes of oil and gas, about 20% of the total resource, are trapped within these sands, large volumes have probably been lost to the basin via their outcrop in the west.

Exploration in the northern North Sea, since 1966, has involved drilling 1750 exploration and appraisal wells and resulted in some 270 discoveries with originally recoverable hydrocarbon reserves and resources of $c.\ 8.5 \times 10^9$ Sm^3 oil equivalent (50×10^9 bbl). The occurrence of these hydrocarbons is intimately associated with the presence of a complex late Jurassic to early Cretaceous rift system buried beneath a Cretaceous and Tertiary cover. Thick, organic-rich, syn-rift mudstones were laid down throughout most of the rift system and provide the main source rocks. Reservoir rocks, principally sandstones, occur in every system from Devonian to Oligocene.

The course of this exploration has been described (Brennand & Van Hoorn 1986; Campbell & Ormaasen 1987) and a wealth of technical information has been made public, especially recently (Brooks & Glennie 1987; Thomas *et al.* 1985; Spencer *et al.* 1986, 1987). It is thus timely to attempt an analysis of the factors which have created this rich hydrocarbon province. We have chosen to do this using the systematic approach advocated by White (1988), namely by first identifying the hydrocarbon source and charting its maturity development in time and in space; and then by describing the hydrocarbon plays grouped according to reservoir (Parsley 1986). The main element which links all of the hydrocarbon geology together is the late Jurassic to early Cretaceous rifting (Fig. 1).

Late Jurassic to early Cretaceous rifting

The key to understanding the Jurassic to Recent evolution of the Northern North Sea is to identify and map the late Jurassic rifts. Their pattern is revealed by a simple isopach map of the Upper Jurassic strata (Fig. 2). It shows three familiar converging rifts — the Viking, Moray Firth and Central Grabens. Also noteworthy is the asymmetry of the faulted troughs and the major erosion of the adjacent highs, which are the footwall blocks to the major boundary faults (Fig. 3).

The Central Graben has an overall NW trend, formed by northerly trending segments progressively offset to the west along WNW trends. The principal rifting phase began close to the end of Callovian times and continued through late Jurassic times, waning during the early Cretaceous. The rifting was transtensional,

From BROOKS, J. (ed.), 1990, *Classic Petroleum Provinces*, Geological Society Special Publication No 50, pp 441–470.

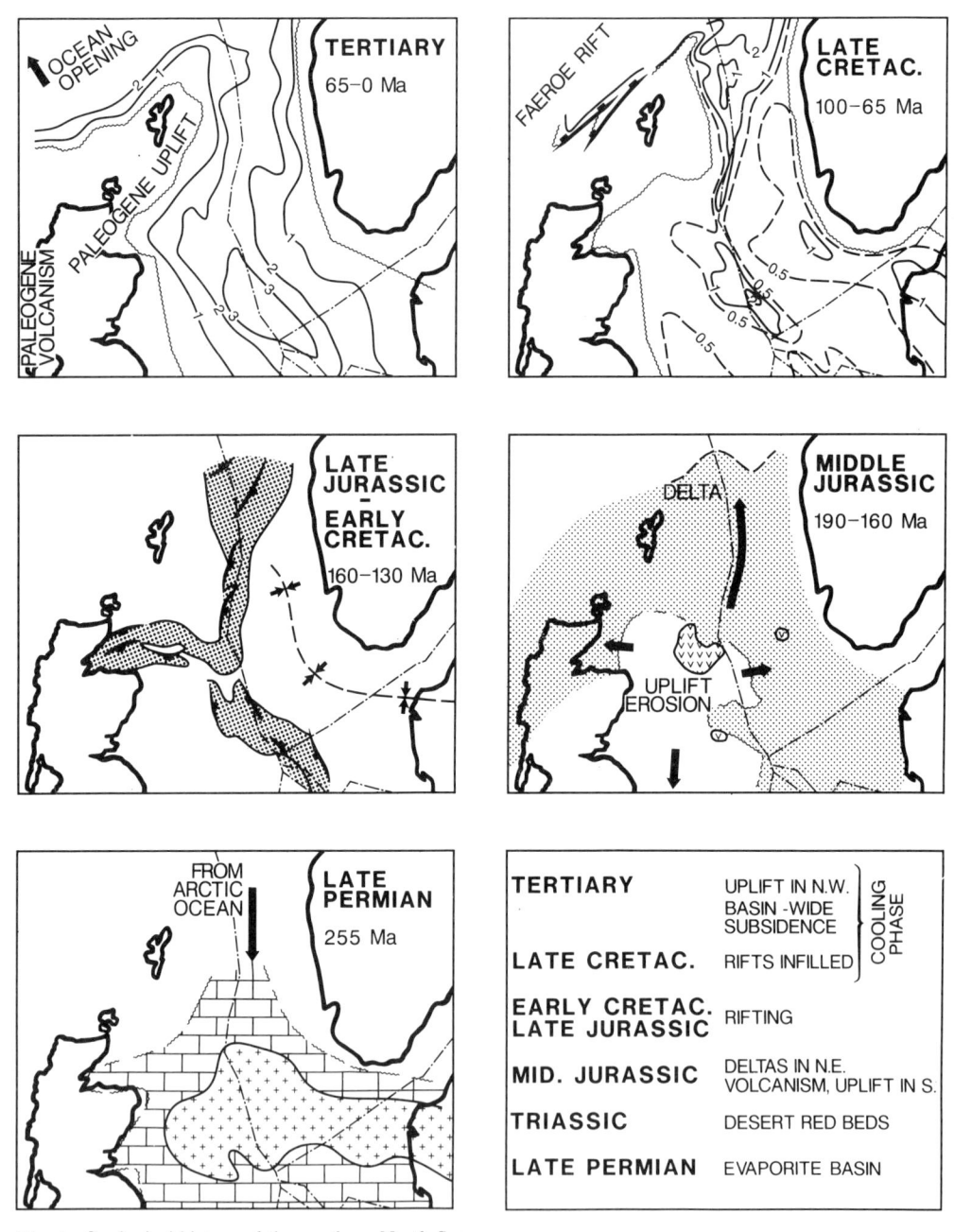

Fig. 1. Geological history of the northern North Sea.

extensional collapse being accompanied by oblique-slip and strike-slip offsets. The lateral component in the rifting was responsible for a complex sequence of structural inversions which began in late Jurassic and continued through Cretaceous times. Another complicating factor is the presence of Zechstein evaporites under much of the Central Graben, resulting in widespread detachment of the Triassic and Jurassic 'cover' from the sub-Zechstein 'basement'. The 'cover' is deformed by gravity driven mechanisms, listric normal faults detaching downwards

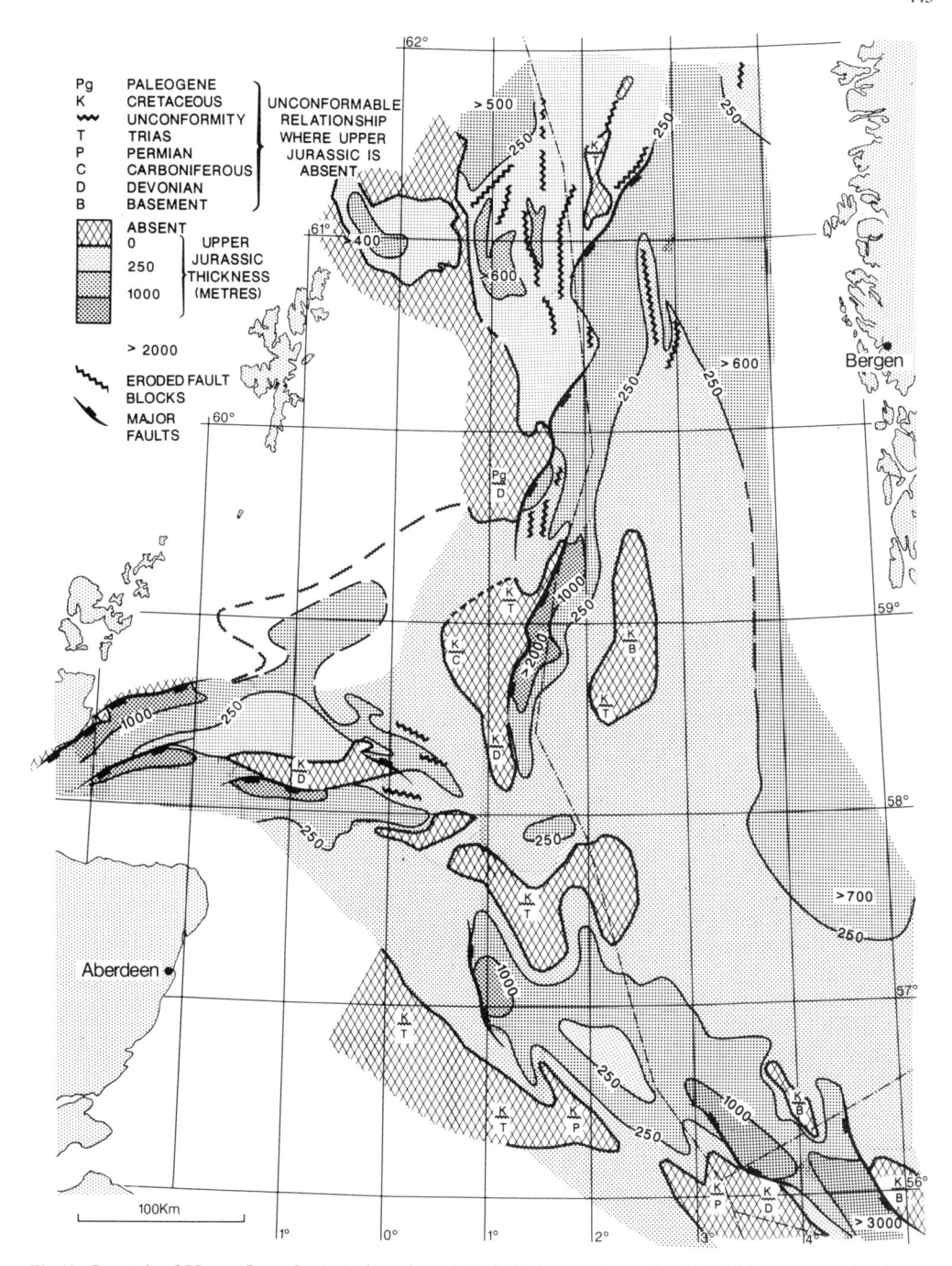

Fig. 2. Isopach of Upper Jurassic strata based on drilled thicknesses in wells. The thicknesses are for the Humber, Viking, Tyne and Boknfjord Groups. The interval includes therefore some Middle Jurassic strata in the north and some thin Ryazanian strata throughout the region.

onto the salt, whereas the 'basement' has rotated fault blocks more characteristic of an extensional regime.

At its northwestern end, in the neighbourhood of the Middle Jurassic Forties volcanic centre, the Central Graben merges with the

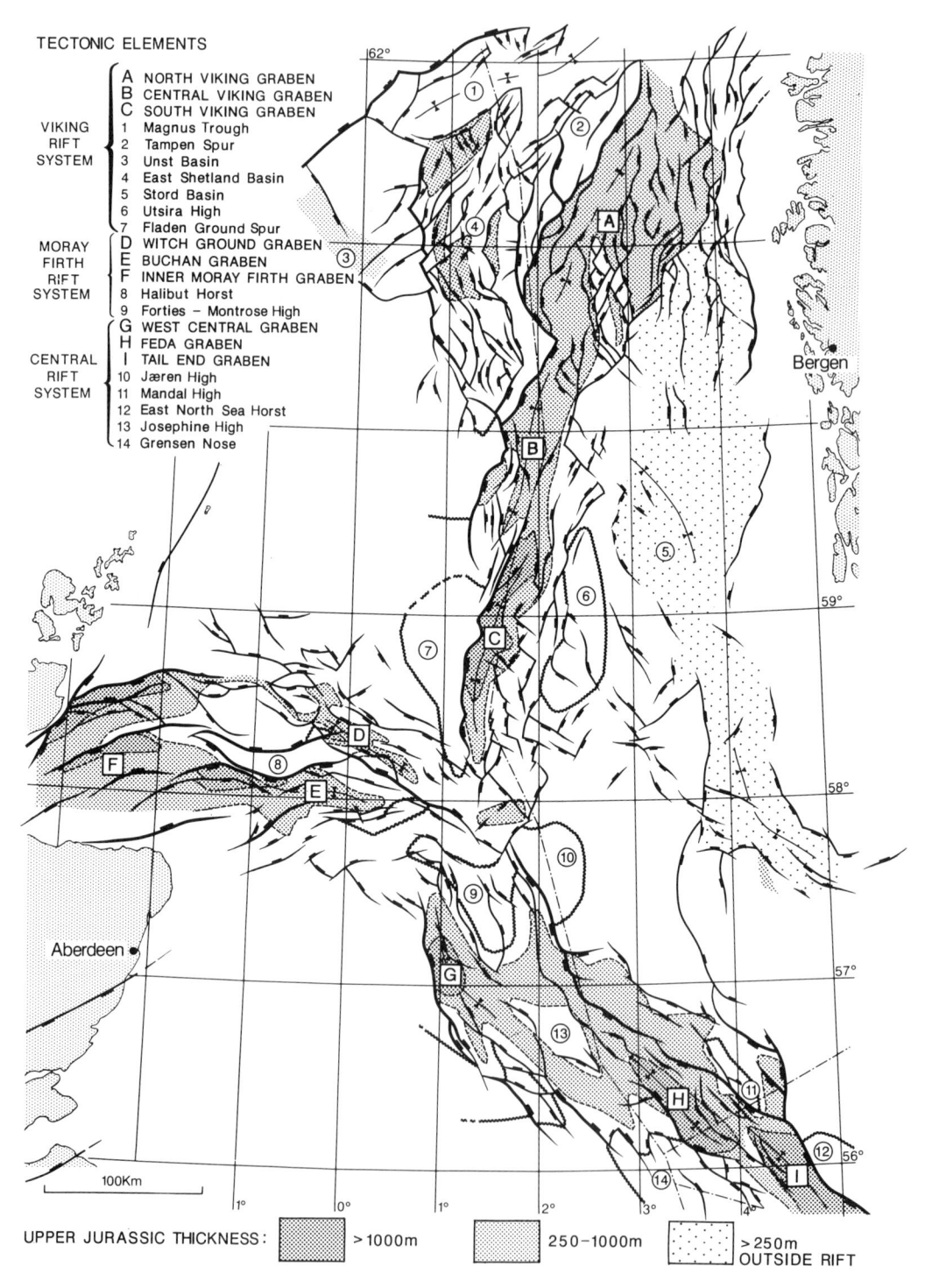

Fig. 3. Tectonic map of the late Jurassic — early Cretaceous rift system, indicating structural elements referred to in the text. The shaded areas show the close relationship between the rifting and the thickness of Upper Jurassic sediments (taken from Fig. 2).

more W-trending Moray Firth rift system. The Zechstein Group evaporites in the Moray Firth consist mostly of carbonates without any particular ability to allow detachment, and many of the faults clearly cut down into the sub-Permian 'basement'. The different sedimentary fill, especially the absence of thick Zechstein evaporites, resulted in a different tectonic style. In both areas the main phase of rift collapse occurred during late Jurassic times, although in the Moray Firth it may have begun later and continued longer into early Cretaceous times.

In the Moray Firth area, complex fault patterns, local compression structures and inversion tectonics, indicate that dip-slip movements were accompanied by oblique-slip and/or strike-slip movements. The Moray Firth rift terminates westward against NNE-trending faults of the Great Glen system. In this zone strike-slip offsets probably compensate for the crustal extension farther east. A dextral strike-slip offset on the Great Glen fault may have been the driving mechanism for the Moray Firth rifting (Bird et al. 1987).

The Viking Graben appears structurally simpler than the Moray Firth and Central Grabens. It has a general northerly trend comprised of three NNE-trending elements arranged in a left-stepping en echelon pattern (Fig. 3A,B,C). Each element is a half-graben bounded in the west by large normal faults which are probably listric at crustal scale. The deposition of syn-rift sequences 1–3 km thick on the east side of these faults, accompanied by major footwall uplift and erosion to the west, indicates that the main period of rifting was during late Jurassic times. There is evidence that faulting migrated outwards from the axial zone of the rift with time, the earliest rift activity occurring in late middle Jurassic times and the latest during early Cretaceous times (Badley et al. 1988).

The Tampen Spur and East Shetland Basin occupy an intermediate position in the rift system, between the deep North Viking Graben and the elevated Shetland Platform. The area suffered widespread extension in late Jurassic times but largely escaped the rapid subsidence typical of the more axial zones. At its northern end the Viking Graben system is truncated by ENE-trending faults of Cretaceous age.

The northern part of the Horda Platform, east of the North Viking Graben, was deformed by the latest stages of the late Jurassic/early Cretaceous rifting phase and subsequently acquired a gentle westerly tilt. Farther south, in the Stord Basin and Egersund Sub-basin, the platform has a thick Jurassic cover and was little affected by rifting but is cut by WNW-trending faults related to the deeply buried Tornquist Zone (Pegrum 1984).

Source rocks

Late Jurassic to earliest Cretaceous

The late Jurassic to early Cretaceous rifting events led to widespread and, in places, great subsidence (Fig. 2), so that marine Upper Jurassic to lowest Cretaceous strata are present throughout the basin. The thickest sequences reach over 3000 m adjacent to major faults, but thin overlapping sequences, sometimes only a few metres thick, cover highs and fault-blocks. The major lithostratigraphic units are the Humber, Viking, Tyne and Boknfjord Groups (Deegan & Scull 1977; Vollset & Doré 1984). Representative wells, including many of the type wells for specific formations, are shown in Figs 4 & 5). These groups commonly rest conformably on Middle Jurassic rocks in the north, the east, the southeast and the west of the northern North Sea. In the central area — the Outer Moray Firth — Humber Group rocks are everywhere unconformable on older rocks.

The shale formations range in age from Callovian at the base, in the north, to Ryazanian (Doré et al. 1985). At many levels there are black shales which show high radioactivity ('hot' shales), with gamma-ray values often above 75°API. They comprise the Kimmeridgian to Volgian Kimmeridge Clay Formation in the Moray Firth, the Volgian to Ryazanian Mandal Formation in the Central Graben, the Kimmeridgian to Ryazanian Draupne Formation in the Viking Graben and the Kimmeridgian Tau Formation in the Norwegian–Danish Basin. There have been few sedimentological studies of these 'hot' shales, for they have rarely been cored. In the Outer Moray Firth — South Viking Graben area, Stow & Atkin (1987) identified fissile-laminated, hemipelagic/turbiditic black shales as having been deposited in quiet, deep, basins. Anoxic conditions were common, due partly to the restricted water circulation in the complicated basin geography. These conditions were finally terminated in late Ryazanian times by a widespread transgression (Rawson & Riley 1982).

The total organic carbon contents (TOC) of the shales are high (> 2%) ranging up to 15% (Stow & Atkin 1987) and, in the case of the 'hot' shales, averaging 5% (Cornford 1986). The typical kerogen types of the 'hot' shales are mixtures of planktonic marine algae and degraded terrigenous humic matter, of sapropelic facies, yielding Type II oil-prone kerogens.

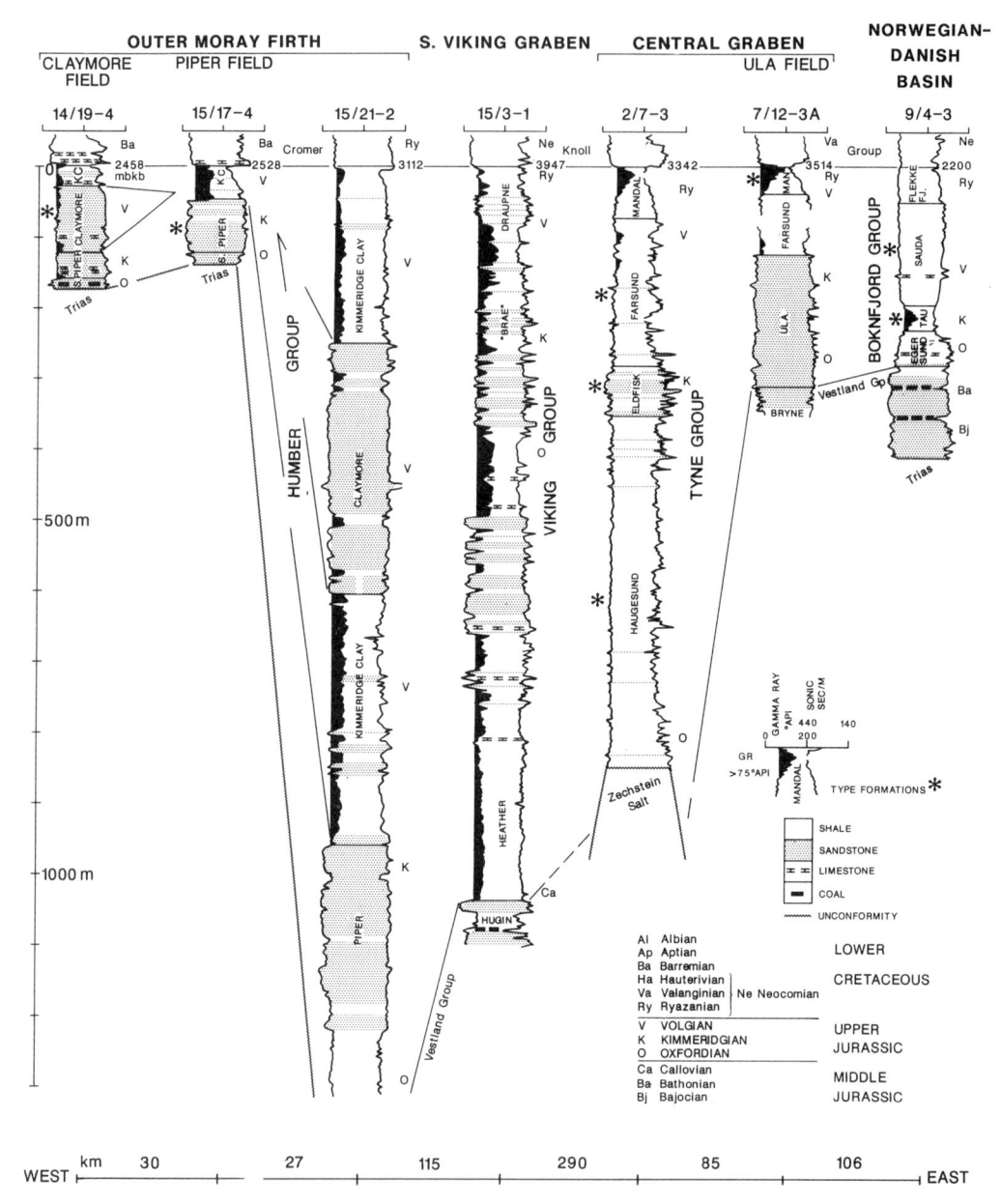

Fig. 4. Stratigraphy of the upper Jurassic to lowest Cretaceous strata of the Moray Firth (Humber Group), South Viking Graben (Viking Group), Central Graben (Tyne Group) and Norwegian–Danish Basin (Boknfjord Group). The terminology is from Deegan & Scull (1977) as modified and extended by Vollset & Dore (1984). Claymore Formation in Well 14/19–4 from Turner *et al.* (1987). s-Sgiath Formation from Harker *et al.* (1987). See Fig. 2 for locations.

The 'cold' shales, of for example the Heather Formation, show mixed organic facies resulting from varying aerobic to mildly anoxic conditions, and contain Type II, III and IV kerogens; these source rocks are gas-prone (Cornford 1986, Thomas *et al.* 1985).

Many studies have strongly suggested that these 'hot' shales are the source rocks from which the hydrocarbons of the northern North Sea have been generated (Van den Bark &

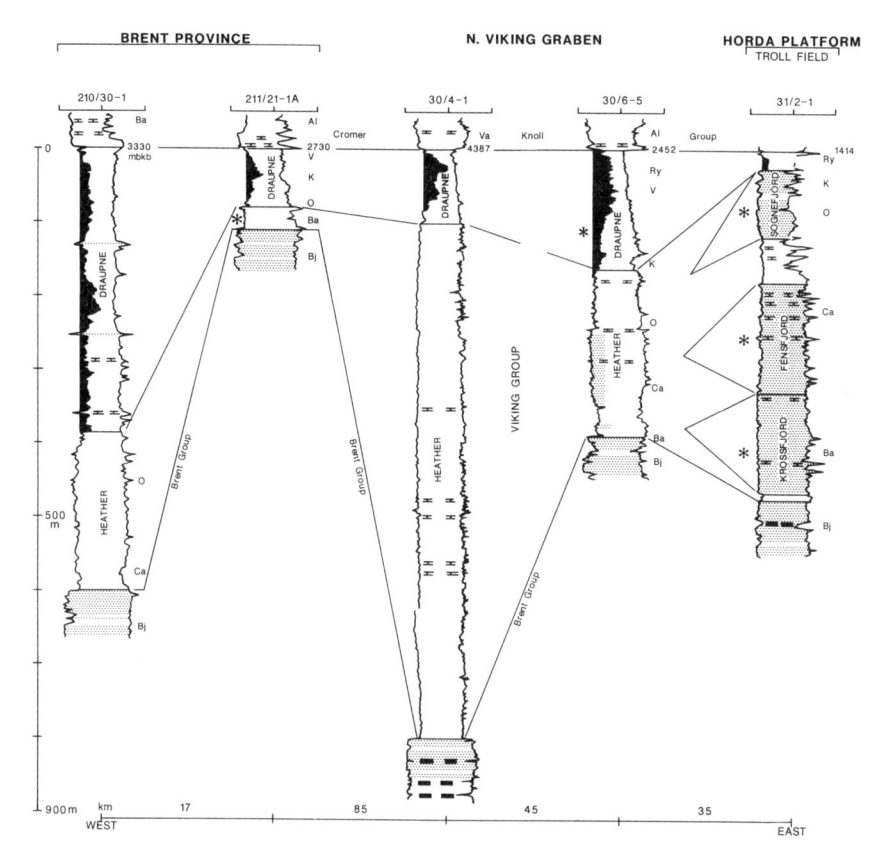

Fig. 5. Stratigraphy of the upper Middle Jurassic to lowest Cretaceous Viking Group in the northern Viking Graben. Key as for Fig. 4. See Fig. 2 for locations.

Thomas 1980; Fisher & Miles 1983; Bissada 1983; Northam 1985; Field 1985; Dahl & Speers 1985; Schou et al. 1985; Østfeldt 1987; Mackenzie et al. 1987).

Middle Jurassic

Middle Jurassic strata are up to 500 m thick in the Viking Graben, Norwegian–Danish Basin and Inner Moray Firth. They occur in the Central Graben, but are absent through much of the Outer Moray Firth. These sequences contain coals and vitrinite-rich carbonaceous shales and have been considered to be dry gas source rocks (Goff 1983). The gas in the Sleipner Vest field may have been generated from a Middle Jurassic source, with the condensate being sourced from the Draupne Formation 'hot' shale (Larsen & Jaarvik 1981; Ranaweera 1987). In the Hild field, gas and condensate may have been sourced from the Middle Jurassic sequence (Rønning et al. 1987).

Lower Jurassic

Lower Jurassic marine shales are present in the Viking Graben (Dunlin Group, up to 200–300 m thick), in the Norwegian–Danish Basin (Fjerritslev Formation) and in the Inner Moray Firth. In the Outer Moray Firth and Central Graben, they are absent or thin. In Western Europe, Lower Jurassic sequences, particularly at Toarcian level (Fleet et al. 1987), frequently contain organic-rich shales which are petroleum source rocks. In the northern Viking Graben the Toarcian shales have TOC values of up to 3%, with limited sapropelic content; they are variously considered to be oil-prone and/or gas prone (Rønnevik et al. 1983; Goff 1983; Field 1985; Schou et al. 1985). In the Norwegian–Danish Basin the Fjerritslev shales show average TOC values of up to 1.6% and have some oil-prone or gas-prone source potential (Thomsen et al. 1987).

Both the Middle and Lower Jurassic intervals

are of secondary importance as sources compared to the Upper Jurassic 'hot' shales.

Maturity

The principal source rock interval of the northern North Sea — the 'hot' shale units of the Kimmeridgian to Ryazanian sequence — has achieved maturity as a result of burial during Cretaceous to Recent times. Throughout most of the region subsidence during this time was continuous. The Tertiary depocentre overlies the late Jurassic graben system, and the regional subsidence is generally assumed to have resulted from cooling of the high thermal anomalies created along the rift zones.

There have been many studies to establish the maturity level of the 'hot' shale units. Direct measurements of vitrinite reflectance in well samples is often difficult, frequently giving spurious values. There is also a general lack of vitrinite in the 'hot shale' intervals (Thomas et al. 1985). Consequently, we have estimated present day maturity using a time/temperature maturity model and have chosen maturity levels equivalent to vitrinite reflectance values of 0.6%; 1.0% and 1.3% as indicators of the main hydrocarbon generation zones. Figure 6 shows the present geographical distribution of these zones and is closely comparable to other published maturity maps (Goff 1983; Fisher & Miles 1983; Bissada 1983; Thomas et al. 1985; Field 1985; Leadholm et al. 1985; Baird 1986; Dahl et al. 1987).

The maturity development of the Kimmeridgian to Ryazanian source interval is simple to understand, because throughout most of the region later subsidence has been continuous (Fig. 7). Oil generation had started over wide areas by Eocene times, whilst gas generation was only achieved in Neogene to Recent times. Throughout most of the region, because of the continuous subsidence, the widest area of generation and the maximum rank of generation occur at the present day. Similar maturation histories have been proposed by numerous workers (Goff 1983; Bissada 1983; Eggen 1984; Field 1985; Leadholm et al. 1985; Baird 1986; Dahl et al. 1987).

The main area where this simple maturity history does not apply is in the Inner Moray Firth. There, thick Upper Jurassic sequences were deposited (Fig. 2) but they only locally reached the oil-window before subsidence was terminated by major Paleogene uplift. In addition, maturity development has been interrupted in parts of the Central Graben by late Cretaceous tectonic inversion (Lindesnes Ridge). In the Egersund sub-basin the Tau Formation has only locally achieved maturity for oil generation.

Play classification

In this review we have analysed the hydrocarbon geology of the Northern North Sea by identifying a series of hydrocarbon plays (White 1988). We have adopted a division by reservoir age, into pre-Jurassic, Lower–Middle Jurassic, Upper Jurassic, Lower Cretaceous, Chalk and Paleogene Plays (Fig. 8). The tectonic history of the area allows these divisions to be grouped into 'Pre-rift', 'Syn-rift' and 'Post-rift' plays. We have found that these two classifications can be applied easily, the only uncertainties being in the equivalence of the syn-rift group with the Upper Jurassic and Lower Cretaceous divisions, for the timings of the onset and close of rifting vary from one area to another.

For each play, the principal controlling parameters have been identified and mapped (Figs 9–14). An essential parameter, the presence of mature source rocks (Fig. 6), is common to all the plays. For each play we describe: the age, distribution, lateral variations, depositional environments and quality of the reservoir rocks; the sealing and trapping mechanisms, quoting typical fields as examples; the relationships between the traps and the thermally mature source rocks; and the effective limitations or controlling factors.

Pre-Jurassic play

The pre-Jurassic play (Fig. 9) includes reservoirs of widely different ages ranging from Devonian (Buchan), to Rotliegendes (Argyll), to Zechstein (Auk, Ettrick), to Triassic (Snorre, Crawford). The common factor in these fields is the stratigraphic relationships of their reservoirs: all lie in areas where the reservoir is unconformably overlain by Upper Jurassic or Cretaceous strata. They are located in the eroded highs which formed during the late Jurassic rifting episode (Fig. 2). The exact age of the reservoir rock containing the hydrocarbons is therefore not relevant in defining the play.

In the Buchan Field the reservoir is a Devonian continental red bed sequence which owes some of its reservoir properties to the fracturing associated with formation of the horst trap. The Auk Field has a reservoir of Zechstein carbonates, the vuggy porosity of which has been enhanced by weathering associated with the base Cretaceous unconformity (Brennand & van Veen 1975). This is also the case at the

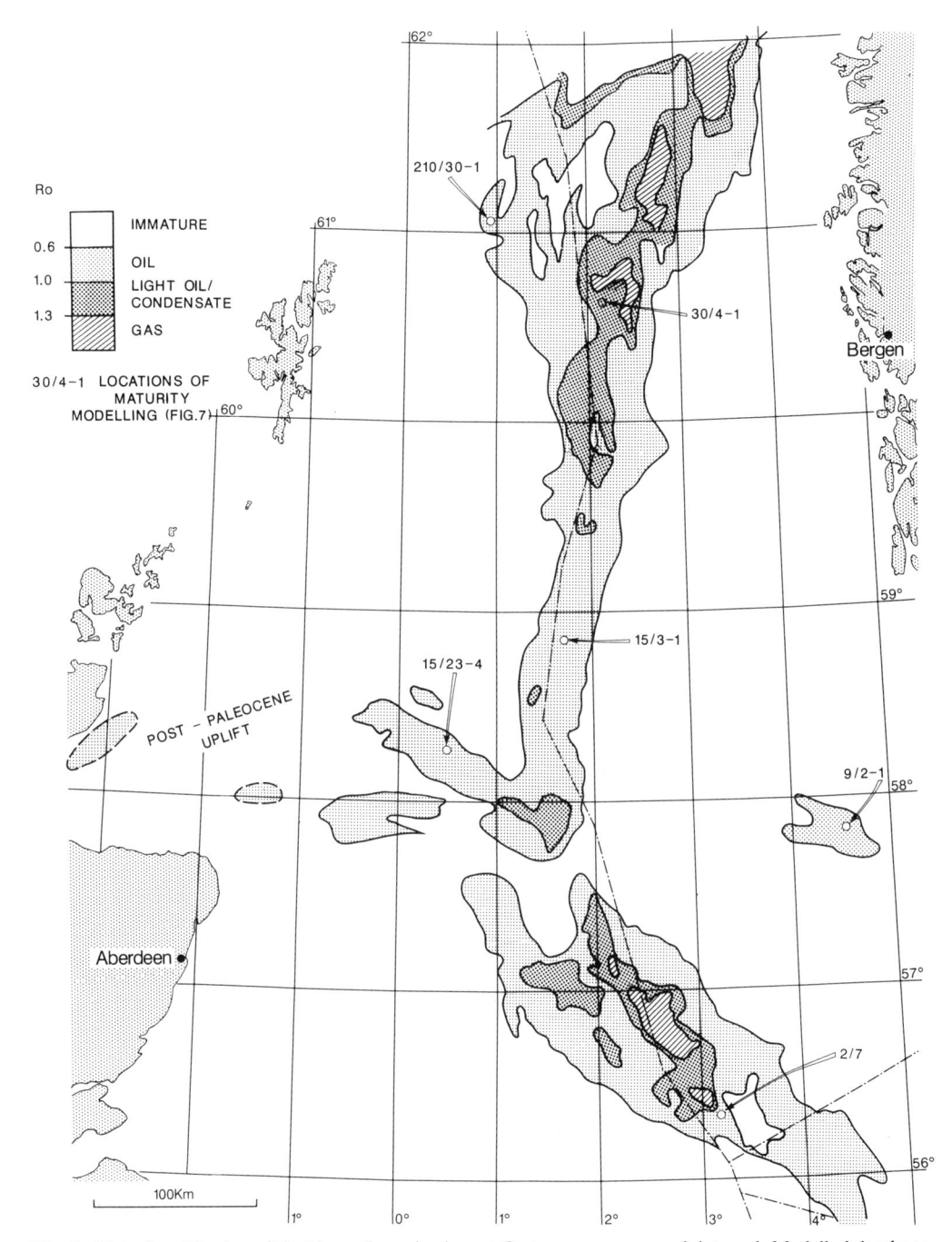

Fig. 6. Maturity of the top of the Upper Jurassic–lowest Cretaceous source rock interval. Modelled depths to the maturity levels shown were calculated area by area and equated with seismic two-way travel time. The maturity zones were then mapped using a regional isochron map of the 'Base Cretaceous' seismic horizon. Finally, areas where Upper Jurassic rocks are absent (Fig. 2) were excluded from the mature area.

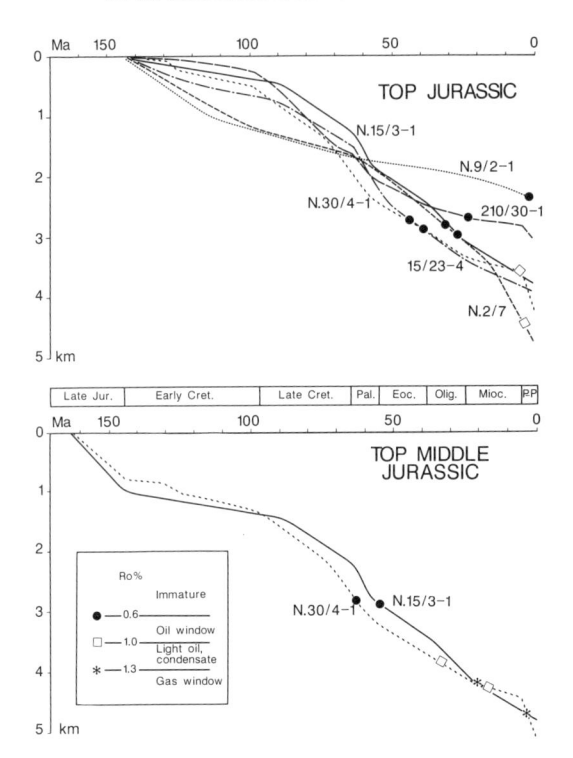

Fig. 7. Subsidence and maturation history of the Upper Jurassic source interval in six deep locations selected for the following reasons: UK Well 210/30−1, deep trough in Brent province; Norwegian Well 30/4−1 — in axis of north Viking Graben; remaining four locations — in deep, but not the deepest parts of the grabens/basins. The curves for Top Middle Jurassic indicate how much earlier in time maturity levels are achieved at the base of thick Upper Jurassic sequences (see Figs 4 & 5). For locations see Fig. 6.

Argyll Field, where Rotliegendes sandstone is also a reservoir, having porosities of 15% and permeabilities of 50 mD (Pennington 1975). The Auk and Argyll fields have varied cap rocks, due to the unconformable relationship, including Upper Cretaceous chalk.

The most important field belonging to the play is Snorre (Hollander 1987). There, erosion has cut so deeply that over large areas Cretaceous shales rest directly on the Triassic reservoir sequence, and only in the west is the important Statfjord Formation reservoir still preserved. The Triassic to Lower Jurassic reservoir sequence is much faulted. Erosion has planed off the tilted fault-blocks so that though the oil-column is less than 300 m high nearly 1000 m of strata contribute to the reservoir. The Triassic sandstones have porosities of 19−29% and permeabilities of 300−500 mD.

All the fields have complicated faulted traps formed during the late Jurassic rifting. Auk and Argyll Fields are in footwall uplifts tilted away from the graben. The fields are all close to areas with mature Upper Jurassic source rocks: indicating the importance of short migration routes for this play.

Lower−Middle Jurassic play

In the North Sea, the Lower−Middle Jurassic sequence of the Northern Viking Graben area is of outstanding importance. These rocks contain many of the largest fields (Fig. 10).

Reservoirs

The oldest reservoirs are the sandstones of the Rhaetian−Sinemurian Statfjord Formation which are over 200 m thick in the vicinity of the Statfjord Field where they lie conformably on Triassic continental red-beds more than 1400 m thick. They record a passage from non-marine to marginal marine conditions: floodplain, sinuous stream, braided stream and coastal plain deposits all being recognized (Kirk 1980). In the Statfjord Field the gross pay is about 120 m, with an average porosity of 22% and an average permeability of 470 mD (Roberts *et al.* 1987). The formation extends westwards across the

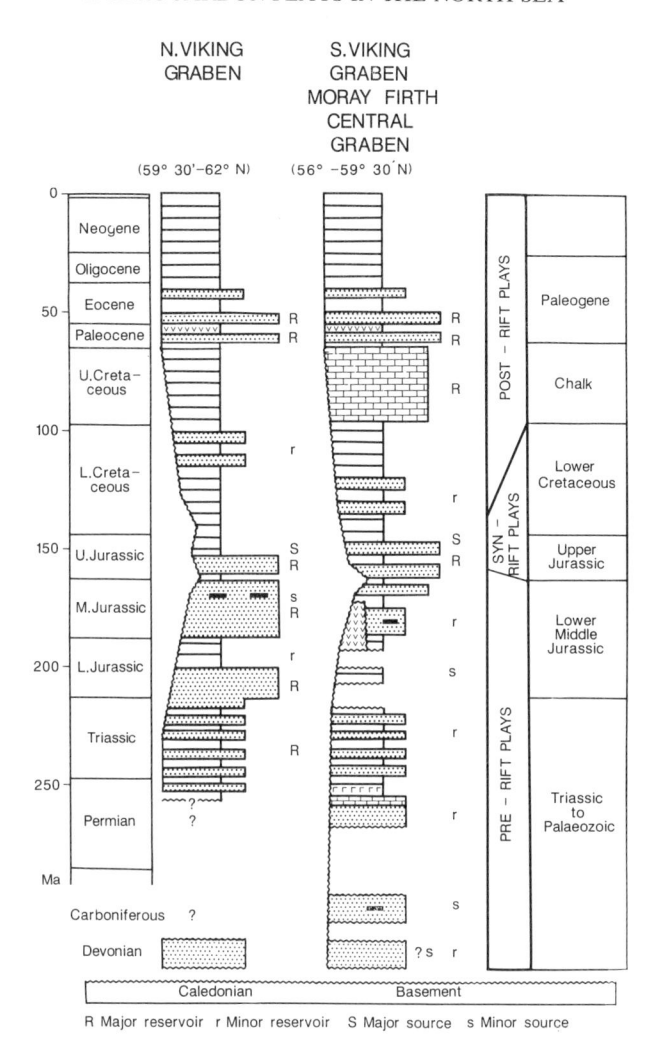

Fig. 8. Classification of the plays in the northern North Sea.

East Shetland Basin, its base becoming increasingly unconformable onto eroded Triassic rocks (Deegan & Scull 1977). It can be recognized eastwards onto the Horda Platform. In the Brage Field it is 150–200 m thick with average porosities of 24% and permeabilities of 500–2000 mD (Hage et al. 1987). Basal Jurassic sandstones, equivalent to the Statfjord Formation, are recognized as far south as the Beryl Field, where they are referred to informally as the Lower Beryl Sands (Marcum et al. 1978).

The Statfjord Formation sandstones in the North Viking Graben are overlain by marine shales of the Dunlin Group. There is a transitional passage from sandstones into shales and the base of the shales is diachronous, ranging in age from Hettangian to Sinemurian. Marine shale deposition continued until terminated by the northward prograding Brent delta in late Toarcian–early Aalenian times. The Dunlin shales are important in providing a top seal to the Statfjord sandstone reservoirs. A regressive sandstone/siltstone interval within the Dunlin shales (Cook Formation, Pliensbachian–Toarcian) is locally a reservoir; for example in the Gullfaks Field where it is up to 150 m thick, with porosities of 24–36% and permeabilities up to 1000 mD (Erichsen et al. 1987).

In middle Jurassic times, the Brent Group delta system dominated the North Viking Graben. The basal sandstones are sub-littoral deposits (Broom Formation). They are followed

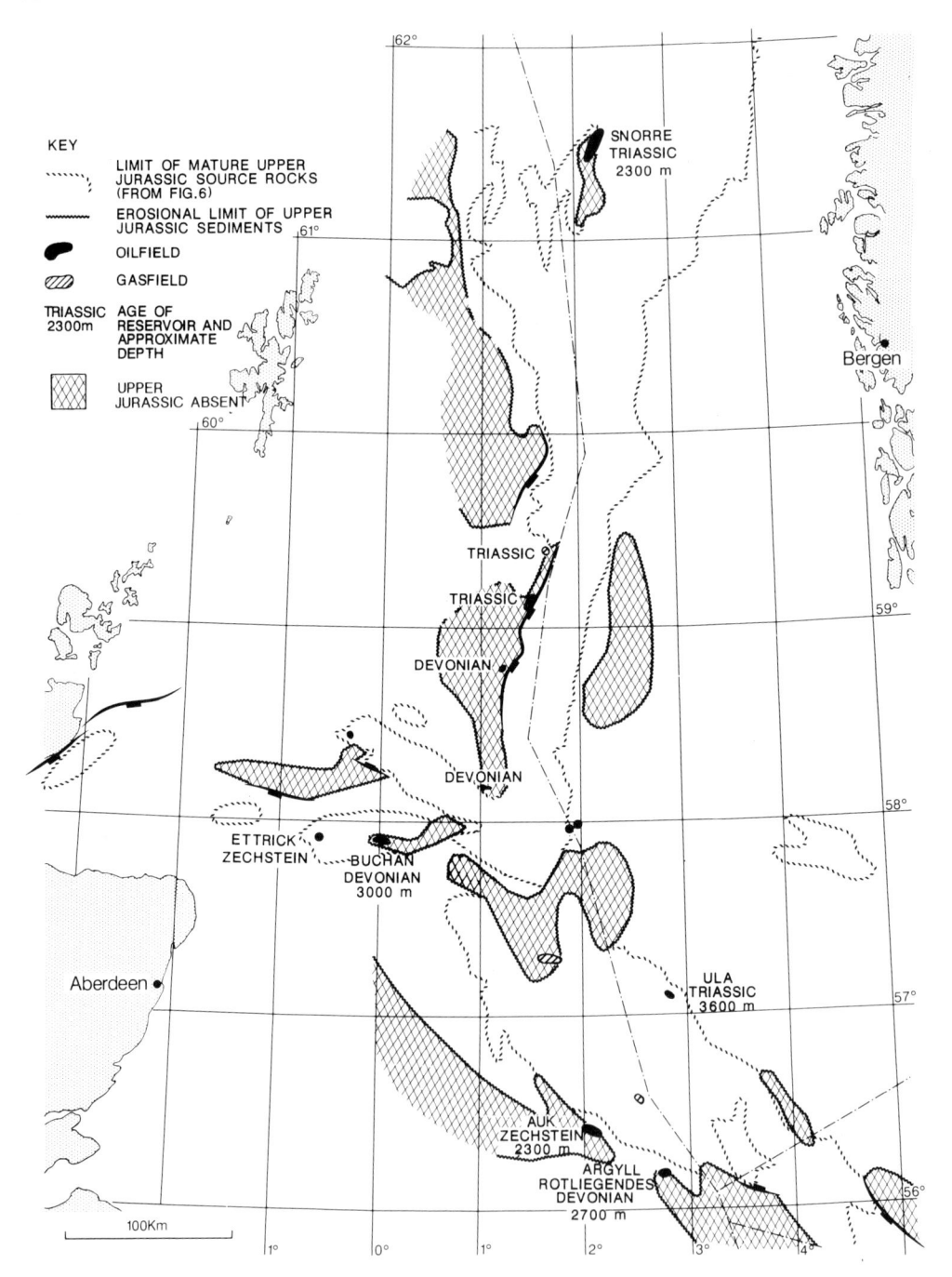

Fig. 9. Pre-Jurassic play map. The cross-hatched areas are the eroded highs of Fig. 2, where Upper Jurassic strata are absent. All the fields lie within these areas or in areas where Jurassic rocks are thin.

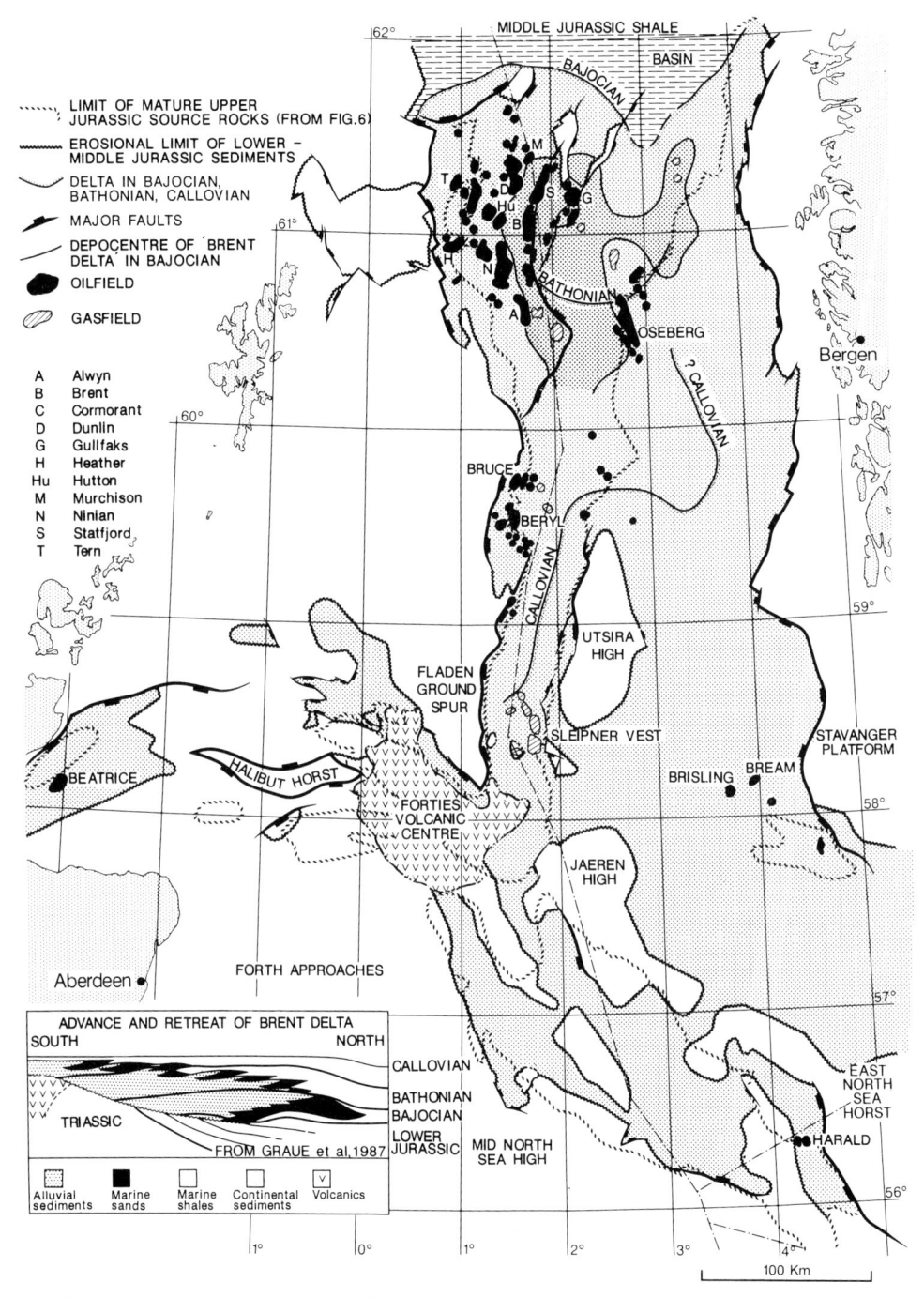

Fig. 10. Lower/Middle Jurassic play map. Shaded areas represent the present distribution of sediments of this interval — the limits are in most cases faulted or erosional and in most cases a direct consequence of the late Jurassic rifting. The approximate depocentre and maximum northerly extent of the Brent delta (in Bajocian times) are indicated.

by prograding, shallow marine, delta-front sandstones which are frequently micaceous and have poor permeabilities (Rannoch Formation). Above lie cleaner and better reservoir sandstones (Etive Formation) considered to be prograding, barrier-bar, delta-top deposits (Budding & Inglin 1981). The succeeding sandstones, shales and coals (Ness Formation), were deposited under back-barrier, alluvial plain and lagoonal conditions. The topmost unit (Tarbert Formation) consists predominantly of sandstones with minor shale interbeds. These sandstones record the start of a transgression which continued with the Heather Formation marine shales drowning out the Brent delta in Callovian–Oxfordian times.

All the subdivisions of the Brent Group are diachronous, for the delta system prograded northwards during early middle Jurassic times, reached its maximum extent in late Bajocian times, and retreated southwards during Bathonian and Callovian times (Graue et al. 1987) (Fig. 10). Towards the north, the delta interfingered with marine clays of the Heather Formation. Within this framework, the thickness, quality and distribution of reservoir sandstone units in individual fields vary. Several factors controlled this variability. Middle Jurassic syn-sedimentary fault movement, caused some local thickness variations (Hallet 1981), but this phenomenon may not be widespread. More important are the thickness variations caused by erosion following late Jurassic rifting and fault-block rotation. Commonly the uppermost divisions, the Ness and Tarbert, are truncated or absent, but in some areas the entire Brent Group has been removed (as in the Snorre Field). Brent sandstones normally have excellent reservoir characteristics, with average porosities approaching 30% and permeabilities c. 2500 mD. Diagenetic changes, especially the growth of fibrous illite, can significantly reduce reservoir quality at depth (Larese et al. 1984).

Middle Jurassic sandstones are important reservoirs to the south. In the Beryl Field, the Middle and Upper Beryl sandstones are broadly equivalent to the Ness and Tarbert Formations (Marcum et al. 1978). In the Sleipner Vest area, Callovian sandstones of the Hugin Formation are the main reservoir (Ranaweera 1987). They are underlain by Bathonian–Bajocian coastal plain sandstones, siltstones, shales and coals (Sleipner Formation) which rest unconformably on Triassic red-beds (Larsen & Jaarvik 1981).

South of this region of important Lower–Middle Jurassic reservoirs, the rock sequence is quite different. In the Outer Moray Firth and in the north of the Central Graben Middle Jurassic strata consist of extrusive basalts (Rattray Formation), locally more than 1000 m thick, which are interbedded with or laterally replaced by fluvial sandstones, siltstones and shales (Pentland Formation). Lower Jurassic rocks are generally absent, and the Pentland strata rest on eroded Permo–Triassic rocks. In the Central Graben, sequences of alluvial, coastal plain to shallow marine sandstones, siltstones and coals of middle Jurassic age (Bryne Formation) occur locally but to date are of limited commercial importance (Harald Field, Danish sector). Lower Jurassic strata are generally missing. The incomplete and localized distribution of these Lower and Middle Jurassic strata in the Outer Moray Firth and Central Graben areas indicates that they were part of an uplifted region. Lower and Middle Jurassic strata are preserved in the Norwegian–Danish Basin and Egersund Sub-basin. The Lower Jurassic is characterized by fine grained alluvial/fluvial clastics (Fjerritslev Formation). They are succeeded by Middle Jurassic sandstones, shales and coals (Bryne Formation) and by Callovian marine sandstones (Sandnes Formation). Middle Jurassic sandstones provide the reservoirs in the Bream and Brisling Fields (D'Heur & de Walque 1987) and in the recent Norwegian Block 9/2 oil discovery. In this area the Middle Jurassic play is restricted by the paucity of mature source rocks, not by a lack of suitable reservoirs.

Traps

The majority of hydrocarbon traps in this play are formed by tilted fault blocks. The traps vary considerably in size and complexity. The Statfjord Field occupies a simple, NE-trending, westerly tilted fault-block trap 24 km in length, 4 km in width and with a productive area of around 80 km^2 (Roberts et al. 1987). In contrast the Gullfaks Field, located only 20 km to the SE, is structurally much more complex, consisting of many small rotated and partly eroded fault-blocks, generally orientated N–S and only 1–2 km in width (Hazeu 1981; Erichsen et al. 1987).

The fault-blocks typical of the Brent province developed due to extension during late Jurassic rifting. On the western flank of the Viking Graben the rotated fault blocks generally dip to the west whereas on the eastern flank they dip to the east. Local exceptions resulted from transfer faulting and intersections of major fault trends — in such areas oblique-slip and strike-slip offsets complicate the structural picture and may result in reversals against the regional dip. The up-dip edges of the rotated fault-blocks are frequently eroded and complicated by 'gravity glide' listric faults, detaching at shallow depths in shale intervals (Statfjord Field: Roberts *et al.* 1987) or by leading edge horst blocks (Gullfaks: Hazeu 1981; Brage: Hage *et al.* 1987). In some instances the erosion which accompanied the fault-block rotation is severe and has resulted in the removal of much or all of the Jurassic sequence, as at Snorre Field. The traps are sealed vertically by Upper Jurassic or Cretaceous shales lying unconformably on them. Lateral seals are provided by the juxtaposition of shales against sandstones on fault planes. Shallow gas pockets have been encountered above some of the traps (Gullfaks: Erichsen *et al.* 1987) and indicate that the seals have locally leaked. The hydrocarbons have been sourced from mature Upper Jurassic shales, present both in the local back-basins of individual rotated fault blocks and more generally in the axial zone of the Viking Graben. Migration distances are probably short (Goff 1983), but the hydrocarbons may have spilled from field to field (Karlsson 1986).

In the Sleipner complex (Larsen & Jaarvik 1981) and in the Bream and Brisling Fields (D'Heur & de Walque 1987) the traps are domal and overlie Zechstein salt, but in Sleipner are complicated by faulting. The Sleipner Vest reservoirs are sealed by late Jurassic shales (Ranaweera 1987). The Bream and Brisling traps were sourced from marginally mature late Jurassic shales and very short migration distances are likely (D'Heur & de Walque 1987).

Limits on play

The Lower–Middle Jurassic play (Fig. 10) is controlled by the location of the Brent delta where thick, porous sandstone reservoirs were deposited prior to the late Jurassic rifting. The rifting segmented sequences deposited on the former delta into major fault-blocks which rotated and subsided simultaneously with the deposition of organic-rich Upper Jurassic shales. These shales drape many of the fault-blocks, infilling the back basins and the axis of the rift. Shale deposition continued through Cretaceous times providing thick seals. Subsidence in Cretaceous and Tertiary times, resulting from post-rift cooling, allowed the Upper Jurassic source rocks to reach thermal maturity in late Eocene times. This subsidence buried the late Jurassic fault blocks without disrupting them. The close physical relationship between the source rocks and the reservoir sandstones permitted early migration into the pre-formed traps.

The play is limited northwards by the shale-out of the sandstone intervals, and westwards by the boundary faults of the rift. Eastwards the play is limited either by the presence of late Jurassic sandstones which have allowed leakage to higher stratigraphic levels (e.g. in the Troll Field) or, as in the Stord Basin, by the presumed absence of mature source rocks. The play continues southwards along the Viking Graben, in progressively younger sandstones deposited by the retreating Brent delta, to the Sleipner area. Farther south the play is variously limited by the absence of Lower–Middle Jurassic sandstones and the widespread presence of stratigraphically younger reservoirs; also the presence of Zechstein salt gave rise to a fundamentally different tectonic style.

Upper Jurassic play

The most active period of rifting in the northern North Sea occurred during late Jurassic times creating the graben systems, the related eroded highs and the multitude of individual fault-blocks (Figs 2, 3 & 11). The clastic marine strata which accumulated during these movements show great lateral variation in thickness and lithology (Figs 4 & 5). However, the change from the pre-rift Middle Jurassic setting with its blanket of deltaic sediments was not instantaneous and the onset of rifting varied in time from area to area.

In the Outer Moray Firth the Upper Jurassic sequence begins with pre-rift deltaic strata (Sgiath Formation, Oxfordian), which are overlain by transgressive marine sandstones (Piper Formation, late Oxfordian–Kimmeridgian) and then by turbiditic sandstones (Claymore Formation, late Kimmeridgian to mid Volgian) which mark the onset of extensional collapse (Harker *et al.* 1987; Boote & Gustav 1987). In the Central Graben, shallow marine sandstones accumulated inside the active graben margins (Ula Formation, Oxfordian to Kimmeridgian) (Spencer *et al.* 1986), but the presence of a thick Zechstein salt sequence prevented many of the basement rooted faults reaching up to Jurassic strata, so there is no clear relationship between footwall uplift, erosion and sand deposition.

The South Viking Graben is at the northern limit of Zechstein salt. There, syn-rift submarine fan and turbiditic sequences, 1–3 km thick accumulated along the giant Brae-trend fault system ('Brae' Formation, Volgian) (Harms *et al.* 1981; Stow *et al.* 1982). Further north, syn-rift fault line sandstones are rare.

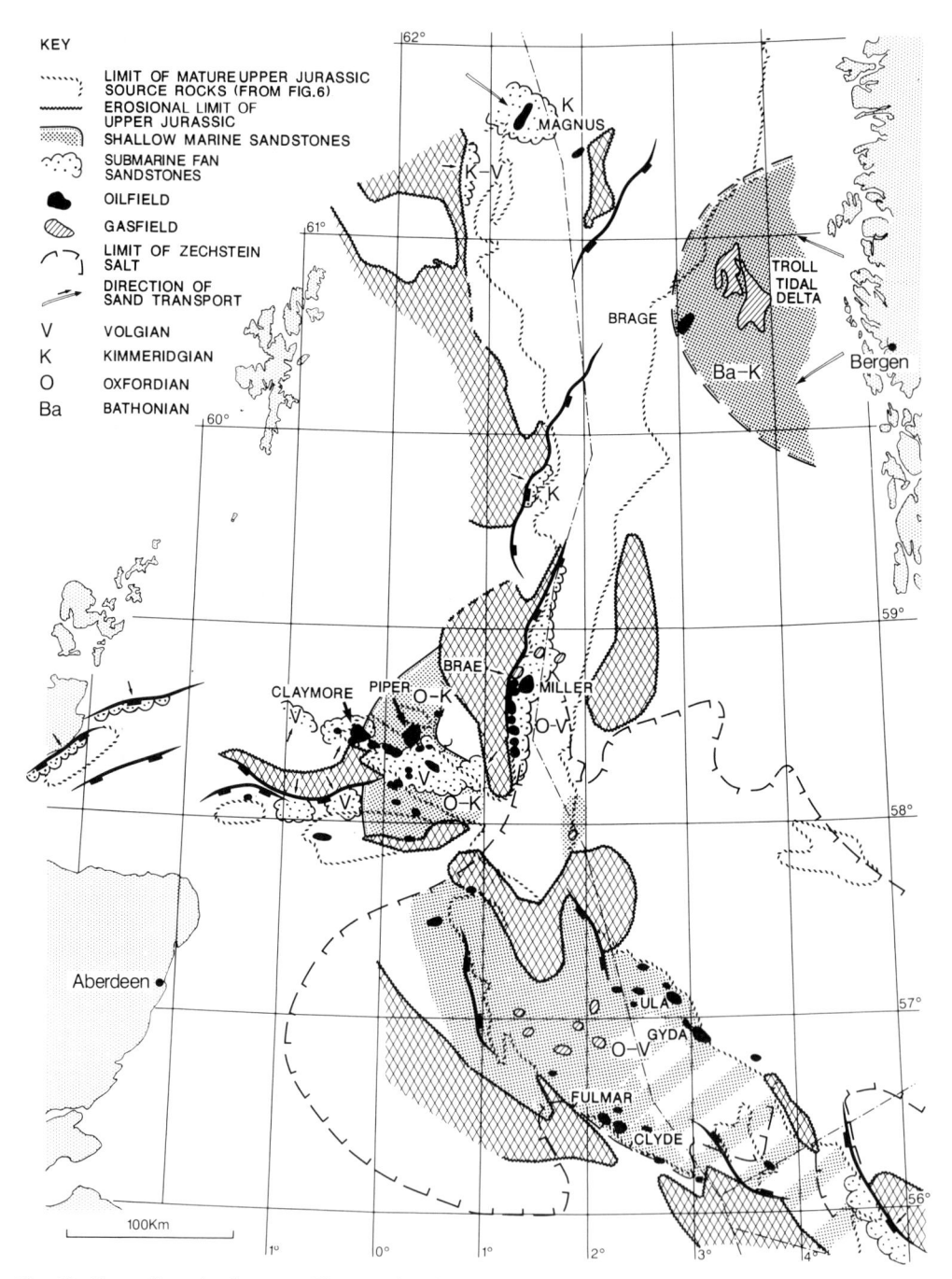

Fig. 11. Upper Jurassic play map. The cross-hatched areas are the eroded highs of Fig. 2, which were the source areas for many of the reservoir sandstones.

The shallow marine sheet sandstones of the Troll area form a sedimentary wedge of Bathonian to Kimmeridgian age, produced by uplift of the Norwegian mainland; the sands were prevented from spilling into the Viking Graben by the east tilt of the Horda Platform; and they were faulted in early Cretaceous times. The Kimmeridgian sandstones of the Magnus

Field were deposited as submarine fans prior to Cretaceous faulting (De'Ath & Schuyleman 1981).

Two groups of sandstones may be recognized: older, pre-rift units (Magnus, Troll, Piper) and later, syn-rift units (Claymore, Brae, Ula). Rather than subdivide this Upper Jurassic play, we believe this complexity is a consequence of the complex geological events in late Jurassic times, and so retain a single grouping.

Reservoirs and traps

The Troll Field reservoir (Fig. 5) is 230 m thick with porosities of up to 34% and permeabilities of up to 10 Darcies. The trap is unique in several respects. It is the shallowest major field (only 1000 m below sea floor) and in places the reservoir is unconformably overlain by Paleocene strata (Gray 1987). It is enormous (770 km^2), the trap being two very gently eastward dipping fault blocks. A remarkably extensive 'flat spot' can be seen on seismic records and is associated with the hydrocarbon-water contact (Birtles 1986).

The Brae 'trend' continues southwards for 100 km on the eastern, downfaulted side of the Fladen Ground Spur (Fig. 3). The fields include the Brae group, Miller, Tiffany, Toni and Thelma, totalling more than ten fields and finds. The reservoirs consist of conglomerates, sandstones and shales deposited as submarine fans (Stow et al. 1982; Turner et al. 1987). The sequence is very variable laterally and the fans are replaced by basinal mudstones only 10–15 km east of the fault line. At the Brae Field the maximum hydrocarbon column is 450 m thick and the trap is caused by eastward dip, reflecting the original depositional geometry, and by lateral seal against the fault to the west. The cap rock is the Kimmeridge Clay.

In the Outer Moray Firth the largest fields are Piper and Claymore. At Piper the reservoir sandstones belong to the deltaic Sgiath and the shallow marine Piper Formations (Fig. 4). They attain a total thickness of about 150 m and have average porosities of 24% and permeabilities of 4 Darcies (Maher 1981). The cap rock is the conformably overlying Kimmeridge shale, but on the crest Upper Cretaceous marls rest unconformably on the reservoir. The trap comprises three parallel, tilted fault-blocks. Gentle movements started during the deposition of the reservoir but the trap was formed principally in early Cretaceous times. The Claymore Field is more complex, and although most of the oil is in Upper Jurassic sandstones (Fig. 4), 32% occurs in unconformably overlying Lower Cretaceous sandstones (Scapa sand), and 3% occurs in unconformably underlying Permian carbonates and Carboniferous sandstones (Maher & Harker 1987). The Claymore Formation is the main Upper Jurassic reservoir and has porosities of 10–29% and permeabilities of 0.2–1300 mD in thin (0.2–0.5 m) turbidite sandstones. Over most of the field the cap rock is the conformable Kimmeridge silt and shale, but on the crest Barremian strata rest unconformably on Carboniferous rocks. The tilted fault-block trap developed in Volgian to Hauterivian times.

In the Central Graben the largest Upper Jurassic oil fields are Fulmar and Ula. The Fulmar Field has a 150–335 m thick sandstone reservoir, which rests unconformably on Triassic rocks. It is mainly of shallow marine origin (Oxfordian to Kimmeridgian), but has a subordinate deepwater sandstone interval (Volgian). Syn-sedimentary deformation occurred. Conformable Kimmeridge Clay forms the cap in the west, but over the crest the reservoir is capped unconformably by chalk. The trap is domal in form, perhaps largely produced by salt withdrawal during late Jurassic times (Johnson et al. 1986). As it lies within 4 km of the main graben boundary fault, half-graben detachment faulting may also have contributed to trap development, as has been suggested along strike at Clyde Field (Gibbs 1984). Recently the Clyde Field structure has been attributed mainly to salt structuring (Smith 1987).

On the northeast flank of the Central Graben the Ula Field lies in a fairway containing ten oil finds (Spencer et al. 1986). At Ula the reservoir sandstone is about 160 m thick, with porosities of 15–22% and permeabilities of 650–850 mD (Home 1987). Vertical sealing is provided by shales of the Mandal Formation (Fig. 4). The trap here is a simple dome above a salt swell, but the other finds along the trend include hanging wall closures involving stratigraphic truncations.

Limits on play

Most of the fields belonging to this play have two features in common. The cap rocks are normally Upper Jurassic shales, and, as these shales are the hydrocarbon source rocks, the fields commonly have very short migration routes (Cayley 1987). This is because the area where thermal maturity has been achieved (Fig. 6) is almost coincident with the graben system (Figs 2 & 3), within which the deposition of the sandstones took place. The most variable feature is the trap type: almost every field is unique, the only fields resembling each other are some adjacent fields along the Brae trend.

Lower Cretaceous play

Rifting continued into early Cretaceous times but was less extensive than in late Jurassic times. Movements on many of the major fault zones had by now ceased (compare Figs 12 & 2), so Lower Cretaceous strata infilled the post-rift relief. Cretaceous strata commonly rest unconformably on older rocks (the 'base Cretaceous unconformity') but in basin centres there is a conformable and usually continuous sequence from Jurassic into Cretaceous strata (Rawson & Riley 1982). The Lower Cretaceous strata belong to the Cromer Knoll Group which everywhere are marine and consist dominantly of mudstones and marls — so that they form cap rocks to Jurassic and older reservoirs.

The Moray Firth was the main area in which rifting movements continued through early Cretaceous times. There, thick sandstone-rich sequences accumulated adjacent to the major faults. Due to Tertiary uplift, however, these rocks are now exposed at the sea floor in the Inner Moray Firth, so only in the graben of the Outer Moray Firth are Lower Cretaceous sandstones in contact with mature source rocks. In the Scapa and Claymore Fields, Valanginian to Hauterivian submarine fan sandstones (Scapa Sandstone, Harker et al. 1987) accumulated as unconformable aprons flanking the eroding faulted highs (Boote & Gustav 1987). On the north flank of Claymore Field, the Scapa Sandstone is 0–140 m thick with porosities of 20–30% and permeabilities of 0.1–4 Darcies. Farther east, Aptian sandstones form the reservoir in the Bosun Field.

In the northeast, only 50 km from the Norwegian mainland, minor gas accumulations occur in Albian sandstones in the Agat finds (Gulbrandsen 1987). The gas is stratigraphically trapped in a westerly thickening Aptian–Albian wedge up to 500 m thick comprising interbedded thin sandstones and shales which were deposited in submarine fans.

Upper Cretaceous play

Rifting had essentially ceased throughout the Northern North Sea by the end of early Cretaceous times. The high heat flows associated with the rifting gradually waned, and the regional pattern was one of cooling subsidence. Subsidence rates were greatest along the axes of the earlier rifts, and the strata are thickest in these areas (Fig. 13), locally exceeding 1000 m. Thick sequences were also deposited in the Norwegian–Danish Trough and the Egersund Sub-basin, associated with subsidence along the Tornquist Zone. An exception to the general pattern of subsidence occurs in the south of the Central Graben where late Cretaceous inversion movements are widespread.

The Upper Cretaceous strata are almost entirely in a chalk facies (the Chalk Group) as far north as 57°N. Northwards there is a gradual increase in clay and north of 59°N clays and shales dominate, being represented by the Shetland Group. Here, chalks are restricted to the uppermost part of the succession or to basin margin positions (Fig. 13). Siltstones and thin sands occur in the north of the Viking Graben and indicate a northerly provenance for the late Cretaceous clastics (Hancock 1975).

Chalk deposition continued into earliest Paleocene (Danian) times in the south of the Central Graben — the Chalk Group therefore includes both late Cretaceous and early Tertiary strata. The Chalk Group is an important oil and gas reservoir only in the south of the Norwegian Central Graben and in the Danish sector. It had not been considered as a potential reservoir until the unexpected discovery of the giant Ekofisk Field in 1969.

Reservoirs

Depositional porosities in chalk can be as high as 70%, after de-watering porosity is reduced to 50–55% and these values are normally greatly reduced during burial — to 15–30% with 0.1–1 mD permeability at 1500–2000 m and to 2–25% with 0–0.5 mD permeability at 2700–3000 m (Hancock & Scholle 1975; Hancock 1986). In the productive fields however, much higher porosities are retained, for example up to 40% in Ekofisk at around 3000 m (Scholle 1977) and more than 30% in Hod Field at 2500–2750 m (Hardman & Kennedy 1980). These high porosities have been retained due to the early migration of oil and the development of high overpressures (Hardman 1982).

Oil migration into the structures began in late Eocene to early Miocene times, when burial depth was in the order of 1500 m. Retention of the oil, and the development of high overpressures with increasing burial was facilitated by the thick Paleocene shale above the Chalk, which formed topseals.

Depositional controls over porosity are particularly important: the most porous intervals are in redeposited debris-flow chalks. They form up to 50% of the thickness (D'Heur & Michaud 1987) in the pay zones, which are up to 250 m in thickness. The debris flows developed on

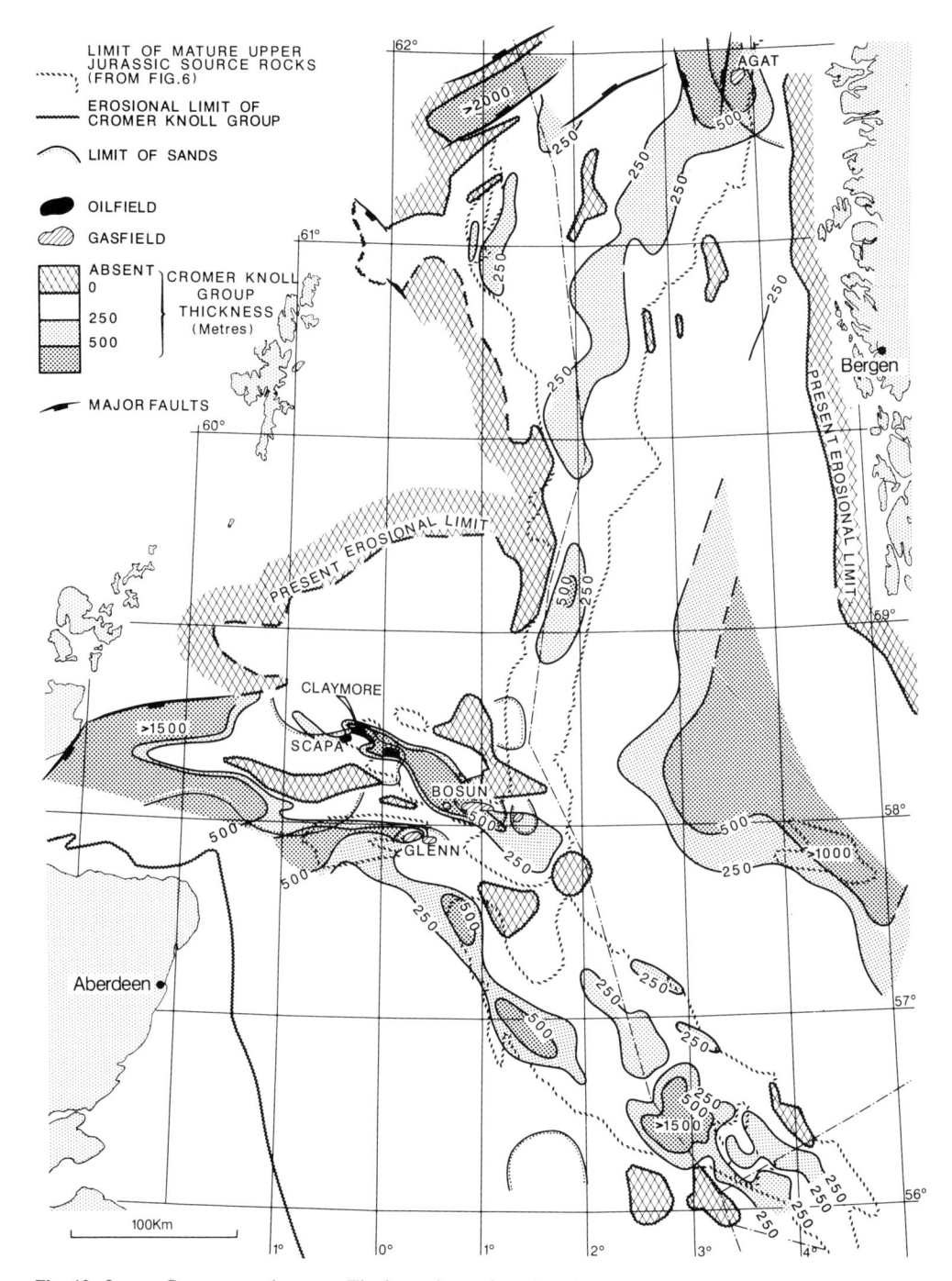

Fig. 12. Lower Cretaceous play map. The isopachs are based on drilled thicknesses in wells. Comparison with Fig. 2 shows that the active faulting was now largely limited to the Moray Firth and this is where sandstones are most developed.

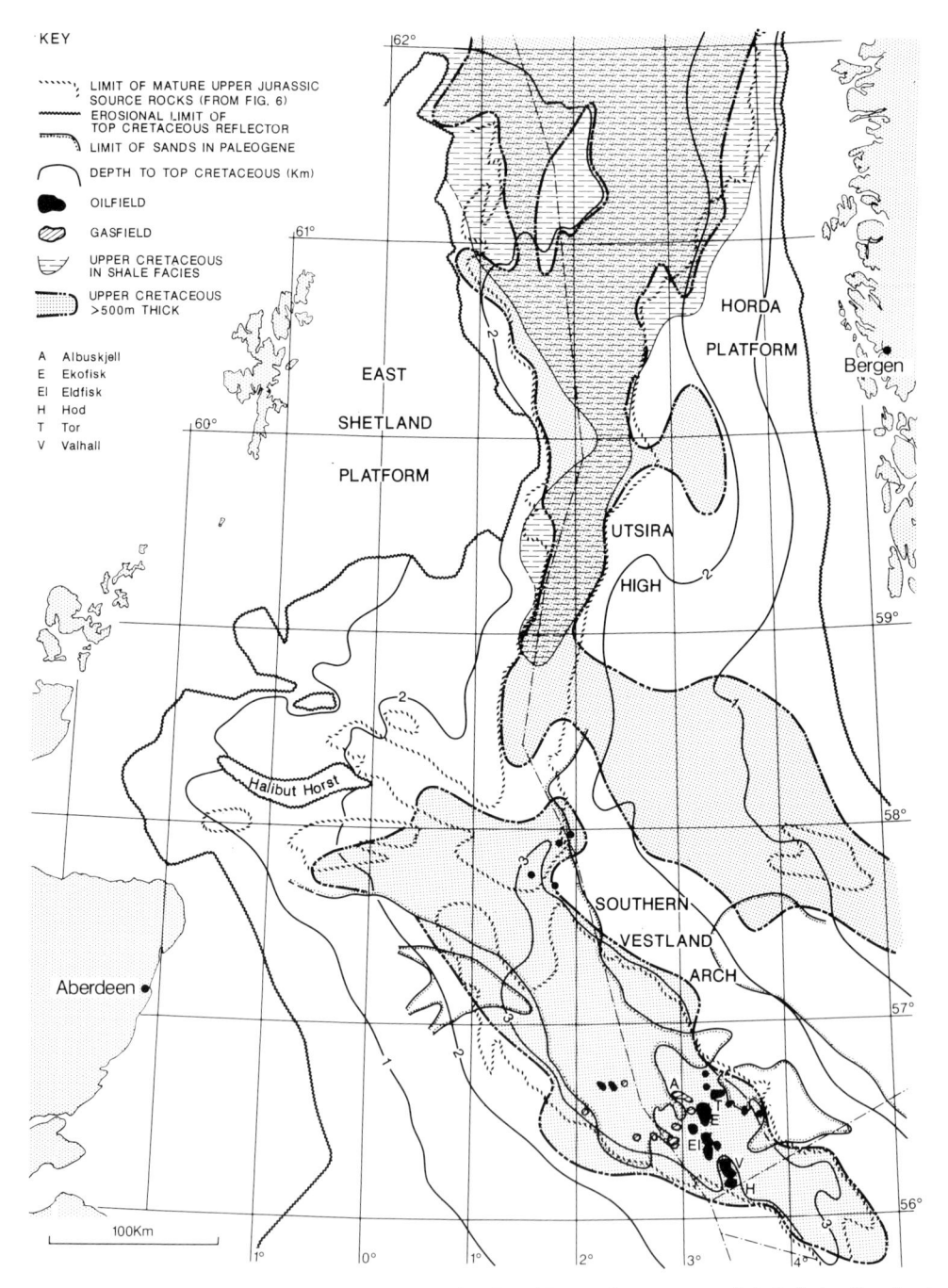

Fig. 13. Upper Cretaceous (Chalk) play map. Despite the widespread occurrence of chalk facies the play is limited to a small area where the chalk is thick, deeply buried, overpressured, structured, underlain by mature source rocks and not overlain by Paleogene sandstones.

submarine slopes above rising salt and inversion structures and locally along the oversteepened flanks of the trough (Kennedy 1987). It is these features which control the locations of the Chalk Group hydrocarbon traps.

The 'normal' pelagic chalk intervals, which have not been redeposited, are generally tight, and form non-productive zones within the fields and even act as cap rocks (as at Auk, Fulmar and Argyll).

Traps

Salt-supported Chalk fields in the Central Graben include Albuskjell, Ekofisk, Edda, Eldfisk, Tor and Tommeliten (D'Heur 1987a, b,c; Pekot & Gersib 1987). In several of these fields the effects of salt diapirism and structural inversion are difficult to separate and both mechanisms may have operated simultaneously. Valhall (Leonard & Munns 1987) and Hod (Norbury 1987) are local culminations on the Lindesnes inversion ridge, in an area where salt diapirism probably played only a minor role. All the chalk traps have four-way dip closure and can be classified as structural traps although stratigraphic variations control the distribution of the productive zones in each field.

The Chalk fields are underlain by thick thermally mature Upper Jurassic source rocks, and migration from source to reservoir has been essentially vertical. This migration may have been assisted by the actively rising salt and by inversion structures causing extensive microfracturing. Several of the Chalk fields have associated gas 'chimneys' in the overlying Tertiary strata. Gas leaking from the reservoir charges the overlying sediments, reducing their bulk density and so their seismic velocity, and disrupting the seismic travel paths. This results in areas of poor data quality and apparent sags or grabens on the tops of the structures (Dangerfield & Brown 1987). The fields can be considered to be 'dynamic', with active migration from below and partial upwards leakage. The predominance of gas condensates in the Albuskjell, West Ekofisk and Tommeliten fields correlates well with the present day maturity map (Fig. 6) and confirms that the migration is essentially vertical.

Limits on play

A clear spatial relationship exists between the location of the Chalk fields and the pinchout line of the Paleogene sandstones (Fig. 13). Sandstones near the base of the Tertiary sequence have provided an escape route for hydrocarbons migrating to the top of the chalk and, possibly more importantly, have prevented the development of overpressure (Cayley 1987) and the retention of porosity in the subjacent chalk. No major Chalk fields have been found in areas where there is a significant sandstone development within the lower Tertiary sequence.

The Chalk play, although stumbled upon early in the exploration history of the Northern North Sea, has proved to be very restricted. Necessary ingredients are the presence of favourable chalk lithologies — redeposited facies associated with unstable submarine slopes; the presence of thick Upper Jurassic source rocks vertically beneath; sufficiently rapid and deep burial to ensure early hydrocarbon generation, migration and overpressuring; and the presence of a contemporary deformation mechanism such as deeply buried salt or structural inversion. These conditions are fulfilled throughout much of the Central Graben but the final, essential, ingredient — the absence of early Tertiary sands — restricts the Chalk play to small areas within the Norwegian and Danish sectors.

Paleogene play

Basinwide subsidence continued from late Cretaceous into Tertiary times, the depocentre approximately coinciding with the buried rift zone. The Tertiary sedimentary sequence of the Northern North Sea is up to 3000 m thick. It is dominated by mudstones, but sandstones are also present especially along the western flank of the basin (Fig. 14). These sandstones are principally Paleocene and Lower Eocene and provide important reservoirs for both oil and gas. The sand input occurred as a result of uplift and erosion associated with the initial opening of the North Atlantic Ocean. Sandstones also occur locally higher in the Paleogene and in the Neogene. Eocene sandstones are of great exploration interest at present.

In the Central North Sea four major Paleocene sandstone units are recognized (Morton 1979; Knox et al. 1981). Heavy mineral assemblages indicate provenance from the Caledonian basement rocks of the Scottish Highlands or the erosion of pre-existing sandstones on the East Shetland Platform. The Paleocene sequence consists of overlapping, prograding submarine fans in part disconformably overlain by prograding shelf-deltaic deposits (Rochow 1981); progradation was eastwards across the basin flank and southeast along the axis of the Central Trough. Distally, the fans thin and pass into siltstones and marine mudstones.

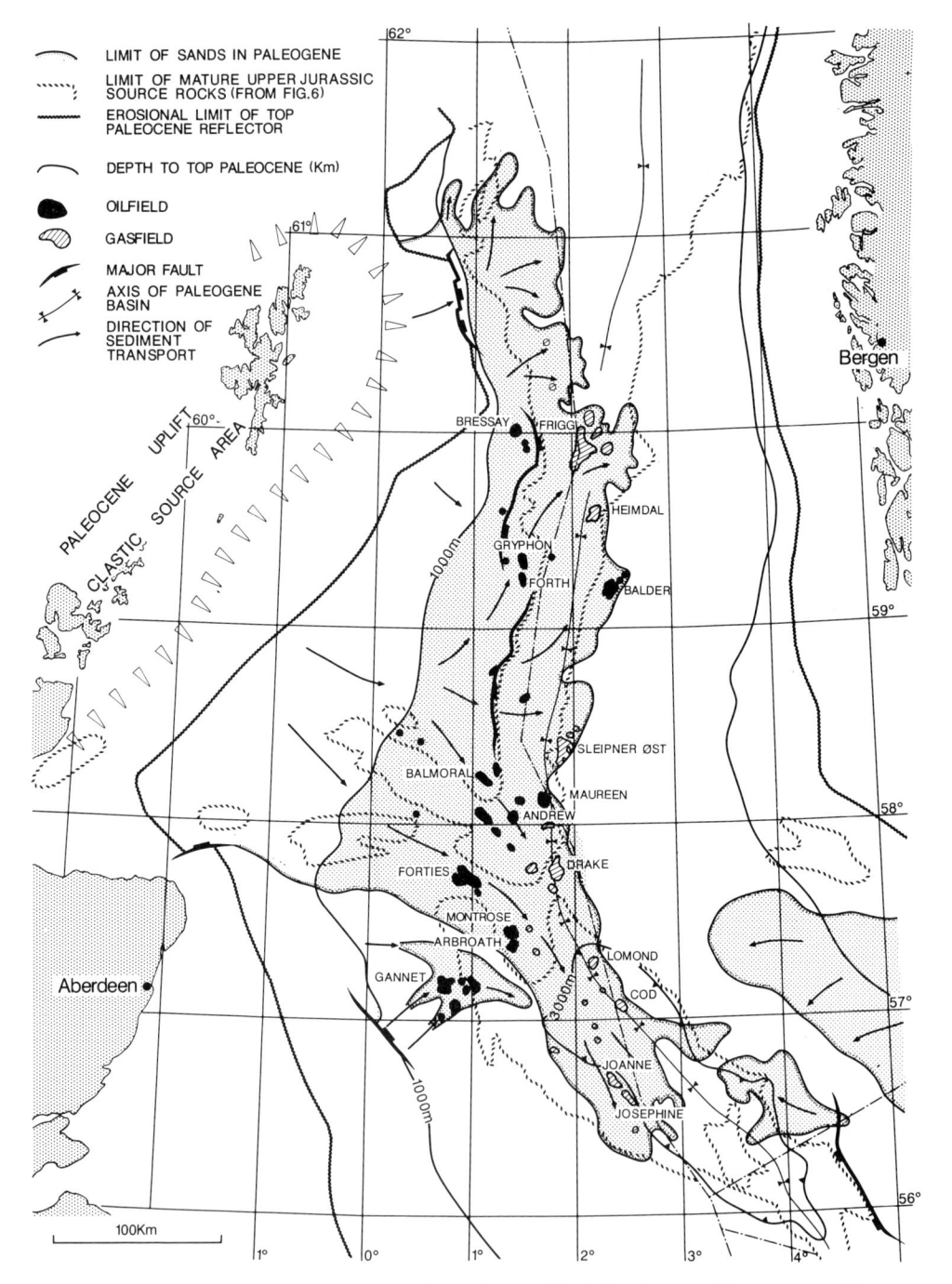

Fig. 14. Paleogene play map. The shaded area shows the combined distribution of Paleogene and Lower Eocene sandstones, mainly derived from the uplifted area of the basin axis. Note that in many cases fields are not directly underlain by mature source rocks.

In the Viking Graben areas sand input was also from the west. Sands shed from the eroding Shetland Platform were first deposited in middle Paleocene times in nearshore locations (Mudge & Bliss 1983). Sand progradation was towards the east. During late Paleocene and early Eocene times sand deposition spread northward along the edge of the East Shetland Platform. Progradation continued into early Eocene times with the local emplacement of submarine fans, as in the Frigg area (Heritier et al. 1979). Morton (1982) recognised four phases of basin subsidence and source area uplift along the western flank of the Viking Graben. Heavy mineral studies again indicate two sources, Caledonian basement rocks of the northern Orkney–Shetland Platform, and reworked sandstones of probable Jurassic age from further South.

Northwards and eastwards the Paleogene sandstones thin and pass into marine siltstones and mudstones. The early Tertiary basin axis was closely related to the position of the late Jurassic rift system. The basin axis (Fig. 14) exerted an important control on the submarine fan sands; the turbidity currents were unable to flow far 'uphill' and were deflected into axial directions. The abrupt retardation of the flows resulted in some areas in massive, clean sands being deposited in distal positions, (Sleipner Øst: Pegrum & Ljones 1984).

Seismic data and limited well data indicate that sands were also transported into the basin from the east (Fig. 14). The western limit of these sandstones is imperfectly known but it seems that they barely reach the Central Graben and, to date, they have not proved important as hydrocarbon reservoirs.

Reservoirs

The Paleogene sandstones may reach several hundred metres in thickness and frequently have excellent reservoir characteristics. Paleocene sandstones have porosities ranging from 17.5% to 35% at depths of 1200 to 2300 m, dependant on the facies. The best reservoirs are in structureless grainflow facies, the turbidite facies being poorer. In the Forties Field mean permeability of the massive sandstones approaches 650 mD, it is 400 mD in the sandy turbidites and < 10 mD in the muddy turbidites (Carman & Young 1981).

Vertical and lateral seals to the Paleogene sandstones are provided by the marine mudstones. Especially towards the top of the Paleogene sandstone sequences, and in distal positions, complex stratigraphic alternations may occur where discrete fan lobes or channel systems are interbedded with impermeable shales and silts (Balder Field: Hanslien 1987).

Traps

Several different trapping mechanisms are recognized. Despite the fact that the sequence has not been affected by regional tectonism most of the traps are 'structural' in that they have four-way dip closure. In some, including Forties (Hill & Wood 1980), the Frigg complex (Mure 1987a), Heimdal (Mure 1987b) and Montrose (Fowler 1975), the closures seem to have been enhanced by differential compaction over the depositional relief of the sandstones. Other fields lie near the limit of the sandstones and may have important stratigraphic components (Balder: Hanslien 1987; Sleipner Øst: Østvedt 1987). In Sleipner Øst the pinchout has been modified by gentle compression associated with rejuvenation of a buried fault system (Pegrum & Ljones 1984). In the Central Graben, Zechstein salt diapirs formed domal uplifts which provide traps in Andrew, Maureen, Lomond, Cod (D'Heur 1987b) and Gannet (Armstrong et al. 1987). The trapping mechanism in some of the recent Eocene sandstone discoveries (e.g. Alba, Gryphon, Forth) has not been published but the channelised nature of those sandstones suggests that stratigraphic controls may be important.

Sources and migration

All the Paleogene fields have been ultimately sourced from Jurassic source rocks. Fields located near the graben axes (Frigg, Heimdal, Andrew, Maureen, Lomond, Cod) overlie deeply buried Upper Jurassic source rocks and 'vertical' migration paths are probable (Goff 1983). Migration may have been focused by buried fault blocks or by salt piercements. In other fields (Forties, Montrose, Arbroath, Balder, Sleipner Øst) mature Upper Jurassic source rocks are not present vertically beneath the accumulations and lateral migration within the sands has to be invoked. The presence of oil, frequently of low gravity and biogenically degraded, in traps located on the East Shetland platform also implies that lateral migration has occurred. Due to the easterly pinchout of the Paleogene sandstones, hydrocarbons which have migrated eastward from the basin axis are trapped in a stratigraphic closure and cannot escape laterally from the basin. By contrast, hydrocarbons which migrated westwards from the basin axis, encounter sandstones which progressively thicken up-dip. Here, the system is open and hydrocarbons have been able, in the absence of intervening traps, to escape to the surface. It is probable that large volumes of hydrocarbons have been lost in this direction.

Limits on play

The principal factors controlling the Paleogene play are the presence of sandstones of reservoir quality in closed traps. Long distance lateral migration allows traps to be located outside the area of mature source rocks. On the eastern flank of the basin the play is limited by the eastward pinchout of the sandstones. On the western flank the sandstones thicken to outcrop and practical limits are thus imposed, the shallowness of the sandstones resulting in oils of low gravity which are biodegraded. This limit is controlled by reservoir temperature and by freshwater influx.

Resources

Table 1 gives an overview of the discoveries made in each play. The two plays which are most important are those which are stratigraphically closest to the source sequence, namely the Lower/Middle Jurassic and the Upper Jurassic. Next in importance is the Tertiary play, with its widespread, excellent reservoirs. The Chalk play is prolific within a small, restricted area. The other two plays are

Table 1. *Discoveries classified by play*

Play	Number of Fields on production Oil	Gas	Approximate total number of fields and discoveries	Original recoverable reserves and resources in main discoveries (x10^6 Sm^3 oil equivalent) Oil	Gas	Total discovered resources (x10^6 Sm^3 oil equivalent)
Tertiary	3	5	60	Forties 380 Maureen 33 Montrose 16	Frigg 191 Heimdal 34 Odin 36	*c.* 1500
Chalk	5	3	25	Ekofisk 363 Eldfisk 87	W. Ekofisk 38 Albuskjell 26	*c.* 700
Lower Cretaceous	1	—	< 10	Scapa 11		100
Upper Jurassic	10	—	60	Piper 151 Magnus 106 Fulmar 67 Claymore 77 Brae 'A' 47 Ula 25	Troll 1252 Brae 'B' 45	*c.* 2500
Lower/ Middle Jurassic	17	—	100	Statfjord 570 Oseberg 292 Brent 287 Gullfaks 210 Ninian 167 Beryl 125 Cormorant 98 Murchison 89 Beatrice 20 Bream < 1	Sleipner V.171	*c.* 3500
Pre-Jurassic	4	—	15	Snorre (Trias) 77 Auk 15 Argyll 11 Buchan 13		< 250
Totals	40	8	270			*c.* 8500

UK discoveries are quoted from the 'Brown Book' for 1987 (Department of Energy 1988).
Norwegian discoveries from Spencer *et al.* (1987), reserves include associated gas.
1000 Sm^3 gas is taken as equivalent to 1 Sm^3 oil.

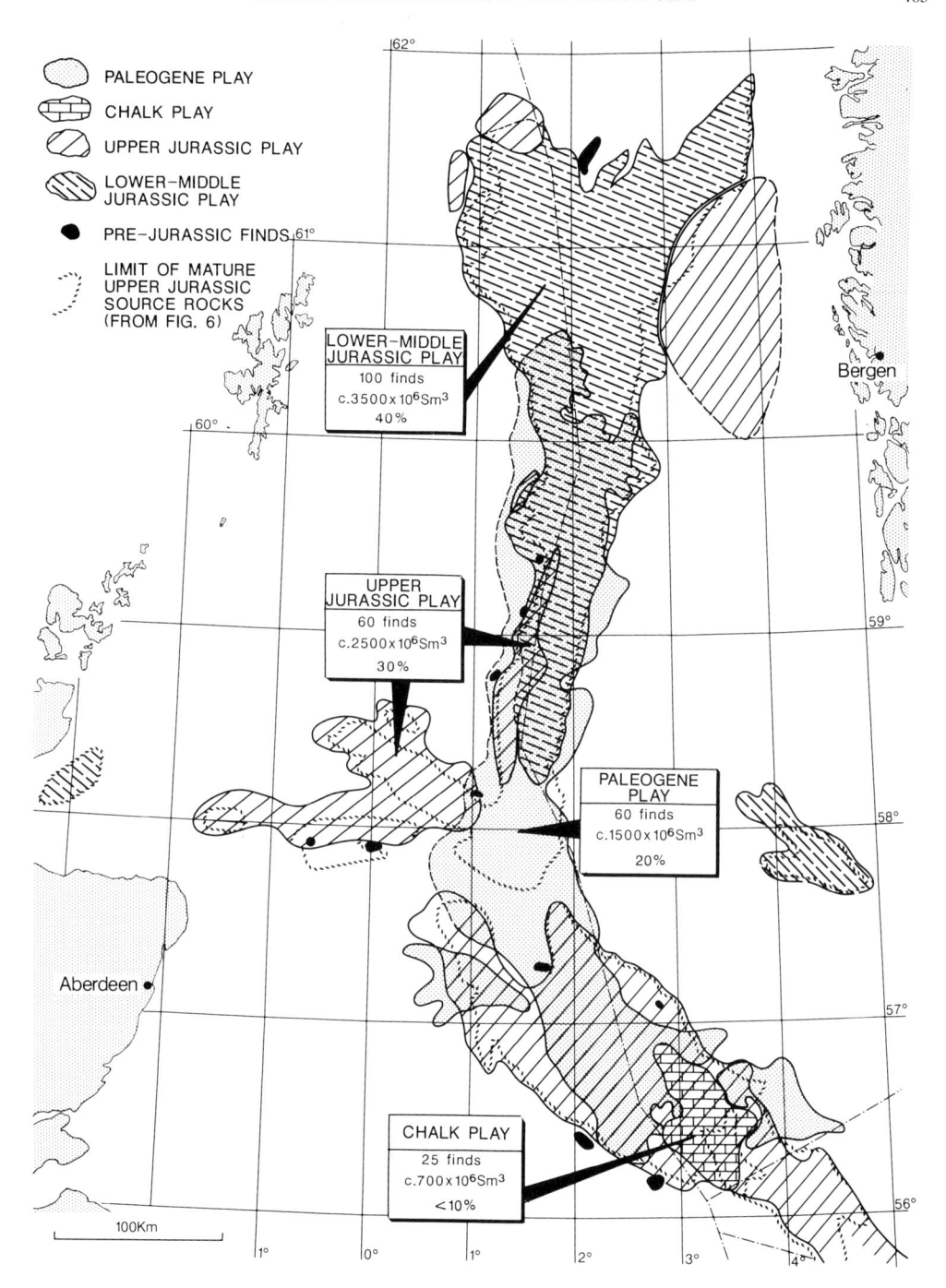

Fig. 15. Map showing compiled plays: the spatial relationship between the four principal plays. Note that the Chalk play is limited by the extent of the Paleogene sand play and that on the northern Horda Platform the Upper Jurassic play may limit the Lower–Middle Jurassic Play. Only in a few areas do more than two plays overlap. The resource figures are derived from Table 1. Play limits derived from Figs 7, 8, 10 & 11.

of minor importance. As is usual with field size distributions, a few giant fields dominate each play, the most extreme example being Troll.

Compiled plays

Stratigraphically the six plays form a sandwich centred on the main Upper Jurassic, source interval (Fig. 8). Thus hydrocarbons have migrated stratigraphically downwards into the pre-Jurassic and Lower–Middle Jurassic plays and stratigraphically upwards into the Cretaceous and Paleogene plays. In terms of depth, however, the order of the plays is different. The deepest hydrocarbon finds are in the Upper Jurassic play in the Central Graben axis (> 5000 m) and in the Lower–Middle Jurassic play in the Viking Graben axis (c. 5000 m). The shallowest finds (1000–2000 m) are in the sandstone sheets which ascend to seabed outcrops in the west (Paleogene play) and in the east (Upper Jurassic play at Troll). The areal relationship between the plays is summarized in Fig. 15. Some of the plays are mutually exclusive: the Chalk play is limited by the superimposed Paleogene play; the Upper Jurassic (Troll sandstone) play may limit the Lower–Middle Jurassic play in an eastward direction. In other areas, e.g. the South Viking Graben, several plays overlap — implying that sealing sequences are present to inhibit vertical migration from one reservoir system to a higher one. Most of the plays are closely related to the distribution of mature Upper Jurassic source rocks. Important exceptions are the Upper Jurassic play in the northeast and the Paleogene play. In both cases long lateral migration has been possible once the hydrocarbons entered the reservoir levels. Also, both these plays are 'open' on the flanks, which has probably allowed large volumes of hydrocarbons to escape from the basin. In Fig. 15 the pre-Jurassic play has been treated differently — individual accumulations have been plotted to emphasise the fact that these fields are largely fortuitous.

Finally, the six plays can be considered from the point-of-view of their genesis: which are direct consequences of the rifting episode and which are not. The 'Syn-rift plays' are directly related to the rifting but the timing in the North Sea can be considered particularly fortunate. Rifting occurred during a period when the latitude and climate were particularly favourable for plant and animal life — the syn-rift sediments are therefore particularly rich in organic material. Had the main rifting occurred under the barren red-bed conditions and desert climate of the Triassic a different story would have to be told. Similarly, although 'Pre-rift plays' can be expected in the exploration of similar extensional basins, the development prior to rifting of the Brent delta — with its thick, extensive and porous reservoirs sands — was a second highly favourable happenstance in the Northern North Sea's history.

The 'Post-rift plays' are not directly connected with the rifting. Although Upper Cretaceous and Danian chalks are extremely widespread throughout northwestern Europe and the North Sea they are only reservoirs under a combination of special circumstances and so only contain oil and gas in extremely limited areas. The Chalk play may have been much more widespread except for the final geological event contributing to the Northern North Sea's prospectivity, the uplift of the Shetland Platform in early Palaeogene times. Uplift, deep erosion and reworking of the earlier sedimentary cover sent a flood of sandstones into the early Tertiary basin. This was clearly an 'external' event, associated with the opening of the North Atlantic. It was the final geological event which has added significantly to the Northern North Sea's hydrocarbon potential by creating the Paleogene play, but perhaps at the cost of a wider Chalk play than that actually encountered.

The authors wish to thank their many colleagues in Statoil who contributed towards the production of this article, by their earlier work and through fruitful discussions and especially E. M. Carlsen, C. Gunnesdal and T. Oliversen. We acknowledge with thanks comments on the article by C. J. Campbell, R. F. P. Hardman, A. Hurst and K. Sørensen. Many geologists and geophysicists have contributed to the present understanding of the geology of the North Sea and a review of this nature necessarily draws heavily on the work of others, through the published literature. We have tried to credit all our sources in the text. Inevitably we will have missed a few — to those we apologize. Finally we wish to thank the management of Statoil (Den norske stats oljeselskap a.s.) for permission to publish this article but wish to state that the opinions expressed are those of the authors and not necessarily those of Statoil.

References

ARMSTRONG, L. A., TEN HAVE, A. & JOHNSON, H. D. 1987. The geology of the Gannet Fields, Central North Sea, UK Sector. In: BROOKS, J. & GLENNIE, K. W. (eds) *Petroleum geology of North West Europe*. Graham & Trotman, London, 533–548.

BADLEY, M. E., PRICE, J. D., RAMBECH DAHL, C. & AGDESTEIN, T. 1988. The structural evolution of the northern Viking Graben and its bearing upon extensional modes of basin formation. *Journal of the Geological Society of London*, **145**, 455–472.

BAIRD, R. A. 1986. Maturation and source rock evaluation of Kimmeridge Clay, Norwegian North Sea. *Bulletin of the American Association of Petroleum Geologists*, **70**, 1–11.

BIRD, T. J., BELL, A., GIBBS, A. D. & NICHOLSON, J. 1987. Aspects of strike-slip tectonics in the Inner Moray Firth Basin, offshore Scotland. *Norsk geologisk Tidsskrift*, **67**, 353–69.

BIRTLES, R. 1986. The seismic flatspot and the discovery and delineation of the Troll Field. *In*: SPENCER, A. M. *et al*. (eds) *Habitat of hydrocarbons on the Norwegian Continental Shelf*. Graham & Trotman, London, 207–215.

BISSADA, K. K. 1983. Petroleum generation in Mesozoic sediments of the Moray Firth Basin, North sea area. *In*: BJORØY, M. (ed.) *Advances in organic geochemistry 1981*. John Wiley, 7–15.

BOOTE, D. R. D. & GUSTAV, S. H. 1987. Evolving depositional systems within an active rift, Witch Ground Graben, North Sea. *In*: BROOKS, J. & GLENNIE, K. W. (eds) *Petroleum geology of North West Europe*. Graham & Trotman, London, 819–834.

BRENNAND, T. P. & VAN HOORN, B. 1986. Historical review of North Sea exploration. *In*: GLENNIE, K. W. (ed.) *Introduction to the petroleum geology of the North sea*. Blackwell, Oxford, 1–24.

—— & VAN VEEN, F. R. 1975. The Auk oil-field. *In*: WOODLAND, A. W. (ed.) *Petroleum and the Continental Shelf of North–West Europe*. Applied Science, Barking, 275–283.

BROOKS, J. & GLENNIE, K. H. 1987. *Petroleum geology of North West Europe*. Graham & Trotman, London.

BUDDING, M. B. & INGLIN, H. F. 1981. A reservoir geological model of the Brent Sands in Southern Cormorant. *In*: ILLING, L. V. & HOBSON, G. D., (eds) *Petroleum geology of the continental shelf of north–west Europe*, Heyden, London, 326–34.

CAMPBELL, C. J. & ORMAASEN, E. 1987. The discovery of oil and gas in Norway: an historical synopsis. *In*: SPENCER, A. M. *et al*. (eds) *Geology of the Norwegian oil and gas fields*. Graham & Trotman, London, 1–37.

CARMAN, G. J. & YOUNG, R. 1981. Reservoir geology of the Forties oilfield. *In*: ILLING, L. V. & HOBSON, G. D. (eds) *Petroleum geology of the continental shelf of north–west Europe*, Heyden, London, 371–379.

CAYLEY, G. T. 1987. Hydrocarbon migration in the Central North Sea. *In*: BROOKS, J. & GLENNIE, K. W. (eds) *Petroleum geology of north west Europe*. Graham & Trotman, London, 1029–1038.

CORNFORD, C. 1986. Source rocks and hydrocarbons of the North Sea. *In*: GLENNIE, K. W. (ed.) *Introduction to the pertroleum geology of the North Sea*. Blackwell, Oxford, 197–236.

DAHL, B., NYSÆTHER, E., SPEERS, G. C. & YUKLER, A. 1987. Oseberg area — integrated basin modelling. *In*: BROOKS, J. & GLENNIE, K. W. (eds) *Petroleum geology of North West Europe*. Graham & Trotman, London, 1029–1038.

—— & SPEERS, G. C. 1985. Organic geochemistry of the Oseberg field (I). *In*: THOMAS, B. M. *et al*. (eds) *Petroleum geochemistry in exploration of the Norwegian Shelf*. Graham and Trotman, London, 185–195.

DANGERFIELD, J. & BROWN, D. A. 1987. The Ekofisk Field. *In*: KLEPPE, J. *et al*. (eds) *North Sea Oil and Gas Reservoirs*, Graham and Trotman, London, 3–22.

DE'ATH, N. G. & SCHUYLEMAN, S. F. 1981. The geology of the Magnus oilfield. *In*: ILLING, L. V. & HOBSON, G. D. (eds) *Petroleum geology of the continental shelf of North–West Europe*. Heyden, London, 342–351.

DEEGAN, C. E. & SCULL, B. J. (Compilers) 1977. A Standard Lithostratigraphic Nomenclature for the Central and Northern North Sea. *Report Institute geological Sciences London*. 77/25. *Norwegian Petroleum Directorate Bulletin 1*.

D'HEUR, M. 1987a. Albuskjell. *In*: SPENCER, A. M. *et al*. (eds) *Geology of the Norwegian Oil and Gas Fields*, London, Graham & Trotman, 51–62.

—— 1987b. Cod. *In*: SPENCER, A. M. *et al*. (eds) *Geology of the Norwegian Oil & Gas Fields*, London, Graham & Trotman, 51–62.

—— 1987c. Tor. *In*: SPENCER, A. M. *et al*. (eds) *Geology of the Norwegian Oil & Gas Fields*, London, Graham & Trotman, 129–42.

—— & DE WALQUE, L. 1987. Bream-Brisling. *In*: SPENCER, A. M. *et al*. (eds) *Geology of the Norwegian Oil and Gas Fields*, London, Graham & Trotman, 63–72.

—— & MICHAUD, F. 1987. Edda. *In*: SPENCER, A. M. *et al*. (eds) *Geology of the Norwegian Oil and Gas Fields*, London Graham & Trotman, 63–72.

DEPARTMENT OF ENERGY 1988. *Development of the oil and gas resources of the United Kingdom 1988*. London, Her Majesty's Stationery office.

DORÉ, A. G., VOLLSET, J. & HAMAR, G. P. 1985. Correlation of the offshore sequences referred to the Kimmeridge Clay Formation — relevance to the norwegian sector. *In*: THOMAS, B. M. *et al*. (eds) *Petroleum geochemistry in exploration of the Norwegian shelf*. Graham & Trotman, London, 27–37.

EGGEN, S. 1984. Modelling of subsidence, hydrocarbon generation, and heat transport in the Norwegian North Sea, *In*: DURAND, B. (ed.) *Thermal Phenomena in Sedimentary Basins*, Technip, Paris, 271–283.

ERICHSEN, T., HELLE, M., HENDEN, J. & ROGNEBAKKE, A. 1987. Gullfaks. *In*: SPENCER, A. M. *et al*. (eds) *Geology of the Norwegian oil and gas fields*. Graham & Trotman, London, 273–286.

FIELD, J. D. 1985. Organic geochemistry in exploration of the northern North Sea. *In*: THOMAS, B. M. *et al*. (eds) *Petroleum Geochemistry and exploration of the Norwegian Shelf*, Graham & Trotman, London, 39–57.

FISHER, J. D. & MILES, J. A. 1983. Kerogen types, organic maturation and hydrocarbon occurrences in the Moray Firth Graben, North Sea Basin. *In*: BROOKS, J. (ed.) *Petroleum Geochemistry and Exploration of Europe*, Geological Society

Special Publication **12**, 195–201.
FLEET, A. J., CLAYTON, C. J., JENKYNS, H. C. & PARKINSON, D. H. 1987. Liassic Source rock Deposition in Western Europe. *In*: BROOKS, J. & GLENNIE, K. W. (eds) *Petroleum geology of North–west Europe*, Graham & Trotman, London, 59–70.

FOWLER, C. 1975. The geology of the Montrose Field. *In*: WOODLAND, W. A. (ed.) *Petroleum and the Continental Shelf of North–West Europe, Vol. 1: Geology*, London, Applied Science, 467–476.

GIBBS, A. D. 1984. Clyde Field growth fault secondary detachment above basement faults in North Sea. *Bulletin of the American Association of Petroleum Geologists*. **68**, 1029–1039.

GOFF, J. C. 1983. Hydrocarbon generation and migration from Jurassic source rocks in the Shetland Basin and Viking Graben of the northern North Sea. *Journal of the geological Society of London* **140** (3), 445–474.

GRAUE, E., HELLAND-HANSEN, W., JOHNSON, J., LØMO, L., NØTTVEDT, A., RØNNING, K., RYSETH, A. & STEEL, R. 1987. Advance and retreat of Brent delta system, Norwegian North Sea. *In*: BROOKS, J. & GLENNIE, K. W. (eds) *Petroleum geology of northwest Europe*. Graham & Trotman, London, 915–937.

GRAY, D. I. 1987. Troll. *In*: SPENCER, A. M. *et al.* (eds) *Geology of the Norwegian oil gas fields*. Graham & Trotman, London, 389–401.

GULBRANDSEN, A. 1987. Agat. *In*: SPENCER, A. M. *et al.* (eds) *Geology of the Norwegian oil and gasfields*. Graham & Trotman, London, 363–370.

HAGE, A., BLOMSTAD, K. & STRAND, J. E. 1987. Brage. *In*: SPENCER, A. M. *et al.* (eds) *Geology of the Norwegian Oil & Gas Fields*. London Graham & Trotman, 371–378.

HALLET, D. 1981. Refinement of the geological model of the Thistle Field. *In*: ILLING, L. V. & HOBSON, G. D. (eds) *Petroleum Geology of the Continental Shelf of North–West Europe*, Heyden; London, 315–325.

HANCOCK, J. M. 1975. The sequence of the facies in the Upper Cretaceous of northern Europe compared with that in the western interior. *In*: Caldwell, W. G. F. (ed.) *Cretaceous system in the Western Interior of North America*, Special Paper Geological Association of Canada **13**, 83–118.

—— 1986. Cretaceous. *In*: GLENNIE, K. W., (ed.) *Introduction to the Petroleum Geology of the North Sea*, Blackwell, Oxford, 161–178.

—— & SCHOLLE, P. A. 1975. Chalk of the North Sea. *In*: WOODLAND, A. W. (ed.) *Petroleum and the continental and the continental shelf of north-west Europe, Vol. 1, Geology*, Applied Science, 413–427.

HANSLIEN, S. 1987. Balder. *In*: SPENCER, A. M. *et al.* (eds) *Geology of the Norwegian oil and gas fields*. Graham & Trotman, London, 193–202.

HARDMAN, R. F. P. 1982. Chalk reservoirs of the North sea. *Bulletin of the geological Society of Denmark* **30**, 119–37.

—— & KENNEDY, W. J. 1980. Chalk reservoirs of the Hod fields, Norway. *In*: *The sedimentation of the North Sea reservoir rocks*. Norwegian Petroleum society. Article XI.

HARKER, S. D., GUSTAV, S. H. & RILEY, L. A. 1987. Triassic to Cenomanian stratigraphy of the Witch Ground Graben. *In*: BROOKS, J. & GLENNIE, K. W. (eds) *Petroleum geology of northwest Europe*. London, Graham & Trotman, 809–818.

HARMS, J. C., TACKENBERG, P., POLLOCK, R. E. & PICKLES, E. 1981. The Brae Oilfield area, *In*: ILLING, L. V. & HOBSON, G. D. (eds) *Petroleum Geology of the Continental Shelf of North–West Europe*, Heyden, London, 352–357.

HAZEU, G. J. A. 1981. 34/10 Delta structure geological evaluation and appraisal. *In*: *Norwegian Symposium on Exploration*, Norwegian Petroleum Society, Article 13.

HERITIER, F. E., LOSSEL, P. & WATHNE, E. 1979. Frigg Field — large submarine-fan trap in Lower Eocene rocks of the Viking Graben, North Sea. *Bulletin of the American Association of Petroleum Geologists* **63**, 1999–2020.

HILL, P. J. & WOOD, G. V. 1980. Geology of the Forties Field UK Continental Shelf (North Sea), *Bulletin of the American Association of Petroleum Geologists*, **64**, 81–93.

HOLLANDER, N. B. 1987. Snorre. *In*: SPENCER, A. M. *et al.* (eds) *Geology of the Norwegian Oil & Gas Fields*, London, Graham & Trotman, 307–317.

HOME, P. C. 1987. Ula. *In*: SPENCER, A. M. *et al.* (eds) *Geology of the Norwegian oil and gas fields*. Graham & Trotman, London, 143–152.

JOHNSON, H. D., MACKAY, T. A. & STEWART, D. J. 1986. The Fulmar Oil-field (Central North Sea): geological aspects of its discovery, appraisal and development. *Marine and Petroleum Geology*, **3**, 99–125.

KARLSSON, W. 1986. The Snorre, Statfjord and Gullfaks oilfields and the habitat of hydrocarbons on the Tampen Spur, offshore Norway. *In*: SPENCER, A. M. *et al.* (eds) *Habitat of hydrocarbons on the Norwegian Continental shelf*. Graham & Trotman, London, 181–197.

KENNEDY, W. J. 1987. Sedimentology of late Cretaceous–Paleocene Chalk reservoirs, North Sea Central Graben. *In*: BROOKS, J. & GLENNIE, K. W. (eds) *Petroleum geology of Northwest Europe*. Graham & Trotman, London, 469–482.

KIRK, R. H. 1980. Statfjord Field – a North Sea giant. *In*: HALBOUTY, M. T. (ed.) *Giant oil and gas fields of the decade 1968–1978. Memoir of the American Association of Petroleum Geologists*, **30**, 95–116.

KNOX, R. W. O'B., MORTON, A. C. & HARLAND, R. 1981. Stratigraphical relationships of Paleocene sands in the UK sector of the central North Sea. *In*: ILLING, L. V. & HOBSON, G. D. (eds) *Petroleum Geology of the Continental Shelf of North–West Europe*. Heyden, London, 267–281.

LARESE, R. E., HASKELL, N. L., PREZBINDOWSKI, D. R. & BEJU, D. 1984. Porosity development in selected Jurassic sandstones from the Norwegian and North Seas, Norway – an overview. *In*: SPENCER, A. M. *et al.* (eds) *Petroleum Geology*

of the North European Margin, London, Graham & Trotman, 81–95.
LARSEN, R. M. & JAARVIK, L. J. 1981. The geology of the Sleipner Field Complex. In: Norwegian Symposium on Exploration, Norwegian Petroleum Society, Article 15.
LEADHOLM, R. H., HO, T. T. Y. & SAHAI, S. K. 1985. Heat flow, geothermal gradients and maturation modelling on the Norwegian continental shelf using computer methods. In: THOMAS, B. M. et al. (eds) Petroleum geochemistry in exploration of the Norwegian shelf. Graham & Trotman, London, 131–143.
LEONARD, R. C. & MUNNS, J. W. 1987. Valhall. In: SPENCER, A. M. et al. (eds) Geology of the Norwegian Oil & Gas Fields, London, Graham & Trotman, 153–164.
MACKENZIE, A. S., LEYTHAUSER, D., MULLER, P., RADKE, M. & SHAEFER, R. G. 1987. The expulsion of petroleum from Kimmeridge clay source rocks in the area of the Brae Oilfield, UK continental shelf. In: BROOKS, J. & GLENNIE, K. W. (eds) Petroleum geology of North west Europe. Graham & Trotman, London, 865–878.
MAHER, C. E. 1981. The Piper Oilfield. In: ILLING, L. V. & HOBSON, G. D. (eds) Petroleum Geology of the Continental Shelf of North–West Europe, Heyden, London, 358–370.
—— & HARKER, S. D. 1987. Claymore Oil Field. In: BROOKS, J. & GLENNIE, K. W. (eds) Petroleum geology of North west Europe. Graham & Trotman, London, 835–846.
MARCUM, B. L., AL-HUSSAINY, R., ADAMS, G. E., CROFT, M. & BLOCK, M. L. 1978. Development of the Beryl 'A' Field, Proceedings of the European Offshore Petroleum Conference, 319–321.
MORTON, A. C. 1979. The provenance and distribution of the Paleocene sands of the Central North Sea. Journal of Petroleum Geology, 2.
MURE, E. 1987a. Frigg. In: SPENCER, A. M. et al. (eds) Geology of the Norwegian Oil & Gas Fields, London, Graham & Trotman, 203–213.
—— 1987b. Heimdal. In: SPENCER, A. M. et al. (eds) Geology of the Norwegian Oil & Gas Fields, London, Graham & Trotman, 229–234.
NORBURY, I. 1987. Hod. In: SPENCER, A. M. et al. (eds) Geology of the Norwegian Oil & Gas Fields, London, Graham & Trotman, 107–116.
NORTHAM, M. A. 1985. Correlation of northern North sea oils: the different facies of their Jurassic source. In: THOMAS, B. M. et al. (eds) Petroleum geochemistry in exploration of the Norwegian shelf. Graham & Trotman, London, 93–99.
PARSLEY, A. J. 1986. North Sea hydrocarbon plays. In: GLENNIE, K. W. (ed.) Introduction to the Petroleum Geology of the North Sea. Blackwell, Oxford, 237–263.
PEGRUM, R. M. 1984. The extension of the Tornquist Zone in the Norwegian North Sea. Norsk Geologisk Tidsskrift, 64, 29–68.
—— & LJONES, T. E. 1984. 15/9 Gamma gas field offshore Norway, new trap type for North Sea basin with regional structural implications. Bulletin American Association of Petroleum Geologists. 68, 874–902.
PEKOT, L. J. & GERSIB, G. A. 1987. Ekofisk. In: SPENCER, A. M. et al. (eds) Geology of the Norwegian Oil and Gas fields. Graham & Trotman, London, 73–87.
PENNINGTON, J. J. 1975. The geology of the Argyll Field. In: WOODLAND, A. W. (ed.) Petroleum and the continental shelf of North West Europe. Applied Science, Barking, 285–294.
RANAWEERA, H. K. A. 1987. Sleipner Vest. In: SPENCER, A. M. et al. (eds) Geology of the Norwegian Oil and Gas fields. Graham & Trotman London, 253–264.
RAWSON, P. F. & RILEY, L. A. 1982. Latest Jurassic–Early Cretaceous events and the 'Late Cimmerian unconformity' in the North Sea area. Bulletin of the American Association of Petroleum Geologists. 66, 2628–2648.
RØNNEVIK, H., EGGEN, S. & VOLLSET, J. 1983. Exploration of the Norwegian Shelf, In: BROOKS, J. (ed.) Petroleum Geochemistry and Exploration of Europe, Special Publication Geological Society of London 12, 71–93.
ROBERTS, J. D., MATHIESON, A. S. & HAMPSON, J. M. 1987. Statfjord. In: SPENCER, A. M. et al. (eds) Geology of the Norwegian Oil and Gas Fields, London, Graham & Trotman, 319–40.
ROCHOW, K. A. 1981. Seismic stratigraphy of the North Sea 'Paleocene' deposits. In: ILLING, L. V. & HOBSON, G. D. (eds) Petroleum Geology of the continental shelf of north–west Europe, London, Heyden, 255–266.
RØNNING, K., JOHNSTON, C. D., JOHNSTAD, S. E. & SONGSTAD, P. 1987. Hild. In: SPENCER, A. M. et al. (eds) Geology of the Norwegian oil and gas fields. Graham & Trotman, London, 287–294.
SCHOLLE, P. A. 1977. Chalk diagenesis and its relation to petroleum exploration — oil from chalks, a modern miracle? Bulletin of the American Association of Petroleum Geologists, 61, 982–1009.
SCHOU, L., EGGEN, S. & SCHOELL, M. 1985. Oil–oil and oil-source rock correlation, northern North Sea. In: THOMAS, B. M. et al. (eds) Petroleum geochemistry in exploration of the Norwegian shelf. Graham & Trotman, London, 101–117.
SMITH, R. L. 1987. The structural development of the Clyde Field. In: BROOKS, J. & GLENNIE, K. W. (eds) Petroleum geology of north west Europe. Graham & Trotman, London, 523–532.
SPENCER, A. M. et al. (eds) 1986. Habitat of hydrocarbons on the Norwegian Continental Shelf. Graham & Trotman, London.
—— et al. (eds) 1987. Geology of the Norwegian Oil and Gas fields. London, Graham & Trotman.
——, HOME, P. C. & WIIK, V. 1986. Habitat of hydrocarbons in the Jurassic Ula Trend, Central Graben, Norway. In: SPENCER, A. M. et al. (eds) Habitat of hydrocarbons on the Norwegian Continental Shelf. Graham & Trotman, London, 111–127.
STOW, D. A. V. & ATKIN, B. P. 1987. Sediment facies and geochemistry of Upper Jurassic mudrocks in the Central North Sea area. In: BROOKS, J. &

GLENNIE, K. W. (eds) *Petroleum geology of North west Europe*. Graham & Trotman, London, 797–808.
—— BISHOP, C. D. & MILLS, S. T. 1982. Sedimentology of the Brae oilfield, North Sea: fan models and controls. *Journal of Petroleum Geology*, **5**, 129–148.
THOMAS, B. M. *et al.* (eds) 1985. *Petroleum geochemistry in exploration of the Norwegian Shelf*. Graham & Trotman, London.
THOMSEN, E., DAMTOFT, K. & ANDERSEN, C. 1987. Hydrocarbon plays in Denmark outside the Central Trough. *In*: BROOKS, J. & GLENNIE, K. W. (eds) *Petroleum geology of north west Europe*. London, Graham & Trotman, 375–388.
TURNER, C. C., COHEN, J. M., CONNELL, E. R. & COOPER, D. M. 1987. A depositional model for the South Brae Oilfield. *In*: BROOKS, J. & GLENNIE, K. W. (eds) *Petroleum geology of North West Europe*. Graham & Trotman, London, 853–864.
VAN DDEN BARK, E. & THOMAS, O. D. 1980. Ekofisk: first of the giant oil fields in Western Europe. *Memoir of the American Association of Petroleum Geologists*, **30**, 195–224.
VOLLSET, J. & DORE, A. G. (eds). 1984. A revised Triassic and Jurassic Lithostratigraphic Nomenclature for the Norwegian North Sea. *Norwegian Petroleum Directorate Bulletin* **3**.
WHITE, D. A. 1988. Oil and gas play maps in exploration and assessment. *Bulletin of the American Association of Petroleum Geologists* **72**, 944–949.
ØSTFELDT, P. 1987. Oil-Source rock correlation in Danish North Sea. *In*: BROOKS, J. & GLENNIE, K. W. (eds) *Petroleum geology of North West Europe*. Graham & Trotman, London, 419–430.
ØSTVEDT, O. J. 1987. Sleipner Øst. *In*: SPENCER, A. M. *et al.* (eds) *Geology of the Norwegian oil and gas fields*. Graham & Trotman, London, 243–252.

Haltenbanken hydrocarbon province (off-shore Mid-Norway)

O. R. HEUM & R. M. LARSEN

Statoil, Staranger, Norway

Abstract: Exploration drilling at Haltenbanken started in 1980 and to date 11 significant hydrocarbon discoveries have been made. The estimated total recoverable discovered reserves are 310 million SM3 of oil and condens and 305 billion SM3 of gas. Only 14 out of 25 proper wildcat wells turned out dry, giving a pure exploration success ratio of 44%. The major structural elements at Haltenbanken are (from East to West) the Troendelag Platform, the Halten Terrace and the Deep Moere Basin. The major hydrocarbon discoveries are located on the Halten Terrace and on the western margin of the Troendelas Platform.

After a long period of regional passive subsidence during Triassic and most of Jurassic the Halten Terrace was formed as a consequence of Late Jurassic east–west rifting and early/middle Cretaceous north–south Dextral wrenching. During late Cretaceous and Tertiary again regional passive subsidence dominated. In Miocene very deep waters had developed at Haltenbanken as well as over most of the Mid-Norwegian Shelf. In Pliocene uplift of the Fennoscandian Shield resulted in rapid westward shelf progradation and deposition of up to 1000 m of sediments.

This Pliocene burial had a great impact on the formation and entrapment of hydrocarbons at Haltenbanken. As the temperatures increased the rich Upper Jurassic marine and Lower Jurassic terrestrial source rocks passed through the oil window and are currently still producing large amounts of oil, condensates and gas.

The rapid Pliocene sediment loading also led to great overpressure in the Upper Jurassic to palaeocene line-grained sequence. In the Eastern areas this overpressure resulted in a perfect seal over the Middle and Lower Jurassic sandstone reservoirs. In the Western areas also the Jurassic reservoir sequence is severely overpressured leading to less favourable sealing conditions. On the other hand, the overpressure in the reservoirs in the Western areas seems to have reduced the rate of compaction, hence preserving the porosity and permeability better at great depths. Seven of the most important discoveries at Haltenbanken are discussed as case examples. The variations in type of hydrocarbons (oil, vs. gas or gas/condensates) entrapped in the individual fields are caused by subtle, but important variations in source rock maturity, migration conditions (including drainage area and storage capacity), trapping conditions (fill-spill or fill-leak), timing and *PVT*-conditions.

Lena–Tunguska Upper Proterozoic–Palaeozoic petroleum superprovince

A. E. KONTOROVICH, M. M. MANDEL'BAUM, V. S. SURKOV, A. A. TROFIMUK & A. N. ZOLOTOV

NPO Sibgeo, Sniiggims, Norosibirsk, USSR

Abstract: The Lena–Tunguska Superprovince is the only region of the world where extensive economic Upper Proterozoic petroleum production is proved. This province extends the stratigraphical range of hydrocarbon accumulations to the deposits older than one billion years.

The Upper Proterozoic deposits are represented mainly by Riphean carbonate formations and terrigene carbonate sulphate-bearing sediments of Vendian (Eocambrian). These deposits cover an area of some two million square kilometres.

Oil and gas potential in the Vendian and Upper Cambrian sediments in the central and southern regions of the province is proven and more than thirty fields have been discovered. Sheet, roof, lithologically and tectonically screened and non-anticlinal pools dominate these discoveries. The Lower Cambrian salts are the regional seal.

Riphean petroleum potential has been proved in the western part of the province where the extensive Yurubcheno–Takhomsk zone of oil and gas accumulations occur within the Baikit anteclise and Katanga Trough. Oils appear to be of one genetical type as low sulphur, naphthenic-rich oils containing n-alkanes, isoprenoids and are enriched in 12,13-monomethylalkanes, ^{12}C carbon isotope.

The main current oil-producing plays are confined to Riphean Formations.

The Lena–Tunguska Petroleum Superprovince (approximately 2.8×10^6 km^2) occupies the major part of the ancient Siberian Platform and is one of the largest on the Euroasian continent. It was first described by Kontorovich in 1975 (Kontorovich *et al*. 1975). The name is derived from four great Eastern Siberia rivers, the Lena River bordering the Province in the east, the Low, Stony and Upper (Angara) Tunguska flowing through the southern and western regions of the Province. The Lena–Tunguska Province boundaries in the west and south are essentially coincident with the platform boundaries, and in the north and east they are in the zone adjacent to the Khatanga–Vilyui Petroleum Province within which main hydrocarbon prospects are associated with the Upper Palaeozoic and Mesozoic Formations. Tectonically the Province is bounded by the Yenisei–Khatanga regional trough, Anabar–Khatanga saddle, Lena–Anabar Trough in the north, and by the Vilyui syncline and Predverkhoyansk trough in the east.

Two features make the Lena–Tunguska Province an object of a special importance to theoretical and applied petroleum geology. The first one is that the Lena–Tunguska Province is the first region in the world where commercial Upper Proterozoic petroleum production has been proved in a vast area of about one million square kilometers. Predicted Upper Proterozoic production was originally described by Trofimuk in 1960 (Trofimuk 1960). Production from such ancient formations is of great importance to oil and gas industry development both in the USSR and in many other countries of the world as it allows for more optimistic evaluation of petroleum potential to be made for Upper Precambrian sedimentary basins. The second important feature is that this Province is a unique region in the world, where regional petroleum potential has been proved for the basins with abundant trappean intrusions in the sections. This should encourage oil and gas exploration in undertrappean sedimentary basins of Eurasia, Africa and South America.

Scientific prediction and discovery of this major Upper Proterozoic–Paleozoic Petroleum Province are proof that the theory of absence of hydrocarbon prospects in the Precambrian ancient sedimentary formations is erroneous. These discoveries provide excellent evidence to encourage hydrocarbon exploration in other worldwide basins and to enhance essential world petroleum resource evaluations in ancient basins. Studies of petroleum geology in the Lena–Tunguska Province contribute substantially to improvement of petroleum generation theory and prediction methods of oil and gas potential and also give knowledge on petroleum genesis evolution in the Earth history.

Drilling history

In the Lena–Tunguska Province deep drilling for oil and gas started during the second half of 1930s. About the same time in the east of the Province, on the southern slope of the Aldan anteclise within the Tolba River basin, a small oil sample was recovered from deposits, which according to modern stratigraphy was Vendian in age. In 1948–1960, the wells drilled and geophysical surveys carried out were relatively low and were mainly in the southern regions of the Province. In 1962, a key well penetrated the small Markovo pool in the Vendian and Lower Cambrian deposits. Exploration in the Province has increased and became more intensive in the mid-1970s when the main petroleum accumulations was confirmed. These accumulations were confined in the Nepa–Botuoba, Baikit anteclises and in the Khatanga saddle which connects them. Currently about 20 oil and gas pools have been discovered within this area.

By early 1988, more than 3 million metres of deep wells had been drilled, and drilling density in the region amounted to some 1.2 m km². Drilling density on prospects is extremely uneven. The Nepa–Botuoba anteclise and Angara–Lena step are better explored, with drilling densities of 6.6 m km². and 3.6 m km² respectively. Commercial development of discoveries in the Province is very complicated because of the geologic and natural-climatic condition. Among the former good seismic information capability of seismic exploration is considered to be the most limiting factor in correctly evaluating the region's hydrocarbon potential, especially those plays with abundant traps.

Tectonic subdivision

The basement of the Lena–Tunguska Province is a complex system of crustal blocks of different age. All these blocks are integrated into two large Aldan–Anabar and Tunguska regions (Antsiferov et al. 1981). The Iengrskii reworked Katarchean massif, Aldan and Anabar fold Systems of Early Archean (3300 ± 200 million years), Batomgskaya Fold System of Late Archean (2600 ± 100 Ma), early Caledonian Olenek and Olekma Fold Systems (2000 ± 100 Ma) are the largest tectonic elements in the basement of the Aldan–Anabar region. In the Tunguska region the Angara–Lena and Kotui Systems have been distinguished and have a folding age of 1700 ± 100 Ma.

These rocks out crop and occur near the surface over much of the Anabar anteclise and within the Aldan anteclise along the southern boundary of the Province. Within the rest of the Province, depth to basement range from 2–3 to 5–8 km. Some areas are suggested to subside to depths of 12–14 km (Fig. 1).

The platform sedimentary cover within the Province consists of the Riphean, Vendian and Phanerozoic sedimentary and volcanogenic–sedimentary complexes. Wide development of unconformities; Vendian, Cambrian and Devonian thick evaporite strata, as well as intrusive and effusive rocks (traps) are characteristic of the platform sedimentary cover. Intrusions occur as thick sheet bodies (varying from a few metres to over 200–300 m thick) within various complexes and form numerous dykes and veins. Most of the trappean bodies is dated as Late Permo–Triassic and in the east of the Province as Devonian.

Sedimentary cover includes five main structural stages, such as Riphean, Vendian–Lower Palaeozoic, Middle Palaeozoic, Upper Palaeozoic and Triassic (Antsiferov et al. 1987; Zolotov 1981). Between theses structural stages there are major structural unconformities which in some parts of the Province are manifested in different ways.

A number of sedimentary basins have been identified in the Riphean structural stage. According to their position in the platform regional structure and evolution history, they are divided into two main types—pericratonic and intracratonic basins (Fig. 2). Pericratonic

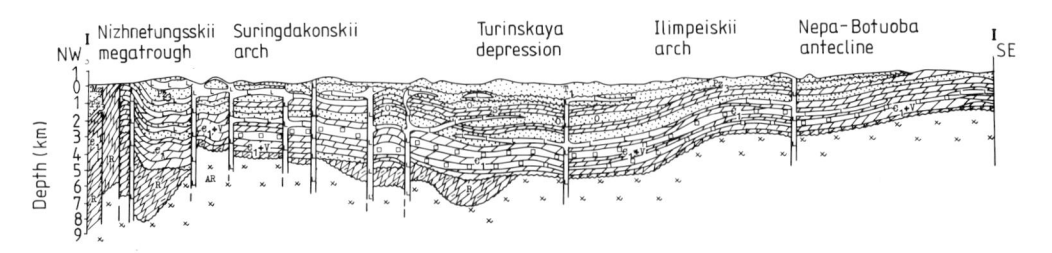

Fig. 1. Geological cross section of the Lena–Tunguska petroleum bearing province. For symbols see Fig. 5.

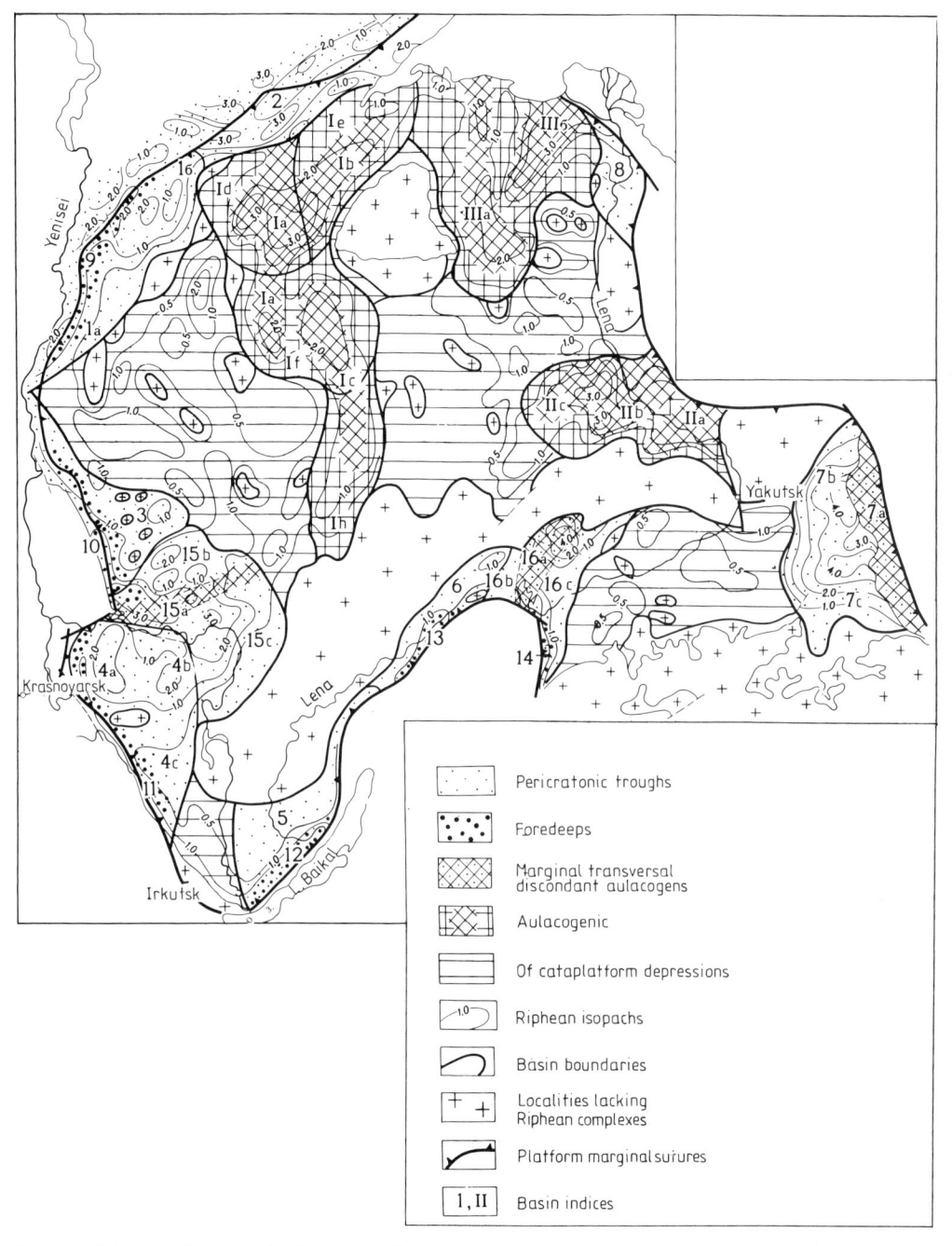

Fig. 2. Riphean sedimentary basins of the Siberian Platform. Marginal cratonic sedimentary basins of: Pericratonic troughs: Turukhan−Khantaiskii (1), depressions: Turukhan−Olenek (a) Lamsko−Khantaiskaya (b); Pre-Taimyr (2), Vel'minskii (3), Chuno−Biryusa (4); depressions: Dolgomostovskaya (a), Mursko−Chunskaya (b), Iiskaya (c); Pre-Baikal (5), Nyuisko−Dzherbinskii (6), Uchuro−Maiskii (7); Sette−Dabanskii 'aulacogen' (a), depressions: Ust−Maiskaya (6), Uchurskaya (c), Chekurovskii (8). Foredeeps: Turukhan−Igarka (9), Pre-Yenisei (10), Pre-Sayany (11), Pre-Baikal (12), Ust−Vitim (13), Berezovskii (14). Marginal cross aulacogens: Irkineevo−Vanavarskii (15), Irkineevskii aulacogen (a), depressions: Taiga (6), Katskaya (c), Urinskii (16), Urinskii aulacogen (a), depressions: Dzherbinskaya (b), Biryukskaya (c). Intracratonic sedimentary basins: Aulacogenic: Kotui−Vovukanskii (1), aulacogens: Kotui (a), Fomichevskii (6), Esseiskii (c), depressions: Maimechinskaya (d), Fomichevsko−Kostrominskaya (e), Moiero−Kotui (f), Vovukanskaya (g); Vilyui II depressions: Lungkhinskaya (a), Ust−Vilyui (b), Lindenskaya (c); III − Khastakhsko Udzhinskii, aulacogens: Udzhinskii (a), Khastakhskii (b).

sedimentary basins during their development were closely associated with the evolution of fold systems adjacent to the platform, that predetermined the petroleum potential of the Upper Proterozoic in the southern regions of the platform, particularly during the miogeosynclinal development stage. The Velminskii and Pre-Baikal basins are the most important basins in the region.

A series of poorly studied intracratonic basins includes aulacogen basins of rift nature and cataplatform depressions.

The Anabar, Aldan, Nepa–Botuoba, Baikit anteclines and Bakhtinskii megasalient, Kureika and PreSayany–Yenisei synclines, Predpatomskii regional trough and Angara–Lena step are distinguished in the Vendian–Lower Paleozoic structural stage (Fig. 3).

Characteristics of these structures is given in Antsiferov et al. (1981).

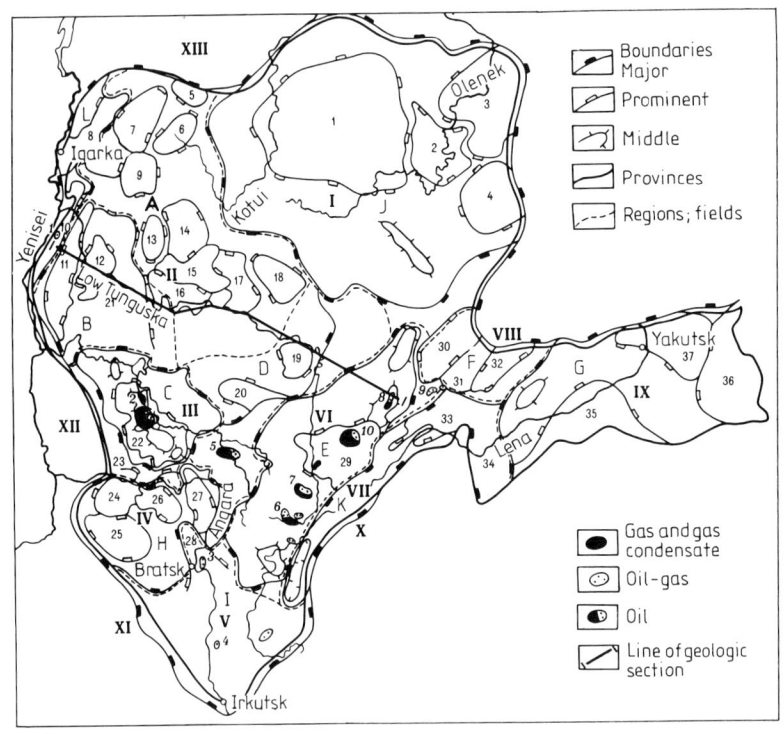

Fig. 3. Tectonic and petroleum geologic regions in the Lena–Tunguska Province. Major structural elements: I, Anabar antecline; II, Kureika syncline; III, Baikit antecline; IV, Pre-Sayany–Yenisei syncline; V, Angara–Lena Stage; VI, Nepa–Botuoba antecline; VII, Predpatomskii regional trough; VIII, Vilyui hemisyncline; IX, Aldan antecline; X, Baikal–Patomskaya folded region; XI, anticlinorium of Eastern Sayan; XII, Yenisei Ridge; XIII, Yenisei–Khatanga regional trough. Prominent structural elements: 1, Anabar arch; 2, Sukhanskaya depression, 3, Olenek arch; 4, Munskii arch; 5, Ledyanskii arch; 6, Ayanskii arch; 7, Lamsko–Khantayskii megatrough; 8, Khantaysko–Rybninskii megaswell; 9, Anamskii arch; 10, Kureika–Baklanikhinskii megaswell; 11, Nizhnetungusskii megatrough; 12, suringdakonskii arch; 13, Yukteliyskii arch; 14, Verkhnekochechumskaya depression; 15, Kochechumskaya depression; 16, Turinskaya depression; 17, Korvunchanskaya depression; 18, Turunskii arch; 19, Ilimpeyskii arch; 20, Chun'skii salient; 21, Bakhtinskii salient; 22, Kamovskii arch; 23, Terinskii megatrough; 24, Boguchano–Manzinskii salient; 25, Dolgomostovskaya depression; 26, Mursko–Chun'skaya depression; 27, Katskaya depression; 28, Bratsk salient; 29, Nepa arch; 30, Yugyattinskaya depression; 31, Suntarskii arch; 32, Kempendyaiskaya depression; 33, Nyuisko–Dzherbinskaya depression; 34, Berezovskaya depression; 35, Tolbinskii salient; 36, Aldan-Maiskaya depression; 37, Yakutsk arch. Geological zoning of oil and gas fields. Perspective petroleum-bearing regions: A, North Tunguska; B, South Tunguska; C, Baikit; D, Khatanga; E, Nepa–Botuoba; F, West Vilui; G, North Aldan; H, Pre-Sayany–Yenisei; I, Angara–Lena; J, Anabarskaya; K, Predpatomskaya; L, Turukhan–Noril'sk petroleum perspective area. Fields: *1*, Podkamennoye; *2*, Yurubchen–Tomoto zone; *3*, Bratskoye; *4*, Atovskoye; *5*, Sobinskoye; *6*, Yaraktinskoye; *7*, Dulisminskoye; *8* Srednebotuobinskoye; *9*, Verhnevilyuchanskoye; *10*, Verkhnechonskoye.

Petroleum-bearing regions: their potential

Eleven petroleum-bearing regions and one individual petroleum-bearing area have been distinguished in the Lena–Tunguska Province (Table 1).

The North Tunguska petroleum-bearing region includes the northern and central parts of the Kureika syncline. The region is composed of Riphean, Vendian and Palaeozoic terrigenous–carbonate deposits with total thickness from 3 to 8 km and Triassic tuffaceous–effusive formations of 1 to 3 km thick. Maximum thicknesses of these sedimentary formations are reported in the northwest of the region.

Within the North Tunguska region the plays with petroleum potential includes the sedimentary cover section from the Riphean to Permian in age. The Riphean, Vendian and Cambrian complexes have not yet been fully explored.

The Ordovician petroleum-bearing play complex within the region consists of a carbonate sequence (Ust–Kut stage and lowermost of Chunskii stage) and terrigene sequence (Baikit sandstones at the top of Chunskii stage). Thickness of carbonate sequence in the complex ranges from 550 m (in the north) to 300–250 m (in the south). Common breaks in this sequence and associated cavern porosity indicates the development of fracture reservoirs in the carbonate complex. The terrigenous sequence of the complex contains some sandstone beds (up to 70 m thick with good average 20–30) reservoir properties. A thick sequence of alternating mudstones, marls and argillaceous limestones of Middle–Upper Ordovician age is the regional seal in the complex.

The Silurian play complex is represented by a sequence of carbonate and clay–carbonate deposits. Zonally developed members of organic-–clastic limestones and dolomites in the Wenlock stage of the Lower Silurian appear to be the most promising reservoirs. Such plays are 60–80 m thick and show satisfactory storage capacity (porosity and permeability). In many areas (Suhhotungusskaya, Gornaya, Sigovo–Podkamennaya) structural holes in this horizon, penetrate remains of destroyed oil pools. The regional seal of the complex is a stratum of the Upper Silurian argillaceous dolomites (140–180 m). Locally distributed (predominantly in the northwest) sandstone beds of 2–5 m thickness (porosity of 7–21% and permeability to 10^{-13} m^2 are reservoirs in the Devonian argillaceous strata.

The Upper Palaeozoic play complex of 400–800 m thick sediments is represented by alternation of thick (15–60 m) sandstone beds and members of sandy–argillaceous rocks and coals. Terrigenous reservoir beds have good storage capacity. Tuffaceous and argillaceous rocks of Lower Triassic Tutonchanskaya and Korvunchanskaya Formations are the seals in the complex.

Structural patterns of different horizons in the cover are poorly understood within the region. Interpretation of geophysical data has identified a number of arches at the basement top, $10-20 \times 10^3$ km^2 each and an amplitude of about a hundred metres. These are considered to be the most promising zones for oil and gas accumulation.

The structure of the North Tunguska petroleum-bearing region is complicated by trappean discordant intrusions and sills being well developed in marginal zones of the Kureika syncline. Trappean magmatism has a marked influence on the petroleum potential of these sediments. Intensive neotectonic movements resulting in destruction and re-migration of hydrocarbon pools have been identified within the area. Formation of the dissected highland relief is also associated with these movements.

Permafrost rocks are distributed throughout the region. The Plutoran Plateau sediments are frozen to depths greater than 600–700 m. In the northwestern Middle Siberian highland the permafrost rock thickness reaches 300–450 m. There is a possibility of hydrocarbon pools occurring locally in hydrate form due to permafrost.

Over most of the area permeable horizons of sedimentary cover are saturated with highly concentrated and metamorphosed brines. This suggests that the Palaeozoic section may have favourable conditions for oil and gas pools preservation.

The Turukhansk–Noril'sk petroleum-bearing region covers an area of the Khantaisko–Rybninskii and Kureika–Baklanikhinskii megaswells. The Riphean, Vendian and Lower Paleozoic strata with total thicknesses of 5–6 km are broadly distributed within the region. The mentioned megaswells have deeply denudated arched zones reducing the petroleum potential of the region. The Kureika–Balkanikhinskii megaswell does, however, produce commercial oil and gas flows from fractured dolomites of the Lower Cambrian Platonovskaya Formation in Sukhotungusskaya, Volodinskaya and Nizhneletninskaya areas. Commercial gas production is also obtained from two wells drilled in the upper Lower–Middle Cambrian Kostinskaya Formation.

Within the Turukhansk–Noril'sk region the Vendian and Cambrian complexes are promising play for petroleum discoveries.

The Anabar petroleum-bearing region en-

Table 1. *Basic indices of petroleum-bearing regions and areas of Lena–Tunguska petroleum Province*

Petroleum-bearing regions and areas	Area (10^3 km^2)	Sedimentary cover maximum thickness (km)	Main productive complexes	Main perspective complexes	Reservoir lithologies	Fields
(a) North–Tunguska	430	8	—		sandstones, limestones, dolomites	
(b) South–Tunguska	45	6			sandstones, limestones,	
(c) Baikit	155	3.5			limestones, dolomites, sandstones	Kuyumbinskoye Omorinskoye, Yurubchenskoye
(d) Katanga	230	5			the same	Sobinskoye
(e) Nepa–Batuoba	250	3			sandstones, carbonates	Markovskoye, Yaraktinskoye, Ayanskoye, Danilovskoye, Verkhnechonskoye, Dulisma, Srednebotuoba, Irelyakhskoye, Tas–Yuryakhskoye, Nizhnekhamakinskoye, Khotogo–Murbaiskoye, Verkhnevilyuchanskoye, Vilyui–Dzherbinskoye, Ozernoye, Iktekhskoye, Machchobinskoye, North–Nelbinskoye
(f) West–Vilyui	80	9			sandstones, carbonates	
(g) North–Aldan	300	6	—		limestones, sandstones	
(h) PreSayany–Yenisei	150	7.5	—		sandstones, carbonates	
(i) Angara–Lena	170	3.5			the same	Batskoye, Atovskoye, Kovyktinskoye
(j) Anabar	540	5			sandstones, dolomites,	
(k) Predpatoma	125	7			sandstones, carbonates	Bysakhtakhskoye, Kehdehrginskoye
(l) Turukhansk-Noril'sk	45	8			sandstones, dolomites	Podkamennoye

compasses an area of the Anabar anteclise. Sedimentary cover is mainly composed of the Riphean, Vendian and Cambrian sediments. Commercial pools have not yet been discovered in this area. In the southern part of the region, Kenelekan oil sources are confined to the zone of the Lower Cambrian exposures. Southwards the Cambrian rocks are saturated with bitumen in the Olenek, Alakit and Siligir river basins.

The Lower and Middle Cambrian oil-and gas-rich sandstones and carbonate rocks rich have been penetrated by wells in the Markhinskaya area.

Within the section the Riphean, Vendian terrigenous and Cambrian carbonate play complexes have been identified. In the Vendian complex (porosity 5–10 %) sandstones of basal horizons with thicknesses from 1–5 to 40 m in the east and from zero to 10 m in the west could serve as reservoir beds. In Cambrian sediments, the thick strata are vuggy, often reef dolomites in the south and west and argillaceous carbonate and siliceous sediments enriched in the organic matter are developed in the northeast.

The South Tunguska petroleum-bearing province covers a territory of the Bakhtinskii region and the southwest margin of the Kureika syncline. The sedimentary cover in this area consists of Riphean, Vendian, Lower and Middle Paleozoic terrigenous–carbonate rocks, Upper Palaeozoic terrigenous and Triassic volcanogenic sediments.

Within the region Cambrian strata containing carbonate horizons with good reservoir properties (the upper Kostinskaya Formation) are the best prospects for oil and gas. The Ordovician complex Baikit sandstones (20–80 m thick) also have some hydrocarbon potential. Petroleum potential is not yet confirmed in the Vendian and Riphean.

The South Tunguska structural pattern is poorly studied from limited seismic survey. The Bakhtinskii projection is the major oil and gas accumulating zone. The sedimentary cover in this area shows abundant intrusive traps (up to 800–1000 m of traps in the section penetrated to depths of 3000 m) which significantly affects the petroleum potential and composition of oil and gas pools.

In the South Tunguska region hydrocarbon–carbon dioxide gas with condensate (Lower Tunguska, Tanachinskaya) occur in the carbonate horizons of the upper Kostinskaya Formation.

Oil and gas with condensate liquids have been produced from intervals in the Upper Usolskaya Formation of Lower Cambrian age. The traps for the accumulations are considered to be of reef origin.

The Baikit petroleum-bearing region covers the area of the Baikit anteclise and adjacent negative structures on the west and south. The area has been reasonably well studied by seismic surveys. The Kamovskii arch (43×10^3 km^2) is the major structure within this section and is complicated by the three dome-shaped uplifts. Currently two of these domes (Kuyumba and Taiga) are targets for exploration drilling. The main petroleum potential is associated with the Riphean (vuggy, cavernous fractured and fractured carbonates), Vendian and Cambrian complexes. The Vendian is mainly represented by mudstones interbedded with sandstones and siltstones. In the central part of the anticline these strata are absent. Within the Cambrian play complex the carbonate horizons (Osinskii, Lower Belskii, Bulaiskii and Litvintsevskii) are considered as reservoirs with total thicknesses of 20–70 m to 170–300 m. Some permeable and high porous (5–18%) beds reach individual thicknesses of 5–10 m.

The number of mapped intrusive traps available for oil and gas accumulations in these Cambrian deposits is insignificant. However, local increase in total thickness appears to be fixed. The best developed intrusive rocks are found on the northern and northeastern slopes of the Baikit anteclise.

Lower sedimentary cover horizons within anteclise appear to have favourable conditions for petroleum accumulation and preservation because of the widely distributed salt beds acting as reliable seals throughout the Lower Cambrian section.

The Baikit petroleum-bearing region is currently of special interest to petroleum geologists since here for the first time in the world Riphean strata are proving to be of commercial hydrocarbon importance over most of the area which is defined as the Yurubchen-Tokhoma oil and gas accumulating zone (Fig. 4). Due to its uniqueness of this geological phenomenon of Riphean zone structure requires further detailed evaluation. Investigations are only commencing and a detailed geological model will be better understood following on from further exploration drilling.

Tectonically the Yurubcheno–Tokhomskaya zone is confined to the Baikit anteclise arch. The Riphean oil and gas-bearing sediments are composed of thick carbonate, terrigenous–carbonate and carbonate–terrigenous strata. Compared with the Yenisey Ridge sections these sediments are referred to as the Middle–Upper Riphean Sukhopitskaya and Tungusikskaya Groups. From seismic studies and deep drilling evidence the Riphean total thicknesses are up to 3000 m.

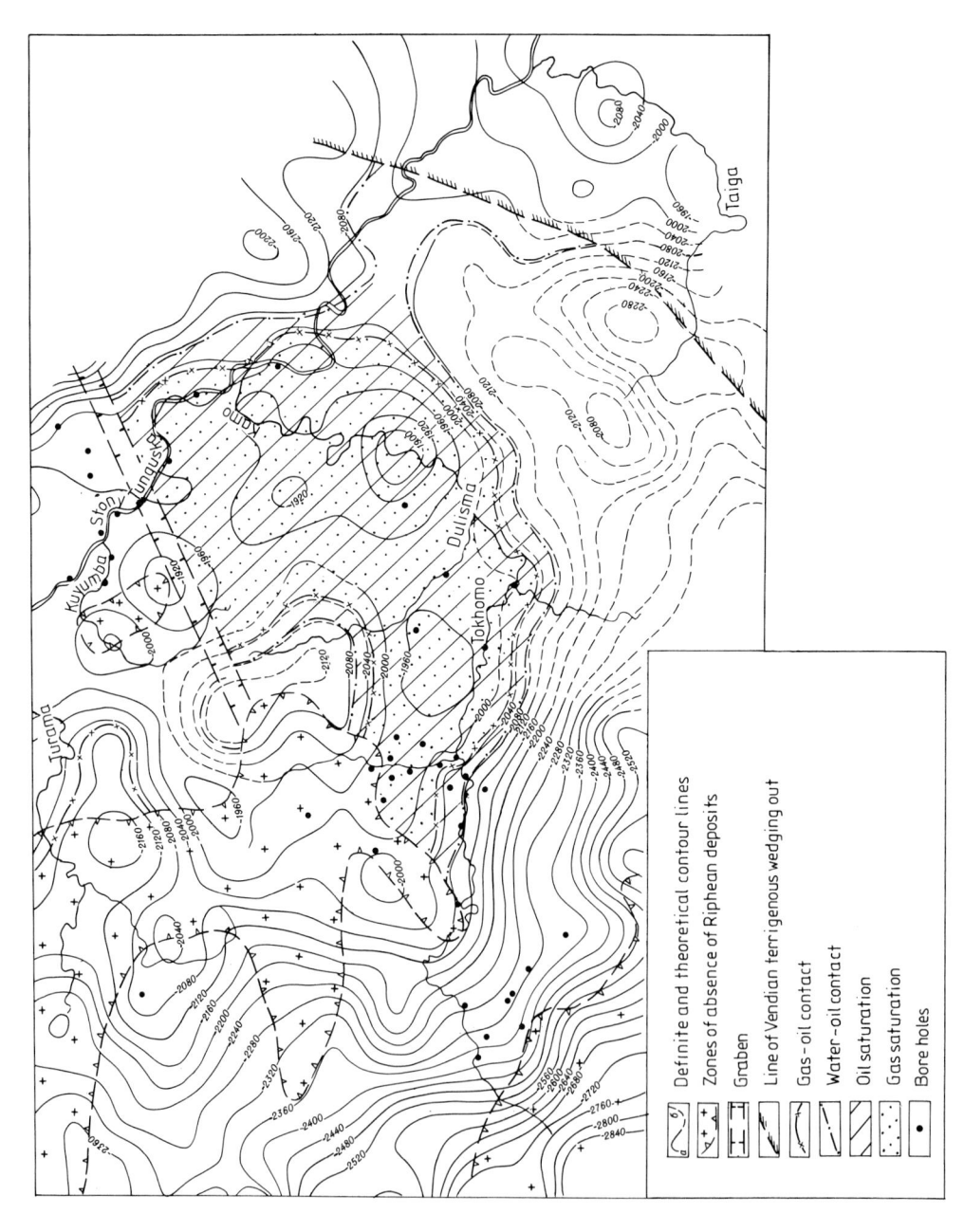

Fig. 4. Structural scheme of the Riphean top for the Yurubchen–Toknomo zone of petroleum accumulation.

Geophysical data and remote-sensing techniques suggest block structure and rock dislocation for the Riphean. Core measured angles of dip within the Riphean rocks vary from 10 to 85° and are frequently from 15 to 25°.

Within the Sukhopitskaya Group most wells penetrated the Alad'in sediments (R_a^1) where commercial petroleum potential is known. The formation consists of grey to light purple, green and brown dolomites. These are medium- and fine-grained with sinuous and lenticular layering, often stromatolithic or organogenous clastic, fractured. Fracturing is rather intensive and is predominantly vertical and subvertical. Fractures are partially or completely filled by dolomite, calcite, siliceous material and naphtides.

The Alad'in dolomites have undergone intensive secondary transformation, silicification and leaching. Recrystallization is most pronounced in rocks containing low non-carbonate material content. Leaching cavities (2–3 mm in diameter) are widely distributed within the rocks. Frequently leaching was developed on fracture walls, which made them crevice-shaped with complex rough surfaces. The major part of non-carbonate material consists mainly of siliceous, epigenetic materials. The latter consists of quartz and chalcedony and often accounts for 25–30% of the rock mass. Siliceous sediments (3–5 cm thick) form spotted and lenticular nodules and interlayers, which sometimes impregnate uniformly recrystallized carbonate material, filled fractures and leached cavities.

In the eastern part of the Yurubchen–Tokhomskaya zone the Sukhopitskaya Group is overlapped by the Tungusikskaya Group. The Group is formed by thick interbedding of carbonate and terrigenous–carbonate rocks with various thicknesses of carbonate-to-terrigenous material. Black clay shale and marlstone sequences of the Shuntarskaya Formation can be distinguished in the upper portion of the Tungusikskaya Group. Reservoir properties of the Tungusikskaya rock Group are poorer than those of the Sukhopitskaya Group. It has been proposed that boundaries of the Yurubchen Tokhomskaya zone may be controlled by the outcrop limit of Pre-Vendian rock formations of the Tungusikskaya Group. This suggestion will be tested by future exploration.

Different levels of Vendian Formations rest on the erosional surfaces of Riphean deposits.

Commercial gas and oil potential of two plays within the Vendian and in the top of the Riphean has been proved, and the pool in the Riphean is stratigraphically confined to the Alad'inskaya Formation. A single massive pool probably coincide with these horizons (Fig. 5), but the actual pool structure is probably more composite (i.e. of block type-structure). Unubiqui-

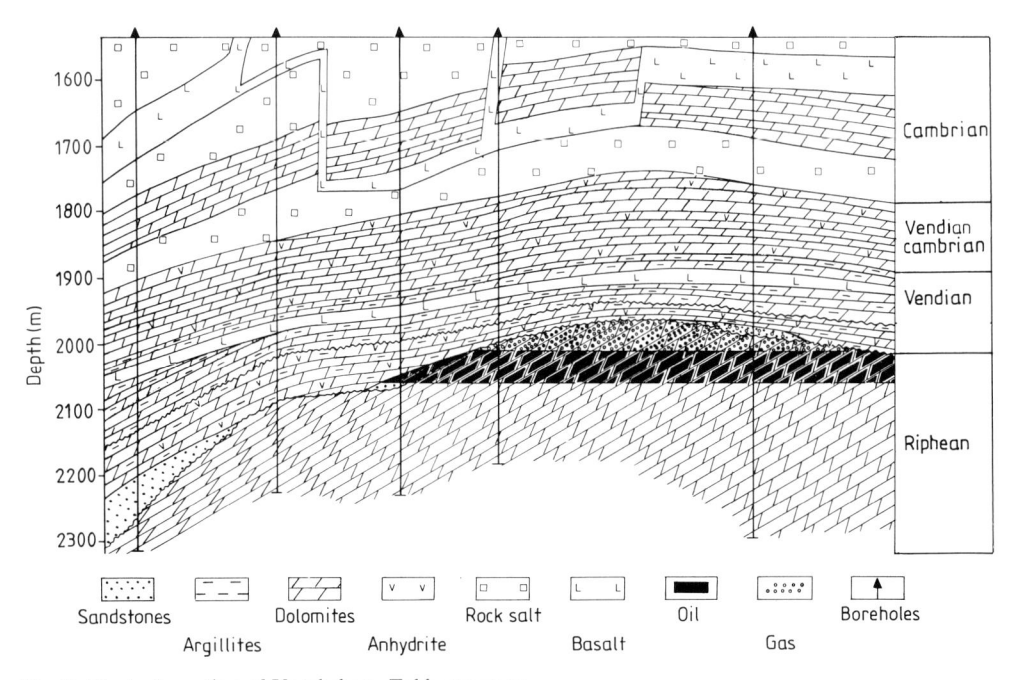

Fig. 5. Geologic section of Yurubchen–Tokhomo zone.

tous reservoir distributions are also possible.

The Katanga petroleum-bearing region occupies the south—eastern part of the Kureika syncline and the Khatanga saddle. Sedimentary cover of the region is composed of the Riphean, Vendian, Lower and Upper Palaeozoic rocks with total thicknesses up to 3—4 km, as well as the Triassic volcanogene formations which are several hundred metres thick. The Vendian complex gives the main oil and gas exploration targets in the region. The Riphean, Cambrian, and Ordovician complexes in the northern part of the region also have good potential.

The structural pattern of the Katanga petroleum-bearing region is poorly studied. The Chun'skii projection and Illimpeiskii arch of 10×10^3 and 14×10^3 km^2 (in area) respectively, have been defined using geological and geophysical data and their amplitude reaches several hundred metres. These structures, together with the Katanga saddle, are the main targets for oil and gas accumulation. Intrusive formations also occur in the section stratigraphically above these prospects.

Oil and gas production from the Vendian has been obtained in the Katanga petroleum-bearing region in a number of areas. The Sobinskoye Field, situated some ten kilometres south from the location of the well known Tunguska meteorite fall, is the best documented discovery. The hydrocarbon pools occur within the Vendian Formation. The field is a multi-horizon, oil-gas condensate discovery (Fig. 6). The accumulations occurs as sheets, with tectonical and lithological barriers.

Riphean and Cambrian plays are also prospective in the region. The Riphean oil shows are reported from one of the parametric wells.

The Nepa—Botuoba petroleum-bearing region corresponds to the anteciline of the same name. The Nepa Arch occurs in the centre of the Nepa—Botuoba anteciline; the Mirninski and Syul'dyukarskii structural Capes and the Verkhnevilyuchanskoye Dome in the northern area, the Ust'—Kut Dome in the southern region.

The Nepa—Botuoba region includes two main petroleum-bearing complexes; the Vendian terrigenous—carbonate and the Cambrian carbonate—saliferous plays. The Vendian complex play is an interbedded sequence of sandstones, siltstones and argillites. The complex was formed in the coastal zone environment of shallow sea. The palaeocrest of anteciline is concentrically bordered by deposits of the complex from the south, southeast, east and northeast. Sand beds with improved reservoir properties are represented by facies of bars, beaches, wash slopes, etc. Numerous gas, gas and oil, sometimes oil pools have been identified in this complex. The pools usually have lithological and tectonical barriers and a great many are confined to lithologic traps within sand beds, monoclinally occurring and wedging out up the dip. Examples are pools of the Verkhnechonskoye, Dulisminskoye, Jaraktinskoye Fields (Figs 7—9). The Srednebotuobinskoye Field is an example of a pool, confined on anticlinal, tectonically dislocated trap (Fig. 10).

Oil production from the Nepa—Botuoba region will commence in 1990. Gas production in small quantities has been already begun.

A number of other prospective horizons with porous, cavernous—porous and fractured reservoir-type are present in the carbonate subsalt deposits, and oil has been found in various of these plays. The Nepa—Botuoba complex is principally an oil province.

The Predpatomskaya petroleum-bearing region is an important area in the Nepa—Botuoba and North Aldan regions. It covers the Predpatomskii Regional Trough and oil and gas occurrences are present in the Riphean, Vendian and Cambrian. The major hydrocarbon prospects are associated with the Vendian complex.

The Predpatomskii Regional Trough is in the southern part of the Siberian Platform. The Trough is of north-east direction and is 35—155 km wide and 1700 km long. It is parallel the Baikal—Patomskoye highland. The Nyuisko—Jerbinskaya and Beryezovskaya depressions as well as Kazachinskii trough, are determined in the north and south of the regional trough, respectively. The trough is characterized by linear, high-amplitude folds in the upper horizons of sedimentary cover with fractures penetrating up to lower horizons of the cover. In the Nyuisko—Jerbinskaya and Beryezovskaya depressions there are a number of swells of the northeast strike.

The Cambrian carbonate deposits (the Pilyndinskoye and Kehdehrginskoye Fields) and the Vendian terrigenous plays (Khotogo—Murbaiskoe and Bysakhtakhskoe Fields) contain the commerical oil and gas of the Predpatomskoye petroleum-bearing region.

In recent years the conception of overthrust structure of the Predpatomskaya petroleum-bearing region has been proposed, and confirmed by field results in some areas. According to this conception the Pilyndinskoye, Kehdehrginskoye and Bysakhtakhskoye Fields are confined to allochthonous plates. The Khotogo—Murbaiskoye Field is connected with autochthonous formation sandstones.

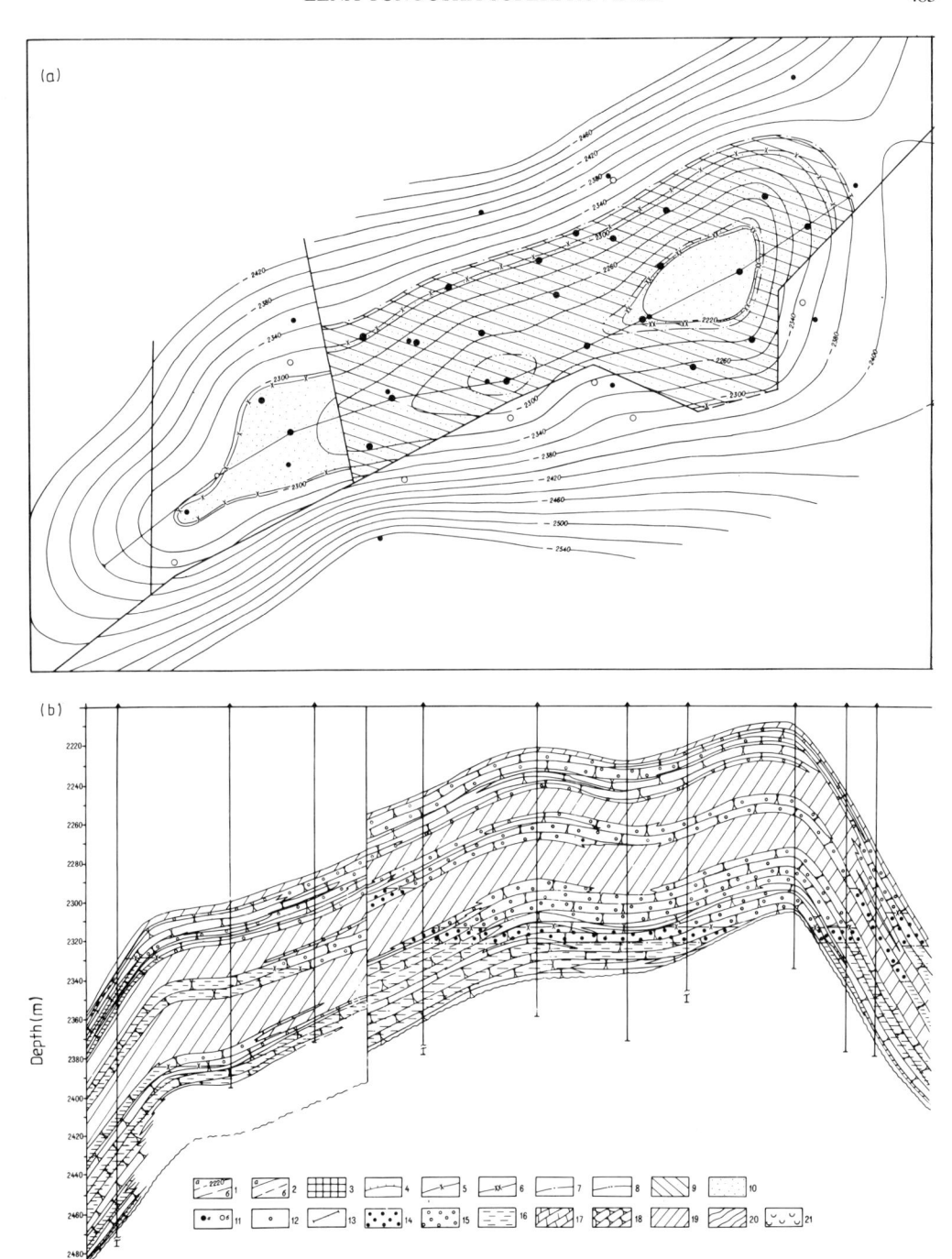

Fig. 6. Structural scheme of BH$_1$ producing formation top and section of Sobinskoye oil–gas-condensate field. 1, a, certain and b, hypothetical countour lines of producing formation top; 2, a, certain and b, hypothetical tectonic dislocations; 3, zones lacking producing formations; 4, line of reservoir wedging out; 7, external and 8, internal gas pool outline; 9, oil and 10, gas field saturation; 11, deep wells: a, with HC inflow, b, without HC inflow; 12, core holes; 13, section lines; 14, oil-saturated; 15, gas-saturated and 16, water-saturated reservoirs; 17, dolomites; 18, sandstones; 19, seals; 20, crystalline basement; 21, crust of weathering.

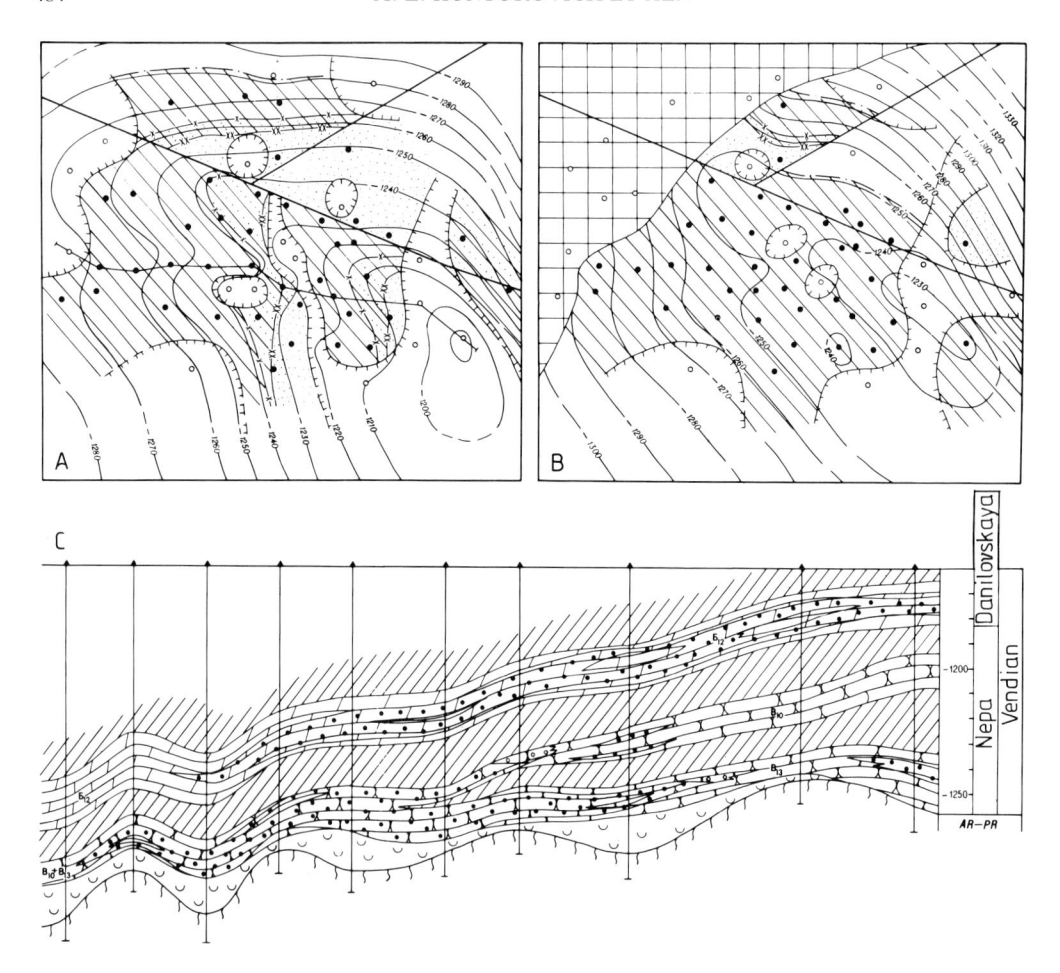

Fig. 7. Structural scheme and profile of Verkhnechonskoye field. A, layer VC$_1$ (B$_{10}$); B, layer VC$_2$ (B$_{12}$). For legend see Fig. 6.

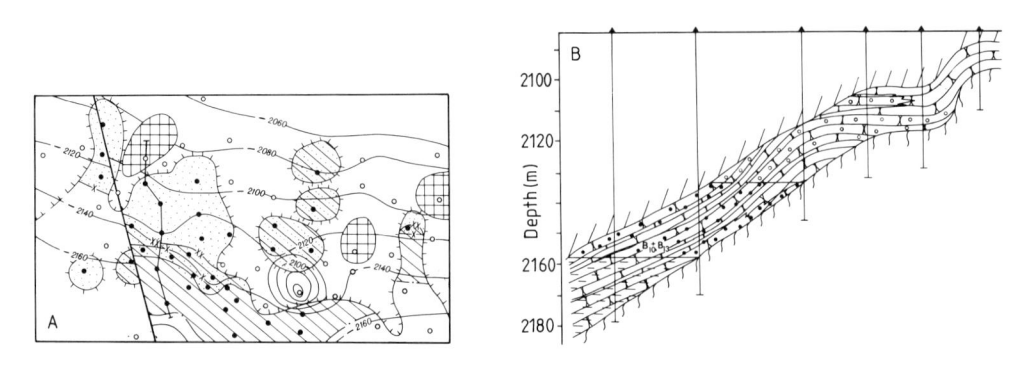

Fig. 8. (A) Structural scheme of B$_{10}$ producing formation top and (B) section of Yaraktinskoye and Ayanskoye fields. For legend see Fig. 6.

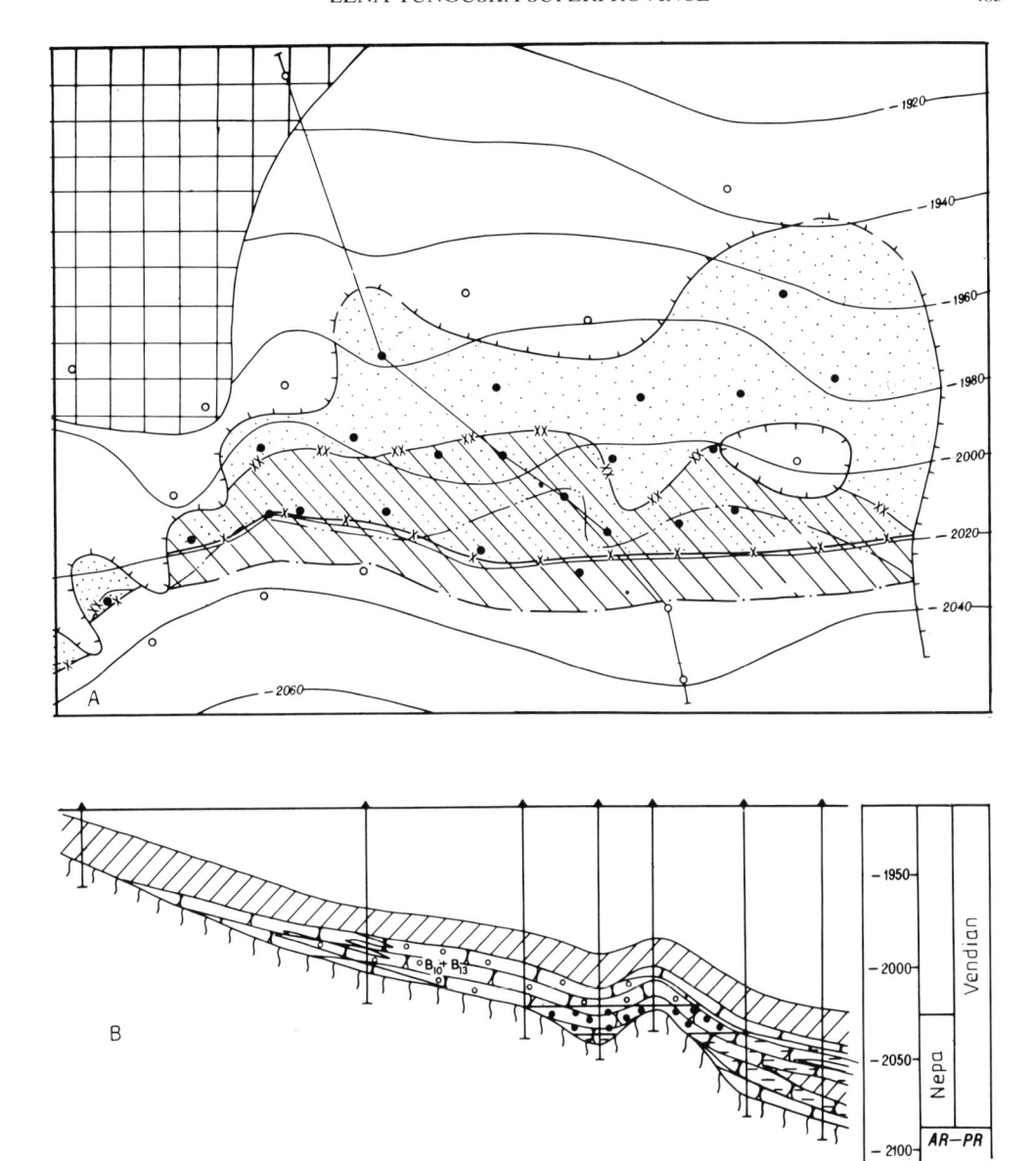

Fig. 9. (A) Structural scheme of B_{10} producing formation top and (B) section of Dulisminskoye field. For legend see Fig. 6.

The West Vilyui petroleum-bearing region is located in the western part of the Vilyii hemisyncline. In terms of tectonics, this region has the most complicated structure when compared with the other regions of the Lena−Tunguska province. The region contains the Ygyattinskaya and Kempendyaiskaya Depressions and the Suntarskii Arch. In addition, some local structures have been identified from seismic studies. Sedimentary cover thicknesses are 6−10 m in the depressions and 0.5−3 km in the Arch. This sedimentary cover is formed by the Riphean terrigenouscarbonate deposits; the Lower and middle Palaeozoic and the Mesozoic terrigenous deposits. A considerable quantity of rock salt is present in the Devonian and Lower Cambrian deposits. Prospects of oil and gas occurrence are associated with these Riphean and Vendian plays. Oil tests have been obtained from the Vendian sandstones in the region.

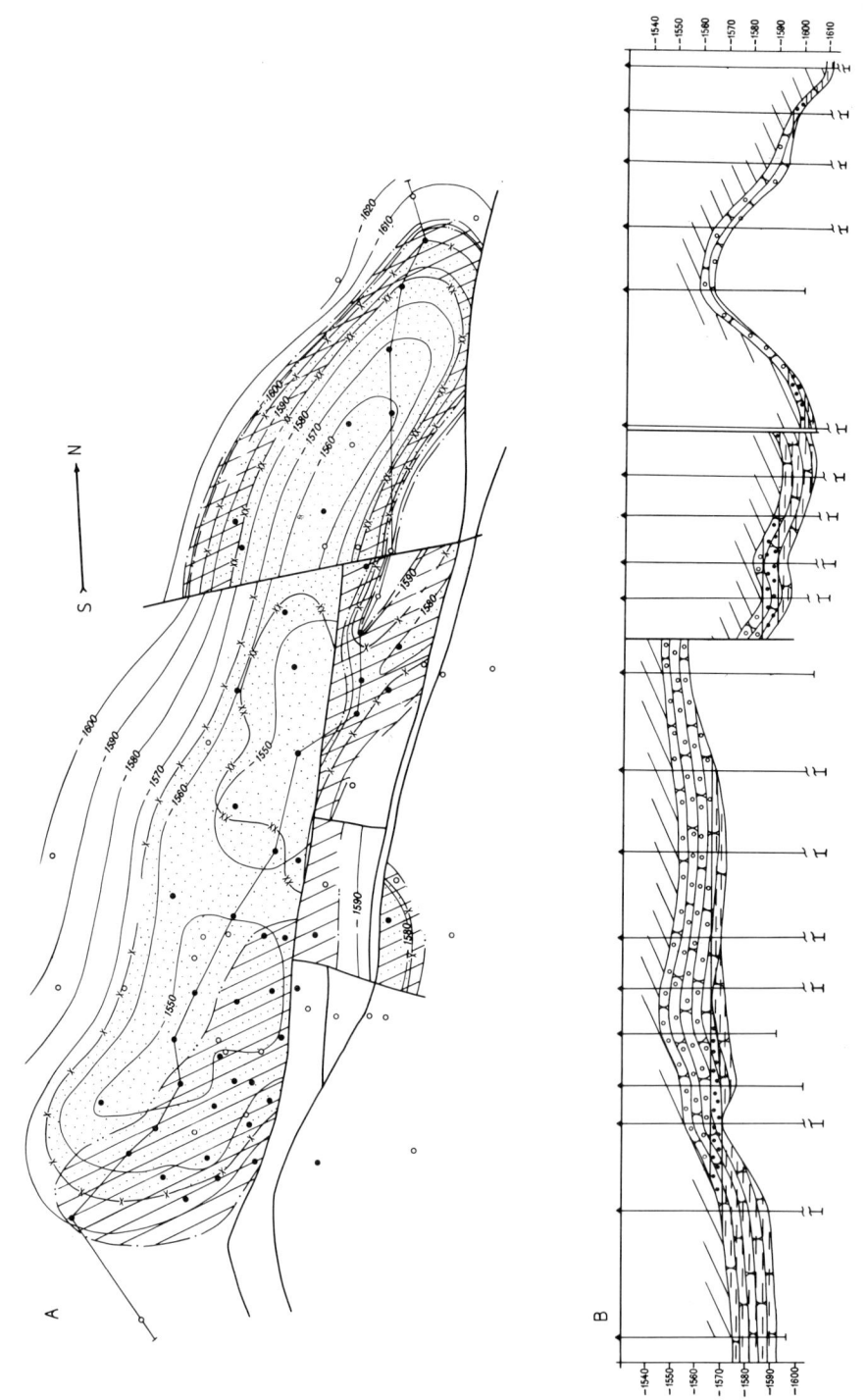

Fig. 10. (A) Structural scheme of B_{10} producing formation top and (B) section of Srednebotuobinskoye field. For legend see Fig. 6.

The North Aldan petroleum-bearing region occupies the northern slope of the Aldan anteclise. The Jurassic, Cambrian, Vendian and Riphean formations out crop in different zones of the region. The Aldan–Maiskaya Depression (58×10^3 km^2) is the largest structure with the most complete section, but the Jakutsk Arch (43×10^3 km^2) and the Tolbinskii Projection (87×10^3 km^2) have a reduced section.

The Riphean, Vendian and Cambrian plays are all prospective for oil and gas. Oil-saturated sandstones (confined to the Cambrian basal levels) have been drilled in some areas. The thickness of the individual sandstone beds varies from 8 to 35 m.

Oil and gas pool preservation is a most complex problem in evaluation of petroleum potential in the North Aldan region. Deep erosion of the Cambrian and Riphean sections; lack of reliable seals (the section is mainly formed by carbonate rocks) allow for subsided anteclise slopes to be regarded as primary targets of petroleum exploration.

The PreSayany–Yenisei petroleum-bearing region covers the syncline of the same name, complicated by a number of anteclise first-order structures (e.g. Dologomostovskaya, Mursho–Chunskaya, Katskaya Depressions 13×10^3 to 28×10^3 km^2) and the Boguchano–Manzinskii Projection. In the west of the region swells developed, the formation of which have been influenced by salt tectonics. Sedimentary cover of the region is mainly composed of the Riphean, Vendian, Cambrian and Ordovician. In restricted areas there are also some Silurian, Devonian, Upper Palaeozoic, Triassic deposits.

Major prospects of oil and gas presence are associated with the Vendian terrigenous and Cambrian saliferous carbonate (the Osinskii Horizon, the Bel'skaya and Bulaiskaya Formations) plays. In the region oil tests were obtained from the lowermost of the Lower Cambrian Belskaya Formation. Reservoir properties of the drilled section of the Cambrian play are not very favourable, nevertheless the presence of salt-bearing caps may allow or good hydrocarbon preservation in prospective horizons.

The Angara–Lena petroleum-bearing region occupies an area of the same name. In this section the Vendian terrigenous complex and carbonate horizons are located within the Lower Cambrian Salt intervals and found to be are petroliferous. Gas pools in Vendian sandstones have been tested in the Bratsk and Atovskoye Fields and there are gas occurrences in the Parfenovo area. Further exploration and appraisal of the Kovyktinskoye Field has recently commenced. Oil tests from the Osinskii horizon of the Lower Cambrian Usol'skaya formation were run in the Atovskaya, Yuzhno–Raduyskaya and Osinskaya areas and gas with condensate was tested in the Ilim area. The upper part of the Usol'skaya Formation contains the Balykhtinskii horizon which tested gas in the Balykhtinskaya area. Commercial oil flow from carbonate horizons of the Lower Cambrian Belskaya Formation have been obtained from the Khristoforovskay area, and gas was also tested in the Birkinskaya area. Gas flows from deposits of the Bil'chirskii horizon in the Khristoforovo and Bil'chirskaya areas were tested from the Angara Formation of the Lower Cambrian.

Thickness of sand horizons in the Vendian terrigenous plays reduces from 50–60 m in the southern region to less than 10–30 m in its central areas.

Petroleum source rocks: genetic relation between organic matter and oils

According to available geochemical data, the Riphean deposits, and to a lesser extent Vendian deposits, were important source rocks for the oil and gas accumulations on the Siberian Platform. Hydrocarbons migrated into the Nepa–Botuoba and Baikit anteclines and the Angara–Lena stage from peri-cratonic miogeosynclinal basins (see Figure 2). Thick organic-rich sediments accumulated within these formations. In folded platform margins there are several horizons where organic matter content reaches TOC values of 5–10%, but the content of organic carbon in these thick strata averages **>0.2%**. The organic matter is aquagenic in origin, ^{12}C enriched, which supports a polymer-lipid nature of kerogen. Continuous uniform subsidence during Upper Proterozoic and Early Paleozoic characterized these peri-cratonic basins, which resulted in generation of huge quantities of oil and hydrocarbon gases. The majority of these hydrocarbons were lost and only the final generation phases during the Vendian and especially in Cambrian, resulted in migration of hydrocarbons into Vendian–Cambrian reservoirs by lateral migration and into the Baikit anteclise by vertical migration.

Petroleum generation in intra-cratonic basins is of different character. In cata-platform basins with relatively thin of Riphean deposits major oil generation occurred only during the late Cambrian–Ordovician–Silurian. During this time sedimentary cover were formed reservoirs of high quality and strongly favoured accumulation and preservation of migrating hydrocarbons. Oil and gas generation in basins of aulacogen-type began in Riphean and continued throughout the Early Palaeozoic.

Organic matter in Riphean and Vendian source rocks are mainly of aliphatic hydrocarbon composition. Normal alkanes are 7–10% of saturated hydrocarbon (HC) fraction; C_{15}–C_{17} compounds are maximum concentrations; Isoprenoids and 12,13-monomethylalkanes are also present, and triterpane hydrocarbons dominate over steranes. Correlation of individual stereoisomers of these steranes and triterpanes supports the high maturity of organic matter and generated oils.

Aromatic hydrocarbons are present in small amounts in bitumoids and are mainly monocyclic compounds. In bitumoid composition porphyrins are almost absent, sulphur content is small, and the materials are characterised by high abundance of ^{12}C, with $\delta^{13}C$ values -30 to $-31‰$

Petroleum-generating potential of Vendian and Lower Paleozoic sedimentary rocks is much lower in the southern Lena–Tunguska Province. They are low in organic matter and have not reached the main oil generation zone.

In eastern parts in Lower–Middle Cambrian sediments a sequence of siliceous clay–carbonate rocks are extremely rich in organic matter and the Kuonamian Formation and its analogues are well developed. The Vilyui syncline has subsided to great depths and generated large volumes of oil and gas (Kontorovich 1984). Migration pathways could be accumulated these hydrocarbons in the Anabar and Aldan anticline slopes and in Cambrian reef traps, which are distributed in a broad extensive belt along shelf zone of the ancient Cambrian East Siberian Seas in Western Yakutiya.

Stratigraphic range and distribution of oil source rocks is significantly higher in northwestern and northern Province, mainly in the North Tunguska region (see Fig. 3). Source rocks enriched in organic matter are present in the Cambrian and Silurian. Deposits of the other Palaeozoic systems also have good–fair source rock characteristics.

Oil from the Lena–Tunguska Petroleum Province ranges in density from 800 to 900 kg m³; resin content varies from traces to 20–25%; asphaltenes are almost absent. Hydrocarbon compositions of oils from Upper Proterozoic and Cambrian Khatanga and Baikit petroleum-bearing regions are very similar (Aref'ev et al. 1980; Antsiferov et al. 1981; Kontorovich et al. 1986). Studies on the composition peculiarities of Riphean Yurubchenskaya oil show that methanic–naphthenic hydrocarbons represent about 80% of the analysed oil fraction. The Naphthenic fraction is about 10% per fraction of isocycloaliphatic hydrocarbons with boiling temperature range >200°C. N-alkanes are insignificant in these oils. Concentration of N-alkanes (with boiling temperature >200°C) is 3.5–5.5% for oil from different samples. Hydrocarbons with C_{Nos} up to C_{37} have been identified. Penta- (C_{15}) and hexa-decanes (C_{16}) are in maximum concentrations (about 0.5%) and Isoprenoid alkanes are about 1% by volume in the oil. A whole spectrum of hydrocarbons from 2,6-dimethylnonane (C_{11}) to 2,6,10,14-tetramethyloctadecane (C_{22}) have been identified with C_{14} and C_{13} in highest concentration, and C_{15} and C_{16} in lowest concentration. Concentration of phytane and pristane, dominating within isoprenanes, is three times lower than C_{14}; the pristane to phytane ratio is about 1.10; the ratio of isoprenanes and normal alkanes is 0.2–0.3. Aliphatic hydrocarbons of two more homologous series, (12- and 13-branched monomethylalkanes) are contained in higher concentrations in the Yurubchenskaya oil. Their total concentration in oil is about 0.25%. Hydrocarbon in the range C_{19} to C_{30} have been identified.

Oils of the Vendian producing horizons, especially within the Nepa–Botuoba anticline, are characterized by higher concentrations of steranes and hopanes biomarkers, the latter being more abundant.

In zones where oil-bearing sulphate-dolomite rocks (the Kostino Formation, the Osinskii horizon of the Usol'e Formation) occur near doleritic intrusions, oils are rich in mercaptan hydrocarbons and often carbon dioxide gas pools are also found. Inclusions (e.g. elemental sulfur carbonic acid) are often found in salt at the same stratigraphic levels. All these 'by-products' being formed by the intense heating of oil-bearing sulphate-dolomite rocks by trappean intrusions which are a consequence of a single complex of paragenetically related events.

Distribution of petroleum resources

Evaluation of petroleum potential of the Siberian Platform (Antsiferov et al. 1981) shows that Riphean to the Upper Paleozoic stratigraphic intervals of the Lena–Tunguska Province are all petroliferous.

The most important controlling factors for oil accumulation are the following.
(i) Prospectivity for oil and gas decreases from the Upper Proterozoic to the younger Middle and Upper Paleozoic (Fig. 11).
(ii) The greatest stratigraphic interval of oil-

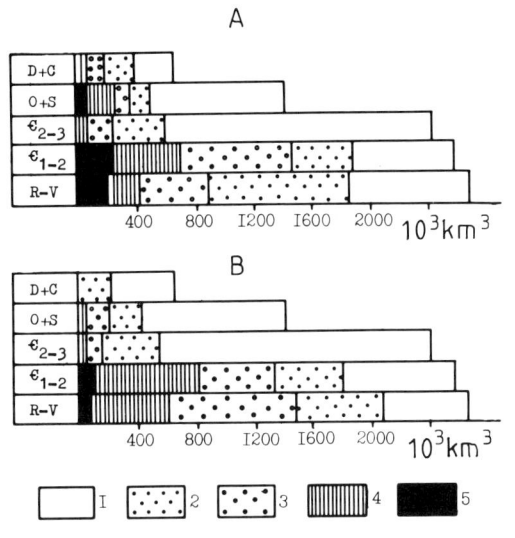

Fig. 11. Correlation of perspective for oil (A) and gas (B) areas of stratigraphic complexes in Lena–Tunguska Province. Areas: 1, with no prospects; 2, with low prospects, 3, perspective of category III; 4, perspective of category II; 5, perspective of category I.

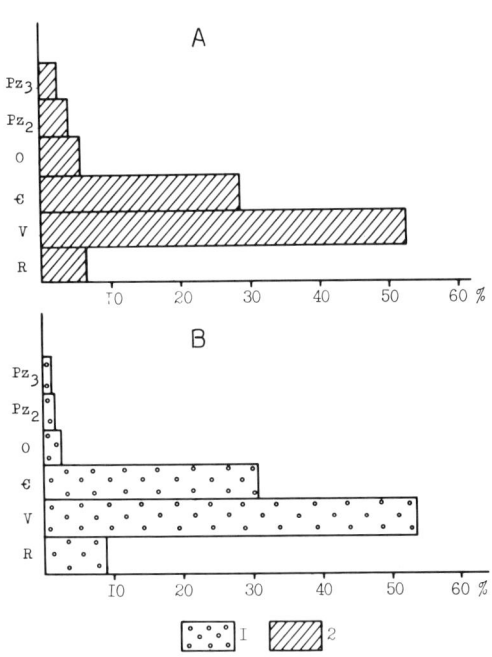

Fig. 12. Distribution into stratigraphic complexes of initial (A) oil- and (B) gas-in-place resources in Lena–Tunguska petroleum-bearing province.

gas bearings occurs in the North-Tunguska petroleum-bearing region, in the central and northern Kureika syncline.

(iii) Zones for the main oil prospects are localized in two different regions of the Lena–Tunguska Province.

The central part of the Province which includes the Baikit and Nepa–Botuoba anteclines and the Khatanga Saddle. The main prospects of this region are associated with the Riphean, Vendian and Lower Cambrian Formations, and it is called the main petroleum belt (Kontorovich et al. 1982). Petroleum has been proved over this whole area.

Confined to the central and northern parts of the Kureika syncline. The main prospectivity is within the Lower and Middle Palaeozoic and the presence of major petroleum resources has yet to be confirmed.

(iv) In the Province more than 50% of oil and gas reserves are found in the Upper Proterozoic Formations (Fig. 12).

Exploration has conclusively revealed presence of commercial petroleum accumulations in the Precambrian of the Lena–Tunguska Province. Their subsequent development of these discoveries will undoubtedly lead to revision of ideas about the petroleum potential of ancient strata in the Earth sedimentary cover.

References

ANTSIFEROV, A. S., BAKIN, V. E., VARLAMOV, I. P., et al. 1981. *Petroleum geology of Siberian Platform.* Nedra, Moscow.

AREF'EV, A. A., ZABRODINA, M. N., MAKUSHINA, V. M. & PETROV, A. A. 1980. Relic tetra- and pentacyclic hydrocarbons in ancient petroleums of the Siberian Platform. *Izv. Academic Nauk SSSR. Series Geology*, **3**, 135–140.

KONTOROVICH, A. E. 1984. Geochemical methods for the quantitative evaluation of the petroleum potential of sedimentary basins. In: DEMAISON, Q. & MURRIS, R. J. (eds) *Petroleum Geochemistry and Basin Evaluation.* AAPG Memoir **35**, Tulsa, Okla. 79–109.

——, SURKOV, V. S. & TROFIMUK, A. A. 1982. Main zones of oil generation in the Lena–Tunguska Province. In: *Progress in Teaching of Academician I. M. Gubkin in Petroleum Geology of Siberia.* Nauka, Novosibirsk. 22–42.

——, —— & —— 1986. Nepa–Botuoba Antecline — a New Perspective Area of Petroleum Production in the East USSR. In: KONTOROVICH, A. E. et al. (eds) Nauka, Novosibirsk.

TROFIMUK, A. A. 1960. Oil and gas bearing of the Siberian Platform. *Geology and Geophysics*, **7**, 3–11.

ZOLOTOV, A. N. 1982. *Tectonics and Petroleum Potential of Ancient Strata*, Nedra, Moscow.

West Siberian oil and gas superprovince

I. I. NESTEROV, F. K. SALVAMANOV, A. E. KONTOROVICH, N. K. KULAKHMETOV, V. S. SURKOV, A. A. TROFIMUK & V. I. SHPILMAN

USSR Ministry of Geology, Moscow, USSR

Abstract: The West Siberian oil and gas superprovince is an area of more than 2.5×10^6 km^2. The province basin is filled with Cenozoic, Cretaceous and Jurassic platform sediments and, in the north and west also with Triassic and Palaeozoic rocks. The volume of sedimentary platform deposits is more than 9×10^6 km^3 which includes 0.75×10^6 km^3 of Triassic deposits and 1.3×10^6 km^3 of Palaeozoic deposits. The composition of the reservoir rock is terrigenous and it is inferred that Palaeozoic deposits can be represented by carbonates. Sediments accumulated in marine conditions are predominant in the region. Oil and gas pools occur both in marine and continental deposits. The total amount of discovered oil and gas in various plays is distributed as follows: 43% of pools are associated with regional caprocks, 40% with zonal caprocks and 17% with local caprocks.

The largest gas pools are mainly accumulated under regional caprocks and oil pools occur under zonal caprocks. There are no known major gas and oil pools under local caprocks. In future, discoveries of large hydrocarbon accumulations under zonal caps and an increase in discoveries of local reservoirs are confidently anticipated.

Regional studies, basin evaluations and calculations confirm the West Siberian basin to be a superprovince according to its known hydrocarbon potential. However, the regional evaluations do not meet current commercial requirements of the petroleum exploration industry, the main task of which is substantiation of potential reserves in concrete regional, zonal and local layers, ranging in thickness from 10 to 20 m instead of thick formations. Such needs call for development of modern methods such as remote-sensing to assist in oil-and-gas potential predictive exploration. Such methods must later be further developed to identify thicker oil-and-gas reservoirs of several square kilometres in size with less than 5–10 m bed thickness.

The West Siberia basin is being developed very rapidly. It took only eight years from the first small gas field discovery (1953) to the first significant oil production. Since the first oil pipeline installation in 1964 an average of 848×10^3 m^3 per day of oil and gas has been produced, and in 1987 production had reached 2.38×10^6 m^3 per day.

Hydrocarbon potential evaluation of the large West Siberian sedimentary basin has been studied using geological, lithological–facial, geochemical, hydrogeological and other similar methods. These have been continually uprated through regional geological structure studies and after oil-and-gas pool discoveries. As early as the 1950s, after limited stratigraphic and parametric well drilling and prospecting drilling, the oil-and-gas potential of the basin was considered to be rather low. Even after the discovery of the first oil-and-gas fields in the pre-Urals region, the USSR oil and gas industry did not pay much attention to these discoveries. The situation drastically changed in 1961 after highly productive oil pool discoveries in Neocomian deposits of the Megion area in the centre of West Siberia. The methods of regional evaluation then began to improve. In the 1970s new methodology (including statistical calculations) were applied, which permitted prediction and calculation of potential hydrocarbon resources independent of prospecting and exploration drilling. Following approval of these estimates by Gosplan (State planning Committee of the USSR), they were used as a basis for long-term planning in West Siberian basin exploration. Forecast evaluation, confirmed by actual new field discoveries, was taken as a basis not only for planning but for the construction of major industrial complexes and transport systems connected with oil and gas production.

However, after a substantial increase of geological exploration activities, the accuracy of regional estimations of potential reserves across the sedimentary mantle section and through oil-and-gas bearing regions and areas became insufficient to determine the annual perspective trends in the search for new pools. In the north of West Siberia there appeared to be a discrepancy between proven oil and gas reserves and prospective forecasts. The number of 'low-informative' wells tended to increase and reached some 50% of total drilling in the region. This caused problems of evaluating and revising basic regional and geological data and re-

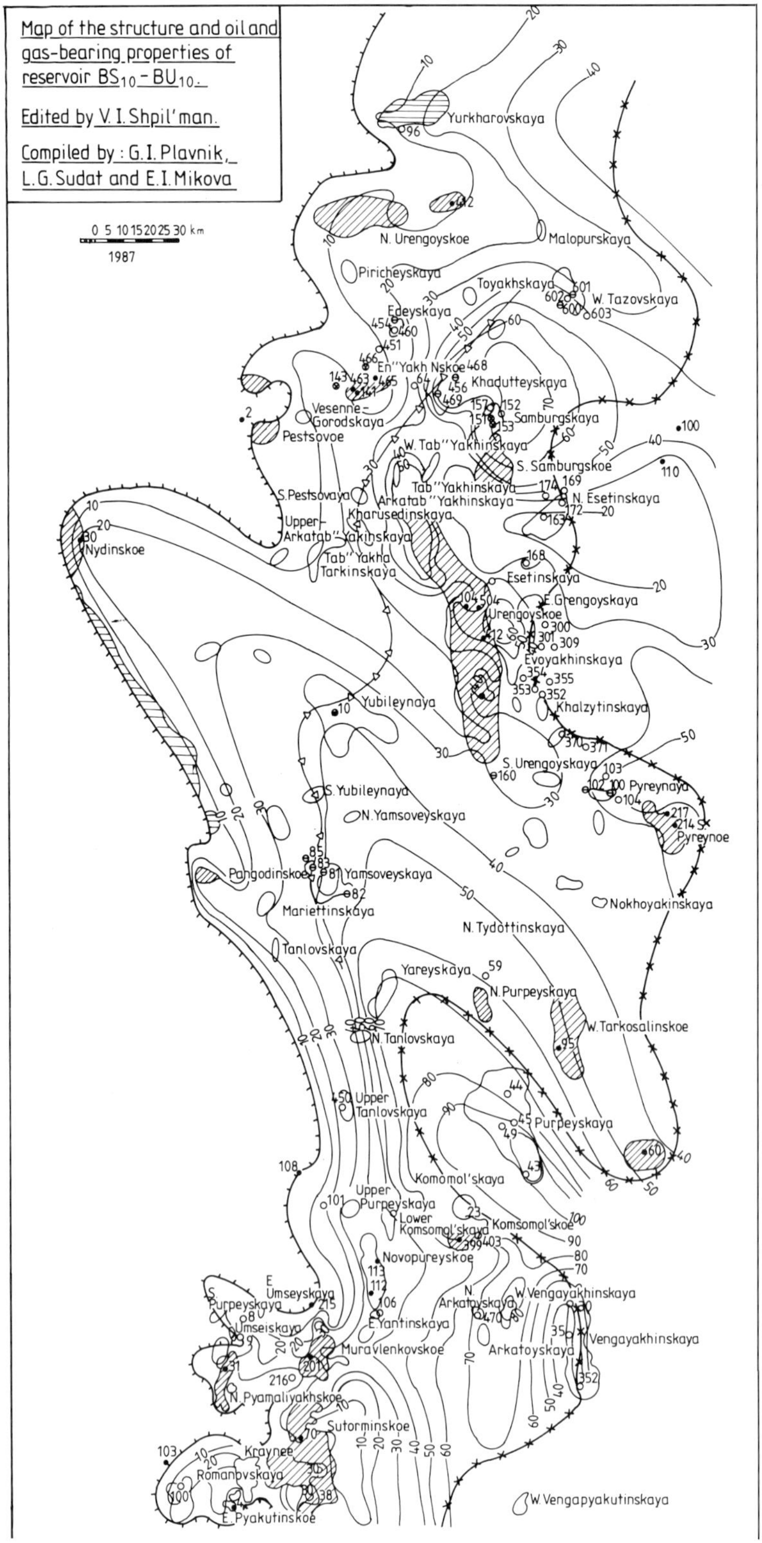
Map of the structure and oil and gas-bearing properties of reservoir BS_{10} – BU_{10}.
Edited by V. I. Shpil'man.
Compiled by : G. I. Plavnik, L. G. Sudat and E. I. Mikova
0 5 10 15 20 25 30 km
1987
Yurkharovskaya
N. Urengoyskoe
Malopurskaya
Piricheyskaya
Toyakhskaya
W. Tazovskaya
Edeyskaya
En''Yakh Nskoe
Khadutteyskaya
Vesenne-Gorodskaya
Pestsovoe
Samburgskaya
W. Tab''Yakhinskaya
S. Samburgskoe
Tab''Yakhinskaya
Arkatab''Yakhinskaya
N. Esetinskaya
S. Pestsovaya
Kharusedinskaya
Upper-Arkatab''Yakinskaya
Nydinskoe
Tab''Yakha Tarkinskaya
Esetinskaya
E. Urengoyskaya
Urengoyskoe
Evoyakhinskaya
Khalzytinskaya
Yubileynaya
S. Urengoyskaya
Pyreynaya
S. Pyreynoe
S. Yubileynaya
N. Yamsoveyskaya
Pangodinskoe
Yamsoveyskaya
Mariettinskaya
Nokhoyakinskaya
N. Tydottinskaya
Tanlovskaya
Yareyskaya
N. Purpeyskaya
W. Tarkosalinskoe
N. Tanlovskaya
Upper Tanlovskaya
Purpeyskaya
Komomol'skaya
Upper Purpeyskaya
Lower Komsomol'skaya
Komsomol'skoe
Novopureyskoe
E Umseyskaya
S. Purpeyskaya
N. Arkatoyskaya
W. Vengayakhinskaya
Umseiskaya
E. Yantinskaya
Vengayakhinskaya
Muravlenkovskoe
Arkatoyskaya
N. Pyamaliyakhskoe
Sutorminskoe
Kraynee
Romanovskaya
E. Pyakutinskoe
W. Vengapyakutinskaya

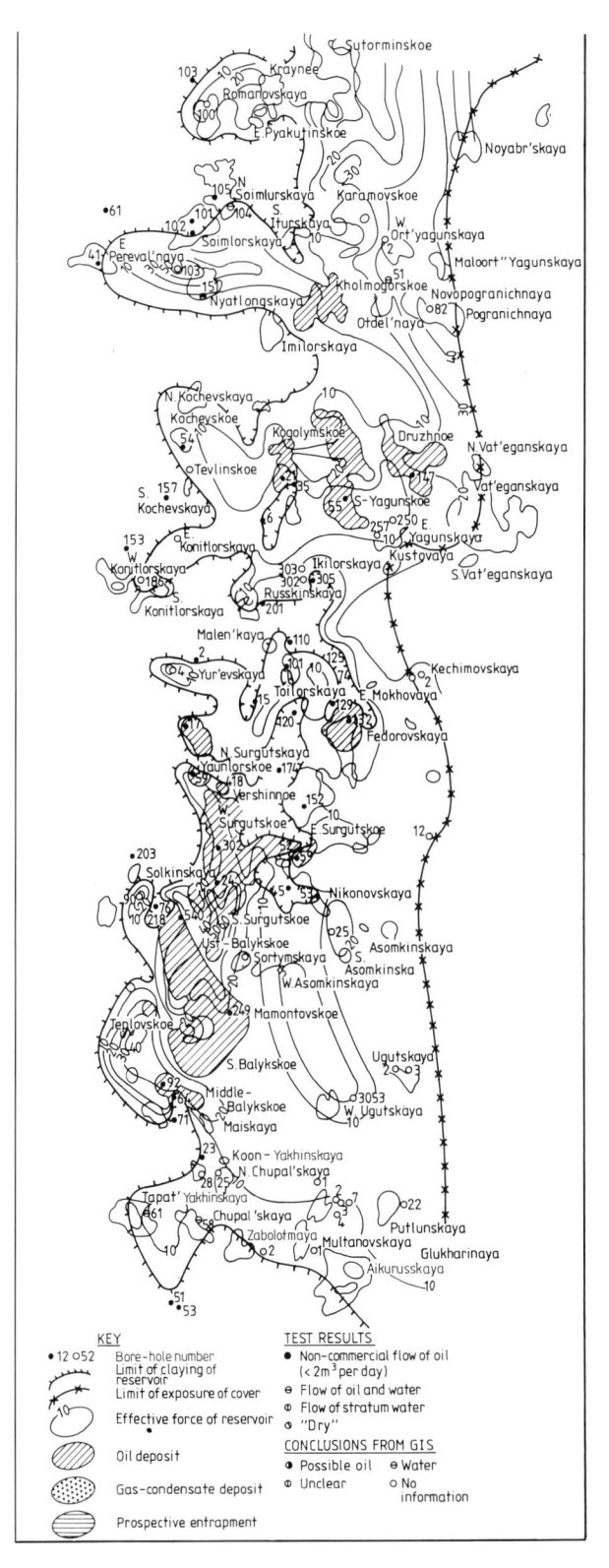

Fig. 1.

visions had to be made on the selection of main prospecting trends and plays.

For regional estimation it is necessary to pass from strata evaluation to forecast of reservoir beds, 10–20 metres in thickness, according to maps of 1:500 000 and 1:200 000 scales. Figure 1 illustrates a section of such a map for the Valanginian trends and its analogues.

Local evaluations show that the main role belongs to interpretative methods of remote operation, based on determination of physical–chemical thermal and geochemical characteristics of sedimentary rock sections above the hydrocarbon pools, which developed during hydrocarbon generation.

It was recognized that studies of oil and gas pool formation processes and mechanisms were needed to understand and assist in these regional studies. Geographical and statistical applications, basin and prospect evaluations of discovered and predicted fields distribution were supplemented with studies of physical–chemical processes. Such applications allowed the study of oil and gas composition, rocks containing oil and gas and organic matter, and correlation of sources for commercial hydrocarbon generation and accumulation.

Estimations of economical region development excluded evaluations determined for fulfilment of plan goals, which did not take into account relative or absolute profit value and the state of social security of people who are developing this region.

Limitations for traditional regional criteria of oil and gas potential

In West Siberia a set of regional geochemical, hydrogeological, lithofacies maps and other criteria has been used to delineate the area's prospectivity for oil and gas.

It was assumed that commercial amounts of oil and gas could not be generated in an environment where the catagenesis of dispersed organic matter did not exceed a 'middle-brown coal stage' of its transformation. For this reason, regions with sedimentary rock thicknesses of less than 1.0 km were excluded from prospecting.

In practice, there is one commercial gas accumulation, found at a depth of 400 m in an area which excludes migration of gas from below. In the north of the West Siberia Basin, gas showings are registered at a depth of only 100–150 m. Naphthenic oil pools are known to occur at a depth of about 900 m. In view of these data, when making regional predictions of oil and gas potential, the parameter of organic matter catagenesis should be used only with more precise characteristics, such as the energy potential of source organic matter for oil and gas which is estimated by the quantity of paramagnetic centres in kerogen.

Application of hydrogeological data suggests that washing-out with fresh percolating waters decreases oil-and-gas bearing potential. Washing-out zone width from basin margins to its centre was estimated to be 50–100 km. It was ssumed that underground waters mineralization values were important parameters of subsurface washing-out with fresh waters. And though it was taken into account that their desalting was caused by waters displaced from clay rocks and by newly formed waters in the process of regional transformations of organic matter, the presence of low salinity waters (e.g. less than 10 g per litre) was not considered to favour a positive evaluation of oil and gas potential.

These assumptions were not confirmed in practice. In the north of West Siberia, underground water mineralization decreases from 15–20 g per litre in upper parts of section (1000 –1500 m) to 5g per litre at depths of 3–4 km. The role of liquid hydrocarbons in relation to gaseous hydrocarbons increases in pools downward through the section. Without obtaining additional mineralization data of underground waters it does not seem that such methods can be applied in evaluating the prospects of oil and gas. The significance of quantity and composition of the underground water gas component is also not yet defined. Moreover, it is necessary to explain pool occurrence (primarily gas pools) in zones with a high deficit of partial pressure (up to 20 MPa) of water-dissolved gases. It should be taken into account that even with gas cap presence in oil pools, oils are mainly undersaturated with gas.

In defining the limits of oil-and-gas bearing potential on the basis of lithofacies characteristics, marine genesis sequences were always preferred. In practice, oil and gas pools, irrespective of their reserves volume, have been discovered in rocks which were accumulating both in marine and continental environments. In regional evaluation of oil-and-gas potential, the previously used indices of marine sediments portion in their total volume were replaced by more objective forecasting such as; source rock facies, source rock character and richness and its potential to generate oil and gas. Indices such as pristane/phytane ratio or other similar parametres in oils and source rocks reflect hydrocarbon generation stages being developed during burial in the basin. The indices of oxidation reactions in rocks are still reliable. In West Siberia to the south and southeast a thick series of mottled rocks have been developed in Jurassic and Noecomian complexes. And though it has been proved that red-coloured rocks grade from the margins to the centre of the West Siberia Basin, into green-coloured forms, due to reduction of ferric into ferrous oxide, it is still considered that the residual organic matter energy potential is insufficient for oil and gas generation because of the great consumption of energy for iron-reducing reactions. In oil-and-gas potential forecasting, much attention has been paid to quantitative geochemical indices of dispersed organic matter in the potential source rock. The increase of organic carbon content and the portion of contained bitumoids in each separate horizon, towards the centre of the West Siberia sedimentary basin, correlates with the prospective oil-and-gas plays, which have been confirmed by actual discoveries of oil and gas pools. Some explorationists still have some doubt concerning the application of these indices in the basin. If different stratigraphic levels are compared it can be seen that sometimes in rocks with low organic matter and bitumoid content many oil & gas pools are found, while in rocks with these indices 2–3 times higher oil and gas are absent. Even allowing for organic matter transformation in rocks does not provide a reasonable explanation of these observations.

Much attention has been given to thermodynamic indices and processes. An isotherm of 70° was considered by many scientists to be critical for the evaluation of oil-bearing generation potential. Also, static relationships seemed to confirm the use of this thermal parametre for regional oil-and-gas generation & hydrocarbon bearing potential. Exploration of the West Siberian basin has made the validity of this individual index doubtful. It has been shown that current unsteady thermal state of rocks in the basin resulted from the effect of climatic factors (glaciation) during the Quaternary. The depth of the unsteady state of rocks reaches 2800 m, and the difference between current and palaeotemperatures varies from 10–15°C to 40°C (Fig. 2). The introduction of these correction factors shows that all 'regularities' concerning the relation between oil occurrence, its properties and the current temperature, are completely erroneous.

The processes of hydrocarbon generation and oil and gas pool formation cannot be explained by any simple individual factor. Integration of the complex geochemical processes are required to provide an explanation of actual oil and gas pool allocation, their phase state and their distribution through the section and region. Current studies using methods based on the complex use of the above listed regional parameters fail to explain the occurrence of oil and gas pool distribution stratigraphically and/or regionally in the West Siberian Basin.

In the West Siberia sedimentary basin approximately 10^8 traps are distributed across the section of sedimentary rocks occupying an area of some 3×10^6 km^2 and not less than 10 000 oil and gas pools are predicted to be present in the basin. Only one oil and gas accumulation occurs every 10 000 traps. This fact may not fully account for any combination of all the above listed and generally accepted indices of oil-and-gas-bearing potential. The present state of knowledge on oil and gas geology more or less reliably explains the regional hydrocarbon potential, but does not fully explain their local and regional distribution.

Major reservoirs for new oil and gas pools

More than 1700 oil and gas pools have been discovered in West Siberia during the last 30 years of routine hydrocarbon exploration (Table 1).

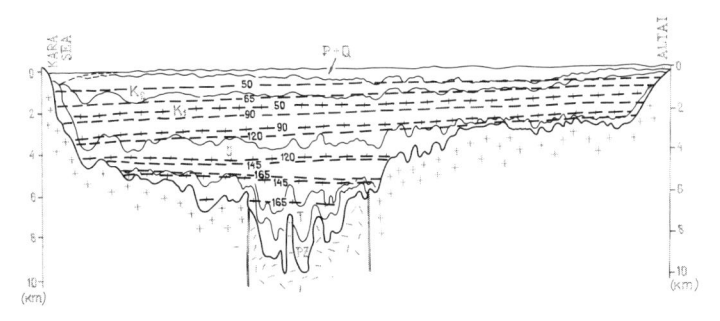

Fig. 2. Current temperatures — and palaeo-temperatures — along the section Novaya Zemlya – Altai, Western Siberia sedimentary mantle.

Table 1. *Distribution and number of oil and gas pools associated with regional, zonal and local caprocks in West Siberia*

Reservoir type	Jurassic			Cretaceous		
(caprocks)	oil	gas & oil	total	oil	gas & oil	total
Regional	501	81	582	73	67	140
Zonal	1	–	1	409	277	686
Local	12	3	15	150	130	280
Total:	514	84	598	632	474	1106

Of these discoveries 582 pools have been found in Jurassic rocks under regional caprocks; 15 pools with local caprocks and 1 pool with a zonal caprock in the Krasnoleninsk region (beds v_{10-11} of the Talin area). In Cretaceous rocks 1106 pools have been discovered; 140 pools under regional caprocks; 686 with zonal caprocks and 280 having local caprocks. Almost 85% of discovered hydrocarbon pools have been found under regional and zonal caprocks and in the near future, it is predicted that further increase of such discoveries will be made, especially those plays under zonal caprocks. This accounts for the transition to oil-and-gas-bearing capacity forecast based on specific reservoir maps which may be constructed with a sufficient degree of accuracy using seismic and drilling data on a scale of 1:500 000 and for separate regions on a 1:200 000 scale.

Figures 3 & 4 show geological profiles with section types outlined, where regional, zonal and local caprocks dominate in the basin.

The lowest target for oil and gas plays is a basement of weathered crust and areas of its

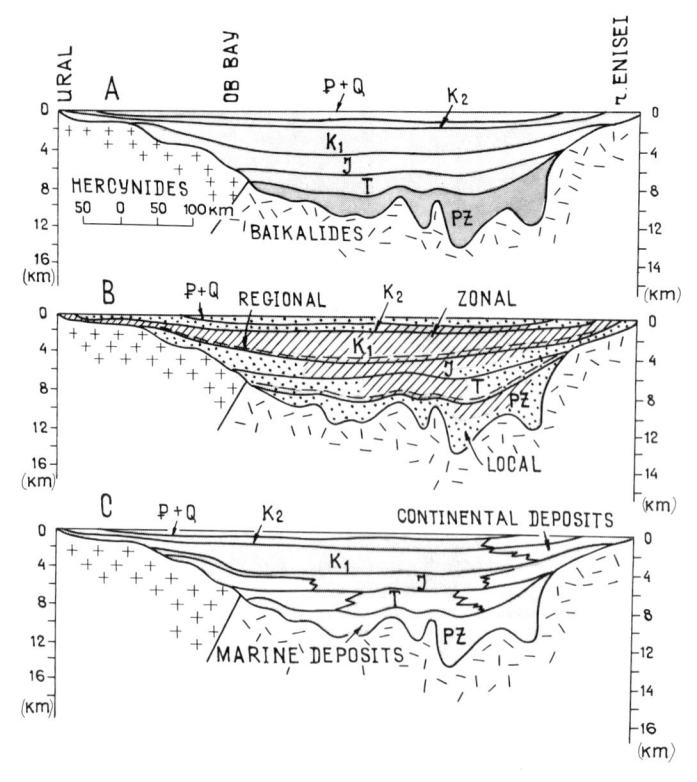

Fig. 3. (A) Geological section, (B) caprock types and (C) facies along the merimian line of the West Siberian polar circle.

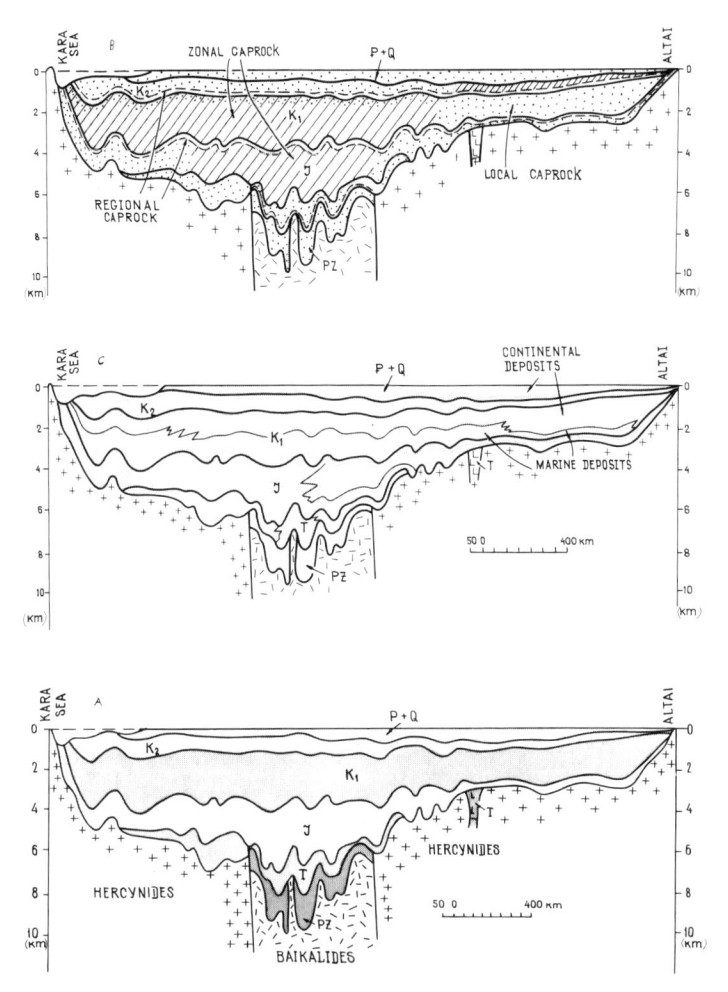

Fig. 4. (A) Geological section, (B) caprock types and (C) facies along the line Novaya Zemlya–Altai.

redeposition in the nearest zone with terrigenous material being transported for a distance of 5–10 km from source rock supply. New regional reservoirs are not expected to be discovered in this oil-and-gas-bearing horizon. The number of zonal caprocks is also expected to be limited. Oil Fields with production rates over 100 tons per day have been registered in a shale weathered basement (Severo–Varyeganskoye Field), and in shale and carbonate rock weathered basements with gas accumulations over 10^6 m^3 (Novoportovskoye Field). However, in similar rock types in the Tomsk region and Shaim area of the Tjumen region, oil and gas production rates were not so high. Oil and gas pools with large reserves have not been revealed in weathering crusts. Figure 5 is an example of oil distribution in a shale weathered basement of the Severo–Vareganskoye Field located in the central part of the West Siberian superprovince.

Reservoir properties, production rates and volume of oil and gas reserves in the nearest weathered basement erosion zone have been predicted rather accurately. And basement erosion in marine or continental environments is of primary importance in a number of cases. On the greater part of the West Siberian region, the basement is overlain by Lower–Middle Jurassic sandy-argillaceous deposits which accumulated in continental environments. In periods of active tectonic movements, clastics with better reservoir properties accumulated in such environments around the protrusions of the basement paleo-relief. Sediment accumulations of Hettangian and Sinemurian, Upper

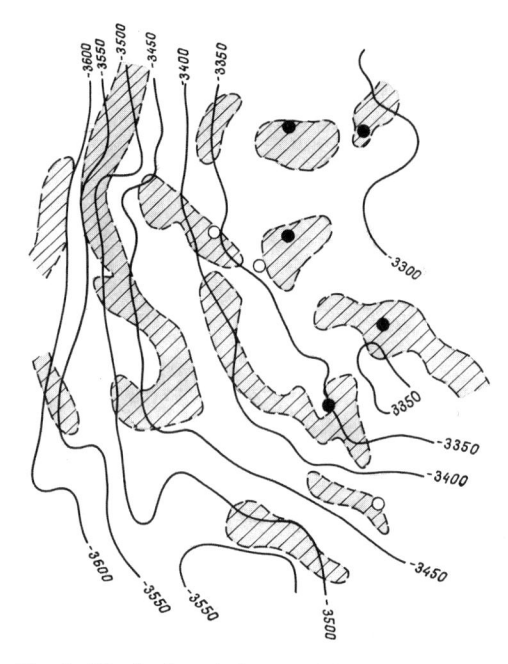

Fig. 5. Distribution of oil pools within the palaeozoic weathering crust of the Severo–Vareganskoye field.

Pliensbachian, Aalenian and Bathonian age date these periods of deposition. If sediments of these time intervals are adjacent to the basement protrusions, then the probability of identifying sediments with better reservoir properties increases. The composition of basement rocks is also important. An example of an oil pool in a section of the nearest erosion zone is the Medvedevskaya area within the Sosninsko–Sovietskoye Field in the Tomsk region where an oil accumulation with flow rate of over 400 m^3 d^{-1} was obtained from Aalenian continental deposits overlying the basement. When the basement protrusion is overlain by sediments which were accumulating in marine conditions, the reservoir properties of sediments in the weathered crust erosion-zone was considerably improved, especially so if granite–gneissic rocks are abundant in the basement. Oil and gas accumulations in such zones have been found in the Berjezovo, Shaim and Krasnoleninsk areas of West Siberia. Oil production rates in these areas can be as high as 100 m^3 per day, and gas production rates exceed 10^6 m^3 per day. General regularities of reservoir formation under these conditions have also been defined. If the basement protrusion, in which granite–gneissic bodies or other rocks capable of providing coarse-clastic terrigenous material occurs, was elevated in paleo-relief above the level of underlying continental sediments for more than 50–70 m, sediments with good reservoir properties were usually formed. If this erocess is absent rocks in the weathered crust erosion-zone tend to show poor reservoir properties, (porosity < 10–15%; density < 10 md). Heat flow maps provide reliable information for prediction the composition and areal development of various basement rock types (Fig. 6). Regional maps indicate the time of basement consolidation while detailed maps illustrate the petrographic composition of rocks.

The Palaeozoic platform sediments are distinguishable only from seismic data in the eastern and northern parts of West Siberia at depths of 5–14 km and cover an area of 800 000 km^2. They occur in the basement of Baikalian or Karelian consolidation. The structure of these rocks and their regional stratification have not been studied yet. A single regional reservoir is considered to be present above the structural and stratigraphic unconformity identified by seismic data. Additional data will be available from the drilling of a superdeep hole which was spudded in December 1987 near Tikhaya station of the road Tyumen–Novy Urengoy. Reservoir formation maps cannot be succesfully constructed until data on these sediments are available. Evaluation of their oil-and-gas content is estimated from the regional parametres, such as volume rate of sedimentation, rock volumes, lithofacies characteristics. Such evaluations suggested that sediments are of marine origin. The rocks are sandy/shales with some, carbonates. Their total thickness is up to 5 km with average 1.6 km.

The Triassic platform sediments are abundant in the north of West Siberia with identified distribution area of about 600 000 km^2. The distribution was obtained from seismic data. These sediments have been penetrated only in the Ust–Yenisei trough and are represented by continental clay–siltstone rocks with interbeds of sandstone. In the Gydan peninsula the sediments are considered to have accumulated in marine environments. The Triassic sediments occur at depths of 5–8 km. They have a total thickness of up to 2.5 km with an average of 1.25 km. It is currently not possible to construct formation-reservoir maps from the limited data, and their hydrocarbon reserves potential has been estimated using simple geological parametres for the entire complex. To the centre and south of West Siberia, Triassic basaltic cover rocks are widely developed. The Triassic sediments also occur in single grabens up to depths of 3–5 km, where they are associated with Carboniferous sandy/shale rocks.

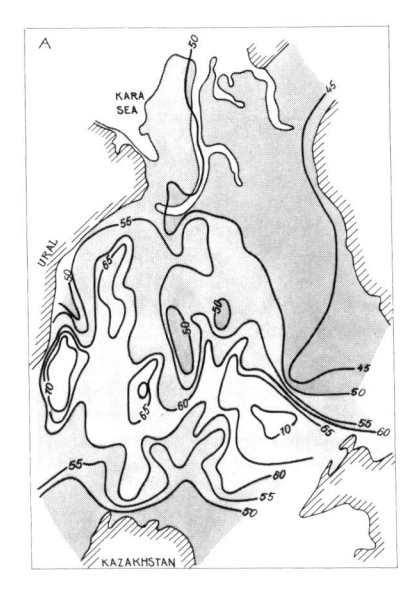

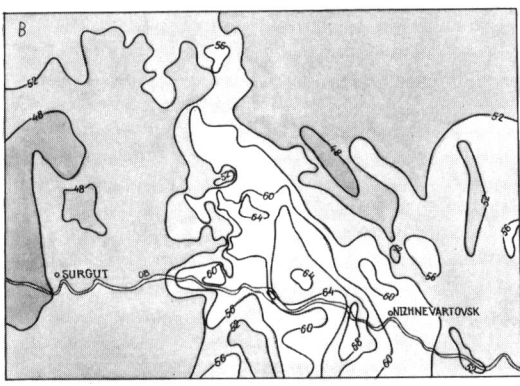

Fig. 6. Heat flow variation (A) north of Kazakhstan and (B) in the Nizhneartovsk region in Western Siberia.

The Lower and Middle Jurassic sediments represented by continental sand/shale rocks occur in the base of the platform mantle in the most of the West Siberian sedimentary basin. The top of these Jurassic sediments is mapped at depth of 2.5–3.0 km and to the north is at depth of up to 4.5–5 km. Locally their thickness reaches 2–2.5 km and do not appear to exceed 200–300 m on the average. In general, local caprocks were developed over the entire strata. In the top locally oil-bearing, variable sand siltstone bed, underlies the regional caprock. These plays low oil production. Toarcian zonal caprock have been identified in the region of Khanty–Mansiysk, with a high production oil pool discovered in the underlying sediments.

In the Polar Circle of West Siberia in the area of Gydan and Yamal Peninsulas; the Lower and Middle Jurassic continental deposits grade into marine facies, where at least three zonal caprocks are expected to control sandy bodies with good reservoir properties. The discovery of high-output oil pools is predicted in these areas.

In Upper Jurassic rocks of marine origin there are two regional horizons whose distribution has been used for reservoir map construction. There are no zonal reservoirs in most of the sediment area deposition. The first regional oil-and-gas bearing horizon in Upper Jurassic marine sediments is confined to the Upper Oxfordian and represented mainly by sandy rocks with greatly variable reservoir properties. The Kimmeridgian Clays serve as caprocks. Regionally the Upper Jurassic horizon (10–50 m thick) contains gas and oil, and is present at depths of 1.5–4 km. To the west along the Surgut–Taz axis sandy deposits are replaced by clays, and along the basement protrusions in Berezovo and Shaim areas Vogulkino sandstones occur with small oil and gas pools that yield up to 300 tons of oil per day and 1.5×10^6 m^3 of gas per day.

Bituminous clay sediments, mainly Volgian (Tithonian), are 5–15 m above the Oxfordian regional reservoir and form a regional oil-and-gas bearing horizon. In West Siberia, for the first time in the history of oil and gas exploration, highly productive oil pools are found in clays. Here the reservoir developed contemporaneously with the formation of oil pools. The main condition controlling the formation of these pools is not organic matter volumes in rock, but its distribution. When organic matter is present in the form of microlenses with density of more than 30–40% of rock volume, then oil and gas pools form at the level of kerogen maturation. Critical energy levels for West Siberia are at least $(200-300) \times 10^{17}$ spins g^{-1} of paramagnetic centres in kerogen. Pools occur in the form of lenticular bodies which can coalesce during organic maturation. In West Siberia there are more than 40 oil pools that have been discovered in bituminous clays. Thickness of Tithonian oil–gas bearing clay horizons is 20–40 m. With increase of the horizon thickness to the margins of the West Siberian sedimentation basin the density of organic microlenses distribution decreases and, correspondingly, the probability of oil pool formation is reduced. In the Salym Field a recovery from bituminous argillaceous rocks was tested but the results were disappointing although flow rates of some wells achieved 6000 m^3 per day. The main reason for ignoring the development of this type of oil play is a traditional approach to their development. When oil was recovered from large depression cones for the purpose of getting large short-time effects the absence of rigid reservoir matrix induced the compression of horizontal supplying fractures, and oil flow was reduced. To restore the initial flow rates wells had to be shut-in for more than one year. These shut-in time intervals were considered to be too long, and producing organizations lost their interest in this type of pools because to fulfil planned oil recovery targets there were enough fields with a traditional type of reservoir and flow rates.

Above the Tithonian argillaceous oil–gas horizon there occurs a series of mainly argillaceous grey-coloured marine sediments of Berriasian–Valanginian–Hauterivian–Barremian and, partially, Aptian age in which number of sandy interbeds increases upwards in the section. The combined total thickness of these horizons reaches 300–500 m and up to 2 km in the northern part. In the lower part of this section local reservoirs are predominant. Zonal reservoirs appear with at least ten of them having a development area of more than $50-100 \times 10^3$ km^2. In the uppermost part of the series, zonal reservoirs again prevail due to widespread caprock development. Along with a vertical regularity of quantitative zonal to local reservoir ratios there is also a horizontal zone. In the western part, zonal reservoirs gradually shift upwards from east to west in the direction of regional replacement of sand bodies with clays, approximately along the meridian line from Tobolsk to Nadym.

In the basement of this Berriasian–Aptian complex there is a sandstone series up to 100 m thick characterized by heterogeneous reservoir properties and lithology. Regionally this series contains oil and gas, but because of its heterogeneity distribution, oil and gas pools and well flow rates vary considerably with values ranging from first hundreds of tons and hundreds of cubic metres. Another peculiarity is characteristic of this series; gorge-shaped erosion zones formed by bottom sea streams parallel to the coastal zone are distinguished in basin. The width of these erosion zones is 2–5 km and their length can be up to 200 km. Studies up to 1st of January 1988 showed ten such zones have been found. Highly productive oil accumulations are expected in these zones.

A thick series mainly of continental Aptian–Albian–Cenomanian sandstones overlies 50–100 metres of clays forming a regional Aptian caprock. At the top is a regional Upper Cretaceous caprock covering an area of 10^6 km^2. Local reservoirs are predominant in Aptian–Cenomanian oil-and-gas bearing horizon, and the largest gas pools are confined to the top of this series. To the west from the latitude of Khanty–Mansiysk, a zonal reservoir appears below the marine Albian sequence. There are several narrow zonal reservoirs along the meridian direction within the Yamal Peninsula and highly productive gas pools appear to be confined to them. The top of Aptian–Cenomanian contains mainly sandy oil-and-gas bearing horizon at depths of 800–1200 m and maximum thickness up to 2000 m.

Almost all of the sandy horizon area (90%) is accumulated in a continental environment as a series of Upper Cretaceous and Paleocene–Eocene marine clay-siliceous rocks with a thickness of 800 m, occurring stratigraphically higher in the section. In the eastern part of the region along the meridian of Nizhnevartovsk–

Tazovskoye, a sandy Turonian–Santonian series (50 m thick) is found where there are local reservoirs and a zonal one. In this horizon, gas pools are encountered at current depths of 400–900 m.

After the Chegan (Tavda) epoch in Middle Eocene a continental mode of sedimentation was established over the whole West Siberia territory. The bottom of these continental rocks lies at depths of 350–400 m. In the north of West Siberia, during the Lower Quaternary period a sea boreal basin advanced sometimes up to 500 km of the modern coast line of Kara sea. In the most part of West Siberia the thickness of Quaternary sediments is not more than 20–30 m, but increases up to 200–300 m in the north and southmost part of the Zaisan trough. Limited gas flows (up to 20 m^3 per day) have been reported at depths of 100–150 m in a zone of perennial-frozen rocks located in the continental Middle Eocene–Holocene sediments.

The problem of deep oil and gas prospecting

Deep drilling at more than 4500 m is well established in hydrocarbon exploration. Below this depth mark there occur Jurassic, Triassic and Palaeozoic platform beds in the region. The problem of oil and gas occurrence, evaluation and probability of risk in oil and gas prospecting at such great depths is dependant on the results of reservoir properties of rocks and the theory of formation of oil and gas pools. During increased burial rocks, especially argillaceous, consolidate and reduction in reservoir properties usually occur. But sand deposits also consolidate. In theory, this process is based on the idea that an increase in both pressure and temperature deteriorates the reservoir properties of sandstones. The influence of these factors on carbonate rocks is not fully evaluated. It is suggested that at high pressures and temperatures, reservoir properties of both terrigenous and carbonate rocks can improve due to leaching. But this is not proved for the natural conditions. By statistical evaluation of reservoir properties of rocks it is shown that they deteriorate with depth and at present we have no scientifically proven alternative to statistical data. In practice each individual case of the location of deep wells must be carefully modelled from the point of view of improvement or deterioration in the reservoir properties of rocks.

From the standpoint of the current theories of formation of oil and gas, the probability of such discoveries at great depths becomes less favourable. Considering temperature and pressure parametres, the latter always appears to be negative while the former contributes to the processes of oil and gas generation. Omitting the theoretical details it is not recommended to carry out deep drilling in search of oil and gas pools without more data and specific modelling.

Principles of local forecast of oil and gas pools

Three main ways of solving this problem have been distinguished. First it is necessary to use direct remote sensing methods for identification of the location of wells to discover oil and gas pools. Direct prospecting reflects information of pools in physical, chemical and other fields, which can be recorded by remote-sensing techniques.

In West Siberia seismic waves reflected from gas–water surfaces are recorded in several gas pools of massive structure. However, currently we cannot record the same reflections from the top of pools. The applicability of this phenomenon to direct prospecting is therefore very poor and the data obtained is more symbolic than practical. With the currently available resolution of geophysical equipment there is however, no other reliable direct method for oil and gas prospecting.

The second method of prospecting for oil and gas pools is to identify geological bodies represented by a structural relief and separate rock lenses by geophysical methods. Proceeding from the regional geological investigations, interpretation of the relation between the boundaries obtained and vertical and horizontal heterogeneity of the geological section caused by alternation of permeable and impermeable rocks is attempted. Drilling of wells for oil and gas prospecting depends on the extent of risk estimated by regional or prospect evaluations. The value of such risk determines the ratio of the number of producing wells to their total number drilled within a certain area or a region. In West Siberia on average, the risk of drilling is considered to be justified when one in three results in discovery of a new oil or gas pool. With the risk increasing to only one in ten, the interest in prospecting within this region decreases. The area of the region and the degree of substantiation of potential hydrocarbon reserves is the decisive factor. In the vast region of the West Siberian basin, approximately 100 wells were drilled before the discovery of the first gas field. Such a high risk was considered to be justifiable, though many specialists who determined the commercial aspects of exploration did not consider the region to be a major hydrocarbon province.

The third method of local forecast of oil and gas bearing potential is based on the study of physical and chemical parametres reflecting the processes of formation and accumulation of oil and gas. Until recently the development of these methods was slow due to the fact that hypotheses of oil and gas origin and formation of oil and gas pools were not fully understood and appreciated.

The most reliable models of formation and accumulation of oil and gas are those that take into account concentrations of paramagnetic centres in kerogen of organic matter. According to these models oil and gas form during fast regional burial of rocks within separate local areas, where pressure and temperature stabilize with different velocities. Favourable conditions form only in those parts of local stratification systems where there is no equilibrium between the speed of formation of new paramagnetic centres in kerogen of organic matter and the velocity of change in temperature and pressure. such conditions occur when temperature increases faster than pressure and absolute concentrations of paramagnetic centres are also significant. In West Siberia the maximum concentrations amounting to 600×10^{17} spin g^{-1} are registered in kerogen of Valanginian and Jurassic deposits of the Middle–Ob area. These sediments are considered to be important source rocks.

When these conditions are available, radical groups with high energies of up to 100 kcal mol^{-1} break off, causing mass breakdown of C–C bonds in the kerogen macromolecules.

The rate of reaction of molecular formation with hydrocarbon chain lengths of 10–35 carbon atoms can increase up to 300°C. This thermal anomaly disperses for 95% in 1.5–2.0 million years and for 98–99% in 20–30 million years. The presence of thermal anomaly results in the decrease of volume density of rocks overlaying oil and gas pools. This change in the density of rocks can be observed by remote-sensing methods based on estimation of acoustic properties of rocks. In West Siberia, acoustic anomalies are distinguished over all the investigated oil and gas pools both in rocks occurring directly above the pools, and in surface rocks. The same physical phenomenon is the basis for local forecast methods of oil and gas pools by interpretation of space photographs and the composition of gases recovered from the rocks in the depth interval 100–300 m. Theoretical grounds for the above-mentioned local forecast of oil and gas bearing potential account for the presence of the anomalies in gas composition of snow cover over hydrocarbon accumulations and beyond their boundaries. The level of theoretical substantiation of local forecast of oil and gas potential in West Siberia allows the use of new methods for oil and gas discovery within other regions of the world, where geological conditions for the formation of hydrocarbon accumulations are similar to the paleoenvironment of West Siberia. In this case economic efficiency is guaranteed due to decrease the extent of risk of investments in oil and gas prospecting even within the regions, where new oil and gas fields are not discovered for the present.

Pakistan: a history of petroleum exploration and future potential

P. DOLAN

Dolar & Associates, Greyhound House, 23–24 George St., Richmond, Surrey TW9 IHY, UK

Historical review of petroleum exploration

The documented history of systematic petroleum exploration in that part of southern Asia now known as Pakistan, see Fig. 1, is here divided into five stages.

(1866–1947) British Companies, pre-Independence

In 1866 some seven or eight holes were dug in and around the seepages near Fatehjang, 25 miles west of Islamabad, yielding a few gallons of oil per day (Pascoe 1920). Then, in 1870, B. S. Lyman, an American from Pennsylvania was hired to investigate the oil prospects of the Punjab. His report is of special interest to a meeting on Classic Petroleum Provinces, since its maps were probably the first ever published on which subsurface structure was delineated by contours, Owen (1975). During the last quarter of the 19th Century there were sporadic attempts to drill shallow boreholes in the northern parts of the Central Fold Belt. This activity culminated in the drilling of thirteen wells at Kattan in the Marri Tribal Territory of northeast Baluchistan near oil seepages and from which 25 000 barrels of oil were produced between 1885 and 1892. The impetus for this drilling was to establish a convenient source of fuel for the railway system which the then Indian Government was constructing to help secure the Afghan/Indian border.

Immediately prior to World War I, various investment syndicates (mainly British financed) were formed to explore in the Punjab and it is from these origins that the first significant company was established on 1 December 1913 to acquire extensive exploration rights; The Attock Oil Company.

In 1915, oil was discovered in quantity at Khaur by Attock and oil exploration commenced in earnest. Burmah Oil Company then undertook geological surveys in the Punjab, thereby broadening its exploration interests outside Burma, where it had been successfully exploring for 30 years, in contrast to Attock which was established with the express intent of exploring only within the Punjab. From 1916 to 1923, Attock, Burmah and a newcomer, Whitehall Petroleum Corporation undertook geological reconnaissance work and drilled several dry holes.

Between 1923 and 1928, geophysical methods were applied for the first time in Pakistan for investigating subsurface structure. This was at Khairpur, immediately south of Sukkur, where Burmah commissioned a torsion balance survey and simultaneous magnetic observations. The resultant Khairpur-1 well of 1925 was a dry hole (although subsequent deeper drilling has established a significant gas accumulation). From 1927 to 1939, gravity surveys were extended to several areas throughout the current area of Pakistan. Early in 1931 Attock made a torsion balance survey near Khaur and carried out some preliminary trials with a seismograph, see Fig. 2.

Between 1937 and 1939, the Lower Sind area was photographed from the air by the Indian Air Survey and Transport Company for Burmah. A geological reconnaissance map with the help of the air photographs was made in 1939 and most of the prominent structures were licensed by Burmah who also planned to conduct extensive seismic surveys in alluvium covered areas. However, this work was forestalled at the start of World War II. The exigences of war conditions resulted in limited exploration effort being focussed on the Potwar area, and the discovery of the Joya Mair Field (1944) and Balkassar Field (1946).

(1947–1961) Post-Independence, British and American companies explore under new regulations

In 1947 Pakistan assumed independence as an Islamic state with two 'wings', West and East. Following the previous Potwar successes, several companies started limited seismic exploration.

The Pakistan Petroleum (Production) Rules were promulgated in 1949, by virtue of which oil prospecting became a subject for the Central Government rather than a provincial matter as in pre-Independence years. Under these new rules concessions could be granted only to companies incorporated in Pakistan. Consequently, Attock transferred its activities to a new local subsidiary, Pakistan Oil Fields Limited (POL), while Burmah formed Pakistan Petroleum Limited (PPL). Both new companies had 30% of the equity held by the Government. Soon afterwards, in 1951, gas was discovered at Sui by PPL. Also in 1951, an agreement was made under the post World War II Colombo Plan for Canada to provide Pakistan with technical assistance. Between 1952 and 1956 aerial survey work and extensive geological field mapping was conducted. This and other work resulted in a major geological report published by the Canadian Government in 1961.

In 1954 the American company Standard Vacuum became the first non-British company to register itself in Pakistan for oil exploration in the post-War period. Standard Vacuum conducted the first aeromagnetic survey in Pakistan covering the Indus Plain.

(1961–1968) Oil & Gas Development Corporation (OGDC) assisted by Soviet technicians

To maintain exploration momentum, the Pakistan Government negotiated a five-year loan (about $US 35 million of the day) from the Soviet Government and established the Oil & Gas Development Corporation (OGDC) in September 1961. Soviet experience and equipment were mobilized to systematically study the whole country. Aerial reconnaissance and detailed aeromagnetic work was undertaken while six geological surface parties, three gravity parties and six seismic parties were mobilized. The resultant OGDC drilling program led to a small, but in retrospect significant, gas discovery at Sari Sing in 1966, the first in the south of the country. In 1968, after OGDC discovered the Toot (Tut) and Kot Sarang oil fields, and POL discovered oil at Meyal-1, several foreign companies began to show interest in the potential of Pakistan.

(1969–1976) Private foreign companies resume exploration

In the aftermath of the war between Pakistan and India, West Pakistan and East Pakistan became the independent states of Pakistan and Bangladesh respectively. Exploration activity continued in the producing regions of Pakistan but also began to impinge more on southern coastal areas, but without success.

(1976–Present) Improved economics and technology prove new exploration plays

In 1976 OGDC announced that the Dhodak-1 well had discovered 4 TCF of gas and 200 (subsequently revised to 34) million barrels of condensate. This was the first report of significant liquid petroleum outside the Potwar region. Then, with modified petroleum regulations, several companies concluded joint venture agreements with OGDC. Most significantly, Union Texas and partners acquired acreage in what is now the productive area southeast of Hyderabad, the Badin Block. Thereafter, the pace of activity and production accelerated (see Figs 3 & 4). In 1981 Union Texas made a significant oil discovery at Khaskeli, thereby disproving the long-held concept that southern Pakistan was principally a gas bearing region. Despite this successes at Khaskeli and numerous satellite structures, as well as renewed offshore seismic activity in 1982 (see Fig. 2) it is noteworthy that several large operators closed down operations. While activity continued to be dominated by Union Texas in the Indus Delta region, Occidental reported the most prolific oil discovery yet, at Dhurnal-1 in the Potwar region, in 1984.

In late 1987, industry interest in Pakistan increased significantly and there were numerous licence applications pending throughout much of the Lower and Upper Indus Basins.

Geological structure

Current structure

Remote sensing and surface data. Aerial photography has been routinely used in Pakistan for cartography since the 1930s. However, these data are generally difficult to obtain, primarily for security reasons. This is not a major hindrance to exploration, since the satellite imagery that can be obtained for Pakistan is of excellent quality and the inherent nature of the terrain yields spectacular results.

At a more detailed level there is, potentially, a wealth of surface data to be considered, particularly from within the Central Fold Belt, see Fig. 5. Gathering these data commenced in the second half of the 19th Century under the auspices of the Geological Survey of India. This work was initially of a strategic survey nature and was restricted mainly to the Potwar area. Surface mapping in other parts of the country was restricted by political instability and security problems in the tribal areas. It was not until

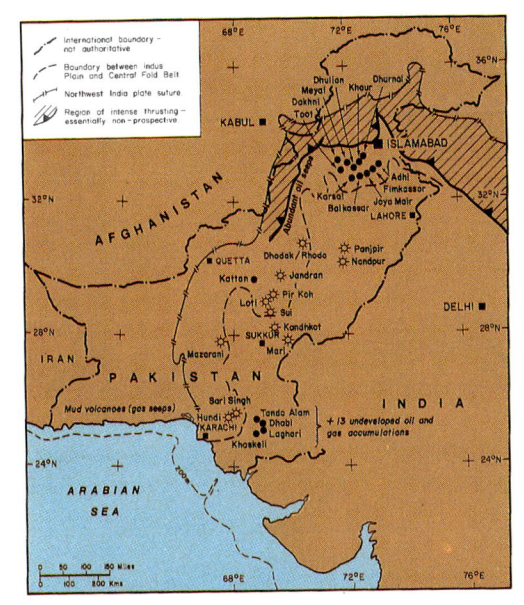

Fig. 1. Location map.

1953 that extensive geological field mapping was undertaken with Canadian technical assistance as part of the post-World War II Colombo Plan.

The geomorphology of Pakistan can be conveniently divided into four areas of discrete surface geology; in broad informal geographic terms they are the Indus Plain, Central Fold Belt, Baluchistan and Northern Mountains, see Fig. 5.

The progressive incision of the Plain by the Indus River and the westerly migration of the main course in the region south of latitude 27° North (about 80 km in 5000 years) are clearly evident. It is interesting to note that the same situation, in a mirror image sense, occurs in Bangladesh where the Ganges River migrates eastwards away from the Indian craton. Snelgrove (1979) described these shifts as a side effect of the northward drift of the Indian plate.

The Central Fold Belt has a conspicious sinusoidal trace which reflects the interaction of the Indian plate and Eurasian 'blocks' as they collided and became sutured during the Early Tertiary.

Along the trace of the Central Fold Belt there are two distinct elements, those that are narrow (generally north/south trending) and those which are broader and pendulous in appearance (generally east/west trending). It is implicit in most published literature that the sinusoidal shape owes its origins to old established promontories and re-entries along the northwestern edge of the Indian plate which left their imprint on the suture zone. Such a model may be correct in whole, or in part, but is the irregular shape of the suture zone dictated by elements exclusive to the Indian craton? There are two other possible mechanisms.

First, similar old established elements may exist on the Afghanistan, Turan and Tibetan cratonic blocks. Secondly, the suture is unlikely to be an unmodified expression of the pre-collision boundaries. New satellite image data suggest that there has been far-reaching transcurrent structural modification during suturing along lines of pre-existing cratonic fractures.

Baluchistan is regarded as unlikely to have anything other than the most speculative long-term potential, and is not considered further here.

The Northern Mountains of Pakistan are a series of complex structural belts with still poorly understood surface geology. For example, in the 'benchmark' publication (Farah & De Jong 1979), most authors who addressed the geology of this area presented more than one model for the evolution of the 'Indian Eurasia Suture Zone'. For further details of the fault blocks, thrust sheets and granite plutons, reference should be made, amongst others to Le Fort (1975), Desio (1979), Tahirkheli *et al.* (1979) and Coward *et al.* (1982).

Magnetic Data: Quadri and Shuaib (1986) have described aeromagnetic surveys conducted by Standard Vacuum in 1955 and by OGDC in 1962 and 1963 in the Lower Indus Basin. It is worthy of note that some of the current reconnaissance permit applications probably include proposals to acquire comprehensive, aeromagnetic data. The existing data are somewhat fragmentary.

Perhaps the most significant features of the currently available magnetic data are inferred northwest-southeast lineaments, see Fig. 6. A good example of this is the lineament evident in Fig. 3 of Quadri & Shuaib (1986) which extends northwestwards directly into the one major offset of the Indus River course. The significance of the northwest–southeast lineaments is that they may be basement related offsets which affect the structural style (and hence deposition patterns) of the overlying sedimentary fill, and may control petroleum migration paths (Dolan *et al.* 1987).

Gravity Data (Bouguer). Considerable amounts of gravity data are available across the Indus

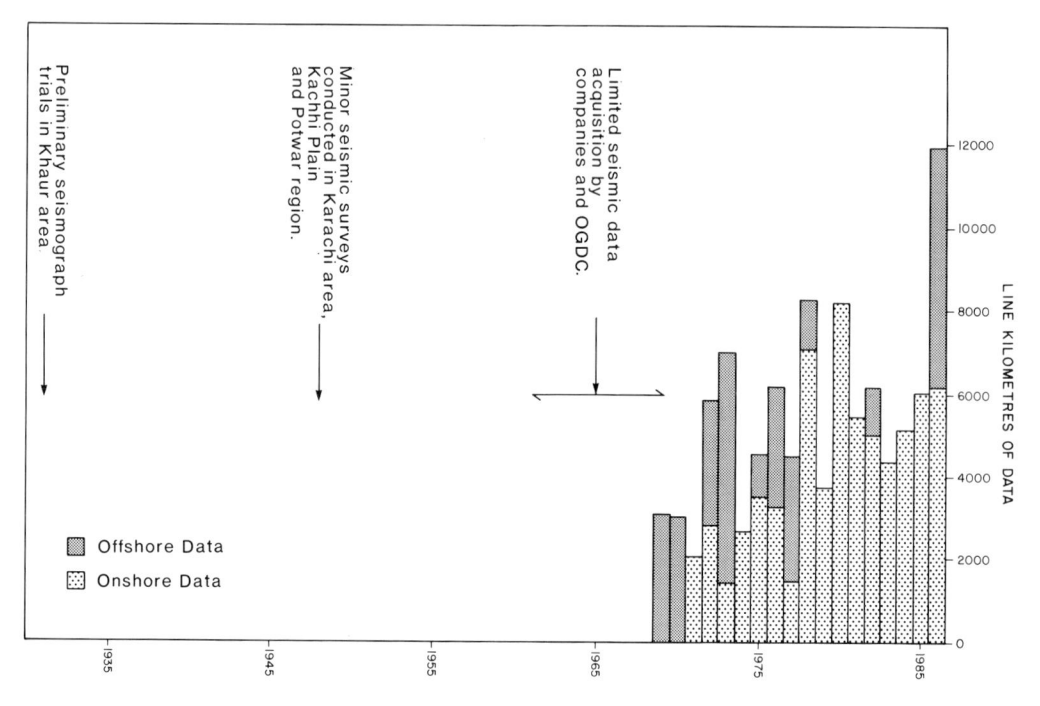

Fig. 2. Seismic data, history of activity.

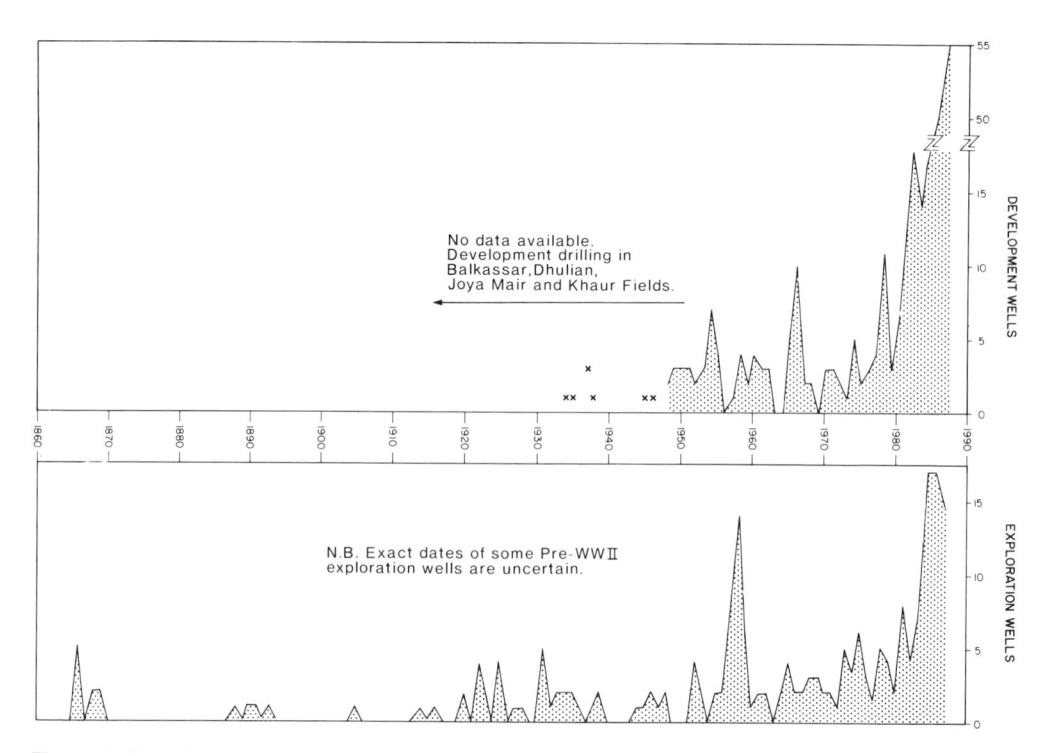

Fig. 3. Drilling, history of activity.

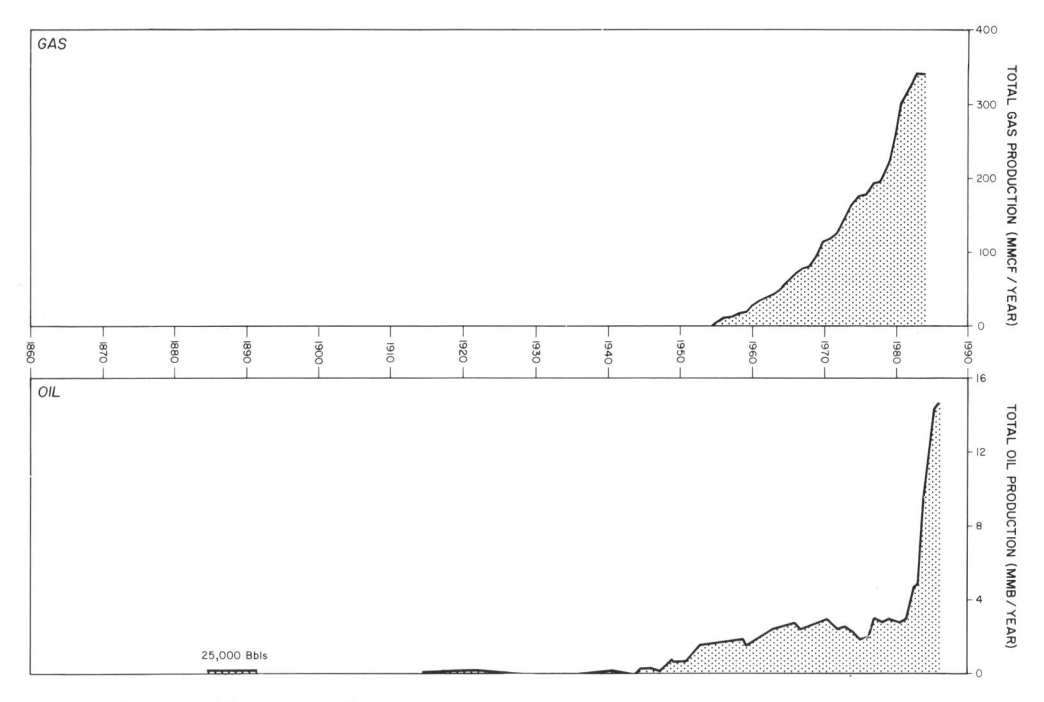

Fig. 4. Oil and gas, history of production.

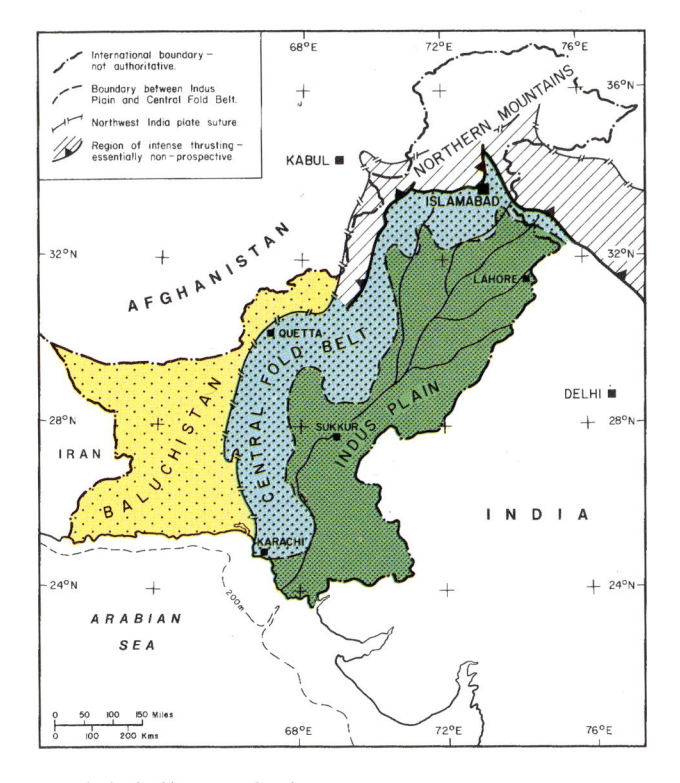

Fig. 5. Principal geomorphological/structural units.

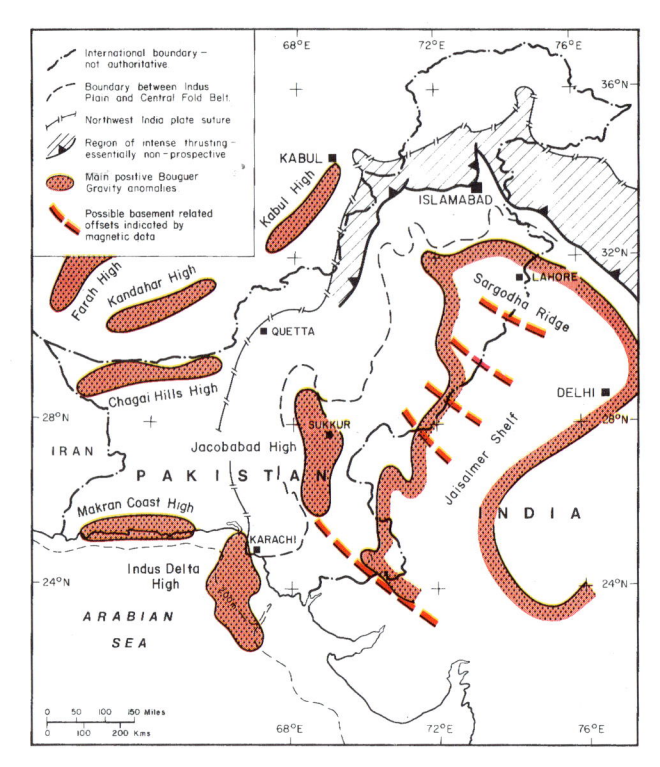

Fig. 6. Principal gravity and magnetic anomalies.

Plain which show the most conspicuous feature to be the Jaisalmer Shelf along the Pakistan/Indian border. This shelf has a major promontory (the Sargodha Ridge) at its northern end which re-inforces the northwest–southeast trend described above. The Jaisalmer Shelf dips off to the west with significant negative anomalies being created by the thick Lower Indus Basin sediment fill. The one major exception is the Jacobabad High which is an anomalous and poorly understood feature. Further west, in the Central Fold Belt, Baluchistan and Afghanistan, data control is sparse and the indicated anomalies should be regarded as poorly defined.

Seismic Data. Relatively large amounts of seismic data exist (see Fig. 2) and are known to confirm the occurrence of thick sedimentary sequences throughout much of Pakistan.

Structural Evolution

The gross basin development is considered in the context of plate tectonic evolution, see Figs 7 & 8, prior to considering the stratigraphy and paleogeography of individual chronostratigraphic units.

Punjab Basin. The Phanerozoic history of this basin commenced with an Interior Sag sedimentary cycle above an igneous and metamorphic basement. The absence of basal conglomerates in wells drilled to date, the presence of evaporitic sequences and the lack of angular discordance at the unconformity all suggest that the basement surface has a very low relief and represents the mature peneplaned surface of Pangaea.

During Early Cambrian times the basin downwarp resulted in deposition of a major sequence of continental red-beds. Within this sequence is a series of dolomites and evaporites and another of glauconitic sandstones. These two series are indicative of ephemeral marine incursions, probably from the east.

The geological history of Pakistan is usually described as having a lengthy period (about 250 million years) of erosion and non-deposition, lasting from the Late Cambrian to Permo-Carboniferous times. However, to the west in Iran and Afghanistan, Paleozoic sequences do occur, and it is conceivable that similar

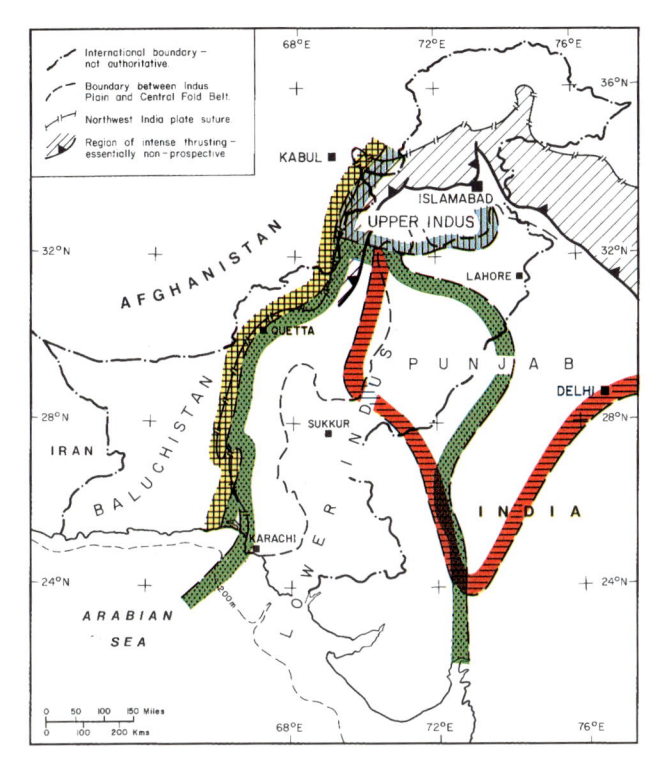

Fig. 7. Pakistan and NW India, basin outlines.

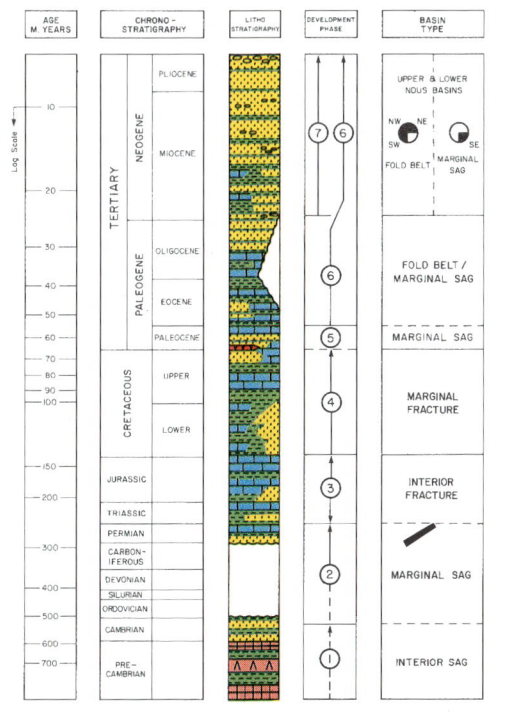

Fig. 8. Basin development.

sequences could exist in Pakistan, particularly in the Upper Indus Basin.

If present in Pakistan, they would probably be the basal units of a Marginal Sag (Phase 2) basin which succeeded the initial Interior Sag (Phase 1) sediments as the 'northern' flank of Pangaea experienced intermittent marine transgressions. Within the Indus Basins, the oldest recorded Marginal Sag basin sediments are periglacial and non-marine clastics of Early Permian age.

Upper and Lower Indus Basins. In Late Permian times, there appears to have been the onset of thermal bulging in the east and south with associated crustal flexure and platform tilt to the west. These movements heralded the beginning of an Interior Fracture basin (Phase 3). Rifting continued to develop, as part of the greater Triassic fragmentation of Gondwana, and a marine incursion became well established, first in the Upper Indus Basin and then, probably, in the northern parts of the Lower Indus Basin.

The Jurassic sequences of the Interior Fracture basin are characterized by an easterly belt of clastic dominated marginal marine sediments, a central zone of pelagic deep-water sediments and a westerly belt of high energy shallow water carbonates. In Early Cretaceous times a regressive phase commenced, as indicated by Lower Cretaceous deltaic facies in the Sind area of the Lower Indus Basin.

These Mesozoic sequences remained undisturbed on the continental margin. To the west, in the Central Fold Belt, volcanic facies developed in Late Jurassic/Early Cretaceous times with associated disturbance of the host sedimentary sequences. This activity was probably related to the pending sea-floor spreading between India and Madagascar which heralded the beginning of another phase of structural development, the Marginal Fracture basin (Phase 4).

Cretaceous rifting between India and Madagascar led to the development of a basin with normal faulting and attendant variations in the thickness of syn-rift sediments. Wrench faulting also probably occurred at this time as the Indian craton moved tangentially away from Africa. If so, then 'inversion' structures may be present within the Mesozoic section.

The Late Cretaceous stage of Marginal Fracture basin development was accompanied by the extensive development of flood basalts; the Decan Trap of northwest India, which extended into the Sind area of southeast Pakistan.

Once the Indian craton was clearly detached from the Gondwana continent a Marginal Sag basin (Phase 5) developed. This basin was the site of widespread marine carbonate deposition which suggests a relatively passive margin regime. However, in the Lower Indus Basin over 4500 feet of Paleocene sandstones are developed, indicating local uplift and erosion of the now northwardly mirgrating Indian craton.

The Marginal Fracture and Marginal Sag basins (Phases 4 and 5) exhibit one particularly anomalous tectono-stratigraphic feature, namely intermittent westerly sediment sources and associated shallow marine environments. This apparent enigma can be explained by invoking one of two mechanisms. First, there may have been micro-continental slivers (allochthonous terranes) spalled from Gondwana ahead of the Indian craton and subsequently caught up in the collision melange developing between India and Eurasia. Secondly, the northwestern margin of India (now partially subducted) may have been isostatically elevated, perhaps by passage across on oceanic thermal anomaly.

Whatever explanation is correct, the Early Eocene was when the penultimate Fold Belt/ Marginal Sag basin (Phase 6) stage of basin development commenced. In the north, there was segmentation of the Mesozoic basin into discrete depocentres which became progressively smaller throughout the Eocene and Oligocene.

A similar depositional style prevailed in the Lower Indus Basin, though sedimentation during the Eocene and Oligocene was restricted to a basin which tapered to the north. Only in the southern part of the Lower Indus Basin did the Marginal Sag regime persist throughout the Paleogene undistributed by the Central Fold Belt. The locus of this essentially quiescent basin migrated progressively further south throughout the Neogene until the present, being restricted to what is now the Indus Delta and Indus Fan.

Elsewhere in Pakistan, the Fold Belt/ Marginal Sag basin (Phase 6) stage ended at the end of the Paleogene when the full impact of the collision between the Indian craton and Eurasian blocks was felt. This collision probably rejuvenated old, intra-basement, fractures and resulted in wrench movements. At this time there was widespread uplift, particularly in the north and east, as the Indian cratonic wedge was driven beneath the Eurasian blocks. Deposition was then almost exclusively terrestrial in a Fold Belt basin (Phase 7). Marine sedimentation was restricted to a narrow, rapidly subsiding trough, in the Kirthar area. Sediment supply exceeded the rate of subsidence within the new foreland basin of the foldbelt and the marine conditions retreated from the Kirthar area to the present coast line.

Baluchistan. This basin, as defined in Fig. 7, has a geological history which is both very different and, for long periods, geographically remote from the rest of Pakistan. Indeed, in most plate tectonic reconstructions the area is considered not to have existed prior to the Early Jurassic. The oldest known stratigraphic record is of Early Cretaceous age (see Fig. 9). The geological history is poorly understood for several reasons including, its structural complexity, a preponderance of Cainozoic and younger deposits at the surface, a lack of obvious natural resource potential and its geographical isolation. However, micro-continental slivers could be involved in the history, see Fig. 10, and hence the Baluchistan area may not be entirely a 'young' development in a formerly oceanic area.

There is, for instance, the Bajgan/Dur-Kan Zone in Iran, a very narrow sliver of continental crust which has remnants of Carboniferous, Permian and Triassic shelf limestones. This is well documented by McCall & Kidd (1982), who include it in their plate tectonic synthesis for southeastern Iran and speculate on its extension into the Makran region of Pakistan. If so, the geological development of the area may

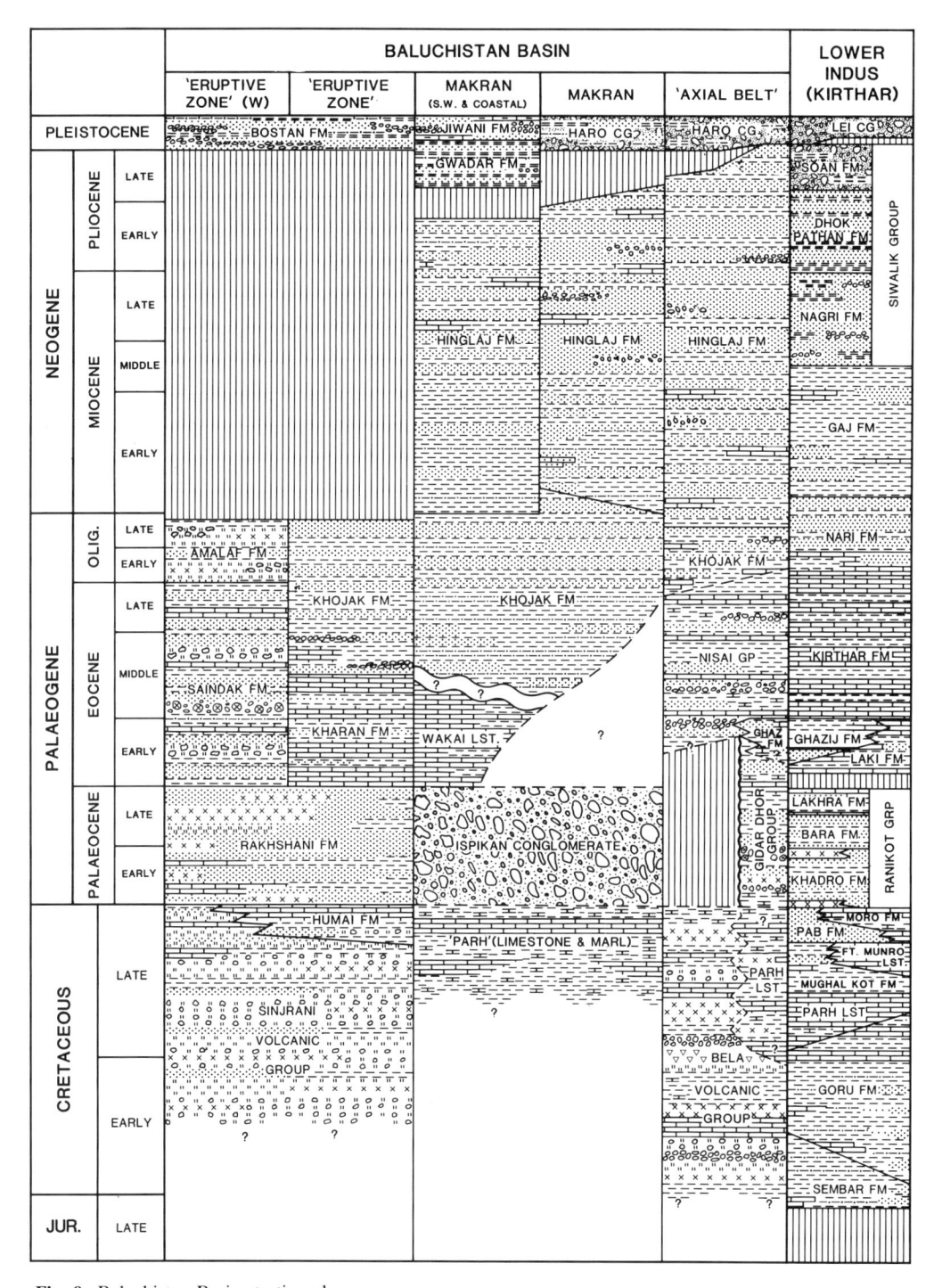

Fig. 9. Baluchistan Basin stratigraphy.

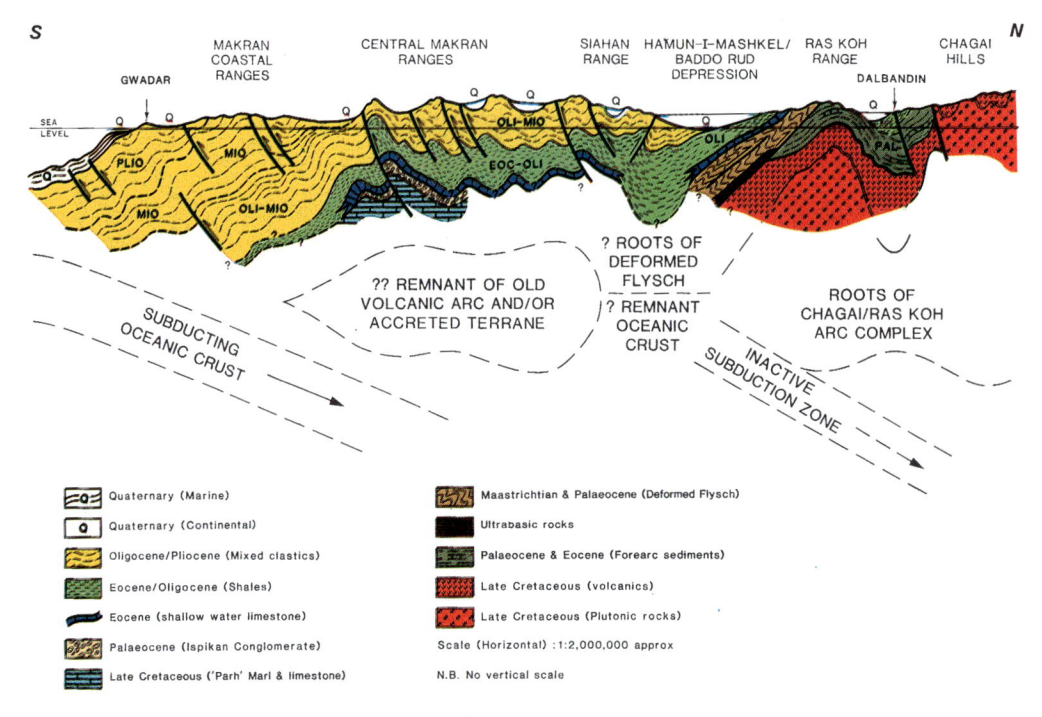

Fig. 10. Baluchistan Basin, structural cross-section.

well have been similar to the rest of Pakistan up to the end of the Triassic.

However, it is the subsequent episodes of island arc volcanism and plutonism which make any petroleum plays associated with a favourable Paleozoic/Mesozoic history both highly speculative and long term.

Petroleum stratigraphy and paleogeography

The generalized petroleum stratigraphy of Pakistan, as illustrated in Fig. 11, is a synthesis of over 330 terms which have been used from time to time.

The first commercial accumulations found were the fractured carbonate oil reservoirs of the Paleogene in the Upper Indus Basin, followed by the Paleogene accumulations of the Lower Indus Basin. As drilling was extended to deeper targets the Mesozoic and Paleozoic mainly clastic, reservoirs of the Upper Indus and Punjab Basins were exploited and finally the Cretaceous clastic reservoirs of the Lower Indus Basin. It is evident from Fig. 11, that petroleum prospects can be envisaged at most stratigraphic levels over wide geographical areas.

Early Paleozoic

The Late Precambrian (Infra-Cambrian) and Cambrian are known from a narrow outcrop belt in the Salt Range and Trans-Indus Ranges (which form the southern margin of the Fold Belt in the Potwar and Kohat areas), and in the subsurface from a number of wells. Most of the wells are in the Potwar area, but a few are more remote to the south. These have established that the Lower Paleozoic probably underlies most of the Punjab Basin, attaining maximum thicknesses of several thousand feet. The eastern Salt Range outcrops are the type occurrences and exhibit a basic twofold division into a lower evaporitic part, the Salt Range Formation and an upper clastic part, the Jhelum Group.

The Punjab Basin (see Fig. 12) was a large desiccating basin with marine replenishment and increasing clastic input. The occurrence of oil shales within the sequence suggests restricted conditions during periods of reduced desiccation and enhanced biological activity. The basin form is speculative to the west, but is controlled to the east by the Vindhyan Group and associated sediments of northwest India.

The known reservoir rocks are sandstones

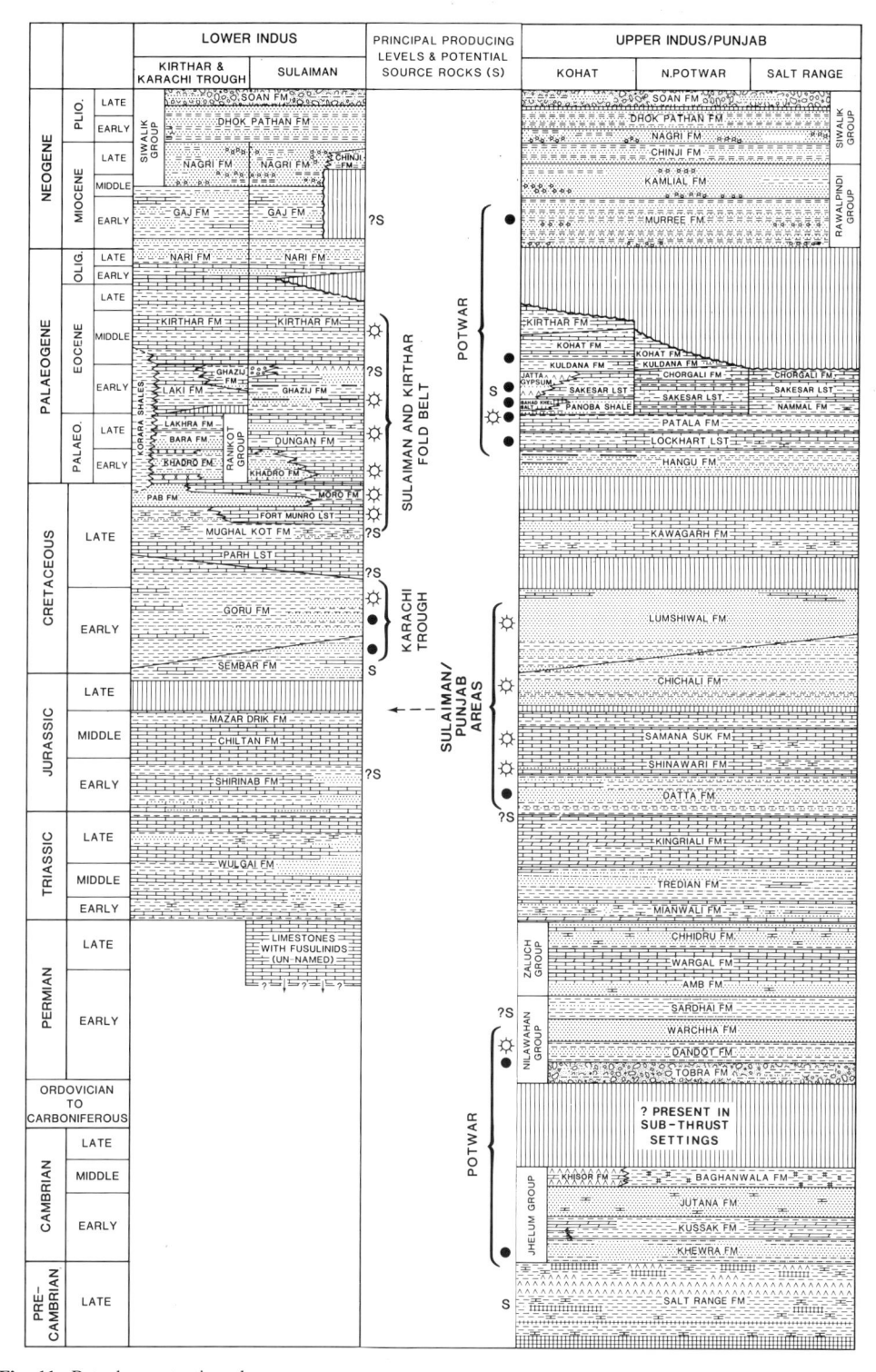

Fig. 11. Petroleum stratigraphy.

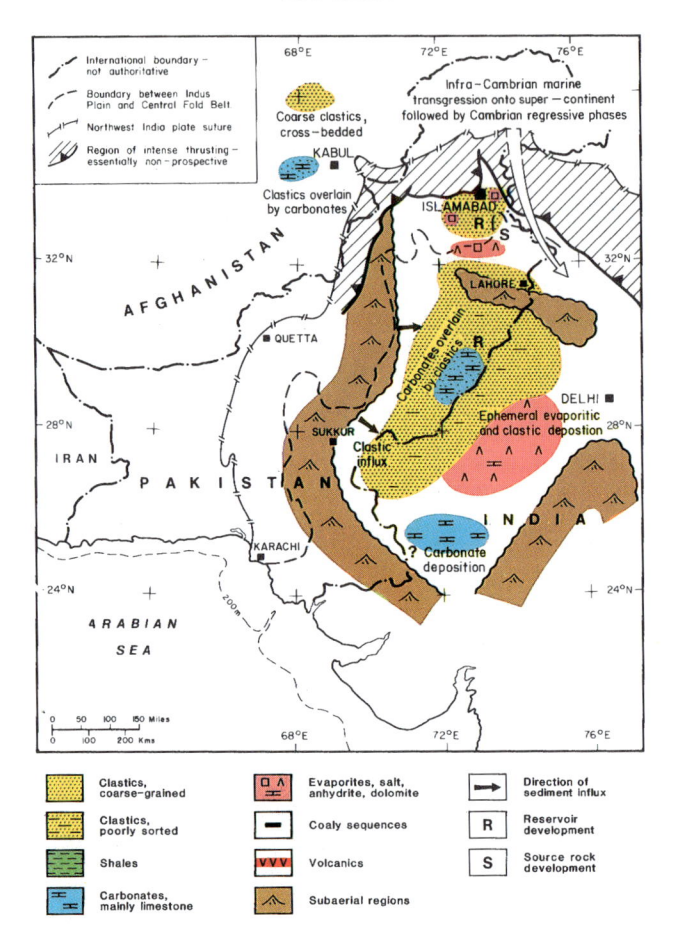

Fig. 12. Early Paleozoic depositional patterns.

which have been productive in the Potwar area. Further to the south, their potential is probably contingent upon a favourable juxtaposition with Mesozoic source rocks.

Ordovician to Carboniferous

Apart from some enigmatic, unfossiliferous dolomites in neighbouring northwest India, there are no sequences suspected to be of this age in the Punjab Basin south of the Main Boundary Thrust. However, to the north of this line (the thrust in Fig. 12) there are the Lower Palaeozoic Attock Slates as well as Silurian to Devonian phyllites, quartzites and crinoidal limestones. In Kashmir, the Syringothris Limestone and Fenestella Shales are confidently assigned a Carboniferous age.

Hence the areas south of the Main Boundary Thrust were in reasonably close proximity to prolonged marine sedimentation, and in a subthrust setting, marginal marine deposits may be present. The potential for source rocks and reservoir development are intriguing if the peripheral elements of the Gondwana jigsaw, inter alia South Tibet, North Tibet, Turan and Tarim Massifs were already separated by rifts or small oceanic basins.

Early Permian

The control points of the Early Permian are similar to those of the Lower Palaeozoic. Deposits of this age are referred to the Nilawahan Group, and typically comprise 500 to 1000 feet of clastic section.

The basin orientation in Early Permian times appears to still have been essentially south to north, as in Cambrian times. However, control points along the eastern flank of the Potwar area and at the Adampur-1 well in India suggest that there was no longer an extension of the

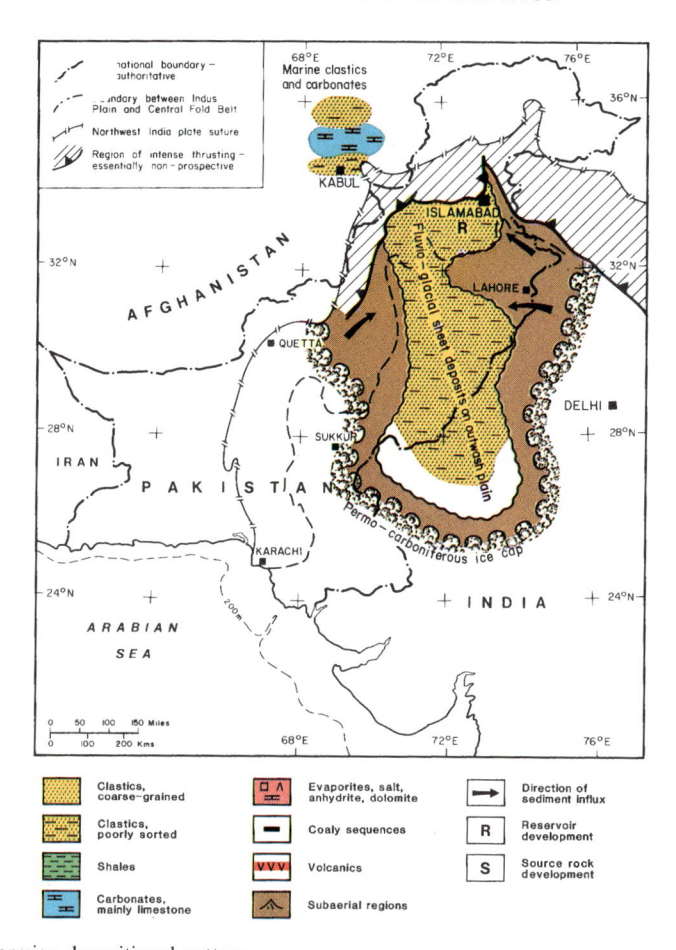

Fig. 13. Early Permian depositional patterns.

basin to the east. At the Pugal-1 well, Krishnan (1982) reported '...100 m of shales which are unfossiliferous but contain pollen indicating Permian age'.

Throughout much of the preceding Silurian, to Carboniferous times, Pakistan was at high southerly latitudes (typically 50° to 60° South) and was, for part of the time, under the influence of glacial conditions (see Fig. 13). North of the present day latitude 28° North there are clastic sequences of Early Permian age which commence with a basal tillite, the Tobra Formation. This sequence is followed by a series of three formations, the Dandot, Warchha and Sardhai (shale, sandstone, shale) which are remarkably uniform in development.

The Tobra Formation is reported to be oil productive in the Potwar area with probably fair to good sandstone reservoir characteristics. The oil source is presumed to be Infra-Cambrian shales, although the Sardhai Formation is known to contain some black shales which may be of interest.

Late Permian/Triassic

At the beginning of Late Permian times there was a distinct shift of the easterly depositional edge towards the west with the basin axis also moving westwards due to regional tilting and incipient rifting (see Fig. 14).

The northeastern margin of deposition is controlled by numerous wells, however, to the south only one well is reported to have encountered the Triassic and even this is a contentious red-bed sequence. To the west, the Upper and Lower Indus Basin margins are also debatable. In the region of Quetta there are dark limestones of possible Late Permian age (Shah 1978). The maximum thickness of Upper Permian and Triassic sediments is unknown but, from evidence in neighbouring Iran and

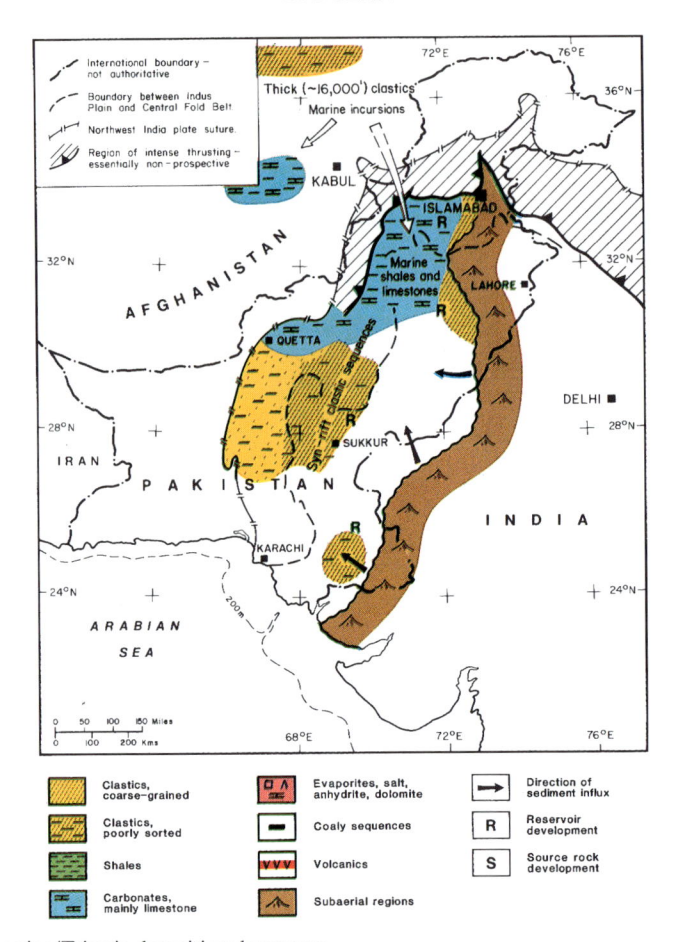

Fig. 14. Late Permian/Triassic depositional patterns.

northern Afghanistan, could attain 5000 to 10 000 feet of syn-rift coarse clastics with reservoir potential.

Evidence of periodic marine incursions in northwestern Pakistan raises the possibility of limited source rock development. Coals have been described within the Triassic of one well west of the Potwar which also indicates some petroleum source potential.

Jurassic

Carbonate sediments predominate in both the Upper and Lower Indus Basins, see Figure 11, but are more ubiquitous in the latter. In the Upper Indus Basin, the Lower Jurassic Datta Formation comprises mainly continental sands, whereas the Chichalli Formation, which follows intervening marine carbonates, is a marine sand. These different sequences combine to form an easterly marginal marine facies, mainly clastic, grading into massive marine carbonates and shales, which in turn give way to a westerly oolitic and shelly carbonate facies, see Fig. 15. Isopach values for the Jurassic sequences typically total 5000 to 7500 feet.

The easterly facies contain reservoir quality sands and carbonates which are oil and gas bearing. Further west, little is known of reservoir quality, though fractured reservoirs might reasonably be expected.

Broad paleogeographic considerations suggest that west of the Potwar area, Jurassic source rocks may be developed. In the Lower Indus Basin, source rocks may be widely developed. It should be noted that between the scattered easterly well control and the outcrop control of the western Central Fold Belt, there is very little subsurface control. Although probably insignificant, Jurassic coals in the northeastern part of the Lower Indus Basin may be a minor source facies.

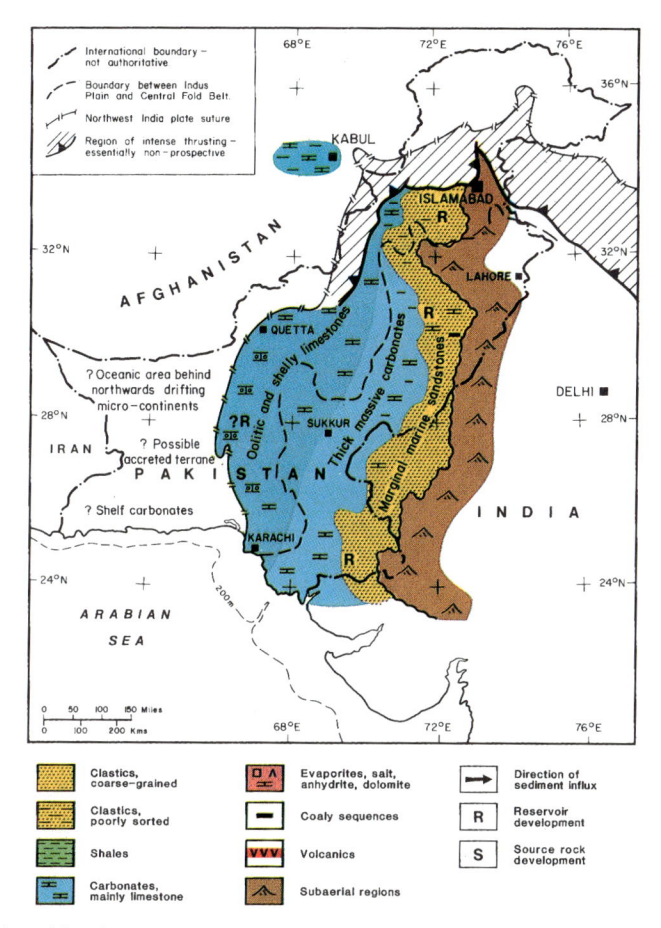

Fig. 15. Jurassic depositional patterns.

Early Cretaceous

The sediments of Early Cretaceous age have a distribution throughout the Lower Indus Basin comparable to that of the preceding Jurassic sediments (see Fig. 16). However, in the Potwar area of the Upper Indus Basin, they are likely to be restricted mainly to the westerly Kohat area; this is probably due to erosion rather than non-deposition. Within the Lower Indus Basin the Lower Cretaceous sediments can, like the preceding Jurassic sequences, be divided into an easterly sandstone dominated zone and a westerly area of mixed fine grained clastics and interbedded limestones. The easterly marginal marine/deltaic facies attain a maximum thickness of about 10 000 feet while the more condensed westerly sequences are typically 1000 to 3000 feet.

The thick sandstone facies of the Goru Formation is developed above the Sembar Formation shales which are a confirmed mature source rock. The Goru Formation sandstones can be expected to extend northwards along strike, as they do to the south into the Kutch area of India, Richter-Bernburg and Schott (1963). Equally interesting, is the possibility of distal members of the sandstone prograding down-slope into the western parts of the Lower Indus Basin, the sediment transport being controlled by pre-existing structural lineaments.

Late Cretaceous

Upper Cretaceous sequences are absent in the Potwar area of the Upper Indus Basin, but may exist to the west in the Kohat area. By contrast, they are widely developed in the Lower Indus Basin. Whereas this basin was asymmetric in Early Cretaceous times, it appears to have developed a central axis during the Late Cretaceous. This apparently resulted in thin

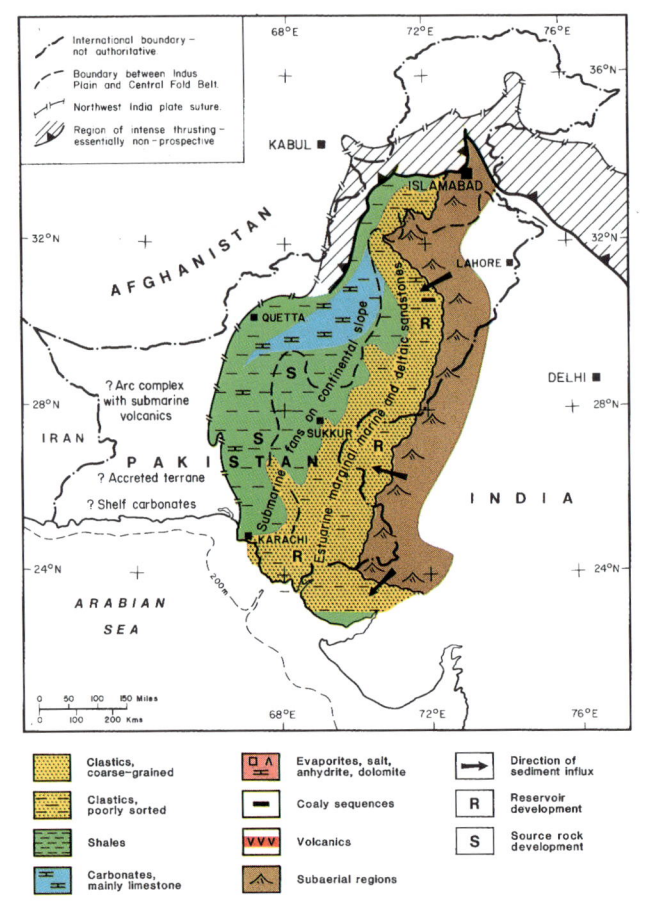

Fig. 16. Early Cretaceous depositional patterns.

marginal marine sandstones in the east and similar thicker sequences in the west, combined with shallow water foraminiferal limestones (see Fig. 17). The axial part of the basin contains deep water shales and limestones, probably sandy in the upper levels.

The easterly Lumshiwal Formation sandstones are gas bearing as are the westerly Pab Formation sandstones throughout much of the Central Fold Belt. The gross reservoir sequences are thought to be typically a few hundred feet thick, but the Upper Cretaceous sequences probably have a cumulative thickness of several thousand feet.

Quantitative data regarding the source potential of Upper Cretaceous sequences are unavailable. However, oil seeps from the Mughal Kot Formation in the northern Central Fold Belt and the presence of several other shale sequences suggest that petroleum may be sourced from this part of the succession.

Paleocene

The Paleocene is well known from many outcrops and from the great majority of wells drilled to date. It was a period of transition from essentially clastic to carbonate deposition and as a result the highly variable lateral facies variations have led to a complex and often unsatisfactory stratigraphic nomenclature. Throughout most of the Lower Indus Basin the Paleocene succeeds the Cretaceous Pab or Moro Formations conformably or after a short hiatus, whereas in the Upper Indus Basin there is a very strong Base Tertiary unconformity, with Paleocene resting directly on rocks as old as the Cambrian.

In the Upper Indus Basin the Paleocene comprises the Hangu Formation, sandstones and shales with some thin coals; the Lockhart Limestone, a massive limestone with minor marls and shales; and the Patala Formation,

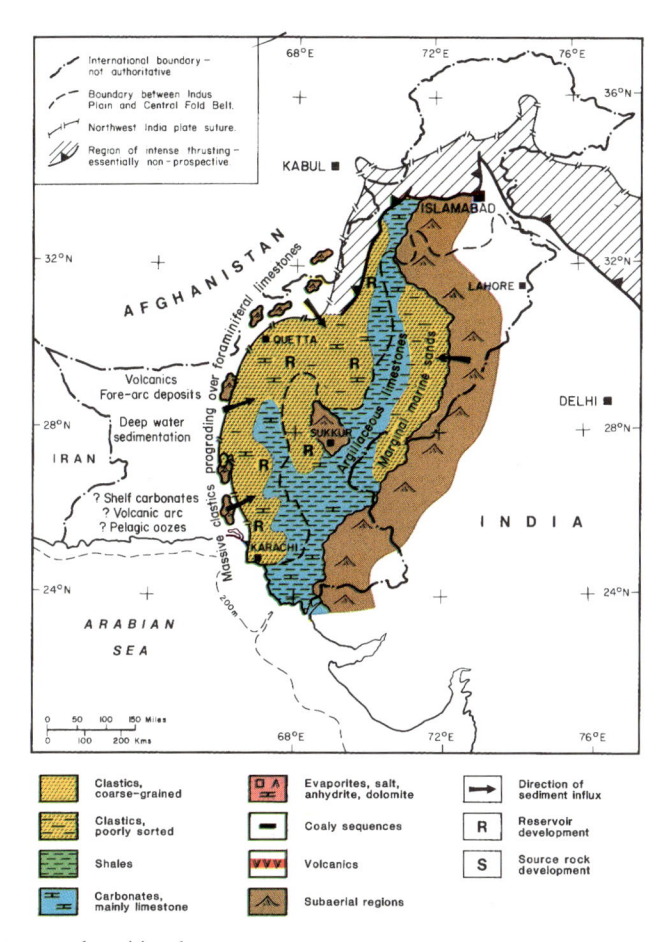

Fig. 17. Late Cretaceous depositional patterns.

shales and marls with minor limestone and sandstone (see Fig. 11). In the Lower Indus Basin the succession is more complicated because different formations are recognised in two sub-basins, the Sulaiman and the Kirthar, albeit with some overlap.

In the Kirthar sub-basin, and including the Karachi Trough, the Paleocene is normally represented by the constituent formations of the Ranikot Group. These are the Khadro Formation, sandstones and shales with varying amounts (sometimes dominant) of volcanic material; the Bara Formation, similar but with sandstones, sometimes calcareous, carbonaceous shales and coals; and the Lakhra Formation, limestone with calcareous sandstone and shale interbeds. The emplacement of the extensive Deccan Trap Volcanics at the beginning of the Paleocene in northwest India affected only the southern part of the Lower Indus Basin (see Fig. 18). In the western part of the Karachi Trough there is an area where the whole of the Paleocene is represented by a monotonous shale section, the Korara Shales. This development also laterally replaces formations up to Middle Eocene in age, and possibly also some of the uppermost Cretaceous Pab Sandstone.

In the Sulaiman sub-basin the bulk of the Paleocene is represented by the Dungan Formation, generally a massive limestone with only minor shale and sandstone interbeds. In a few localities the Dungan is underlain by sandy units attributable to the Khadro and/or Bara Formations.

In the Upper Indus Basin, sedimentation along the southern flank of the basin was dominated by sandstone sequences, the Hangu Formation. The greater part of the Potwar area was the site of thin shelfal limestone deposition, the Lockhart and Patala Formations, throughout Paleocene times with a shale component

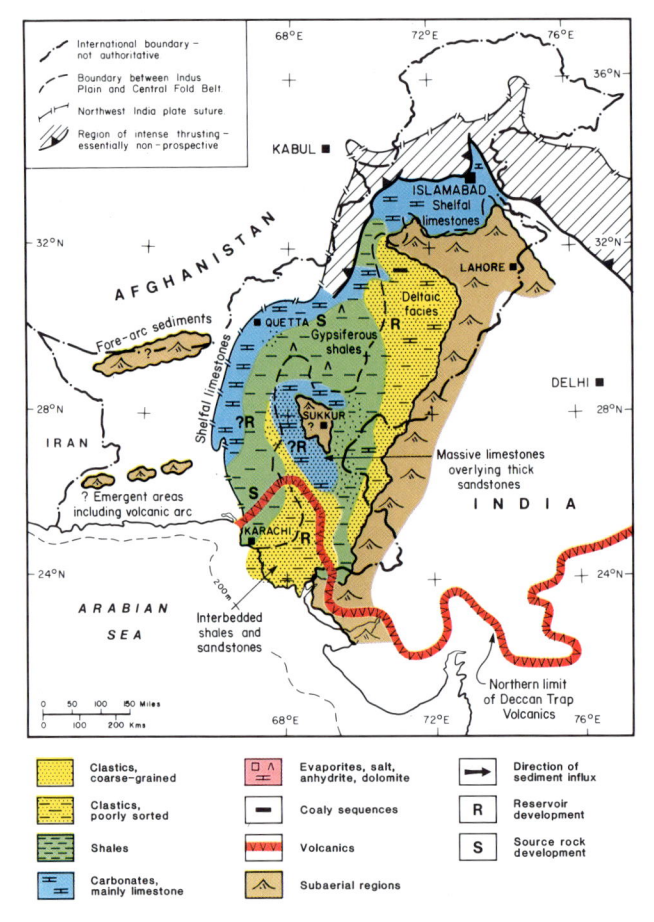

Fig. 18. Paleocene depositional patterns.

which becomes progressively more prevalent into the westerly Kohat area. Although no data are available, it is assumed that the Kohat area was the site of relatively low energy, deep water shale deposition.

To the south, in the Lower Indus Basin, the interpreted sedimentary patterns are more complex, comprising an eastern area of mixed lithologies that are dominated by sandstone, an axial area where mostly shales are developed, and a westerly area of mixed lithologies in which limestone is the dominant component. These facies contain numerous local reservoirs, the most notable of which are the Ranikot Group sandstones and limestones. There may be substantial reservoir potential still to be explored in the limestones around the margins of the Lower Indus Basin and particularly around the Jacobabad High. The possibility of carbonate reservoirs being present offshore has been noted in earlier publications, but has attracted little industry interest to date.

Both basins also probably contain source rock intervals, though data are scarce. In the Lower Indus Basin there are ubiquitous subsurface gas shows and reports of surface oil seeps and bituminous shales in the Ranikot Group. Although there is presently no evidence to indicate that the laterally equivalent Korara Shales are organic rich, it is worth noting that the time equivalent shales in the Cambay Basin, see Fig. 19 for location, are good source rocks, Bhandari and Chowdhary (1975), Gambhir (1976). In the Upper Indus Basin, the Patala Formation shales are likely to be the principal source rock.

Eocene

Apart from the previously mentioned Korara Shales which straddle the Paleocene/Eocene

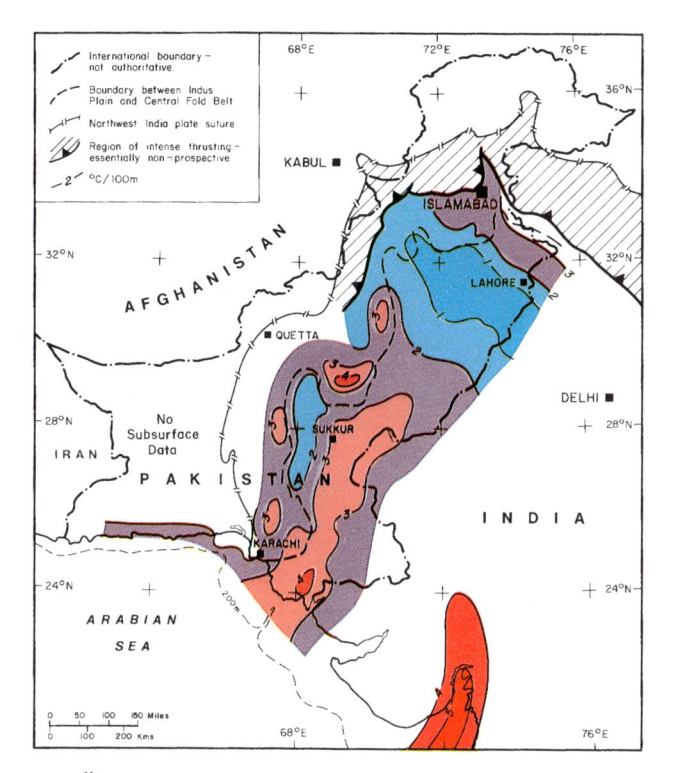

Fig. 19. Temperature gradients.

boundary, the Eocene of the Lower Indus Basin is represented by three formations. The Laki Formation is a limestone unit (often an important 'reefoidal' reservoir with subordinate marls, shales and sandstones. The Ghazij Formation is developed as a lateral equivalent of the Laki Formation and has a very mixed lithology of shales (some gypsiferous) in the main depocentres and sandstones and coals or limestones around the western margin where plate suturing occurred in the Early Eocene and fluviatile/estuarine sedimentation developed, see Fig. 20. A particularly exotic facies is the Gwanik Formation conglomerates which reportedly prograde from the west across the suture south of Quetta.

The Laki and Ghazij Formations are succeeded by the Kirthar Formation, a limestone and shale unit similar to the Laki, of Middle to Late Eocene age. Although not normally described as a reservoir, the presence of the Mari Gas Field (approximately 4 TCF recoverable reserves) in the Kirthar Formation makes it worthy of further study, like the preceeding Paleocene sequences around the basin margin and intra-basin 'highs'. The source potential of inter-bedded shales is speculative.

The Eocene succession in the Upper Indus Basin shows appreciable variation between different areas, see Fig. 11, although most of the units recognized comprise varying amounts of limestone and shale, and there is a strong impression that the current published terminology is more finely divided at formation level than really necessary. Consequently, the main limestone unit deposited during the Early Eocene is referred to here as the Sakesar Limestone.

The reservoir potential of the Eocene succession in the Upper Indus Basin is dependent more upon post-depositional fracturing than primary porosity. This secondary porosity has been developed in every Potwar area field which has produced to date. Like the Upper Indus Basin, Eocene source rocks are speculative.

Post-Eocene

Major uplift occurred during the Oligocene with deposition continuing probably only in the current offshore area. In Early Miocene times, deposition resumed onshore and a thick sequence of mainly coarse clastics developed (see Fig. 11). These sediments contain excellent

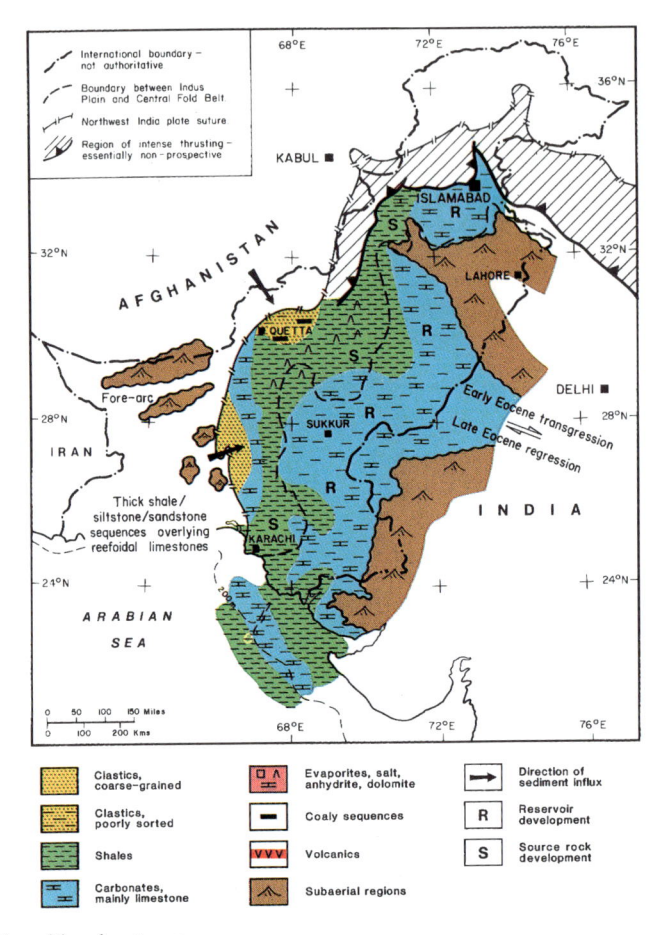

Fig. 20. Eocene depositional patterns.

reservoirs, but they are not favourably positioned with respect to source rocks and are, in any event, often charged with high pressure meteoric water. Apart from isolated, small oil accumulations created by secondary migration from deeper reservoirs, they are not prospective.

In the offshore, there may well be significant Oligocene to Pliocene submarine channel sand prospects developed high up on the present day continental slope (see McHargue & Webb 1986).

Petroleum potential

Reservoirs

The general reservoir potential of the various stratigraphic units has been summarized above and, despite a scarcity of data, they appear to be widespread.

Source rocks

Very little geochemical information is currently available in the public domain but the possibility of several petroleum source rock intervals being developed appears to be good. The likely maturity level of any organic rich intervals can be gauged with the assistance of the summary geothermal data in Fig. 19; based on information published by Khan & Raza (1986). Salient points are the low values over the Sargodha Ridge, the belt of medium values along the eastern side of the Central Fold Belt which are comparable to those of the productive Potwar and Badin Block areas and the trend of apparently low values (typical of a foreland basin) infront of the Central Fold Belt. Finally, apprehension about high geothermal values should be tempered by recognizing the much higher values of the Cambay Basin where there are oil reserves of several billion barrels.

Traps

Stratigraphic traps almost certainly abound in Pakistan though, apart from in a few local circumstances, there are inadequate subsurface data to predict them with confidence. In a very general sense, exploration risk may well increase to the east where many sequences become sandier or where seals may be breached.

Structural traps are of three types, those of a tensional nature which are mostly developed beneath the Indus Plain; those that are compressional, restricted for the most part to the Central Fold Belt; and those related to transcurrent motions which may be in either province.

The tensional traps are mostly related to post-Late Permian/Triassic rifting of the Lower and Upper Indus Basins. The normal faulting should be interpreted in the context of one side of a proto-rift which became a passive margin and was then involved in a collision phase only late in its history. It might be illuminating to interpret this structural system in terms of the rifting processes and accomodation zones described by Rosendahl (1987).

Compressional tectonics have generated hundreds of structures in the Central Fold Belt. These folds can be tens of kilometres long, are normally asymmetric and are probably thrust related. The anticlinal axes are likely to vary in pitch with depth due to complications caused by transcurrent motion. In the Potwar area the structure is further complicated by the Infra-Cambrian halokinetic structures. The deep structure of the Central Fold Belt is essentially unknown, as is exemplified by the relatively well explored Potwar area. Those that would have it created by intra-basement thrusting quote the evidence of the orthogonal symmetry of the area and its general alignment with intracratonic trends. Adherents of the Infra-Cambrian salt detachment model point to the lack of any gravity anomaly along the southern Salt Range margin. However, newly emerging ideas about second-order thrust features being developed in what are fundamentally tensional regimes, may suggest that even more radical salt-related concepts of deformation may need to be considered.

Transcurrent lineaments are likely to be significant in creating traps and possibly in influencing preferred zones of carbonate build-up and the location of clastic runnels. Beneath the Indus Plain, transcurrent lineaments will be difficult to map. However, where cross-trends are manifest on satellite images they may represent re-activated transcurrent trends which can be projected beneath the Indus Plain.

Petroleum types

In recent years it has become apparent that oil, condensate and gas are present and producable in both the Upper and Lower Indus Basins.

The distribution of oil is currently restricted to two specific areas. However, the 19th Century oil production at the isolated Kattan wells east of Quetta must be considered as very significant for future exploration. Having regard for the tectonic regimes present in Pakistan it is also necessary to anticipate a widespread distribution of condensate reserves.

Outside the Potwar area, gas has dominated discoveries to date. A feature of this gas has been the locally high percentage of inert components, on an apparently random basis. Possibly, there is a simple relationship of high inert components in close proximity to the basement, from which they are derived. Another exotic aspect of gas occurrence in Pakistan is the series of mud volcanoes along the Makran coast. These features (40–300 feet high) are created by gas seepages transporting unconsolidated sediments to the surface. The cones are situated along structural lineaments and periodically erupt, in association with earthquakes.

Conclusions

Pakistan contains at least three sedimentary basins which contain thick Lower Palaeozoic, Mesozoic and Cenozoic sediments. All of the ingredients for successful petroleum exploration are present, including reservoirs, source rocks, traps and favourable migration history. Only about 200 millions barrels of oil have been produced in the last 73 years, and yet the basin configuration and geological history, as well as the volume of contained sediments is not much different to that of the Arabian/Persian Gulf and coastal fold belt of Iran. While the ultimate potential of Pakistan may be only a fraction of the Gulf area, it still affords a very significant opportunity in a country which is likely to encourage domestic exploration for the foreseeable future.

References

BHANDARI, L. L. & CHOWDHARY, L. R. 1975. Stratigraphic Analysis of Kadi and Kalol Formations, Cambay Basin, India. *Bulletin American Association of Petroleum Geologists*, **59**, 856–871.

COWARD, M. P., JAN, M. Q., REX, D., TARNEY, J., THIRLWALL, M. & WINDLEY, B. F. 1982. Geotectonic framework of the Himalaya of N. Pakistan. *Journal of the Geological Society, London*, **139**, 229–308.

DESIO, A. 1979. Geological Evolution of the Karakorum. *In*: FARAH, A. & DE JONG, K. A.

(eds) *Geodynamics of Pakistan*, Geological Survey of Pakistan, Quetta, 111–124.

DOLAN, P., EDGAR, D. C., BRAND, C. & SHEVLIN, B. J. 1987. Pakistan, Regional Geology and Petroleum Exploration Potential, Dolan & Associates, Non-Exclusive Report, London.

FARAH & DE JONG 1979. *Geodynamics of Pakistan*. Geological Survey of Pakistan, Quetta.

FARAH & DE JONG 1979. *Geodynamics of Pakistan*.

GAMBHIR, S. C. 1976. Pattern of accumulation of oils in Mehsana Area of Cambay Basin, India. *Bulletin of American Association of Petroleum Geologists*, **60**, 1955–1962.

KHAN, M. A. & RAZA, H. A. 1986. The Role of Geothermal Gradients in Hydrocarbon Exploration in Pakistan. *Journal of Petroleum Geology*, **9**, 245–258.

KRISHNAN, M. S. 1982. Geology of India and Burma, 6th Edition, CBS Publishers and Distributors, Delhi.

LE FORT, P. 1975. Himalayas: The Collided Range, Present Knowledge of the Continental arc., *American Journal of Science*, **275**, 1–44.

MCCALL, G. J. H. & KIDD, R. G. W. 1982. The Makran, Southeastern Iran. *In*: LEGGETT, J. K. (ed.), *Trench-Forearc Geology: Sedimentation and Tectonics on Modern and Ancient Active Plate Margins*. Geological Society, London, Special Publication, **10**, 387–397.

MCHARGUE, T. R. & WEBB, J. E. 1986. Internal geometry, seismic facies and petroleum potential of canyon and inner fan channels of the Indus submarine fan. *Bulletin of the American Association of Petroleum Geologists*, **70**, 161–180.

OWEN, E. W. 1975. Trek of the Oil Finders: A history of exploration for petroleum, *American Association of Petroleum Geologists Memoir* **6**, 87.

PASCOE, E. H. 1920. Petroleum in the Punjab and North–West Frontier Province. *Memoirs of the Geological Survey of India*, **40**, 330–489.

QUADRI, V. N. & SHUAIB, S. M. 1987. Geology and hydrocarbon prospects of Pakistan's offshore Indus Basin, *Oil & Gas Journal*, **85**, no 35, 65–67.

RICHTER-BERNBERG, G. & SCHOTT, W. 1963. Jurassic and Cretaceous at the western border of Gondwana Shield in India, and the stratigraphy of oil possibilities. *In*: Proceedings of the Second Symposium on the Development of Petroleum Resources of Asia and the Far East, Mineral Resource Development Series, U.N. Publication, New York: No. 18, 1, 2/aa-XII, 230–236.

ROSENDAHL, B. R. 1987. Architecture of continental rifts with special reference to East Africa, *Annual Review of Earth & Planetary Science*, **15**, 445–503.

SHAH, S. M. L. 1978. Stratigraphy of Pakistan, *Memoir of the Geological Survey of Pakistan*, **12**.

SNELGROVE, A. K. 1979. Migrations of the Indus River, Pakistan in response to plate tectonic motions. *Journal of the Geological Society of India*, **20**, 392–403.

TAHIRKHELI, R. A. K., MATTAUER, M., PROUST, F. & TAPPONNIER, P. 1979. The India Eurasia Suture Zone in Northern Pakistan: synthesis and interpretation of recent data at plate scale. *In*: FARAH, A. & DE JONG, K. A. (eds), *Geodynamics of Pakistan*, Geological Survey of Pakistan, Quetta, 125–130.

Sequence stratigraphy and the habitat of hydrocarbons, Gippsland Basin, Australia

V. D. RAHMANIAN[1], P. S. MOORE[2], W. J. MUDGE[2] & D. E. SPRING[2]

[1] *Exxon Production Research Company, Houston, TX 77252, USA*
[2] *Esso Australia Ltd., 127 Kent Street, Sydney NSW 2000, Australia*

Abstract: The Gippsland Basin is Australia's most prolific oil and gas province, with initial reserves estimated at 3.6 billion barrels crude/condensate, 0.6 billion barrels liquid petroleum gas (LPG) and 8.3 TCF dry gas. It is located in southeastern Australia, mainly offshore. Significant hydrocarbons were first discovered in 1964, in large structures at the top of the Late Cretaceous–Eocene Latrobe Group. All of the large structures have now been drilled, with current exploration concentrating on fault-dependent and combined stratigraphic structural traps within the Latrobe Group. Hydrocarbon, source rock and maturation data suggest that there is a widespread and uniform non-marine source of Late Cretaceous age. The concentration of hydrocarbons at the top of the Latrobe Group is due to the presence of large structures, an excellent cap seal, and the relatively sandy nature of the underlying, intra-Latrobe section. The distribution of oil versus gas at this level is largely explained by maturation variations and migration pathways within the basin. Vertical migration of up to 2 km is required in most cases. The distribution of oil versus gas is also influenced by minor source rock variations, fresh water washing, biodegradation, and the variable quality of fault seal. Recent exploration studies have concentrated on locating stratigraphic and combination traps, especially within the Latrobe Group. Prediction and subsequent discovery of these traps is a challenging task that has been aided by sequence-stratigraphic and seismic-stratigraphic analysis of the rock record. This has allowed the prediction of reservoir, source and seal geometries and distributions. The techniques used to construct this chronostratigraphic framework are presented, together with examples of its application within fields and on a regional basis.

The Gippsland Basin is located mainly offshore, in southeastern Australia (Fig. 1). Discovered hydrocarbon reserves comprise 3.6 billion barrels of crude/condensate, 0.6 billion barrels of liquid petroleum gas (LPG) and 8.3 trillion cubic feet (TCF) of dry gas. The basin has produced over two billion barrels of oil and two trillion cubic feet of gas to date, and in 1987 supplied 70% of Australia's crude oil needs. In addition, it supplies virtually all of Victoria's natural gas requirements, and substantial volumes of LPG for domestic and export consumption. The positive impact on Australia's balance of payments for 1987 was A$4 billion.

All significant discoveries to date have been contained within the Late Cretaceous–Early Tertiary Latrobe Group (Fig. 2). Most of these discoveries are well documented in the literature, with papers by Threlfall *et al.* (1976), Marlow (1978), Thornton *et al.* (1980), Limbert *et al.* (1983), O'Byrne & Henderson (1983), Henzell *et al.* (1985), Battrick (1986), Brown (1986), Roder & Sloan (1986), Young & Coenraads (1986), Sloan (1987), Ozimic *et al.* (1987) Esso Australia Ltd. (1988) and Glenton (1988).

The timing and magnitude of hydrocarbon discoveries in the Gippsland Basin is shown in Fig. 3. Outstanding success in the 1960s is attributed to the discovery of major hydrocarbon pools associated with large structural traps. Examples include Barracouta (discovered 1964; with reserves of 1.5 TCF gas), Marlin (1966; 2.3 TCF gas), Kingfish (1967; 1.3 BB oil), Halibut (1967; 790 MB oil), Snapper (1968; 1.8 TCF gas) and Mackerel (1969; 530 MB oil). This trend slowed in the 1970s and tapered off during the 1980s as the inventory of large structural traps became depleted.

Prior to 1978 most of the hydrocarbons discovered were in traps at the top of the Latrobe Group. With depletion of this prospect inventory, attention has become increasingly focussed on intra-Latrobe plays, in which 95% of oil discovered since 1978 is contained. Some of these discoveries are large. The Flounder Field, for example, contains over 100 million barrels of oil reserves, while the Kipper discovery has in excess of 0.5 TCF recoverable dry gas (Figs 1, 4, 5). Thus, the intra-Latrobe interval is believed to contain the majority of undiscovered hydrocarbons in the basin, despite the fact that, at present, it accounts for less than 10% of the basin's hydrocarbon reserves.

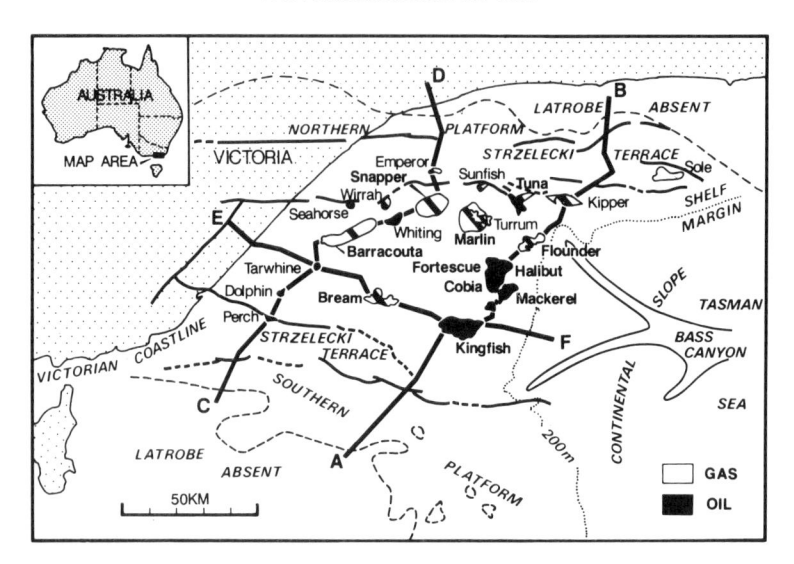

Fig. 1. Location map, Gippsland Basin, showing producing fields (heavy typescript) and other major discoveries. Cross-sections AB, CD and EF are shown in Figs 4, 5 and 11 respectively.

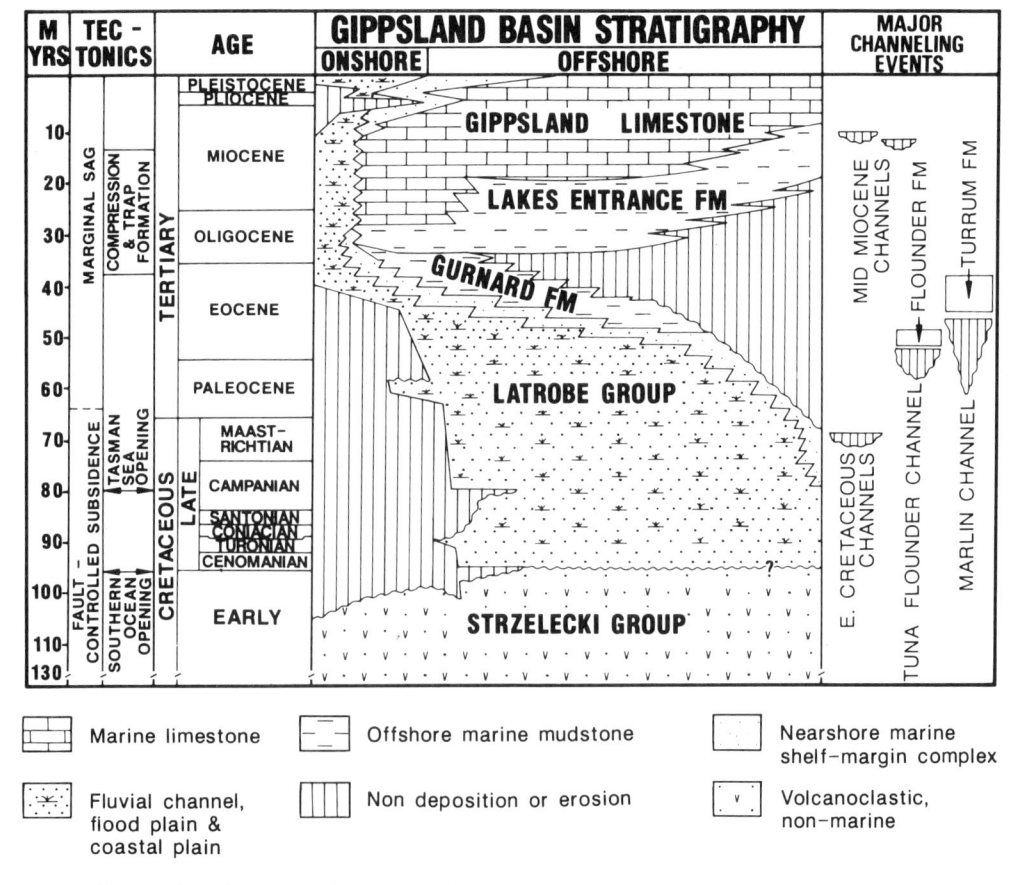

Fig. 2. Gippsland Basin stratigraphy and tectonics.

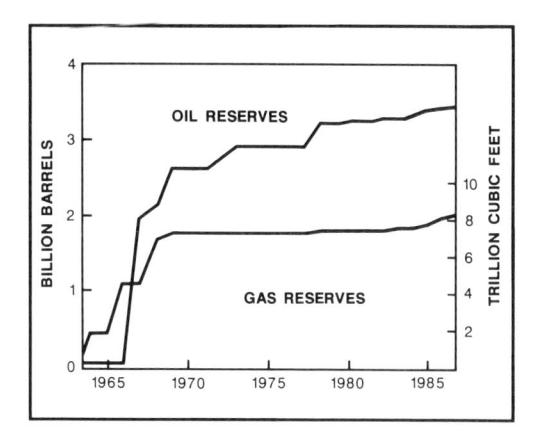

Fig. 3. Gippsland Basin additions to reserves, 1964–1987.

Exploration for structural traps has, in the past, led to the discovery of several large stratigraphic and combination traps. However, with the depletion of the structural prospect inventory, a systematic search for subtle traps has become imperative. With this in mind, a comprehensive, integrated regional study was recently carried out in the Gippsland Basin. The objectives of this study were threefold: (1) to develop a better understanding of the interrelationship between tectonic evolution, structure and stratigraphy within the basin; (2) to better understand the distribution of oil and gas within the basin; and (3) to develop play concepts to search for new stratigraphic and combination traps, both in wildcat areas and adjacent to existing production. Aspects of this study are presented here.

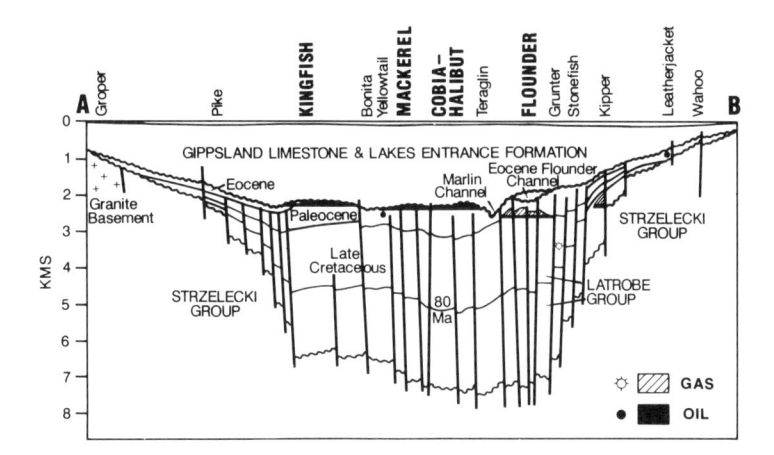

Fig. 4. Regional cross-section AB. Location shown in Fig. 1.

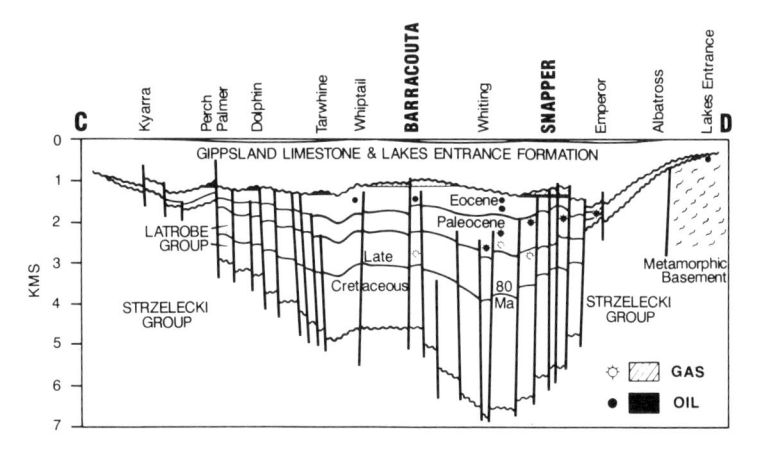

Fig. 5. Regional cross-section CD. Location shown in Fig. 1.

Tectonic evolution

The Gippsland Basin is located on the continental shelf, between mainland Australia and Tasmania. The origin and evolution of the basin are related to the progressive fragmentation and dispersal of Gondwanaland.

Early Cretaceous rifting (130–95 Ma)

Sedimentation commenced in the earliest Cretaceous (approximately 130 Ma) (Haq *et al.* 1987), in a pre-break-up depression which extended along the entire southern margin of mainland Australia (Fig. 6(a)). This Early Cretaceous depression developed into a complex rift system, up to several hundred kilometres wide (Veevers 1984). Deposition during this phase was continuous between the Otway, Bass and Gippsland Basins. In the Gippsland Basin, these initial deposits make up the Early Cretaceous Strzelecki Group (Fig. 2), a thick sequence of volcano-clastic sediments of fluviatile origin regarded as economic basement.

Early Cretaceous tectonism (approximately 95 Ma)

At the end of the Early Cretaceous, Australia separated from Antarctica (Fig. 6(b)). Break-up did not extend into the Gippsland Basin, but instead, continued down the western side of Tasmania. Veevers (1986) proposed an age of 95 ±5 Ma for this event. The break-up created an end of Early Cretaceous unconformity, not only in those basins where new oceanic crust formed, but also further east in the Bass and Gippsland Basins (Fig. 2; Williamson *et al.* 1987). In Gippsland, the unconformity is associated with considerable uplift around the margins of the old rift, with at least 1 km of sediment eroded from parts of the Southern Platform (Fig. 1; Duddy, pers. comm., 1985). The major ridges which separate the Otway, Bass and Gippsland Basins became prominent structural features and barriers to deposition at this time. High heatflows associated with this tectonism are interpreted from apatite fission track analysis of granites derived from these ridges.

In the Gippsland Basin, this tectonic readjustment had two additional effects: firstly there was a major provenance change, from lithic and volcano-clastic deposits of the Strzelecki Group into quartzose deposits of the Latrobe Group; secondly, the area of deposition became much more confined.

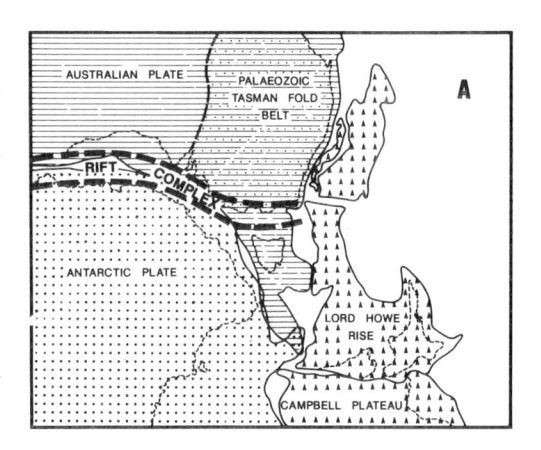

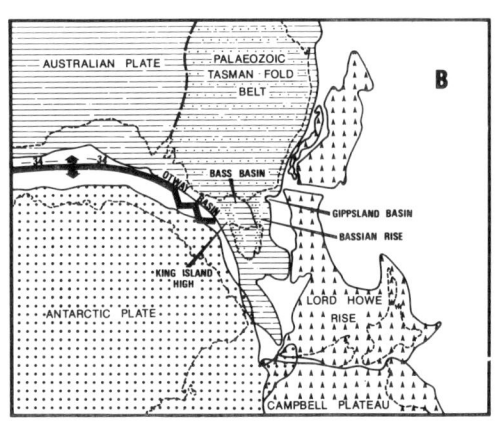

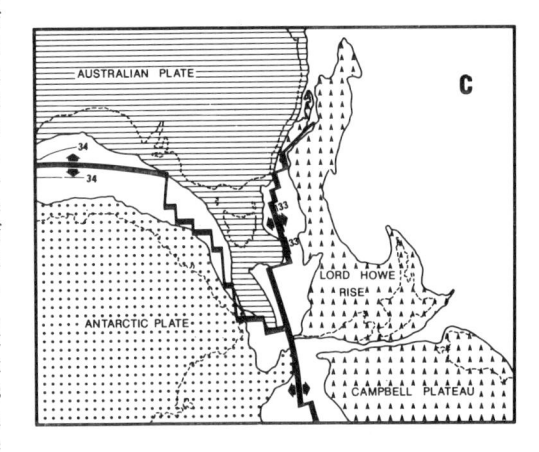

Fig. 6. Plate tectonic evolution of the Gippsland Basin. A: Early Cretaceous. Broad rift complex exists along the southern margin of mainland Australia. B: Mid Late Cretaceous. The Southern Ocean has formed (at 95 Ma) and the Gippsland, Bass and Otway Basins are separate entities. C: Latest Cretaceous. Breakup has occurred along the eastern margin of Australia (at 80 Ma) and the Gippsland Basin is left as a failed rift.

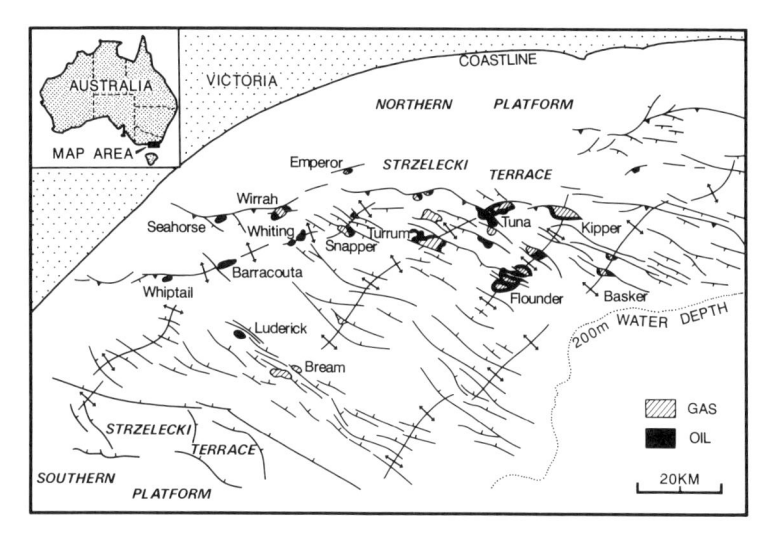

Fig. 7. Major fault patterns, discoveries and anticlinal trends, *within* the Latrobe Group. Illustration is a composite of several horizons, but excludes discoveries at the top of the Latrobe Group.

Late Cretaceous rifting (95–80 Ma)

The Latrobe Group was deposited initially in a well developed rift valley, 40–80 km wide. The northern and southern margins of the rift correspond with the basinward margins of the Strzelecki Terraces (Fig. 1), which trend roughly east–west, parallel to the underlying Early Cretaceous depression. Extension occurred in a NE–SW direction, resulting in fault-controlled subsidence with faults orientated NW–SE (Fig. 7). The rift is asymmetrical, with maximum downfaulting occurring adjacent to the northern margin.

These early Latrobe Group deposits, of early to mid Late Cretaceous age, have only been intersected around the margins of the basin. In the most proximal of these areas, they consist of sandstones and conglomerates of braided-stream and alluvial-fan origin. In the more basinal locations, they include finer grained fluvial deposits. A thick lacustrine unit of Coniacian–Turonian age (*P. mawsonii* Zone of Helby *et al.* 1987) occurs in several wells (Marshall 1989a). The oldest marine fossils identified to date are of Santonian age (Marshall in press 1989b), indicating that marine conditions were present within the rift during the later stages in its evolution. Basaltic lava and minor related intrusives are also present, especially in the upper part of this section.

Late Cretaceous rifting in the Gippsland Basin is interpreted as being accompanied by rifting along the eastern margin of Australia, with a triple junction connecting the two rifts.

Breakup along the eastern edge of Australia occurred in the Campanian, with the oldest magnetic anomaly (A33) giving a break-up age of approximately 80 Ma (Fig. 6c). This event is recorded in the Gippsland Basin by a mild unconformity and a peak in volcanism. The unconformity is best illustrated by well and seismic data in the northeast of the basin, where it lies within the *N. senectus* Zone (age range approx. 80–83 Ma). This is the same unconformity as recorded by Lowry (1987) from the northwest of the basin.

Latest Cretaceous–Eocene: transition from fault-controlled subsidence to marginal sag (80–39 Ma)

Break-up along the eastern margin of Australia left the Gippsland Basin as a failed rift. Subsidence and sedimentation rates initially remained high, but eventually decreased as the basin evolved from a fault-controlled depression into a marginal sag. The upper Latrobe Group was thus deposited under the progressive influence of the Tasman Sea, which encroached from a southeasterly direction. Deposits of this age are more marine in origin than underlying Latrobe Group sediments. They typically formed along wave-dominated shorelines in a suite of environments ranging from deep offshore, through shoreface, foreshore, beach-barrier, to coastal plain and alluvial plain. Deep offshore deposits and glauconitic condensed sections at the top the Latrobe Group are termed the 'Gurnard Formation' (Fig. 2), despite a wide age range.

Submarine erosion, canyon formation and infill occurred during this period, particularly in the east of the basin. The location of the major Eocene channels is shown in Fig. 8.

Late Eocene—Recent: structuring and late sag-phase deposition

Major compression was initiated during latest Latrobe deposition and continued in some areas until the mid Miocene, with a period of reactivation in the Pleistocene—Recent. The compression produced a series of NE—SW trending anticlinal axes, and minor E—W folds. The structures that formed along these axes include simple anticlines, faulted anticlines and highside and lowside fault closures (Figs 4, 5). Compression against the southern and particularly the northern margins of the basin led to partial inversion of E—W trending normal faults, with the formation of compressional rolls (e.g. Snapper, Figs 1 & 5). Uplift associated with structuring resulted in widespread erosion across the basin, forming the top Latrobe Group unconformity. Several of the largest traps in the basin are the result of compressional folding combined with erosional topography (e.g. Marlin, Kingfish, Fortescue, Cobia, Halibut; Figs 1, 4 & 5).

After formation of the top Latrobe unconformity, but during this compressional phase, the basin continued to slowly subside. Shales and marls of the Oligocene Lakes Entrance Formation (Fig. 2) were deposited over the top of the Latrobe Group, in some areas onlapping the newly forming structures. Deposition occurred in shelf, slope and basinal environments. The resulting fine-grained strata constitute a basin-wide seal for potential hydrocarbon reservoirs at the top of the Latrobe Group. Lakes Entrance sedimentation was followed by extensive marine channelling, which accompanied the deposition of a thick bioclastic wedge-known as the Gippsland Limestone. Carbonate formation was probably aided by upwelling of nutrient-rich cold water via submarine canyons on the continental shelf (cf. modern Bass canyon, Fig. 1). Rapid carbonate deposition during the Plio—Pleistocene produced the final easterly progradation, and extended the shelf to its present position in the basin.

Thermal history and maturation levels

The mechanisms which create a sedimentary basin and control its subsequent evolution must be understood in order to model heatflow history and thus assess past and present-day source rock maturation levels. The first step in this process is reconstruction of the tectonic subsidence history of a basin.

Tectonic subsidence curves, extending from the onset of Latrobe Group deposition, have been constructed for many locations in the Gippsland Basin. These curves show a similar pattern, with rapid, fault-controlled subsidence in the Late Cretaceous being succeeded by a progressive decrease in sedimentation rates and basinal sag in the Palaeocene to Recent (50 Ma to present day) (Fig. 9). This pattern is typical of passive continental margins, and is attributed to heating of the lithosphere during rifting followed by cooling and subsidence.

The reason for increased heatflow during rifting of the southern margin of Australia is uncertain (Houseman & Hegarty 1987; Hegarty et al. 1989). However, in this tectonic setting, rifting is most commonly attributed to crustal stretching, either of short duration ('instantaneous'; McKenzie 1978) or over a long period of time (Jarvis & McKenzie 1980). The difference between these two models is the amount of post-rift subsidence that is produced. Instantaneous stretching results in rapid subsidence, whereas extended stretching produces lesser amounts of post-rift downwarp (Cochran 1983).

In the Gippsland Basin, short duration rifting (less than about 15 Ma) is required to successfully model the high subsidence rates that occur in the centre of the graben. Rifting associated with the Latrobe Group is interpreted to extend

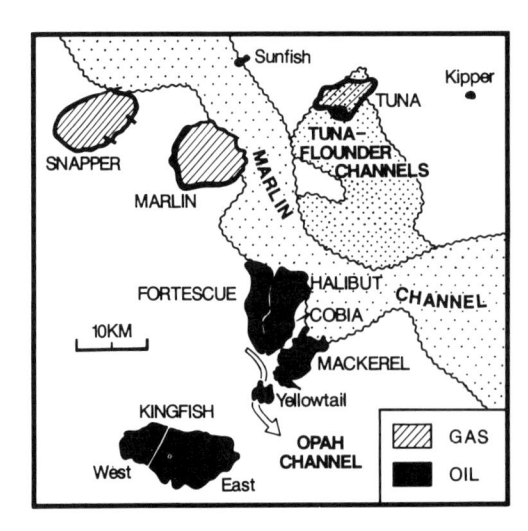

Fig. 8. Location of major Eocene channels and adjacent top Latrobe Group discoveries.

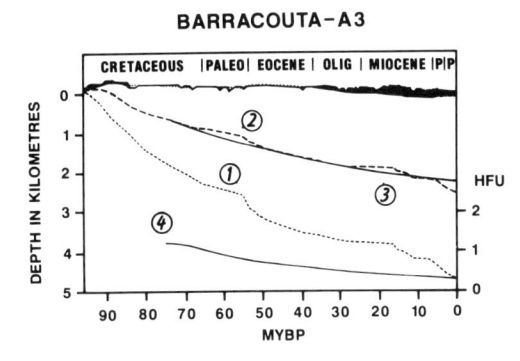

Fig. 9. Subsidence and heatflow history, Barracouta field. Blackened area is modelled water depth. Curves are 1: Total subsidence, corrected for compaction. 2: Actual thermal subsidence. 3: Modelled thermal subsidence based on 15 Ma rifting event from 95–80 Ma. 4: Heatflow anomaly resulting from post-rift lithospheric cooling.

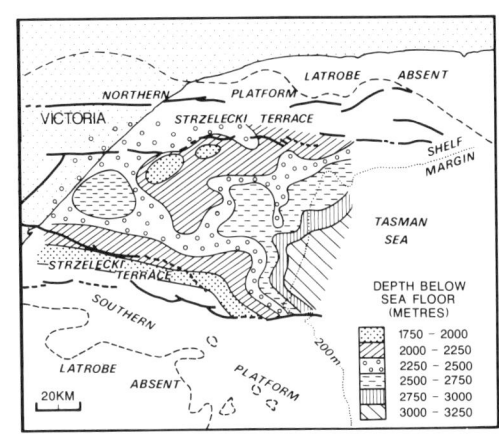

Fig. 10. Depth to 100°C isotherm, showing effects of decreasing geothermal gradient from basin margin to centre, and from west to east.

from the onset of deposition (maximum age 95 Ma) until the opening of the Tasman Sea at 80 Ma. Fault controlled subsidence continued after this time in many areas, but is interpreted to be a response to high sedimentation rates rather than lithospheric stretching. Similar interpretations have been made by other workers in the Gippsland Basin (Hegarty *et al.* 1986), for the near-by Otway Basin (Williamson *et al.* 1987; Hegarty 1985), and also in the North Sea (Sclater & Christie 1980; Goff 1983).

The subsidence model and heatflow anomaly that result from the above interpretation are shown for the Barracouta field in Fig. 9. This interpretation is not a unique one for the data; however, alternative models, involving dyke injection or sub-crustal erosion, produce very similar heatflow histories.

Total heatflow is calculated by adding heatflow due to rifting to a background heat flow of between 0.8 and 1.4 HFU. Thus, large variations occur in the magnitude of background heatflow in the Gippsland Basin. These variations are reflected by present-day geothermal gradients, which show a marked decrease, passing from the basin margins into the central graben (Fig. 10). Model studies show that these base level heatflow variations have probably been present since the onset of Latrobe Group deposition. They are clearly reflected in maturation profiles of wells in the basin, and have had a profound effect on determining the ratio of oil to gas in large top Latrobe Group structures. The variations are attributed primarily to changes in the thickness and conductivity of crustal material underlying the Latrobe Group.

The maturation profile that results from vitrinite reflectance measurements and heatflow modelling is illustrated in cross-section in Fig. 11. The effect of the decreasing heatflow from west to east is clearly seen, as the oil window expands and becomes more deeply buried. East of the 200 m isobath, water depth increases dramatically, so that the Latrobe Group has little sedimentary cover and is immature for oil generation. In plan view, the pattern is more complex. The combination of high heatflow and fast subsidence in the northern, axial part of the graben makes this area the most gas prone. By comparison, the Latrobe Group in the east and southeast, near Fortescue, Cobia–Halibut and Kingfish, is ideally situated for oil generation.

Hydrocarbon source

Very few wells in the Gippsland Basin have been drilled deep enough to intersect rocks which are at peak maturity for oil generation, and all of these wells are located near the basin margins where the strata are relatively thin and sandy. As a result, understanding of the source of hydrocarbons in the basin is derived largely from a study of oils and gases, rather than source rocks.

All studies to date indicate a terrestrial, higher-plant origin for Gippsland Basin hydrocarbons (Brooks & Smith 1969; Philp & Gilbert 1982; Smith & Cook 1984; Alexander *et al.* 1983, 1987; Burns *et al.* 1984, 1987; Powell 1987). Oils are typically paraffinic, with early mature samples exhibiting a waxy character and

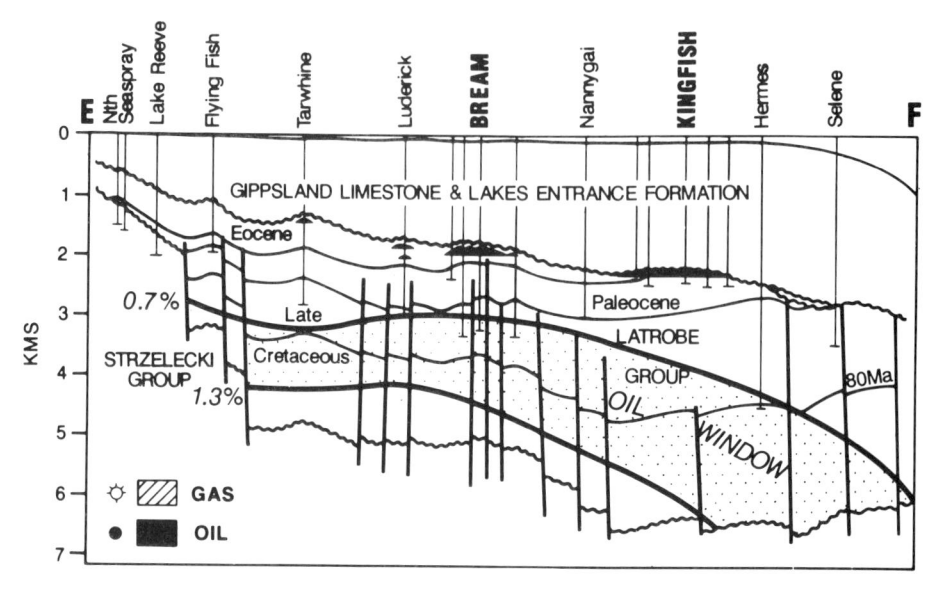

Fig. 11. Regional cross-section EF, showing estimated position of oil window. Location shown in Fig. 1.

slight odd-over-even preference in C_{25+} n-alkanes. Gas chromatography (GC) and mass spectrometry (MS) studies indicate that all of the oils have similar characteristics, suggesting a widespread source of rather uniform character.

Variations exist in the gross composition of the oils, but these are mostly related to maturity differences, with the more mature oils being dominated by short chain n-alkanes, especially in the kerosene and gasoline ranges. These light, condensate-like oils have an API gravity of about 60°, compared with the usual 35–45° recorded for the mature and early-mature oils.

Oil maturity levels have been estimated from gas chromatograms and GC–MS traces, isotope data for dissolved gases, and key maturity indices (Burns et al. 1984, 1987). Most of the oils have been generated at maturity levels corresponding to a vitrinite reflectance range of $R_0 = 0.9–1.1\%$. Source rocks at this level of maturity are of Late Cretaceous age, typically Campanian or older. As outlined above, Latrobe Group sediments of this age are fluviatile to lacustrine in origin, and were deposited in a well developed rift-valley setting.

Gas maturity levels in the Gippsland Basin have also been estimated (Burns et al. 1987) and correspond to the overmature state of hydrocarbon generation. Thus, a similar source is proposed for both gas and oil in the basin, with maturity being the main control over the types of hydrocarbons generated.

Source rock data from the Latrobe Group confirm that it contains significant oil prone material. Oil prone kerogen typically constitutes about 50% of the organic matter, and is dominated by biodegraded terrestrial material plus minor spores and pollen. Coals and dispersed organic matter are dominated by conifer remains and have variable but often high (up to 30%) exinite contents (Smith 1981; Smith & Cook 1984).

Geochemical analyses of Latrobe Group shales indicate only minor lateral and vertical variation. However, the sequence does become shalier and the shales slightly richer in dispersed organic matter passing from the basin margins into the axis of the rift. Total Organic Carbon (TOC) values of shales average 2.4%. Most source rocks are Type II/III and Type III, with Campanian intervals characterized by Type II organic matter. Extractable hydrocarbon yields range up to and exceed 50 mg per gram organic carbon, placing these units in the regime of good petroleum source rocks (Powell 1987). Overall, there is a good correlation between GC–MS traces of Late Cretaceous source rocks and oils reservoired within the Latrobe Group (Alexander et al. 1987). Detailed GC–MS analyses of shales from selected deep wells suggests that the best oil-source environments were coastal-plain swamps, lagoons and possibly larger lakes where quiet water deposition and reducing conditions were best developed.

Hydrocarbon generation, migration and entrapment

Most of the hydrocarbons in the Gippsland Basin are contained in reservoir rocks with maturities in the range $R_0 = 0.4-0.6\%$, thus indicating significant secondary migration. For example, oil is reservoired at approximately 2300 m in Kingfish, whereas the depth to $R_0 = 0.9\%$ in the drainage area varies from about 4500 m to 5500 m, thereby implying vertical hydrocarbon migration of at least 2 km.

The timing of hydrocarbon generation and migration have been estimated by using the heatflow and burial history models discussed above, in combination with standard algorithms for maturation of organic matter. The results at two key locations in the basin are presented in Fig. 12 and are discussed below.

In the Kingfish area (Fig. 12(a)), the basal Latrobe Group entered the oil window at 70 Ma and reached optimum maturity for oil generation at about 50 Ma. Trap formation occurred approximately 20 million years later, when the Latrobe Group was still mostly immature for hydrocarbon generation. The Upper Cretaceous section passed rapidly through the level of peak oil generation about 10–15 Ma, so that today, a 2 km thick oil window exists, with only a small proportion of the Latrobe Group being overmature for oil generation. The Kingfish field contains a modest amount of light hydrocarbons in solution with the 1.6 billion barrels crude/condensate, with both the oil and dissolved gas suggesting maturation levels in the range $R_0 = 0.9-1.1\%$ (Burns et al. 1984, 1987 and more recent data). Thus, migration must have occurred from Upper Cretaceous source rocks buried below 4500 m. Vertical migration may have been assisted by faulting in the deep section and an abundance of sandstone in the upper part of the Latrobe Group.

In the Barracouta field (Fig. 12(b)) the top Latrobe accumulation contains 1.5 TCF wet gas, generated mostly from the overmature zone. In this area, higher heatflows result in a thinner oil window, and thus a greater volume of overmature source rock. In addition, the section has undergone only a limited increase in maturation since trap formation, further favouring gas generation. Model studies indicate that the main generative centre is a syncline which lies to the southeast of the field. Vertical migration in the order of 2 km and lateral migration of up to 15 km are suggested.

Simple migration pathways, combined with maturation estimates of the deep Latrobe section, can explain the presence and absence of hydrocarbons at the top of the Latrobe Group and provide an insight into the oil/gas ratios of those accumulations. This is illustrated in Fig. 13, where top Latrobe Group hydrocarbon accumulations are plotted, along with their gas/oil ratios, on a map showing drainage areas at top Latrobe. Also included on this map is the predicted ratio of mature versus overmature Latrobe Group. Fields which drain predominantly oil mature areas generally contain predominantly oil, and a similar principle applies for gas. Several small oilfields which occur in the northwestern part of the basin near the large Barracouta and Snapper gas fields can be explained by limited drainage from oil-mature areas around the basin margins. Dry structures, such as Sweep (Fig. 13) can be attributed to their location in migration shadows. Thus, while the model presented in Fig. 13 is very simplified, it does provide a reasonable explanation of the distribution of hydrocarbon types in top Latrobe structures.

The pattern of gas/oil ratios presented in Fig. 13 is not repeated for intra-Latrobe accumulations. Most intra-Latrobe accumulations are

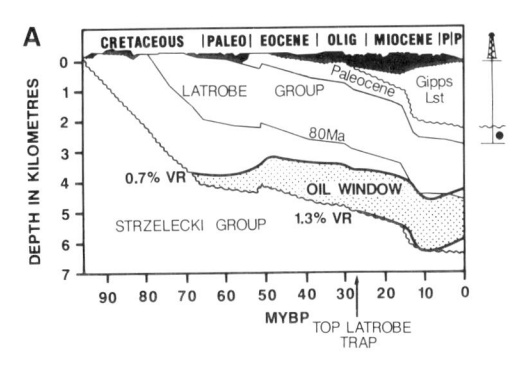

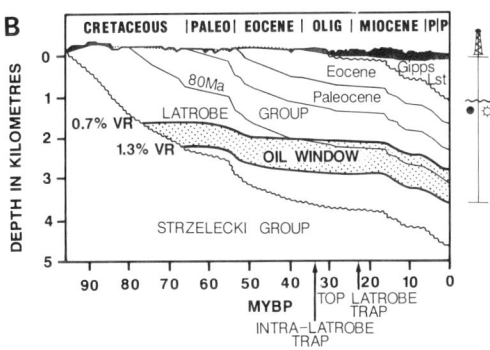

Fig. 12. Timing charts for hydrocarbon generation. A: Kingfish field. B: Barracouta field. Estimated time of trap formation is also shown.

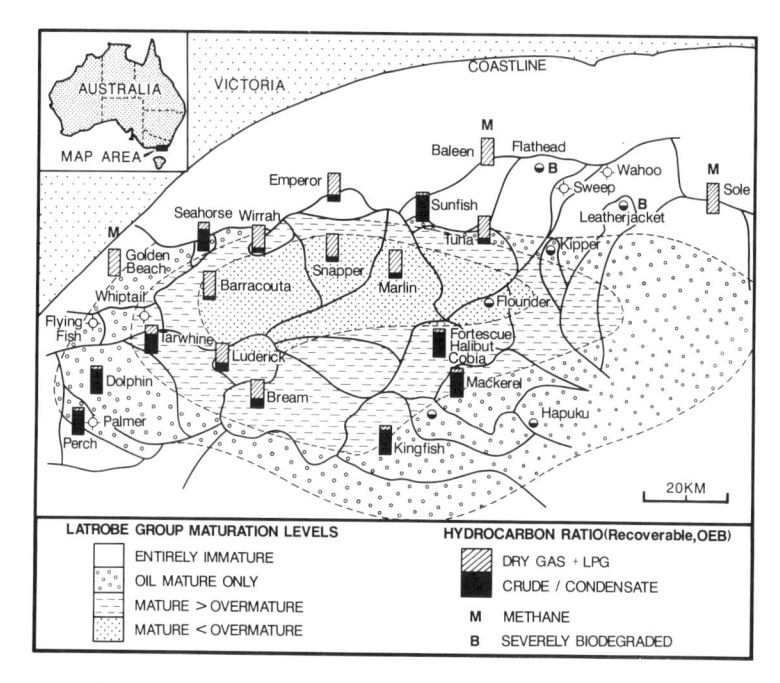

Fig. 13. Relationship between top Latrobe Group hydrocarbons, top Latrobe Group drainage areas, and maturity levels within the Latrode Group.

fault-dependent, and the ratio of gas to oil in these traps is strongly influenced by the ability of cap rocks and fault zones to seal gas and/or oil.

A further complication is the modification of hydrocarbons after entrapment. Kuttan *et al.* (1985) have shown that a fresh water wedge occurs within the Latrobe Group reservoirs in the northwest of the basin. The wedge first developed 2–5 million year ago, as coastal outcrops were uplifted, eroded, and exposed to meteoric waters. Small oil pools which occur in this area are grossly undersaturated, and show evidence of gas stripping by water washing. Where the reservoir temperature is below 30°C, biodegradation is also apparent. Thus, the juxtaposition of small, undersaturated oil pools (e.g. Whiting, Barracouta intra-Latrobe) and large gas fields (Barracouta top Latrobe) can be explained by modest gas stripping by water washing.

Sequence stratigraphic analysis of the Latrobe Group

Introduction

The discussions presented above have concentrated on regional aspects of the Gippsland Basin, in order to show relationships between basin evolution, sedimentation, source development, maturation, migration and entrapment of hydrocarbons. These relationships explain the overall distribution of oil and gas found in the basin to date, and are valuable conceptual tools when risking future drilling targets. However, as targets become smaller and riskier and as the search for stratigraphic traps develops, a detailed model of the basin's stratigraphy and facies architecture is required. Sequence stratigraphy is the best way to construct such a model. This approach represents a marked departure from the more traditional lithostratigraphic approach, because it provides significant additional guidelines for, and constraints to, the way strata are correlated.

Sequence stratigraphic analysis has involved subdivision of the Latrobe Group into a heirarchy of genetically related and time-bounded units (parasequences, systems tracts, and sequences). This approach resulted in the development of a series of cross sections and maps which can be used to predict the distribution of various environments of deposition within the basin. The study utilized a data base consisting of more than 3600 m of core, well logs, seismic, and spore-pollen and dinoflagellate data. Analysis techniques included facies analysis in cores, calibration of these facies with well-logs, construction of well-log chronostratigraphic cross sections, seismic stratigraphic analysis, and construction of integrated seismic facies and paleogeographic maps.

This analysis has allowed more confident prediction of reservoir and seal distribution throughout the basin, resulting in identification of new plays, prospects, and leads. The approach to this study, and illustrative results at both the local and regional scale, are presented below.

Sequence model and terminologies

A depositional sequence is defined as a relatively conformable succession of genetically related strata, bounded by unconformities and their correlative conformities (Mitchum 1977). Sequences are interpreted to be deposited during a eustatic rise and fall. The Latrobe Group is made up of more than 25 such depositional sequences.

Sequences are composed of a predictable succession of genetically linked stratal packages called systems tracts (Van Wagoner et al. 1987). Sequences referred to in this paper mainly correspond with Type-1 sequences defined by Van Wagoner et al. (1987) and Vail (1987), and only contain lowstand, transgressive and highstand systems tracts (Fig. 14).

The principal stratal building blocks of systems tracts are parasequences. A parasequence is defined as a relatively conformable succession of genetically related beds or bed sets, bounded by marine flooding surfaces and their correlative surfaces (Van Wagoner 1985). Parasequences form in response to small-scale perturbations in sediment supply and/or relative changes of sea level which frequently occur during the evolution of a sequence. They consist of a shoaling-upward, progradational lithofacies association of nearshore marine origin.

Stratal patterns and facies geometries of an individual depositional sequence, including the stacking arrangement of its parasequences, are mainly controlled by the interplay of short term eustatic fluctuations, basin tectonics, sediment supply, and basin physiography. Among these variables, the interaction between eustasy and basin tectonics, within each eustatic cycle, leads to a systematic shift in the relative position of sea level. This results in the development of a predictable succession of parasequences and their fluvial counterparts, and a predictable distribution of reservoir and seal facies. For a complete discussion of depositional sequences, the reader is referred to Van Wagoner et al. (1987), Vail (1987) and Posamentier et al. (1989).

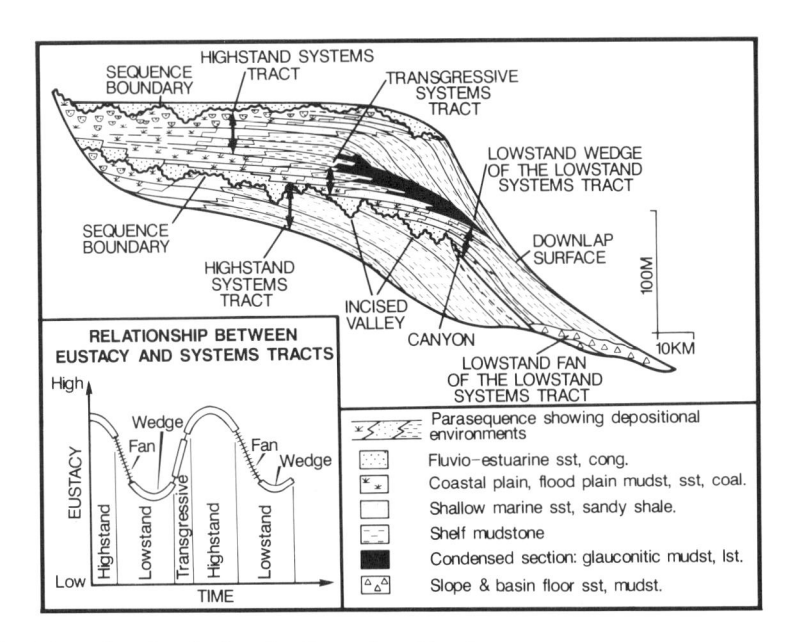

Fig. 14. Physical stratigraphy of a Latrobe Group depositional sequence.

Anatomy and evolution of a typical Latrobe sequence

Figure 14 depicts the physical stratigraphy of a typical depositional sequence in the Gippsland Basin. For any particular sequence, slope and basin-floor deposits of the lowstand systems tract are interpreted to have developed during a eustatic fall, as relative sea level dropped to below the depositional shelf-slope break. At this time, areas landward of the lowstand shoreline were subjected to erosion, sediment bypass, and the formation of incised valleys. Sedimentation in the lowstand wedge occurred in response to a slow relative rise in sea level during the final stages or at the end of the eustatic fall. Initially, the lowstand shorelines prograded basinward as the limited accommodation rate was overwhelmed by sediment supply. The early parts of the lowstand systems tract is thus represented by progradationally stacked marine parasequences and/or fluvial to estuarine incised valley fills (Fig. 14).

Sediment accommodation progressively increased in response to an acceleration of the rise in relative sea level during the late lowstand and transgressive systems tracts. Sediment supply was first balanced and then overwhelmed by the increasing rate of accommodation. Therefore, the depositional records of the late lowstand and transgressive systems tracts are represented by aggradationally and retrogradationally stacked fluvial estuarine incised valley fills as well as near-shore marine parasequences, respectively. During the peak of the transgression, offshore areas were commonly starved of sediments. This condition led to deposition of a condensed section, comprising hemipelagic and pelagic sediments starved of coarse terrigenous material.

As the rate of the relative rise in sea level slowed during the early stages of the highstand systems tract, a balance between basin accommodation and rate of sediment supply was reached once again. This relationship was then altered in favour of sediment supply as a relative fall in sea level began, during the latter part of the highstand. Sediments prograded into the basin, in many cases downlapping onto the transgressive systems tract and condensed section (Fig. 14). The depositional record of the early and late segments of the highstand systems tract in marginal marine areas is, therefore, represented by aggradationally and progradationally stacked parasequences, respectively. In many areas of the Gippsland Basin, these highstand deposits are partially or completely removed by erosion associated with the next sequence boundary.

Distribution of depositional sequences

Stratal patterns in the lower Latrobe Group (Late Cretaceous age) are typical of a late rift depositional history. Large-scale accommodation, generated by fault-controlled subsidence, was balanced by a high rate of sediment supply. This resulted in aggradational stacking of the early Latrobe fluvial and marginal marine sequences (Fig. 15).

Fault-controlled subsidence continued during the early stages of the post-rift phase. This effect coincided with a steady decrease in the rate of sediment supply and a eustatic rise during the Paleocene–Early Eocene (Haq et al. 1987). This process led to rapid transgression and retrogradational stacking of Paleocene- to-Eocene (68–36 Ma) depositional sequences (Fig. 16).

Through the Middle and Upper Eocene (49.5–36 Ma), the eastern and central areas of the basin were starved of siliciclastics as the Latrobe shoreline retrograded to the west and northwest. The depositional record of this interval is represented by the Gurnard Formation

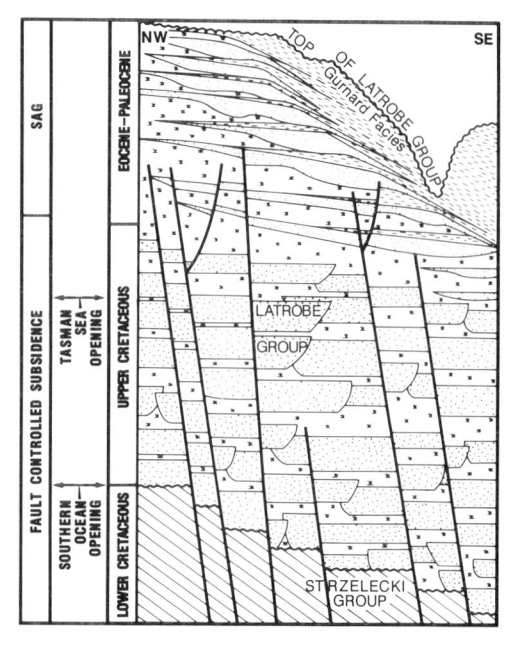

Fig. 15. Schematic diagram showing the distribution of facies within the Latrobe Group and their relationship to regional tectonic events. Facies symbols as for Fig. 14.

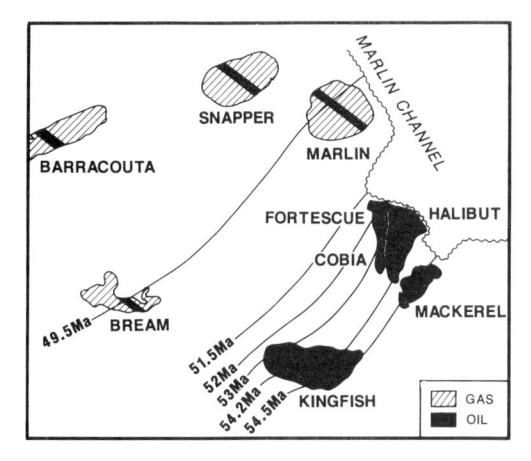

Fig. 16. Maximum landward position of shorelines associated with transgressive system tract deposits of the 54.5 Ma to 49.5 Ma sequences.

(Figs 2 & 15), a glauconitic, sandy mudstone facies. This formation is, therefore, a time-transgressive facies representing condensed sections of Middle to Late Eocene depositional sequences. The Gurnard Formation (up to 50 m thick) acts as a local to semi-regional seal for several of the top Latrobe hydrocarbon accumulations. However, this interval is generally absent in the east of the basin where erosion by channel complexes have produced a marked unconformity at the top of the Latrobe Group (Figs 2, 4 & 5).

The Latrobe Group is also internally characterized by several episodes of canyon formation and fill in the eastern parts of the basin. The oldest major channel system identified to date is located in the southeast of the basin associated with the 68 Ma and 71 Ma sea-level drops at the end of the Cretaceous (Fig. 2). Due to limited well data and structural complexities, the geologic history of these channels is not well understood. Another major canyon formation, formed in the early Eocene, occurs in the eastern portion of the basin, and is known as the Tuna–Flounder Channel (Figs 2, 4 & 8). Its formation south of the Flounder field began with the 53 Ma sea-level drop. Affected by subsequent Eocene sea-level drops (53–50 Ma), the Tuna–Flounder Channel complex evolved in a northerly direction toward the Tuna field, mainly by headward erosion of its canyon.

Sequence framework of hydrocarbon occurrences

Stratigraphic analysis of the Latrobe Group initially involved the detailed study of 14 hydrocarbon fields. This study was conducted to: (i) understand the nature and degree of stratigraphic control on hydrocarbon occurrences, and (ii) develop play concepts that can serve to delineate additional hydrocarbon potential within fields, adjacent to discoveries and in unexplored areas. In order to illustrate this approach and demonstrate the results that were achieved, three fields have been chosen for discussion. They are Flounder, Cobia–Fortescue, and Kingfish (Fig. 1). Together, they represent examples of hydrocarbon entrapment at both the top of the Latrobe Group and within the Latrobe section. These case studies detail hydrocarbon occurrences within the lowstand, transgressive, and highstand systems tracts.

Flounder field

The Flounder field is an intra-Latrobe accumulation, containing reserves in excess of 115 MB oil and 155 BCF wet gas. The main pool occurs approximately 600 m below the top of the Latrobe Group, in a northeast-trending faulted anticline (Figs 1, 4 & 7). The petroleum geology of the field has been described in detail by Young & Coenraads (1986) and Sloan (1987). Our aim is to show the relationships between the sequence framework and the distribution of hydrocarbons.

Figure 17 summarizes the sequence stratigraphic framework of the Flounder field. This framework is based on facies analysis of more than 150 m of core, combined with well-log interpretations.

The 68 and 67 Ma sequence boundaries are both abruptly overlain by massive, heavily bioturbated and slightly glauconitic sandstones of tidal channel origin, identified as lowstand systems tract. Seismic stratigraphic analysis suggests that these sands are associated with broad, incised valleys. Lowstand wedge sandstones of the 68 Ma sequence are overlain by lower shoreface parasequences in the westernmost part of the field (F1 and F6, Fig. 17). These lower shoreface deposits represent the erosional remnant of the transgressive systems tract of this sequence.

Lowstand sandstones of the 67 Ma sequence are overlain by a succession of offshore marine, silty shales and glauconitic mudstones. This unit, below the datum in Fig. 17, represents the transgressive systems tract. The lack of intervening nearshore marine sediment is attributed to a rapid relative rise in sea level resulting in bypass of shoreline sedimentation. The datum is interpreted as the downlap surface of the 67 Ma sequence. The highstand systems tract

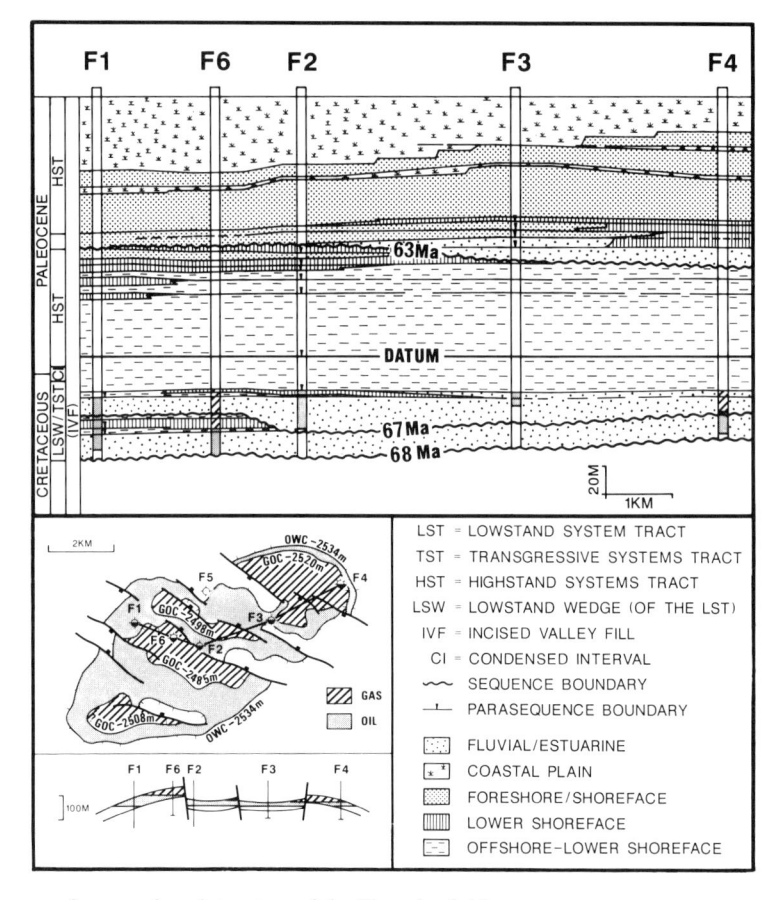

Fig. 17. Sequence framework and structure of the Flounder field.

above the downlap surface is represented by aggradationally and progradationally stacked nearshore marine parasequences.

The 63 Ma sequence contains a thin lowstand to transgressive systems tract, overlain by nearshore marine parasequences of the early and late highstand systems tract.

The principal hydrocarbon accumulation of the Flounder field occurs in lowstand wedge sandstones of the 68 Ma and 67 Ma sequences. Porosities and permeabilities in these units average 20% and 1200 md. Production rates for development wells vary from 1000–4000 barrels per day, with a good water drive. Recovery factors for oil and gas are currently estimated at 55 and 70 percent respectively (Sloan 1987).

Top seal is provided by offshore marine mudstones of the transgressive systems tract and condensed section. The trap is therefore due to the combination of (a) relatively simple anticlinal structure, (b) laterally continuous, high quality reservoir, and (c) thick marine mudstone seal. Faults which dissect the anticline generally have insufficient throw to breach the thick seal provided by the transgressive and early highstand systems tracts.

Fortescue and Cobia–Halibut fields

The Fortescue and Cobia–Halibut fields are large stratigraphic traps at the top of the Latrobe Group (Figs 1, 4, 8 & 18; Thornton et al. 1980; Henzell et al. 1985). Entrapment is provided by westerly dipping strata being truncated by the erosional unconformity at the top of the Latrobe Group. Hydrocarbons occur in several pools, isolated by relatively thin, laterally continuous seals.

The stratigraphic cross section from Fortescue to Cobia (Fig. 18) shows the depositional record of late Paleocene (54.5 Ma and 54.2 Ma) and early Eocene (53 Ma through 51.5 Ma) sequences. This correlation is based on facies analysis of more than 200 m of core, well logs

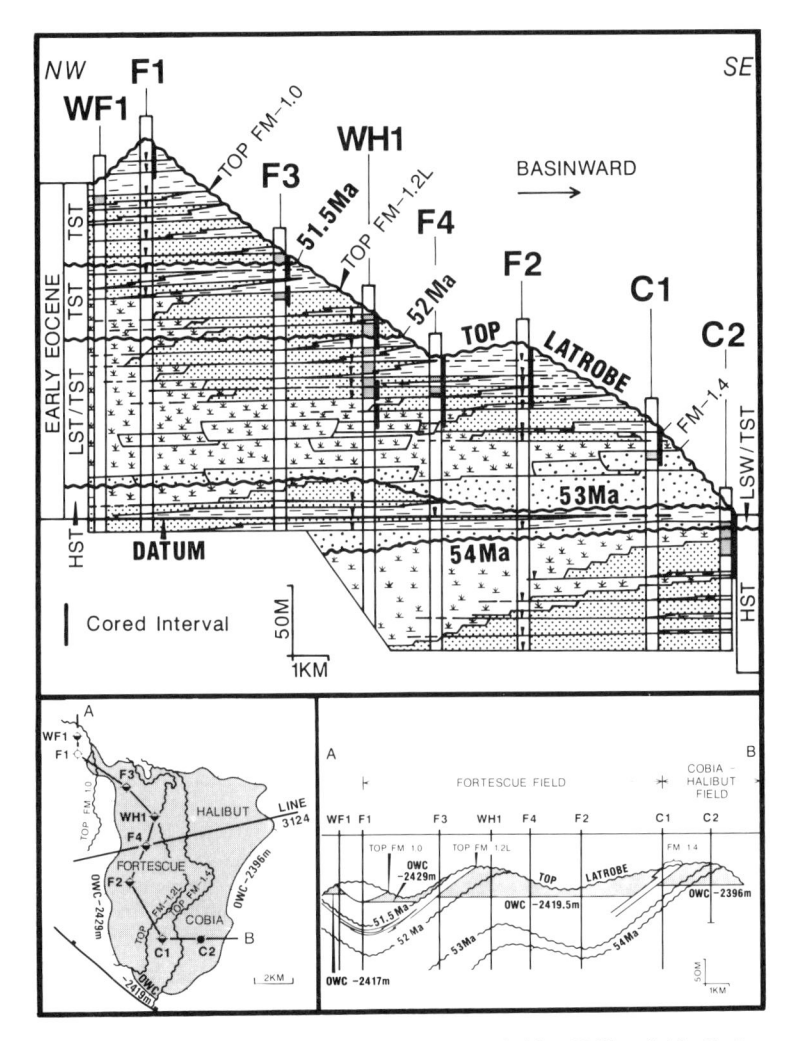

Fig. 18. Sequence framework and structure of the Fortescue and Cobia–Halibut fields. Facies symbols as for Fig. 14.

and seismic data.

Progradationally stacked parasequences (Fig. 18, lower right of stratigraphic section) represent the late highstand systems tract of the 54.5 Ma sequence. Seismically, this systems tract is represented by a series of progradational geometries (Fig. 19).

The 54.2 Ma sequence boundary is identified from seismic and well logs by its highly erosive character. Sediments of the 54.2 Ma sequence comprise blocky, fluvial/estuarine sandstones filling incised valleys of the lowstand systems tract, overlain by a thin transgressive systems tract. The cross section is datumed on a flooding surface that marks the Eocene–Palaeocene boundary. This surface also represents the top of the transgressive systems tract and is interpreted as the downlap surface of the 54.2 Ma sequence. Progradational high-stand parasequences are preserved to the west of Fortescue-4, but are truncated by the 53 Ma sequence boundary to the east.

Braided stream sediments characterize lowstand strata of the 53 Ma sequence. The overlying transgressive systems tract is well developed and from base to top, includes fluvial, coastal plain, nearshore marine, and offshore strata. These parasequences are retrogradationally stacked. The 52 Ma and 51.5 Ma sequences are also dominated by transgressive deposits.

The illustrated depositional sequences of

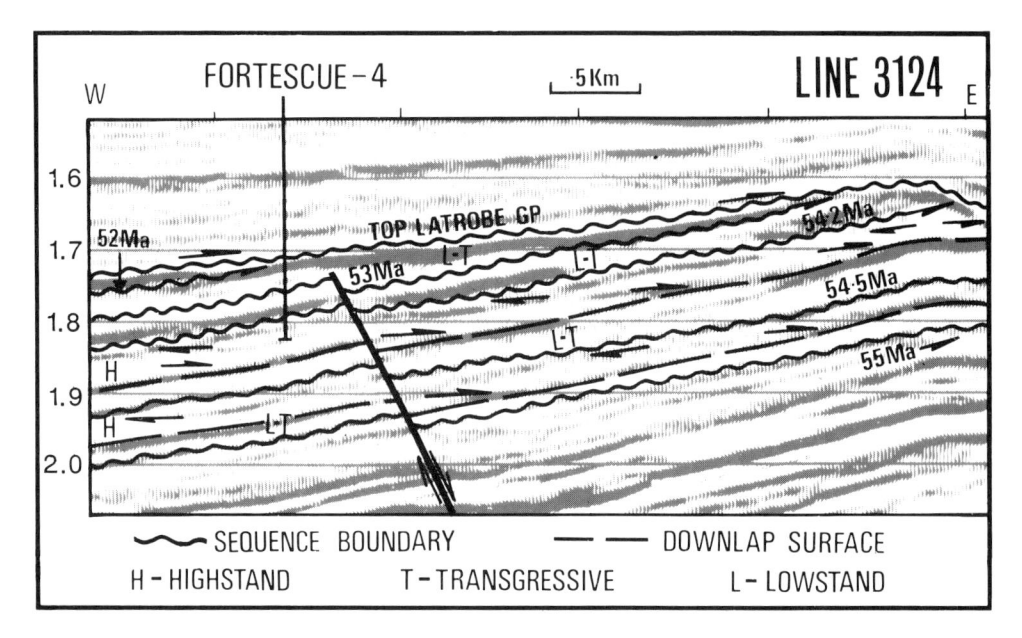

Fig. 19. Seismic section across Fortescue and Cobia–Halibut fields. Location of line shown on Fig. 18.

Fortescue and Cobia display an overall transgressive arrangement, attributed to large, basin-wide accommodation that resulted from the combined effect of tectonic subsidence and eustatic rise. The Fortescue field, like many other examples in the Gippsland Basin, contains hydrocarbons in the lowstand wedge and early transgressive systems tracts. The main reservoirs are sandstones of foreshore and shoreface origin, belonging to the 53 Ma and 52 Ma sequences. Porosities in the field average 20%, with permeabilities up to 5000 millidarcies. Production is supported by a very strong water drive, with a recovery factor estimated at 67% (Henzell 1987). Seals between the various oil pools comprise offshore marine mudstones and minor coastal plain deposits of the transgressive systems tract.

The lateral seal between the Fortescue and Cobia–Halibut pools is a thin succession of coastal plain strata belonging to the 53 Ma transgressive systems tract. Hydrocarbons in Cobia–Halibut are mainly reservoired in fluvio-estuarine, lowstand wedge sandstones of the 53 Ma sequence. However, they also occur in highstand deposits of the 54.5 Ma sequence (Fig. 18), making this field one of the few examples where a significant volume of hydrocarbons is trapped within the highstand systems tract. Porosities for Cobia–Halibut average 22%, permeabilities range up to 7000 md, and the recovery factor is estimated at 71%. Cap seal for both Cobia-Halibut and Fortescue is provided by marls and offshore marine shales of the Lakes Entrance Formation.

The Fortescue field, with 280 MB reserves, was discovered in 1978, eleven years after Halibut. This delay was due largely to the unknown stratigraphic component of the Fortescue trap. After the discovery, some uncertainty existed regarding the precise correlation of reservoirs within the Fortescue field. Sequence stratigraphy has been a valuable tool in resolving these problems.

Kingfish field

The Kingfish field is a large and complex hydrocarbon accumulation, located in the southeast of the basin (Figs 1, 4, 8 & 11). Initial reserves are estimated at 1.3 billion barrels crude/condensate, 170 MB LPG and 340 GSCF dry gas. Two separate oil pools are present, and are known as East and West Kingfish (Figs 20 & 21). They are separated by westerly dipping, coastal plain deposits belonging to the highstand systems tract of the 54.2 Ma sequence.

The East Kingfish accumulation is a sub-unconformity stratigraphic trap, showing similarity to Cobia–Halibut, with westerly dipping strata truncated by a major erosional unconformity at the top of the Latrobe Group. Hydrocarbons are mainly trapped within shoreface to foreshore sandstones of the transgressive and

for oil is currently estimated at 68%.

The West Kingfish has an additional stratigraphic component, since the top of the reservoir does not coincide with the top of the Latrobe Group (Fig. 20). Instead, a favourable stratigraphic trap geometry is generated as westerly dipping 53 Ma reservoir units pinchout updip into their correlative offshore marine mudstones. Thus oil in West Kingfish accumulated in a stratigraphic trap, within nearshore marine sandstones of the 53 Ma transgressive systems tract.

Discussion

The Latrobe Group contains abundant reservoirs, of both marginal marine and fluvial origin. Thus, the main challenge lies primarily in predicting the occurrence and distribution of sealing units. The results of detailed sequence stratigraphic analysis offer significant clues to this problem.

Hydrocarbon occurrences, in more than 80% of the major fields, are associated with sediments of the lowstand and transgressive systems tracts. These hydrocarbons are sealed by offshore marine, coastal-plain, and flood-plain shales of the transgressive systems tract, both at the top and within the Latrobe section. At the top of the Latrobe Group, these shales are the Gurnard and basal Lakes Entrance Formations, while within the Latrobe Group, they are interstratified seals.

Sediments of the highstand systems tract are characterized by a poor preservation potential, because they are often subjected to erosion associated with the overlying sequence boundary. When present, these strata, relative to their lowstand and transgressive counterparts, are often characterized by poorer internal seals. It is largely for this reason that most of the hydrocarbon occurrences found within the highstand systems tracts (e.g. East Kingfish and Cobia–Halibut fields) are in sub-unconformity traps and are thus dependant on non-indigenous seals (provided by the transgressive systems tract of an overlying sequence).

Integrated palaeogeographic and seismic facies maps of the 55 Ma sequence (Fig. 22) depict this evolution of reservoir and seal units during the course of lowstand and transgressive systems tract development. The lowstand shoreline of the 55 Ma sequence was developed as an estuarine coastal complex flanked by a network of incised valleys (Fig. 22(a)). The incised valley complex was carved into the underlying sequence during the preceding eustatic fall.

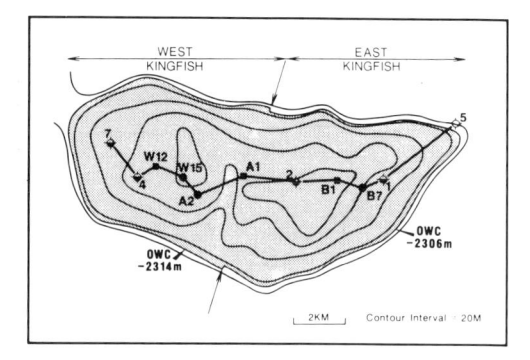

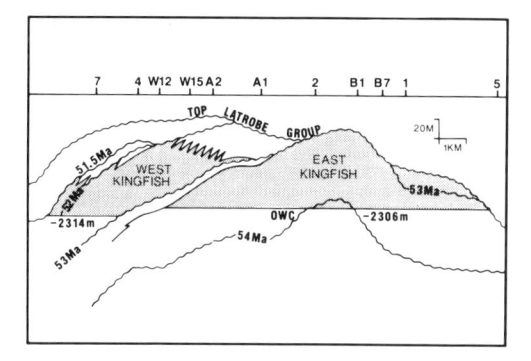

Fig. 20. Structure of the Kingfish field. Map shows structure on top of reservoir. Cross-section shows how this relates to various sequence boundaries and sealing lithologies.

highstand systems tracts of the 54.2 Ma sequence, which provide excellent reservoirs. Field wide porosities are estimated at 21 percent, with permeabilities in some areas exceeding 5000 millidarcies. The recovery factor

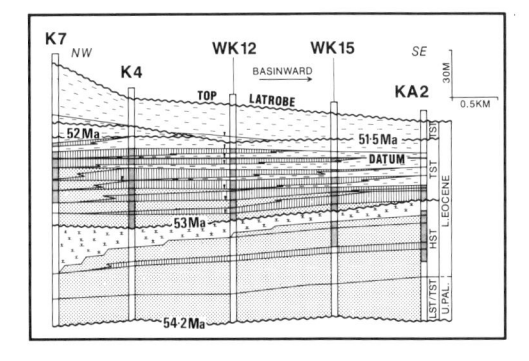

Fig. 21. Sequence framework of West Kingfish. Location of cross-section shown in Fig. 20. Facies symbols as for Fig. 17.

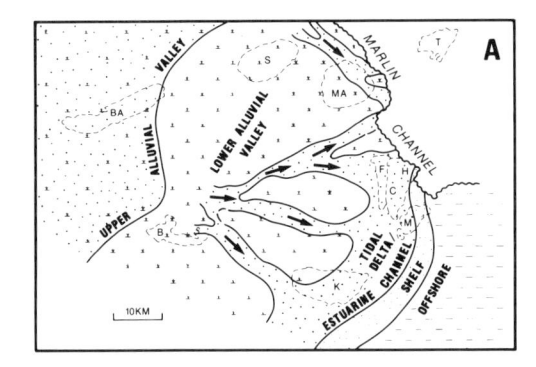

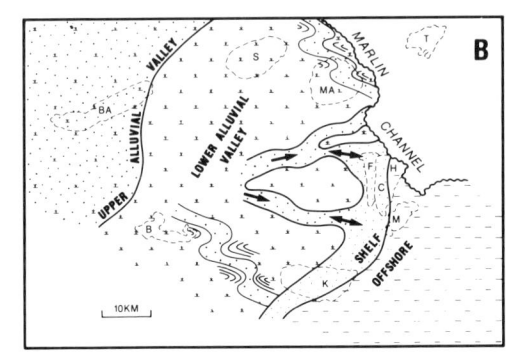

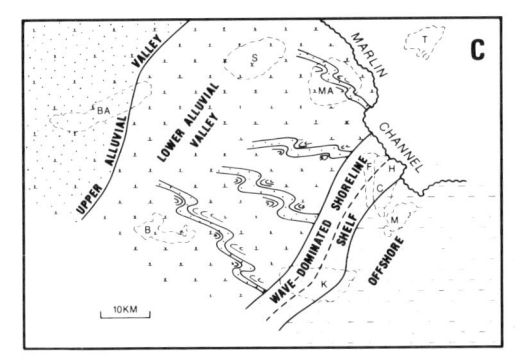

Fig. 22. Evolution of the 55 Ma sequence.
A: lowstand systems tract, B: early transgressive systems tract, C: late transgressive systems tract. Facies symbols as for Fig. 14.

During late lowstand wedge time, segments of the river valleys that were affected by the relative rise in sea level experienced a change from stream-bed erosion and sediment bypass to aggradational, sand-prone, estuarine-to-fluvial sedimentation. These standstones commonly have excellent reservoir characteristics.

Rapid sea-level rise during the transgressive systems tract caused widespread deposition of offshore marine seals on top of lowstand wedge estuarine reservoirs (Fig. 22(b)). Nearshore marine sandstones became interbedded with offshore mudstones as they backstepped under the influence of rising relative sea level. This process led to a retrogradational stacking arrangement of nearshore marine parasequences, with the development of multiple reservoir-seal units (Fig. 22(b, c)).

Landward of the retrograding shoreline, sand-prone alluvial aggradation gave way to fine-grained, flood-plain aggradation in response to a rapid rise in stream-base level during the transgressive systems tract. Vast areas of the alluvial and coastal plain were covered by lakes and swamps, leading to a predominance of fine-grained, flood-plain deposits, extensive coal beds, and a paucity, of major fluvial channel deposits. Consequently, these deposits represent potential seals for underlying fluvial lowstand wedge reservoirs (Fig. 22(b, c)).

The usefulness of sequence stratigraphy as a tool for explaining the distribution of reservoir and seal facies is well documented in the Gippsland Basin. However, the ability to extrapolate the framework away from well control and to high-grade leads inferred by well correlations critically depends on our ability to accurately map sequences, system tracts and parasequences on seismic data. Seismic stratigraphy can achieve these objectives: however, its worth is constrained by the limitations imposed by data quality, resolution and data density. In the Gippsland Basin the mapping of individual parasequences and subtle facies changes have been, to a large extent, beyond seismic resolution (bandwidth:8–40 Hz; minimum resolvable bed thickness 20 m). Recent advances in offshore acquisition technology have provided data of improved quality and bandwidth, leading to optimism regarding the future applications of this technique.

Conclusions

Exploration in the Gippsland Basin is now at a reasonably mature stage. Important technical challenges facing explorers include (a) adequate prediction of risk associated with small, intra-Latrobe structural prospects, and (b) identification of prospects which are partially or fully stratigraphic in nature. With this in mind, a major regional study of Esso's acreage was recently carried out. The results emphasise that the main controls on the distribution of hydrocarbons at the top of the Latrobe Group are maturation, migration, fresh-water flushing and biodegradation, in that order of importance. Source variations are difficult to quantify, due to the fact that few wells penetrate to the level of peak maturation. However, geochemical data on the oils, gases and source rocks suggest that there was a widely distributed and uniform non-marine source for the hydrocarbons.

The distribution of oil and gas within the Latrobe Group is more complicated, and depends to a greater degree on the presence or absence of cap seals and the adequacy of fault seals. Thus, in order to be able to search economically for these traps, a detailed, reliable model of the distribution of reservoir seal units was required. This model was constructed using Exxon's latest techniques in sequence stratigraphy. It has already proved to be valuable in field evaluation, and will be fully tested in the exploration arena in the next few years, as the focus in the Gippsland Basin moves increasingly towards intra-Latrobe and non-structural plays.

References

ALEXANDER, R., KAGI, R. I., WOODHOUSE, G. W. & VOLKMAN, J. K. 1983. The geochemistry of some biodegraded Australian oils. *Australian Petroleum Exploration Association Journal*, 23(1), 53–63.

——, NOBLE, R. A. & KAGI, R. I. 1987. Fossil resin biomarkers and their application in oil to source-rock correlation, Gippsland Basin, Australia. *Australian Petroleum Exploration Association Journal*, 27(1), 63–72.

BATTRICK, M. A. 1986. The West Seahorse and Seahorse Fields: fluvial, lacustrine and deltaic deposition in the Eocene Latrobe Group, Gippsland Basin. *In*: GLENIE, R. C. (ed.) *Second South-Eastern Australian Oil Exploration Symposium*, Petroleum Exploration Society of Australia, 115–158.

BEIN, J., GRIFFITH, B. R. & SVALBE, A. K. 1973. The Kingfish field—offshore Gippsland Basin. *Australian Petroleum Exploration Association Journal*, 13(1), 68–72.

BROOKS, J. D. & SMITH, J. W. 1969. Coalification and the formation of oil and gas in the Gippsland Basin. *Geochimica et Cosmochimica Acta.*, 33, 1183–94.

BROWN, B. R. 1986. Offshore Gippsland Silver Jubilee. *In*: GLENIE, R. C. (ed.) *Second South-Eastern Australian Oil Exploration Symposium*, Petroleum Exploration Society of Australia, 29–56.

BURNS, B. J., BOSTWICK, T. R. & EMMETT, J. K. 1987. Gippsland terrestrial oils—recognition and biodegradational effects. *Australian Petroleum Exploration Association Journal*, 27(1), 73–84.

——, JAMES, A. T. & EMMETT, J. K. 1984. The use of gas isotopes in determining the source of some Gippsland Basin oils. *Australian Petroleum Exploration Association Journal*, 24(1), 217–21.

COCHRAN, J. R. 1983. Effects of the finite rifting times on the development of sedimentary basins. *Earth & Planetary Science Letters*, 66, 289–302.

ESSO AUSTRALIA LTD. 1988. The Gippsland Basin. *In*: *Petroleum in Australia: the first century*, Australian Petroleum Exploration Association, 174–90.

GLENTON, 1988. The Snapper development, Gippsland Basin. *Australian Petroleum Exploration Association Journal*, 28 (1), 29–40.

GOFF, J. C. 1983. Hydrocarbon generation and migration from Jurassic source rocks in the East Shetland Basin and Viking Graben of the Northern North Sea. *Journal of the Geological Society London*, 140, 445–74.

HAQ, B. V., HARDENBOL, J. R. & VAIL, P. R. 1987. Chronology of fluctuating sea levels since the Triassic. *Science*, 235, 1156–67.

HEGARTY, K. A. 1985. Origin and evolution of selected plate boundaries. Ph.D Thesis, Lamont-Doherty Geological Observatory, Columbia University, NY.

——, DUDDY, I. R., GREEN, P. F., GLEADOW, A. J. W., FRASER, I. & WEISSEL, J. K. 1986. Regional evaluation of the tectonic and thermal history of the Gippsland Basin. *In*: GLENIE, R. C. (ed.) *Second South-Eastern Australia Oil Exploration Symposium*, Petroleum Exploration Society of Australia, 65–74.

——, WEISSEL, J. K. & MUTTER, J. C. 1989. Subsidence history of Australia's southern margin: constraints on basin models. *American Association of Petroleum Geology Bulletin*.

HELBY, R., MORGAN, R. & PARTRIDGE, A. D. 1987. A palynological zonation of the Australian Mesozoic. *Memoir Association Australasian Palaeontologists*, 4, 1–94.

HENZELL, S. T., IRRGANG, H. R., JENSSEN, E. J., MITCHELL, R. A. M., MORRELL, G. O., PALMER, I. D., SEAGE, N. W., HICKS, G. J., HORDERN, M. J. & KABLE, C. W. 1985. Fortescue reservoir development and reservoir studies. *Australian Petroleum Exploration Association Journal*, 25 (1), 95–106.

HOUSEMAN, G. A. & HEGARTY, K. A. 1987. Did rifting on Australia's southern margin result from tectonic uplift? *Tectonics*, 6, 515–27.

JARVIS, G. T. & MCKENZIE, D. P. 1980. Sedimentary basin formation with finite extension rates. *Earth & Planetary Science Letters*, 48, 42–52.

KUTTAN, K., KULLA, J. B. & NEWMAN, R. G. 1986. Freshwater influx in the Gippsland basin: impact on formation evaluation, hydrocarbon volumes, and hydrocarbon migration. *Australian Petroleum Exploration Association Journal*, 26(1), 242–9.

LIMBERT, A. R., GLENTON, P. N. & VOLARIC, J. 1983. The Yellowtail oil discovery. *APEA Journal*, 23(1), 170–81.

LOWRY, D. C. 1987. A new play in the Gippsland Basin. *Australian Petroleum Exploration Association Journal*, 27(1), 164–172.

MARLOW, R. B. 1978. Golden Beach: a bright spot. *Australian Petroleum Exploration Association Journal*, 18(1), 109–15.

MARSHALL, N. G. 1989a. An unusual assemblage of algal cysts from the Late Cretaceous of the

Gippsland Basin, southeastern Australia. *Earth Resources Foundation Report* 1988/1, 1−51.

—— 1989b. A Santonian dinoflagellate assemblage from the Gippsland Basin, southeastern Australia. *Memoir Association Australas Palaeontologists*, 5.

McKenzie, D. P. 1978. Some remarks on the development of sedimentary basins. *Earth & Planetary Science Letters*, 40, 25−32.

Mitchum, R. M. 1977. Seismic stratigraphy and global changes of sea level Part 1: glossary of terms used in seismic stratigraphy. *In*: Payton, C. E. (ed.) *Seismic stratigraphy. Applications to hydrocarbon exploration*, American Association Petroleum Geology Memoir 26, 205−12.

O'Byrne, M. J. & Henderson, D. J. 1983. The Tuna Field: a recent development. *APEA Journal*, 23(1), 44−52.

Ozimic, S., Nicholas, E., Pain, L. & Vuckovic, V. 1987. Gippsland Basin, Victoria. *Bureau of Mineral Resources, Geological Geophysics, Australian Petroleum Accumulations Report*, 3.

Philp, P. R. & Gilbert, T. D. 1982. Unusual distribution of biological markers in an Australia crude oil. *Nature*, 299, 244−7.

Posamentier, H. W., Jervey, M. T. & Vail, P. R. 1989. Eustatic controls on clastic deposition. *In*: Van Wagoner, J. C. (ed.). *Sea level change−an integrated approaches*. Society of Economic Paleontology & Mineralogy Special Publication.

Powell, T. G. 1987. Depositional controls on source rock character and crude oil composition. *Proceedings 12th World Petroleum Congress*, 1(4), 31−42.

Roder, & Sloan, M. W. 1986. Barracouta: history of exploration and development, and geology of the field. *In*: Glenie, R. C. (ed.). *Second South−Eastern Australian Oil Exploration Symposium*, Petroleum Exploration Society of Australia, 75−90.

Sclater, J. G. & Christie, R. A. F. 1980. Continental stretching: an explanation of the post-mid-Cretaceous subsidence in the central North Sea Basin. *Journal of Geophysical Research*, 85, 3711−39.

Sloan, M. W. 1987. Flounder − a complete intra-Latrobe oil and gas field. *Australian Petroleum Exploration Association Journal*, 27(1), 308−17.

Smith, G. C. 1981. Tertiary and Upper Cretaceous coals and coal measure sediments in the Bass and Gippsland Basins. PhD. Thesis, University of Wollongong, NSW.

—— & Cook, A. C. 1984. Petroleum occurrence in the Gippsland Basin and its relationship to rank and organic matter type. *Australian Petroleum Exploration Association Journal*, 24(1), 196−216.

Thornton, R. C. N., Burns, B. J., Khurana, A. K. & Rigg, A. J. 1980. The Fortescue Field − new oil in the Gippsland Basin. *Australian Petroleum Exploration Association Journal*, 20(1), 130−42.

Threlfall, W. F., Brown, B. R. & Griffiths, B. R. 1976. Gippsland Basin, offshore. *In*: Leslie, R. B., Evans, M. J. & Knight, C. L. (eds) *Economics Geology of Australia and Papua New Guinea. 3. Petroleum* Australian Institute of Mining & Metallurgy.

Vail, P. R. 1987. Seismic stratigraphy interpretation using sequence stratigraphy. Part 1: seismic stratigraphy interpretation procedure. *In*: Bally, A. W. (ed.) *Atlas of seismic stratigraphy volume 1*. American Association of Petroleum Geology, Studies in Geology 27.

Van Wagoner, J. C. 1985. Reservoir facies distribution as controlled by sea-level change. *Society Economic Paleontology & Mineralogy, Mid-year Meeting, Golden, Colorado, Abstracts*, 91−2.

——, Mitchum, J. R. & Posamentier, H. W. 1987. Seismic stratigraphy interpretation using sequence stratigraphy. Part 2: key definitions of sequence stratigraphy. *In*: Bally, A. W. (ed.) *Atlas of seismic stratigraphy, volume 1*. American Association of Petroleum Geology, Studies in Geology 27.

Veevers, J. J. 1984. *Phanerozoic Earth History of Australia*. Clarendon, Oxford.

——, 1986. Breakup of Australia and Antarctica estimated as mid-Cretaceous (95 ±5 Ma) from magnetic and seismic data at the continental margin. *Earth & Planetary Science Letters*, 77, 91−9.

Williamson, P. E., O'Brien, G. W., Swift, M. G., Felton, E. A., Scherl, A. S., Lock, J., Exon, N. F., Falvey, D. A. & Marlow, M. 1987. Hydrocarbon potential of the offshore Otway Basin. *Australian Petroleum Exploration Association Journal*, 27(1), 173−94.

Young, A. J. & Coenraads, R. R. 1986. A 3D seismic interpretation − Flounder Field, Gippsland Basin. *In*: Glenie, R. C. (ed.) *Second South−Eastern Australian Oil Exploration Symposium*, Petroleum Exploration Society of Australia, 97−108.

Aspects of the petroleum geology of the Junggar Basin, Northwest China

S. R. LAWRENCE

Exploration Consultants Limited, Highlands Farm, Greys Rd, Henley-on-Thames, Oxfordshire RG9 4PS, UK (Present address: Quad Consulting, Churchill House, St Aldates, Oxford OX1 1BN, UK)

Abstract: The Junggar Basin is one of a family of productive interior basins in Northwest China. These are generally referred to as 'flexural' or 'orogenic' basins nucleated by Variscan orogenic episodes. Their uniqueness is enhanced by later rejuvenation and re-activation in the successive orogenic phases affecting Central Asia.

The northwest margin of the Junggar Basin has been productive since the discovery of the Karamay–Wuerhe group of fields in the 1950s. The source rocks are generally considered to be Permian lacustrine sediments which provide mostly liquid hydrocarbons to basin margin alluvial fan reservoirs of Permian, Triassic and Jurassic age. The Karamay Thrust Belt, which forms the northwestern margin of the Junggar Basin, was initially formed during Variscan movements and re-activated during several subsequent structural episodes. According to KOC work there are many trap permutations associated with the Thrust Belt in 'hanging-wall' and 'footwall' situations.

Recent studies have focused on the southwestern part of the Junggar Basin. Here marginal production from Himalayan periclinal structures adjacent to the Tien Shan, dates back to the beginning of the century. More recently Carboniferous 'basement' rocks on the southward extension of the Karamay Thrust Belt, the Hong Che Fault Belt, have proved productive.

The stratigraphy shows the fairly continuous existence of lacustrine conditions in the mid basin since the Late Palaeozoic. Other studies have shown that 'black mudstone' conditions are prevalent in the Permian, Triassic, Jurassic and Early Tertiary and that oil-prone algal-rich intervals, indicating permanent or semi-permanent anoxia, are most likely in the Permian. The stratigraphic development also demonstrates the occurrence of suitable reservoir units in marginal alluvial fan–fluviatile clastics at most stratigraphic levels. The source–reservoir spatial relationships demonstrate favourable conditions for primary and secondary migration. The maturation history shows 'deep' Permian source rocks becoming mature for oil generation in the Triassic and for gas-generation from Cretaceous times. More gas-prone higher level source rocks generated hydrocarbons progressively from the Cretaceous through the Tertiary.

The structural history shows migration routes controlled by regional tilting into the Tien Shan foredeep, and local migration and entrapment by re-activation of the Karamay Thrust Belt along the northwestern basin margin, and of intra-basinal late Variscan transpressional trends. Late Himalayan structuring along the southern margin and basin-wide tilt has effected a phase of re-migration and further funnelling of hydrocarbons to the productive zones of the northwest.

The Junggar Basin is one of a family of productive basins in Northwest China (Fig. 1). In Junggar, major production dates back to the mid-1950s with the development of the Karamay oilfield which lies in the extreme northwestern part of the basin. However, commercial production was first established from the Dushanzi oilfield in the southwest in the early 1900s. This history, and the fact that the Karamay area was until recently the most important production area in China, I believe adequately qualifies it as a 'Classic Petroleum Province'. Exploration and development in the Karamay oilfield and the Junggar Basin is administered by the Karamay Oil Company (KOC). The work which forms the basis for this paper was commissioned by KOC.

Plate tectonics and basin formation

The nature and origin of the so-called 'flexural' or 'orogenic' basins of Northwest China is still a subject of some debate. A distinction can be made between '*inter*-sutural' basins such as Junggar and Tarim ostensibly located upon depressed continental crust, and 'intra-sutural' basins such as Tulufan and Tsaidam which are set within the suture zones or mountain belts.

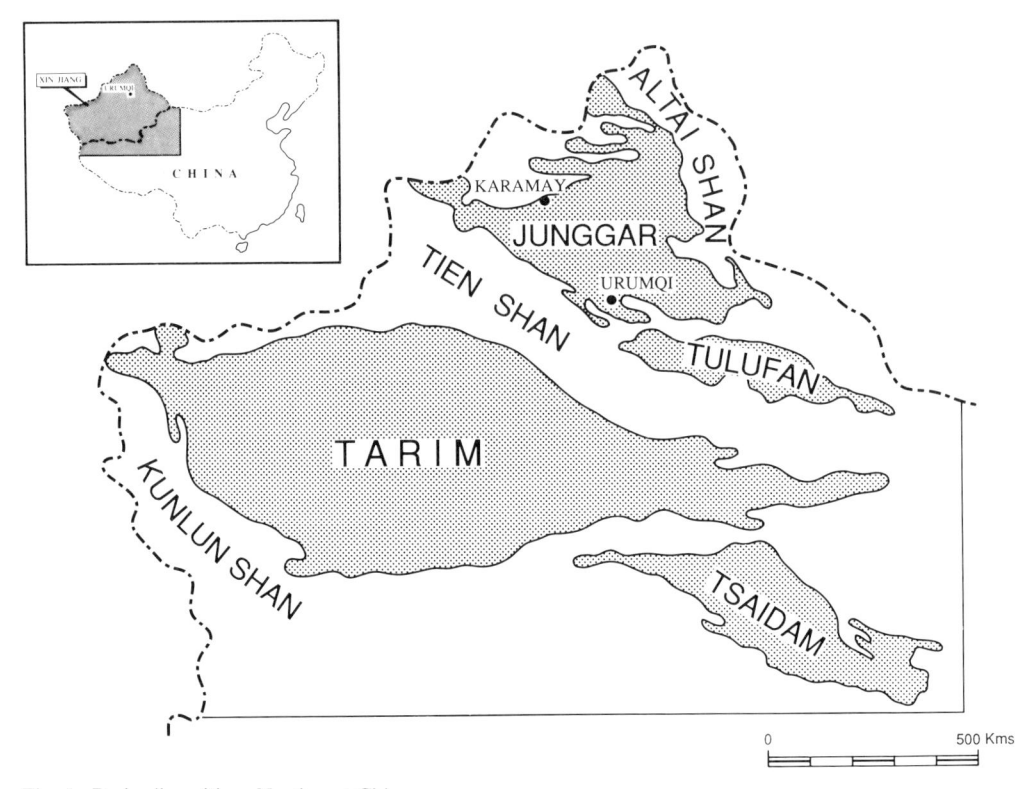

Fig. 1. Basin disposition: Northwest China.

There is little doubt that these *inter*-sutural basins originated in the Late Carboniferous–Early Permian as foredeep–foreland basins with respect to Variscan collisional episodes in Central Asia (Fig. 2). However, their subsequent development is linked to re-activation of suture zones during Cimmerian (Indosinian and Yanshanian) and Himalayan orogenic episodes. The subsidence is thought to be in the form of flexural downbows depressed under the tectonic load of thrust sheets associated with the uplifted mountain ranges (see Watson *et al.* 1987).

The Junggar Basin, the second largest of this group of basins at 130 000 km^2, exemplifies this complexity in its structural architecture and history (Fig. 3). The basin is dominated in the northwest by the Karamay Thrust Belt where thrusted sequences of Carboniferous rocks form part of the Junggar Border Mountains. The fabric of the basin itself is formed by a series of northwest–southeast oriented uplifts and depressions which broadly parallel the bounding suture zones of the Altai Shan and Tien Shan to the north and south. Thus the conventional structural picture shows Carboniferous and Permian rocks involved in large-scale overthrusting along the northwest margin, Permian compression-related uplift within the basin itself and the effects of Himalayan compressional folding and high-angle reverse faulting along the southern margin in front of the Tien Shan.

Published plate tectonic accounts have largely ignored the origin and evolution of the Junggar Border Mountains (see Zhang *et al.* 1984). The Kazakhstan Block is usually depicted as a single continental block underlying the Junggar Basin and stretching beyond its limits to the northwest. However, the Junggar Border Mountains appear to display all the features of B-subduction notably ophiolites, calc-alkaline volcanics and accretionary sediments. Unless a wide suture zone extends under the Junggar Basin, to form a link between the Tien Shan and Altai Shan orogenic systems, then the Junggar Border Mountains (including the Karamay Thrust Belt) must represent a collision zone between two previously separate parts of the Kazakhstan Block. This is the preferred interpretation for this paper. The nature of the Junggar Collision Zone (Fig. 4) depicts a history of oceanic consumption and continental impingement pro-

reserves are estimated unofficially between 200 and 300 million barrels. As of 1984 Karamay was producing 75 000 barrels per day. Up to this date a total of eight oilfields have been discovered in the Karamay–Wuerhe group along the Karamay Thrust Belt, producing from reservoirs at a variety of stratigraphic levels (see Ulmishek 1984).

Traps for Permian and Carboniferous reservoirs occur in 'footwall' positions either in pro-thrust structures or against thrust slices. Triassic reservoirs occur in traps abutting the thrust blocks, whilst Jurassic production comes from stratigraphic traps formed by onlap or wedge-out in the section overstepping the thrust-belt (see Fig. 5 which is modified after Wang Hansheng 1986; see also Xie Hang et al. 1984; Zhang 1984). Reservoirs are nearly always provided by proximal alluvial fan clastics or fluvia-tile sandstones. The major source rocks are considered to be Permian lacustrine shales according to KOC information.

Recent studies have focused on the southwestern part of the Junggar Basin. Here marginal production from one of the Himalayan periclinal structures adjacent to the Tien Shan, the Dushanzi oilfield, dates back to the beginning of the century. More recently Carboniferous so-called 'basement' rocks on the southward extension of the Karamay Thrust Belt, the Hong Che Fault Belt, have proved productive. Studies in this part of the basin have served to highlight some of the important aspects of the petroleum geology of the western half of the basin as a whole.

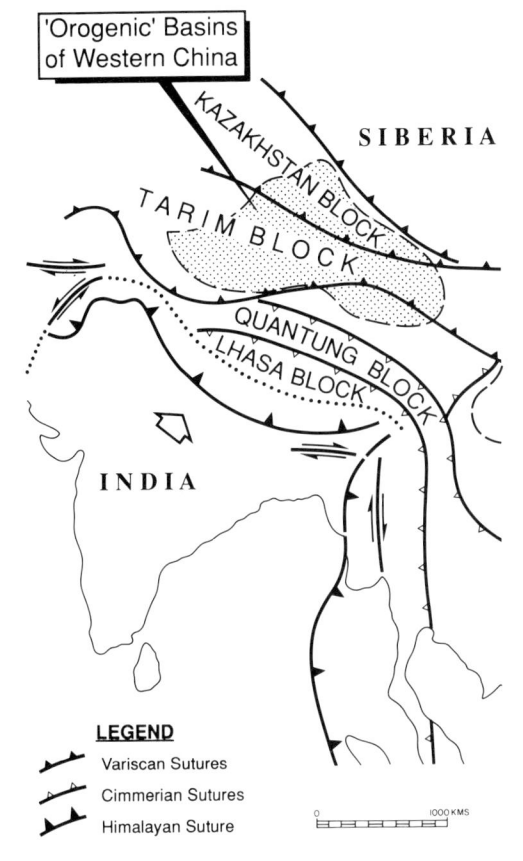

Fig. 2. Regional structural setting: Central Asia.

ceeding from the Early Palaeozoic to the Late Carboniferous.

The Junggar Basin was created therefore by subsidence related to foreland depression on three sides, with Altai Shan formation to the north, Tien Shan formation to the south and the Junggar Border Mountains to the northwest. Although unlikely to be synchronous, all collisional episodes broadly fall within the time range Mid-Carboniferous to Mid-Permian and can be categorised as 'Variscan'.

Outline of petroleum geology of Karamay

With the exception of the Dushanzi Field, the major commercial success has been along the Karamay Thrust Belt. Early drilling was based on the occurrence of numerous seeps and tar sands around Karamay. The Karamay oilfield was discovered in 1955. Original production was from the Jurassic and Triassic sections, but is now dominated by Triassic reservoirs. Oil

Conditions for petroleum generation and expulsion

Stratigraphic development

The stratigraphic development of the northwestern Junggar Basin can be rationalized into 'orogenic', that is leading up to Variscan collisional events, and 'epeirogenic', that is development as a 'flexural' basin following Variscan consolidation (Fig. 6).

The 'orogenic' phases are explained in terms of the formation of a foredeep in the future Junggar Collision Zone. Early and Mid Carboniferous sediments and volcanics form the western part of the Karamay Thrust Belt, formed in an early marine foredeep which developed on the old passive margin of the Junggar Block, and in front of the impinging active margin of the Kazakhstan Block. Foredeep development continued into the Late Carboniferous–Early Permian with subsidence induced

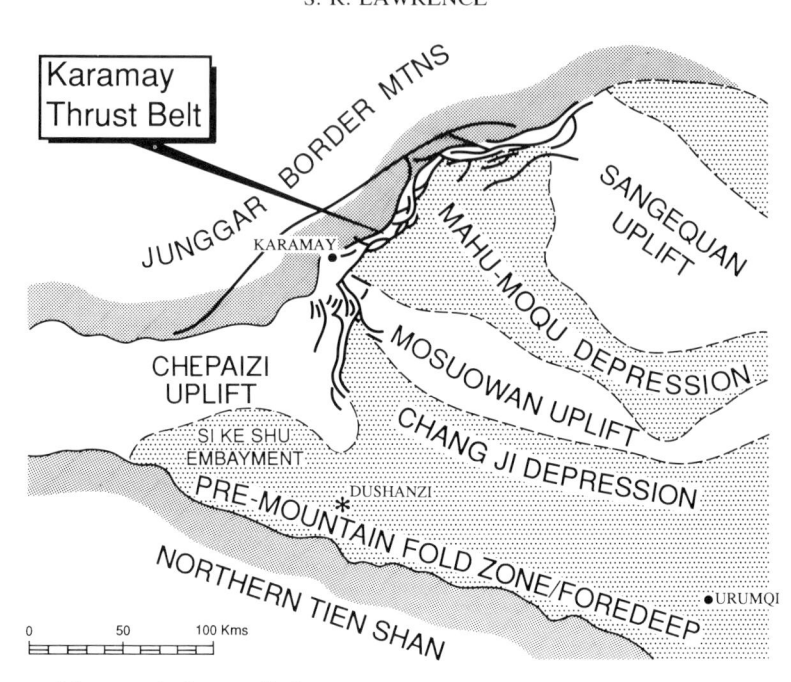

Fig. 3. Structural framework: Junggar Basin.

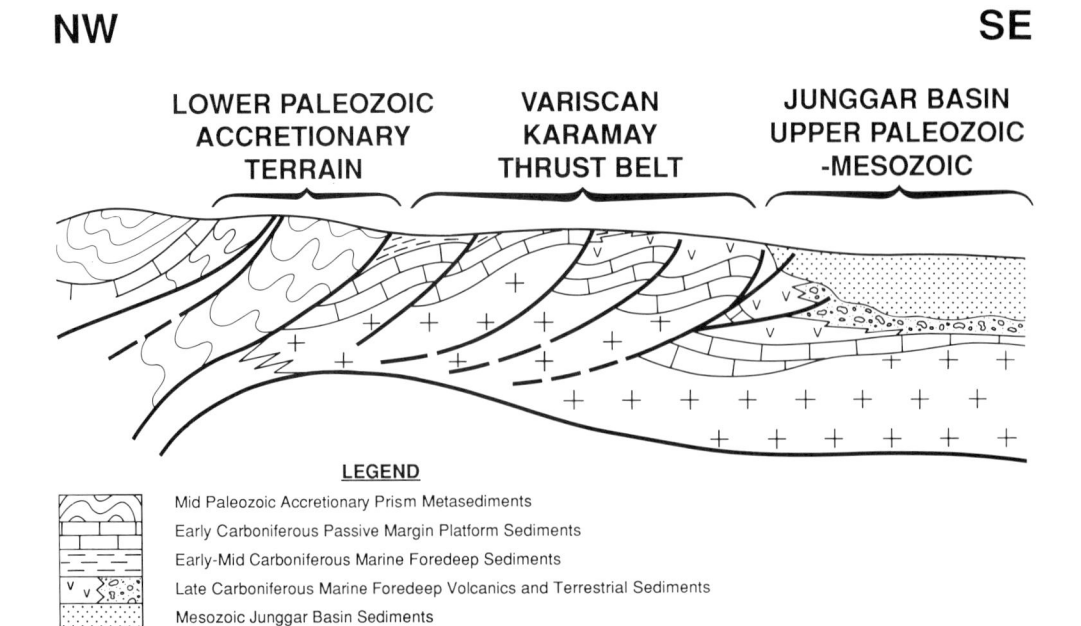

Fig. 4. Architecture: Junggar Collision Zone.

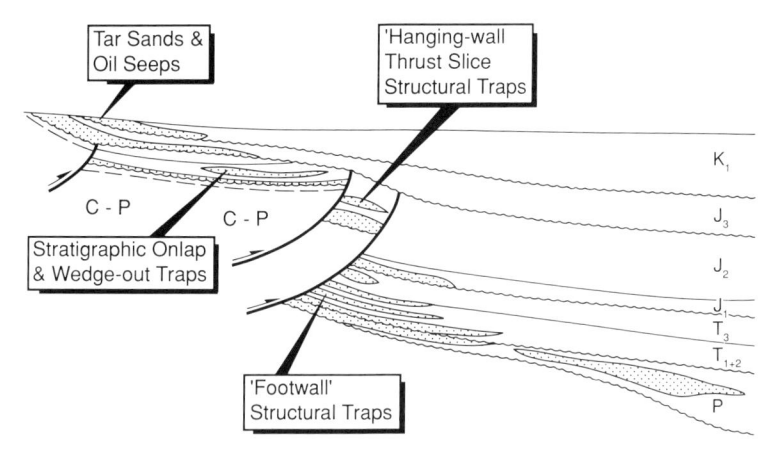

Fig. 5. Petroleum geology: Karamay Thrust Belt.

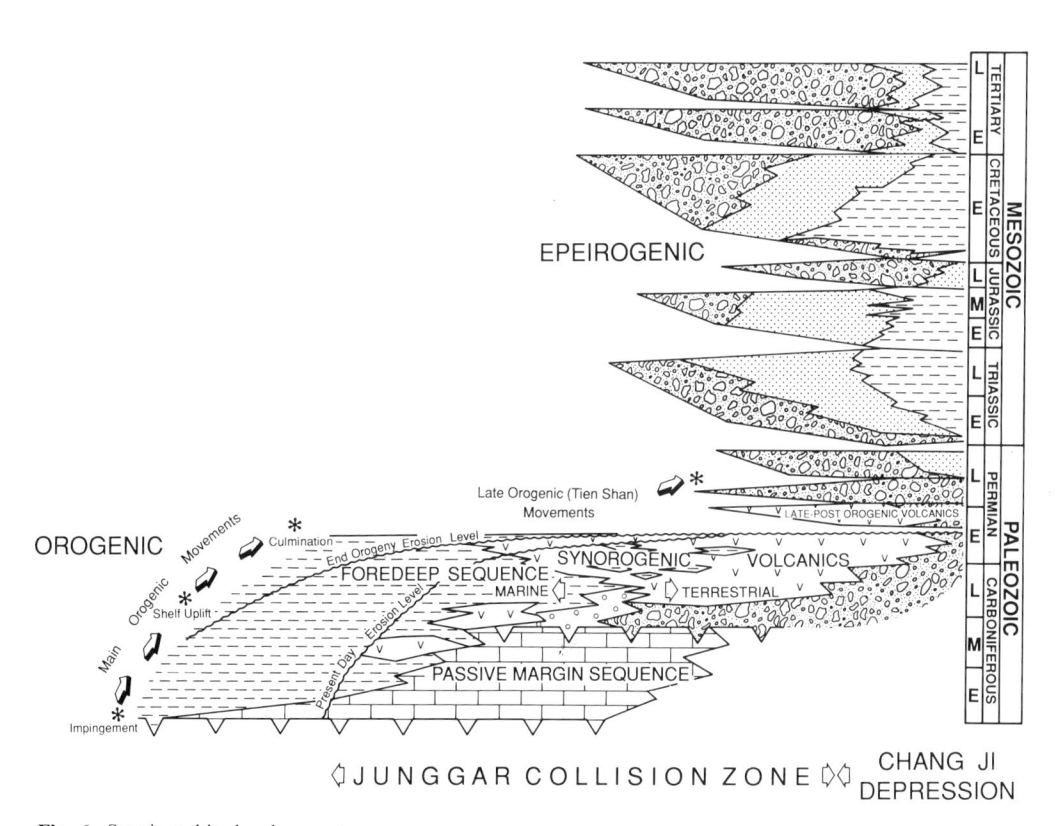

Fig. 6. Stratigraphic development.

by the flexure of the continental margin under the influence of the collision. A progressive shoaling of the sequence is seen in the increasing dominance of terrestrial sedimentation. Carboniferous and Permian volcanics are thought to be the product of syn-orogenic magmatism. The main phase of uplift of the foredeep sequence appears to have been over by the Early Permian, and the products of the erosional stripping ('molasse') collected in the adjacent 'foreland' basin, by now entirely terrestrial. Late and post-orogenic magmatism is indicated by an important volcanic component in these early sequences. Also strong unconformities suggest that tectonism continued probably as a result of the final orogenic phases of the Tien Shan to the south. This is thought to be the generator for the northwest−southeast structural fabric of the basin, which provided the main control for depositional patterns during the Permian.

The whole of the Triassic represents one major tectono−stratigraphic cycle following final uplift of the Tien Shan along the southern edge of the basin. Early regression was quickly followed by transgression as flexural subsidence outpaced sediment supply. A distinct foredeep developed in front of the Tien Shan. The Late Triassic saw the first major re-activation of the Karamay Thrust Belt during Indosinian orogenic movements. The marginal Triassic sediments were stripped-off and re-deposited in a regressive Early Jurassic depositional cycle. The stratigraphic relationships observed in the Lower and Middle Jurassic section along the northwestern basin margin, which includes evidence for minor faulting, are attributed to relaxational 'forebulge' development and lithospheric adjustments induced by Indosinian and Early Yanshanian orogenic episodes.

The end of the Triassic−Jurassic depositional cycle, and the beginnings of the first major phase of Tien Shanian re-activation is marked by the truncational unconformities recognised at the base of the Upper Jurassic and the Lower Cretaceous sections. During this structural episode wrench-controlled re-activation of old northwest−southeast uplift zones in the basin was responsible for the regressive depositional sequences of the Upper Jurassic. The unconformity at the base of the Cretaceous marks the culmination of this structural episode. Cretaceous subsidence which followed was again 'flexural' as a result of the Late Jurassic Tien Shanian rejuvenation and thrust-loading.

The unconformity at the base of the Tertiary section marks the end of Cretaceous flexural subsidence and the beginnings of the Himalayan re-activation of the Tien Shan. The earlier Himalayan orogenic events were represented by an effective cessation of sedimentation in at least the western part of the Junggar Basin. Flexural subsidence was re-established in the Oligocene but the effect of progressive Tien Shanian uplift is represented by major unconformities in the Late Miocene and Late Pliocene.

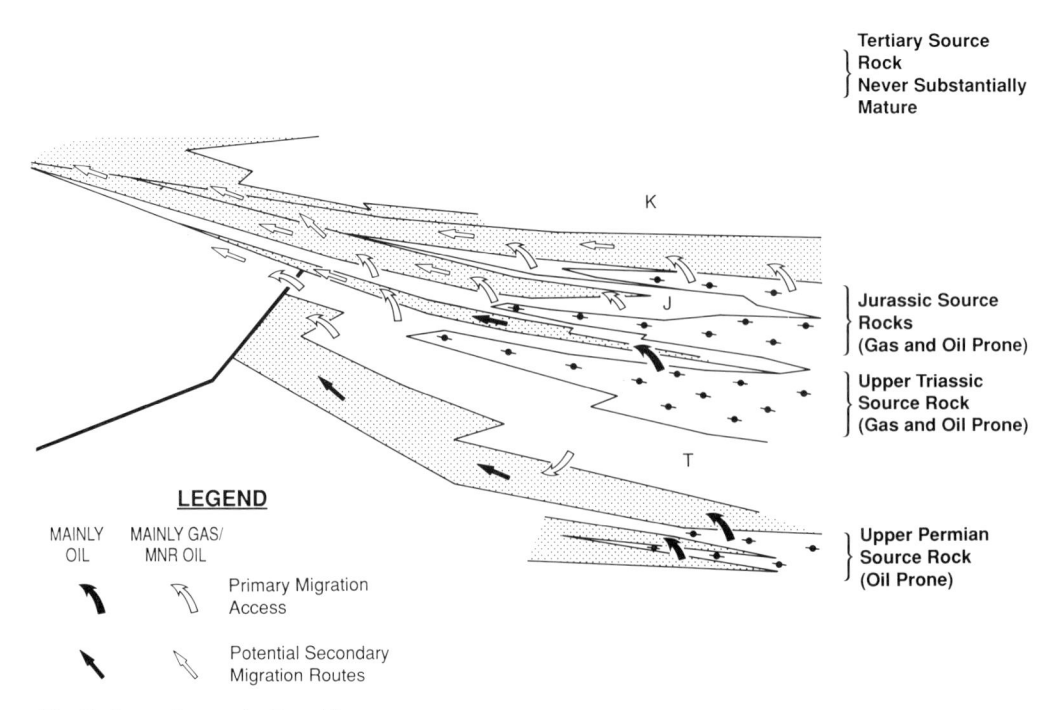

Fig. 7. Source/reservoir disposition.

Reservoirs and source rocks

The spatial stratigraphic relationships across the Karamay Thrust Belt into the basin show a favourable disposition of potential source and reservoir rocks (Fig. 7). Reservoirs are provided by alluvial fan and fluviatile clastics preferentially developed around the basin margin at nearly every stratigraphic level. Similarly, source rocks are developed in lacustrine sediments deposited in 'Lake Junggar' which existed throughout most of geological history. As is very often the case, the source rocks have only been examined and analysed in basin margin sections. Thus mid-basin prediction is based on depositional and climatic models.

A glimpse of the prospective Mid to Late Permian source rocks is obtained from an outcrop section near Urumqi along the southern edge of the basin. The Yao section displays about 600 m of the lower part of the Upper Permian which contains prominent 'oil shale' horizons (Fig. 8). Three samples examined have TOC values increasing from 3% at the bottom to 16–28%, and H/C ratios correspondingly rise from 0.9 to 1.4 (Equiv. to HI of 200–800 mg/g of TOC). Kerogen quality also improves towards the top. It appears that about 50% of the top 200 m contains the rich source rock lithology. Palaeo-latitude during the Permian appears to have been appropriate for accumulation of algal-rich lacustrine source rocks. The oil shales are not hypersaline but stratification of water is highly likely, since the presence of dolomitic mudstones suggests a broad shallow-water playa lake and a high degree of salinity. In this situation prolonged development of anoxic conditions can occur.

The Late Triassic saw the highest lake water levels but mudstones of this age appear to be part of a pro-delta system. Two mid-basinal models have been considered. The high degree of lamination seen in Upper Triassic mudstones suggests sedimentation in a quiet water body ideally suited to the sporadic development of stratification and anoxia. In addition the relatively high rainfall indicated by the river–delta systems would mean that the lake contained fresh water. Under these conditions of 'transient anoxia' a rich source rock will not develop. It is predicted that TOC values will remain under 2% and the kerogen type will be II or II/III. It is possible that the lake established permanent stratification for significant periods, say up to 100 000 years. This would result from a lack of seasonality, a lack of strong winds, deep water and stratification provided by salinity or temperature. In eutrophic freshwater lakes of this nature with 'permanent anoxia', mid-basinal TOC values from 2 to 10% might be expected. Kerogen type would be of a mixed Type II composition, with significant gas potential derived from the terrigenous input.

The basal transgressive shales of Early and Mid Jurassic cycles might improve in source rock quality basinwards. The marginal facies indicate a coastal plain–delta environment. The deeper water that must have existed in the lake is not likely to have developed permanent anoxia at depth. However, local seasonal anoxia could have developed as a result of the organic (land plant) input from the river–delta systems. A mixed algal/terrestrial Type II or Type II/III kerogen would have accumulated with predicted TOC values in the 1 to 2% range.

The varied nature of reservoired and seepage oils from the Karamay Thrust Belt and along the southern margin of the basin can be accounted for by two possible source rock models:
(a) A single source rock with laterally variable facies. The favoured choice in this case is the Mid and Late Permian section. The character of the sediments at Yao and the tectonic setting of the Mid to Late Permian deposition has led to the construction of a Permian lacustrine model (Fig. 9). (The presence of gammacerane in some oils indi-

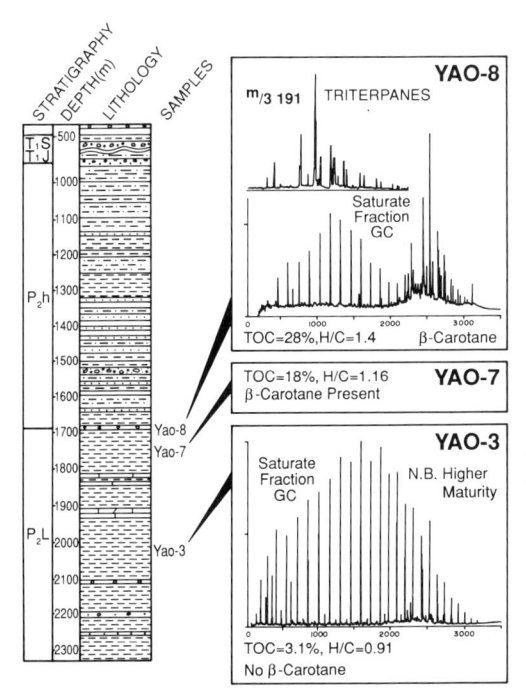

Fig. 8. Permian source rock analysis.

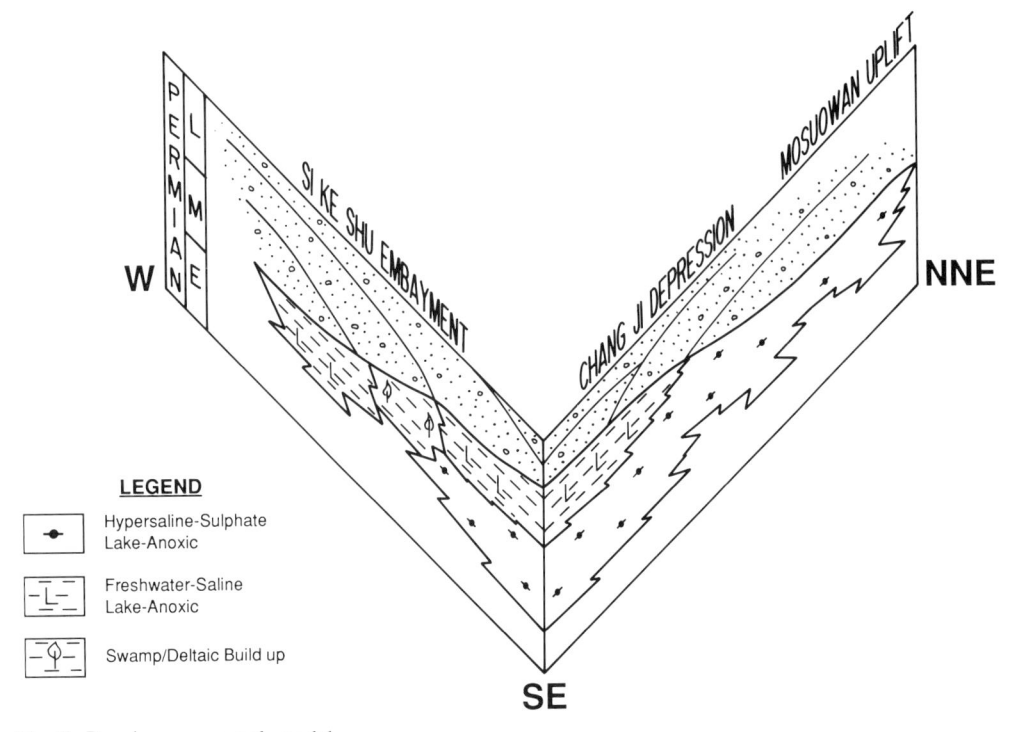

Fig. 9. Permian source rock model.

cates a saline–hypersaline source with a high-level of anoxia. Stratigraphic and depositional models suggests that this can only be Permian).

(b) Several source rocks developed at different stratigraphic levels. The indicated options for this are Permian, Late Triassic, Early–Mid Jurassic and Early Tertiary lacustrine sediments.

Model (b) is used for predictive purposes in the subsequent discussion of hydrocarbon expulsion and migration. In this way most petroleum occurrences and their characteristics can be accounted for. In this scheme the Permian source rocks are considered to be varying in depositional setting according to Model (a) but are basically oil-prone. Triassic and Jurassic source rocks are considered to be of the 'transient anoxia' type with overall poor source rock quality and with a dominant gas-prone character.

Conditions for migration and entrapment

The thrust structures of the Karamay Thrust Zone were already in position by the Early Permian, thereby providing structural traps for early-generated hydrocarbons. Maturation studies have demonstrated the timing of the different phases of hydrocarbon generation for the prospective source rock intervals. The earliest generation, from deep Permian source rocks, probably began in the Triassic. The primary control, therefore, on hydrocarbon migration and entrapment, during the subsequent Mesozoic and Tertiary time period, is provided by the 'epeirogenic' structural evolution of the basin.

Structural development

Generally the formation and subsidence of 'flexural' or 'orogenic' basins is explained by the thrusted re-activation of mountain uplifts in the intervening suture zones (Watson *et al.* 1987). In this way an analogy is drawn with the formation of foreland basins which form as a consequence of lithospheric flexure beneath thrust loads (Beaumont 1981). In the application of this simplistic model to the Junggar Basin, allowance has been made whenever possible for the complicating factors presented by inherited structural fabric in the lithosphere, the varying direction of push from the re-activated mountain system; and the effect of strike-slip movement

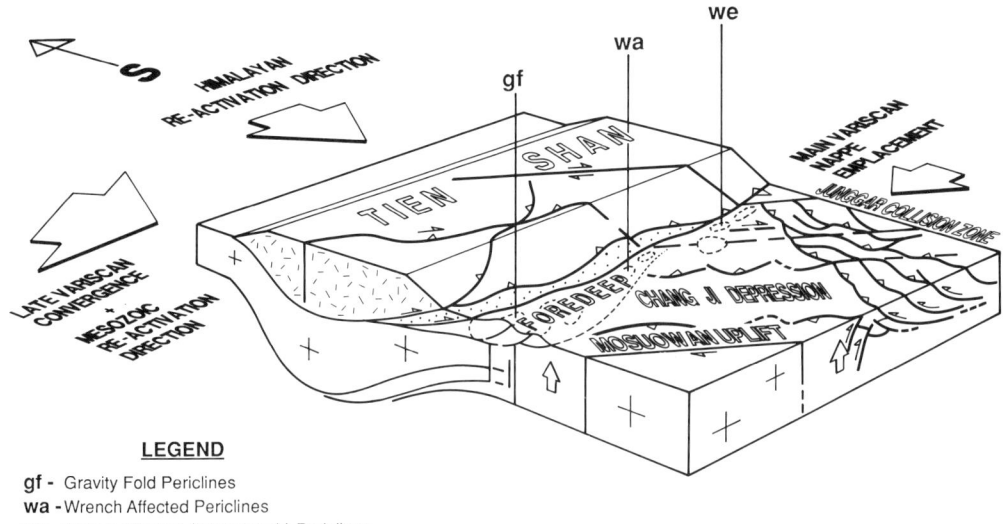

PETROLEUM GEOLOGY: JUNGGAR BASIN

LEGEND
gf - Gravity Fold Periclines
wa - Wrench Affected Periclines
we - Wench Effected (Interwrench) Periclines

Fig. 10. Regional structural model.

in the suture belts and the foreland lithosphere.

The effects of the Indosinian orogenic episodes are seen as the reactivation of the Karamay thrust structures in the Late Triassic–Early Jurassic (Fig. 10). This is envisaged as 'forebulge' uplift of old thrust fault-blocks and the re-emergence of an imbricate fan thrust system. The next major phase of regional structural development is in the Late Jurassic and can be attributed to hinterland epeirogenic movements related to Early Yanshanian orogenesis. At this stage there appears to have been major uplift along northwest–southeast structural zones (see Fig. 10). It is envisaged that strike-slip movements played a role in the re-activation of these late 'Variscan' transpressional trends.

The Himalayan uplift of the Tien Shan appears to have had its main effect in the Quaternary. Reverse faulting is dominant in the Tien Shan but it is widely acknowledged that this episode of tectonic activity was accompanied by major dextral wrench faulting on Palaeozoic faults (Fig. 10). With the recognition of gravity folding and broadly contemporaneous strike-slip tectonics, a model is presented for the development of the fold structures in the foldbelt–foredeep zone of the Tien Shan (Fig. 11). The major periclines in the deepest part of the foredeep are attributed to gravity folding modified by 'ramping' over differential fault-blocks which are moving in response to strike-slip movements in the basement. These are designated 'gf' structures (Fig. 10). Further west

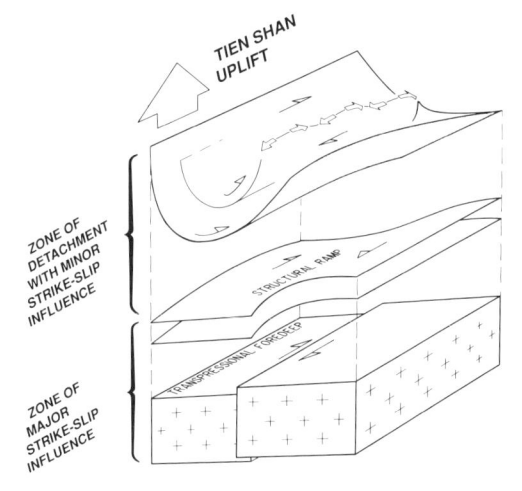

Fig. 11. Mechanics for wrench/ramp model.

the fold structures, which include the Dushanzi oilfield, have been formed in a thinner part of the foredeep section. The geometry is characterised by smaller more restricted and tighter structure. They are considered to be 'wrench-affected' or 'wa' structures formed more directly in response to positive wrenching above a major wrench zone. The westernmost fold structures are gentler in expression and do not have penetrative faulting. They are presented as inter-wrench or 'wrench-affected' or 'we' structures (Fig. 10).

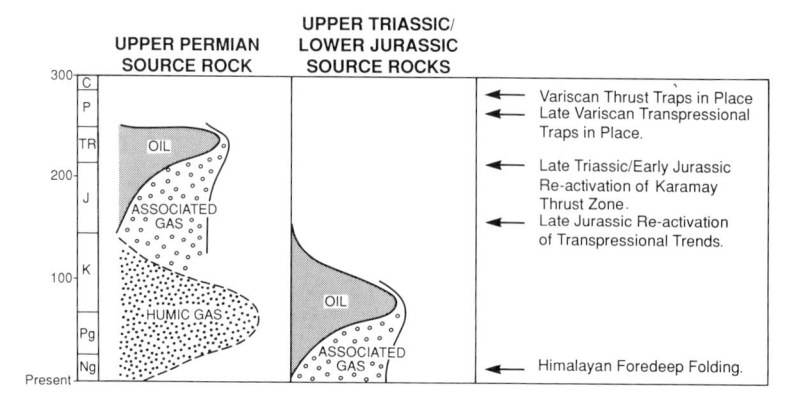

Fig. 12. Timing of hydrocarbon generation.

Migration and entrapment

A consideration of the maturation–expulsion history of the principal source rocks, together with the main phases of structural activity outline above, provides a guide to migration and entrapment in the basin (Fig. 12).

The deep Permian source rocks achieved oil maturity in depression areas, and in the Tien Shan foredeep, prior to Late Jurassic uplift. Early generated oil would have found Permian and Triassic reservoirs and followed regional migration routes out of the sub-basin deeps to the marginal uplifts (Fig. 13a). It is anticipated that palaeo-entrapment could have taken place along the northwest–southeast directed uplift zones in deep Permian reservoirs, and also along the Karamay Thrust Zone in Triassic reservoirs.

Late Jurassic tectonics re-emphasised the northwest–southeast structural trends and caused a partial re-alignment of regional migration patterns (Fig. 13b). Hydrocarbons still moving out of the Permian source rocks in the sub-basins and foredeep areas will have moved dominantly northwestwards along the uplift zones towards the Karamay Thrust Belt and Chepaizi Uplift. Palaeo-entrapment in Permian and Triassic reservoirs was probable along rejuvenated northwest–southeast oriented uplifts. Source rocks in the lower part of the Triassic section in the Tien Shan foredeep were probably contributing gaseous hydrocarbons to Triassic reservoirs at the same time.

By Cretaceous times deep Permian source rocks would have become overmature for oil and were contributing large amounts of gas to Permian and Triassic reservoirs (Fig. 13c). The gas would have displaced previously-trapped oil. It is envisaged that the main leakage of oil into the 'basement' rocks of the Chepaizi Uplift probably occurred at this time. During the Cretaceous and Early Tertiary the upper level Triassic source rocks would have become mature in the Tien Shan foredeep. For the first time Lower Jurassic reservoirs would have become charged with mostly gaseous hydrocarbons. Regional migration was towards the northwest and palaeo-entrapment in Lower Jurassic reservoirs was possible in structural traps along the northwest–southeast orientated uplift zones, and in stratigraphic traps over the Karamay Thrust Belt. It is envisaged that deep structural traps in the Tien Shan foredeep, particularly along the line of major wrench-related fault-blocks, would have retained their early oil in deep reservoirs but would be more likely to have been charged by gas in Jurassic reservoirs.

During Tertiary burial, Jurassic source rocks began contributing gas and supplemental oil from the Tien Shan foredeep to even higher level reservoirs in the Upper Jurassic and Lower Cretaceous section (Fig. 14a). Regional migration was again updip towards the northwest. Palaeo-entrapment is possible in some structural traps along uplift zones and in stratigraphic traps along the Karamay Thrust Belt. Also during this time, deeper Permian source rocks in the Si Ke Shu Embayment would have become mature for oil generation and charged Permian and Triassic reservoirs in deep structural traps.

The Himalayan movements had two main effects which induced re-migration of previously-trapped hydrocarbons both vertically and laterally. In the Tien Shan foredeep the uplift and folding caused exhumation of deep Triassic and Jurassic reservoirs, vertical leakage of hydrocarbons from deep to shallow reservoirs in Himalayan structures and updip spillage of

13a. Mid Jurassic

13b. End Jurassic

13c. End Cretaceous/Early Tertiary

LEGEND

MAINLY OIL MAINLY GAS/
            MNR OIL

Secondary Migration

Re-migration

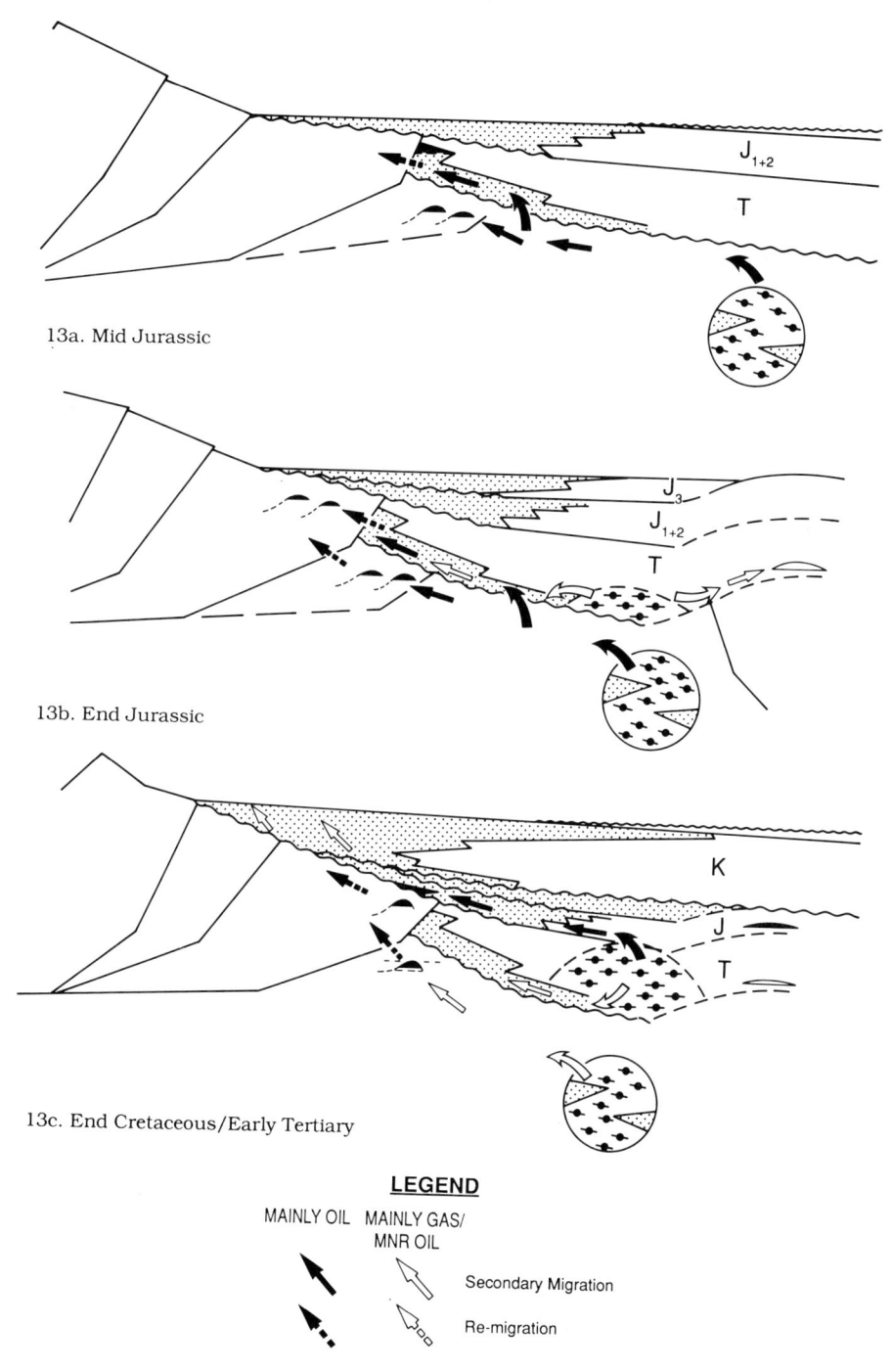

Fig. 13. Migration and entrapment.

14a. Lateral in N.W. Basin Margin

Wrench-Effected or Inter-Wrench   **Wrench-Affected e.g. Dushanzi**   **Gravity Fold e.g. Her Gos**

Exhumed Foredeep

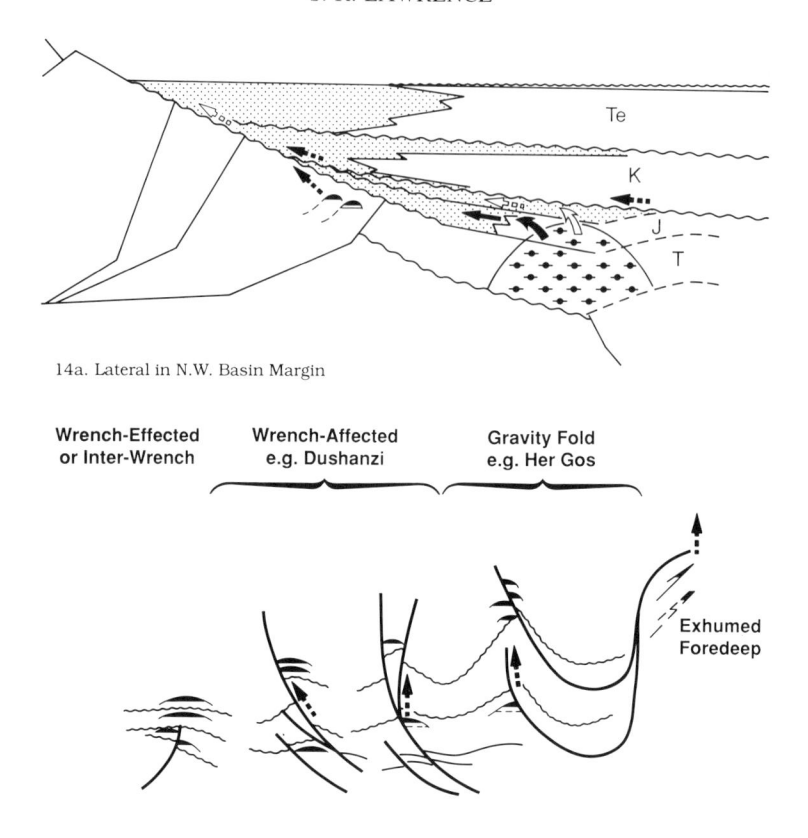

14b. Vertical in Tien Shan Foredeep

Fig. 14. Re-migration following Himalayan tectonics.

hydrocarbons from deeper reservoirs.

The up-to-the-north tilt imposed by the late Himalayan movements provided a regional northerly migration route to re-migrated hydrocarbons. Structural and stratigraphic traps along the southern end of the Karamay Thrust Belt (Hong Che Fault Zone) would have been subject to spillage (Fig. 14a). This late tilt is tending to funnel hydrocarbons to the north, towards the prospective Karamay tract of the Karamay Thrust Belt, but some hydrocarbons have been retained in 'basement' reservoirs and Jurassic stratigraphic traps along the Hong Che Fault Belt to the south.

In the outline of structural development, a distinction was made between types of Himalayan fold structures in the foredeep zone. The nature of their development has a bearing on how hydrocarbons can be retained or migrate vertically to gain access to higher level reservoirs (Fig. 14b). In the extreme case, reservoirs subject to folding and uplift in the 'exhumed foredeep' zone have been subject to extensive breaching and leakage as attested by the numerous seepages encountered in the Tien Shan foothills. Nevertheless hydrocarbons can be retained or re-trapped in this zone as shown by the production from the six wells drilled on the Qi Gou anticline from Jurassic reservoirs.

The nature of the 'gf' structures with the more vertically-biased 'wrench-ramp' mechanics would suggest that deep-trapped hydrocarbons could gain access to higher level reservoirs during Himalayan folding and faulting. This appears to have been the case in the Her Gos anticline although production proved to be sub-commercial.

The commercially-productive Dushanzi anticline is classified as a 'wa' structure. Production was established from Neogene–Paleogene reservoirs in ten pay zones and total reserves have been unofficially estimated at 30 to 50 × 10^6 barrels. This serves to demonstrate that these types of structures are probably more prone to charging by vertical leakage. The origin of the hydrocarbons is thought to be deep traps associated with an underlying major wrench zone. For the 'we' structures, there is a stronger chance of retention of hydrocarbons in deep traps but less likelihood of deep traps actually existing.

Conclusions

The northwestern part of the Junggar Basin is a long-established, indeed classic petroleum province in most senses of the description. Its success was founded upon the drilling of seeps in the first instance but 'development' was firmly based on the application of modern exploration techniques adopted by KOC during the 1970s and 1980s. Recent studies applying a modern integrated approach to exploration provides a better insight into the conditions controlling petroleum accumulation.

We are confident that new exploration philosophies will improve the success ratio in the Junggar Basin which will re-emphasize its historical importance with respect to China's petroleum resources. This bodes extremely well for the massive Tarim Basin which is ostensibly of a similar origin and evolution.

Grateful thanks are extended to the Karamay Oil Company for permission to present and publish this paper, and also to several KOC staff who contributed through participation and discussion. Zhang Ji Yi deserves special mention in this respect. In addition the assistance of C. Cornford and P. Jones in the respective interpretation of source rock geochemistry and structural geology is gratefully acknowledged.

References

BEAUMONT, C. 1981. Foreland Basins. *Geophysical Journal of the Royal Astronomical Society*, **65**, 291–325.

ULMISHEK, G. 1984. *Geology and Petroleum Resources of Basins in Western China*. Publication of US Department of Commerce, National Technical Information Service.

WANG HANSHENG, 1986. Tectonic evolution and oil pools of Junggar Basin, Xinjiang. *Spring Issue, China Oil Magazine*, 26–33.

WATSON, M. P., HAYWARD, A. B., PARKINSON, D. N. & ZHANG, Zh. M. 1987. Plate Tectonic history, basin development and petroleum source rock deposition onshore China. *Marine and Petroleum Geology*, **4**, 205–225.

XIE HONG, ZHAO BAI., LIN LONG DONG & YOU QI-MEI. 1984. Oil-bearing features along the Overthrust Belt in the Northwest Margin of Junggar Basin. Preprint for 'International Petroleum Geology Symposium', Beijing, September 1984.

ZHANG JI-YI, 1984. Petroleum accumulation in the Northwest Margin of Junggar Basin. Preprint for 'International Petroleum Geology Symposium', Beijing, September 1984.

ZHANG, Zh. M., LIOU, J. G. & COLEMAN, R. G. 1984. An outline of the plate tectonics of China. *Bulletin of the Geological Society of America*, **95**, 295–312.

Index

Abu Dhabi (Zakum field)
  diagenesis 309–13
  exploration history 299
  facies analysis 308–9
  petrography 303–8
  reservoir classification 301–3
  reservoir properties 313–5
  stratigraphy 300–1
  structure 299–300
Abu Mahara Formation 286, 289
Aden-Abyan Basin 330, 337
Aguardiente Formation 23
Alad' inskaya Formation 481
Alaska
  oil generation 179–81
  plate sequences 148–9
    Beaufortian 160–8
    Brookian 168–77
    Ellesmerian 149–60
  stratigraphy 144–8
Alaska (North Slope) petroleum province 2, 7
Albacora field 125–6, 135
Alborz Mountains 288–9
Albuskjell field 464
Algerian petroleum province 2, 7
Amal field 320–1
Ameland field 410
Amethyst field 410
Anabar province 477–9
Anahuac Formation 237–8
Anequim field 126
Angara-Lena province 487
Angayucham Terrane 168
Angelina-Caldwell Flexure 266
anhydrite
  cements 259–60, 409
  crystallization 382
  seals 339
Apalachicola Embayment 226
Apennine Thrust Belt *see* Fossa Bradanica
Apulia Plate 369
Apure Basin *see* Barinas-Apure
Ara Formation 287, 289
Arabian Gulf 287–8, 294
Arajuno Formation 96
Areo Formation 44
Argyll field 450, 464
Arkana field 271
Asmari Formation 296
Auca field 97
Audrey field 410
Auk field 448, 450, 464
Austin Group 234
Australia (Gippsland Basin)
  exploration history 525–7
  field characteristics
    Flounder 535–6
    Fortescue-Cobia-Halibut 536–7
    Kingfisher 537–8
  maturation 529–30
  sequence stratigraphy 532–5
  source rocks 530–1
  structure 527–9
  trapping 531–2

backarc basin type 4, 5, 6
Badayi Basalt Formation 285
Badejo field 126
Baghanwala Formation 288
Bagre field 126
Baikit province, 479
Bakken shale 193, 200
Baluchistan Basin 510–2
Bangestan Group 296
Bani Ghayy Group 281
Bara Formation 519
Barinas-Apure Basin 13
  exploration history 55
  field analysis 69–75
  generation capacity 29–30
  geochemistry 63–9
  maturation 27
  production statistics 9
  source rocks 24, 63
  stratigraphy 55–61
  structure 22–3, 61–3
Barnett field 270
Barque/Clipper field 410
Barranquin Formation 23
Barrow Arch 147, 149, 166, 172
Barut Formation 288–90
basin classifications 3–5
Baxterville field 271
Bay Springs field 271
Bayandor Formation 288–90
Beatrice field 464
Beaufort Sea 149
Beaufortian plate sequence 148–9
  resource potential 166
  sequences
    Lower 161–2
    Upper 162–6
  tectonic regime 160–1
Beaverhill Lake Group 192–3
Bedinan Formation 284
Beechwood field 271
Belayim Formation 335
Belly River sands 189, 198
Bel'skaya Formation 487
Bergen field 410
Bermejo field 97
Beryl field 454, 464
Bicudo field 126
Big Island field 271

559

Black Creek field 274
Black Jack Creek field 270
Boknfjord Group 445
Bolton field 271
Bonaire Basin 9, 29
Bone Island Formation 232
Bonito field 125–6
Bonoco Fault 17–18, 79
Borracha Formation 23–4
Bossier Formation 230
Boundary Creek Formation 174, 176
Bowland shales 432
Brae field 457, 464
Brazil (Campos Basin)
  exploration history 125–7
  geochemistry 127
  oil chemistry 130–3, 138
  reservoirs 134–9
  source rocks 127–34
  stratigraphy 121
  tectonic evolution 119–25
Bream field 464
Brent field 464
Brent Group 451, 454
Brookian plate sequence, 149
  basin initiation 168–9
  oil generation 179–81
  resource potential 175–7
  sequences
    Lower 169–73
    Middle 173–4
    Upper 175
Broom Formation 451
Bryne Formation 454
Buah Formation 286, 289
Buchan field 448, 464
Buckner field 270
Buda Formation 230
Bulaiskaya Formation 487
Bure field 410
Burguita Formation 23
burial graphs
  Australia 530
  Brazil 128–9
  Canada 200
  Ecuador 100
  England (North) 434–5
  Venezuela 67
Burj Formation 284
Buzbee field 270

Caledonian structures in N England 417–20
California petroleum province 2, 6
Calow gasfield 432
Cambrian reservoir rocks 477–9, 482, 487
Campos Basin
  exploration history 125–7
  geochemistry 127
  oil chemistry 130–3, 138
  reservoirs 134–9
  source rocks 127–34
  stratigraphy 121
  tectonic evolution 119–25
Campos Formation 120–2

Canada Basin opening 166–8, 172
Canada (Western)
  exploration history 189–91
  migration modelling 198–200
  production statistics 2, 6
  source rocks 191–8
  stratigraphy 190
Candela field 375, 378
Cane River Formation 235
cap rock types
  Australia 536–7
  England (North) 432
  Middle East Salt Basin 296
  North Sea 455, 457, 461
  Suez Gulf 363
  US W Texas 260
  USSR Siberia 496–7
  Venezuela 45
  Yemen (PDR) 333, 338–9
Capacho Formation 23–4
Carapeba field 126
Carapita Formation 44–6
Caratas Formation 44
carbon content *see* total organic carbon
carbon dioxide occurrence 378
carbon isotope values for oils
  Ecuador 102–4
  England (North) 433
  Netherlands 394
  US Gulf Coast 270–1
  US W Texas 260, 267–8
carbon preference index (CPI) 267–8
carbonate diagenesis
  Abu Dhabi 309–13
  Netherlands 389–97
  North Sea 409–10
  US W Texas 275
carbonate platform development 382, 385–9
Carbonera Formation 28–9, 59
Carboniferous rocks
  reservoirs
    Alaska 157
    China 545
    England (North) 430–4
    USSR Siberia 477
  sedimentary history
    Alaska 154–7
    England (North) 421–7
    Pakistan 514
  sources
    Alaska 157
    Canada 193
    England (North) 417, 433
    Netherlands 379
    North Sea 399, 400
Cardium sands 189, 198
Cariaco Basin 9, 18, 28–9
Cariaquito Formation 25
Caribbean Plate 13–14, 21
Carupano Basin 9, 18, 25, 28
Catahoula Group 237
Caucasus (North) petroleum province 2, 6
cements *see* anhydrite; carbonate diagenesis; gypsum
Central Graben of North Sea

sediments 447, 454, 457–8, 466
structure 441–5
Ceuta field 78–88
Chaguaramos Formation 28, 29
Chalcana Formation 96
Chalk Group 458, 461
Chambira Formation 94
Chapel Hill field 271
Chapiza Formation 91
chemical analyses of oils
  Brazil 130–3, 138
  Canada 200
  Ecuador 101–4
  Oman 320
  US W Texas 267–72
  USSR Siberia 488
Cherne field 125–6
Chichalli Formation 516
China (Junggar Basin)
  exploration history 545
  migration and trapping 552–6
  reservoirs 550–2
  source rocks 551
  stratigraphy 547–50
  tectonic setting 545–7
China (North) petroleum province 2
Chinese type basin 4–6
Chukchi Sea 149, 171, 177, 180
Chunchula field 270
Claiborne Group 235–7
Clarksville field 271
Claymore field 464
Claymore Formation 457–8
Clayton Formation 235
Clear Springs field 271
Cleeton field 410
Cobia-Halibut field 536–7
Coca field 97
Cockfield Formation 235–7
Cocos Plate 18
Coevorden gasfield 384
Cogollo Group 23–5, 77–8
Colon Formation 23–4, 78
Colorado Shale Group 196–8
Colville Group 146, 173
Colville Trough 149, 172–3, 178–9
Cononaco field 97
Cormorant field 464
Corvina field 126
Corwin delta 171
Cotton Valley Group 230
Cowden-Foster field 250
cratonic type basin 4–6
Cretaceous rocks
  reservoirs
    Abu Dhabi 301–3
    Alaska 166
    Brazil 134–9
    Canada 201
    Ecuador 104–9
    Middle East Salt Basin 296
    US Gulf Coast Basin 230–5, 273–4
    Venezuela 45, 46
    Yemen (PDR) 332–3

  sedimentary history
    Alaska 162–6, 171–3
    Australia 527–8
    Ecuador 91–4
    Italy 369
    North Sea 404
    Pakistan 509–10, 517–18
    Suez Gulf 354
    Venezuela 13, 43–4, 56
  sources
    Australia 530–1
    Brazil 127–34
    Canada 195–8
    Ecuador 97–8, 104
    Middle East Salt Basin 295
    North Sea 445
    Oman 317
    Pakistan 513, 517–18
    US Gulf Coast 274
    Venezuela 9–12, 23–9, 45, 63, 79
    Yemen (PDR) 333
Cromer Knoll Group 458
Cypress Lake West field 270

Dahu Formation 287
Datta Formation 516
Deccan Trap basalts 510, 519
Derik Formation 284, 289
Desu Formation 287
Devonian rocks
  reservoirs
    Canada 201
    North Sea 448
  sedimentary history
    Alaska 154
    Pakistan 514
  sources
    Canada 191–3
    Yemen (PDR) 333
Dhruma Formation 295
diagenesis *see* carbonate diagenesis
Diyab Formation 313, 317, 320, 325
Dokhan volcanics 283
Dolomite Formation 284
dolomitization
  Abu Dhabi 310
  Netherlands 392
  North Sea 409
Doran Granite 280–1, 288
Dorcheat-Macedonia field 270
Draupne Formation 447
Dukhan Formation 295
Dune field 250
Dungan Formation 519
Dunlin Group 451
Dunlin Group 451
Duvernay Shale Formation 198, 200

Eagle Ford Group 234, 274
Eakring oilfield 420
East Schuler field 270
East Shetland Basin 445, 451
East Texas field 271
East Texas Salt Basin 2, 224, 266

INDEX

East Venezuela Basin 12–13
  exploration history 37–9
  field analysis 345–53
  geochemistry 45
  maturation 27
  production statistics 6, 9
  source rocks 24, 28
  stratigraphy 42–4
  structure 22–3, 39–42
East Zeit field 354–5, 358–60
Echooka Formation 146, 157
Ecuador (Oriente Basin)
  exploration history 89
  geochemistry 97
  oil chemistry 101–4
  oil generation 98–101
  oil migration modelling 109–10, 116
  reservoirs 104–9
  source rocks 97–8
  stratigraphy 93–6
  structure 96–7
  tectonics 89–91
Edale shales 432
Edwards Limestone Formation 230
Egersund Sub-Basin 445, 448, 454
Egypt
  Infracambrian 283–4
  see also Suez Gulf
Eileen Formation 146, 158
Ekofisk field 464
El Cantil Formation 23, 24
El Carito field 38, 39, 49–51
El Furrial field 12, 45–9
El Pilar Fault 17, 20, 39
El Zeit, Gebel 356–7
Elba field 271
Eldfisk field 464
Elk Point Group 191–2
Ellesmerian plate sequence 148
  basin development 154–60
  basin initiation 149–54
  resource potential 160
Embore Formation 120, 121, 122
Ems Low 382–4
Enchova field 126
Endicott Group 146–7, 155–6
England (North)
  exploration history 418
  hydrocarbon distribution 429–30
  maturation 434
  oil field evolution 434–7
  reservoir potential 430–1
  seal rocks, 432
  structural factors
    Caledonian 417–20
    Cenozoic events 428–9
    Late Mesozoic events 428
    Permo-Triassic events 428
    Variscan 420–7
  trap mechanisms 432–3
Enterprise field 271
episutural basin type 3
Escandalosa Formation 23–4
Esna shale 354

Etive Formation 454
Eunice-Monument field 250
Exshaw/Bakken shale 193, 198, 200

facies distribution maps
  Abu Dhabi 308–9
  Australia 590
  Brazil 134
  Ecuador 116
  England (North) 425–6
  Netherlands 383
  US W Texas 251–2
  Venezuela 26, 60
Falcon Basin 9, 25, 28–30
Fara Sandstone Formation 287, 290
Farallon Plate 13, 18
Fatima Formation 285, 289
faults see growth faults
Fayette field 271
Fish Scale Zone 196, 200
Fjerritslev Formation 454
Flomaton field 271
Flora field 271
Flounder field 536–7
folded belt type basin 3
Fordoche field 271
foredeep type basin 4–6
Fortescue field 537–8
Forties field 464
Fossa Bradanica
  exploration history 370–1
  geochemistry 377–8
  reservoir properties 377
Fouke field 270
Fredericksburg Group 230–1
Frigg field 464
Fringe Zechstein Group 379
Frio Formation 237
Fulmar field 457, 464

Gach Saran Formation 296
Gadvan Formation 295
Garau Formation 295
Garoupa field 126
Garoupinha field 126
gas compositions
  Italy 378
  Netherlands 379
  US Gulf Coast 274–5
gasfields
  Canada 199
  England (North) 432
  Italy 373, 375, 377–8
  Netherlands 379
  North Sea 399, 400, 410, 464
  Pakistan 523
  US Gulf Coast 231
  USSR Siberia 482, 494
generating capacity 29
generation timing
  Alaska 179–81
  Australia 532
  China 554
  Ecuador 98–101

Middle East Salt Basin 296–7
US Gulf Coast 269
geochemistry of oils
  Brazil 127
  Ecuador 97
  Italy 377–8
  Oman 320–6
  US Gulf Coast 270–1
  Venezuela 45, 63
geothermal gradient estimates 63–4, 127
Germany 410
Ghabar Group 286
Ghadyah Formation 339
Gharish Group 286
Ghazij Formation 521
Gippsland Basin
  exploration history 525–7
  field characteristics
    Flounder 536–7
    Fortescue-Cobia-Halibut 537–8
    Kingfisher 538–9
  maturation 530–1
  sequence stratigraphy 533–6
  source rocks 531–2
  structure 528–30
  trapping 532–3
Glen Rose Formation 230
Goldsmith field 250
Goru Formation 517
gravity anomaly data 507–8
gravity values (API)
  Abu Dhabi 314
  Ecuador 108, 110–3
  Italy 377
  Oman 323–4, 326
  US Gulf Coast 233, 269
  US W Texas 256–7
  Venezuela 79–80, 101, 105, 107
Grayburg Formation 251, 255–7, 259–62
Groningen field 408, 410–1
Grosmont Formation 198–200
Grotolle-Ferrandina field 373, 376
growth faults
  characteristics 203–4
  classification 206–7
    ductile shale detachment 208–10
    non-detached 214–6
    salt withdrawal 210–4
    salt-sill detachment 214
    stratigraphic detachment 206–8
  interrelationships 216–8
Guafita Formation 58–9, 71
Guarico Formation 25, 29
Guasare Formation 78
Guayuta Group 43
Gudgeon field 410
Gulf Coast Basin (US)
  exploration history 221
  future development 243–4
  geological setting 221–4, 265–6
  growth fault studies 203–4, 206–16
  production statistics 2
  reservoirs 229–43
  salt diapirism 227–9

salt distribution 205
source rocks 205–6
structural setting 224–7
Gulf of Venezuela Basin 9
Gullfaks field 454
Gurnard Formation 529, 535, 539
Gurpi Formation 295
gypsum cements
  North Sea 382, 409
  US W Texas 259–60

Habshan Formation 301
Habshiya Formation 339
Hadramaut-Jeza Trough 329, 334, 338
Hajir Formation 286
Halten Terrace 471
Haltenbanken province 471
Hammamat Formation 283–4, 289
Hangu Formation 518–9
Hanifa Formation 295
Hanna Trough 149, 180
Harut Formation 286
Hassi Messaoud field 7
Hathern Shelf 42
Hatter's Pond field 270
Hauptsandstein 420, 410
Haynesville Formation 267
Heather Formation 446, 454
Heidelberg field 271
Heimdal field 464
Highly Radioactive Zone (HRZ) 148, 171, 176
Hilal field 354–5, 358–60
Hobbs field 250
Hollin Formation 93–4, 104–6
Hominy field 271
Horda Platform 445, 451
Hormuz Formation 288, 290
Hosston Formation 231–2
Humber Group 445
Hunt Fork Formation 145–6, 154
Huqf Group 286, 290, 317, 320–1
hydrogen index 63
hydrogen sulphide occurrence 193

I & L field 270
Idsas orogeny 280
illitization 409
Indefatigable field 410
Indus Basin 509–10, 516–8
Infracambrian
  sediments of Middle East 282–9
  source rocks of Pakistan 512
Interior Range (Venezuela) 39–41
Interior Salt Basin (US) 224, 231
Iran 287–9
Iraq 295
Italy (Fossa Bradanica)
  exploration history 370–1
  geochemistry 377–8
  reservoir properties 377
  stratigraphy 369–70
  structure 373–7
Itkilyariak Formation 157
Ivishak Formation 158, 160

Izhara Formation 295

Jackson Dome 224, 233, 235, 266
James Formation 230
Jay field 270
J'Balah Group 285, 290
Jhelum Group 512
Jordan field 250
Jordan Rift Valley 284
Juan Griego Group 25
Junggar Basin 543–54
Jurassic rocks
 reservoirs
  China 547, 554
  Middle East Salt Basin 296
  North Sea 450–8
  Norway offshore 471
  Pakistan 513, 516
  US Gulf Coast 229–30
  USSR Siberia 496–500
  Yemen (PDR) 332–3
 sedimentary history
  Alaska 161–2, 169–71
  China 550
  Italy 369
  North Sea 403–4
  Pakistan 509–10, 516
  US Louann Salt Basin 205
  Venezuela 13
 sources
  Abu Dhabi 313
  Alaska 162, 166
  Canada 194–5
  China 550–1
  Middle East Salt Basin 294–5
  North Sea 445–8, 463
  Norway offshore 471
  Oman 317
  Pakistan 513, 516
  US Gulf Coast 266–7
  Yemen (PDR) 333, 338
Jusepin field 12, 37, 38
Jutana dolomite 288

Kahar Formation 280, 288, 290
Kalubik Formation 146, 148, 166
Kanayut Formation 145–6, 154
Karamay Thrust Belt 547, 549
Kareem Formation 354–5
Katanga province 482
Kavik Formation 146, 157
Kayak Formation 146, 157
Kazhdumi Formation 295, 297
Keg River Formation 191, 200
Kekiktuk Formation 146, 155, 157
Keoun Creek field 270
kerogen assessments
 China 549
 North Sea 445–7
 US Gulf Coast 267–8
Khabla Formation 286, 289
Khadro Formation 519
Kharaib Formation 301, 303–8
Kharus Formation 286

Khewra Sandstone Formation 288
Khisor Formation 288
Khufai Formation 286
Khuzestan 295
Kingak Formation 146–7, 161–2, 166
Kingfisher field 538–9
Kirthar Formation 521
Kopet Dag petroleum province 2, 6
Koruk Formation 284
Korvunchanskaya Formation 477
Kostinskaya Formation 479
Kot Formation 518
Kuhbanan Formation 287
Kuna Formation 157
Kuonamian Formation 488
Kuparuk River Formation 146–7, 165–6
Kussak Sandstone Formation 288
Kuwait 295

La Ceiba field 73–5
La Luna Formation 23–5, 78
 generating capacity 29
 geochemistry 79
 maturation 25–7
La Morita Formation 23–4
La Pascua Formation 28, 59
La Pica Formation 44
La Quinta Formation 77
La Rosa Formation 28–9, 78
La Vela Bay Basin 9, 18, 29
La Victoria field 70–2
Laffan Formation 295
Lagoa Feia Formation 121–2
Lagunillas Formation 28–9, 78, 80
Lakes Entrance Formation 530, 539
Lakhra Formation 519
Laki Formation 521
Lalun Sandstone Formation 288–90
Las Mercedes Formation 25, 28
Las Piedras Formation 44
Latrobe Group 528–9
 fields
  Flounder 536–7
  Fortescue-Cobia-Halibut 537–8
  Kingfisher 538–9
 sequence stratigraphy 533–6
Leduc Group 189
Lekhwair field 325
Lekhwair Formation 301
Leman Sandstone Formation 402, 405–6
Lena-Tunguska province
 exploration history 473–4
 productive regions 477–87
 reserves 488–9
 source rocks 487–8
 structure 474–6
Leon Formation 59
Linguado field 59
Linguado field 126
Lisburn Group 146–7, 156–7
Livingston field 271
Lockhart Formation 519
Los Jabillos Formation 44
Los Robles Group 25

Louann Salt 205, 222
Louisiana Salt Basin 224, 266
Lovett's Creek field 270
Lumshiwal Formation 518
Lurestan 295

Macae Formation 120–1
McElroy field 250
Mackenzie delta province 149
Macuma Formation 91
magnetic survey data 360, 507
Mahatta-Humaid Formation 286–7, 290
Maidin Formation 286, 289
Malhado field 126
Maljamar field 250
Mandal Formation 457
Mannville Group 195–6, 198, 200
Maracaibo Basin 10–2
  maturation 27
  potential 85–8
  production statistics 2, 6, 9
  reserves 83–5
  source rocks 23–4, 28
  stratigraphy 77–9
  structure 22–3, 80–3
Marcelina Formation 28–9
Mariann field 97
Marimba field 125–6
Marlim field 125–6, 135
Masseria Pepe field 373
Matatere Formation 25
Matulla Formation 354
maturation
  Ecuador 98
  Middle East Salt Basin 295
  Venezuela 25–7, 45
Maturin Sub-Basin 41–2
maturity measurements
  Australia 530–1
  England (North) 434–5
  North Sea 448
  US Gulf Coast 268–9, 273–4
Maureen field 464
Means field 250
Megan field 270
Melanico field 373
Merecure Group 28–9, 44
Mesa Formation 44, 94
Mezzarelle field 373
micritization 309–10
Middle East Salt Basin
  maturation 295
  oil migration 296–7
  production statistics 2, 6
  reservoirs 296
  seals 296
  source rocks 294–5
  stratigraphy
    Permian to present 293–4
    Precambrian 279–82
    regional correlation 282–9
  traps 296
Midlands Farms field 250
Midway Group 235

migration studies and models
  Brazil 134
  Canada 198–200
  China 552–5
  Middle East Salt Basin 296
  US Gulf Coast 269, 273
  US W Texas 260
  Venezuela 45, 69
Mila Formation 288, 289, 290
Miluveach Formation 146–7, 165
Minewah volcanics 283
Minhamir Formation 286, 289
Mirador Formation 28–9
Misoa Formation 78
Missionary field 271
Mississippi Salt Basin 2, 224, 266
Mistal Formation 286, 289
Mito Juan Formation 23–4
Monagas *see* East Venezuela Basin
Monroe Uplift 224, 233, 235, 266
Monte Strombone field 373
Monte Taverna field 373
Montney shale 193–4, 200
Montrose field 464
Moon Rock field 271
Morad Series 280
Moray Firth Graben 441, 445
Moreia field 126
Morgan field 355
Moro Formation 518
Mount Carmel field 271
Muraykhah Formation 285, 289
Murchison field 464
Musipan field 39, 45, 51–2
Muskeg Formation 191

Nabitah ophiolitic suture 285
Nahr Umr shale 295, 301
Najd Fault System 280–1, 285–6
Najmah Formation 295
Namorado field 126
Nancy field 270
Nanushuk Group 146, 148, 171
Napo Formation 96–8, 106–9
Naricual Formation 44, 48–9, 50
Natih Formation 317, 325
Navarro Group 234
Nazca Plate 18, 79
Nebo field 271
neomorphism 310
Nepa-Botuoba province 482
Ness Formation 454
Netherlands (Zechstein Basin)
  depositional model 385–9
  facies distribution 381–2
  gas reserves 408, 411
  stratigraphy 379–81
  structure 382–5
Niger delta petroleum province
  production statistics 2, 6
  reservoir studies 365, 367
Nilawahan Group 514
Nimra Formation 284
Ninian field 464

Nisku Group 189
Nordegg Formation 194–5, 198–200
Norphlet Formation 229, 267
North Aldan province 487
North Caucasus petroleum province 2, 6
North China petroleum province *see* China
North Louisiana Salt Basin 224, 266
North Mount Holly field 270
North New London field 270, 272
North Sea petroleum province 2, 6
  North
    exploration history 441
    maturation 448
    reservoirs 448–64
    resource evaluation 464–6
    source rocks 445–8
    structure 441–5
  South
    diagenetic effects 409–10
    reservoirs 406–10
    resource evaluation 414, 441
    Rotliegend history 405–6
    structure 399–405
North Slope petroleum province *see* Alaska
North Tunguska province 477
Norway (offshore) hydrocarbon province 471
Nova Siri Scalo field 377
Novaya Zemlya-Altai section 494, 496
Nubia sandstone 284, 353, 361–3
Nuka Formation 157
Nukhul Formation 354

Oca-Ancon Fault System 17, 79
Odin field 464
Oficina Formation 28, 29
Okpikruak Formation 146, 148, 171
Oman
  exploration history 317
  oil geochemistry 320–5
  stratigraphy 317–9
    Infracambrian 286–7
Ordovician reservoir rocks 447
Oriente Basin
  exploration history 89
  geochemistry 97
  oil chemistry 101–4
  oil generation 98–101
  oil migration modelling 109–10, 116
  reservoirs 104–9
  source rocks 97–8
  stratigraphy 93–6
  structure 96–7
  tectonics 89–93
Orinoco Heavy Oil Belt 9, 11–2, 30
Orinoco and Trinidad petroleum province 2
Orocual field 37, 39, 45–6
Orocue Formation 28, 29
Orsino field 373
Orteguaza Formation 94
Oseberg field 464
Otuk Formation 160–2
oxygen isotope values 260

Pab Formation 518–9

Pabdeh Formation 295
Pachuta Creek field 270
Pakistan
  basin development 508–12
    Baluchistan 510–12
    Indus 509–10
    Punjab 508–9
  exploration history 503–5
  oil geochemistry 323
  reservoirs 522
  source rocks 522
  stratigraphy 512–22
    Infracambrian 288
  structure 506–8
  traps 523
Palaeofacies maps
  Abu Dhabi 308–9
  Australia 539
  Brazil 134
  Ecuador 116
  England (North) 425–6
  Netherlands 383
  US W Texas 251–2
  Venezuela 26, 60
palaeogeography *see* palaeofacies maps
Paluxy Formation 232
Pampo field 125
Paraguana Peninsula 18–9
Parangula Formation 58
Parati field 126
Pargo field 126
Parsons sandstone 166
Patala Formation 518–20
Pauji Formation 80
Pearsall Formation 232
Pebble Shale Gamma Ray Zone 171
Peninsular Arch 226
Pentland Formation 454
People's Democratic Republic of Yemen *see* Yemen
Perija Fault 79
perisutural type basin 3
permeability measurements
  Abu Dhabi 302–3, 313–15
  Australia 537–9
  Netherlands 396, 408
  North Sea 450–1, 454, 457–8, 463
  US Gulf Coast 233
  Venezuela 72
  West Germany 410
Permo-Triassic rocks
  reservoirs
    Alaska 160
    China 547, 554
    Netherlands 389–97
    North Sea 406–9, 448
    Pakistan 513, 515
    US W Texas 255–6
  sedimentary history
    Alaska 157–60
    China 550
    Netherlands 382–5
    North Sea 402–3, 405–6
    Pakistan 514–16
    US W Texas 251–4

sources
　Alaska 160
　Canada 193—4
　China 550—1
　US W Texas 256—7
petroleum provinces summarized 1—4
Pettet Formation 232
Phosphate Zone 193—4
Pine Tree field 270
Piper field 457, 464
Pirauna field 126
Pisticci field 373, 376, 378
Plate tectonic influence on hydrocarbons
　Apulia 369
　Arctic Alaska, 148—9
　Australia 528
　China 545—7
　Cocos 18
　Farallon 13, 18
　Nazca 18, 79
　South America 14—18
　Venezuela 13—18
　　causes 18—20
　　results 20—1
plateau basalt type basin 3, 4
Platonovskaya Formation 477
Pleistocene reservoir rocks
　Italy 376—7
　US Gulf Coast 241—3
porosity measurements
　Abu Dhabi 302—3, 313—5
　Australia 537—9
　England (North) 431
　Netherlands 396, 408
　North Sea 409, 450—1, 454—8, 463
　US Gulf Coast 233
　US W Texas 258—9
　Venezuela 72
　West Germany 410
Porters Creek Formation 235
Porto Cannone field 375
Portuguesa Basin 25
Precambrian rocks
　reservoirs, USSR 477, 479, 481—2, 487
　sedimentary history, Middle East 282—9
　sources
　　Oman 317
　　Pakistan 512
Predpatomskaya province 482
PreSayany-Yenisei province 487
ProtoTexel-IJsselmeer High 382—4
Prudhoe Bay field 7, 143
Pumbuiza Formation 93
Pumpkin Bay Formation 232
Punjab Basin 508—9, 512

Q crudes of Oman 321—2
Quaternary rocks
　reservoirs
　　Italy 376—7
　　US Gulf Coast 241—3
　sources 9
Queen Formation 251
Querecual Formation 23—4, 27, 43

Quevedo Formation 23—4
Quiriquire field 37—8, 41
Quweira Formation 284
Ram Sandstone Formation 284
Ranikot Group 519—20
Rannoch Formation 454
Rattray Formation 454
Ravar Formation 287
Ravenspurn field 410
Rawson field 271
Recent source rocks 9
Reindeer Formation 174
reservoir rock studies *see under each Period*
Rift Zone type basin 5, 6
Rio Grande Embayment 224, 230—1, 234—6
Rio Guacha Formation 25
Rio Negro Formation 23, 77
Rizu Formation 280, 287, 289
Roblecito Formation 28—9, 59
Robutain Formation 285, 289
Rodessa Formation 232
Rodney field 271
Romashkino field 7
Roseto-Monte Stillo field 375
Rotliegend of North Sea
　reserves 411, 414
　reservoir quality 406—10, 448, 450
　stratigraphy 405—6, 409—10
Rotondella field 377
Rough field 410
Rub al Khali Basin 329, 332—3, 4
Rudeis Formation 354

Sabatain Formation 337
Sabden shales 432
Sabine Uplift 224, 234, 266
Sacha field 97
Sadan Formation 284—5
Sadlerochit Group 146—7, 158
Safiq Formation 317, 322—5
Sag River Formation 147—7, 158
Sahara petroleum province 2, 7
Sahmah field 322—5
salinity in oilfields 106, 109, 114
salt basins *see* Middle East; Suez Gulf; US Gulf Coast; Zechstein
salt diapirs and seismic interpretation 341
Salt Range Formation 288, 512
Samotlor field 7
San Andres Formation 251, 255—62
San Antonio Formation 23—4, 43—4
San Francisco Fault 39—41
San Juan Formation 44, 46
San Marcos Arch 224, 230—1, 234, 266
Sand Hills field 250
Sandres Formation 454
Santa Anita Group 44
Santa Marta Fault 79
Santiago Formation 93, 97
Saq sandstone 285, 289—90
Saramuj Formation 284, 289
Sargelu Formation 295
Saudi Arabia 285, 295
Sayhut Basin 329

Sayyala field 321–2
Scapa field 458, 464
Schneverdingen sandstone 410
Schuler Formation 230
Scram field 410
seals of reservoirs
   Australia 537, 538
   England (North) 432
   Middle East Salt Basin 296
   North Sea 455, 457, 461
   Suez Gulf 363
   US W Texas 260
   USSR Siberia 496–7
   Venezuela 45
   Yemen (PDR) 333, 338–9
Sean field 410
seismic sections
   Alaska 150, 151
   Brazil 122
   England (North) 420
   Suez Gulf 343–51, 360–1
   US Gulf Coast 208–9, 212, 215–7
   Venezuela 47–8, 62, 70–4, 81, 4
seismic techniques in salt basins
   methods 341–2
   modelling 342
   processing data 343–4
   results 344–50
selenite crystallization 382
Sembar Formation 517
Seminole field 250
sequence stratigraphy
   Australia 533–6
   England (North) 422
Seven Rivers Formation 251, 259–60
Severo-Vareganskoye field 498
Shabb Formation 286
Shabwa-Balhaf Graben 329–30
Shale Wall Formation 146, 173
Shammar rhyolite 281
Shear Zone type basin 5–6
Shelton field 271
Shetland Group 458
Shibkah field 325
Shilaif Formation 295
Shuaiba Formation 295, 301
Shublik Formation 146–7, 158–60
Shuntarskaya Formation 481
Shuram Formation 286, 289
Shushu Findi field 97
Siberian Platform petroleum provinces see Lena-Tunguska; West Siberia
Sidki field 354–5, 358–60
Siksikpuk Formation 157–8
Silurian rocks
   reservoirs in USSR 477
   sources in Oman 317
Silverpit Formation 402
Silvestre-Sinco field 13
Sinai Peninsula 283–4
Siq Formation 285, 289–90
Sirte-Libya petroleum province 2, 6
Slaughter Levelland field 250
Sleipner Formation 454–5

Sligo Formation 232
Slochteren Formation 402, 406
Smackover field 271
Smackover Formation 229–30, 266–75
Smithdale field 271
Snorre field 450, 455, 464
Socotra Basin 330
Sohlingen gasfield 410
Soltanieh Mountains 288–9
Soriano field 375
source rock studies see under each Period
South America see Brazil; Ecuador; Venezuela
South American Plate 14–18
South Florida Basin 226, 232–3
South Garib Formation 341, 355–6
South Paulding field 270
South Summerland field 271
South Tunguska province 479
Sparta Formation 235
Spruce Tree Formation 146, 173
Statfjord field 464
Statfjord Formation 450–1
Stord Basin 445, 455
Strzelecki Group 528
stylolitization 312
Sucre Group 43
Sudr Formation 354
Suez Gulf
   geophysical features 360–1
   reservoir distribution 361–3
   sequence thicknesses 358–60
   stratigraphy 353–6
   structure 341, 353
   unconformities 357–8
Sukhopitskaya Group 479, 481
Sulaiy Formation 295
sulphates see anhydrite; gypsum; selenite
sulphur occurrence 379
Sumatra petroleum province 2
Sunniland Formation 232–3

Talco Graben field 270
Tampen Spur 445
Tampico-Reforma-Campeche 2, 6
Tarbert Formation 454
Taylor Group 234
Telbismi Formation 284–5
Ten Boer Formation 402, 406
Tena Formation 94, 98–100
Tertiary rocks
   reservoirs
      Australia 536–7
      Brazil 134–9
      China 554
      Italy 373, 375–6
      Middle East Salt Basin 296
      Niger delta 365
      North Sea 461–4
      Pakistan 512–3, 520–1
      US Gulf Coast 205–6, 235–41
      Venezuela 45–6, 49, 83–5
      Yemen (PDR) 339
   sedimentary history
      Alaska 173–5

INDEX

   Australia 529
   Ecuador 94
   Italy 369–70
   North Sea 404–5
   Pakistan 510, 518–22
   Suez Gulf 354–6
   US Gulf Coast 226–7
   Venezuela 13, 44, 56–61
  sources
   China 550
   Pakistan 513, 520–1
   US Gulf Coast 274
   Venezuela 9–10, 12–13, 28
Texas
  East Salt Basin 224, 266
  West Permian Basin
   exploration history 257–8
   reservoirs 255–6
   sedimentary history 251–4
   source rocks 256–7
   structure 254–5
Thamama Group 301
Thames field 410
Thebes Formation 354
thrust belt type basin 5–6
Tiguino field 97
Timbalier Trough 212–13
time-temperature index
  Brazil 127
  Ecuador 98–100
Tiyuyaca Formation 94, 98–100
Tobosa Basin 251
Tobra Formation 288, 515
Torok Formation 146, 148, 171
Torrente Saccione field 375
Torrente Tona field 373, 375
total organic carbon (TOC) measures
  Australia 531
  China 551
  England (North) 433
  North Sea 445, 447
  US Gulf Coast 266, 274
  USSR Siberia 487
  Venezuela 63–4
trapping mechanisms
  Abu Dhabi 313
  Australia 536–7
  Brazil 134–9
  Canada 201
  China 552–5
  England (North) 432–3, 435
  Italy 373–7
  Middle East Salt Basin 296–7
  North Sea 454–5, 457, 461, 463
  Pakistan 523
  US W Texas 255–6
  Venezuela 69, 83, 85–6
Travis Peak Formation 231–2
Triassic rocks see Permo-Triassic rocks
Trilha field 126
Trix Liz field 270
Troendelas Platform 471
Troll field 457, 464
Trujillo Formation 82

Tungusikskaya Group 479
Tunguska province 477, 479
Turkey 284–5
Turnbull field 271
Turukhansk-Noril'sk province 477
Tuscaloosa Group 233–4, 274
Tutonchanskaya Formation 477
Tuwaiq Mountain Formation 295
Tyne Group 445

Ugnu Formation 146, 173–4
Ula field 457, 464
Umiat delta 171, 173
United Arab Emirates see Abu Dhabi
Urica Fault 39
Urumaco Trough 18
USA
  also see Alaska
  Gulf Coast Basin
   exploration history 221
   future development 243–4
   geological setting 221–4, 265–6
   growth fault studies 203–4, 206–16
   production statistics 2
   reservoirs 229–43
   salt diapirism 227–9
   salt distribution 205
   source rocks 205–6
   structural setting 224–7
  W Texas
   exploration history 257–8
   production statistics 2
   reservoirs 255–6
   sedimentary history 251–4
   source rocks 256–7
   structure 254–5
Usolskaya Formation 479, 487
USSR
  Lena-Tunguska
   exploration history 473–4
   productive regions 477–87
   reserves 488–9
   source rocks 487–8
   structure 474–6
  West Siberia
   exploration history 493–4
   future prospect 501–2
   production statistics 2, 6–7
   regional potential 494–5
   reservoirs 495–501

Vacuum field 250
Valera Fault 23
Valiant field 410
Vanguard field 410
Variscan structures
  England (North) 420–27
  North Sea 399–402
Venezuela
  also see Barinas-Apure; East Venezuela;
    Maracaibo
  basin structure development 22–3
  generating capacity 29
  palaeogeography 26, 60

plate tectonic setting 13–22
provinces outlined 9
sedimentary history 13
    Cretacous 23–7
    Tertiary 28–9
Vermelho field 126
Vicksburg Group 237–8
Victor field 410
Vidona Formation 44
Viking field 410
Viking Graben 441, 445–7, 450, 454–5, 463
Viking sands 189, 197–8
Villeta Formation 25
Viola field 126
vitrinite reflectance measurements
    Australia 532
    Brazil 127
    Ecuador 98
    US Gulf Coast 269
    Venezuela 68
Volga-Ural petroleum province 2, 6–7
Vulcan field 410

Waddell field 250
Washita Group 231
Wasson field 250
Wata formation 354
Welbeck Basin 433
Welland field 410
West Canada petroleum province *see* Canada
West Germany 410
West Sak Formation 146, 173, 174
West Siberia petroleum province
    exploration history 493–4
    future prospect 501–2
    production statistics 2, 6–7
    regional potential 494–5
    reservoirs 495–501
West Sole field 410
West Texarkana field 270
West Texas Permian Basin
    exploration history 257–8
    production statistics 2
    reservoirs 255–6
    sedimentary history 251–4
    source rocks 256–7
    structure 254–5
West Vilyui province 485
White Speckled shale 196–7, 200
Widmerpool Gulf 421

Wiggins Uplift (Arch) 226, 266
Wilcox Group 235, 274
Willis Branch field 271
Winterburn Group 193
Wissey field 410
Woodbend Group 192–3
Woodbine Group 233–4

Yare field 410
Yates field
    exploration history 257–8
    porosity studies 258–9
    production statistics 261–2
    reservoirs 255–6, 260–1
    seal rocks 260
    sedimentary history 251–4
    source rocks 256–7, 260
    structure 254–5, 259
Yegua Formation 235–7
Yemen (PDR)
    basin development 329–30
    geophysical studies 334
    oilfield potential 334–9
    sedimentary history 286, 329
    structure 331–3

Zagros foldbelt 6, 293
Zagros Mountains 287–8
Zaigun Formation 288–90
Zakum field
    diagenesis 309–13
    exploration history 299
    facies analysis 308–9
    petrography 303–8
    reservoir classification 301–3
    reservoir properties 313–5
    stratigraphy 300–1
    structure 299–300
Zebra sandstone 284
Zechstein Basin
    Netherlands
        depositional model 385–9
        diagenetic studies 389–97
        facies distribution 381–2
        stratigraphy 379–81
        structure 382–5
    North Sea, sedimentary history 402–4
Zechstein Group 445–8
Zeit Formation 341, 356